WITHDRAWN

# LIFE OF INVERTEBRATES

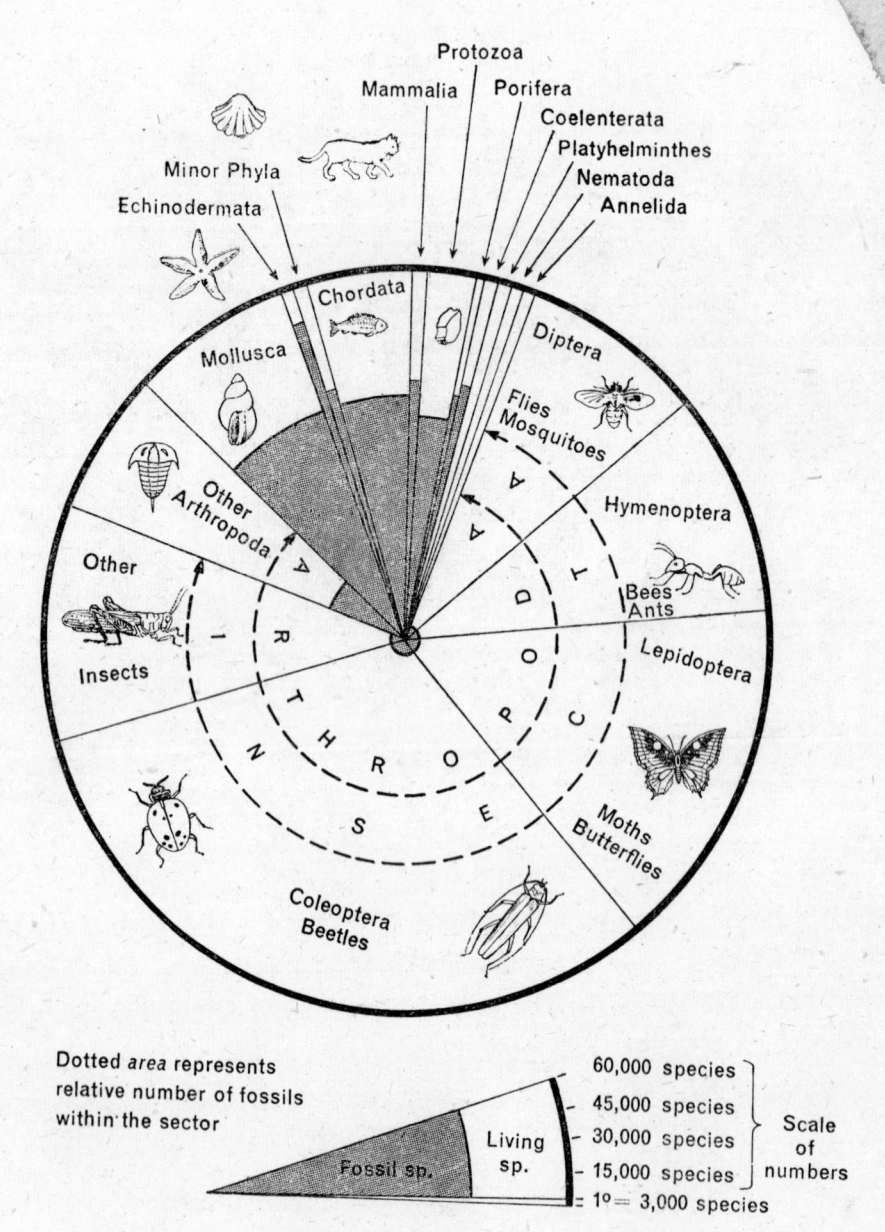

Relative numbers of known species, living and fossil, of various animal phyla.

# LIFE OF INVERTEBRATES

[FOR B.Sc. (PASS AND HONS.) STUDENTS OF INDIAN UNIVERSITIES]

**S. N. PRASAD**
*Professor, Department of Zoology*
*Allahabad University, Allahabad*

Exclusive Distributor
for USA and CANADA
Advent Books, Inc.
141 East 44th Street
New York, NY 10017

VIKAS PUBLISHING HOUSE PVT LTD

VIKAS PUBLISHING HOUSE PVT LTD
*Regd. Office*: 5 Ansari Road, New Delhi 110002
*Head Office*: VIKAS HOUSE, 20/4 Industrial Area, Sahibabad
Distt. Ghaziabad, U.P. (India)

COPYRIGHT © S.N. PRASAD, 1980

1V2P3901

ISBN 0-7069-1042-7

Printed by Typographers India at Rashtravani Printers,
A-49/1 Mayapuri, Phase 1, New Delhi

# PREFACE

The Invertebrates are those animals that have no internal bony skeleton, no vertebrae or backbone. They include 95 per cent of the entire Animal Kingdom. Aristotle was the first to identify the invertebrates from the vertebrates or those animals that possess a vertebral column and internal bony skeleton. But his classification was different. According to his definition the vertebrates were the animals with blood and the invertebrates were those without blood. The correct distinction was finally recognized by Lamarck and Cuvier. Many modern writers feel that the division of the Animal Kingdom into vertebrates and invertebrates is artificial. There is not a single positive characteristic that invertebrates hold in common. The subdivision is a vast heterogeneous assemblage of groups of animals. The range in size, in structural diversity and in adaptations to different modes is enormous. It is, however, true that 95 per cent of these animals are without a vertebral column (invertebrates). And it is these animals without backbone that this book describes.

During the last fifteen years there has been a virtual revolution in the study of morphology of tissues, sensory structures and even in the field of cytoplasmic and nuclear structures. New findings in the finer structure of membranes have enabled scientist to interpret their functions correctly. The main aim of the present book is to give new insights into cell functioning and ultrastructure as they apply to various invertebrate forms. It illustrates the wealth of diversity in organization and physiology that has been found in the various groups of protozoans and other forms.

The study of movement of cilia and flagella, the role of hydrostatic skeleton in the locomotion of many invertebrates, the activities of muscle fibres and similar other features have been discussed in the light of recent researches.

The latest trend in the taxonomy of various groups has been incorporated. The systematic classification of Protozoa, Arthropoda and Echinodermata appears to have been completely changed. Following the opinion of all recent students the group Hemichordata has been removed from the Chordata and made an independent phylum of the Invertebrates. It has been placed just after Echinodermata. In most of the contemporary textbooks many smaller groups of animals called "Minor Phyla" have been given as appendices of certain chapters or have been described at the end. This is not so in the present book. The phylogeny has been properly discussed and the groups have been placed correctly and appropriately.

The material of the volume has been compiled from the literature by perusing a large number of original articles. The great resources of the university library have enabled an exhaustive study of literature. Writing such book needs inspiration and I have no dearth of it. During my entire teaching career spanning almost four decades I had hundreds and hundreds of students good, bad or indifferent to inspire me through their super-brilliance, intelligence and utter ignorance of a correct approach to the subject. The students of the last category proved better as they gave me an opportunity to discuss the problems more intimately and analytically. I have no words to thank them.

It is difficult to acknowledge adequately the assistance of all who have contributed to the preparation of this book. First of all I am highly indebted to my colleague, Dr (Miss) Prabha Grover for carefully reading certain chapters and for suggesting material which she believed should be included in the book. Some figures have been drawn by Mr Anis Rizvi, a professional artist, who has done an excellent job. Many figures in the later chapters have been drawn by Km. Nivedita Kashyapa, who, as a beginner, has certainly left an indelible impression. Proper encouragement is bound to make her a first class artist. Km. Praveen Ara, Nivedita and Vasantika have pasted all the labels in the diagrams. I must acknowledge that without their help the figures would never have been completed.

The secretarial assistance of Mr K.M. Chacko and Km. K. Kunjamma is gratefully acknowledged, and special thanks are also due to the latter for help in typing the manuscript.

*S.N. PRASAD*

# CONTENTS

Introduction — xxi—xxix

## 1. PHYLUM PROTOZOA — 1—150

| | |
|---|---|
| *Phylum* Protozoa | 1 |
|   *Class* Mastigophora (Flagellata) | 17 |
|     *Sub-phylum* Sarcomastigophora | 44 |
|       *Superclass* Mastigophora or Flagellata | 44 |
|         *Class* Phytomastigophora | 44 |
|           *Order* Chrysomonadida | 44 |
|           *Order* Silicoflagellida | 45 |
|           *Order* Cocolithophorida | 45 |
|           *Order* Heterochlorida | 45 |
|           *Order* Cryptomonadida | 45 |
|           *Order* Dinoflagellida | 45 |
|           *Order* Ebriida | 47 |
|           *Order* Euglenida | 47 |
|           *Order* Chloromonadida | 47 |
|           *Order* Phytomonadida or Volvocida (Volvocales) | 47 |
|         *Class* Zoomastigophora | 48 |
|           *Order* Choanoflagellida | 48 |
|           *Order* Bicosoecida | 48 |
|           *Order* Rhizomastigida | 48 |
|           *Order* Kinetoplastida | 49 |
|           *Order* Retortamonadida | 49 |
|           *Order* Diplomonadida | 49 |
|           *Order* Oxymonadida | 49 |
|           *Order* Trichomonadida | 49 |

|   |   |
|---|---|
| *Order* Hypermastigida | 50 |
| *Superclass* Opalinatea | 50 |
| *Superclass* Sarcodina | 81 |
|   *Class* Rhizopodea | 81 |
|     *Subclass* Lobosia or Amoebozoa or Amoebina | 81 |
|       *Order* Amoebida (Gymnoamoeba or Nuda) | 81 |
|       *Order* Arcellinida (Thecamoeba or Testacea) | 81 |
|     *Subclass* Filosia | 82 |
|     *Subclass* Granuloreticulosia | 82 |
|       *Order* Athalamida | 82 |
|       *Order* Foraminifera or Thalamophora | 82 |
|   *Class* Piroplasmea | 83 |
|   *Class* Actinopodea | 83 |
|     *Subclass* Acantharia | 83 |
|     *Subclass* Heliozoa | 83 |
|     *Subclass* Radiolaria | 84 |
|     *Subclass* Proteromyxidia | 85 |
|     *Subclass* Mycetozoa | 85 |
| *Subphylum* Sporozoa | 86 |
|   *Class* Telosporea | 103 |
|     *Subclass* Gregarinia | 103 |
|     *Subclass* Coccidia | 104 |
|     *Subclass* Haemosporidia | 105 |
| *Subphylum* Cnidospora | 105 |
|   *Class* Myxosporea | 106 |
|     *Order* Myxosporidia | 107 |
|     *Order* Actinomyxida | 107 |
|     *Order* Helicosporida | 107 |
|   *Class* Microsporea | 107 |
|     *Order* Microsporidia | 108 |
|     *Order* Sarcosporidia | 108 |
|     *Order* Haplosporidia | 108 |
| *Subphylum* Ciliophora | 108 |
|   *Class* Ciliata | 108 |
|     *Subclass* Holotrichia | 144 |
|       *Order* Gymnoblastida | 144 |
|       *Order* Trichostomatida | 145 |
|       *Order* Chonotrichida | 145 |
|       *Order* Apostomatida | 145 |
|       *Order* Astomatida | 145 |
|       *Order* Hymenostomatida | 146 |
|       *Order* Thigmotrichida | 146 |
|     *Subclass* Peritrichia | 146 |
|       *Order* Peritrichida | 147 |
|     *Subclass* Suctoria | 147 |
|       *Order* Suctorida | 147 |
|     *Subclass* Spirotrichia | 147 |
|       *Order* Heterotrichida | 147 |
|       *Order* Oligotrichida | 147 |
|       *Order* Tintinnida | 147 |
|       *Order* Entodiniomorphida | 148 |
|       *Order* Odontostomatida | 148 |
|       *Order* Hypotrichida | 150 |

## 2. PHYLUM MESOZOA     151—153

*Phylum Mesozoa*     151

## Contents

| | |
|---|---|
| *Order* Dicyemida | 152 |
| *Order* Orthonectida | 153 |

## 3. PHYLUM PORIFERA—SPONGES 154—181

| | |
|---|---|
| *Phylum* Porifera | 175 |
|   *Class* Calcarea or Calcispongiae | 175 |
|     *Subclass* Calcaronea | 176 |
|       *Order* Leucosolenida (Homocoela) | 176 |
|       *Order* Sycettida (Heterocoela) | 176 |
|     *Subclass* Calcinea | 177 |
|       *Order* Clathrinida | 177 |
|       *Order* Leucettida | 177 |
|       *Order* Pharetronida | 177 |
|   *Class* Hexactinellida or Hyalospongiae | 177 |
|     *Order* Hexasterophora | 177 |
|     *Order* Amphidiscophora | 178 |
|   *Class* Demospongiae | 178 |
|     *Subclass* Tetractinomorpha | 178 |
|       *Order* Homosclerophorida | 178 |
|       *Order* Choristida | 178 |
|       *Order* Epipolasida | 179 |
|       *Order* Hadromerida | 179 |
|       *Order* Axinellida | 179 |
|       *Order* Lithistida | 179 |
|     *Subclass* Ceractinomorpha | 179 |
|       *Order* Halichondrida | 179 |
|       *Order* Poecilosclerida | 180 |
|       *Order* Haplosclerida | 180 |
|       *Order* Dictyoceratida | 181 |
|       *Order* Dendroceratida | 181 |

## 4. THE METAZOA 182—184

## 5. PHYLUM CNIDARIA (COELENTERATA) 185—241

| | |
|---|---|
| *Phylum* Coelenterata | 188 |
|   *Class* Hydrozoa | 188 |
|     *Order* Hydroida | 216 |
|     *Order* Hydrocorallina | 216 |
|     *Order* Trachylina | 217 |
|     *Order* Siphonophora | 219 |
|     *Order* Graptolithina | 219 |
|   *Class* Scyphozoa | 219 |
|     *Order* Stauromedusae | 225 |
|     *Order* Cubomedusae or Carybdeida | 226 |
|     *Order* Coronata | 226 |
|     *Order* Discomedusae | 227 |
|     *Order* Rhizostomeae | 227 |
|   *Class* Anthozoa | 228 |
|     *Subclass* Octocorallia (Alcyonaria) | 232 |
|       *Order* Stolonifera | 234 |
|       *Order* Telestacea | 234 |
|       *Order* Alcyonacea | 235 |
|       *Order* Coenothecalia | 235 |
|       *Order* Gorgonacea | 235 |
|       *Order* Pennatulacea | 235 |

| | |
|---|---|
| *Subclass* Hexacorallia (Zooantharia) | 236 |
| *Order* Actiniaria | 236 |
| *Order* Madreporaria | 237 |
| *Order* Zoanthidea | 240 |
| *Order* Antipatharia | 240 |
| *Order* Ceriantharia | 240 |
| *Order* Rugosa or Tetracoralla | 240 |
| *Order* Corallimorpharia | 240 |
| *Subclass* Tabulata | 241 |

## 6. PHYLUM CTENOPHORA 242—248

| | |
|---|---|
| *Phylum* Ctenophora | 242 |
| *Class* Ctenophora | 242 |
| *Subclass* Tentaculata | 247 |
| *Order* Cydippida | 247 |
| *Order* Lobata | 247 |
| *Order* Cestida | 247 |
| *Order* Platyctenea | 248 |
| *Subclass* Nuda | 248 |
| *Order* Beroida | 248 |

## 7. PHYLUM PLATYHELMINTHES 249—305

| | |
|---|---|
| *Phylum* Platyhelminthes | 249 |
| *Class* Turbellaria | 252 |
| *Order* Acoela | 261 |
| *Order* Rhabdocoela | 262 |
| *Order* Alloeocoela | 263 |
| *Order* Tricladida | 263 |
| *Order* Polycladida | 263 |
| *Class* Trematoda | 265 |
| *Order* Monogenea (Heterocotylea) | 283 |
| *Order* Aspidobothria | 283 |
| *Order* Digenea (Malacocotylea) | 283 |
| *Class* Cestoda | 284 |
| *Subclass* Cestodaria (Monozoa) | 295 |
| *Order* Amphilinidea | 295 |
| *Order* Gyrocotylidea | 296 |
| *Subclass* Eucestoda (Merozoa) | 296 |
| *Order* Tetraphyllidea or Phyllobothrioidea | 297 |
| *Order* Lecanicephaloidea | 297 |
| *Order* Proteocephaloidea | 298 |
| *Order* Diphyllidea | 298 |
| *Order* Tetrarhynchoidea or Trypanorhyncha | 298 |
| *Order* Pseudophyllidea or Dibothriocephaloidea | 298 |
| *Order* Cyclophyllidea or Taenioidea | 300 |

## 8. PHYLUM NEMERTINA 306—314

| | |
|---|---|
| *Phylum* Nemertina | 306 |
| *Subclass* Anopla | 313 |
| *Order* Palaeonemertini | 314 |
| *Order* Heteronemertini or Heteronemertea | 314 |
| *Subclass* Enopla | 314 |
| *Order* Hoplonemertini or Hoplonemertea | 314 |
| *Order* Bdellomorpha or Bdellonemertini | 314 |

## Contents

**9. PHYLUM ACANTHOCEPHALA** — 315—323

**10. PHYLUM NEMATHELMINTHES (ASCHELMINTHES)** — 324—387

- *Phylum* Nemathelminthes — 324
  - *Class* Rotifera — 325
    - *Order* Seisonacea or Seisonida — 340
    - *Order* Bdelloidea — 340
    - *Order* Monogononta — 340
      - *Suborder* Ploima — 340
      - *Suborder* Flosculariacea — 341
      - *Suborder* Collothecaca — 341
  - *Class* Gastrotricha — 342
    - *Order* Chaetonotoidea — 345
    - *Order* Macrodasyoidea — 345
  - *Class* Kinorhyncha — 346
  - *Class* Nematoda — 348
    - *Order* Enoploidea — 381
    - *Order* Dorylaimoidea — 381
    - *Order* Memithoidea — 381
    - *Order* Chromadoroidea — 381
    - *Order* Araeolaimoidea — 382
    - *Order* Monhysteroidea — 382
    - *Order* Desmoscolecoidas — 382
    - *Order* Rhabditoidea or Anguilluloidea — 382
    - *Order* Rhabdiasoidea — 383
    - *Order* Oxyuroidea — 383
    - *Order* Ascaroidea — 383
    - *Order* Strongyloidea — 383
    - *Order* Spiruroidea — 384
    - *Order* Dracunculoidea — 384
    - *Order* Filarioidea — 384
    - *Order* Trichuroidea or Trichinelloidea — 384
    - *Order* Dioctophymoidea — 384
  - *Class* Nematomorpha or Gordiacea — 385
    - *Order* Gordioidea — 386
    - *Order* Nectonematoidea — 386

**11. PHYLUM ENTOPROCTA** — 388—397

**12. PHYLUM ANNELIDA** — 398—500

- *Phylum* Annelida — 398
  - *Class* Polychaeta — 402
    - *Subclass* Errantia — 420
      - *Family* Aphroditidae — 420
      - *Family* Polynoidae — 421
      - *Family* Phyllodocidae — 421
      - *Family* Alciopidae — 422
      - *Family* Tomopteridae — 422
      - *Family* Glyceridae — 422
      - *Family* Nephthyidae — 422
      - *Family* Syllidae — 422
      - *Family* Hesionidae — 422
      - *Family* Nereidae — 422
      - *Family* Eunicidae — 422

   *Family* Histriobdellidae    422
   *Family* Ichthyotomidae    422
   *Family* Myzostomidae    422
  *Subclass* Sedentaria    422
   *Family* Orbiniidae    422
   *Family* Spionidae    422
   *Family* Chaetopteridae    422
   *Family* Sabellariidae    422
   *Family* Capitellidae    422
   *Family* Arenicolidae    423
   *Family* Opheliidae    423
   *Family* Maldanidae    423
   *Family* Cirratulidae    424
   *Family* Oweniidae    424
   *Family* Sabellariidae    424
   *Family* Amphictenidae    424
   *Family* Ampharetidae    424
   *Family* Terebellidae    424
   *Family* Sabellidae    424
   *Family* Serpullidae    424
 *Class* Oligochaeta    424
  *Order* Plesiopora    452
   *Family* Aeolosomatidae    452
   *Family* Naididae    452
   *Family* Tubificidae    453
   *Family* Phreodrilidae    453
  *Order* Plesiopora Prosotheca    453
   *Family* Enchytraeidae    453
  *Order* Prosopora    453
   *Family* Lumbriculidae    453
   *Family* Branchiobdellidae    453
  *Order* Opisthopora    453
   *Family* Haplotaxidae    453
   *Family* Allurididae    453
   *Family* Moniligastridae    453
   *Family* Glossoscolecidae    453
   *Family* Lumbricidae    453
   *Family* Megascolecidae    453
   *Family* Eudrilidae    453
 *Class* Hirudinea    453
  *Order* Acanthobdellida    474
  *Order* Rhynchobdellida    474
   *Family* Glossiphoniidae    474
   *Family* Piscicolidae    474
  *Order* Arhynchobdellida or Gnathobdellida    474
   *Family* Hirudidae    474
   *Family* Haemadipsidae    474
   *Family* Expobdellidae    475
 *Class* Archiannelida    475
 *Class* Echiuroidea    477
 *Class* Sipunculoidea    478
 *Class* Priapulida    480

## 13. PHYLUM ECTOPROCTA (BRYOZOA)    501—511

*Phylum* Ectoprocta    501

|   |   |
|---|---|
| *Class* Phylactolaemata | 510 |
| *Class* Gymnolaemata | 510 |
| *Order* Ctenostomata | 510 |
| *Order* Cyclostomata | 510 |
| *Order* Cheilostomata | 510 |
| *Order* Treptostomata | 510 |
| *Order* Cryptostomata | 510 |

## 14. PHYLUM PHORONIDEA — 512—517

## 15. PHYLUM BRACHIOPODA — 518—524

|   |   |
|---|---|
| *Phylum* Brachiopoda | 518 |
| *Class* Inarticulata or Ecardines | 524 |
| *Class* Articulata or Testicardines | 524 |

## 16. PHYLUM ARTHROPODA — 525—661

|   |   |
|---|---|
| *Phylum* Arthropoda | 525 |
| *Subphylum* Onychophora | 535 |
| *Subphylum* Pentastomida | 543 |
| *Subphylum* Tardigrada | 345 |
| *Subphylum* Trilobitomorpha | 546 |
| *Subphylum* Chelicerata | 548 |
| *Class* Merostomata | 549 |
| *Subclass* Xiphosura | 550 |
| *Subclass* Eurypterida | 555 |
| *Class* Arachnida | 556 |
| *Order* Scorpionida | 586 |
| *Order* Pseudoscorpionida (Chelonethi) | 587 |
| *Order* Solifugae | 587 |
| *Order* Palpigrada (Palpigradi) | 588 |
| *Order* Uropygi (Pedipalpi) | 588 |
| *Order* Amblypygi (Phrynichida) | 589 |
| *Order* Araneae (Araneida) | 590 |
| *Order* Ricinulei (Podogona) | 591 |
| *Order* Opiliones (Phalangida) | 592 |
| *Order* Acarina (Acari) | 593 |
| *Suborder* Orthognatha (Mygalomorphs) | 593 |
| *Suborder* Labidognatha (Araneomorphs) | 594 |
| *Family* Pholcidae | 595 |
| *Family* Theridiidae | 595 |
| *Family* Araneidae | 595 |
| *Family* Linyphiidae | 595 |
| *Family* Agelenidae | 595 |
| *Family* Lycosidae | 595 |
| *Family* Pisauridae | 595 |
| *Family* Thomisidae | 595 |
| *Family* Salticidae | 595 |
| *Subphylum* Mandibulata | 595 |
| *Class* Crustacea | 596 |
| *Subclass* Cephalocarida | 634 |
| *Subclass* Branchiopoda | 635 |
| *Order* Anostraca | 635 |
| *Order* Notostraca | 636 |
| *Order* Conchostraca | 636 |

| | |
|---|---|
| *Order* Cladocera | 636 |
| *Subclass* Ostracoda | 636 |
| *Order* Mydocopa | 637 |
| *Order* Cladocopa | 638 |
| *Order* Podocopa | 638 |
| *Order* Platycopa | 639 |
| *Subclass* Mystacocarida | 639 |
| *Subclass* Copepoda | 639 |
| *Subclass* Branchiura | 640 |
| *Subclass* Cirripeda | 641 |
| *Order* Thoracica | 641 |
| *Order* Acrothoracica | 642 |
| *Order* Apoda | 642 |
| *Order* Rhizocephala | 643 |
| *Order* Ascothoracica | 644 |
| *Subclass* Malacostraca | 644 |
| *Series* Leptostraca | 645 |
| *Order* Phyllocarida | 645 |
| *Order* Nebaliacea | 645 |
| *Series* Eumalastraca | 645 |
| *Series* Syncarida | 645 |
| *Order* Palaeocaridacea | 646 |
| *Order* Anaspidacea | 646 |
| *Order* Bathynellacea | 646 |
| *Superorder* Hoplocarida | 646 |
| *Order* Stomatopoda | 646 |
| *Superorder* Peracarida | 647 |
| *Order* Thermosbaenacea | 647 |
| *Order* Spelaeogriphacea | 648 |
| *Order* Mysidacea | 648 |
| *Order* Cumacea | 648 |
| *Order* Tanaidacea | 649 |
| *Order* Isopoda | 650 |
| *Order* Amphipoda | 651 |
| *Superorder* Eucarida | 651 |
| *Order* Euphausiacea | 651 |
| *Order* Decapoda | 651 |
| *Class* Diplopoda—Millipedes | 652 |
| *Order* Pselaphognatha | 654 |
| *Order* Chilongnatha | 654 |
| *Class* Chilopoda—Centipedes | 655 |
| *Order* Pleurostigma | 658 |
| *Order* Notostigma | 658 |
| *Class* Symphyla | 659 |
| *Class* Pauropoda | 660 |

## 17. CLASS INSECTA 662—761

| | |
|---|---|
| *Class* Insecta | 662 |
| *Subclass* Oligoentomata | 723 |
| *Subclass* Myrientomata | 724 |
| *Subclass* Apterygota | 724 |
| *Order* Thysanura | 724 |
| *Order* Aptera (Diplura) | 724 |
| *Subclass* Pterygota | 725 |
| *Division* Exopterygota (Hemimetabola) | 725 |

Contents    xvii

    *Order* Orthoptera    725
      *Family* Acrididae    726
      *Family* Tettigonidae    726
      *Family* Gryllidae    726
    *Order* Grylloblattodea    726
    *Order* Phasmida    727
    *Order* Dictyoptera    727
      *Suborder* Blattaria    728
      *Suborder* Mantodea    730
    *Order* Dermaptera    728
    *Order* Plecoptera    728
    *Order* Isoptera    728
    *Order* Zoraptera    729
    *Order* Embioptera    730
    *Order* Psocoptera    730
      *Suborder* Psocida    730
    *Order* Mallophaga    730
    *Order* Siphunculata    731
    *Order* Ephemeroptera    732
    *Order* Odonata    732
      *Suborder* Anisoptera    733
      *Suborder* Zygoptera    733
    *Order* Thysanoptera    733
    *Order* Hemiptera    733
      *Suborder* Heteroptera    733
      *Suborder* Homoptera    735
  *Division* Endopterygota    737
    *Order* Neuroptera    737
      *Suborder* Megaloptera    737
      *Suborder* Planipennia    737
    *Order* Mecoptera    737
    *Order* Trichoptera    738
    *Order* Lepidoptera    739
      *Suborder* Zeugloptera    739
      *Suborder* Monotrysia    740
      *Suborder* Ditrysia    740
    *Order* Coleoptera    741
      *Suborder* Adephaga    741
      *Suborder* Archostemmata    742
    *Order* Strepsiptera    743
    *Order* Hymenoptera    743
      *Suborder* Symphyta    744
      *Suborder* Apocrita    744
    *Order* Diptera    744
      *Suborder* Nematocera    745
      *Suborder* Brachycera    745
      *Suborder* Cyclorrhapha    745
    *Order* Aphaniptera    745

## 18. PARASITISM    762—768

## 19. PHYLUM MOLLUSCA    769—857

  *Phylum* Mollusca    769
    *Class* Monoplacophora    778
    *Class* Amphineura    779

# INTRODUCTION

The invertebrates are animals without vertebral column or backbone. Though they include 95 per cent of the entire animal kingdom yet the animals with backbones or vertebral columns (including man and his allies), comprising only 5 per cent of population, seem to dominate the world. However, the life of invertebrates is as fascinating, revealing and complicated a subject as that of vertebrates. Without a careful and thorough study of invertebrates it is hardly possible to peep into the secrets of life of animals on the whole.

Arising from primitive flagellate animals the invertebrates have evolved into a bewildering variety of forms of ever-increasing complexity of structure. The number and kinds of animals are diverse. Some are represented by enormous numbers and others are extremely scarce. They occur all over the earth in every conceivable environment yet they are fundamentally alike in structure and function. These principles of fundamental unity in structure and function have been discovered through a comprehensive study of animal life in general, which evidently leads one to realise that in spite of great variation in adaptations, with which animals meet different environments, all the animals have originated from some common stock.

An organism, whether one-celled or many-celled, may exist as a separate individual or as one of an aggregate or group, more or less, dependent upon each other. Such a group is called a colony. It should be noted that a colony of unicellular animals is comparable to a colony of multicellular ones, and not to a single multicellular organism. It may be true that in some colonies, however, the subservience of individuals to the needs of the whole colony is so pronounced that the whole colony may reasonably be considered one organism.

The evolutionary history of the invertebrates is incomplete. Many soft-bodied forms have left no fossil records. There are only eight phyla that have many representatives with mineralized skeletons. These are Sarcodina, Porifera, Coelenterata, Bryozoa, Brachiopoda, Mollusca, Arthropoda and Echinodermata. These have an excellent fossil record, and include the dominant invertebrates in the sea today.

Most of the soft-bodied invertebrate phyla, which play minor roles in modern oceans, probably were also of relatively minor importance in ancient seas. Only one soft-bodied phylum, the Annelida, is a significant contributor to modern marine invertebrate faunas. The seas have probably been dominated by the eight phyla with mineralized skeletons and the Annelida since Early Palaeozoic time.

Unfortunately, there is no fossil record of the origin of these phyla, for they were already clearly separate and distinct when they first appeared as fossils. Apparently, the common ancestors from which they arose either lacked preservable hard parts or were so rare and local in occurrence that they have not been preserved in known Cambrian or Late Precambrian rocks. Because there are no intermediate fossils between phyla, speculations about their evolutionary relations must depend on indirect evidence, particularly on comparative studies of the anatomy, embryology, and biochemistry of present-day representatives. Such studies have led to conflicting interpretations, but there is reasonable agreement that the Bryozoa and Brachiopoda are more closely related to each other than to the other phyla. Likewise, the Mollusca, Arthropoda, and Annelida appear to be related as do the Echinodermata and Chordata. The Sarcodina, Porifera, and Coelenterata do not appear to be closely related to any of the other phyla.

Within the major phyla of invertebrate animals, the Cambrian Period was the time of experimentation, and the Early Ordovician Period the time of secondary radiation and modernization. By late Ordovician time most of the invertebrate classes, that dominate the seas today were well established. Since that time, changes in invertebrates life have been mostly smaller-scale evolutionary radiations and extinctions. Throughout the entire history of life relatively few classes of organisms have ever become extinct. Most exceptions to this generalization are invertebrate classes that originated in Early Cambrian time, never became very common, and were extinct by the close of the Palaeozoic Era.

A closely related and equally neglected problem concerns the adaptations and life habits of the extinct Cambrian classes and the changes in habits that accompanied the rise of the modern classes. Similarities in shell form between Early Palaeozoic and present-day representatives of the dominant modern classes make it reasonably certain that their adaptations have changed very little since they first arose. Sponges and bryozoans have probably always been principally epifaunal filterfeeders, cephalopods have been swimming carnivores, and so on. For this reason it is often possible to make detailed reconstructions of the life habits of fossil marine communities from Ordovician time to the present. This situation changes sharply in the Cambrian record where so many classes lack living representatives.

A fundamental characteristic of living systems is that they carry on a continuous exchange of energy and materials with their environment, *i.e.*, they are involved in exchanges that are the driving force of the complex systems of chemical reactions called **metabolism**. One result of their metabolic activity is that they

## Introduction

are able to build up some of the products of metabolism into the substance of their bodies, thereby providing for growth (**constructive metabolism** or **anabolism**) and the replacement of worn-out material. The other phase of metabolism concerns with breaking down or **destructive metabolism** or **katabolism**. Constructive phase overbalances the destructive phase during periods of active growth and repair. On the contrary, during old age and disease, destructive phase predominates.

**Assimilation** is an anabolic activity by which materials derived from digestion are incorporated into the living protoplasm. **Respiration** is a destructive chemical process by which food is burned releasing energy. **Excretion** is the separation from the living protoplasm of the waste products of assimilation and other chemical activities. Indeed, no part of a living body escapes the consequences of this continuous flux. Studies with radioactive tracers have shown that even the molecules of apparently permanent, inert material, such as supporting skeletal structures, are steadily replaced by corresponding molecules taken into the body from outside. A further result of metabolic activity is the capacity for irritability and for adaptive response to stimulation. This controls various activities of life leading ultimately to the **reproduction** of the individual and hence the perpetuation of its species.

Reproduction depends upon the capacity of living systems for making copies of themselves the process that is called replication. The perpetuation of the species, however, depends in the long run upon occasional imperfections in the replication, and as a result of these the copy may differ from the parental form in certain respects. These differences, called mutations are likely either to aid or to impede the adjustments of a particular organism to its environment. But the resources of the environment are not limitless, so that the maintenance and growth of organisms involves competition between them for limited supplies of materials. Organisms, by their own activities, tend to extend the range of their distribution and thus to exploit their environment to the limits of their capacities. This tendency has probably been of immense evolutionary importance. Populations, which develop mutations that aid such extension, will probably be more successful in this competitive exploitation. They will tend to survive and reproduce at the expense of other populations, a consequence that is the basis of the process called **natural selection**. It is apparent that the relations between living material and its environment are continuously moulded, with the resulting production of organisms that are ever more complex and ever more efficient in the exploitation of the environment. This is evolution, a continuous sequence of change leading from the simplest forms of life to the most complex.

Nature proceeds little by little from things lifeless to animal life in such a way that it is impossible to determine the exact line of demarcation, nor on which side thereof an intermediate form should lie. Thus, next after lifeless things in the upward scale comes the plant. Although one plant may differ from another as to its amount of apparent vitality, it is devoid of life as compared with an animal. Seen from the point of view of an evolutionist among plants there exists a continuous scale of ascent towards the animal. In regard to sensibility, some animals give no indication whatsoever of it, whilst others indicate it but indistinctly. Further, the substance of some of these intermediate creatures is flesh-like, as is the case with the so-called *Tethya* (ascidians) and *Acalephae* (sea-anemones) but the sponge is in every respect like a vegetable. And so throughout the entire animal scale there is a graduated differentiation in amount of vitality and in capacity for motion.

We generally speak of "lower" and "higher" animals. What it means is not easy to explain. Animals must be organized so as to function and behave in a manner best calculated to ensure survival and reproduction. From this point of view some environments are more "difficult" than others. For instance the littoral zone of the sea, particularly that part of it below the tidemarks, is easier to occupy than either dry land or the air. Exploitation of the more difficult environments has required the development of new devices that are not needed by animals living in the easier habitats. Examples of such devices include waterproofing for land, and wings for the air. From this point of view animals living in more difficult environments may be regarded as higher animals. The possession by them of new and specialized devices is an objective criterion by which their rightful place on the ladder of life may be defined. However, this analysis is not sufficient. For it is known that many truly lower organisms may inhabit very difficult environment. For example, life within the alimentary canal of another animal presents many problems, but intestinal parasites reside in that environment only, yet this seems an inadequate reason for calling them higher animals. The important consideration here is that the possibilities of life on this planet may be exploited in many ways. One species may survive because it possesses a narrow and inflexible range of responses, allowing it to sample only a small fraction of the potential resources by which it is surrounded. Such an animal is *Peripatus*, which has reacted to the danger of desiccation on land by restricting itself to damp and concealed niches. Other animals may exploit their environment much more fully, perhaps because they possess devices that enable them to resist a wider range of stresses, or perhaps because they can sample a wider choice of food. These animals may be regarded as higher than those that lead more restricted lives.

The biological significance of more complex organization can be explained by taking an objective example of man. The dominant status of man in the world is too well-known. It can be justified by his ability to manipulate his environment to his own purpose. It can be justified equally by the flexibility of his behaviour, and by the unique capacity of his nervous system, which results, among many other things, in making him the only animal that can scrutinize the rest of the animal kingdom in sufficient depth to be able to write books on it.

Two other concepts may conveniently be mentioned here, since they are closely associated with this matter of status. While comparing animals they are referred to either "primitive" or "specialized." By specialized it is meant that they possess characters that tend to debar them from further evolutionary change. Primitive groups or primitive animals, by contrast, possess many characters that are theoretically capable of further change. For example, the nerve-net is described as a primitive type of nervous system, because it is believed to be the forerunner of the polarized and centralized type of nervous system of higher animals.

Finally, it is a common practice to speak of animals or groups as being "successful" or "unsuccessful." These terms, like "higher" and "lower" are relative and can only be usefully employed if certain objective standard is developed.

# Introduction

Since life is always a struggle, and the environment fundamentally hostile to its maintenance, it is fair to say that any group of animals that has survived at all is a successful one. But it can be reasonably asserted that the more successful ones are those that have not merely survived, but have made the fullest use of the potentialities of the environment. In this sense the successful animals are the ones that have just been defined as higher ones. But this is not all, for at any particular level of evolution there will be some groups that may be judged more successful than others. A useful objective criterion here is to consider relative abundance. Groups that have exploited most successfully a particular level of organization will tend to be more abundant than the less successful ones. This abundance will be reflected in the number of individuals representing the group at any given moment, and in the gross mass of their material.

## FORM OF ANIMALS

**Symmetry**. Some animals are asymmetrical, but others may possess spherical, radial, biradial or bilateral symmetry to a greater or lesser degree. This refers mainly to the external form. Many internal organ-system of an externally symmetrical animal may be asymmetrical. **Spherical symmetry** is the symmetry of a ball. Any section through the centre divides the organism into symmetrical halves, e.g., certain marine **rhizopods**. **Radial symmetry** is the symmetry of a wheel as found in jelly-fish and other coelenterates. **Biradial symmetry** is the symmetry of an oval. Only two sections, vertical ones at right angles to each other, divide the organism into symmetrical halves, there is only one axis —the oral-aboral axis—whose poles differ from each other. Ctenophores like sea-walnut or combjelly furnish examples of the type. **Bilateral symmetry** is the symmetry of plank in which the two ends differ. Only one section, a vertical one in the longitudinal axis, divides the organism into symmetrical halves, there are two, dorsoventral and anteroposterior, axes whose poles differ from each other. Such a symmetry is exhibited by animals from earthworm to man.

**Metamerism**. Several groups of animals are characterized by, more or less, complete segmentation of the body, in other words, there is a linear repetition of body parts. This condition is known as **metamerism**, each segment being a **somite** or **metamere**. The segments may be essentially alike (**homonomous**, e.g., in earthworm) or dissimilar (**heteronomous**, e.g., in insect). Segmentation may be both external and internal (earthworm) or only external (arthropods).

**Polymorphism**. Social animals in a colony may be specialized for different work, such as reproduction, protection and work, as in the colony of ants or termites. The condition in which different individuals of the same species are specialized for different functions is **polymorphism**, having different forms. Even in non-colonial animals polymorphism occurs e.g., within some species of butterflies there co-exist individuals which mimic other species and other individuals which do not. **Dimorphism** of two sexes is exactly analogous.

### Reproduction

Animals reproduce either by **asexual** or **sexual** means. In asexual reproduction the unicellular animals simply divide into two equal offsprings by a simple process called **fission**. In some cases one individual may give rise to numerous offsprings either by dividing into many spores or by producing numerous **buds**, simple outgrowths that grow into complete individuals. Many-celled animals also reproduce asexually generally by forming buds. In others individuals may develop from the fragments of a parent that has broken into pieces; they may develop resistant structures formed in an animal body before the onset of unfavourable conditions, **gemmules** of freshwater sponges or statoblasts of freshwater **Bryozoa**.

In sexual reproduction two mature reproductive cells, **gametes**, fuse to form a single cell, the **zygote**. The gametes develop in parents of different sex, male and female. The male gamete is a **sperm cell** or **spermatozoan**, and the female gamete is an **egg cell** or **ovum**. Each gamete develops by a process called **maturation** or **gametogenesis** in an organ called the gonad. The male gonad is **testis** and its maturation is **spermatogenesis**, while the female gonad is **ovary** and its maturation is **oogenesis**. The process of fusion between the two is called **fertilization**, the end product of which is **zygote** that matures through a process called **embryology (embryogeny)**. Usually in higher forms male and female sexes are separate (**dioecious**), but in some cases the male and female organs are contained in one animal which is then said to be **monoecious** or **hermaphroditic**. Analogies of sexual reproduction are found in certain reproductive processes of **Protozoa** as for instance, **conjugation**.

**Variation in reproduction.** Alternation of generations or metagenesis is a method quite characteristic of plants. In this, an alternation of a sexually reproducing generation with an asexually reproducing generation takes place. Metagenesis occurs in some parasitic flatworms.

**Parthenogenesis** is reproduction through the development of unfertilized egg. In some insects and rotifers it is a constant and normal method of reproduction, or it may alternate with sexual reproduction. Scientists have succeeded in bringing about **artificial parthenogenesis** by chemical or physical methods in many species of animals.

**Paedogenesis** is reproduction by larval or immature forms. In a few instances larvae or pupae are capable or parthenogenetic reproduction. Examples of this type are found in insects like *Miastor*, liverfluke and axolotl or tiger salamander (vertebrates).

**Polyembryony** is the production of two or more embryos from a single egg by gemmation. This process is only known in a few groups of animals, it occurs in parasitic hymenoptera.

**Egg-cell and cleavage.** The nucleus of egg-cell is usually eccentric being on the side which gives off polar bodies. This is the **animal pole**. The cytoplasm is concentrated towards this side and the yolk nearer the opposite pole, the **vegetative pole**. Egg with small quantity of evenly distributed yolk is **homolecithal**. If this is large in amount and concentrated toward one pole the egg is **telolecithal**. In insect eggs the nucleus, with a little cytoplasm, lies in the yolk which is central, the whole being surrounded by a layer of cytoplasm. Such an egg is **centrolecithal**. The **cleavage** (cell division) pattern of the egg depends on the quantity of yolk. In the homolecithal egg the cleavage may be **equal** and **total** or **entire**, the cells formed being completely divided and of about equal size. In the second case (telolecithal) the cleavage may be confined to a disc-like region at the animal pole and is called **discoidal**. In the centrolecithal eggs the cleavage is **superficial**, that is, it is confined to the surface extending all over the egg. Entire cleavage is termed **holoblastic**, whereas, the discoidal, or superficial cleavage is **meroblastic**. In some the fate of the cells can be determined at a very early stage and the germ cell is segregated from the beginning. Such a development is of **determinate** type.

## Embryology

Ordinarily embryology is considered the process of development from the fertilized egg to a stage approximating the adult condition, *e.g.*, to hatching in birds or birth in mammals Fertilization awakens the egg into a wonderful state of activity. Carrying with it the combined hereditary factors present in the conjoined nuclei it starts to divide actively and with great rapidity into daughter cells, each of which in turn repeats the process and hands on the stimulus to its descendants. Consequently, a cluster of cells resembling a mulberry is formed. This is the **morula**. Continued multiplication or cleavage results in the formation of a hollow ball of cells the **blastula**, its cavity being **blastocoel** or **segmentation cavity**.

More rapid division of cells at animal than at vegetative pole results in an inpushing or **invagination** of cells near the vegetative pole and consequently the blastocoel is reduced and a double-layered ball of cells is formed. The cavity of this is the **gastrocoel** or **archenteron** that opens to the outside by **blastopore**. The two surrounding layers are **ectoderm** (outer) and **endoderm** (inner). In some animals invagination does not occur but the endoderm is formed by **delamination**, an internal layering off from the cell of the blastula. In others the actively dividing **micromeres** encroach upon the megameres to form layered condition. This process is **epiboly**. The third germ layer of **mesoderm** appears between ecotderm and endoderm where some cells proliferate to form a layer of cells, occupying the segmentation cavity.

## Geological Time Scale

The solid crust of the earth is about fifty miles thick. Whether in its earlier stages this was smooth or rough is not known, but at one period the earth was an unfurnished world devoid of stratified rocks, water, plants and animals. The materials now composing this crust are rocks of mineral matters of various kinds, as granite, coal sandstone, chalk and clay. Some are hard, others soft, but geologically all are rocks. Some, such as sandstones, and lime-stones always occur in parallel layers termed strata which may lie flat, may slope or be bent into arch-like folds. These stratified rocks contain fossils, and thus, are important for our purpose. William Smith (1799) first noticed that the series of animal and plant fossils succeed one another in regular order.

The geological time has been divided into six **eras**, the end of each marking the time of some significant geological change. Each of the eras is divided into shorter **periods** or **epochs**. The first era or **azoic** is the one with no evidence of living organisms. It marks the origin of the earth from the sun and lasted for over 1,000,000,000 years (an estimate based on several lines of evidence). Primitive life probably appeared in the next era called **archeozoic**, but left no fossil record as the organisms were simple one-celled plants and animals. The third era is known as the **proterozoic**, in which also fossils are rare and poorly preserved, but it is presumed that most of the invertebrate phyla must have evolved during this period. This presumption is further strengthened by the fact that in the periods following this (*i.e.*, in **Palaeozoic**) era there is an abundance of fossils of nearly all invertebrate phyla. Trilobites and brachiopods are numerous. This period is known as **Cambrian**, the first epoch of the palaeozoic era. In the Palaeozoic the Cambrian is followed by **Ordovician** having peak of invertebrate dominance. **Silurian** comes next and is known for first great coral reefs, brachiopods at peak, trilobites begin to decline and first land invertebrates begin to appear. **Devonian** is the next period of Palaeozoic followed by carboniferous and lastly Permian. Devonian shows decline of trilobites, abundance of molluscs and first crabs and land snails. Carboniferous is known for insects and decline of brachiopods and Permian presents modern insects, last trilobites and eurypterids. The **Mesozoic** era includes three periods, **Triassic** showing decline in number and importance of marine invertebrates; **Jurassic** in which modern crustaceans appear and ammonoids are abundant, and **Cretaceous** known for extinction of ammonoids. Lastly comes the **Caenozoic** era of recent life. It includes two periods, **Tertiary** noted for the appearance of modern invertebrate types and **Quarternary** with arthropods and molluscs most abundant and all other phyla are well represented.

## Ecological Distribution

Obvious major divisions of environments are **salt-water, fresh-water** and **land**. Salt waters include **oceans, seas** and **bays** and cover approximately 71 per cent of the earth providing extensive and stable habitats. Some marine animals are **planktons**, *i.e.*, those that float on the surface and are moved passively by winds, waves or currents and are naturally of small to microscopic size. Others are **nektons**, *i.e.*, they swim freely by their own efforts. Both these types are referred to as **pelagic**. Strict bottom dwellers are known as **benthos**. Some animals may be found between the tide lines and are termed **littoral** or **shore forms**; others are found below tide lines to depths of about 100 fathoms (1825 metres) and are called **neritic**; those that live at depths of 1825 to 18,250 metres are called **bathyal** and **abyssal** is the name given to those found below, 18,250 metres. Fresh waters are scattered, of lesser depths and volume and have more variable temperature. Fresh waters may be running water such as rivers or standing waters such as lakes.

Land is the third habitat of animals. The life on land is **terrestrial**. The land presents interaction on many physical, climatic and biological factors which produce a variety of ecological conditions for the inhabitants on islands or continents. Forests including evergreen coniferous forests, broad-leaved deciduous forests, tropical rain forests, tundras, grasslands (prairies), sagebrush, deserts, mountains, etc., are various land habitats which influence the life of the animals in various ways.

## CLASSIFICATION OF ANIMALS

For a proper study of the enormous number of animals inhabiting the world some means of grouping them scientifically is necessary. In fact biological science made little progress until it was realized that animals must be grouped accord-

*Introduction*

ing to their fundamental similarities in structure. The branch of zoology that devotes itself to such a study is called **Classification** or **Taxonomy** or **Systematics**.

The scheme of classification begins with the conception of the **species** (plural also species) which may be defined as natural population of organism which has a heredity distinct from that of any other group, and the members of which breed only with one another to produce fertile offsprings, moreover, the species usually has a geographic range separate from related groups. Two or more species having certain characters in common are grouped together as a **genus** (plural **genera**). Each species is then named by a combination of a generic and a specific name such as *Ameoba proteus* and *Amoeba dubia*, etc., *Amoeba* being generic name and *proteus* and *dubia* specific names of two different species. Now no other kind of protozoa, or indeed no other animal, has these two names applied to it, so that they distinguish it from all other animals. This system of giving two names to an individual is called the **Binominal system** and owes its inception to **Linnaeus**.

Genera having certain common characters are united into a family, similar families with certain degree of resemblance are grouped into an order, orders into a class, and classes into **phyla** (singular **phylum**), or **divisions**. Thus, the taxonomic divisions for a natural grouping are species, genus, family, order, class and phylum, but often many other divisions are employed usually formed by prefixing "**sub**" for a lower division and "**super**" for a higher division. Biological classification reflects degrees of evolutionary relationship, since the more closely the organisms are related by descent, the more features they have in common. The system of classification is always subject to modification as more and more is learnt about organism, and about their fossil history. Up to now the criteria used for classification of animals were entirely structural, but now other criteria are springing up as more biochemical and physiological differences are being discovered.

# 1 PHYLUM PROTOZOA

## INTRODUCTION

The **Protozoa**[1] are the "first animals" (Greek, *protos*—first+*zoon*—animal) because as far as known they comprise the first animals that have appeared in evolution. The Protozoa were simple animals, but now they comprise considerably diversified organisms, and all multicellular animals are believed to have originated from them. They are often defined as **unicellular,** but this name appears to convey a wrong sense because the Protozoa are not loose cells moving about. They are complete organisms that may be more complicated in construction than the simplest Metazoa. It is preferable, therefore, to call the Protozoa as **non-cellular** or **acellular** animals rather than unicellular. The group comprises four quite distinct classes of organisms, three of which, the **Ciliophora, Sporozoa** and **Cnidospora,** consist of animals only, while the **Sarcomastigophora** contain both animals and plants. Many taxonomists feel that the organisation of the members of these four classes is sufficiently different so much so that each of them can be given the status of phylum. However in the majority of the schemes of classification they appear as sub-phyla of the phylum Protozoa.

[1] The Protozoa are "so small in size that a dozen may inhabit one red blood cell (e g., *Babesia*—a Syorzoan) or several hundred may inhabit a single tissue cell (e.g., *Leishmania*—a flagellate). For this reason they are measured in microns or µ (1µ=1/1000 mm or 1/25000 of an inch). A millimicron i.e., one thousandth of a micron is written µm, and is 10 Angstroms (the physicists' unit of small distance)".

## A. FEATURES OF THE FOUR SUB-PHYLA

The simplest group is the **Phytomastigophorea** (plant, flagellates). This group includes the photosynthetic flagellates and some colourless flagellate Protozoa which do not have features indicating a close relationship with any algal group. They are placed in the Zoomastigophorea (animal flagellates). This group may thus be polyphyletic assemblage. Several of these groups of animal flagellates include specialised symbionts, many of them parasites in animals, and in some groups the organisms carry large numbers of flagella, while the plant flagellates commonly have two and seldom more than four flagella.

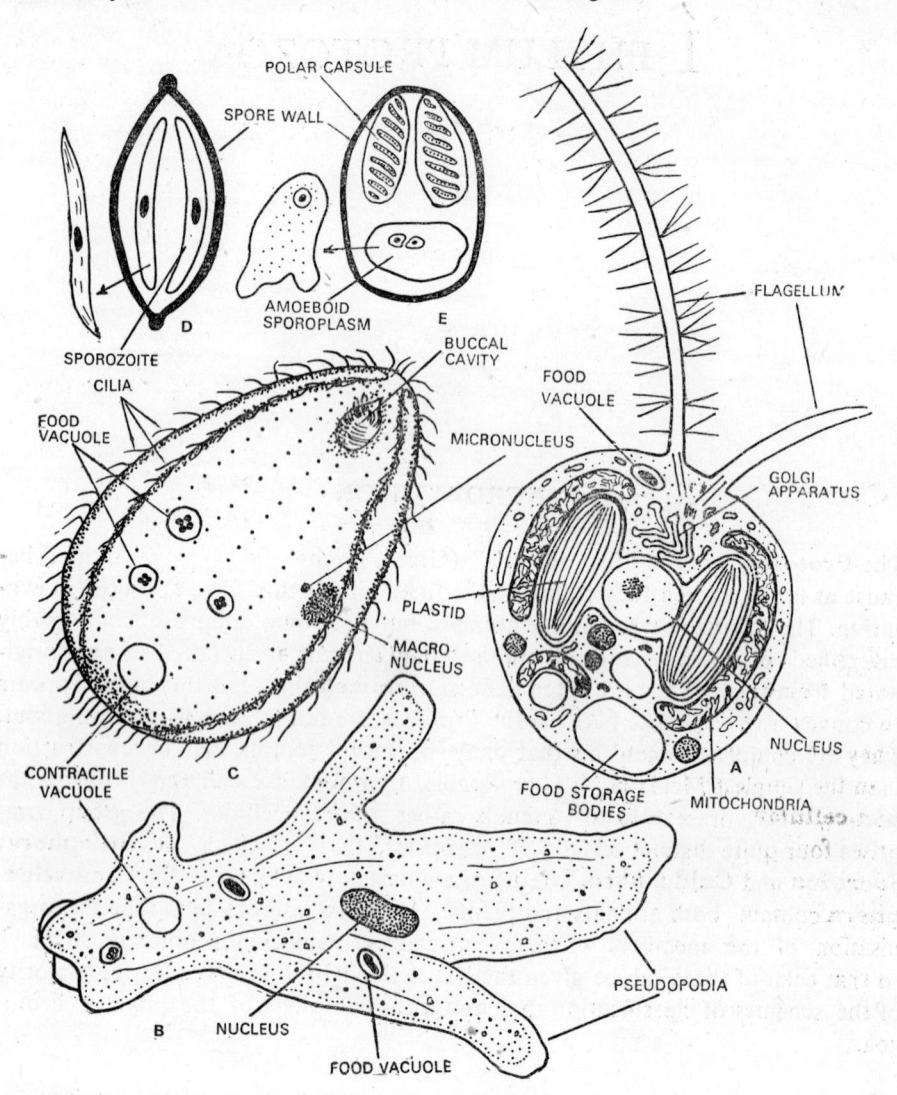

FIG. 1.1. Different types of acellular animals (Protozoa). A—Flagellate, *Ochromonas*; B—*Amoeba*; C—Ciliate; D—Sporozoan; and E—Cnidosporan.

Figure 1.1A gives the structure of a simple chrysomonad flagellate. The cell possesses the basic components of a plant flagellate, including flagella, nucleus, plastid, food storage bodies, mitochondria, vacuoles, vesicles and other membranous and fibrous inclusions, the cell surface is not structurally specialised. Various structures tend to occur in characteristic positions within the cell of a particular species of flagellate.

The amoeboid Protozoa (Sarcodina) show both flagella (or non-gametic cells) and amoeboid pseudopodia, together or at different phases of the life cycle. Therefore, they are now placed in the sub-phylum Sarcomastigophora close to flagellate organisms. It seems that several parallel lines of evolution have led from pigmented flagellates to the various groups of amoeboid forms in the Chrysophyceae, but generally amoebae are phagotrophs and clearly animals. The primary characteristic feature of sarcodine protozoa is the possession of some form of **pseudopodium,** but there is considerable diversity of structure, notably in the character of test, shell or skeletal material when present, and in the type of pseudopodium, which may be broadly lobed, needle-like or reticulate. Figure 1.1B shows how little structural organisation an amoeba (*Amoeba proteus*) may have in comparison with a flagellate or ciliate protozoan. The cytoplasm is differentiated into the outer ectoplasm and the inner endoplasm which flows forward through the body into the cylindrical pseudopodia. The nucleus, food storage granules, mitochondria, vacuoles and other membranous or crystalline inclusions lie within the endoplasm.

The Sporozoa are a group of specialised parasitic Protozoa with complex life cycles, initially classed together because they showed a common mode of life and were different from other types of single-celled organisms. A diversity in certain features led to the suggestion that the group is polyphyletic in origin. But recent studies of fine structure have revealed the presence of certain unique organelles in members of several groups, indicating that the Sporozoa are really phylogenetically compact. They are haploid organisms with a zygotic meiosis, a character that they share only with certain flagellates. Many Sporozoa have flagellated gametes.

Members of the Cnidospora have distinctive spores whose infective contents are amoeboid sporoplasms, not regular-shaped sporozoites as in Sporozoa (Fig. 1.1D). The elaborate structure of Cnidospora is, in many forms, produced by differentiation of several cells derived from a multinucleate plasmodium. They are clearly distinct from the Sporozoa, and some authorities doubt their inclusion under Protozoa because of their multicellular spores. All members of the sub-phylum are parasitic, occurring in cold-blooded vertebrates and in invertebrates. It has been suggested that they have been derived from as diverse Protozoa as amoebae, dinoflagellates and even several metozoan groups, but their structure is so specialised that it has not revealed any likely relationships. The polar capsules of the Cnidospora are reminiscent of nematocysts, and it will be interesting if they are found to show ultrastructural details comparable with the nematocysts of coelenterates or dinoflagellates.

The ciliates (Ciliophora) clearly form a natural group isolated from other Protozoa by a number of specialised features, such as the pellicular organisation based primitively on rows of cilia, the unique nuclear dimorphism and the sexual process of conjugation. It is likely that the ciliates have evolved from a flagellate stock by multiplication of the locomotor structure, and that the nuclear dimorphism occurred later since some simpler ciliates have only one unspecialised nucleus. Ciliates are diploid with gametic meiosis, a pattern shared among Protozoa with opalinids, heliozoans and some zooflagellates.

Figure 1.1C shows some common features of ciliates. The body surface is covered with cilia which are mostly aligned in rows called **kineties**. This particular kinety pattern is interrupted in the region of the **cell mouth** (**cytostome**) where there may be specialised compound cilia which are used in feeding; the depression in which these compound cilia lie, and which leads into the cytostome, is called the **buccal cavity**. The pellicle is a specialised superficial zone of cytoplasm including the ciliary bases and attached fibres ('infraciliature') and often membranous structures, some mitochondria and other organelles. The cytoplasm in the interior is more fluid and mobile; it contains the macronucleus, micronucleus, food vacuoles, mitochondria, and many other inclusions. Contractile vacuoles are often closely associated with pellicular structure, although they protrude into the interior cytoplasm.

In the light of the newer trends the opalinid Protozoa are not classified with Ciliophora but with Sarcomastigophora. They possess rows of cilia—a character of ciliates. But they appear to resemble flagellates in nuclear and reproductive features. Ecologically they form a very restricted group with no close relations with any other protozoan group. It is possible that they may have evolved in parallel with the ciliates from a common flagellate ancestor.

## B. RELATIONSHIPS OF PROTOZOAN GROUPS

It is believed that the first cellular organisms were without membrane-enclosed organelles in the cytoplasm, and, therefore, did not have true nuclei, mitochondria and plastids. Such organisms are called **procaryotic**. Organisms with true nuclei, mitohcondria and plastids are known as **caryotic** organisms. Moreover, the flagella found on bacteria (procaryotes) are simpler fibrous organelles than the complex flagella built of a bundle of 9+2 longitudinal fibrils, which are found on **eucaryotic** cells. Flagellar organelles of both types are locomotory structures, but it is unfortunate that organelles of totally different construction, associated with organisms at two different evolutionary levels, should be called by the same name. Bacteria and blue green algae are included in the **Cyanophyta**, the photosynthetic pigment of which is chlorophyll $a$.

The red algae (**Rhodophyta**) possess genuine plastids in which at least two unit membranes enclose the chlorophyll-bearing lamellae with single separate **thylacoids**. These are **eucaryotes**, but lack the 9+2 flagella found in all the other major groups of eucaryotes. This and the form of their plastids suggest that the Rhodophyta is the most ancient eucaryote group.

The remaining algae (**Contophora**) have 9+2 flagella, at least at some stage in members of each group, and have thylakoids adhering in groups within the plastids. A major distinction between two types of these algae is the appearance of chlorophyll $b$ as one of the photosynthetic pigments in a few groups, collectively called the Chlorophyta, while the pigments of the Chromophyta do not include chlorophyll $b$.

Included among the Chromophyta are several groups, notably the diatoms (**Bacillariophyceae**) and the brown algae (**Phaeophyceae**), which do not have a dominant flagellate stage and which are never included in the Protozoa.

Most of the other groups are frequently classified with the Protozoa as well as with the Algae, and this is particularly true of those groups in which some members have lost their photosynthetic pigments and have taken up saprophytic or holozoic nutrition. Many dinoflagellates, for example, have so many animal characteristics that they are usually described with protozoans and never with Algae.

These early organisms stand at the level of the evolutionary separation between animals and plants. The evolution of animals from plants has taken place many times among these groups. Many authors, therefore, include the whole assemblage of algal groups in a single phylogenetic scheme with protozoa in order to understand the true relationships of many holozoic flagellates.

The specialised features of such protozoan groups as the Ciliophora or the Sporozoa, on one hand, and of the pigmented flagellates, on the other, make it difficult to suggest probable relationships between these protozoa and any of the algal groups. However, it seems likely that Algae of the Chromophyta, which is probably a more ancient group than the Chlorophyta, are closer to the animals. Among the Chromophytes the possible ancestors of the holozoic protozoa seem to be the Chlorophyceae. It is thought that ciliates, sporozoans and various groups of amoeboid and flagellate protozoa may have evolved independently different stages of the evolution of the chromophyte Algae. Some writers suspect that electron microscope studies of some small groups of colourless flagellates (which at present occur almost exclusively in protozoan classification) may soon show that they are more closely related to a group of chromophyte algae than to other colourless "zooflagellates"

It is widely believed that multicellular animals evolved from protozoan organisms. It is not quite clear as to which protozoan group is involved. Most workers agree that the collar flagellate protozoa (among Zoomastigophorea) have given rise to the sponges, since both groups contain flagellate cells of the same unusual type. Now sponges are not considered to be ancestor to any of the metazoan phyla. Some authorities feel that the first coelenterates have evolved from a colonial phytoflagellate, perhaps through an intermediary planuloid form. Another theory is that the first metazoan organisms were the turbellarians, and that these appeared as a result of the cellularisation of primitive, multinucleate, ciliated protozoa. Yet other possibilities had been suggested including the hypothesis of a polyphyletic origin of the lower metazoan phyla from the several groups of protozoa.

The divergence between the major groups of lower organisms is extremely ancient; there are reports of fossil flagellates from well over 2000 million years old deposits, but the remains of pre-Cambrian rocks are generally fragmentary. Most of the animal Phyla had evolved, nearly 600 million years ago, by the beginning of the Cambrian era. It is clear that evolution has been continuing at a faster or slower rate in all groups of protozoa, and it is likely that any organisms which would have provided valuable phylogenetic links may have been eliminated long ago by more successful species.

## C. PROTOZOA AS CELLS

The **Protozoa** are the simplest organisms, clearly unicellular. The ultrastructure of a protozoan shows many of the *same* characteristic feature as cells of metazoa—the *same* membrane formations and organelles made from the *same* type of components. In several protozoan groups organisms have evolved extremely elaborate organelles of such a diversity of types within the cell membrane that these cells are the most complex known. Many protozoan cells are very large. This increase in size apparently demands more nuclear material in the form of duplicate sets of chromosomes, either within a polyploid nucleus or in multiple nuclei. The formation of colonies of protozoans is common in many groups, and probably originated by incomplete separation following division of the organism. Protozoans of several groups show an approach to genuine multicellular organisation, including a division of labour of partially differentiated cells, within the most integrated types of colony.

A protozoan cell is a complex, highly integrated unit of life. A turnover of chemical materials, a store of bio-chemical information and a through-flow of energy, are required for its functioning, maintenance and growth and for other activities of the organism. Organic molecules and many vital components of the cell may be built up within autotrophic organisms from such simple molecules as carbon dioxide, water and salts containing nitrogen, sulphur, phosphorus and other elements. The synthesis of organic compound from such simple precursors occurs in association with the fixation of light energy in the plastids of photosynthetic cells. Energy derived from the breakdown of simple organic compound is also utilised for similar synthesis. Heterotrophic protozoa take in large molecules of raw materials for cell construction and as a source of energy. Their powers of synthesis of small organic molecules are limited. The small molecules required by autotrophic organisms may enter the cell by diffusion. The uptake of complex molecules requires an active process and usually involves taking the food substance into vacuoles in the cell, and the subsequent enzymic breakdown of macromolecules into smaller units which may be passed through the vacuole membrane. The cell membrane forms the zone of contact between a protozoan cell and the environment. It is a vital controlling region exerting an important influence on intracellular concentrations. Everything that enters or leaves the organism must pass through this membrane. Agents which act on the protozoan must either act at the cell membrane or pass through the membrane to an intracellular site of action.

Energy obtained from light or the breakdown of organic molecules is stored in the protozoan cells in the chemical bonds of organic compounds. Some of these compounds are changeable and carry a relatively large quantity of readily available energy. The best known example is **adenosine triphosphate** (ATP), which carries the energy in "high energy" phosphate bonds. Many other compounds which serve as reservoirs of energy are inert. Their energy is mobilised by the enzymic oxidation processes associated with respiration and is transferred to energy carriers like ATP. Aerobic respiration involves the utilisation of oxygen and is generally associated with mitochondria in the cytoplasm. Energy given up by ATP and similar carrier molecules is used in protozoan cells for synthesis and for movement. It is possible that energy is normally used in the formation of chemical bonds; this is obvious in the case of molecular synthesis, and in movement it is thought that the changes in shape or position of molecules which lead to a diversity of movements are also the result of the formation of

*Phylum Protozoa*

temporary intermolecular or intramolecular chemical bonds. Most reactions in protozoan cells involve loss of some energy and the energy released on the breakage of many bonds may not be recoverable.

Molecular synthesis within protozoa is the responsibility of DNA (**deoxyribonucleic acid**) present in the genes. Activities of DNA involve RNA (**ribonucleic acid**) and produce ribonucleic protein bodies or **ribosomes**, the majority of which occur in cytoplasm. Some proteins synthesised are structural proteins, but the most of them are enzymes responsible for the promotion of various cell functions. Molecules synthesised in protozoan cells are used within the cells but most protozoa produce extracellular products such as envelope cyst walls, extrusive organelles and mucous materials.

The character of protozoan cells depends on the types of proteins it contains. Individuals within a species may show minor variations in their total protein constitution; if the differences between the protein constitutions of two populations of a protozoan species become extensive, it is likely that the two populations will no longer interbreed and, therefore, that they become separated as two distinct species. The differences between the protein constitutions of two species will be greater or less according to the remoteness or closeness of their phylogenetic relationship.

## D. STRUCTURE OF PROTOZOA

**Cell-membrane.** The outer surface of the Protozoans, even naked protozoans like *Amoeba*, is bounded by an exceptionally thin, elastic and permeable membrane which is known as **cell membrane**. The cell membrane has a tripartite structure detectable with electron microscopy as a central light zone separating two dark ones. It is about 75 Å thick, each, zone, having a thickness within the range of 20-30 Å (an angstrom unit, Å is $10^{-7}$ mm). Such a membrane has been conveniently referred to **unit membrane**. According to older view this membrane is made up of two layers of lipid molecules. The outer surface of these is electrically charged (hence called polar surface). It is covered by a relatively solid layer of protein molecules. The electron micrographs subsequently confirmed this interpretation, the central layer corresponds well with the lipid zone and the outer layers with the protein components.

Unit membrane is an example of precisely oriented molecular structure. The unit membrane controls the amount of substance passing out or entering the cell. It permits the passage of some substances (**permeable**) and does not allow other substances to pass (**impermeable**). The lipid component accounts for the ready passage of fat-soluble substances, while the protein gives elasticity and strength, and is responsible for the observed low surface tension. The presence of pores in the plasma membrane explains to some extent why molecules of different sizes pass through with different degrees of ease. There are certain substances which may be able to pass through the membrane by use of metabolic energy. Due to this selecting permeability of the cell membrane an electrical potential is maintained across it and is used in many physiological activities.

The unit membrane separates the outer environment from the inner and is thus protective, but it cannot protect from electromagnetic or other radiations. Since the protozoans depend upon the membrane for the protection of its body surface there are necessarily various modifications and elaborations of its structure.

It may be simple surface membrane (**plasmalemma**) in amoeboid Sarcodina (Rhizopoda), although even in some amoebae it may bear a continuous covering of filamentous molecules, where the filaments are thought to be microproteins meant for adhesion to the substratum. In many other protozoans the body surface is much more complex, as for instance, the longitudinally striated pellicle of *Euglena* or the body surface of the ciliate *Paramecium* (to be considered later).

There are some membranes inside the cell. **Endoplasmic reticulum** (Fig. 1.2) is one such example. Its existence, suspected previously was later confirmed

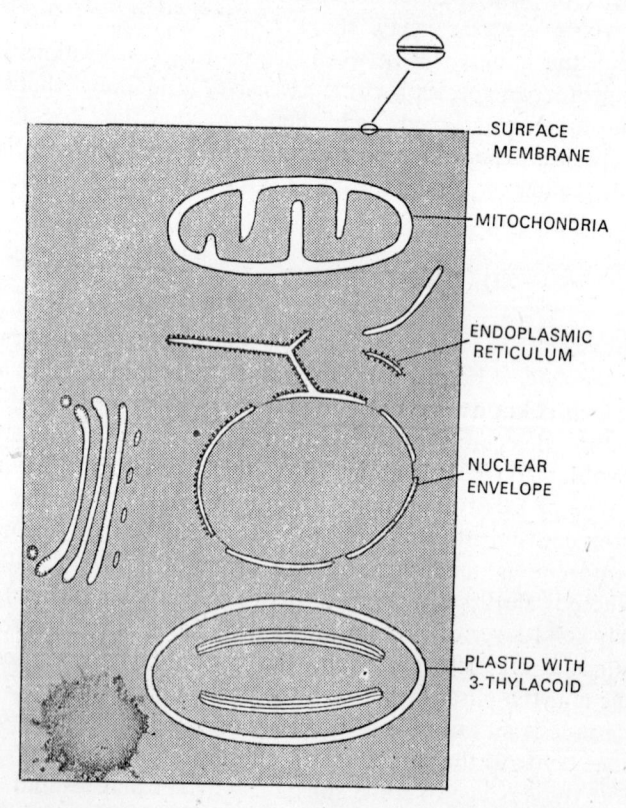

**FIG. 1.2.** Diagrams to show the arrangements of membranes in some cell organelles. Every membrane shown as a three-unit membrane as indicated at the top. Ribosome granules are shown attached to parts of endoplasmic reticulum and nuclear membrane. The membrane of the golgi body forms cisternae.

by electron microscopy. It is found in all the cells to some extent. It encloses a channel which is not in continuity with the cytoplasm. It may divide the cell in two parts, each maintained under two conditions. It is usually covered on the outside with small bodies (**ribosomes**) concerned with protein synthesis. The endoplasmic reticulum without ribosome is termed **smooth surfaced** and with ribosomes **rough surfaced.**

## Phylum Protozoa

**Nuclear envelope.** The nuclear envelope (Fig. 1.2) consists of two unit membranes. It is often connected with the endoplasmic membrane. It bears, sometimes, ribosomes on its outer surface. Following nuclear division the nuclear envelope is reformed from the elements of the endoplasmic reticulum. A space of 20 to 30 mm. separates the two unit membranes of the nuclear envelope. Large pores present in this envelope allow various materials to pass from cytoplasm into the nucleoplasm and *vice versa*.

**Golgi bodies.** The **golgi bodies** or **dictyosomes** are membranous structures associated with endoplasmic reticulum and are in the form of vesicles or cisternae. They are believed to be concerned with the elaboration or storage of products of cell synthesis. In flagellated protozoan these may be numerous while in most ciliates they may be poorly developed. Sometimes it is possible to differentiate the forming face of a dictyosome at which the cisternae are formed (*i*) by vesicles derived from the endoplasmic reticulum or nuclear envelope, and (*ii*) the opposite (mature) face at which vesicles are formed from dictyosome cisternae and emigrate to the plasma membrane. The materials formed at the ribosomes are transported to the forming face of golgi bodies and then are transported away at the mature face after the materials have been changed into the ultimate form. Formation of complex scales within the Golgi vesicles in some protozoans is an example.

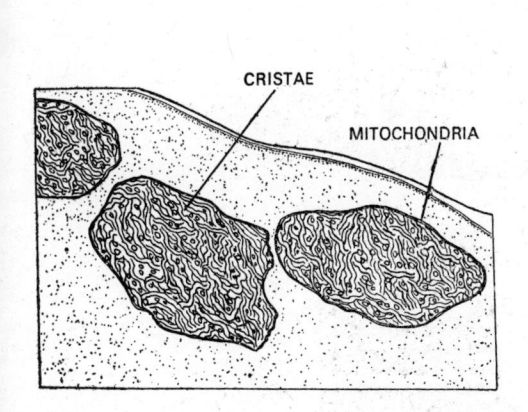

**FIG. 1.3.** Electron micrograph of section through mitochondria to show the cristae in the ciliate *Euplotes* with numerous cristae.

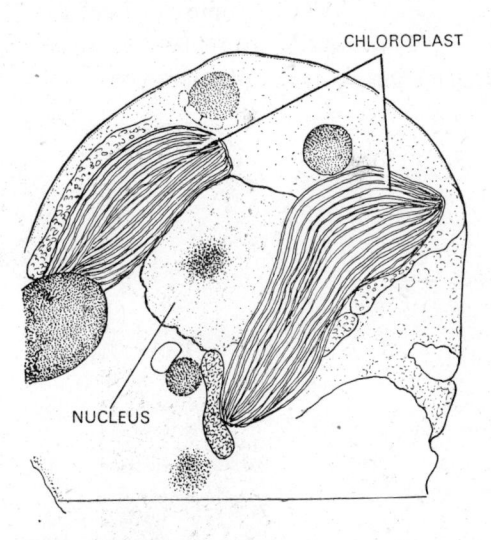

**FIG. 1.4.** Electron micrograph of a section through a flagellate of the Chrysophyceae to show the typical appearance of the chloroplast and the nucleus.

**Mitochondria.** The cytoplasm also contains **plastids** and **mitochondria** (Fig. 1.3). Of these mitochondria are one $\mu$m across to several $\mu$m long. They are made up of two membranes, the outer forms its surface while the inner is continuous with projections that exist inside. These lamellar projections may be reduced or may form tubular or vesicular cristae (Fig. 1.3). The membrane system forms two compartments, the inner membrane encloses a mitochondrial matrix compartment and the space between the two membranes and within the projecting cristae forms the second compartment. Tubular cristae occur in most protozoa but lamellar or discoidal cristae occur in flagellates. Amoebae and gregarines have vesicular cristae. In protozoa that live in the gut of vertebrates (e.g., *Balantidium*) the mitochondria are without cristae. Mitochondria contain enzymes associated with the Kreb's cycle and with oxidative phosphorylation. Mitochondria are the main sites of formation of energy rich phosphate compounds which contribute to energy requiring reactions in cells. Mitochondria contain DNA and are able to carry on independent synthesis of their own ribosomes. In some protozoans the cytoplasm has **peroxysomes** which contain oxidase enzymes which release hydrogen peroxide. These particles vary from 0.5 to 1.0 $\mu$m in diameter and are bound by a single membrane. They are found in *Acanthamoeba* and *Tetrahymena*. The primary function of these granules is the synthesis of carbohydrates from fats. In *Tetrahymena* mitochondria and peroxysomes work together.

The **plastid** (Fig 1.2) is found in autotrophic flagellates. Each plastid is bounded by two membranes which enclose numerous lamellae inside. The photosynthetic pigments are enclosed in the lamellae. In red algae the lamellae are separate, but in flagellates they are packed together. In some flagellates the plastids contain a pyrenoid, which forms the centre for the production of reserved carbohydrates. Plastids have been found to contain DNA and small ribosomes capable of some small protein synthesis.

**Contractile vacuoles.** Contractile vacuoles (Fig. 1.5) are present in most ciliates, in many flagellates and in amoebae. Many protozoa have several con-

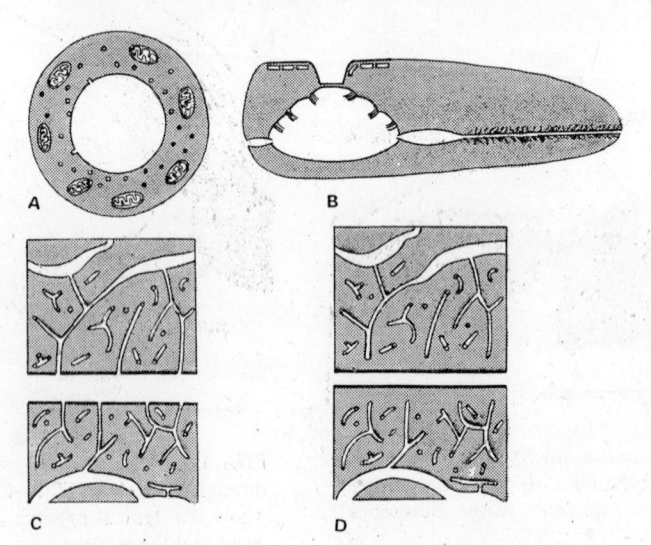

FIG. 1.5. Diagrams to show the structure of contractile vacuole. A—*Amoeba* surrounded by vesicle and Mitochondria; B—A section through contractile vacuole and parts of two supply canals of *Paramecium*; C—Connection between the fine tubules in *Paramecium* during diastole of the canals; and D—The same during systole of the canal.

tractile vacuoles, others have only one. They occur close to the plasma membrane and are seen to swell slowly **(diastole)** before suddenly collapsing

## Phylum Protozoa

(**systole**) and releasing their fluid contents to the outside medium. In some protozoa the position of the contractile vacuole is fixed. They are filled from system of canals all round them and their opening to the outside is permanent. In some amoeba the simple contractile vacuoles consist of a unit membrane enclosing the clear vacuole surrounded by a region containing many small vesicles and mitochondria. A much more complex structure has been found in the contractile vacuole of *Paramecium* where the main vacuole is fed by about six radiating canals. The primary function of contractile vacuole is regulation of water contents. There is evidence that the osmotic pressure of the cytoplasm of fresh water protozoa is well above that of the surrounding medium while the marine protozoa are likely to be slightly hypertonic to sea water, as a result water will flow by osmosis into the body of protozoa. Water will also enter by food taken in by phagocytosis and may be produced in metabolism. The rate of pulsation of vacuoles differs in different protozoa. It may contract only once or twice in an hour in some ciliates, while in some smaller fresh water ciliates it may contract every few seconds. The relative rate of removal of water appears to be greatest in the smallest forms. A small ciliate may expel a water volume equal to its body volume in a few minutes while a large ciliate may take several hours to pump out the equivalent of its body volume.

**Food vacuoles.** The food vacuoles are concerned with the intake and digestion of food, and are present in the protozoans. There are two types of food vacuoles: **phagocytic** and **pinocytotic** vacuoles. While the former enclose large food particles the latter enclose invisible food materials either in solution or absorbed on the surface membrane. It is not possible to make sharp distinction between these two types of vacuoles. In those Protozoa that have well-formed cytostomes the intake of food may only involve phagocytic vacuoles and in those forms that have no permanent cytostome pinocytosis may be the main mode of feeding. It is likely that both the types of feeding occurs. It is also possible that pinocytosis may occur at the surface of phagocytic vacuoles.

The organic contents of the food (phagocytic or pinocytotic) vacuoles are digested by the activity of enzymes contributed by **lysosomes**. Lysosomes are minute bodies (about 0.2-0.8 $\mu$m in diameter) which are believed to be formed in the Golgi region. Presumably they are the neutral red granules of earlier Protozoologists. Each lysosome has a single unit membrane containing hydrolytic enzymes. They are capable of breaking major classes of organic molecules to simple soluble molecules of the order of size of amino acids and monosaccharides. Generally, soon after formation, the food vacuoles shrink slightly and then become acid at about the time that the prey organism is killed. Later, the contents of the vacuole become alkaline and the vacuole swells again, presumably due to the presence of digestive products of the prey. After the absorption of the digested food materials into the cytoplasm, the food vacuole shrinks again to a bag containing indigestible residue which leaves the body through the unspecialised surface membrane or at a **cytopyge**.

**Fibrous structure.** The cytoplasm of Protozoa also contains fibrous structures. There are two classes of structures: (*i*) those built of **filaments** 4-10 $\mu$m thick, or (*ii*) those built of **microtubular fibrils** about 20 $\mu$m in outside diameter. Filaments are made up of globular protein molecules in a single row (Fig 1.6 A) or two or more of these molecular chains woven together. In some fibres the component filaments are loosely aggregated into bundles. Such fibres are capable of changing length. In other fibres the filaments are accurately lined up and cross-linked with regular periodicity to form the striated fibres that are often seen in electron micrographs. There are rigid fibres that are not capable of changing length. Loose bundles are found in the contractile myonemes of the ciliates *Spirostomum* and *Stentor* and also in the contractile stalk of *Vorticella*. *Paramecium*, on the other hand, possesses striated fibres.

Sometimes, single rows of globular protein molecules join together to form walls of a cylindrical fibril (Fig. 1.6 C), the **microtubular fibril**. It is evident from the studies of several fibrils of about 20 $\mu$m that each is made up of 12-13 rows of molecules. The number of rows may vary in larger or smaller microtubular fibrils. Such microtubular fibrils form the basis of a number of fibrous structures in Protozoa of all types. They form the longitudinal elements of flagella and cilia and occur in other situations which are mentioned elsewhere in this book.

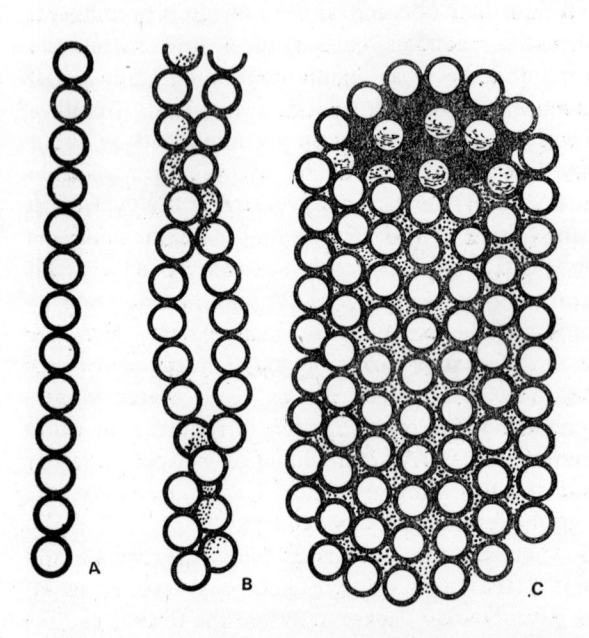

FIG. 1.6. The arrangement of globular molecules in filaments and microtubular fibrils. A—A single row of molecule; B—Two strands of molecule; and C—The walls of a tubular microtubule.

**Nucleus.** All protozoans have at least one nucleus and in many species several or even many nuclei may be enclosed within the same body of cytoplasm by a single plasma membrane. Some Protozoa such as ciliates have characteristically two nuclei (**macronuclei** and **micronuclei**) and at a certain stage in the life cycle of some foraminiferans two types of nuclei are known to be present within the same organism. All nuclei contain the genetic material **deoxyribonucleic acids** (DNA) in the chromosomes within the nucleoplasm. The form of nuclei and their manner of multiplication are more diverse in Protozoa than in higher animals. Most common protozoan nuclei are vesicular and spherical or oval, the best known exceptions being the macronuclei of ciliates which are more dense and have complex shapes. The nucleus is bounded by a double unit membrane (about 7 $\mu$m thick separated by a space of 20 $\mu$m or so) which is, in places, continuous with the endoplasmic reticulum. The nuclear envelope has many pores which allow continuity between the cytoplasm and the interior of the nucleus. The inner and outer membranes of the nuclear envelope come together to form the margin of the pore. In surface view the pores appear as dense rings.

The protozoan nuclei contain deoxyribonucleoprotein and sometimes at least ribonucleoprotein which is often concentrated into one or more intranuclear masses called **nucleoli**. **Chromosomes** have been reported from the nuclei of some Protozoa. Perhaps, Chromosomes may be found in all, but the small size of the nuclei makes their demonstration difficult. The (haploid) chromosome numbers recorded for most parasitic Protozoa are low (three in several species of *Trypanosoma* and *Leishmania tropica*, two in most species of *Plasmodium*, or in some Cnidospora and four in some Gregarinia and *Giardia* species). The largest numbers recorded for the parasitic Protozoa are in the hypermastigid flagellates living in the guts of termites and some orthopterous insects. They have from 6-48 chromosomes (Garnham, 1966c). The deoxyribonucleic acid (DNA) content of the nuclei of parasitic Protozoa is also low in a few organisms in which it has been determined. In *Trypanosoma evansi* the nuclear DNA content is only about three per cent of that found in human diploid nuclei (Baker, 1961). Some protozoans (e.g., the Opalinata) have many nuclei. In Protozoa sexual processes are known only among the Ciliophora, Mastigophora and Sporozoa. In the Ciliophora the micronuclei is diploid throughout most of the organism's life history, while in some (perhaps, all) Sporozoa the nucleus is haploid for all but a brief period of time after fertilisation.

The nucleus has two major functions to perform, the replication of the genetic material of the cell and the release of genetic information to the synthetic machinery of the cell. Replication involves the synthesis of new DNA to duplicate the chromosomes, and, subsequently, the separation of the chromosomes into daughter nuclei, normally immediately before cell division. The synthesis of DNA in a nucleus normally occurs at a characteristic time in the cell cycle, and this is usually at a different time in the cycle from the division of the nucleus.

Division of the protozoan cell is usually immediately preceded by the division of the nucleus, except in some multinucleate forms. The genetic function of the nucleus demands that at division each daughter nucleus should receive a complete set of chromosomes identical with that passed by the present cell. This is achieved by mitosis, which in certain Protozoa, according to some reports, is not of the conventional type, but most of these reports have since been found to be incorrect and the process normally differs only in the minor ways from mitosis in cells of higher plants and animals. Haploid and diploid nuclei undergo a mitotic process in which each chromosome divides longitudinally into two identical parts (the replicate chromosome being found as a result of DNA synthesis). Then the twin products of each chromosome division move apart so that identical groups of chromosomes come together in each of the two daughter nuclei.

Mitotic division in Protozoa takes place in very simple way. First, the centriole or its equivalent divides into two, the daughter centrioles move to opposite sides of the nucleus, the **mitotic spindle** of microtubular fibrils forms between the centrioles. The chromosomes, which are long filamentous before mitosis, shorten and thicken progressively at prophase. When shortened they appear to be double, each chromosome consists of two chromosomes (or chromatids). The chromosomes may be banded or beaded with aggregations of DNA and are constricted at the centromere. At the **metaphase**, the centromeres form attachments with fibrils of the spindle near its equatorial plane and typically become arrayed across the equator of the spindle. By this time the nuclei and the nuclear membrane disappear. During anaphase the centromere of each chromosome splits and the two chromatids are separated as the centromeres move towards the poles of the nuclear spindle. The mechanism of movement is not known, but may depend on changes in the length of spindle fibres, some of which are continuous throughout the length of the spindle, while those attached to the chromosomes are shorter. Finally, the nuclei are reconstructed, the chromosomes lengthen, and the nucleoli and the nuclear envelope are reformed. This stage is the **telophase**.

This is the simplest way of mitotic division, but there are a number of ways in which mitotic division in various species differs.

**Skeletal structures.** There are some free-living Protozoa that form elaborate exo-skeleton (e.g., Radiolaria, Foraminifera). Some people consider the cysts of parasitic forms as exo-skeletal structures while others regard the tough wall surrounding the spore of the Cnidospora (Fig. 1.1) as exo-skeleton. In some parasitic Protozoa, particularly the Mastigophora, certain endo-skeletal structures are formed. Axostyle *Trichomonas* is an example. In the same way, the sub-pellicular microtubules of trypanosomatids and malaria parasites may be considered endo-skeletal structures.

## E. NUTRITION

Nutrition in Protozoa may be of two types **holophytic** or **holozoic**. The holophytic nutrition is plant-like or synthetic, i.e., the food is synthesised as in plants, based on cholorophyll. In this case the energy derived from light is used to build up organnic molecules from inorganic materials. This type is also called **photoautotrophic**[2] nutrition and is found in flagellates which contain chlorophyll pigments. Holozoic is that type of nutrition in which organisms obtain energy from the breakdown of organic materials derived from other organisms and is based on predation. This has also been called **heterotrophic** nutrition. This consists essentially of the ingestion of the form of other animals or plants (either whole or in pieces), breaking these molecules down into simpler units (amino acids, simple sugars, fatty acids) and rebuilding them into animal's own proteins, carbohydrates and fats.

When holozoic Protozoa obtain their food by actively ingesting large particles, whole cells or large chunks of them, the feeding process is said to be **phagotrophic**. Some protozoans were, however, known to have **saprozoic** nutrition, in which simpler organic molecules (amino acids, simple sugars and fatty acids) produced as a result of decay or (as in the case of parasites) the host's digestive process, are absorbed by the protozoan in solution by diffusion through the pellicle. The trypanosomes were thought to be saprozoic feeders, but they are now known to feed by **pinocytosis,** a process which consists of the ingestion of minute droplets of the surrounding fluid by their incorporation into tiny vacuoles formed by invagination of the surface (unit) membrane of the

---

[2]In bacteria chemoautotropic nutrition occurs, in this energy is derived from the oxidation of inorganic substances to build up organic molecules from inorganic materials.

cell. These vacuoles then break off and move inward into the cytoplasm. This process is similar to phagotrophy with only minute differences.

Nutrition in amoeba is **heterotrophic**. It feeds on the tissues of other animals or plants. It takes in complete complex molecules of many types used in metabolism. Such animals were also called **holozoic**. Synthetic powers of such animals (**heterotrophs**) are often limited. Such a heterotroph may require not only an organic energy source but also one or more classes of organic molecules as organic carbon compounds and organic nitrogen compounds. It may have limited ability to interconvert amino acids, so that it may require many or most of about 20 amino acids as well as a source of **purine** and **pyrimidine** compounds. Many organisms need growth factors (vitamins) which are generally complex molecules necessary for coenzyme functions.

An organism is made of organic molecules which contain a number of important elements which must be present in the nutrients that they take (or absorb). Carbohydrates contain carbon, hydrogen and oxygen, proteins contain also nitrogen and often sulphur and lipids often contain phosphorus. The nucleotides which form the nucleic acid and such vital compounds as ATP contain nitrogen in purine and pyrimidine bases and phosphorus. An autotrophic organism is able to synthesise all the complex molecules it requires from simple inorganic compounds containing these elements, together with a few traces of such elements as copper, iron and magnesium.

## F. CLASSIFICATION

The classification of the animal kingdom is under continuous review so that the treatment of particular groups is liable to be modified from time to time. The Protozoa are exceedingly difficult to classify because of the many transitional types and forms of uncertain relationship. It is impossible to erect exclusive definitions, and consequently the definitions centre about typical members. The scheme given below is a simplified and conservative outline following that included by Barnes (1963) and Sleigh (1973).

## PHYLUM PROTOZOA

| | |
|---|---|
| **Sub-phylum:** | SARCOMASTIGOPHORA |
| **Superclass:** | MASTIGOPHORA |
| **Class:** | PHYTOMASTIGOPHORA |
| **Orders:** | Chrysomonadida |
| | Silicoflagellida |
| | Coccolithophorida |
| | Heterochlorida |
| | Cryptomonadida |
| | Dinoflagellida |
| | Ebriida |
| | Euglenida |
| | Chloromonadida |
| | Volvocida |

| | |
|---|---|
| Class: | ZOOMASTIGOPHOREA |
| Orders: | Choanoflagellida |
| | Bicosoecida |
| | Rhizomastigida |
| | Kinetoplastida |
| | Retortamonadida |
| | Diplomonadida |
| | Oxymonadida |
| | Trichomonadida |
| | Hypermastigida |
| Superclass: | OPALINATA |
| Order: | Opalinida |
| Superclass: | SARCODINA |
| Class: | RHIZOPODEA |
| Subclass: | LOBOSIA |
| Orders: | Amoebida |
| | Arcellinida(=Testacida) |
| Subclasses: | FILOSIA |
| | GRANULORETICULOSIA |
| Orders: | Athalaniida |
| | Forminiferida |
| Class: | PIROPLASMEA |
| Order: | Piroplasmida |
| Class: | ACTINOPODEA |
| Subclasses: | ACANTHARIA |
| | HELIOZOIA |
| | RADIOLARIA |
| | PROTEOMYXIDIA |
| | MYCETOZOA |

| | |
|---|---|
| Sub-phylum: | **SPOROZOA** |
| Class I: | TELOSPOREA |
| Subclass (i): | GREGARINIA |
| Order: | Archigregarinida |
| | Eugregarinida |
| | Neogregarinida |
| Subclass (ii): | COCCIDIA |
| Orders: | Protococcida |
| | Eucoccida |

| | |
|---|---|
| Sub-phylum: | **CNIDOSPORA** |
| Class: | MYXOSPOREA |
| Orders: | Myxosporida |
| | Actinomyxida |
| | Helicosporida |

| | |
|---|---|
| Class: | MICROSPOREA |
| Orders: | Microsporida |
| | Sarcosporida |
| | Haplosporida |
| | |
| Sub-phylum: | **CILIOPHORA** |
| Class: | CILIATA |
| Subclass: | HOLOTRICHIA |
| Orders: | Gymnostomatida |
| | Trichostomatida |
| | Chonotrichida |
| | Apostomatida |
| | Astomatida |
| | Hymenostomatida |
| | Thigmotrichida |
| Subclass: | PERITRICHIA |
| Order: | Peritrichida |
| Subclass: | SUCTORIA |
| Order: | Suctorida |
| Subclass: | SPIROTRICHIA |
| Orders: | Heterotrichida |
| | Oligotrichida |
| | Tintinnida |
| | Entodiniomorphida |
| | Odontostomatida |
| | Hypotrichida |

## CLASS MASTIGOPHORA (FLAGELLATA)

The **Mastigophora** (Gr. *mastix*, whip + *phora*, bearing) are flagellate Protozoa and bear one or more flagella (singular flagellum) permanently or temporarily in the adult state for locomotion and food caputre. The flagellates are extremely difficult to separate from rhizopods, algae and sporozoans. Symmetry is generally radial, but bilateral and irregular types also occur. In many flagellates a red eyespot or **stigma** occurs near the anterior end. It consists of a cup of carotin pigment called **hematochrome** and is believed to shade a light-sensitive substance located in the cup and overlain in some forms by a lens-like thickening. The whole constitutes a sensory organelle for light perception and makes detection of the animalcule easy. The flagellates are small protozoans which in the adult stage are motile, swimming by means of flagella. The cell-body is usually of definite form, oval, elongate or spherical with differentiated anterior end. A firm **pellicle,** often ringed spirally or longitudinally, maintains the form of the cell-body. In some groups the pellicle is provided with shells, cases and armours which may adhere closely (as the cellulose plates of dinoflagellates) or may consist of loose encasements of jelly, silica or cellulose. In some cases the pellicle is thin or absent with the result that amoeboid changes in shape take place and some species become wholly amoeboid at times, even losing the flagellum. Generally the forms are radially symmetrical, but bilateral and irregular types also occur. Some chrysomonads have stiff pseudopods like those of Heliozoa.

The cytoplasm is usually not differentiated into ectoplasm and endoplasm. The rounded or pointed anterior end bears a gullet-like depression that generally serves for the insertion of the flagella rather than for food intake. Many species contain **plastids (chromoplasts)** of various shapes. They may consist entirely of chlorophyll lending the animals a pure green colour or may be coloured red, yellow or brown. The chromoplasts are usually accompanied by **pyrenoids**. Fresh water forms have contractile vacuoles. The **nucleus** is generally single and vesicular with a single **nucleolus (endosome)**. The nutrition is variable—holozoic, saprophytic or holophytic. Some individuals indulge in worm-like contractions and expansions and are said to perform "**euglenoid movements.**" Reproduction is by longitudinal fission, but some undergo multiple fission and one group presents example of sexual reproduction also.

The flagellates are found everywhere in fresh or salt water. Many are freeliving and solitary, some are sessile and some form colonies of a few to thousands of individuals. Many are parasites of human beings and other animals and some of them cause important diseases. Free-living forms may encyst to avoid unfavourable conditions.

## TYPE EUGLENA

*Euglena* is a common, solitary and free-living flagellate inhabiting stagnant water containing nitrogenous organic matter. Such water with high organic content is called **polysaprobic**, while water with medium organic contents are described as **mesosaprobic** and with low organic contents as **oligosaprobic**. Extremely pure waters are described as **katharobic**. Largest number of protozoan organisms occur in polysaprobic conditions, principally bacteria-consuming forms tolerant of very low oxygen concentration such as the flagellates *Oikomonas mutabilis* and *Bodo* sp. and even photosynthetic forms like *Euglena*. They are often found in such a large number that they produce a green scum on the surface of ponds. Such ponds and ditches usually contain euglenoids of more than one species. Of these the most common from is *Euglena viridis* (Fig. 1.7).

### Structure

The euglenoid is a minute spindle-shaped organism measuring up to 0.1 mm. in length. The shape of the cell-body is constant and the end which is foremost when the organism moves in blunt and the opposite (posterior) end tapers to a point called **tailpiece** (Fig. 1.7A). The shape of the body is maintained by the **pellicle** which presents spiral or parallel striations. The striations may or may not be present in the pellicle of tailpiece. The pellicle of *E. spirogyra* is tuberculate, i.e., bears minute tubercles. Although firm and strong the pellicle is very thin and sufficiently flexible and elastic to permit considerable and characteristic changes of form which the body (e.g., during euglenoid movements) occasionally exhibits.

## Phylum Protozoa

As mentioned earlier the outer surface of every protozoan consists of **unit membrane (plasma membrane)**, which is permeable to certain substance

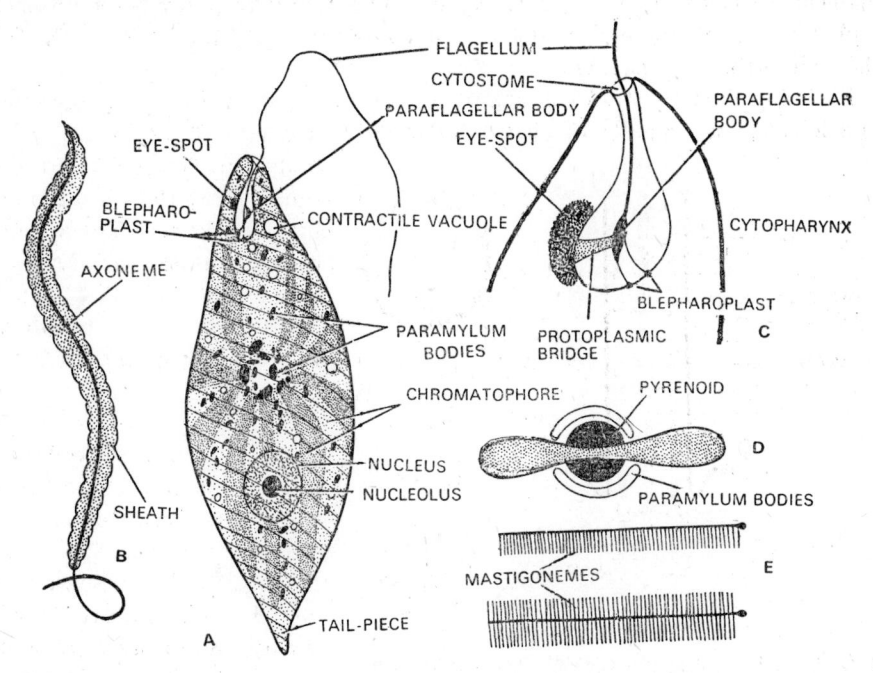

**FIG. 1.7.** A—*Euglena viridis*; B—Flagellum; C—Anterior part of *Euglena* showing attachment of the flagellum; D—Chromatophore; and E—Mastigonemes of stichonematic and acronematic flagellum.

and impermeable to others. The membrane consists of lipoprotein structures about 7.5 $\mu$m thick. In an electron micrograph of section membrane it appears as two dense lines, each about 2 $\mu$m thick, separated by a light space 3-4 $\mu$m thick. The dark lines represent lipid layers separated by a layer of protein molecules. Impermeability is provided by the lipid molecules, and strength with elasticity is provided by the protein. The pellicle of *Euglena* may be strengthened or stiffened with protein fibre or similar other structures (skeletal rods, plates, thecal plates) within the cell membrane (by contrast *Chlamydomonas* and related flagellates have a cellulose cell wall outside the membrane as found in many higher plants.)

At the end of the body there is a single long delicate protoplasmic filament, the flagellum (Fig. 1.7). This whip-like process, by its lashing movements, causes the organism to move in water. It also acts as a sensory organelle for exploring the surroundings. The flagellum of the euglena is also called **tractellum** because it does not propel the body from behind but draws it forward.

The flagellum springs from the anterior tip from a flask-shaped groove, the cytopharynx or "gullet" (Fig. 1.7A) that opens to the exterior by a **cell-mouth** or **cystostome**. The flagellum arises from the bottom of the cytopharynx, is forked at the base and is connected by two **rhizoplasts** to basal granules, the **blepharoplasts,** behind the nucleus. Each blepharoplast is a compact granule that gives rise to a separate filament (axoneme), and in most of the species the two filaments soon unite to form a definitive flagellum. In some species of *Euglena* (*E. viridis*) the flagellum is enlarged at the junction of the two filaments or just behind it (Fig. 1.1A) to form the **paraflagellar** body, probably, a photosensitive structure. A delicate **protoplasmic bridge** connects this with the wall of the cytopharynx.

Each flagellum is cylindrical organelle about .25 µm in diameter, composed of a longitudinal bundle of microtubular fibres (the axoneme) enclosed within a unit membrane which is continuous with the plasma membrane of the cell. The axoneme extends into the surface region of the cell as the flagellar basal body, the **blepharoplast** (**kinetoplast** in ciliates). In some flagellates, (e.g., *Trypanosoma lewisi*) close to the basal body a **kinetoplast** is located (Fig. 1.9). The kinetoplast has been shown to be self reproducing, DNA containing and double-membrane bound and to have tubular extensions identical to mitochondria. Thus kinetoplast and mitochondria are developmentally interrelated. The microtubular fibrils of the axoneme are arranged in a precise pattern, best seen in cross section (Fig. 1.8). Two microtubules 24 µm in diameter occupy the centre of the bundles. These are surrounded by nine double microtubular fibrils, arranged on the circumference of a circle about 0.2 µm in overall diameter (Fig. 1.8A). Each double set consists of two sub-fibrils A and B. The sub-fibril A of each doublet carries two rows of lateral projections along its long axis. The "9+2" cylinder (the axoneme) is surrounded by a hollow cylindrical extension of the uniting membrane of the organism (or cell) and is called the **sheath**. Flagella often (perhaps always) have a band of material within the sheath, parallel to the axoneme, which supports the microtubules. This is needed more because the flagellum is relatively a long structure. The basal body (blepharoplast) essentially of a relatively short extension of the nine outer axoneme fibrils (which may here be triple instead of double) without the two central ones. The central microtubules appear to originate at a **basal plate** which makes the junction of the axoneme and the basal body.

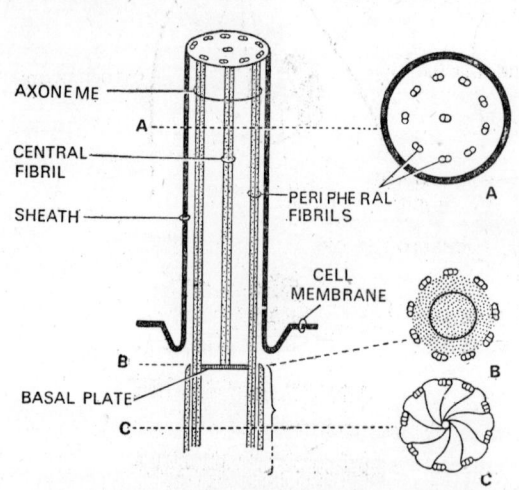

**FIG. 1.8.** The ultra-structure of the flagellum (Diagrammatic). A, B and C—Cross sections through the flagellum at the levels indicated (*i.e.* A, B, C).

The terminal region of the flagellum is narrow because the central fibril is longer than the peripheral fibrils. Commonly the peripheral doublets become single near the tip by the termination of one of such fibrils (*B* sub-fibril).

At the transition between the flagellum and the basal body the radial strands on the peripheral doublets are missing and frequently each is connected to the flagellar membrane. The central fibrils end at about the level of the cell-surface and may be embedded at their inner ends in a transverse plate which may occupy much of the area within the peripheral ring. The basal body (blepharoplast) is without central fibril and consists of relatively short extensions of the nine outer axoneme fibrils, which may here become triple instead of double as a third sub-fibril C is added. Thin strands interconnect the triplets in various ways, the commonest being the cartwheel arrangement (Fig. 1.8 C). The central area is sometimes occupied by granules or other structures.

Flagellum of *Euglena* is frequently provided with one row of slender hairs about 5 μm thick. Other flagellates may have two rows of lateral hairs of the same type. Some other flagellates have thicker, stiffer projections about 20 μm thick and 1 μm or so long. It is usual to refer all types of flagellar hairs as **flimmer filaments** and to reserve the term **mastigoneme** for thicker projections. Mastigonemes are believed to project rigidly from the flagellum in two rows arranged in the plane of flagellar undulation.

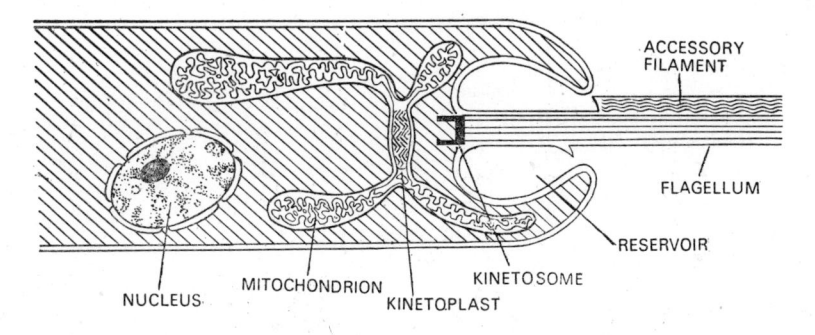

FIG. 1.9. A protozoan mitochondria—Diagrammatic representation of the relationship of the kinetoplast to the mitochondria as seen with the electron microscope.

The structure of the flagellar basal body is similar to that of **centriole** found in animal cells. In some flagellates the basal bodies act as the organising centre for the mitotic spindle of the dividing nucleus functioning both as centrioles and bases of motile flagella. It is likely that the centriole has been derived from the flagellar basal body of ancestral flagellates. The basal body is also the structure from which flagella (and cilia) develop and an organising centre for many cytoplasmic fibril systems. There is some evidence that basal bodies contain DNA and may have some powers of self replication. Earlier workers (Woodcock, Minchin) called it **kinetonucleus**. In some species of *Euglena* one of the blepharoplasts gives rise to a fibre-like structure, the **rhizoplast**, which is connected to a small granule situated near the nuclear membrane. Hall calls the small granule the **extranuclear centrosome** and Baker calls it the **parabasal homologue**.

**Flagellar movement** (Fig. 1.10). Electron microscopy has shown that flagella and cilia have the same fundamental structure. The difference lies in their mode of beating, and in the type of movement that is produced by the beats. Both are adapted for action in a fluid medium. If they are attached to a fixed surface they set up motion in the medium relative to the surface. They, however, arise from the surface of movable object which is caused to move in relation to the medium by their movement. The flagellar and ciliary actions differ from each other in as much as the beat of a flagellum is often symmetrical, with several waves included in it at any moment, whereas, the beat of a cilium is asymmetrical and includes only one wave. The result is that the flagellum is able to move the fluid medium continuously throughout its beat, and in such a way that the movement is at right angles to the surface of attachment. The cilium, by contrast, has an

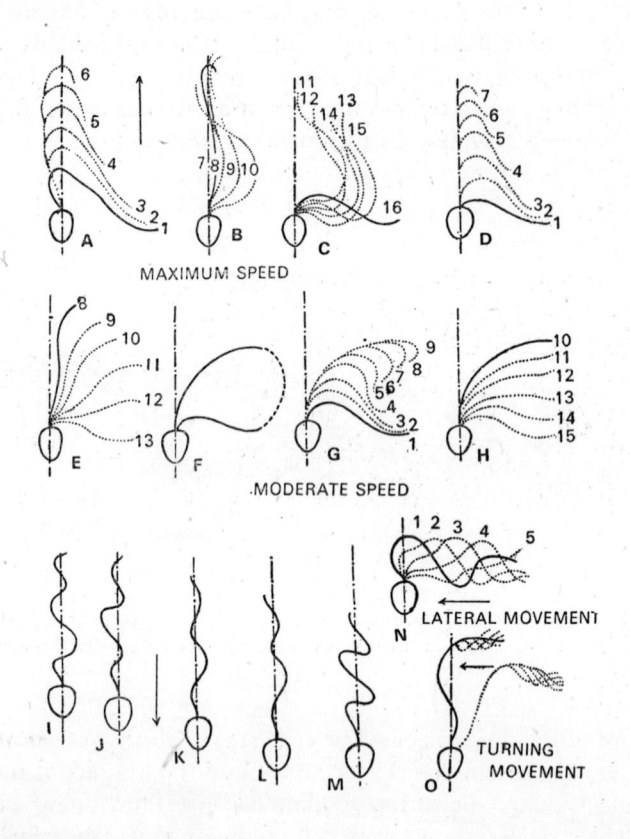

FIG. 1.10. Flagellar movements of *Monas* sp. (*after* Krijasman)

**active phase,** during which movement is brought about, and a **recovery phase,** during which there is no significant movement. The movement of the fluid is parallel to the surface of attachment. Moreover, as compared to flagella, cilia are typically much more densely set and their beats are coordinated in the well-known pattern of **metachronal rhythm**, conventionally compared to the passage of wind over a field of wheat.

In earlier days it was natural to presume that the apparently simple amoeboid movement of sarcodine (rhizopod) Protozoa was the most primitive type of animal locomotion. This is not correct. Flagellar movement is the most primitive. It is the movement of the most primitive protozoan group of Mastigophora. Moreover, it is now known that the underlying mechanism of amoeboid movement, in principle, is similar to flagellar and ciliary movements. It is reasonable, therefore, to conclude that cilia were evolved from flagelia.

Pseudopodial movement must also have evolved in a stock with a flagellate ancestry.

The close relationship with mastigophoran and sarcodine protozoans is better shown by the existence of flagellate reproductive stages in some of the latter e.g., *Elphidium* (*Polystomella*). There are also adult forms that have both flagella and pseudopodia either simultaneously or in two separate phases. Examples are seen in the phytomastigine order Chrisomonadina; many members of this group can assume an amoeboid form, and may even lose their flagella and chloroplasts so that they then become virtually indistinguishable from typical sarcodinians. An example of the reverse transformation is seen in the sarcodine *Naegleria gruberi*. This organism normally lives in the soil, where it ingests bacteria. Transformation of a flagellate phase takes place in the laboratory cultures when the medium is diluted with pure water. At first the body becomes polarised in organisation with lobose pseudopodia at one end and filiform pseudopodia at the other; flagella then appear among the latter, and the organism passes into the flagellate phase. This phase, which is freely motile, is a transient one; it has been suggested by Willmer that it may be a response enabling the organism to obtain fresh food supplies or to reach an environment which is in other respects more suitable. In short, it can get the best of both worlds.

The forces set up by flagella and cilia are small, and are totally indequate for the production of lively movement as found in larger animals (due to muscle cell).

We must now consider briefly how far the mode of functioning of flagella and cilia depends upon properties that they share with the contractile myofibrils of muscle cells. To what extent, in other words, is muscular contraction a development of biochemical mechanisms that were already well established before the emergence of metazoan organisation? The interpretation of the action of flagella and cilia presents problems that they may be approached in more than one way. It is possible, for example, to analyse the form of their movement in the organism that depends upon it. Alternatively, we can face more difficult problems of the nature of the cellular mechanism that is responsible for the movement of these delicate threads, and of the ways in which the necessary supplies of energy are made available. These problems are, of course, related.

The movement of flagella commonly involves the generation of waves that are transmitted along it, either in a single plane, or in a corkscrew pattern. The effect of this upon the movement of a protozoan is well exemplified by *Euglena*. The waves arise at the base of the flagellum, and pass to its tip. The flagellum beats at a rate of about 12 beats per second. The result is that the forces are created in two directions, one parallel to the main axis of the body driving the animal forward and the other at right angles to the main body axis tending to rotate the animal on its long axis. Because of this, *Euglena* rotates (Fig. 1.11) as it swims (at a rate of about one turn per second), and it also follows a corkscrew course. The movement of its body is thus comparable with that of a propeller, for it sets up courses on the water that bring about forward displacement. It is not essential, therefore, for the flagellum itself to provide a forward component in such circumstances; it may well do so, however, in the particular case of *Euglena*, since in that organism the flagellum is directed backward along the side of the body. (one reason incidentally, why students find it so difficult to observe). Lateral hairs on the flagellum modify the movement of water that is

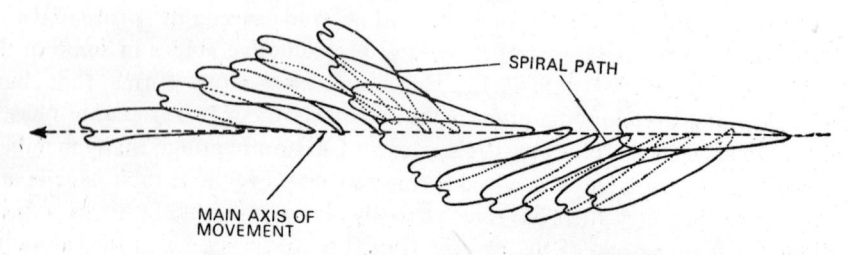

FIG. 1.11. Movement (spiral path) or *Euglena*.

produced by flagellar undulations. They cause a reversal of the flow of water produced by the undulations of the flagellum so that the propagation of bending waves along the flagellar axis from base to tip results in a flow of water towards the flagellar waves.

While the flagellum is moving it must be dissipating energy[3] as a result of work done against the surrounding water; this means that if it were solely dependent upon energy propagated along it from some force in the organism's body, its pulse would necessarily become diminished towards its free end. High speed cinematography reveals that this does not happen. On the contrary, as the wave passes along the flagellum it actually develops an increase both in velocity and amplitude. From this it is concluded that the flagellum must itself contribute energy to its movement. It is not yet clearly understood how this energy is produced, but there is at least some evidence that the underlying biochemical material is similar in principle to that of muscular contraction, in that ATP is the source of the required energy.

The evidence for this conclusion depends in part upon studies of the responses of intact cilia to the presence of ATP in the medium. Cilia from both the oyster and the frog show increased activity in the presence of ATP, the effect disappears if the ATP is destroyed by hydrolysis. Such evidence has been supplemented by the use of glycerol extraction method. Cilia of frog that have been extracted with glycerol for several days are inactive. But they can be activated if ATP is added to the medium. Cilia will beat in such circumstances even if they have been separated from any protoplasmic connection. The presence of ATP-ase in ciliated cells has, in fact, been demonstrated by both direct enzymatic studies and histochemical procedures. There is some evidence that it may be localised within the cilia themselves in the ciliate protozoan *Tetrahymena*, and it is known also to be present in the tails of sperm. Finally, ciliary activity can be arrested by the use of reagents such as—sodium fluoride, sodium azide, and sodium cyanide, which selectively inhibit specific stages in the metabolic path-ways of anaerobic glycolysis, the citric acid cycle, and the cytochrome system respectively. It may be concluded, then, therefore, that the flagellar and ciliary activity, like muscular contraction, depends upon the pattern

---

[3] C. Brokaw calculated that the total energy dissipation of a 40 $\mu$m long flagellum beating about 30 times a second with a wave-length of about 30 $\mu$m is about $3 \times 10^{-7}$ erg/sec. at 16°C.

# Phylum Protozoa

of biochemical organisation that plays fundamentally an important part in the energetics of living system.

During swimming the flagellum is always directed backwards and a series of waves pass along it from base to the tip in a spiral manner, with increasing amplitude. This produces two components of force on the anterior end of the body which not only rotate the body on its long axis but gyrate it along the general direction of locomotion. Thus as the *Euglena* moves forward it takes its spiral course along the axis of the general direction of movement, with its body always inclining at the angle of 30° to this axis (Fig. 1.11). Evidently the posterior end of the body makes a smaller circle than the anterior end and keeps quite close to the axis. The observer particularly noticed that the waves passed along the flagellum each second, and each time, a wave beat out from its base, the anterior end of the animal was rebounded. This made the anterior end vibrate rapidly, but the vibrations were hardly visible. The animal, according to his findings, completed a spiral in one second, covering a distance of about 0.17 mm.

**Contractile vacuole.** At one side of the cytopharynx is found a large **contractile vacuole** (Fig. 1.12) that discharges into the expanded portion of the cytopharynx. Just before collapse of the vacuole there appear around it many small vacuoles which fuse together to form the next vacuole. The vacuolar system, therefore, is like that of the protozoa and not "complex" as described

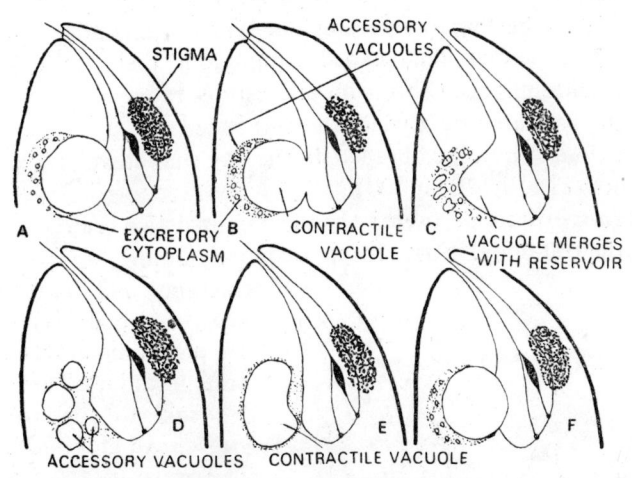

FIG. 1.12. Formation of the contractile vacuole in *Euglena*. A—Wellformed contractile vacuole; B—Contractile vacuole opening into the *cytopharynx*; C—Vacuole merges with the reservoir; D—Accessory vacuoles appear in excretory cytoplasm; and E—New vacuole formed.

in many text-books. The wall of the reservoir adjacent to the contractile vacuole is unstable being without cuticular lining. At the cystole of the vacuole the walls of the reservoir and of the vacuole burst pouring the contents of the vacuole in the reservoir. The vacuole thus serves to regulate the water content of the organism, eliminating excess water that enters through the surface or with food. Associated with the vacuole of *Euglena* Gatenby and Singh found structures (Golgi apparatus) meant to collect water for the vacuole. The pulsation of the contractile vacuole is rhythmic, but its rate varies with temperature.

On one side of the cytopharynx lies the red pigment spot consisting of clusters of lipid globules containing orange or red carotinoid pigments. This is known as the **stigma** or **eye-spot** and is photoreceptor (Fig. 1.12). They are common in pigmented flagellates and may show several patterns of relationship with flagella and chloroplasts. In most groups where it is present the eye-spot is found within a chloroplast and may be associated with a short or backwardly directed flagellum. In *Euglena* the eye-spot is a hyaline enlargement near the base of the flagellum. This is shaped from one side by a yellowish-red cup lying in the cytoplasm, thus, producing the masking essential in a light receptor which is to be used to obtain direction from light rays.

The **nucleus** is usually large situated near the pointed end of the organism and contains one **endosome** (nucleolus). The cytoplasm contains many green chloroplasts of various shapes (discs, ovals, stars and bands). In *Euglena viridis* is present a single group of slender chloroplasts radiating from a central point (Fig. 1.7), imparting a stellate (star-shaped) appearance to it. Typically, there is a **pyrenoid** in the centre of chloroplasts. Pyrenoids are areas of dense matrices found within chloroplasts and associated with the formation of polysaccharides. They are often surrounded or capped by large grains of starch or some other polysaccharide, such as **paramylum**, a carbohydrate similar to glycogen. Paramylum is also found in general cytoplasm in the shape of large ellipsoidal granules. The shape of paramylum varies generally. *E. oxyuris*, for instance, has a conspicuous link-shaped paramylum body in each end. The fine structure of chromoplasm shows an interesting range of variations. Chloroplasts consist basically of thylacoids embedded in a matrix and enclosed ordinarily by a double unit membrane. In the euglenas there are three chloroplast membranes (Figs. 1.12G and H).

**Euglenoid movement or metaboly.** The animal also crawls by the spiral movement of the cell-body. Some forms frequently indulge in worm-like contractions and expansions of the superficial body layer (simulating the peristaltic movement of vertebrate intestine). These movements are said to be **metabolic** or **euglenoid** movements (Fig. 1.13). The wave passes over the entire body. There are many explanations given for these movements or metaboly. During classroom study they are noted when the water dries up or the organisms are about to die.

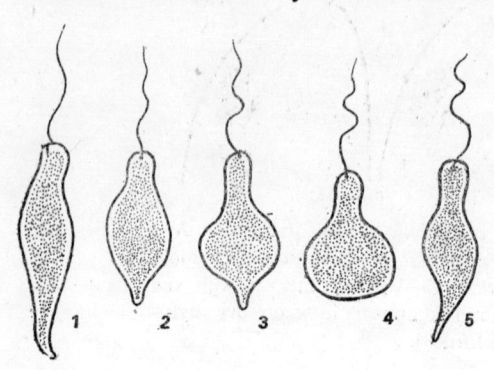

FIG. 1.13. Euglenoid movement in *Euglena*.

## Nutrition

The euglenas take no solid food, in other words, they are not holozoic. They subsist entirely by holophytic nutrition whereby some food is synthesised within the body with the help of chloroplasts. They thus take inorganic materials and with light energy synthesise food (photosynthesis). Such a mode of nutrition is called **phototrophic**. The genus *Euglena* is typically phototrophic, but *E. gracilis* can survive and grow in the dark, in conditions in which it cannot possibly be deriving energy from photosynthesis. Its survival is due to **osmotrophic heterotrophy.** In the absence of solar radiation it can obtain energy by oxidising acetic acid. Indeed many flagellates flourish in the presence of this substance. They seem to have little capacity for utilising carbohydrate, and the acetate provides them with what is probably the sole source of carbon.

The osmotrophy of *E. gracilis* provides the organism with a source of energy but it also meets other needs, for euglenids despite being phototrophic have certain specific vitamin demands. This was first demonstrated for *E. pisiformis* which cannot grow without an exogenous supply of thiamine. This dependence seems to be a common situation in euglenids, nor is thiamine the only substance so required, as *E. gracilis* and *E. viridis* together with many other phytoflagellates, need exogenous supply of vitamin $B_{12}$, although closely related cobalt compounds can sometimes be substituted for this.

Some euglenids lose their chlorophyll after prolonged culture in dark. *E. deses* does so. *E. gracilis* may spontaneously give rise to colourless form, which is identical with a flagellate that was known at one time as *Astasia longa*. The appearance of colourless forms can be introduced by treatment with streptomycin. *E. gracilis* responds to this and produces colourless forms. *E. paciformis* on the other hand, does not keep up photosynthesis, and when maintained in dark it dies. Thus *E. paciformis* is **obligatory phototroph** whereas *E. gracilis* is **facultative** one.

The survival of species without chlorophyll, of course, depends upon their capacity for osmotrophic nutrition, but we might expect that during evolution the establishment of colourless forms through mutation would be followed by the development of phagotrophy. Truly, animal-like forms, colourless and phagotrophic have originated in this way. *Peranema*, which swallows even *Euglena*, is one such animal. The euglenas are sometimes holophytic and sometimes heterotrophic depending upon external conditions. Some authors call this double forms of nutrition as **mixotrophic mode of nutrition**. Food is generally stored as carbohydrate granules or lipid droplets. Carbohydrate occurs in the form of starch and glycogen or in the form of leucosin, laminarin or paramylon. In some cases starch is stored in chloroplasts. Lipid droplets are formed in the cells of most groups and in some cases form the only food reserve.

**Respiration and Excretion**

Respiration takes place through diffusion. Oxygen diffuses into the cytoplasm through the pellicle and carbon dioxide diffuses outwards. During the active photosynthesis in the organism diffusion of carbon dioxide inwards and the output of oxygen mask the respiratory diffusions. Excretory products pass out by diffusion.

Some workers have assigned the excretory function to the contractile vacuole also. Working on *Euglena pseudoviridis* Chadefaud observed that the contractile vacuole is surrounded by a specialised granular cytoplasm which he has called **excretory cytoplasm**. In this appear small accessory vacuoles which merge to form larger vacuoles, which ultimately form the contractile vacuoles. There is nothing new about the observations of Chadefaud who has named the area of the formation of accessory vacuoles as excretory cytoplasm. He has not mentioned whether the vacuoles forming in the cytoplasm contain any nitrogenous waste or not. It must be mentioned, however, that regulation of water content is an important function of the excretory systems throughout the animal kingdom. Thus it can be concluded that the vacuolar system of the protozoans is the most primitive form of excretory system.

**Phototaxis.** The rate and direction of swimming of flagellate Protozoas influenced by light (phototaxis). *Euglena* shows positive phototaxis. Near the base of the locomotor flagellum is a swelling (paraflagellar body) which is believed to be light sensitive, and the manner in which light falls on the swelling is held to determine flagellar activities. The swimming response is photopositive when the flagellar swelling is periodically shaded by the adjacent red-pigmented stigma, but photonegative when the flagellar swelling is continuously illuminated.

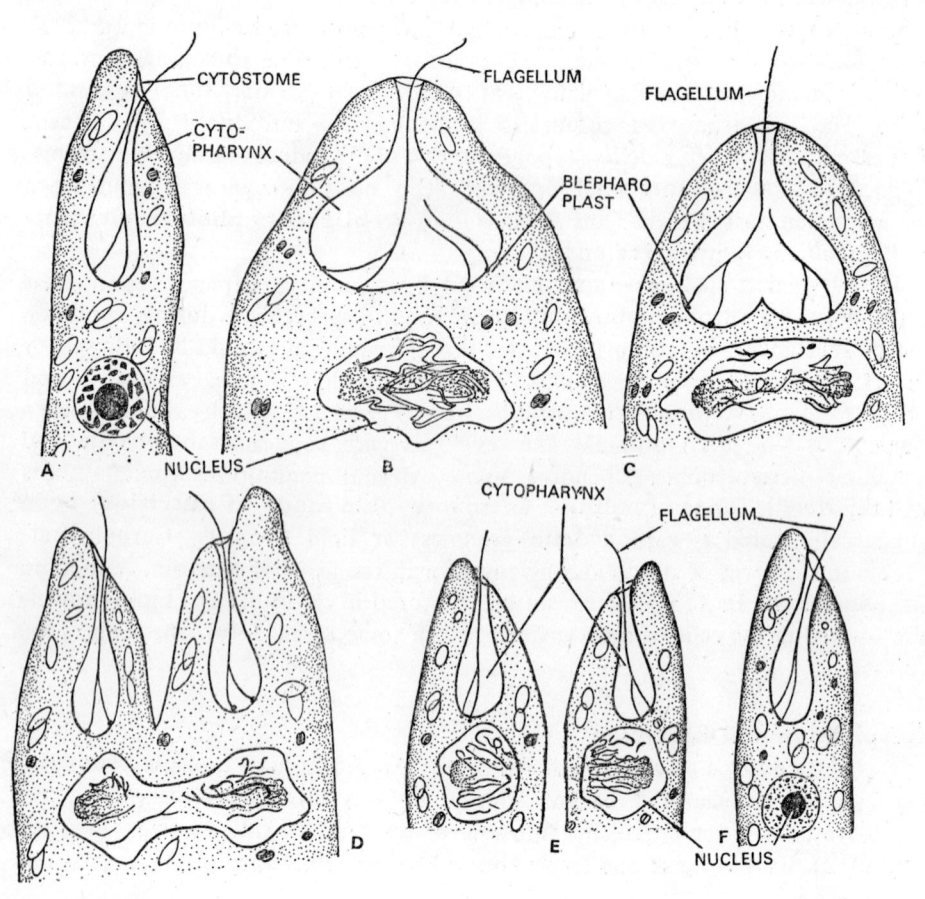

FIG. 1.14. Binary fission in *Euglena*. A—Individual ready to divide; B—Development of chromosomes, division of blepharoplasts (metaphase); C—Separation of chromosomes (anaphase); D—Division of the body begins from the anterior end (telophase) (*after* A. Hollande); E—Two separate individuals formed; and F—Nucleus assumes the same form as in A.

Pigmented Protozoa are most sensitive to light since they absorb a greater part of its energy. *Euglena*, for instance, moves towards light of moderate intensity. Exceedingly strong light is damaging. *Euglena* responds to high light intensities by negative phototaxis. In some species of *Euglena* (*E. sanguinea*, and *E. haematodes*) a protective red-pigment is found to migrate to the body circle in bright light and to migrate inwards in weak light.

**Behaviour**

While swimming forward rotating and gyrating on its long axis *Euglena* may reach an unfavourable area. When it does it turns violently usually heading into favourable area again. For this the gyration of the anterior end is suddenly widened (Fig. 1.15A) and a new spiral path is taken. Should it still encounter the unfavourable area, it will halt and turn again towards favourable area. This behaviour gives the appearance of an animal trying a particular direction and correcting its error, if misled. Thus, such a behaviour is called **trial** and **error**, behaviour but it is better to call it **avoiding reaction**. *Euglena* reacts positively to light of favourable intensity and swims towards the source, but is negative to strong light and hence tries to avoid it (colourless flagellates are indifferent to light). The light is perceived only at the anterior end by the specialised eye-spot. This specialisation is important since they depend upon photosynthesis for their nourishment. Hence it is essential that they should expose themselves to as much light as possible. When placed in a dish euglenas quickly aggregate on the side of the dish nearest to the light. When exposed to darkness the animals become scattered throughout the container or become inactive and sometimes even encyst. Euglenas also react to mechanical shock,

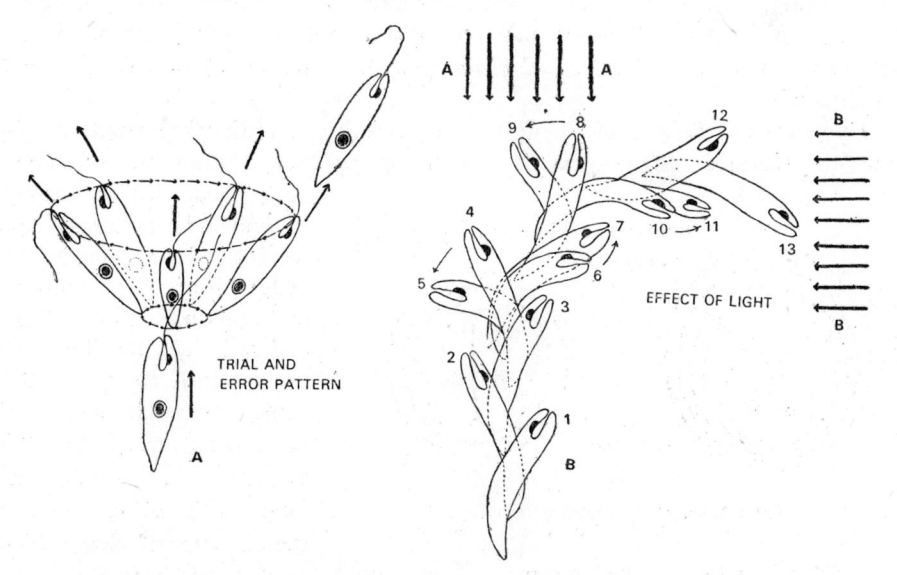

**FIG. 1.15.** A—Avoiding reaction of *Euglena*; and B—Effect of light on *Euglena*.

temperature change and chemicals. As a reaction to such a stimulus the organism slows down or stops, traces a wider spiral by swinging the anterior end farther out and starts off in a new direction (Fig. 1.15B).

## Reproduction

In the active stage the euglenas reproduce by longitudinal binary fission. The nucleus divides into two mitotically. The endosome also persists, it elongates and then constricts into two. The cause for persistence of endosomes is not known. Some authors have called it intranuclear spindle, for which there is no evidential support. Following the nuclear division all the anterior extranuclear organelles such as the blepharoplasts, reservoir, cytopharynx and stigma, etc., are duplicated. This is accompanied by the longitudinal splitting of the cytoplasm which begins from the anterior end of the body. In some species, it has been found that the paraflagellar body disappears before fission begins. In some cases the stigma breaks into component granules. Ordinarily, the original flagellum of the parent is retained by one daughter euglena and the other develops a new one. On the other hand, some observers have reported complete disappearance of the entire locomotory apparatus during division, and each daughter cell constructs a new set (Fig. 1.14). Euglenoids also reproduce by longitudinal fission in resting stage. The flagellum is withdrawn, the animal ceases to swim and encloses itself in an envelop of mucilage. The nucleus and chloroplasts divide followed by the division of the cytoplasm. Cases of multiplication, though rare, have also been reported in this condition. Multiplication also takes place in encysted condition. *Euglena* very rapidly encysts forming both thick and thin-walled cysts within which it divides into several (sixteen or thirty-two) daughter euglenas. Sometimes, the flagellate loses its flagellum and rounds up into an alga-like cell in which metabolism continues and reproduction occurs by fission, thus forming extensive green scums on the surface of ponds. In this condition they are said to assume the **palmella state**. Such a palmella state is of regular occurrence in some species. There is no evidence of sexual reproduction in *Euglena*. Biencler has recently described sexual reproduction by syngamy in some species of *Euglena*, but his findings do not find general acceptance.

**Encystment** (Fig. 1.16). Under certain conditions ordinary protective cysts are also formed. Encystment is stimulated by lack of food, lack of oxygen, drying heat (as in strongly illuminated cultures) and fouling of the medium. The yellowish-brown coloured cysts is composed of a special carbohydrate. The cysts are generally rounded, their walls are made of two or three concentric layers. The cysts are usually small, their total width being equal to the diameter of the animal, though it may be larger sometimes. Thin and stalked cysts have also been reported in some species and in others each may be provided with an operculum. The organisms may lie in the centre of the cysts or towards one side (eccentrically). The cysts are protective structures which help the organisms to withstand unfavourable circumstances and also aid in their dispersal. On the return

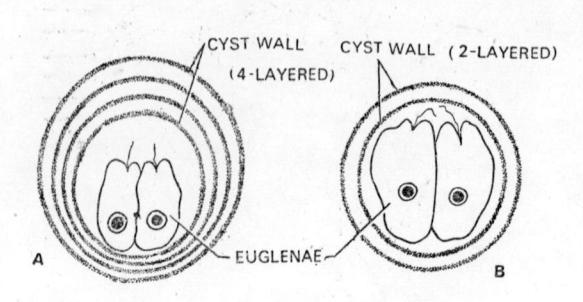

FIG. 1.16. Encystment of *Euglena*.

# Phylum Protozoa

of favourable conditions the cysts dissolve and the organism come out and begin normal life.

## TYPE VOLVOX

*Volvox* is a colony of flagellated individuals which keep on swimming in temporary or permanent fresh water ponds. In the shape of tiny green balls it keeps on rolling on water surface, hence it is called *Volvox* (from *volvere*, to roll or revolve). It multiplies so rapidly that the water surface of the ponds in which it occurs may appear green. The most common species of fresh-water ponds or ditches is *Volvox globator* (Fig. 1.17). About a dozen other species may also be found along with it. It usually appears in the spring, increases in

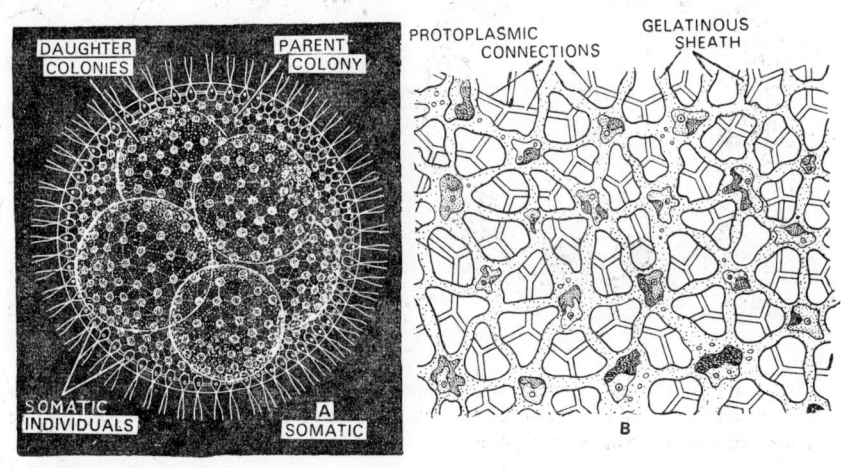

**FIG. 1.17.** *Volvox aureus.* A—Entire colony with daughter colonies (*after* Klein); and B—Surface view of the same.

abundance and then disappears abruptly early in summer. It passes the remaining period in the zygote stage.

## Structure

*Volvox* occurs in the form of hollow spheres (Fig. 1.17) or ovoid balls of mucilage, rarely excluding the size of a pin's head (one species has spheres three millimetres in diameter). The mucilage is transparent and into it are embedded some 500 to 50,000 very small pear-shaped pale green cells. These cells lie towards the surface where the consistency of the mucilage is more compact than at the centre. The soft mucilage at the centre is watery and in some species consists of water only. In some species the individual cells of *Volvox* are connected to their neighbours by stout standards of protoplasm (Fig. 1.17B) and the whole colony appears as a spherical network bounded by a transparent layer. In other species these strands are reduced or absent and the colony looks like a transparent ball into which are embedded numerous green dots. A suitably stained preparation of *Volvox* reveals a pattern of hexagonal or more or less circular areas (Fig. 1.18D) each with a cell or individual in the centre. Really speaking each cell is surrounded by a thick sheath of mucilage (Fig. 1.18C), bounded by a thinner and presumably firmer layer. the wall (or middle lamella). Each cell consists of a nucleus, a large chromatophore in which is located the green pigment (chlorophyll), a red eye-spot, two or more contractile vacuoles and one or more pyrenoids. The pyrenoids are of proteinaceous nature and perhaps concerned with nutrition. From each cell pass out two flagella whose movements are responsible for the propulsion of the colony in water.

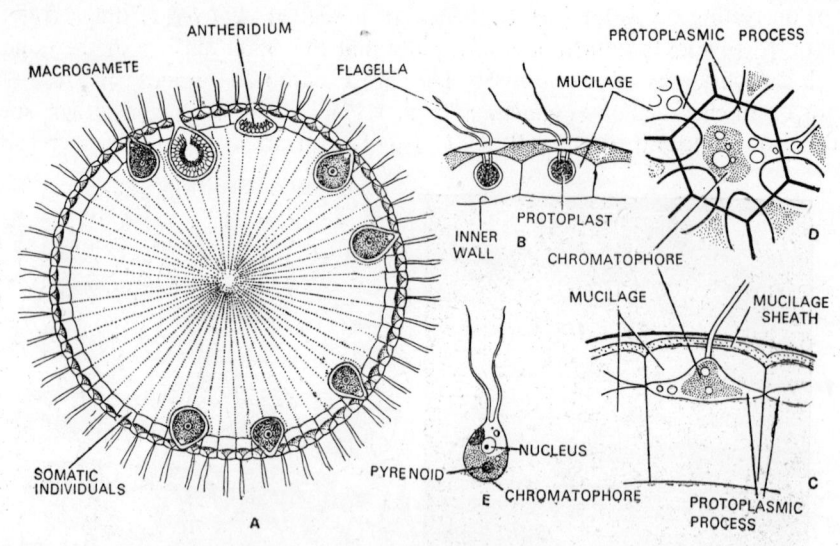

FIG. 1.18. A—*Volvox globator* (*after* Janet); B—Part of the surface of *V. tertius* in section; C—Vertical sections of the cell of *V. globator*; D—Surface view of the same cell; and E—Single protoplast.

*Volvox* does not roll about in the water in a haphazard manner but while rolling it indicates that it has a polarity—a particular part of the colony is always kept in front. This part has been called **anterior pole**. The cells around the anterior pole have larger eye-spots than their most posteriorly placed fellows, and they are more crowded on the anterior side. In some species the posterior pole of the colony possesses a small opening. As the colony moves forward it spins round its axis. Recent observations indicate that the spinning is always to the left.

The individuals are all alike in a young colony but as it matures, some cells in the posterior half become larger until they are, perhaps, ten times as big as their neighbours. As they grow they project towards the centre of the sphere (Fig. 1.18A) contained in a mucilaginous sac. The number of pyrenoids in these cells increases. These enlarged cells are reproductive cells, either asexual or sexual. It is evident from the above that a mature colony of *Volvox* can be regarded as a multicellular type composed of cells set apart to perform different functions in life. Some are vegetative cells responsible for the manufacture of food and locomotion, some are asexually reproducing cells and others are sexually reproducing cells, forming ova and sperms.

### Reproduction

*Volvox* reproduces in two ways, sexual and asexual. At the beginning of the

## Phylum Protozoa

growing season a colony reproduces exclusively by asexual method while at the end of the season reproduction is exclusively sexual.

Asexual reproduction proceeds rapidly during summer producing generations after generations. As the colony matures some cells at the posterior end of the body become enlarged until they are about ten times as large as the vegetative cells. These enlarged cells are called **parthenogonidia** (Fig. 1.19A). These

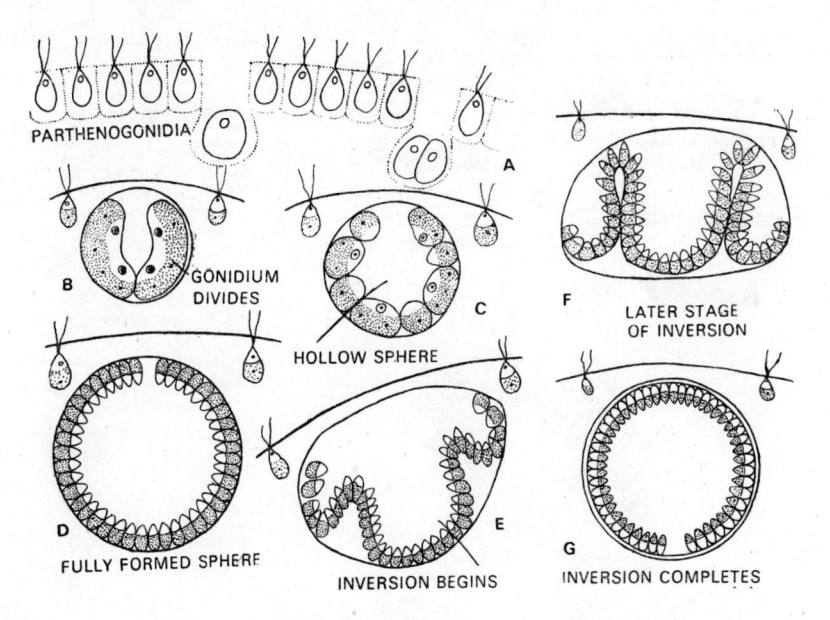

FIG. 1.19. Development of reproductive cells in *Volvox*.

cells lose their flagella and their pyrenoids increase in number, they have well-defined nuclei and denser granular protoplasm. As the gonidia develop, each of them divides into two by a wall perpendicular to the surface of the sphere (Fig. 1.20A). These two cells divide into four in the same way and form a little plate of eight cells. The cells continue multiplication and finally form hollow sphere perforated by a hole at the outer pole (Fig. 1.20C). It is a very small pore called the Phialopore, which in some species persists in mature *Volvox*. A very remarkable change now occurs and the sphere turns completely inside out (Figs. 1.19D—G). In the beginning the pointed ends of the cells are towards the centre of the sphere. When the cell-division stops the young colony curves inwards in the middle like a saucer and then turns inside out through the phialopore. As the inversion is completed the cells develop flagella and form a young colony. Several such young colonies may be formed inside a single sphere (Fig. 1.17A). These colonies may again reproduce asexually and may form as many as four generations. These newly formed spheres are released by the breaking up and death of the parent colony. Asexual reproduction is a means of propagating the species rapidly, as such it takes place when conditions are favourable for *Volvox*.

Sexual reproduction involves fusion of two cells or gametes and occurs when conditions are becoming less favourable. Under such condition a zygote is formed (Fig. 1.20C). It is a resting stage to pass through the unfavourable period in a state of suspended animation. When the favourable conditions

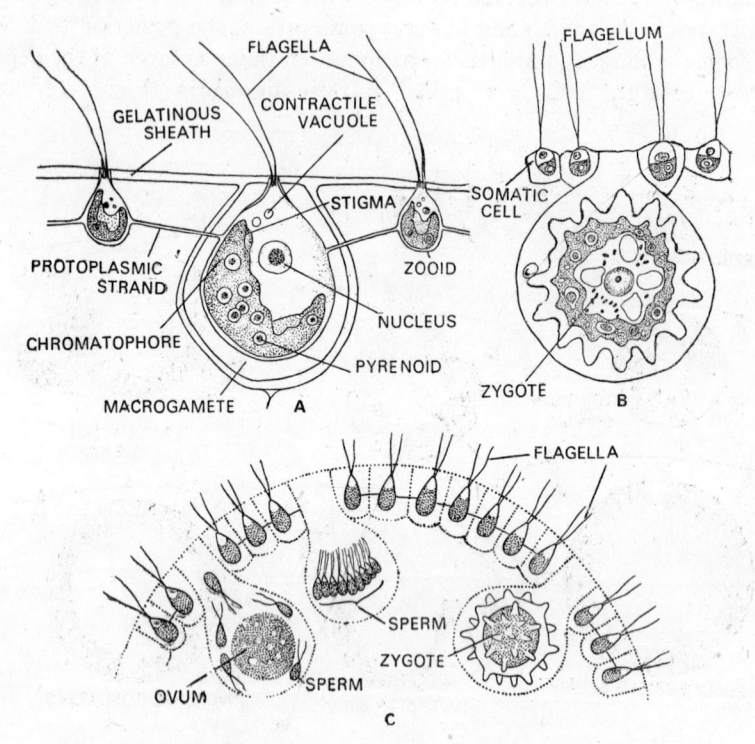

FIG. 1.20. Individuals of *V. globator*. A—Macrogamete; B—Zygote; and C—Development of gonidium in *V. aureus*. (*after* Zimmerman).

return the zygote germinates to form a motile *Volvox*. The zygote is formed by the fusion of the male and female gametes. The female gamete is a relatively large motionless ovum, while the male gamete is a small and motile spermatozoid (Fig. 1.20C). The cells which form eggs are similar to those reproducing asexually and are called oogonia. They increase in size until they are 5-10 times as large as the normal cells and acquire many pyrenoids. These eggs are surrounded by a mucilage sheath (Fig. 1.20). They withdraw their flagella and become large, they contain many vacuoles and stored food material and are non-motile.

The cells that produce the male gametes are called antheridia and like oogonia are simply enlarged vegetative cells. Usually a few antheridia occur in a sphere but in certain species a majority of or all cells of a colony may produce gametes. The cell divides into a bunch of 16, 32, 64 or 128 needle-like cells, each with two flagella. They are set free as a plate of cells which moves through the water. When it reaches the vicinity of an egg it breaks up releasing spermatozoids. The sperms swim around the egg (Fig. 1.20C), which remains in its original position in the colony. One of the sperms, that reaches the egg, fuses with the ovum forming the zygote, the first cell of the new generation. The zygote soon secretes a thick firm three-layered wall which may be smooth or

## Phylum Protozoa

spiny and develops a pigment (**haematochrome**) imparting it orange-red colour. The zygote remains dormant in the parent colony till it dies and disintegrates. Then it sinks to the bottom of the water and passes through a resting stage during unfavourable conditions. When favourable conditions return the thick wall of the zygote splits liberating the little individual in a little vesicle formed of the inner wall of the zygote (**endospore**). The cell in the vesicle undergoes similar divisions as in the case of parthenogonidia and forms a new colony in the same way. This happens in *V. aureus*. In *V. capensis* the cell within the vesicle develops flagella forming a biflagellate zoospore that escapes from the vesicle. This keeps on swimming freely for some time and then begins to divide. It forms a young sphere of about 500 cells which finally undergoes inversion forming a normal sphere.

In some species of *Volvox* the asexual and sexual reproductions do not occur in the same sphere. Likewise there are some monoecious species (*V. aureus*) and others dioecious (*V. globator*).

*Volvox* is now not regarded as an aggregation of cells such as found in *Synura* (Fig. 1.21) but as a multicellular individual showing division of labour and

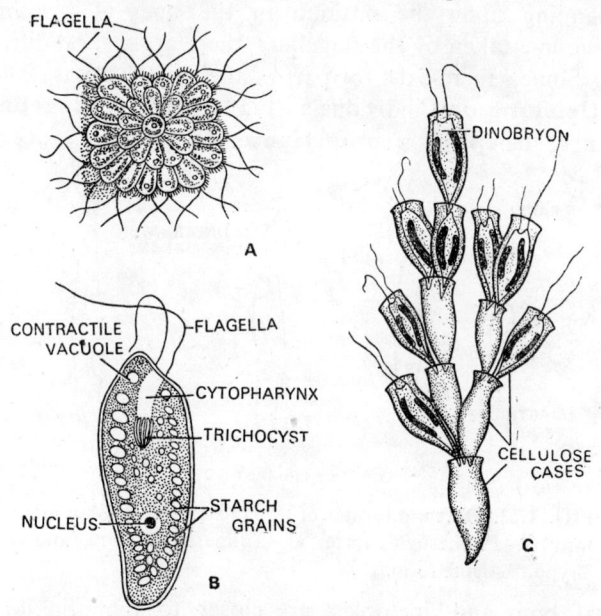

FIG. 1.21. Flagellates in drinking water. A—*Synura uvella*; B—*Chilomonas paramecium*; and C—*Dinobryon sertularia*.

exhibiting marked polarity. These were formerly regarded as individual ancestral to the Metazoa, but that view does not find much support now. The colonial Phytomonads are provided with chlorophyll and contain cellulose walls. This alone precludes the possibility that any of these could be ancestral to animal groups. However, it seems quite reasonable to consider that the real ancestors of the Metazoa might have passed through similar stages.

### PARASITIC FLAGELLATES

A number of flagellates are important from human standpoint. Some **inhabit** the intestine or other parts of the body and others live in blood and invade various body tissues causing definite pathological conditions. There are some flagellates that are commonly called **trypanosomes**. Of these *Leishmania donovani* causes **kala-azar**, a deadly oriental disease and *L. tropica* is responsible for oriental sores or boils. The chief pathogenic trypansomes of man are *Trypanosoma gambiense* and *T. rhodesiense* organisms that cause African sleeping sickness in man. The trypanosomes are characterised by the possession of DNA-rich organelle called the **kinetoplast** which is associated with the mitochondrion (Fig. 1.9). The **kinetoplast** is also present in another flagellate *Bodo* (Fig. 1.23A). The biflagellate *Bodo* and its relatives and the uniflagellate *Trypanosoma* and related forms are now grouped in an order **Kinetoplastida**.

All trypanosomes are elongate, slender Protozoa, at least at some stage of their life-cycle, with a single nucleus and a kinetoplast situated near the origin of the single anterior flagellum, by means of which they swim actively. The electron microscope has recently revealed that the kinetoplast is a mass of DNA contained within a very large mitochondrion[4] (Fig. 1.9) (prior to this discovery people knew that DNA occurred only in nuclei). Depending upon the position in the body of kinetoplast and basal body and the course taken by the flagellum there are several different forms of the organism. Some genera exist for part of their life-cycle as *non-***flagellate** or **amastigote (leishmanial)** individuals (Fig. 1.22). The **flagellate** or **mastigote** forms may be **epi-** or **promastigote**. In **promastigote (leptomonad)**

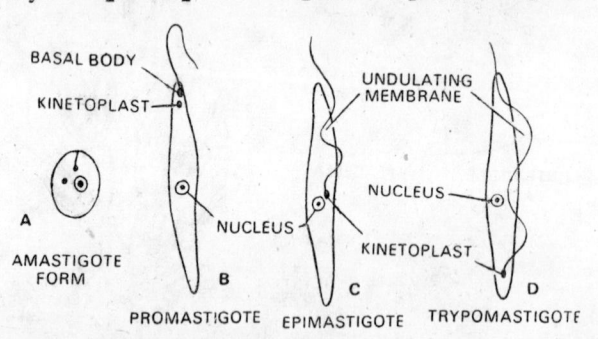

FIG. 1.22. Different forms of trypanosomes. A—Amastigote form; B—Promastigote form; C—Epimastigote form; and D—Trypomastigote form.

forms the basal body and kinetoplast are closer to the anterior end and the flagellum emerges through a short intucking of the body surface the **"reservoir"** or **flagellar pocket**. In the **epimastigote (critidial)** individual the kinetoplast and basal body lie close to, usually, besides the nucleus, while the **trypomastigote (trypanosomal)** forms are slender, flagellate individuals in which the kinetoplast and basal body are near the posterior end and the flagellum emerges through a short pocket running anterolaterally. The flagellum then passes forwards along the surface of the body of the organism to which it is apparently attached in such a way that as the flagellum undulates, it draws out a fin-like expansion of the body to form the **undulating membrane** (Fig. 1.22D).

[4] It has subsequently been recognised that DNA occurs in mitochondria of many other organisms.

*Phylum Protozoa*     37

Genera *Leptomonas, Herpetomonas, Crithidia* (Fig. 1.23), *Blastocrithidia* and *Rhynchoidomonas* are exclusively parasites of insects (and a few other invertebrates). They all inhabit the gut, and are presumably transmitted via the faeces, sometimes at least as encysted amastigote forms. The genus *Phytomonas* parasitises plants, especially succulents, and is transmitted by Hemiptera which feeds on the plant juices. Three genera are parasites of vertebrates. These are *Leishmania, Trypanosoma* and *Endotrypanum*. Almost all species of these three genera parasitise blood sucking invertebrates (usually insects or leeches) by means of which they are transmitted from one vertebrate to another. Some of them are pathogenic to their vertebrate hosts, but with one possible exception (*T. rangeli*) there is no evidence that any harms its invertebrate host. *Endotrypanum* lives inside the erythrocytes of its host (Sloths of South and Central Africa). *Leishmania* and *Trypanosoma* are parasites of vertebrates.

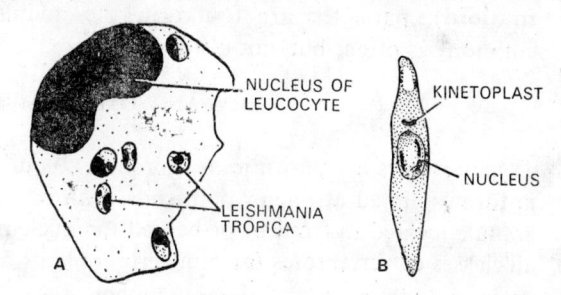

FIG. 1.23. *Leishmania* of oriental sore. A—*Leishmania* in white blood cell; and B—*Leishmania* without flagellum.

## LEISHMANIA DONOVANI

In the vertebrate host the parasite exists as amastigote form, usually 2-4 $\mu$ in diameter often referred to as "Leishman Donovan" or "L. D." after the names of their first observers. These are ingested by the macrophages as part of the latter's phagocytic activity. The parasites thrive and multiply by binary fission within the microphages, they are not destroyed. If the microphage divides the parasites pass on to two daughter cells, if the infected cell dies the liberated parasites are presumably ingested by other macrophages, thus infecting most cells. Infected macrophages in the blood or skin are ingested by sand flies (*Phlebotomus* spp.) when the insects (females) feed on blood. In the mid-gut of the sand fly, the parasites emerge from the macrophages, transform into promastigote forms by elongating and developing the flagellum and begin to multiply by binary fission. The flagellates spread forward to the gut of the insect and many of them become attached to the mucosa by their flagella. They become established in the fore-gut and by their multiplication eventually block the cavity of the proventriculus, pharynx and proboscis. About ten days after ingesting the parasites when the insect again attempts to feed (insect feeds about every five days) it injects the flagellates in the skin of the host. Here they are phagocytised by macrophages inside which they become amastigote and begin multiplying.

*Leishmania donovani* causes **visceral leishmaniasis** or **kala-azar** of man throughout the warmer parts of Asia, the Mediterranean coasts, North and East Africa and South America. In some areas (the Mediterranean, China, Central Asia, South America) dogs are often infected and serve as reservoirs of the parasites. In East Africa and Sudan this role is played by wild rodents, while in India man is the only vertebrate host. Leishmaniasis is a slow disease and is usually fatal if not treated. It is sometimes followed by a nodular eruption of the skin which may be allergic in origin (**post-kala-azar dermal leishmanoid**), parasites are found in the nodules. Treatment with compound of antimony is often, but not always, effective.

## TRYPANOSOMA GAMBIENSE

Trypanosomes are parasitic or saprozoic colourless flagellates typically elongated in form, pointed at one or both ends, with one flagellum springing from a basal granule located in front of or behind the nucleus. They occur as parasites in all classes of vertebrates (also in certain plants and other animals) and are pathogenic only to man and domestic mammals. The pathogenic trypanosomes are confined to tropical courtries.

**Structure**

The adult trypanosome has typical fusiform body pointed at both ends (Fig. 1.24A). A firm membranous **pellicle** maintains the form of the body, the cytoplasm of which contains a large, vesicular **nucleus**. The flagellum arises from the posterior end (wrongly regarded as anterior end by some people).

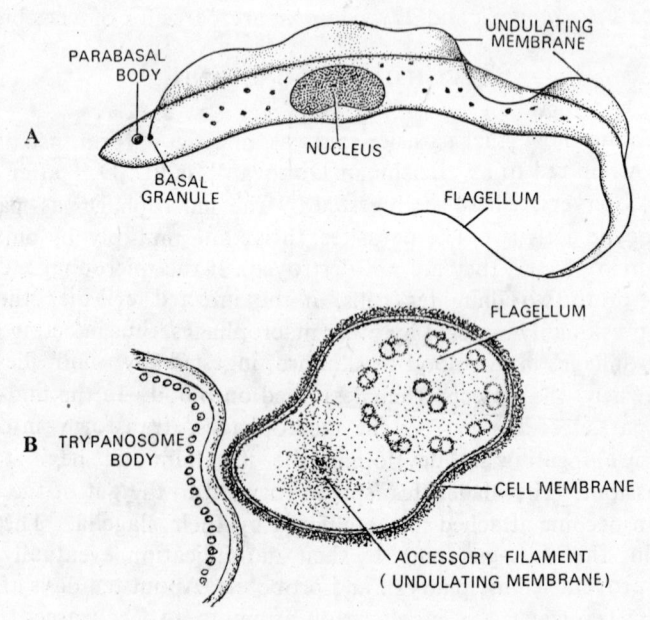

FIG. 1.24. A—*Trypanosoma gambiense*; and B—Transverse section through undulating membrane.

At the base of flagellum is a darkly staining granule, the **blepharoplast**, and there may be present an additional structure, the **parabasal body**. The blepharoplast and the parabasal body are said to form the **kinetic elements** or **kinetoplast**. The flagellum runs along the entire length of the body on the edge of an **undulating membrane** which represents, so to say, pulled out cytoplasm of the body forming a "membrane" connecting the flagellum with the

*Phylum Protozoa*

body. The flagellum becomes free at the anterior end. The trypanosomes swim freely in the blood plasma by vibratile movements of the undulating membrane and flagellum. The free end of latter points forwards.

**Nutrition**

Nutrition is effective by the absorption of nutrient material from the blood plasma. The gaseous exchanges in respiration and the elimination of nitrogenous excretory products take place by diffusion through the pellicle. There is no contractile vacuole as no mechanism for osmo-regulation is needed.

**Reproduction**

Usually reproduction in the trypanosomes takes place by longitudinal binary fission. First, the kinetic elements divide and then the nucleus, which is followed by the longitudinal division of cytoplasm. In some cases the flagellum may also divide, but usually one-half of the protoplasm takes the old flagellum and the other forms a new one, this being done before the complete separation of the cytoplasm.

*Trypanosoma gambiense* causes African sleeping sickness in man, and another sleeping sickness, known as "encephalitis" is caused by virus. The characteristics of the disease include fever, in the early stages, enlarged lymph-glands, anemia and lethargy, etc. probably, produced by the poisonous by-products of metabolism, toxins, liberated in the blood. Finally, the parasites invade the fluid surrounding the brain and spinal cord, the victim loses consciousness, and the "sleeping sickness," as it is called due to this stage, goes to a fatal end.

The disease is transmitted from man to man or from mammals to man by blood-sucking **tsetse-fly** (Fig. 1.25) of the genus *Glossina* in whose digestive

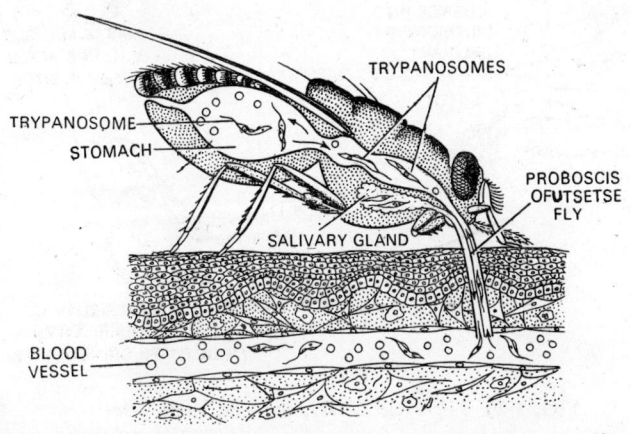

**FIG. 1.25.** Life cycle of *Trypanosoma* that causes African sleeping sickness. The figure shows an infected fly biting an infected man; principal stages in the life cycle are shown; trypanosomes of salivary gland being injected into blood are entering the proboscis of tstse fly.

tract several changes take place. When the tsetse-fly sucks the blood of any wild mammal (almost all wild animals in Africa lodge these in non-pathogenetic condition) or infected man, the blood carries trypanosomes to the intestine, where they survive the action of digestive juices and undergo multiplication by binary fission, and form numerous slender forms. These invade the interior part of the gut from where they work their way into the salivary glands through the labial cavity and hypopharynx. Here they become attached to the cells in the glandular part and a further phase of multiplication occurs, during which, among many other forms similar to those found in the vertebrate blood, are produced. Such forms fill the cavities of the salivary glands. At this stage the

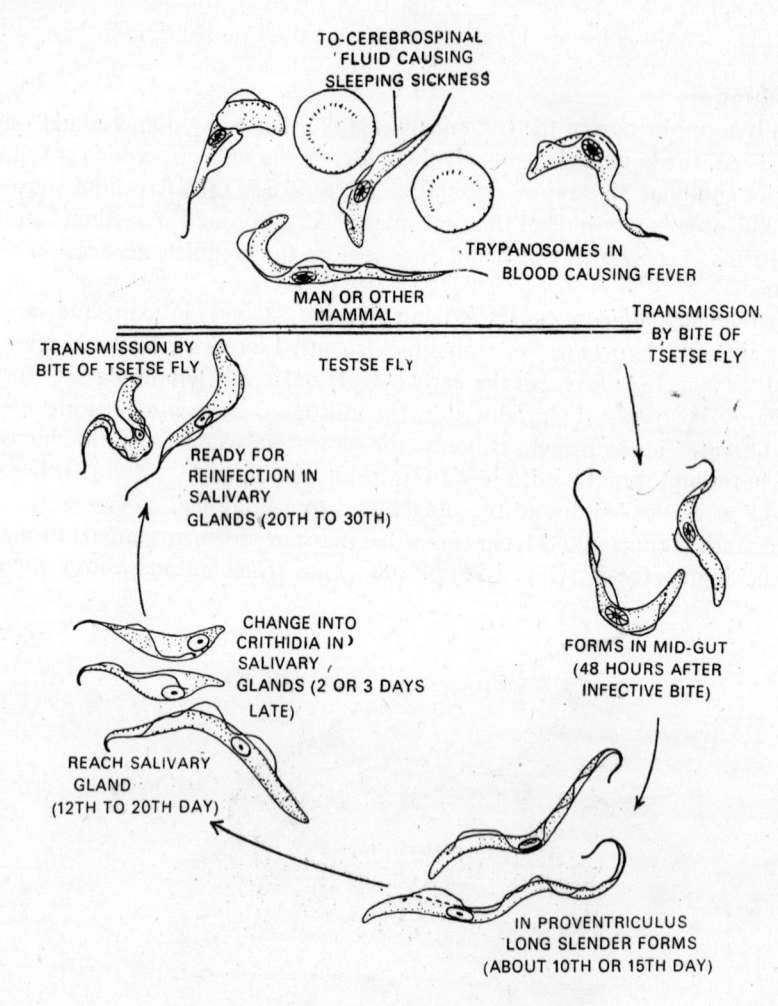

FIG. 1.26. Life-cycle of *Trypanosoma gambiense*.

fly is said to be infective which will release trypanosomes in the blood of another man or mammal along with saliva while sucking blood. *Glossina* becomes infective 20-34 days after sucking the blood, i.e., the parasites complete the reproductive cycle in the fly in this period before they enter the salivary glands. The incubation period of the disease is 2-3 weeks. Between two and five days after the bite of the infected *Glossina*, there occurs at the site an inflammatory swelling known as the **trypanosome chancre**. This is the localised primary effect, the disease will supervene after the incubation period.

# Phylum Protozoa

The ravages of the murderous pair of the tsetse-fly and the trypanosome are great with the result that large regions in Africa are uninhabitable for man and their stock. Control measures for the disease are complicated because of the prevalence of the parasites in the wild mammals. They act as vast reservoirs from which flies are easily reinfected. An adequate solution for this problem seems to lie in the isolation of harmful toxins with the help of which human beings can be made immune to the parasite. The parasite will then be able to live in man without producing an incapacitating and usually fatal disease.

### Other Pathogenetic Trypanosomes

Chagas' disease of Central and South America, in which the parasites multiply inside the muscles, the heart and the nervous system, causing dangerous swellings is caused by *Trypanosoma cruzi* and is transmitted by a bug (*Triatoma megista*). *T. cruzi* is smaller than *T. gambiense,* its large blepharoplast, however, is especially noticeable. The parasites are not plentiful in the blood. On entering the tissue the trypanosomes lose their flagella and assume the form of leishmania, in which form they reproduce by mitotic division which was once considered to be schizogony. Clusters of parasites develop intracellularly. These clusters contain hundreds of parasites in a characteristic grouping. The leishmania-like parasites before entry into circulation, are transformed into flagellate forms. This change between the blood and tissue phase of the parasite is continually repeated. The Chagas disease is accompanied by fever, diarrhoea, swelling of the lymph glands, and in the chronic stages a characteristic myocarditis occurs.

The natural reservoirs of *T. cruzi* are armadillos, cats, dogs and marsupials. That means these animals carry hordes of parasites, and it is from these that the bug *Triatoma megista* sucks and transfers them to the definite host. They are harmless to these animals. *T. brucei* causes nagana fever of a variety of African domestic mammals and is transmitted by some species of *Glossina*. The **surra** disease common among horses, mules, cattle and camels in our country is caused by *T. evansi* which is carried by tabanid flies; *T. lewisi* occurs in the rat and is transmitted by a flea. Several non-pathogenetic forms also occur in mammals. Common among these are *T. primatum* in anthropoid apes, *T. theileri* in cattle and *T. melophagium* in sheep.

## FLAGELLATE PARASITES OF THE ALIMENTARY CANAL AND URINOGENITAL TRACT

Many flagellates live as parasites in the gut and urinary tract. Those that live in the gut of man include *Chilomastix mesnili, Enteromonas hominis, Retortamonas intestinalis, Giardia lamblia* and *Trichomonas hominis*. Of these only *Giardia* is sometimes pathogenic. *Trichomonas vaginalis* lives in female vagina and the male urethra or prostate, and is common throughout the world specially in women.

### GIARDIA LAMBLIA

Species of the genus *Giardia* (Fig. 1.27 B) have been described from the small intestine of man, dogs, cats, cattle, various rodents, rabbits and other mammals; a few species have been reported from amphibia and reptiles. The species *Giardia lamblia* has also been named as *G. intestinalis*, *G. duodenalis* and even as *Lamblia intestinalis*. Antony Van Leeuwenhoeck in 1861 saw this protozoan in his own faeces.

### Pathogenesis

*G. lamblia* inhabits the lumen of the duodenum and upper ileum of man, monkeys and pigs in all parts of the world. It is common in human children. It causes a disease known as **giardiasis** or **lambliasis**. It does not invade tissues. When the number of parasites increases they produce acute diarrhoea, especially in children. This is accompanied by epigastric pain. It is believed that sometimes the parasite swims into the gall bladder and causes jaundice, nausea and vomiting. If the infection continues to be heavy it may interfere with the absorption of fat and fat soluble vitamins.

### Structure

*Giardia* trophozoite is shaped like a pear bisected longitudinally (Fig. 1.27).

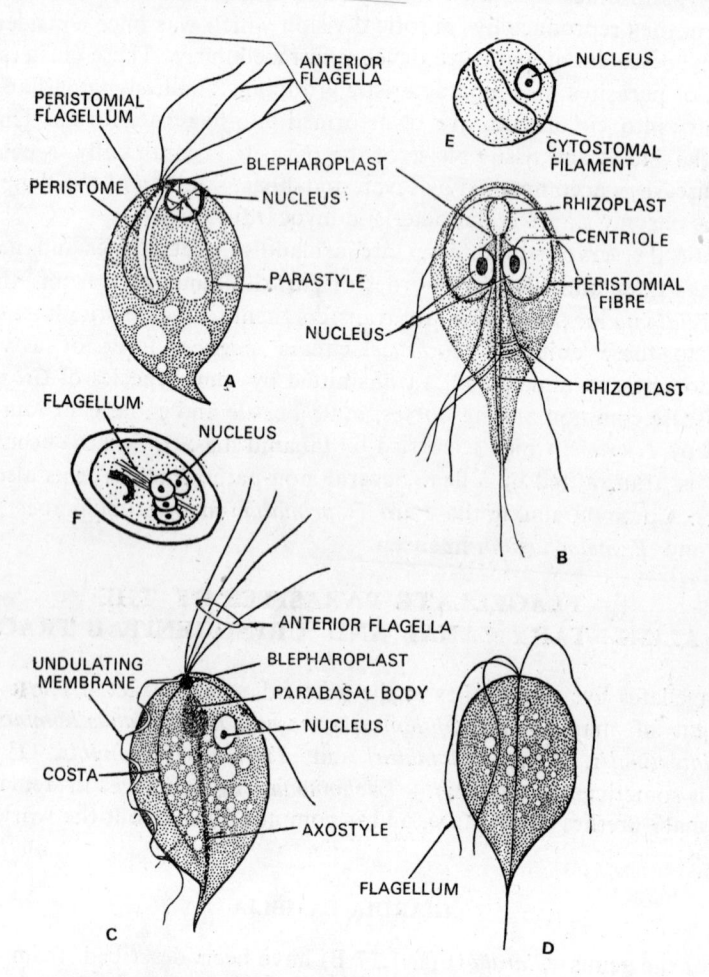

FIG. 1.27. Intestinal flagellates of man. **A**—*Chilomastix mesnili*; **B**—*Giardia*; **C**—*Pentatrichomonas hominis*; **D**—*Enteromonas hominis*; **E**—Cyst of *Chilomastix mesnili*; and **F**—Cyst of *Giardia*.

# Phylum Protozoa

The flat ventral surface of the broad anterior end forms a sucker, with a thickened anterior rim. The protozoan attaches itself to the intestinal mucosa with this end. There are two nuclei above the surface. Between the nuclei eight basal bodies are located. The basal bodies give rise to eight flagella. Two of the flagella emerge directly from the body, two cross over and follow the front edge of the sucker before emerging laterally, and remaining four are directed backwards on the body (Fig. 1.27B). Of these four, two emerge at the hind end of the sucker and the remaining two at the extreme hind end of the body. The flagella help the protozoan in swimming. In the posterior half of the body lie one or usually two curved **"median bodies"** of unknown function. Some authors have called them parabasal bodies which they are not. True parabasal bodies are clear under electron microscope. The protozoan measures 10-20 $\mu$ long and 5-10 $\mu$ broad and 2-4 $\mu$ thick. Reproduction is not known.

The cyst of *G. lamblia* contains four nuclei grouped at one end. The remains of the median body flagella and anterior rim of the sucker appear as a collection of fibrils (Fig. 1.27E) within the cyst; with these characters it is easy to recognise *Giardia* in a preparation of the faeces. When a cyst is swallowed by a susceptible host it presumably hatches in the abdomen. The organism that emerges is quadrinucleate. It divides into binucleate trophozoites.

**Treatment and prevention.** Giardiasis is easily cured by administration of antimalarial drugs **mepacrine** or **chloroquine**. Prevention depends solely on hygiene with regard to food so that cysts may not be ingested.

## TRICHOMONAS VAGINALIS

Species of this genus occur in the intestines of mammals including man (Fig. 1.28), birds, reptiles, amphibia, molluscs (slugs) and termites; in the mouth of man and monkeys (*T. tenax*) and in the urinogenital tract of man and cattle (*T. vaginalis* and *T. foetus*). They reproduce by binary fission, no sexual reproduction is known and no cyst is produced.

*Trichomonas vaginalis* is ovoid narrower at the hind end ranging in size from 10-30 $\mu$ long and 5-15 $\mu$ broad. It has four free anterior flagella and one recurved posterior flagellum and undulating membrane end about half way along the body. All the flagella arise from basal bodies grouped at the anterior end just in front of the simple nucleus. The flagella help the animal swim actively. According to electron microscopic

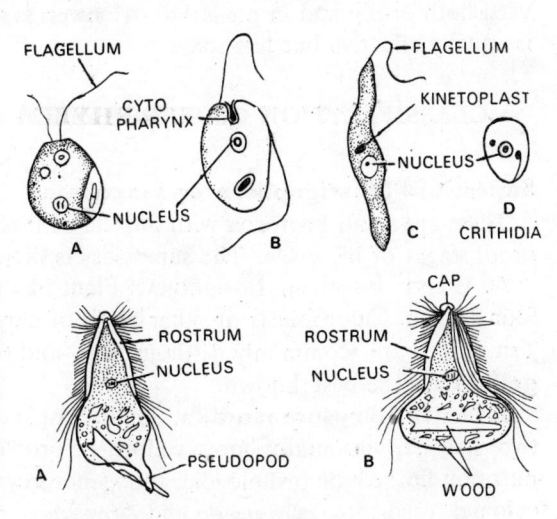

FIG. 1.28. Some flagellates. A—*Monas vestita*; B—*Bodo caudatus*; C and D—Stage of *Crithidia* from the intestine of horse fly; and E and F—*Trichonympha* from termite.

studies by Smith and Stewart (1966) the flagellate also possesses a prominent skeletal axostyle of longitudinally arranged parallel microtubercles, which may protrude from the hind end of the body, a **cystosomal groove,** a supporting **costa** at the base of the undulating membrane, and a true parabasal body beside the nucleus. In the living state the parasite shows beautiful progression of waves backward along the undulating membrane.

**Pathogenesis**

The parasite is non-pathogenic mostly. In males also it is non-pathogenic, although occasionally it has been known to cause mild inflammation of the urethra. In women the parasite causes mild vaginal inflammation associated with a copious foul-smelling discharge. As far as known *T. vaginalis* does not invade the tissues. It has been suggested that the parasite's pathogenecity is associated with endocrinal or other changes resulting in variations in normal bacterial flora of the vagina, leading to a reduction in the acidity of its contents from the usual pH 4-4.5 to pH 5.5. However, the parasite cannot survive at neutrality (pH 7).

*T. vaginalis* is probably transmitted via an infected male, but may also occur by contamination of the vaginal orifice with infected material.

**Diagnosis.** Microscopic examination of fresh preparation of the vaginal discharge showing motile organisms is quite reliable; dried smear stained in Giemsa's stain (or other stain) is not so reliable. The best result is obtained by the *in vitro* cultivation of the vaginal discharge to isolate the parasites.

**Treatment.** Radical cure from the infection is sometimes not easy although good results are obtained in proper treatment. The antibiotic **Trichomycin** is effective both orally and topically. Other treatments include use of organic arsenicals such as **carbasone** (4-carbamylaminophenylarsonic acid) have been used both orally and in pessaries. A newer arsenic compound, **metronidazole**, is equally effective but less toxic.

## CLASSIFICATION OF SUB-PHYLUM SARCOMASTIGOPHORA

**Superclass Mastigophora or Flagellata**

These are small Protozoa with flagella as locomotor organs, present in some or all stages of life-cycle. The superclass is divided into two classes:

A. CLASS PHYTOMASTIGOPHORA. Plant-like flagellates with never more than four flagella. Chloroplasts or other kinds of chromatophores are usually present. Ten orders are commonly distinguished and of these **Euglenida** and **Dinoflagellida** are the best known.

1. **Order Chrysomonadida.** Small, simple often amoeboid forms with one or two flagella, chromatophores yellow or brown one to several; gullet wanting, nutrition holophytic or holozoic, endogenous cysts of silica are produced. Some colonial forms (*Synura uvella* and *Uroglenopsis americana*) inhabit fresh waters, causing bad flavour in water supplies by aromatic oils released at death.

Examples: *Chromulina* (Fig. 1.29), *Chrysamoeba radians*, *Synura uvella* (Fig. 1.21A), *Ochromonas*, etc.

2. **Order Silicoflagellida.** Flagellum single or absent, with brown chromoplast, internal siliceous skeleton. Marine, known mostly from fossil forms.
Example: *Dictyocha*.

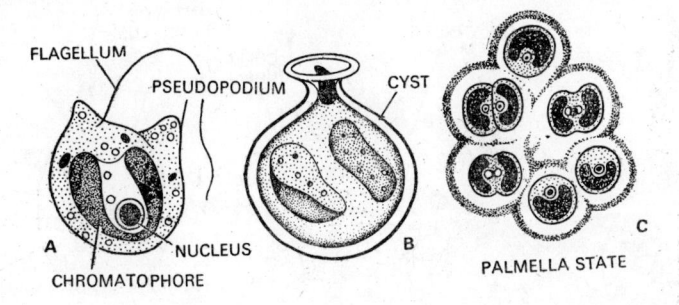

FIG. 1.29 *Chromulina*. A—The flagellate; B—Cyst and C—Palmella stage.

3. **Order Cocolithophorida.** Tiny marine flagellates covered by calcareous platelets (**coccoliths**), with two flagella and yellow or brown chromoplasts. No endogenous siliceous cyst.
Examples: *Cocolithus, Rhabdodosphoera*.

4. **Order Heterochlorida.** Two unequal flagella and yellow-green chromoplasts, siliceous cysts.
Examples: *Heterchloris, Myxochloris*.

5. **Order Cryptomonadida.** Small, oval flagellates of constant forms (not amoeboid), flagella two arising from a gullet-like depression, the walls of which bear small rods resembling trichocysts; nutrition holophytic or holozoic; holophytic types bear chloroplasts of various colours, some saprophytic forms are colourless, some (*Cryptomonas*) live as zooxanthelle in other Protozoa and Metazoa.
Examples: *Chilomonas* (colourless), *Cryptomonas* (coloured).

6. **Order Dinoflagellida.** Plankton protozoans. There is a definite capsule made of cellulose (Fig. 1.30). The capsule is in two parts and has an equatorial groove, the sulcus; with two flagella, one directed backwards (trailing) and usually in the longitudinal groove, and the other transverse flagellum, usually in a more or less spiral or transverse groove, (annulus). A stigma and numerous small yellow brown or infrequently greenish chloroplasts are usually present, Nucleus is single and large. Characteristic of the order is the presence of large sacs connected to the exterior by canals. These non-contractile spaces form the so-called **pusule** system which unlike the vacuolar system serves for the intake of the water, exact function of which is not understood, probably it is hydrostatic. Some through loss of chloroplasts have become entirely holozoic and partly or completely amoeboid in form and behaviour.

Reproduction is by fission the plane of which is oblique but it resembles longitudinal fission of the flagellates as it passes between two flagella. Fission may be within or without a cyst. It may be simple, binary or in the encysted condition it may be multiple. The products of repeated binary fission of pelagic forms sometimes hang together for a considerable time as a chain. Syngamy is suspected but sufficient evidences are not available.

**Example 1:** *Ceratium* (Fig. 1.33) is a typical armoured, holotypic form with three long spines. It bears two flagella. The thick and rigid theca is formed of

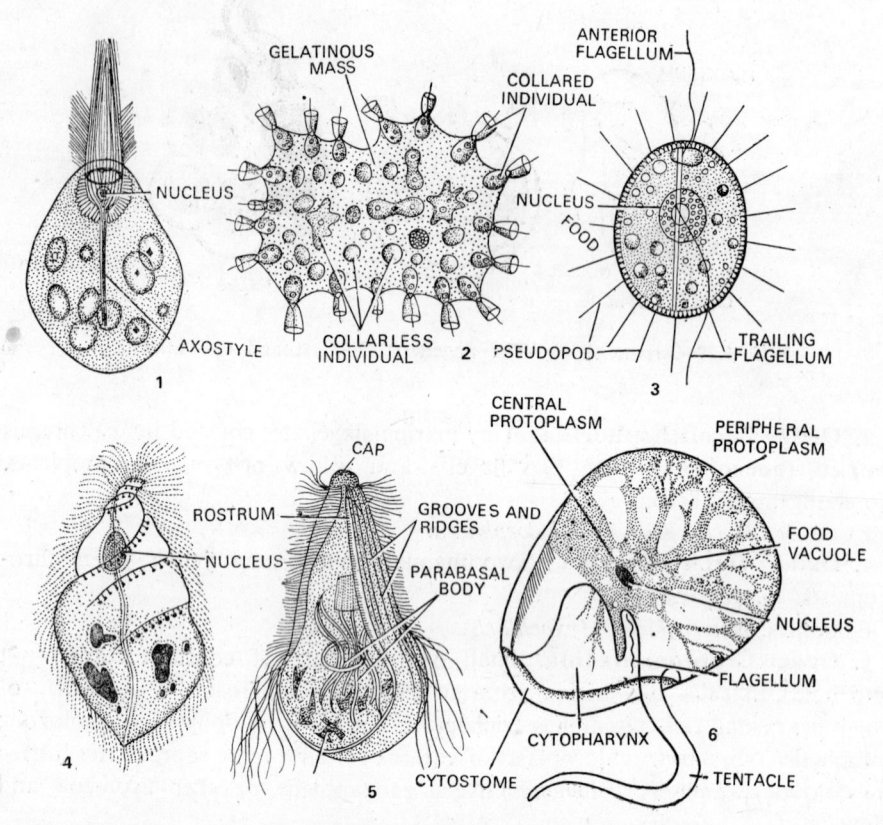

**FIG. 1.30.** Some well-known flagellates. 1—*Lophomonas*; 2—*Proterosponga*; 3—*Thaumatomastix*; 4—*Spirotrichonympha*; 5—*Trichonympha*; and 6—*Noctiluca*.

a variable number of distinct plates and possesses two usual grooves. The plates forming the theca may be differently shaped in different species and bear fine perforations. The long spines are simple elongations of some of these plates. These spines may be curved or straight; one of them is anterior (**apical spine**) and one to four posterior (**antapical spines**). The annulus, which is mostly equatorial, is covered by a single plate, the **cingulum**. Through the pores of the theca protrude out numerous cytoplasmic papillae which give rise to network of **pseudopodia** (Fig 1.33C) which catch micro-organisms for food. Nutrition is holophytic, sometimes holozoic. They reproduce by oblique binary fission (Fig. A.33D) which, it is reported, takes place at night and in the early morning. Chloroplasts are green in fresh water forms, yellow or brown in marine forms.

Example 2. *Noctiluca* (Fig. 1.30) is a large gelatinous sphere often 1 mm. or more in diameter floating on the sea, usually near shores. It is colourless and ingests solid food, one flagellum is short and thick and the other is delicate. *Noctiluca* is a luminescent dinoflagellate and is responsible for much of the phosphorescence of the ocean shore water. *Noctiluca* reproduces by binary fission and by the formation of isogametes each with one long trailing flagellum. These fuse in pairs, but the fate of the resulting zygote is not fully known.

7. **Order Ebriida.** Biflagellate without chromoplasts. Mainly fossil forms. Internal skeleton siliceous.

Example: *Ebria*.

8. **Order Euglenida.** Elongated, spindle-shaped forms with pellicle usually ridged or striated; the anterior end bears a red stigma in green forms and is provided with a depression or cytopharynx from which one or two flagella emerge and which serve for food intake in holozoic type; chromatophores green or none; nutrition holophytic, holozoic or saprophytic; mostly inhabit fresh water.

Examples: *Euglena* (Fig. 1.31), *Astasia* (colourless), *Trachelomonas* (in shell), *Phacus torta* (Fig. 1.31).

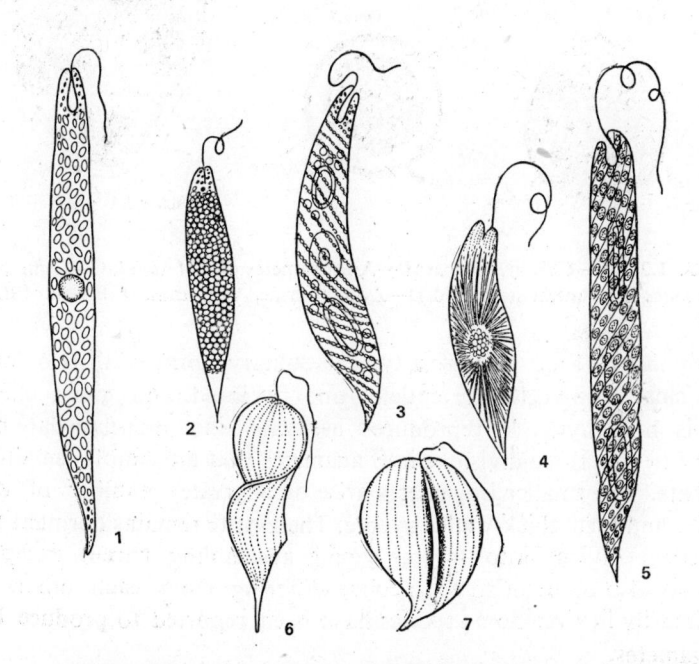

FIG. 1.31. Common species of Euglenoid forms (*after* various sources) 1—*Euglena deses*; 2—*E. proxima*; 3—*E. spirogyra*; 4—*E. viridis*; 5—*E. oxuris*; 6—*Phacus torta*; and 7—*P. pleuronectes*.

9. **Order Chloromonadida.** Small amoeboid bodies with two flagella and pale green chloroplasts; assimilation product is oil and not carbohydrate.

Example: *Coelomonas, Goniostomum*.

10. **Order Phytomonadida or Volvacida (Volvocales).** Small oval or elongated flagellates of fixed form, enclosed in a cellulose membrane usually with two rather short equal flagella; one large cup-shaped chloroplast, red stigma, two contractile vacuoles and a simple vesicular nucleus. Single or colonial forms chiefly inhabiting fresh water; nutrition is holophytic or saprophytic. Asexual reproduction by division within the membrane and sexual repro-

duction with all grades of isogamous gametes. Cyst formation occurs. Palmella formation is also frequent.

Examples: *Chlamydomonas* (Fig. 1.32), *Chlorogonium* typical single forms, and *Pleodorina* and *Volvox* (Fig. 1.17) typical colonial forms.

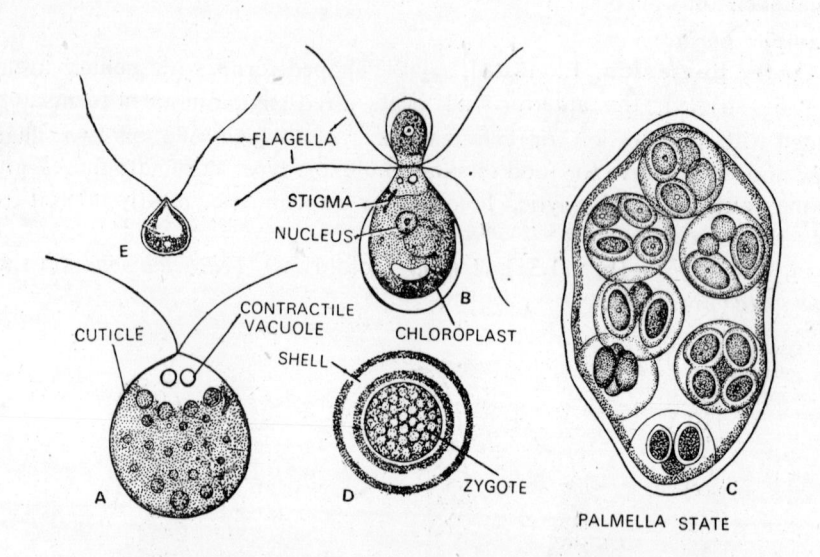

FIG. 1.32. A—*Chlamydomonas*; B—Anisogametes of *C. braunii*; C—*Chlamydomonas* in palmella state; and D—Zygote formed as a result of fusion of B.

*Chlamydomonas* (Fig. 1.32) is a typical solitary form with two flagella, an eye-spot, a close fitting cellulose cuticle, one pyrenoid and green chloroplast. Nutrition is holophytic. It reproduces asexually in a non-flagellate palmella-like state by two or three divisions into agametes that are simply small replicas of the parents. The smaller isogametes arise by a greater number of divisions, fuse in pairs and form thick-walled zygote. The zygote remains dormant for some time and then divides into several young resembling parent except in size. Anisogametes also occur in some species, differing from each other and the parent primarily in size. Some species have been reported to produce both iso- and anisogametes.

B. CLASS ZOOMASTIGOPHORA. Flagellates of distinctly animal nature without chloroplasts and with definitely holozoic nutrition. They have often more than one flagellum and sometimes more than four. Some are free-living and some are well-known parasites. This is divided into four orders.

1. **Order Choanoflagellida.** Fresh water flagellates with a single flagellum surrounded by a collar, sessile, stalked solitary or colonial.

Example: *Codosiga* and *Proterospongia* (Fig. 1.30)

2. **Order Bicosoecida.** Largely fresh water flagellates enclosed in a lorica. Two flagella, one free, the other attaching posterior end of body to shell.

Example: *Salpingoeca*, *Poteriodendron*.

3. **Order Rhizomastigida.** Amoeboid forms with one to many flagella, chiefly fresh water.

*Phylum Protozoa*

Examples: *Mastigamoeba, Dimorpha*.

4. **Order Kinetoplastida**. Basically one, but up to four flagella. Mostly parasitic.

Examples: *Bodo* (Fig. 1.28), *Leishmania, Trypanosoma*.

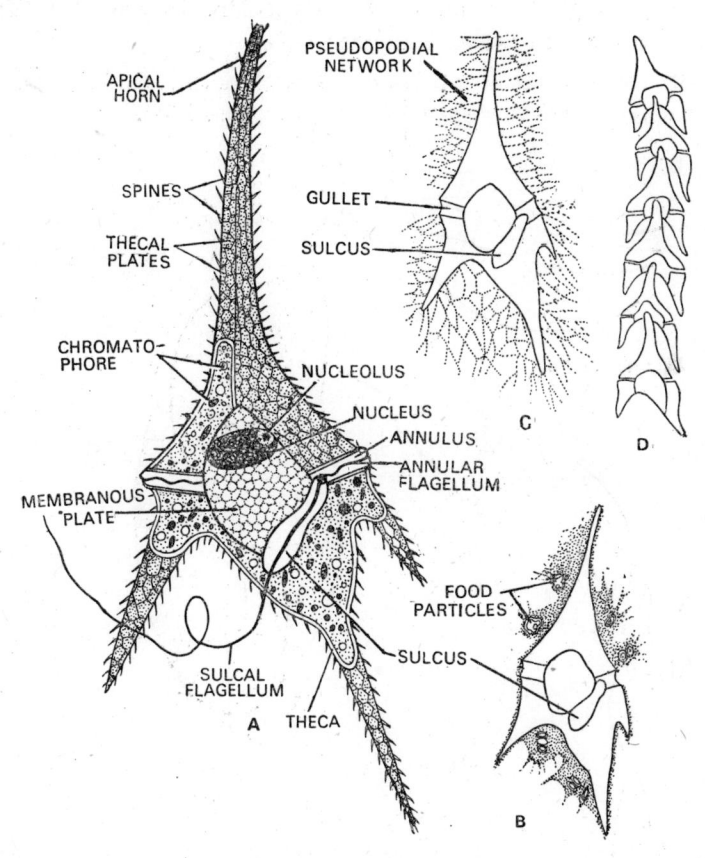

FIG. 1.33. **A**—*Ceratium hirudinella*; **B**—Showing pseudopodial network; **C**—Food particles captured by pseudopodial network; and **D**—Colony formation.

5. **Order Retortomonadida**. Gut parasites of insects or vertebrates with one to four flagella. One flagellum associated with ventrally located cytostome.

Example: *Chilomastix* (Fig. 1.27).

6. **Order Diplomonadida**. Bilaterally symmetrical flagellates with two nuclei, each nucleus associated with four flagella, Mostly parasitic.

Examples: *Hexamita, Giardia* (Fig. 1.27B).

7. **Order Oxymonadida**. Parasitic multinucleate flagellates. each nucleus with four flagella some of which are found posteriorly and adhere to body surface.

Examples: *Oxymonas, Pyronympha*.

8. **Order Trichomonadida**. Parasitic flagellates with four to six flagella, one of which is trailing.

Example: *Trichomonas*.

9. **Order Hypermastigida.** Small, complex flagellates with numerous flagella living as symbionts in the gut of insects. The flagella occur in various ways in a terminal tuft (*Lophomonas*), in two or four elongated groups at the anterior end (*Barbulonympha*) or over all of the body in longitudinal grooves (*Trichonympha*). Reproduction takes place by fission. They assist in the digestion of wood in the gut of wood-eating termites.

Examples: *Lophomonas* (from domestic cockroach), *Barbulonympha* (from wild cockroach) and *Trichonympha* from termite (Fig. 1.34).

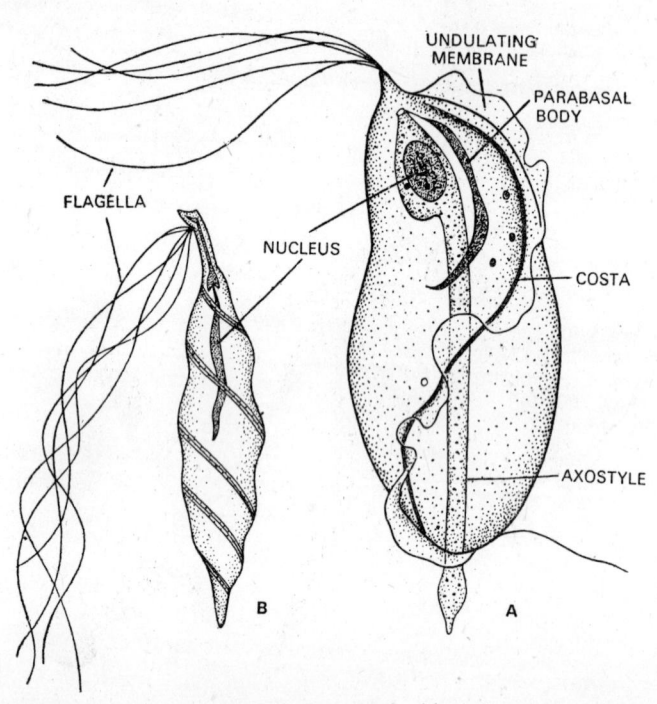

FIG. 1.34. Flagellates of the termite gut. A—*Trichomanas termopsidis*; and B—*Streblomastix strix*.

## THE OPALINID PROTOZOA

The subclass Protociliata of the earlier authors is now renamed class Opalinatea. This includes Opalinid (*Opalina* like) Protozoa found in the rectum of frog and toad. Opalinid Protozoa were found in the faeces of frog by Leeuwenhoek in the days when microscopy was just taking shape. They look like ciliates with which they were classified hitherto, but now it has been found that they differ from Ciliata in their morphology, form of division and life-cycle. It seems likely that ciliates and opalinids evolved along somewhat divergent lines from the same or similar ancestral flagellates. They are now considered to be distinctly separate from both (ciliates and flagellates) groups although they exhibit some features of both flagellates and ciliates.

Opalinids are found in the rectum of frogs and toads. The one recovered from the rectum of *Rana temporaria* is called *Opalina ranarum* (Fig. 1.35). It is a leaf-shaped structure measuring 300 to 800 μm long, 300 μm wide and 30 μm thick. The animal is laterally flattened and the side on which it usually lies on

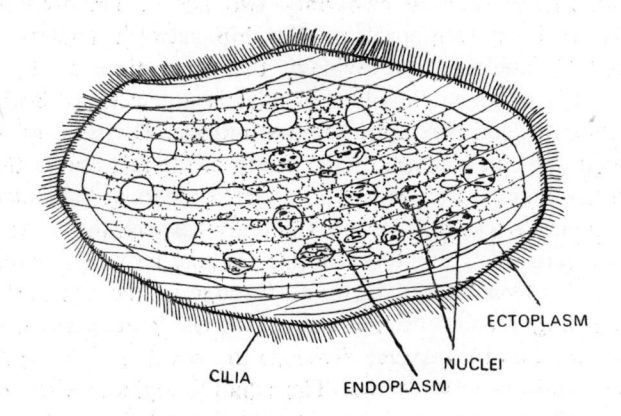

FIG. 1.35. *Opalina ranarum* (*after* Bhatia and Gulati).

a microscope slide is the left side. Both surfaces of the body are densely covered with cilia, which are about 15 μm long and are borne in rows extending backwards. The cilia move in close waves which can be seen sweeping back over the surface of the body as the animal swims forwards. The direction of swimming is changed by alteration in the direction of beat of the cilia.

Between the rows of cilia the pellicle is thrown into narrow longitudinal folds, each supported by a single longitudinal ribbon of 12-24 microtubular fibrils. About six of these folds occur between adjacent ciliary rows, and in between the folds there are signs of pinocytotic activity. *Opalina* has no cytostome, and appears to feed by pinocytosis over much of the body surface. Although the opalinids are found in the rectum of various hosts there is no evidence that any of them harms its host at any stage of its life cycle. It may derive nourishment from mucous secretions as well as from gut fluids in the rectum.

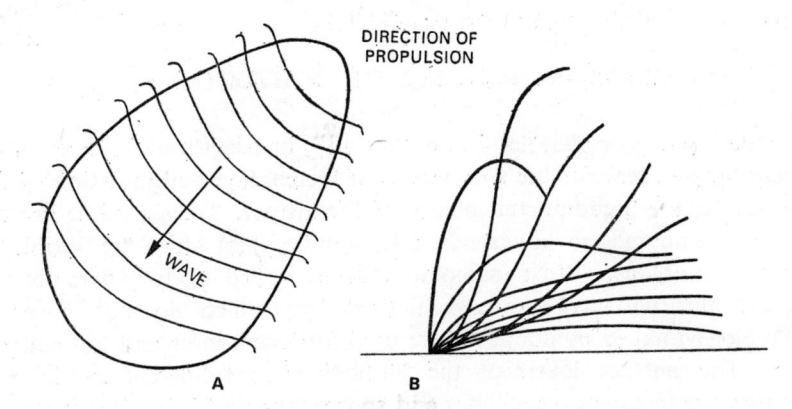

FIG. 1.36. Features of *Opalina*. A—Metachronal wave patterns—viewed from above (Sleigh, 1966); and B—Beat of cilium (Sleigh, 1968).

At the anterior margin of the body there is a zone of cilia, one cilium wide at the ends and four to six cilia wide for most of its length. This region is called the *falx*, a region of some morphogenetic importance. The other ciliary rows arise on either side of the falx. *Opalina* has large number of similar nuclei. Some genera like *Zelleriella* may have only two nuclei. The division of *Opalina* is peculiar. There is no temporal relationship between nuclear division and cell division. All the nuclei do not divide at the same time so that a variable number of nuclei may be involved in mitosis at any time. Both the nuclear membrane and the nucleoli persist throughout mitosis. Division of the body is usually preceded by an increase in the length of the falx and in the number of kinetosomes in this region. Following this a cleft appears at the anterior end of the body and extends backwards across the body between the rows of cilia to divide the animal into two daughters whose sizes may be equal or quite unequal. The trophic forms of *Opalina* may be found throughout the year in the adult host. During the breeding season of the amphibian the protozoans divide in rapid series of longitudinal and transverse divisions to provide many small daughters with a few nuclei and few ciliary rows. The small daughters with about three to six nuclei form approximately spherical cyst about 30 $\mu$m across in the faecal mass of the host and are passed out of the amphibian. The cysts survive for some weeks in water but if they are eaten by tadpoles they hatch quickly in the gut. The protozoans which emerge from the cysts are gamonts. They multiply by a series of longitudinal fissions to produce uninucleate fusiform gametes. The gametes are of two types, slender microgametes and fatter macrogametes. It is assumed that gametes are produced after meiosis. Fertilization takes place by the attachment of microgamete to macrogamete and the subsequent absorption of the former by the latter. The zygote, later, becomes rounded, secretes a cyst wall and is seen to be uninucleate. The zygocysts are shed into water with the faeces of the tadpole. On being eaten by another tadpole they may hatch to produce another generation of gamonts and gametes or may develop into multinucleate forms which are cylindrical in the beginning but become fattened later on.

*Opalina* differs from typical ciliates in (*i*) having many nuclei of only one type, (*ii*) occurrence of a division plane running across the ciliary row, (*iii*) the form of infraciliature, and (*iv*) the process of syngamy.

## CLASS—SARCODINA OR RHIZOPODA

The Sarcodina (Gr. *sarcodes*, fleshy) are "naked" or shell-bearing Protozoa in which pseudopodia serve as the sole means of locomotion and ingestion of food at least during the predominant phase of life-history. The sarcodine Protozoa normally possess pseudopodia (*pseudo*, false + *podos*, feet) which are flexible type of locomotor organs. Most sarcodines have naked body. There are some amoeboid groups that secrete tests, shells or skeleton which do not cover the body completely; naked pseudopodia are used for locomotion and for collection of food. The surface layers of the cell body are not differentiated in special ways as they are in flagellates, ciliates and sporozoans and it is felt that this lack of cortical differentiation is a result of cytoplasmic mobility necessary for pseudopodial movement. Although the surface layer does not show any specialisation, many sarcodines show some remarkably complex structures, particularly in external shells and in internal skeletons, both living and non-

# Phylum Protozoa

living. The mode of nutrition is holozoic. Ready food is captured with the help of pseudopodia and engulfed at any one point of the body. Members reproduce predominantly asexually mostly by binary fission, sometimes by multiple fission. Some members of the group reproduce sexually. Major features of interest in Sarcodina centre around the pseudopodia and their use in locomotion, the composition, organisation and construction of the shells and skeletons, the specialised life cycles and reproduction of members of some groups, and relations with other organisms in various symbiotic relationship.

## TYPE AMOEBA

The common *Amoeba* (Fig. 1.37) of fresh water ponds and ditches, etc. is a simple looking microscopic animal. An attempt to understand how it moves and

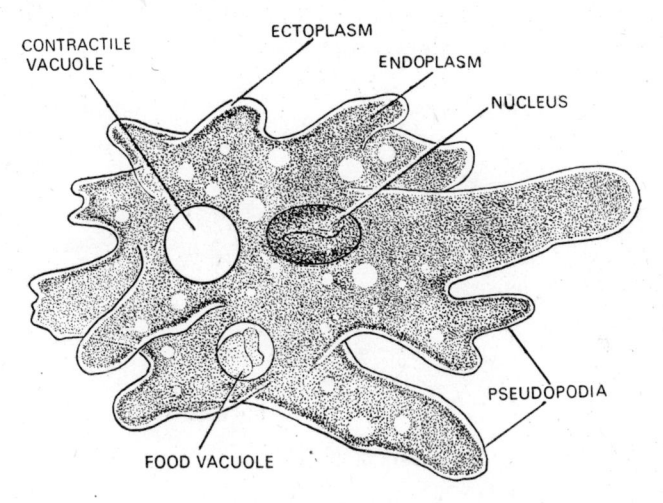

FIG. 1.37. Structure of *Amoeba proteus*.

behaves, how it feeds and how it reproduces makes it clear that this seemingly simple *Amoeba* is rather a complex animal. It is difficult to study it the way the other animals are studied because all the capacities that large animals delegate to their different organs are inextricably mixed up in a shapeless and almost structureless drop of protoplasm. *Amoeba* presents naked protoplasm actively living.

It is not difficult to find amoebae. Damp moss scraped from a wall, or a little ordinary soil will usually produce hundreds of them if it is left for a few days in a bowl of water. But they will be mostly the smaller species, less than 1/20th mm. across and, therefore, quite invisible with the naked eye. There are several large species of amoeba up to 1 mm or so in size, but they are less easy to find. The best known of them is *Amoeba proteus*, which is normally used for teaching, but is comparatively rare except in laboratories.

**Structure**

The most striking feature of the animalcule is its continually changing shape, a characteristic that led the early workers to name it "the **proteus animalcule**," after the Greek god Proteus known for his capacity to change shape. An active amoeba consists of a number of finger-like pseudopodia emerging in various directions. The substance of a pseudopodium consists of two separate phases. Outside, seemingly fairly firm and perhaps a third of the width of the whole pseudopodium, is a layer of jelly, the **plasmagel,** inside there is a fluid **plasmasol**, moving about. Each pseudopodium, in fact, is a tube, the walls of which are made up of plasmagel while the cavity is filled with plasmasol. Some authors (Shaeffer, 1920) call the fluid substance **endosarc (endoplasm)** and the jelly-like substance surrounding it **ectosarc (ectoplasm)**. In some pseudopodia the plasmasol flows outwards and forms fresh pseudopodium at the tip in others it flows inwards and the tube gradually shortens (receding pseudopodium). In *Amoeba proteus* individual pseudopodia may measure as much as 200 microns long and 50 microns across, and there may be as many as a dozen altogether. Most other species of amoeba are smaller, some with innumerable little pseudopodia, some with a single large one.

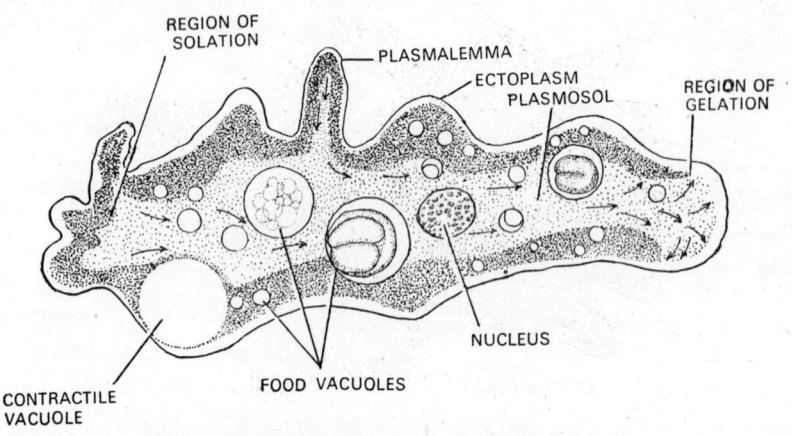

FIG. 1.38. Locomotion in *Amoeba* (*after* Mast).

According to Mast (1925) the **ectoplasm** or **ectosarc** consists of (*i*) the outer surface layer or thin hyaline membrane, the **plasmalemma,** and (*ii*) the inner finely granular layer, while the **endoplasm** or **endosarc** consists of (*a*) a coarsely granular layer surrounding, (*b*) the central fluid mass of substance. The finely granular layer of the ectoplasm is followed by the coarsely granular layer of the endoplasm. These two layers are not sharply differentiated and both appear to be relatively solid. The granules in them move very little or not at all in relation to each other and they do not move forward. Immediately below them there is a mass of substance which appears to be precisely like substances in the coarsely granular layer, except that the granules in it actively tumble over each other and move rapidly forward, they appear to be suspended and carried in a fluid stream. Thus, the central mass of substance is elongated, and it flows within a tube formed by the coarsely granular substance. This tube is open at one end. The finely granular layer not only covers the coarsely granular layer but also extends over the anterior end of the central mass, forming a closed sac. Finally, thin hyaline membrane covers this sac. According to most modern views, the coarsely granular and finely granular layers comprise the plasmagel.

Closely enveloping the whole amoeba there is a delicate, clear, relatively tough, elastic surface layer, the **plasmalemma**[5], approximately a quarter of a micron (0.25 $\mu$) thick. If a drop of water is injected beneath the plasmalemma a blister appears at the place lifting it from the layer below. A slight puncture will burst the blister and the plasmalemma will collapse proving its elasticity. It is insoluble in water and in certain regions it is separated from the plasmagel by a layer of fluid. It is not merely a surface film continuously formed at one end and destroyed at the other, it is a relatively permanent structure and is able to slide freely over plasmagel. However, it is destroyed in parts and reformed from time to time, for a portion is taken in with every food vacuole and destroyed, and if this were not replaced the whole membrane would soon be used up. It is a permeable membrane through which water, oxygen and carbon dioxide pass in and out, but proteins, fats and salts of the protoplasm are not allowed to escape into the surrounding water by the same membrane (*see* page 8).

**Endoplasm (plasmasol).** The endoplasm consists of a fluid in which are suspended structures of various types. Numerous minute spherical granules (0.25 $\mu$ in diameter), about an equal number of slightly irregular granules (1 $\mu$ in diameter) relatively more translucent, hexagonal and rectangular crystals contained in distinct vacuoles, sometimes in a very large number and varying size (the largest being 6.5 $\mu$ in diameter), various food vacuoles, usually the nucleus and the contractile vacuoles constitute the inclusions of the plasmasol. The granules appear to be in suspension in the liquid. Likewise the vacuoles containing the crystals roll about freely in the plasmasol, and they always remain intact no matter what the consistency of the substance in them may be. They also remain intact for a considerable time if the amoeba is broken open and the plasmasol flows in the surface water. The refractive spherical bodies evidently represent a step in the process of digestion. They are found in all sizes in the food vacuoles which apparently break up and discharge them. They are insoluble in water and are abundant some time and scarce or absent at others. Some believe them to be protein (Von Willer). The food vacuoles float freely in the plasmasol. They contain water that the amoeba takes along with the food. The food vacuoles may be small or large and may contain much or little food. They are apparently surrounded by an elastic membrane, not merely a surface film, for they remain intact in water as well as in plasmasol and they can be greatly distorted if made to pass through a small opening. In the formation of each food vacuole a sheet of plasmalemma is taken in. This forms a limiting membrane which probably persists until the vacuole disintegrates. The content is usually slightly acidic for some time after they are formed, and then very slightly alkaline or neutral until they disintegrate. The plasmasol looks plainly like an emulsion.

The **nucleus** in *Amoeba proteus* is practically always found in the plasmasol. Usually it lies somewhere between the middle and the posterior end, but sometimes it may be seen in the very tip of an actively extending pseudopod. Now and then it becomes attached to the inner surface of the plasmagel and remains stationary for a time, while the posterior end moves forward. It is set free due

---

[5] Called "Haut" by Rhumbler; "Häutchen" by Bütschli; "third layer" by Schaeffer and "pellicular layer" or "pellicle" by Doflein.

to the liquefaction of the plasmagel when it reaches the posterior end or is separated off by the action of the streaming plasmasol. Thus, being continuously carried back and forth, it occupies no fixed position. It differs in shape in different species. It is either biconcave, biconvex, oval or discoid. It is a structure of utmost importance to the animal. If it is removed by micro-dissection, the protoplasm may live without it for sometime, but shall remain incapable of carrying out all its vital activities. If the animal is cut into two pieces, one with and the other without the nucleus, the former behaves like an entire animal, soon grows to its previous size and finally reproduces. The piece without the nucleus may live for some time, but is not able to grow or reproduce. But an isolated nucleus also cannot survive. The nucleus and cytoplasm are interdependent.

In active **monopodal** specimens, that is, the specimens that throw out one pseudopodium and move in one direction, the plasmasol is in the form of an elongated fluid column with a knob-like enlargement at the anterior end. It is located in the middle of the body extending nearly its entire length. In **multipodal** specimens a branch of plasmasol extends in each pseudopod. Besides the nucleus the plasmasol also contains a spherical fluid filled contractile vacuole, that regulates the water contents (*see* Osmo-regulation).

**Ectoplasm (plasmagel)**. The plasmasol is usually completely surrounded by a layer of relatively solid substance the **plasmagel (ectoplasm),** which consequently forms a closed tubular sac with branches extending into the pseudopods. The plasmagel is very thin at the tip of extending pseudopods and it varies greatly in thickness elsewhere, in different individuals and also in the same individual under different conditions. It may be so thin that it forms merely a thin membrane under the plasmalemma leaving a larger space occupied by plasmasol or it may be so thick that a very little space, if any, is left. The plasmagel is an alveolar system in general, fairly uniform in thickness throughout, there being merely a local thickening here and there. At the tip of an extending pseudopod it is always extremely thin and probably sometimes absent. The plasmagel changes rapidly into plasmasol, and *vice-versa*. Precisely what changes take place in the transformation of plasmagel into plasmasol and *vice-versa* are not known. Thus, the same kinds of granules, crystals and vacuoles occur in both. The nucleus, the contractile vacuole and the larger food vacuoles are, however, usually confined to the plasmasol. As a whole the plasmagel is essentially like the plasmasol in composition. These two structures are, however, sharply differentiated from each other in the physical state of a portion of the substance of which they are composed. The plasmagel is relatively a rigid structure. The extension of long free pseudopods into the water without external support and the retention of a given form in a specimen, with numerous pseudopods, as it is rolled about prove this point. The granules suspended in the plasmagel show characteristic Brownian movements as they occur in the plasmasol. Such movements are most rapid and extensive in the case of the smallest granules. The movements of these granules is restricted to smaller areas because the substance is relatively rigid.

**Hyaline cap**. At the tip of active pseudopods there is usually a definite **hyaline cap** of varying thickness. Sometimes it is as thick as half the diameter of the pseudopod and sometimes so thin that it is barely visible if it is not

# Phylum Protozoa

absent altogether. The substance in this cap is as fluid as the plasmasol.

Immediately below the plasmalemma over the entire surface of the amoeba, except perhaps where it is in contact with the substratum, there is a relatively thin hyaline layer which is usually free from granules. Usually this layer is not

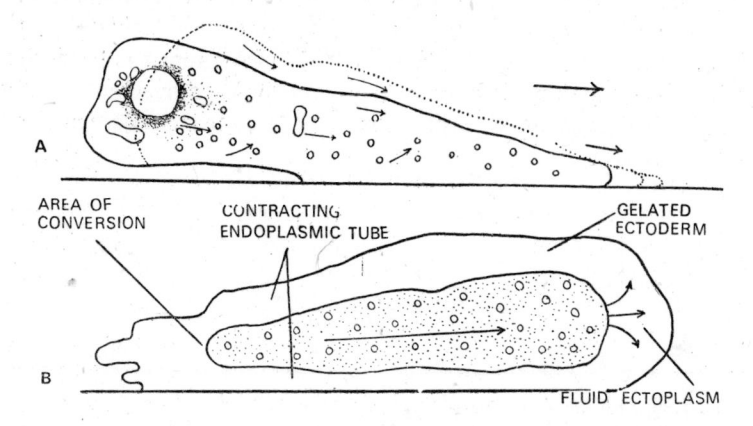

FIG. 1.39. A—Movement of *Amoeba verrucosa* seen from side (*after* Jennings); and B—A marine limax amoeba in locomotion showing changes in consistency of cytoplasm during formation of a pseudopod (*after* Pantin).

very uniform in thickness. In any given region it may be practically absent and in an adjoining region very prominent, and it may be thin in any region at one time and thick at another. In all probability it is a layer of liquid which separates the plasmalemma from the plasmagel, and this layer is doubtless, formed by the spreading of the liquid in the hyaline cap over the plasmagel as the pseudopods extend. This liquid hyaline layer gelates at times, especially that portion which adjoins the region of the plasmalemma which is in contact with the substratum.

## Locomotion

The locomotion of amoeba is affected by the formation of temporary finger-like processess or **pseudopodia** (false feet—Greek, *pseudo*, false; *podos*, foot). Amoebas have no distinct head or tail ends but have a surface which is everywhere the same, and any one point of this surface may flow out as a pseudopod (Fig. 1.38). The protoplasm that enters into it is withdrawn from other parts of the body, and, therefore, if the formation of pseudopods is mainly in one direction the amoeba moves to that side. But sooner or later another similar pseudopod forms at an adjacent point and the cytoplasm flows into it. In this manner the animal progresses in an irregular fashion, flowing first to one side, then to the other. It often alters its course by putting out pseudopod on the side opposite the previous advance. As new pseudopods appear the old ones flow back in the general mass. Such movements are known as **"amoeboid movements."** Amoeboid movement is very slow, and the animal does not proceed for long in any one direction.

Amoeboid movement occurs in other Protozoa, and also in the amoebocytes of sponges and in white blood corpuscles of the vertebrates. Amoeboid movement has always aroused great interest because it is presumed to be one of the most primitive types of animal locomotion. Apparently it is totally different from the muscular movement of complex animals. But it is probable that a thorough understanding of the mechanism of amoeboid movement may throw some light on the general nature of contractility, and thus elucidate the nature of muscle contraction. Consequently the amoeboid movement has been the object of intensive investigation. Probably the amoeboid movement is the basic characteristic of unspecialised protoplasm and like most fundamental processes is difficult to explain.

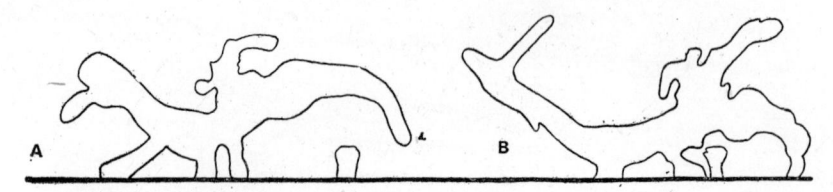

FIG. 1.40. Outline sketches of photomicrographs of *Amoeba proteus* during locomotion as seen from side (*after* Dellinger).

Speculation about amoeboid movement goes back a hundred and fifty years to Ehrenberg, who first suggested that pseudopodia were hernia-like protrusions forced out by muscular contractions of hind part of the animal. Within a year, the alternative view was put forward that the pseudopodia were themselves inherently extensible, and pulled the rest of the animal after them, and ever since biologists have held essentially one or the other of these views.

The elements of the amoeboid movement are not difficult to grasp. At the tip of the advancing pseudopodium the plasmagel seems to spread out, while in the body of the pseudopodium the plasmasol flows forward. When the plasmasol reaches the very tip it fans out to form fresh pseudopodial wall. At the hind end of the animal, the plasmagel seems to draw itself together and get thicker while its surface becomes wrinkled.

It is evident from this that at the anterior end the forward flowing plasmasol changes into rigid plasmagel (gelates), while at the posterior end the plasmagel changes into plasmasol (solates) causing a forward streaming of the more fluid plasmasol. This is the **"change of viscosity"** or **"sol-gel"** theory first advocated by Hyman (1917) and adopted by Pantin (1923-1926) and Mast (1925). In an actively progressing specimen the plasmasol is continuously rapidly streaming forward, while the plasmagel is practically everywhere at rest forming, so to say, a tube within which the plasmasol flows. Forward locomotion occurs because of occasional attachments of the pseudopodium to the substratum. For many years it was accepted that the forward flow of endoplasm is achieved by the contraction of the rear end of the ectoplasmic tube, but recently R.D. Allen has collected substantial evidence in support of a frontal contraction (Fig. 1.41) hypothesis according to which the source of motile forces is a contractile activity in the transformation of endoplasm to ectoplasm near the advancing tip of the pseudopods. This has been called the **"fountain-zone" hypothesis**.

# Phylum Protozoa

Another theory—**"surface tension theory"**—advocated by Butschli was widely accepted for a long time. According to this view locomotion in *Amoeba* is

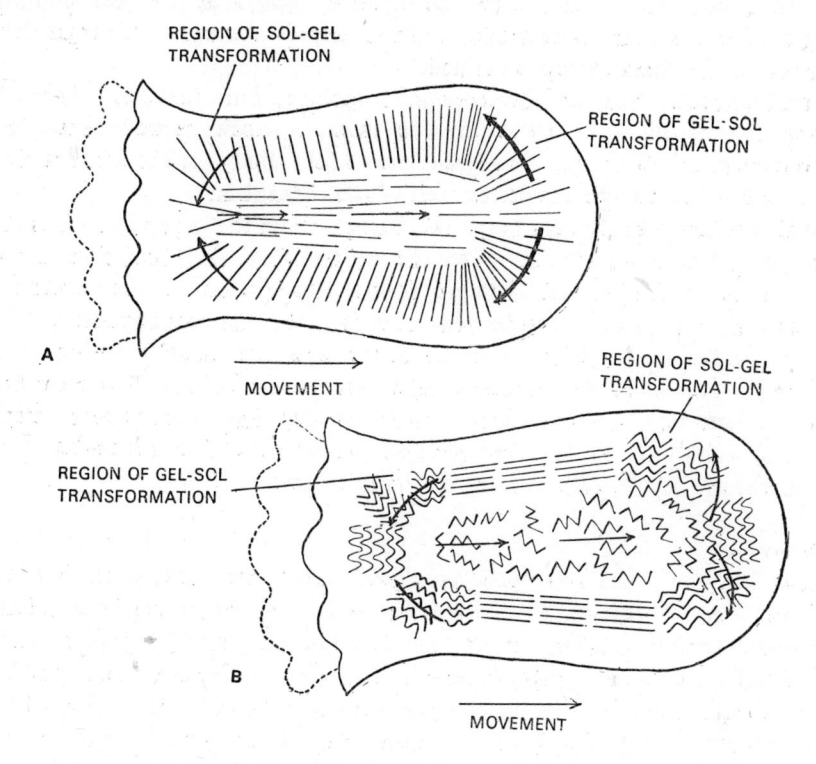

FIG. 1.41. Illustration of "sol-gel" theory of *Amoeba*. A—*after* R D. Allen and B—*after* Mast-Jahn.

essentially like the movement of a globule of mercury or other liquid produced by local reduction of surface tension. This theory is not applicable because when an amoeba moves the upper surface in many species moves forward instead of backwards (opposite to that produced in a globule of liquid); many amoebae are too rigid to produce surface tension, and according to this view the surface is assumed to be liquid, whereas, in most amoeboid forms it is gelatinised.

In multipodal specimens the pseudopods usually become attached to the substratum and then contract pulling the organism forward. Sometimes the pseudopods do not contract, then the amoeba merely flows over the point of attachment due to the contraction of the plasmagel at the posterior end (Fig. 1.38). Such a type of locomotion has been called the **"walking movement"** (by Dellinger).

Four major types of pseudopodium are recognised: (*i*) the broad blunt **lobopodium** characteristic of *Amoeba*; (*ii*) the **filopodia**, needle-shaped branched pseudopods that do not form a reticulum; (*iii*) the **axopodia**, needle-shaped unbranched pseudopods supported by some central skeletal structure; and (*iv*) the **reticulopodia**, slender repeatedly branching and anastomosing pseudopodia.

*Amoeba* has **lobose** pseudopods. The giant herbivorous amoeba *Pelomyxa palustris* does not form tubular pseudopods but retains slug-like or sac-like shape while moving. In this type the hyaline cap is at the rear end and feeding also occurs near the rear end, no food cup is formed. A single large pseudopod of the **limax type** is formed.

In small amoeba such as *Harmanella, Nageleria,* and parasitic *Entamoeba* only one pseudopodium erupts at any one instant. Such pseudopodia are of **eruptive type**, are clear and lack granules. Main body of the amoeba flows more smoothly behind the advancing part of the organism.

Several tubular pseudopodia may also extend from the apertures of shelled amoeba as *Arcella* and *Difflugia*. Since the weight of the shell has to be carried in this case the pseudopod of *Difflugia* forms attachments to the substratum following the appearance of birefringent strands within the pseudopods.

Fine **filopodia** are found in small amoebae and are usually rather transparent with few granules. Reticulopodia are found in the **Foraminifera** and are characterised by the network formed by very fine pseudopodial strands (Fig. 1.33B.) with more or less independent movements. The **axopodia** occur in the heliozoans like *Actinophrys* or *Actinosphaerium*.

**Nutrition**

*Amoeba* feeds on organisms such as algae, other protozoans, rotifers and nematodes, dead protoplasm, preferring small live flagellates and ciliates. It may engulf several paramecia or several hundred small flagellates daily. It also exhibits choice in selecting food and is able to distinguish inert particles from the minute plants and animals upon which it feeds. Movement of the prey or substances diffusing from it attract the amoeba that always avoids highly active organisms.

There are no definite regions or organelles for food intake or ingestion. The food is captured by pseudopodia. A food cup is formed and pseudopodia are thrown around the sides and over the top of the object (Fig. 1.42). In this way the food is held against the substratum, then completely surrounded by cytoplasm and finally incorporated into the main mass of the body. Such a method is sometimes called **circumfluence** and is employed in the ingestion of less active prey. If the prey is active the pseudopodia are thrown out widely so that a wide food cup (Fig. 1.42C) is formed and encloses the prey without touching or irritating it. This method of food capture is known as **circumvallation**.

In some cases the food is taken into the body upon contact with very little movement by the amoeba. This method is described as feeding by **import**. Amoebae are also known to feed by **invagination**. The amoeba touches and adheres to the food, and the ectoplasm in contact with it invaginates into the endoplasm as a tube carrying the food particles with it (Fig. 1.42). The tube then dissolves leaving the food into the interior. Usually the prey is paralysed on contact with the body. Sometimes active animals are ingested but they become inactive in the food vacuole.

The drop of water taken along with the food forms a food vacuole in the endoplasm which secretes enzymes that enter the vacuole. In an actively feeding amoeba several such food vacuoles may be seen. The food vacuole is the

*Phylum Protozoa* 61

miniature **"digestive tract"** of the amoeba. Enzymes act best at a definite acidity or alkalinity, consequently the amoeba furnishes proper conditions

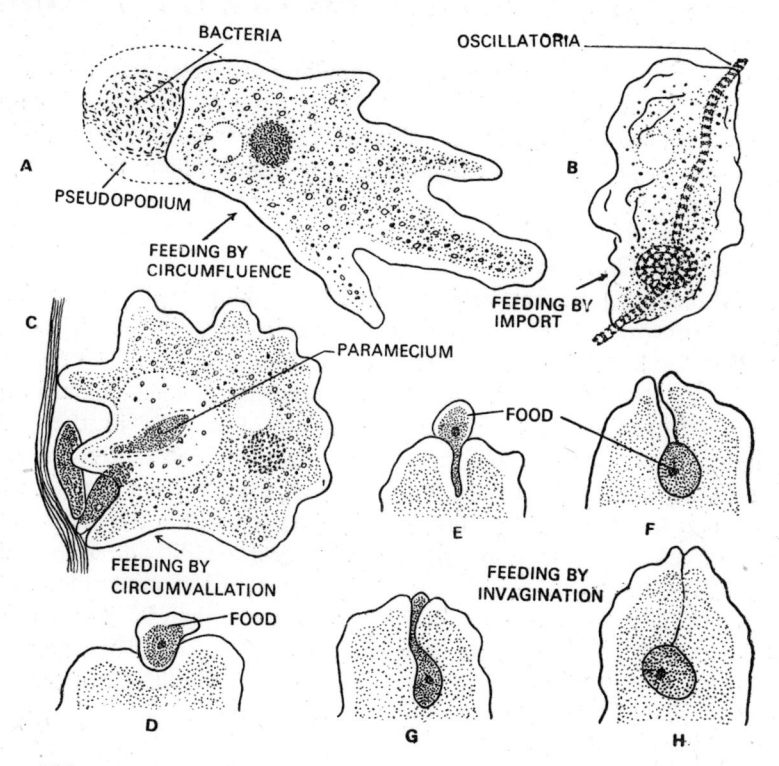

FIG. 1.42. Ingestion or feeding in *Amoeba*. A—Feeding by circumfluence; B—By import; C—By circumvallation; and D to H—By invagination.

for the proper functioning of the enzymes. Thus in the food vacuole the reaction is at first acid, the acidity serving to kill the prey which often struggles for some time after being ingested. The reaction later becomes alkaline because the proteolytic (protein digesting) enzyme (like **trypsin**) of the amoeba is active only in an alkaline medium. Later, the food particles lose their sharp outlines, swell, become more transparent and reduce in amount as the products

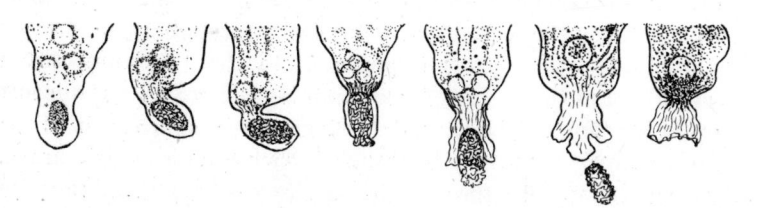

FIG. 1.43. Stages in the egestion of waste matter (*Amoeba verrucosa*).

of digestion are assimilated by the general cytoplasm. Direct evidences showing the digestion of proteins and fats are available. For this purpose ferments like peptidase and proteinase are poured into the food vacuole by the endoplasm. Holter and Doyle have recently demonstrated the presence of amylase in *A. proteus*, which means that the power to convert starch into soluble sugar exists in amoeba. According to Mast, *Amoeba proteus* is also able to digest the neutral fat globules of the food and break into fatty acids and glycerine.

The food so digested passes into the surrounding protoplasm from the food vacuoles. The digested food is distributed to all parts of the body because of the movements of the food vacuoles. The indigestible fragments are eliminated in a very simple way. They are extruded at any point where they happen to reach the surface (Fig. 1.43).

**Respiration and Excretion**

No special breathing mechanism is required in a minute creature like *Amoeba*, and yet it carries on all the essentials of respiration, in which energy is liberated from food and made available for other life processes. All the oxygen necessary for respiratory activities passes through the plasmalemma into the protoplasm by diffusion, since the concentration of oxygen in the water is higher than that in the amoba's cytpolasm. Oxygen constantly enters and is used up in the burning of foods. This is almost a continuous process, for the burning of food consumes oxygen and its concentration within the animal always remains lower than that in the outside water. The carbon dioxide produced as a result of the breaking down of substance diffuses out for the same reasons. It is always at a higher concentration within the amoeba than in the surrounding water, thus the same factors control its passage to the exterior. The water that is produced as a result of breaking down is a normal constituent of the animal body, and its rapid disposal is not necessary. Burning protein yields not only carbon dioxide and water, but also nitrogenous substances that are poisonous and must be excreted rapidly. Such waste products are eliminated by diffusion.

**Osmo-regulation**

The contractile vacuole is a mechanism for osmo-regulation, i.e., control of water content of the cytoplasm. When perfectly formed a contractile vacuole is a spherical cavity containing liquid. It appears as a clear circular area in the cytoplasm, slowly increases in size by the collection of more liquid in it and finally contracts suddenly discharging its contents to the exterior (Fig. 1.44). Soon it appears again from one or more minute droplets, grows in size, as before until it collapses again. From this it is clear that some liquid is collected in it and expelled from the cytoplasm periodically through a temporary pore on the surface (Fig. 1.44). The growing phase up to the maximum size reached is called **diastole**, and the collapsed condition is referred to as **systole**. The water comes in the animal body mainly from three sources: it is produced as a result of respiration; it may be included when food particles are engulfed, and it may enter osmotically through the semi-permeable membrane, the plasmalemma. Thus, there is a continual influx of water which may lead to the rupture of the animal if not expelled. It is this excess of water that is collected in the contractile vacuole and expelled periodically. How the water is forced out of the protoplasm into the vacuole is not exactly known. There are three theories describing the growth of the contractile vacuole. These include the **osmotic** theory, the **filtration** theory, and the **secretion** theory of the

# Phylum Protozoa

diastole. According to the first the water passes into the growing vacuole by the simple process of osmosis. The second (filtration theory) is of the view

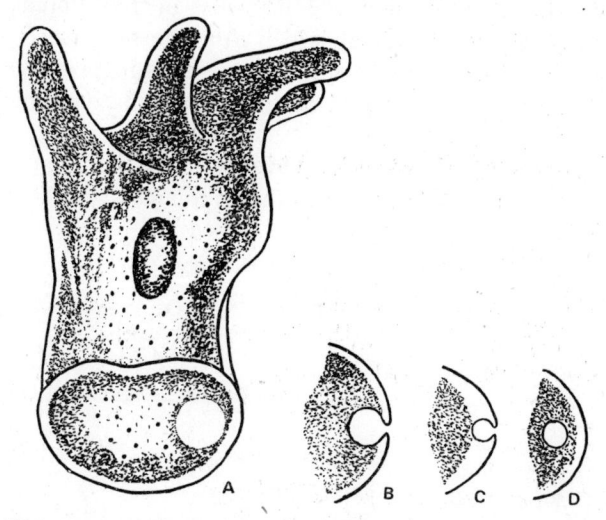

FIG. 1.44. Section of *Amoeba* showing the formation of the contractile vacuole.

that the water filters through the vacuolar membrane into the vacuole because of the hydrostatic pressure of the endoplasm. The third (secretion theory) suggests that the water is first absorbed in the vacuolar membrane and then secreted into the vacuole. It has been recently found in *A. proteus* that the growth of the vacuole takes place in spurts and not steadily. From this it is inferred that the fluid enters the vacuole by the fusion of similar smaller vacuoles. On attaining maximum size the vacuole migrates to the surface until it reaches the plasma-membrane which subsequently bursts discharging the contents. The location of the vacuole is not definite in such forms and it moves about with the cytoplasmic movements. As a rule it is confined to the temporary posterior region of the body.

It is possible that the expelled water may contain some excretory products, but there is no experimental evidence to show that the vacuole serves as an excretory organ. On the contrary, there is evidence against the excretory role of the contractile vacuole. Marine animals similar to *Amoeba* in constitution do not possess a contractile vacuole and yet excrete. Besides, experimentally increasing the concentration of salts in the water surrounding the amoeba causes the vacuole to contract less and less frequently and finally to vanish altogether. Conversely, some marine relatives of amoeba develop contractile vacuoles when placed in fresh water. It is, therefore, concluded that the chief function of the contractile vacuole is to regulate the water content of the amoeba. In the parasitic protozoa with contractile vacuoles, most of the water expelled probably enters the feeding process.

## Behaviour

Amoeba responds to both external and internal stimuli, *i.e.*, it is sensitive to touch, or nearness of food, or some other influence. Responses of an amoeba to stimuli constitute **behaviour,** the old word for which is **irritability**. There are many different types of external stimuli to which an amoeba reacts and the nature of the response depends upon the intensity of the stimulus. A response involves the entire body of the animal, which either moves away from the stimulus **(negative response)** or continues moving towards it **(positive response)**. Hunger, on the other hand, is internal stimulus in response to which an amoeba searches food.

**Reaction to mechanical stimuli (Thigmotropism).** Where an amoeba is pricked with needle (Fig. 1.45), the part that is touched shrinks away and the

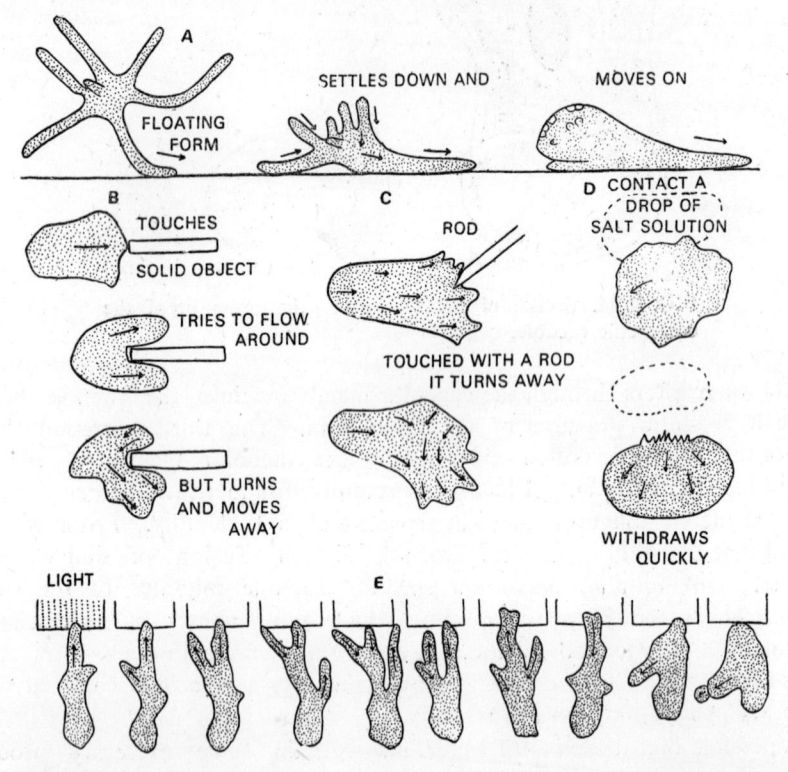

FIG. 1.45. Reaction to stimuli in *Amoeba*. A to D—(*after* Jennings), arrows indicate direction of movement; and E—Successive stages at about 30 second intervals in reaction to strong light focussed continuously on a microscope slide (*after* Mast).

opposite side thrusts out pseudopodia in an effort to escape. When the amoeba is already on the move, and the needle is applied to the leading end, the reaction is particularly strong. This is an example of **negative response** to touch. Jenning's observations provide another similar example. When an amoeba contacts a dead algal filament, its pseudopodia first moves on both the sides of the filament and then the animal stops and moves back.

A floating amoeba, on the other hand, responds to touch in a positive way. As soon as any of its pseudopodia comes in contact with a floating surface it spreads and adheres to the surface, and soon starts creeping on the surface.

**Reaction to light (Phototropism).** When a beam of light is shone on the advancing pseudopodium of an amoeba, the movement is invariably slowed down, and if the light is strong enough, it is completely stopped. Movement later starts up in a new direction and in consequence the amoeba ultimately moves away from the light.

It must be noted that in its natural surrounding an amoeba is exposed to light of varying intensity. For this reason it does not respond to gradual changes in illumination. For any reaction it must be subjected to strong beam of light, Observations of some authors suggest that if *A. proteus* is exposed to light on any side, the plasmagel of that side thickens due to gelation of the adjacent plasmasol. On this side, therefore, the plasmagel contracts and its elastic strength increases consequently forcing the pseudopodia to sprout on the opposite side.

**Reaction to chemical stimuli (Chemotropism).** The effect of stimuli other than light is more complicated, and the amoeba's reaction to them takes two forms depending on their strength. Amoebae are indifferent to normal chemical constituents of water. Most dilute chemicals and gentle contact with an object always cause a pseudopodium to be pushed out at the point where the stimulus is applied. Movement generally continues in the same direction. Stronger stimulation with touch or chemicals, however, has the reverse effect; movement stops completely, and the amoeba becomes rounded and later sets off in a new direction. The fact that strong stimuli produce the opposite effect to weak ones remains something of a puzzle. It is probable that they (strong stimuli) produce such widespread effects that the delicate balance needed for continued movement is quite upset.

**Reaction to heat (Thermotropism).** Change of temperature of the surrounding water has a direct effect on locomotion and rate of feeding. If the temperature of the water is reduced, locomotion and rate of feeding retards, on the other hand, if the temperature is increased the locomotion is speeded up, but ceases above 30°C.

**Reaction to gravity (Geotropism).** It has been suggested that amoebae are positive to the force of gravity. It is for this reason that they are found attached to the bottom of the container.

**Reaction to electric current (Galvanotropism).** If an electric current is passed into water containing amoebae, the amoebae move towards the cathode thus showing a negative response to the anode. This is brought about by the reversal of cytoplasmic streaming, and by the solation and consequent decrease in the elastic strength of plasmagel at the surface of body towards the cathode. If the intensity of the current is increased disintegration of the body begins at the surface towards the anode.

**Reaction to water current (Rheotropism).** When floating freely in water the amoebae place in line with the water current, i.e., they are positive to it. It is evident from the above that the responses by the amoeba to the stimuli are such as to benefit the individual and the species by avoiding unfavourable conditions and seeking those useful to it.

The ability of an amoeba to select food has been used by some as evidence that it exhibits **conscious behaviour** or possesses some trace of those powers which in higher animals are vested in the brain and which have been called the **psychic property** of protoplasm. Here is an experiment that supports this view. If the tip of a pseudopodium is repeatedly illuminated with a very bright light, the amoeba first reacts by stopping and putting out a lot of little pseudopodia to either side; but after 10 or 20 repetitions, it responds by putting out a single pseudopodium in the reverse direction. A day later, though it has not been illuminated in the interval, it still responds by putting out a single pseudopodium. No doubt this is an effect very different in degree from adaptive behaviour and memory in higher animals, but it may not be altogether different in kind. However, others maintain that the simple activities of an amoeba imply no psychic attributes.

### Reproduction

When an amoeba feeds freely it grows gradually to an optimum size. Beyond this size it cannot grow and soon reproduction sets in. The amoeba reproduces asexually by the simple process of binary fission (Fig. 1.46). For long such a

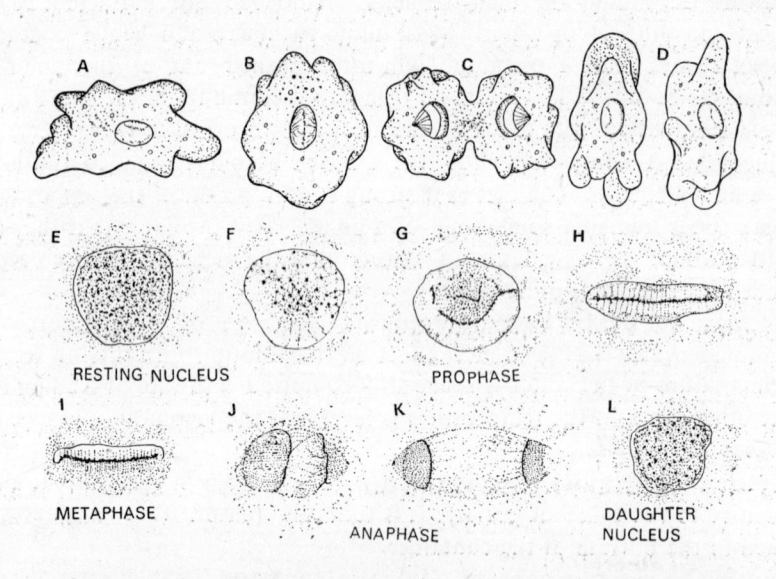

FIG. 1.46. Binary fission in *Amoeba*. A—Ready to undergo fission; B—Withdrawing pseudopodia, and nucleus beginning to divide; C—Nucleus divides into two and cytoplasm constricts; D—Two daughter amoebae separate; and E to L—Nuclear behaviour during division.

division was thought to be amitotic, but it has now been shown that the amoeba divides mitotically. Before mitotic division sets in the chromatin is lodged in the central area of the nucleus and the conspicuous granules present under the membrane contain very little chromatin. As the prophase begins several changes take place: (*i*) the peripheral granules give rise to achromatic figure and the chromatin becomes aggregated in centre (Fig. 1.46 B, F), (*ii*) then they assume a ring-form along the periphery of the central mass of network. At this stage the cytoplasm around the nucleus is much vacuolated (Fig. 1.46 G). (*iii*) A little later the nucleus becomes markedly flattened with its membrane still intact and in it appears a discoid equatorial plate, connected with the nuclear membrane by numerous fibrils (Fig. 1.46 H). This marks the close of prophase. (*iv*) With the beginning of metaphase the nuclear membrane thins down extremely and even disappears over one side of the plate (Fig. 1.46 I). (v) In the next stage (anaphase) the nuclear membrane disappears completely, the equatorial plate splits and each half contracts in the plane of the plate, producing two daughter plates. In some specimens even a faint spindle has been noticed. The vacuoles round the nucleus disappear (Fig. 1.46 J). The plates separate more and more and become smaller in size. Finally, they reach the poles and form two daughter nuclei. These observations are based upon the study of Chalkley and Daniel (1933) who have also suggested that the chromatin granules in the equatorial dlate are chromosomes. If so, the number of chromosomes is enormous (500-600) and they are very small in size. Chromosome numbers vary widely even between closely related species. The separation of two daughter groups of chromosomes takes by the elongation of the region between them. As the chromosome groups are pushed apart the protoplasm gets constricted more and more until the strand connecting the two halves breaks.

According to Chalkley and Daniel there is a definite correlation between the stages of the nuclear division and external morphological changes. During the prophase the animal is rounded, studded with fine pseudopodia, with a distinct hyaline area in the centre, which disappears during the metaphase. In the anaphase the pseudopodia become coarse and during the telophase the body elongates, cleft appears and the pseudopodia return to normal condition. Finally, the two daughter amoebae separate. Each daughter amoeba soon increases in size and behaves just like the parent. Since an amoeba continues to exist in its offspring it may be said to be immortal, that is, it has no dead ancestors.

Earlier workers have reported multiple fission as well, where the parent amoeba divides simultaneously into a large number of small offspring, but it

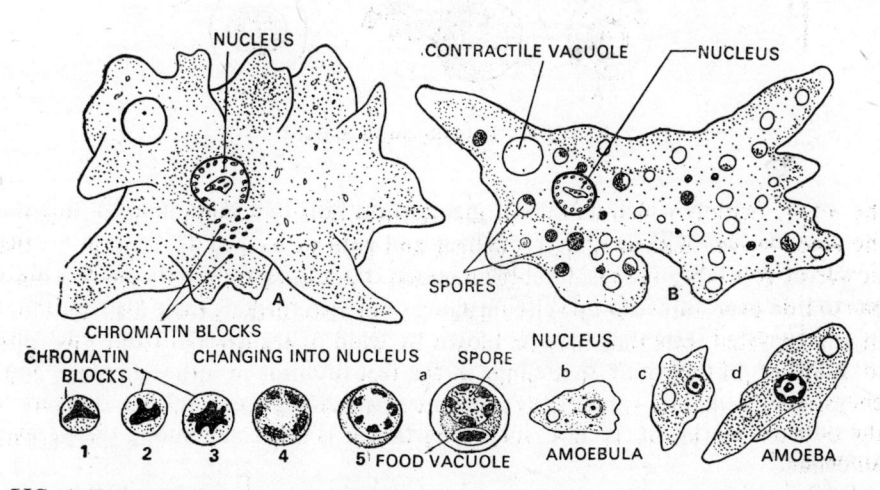

FIG. 1.47. Sporulation without encystment in *Amoeba proteus* (*after* Taylor). A—Shows escape of chromatin blocks that change into spores; and B—1-5, chromation blocks changing into nucleus; a-d, spore producing amoeba.

now seems almost certain that these observations were wrong. Recently, however, Taylor has described a process of sporulation without encystment for *Amoeba proteus*. In the beginning the nuclear membrane dissolves at one or more places and chromatin granules (blocks) escape into the endoplasm (Fig. 1.47). Each chromatin granule develops into a little nucleus, forms a nuclear membrane while moving with the cytoplasm. Each nucleus at this stage can be regarded as an amoebula. Soon each amoebula develops a spore-case around it. Hundreds of such spores are present in the body of the parent, that becomes somewhat shrivelled, the nucleus of the parent is also completely exhausted. When set free these spores sink down to the bottom. They are quite tough and are able to withstand desiccation. On the return of favourable condition from each hatches out a small amoebula which grows into a normal form.

In this case a full-fledged nucleus (secondary nucleus of some) develops from a chromatin block of the parent nucleus. Thus the nucleus of the parent has been called a **multiple nucleus** or **polysenergid** nucleus. In the light of this view the phenomenon of sporulation, as described above, can be considered as the resolution of the multiple nucleus into its constituent units. Sexual process, involving the formation and fusion of gametes, has been described in various species of amoeba, but many of the accounts are open to a good deal of doubt. Conjugation has been described from time to time, but is even less well substantiated than the sexual process. It is, however, evident that these various methods of reproduction are very much less common than ordinary division, if they occur at all.

**Encystment.** If conditions of life become unfavourable, as when the pond dries up, or when the food supply runs low the animal forms a membranous protective shell or cyst (Fig. 1.48 A) around it to tide over the difficult period. The animal rounds up and secretes over itself a hard and impervious covering,

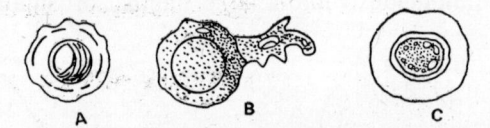

FIG. 1.48. Encystment in *Amoeba*.

the **cyst**, within which its rate of metabolism falls to a minimum. In this state the amoeba can withstand extreme heat and cold as well as desiccation, as their power of resistance is considerably increased. Encystment not only enables amoebae to tide over unfavourable circumstances but also furthers their distribution, as in the encysted state they may be blown by wind or transferred from one pond to another in the mud that clings to the feet of birds or other animals. Since encystment is mainly valuable as a means of surviving some of the dangers of the outside world, it is not surprising that it is universal among the parasitic amoebae.

Replaced in a suitable environment the cyst ruptures and the enclosed animal emerges and resumes its usual activities (Fig. 1.48). According to some the amoeba does not emerge from encystment as its old self; instead a cluster of swarm spores come out, produced by repeated divisions within the cysts. In some species such as *Paramoeba*, there are a dozen or so of these swarm spores, each with two whip-like flagella. In the course of time the flagella are lost and the swarm spore comes to look like ordinary amoeba.

## TYPE ELPHIDIUM (POLYSTOMELLA)

*Elphidium* (*Polystomella*) is a close relative of *Amoeba*. It is a Foraminifera and possesses a shell. From the shell protrude out a number of temporary thread-like structures called pseudopodia (Fig. 1.49). The pseudopodia of amoeba are short and blunt but here they are long and root-like. The name Rhizopoda (the root-feet) is much more appropriate here than to amoeba itself, for when seen alive with the help of a hand lens the pseudopodia of the foraminiferans appear as tufts of tiny root-hairs growing around the body of the animal.

The foraminiferans are typically creeping forms moving slowly and using their net-like pseudopodia chiefly for food capture. There are some genera, on the other hand, that have taken to a pelagic existence and float on the surface of the ocean, spreading their nets in all directions around them, e.g., *Globigerina*. Some forms (*Haliphysema*) attach themselves to submerged objects and lead a sedentary life.

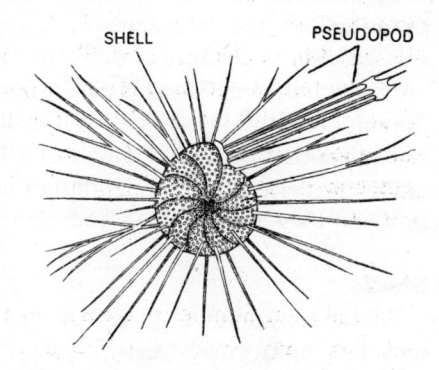

FIG. 1.49. *Polystomella* (living) showing extended pseudopods emerging out of the shell.

### Structure

The body of the animal consists of fluid protoplasm as that of *Amoeba*, but does not show marked distinction of ectoplasm and endoplasm. The nuclear apparatus varies in different forms and even in the same species as will be evident from the study of the reproductive system. In some fresh water forms contractile vacuoles are present, but they are not found in marine animals. The protoplasm contains metaplastic bodies of various kinds, and may become loaded with faecal matter in the form of masses of brown granules (called **stercome** by Schaudinn). Periodically this material is got rid off often as a prelude to the formation of a new chamber.

### Pseudopodia

Pseudopodia are temporary protrusions of the body fluid by means of which these animals crawl about. But crawling is not the only function. By means of pseudopodia they can touch or taste or feel, they can feed, collect things, defecate, build cysts, in some cases even their shells. The pseudopodia of *Polystomella* are exceedingly long, thin, straight threads which radiate outwards from the shell. They vary greatly in thickness. In some cases they take the form of broad bands and in other they are extremely fine threads. A careful examination reveals that a number of fine granules of fairly constant diameter are embedded in these threads. They are in constant movement. There is a double stream of these granules, some moving inwards towards the shell and others outwards. According to many observers there is more solid axial core in each pseudopodium, but others have suggested that this is not always present. It seems to be formed under certain conditions which are not clearly understood. The pseudopodia are covered by some sort of pellicle like the plasmalemma of *Amoeba*. The **pellicle,** it seems, has a movement of its own, independent both of that of the granules and of the more fluid protoplasm. If a small organism like a flagellate comes to rest in contact with the pellicle it may be carried along passively for some distance by the movement of the pellicle. The reaction of pseudopodia to foreign objects changes at different phases of their activity. They are not sticky normally but sometimes (as during feeding) when a copepod touches a pseudopodium it falls motionless as if paralysed by some toxic secretion. It is apparent from the above that the pseudopodia of a Foraminifera are different from those of an amoeba—there is no differentiation into gel or sol, streaming is due to contraction and they are not small and blunt.

**Shell**

All the Foraminifera possess a shell. Although the body of the animal is fluid and has no formed parts to give it any sort of permanent shape, the forms of the shells are as well-defined and characteristic as those of any higher animals. On the basis of their structure there are two types of shells: (*i*) those composed of secreted materials, and (*ii*) **arenaceous shells** composed mainly of foreign materials held together by a secreted cement. In all foraminiferans, there first appears primitive test of the adult. Among the modern types mostly calcareous tests are found. Some may have siliceous test which differs from arenaceous test in the absence of foreign particles In some interesting cases the first few chambers are arenaceous while the later ones strictly calcareous. The shell of *Polystomella* is made up of many chambers (**polythalamous**) although in some foraminiferans the shell may consist of a single chamber (**monothalamous**). A typical form of the shells is a chamber with a wide aperture (sometimes more than one) through which the pseudopodia pass out. Such shells occur in the **Testacea** (*Arcella, Gromia, Difflugia*). In addition to the wide aperture there may be numerous minute pores through which the protoplasm streams out as delicate pseudopodia. Thus there are two types of shells **perforate**, that have numerous minute pores all over the surface of the shell in addition to the aperture, and **imperforate** in which there are no perforation on the surface.

The chamber that is first formed (initial chamber) is called **proloculum**. In multilocular species there are two types of shells differing in relative sizes of the proloculum. In some the proloculum is smaller and is produced by an organism developing from a zygote. Such shells are called **microspheric**. In other the initial chamber is proportionately larger and they are produced by individuals resulting from schizogony. Such shells are called **megalospheric** type. As successive chambers are added their limits are marked externally **sutures** and internally by **septa** (partitions). As each new chamber is formed the anterior wall of the preceding chamber becomes a septum in the simple cases and the old aperture becomes **foramen** joining the two chambers. In *Elphidium* (*Polystomella*) the foramina are gradually closed by chitinous deposits which first appear as rings and finally form plugs, the function of which is unknown. In many foraminiferans the septa are double. The structure may be further complicated by a system of canals, which runs through the wall of the test and within the double septa. The canals communicate with the chambers and also open to outside independent of the pores.

*Phylum Protozoa*

**Formation of shell.** Whether perforate or imperforate the shell remains single in simpler forms (**Testacea**). In such forms when the animal divides by binary fission, the protoplasm streams out through the principal aperture to give rise to the body of the daughter individual. This protoplasm forms a shell for itself and when the division is complete the two separate, the mother retaining the old shell. In many species, on the other hand, when the animal outgrows the original single-chambered shell, the protoplasm flows out and forms another chamber, which is not separated as a distinct animal but remains continuous with the shell. Thus, the single individual develops a two-chambered shell. In the same way many more chambers may be added successively, each newly formed chamber being as a rule slightly larger than that formed just before (Fig. 1.50).

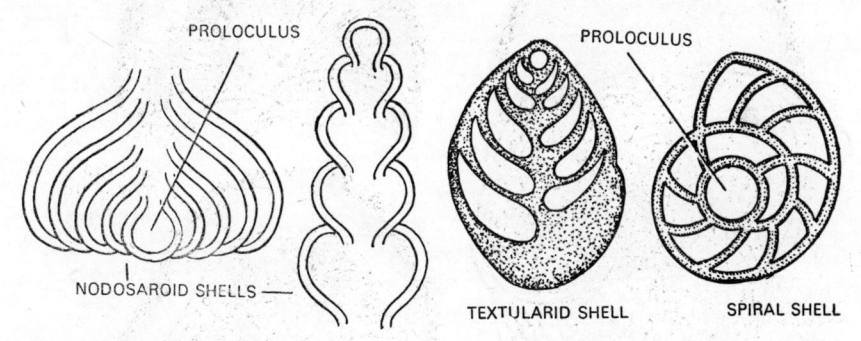

FIG. 1.50. Diagrammatic section of different types of shells to show the arrangement of chambers.

When new chambers are added in a straight line-like beads on a string the shell is said to be **nodosaroid**; when the chambers are coiled in a flat or conical spiral the shell is said to be **spiral**; when the chambers are concentric the shell is called **cyclic**; when the chambers are arranged in two or three alternating rows the shells are called **textularid** type (Fig. 1.50). The chambers may be arranged irregularly or may present a mixture of the two above types. The shells may also be ornamented or may bear different types of markings and spines, etc.

The basic structure of the arenaceous shells is a thin chitinous layer on which are cemented sand grains, sponge spicules, ambulacral plates of echinoderms, fragments of other tests, and the like. The animalcule (little speck of jelly) secretes a delicate envelope and then selects different sized grains which are encrushed round the shell. Curiously enough one particular species will select grains only of one size. One cannot believe that such minute animals have enough mental power to do it, and how such structures can attain this result is beyond one's understanding. This, however, has been experimentally confirmed by Heron-Allen. He kept the foraminiferan *Verneuilina polystropha* in his laboratory in bowls containing artificial mud made by crushing such gems as ruby, sapphire, topaz, etc., and found that the animalcule preferred only grains of higher specific gravity. Some foraminiferans select flakes of mica for their shells. They may further select spicules of sponges and sea-cucumber, etc. Members of the genus *Techinella* (the little craftsman) select spicules of particular length only and arrange them in a particular manner.

In a multi-chambered Foraminifera each new chamber is formed, as mentioned above, by the extrusion of a droplet of fluid protoplasm from the old shell aperture and that the wall of the new chamber is laid down around this. So far as known from the available information it seems that the construction of a new chamber does not take place in the same way in all the species. It is also likely that in the same species the shell may be in different forms at different stages of growth. La Calvez reports that the earliest chambers are formed around

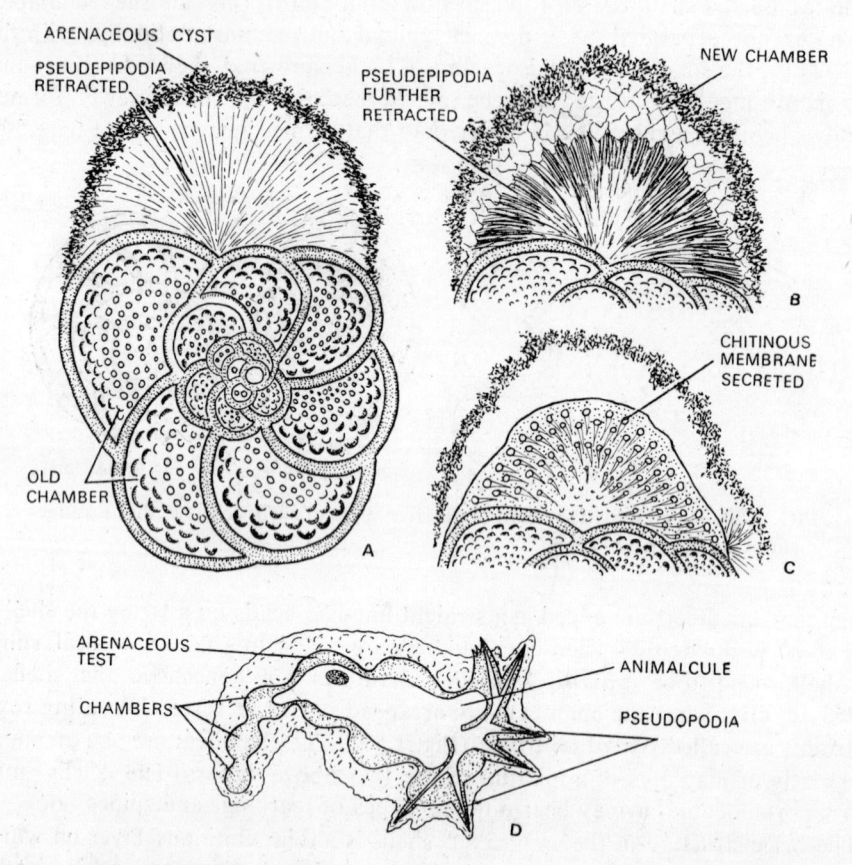

FIG. 1.51. A to C—Formation of a new chamber of *Discorbis bertheloti*; and D—Formation of arenaceous shell of *Iridia serialis* (after Calvez).

fluid masses of the cell material. But in the later (post-embryonic) stages the new chambers are formed differently. In the first place the pseudopodia, which normally keep extended outwards, shorten to form a regular fan-like bundle (Fig. 1.51 A). Around the periphery of this an accumulation of extraneous particles forms a cyst-like structure. After some time the pseudopodial fan shortens again (Fig. 1.51 B). At the periphery of the fan now appears a chitinoid membrane. On this membrane appear a number of little refringent dots which enlarge and coalesce forming calcareous wall. These observations were made on *Discorbis bertheloti*, but Jepps came to similar conclusions while studying on *Polystomella* (*Elphidium*). Thus it is evident that forces that determine the form of the chambers are those that determine the length, disposition and various activities of the pseudopodia.

# Phylum Protozoa

## Behaviour

The most wonderful aspect of the behaviour of the Foraminifera is their power to select particular type for their shells, which has been mentioned above. How are they able to perform the miracle is not known. Jensen in 1901 gave an account of a series of experiments on the reaction of the foraminiferan pseudopodia to particles of various kinds, but that too cannot explain this power of selection of grains. More work on this aspect of the life of foraminiferans is needed to elucidate their behaviour.

## Life Cycle

Typically there are two distinct generations in the life-cycle (Fig. 1.52) of *Polystomella*—a sexual one (the **gamont**) which alternates with an asexual one (or

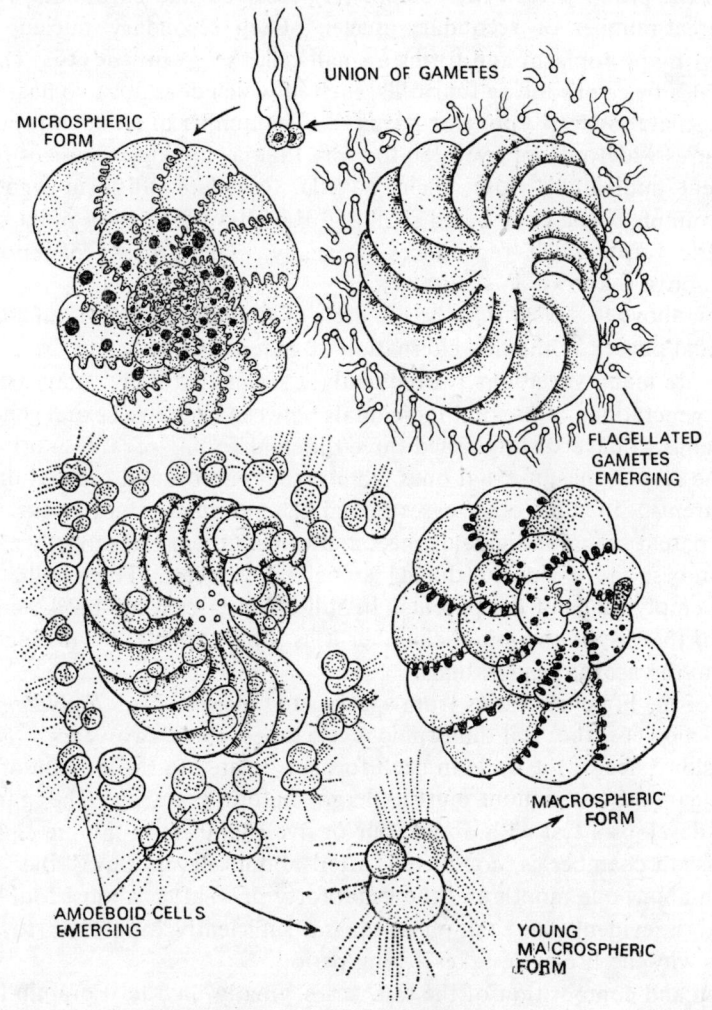

FIG. 1.52. Life cycle of *Polystomella*.

**schizont**). This alternation of generation is further combined with dimorphism in the adult condition. The initial chamber of the asexual individuals is smaller than that of the sexual individuals, i.e., they are **microspheric**. The microspheric form has many nuclei which multiply by fission as the animal grows. During schizogony the nuclei become resolved entirely into chromidia and the protoplasm streams out of the shell, which is abandoned. The chromidia form secondary nuclei and the protoplasmic mass divides into swarms of (about 200 or more) individuals which are all alike, and have been called **amoebulae**. Each amoebula contains a nucleus and chromidia and secretes a single-chambered shell which is the initial chamber of the megalospheric form. The amoebulae separate, feed and grow adding subsequent chambers to the initial chambers already formed and become the megalospheric (sexual) adult. When full grown the sexual form has large nucleus and numerous chromidia. As it grows the nucleus passes from chamber to chamber and gives rise to a number of chromidia into the general cytoplasm. In the end the primary nucleus is completely resolved into chromidia which give rise to a great number of secondary nuclei. Each secondary nucleus becomes surrounded by protoplasm and forms a small cell, the **gametocyte**. The gametocyte divides twice producing four cells, each of which develops two flagella forming a biflagellate **swarm spore** or **gamete**. The number of gametes produced by gamont varies widely, from over 250 to more than a million. Gametes produced by different individuals lose their flagella, copulate and form zygote which secretes a minute single chambered shell and becomes the starting point of a new microspheric form which reproduces asexually. The process of reproduction described above is known as **syngamy**.

From the above it is clear that the life-cycle of *Elphidium* involves an alternation of an asexual phase. While this alternation of generations seems to be the general rule, there are many variations. Frequently it seems that there may be several successive generations of asexual individuals between each two sexual generations. Hofker claims to have distinguished three different forms of the most species, namely, the sexual ones, asexual ones from sexual parents and asexual ones from asexual parents. In some other cases sexual generation does not occur. In some cases two parent sexual individuals meet and cement themselves together within a fertilization cyst and divide producing amoeboid gametes. They unite together within the empty shells of the parents. In still other cases it has been seen that on the basis of the size of the initial chamber it is not possible to distinguish the sexual from the asexual individuals.

Length of the life-cycle varies from species to species. In *Elphidium crispa*, the completion of the full dimorphic cycle takes about two years. Now, the relatively slow pace of the cycle in such forms is related to the amount of growth, the young gamont or agamont must undergo before maturity. The gamont of *Elphidium* develops a test with about four or five chambers in about eight days' time. The sixth chamber is completed after about eleven days, the fifteenth chamber in about one month and the usual forty or so after almost four months. From this it is evident that each phase takes a sufficiently long time to mature, and that is why the life-cycle takes a long period.

The form and composition of the test varies greatly in the Foraminifera. On this basis about fifty families have been recognised. It is not possible to describe all here, nor is it possible even to discuss all the different genera of Foraminifera. Only those forms are described here which are usually given in our laboratory for study (Figs. 1.53 and 1.54).

## Phylum Protozoa

Genus *Rhabdammina*—Test entirely or partly arenaceous with a central chamber and two or more arms [Fig. 1.53(1)].

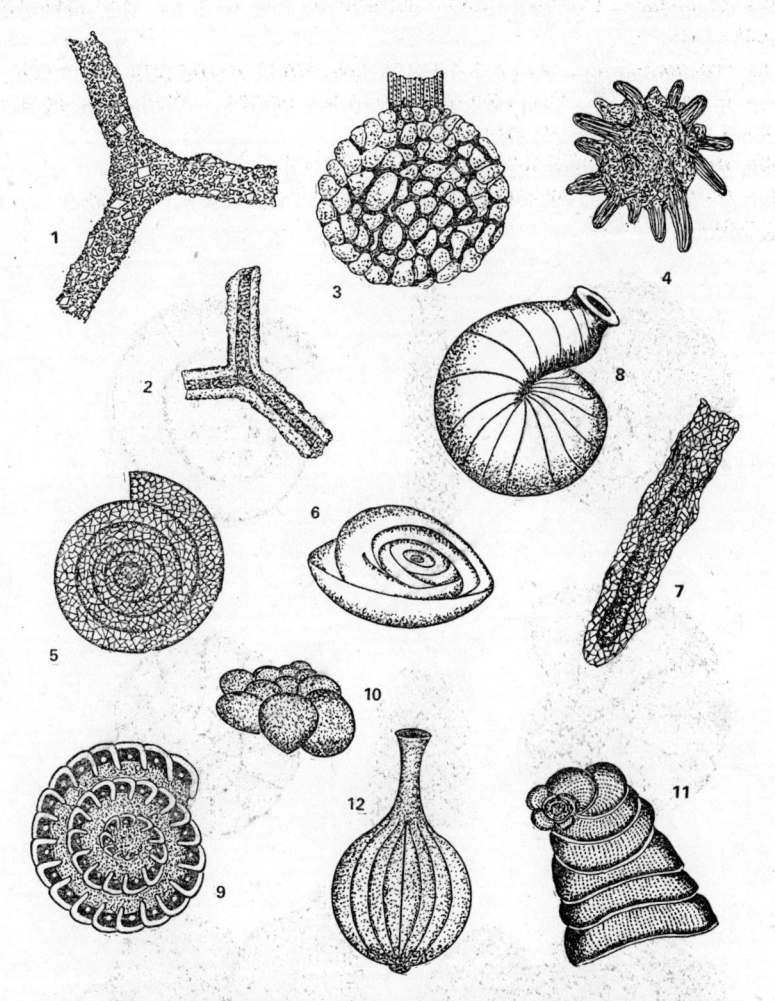

**FIG. 1.53.** Different types of Foraminifera shells. **1**—*Rhabdammina abyssorum*; **2**—*Rhizammina algaeformis*; **3**—*Saccammina sphaerica*; **4**—*Calcarina defrancei*; **5**—*Ammodiscus incertus*; **6**—*Meliola*; **7**—*Hyperammina subnodosa*; **8**—*Lituola nautiloidea*; **9**—*Fusulina*; sp. **10**—*Trochammina inflata*; **11**—*Vertebralina striata* (from various sources); and **12**—*Lagena striata*.

Genus *Saccammina*—Arenaceous test with a chamber or rarely a series of similar chambers loosely attached, with normally a single opening [Fig. 1.53(3)].

Genus *Hyperammina*—Arenaceous two-chambered test with a proloculum and long undivided tubular second chamber which is simple or branching but not coiled [Fig. 1.53(7)].

Genus *Ammodiscus*—Arenaceous test with the second chamber usually coiled at least in the young. [Fig 1.53(5)].

Genus *Silicina*—Test of siliceous material with partially divided second chamber.

*Textularia*—Typically many-chambered shell showing biserial arrangement at least in young microspheric forms [Fig. 1.54(1)].

Genus *Rheophax*—Typically many-chambered test with all the chambers in a rectilineal series.

Genus *Trochammina*—Test trochoid at least while young [Fig. 1.53(10)].

Genus *Spiroloculina*—Test coiled in varying planes. Wall imperforate with arenaceous portion only on the exterior.

Genus *Meliola*—Similar structure [Fig. 1.53(6)].

Genus *Triloculina*—Test **calcareous**, imperforate with chamber coiled in varying planes.

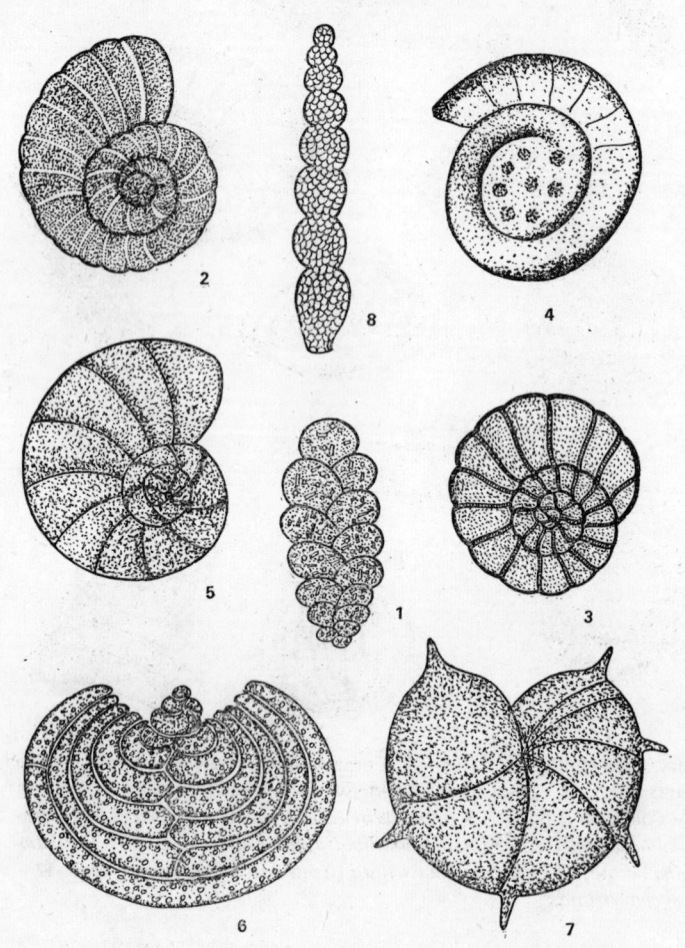

**FIG. 1.54.** Different types of Foraminifera shells. **1**—*Textularia agglutinaus*; **2**—*Operculina ammonoides*; **3**—*Rotalia beccrarii*; **4**—*Silicina limitata*; **5**—*Asterigerina carinatna*; **6**—*Paronina flabelliformis*; **7**—*Hantikenina atapamenis*; and **8**—*Reophax nodulosus*.

Genus *Peneroplis*—Test calcareous typically compressed and often discoid. Chambers mostly subdivided.

Genus *Lagena*—Test calcareous, perforate, plain, spirally coiled or becoming straight or single chambered [Fig. 1.53(12)].

# Phylum Protozoa

Genus *Elphidium*—Test not vitreous. Plain spiral, septa single, no canal system.

Genus *Operculina*—Test plain, spiral, septa double and canal system in higher forms [Fig. 1.54(2)].

Genus *Pavonina*—Test generally biserial in at least microspheric form. Aperture usually large without teeth [Fig. 1.54(8)].

Genus *Hanikenina*—The plain spiral bi- or triserial with elongate spine and lobed aperture [Fig. 1.54(7)].

Genus *Rotalia*—Test trochoid, usually coarsely perforate with equatorial and lateral chamber. No alternative supplementary chamber on ventral side [Fig. 1.54(3)].

Genus *Calcarina*—Test trochoid and aperture ventral in young with supplementary large spine independent of chamber. [Fig. 1.53(4)].

Genus *Globigerina*—With chambers mostly finely spinous and wall cancellated, adapted for pelagic life. Globular form with last chamber completely involute.

## ENTAMOEBA HISTOLYTICA

Only two species of *Entamoeba* are harmful—*E. histolytica* in man and mammals and *E. invadens* in reptiles. Members of the genus *Entamoeba* are distinguished

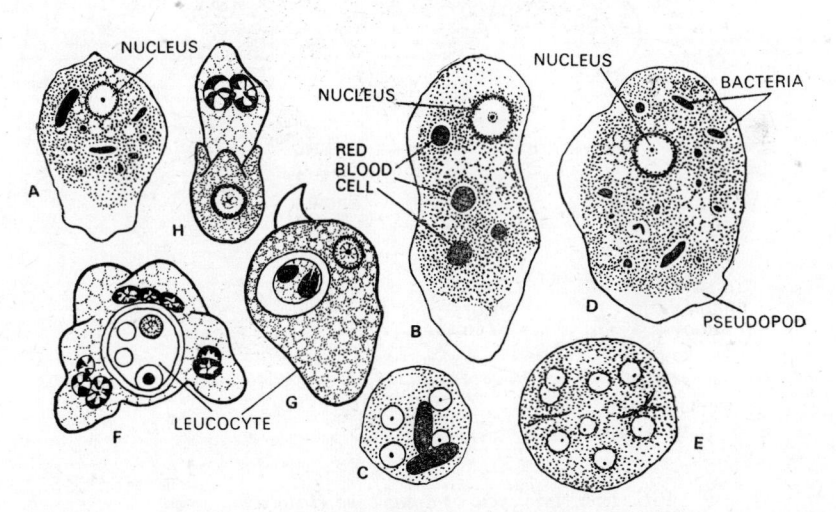

FIG. 1.55. A—Trophozpite of *Entamoeba hartmanni*; B—Trophozoite of *Entamoeva histolytica* from smear of human faeces; C—Mature cyst of the same; D—*Entamoeba coli*; E—Mature cyst of the same; F—Three leucocytes engulfing an amoeba; G—*E. gingivalis* with an engulfed leucocyte; and H—*E gingivalis* engulfing a leucocyte.

by the nuclear structure. The nucleoprotein is arranged in a peripheral ring lining the nuclear membrane. There is more or less a central karyosome, the rest of the space within the nuclear membrane appears empty and is, perhaps, filled with fluid. The species of the genus which infect man and domestic animals have been divided into four groups based mainly on the number of nuclei present in the mature cyst. The trophozoite of all species is uninucleate when it encysts but in many species it undergoes two or three nuclear divisions forming four or eight nuclei. The cysts often possess, when young, a vacuole containing glycogen which disappears by the time nuclear division completes. Cysts also contain structures called chromatoid bodies which appear transparent in fresh preparations but stain intensely with various haematoxylins. The chromatoid body also disappears as the cyst becomes old. Each chromatoid body is made up of ribonucleo-protein arranged as ribosome-like particles in a regular almost crystalline array.

The healthy trophozoite of *E. histolytica* is large and moves with a forward flowing movement. This is called **"limax type"** movement because it resembles the movement of a **garden slug** (*Limax* species). The entire body is elongated and of fairly regular in outline; it flows smoothly forward, all of its anterior end functioning as a single broad pseudopodium. The trophozoite contains a single nucleus with a fine karyosome and peripheral granules. There is no contractile vacuole. Food vacuoles, however, occur in the endoplasm. They contain various objects such as bacteria, host cell nuclei and sometimes even host erythrocytes. Trophozoites usually live in the lumen of the host intestine where each measures 10-20 $\mu$ in diameter. Sometimes trophozoites invade the mucosa and submucosa and many spread to other tissues (chiefly the liver). In these richer environments

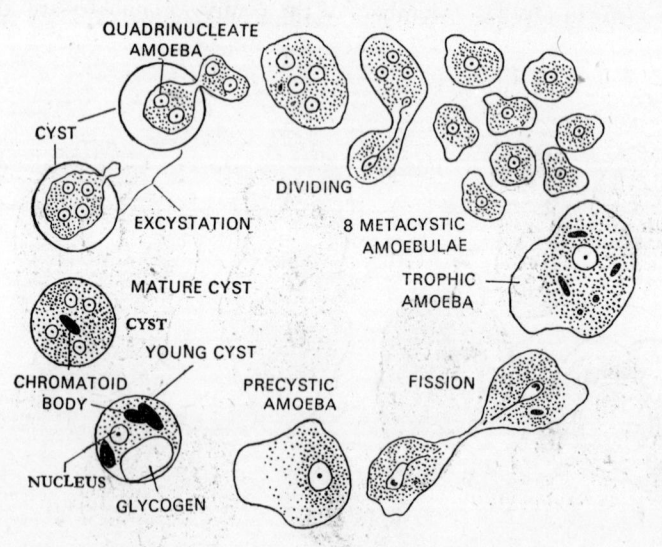

FIG. 1.56. Life cycle of *Entamoeba Histolytica*.

they become large. The trophozoites encyst only in gut lumen and not in other tissues. They round up, become quiescent and secrete a thin cyst wall. A diffuse mass of glycogen may be seen in the young cyst. Chromatoid bodies are also seen in young cyst. Soon the nucleus of trophozoite undergoes two divisions and forms four characteristic nuclei of the mature cyst. The cyst is the transmissive stage. It does not further develop in the original host and is passed out with faeces. It can survive outside the host for some time only. When swallowed by another susceptive mammal the cyst passes unharmed through the stomach. In the small intestine the parasite emerges from the cyst as four nucleate amoeba. These nuclei divide once more and then cytoplasmic division takes place. Thus, eight small uninucleate amoebae are formed. These pass into the large intestine and grow to full size.

In olden days *E. histolytica* was reported to exist in two races of *minuta* and *magna* forms. The minuta forms are small and live in the lumen of intestine dividing by binary fission. The small forms live in the lumen of intestine and reproduce by binary fission. They show a vacuole whose contents are homogeneous because this form obtains its nourishment primarily from dissolved substances. Occasionally bacteria are also taken up. They are not responsible for any special clinical symptoms, i.e., they are harmless. Now, there are evidences available that the minuta forms do not form a biological race of *E. histolytica*, but they are an independent species *E. hartmani* that lives in the intestine of man and mammals as harmless individuals. Only the large trophozoites of *E. histolytica* are parasitic.

**Pathogenesis**

*Entamoeba histolytica* usually lives in the gut lumen of man as a harmless commensal. Occasionally, for reasons unknown, the trophozoites penetrate the mucosa and the muscularis mucosae and invade the submucosa. Here they multiply and spread radially outwards below the mucosa to form a characteristically flask-shaped lesion or ulcer the centre of which is filled with cellular debris, lymphocytes, plasma cells and macrophages. Secondary bacterial infection may occur and then polymorphonuclear leucocytes will also be found. As the submucosa is eroded by the amoeba many blood vessels are broken and the typical blood dysentery results. Amoebae are found chiefly at the advancing edges of the ulcer. Rarely the ulcers perforate the gut wall entirely and cause peritonitis. Sometimes the amoeba spread via blood vessels to other organs where they invade and destroy the tissue causing amoebic abscesses (without bacteria) which may become large (several centimetres in diameter). The most common site for their development is the liver, but they may form such abscesses in lung and even brain. Untreated amoebic dysentery may result in death due to loss in fluid and blood. Abscesses in the liver or elsewhere may rupture through the body wall or into the peritoneal cavity with serious or fatal results.

**Epidemiology.** Infection with *E. histolytica* takes place by direct ingestion of faecal material containing cysts. The source of infection is usually man but dogs and rats living around him also get infected. The infection is common in those areas where food hygiene is less effective. There are some people who are chronically infected without being ill. Such persons are called **cyst-passers**. If the standard of their cleanliness is not ideal they become a regular source of infection. Sometimes, such persons are employed as food handlers and the matter becomes worse. The use of untreated human faeces as a fertiliser is another common source of infection. It often contaminates water supplies. Coprophagous insects such as flies and cockroaches may carry viable cysts, either on their legs or in their intestines, to food stuffs. The cysts survive for several weeks outside the body if not desiccated. They are not killed by aqueous potassium permanganate often used for washing salads. They are readily killed by heat (50°C and above).

**Diagnosis. Amoebiasis** is usually diagnosed by the presence of parasites in faeces. Active trophozoites are visible in fresh faeces mixed with saline and examined microscopically at 37°C. In diarrhoeic and normal faeces cysts may be found either by direct microscopical examination or after their concentration from the sample. In case of doubt attempts may be made to cultivate the amoebae in suitable artificial media.

**Treatment and prevention.** It is not easy to control amoebiasis. There are various drugs to curb the infection of *E. histolytica*. The alkaloid emetine has been used for many years (in various compounds) and is quite effective, but it is toxic. **Dehydro-emetine**, a synthetic derivative, is easily effective and less toxic. Chloroquine, an antimalarial drug, cures amoebic abscesses in liver only. Larger abscesses have to be treated surgically.

Prevention of infection is entirely a matter of hygiene, both personal and municipal. Personal hygiene involves washing of hands, avoiding the intake of raw vegetables and salads in dangerous areas and protection of food from coprophilic insects. The municipal hygiene calls for proper sewage disposal and water purification. Food-handlers in endemic areas should be examined for infection and treated if necessary.

**Other Intestinal Amoebae of Man**

There are five species of amoebae that live in human intestine. All are non-pathogenic. They, however, create confusion when attempt is made to diagnose amoebiasis. Therefore, it is necessary to know them. *Entamoeba hartmanni* closely resembles *E. histolytica*. The trophozoites are about the same size as the smaller individuals of *E. histolytica*, while the cysts which have four nuclei and

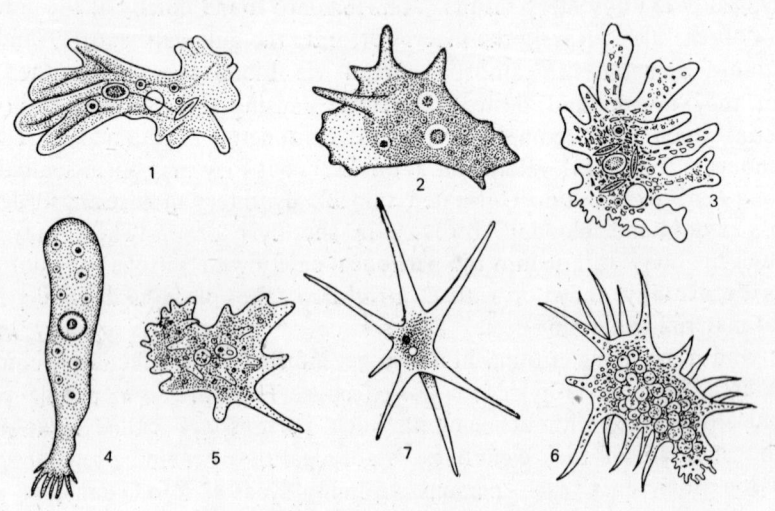

FIG. 1.57. Different types of amoeba.

chromatoid bodies exactly like those of *E. histolytica* are definitely smaller (4.0 to 10 5 $\mu$ average 7.4 $\mu$ in diameter). Thus, cysts of this type in human faeces which are less than 10 $\mu$ in diameter are almost certainly not *E. histolytica*. *E. hartmanni* have often been referred to as the "small race" of the former. but the evidence that they are separate species seems to be good (Baker, 1969). It is probable that many records of *E. histolytica* indigenous to the temperate parts of the world refer to *E. hartmanni*.

*Phylum Protozoa* 81

*Entamoeba coli* is another species living in the gut. The trophozoite measures 15-30 $\mu$, while the nuclear structure is *Entamoeba*-type, coarser. The cyst is smaller being 10-30 $\mu$ and possesses eight nuclei. The chromatoid bodies have thin sharp ends.

*Endolimax nana* has smaller (6-12$\mu$) trophozoite with nucleus that is not of *Entamoeba*-type. The cyst measures 6-9 $\mu$ and possesses four nuclei while the chromatoid bodies are absent.

*Iodamoeba buetschlii* varies in size from 5-20$\mu$. The nuclear structure is not of *Entamoeba*-type. The cyst is 9-15$\mu$ with only one nucleus and is without chromatoid bodies. Persistent glycogen vacuoles are present.

*Dientamoeba fragilis* is another species the trophozoite of which varies from 7-12$\mu$ in size, and is usually binucleate. Cyst of this species has not been reported.

## CLASSIFICATION

**Superclass Sarcodina**

A. CLASS RHIZOPODEA. It consists of Protozoa which are usually free-living and which move and ingest their food by means of pseudopodia, at least during the predominant phase of life-history; asexual reproduction is usually by binary fission. Differentiation of the body is rarely carried to an advanced degree.

(*i*) **Subclass Lobosia** (or **Amoebozoa** or **Amoebina**) comprises the Amoeboid rhizopods. Pseudopods relatively short and blunt changing, not stiff or ray-like; clear distinction between ectoplasm and endoplasm. They may be naked or shelled.

**1. Order Amoebida (Gymnoamoeba** or **Nuda)**, the naked Lobosia. Body naked or enclosed in a thin membrane; pseudopods arising at any point.

Example: *Amoeba* (Fig. 1.37).

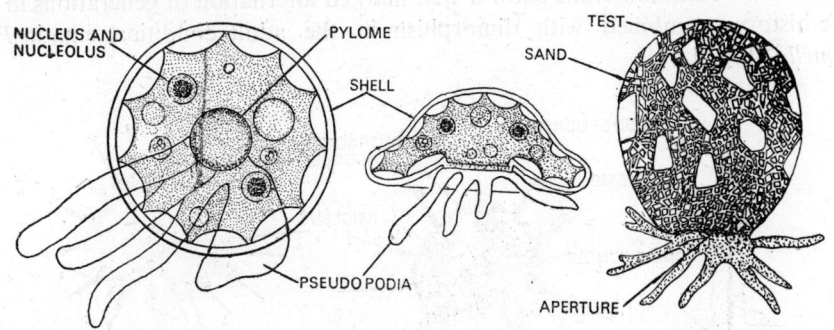

FIG. 1.58. A—*Arcella vulgario* horizontal section; B—Vertical section of the same; and C—*Cucurbitella mespiliformis* (*Difflugia*-like form).

**2. Order Arcellinida (Thecamoeba** or **Testacea)**, the shelled Lobosia. Body enclosed in a shell with one aperture through which alone the pseudopods protrude. The protective covering may be in the form of gelatinous or membranous encasement or shell composed of siliceous plates or of foreign particles embedded in a gelatinous matrix secreted by the animal. The shells are of simple shape, mostly oval, urn or bowl-like with a single opening or **pylome** through which the pseudopods protrude.

Examples: 1. *Difflugia,* a free living form which gathers sand grains, cements them together with sticky secretions to form a protective shell.

2. *Arcella,* the common fresh water form found in ponds, secretes a hard protective shell about itself. When the animal divides, one daughter cell retains the shell whereas the other has to construct one afresh (Fig. 1.59).

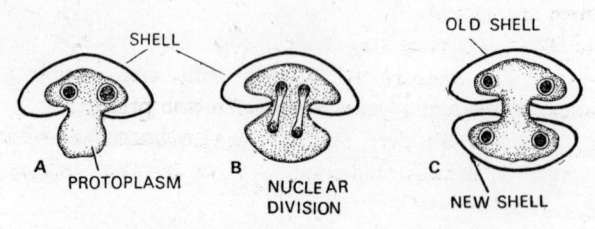

FIG. 1.59. Division in *Arcella.*

(*ii*) **Subclass Filosia**. Marine and fresh water Sarcodina with tapering or branching filopodia. Naked or test with a single aperture.

Examples: *Tenardia* (naked), *Allogromia*.

(*iii*) **Subclass Granuloreticulasia**. Sarcodina with delicate and granular reticulopodia.

1. **Order Athalamida**. Naked.

Example: *Biomyxa.*

2. **Order Foraminifera or Thalamophora**. Amoeboid forms, with slender pseudopods, often anastomosing enclosed in a simple or chambered shell with one opening or with numerous pores for pseudopodia. The body-protoplasm exhibits no marked distinction of ectoplasm and endoplasm, contractile vacuoles are present in some of the fresh-water genera, but are absent in marine forms. The marine foraminiferans show a well-marked alternation of generations in their life-history, combined with dimorphism in the adult condition. e.g., in *Polystomella* (Fig. 1.49).

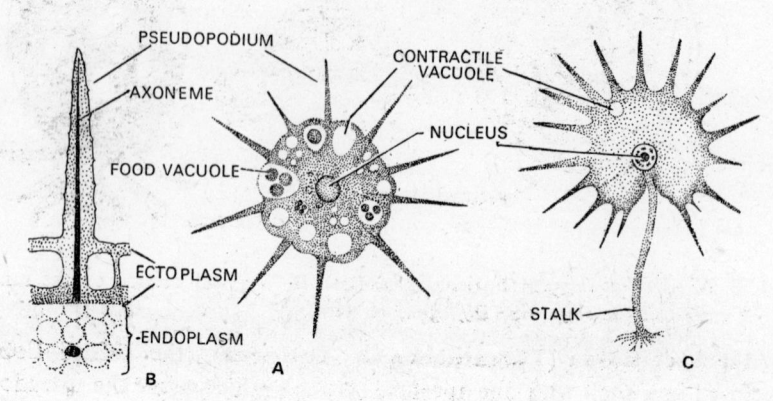

FIG. 1.60. A—*Actinophrys sol*; B—Portion of its peripheral cytoplasm; and C—*Clathrulina elegans.*

Examples: *Polystomella* (Fig. 1.49), *Globigerina, Nodosaria, Lagena, Discospirulina, Frondicularia. Discorbina* and *Planorbulina,* etc.

B. CLASS PIROPLASMEA. Small, round, rod-shaped or amoeboid parasites in vertebrate red blood cells. Formerly these were included with sporozoans but spores are absent. *Babesia* which is transmitted by ticks, is cause of cattle-tick fever.

C. CLASS ACTINOPODEA. Primarily floating or sessile Sarcodina with pseudopodia radiating from a spherical body.

(*i*) **Subclass Acantharia**. Actinopodea with imperforate non-chitinoid central capsule without pores, anisotropic skeleton of strontium sulphate. Marine.

Example: *Acanthometra* (Fig. 1.63A).

(*ii*) **Subclass Heliozoa**. Predominantly fresh-water rhizopods of spherical shape with radiating pseudopods or axopods (Figs. 1.60 and 1.61). The body is radially symmetrical. The ectoplasm is highly vacuolated and is termed cortex, that bears one to several contractile vacuoles at its surface. Endoplasm is denser granular mass in the centre containing matted needles of silica. The radiating pseudopodia project through the skeletal covering, when present. Each consists of a stiff axial rod clothed with a layer of streaming protoplasm (Fig. 1.61B). The axial rods terminate free in the medulla or on a nucleus in multinucleate forms. In good many the axial rods converge to a central granule. Movement is very slow, the pseudopods serve for food capture. Nutrition is holozoic. The axopods adhere to the prey, paralyse it probably by some toxic secretion and engulf the food in one of the following ways; the axopods may shorten conveying the prey to the main mass, shortening takes place by the absorption of the

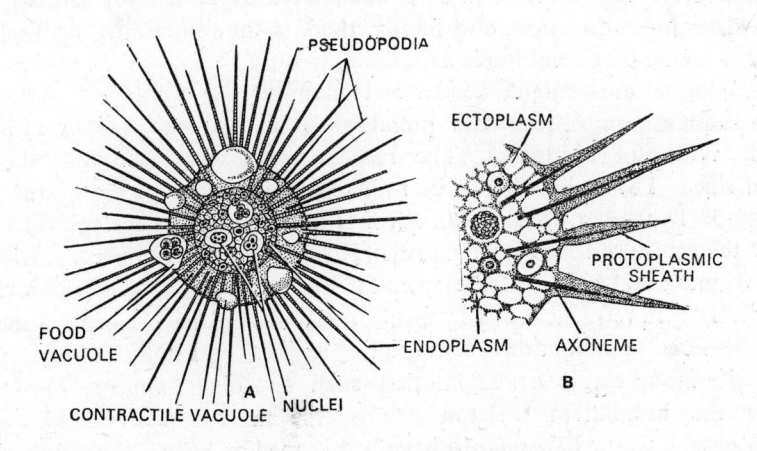

FIG. 1.61. *Actinosphaerium eichhorni*. A—Entire; and B—A part showing structure in details.

axopods at the base; or the axopods may melt around the prey, including it in a food vacuole, which later retracts into the animal. Many species harbour **zoochlorellae**. Uninucleate forms reproduce asexually by binary fission involving a regular mitotic division of the nucleus. Gemmation has also been frequently reported. Sexual phases are only known in a few cases and are peculiar, a form of **hologamy** called **autogamy** has been described for *Actinophrys* and *Actinosphaerium*. The animal withdraws its pseudopodia, secretes a gelatinous cyst and finally divides into two daughter cells that undergo two typical maturation divisions (like the metazoan sex cells). Later they fuse and form an encysted zygote that hatches into a young heliozoan.

Examples: *Actinophrys* (small uninucleate form); *Actinosphaerium* (large multinucleate form) and *Clathrulina* (beautiful form fastened by stalk).

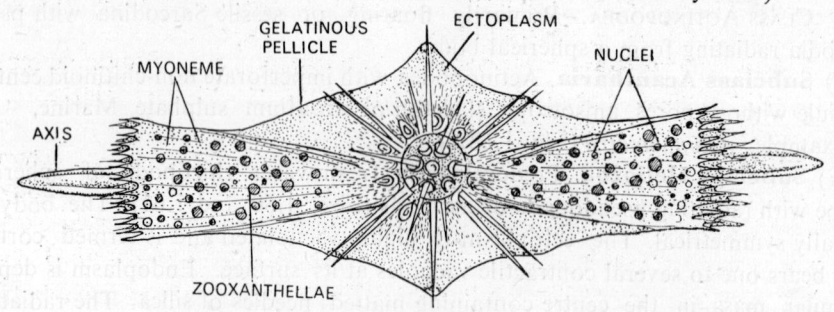

FIG. 1.62. *Diploconus cylindricus* (*after* Schewiakoff).

(*iii*) **Subclass Radiolaria**. Radially symmetrical rhizopods living in the surface layer of the sea. The cytoplasm is differentiated in two distinct layers separated by a membranous capsule, thus the two layers are sometimes referred to as extracapsular and intracapsular cytoplasm. The capsule membrane is provided by pores that allow protoplasmic communications. The extracapsular protoplasm can be divided into three zones. Immediately outside the capsule is the assimilative layer. This is followed by a vacuolated layer known as the **calymma**, hydrostatic in function, and finally, there is the outermost protoplasmic layer from which the pseudopods arise.

The pseudopods are straight, slender and filamentous, composed entirely of motile protoplasm; but in some the pseudopods are axopods supported by stiff axial rods. Nutrition is holozoic. The radiolarians secrete elaborate skeletons mostly of silica. The skeletons may be radiate or concentric or of both types intermingled. The radiating skeleton consists of long needles (or spines) radiating from the centre of the central capsule (*Thalassicola*). In the other variety the skeleton forming material (separate spicules or lattice spheres) are arranged concentrically with reference to the capsule membrane (found in order Spumellaria). All possible combinations of radiating spines and lattice spheres occur, together with innumerable ornamentation such as spines, thorns, hooks, etc. rendering the radiolarian skeleton one of the most wonderful and exquisite objects in nature. Some (*Nasselaria*) have bell-shaped or helmet-shaped skeletons. Radiolarian skeletons can be found in marine deposits in shallow water but it is only in very deep region that they occur in a concentration of at least 20 per cent.

Reproduction is effected by binary fission. The mitotic division of the nucleus is remarkable inasmuch as it produces a large number of chromosomes (over a thousand) and a centrosome is lacking. The central capsule divides later, and finally, the calymma divides. The skeleton also divides when possible and each daughter regenerates the missing half, but when the skeleton is indivisible one daughter cell separates and secretes its own skeleton. Many Radiolaria produce biflagellate swarm spores by multiple fission within the central capsule. Two types of swarmers have been reported: (*i*) **isospores,** all similar in size and appearance, (ii) **anisopores** of two types smaller microspores and larger macrospores. Some authors (Hyman) regard the anisopores as escaped zooxanthellae, symbiotic plant bodies ("yellow cells") inhabiting the extracapsular protoplasm. The complete life history is unknown in any case.

Examples: *Thalassicola* (Fig. 1.63), *Collozoum, Sphaerozoum, Collosphaera*, etc.

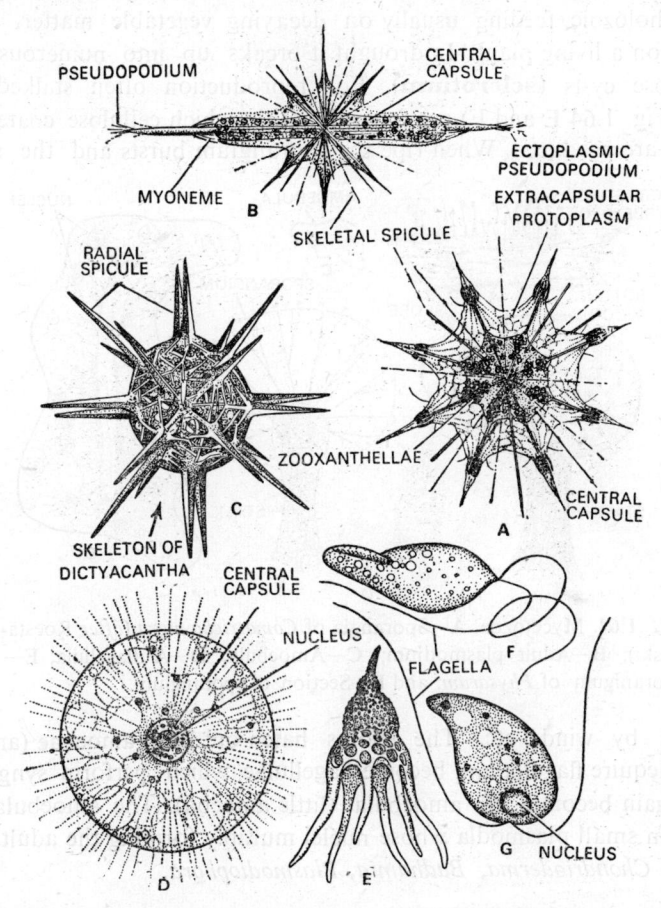

FIG. 1.63. Different types of Radiolaria. **A**—*Acanthometra* (*after* Morroff and Satiasny); **B**—*Amphilonche elongata* (*after* Schewiakoff); **C**—*Dictyacantha tetrogonopa* (*after* Schewiakoff); **D**—*Thalassicola pellucida* (*after* Grasse); **E**—Helmet-shaped skeleton of *Nassellaria*; **F**—Biflagellate isogamete of *Collozoum fulvum* (*after* Brandt); and **G**—Isogamete of *Coclodandrum*.

(iv) **Subclass Proteromyxidia.** Largely marine and freshwater parasites of algae and higher plants. Filopodia and reticulopodia in some species.

Example: *Vampyrella*.

(v) **Subclass Mycetozoa.** Semi-terrestial rhizopods which form encrusting masses on rotten wood or decaying vegetable matter of various kinds. The masses are really colonies of amoeba-like rhizopods, but the colony has the power to move as a unit. The individual is devoid of shell, skeleton or central capsule in the active phase, but in the quiescent stage a cyst of cellulose is formed. The most characteristic feature is the formation of plasmodia which represent the adult vegetative phase of the life-history. Reproduction takes place by the formation of resistant spores very similar to those of fungi. That is the reason why botanist include this under Myxomycetes.

The adult plasmodium is a sheet of protoplasm containing thousands of nuclear pieces and numerous contractile vacuoles (Fig. 1.64B). Its mode of nutrition is holozoic, feeding usually on decaying vegetable matter, sometimes (*Badhamia*) on a living plant. In drought it breaks up into numerous multinucleate cellulose cysts (**sclerotium**). For reproduction often stalked cellulose sporangia (Fig. 1.64 E and F) are formed, within which cellulose coated uninucleate spores are produced. When ripe the sporangium bursts and the spores are

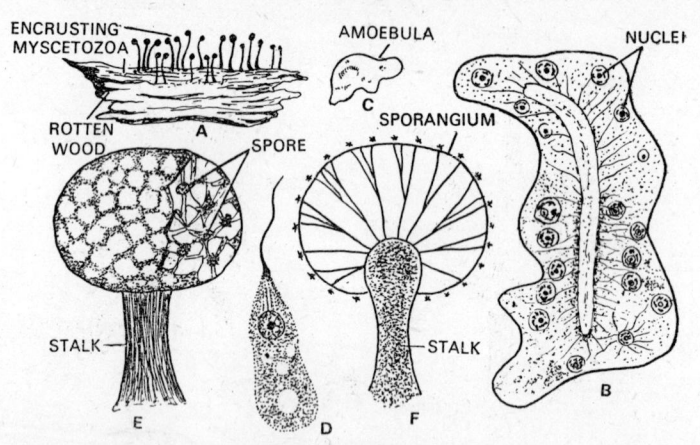

FIG. 1.64. Mycetozoa. A—Sporangia of *Comatricha nigra* (*after* Roestafinski); B—Adult plasmodium; C—Amoebula; [D—A flagellula; E—Sporanigum of *Physarum*; and F—Section of sporangium.

disseminated by wind, etc. The spores hatch out little amoebae (amoebulae) which may acquire flagella and become flagellulae which perform syngamy and the zygote again becomes all amoebula (little amoeba). The amoebulae tend to fuse and form small plasmodia whose nuclei multiply forming the adult.

Examples: *Chondrioderma, Badhamia, Plasmodiophora*.

## SUBPHYLUM SPOROZOA

The **Sporozoa** are characterised by the possession of typical simple spores, which were originally a resistant stage meant for dispersion and transmission of the species. In more sophisticated members of the subphylum (*e.g.*, **Haemosporina**) the spore itself may be lacking, since with the adoption of an insect vector it is no longer necessary to protect the parasite while outside a host. The Sporozoa lack cilia or flagella (except on the microgametes of some genera). In almost all Sporozoa, however, the infective stage is **sporozoite**, which develops within the spore of the more primitive members. The growing vegetative parasite is called a **trophozoite,** and in the majority of groups the trophozoite attains full size and divides by a type of multiple fission called **Schizogony**. What was regarded as multiple fission originally is now considered a type of multiple budding as revealed by electron microscopy. The trophozoite at this stage is called a **schizont** (or a **gamont**). After two or more nuclear divisions,

# Phylum Protozoa

the "daughter" nuclei move to the periphery of the organism (**schizont**) and daughter individuals or **merozoite**s appear as buds, one related to each nucleus. Finally, the merozoites break off, the schizont is destroyed in the process, only residual body of cytoplasm remains (and in the case of erythrocytic schizonts of *Plasmodium* the malarial pigment remains). But formation begins only after nuclear division. The number of merozoites produced ranges from four to many thousands. In very large schizonts the area available for the production of merozoites is increased by complex invagination of the schizonts called **cytomeres**. Schizogony is found only in Sporozoa.

The merozoites reinfect the host and may repeat the schizogonic cycle several times and finally they give rise to **gamonts** or **gametocytes** which by multiple fission give rise to **gametes** (**micro-** and **mega-**); this process is called **gamogony**. The gametes fuse in pairs and form the zygote which is called **oocyst** if non-motile, or **ookinete**, if motile. The zygote again undergoes multiple fission,

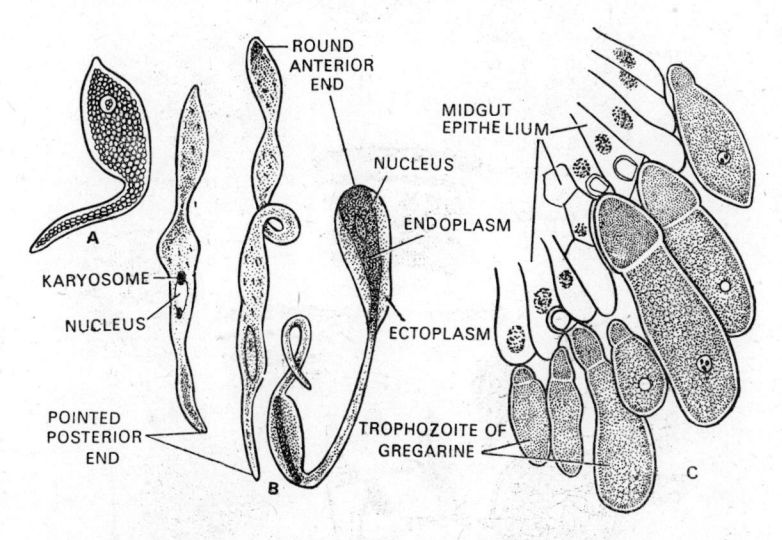

FIG. 1.65. A—Adult trophozoite; B—Trophozoite of *Nematocystis* from *Eutyphoeus* (*after* Bhatia and Chatterjee); and C—Gregarine trophozoites.

known as **sporogony**, either forming naked young, **sporozoites** which, infect directly, or resistant spores from which sporozoites are formed under proper conditions. In the life-cycle of many sporozoans an alternation of schizogony and sporogony takes place, often in relation to change of hosts.

## MONOCYSTIS

*Monocystis* is an inhabitant of a part of the reproductive apparatus of the common earthworm. Various stages in the life-history of this organism are invariably found in the seminal vesicles of the earth-worm.[6] The common spe-

---

[6]The genus *Eutyphoeus* usually dissected here has been noticed to be normally infected with *Nematocystis*, which is relatively very long tapering at one or both ends, generally with anterior end rounded and posterior pointed, showing numerous constrictions and bulgings probably for the progression of the parasite. It appears as if the granular cytoplasm, together with the nucleus, flows from one pole to the other. There are, in addition, numerous specimens referable to *Dirhybocystis globosa* (Bhatia and Chatterjee).

cies usually met with include *Monocystis beddardi* (Ghosh), *M. pheretimi* (Bhatia and Chatterjee) and *M. Iloidi* (Ghosh).

**Structure**

The young and growing trophozoite is a minute nucleated body found in the central protoplasm of one of the collections of the developing spermatozoa or sperm morula. It feeds and grows at the expense of the protoplasm around it until the whole protoplasm has been consumed. Thereupon, the trophozoite is found surrounded by the tails of dead **spermatozoa** of the earthworm, for a time, and appears like a ciliated organism. Ultimately these spermatozoa are detached and the trophozoite becomes free.

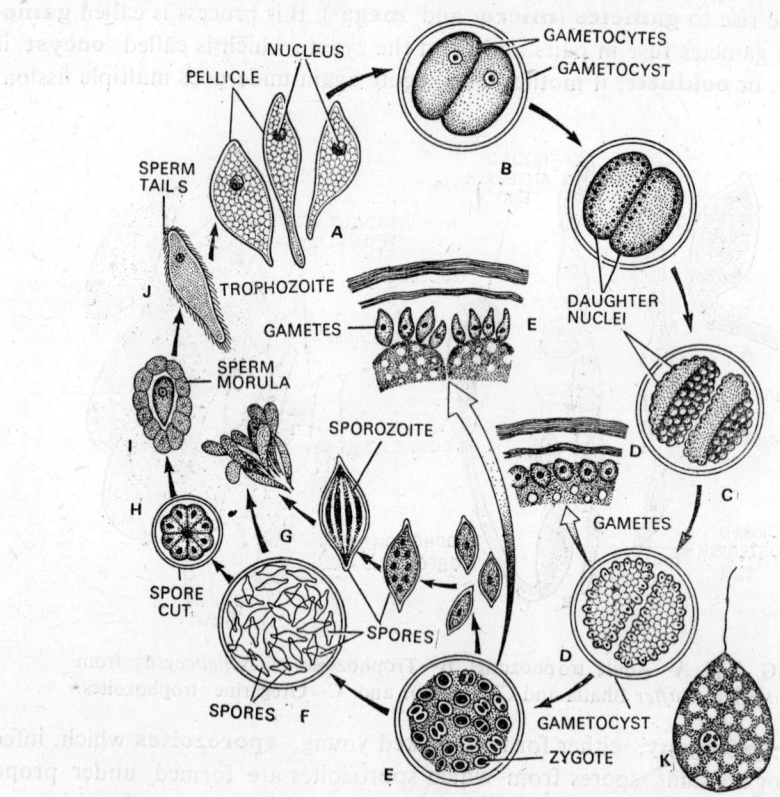

FIG. 1.66. Life-cycle of *Monocystis*. A—Adult trophozoites; B—Syzygy; C—Nuclear division and migration of daughter nuclei towards periphery; D—Formation of gametes; E—Conjugation; F—Spore formation; G—Spore; H—Cross section of a spore; I—Sporozoite in sperm morula; and K—Male gamete of *Monocystis marazeki* (enlarged).

The mature trophozoite is fusiform (Fig. 1.66) limited externally by a thick smooth and porous pellicle (also called **ectocyte**). Its cytoplasm has an outer layer, **ectoplasm** (or **cortex**), and an inner layer **endoplasm (medulla)**. The ectoplasm may be differentiated into an outer homogeneous **sarcocyte** and an inner **myocyte**. In the myocyte of the ectoplasm are found longitudinally running tracts of cytoplasm called **myonemes**, specially developed contractile structures. In some species circular myonemes are also present. Within the endoplasm in the **nucleus**, which is surrounded by a delicate nuclear membrane, and contains a clear liquid in which are suspended denser bodies called **nucleoli**. Chromatin is distributed throughout the **nucleoplasm**. The endoplasm contains other inclusions including **golgi material, chondriosome (mitochondria), paraglycogen** or **zoomylum**, the refractile carbohydrate granules, etc.

## Locomotion

The trophozoites show only slow wriggling movement brought about by the rhythmical contractions of the myonemes. Such movement is characteristic of the group (Gregarinida) to which *Monocystis* belongs, and is consequently called **gregarine movement**. In fact, the parasitic mode of life reduces the need of active movement.

## Nutrition

In the early stages the trophozoites feed upon the protoplasm of the sperm morula. Digestive enzymes exude out of its body and render the protoplasm around it assimilable. The digested products later are absorbed through the pellicle. After the sperm tails have been shed, the trophozoite lies freely in the fluid of the seminal vesicle and doubtless absorbs nutriment from it. Reserve food material is stored in the endoplasm in the form of granules of **paramylum (paraglycogen)**.

## Respiration and Excretion

Respiration is brought about by the usual method of diffusion. Being parasitic, the organism lives within the seminal vesicle of the earthworm surrounded by fluid. The oxygen from this fluid diffuses in its body and carbon dioxide likewise diffuses out. The fluid in the seminal vesicle receives supplies of oxygen through the blood of the earthworm, and in the same way the excess of carbon dioxide is handed over to the fluid whence it goes out in the same way (*i.e.*, through the blood). The nitrogenous products are also disposed of in the same way.

## Reproduction

Before entering the actual reproductive phase (Fig. 1.66) each fully matured trophozoite becomes a **gametocyte (gamont)**. Then, two such gametocytes become associated with each other and each such pair secretes around itself double-walled cyst. Since the gametocytes do not fuse, but are merely associated together, the cyst is commonly called an **association cyst**, but can equally well be called a **gametocyst**. The process is called **syzygy** while some prefer to call it **pseudoconjugation** as they come close to each other but there is no exchange of nuclear material. Within the gametocyst each gametocyte produces by simple binary fission a number of small nucleate bodies, the **gametes**. The process of gamete formation is known as **gametogony**. The last division of the nucleus leading to the gamete formation is a reduction division resulting in the formation of haploid gametes. During gametogony first the parent nucleus divides mitotically several times, finally the daughter nuclei divide meiotically forming haploid nuclei which migrate to the periphery of the parent gametocytes forming buds in the surface imparting to it a mulberry-like appearance.

Finally, each of these buds separates forming the gametes which are released in the cavity of the gametocyst. The gametes produced by the two gametocytes are equal in the number and similar in shape (**isogametes**). It is, therefore, difficult to distinguish the gametes from the two gametocytes, but it is presumed from other evidence that when fusion takes place in pairs one gamete comes from one gametocyte and the other from the other. The fusion of the gamete is complete involving both nucleus and cytoplasm. Recently it has been suggested, on the basis of the new findings of some workers, that sex differentiation occurs between the gametes of two gametocytes. These may be in the form of cytoplasmic inclusions or in shape. For instance, it has been pointed out that the male gametes have a pointed tail or are flagellated. Flagellated microgametes of *Monocystis marazeki* as found by Hahn are cited as example.

As a result of fusion (**syngamy**) large diploid **zygotes** are formed. The zygotes are also termed **sporoblasts** because of their later development. Each sporoblast secretes a tough resistant chitinoid covering, in external appearance resembling diatom of the genus *Navicella*. For this reason the sporocyst, at this stage, is called a **pseudonavicella**. In this stage they are commonly called **spores**. Within the sporocyst, the sporoblast undergoes three successive divisions producing eight sickle-shaped or fusiform, nucleated protoplasmic bodies, the **sporozoites**. By now the original gametocyst bursts releasing the spores in the cavity of the seminal vesicles of the earthworm. Further development of the sporozoite is possible only if it is transferred to another earthworm. The sporocysts (spores or pseudonaviscellae) are ingested by fresh (uninfested) worms, then their walls are dissolved by the digestive secretion of the worm and the sporozoites are set free. Being actively motile the sporozoites pass out through the tissues of the gut into the coelom and then migrate into the seminal vesicles. On reaching the seminal vesicles of a new worm each penetrates into a new sperm morula, becomes a trophozoite, and the cycle begins again.

*Monocystis* is a parasitic form, and in response to its parasitic habit various vegetative organelles such as the contractile vacuole, mouth, gullet and locomotory structures are reduced or lost. There is a definite increase in the complexity and efficiency of the reproductive system. The potential reproductive units (sporozoites) have to pass through the outside world and there are chances of destruction of the individuals. Therefore, these are produced in large numbers so that at least some of them may survive and may enable the organism to continue the race.

**Mode of infection.** Nothing is definitely known about the propagation of the parasite. Almost every worm examined shows the presence of the parasite, it may be supposed that the sporocyst may be transferred from one host to another in the process of copulation when the spermatozoa are conveyed from the seminal vesicles of one worm to the spermathecae of another. But the pseudonavicellae have never been seen in the spermathecae or in the cocoon, it is, therefore, presumed that the sporocysts are set free when the host worm dies and decays in the normal course of events or when it is devoured by some animal, such as a bird, and is digested within its gut. The latter is evidenced by the fact that the sporocysts have been actually seen in the contents of the intestines of various birds. From the gut the sporocysts pass with the excrement of birds into the soil with which they are swallowed by other earthworms.

## EIMERIA

Two genera *Eimeria* (Fig. 1.67) and *Isopora* of the family Eimeridae are quite well-known among the Coccidia. *Eimeria* includes many important parasites of domestic mammals and birds, while *Isopora* parasitises man (inhabits human intestine as mildly pathogenic parasites). *Eimeria* contains a large number of species parasitising vertebrate animals of all classes. Some species have been reported from annelids, arthropods (mostly centipedes) and Protochordates. None infects man.

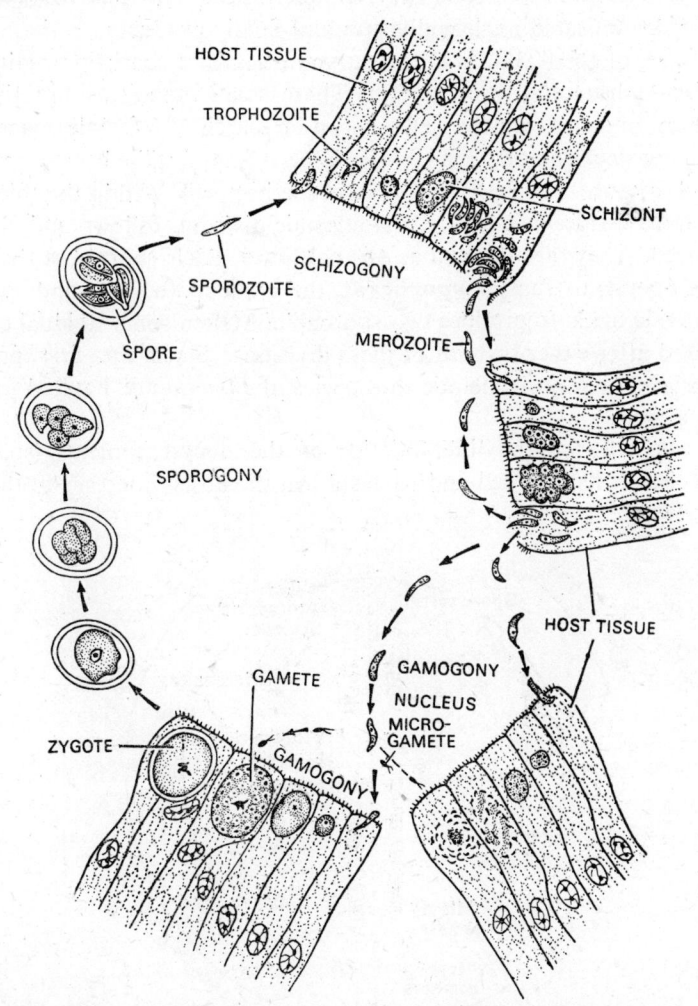

**FIG. 1.67.** Life-cycle of an eimeriid coceidian (*after* Morgan and Hawkins).

Parasite enters the host in the form of oocyst. When swallowed the oocyst hatches in the small intestine of the host (probably under the influence of mechanical pressure, pepsin and trypsin). The sporozoites emerge from the sporocysts and penetrate the cells of the intestinal mucosa. In some species they round up and grow in these cells, while in others they are carried by microphages, elsewhere in the body. The growing forms are called **trophozoites**. Most of them begin nuclear division and thus become schizonts. Then by multiple budding merozoites are produced (by asexual process). The size of schizonts varies widely with different species from about 10 to several 100 microns in diameter. The number of merozoites varies from about 16 to thousands. The cell and schizont rupture releasing the merozoites, which in most species re-enter other cells (either nearby or in some species in more distant tissue) and recommence schizogony. This continues for a varying but limited number of generations.

Sooner or later, for unknown reasons, the merozoites enter usually host cells of the intestinal mucosa and develop into sexual individuals or **gametocytes**. The female or **macrogametocyte** remains uninuclear, but the male **microgametocyte** undergoes repeated nuclear division and finally produces as its surface a large number of small (2-3 $\mu$ long) curved organisms consisting mainly of nucleus, mitochondrion and three flagella. These small organisms are the motile male gamete or microgamete. They swim in search of a female macrogamete, and on finding one, one of them fertilises her. The fertilised macrogametes is now called **zygote**. It encysts still within the host cell. Within the thick cyst its diploid nucleus divides meiotically, cytoplasmic division follows and four cells are produced. They are known as **sporoblasts**. Each sporoblast then encysts (within the oocyst) to form a **sporocyst**, the contents (nucleus and cytoplasm) of which divide again to produce two sporozoites. Often some residual cytoplasm is left unused after either or both of these divisions. Since the first post-fertilization nuclear division is meiotic the species of *Eimeria* are haploid for almost all their life cycle.

At some stage during the differentiation of the oocyst contents (sporogony), the oocyst leaves its host cell and passes down the hosts' intestine to the outside

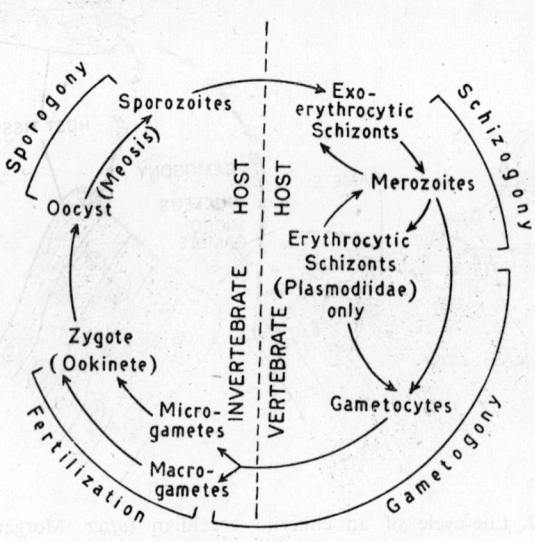

FIG. 1.68. Diagrammatic representation of life-cycle in the suborders Haemosporina.

world. There it can survive for a considerable time, until eaten by another susceptible host. The development cycle of *Eimeria* resembles that of malarial parasite except that in the latter fertilization and sporogony occur in a second host (**vector**) and the parasite is not exposed to the dangers of the outside world at any time.

## MALARIAL PARASITE

**Malaria.** Of all the tropical diseases malaria was one of the biggest health problems confronting the human race. It was responsible for roughly one-third of all the attendance at hospitals in the tropics. About one-third of the entire population of many hot countries suffered from it every year. Although only one case in several hundreds proved fatal, the diseases was so prevalent that the total number of deaths due to it was colossal. It had been officially estimated that in India alone something like 1,3,00,000 deaths were caused by it in an average year. In Uttar Pradesh, it was estimated that one-quarter of the population was totally incapacitated for two months in the year. The loss of working ability which even mild attacks of malaria cause was very serious. A couple of years back the disease was believed to have been eliminated, but it has staged a comeback now, that too in a relatively dreaded manner.

In 1952, when the government first began its campaign against malaria, the disease afflicted some 75 million Indians every year, killing nearly 800,000. Regular spraying of mosquito-infected areas with insecticides and mass distribution of Chloroquin, an anti-malaria drug, dramatically reduced the total number of cases by 1964 to 100,000 of which none was fatal. Soon afterwards, the figures started rising, and by 1975 the number of malaria cases had skyrocketed to five million, with one hundred deaths.

The Indo-Pakistan war of 1965, which disrupted the procurement and distribution of insecticides, caused the initial setback to the anti-malaria campaign. But there were other reasons too. The mosquitoes in many areas of India had become immune to insecticides and learnt to avoid DDT-sprayed walls. In one village in Assam, even the malaria parasite has shown resistance to Chloroquin. What is more, the rise in oil prices has resulted in the short supply of petroleum based insecticides. Hence other methods of fighting malaria are now being studied, and a malaria vaccine may be developed in future. But the World Health Organization has predicted that unless a major breakthrough occurs, there will be nearly 12 million cases in India by 1980, 400,000 of them fatal.

Though characteristically a tropical disease malaria has affected Europe as far north as Holland and England. Until the later part of nineteenth century it was quite common in the marshes of Essex, and Kent. Falstaff was "shaken of a burning quotidian tertian" (Shakespeare in **Henry V**), and Dickens refers to it in **Great Expectations**. Malaria has been practically wiped out in England, but these districts remain potentially malarious. In Greece and Rome the disease was until recently a curse. It is said that the disease has protected the city more efficaciously than armies against foreign invaders.

Malaria has played a significant role in military history. It has decided the fate of many armies, notoriously that of the Walcheren Expedition of 1809. In this highly malarious district of Holland 10,000 out of 15,000 British troops had malaria at one time, and they were dying at the rate of 25 to 30 per day. During the war of 1914-18 malaria was probably responsible for more casualties than any other disease. Among British troops there were 160,000 admissions of malaria cases to hospitals in Macedonia, 35,000 in Egypt, 107,000 in East Africa, and 120,000 in Mesopotamia. As all soldiers suffering from it did not receive treatment in hospitals, these figures under-estimate the total loss of man power.

During 1918 the British forces in Macedonia alone lost 2,000,000 service days.

**Historical account**. For long it was believed that malaria was caused by harmful vapours produced in marshy land (Gr. *malo*, bad+air). Laveran, a French military surgeon, for the first time, noticed *Plasmodium* in the blood of a malaria patient in 1880. In 1885 Golgi, an Italian worker, further confirmed that *Plasmodium* always occurred in the blood of persons suffering from malaria. This posed a problem before the scientist—how *Plasmodium* gets into the blood of a man?

Meanwhile, Sir Patrick Manson, a Scottish doctor at a Chinese hospital in Formosa, had discovered that mosquito carried the germs which caused elephantiasis. A German doctor, Richard Pfeiffer, suggested in 1892 that some bloodsucking insect carried malaria also. Laveran, working in Algeria, definitely suggested that the mosquito might spread malaria.

In 1894 Major Ronald Ross of the Indian Medical Service, a doctor who had long been interested in the study of malaria, returned home on leave, and while in London called on Manson, who had retired by then. Manson explained his theories to Ross. Back in India Ross settled to work in earnest and in 1897 he succeeded in his efforts. He established that the germs of malaria are sucked by the mosquitoes of one type, i.e., female *Anopheles* from the body of the infected human being and develop within its body to infective stage. When this mosquito bites a healthy man the germs are inoculated in his blood ultimately resulting in his sickness. For this discovery Sir Ronald Ross was awarded the Nobel Prize for medicine in 1902.

The term malarial parasites is generally restricted to species of the genus *Plasmodium* which parasitises reptiles, birds and mammals. They are obligate intracellular parasites for almost all of their life-cycle, and have two hosts: a vertebrate in which asexual reproduction or schizogony occurs, and an invertebrate (always a blood sucking dipterous insect) in which sexual reproduction occurs, the insect being regarded as the vector.

**Systematic position**. The genus *Plasmodium* belongs to the family Plasmodiidae. Other closely related genera, which do not infect man, are usually grouped into two separate families Haemoproteidae and Leucocytozoidae. All these families belong to the suborder Haemosporina under the sub-phylum Sporozoa. The family Plasmodiidae comprises a single genus *Plasmodium* Marchiafava and Celli 1885. This genus of Sporozoa undergoes sexual reproduction in mosquitoes (Diptera, Culicidae) and asexual reproduction (schizogony) in vertebrates. Schizogony occurs partly in certain fixed tissue cells and partly in erythrocytes. Gametocytes develop in erythrocytes. The intraerythrocytic forms metabolise haemoglobin, producing a characteristic yellow, brown or black **malarial pigment** or **haemozoin** (a compound containing **haeme**) in vacuoles within their cytoplasm. The genus plasmodium has been divided into following subgenera:

(1) In primates: *Plasmodium* and *Laverania* (transmitted by mosquitoes of the genus *Anopheles*).

(2) In lemurs and lower mammals: *Vinckeia* (transmitted by *Anopheles*).

(3) In birds: *Haemamoeba, Huffia, Giovannolaia,* and *Novyella* (transmitted by various genera of mosquitoes, anopheline and culicine).

(4) In reptiles: *Sauramoeba, Carinamoeba* and *Ophidiella* (transmission unknown).

# Phylum Protozoa

## Life-cycle in Vertebrate Host

The life-cycles of the parasites of all the subgenera listed above are similar, and also resemble the life-cycles of the members of the suborder Eimeriina. The vertebrate host is infected by means of small $(10\text{-}15 \times 0.5\text{-}1\mu)$ fusiform, uninucleate **sporozoites** injected with the saliva of an infected female mosquito when the latter feeds. It may be mentioned here that only female mosquitoes can feed on blood (only female can serve as vector). The sporozoites penetrate fixed tissue cells and commence **schizogony** (a type of asexual reproduction resulting in the production of four or more progeny i.e., merozoites, during which all or almost all the nuclear divisions are completed before cytoplasmic cleavage begins). The cells in which schizogony takes place are different in different subgenera. In those parasitising mammals schizogony always takes place in liver parenchyma cells. In those parasitising birds and reptiles the first generation (**primary exoerythrocytic schizonts**) takes place in lymphoid-macrophage cells in the skin near the site of the mosquitoes bite and subsequent generation (**secondary**

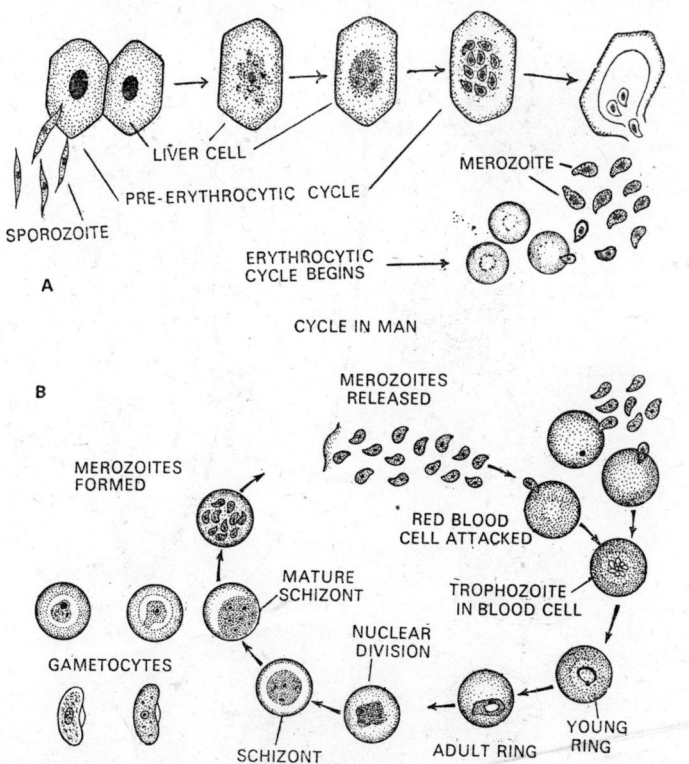

FIG. 1.69. A—Exoerythrocytic cycle in man. Sporozoites enter liver cells and multiply. The cells produced are merozoites. They attack red blood cells but do not enter liver cells again as hitherto believed; and B—The erythrocytic cycle in man.

**exoerythrocytic schizonts**) in various lymphoid-macrophage cells throughout the body. There are certain other terms that have been used for primary and secondary exoerythrocytic-schizonts. The primary exoerythrocytic schizonts have been called **pre-erythrocytic** or **cryptozoic schizonts**, while those of the subsequent generation have been called **secondary exoerytbrocytic** or **exo-**

**erythrocytic** or **metacryptozoic** or **phanerozoic** schizonts.

In the former group most or all merozoites produced by primary exoerythrocytic schizonts enter erythrocytes (red blood cells) and form small rings. They then grow and form what are known as trophozoites. Finally nuclear division begins in the trophozoites, as in the beginning the erythrocytic schizogony. The classical view is that in all species except *P. falciparum*, merozoites produced in the liver cells re-enter liver cells on being released and develop as successive generations of secondary exoerythrocytic schizonts, but this has never been conclusively demonstrated. Authorities like Garnham (1967) doubt this. In the subgenera infecting birds and reptiles, there is no doubt, that some of the merozoites arising from the primary exoerythrocytic schizonts invade other lymphoid macrophage cells to produce secondary exoerythrocytic schizonts. Some merozoites produced by these secondary schizonts invade other lymphoid macrophage cells, while others enter red blood cells (or their precursors) to initiate

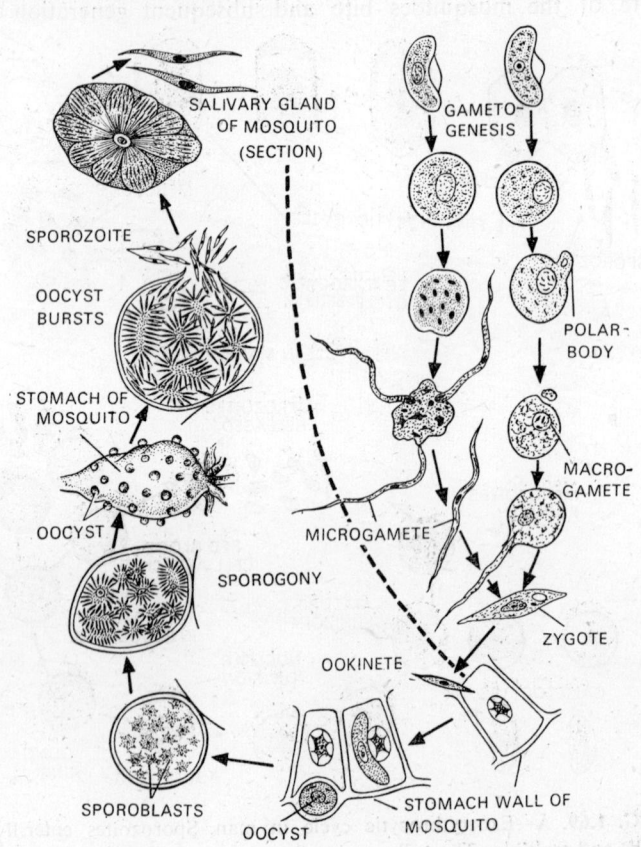

FIG. 1.70. Sexual cycle of the malarial parasite in the mosquito.

erythrocytic schizogony. Both types of schizogony exoerythrocytic and erythrocytic) continue side by side. But in mammalian malaria the merozoites produced by the primary erythrocytic schizonts do not re-attack liver cells. Sometimes merozoites from these schizonts appear more than one time, that is because some primary exoerythrocytic schizonts grow more slowly or have a dormant phase.

Hence merozoites are released at different times.

In the mammalian malaria parasites merozoites from erythrocytic schizonts invade only other erythrocytes, but those of avian and reptilian malaria may enter either other red blood cells or lymphoid-macrophage cells, in the latter they develop as secondary exoerythrocytic schizonts.

## Life-Cycle in Mosquito

Some of the merozoites entering erythrocytes (in both groups) develop not as erythrocytic schizonts but as gametocytes (sexual individuals). The male and female gametocytes (which can be differentiated in stained blood films as the latter have a more compact nucleus and very basophilic cytoplasm) do not divide, but remain within their last erythrocytes until they either die or are ingested by a mosquito in which they continue their development in the insect's stomach.

The female emerges from the red cells, rounds up (if not already spherical) and, now a mature gamete awaits the arrival of a male gamete. The male gametocyte meanwhile becomes very active. On emerging from the host cell it undergoes three nuclear divisions and develops eight flagella all within 15-20 minutes. The flagella emerge from the surface of the gametocyte (this process is termed "**exflagellation**"), one nucleus passes inside the membrane surrounding the axoneme of each flagellum, and the resulting male gametes break off and swim away rapidly. Those that meet the female gametes penetrate and fertilize them. The process of gamete formation is termed **gametogony**. The female gametocyte is termed **macrogametocyte** and **gamete** is called **macrogamete**, while the male gametocyte is called the **microgametocyte** and the gamete is the **microgomete**.

The fertilized female gamete elongates in a motile **ookinete**, a stage in which the male and female nuclei fuse sooner or later. The ookinete burrows into and usually through a cell of the single-layered epithelium forming the wall of the mosquito's stomach. On the outer surface of the stomach it encysts and becomes an **oocyst**. The contents of the oocyst undergo **sporogony**, a process essentially like schizogony, which results in the formation of thousands of uninucleate **sporozoites**. The first nuclear division of sporogony is a reduction division and all stages except the ookinete are haploid. The oocyst increases in size throughout sporogony and finally bursts liberating the sporozoites in the haemocoel of mosquito. They spread throughout the haemocoel and

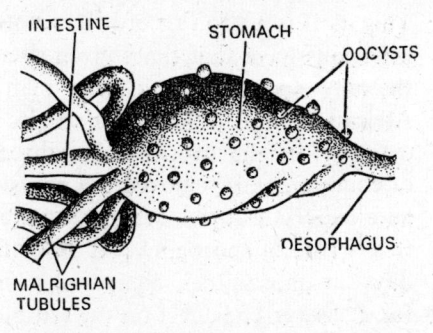

FIG. 1.71. Stomach of an infected mosquito.

eventually penetrate the salivary glands, where they remain until the insect has its next blood meal when they are injected with the saliva into the animal on which it is feeding. If it happens to be a susceptible host then sporozoites enter the appropriate tissue cells and commence primary erythrocytic schizogony.

## MALARIA IN MAN

There are four species of *Plasmodium* that attack man. All are transmitted by mosquitoes of the genus *Anopheles*. The description of the species is given below:

*Plasmodium* (*Plasmodium*) *vivax*. This is the most wide-spread species causing malaria in man. It produces a relatively mild malaria called **benign tertian malaria**. This species is called **vivax** (the lively *Plasmodium*) because of active amoeboid movement shown by the growing trophozoite. It is because of this that it appears irregular in form in dried stained preparations of blood films. It is called **tertian** because fever occurs at intervals of 48 hours, thus if the day of the first fever is numbered one, the next fever will come on third day. The adjective "**benign**" carries the suggestion of a mild and favourable disease, only because it is not fatal. Otherwise those who have suffered from malaria will never agree that the attack is mild as it makes the patient very weak. If untreated, it persists for many years, relapses eight years after infection has been recorded, presumably because of successful re-invasion of erythrocytes by merozoites produced by persisting exoerythrocytic schizonts in the liver. Study of a thin blood film taken from a patient shows that as a parasite grows the red cell becomes large because of mechanical pressure from the growing parasite, and becomes pale in colour perhaps because of ingestion and digestion of haemoglobin by the parasite. Moreover, the surface membrane becomes covered with closely packed fine dots (that appear pink with **giemsa's stain**), known as **Schüffner's dots** (after the discoverer). The Schüffner's dots are not visible in unstained parasitised cells. Even electron miscroscope has not revealed their nature. Work with fluorescent antibodies on other species of malarial parasites has suggested that they may represent aggregations of some antigenic material, perhaps excreted by the parasite and deposited on the red cell's plasmalemma, but there is no evidence.

*Plasmodium* (*Plasmodium*) *malariae*. This species is less common than *P. vivax*. The infection of this species, in untreated persons, may continue for 40 years. This itself is a good record. It is not very pathogenic organism although chronic infections have sometimes given rise to a lethal kidney condition. *P. malariae* is the only species that attacks man as well as other primates. It attacks West African Chimpanzees. Other species which infect man can develop exoerythrocytically in the liver of chimpanzees, but the blood resists the attack. If spleen of chimpanzee is removed the parasites can attack its blood also. Both intra- and exoerythrocytic schizogony are slower in this species than in others, and this is true of sporogony. At 24°C the latter cycle (sporogony) takes about 21 days in mosquitoes compared with 16 for *P. ovale*, 11 for *P. falciparum* and 9 for *P. vivax*. The effect on the red blood cells by *P. malariae* is not as prominent as by *P. vivax*. The erythrocyte is not enlarged. No dots are seen on the surface. However, prolonged staining shows very fine dots scattered irregularly over the plasmalemma. These have been called **Ziemann's dots**.

*Plasmodium* (*Plasmodium*) *ovale*. This is the rarest of the four species of parasites which infect man. It was finally recognized as a distinct species in 1922. Morphologically it looks like *P. malariae* although the duration of the erythrocytic cycle of schizogony is different. The effect produced by this species on the host cell is similar to that of *P. vivax*. Some authors have mentioned it as

"*P. malariae* in a *P. vivax* type cells." The host cell shows prominent Schüffner's dots and is enlarged (slightly less than that infected with *P. vivax*). While a blood film is prepared for study the red cell is drawn out into an elongated oval shape (hence *"ovale"*) and even shows a tattered appearance at one end. *P. ovale* is not highly pathogenic.

*Plasmodium* (*Laverania*) *falciparum*. This is the most common species of *Plasmodium* attacking man. According to Garnham (1966) it is "almost unchallenged in the supremacy as the greatest killer of the human race over most parts of Africa and elsewhere in the Tropics." The specific name (*falciparum*) is derived from the fact that, in this species, the gametocytes are crescentic. *P. falciparum* is the species which almost certainly does not have secondary exoerythrocytic schizonts (Garnham, 1966C).[7] The growing trophozoites, schizonts and immature gametocytes of the species are very rarely seen in the circulating blood, as they are concentrated within the capillaries and blood sinuses of internal organs such as the brain, liver, kidneys, spleen, bone marrow, and specially in placenta (in pregnant women). The infected erythrocytes become sticky, and tend to adhere to each other and to the walls of smaller blood vessels. Thus, blood films made from person infected with *falciparum* usually contain only very young trophozoites (signet ring stage) and mature gametocytes. The red cells containing asexual parasites, except for those inhabited by the very young ring forms, show when stained a few relatively large dots or lines, called **Maurer's clefts,** but there are no Schüffner's or Ziemann's dots. It is evident from electron microscope studies that these clefts are within the cytoplasm of the erythrocyte and are probably cast off portions of the outer membrane of the parasite and it is likely that Schüffner's and Ziemann's dots have a similar origin.

## PATHOLOGY OF MALARIA IN MAN

As mentioned above *P. falciparum* causes fatal disease, the other three species are mild and not fatal. However, they cause recurrent fevers which are very debilitating and may lower the patient's resistance to other infection.

Malaria is characterised by fevers which occur when the schizonts in the erythrocytes burst, setting free their merozoites and also liberating in the blood various excretory products including the malarial pigment and the remains of the cytoplasm of the red cell. It is believed that some of these liberated substances (not the merozoites themselves) are toxic and induce a high fever followed by **rigor, (violent shivering)**. The exact mechanism responsible for these effects is not fully understood. The schizonts develop simultaneously so that they all burst at the same time. Therefore, the time elapsing between successive bouts of fevers is often the same as the duration of the erythrocytic schizogony i.e., 72 hours with *P. malariae* and 48 hours with the other three species. The fevers thus produced are called **"quartan"** or **"tertian."** In some cases in the early phase of attack schizogony is less synchronized and fevers may occur daily, but the rhythm, characteristic of the species, may be soon adopted. Such a fever is called **Quotidian malaria**. Sometimes this type of fever is caused

---

[7]Garnham P.C C. (1966C)., *Malaria Parasites* Oxford, Blackwell Scientific Publications.

by mixed infection.

The exoerythrocytic schizonts do not produce any pathogenic symptoms, hence no symptoms develop during early part of the infection, before the erythrocytes are attacked. This is called the **pre-patent** period (the period before the parasites can be detected in the blood). The period between infection and the appearance of the first symptoms is the incubation period, which is longer than the pre-patent period, must be longer by the duration of at least one erythrocytic cycle, since symptoms do not develop until the first erythrocytic schizonts burst. The rupturing of only a few schizonts in the blood does not produce noticeable symptoms, therefore, if the infection is slight the incubation period becomes longer.

The exoerythrocytic schizonts are not affected by antibodies, but the erythrocytic schizonts are. If the infection is not treated and if death does not result rapidly (as it may with *P. falciparum*), after a variable number of erythrocytic schizogony cycles the host's antibodies will destroy most or all of the erythrocytic schizonts and all symptoms will disappear. The infection becomes latent and will no longer be detectable in the peripheral blood. However, the cure will not be complete, either a few erythrocytic schizonts survive in the capillaries of various viscera, or exo-erythrocytic schizonts persist in some way in the liver or both (except *P. falciparum* which has no persistent exo-erythrocytic schizonts). Eventually when either the level of circulating antibody has dropped or when the parasite itself has changed its antigenic structure (possibly by mutation) and is no longer affected by the existing antibody, the number of parasites in erythrocytes increases and a second clinical attack develops. In the absence of treatment this process may continue for years. Persons living in areas where malaria is common are regularly re-infected throughout their lives by the bite of infected mosquitoes and may, if they survive, develop incomplete resistance to or partial tolerance of the parasites. The infant death rate in such areas is appallingly high unless medical facilities are adequate to cope with rapid diagnosis and treatment.

The malarial infection (in man and other animals) always leads to a massive increase in the number of phagocytic cells of the lymphoid macrophage system. The spleen is the largest agglomeration of these cells and so it becomes grossly enlarged in chronic malaria. The spleen seems to be of great importance in the body's defence against malaria. In experimentally infected animals, in which the infection has become intense, the surgical removal of spleen leads to the rapid reappearance of parasites in the blood.

*P. falciparum* blocks the delicate capillaries in the viscera. This is responsible for its lethal effect. The precise mechanism of this effect is not fully understood. It is assumed that parasites secrete a substance which constricts the capillaries. Coupled with this the surface membrane of the infected erythrocytes becomes sticky and tends to adhere to the walls of capillaries and to each other. This causes two symptoms: interference with the tissue's oxygen supply and eventually rupture of the blocked capillaries and bleeding into the surrounding tissue. This may occur in any or all of the internal organs, but is most serious in the brain. Damage to brain causes death of most persons. It is for this reason that this is called **malignant tertian malaria** or **cerebral malaria**. Black-water fever is another very serious complication of *P. falcipa-*

*rum*. This was associated with inadequate treatment with quinine, and is now rare. It consisted of wholesale destruction (lysis) of the patient's erythrocytes (for reasons not clearly understood), with excretion of the liberated haemoglobin in the urine (hence the term black water).

**Diagnosis of Malaria in Man**

Recurrent fevers with enlarged spleen constitute the clinical findings to diagnose malaria. The occurrence of the parasites in thick or thin blood films is another dependable method. There is no serological test for the detection of scanty infection, nor is there any experimental animal into which blood may be injected for culturing the parasite. In doubtful cases fluorescent antibody tests have been found useful.

**Treatment and prevention of malaria**. The control measures fall under three headings: (*a*) treatment of the infection in the patient; (*b*) the prevention of the infection-prophylaxis; (*c*) control of the vector.

(*a*) It is clear that like the germs of other infectious diseases malarial parasites do not produce antitoxins or antibodies, or if produced at all the antibodies are not efficacious in controlling the disease, Malaria, therefore, cannot be treated by vaccination or inoculation with immune sera. It can only be treated with drugs that may kill all stages of the parasite without poisoning the patient. Quinine and other synthetic drugs are doing this, though none of our present day drugs has reached perfection.

The list of the synthetic drugs include *Paludrine, Atebrin, Plasmochin, Resochin, Pamaquin* and *Chaemoquin*, etc. Quinine is extracted from the bark of cinchona, a tree which is grown successfully in Peru, India, Ceylon and Java. The Dutch East Indies have a virtually monopolised quinine production, providing 90 per cent of the total. It rapidly kills the schizonts but the gamonts are resistant.

All clinical symptoms of malaria are caused exclusively by the schizonts of the erythrocytic phase. The most effective treatment of an attack of malaria is a schizonticide (schizont killer). Quinine is one but is unable to kill gametocytes. Because of the association of quinine with blackwater fever general treatment is usually given by synthetic drugs. Atebrin has the same effect. Paludrine is reported to kill the parasite in all its forms. But the drug now in a wider use is Resochin. In *falciparum* malaria after the first attack, no exoerythrocytic parasitic forms persist, and therefore, a thorough treatment with a schizonticide radically cures this type with no fear of relapse. In *vivax, ovale* and quartan malaria, after the erythrocytic schizonts, which are responsible for the clinical symptoms, are destroyed, there are still present the persistent exoerythrocytic parasites which cause the relapses. To avoid these relapses the best is to administer either Pamaquin or Plasmochin after the conclusion of Resochin treatment. Pamaquin is a rather toxic drug which should be used cautiously and preferably only in hospitals.

A casual prophylactic which destroys the inoculated sporozoites before they can develop into pre-erythrocytic or exoerythrocytic forms is at present unknown. The most practical method of prophylaxis available at the moment is suppressive treatment (chemoprophylaxis). For suppressive treatment a drug is necessary which will destroy the schizonts growing from either the exo- or erythrocytic parasites immediately they are formed, and should be capable of achieving this when regular small doses are administered. Paludrine or Daraprim in small weekly doses has proved perfectly reliable.

(b) The prevention of infection can be effected in two ways: (i) by using protective measures such as mosquito nets, etc., by which the mosquito is prevented from biting, and (ii) the use of prophylactic drugs, small daily doses of antimalarial drugs which will kill the parasite either in the sporozoite or merozoite stage.

(c) It is perfectly clear that if the vector is completely exterminated then the infection cannot be transmitted from one person to another. In fact, it is this approach to the control of the disease which has proved extremely effective. The dependence of mosquitoes on water for breeding makes the control measures easier, for it is easier to kill the aquatic larvae than the winged insect. Water can be poisoned with oil or other insecticides. Large sheets of water can be air sprayed with these poisons. Natural enemies like fish or ducks that feed on the larvae can be introduced. Insectivorous water plants such as the bladderwort (*Utricularia*) also catch a lot of mosquito larvae. Further, useless swampy places can be drained and the breeding space for the mosquitoes can be reduced. By all these methods the mosquito population will be considerably reduced and may even be completely eliminated.

**Other Species of Plasmodium**

There are some other species of *Plasmodium* that attack animals other than man. They are used in experimental work regarding the parasites. They are listed below:

1. *Plasmodium cynomolgi*. This species attacks monkeys mainly *Macaca* spp. in India, Ceylon, and Far East. Close to *P. vivax* it has several subspecies some of which may infect man.

2. *Plasmodium knowlesi*. This species also attacks monkeys mainly *Macaca* in Asia. It has 24-hour erythrocytic schizogony cycle. It is lethal to rhesus monkeys and may infect man causing only a mild disease.

3. *Plasmodium gonderi*. This species occurs in monkeys only in Africa, usually it is not markedly pathogenic. Some other less pathogenic species have been reported from chimpanzees and gorilla.

4. *Plasmodium (Vinckeia) berghei*. This species, discovered in the Congo, is one of the two species known to attack marine rodents. It can be maintained in the laboratory on rats and mice and is fairly pathogenic to them.

5. *Plasmodium (Haemamoeba) gallinaceum*. This species is a natural parasite of the jungle fowl in Asia. It can be maintained in the laboratory on chicken and is used in testing new malaria drugs. It can cause outbreaks of disease in flock of domestic hens and often kills the younger birds by blocking brain capillaries by large secondary exoerythrocytic schizonts, which develops in the endothelial cells.

There are many species of *Plasmodium* that attack birds and reptiles but they do not appear to be very pathogenic.

## CLASSIFICATION OF SUB-PHYLUM SPOROZOA

These are parasitic Protozoa which are usually propagated by spores. The spores usually have a resistant envelope. Trophic organelles are lacking in the organism. The subphylum **Sporozoa** is divided into four classes.

I. CLASS TELOSPOREA. Elongated sporozoites without polar capsules in the spores. Reproduction sexual and asexual. The Telosporea falls into three subclasses—Gregarinia, Coccidia, Haemosporidia.

(i) **Subclass Gregarinia**. The Gregarinia are chiefly lumen dwelling (**coelozoic**) parasites of invertebrates, especially arthropods and annelids, usually inhabiting the digestive tract, less frequently the coelome or the vascular system. They are typically intracellular only in the early part of their growth, i e., in the **trophozoite** stage. Later they leave the epithelial cells and develop into more or less elongate motile adults, usually referred to as **gregarines.** Majority of these do not show asexual reproduction or schizogony and multiplication takes place solely by sporogony following upon gametogony. The adult sporozoites themselves are gametocytes but do not show any differentiation into male or female as in others (Coccidia and Haemosporidia). The gametocytes tend to adhere in chains of two or more individuals. The adherence of two gregarines, known as **syzygy,** anticipates gamogony, for the later history indicates that the anterior member is female and the posterior male. Gamogony follows the same course in all members of the order. Both the pairing individuals produce an equal number of gametes and are usually equal in size though always similar in character. Conjugation takes place between similar or dissimilar gametes (isogamy or anisogamy).

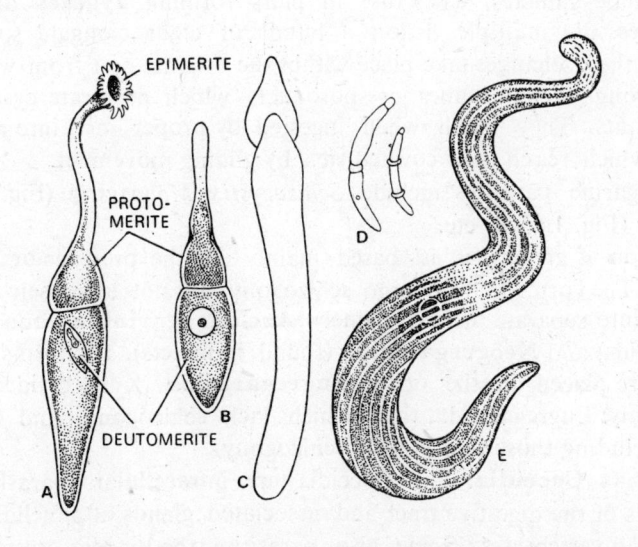

FIG. 1.72. Gregarina. A—*Menospora polyacantha*; B—*Gregarina longirostris*; C—*Didymophyes gigantea*; D—*Hermocystis harpail* in syzygy; and E—*Selenidum*, and schizogregarina.

*Gregarina* (Fig. 1.72) is a sporozoan in which the cell body is divided into two regions by a partition. The substance of the body is divided into a firmer outer layer, the **ectoplasm**, and a more fluid central substance, the endoplasm. The ectoplasm may be further differentiated into three layers: an external cuticle called **epicyte**, a middle layer, the **sarcocyte**, and a deeper contractile layer containing myonemes called the **myocyte**. Ectoplasm is thin in motionless forms. The cuticle is firm and grows inwards to divide the body into a larger nucleated posterior region or **deutomerite**, and a smaller anterior region, the **protomerite**. In the young stage the cuticle forms a small anterior cap or **epimerite**. When young they are intracellular parasites, but later they become free in the gut. They feed by absorption of foodstuffs and store glycogen. On an average many are about one-tenth of an inch in size but the size of the adult *G. gigantea* is sometimes three-quarters of an inch.

The young *Gregarina* is parasitic in one of the lining cells of the gut, where it grows. It leaves the cell but remains for a time attached to it by the cap, this is cast off later, and the individual becomes free in the gut, while growth still continues. Many species of these individuals possess a remarkable power of gliding forward. The spore formation in *Gregarina* is preceded by a close apposition of two trophozoites and their enclosure in a common cyst or **association cyst**. As they represent cells in which gametes are produced they are called, **gametocytes**. True gametes are formed within these. The gametes are quite alike in most but in some they can be distinguished into male and female gametes. They fuse in pairs forming **zygotes** or spores of various shapes. By multiple fission a bundle of eight elongate sporozoites is formed. All these changes take place within the original cyst from which spores come out through definite tubes or sporoducts which penetrate cyst wall and act as conduits. They hatch when ingested by proper hosts into a worm-like sporozoites which reach their coveted sites by gliding movement.

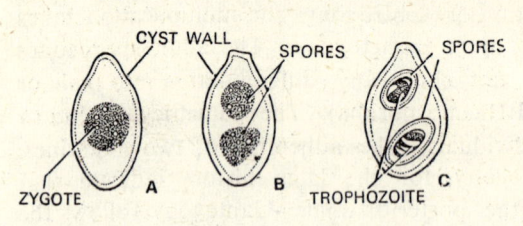

FIG. 1.73. Human coccidia *Isopora hominis*—oocyst.

Other gregarine parasites include *Schizocystis, Monocystis* (Fig. 1.66) and *Nematocystis* (Fig. 1.65 B) etc.

Classification of gregarines is based mainly on the presence or absence of schizogony. The forms that undergo schizogony are not all closely related and are grouped into separate orders, namely **Archigregarinida** (found mainly in marine annelids) and **Neogregarinida** (found in insects). The forms which lack schizogony are placed in the order **Eugregarinida**. Kudo divided them into two orders only: Eugregarinida, those which lack schizogony, and **Schizogregarinida** including those that possess schizogony.

(ii) **Subclass Coccidia**. The Coccidia are intracellular parasites of the epithelial cells of the digestive tract and associated glands of annelids, molluscs, arthropods and vertebrates. Some also parasitise the kidneys, testes, lining of blood vessels and coelomic spaces. Life-cycles are complex involving alternation of schizogony and sporogony, sometime with change of host (*Hepatozoon*).

*Eimeria* (Fig. 1.67) is an example of typical coccidians. It is a tissue cell parasite. It gains entrance into the host as oocyst through the mouth. The oocyst bursts releasing the sporocysts from which escape the sporozoites and move about in the lumen of the gut till they enter the epithelial cells of the gut wall, where they grow into schizonts. Schizonts are large rounded bodies, which

*Phylum Protozoa*

give rise to merozoites, escape in the lumen of the gut and enter new host cells and repeat the process. Some merozoites develop into macrogametocytes. The macrogametocyte produces a single **macrogamete** after extruding part of its nuclear material. The microgametocyte produces a number of biflagellated microgametes. Syngamy takes place and the zygote produces a cyst around itself forming an oocyst. The nucleus divides twice and four sporoblasts are developed inside the oocyst. Each sporoblast secretes a membrane and becomes a sporocyst, and two sporozoites are developed inside each sporocyst. Oocysts pass out in the faecal matter of the host and infect host by being ingested. Numerous species of *Eimeria* are known inhabiting arthropods and all classes of vertebrates, particularly domestic birds and mammals.

2. *Hepatozoon*. This genus parasitises cells of the lung, liver, spleen, bone marrow, etc. of mammals and passes through schizogonic cycle in such tissues (Fig. 1.74). The gamonts penetrate red and white blood corpuscles and do not develop further unless ingested by bloodsucking invertebrates such as ticks, mites or leeches. Gamogony and sporogony begin in the invertebrate host and the reinfection of the primary vertebrate host is accomplished either by ingestion of infected mites and ticks or by the bite of an infected animal. *Hepatozoon muris* is a good example.

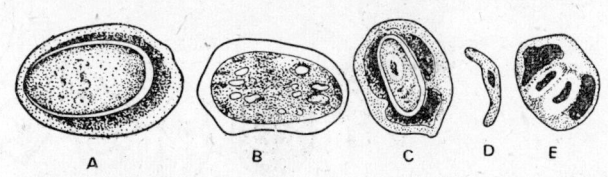

FIG. 1.74. *Hepatozoon*. A to C—*Hepatozoonadieli*; D—*H. canis*; free vermicule; E—*H. funambuli* in leucocyte (*from* Fauna of British India).

(*iii*) **Subclass Haemosporidia**. The haemosporidia are specially modified for parasitic life in the blood. There is alternation of hosts, asexual reproduction or schizogony takes place in the blood of vertebrates and sexual reproduction or sporogony takes place in the alimentary canal of some blood sucking invertebrates. They are minute, usually intracellular parasites of red blood corpuscles, showing motile amoeboid forms in their schizogonous cycle in the vertebrate host. Gametocytes and dimorphic gametes are formed but the microgametes have no flagella, as a rule, and move by lashing movement. The motile **zygote** is ookinete. After becoming encysted the zygote gives rise to a large number of sporozoites which are introduced into the blood of the vertebrate host. As they do not pass any stage of their life-history outside the body of a host, the sporozoites are not enclosed within a resistant cyst. They parasitise all kinds of vertebrates—mammals, birds, reptiles, amphibians and fish. A number of them occur in man and cause malaria.

Examples: *Haemoproteus*, *Plasmodium* (Fig. 1.70.) etc.

## SUB-PHYLUM CNIDOSPORA

These are Amoeboid trophozoites in which dissemination takes place through resistant spores which are provided with one to four polar capsules. Each spore contains one to many generative cells called **sporoplasm** and is covered by a shell of one piece or bivalved or trivalved. Each polar capsule is oval (resembling nematocyst) enclosing a spirally coiled filament that, under the action of digestive juices, discharges by turning inside out and serves to attach the spore

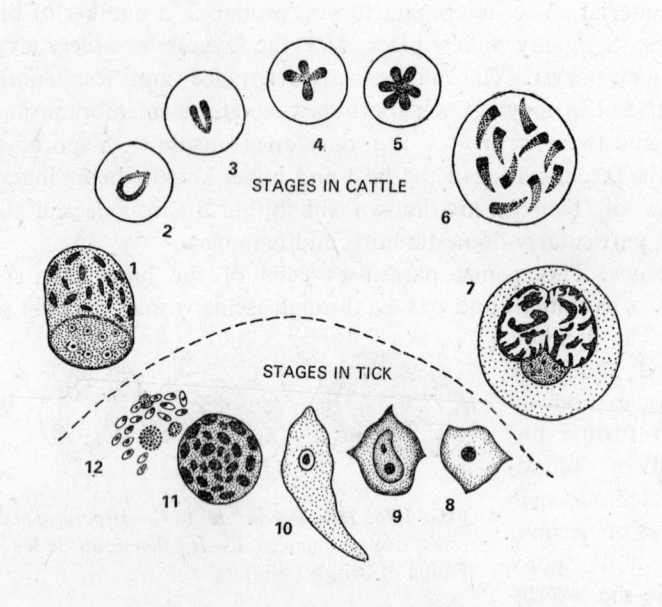

FIG. 1.75. Life-cycle of the parasite, *Theileria Parva*.

to the intestinal wall till the amoeboid body escapes from the spore and infects the tissues of the new host. The group is divided into two classes comprising three orders each.

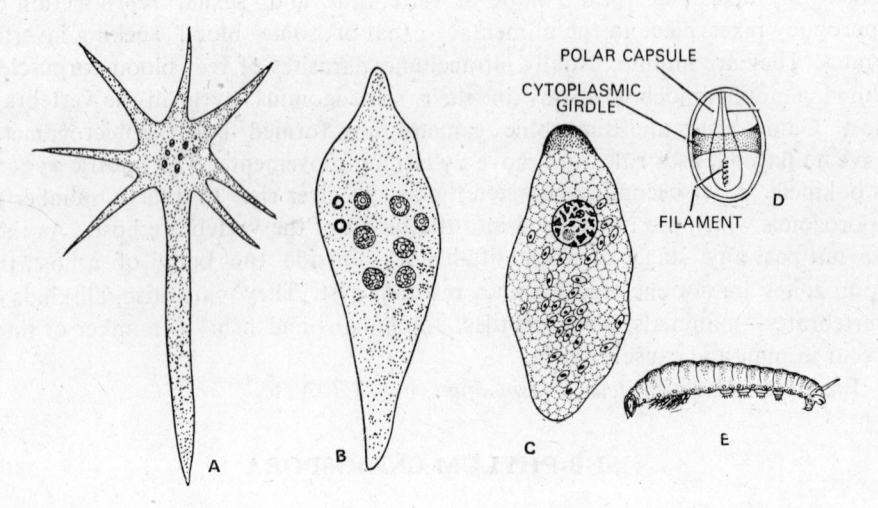

FIG. 1.76. The trophozoite of **A**—*Leptotheca* and **B**—*Ceratomyxa* (*after* Davis); **C**—*Nosema* in a cephaline gregarine; **D**—Spore of *Nosema bombycis*; and **E**—Silkworm larva infected with *Nosema*.

I. CLASS MYXOSPOREA. The Myxosporea are parasites of cold-blooded vertebrates particularly fish. Each species attacks one or several species of fish. They inhabit hollow organs, such as the gall and urinary bladders, where they are actively amoeboid; or they inhabit connective tissue or the cells of the gills, kidneys, liver, spleen, etc. They are mostly harmless to the host but may cause damaging tumour-like masses. Each large spore is enclosed in a shell made of two valves and contains two or four polar capsules.

1. **Order Myxosporidia** includes *Myxidium* which occurs abundantly in the gall and urinary bladders and kidneys of fish (sometimes reptiles). It spores have the polar capsules pointed at opposite direction.

Other genera include *Leptotheca* (Fig. 1.76) with elongated pseudopodia at one end, occurs chiefly in the coelom of fish and has rounded spores; *Ceratomyxa* (Fig. 1.76) attaches to the gall bladder of fish and *Myxobolus* to the solid organs of fresh water fish.

2. **Order Actinomyxida**. The *Actinomyxida* are inhabitants of the coelom of fresh-water oligochaetes and marine sipunculids but have not been reported in India. Each spore has its shell composed of three valves, which may be drawn out into simple or bifurcated processes. Each spore has three polar capsules and one to many amoeboid young or multinucleate mass.

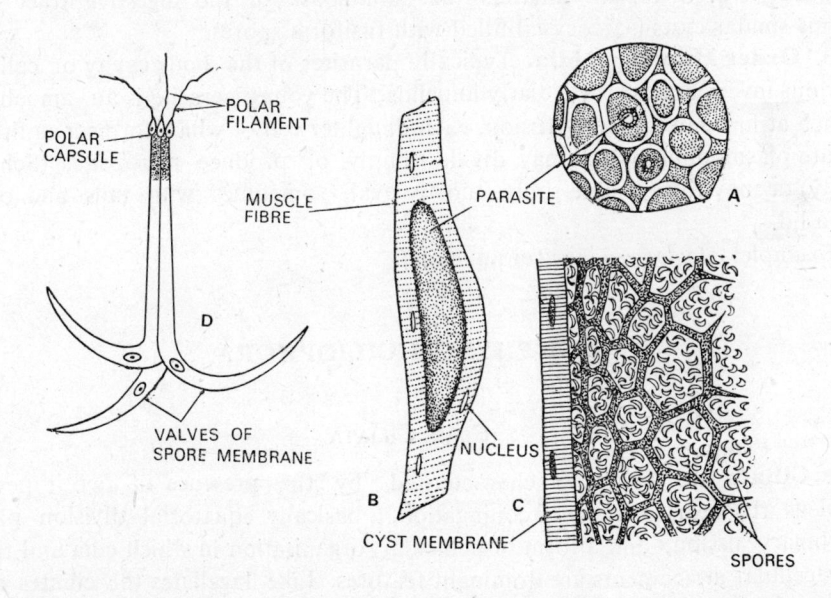

FIG. 1.77. *Sarcocystis blanchardi*. **A**—Transverse section; **B**—Longitudinal section of the muscle fibre showing the parasite; **C**—Longitudinal section of sarcocyst (*from* Fauna of British India); and **D**—Spore of *Triactinomyxon*.

Examples: *Triactinomyxon* (found in *Tubifex* (Fig. 1.77); *Tetractinomyxon* and *Hexactinomyxon*.

3. **Order Helicosporida** contains a single species parasitic in arthropods, example is *Helicosporidium parasiticum*.

II. CLASS MICROSPOREA. The Microsporea are intracellular parasites typically of arthropods and fishes inhabiting almost all the types of tissues of the host and destroying them on an extensive scale. The spores are minute, of simple form and structure, containing a long coiled filament, representing a polar capsule of which the wall is missing. Some Microsporea are responsible for causing devastating epidemics in silkworm, honeybees and certain fishes.

1. **Order Microsporidia** includes *Nosema bombycis* (parasitic on silkworms, Fig. 1.76); *N. apis* (on honey bee); *Glugea* (on fresh water fish).

2. **Order Sarcosporidia.** Typically parasites of the muscles and connective tissue of mammals, although birds and reptiles are also known to harbour them. They usually affect the striped muscle fibres of the skeleton, tongue, larynx, and diaphragm. Lying between the muscle fibres are thin-walled large cysts called sarcocysts (Miescher's tubes). Each is made up of many chambers of which the peripheral ones are filled with crescentic spores. The life-history and mode of transmission are unknown.

Example 1. *Sarcocystis* (Fig. 1.77), the chief genus that parasitises the muscles and develops sarcocysts or cysts made of many chambers, the peripheral ones of which are filled with crescentic spores. It produces a toxin resembling a bacterial toxin, which is fatal when injected in small mammals (mice, guniea pigs) but not so harmful to the host, it seems.

Example 2. *Globidium* inhabits the submucosa of the digestive tract and forms similar cysts (sarcocysts) filled with fusiform spores.

3. **Order Haplosporidia.** Typically parasites of the body cavity or cells of various invertebrates, particularly annelids. The young parasite is an amoebula, which at first multiplies by fission, each daughter cell of which forms a multinucleate plasmodium, which may divide simply or produce merozoites (Schizogony) or may form simple spores mostly oval, sometimes with tails and open by a lid.

Example: *Haplosporidium* (in annelids).

## SUB-PHYLUM CILIOPHORA

### CLASS CILIATA

The **Ciliata** are Protozoa characterised by the presence of two types of nucleus, the sexual process of conjugation, a basically equatorial division plane in binary fission, and a form of pellicular organisation in which cilia and their intracellular attachments are dominant features. Like flagellates the ciliates also have a definte permanent shape in most of which anterior end can be differentiated from the posterior. Most ciliates possess nuclei of two types, namely, **macronuclei** and **micronuclei**. Viable strains of ciliates without micronuclei are known, but with the exception of *Stephanopogon*, all ciliates are thought to possess at least one macronucleus (many micronuclei in some); the number of micronuclei is often greater than the number of macronuclei. The micronuclei are small and compact and are normally diploid with large amounts of histone protein, but little or no RNA (no nucleoli). The micronuclei divide mitotically, at the time of binary fission, within persistent nuclear membranes, and in the sexual processes of conjugation and autogamy a micronucleus undergoes meiosis before giving rise to the gametic pronuclei. Normally, the macronucleus degenerates following sexual process and a new macronucleus is formed from division products of the micronuclear syncaryon. Macronucleus is larger and normally highly polyploid. In the fully formed macronucleus the chromatin is confined to many small dense bodies containing numerous microfibrils twisted together. The numerous larger less dense nucleoli contained in the cytoplasm depends on RNA of macronuclear origin. Formation of DNA as well as RNA occurs in most macronuclei.

The behavioural pattern of the nuclei varies in different ciliates. In a *Stephanopogon*, an ancestral ciliate, young individuals have two identical nuclei, each with a central nucleolus and peripheral chromatin. During growth both the nuclei undergo mitosis and several (up to 16 in larger form) nuclei are found. Following encystment the cytoplasm divides into uninucleate masses, within each of which mitosis takes place so that at excystment a number of binucleate ciliates emerge (Fig. 1.99). In *Loxodes striatus* both the macronuclei and micronuclei are diploid, but macronuclei never divide, so that new macronuclei are always formed from micronuclei and additional micronuclear divisions are necessary to maintain the full number of nuclei. In a typical higher ciliate (e.g., *Paramecium*) the macronucleus disappears during conjugation, in the course of which the micronucleus undergoes meiosis and fertilisation and then a new macronucleus is formed from the micronuclear syncaryon. During binary fission both macronuclei and micronuclei divide.

They are the most specialised of protozoa in having various organelles to perform particular vital processes. This results in a division of labour between different parts of the organism, which can be compared to that between organ-systems in multicellular animal. The body cytoplasm is distinctly marked out forming **ectoplasm** and **endoplasm**. The ectoplasm is often highly differentiated into the **pellicle** and underlying layers. The **cilia** are short filaments that spring from the **basal bodies** and pierce the pellicle to come out. The cilia occur in longitudinal or diagonal rows and have the same structure and mechanism of movement as flagella, the difference being that they are shorter in length, greater in number and their basal bodies are not related to mitotic division. Contractile vacuoles are present in all ciliates. They are mostly solitary free swimming forms found in fresh and salt water. Some are commensal (*Nyctotherus*, an intestinal commensal) or parasitic (*Balantidium coli* in pig) in other animals and a few form colonies (many vorticellids). The nutrition of ciliates is almost wholly **holozoic** and there are generally two types of forms so far as their feeding habits are concerned (*i*) the **raptorial** that hunt and ingest large prey, and (*ii*) the **current producing** forms that obtain their food by means of ciliary current. They reproduce asexually and the sexual process takes a peculiar form known as conjugation. Encystment is widespread.

## TYPE PARAMECIUM

*Paramecium* is a common fresh-water ciliate found in decaying vegetables abounding with bacteria. It can be easily "grown" in the laboratory in an infusion made by boiling a little hay or some wheat in water. They are visible to the naked eye as elongated whitish spots. In general appearance, *Paramecium* (Fig. 1.78) appears like the sole of a slipper, hence it is popularly known as the "**slipper-animacule.**" One species was named *Paramecium caudatum* because of its resemblance to an imprint of human foot. Due to ease of culture and rapid rate of reproduction they make ideal material for use in the study of many fundamental problems in general biology.

The first description of *Paramecium* comes from a letter of Christian Huygens, the physicist, written in 1678. The name *Paramecium* was coined for the first time by John Hill in 1752 from the Greek *Paramekos*, meaning oblong and refers to the shape of the animal. Several species of *Paramecium* are known, most of them living in pools or ponds of fresh-water, though a few live in sea water pools. All the species may be found in all the parts of the world. The small bodies of water in which they live must often dry up, but whether they can escape desiccation, as do many other related organisms by secreting a resistant cyst about themselves is not known. They are, however, transferred to new ponds or pools by some means, perhaps in cysts, or naked in drops of water, carried possibly by wind or on the feet of birds. They themselves provide food for various animals such as *Didinium*, another ciliate and *Podophyra*, a suctorian.

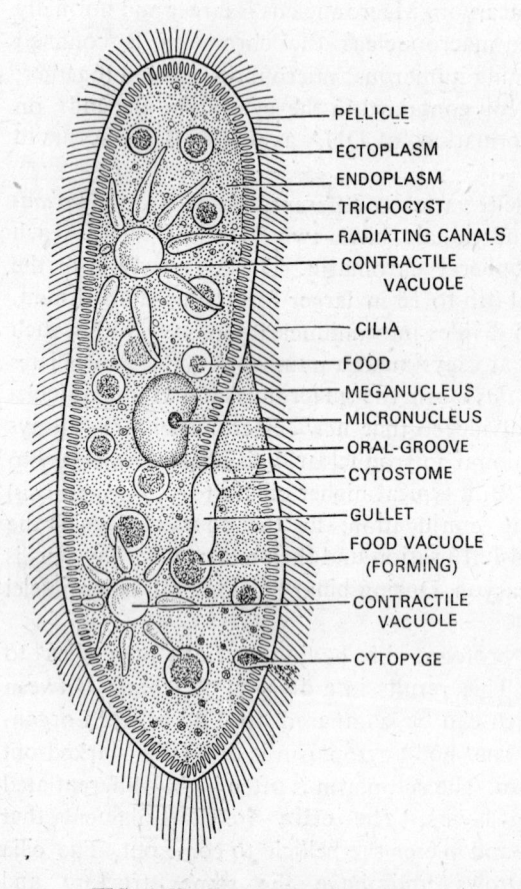

FIG. 1.78. *Paramecium*, from life.

They vary in size. The largest species, *Paramecium caudatum* varies in length from 180 to 350 microns, while *P. trichium* may be only 60 microns long. The different species differ slightly in shape and structure. Certain environmental factors markedly influence size and shape. Food is one of them, there may be a fluctuation of as much as 62 microns between starved and normal individuals. Temperature and pH of the culture medium also influence the size of individuals.

## Colour

Some species are somewhat translucent and colourless, others vary from light gray, white to pale yellow. One species *Paramecium bursaria* is green because of

# Phylum Protozoa

the presence of *Zoochlorellae* in the endoplasm. Under certain conditions *Zoochlorellae* free colourless varieties (clones) of this species also appear.

## Structure

The shape of the body is definite (Fig. 1.78) and constant because the body is covered by a thin but firm flexible outer covering, the **pellicle,** secreted by its outer surface. It is elongated, more or less flattened, rounded at one end and bluntly pointed at the other presenting a good example of a streamlined form. The rounded end is directed forwards during locomotion and is, therefore, termed anterior, whereas, the pointed end is posterior. The terms **anterior** and **posterior** are used for convenience only because occasionally the animal reverses its direction of movement. The surface carrying a shallow depression or oral groove is the **ventral** or **oral surface,** while that opposite to this is the **dorsal** or **aboral** surface.

A closer examination will reveal that the body cytoplasm is divisible into a semifluid granular interior, the **endoplasm**, and a more dense, clear outer layer, the **ectoplasm.** The ectoplasm is a distinctly permanent part of the body clearly differentiated from the endoplasm. It is the part of the ciliate that comes in direct contact with the environment. It contains **trichocysts, cilia** and **fibrillar structures,** and is bounded externally by a body covering envelope called **pellicle**.

The pellicle appears homogeneous in the living conditions under low power, but electron microscopy has revealed that it is very complicated in structure. The

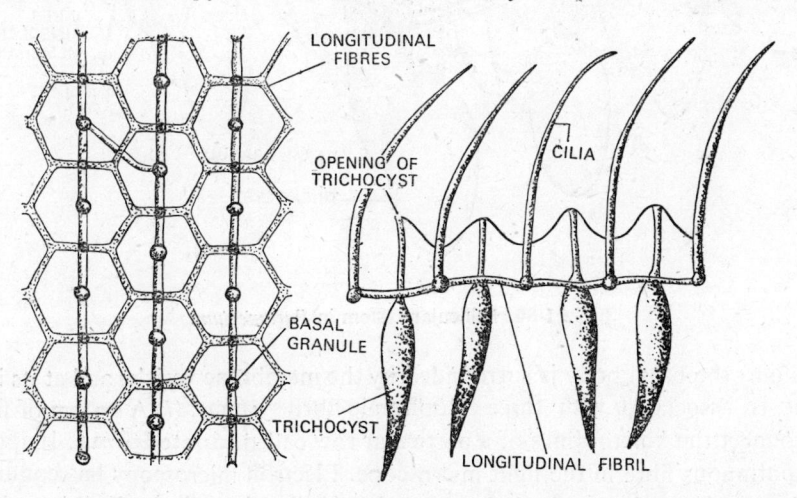

FIG. 1.79. *Paramecium multimicronucleatum.* A—Semidiagrammatic sketch of pellicular structure showing basal granules, and longitudinal fibrils; and B—Surface view of the pellicle.

pellicle (Fig. 1.79) has inflated kidney-shaped **alveoli**. The inflated condition and the shape of alveoli produce a polygonal space (Fig. 1.80) around one or two cilia which arise between them. Alternating with the alveoli are bottle-shaped organelle, the **trichocysts,** which form a second deeper compact layer of the pellicular system. The cilia covering the body are arranged in longitudinal rows called **kineties** (singular **kinety**). The base of each cilium is associated with several intracellular fibrils, and a pair of membranous vesicles to form the basic unit of pellicular organisation which is called the **kinetid**. A surface membrane covers the whole organism including the cilia. Beneath this membrane, in most places, there are two further membranes, which form the **outer** and **inner walls** of flattened vesicles or **alveoli** extending over the body (Fig. 1.80) surface and arranged in pairs around the base of each cilium. The outer wall of the vesicle lies close to the cell membrane. The membranes between adjacent vesicles are seldom complete and the aveolar cavities of a whole kinety may be continuous. Below the innermost membrane lies a dense layer of cytoplasm called the **epiplasm**.

Each cilium arises from a **basal body** or **kinetosome** located in the ectoplasm (Figs. 1.79 B, and 1.80). At its outer end (near the place where it gives off

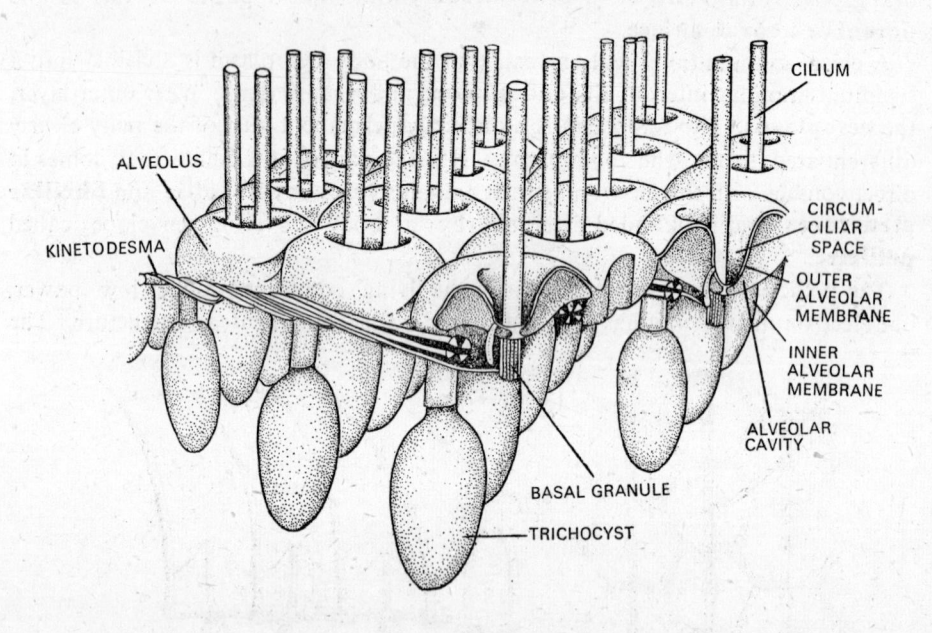

FIG. 1 80. Pellicular system in *Paramecium*.

the cilium) the basal body is surrounded by the membrane vesicle, and at its inner end it is associated with three ectoplasmic fibril systems: (*i*) A system of fibrils that connect the kinetosomes of a particular row called **kinetodesma**. It appears as a continuous fibre in the light microscope. Electron microscopy has confirmed that it consists of a series of overlapping fibril units. Each fibril is striated. Electron microscopy has also shown that the kinetodesmal fibril runs along the right side of the row of kinetosomes of a kinety; (*ii*) A set of 7-12 longitudinal microtubular fibrils is found between the innermost pellicular membrane and the epiplasm; and (*iii*) A group of five or six transverse microtubules that run outwards towards the pellicle and transversely towards the left so that they terminate close to the longitudinal microtubular fibrils near the next kinetodesmos on the left.

The fibrillar system or **neuromotor apparatus** is visible with the help of nuclear stains, especially Heidenhain's. The study with such stains also shows that the surface of paramecium consists of a series of polygons (Fig. 1.79A). The basal granules are always at or near the centres of the polygons forming

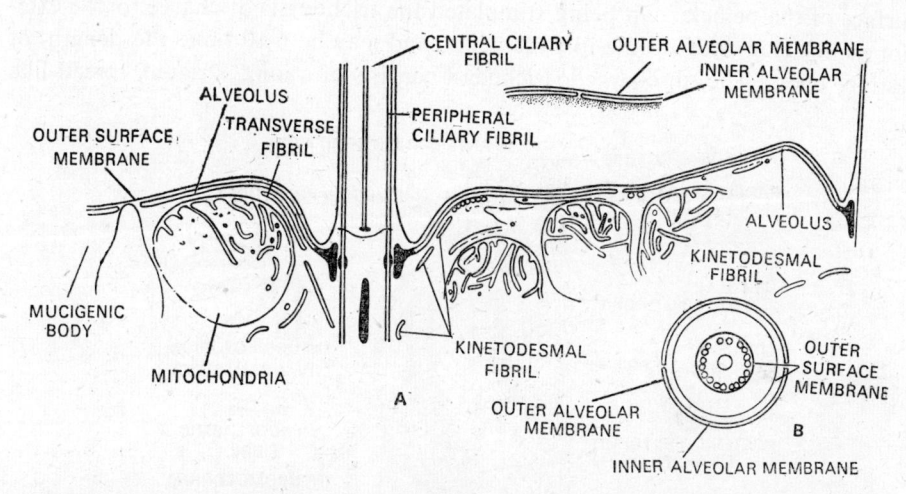

FIG. 1.81. Section through cilium and pellicle of *Colpidium*. The alveoli are greatly flattened and their inner and outer members are found at the base of the cilium. **A**—Enlarged view of surface and alveolar membrane; and **B**—Cross section of a cilium and surrounding pellicle. There are nine doubled peripheral ciliary fibrils.

longitudinal chains of the body fibrils. These run nearly parallel to the long axis of the body except in the regions of the sutures and the cytostome, where they curve sharply inward. The fibrils in the region of cytopharynx and oesophagus are many and will be studied later.

**Cilia.** The **cilia** (Lat. *cilium*, an eyelash) are short fine protoplasmic processes, completely covering the body of a paramecium and serving as organs of locomotion and food capture. The cilia are arranged on the body in parallel rows (**kineties**). On the body proper the cilia are of moderate and fairly uniform length, but at the extreme posterior tip they become longer forming the caudal tuft. The cilia found in the cytopharynx and oesophagus show a wide range of variation in length. They attain a length of 16 microns approximately and have a diameter of 0.1 to 0.3 microns. The number of cilia on paramecia varies from 350 (small paramecia) to 2,500 according to various estimates. Gelei (1925) has estimated about 18,000 cilia in *P. nephridiatum*. In the living stage, observed by transmitted or polarised light, the cilia appear optically homogeneous, but earlier electron microscope studies have revealed that a cilium of paramecium consists of a bundle of about eleven fibrils extending its full length. These are surrounded by an external membrane. Only one fibre reaches the tip making the cilium a tapering structure. The structure of a cilium is similar to that of a flagellum (*see* p. 20) except that it is smaller and shows certain minor functional differences. It is reasonable to describe cilia as specialised class of flagella.

**Trichocysts.** The **trichocysts** (Gr. *thrix, trichos*, hair and *kystis*, a bladder) are small fusiform or carrot-shaped retractile elements embedded in the ectoplasm of paramecium. They are suspended at right angles to the surface of the body (Fig. 1.80) being located at the centres of the anterior and posterior walls of the pellicular polygons. The size of the trichocysts is small. In the resting stage a trichocyst measures four microns in length and extends 1-2 microns below the surface of the pellicle. On being stimulated the trichocysts discharge to the exterior as fine elongated needle-like filaments, and may be 6-10 times the length of resting ones. The discharged trichocyst consists of a long, striated, thread-like

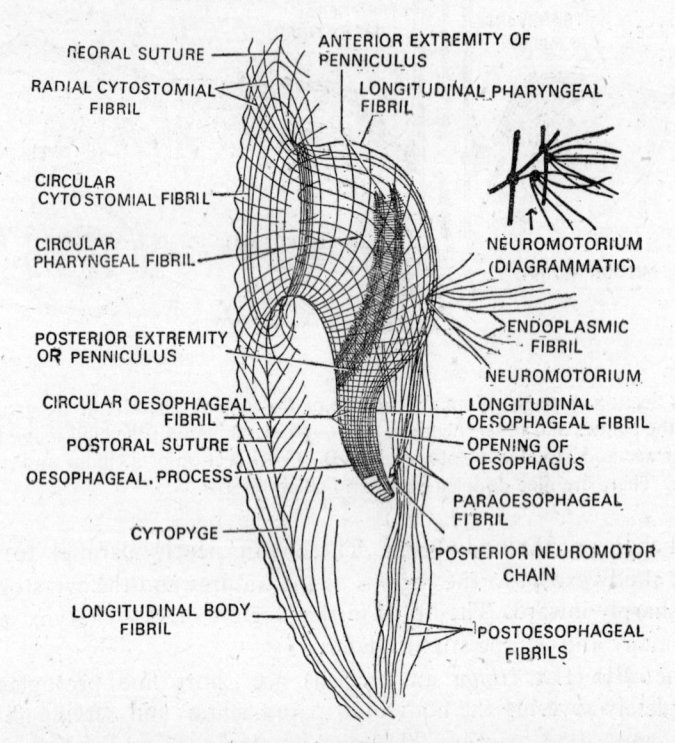

FIG. 1.82. Diagrammatic sketch of the pharyngo-oesophageal network; with its associated fibrils of the body surface and the endoplasm in *P. multimicronucleatum*. Neuromotorium is shown on the top left. (*after* Lund).

shaft surmounted by a barb, shaped somewhat like a golf tee. The shaft is not evident in the undischarged state and is probably formed in the process of discharge. Many different kinds of stimuli can evoke discharge. They include chemicals, electric shocks, mechanical injury as in cutting or with pressure and desiccation. The trichocysts are discharged very rapidly (within several milliseconds). The degree of extrusion varies, some may discharge partly with a given stimulus, others fully, while still others may be discharged and set free from the body.

In *Dileptus anser* the trichocysts are apparently turned inside out at discharge. The expelled trichocyst consists of a bulbous base that tapers into a long thread. In *D. gigas* the trichocysts are fluid-filled vesicles the contents of which can be expelled. Such trichocysts (called **toxicyst**) have a definite paralysing effect on rotifers and other protozoa. The trichocysts have been compared with the rhabdites of the Turbellaria by some authors who considered the extruded forms to be fibrous slime of poisonous nature. Others compared them with nematocysts of

Phylum Protozoa 115

the coelenterates. But its structure differs from both of these. A trichocyst is made up of a large, roughly oval and cross-striated shaft, continued into a small elong-

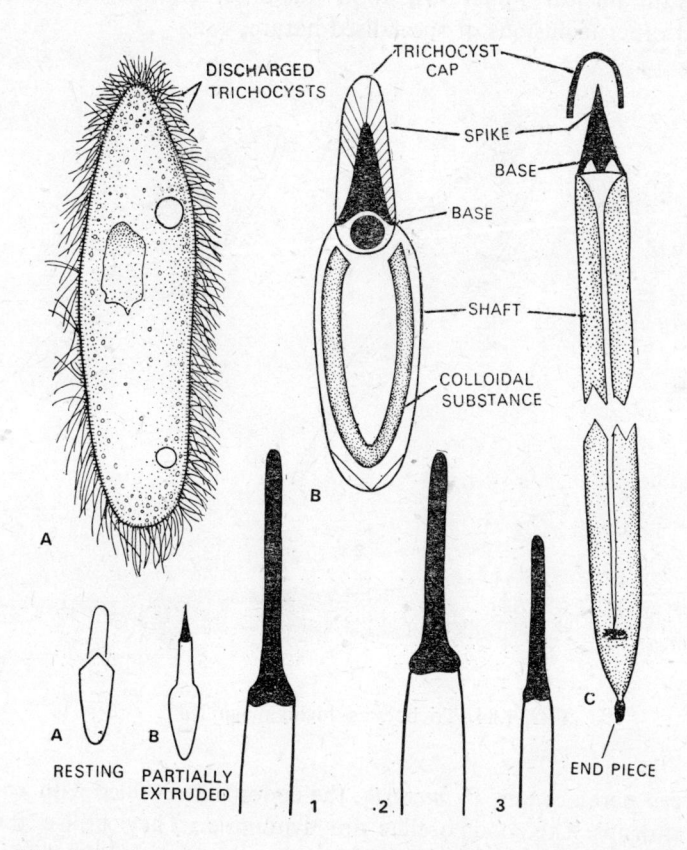

FIG. 1.83. A—*Paramecium* with discharged trichocyst after the application of picric acid; B—Hypothetical longitudinal section of the resting trichocyst; C—The same of a discharged trichocyst; D—Diagrams show how a resting (a) and partially extruded (b) trichocysts look under dark field illumination of a microscope. The tips of the trichocysts under electron microscope reveal differences in shape. 1—*Paramecium caudatum*; 2—*P. multimicronucleatum*; and 3—*P. bursaria* (*after* Wichterman).

ated and pointed tip at its broader end. The tip is covered over by an elongated structure, the **trichocyst cap**. Before the trichocyst is extruded the cap is lifted and enough quantity of water enters in through a minute opening at the base of the tip. Within the extensile membrane of the shaft there is a substance which can absorb enough water and swell. The substance has been called **"quellkorper"** in German which means substance of the body. Thus, as soon as the water enters in, it is absorbed by this substance with the result that the shaft-membrane becomes elongated (forms the body). This pushes the trichocyst through the pore in the pellicle and it projects out. As the trichocysts are discharged when they are touched by injurious chemicals (Fig. 1.83A) or when attacked by an enemy it is concluded that they are defensive in function. It has also been suggested that the trichocysts serve to anchor the ciliate while feeding on bacteria.

**Endoplasm.** The endoplasm is the more fluid portion of the cytoplasm that shows characteristic streaming movements (cyclosis). It is a relatively larger part and includes the nuclear apparatus, food vacuoles, contractile vacuoles, food reserves and other inclusions of specialised nature.

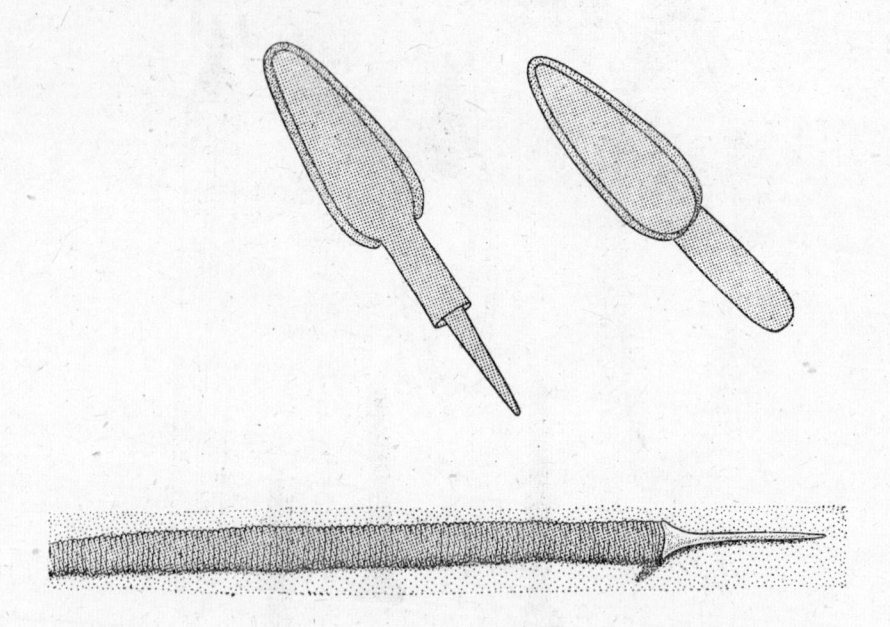

**FIG. 1.84.** Trichocysts, highly magnified.

In the green paramecium, *P. bursaria*, the endoplasm is filled with **zoochlorellae**, and fat drops. The zoochlorellae are symbiotic. They utilise the carbon dioxide and nitrogenous and phosphorus wastes of the host and in return furnish oxygen and synthesised food and may even be digested by the host in times of hunger.

**Food passage-way.** On the oral surface of the animal is found a broad shallow obliquely directed groove called the **oral groove**. It is widest at the extreme anterior end and extends from it to the region two-thirds of the body length. Its basal portion is known as **vestibulum** consisting of invaginated body pellicle. The vestibulum leads directly into the **mouth (cytostome)** which is a fixed oval opening. The mouth directly opens into the wide **pharynx (cytopharynx)** which extends toward the centre of the body. Then the food passage-way becomes slender and turns to the posterior side forming the tapering **oesophagus**. The oesophagus runs parallel to the surface and then suddenly turns toward the centre of the body and ends in a blind pouch, the oesophageal sac or process which is heavily ciliated.

The vestibulum bears shorter cilia than those of the body and is free of trichocysts. The pharynx bears four rows of long powerful cilia (called **membrana quadripartita** by Gelei). On the left wall of the pharynx is situated a ciliary structure called the **penniculus**. It consists of eight rows of cilia arranged in two closely-set blocks of four each. It spirals through approximately

*Phylum Protozoa*

90 degrees so that its posterior extremity is on the oral (ventral) surface of the oesophagus. It drives the food in the body. Just behind the mouth on the oral

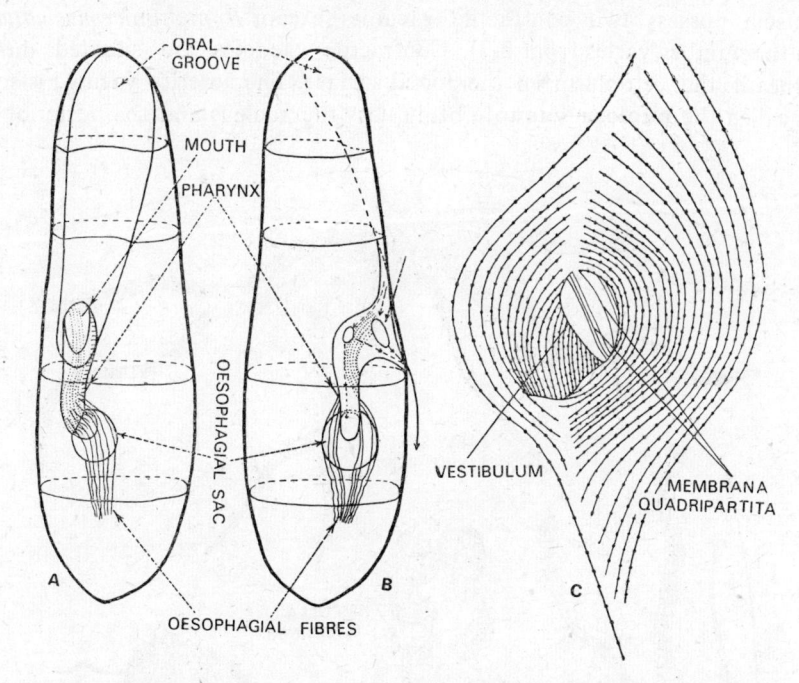

FIG. 1.85. Feeding apparatus of *P. aurelia*. A—Oral groove and the mouth facing upward; B—Oral groove and mouth facing to the right (*after* Mast); and C—Ciliary fibrils pattern in relation to the mouth (*after* Gelei).

surface there is a minute pore through which the unabsorbed remains of food are passed out. This is called **cell-anus** or **cytopyge**. The anus is visible when the animal is in the process of egestion. The position of the anus varies in different paramecia. In some it is on the posterior extremity and in others between the oral groove and posterior end.

**Nuclei.** In the endoplasm are embedded two nuclei. (Fig. 1.78). They differ in size and appearance, and also in function. The smaller one or the **micronucleus** is lodged in a depression at one side of the larger nucleus called **meganucleus** or **macronucleus**. The meganucleus is a conspicuous kidney-shaped body lying approximately in the centre. It is smooth and generally regular in outline. It is a compact structure containing fine threads and tightly packed discrete chromatin granules of variable sizes. It usually divides amitotically. The micronuclei are smaller structures and may be compact (*P. caudatum*) or vesicular (*P. aurelia*). Their number varies in different species. *P. caudatum* and *P. bursaria*, etc., have one compact ellipsoid micronucleus, *P. aurelia* two spherical vesicular micronuclei and *P. multimicronucleatum*, has 3-9 spherical vesicular micronuclei. Because two types of nuclei are present the paramecia are said to possess dimorphic nuclei. The meganucleus seems to control the ordinary "vegetative" activities of the animal while the micronucleus is concerned with reproduction.

**Contractile vacuole.** The endoplasm includes food vacuoles of various sizes containing material undergoing digestion, whereas, towards each end of the cell-body there is a large clear contractile vacuole (Fig. 1.78). Normally paramecia possess two contractile vacuoles (except *P. multimicronucleatum* in which the number varies from 2-7). Contractile vacuoles are situated directly underneath the ectoplasm on the dorsal surface. The anterior vacuole is sometimes called the **nuclear vacuole** being near the nucleus and the posterior one

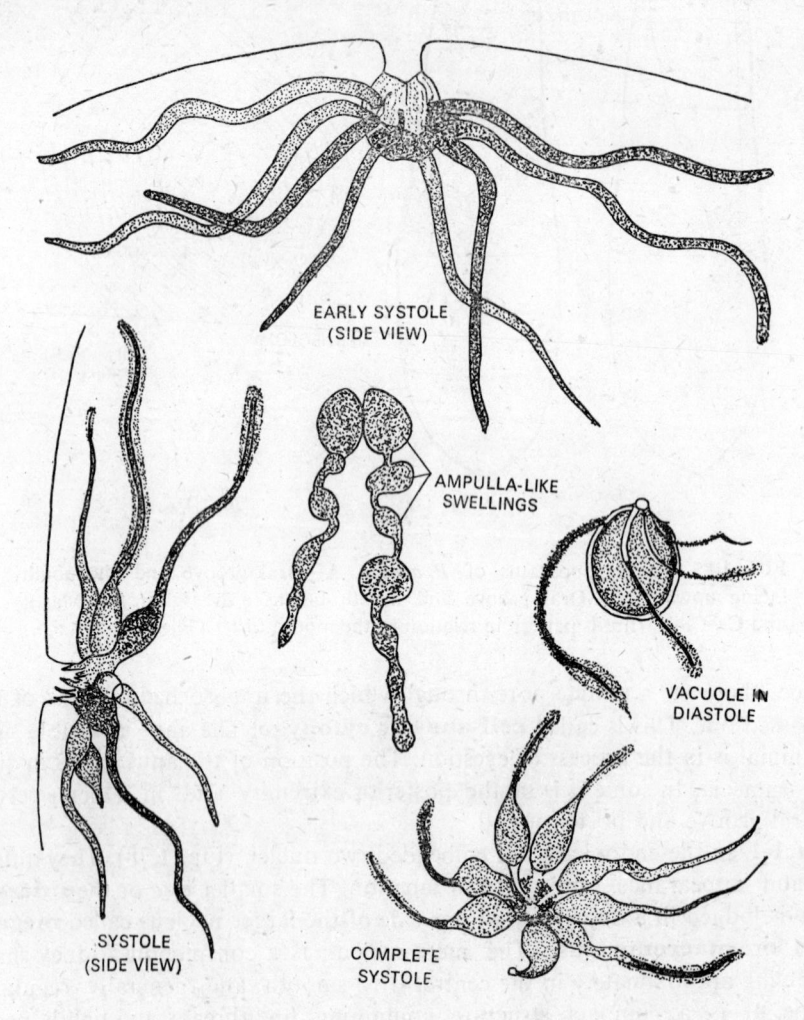

FIG. 1.86. Contractile vacuole of *P. multimicronucleatum* (*after* King).

is called the **peristomial vacuole** being in the vicinity of the peristome. In the vacuoles of *Amoeba* and *Euglena* smaller accessory vacuoles coalesce to form a definitive contractile vacuole, therefore, each is termed **vesicle-fed vacuole**. In paramecia the vacuole is fed by canals and is called the **canal-fed vacuole**. The number of canals varies from 1 to 10 although the common number is 5 to 7. They are slender and radiating structures found in one plane, and form a characteristic rosette about each larger, roughly spherical vacuole and empty into it (Fig. 1.86). They are called feeders or feeding canals. The

*Phylum Protozoa* 119

functional cycle of the vacuole is simple. A vacuole becomes enlarged with fluid (mainly water) until it reaches the maximum size (**diastole**), then it suddenly collapses (**systole**) discharging the vacuolar contents to the outside, through a pore which is located directly above each vacuole and is fixed in position.

Each radial canal or feeder consists of (*i*) a tubular terminal part, (*ii*) an ampulla and (*iii*) an injector (Fig. 1.87). The fluid from the protoplasm surrounding the radial canal passes into the terminal part which is also able to secrete hypertonic fluid into the lumen. This fluid is collected in the ampulla which becomes **bulb** or **flask-shaped** when distended (diastole of the ampulla), but when the fluid is passed out (systole of the ampulla) it is of the same diameter as the terminal portion (Fig. 1.87). Instead of one flask-shaped enlargement sometimes the ampulla consists of a series of bulb-like swellings (Fig. 1.86). When the ampulla is fully distended (diastole) it collapses and the fluid is passed

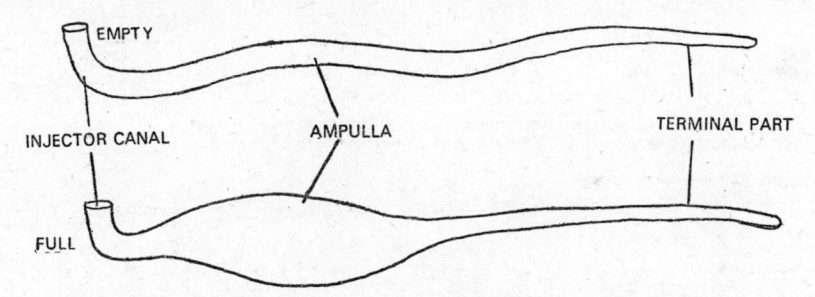

FIG. 1.87. Radial canals of contractile vacuole (*after* Wichterman).

through the injector to form the contractile vacuole. Thus it is apparent that the systole of the ampulla becomes the diastole of the vacuole. In some species of paramecium (*P. trichium*) the contractile vacuole is without radial canals.

**Locomotion**

*Paramecium* swims in water by beating its **cilia** and has put on speed because of the development of a large number of these locomotory organs. Watched under a microscope they seem to move at a great speed, but remembering that each cilium is in fact only 1/1000th of an inch long, it may be realised that their movement is really very little. Hence the speed of the animal is very slow, perhaps 5 metres per hour or 110 metres per day.

A cilium performs whiplash movements (Fig. 1.89) by which water is propelled. The collaborative action of cilia causes the organism to swim or to maintain a current of fluid over the ciliated surface. The beat of a cilium may be separated into two distinct phases, one phase having **effective stroke** (when the cilium moves in the direction of the main flow of fluid) and the other the **recovery stroke**. In the classical form of effective stroke the cilium bends near-rigidly, protruding full length from the cell (Fig. 1.89), while a flexure at the base produces a swing of the ciliary axis. This is followed by a sharp movement back to the other side of the base, while the distal part of the cilium continues to swing in the effective direction. The bend in the cilium produced by these two angular movements at the base is propagated up to the tip of the cilium during the recovery stroke, in such a way that the cilium moves back close to the cell surface in preparation of the next effective stroke. Generally, the recovery stroke takes longer than the effective stroke; both the strokes are active processes. The effective beat, therefore, appears to pass as a wave forwards in the series of cilia, though the surrounding fluid is driven backwards. The cilia are very close together and because of fluid mechanical interaction they tend to beat in phase (i.e., they are synchronized). While the adjacent

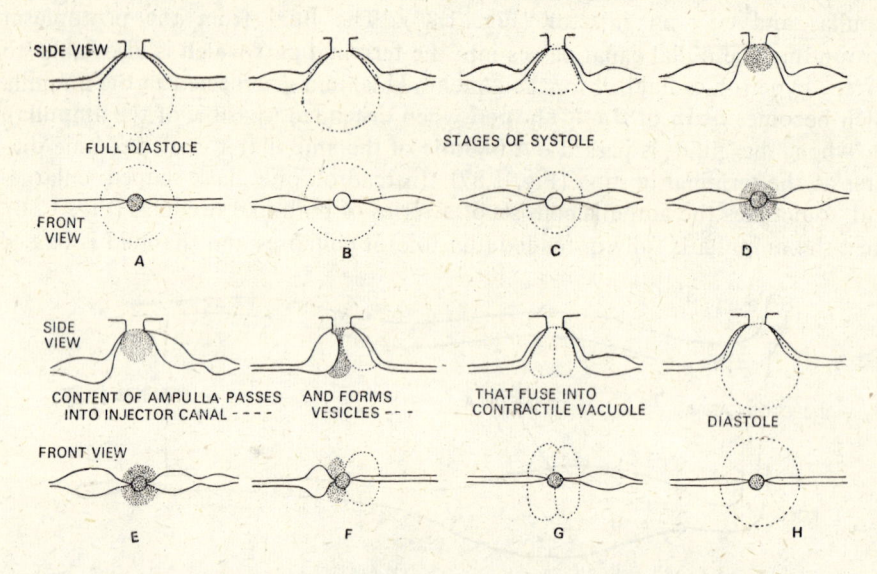

FIG. 1.88. Successive stages in the life-cycle of the contractile vacuole (*after* King).

cilia move in phase those along other rows show metachronism. Waves of activity pass across the ciliated surface in the direction of the main line metachrony and give rise to an appearance comparable with the waves of motion of corn-stalks when wind blows over corn-fields. However, the relationship between the direction of the effective stroke of the ciliary beat (which is approximately the direction of propulsion of water) and the direction of propagation of the metachronal waves varies in different ciliated systems.

The patterns of metachronal co-ordination are diverse. In long rows of cilia the phase of effective stroke of the beat is usually approximately at right angles to the row along which the metachronal waves pass. Such metachronism is called **diaplectic** (Fig. 1.89D). These may be of two types: (*i*) **dexioplectic** when the beat is towards the right; (*ii*) **laeoplectic** pattern when the beat is towards the left of the propagation of the waves. When the metachronal waves of cilia pass over a field of cilia in the same direction as the effective stroke of the ciliary beat the metachronism is called **symplectic** (Fig. 1.89B). When the metachronal waves of the cilia pass over a field of cilia in a direction more or less opposite to the effective stroke the metachronism is called **antiplectic**. The metachronism of paramecium is approximately dexioplectic but if a viscous agent such as methyl cellulose is added to the medium around the ciliate near-symplectic metachronism is induced.

It was formerly believed that the co-ordination of ciliary beating is brought about by fibre system associated with the ciliary bases. There is now substantial evidence that neither the co-ordination of the ciliary-metachronism nor co-ordination of reversed beating (*e. g.*, *Opalina*, *Paramecium*, *Euplotes*) are dependent

*Phylum Protozoa*

on the fibre system. When a cilium moves it carries with it a surrounding layer of fluid, the extent of which depends on the viscosity of the fluid and velocity of the cilium. In the normal pattern of ciliary beat more water is carried by the

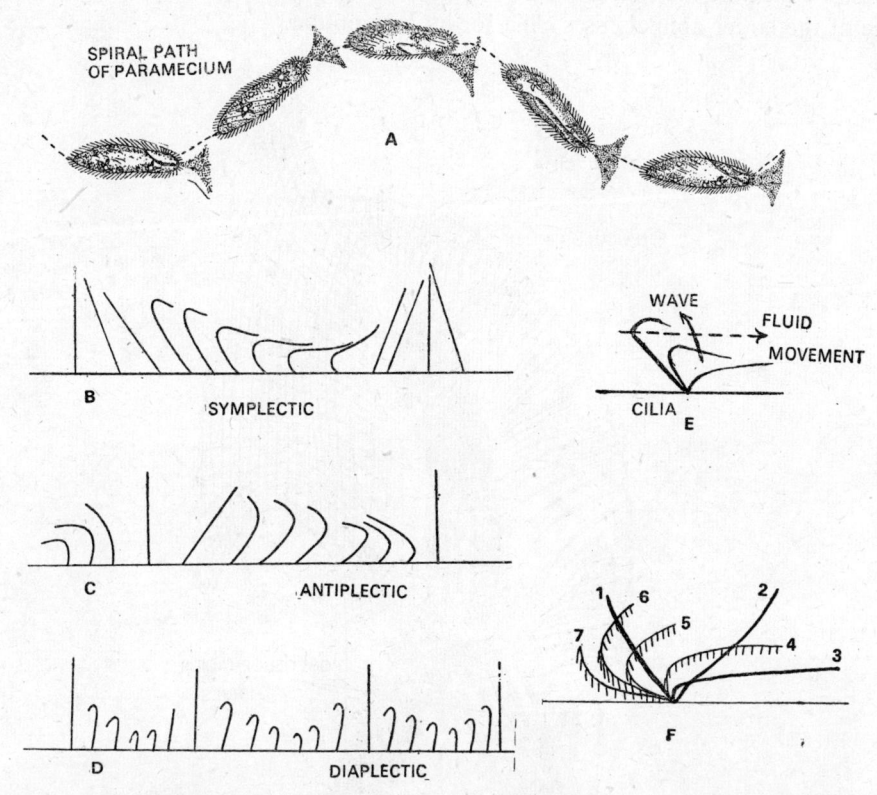

FIG. 1.89. Ciliary movements. A—Spiral path of *Paramecium*; B—Symplectic movement of cilia; C—Antiplectic movement; D—Diaplectic movement; E—Movement of fluid relative to the recovery stroke; and F—Recovery stroke, the cilia in positions 4—7 are bent to the side away from the observer.

faster effective stroke than by the slow recovery stroke so that a throbbing flow of water is caused from one side of the cilium towards the other. When two cilia lie close enough for their transported water layers to overlap, the activity of the two cilia becomes hydrodynamically linked. This **viscous mechanical coupling** between adjacent active cilia results in the formation of metachronal waves. The co-ordination between the adjacent cilia is thus brought about by the interaction between the beating cilia.

The animal moves backward or forward depending upon the direction of the effective strokes of the cilia. The stroke is generally oblique, and, therefore the paramecium appears to rotate on its long axis in the manner of a left spiral. Formerly, it was believed that the spiralling is brought about by the obliquely placed oral groove but this has been proved by experiments to be incorrect. The cilia of the oral groove do help the movement of the paramecium but in a different way. The effective strokes of the cilia of the oral groove, which strike more directly backward, turn the body of the animal towards the aboral surface.

The rolling of the animal over to the left coupled with the swerving (turning) of the body to the aboral surface enable this asymmetric animal to follow a more or less straight course while forming large spirals (Fig. 1.89A). The forces exerted by a cilium are minute and it is, therefore, not surprising to find that none of the larger animals uses cilia for its locomotion.

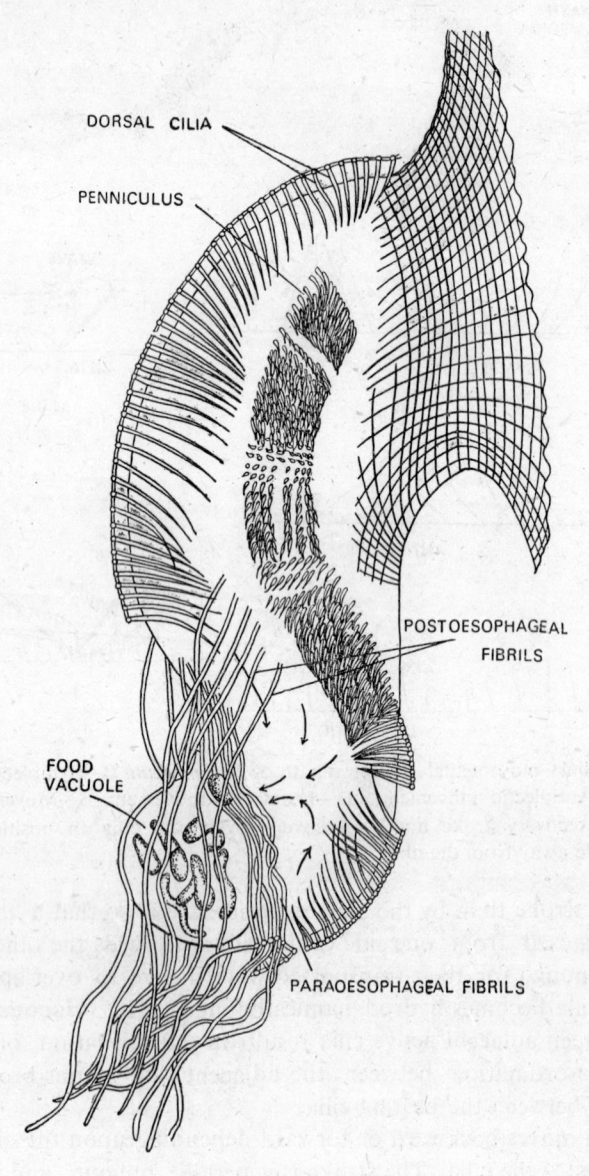

FIG. 1.90. Side view of the cytopharynx showing penniculus and other fibrils associated with feeding (*after* Lund).

## Nutrition

The nutrition of *Paramecium* is holozoic. It feeds mainly upon bacteria, which abound in water in which it lives, although it may also feed upon small protozoans, algae and yeast. While actively feeding, the paramecia move rather slowly. Food is taken in only at a definite place on the surface, the **cell-mouth**, situated at the bottom of the **vestibulum**. The **oral groove** is provided with relatively larger cilia (Fig. 1.90) which are arranged in definite tracts and direct the food down to the mouth. As the food particles are driven in by ciliary action all sorts of particles from the neighbourhood reach the vestibulum. From here only smaller particles of one particular size move in. Thus, a selection of smaller particles takes place. Other particles finally pass out. These ciliary tracts were formerly called **undulating membranes**. The food particles (bacteria) are whirled around by the special ciliary tracts and are concentrated into balls at the bottom of the food tract. The finished ball then passes into the endoplasm as a food vacuole (Fig. 1.91). The food vacuoles are formed in about 1-5 minutes depending upon the abundance of material and rate of

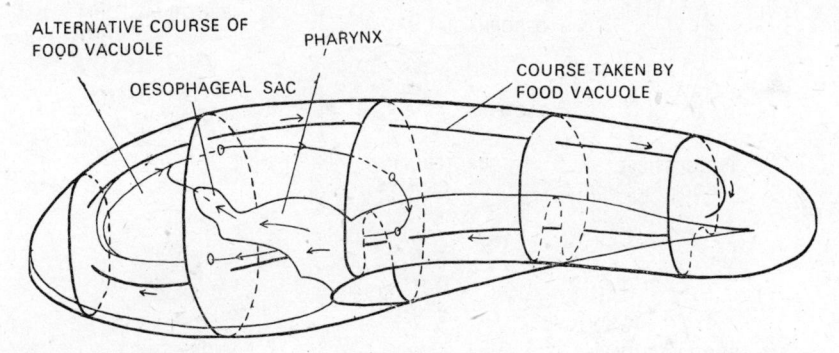

FIG. 1.91. Cyclosis in *Paramecium* and paths taken by the food in the endoplasm (*after* Kalmus).

feeding. The size of the vacuoles depends upon the quantity and quality of food, the viscosity of the surrounding medium and the physiological state of the organism. They are generally spherical when formed first and their membrane is semipermeable. Within the endoplasm the food vacuole moves more or less in a definite course (**cyclosis**) by a slow circulation of the semifluid cytoplasm, and in the meantime, its contents undergo digestion. Soon after it enters the endoplasm the food vacuole moves, posteriorly up to the end where it stops momentarily and then turns dorsally and turning anteriorly it moves forwards. On reaching the anterior end of the body it moves back again along the oral side casting off unabsorbed remains. The food vacuole remains in the body from one to three hours at the room temperature depending upon the type of food, and physiological state of the animal. On leaving the cytopharynx the vacuole first decreases in size (Fig. 1.92) and finally increases. The decrease in size takes place because the osmotic concentration of the endoplasm is higher than that of the vacuole, which, therefore, loses water and becomes small. Gradually the contents of the vacuole become acidic and increase the osmotic pressure inside. This enables the vacuole to absorb water and attain its normal size at the cell-anus which is sometimes called the potential cell-anus because protoplasm is exposed here only at the time of defecation and not permanently as at the cell-mouth.

**Digestion**

The food vacuole first becomes acidic, the acid being secreted by the cyto plasm adjoining the vestibulum and the pharynx. Increase in acidity probably causes hydrolysis thereby increasing the osmotic concentration which results in an inflow of fluid containing digestive enzymes. The enzymes are produced by certain discrete granules or globules or other fine structures in the protoplasm. These have been called neutral red granules or globules, digestive granules mitochondria, chondriosome, etc. Acid in the food vacuole marks the beginning of the digestive phase and protein digestion takes place in the pre-

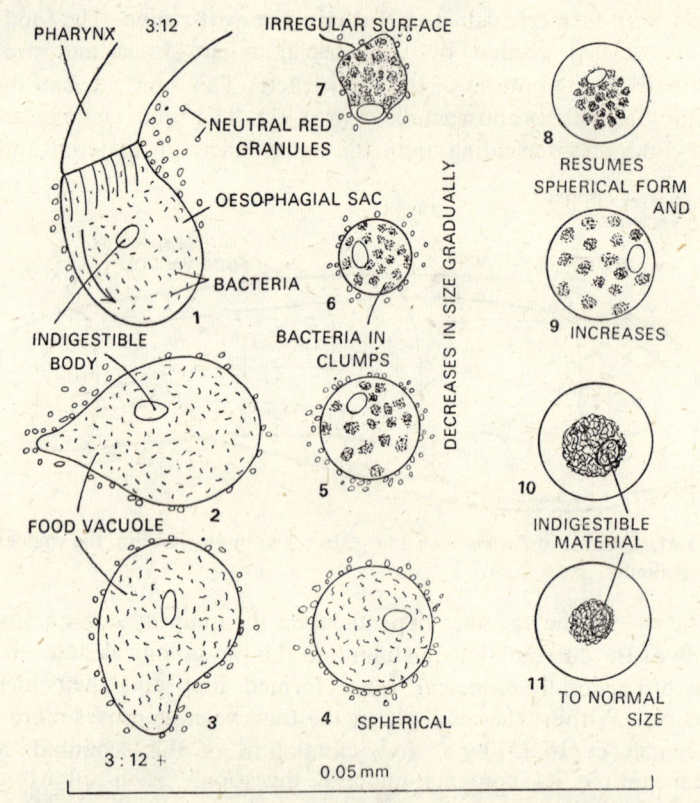

**FIG. 1.92.** Fate of food in *Paramecium aurelia* (*after* Mast.) Note that the vacuole rapidly becomes spherical (1—4), and then decreases greatly in size (5—7) and again increases greatly (8—11). Neutral red granules on the surface disappear with increase in size (8—11), and bacteria die and become agglutinated during the decrease (5—7).

sence of acid. Complete digestion takes place mainly during the later alkaline phase of the food vacuole. Proteins are acted upon by certain proteolytic enzymes resembling trypsin and are changed into amino acids. There are many evidences that show that starch and other carbohydrates are digested in the food vacuoles and changed to dextrins which are utilised by the animal. It is reserved in the form of glycogen, which is utilised during starvation. Difference of opinion about the digestion of fat still exists. Many observations indicated that there is a close relationship between carbohydrate and fat metabolism. Thus, it is evident that the substances in the food vacuoles are digested to form monosaccharides, amino-acids and other breakdown products and proteins are converted into peptones only in many cases.

*Phylum Protozoa* 125

If paramecia do not get food in the environment they feed upon the reserve material in the cytoplasm. When even these are exhausted the macronucleus becomes smaller, the animal becomes transparent, vacuoles appear in the cytoplasm, cilia disappear, and the degeneration of specialised structure sets in. This phenomenon has been called **inanition**. When rich food such as bacteria are supplied the paramecia may regain the normal form. The unused material in the food vacuole, after the digestion is over, is egested rapidly at the cell-anus. The process is called **egestion** or **defecation**.

**Respiration and Excretion**

Respiration takes place by diffusion through the surface and is essentially the same as in all other animals, *i.e.*, oxygen is taken and used for the burning of food; and carbon dioxide, water and nitrogenous wastes are given off. Some authors consider the contractile vacuole to be a respiratory structure. Oxygen needed for respiration is obtained from water that enters the body and reaches the contractile vacuole and carbon dioxide is removed by the latter. The pellicle is also believed to play some part in respiration for most of the oxygen enters through it. Active paramecia utilise more oxygen that is why the rate of oxygen consumption in a freshly made culture containing young and growing paramecia is higher than in culture containing older non-dividing paramecia. If the temperature is increased the rate of respiration is markedly accelerated. It is a fact that, like all other organisms, paramecia require oxygen to carry on the functions of life. It is not surprising, therefore, to see paramecia gather at the surface of the culture media where the oxygen concentration is maximum. If oxygen is removed completely from the medium in which paramecia live they will die in twelve hours.

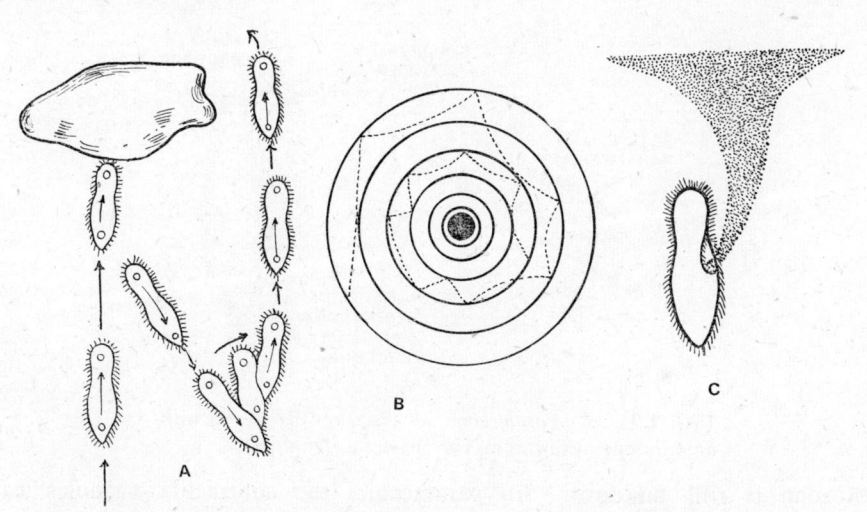

FIG. 1.93. A—Avoiding reaction; B—'Trial and error' in relation to chemical. A little chemical is placed in the centre of a drop of water, and it slowly diffuses outwards. The concentric circles indicate zones of diminishing concentration. As the animals swims, it gives an avoiding reaction whenever it enters a zone less favourable than the one it is in, but soon becomes accustomed to it (*after* Kühn); C—*Paramecium* 'get advanced information' about the material ahead by drawing the current of water in the form of a cone (*after* Jennings).

Waste products of cellular metabolism are eliminated from the protoplasm by diffusion. Such waste products usually include nitrogenous substances only. The other most widely accepted view held today with regard to the removal of nitrogenous wastes is that they are eliminated by the contractile vacuoles, which are primarily organelles for the removal of excess of water (osmo-regulation). As the water from the environment enters the cytoplasm it becomes laden with nitrogenous waste products of metabolism that are excreted through the contractile vacuole. Excretion also functions in order to maintain osmotic equilibrium (*see* below). The cytoplasm of paramecia contains crystals and crystalline granules. They are catabolic products of metabolism. They are first dissolved and then got rid of with the water of the contractile vacuole. How the insoluble crystals such as those of calcium phosphate are got rid of is not known yet.

**Osmo-regulation.** Two contractile vacuoles regulate the water content of the body of paramecium. They occupy rather fixed positions (unlike the roving type of *Amoeba*), one being anterior and the other posterior. The contractile vacuoles of *Paramecium* have more complex structure (Fig. 1.86) than those of the amoebas. The contractile vacuole of *Amoeba* is apparently a temporary structure which reforms before each contraction; but in *Paramecium* the vacuole canals and the pore through which the vacuole discharges are probably permanent structures even though the vacuole is probably not. A permanent pore presumably constitutes a weak spot more liable to form an opening than the rest of the body surface, and not an actual opening. Contractile vacuoles also appear in a large number of marine and parasitic ciliates, but the rate of output of fluid is much lower in marine than in fresh water ciliates of the same size. Exact function of the contractile vacuoles of the marine and parasitic

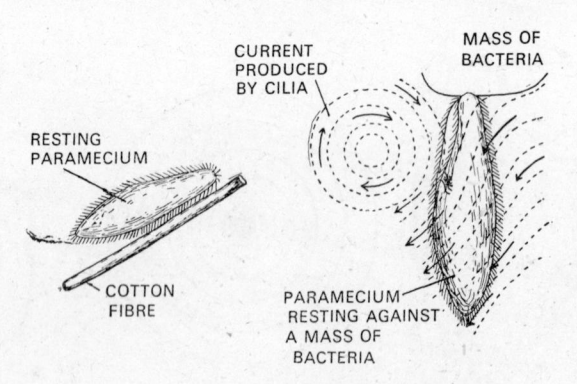

FIG. 1.94. A—*Paramecium* at rest; and B—at rest with anterior end against a mass of bacteria (Jennings).

protozoan is still unknown.[8] In paramecium the contractile vacuoles can eliminate a volume of water equivalent to its body volume in about half an hour, as compared with 4-30 hours required by an amoeba, though the rate of discharge depends upon several other factors. It varies with temperature, rise of temperature leads to an increase of frequency of discharge. The rate of discharge is also higher in an inactive animal than in one that is swimming

[8] J.A. Kitching (1938), *Biol. Reviews*, Vol. 13.

*Phylum Protozoa*

about; and is higher in water with a scant supply of dissolved salts than with stronger concentration.

**Behaviour**

The reactions of *Paramecium* to conditions in its environments are interesting. It is certainly sensitive to conditions of light intensity, concentration of oxygen and carbon dioxide, and to other chemical substances in the water.

**Reaction to mechanical stimuli (Thigmotaxis).** Its responses to tactile stimuli are perhaps worthy of special mention. If during its free swimming activities it bumps into obstacles in its path, the action of its cilia are immediately reversed, it turns to a side and goes off in a new direction. The set of movements in which the animal turns back, and swims off in a new direction is called avoiding reaction (Fig. 1.93). In case of a second obstacle the same set of movements is repeated. Mechanical obstacles, excessive heat, excessive cold, irritating chemicals, unsuitable food, a predaceous enemy all elicit the avoiding reaction, which may be said to constitute most important behaviour of a paramecium.

*Paramecium* is also known to learn by **trial and error** reactions. In its constant movements *Paramecium* may swim by chance into a region rich in bacteria. Each time it crosses the boundary of this region into a less favourable area, it gives the avoiding reaction, in other words it avoids getting out of

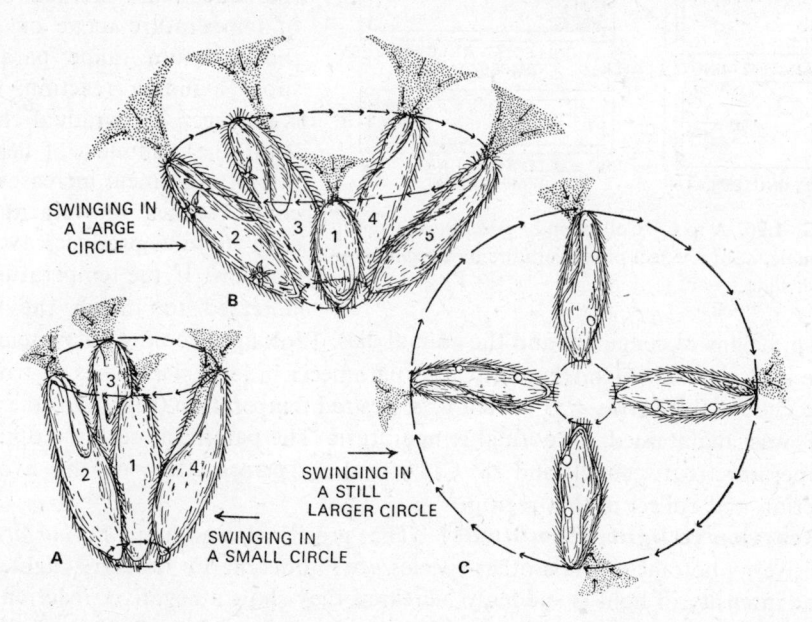

FIG. 1.95. Avoiding reaction. A—Swinging a small circle; B—Larger circle; and C—Still larger.

the favourable area. Often *Paramecium* begins to react negatively before getting into the unfavourable surrounding. The beating of cilia in the oral groove draws a constant current of water towards the oral groove in the form of a cone. (Fig. 1.93). If there is an irritating chemical in the water ahead, or if the temperature of water is disagreeable a portion will be drawn to the oral groove in advance. Such advance information about the environment ahead enables the animal to avoid the unfavourable region without actually entering it.

The ability to discriminate between foods is very poorly developed in paramecia. They readily take in and make food balls of almost any minute particles, such as carbon grains, dye suspensions and the like. After a time, however, they start rejecting such inert particles while keeping on accepting bacteria.

**Reaction to chemical stimuli (Chemotaxis).** Paramecia avoid strong acids, but they give the avoiding reaction when passing from dilute acids to ordinary water, and therefore, tend to aggregate in regions of low acidity. This behaviour aids the animal in feeding, because bacteria are most likely to be present near decaying organic matter which makes the water slightly acid. Alkalies repel these ciliates powerfully.

**Reaction to temperature (Thermotaxis).** The optimum range of temperature for paramecia is around 24-28°C, but they can withstand considerably lower and higher temperature than the optimum range. Paramecia have a great power of adaptability which helps them to acclimatize or adjust to the environmental temperature. But if there is a sudden and marked change of temperature above or below the optimum range paramecia show avoiding reaction, which is not seen in a gradual change. With the stimulus of heat the rate of movement increases and cold reduces movement and each stimulus produces avoiding reaction. If the temperatures is increased too much the cytoplasm begins to coagulate and the animal dies. Prolonged cooling at low temperature is also lethal. Mendelssohn placed paramecia in long glass tubes or troughs. One end of the device was heated to a desired temperature (Fig. 1.96) the other end was maintained at normal temperature. The paramecia seek the optimum temperature (between 24° and 28° C) by trial and error pattern of the avoiding reaction and collect in that region.

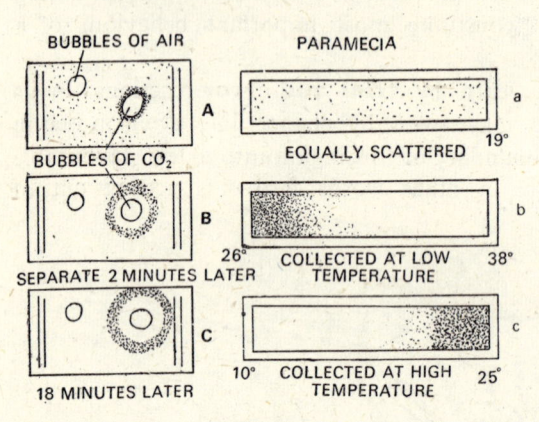

FIG. 1.96. A to C—Collection of *Paramecium* about a buble. a, b, c, effect of temperature on *Paramecium* on a slide.

**Reaction to light (Phototaxis).** The green paramecium, *P. bursaria*, is positively phototactic while other species are indifferent to ordinary light. But if the intensity of light is suddenly increased they show a negative reaction *i.e.*, avoid it. They give a negative reaction to ultraviolet light. Ordinarily X-rays have no apparent effect on the ciliates. If the dividing forms are exposed to X-rays the division is temporarily inhibited. The greater the dosage the greater the delay in fission.

**Reaction to electric current (Galvanotaxis).** When two electrodes are placed opposite each other in a shallow dish containing paramecia (Fig. 1.97) and a constant current applied all the organisms swim in the same direction towards the cathode or negative electrode, where they concentrate in large

## Phylum Protozoa

number. If the direction of the current is reversed, while the paramecia are swimming, the organisms reverse their direction and swim toward the new cathode. Why they do so is not understood. A weak current has no reaction.

**Reaction to gravity (Geotaxis).** The paramecia show a negative response to the force of gravity and keep on swimming upwards with their anterior end directed away from the force of gravity.

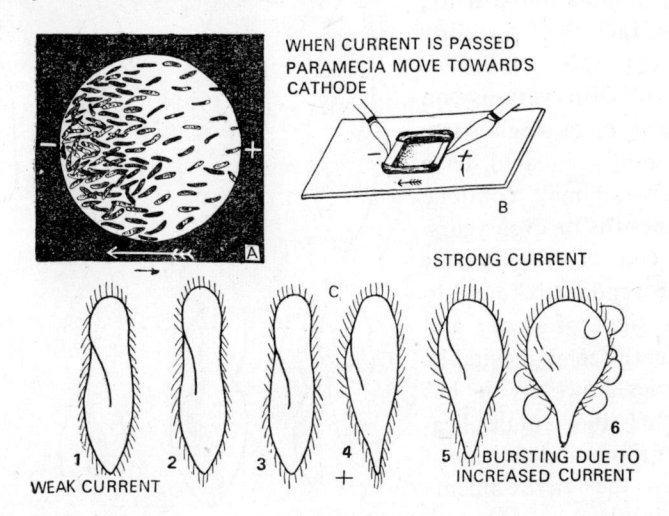

FIG. 1.97. When current is passed by means of unpolarised brush electrodes through a cell with porous walls the organisms gather at the cathodic side (B); A—Magnified view of the same (*after* Verworn); C—Successive stages in paramecia if the current is gradually increased. In weak current (as in 1) only a few cilia are reversed but gradually the cilia reverse and animal bursts (*after* Jennings).

**Reaction to water currents (Rheotaxis).** Normally the paramecia swim with water currents, but it has been observed that, with a certain velocity of the water current, they turn their anterior ends upstream (a direction against the stream).

**Reaction to hydrostatic pressure (Barotaxis).** When the hydrostatic pressure is increased, the rate of movement is decreased. Further increase forces them to stop their movement completely.

For a small animal devoid of any sensory mechanisms as it is, *Paramecium* is remarkably adaptive.

## Reproduction

Commonly *Paramecium* reproduces by transverse binary fission (Fig. 1.98). The micronucleus divides first mitotically and the two daughter nuclei move towards opposite ends of the cell and then the meganucleus divides probably amitotically. Then the cell body divides by a transverse constriction into two daughters. The front and the rear halves of *Paramecium* are not exactly alike, but even before separation occurs, each half forms the parts necessary for a complete individual. Thus, the gullet, which is behind the middle, falls to the rear daughter, the front daughter, early in the process of division, forms a new gullet. The resulting daughter individuals are equal-sized each carrying a set of organelles, and grow to full size before fission sets in again. Fission takes about two hours to complete and may occur one to four times per day producing two to sixteen individuals. All paramecia produced by fission (uniparental reproduction) from a single individual are known collectively as a **clone**. The rate of multiplication depends on several factors important of which are temperature, availability of nutrient, age of the culture, population density, etc. Internal factors of heredity and physiology also influence multiplication. Observations on the descendant of a single paramecium, carefully isolated, have shown that fission may continue for several months or even years. But by the end of that time the individual descendant is found to be showing signs of decay and eventually death seems inevitable. In some species, however, it has been suggested that multiplication may continue indefinitely if conditions in the environments remain favourable.

**Conjugation in** *Paramecium aurelia*. Sexual reproduction in ciliates involves exchange of nuclear material. *Paramecium aurelia* has one macronucleus and two micronuclei. At the beginning of conjugation two ciliates come

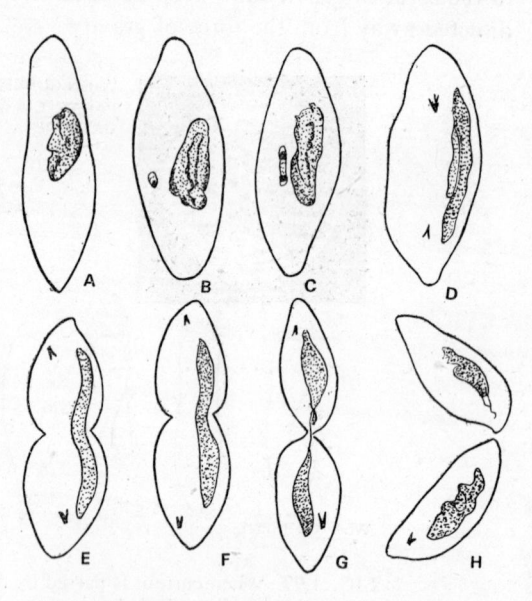

FIG. 1.98. Binary fission in *Paramecium*. A—Mature adult ready to divide; B to D—The micronucleus divides mitotically and the daughter nuclei move towards apposite ends; E—Meganucleus prepares to divide; F-G—Meganucleus divides amitotically and also the cell body; H—Two daughter paramecia separate.

together and become attached first at deciliated areas near the anterior end of their ventral surfaces, and later by attachments formed in the gullet region where membrane fusion leads to the formation of a number of pores providing cytoplasmic bridges between one paramecium (now called **gamont**) and the other. Adhesion probably takes place by the secretion of a sticky substance by the cilia. Attachment lasts for several hours. Both micronuclei in each conjugant undergo two division meiosis, so that eight haploid nuclei are formed in each cell. Seven of these degenerate. The remaining micronuclei undergo mitosis to produce two gamete nuclei in each cell. One of these is stationary and can be considered **female nucleus**, the other is called **wandering** or **male nucleus**. The wandering nucleus from each gamont migrates across a cytoplasmic bridge into the other cell and fuses with the stationary nucleus and a **synkaryon** (a **zygote nucleus**) is formed. Subsequently, the cells separate, each is now called an **exconjugant**. By this time the original macronuclei of the two cells have more or less disintegrated. The micro-nuclear synkaryon divides twice mitotically and two of the four products become macronuclei and two remain as diploid micronuclei. At the first binary fission of the exconjugant one macronucleus passes to each daughter cell and both micronuclei divide mitotically, so that the original nuclear condition is

# Phylum Protozoa

restored.

**Autogamy.** The entire process of autogamy takes place inside the body of a single animal, without the co-operation of another individual. The end result of autogamy is the establishment of a new nuclear apparatus from products of the division of the micronuclei, two of which fuse to form a synkaryon. The old

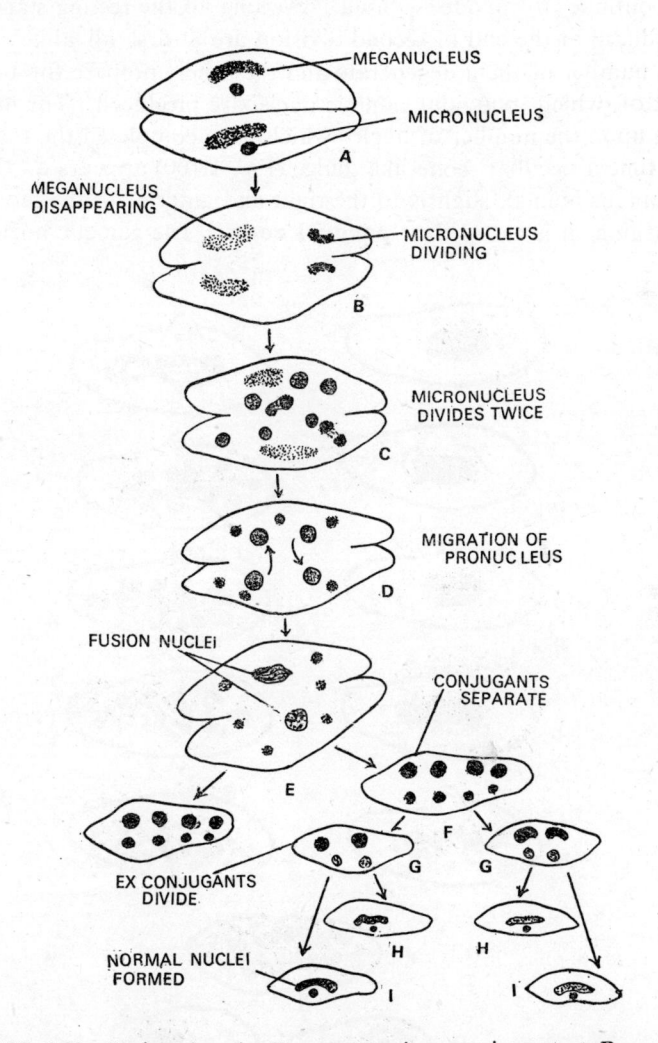

FIG. 1.99. Conjugation. A—Two paramecia come in contact; B—The meganucleus tends to disappear and the micronucleus tends to divide; C—Micronucleus divides twice producing 4 daughter nuclei of which three degenerate and the fourth divides again; D—The migratory pronucleus of one conjugant passes into body of the other; E—Fusion nuclei formed, the two conjugants now separate; F—Synkaryon undergoes repeated division forming four pairs of nuclei; and G to I—The exconjugants divides into four daughter cells, each forming a fresh individual.

macronucleus fragments and disappears during the process. Two divisions of the synkaryon produce the new nuclear complex.

132                                              LIFE OF INVERTEBRATES

The two characteristic vesicular micronuclei of the species become larger—the material between the endosome and the membrane increases and the two look a good deal like small food vacuoles. At the next stage these two change to form thin thread-like crescents (U-shaped), some may even be straight. The mitotic division of the nuclei then begins. The four nuclei produced by the first division continue to divide without reverting to the resting stage. The eight nuclei produced at the end of second division are at first all alike. After a time a variable number of them degenerate and the others prepare for third division as a result of which potential gametic nuclei are produced. The number varies depending upon the number of nuclei which have completed the third division.

By this time a peculiar cone-like bulge (Fig. 1.100) appears on the surface of the autogamous animal, slightly to the right and anterior to the mouth. Because of its position it is called the **"paroral cone."** The gametic nuclei are at first

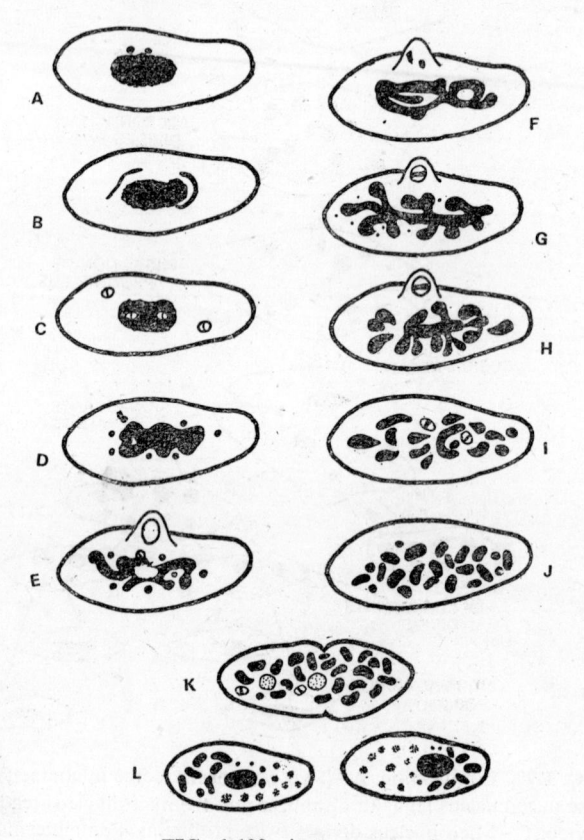

**FIG. 1.100.** Autogamy.

elongated spindle-like bodies that stain faintly. Often there is slight size difference between them so that they may be referred to as **"male"** (smaller) and **"female"** (larger). They approach each other and move into the paroral cone and fuse to form a synkaryon. Later it withdraws from the paroral cone into the body and divides. By this time the old macronucleus is pretty well disintegrated and its fragments choke up the interior of the cell. The synkaryon decides twice forming four nuclei; all of the same size. Of these two increase in size and become macronuclear primordia. Their growth is quite rapid. The other two become functional micronuclei and condense and decrease in size. Then the autogamous protozoan divides, one of the macronuclear primordium passes to each cell, the two micronuclei also divide and pass two to each daughter cell and the old macronuclear fragments are distributed in approximately equal numbers to the two daughters. In the beginning the young macronucleus has a smooth and regular form and its chromatins are very finely granular—almost homogeneous. Gradually, its substance increases and normal form is attained. Sometimes two paramecia come together as if to conjugate, but do not exchange nuclear material, and end up by performing autogamous self-fertilisation. This process is referred to as **selfing** or **cytogamy**.

Throughout the process of reorganisation the macronuclear and micronuclear behaviour is correlated fairly closely, so that it is often possible to predict what the micronuclear condition will be, from an inspection of the macronucleus. An exceedingly close parallelism exists between the events of autogamy and those of conjugation (which however, involves two animals, not one as in autogamy). Whenever conjugation is found in a culture, autogamy is likely to be taking place. Simultaneously and conversely, when no conjugation is occurring in a culture autogamy is unlikely to be seen. Apparently, the same conditions, be they environmental or intrinsic, which induce conjugation likewise induce autogamy. Increase in the number of individuals does not take place by any of the sexual processes, except by mitosis and binary fission following nuclear reorganisation.

**Conditions for conjugations**. It is known that two individual ciliates will conjugate only under certain conditions. The individuals must be of the correct type and the age of the clone to which they belong must be within certain limits. Members of the species of *Paramecium aurelia* have been separated into 12 varieties. Members of one kind will not normally conjugate with members of another kind. This indicates that each variety is effectively and genetically isolated species. Varieties have been called **syngens** (from the fact that they have the same **gene pool**). With each syngen of *P. aurelia* there are two mating types (in one syngen only one mating type has been found), and conjugation will only take place between individuals of different mating type belonging to the same syngen. Mating type specificity in *Paramecium* seems to be expressed in chemical characteristic of the surface membrane.

**Mating types**. In *Paramecium bursaria* six syngens have been recognised. Most of them contain more than two mating types. In one syngen there are eight mating types and a mature member of one mating type will conjugate with a mature member of any of the seven mating types of that syngen, but not normally with any other member of other syngen. In *P. bursaria* the immature period lasts several days to many weeks and the period of maturity may last for many years, so that the complete life span of the clone may exceed ten years. Control of the mating types resides in the macronucleus. The mating type of a clone is established at the formation of the new macronucleus following conjugation and is transmitted at macronuclear fission. The character of cytoplasm in which the macronucleus develops is also known to influence mating types.

**Process of rejuvenation**. Each exconjugant gives rise to a clone of cells by successive binary fission. The ability of individuals to take part in sexual processes depends upon the age of the clone. Members of a clone are not capable of conjugation until a certain number of fission cycles have been completed since the previous conjugation. Following this period of immaturity lasting 2-10 days the ciliates are mature for a month or so and will conjugate successfully with ciliates that are ready to mate.

As clone grows old and enters a period of old age (senescence), fertility decreases, and, as a result, conjugations become rare. However, some ex-conjugants that have remained non-viable show an increasing proportion of conjugations. Meanwhile, autogamy becomes more common in separate individuals. As senescence progresses, autogamy no longer produces viable clones, so that eventually after some months no further sexual processes of any type are successful and the ageing clone reaches a state of genetic death. Its member can no longer participate in the formation of a new clone. Ciliates in the senescent clone may still divide asexually and the clone may persist for some months in a state of reduced vigour before the entire population dies and the clone suffers somatic death by ultimate loss of macronuclear function. It is important to appreciate that both forms of sexual processes (conjugation and autogamy) can result in rejuvenation and the foundation of a new clone. The successful renewal of the macronucleus is the vital requirement in these processes.

In 1914, Woodruff and Erdmann described a complicated process of nuclear rearrangement named **Endomixis**. Diller (1936) was not able to confirm such a micronuclear behaviour in *P. aurelia*. He thus denied the existence of **Endomixis**. In the same year, Diller described certain methods of nuclear reorganisation in mass cultures of *P. aurelia* which cannot be explained under conjugation or autogamy. This process was called by him **Hemixis**. Modern workers (Sleigh, 1973) do not include Hemixis among the methods of nuclear reorganisation found in ciliates.

**Killer**. Sonneborn discovered that certain strains of *Paramecium aurelia* liberate into the medium, in which they live, a poison which kills other strains of *Paramecium*. He called the former "**killers**" and the latter "**sensitives**." The killers are resistant to their own poision, which was named **paramecin**. Killing takes placeslowly, the sensitives taking about two days to die. The killers are also found in nature but perhaps they are not effective. Probably, the paramecin set free is diluted very much to have any effect.

Because the process of killing is slow it is possible to get conjugation between killers and sensitives. If each exconjugant is cultured separately, the killer exconjugant is found to give rise to killer cultures and the sensitive to sensitive cultures. Usually, there is no exchange of "killer" properties (killer factor), but in prolonged conjugation exchange of cytoplasm takes place, and the killer factor gets carried into the sensitive which then produces a killer culture. This has been worked out successfully by Sonneborn.

In *Paramecium aurelia* certain symbiotic organisms have been found within the cytoplasm. These are believed to be bacteria and have been given such names as **kappa, mu, pi, lambda**, etc. Lambda particles carry appendages that look like bacterial flagella, and several types of particles have cell walls of bacterial type. Killer paramecia were found to contain kappa particles. Only kappa particles that contained refractive inclusions were able to kill, but other kappa particles could develop these inclusions. Paramecia contain **mu** kill sensitives

# Phylum Protozoa

at conjugation (they are mate killers), while **pi** may be an inactive kappa. Paramecia containing lambda produce large lethal granules which are released into the medium. These symbiotic organisms derive nutrients from the paramecia, but a killer stock has an advantage over sensitive stock in ecological competitions.

### TYPE VORTICELLA

*Vorticella* (Fig. 1.101) is the common type of stalked ciliates. It is found in abundance in fresh water. It is hardly possible to examine the water of a pond without finding the vorticellids. It also abounds almost in all laboratory cultures. The vorticellids are attached to any aquatic object, plant or animal. They may be found attached to various aquatic plant leaves or stems, or to mollusc shells, or even to several of the living oligochaetes. Some species (*V. convallaria*) are present only in stagnant water or infusions. Some (*V. microstoma*) are common in oils and sour infusions.

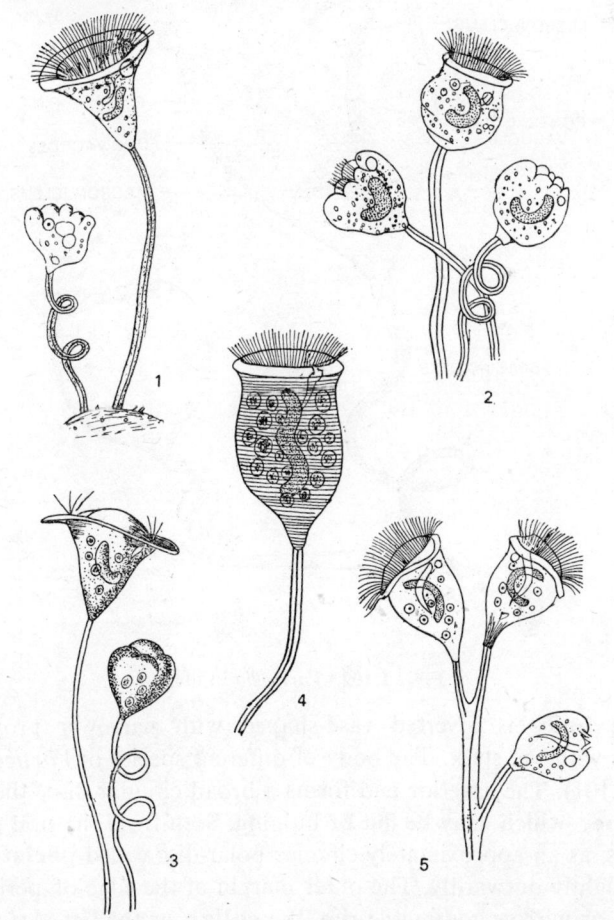

FIG. 1.101. Different types of vorticellids found in India. 1—*Vorticella citrina*; 2—*V. campanula*; 3—*V. patellina*; 4—*V. convallaria*; 5—*Carchesium epistyles*.

### Structure

The animalcules have a small bell-shaped, oval or companulate body attached to the submerged objects by a slender stalk enclosing an elastic, spirally disposed contractile axial fibre or **spasmoneme**. The axial fibre of the stalk consists of a band of myonemes enclosed in a larger sheath, continuous with the pellicle of the bell, and fastened in a spiral course to the inner surface of the wall of the stalk. The stalk is secreted by the body and comprises a pseudochitinous tube filled with granulated **thecoplasm**, in which is embedded a central rod of **kinoplasm**. Numerous elastic skeletal fibrillae or myonemes are arranged on the surface of the central rod. When undisturbed the bell is poised fully expanded on the end of the long, straight stalk with cilia in rapid action. The least disturbance causes the stalk to contract like a coiled spring, while the bell also contracts, folding its edges over the circlet of cilia (Fig. 1.102).

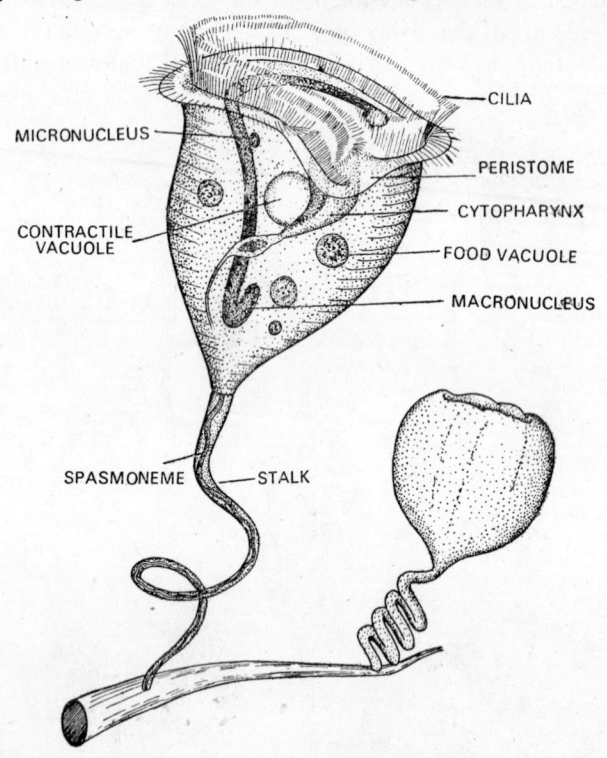

FIG. 1.102. *Vorticella* in life.

**Body.** The **body** is inverted vase-shaped with narrower proximal region which merges with the stalk. The body of different species of *Vorticella* differs in shape (Fig. 1.101). The anterior end forms a broad circular disc, the **epistome** or **peristome,** which may be flat or bulging. Seen from the oral end the epistome appears as an approximately circular polar disc called **peristomial disc**. This bulges slightly outwardly. The outer margin of the edge of peristome forms a thick broad or narrow contractile rim, the **collar** or the **lip**. At the time of contraction the peristomial disc is withdrawn and the collar encloses the vase from above. In between the colar and elevated peristomial disc there is a distinct **perstomial groove**. At the posterior end of the vase a clear area, the **scopula**, is visible. This secretes the inert matrix of the stalk. Sometimes this appears as a small invagination of the wall showing distinct fibrillae (Fig. 1.103A).

# Phylum Protozoa

The cilia are restricted to a certain area that is known as the **adoral zone** (conducting to the oral aperture). On the body there are no cilia, although basal bodies have been demonstrated all over the body. The cilia are extremely fine and short processes of the ectoplasm. Each cilium which originates from a basal body (**kinetosome**) embedded in the ectoplasm, consists of a central axis or axoneme surrounded by a contractile sheath. In the stalked forms cilia help in food capture but in the stalkless stages called **telotrochs** the cilia help in locomotion. The adoral zone of cilia consists of two parallel cilliary girdles which run round spirally to the left (contra-clockwise), one of which descends down into the vestibule or the gullet. The outer girdle consists of a single row of cilia (Fig. 1.103) fused to form the outer peristomial membrane which projects out radially over the peristomial border forming a kind of circular shelf. The inner circle stands erect and is double (consisting of two rows) in some species and keeps up a constant undulation. At the right edge of the vestibule the inner and outer circles separate, the former passes down in the vestibule along its inner wall, the latter travels along the outer wall becoming a strong undulating membrane. Between the disc and the collar at one side is a large cavity, the **vestibule**, where the disc is raised considerably above the collar, with the result that the beginning of the adoral zone is higher than its end. The vestibule continues in the interior as a funnel-shaped cytopharynx. The arrangement of the ciliary girdles is such that their constant beating, guided by the actions of peristomial membrane, drives food particles into the vestibule through which they pass into the interior to form food vacuoles, which take a definite course in the bell (Fig. 1.103).

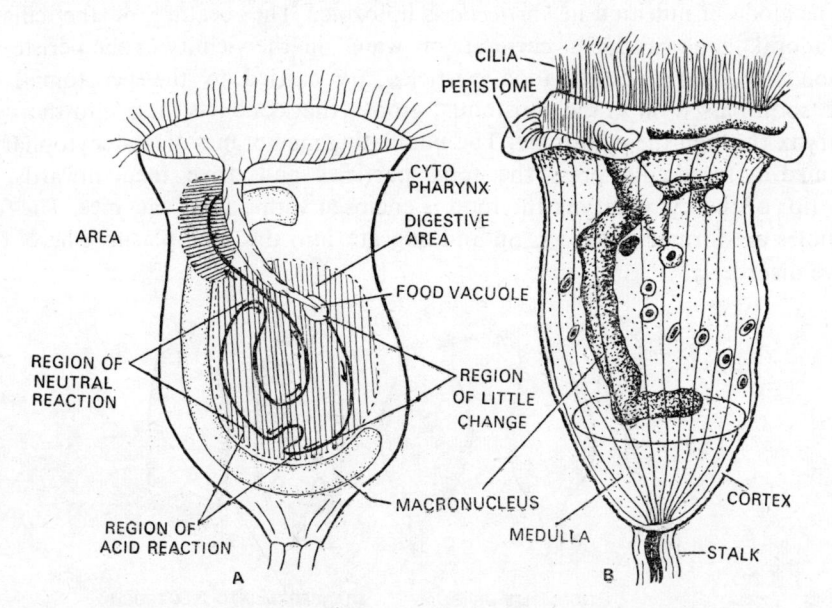

FIG. 1.103. A—Path of food vacuole and; B—Bell of *Vorticella* showing internal structure.

The pellicle of the bell is uniformly thin and delicate (thicker than that of *Paramecium*). It is often marked with circular parallel striations and in some species of *Vorticella* it is tuberculate. It covers the body so closely that it becomes difficult to distinguish it.

The interior of the cell-body contains various food vacuoles. There are two nuclei. The larger **macronucleus** is elongate band-shaped or horse shoe-shaped structure, whereas, the **micronucleus** is minute situated close to the macronucleus. The macronucleus is rather compact and is the centre of all metabolic activities of the body. If it is removed or destroyed the animal dies. The micronucleus is associated with reproductive activities. The cytoplasm of the body is differentiated into a thin cortical layer of **ectoplasm** and a central mass of **endoplasm**. The ectoplasm contains basal bodies (**kinetonuclei**) from which arise the axonemes of cilia. Contractile elements called myonemes are also present in the ectoplasm. These are like longitudinal fibrillae chiefly visible in the scopula where they converge to form a sphincter for closure. The contractility of the body is due to the myonemes.

The endoplasm is a finely granulated fluid showing distinct streaming movement or **cyclosis**. The nuclear apparatus, numerous food vacuoles, pigment granules, reserve food material in the form of numerous refractile granules, crystals and similar other materials lie in the endoplasm. A single contractile vacuole is situated to one side of the vestibule and communicates with a small reservoir which opens into the vestibule. On the side of the vestibule there is a weak spot, the potential **cell-anus** or **cytopyge**, through which useless debris is discharged into the vestibule.

**Nutrition**

The mode of nutrition in *Vorticella* is holozoic. The beating of the cilia of the adoral zone produces currents of water in the vicinity of the peristomial region, with the result that food particles are carried to the peristomial disc and soon passed on to the peristomial groove that conducts them into the cytopharynx through the cytostome. The undulating membranes of the cytopharynx (formed by fused cilia) press the food particles and drive them inwards. At the tip of the cytopharynx the food is enclosed within food vacuoles. The food vacuoles are formed in succession and migrate into the endoplasm where they move disorderly.

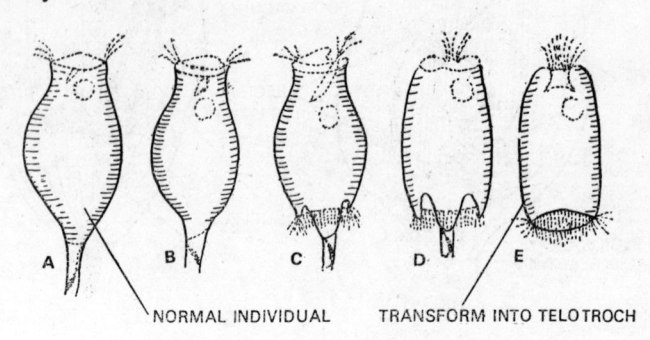

FIG. 1.104. Formation of telotroch.

In *Carchesium*, according to Greenwood, the food vacuoles pass down to one end of the macronucleus and then move close along its concave surface to the anterior end of the nucleus where defecation to the vestibule takes place. A continuous "digestive tract" (Fig. 1.103A) has been described for other vorticellids

# Phylum Protozoa

also. According to Kochring disorderly cyclosis of the food is merely an optical illusion caused by the food vacuoles along this tract.

The digestion of food particles is completed in the food vacuoles. The contents of the vacuoles become first acidic and remains so during entire protein digestion. Then the contents become neutral and finally alkaline when the digestion is completed. Some authors have suggested that certain digestive enzymes pass from the endoplasm into the food vacuoles and complete digestion. These enzymes have been stained by neutral red and have been called **neutral red granules** or **digestive granules**. The undigested material of the food vacuoles is passed out at the potential **cell-anus** or **cytopyge** into the cytopharynx on its lower wall. From here they are passed out. Some authors have even suggested the existence of an **incurrent channel** for incoming food particles and an **excurrent channel** for the exit of waste material, in the cytopharynx. The excess of material is reserved in the form of refractile glycogen granules usually present in the endoplasm.

**Osmo-regulation.** The more or less fixed types of spherical **contractile vacuole** of *Vorticella* regulates the water contents of the animal as in *Amoeba* and *Paramecium*. The water discharged to the exterior by the vacuole may contain a small amount of metabolic wastes and respiratory gases, but the specific role of the contractile vacuole has been established to be the regulation of water content.

## Locomotion

The question of locomotion does not arise in the sessile animals and *Vorticella* is one. Contractions of the body and stalk are the only body movements perfor-

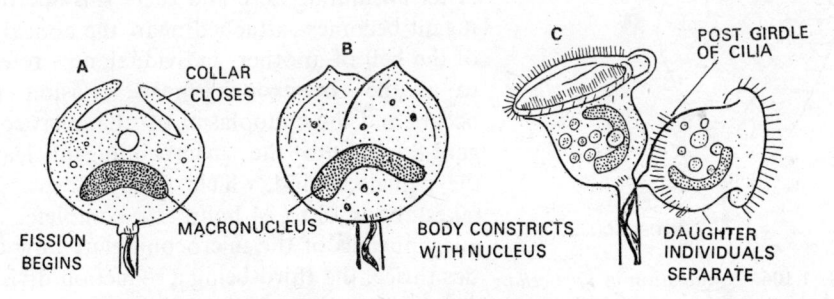

FIG. 1.105. Fission in *Vorticella*.

med by vorticellids. Sometimes a vorticellid gets detached from the stalk, develops an aboral circlet of cilia and is seen swimming. Such an individual is referred to as telotroch, (Fig. 1.105) which always moves by the beat of cilia taking a spiral course.

## Reproduction

Reproduction takes place by simple binary fission which is apparently longitudinal (in other ciliates fission is transverse). But this appears so because of the special modification of the body. The oral side represents the morphological dorsal surface, and the aboral side is the ventral surface. At the onset of fission the collar closes over the peristome and the elongated meganucleus becomes oval. The macronucleus divides amitotically following which the micronucleus divides mitotically. The peristome then shows a vertical constriction which rapidly extends to the stalk. The plane is such that one daughter becomes larger and retains the stalk and the other is smaller without stalk. This daughter develops a posterior girdle of cilia and swims away as a telotroch. Later it attaches to the substratum by means of its scopula and develops a new stalk. When this small daughter cell attaches to the substratum by means of the scopula it resembles another ciliate *Scyphidia* in which the scopula functions as the hold-fast organ. Therefore, this stage is often referred to as the scyphidia stage of *Vorticella*. This is a short-lived stage. Vorticellids also put out such a posterior girdle of cilia, under unfavourable conditions, cut off from the stalks and swim away. They develop stalks on the return of favourable conditions. This process also helps the dispersal of the non-locomotive species. In many species of *Zoothamnium* (colonial stalked ciliate) the zooids are dimorphic. The ordinary bell-shaped forms divide in the usual way and remain attached to the parent stalk forming complex colonies. Besides there are large globular and mouthless zooids that develop posterior girdle of cilia, detach from the parent and swim off. They settle down at new places, develop stalk and produce new colonies by their multiplication.

The vorticellids also reproduce sexually but the sexual mode in this case is very characteristic. Any sessile vegetative individual produces by budding a small individual called **microconjugant** (Fig. 1.106). This cuts off from the parent bell and swims away as a small telotroch. Although the individuals thus produced are dissimilar in size the micronucleus of the parent divides equally. After swimming here and there this microconjugant becomes attached near the aboral end of the bell of another individual now referred to as the **macroconjugant**. Fusion then occurs and the endoplasm of the microconjugant flows into the macroconjugant leaving the pellicle behind, which finally drops. This takes about 18 to 24 hours to complete. The micronucleus of the microconjugant now divides thrice, the third being a reduction division, while the micronucleus of the macroconjugant divides only twice, the second being the reduction division. Of these only two nuclei, one from each conjugant, fuse forming a synkaryon. The synkaryon divides three times forming eight daughter nuclei, of which one becomes a micronucleus and the remaining seven form macronuclei. Now the exconjugant divides repeatedly forming several daughter individuals. The micronuclei of these are derived by the division of the first micronucleus while the macronucleus passes as such into each. According to another description the small microconjugants are produced through several divisions (multiple fission) of the fixed bell-shaped individuals (Fig. 1.107). In this case too the daugher cells are smaller and they separate as telotrochs. Rest of the changes are similar to those described above. Some people have suggested that some species of *Vorticella* seem to exert some sort of attraction for the motile microconjugants within a

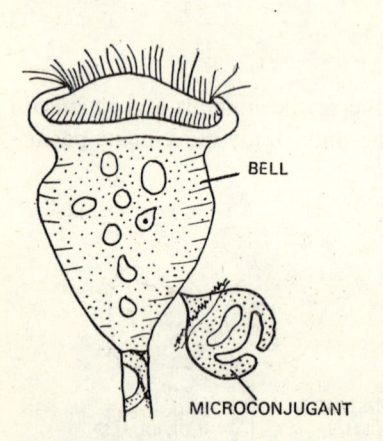

FIG. 1.106. Conjugation in *Vorticella*.

limited distance of a millimetre.

The conjugation of *Vorticella* differs from that of *Paramecium* in two important ways:

(1) The two conjugants are dissimilar morphologically, one being larger (macrogamete) and the other smaller (microgamete),

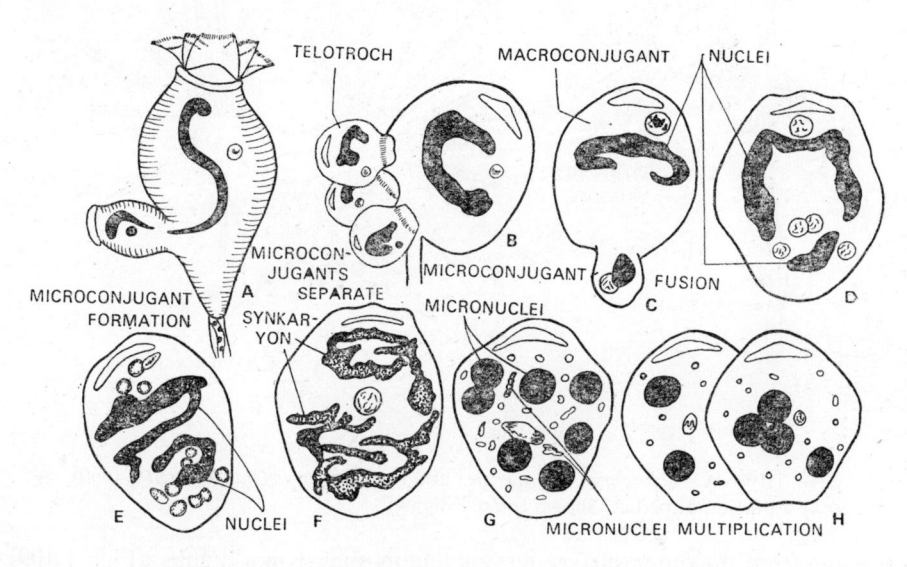

FIG. 1.107. Nuclear changes during conjugation (*after* Finlay).

(2) There is no reciprocal exchange of nuclear material, the endoplasm of the microconjugant containing the nuclei flows into the macroconjugant. Thus the fusion of the material is permanent and only one exconjugant is formed. It is likely that such a characteristic conjugation is an adaptation to sessile mode of life. However, the purpose of the conjugation is renewal of vitality lost because of continuous fission.

**Encystment.** Under certain conditions the vorticellids undergo encystment as has been studied by Brand for *Vorticella microstoma*. An individual secretes a cyst enclosing it and then divides into two, four and finally eight cells (spores). On being liberated each of these cells gives rise to an individual. According to certain reports encystment also occurs when there is dearth of oxygen and food in the medium. The individuals that emerge out of the cysts (excysted individuals) are always telotrochs.

## THE SUCTORIA

The **Suctoria** (also called **Acineta**) possess cilia only during the larval stages. The adults develop suctorial tentacles which serve as food-acquiring mechanism. The young stages are budded off from the adults, which are sedentary and without cytostome. Typically, the form of the body is vase-like but rounded or inverted conical forms are also common (Fig. 1.108). They are attached directly or by a stalk to objects or to other animals such as small Crustacea and acquatic larvae, etc. A few are unattached and are parasitic on free living or parasitic ciliates. The body is often protected by a secreted theca, continuous with the stalk in pedunculate forms. The tentacles are stiff processes

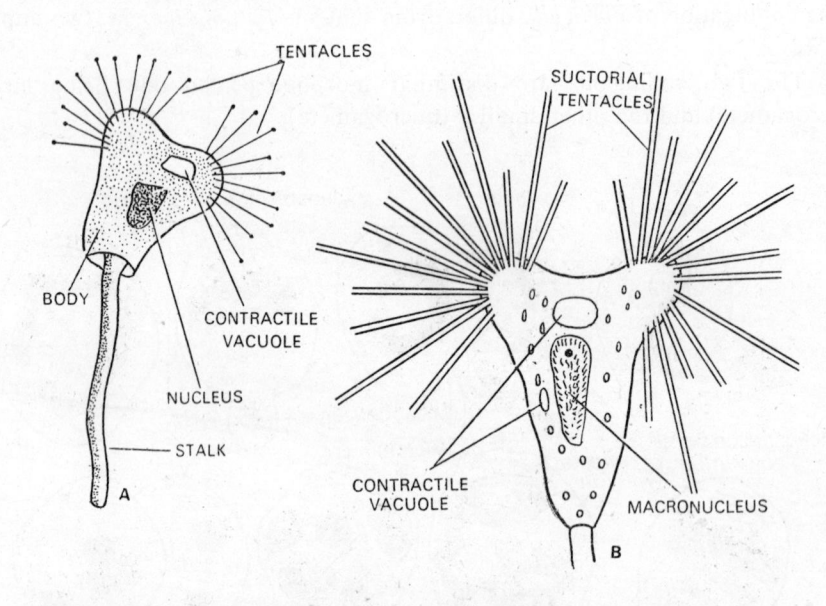

**FIG. 1.108.** A—*Tokophrya bengalensis*; and B—*Tokophrya quadripartita* (with only a portion of pedicel shown (*after* Filipjev).

springing from the outer surface terminating in sucker-like knobs (Fig. 1.109). Each tentacle consists of an outer contractile sheath and an inner stiff tube which extends deeply in the general endoplasm. Such sucking tentacles are always present in the adult organisms. In addition, in *Ephelota* there are pointed prehensile tentacles devoid of sucker-like knobs. There are one or more contractile vacuoles usually present. Nuclei are two. The meganucleus exhibits a great variety of form, it may be oval, elongated or branched. Micronuclei are one to many.

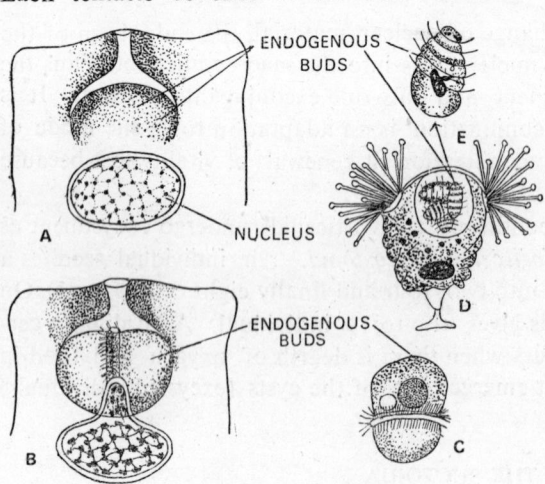

**FIG. 1.109.** Endogeneous budding in Suctoria (*after* Butschilli).

The food consists of small animals such as rotifers and protozoans which adhere to the suckers and are paralysed by a toxic secretion from the tentacles. Their contents are then sucked in by way of the tentacular canal, but how it is done is not understood. When food is scarce the individuals encyst.

Reproduction may be by binary fission but the usual method is budding. Budding may be external, one or more buds being formed on the surface of the organism, or internal (endogeneous), in which case the buds are formed in a

*Phylum Protozoa* 143

deep depression of the surface of the body called a brood-pouch. These buds develop cilia, move about within the pouch and finally come out through a pore

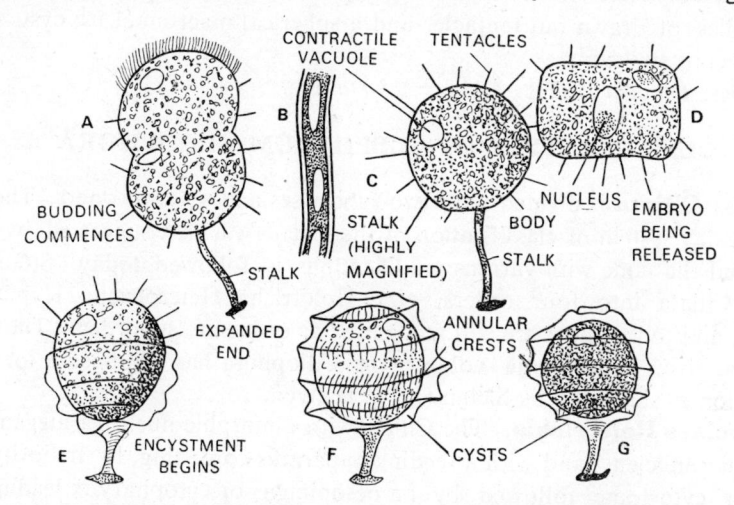

FIG. 1.110. Budding and encystment in *Podophrya fixa* (*after* Collin).

and swim away. After sometime the embryo attaches to some suitable object, loses cilia, secrets a stalk, develops tentacles and grows into the adult pedunculate form. Conjugation also occurs as in Ciliates. When two organisms associate the meganuclei degenerate, micronuclei divide a number of times and exchange nuclear material, fusion of micronuclei takes place and, finally, the organisms separate. Meganuclei and micronuclei are reformed from the products of division of the combination nucleus (synkaryon).

External budding has been observed in *Podophrya fixa*, a form reported from fresh and salt water from Bombay. The spherical body divides by an equatorial groove (Fig. 1.110A) and the individual released has a crown of cilia (Fig. 1.111B) on one side. Encystment has also been reported in the same form as

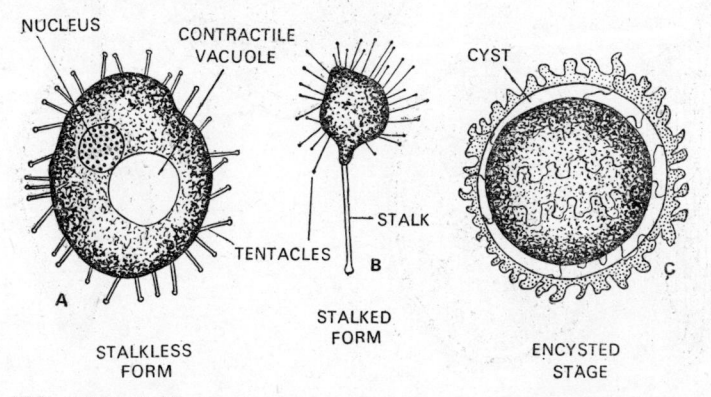

FIG. 1.111. Stalked and stalkless stages of *Podophrya sandi* (*after* Sand).

represented in Fig. 1.111C. Another species of *Podophrya* found in Bengal has a straight pedicle inserted on a conical prolongation of the body. It has numerous unequal knobbed tentacles, uniformly distributed or in one or three bundles. Contractile vacuole is eccentric, macronucleus central, spherical or oval (Fig. 1.111). The motile stage of this species has a complete girdle of cilia, two bundles of drawn out tentacles and a spherical macronucleus; cysts of this species are not stalked.

Examples: *Tokophrya* (Fig. 1.111), *Acineta*, etc.

## CLASSIFICATION OF SUBPHYLUM CILIOPHORA

THE GLASS **Ciliata** is divided into two subclasses and several orders. The basis of present-day system of classification of the Ciliata was first suggested by Stein in 1857 and the same with various modifications is followed today. Stein divided the Ciliata into four orders, viz., Holotricha, Heterotricha, Hypotricha, Peritricha and today these are still recognised as orders or suborders. The classification as given in the fauna volume on Ciliophora has changed a lot. Here classification as suggested by Sleigh (1973) is given.

(*i*) **Subclass Holotrichia.** The Ciliata with dimorphic nuclei (a meganucleus and a micronucleus) and with a feeding apparatus consisting, in majority, of a mouth or cytostome followed by an oesophagus or cytopharynx leading into the endoplasm. Conjugation is quite frequent. Forms, usually free-living, some are commensal and a few others are parasitic. The subclass is divided into several orders.

1. **Order Gymnoblastida.** Large ciliates devoid of oral ciliature, cytostome opening directly to outside.

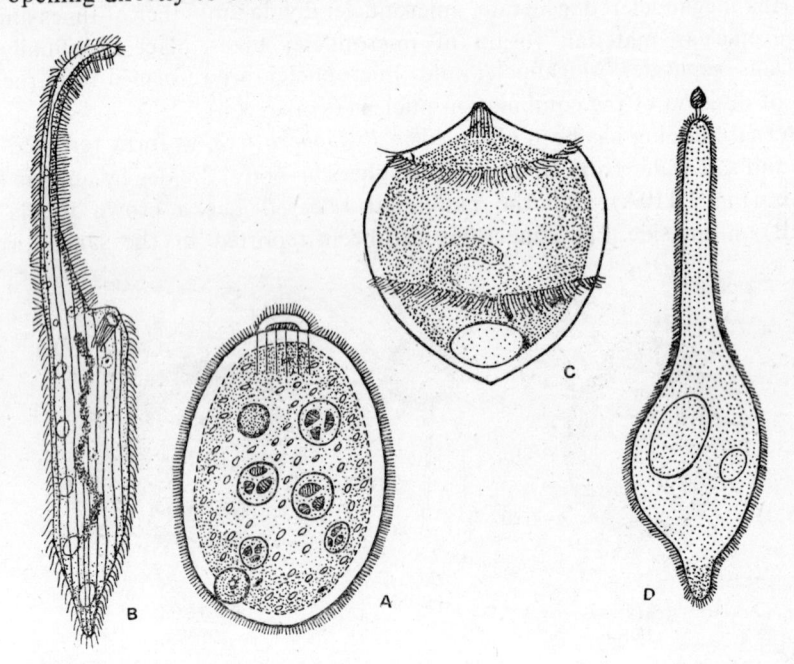

FIG. 1.112. Some common ciliates A—*Prorodon*; B—*Dileptus*; C—*Didinium*; and D—*Lacrymaria*.

Examples: *Prorodon* (Fig. 1.112 A), *Lacrymaria*, *Dileptus* (Fig. 1.112 B-D), *Didinium* (Fig. 1.112 C), *Coleps*, *Nassula*.

# Phylum Protozoa

2. **Order Trichostomatida.** Large ciliates with vestibular but no buccal ciliature.

Examples: *Colpoda* (Fig. 1.113A), *Balantidium* (Fig. 1.113B), *Tillina*, *Nyctotherus* (Fig. 1.113C), *Loxophyllium* (Fig. 1.115F).

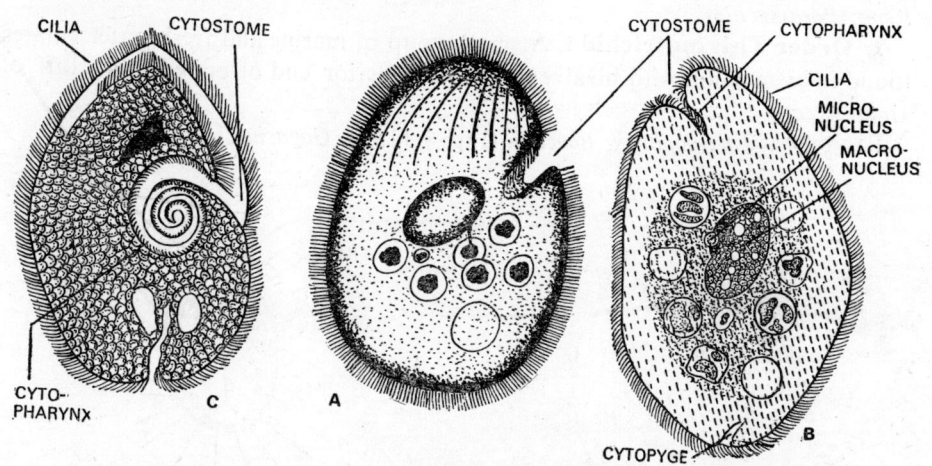

FIG. 1.113. A—*Colpoda*; B—*Balantidium*; C—*Nyctotherus macropharyngeus*.

*Balantidium* (Fig. 1.113B), an intestinal inhabitant of various animals chiefly of frogs and mammals, has a large peristomial depression at the anterior end of the oval ciliated body. It possesses a contractile vacuole also. This is perhaps related to the fact that *Balantidium*, unlike many parasitic forms, feeds by means of a mouth. *B. coli*, whose natural host is pig, has also been found in man. In man it is occasionally harmful. It is transmitted in encysted state. Man probably becomes infested through too loose association with infected animals.

3. **Order Chonotrichida.** Vase-shaped ciliates lacking body cilia. A funnel at the free end of the body bears vestibular cilia. Chiefly marine or ectocommensal on crustaceans.

Examples: *Spirochona* (Fig. 1.114A), *Lobochona*, *Chilodochona*.

4. **Order Apostomatida.** Ciliates with spirally arranged ciliation on the body. Midventral cytostome lies in the vicinity of a peculiar rosette. Marine parasites or commensals with complex life-cycles usually involving two hosts, one of which is commonly a crustacean.

Examples: *Hyalophysa*, *Foettingeria*.

5. **Order Astomatida.** Commensals or endoparasites living chiefly in the gut and coelom of Oligochaete worms. Cytostome absent. Body ciliation uniform.

Examples: *Hoplitophrya*, *Anoplophrya*.

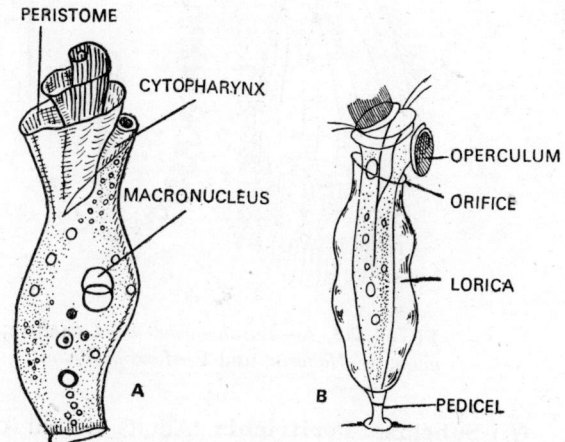

FIG. 1.114. A—*Spirochona gemmipara*; and (*after* Hertwig); and B—*Pyxicola casteri*; a loricate peritricha.

6. **Order Hymenostomatida.** Small ciliates having a uniform body ciliation but possessing a buccal cavity. Buccal ciliature comprises undulating membrane and an aboral zone of membranelles.

Examples: *Tetrahymena* (Fig. 1.115 A), *Cyclidium* (Fig. 1.115 B), *Blepharostoma, Paramecium.*

7. **Order Thigmotrichida.** A small group of marine and fresh water ciliates found in association with bivalve molluscs. Anterior end of body bears a tuft of thignotractic cilia.

Examples: *Thigmophrya, Boveria,* (Fig. 1.115C), *Gagarius.*

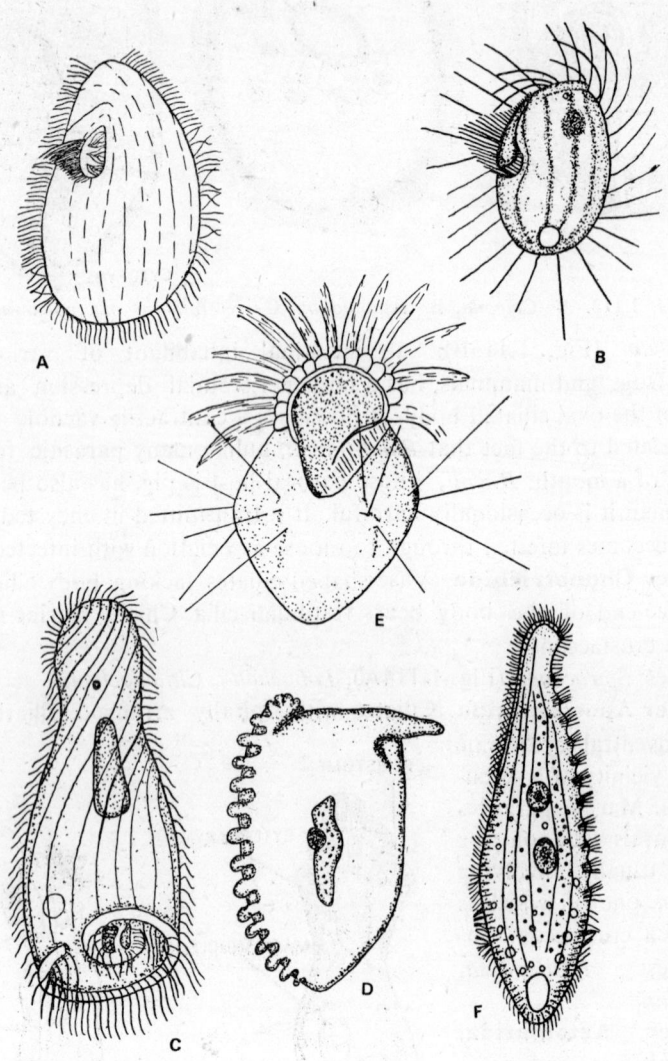

FIG. 1.115. A—*Tetrahymena*; B—*Cyclidium*; C—*Boveria*; D—*Entodinium*; E—*Halteria*; and F—*Loxophyllium.*

(*ii*) **Subclass Peritrichia.** Adults devoid of body cilia, but the apical end of the body typically bears a conspicuous buccal ciliature. Mostly stalked ciliates.

# Phylum Protozoa

**Order Peritrichida.** The Peritrichia have three extensively developed rows of cilia (one row forming an undulating membrane) wind clockwise into the buccal cavity. Some peritrichs are stalked, sessile organisms.

Examples: *Vorticella* (Fig. 1.102), *Carchesium*, *Zoothamnium*, *Trichodina*, *Epistylis* (Fig. 1.116), etc.

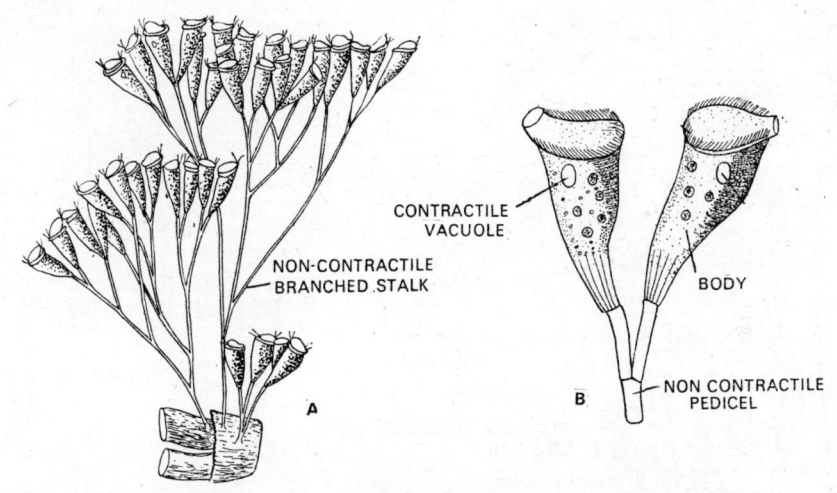

FIG. 1.116. *Epistylis*. A—*Epistylis anastatica*; and B—*E. galea*.

(*iii*) **Subclass Suctoria.** A typical ciliate whose relationship with true ciliates is not fully apparent. The distal end of the body bearing few to many tentacles. Adults without cilia, larval forms ciliated, double nuclei present.

**Order Suctorida.** Sessile, stalked ciliates with the distal end of the body bearing tentacles provided with structures called haptocysts which help in sucking out protoplasm of the prey (usually a ciliate).

Examples: *Acineta*, *Podophrya* (Fig. 1.108), *Ephelota*, etc.

(*iv*) **Subclass Spirotrichia.** The ciliates with generally reduced body cilia and well developed buccal ciliature.

1. **Order Heterotrichida.** With uniform body cilia or body encased in a lorica. Body cilia absent in loricate forms.

Examples; *Bursaria*, *Stentor* (Fig. 1.117), *Blepharisma*, *Spirostomum* (Fig. 1.118B), etc.

*Stentor* has an elegant trumpet-shaped body with a broad peristome at the anterior end. The peristome is encircled by a row of menbranelles, which leads by way of vestibule into the spirally coiled cytopharynx. *S. coeruleus*, the blue species is characterised by alternating blue and white ectoplasmic stripes. The stentors are generally attached but often break loose and swim in a semi-contracted condition. They feed chiefly on small ciliates. The macronucleus is very elongated and moniliform. The number of micronucleus is numerous (80 in *S. coeruleus*). There is single contractile vacuole fed by many canals.

2. **Order Oligotrichida.** Small ciliates with body cilia reduced or absent. Conspicuous buccal membranenelles commonly extending around apical end of body.

Examples: *Halteria* [Fig. 1.115(E)].

3. **Order Tintinnida.** Loricate mostly free swimming ciliates with conspi-

cuous oral membranenelles when extended. Chiefly marine.
Examples: *Codonella, Tintinnus, Favella, Tintinnopsis* (Fig. 1.118B).

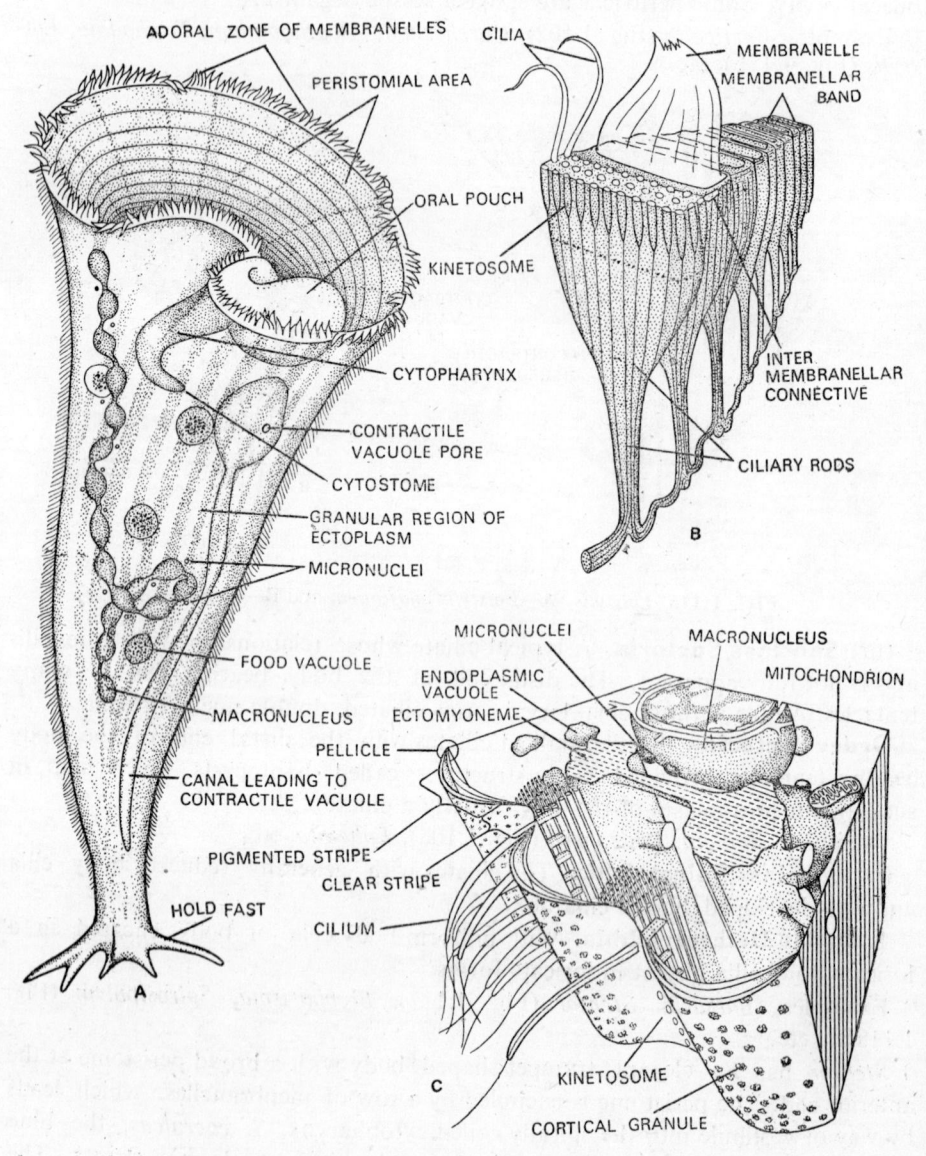

FIG. 1.117. A—*Stentor*; B—A few membranellae under electron microscope; and C—An electron microphotograph of a section passing through a portion of *Stentor*.

4. **Order Entodiniomorphida**. Ciliates that occur as entocommensals in the digestive tract of herbivorous mammals. Body cilia reduced or absent, buccal ciliature prominent, posterior end may be drawn out into spines.
Examples: *Endodinium, Elephantophilus, Cycloposthium*.

5. **Order Odontostomatida**. A small group of laterally compressed and wedge-shaped ciliates with carapace and reduced body and buccal cilia.
Examples: *Saprodinium* (Fig. 1.119B).

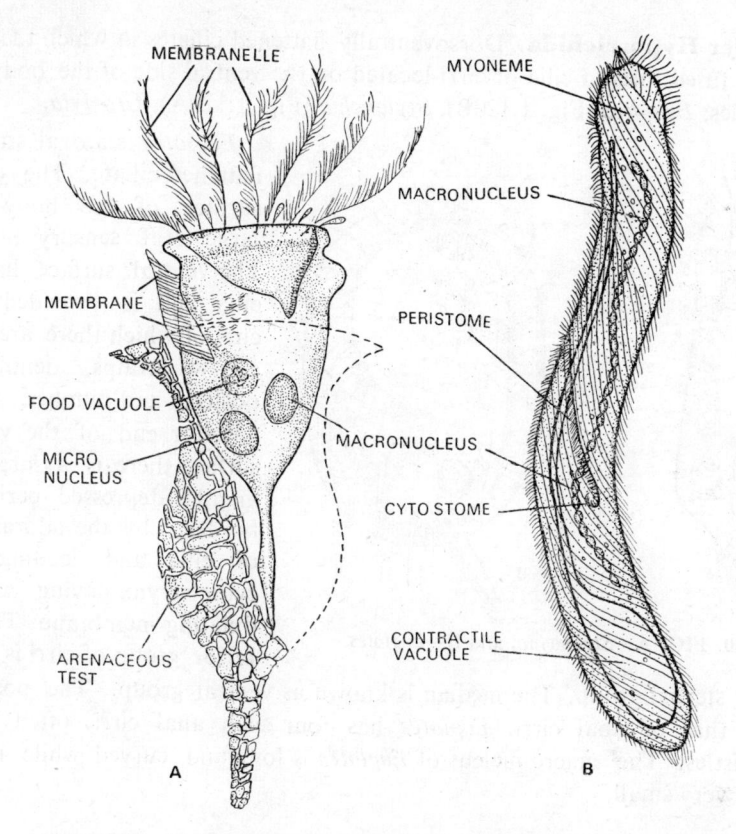

FIG. 1.118. A—*Tintinnopsis campanula*—a ciliate which forms a lorica by cementing sand grains together; and B—*Spirostomum*.

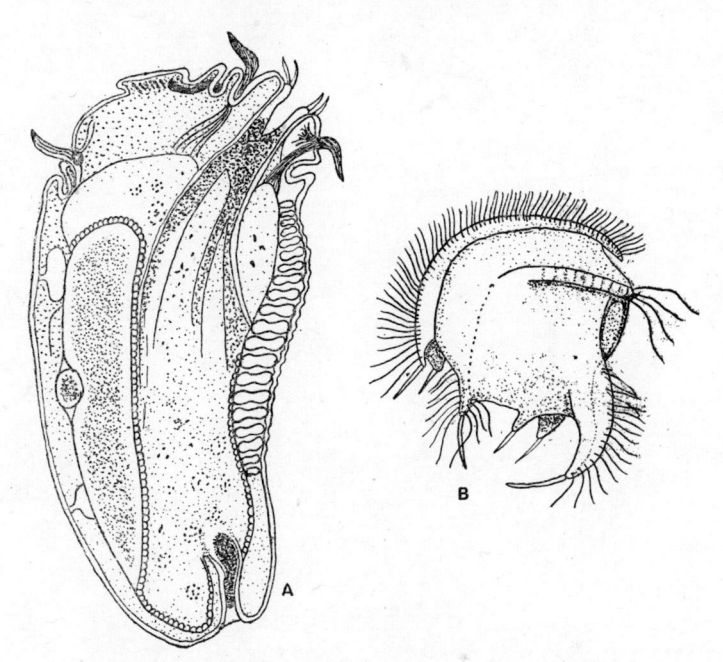

FIG. 1.119. A—*Diplodinium caudatum*, a commensal in the rumen of cattle; and B—*Saprodinium dentatum*.

6. **Order Hypotrichida.** Dorsoventrally flattened ciliates in which the body cilia form fused tufts of cilia or cirri located on the ventral side of the body.

Examples: *Euplotes* (Fig. 1.120B), *Stylonchia* (Fig. 1.120A), *Urostyla*.

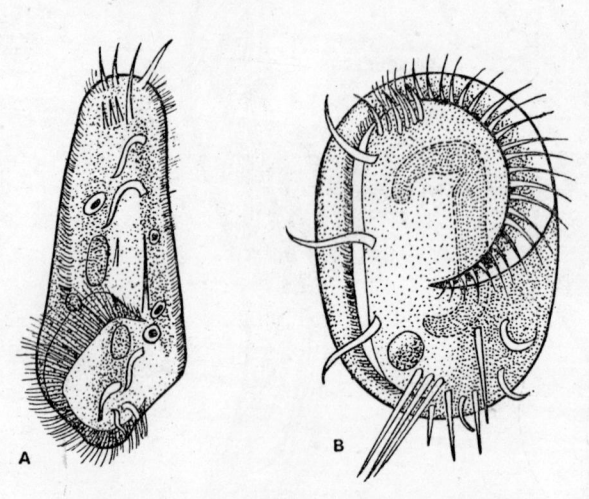

*Euplotes* is a oval strongly flattened ciliate. The dorsal surface of the body has large stiff sensory bristles. The ventral surface has no cilia and is provided with cirri of which there are only a few groups, definite in position and number. At the anterior end of the ventral surface there is a large triangular depressed peristome bordered by the adoral zone of cilia and leading into cytopharynx having an undulating membrane. The anterior group of cirri is called frontal or sternal group. The median is known as ventral group. The posterior group is that of anal cirri. *Euplotes* has four such anal cirri, often called caudal bristles. The macronucleus of *Euplotes* is long and curved while micronucleus is very small.

1.120. FIG. A—*Stylonchia*; and B—*Euplotes*.

# 2 PHYLUM MESOZOA

The **Mesozoa** is a minor but puzzling (enigmatic) phylum of many-celled animals. There are about fifty known species: all are minute parasites found in the body-cavities of various higher invertebrates. While they could be regarded as the simplest organised among many-celled animals, they are thought by some to be extremely degenerate derivatives of flatworms. They are not unlike the planula larva of coelenterates in general form, but although they have a solid two-layered construction, the inner cells are all reproductive cells, with a single type of ciliated cells forming an outer layer around them. Their life-cycle is always complicated and usually seems to involve an alternation of asexual and sexual generations. Some distinguished workers have suggested that the similarities to protistan structure and function in Mesozoa are real, and that it is improbable that they could arise by degeneration of the flatworm body form. If this could be demonstrated to be so, then the question would arise whether the Mesozoa are (like the Parazoa) independently evolved from protistans, or whether they are in some way related to intermediate forms between Protista and true Metazoa. If the latter is true, then the group is of great phylogenetic significance. In any case, development and life-cycle in the group merit more modern investigation.

The Mesozoa, generally, have a solid two-layered construction, but their two layers do not correspond to ectoderm and endoderm of the Metazoa, because the inner layer (of one or more cells) is reproductive and not digestive. The phylum is divisible into two orders: (1) the **Dicyemida** or **Rhombozoa** and (2) **Orthonectida**.

## ORDER DICYEMIDA

The Dicyemida occur as parasites in the cephalopod molluscs (squids and octopuses). The body of the parasite is worm-like and up to 6-7 mm long. It consists of an outer layer of ciliated somatic cells or **somatoderm** enclosing a single inner cell, the **axial cell** (Fig. 2.1). A slightly distinct anterior **head** can be distinguished from an elongated trunk. The number of cells is almost constant for a species. The head region has eight or nine cells, the **collar cells**, and are

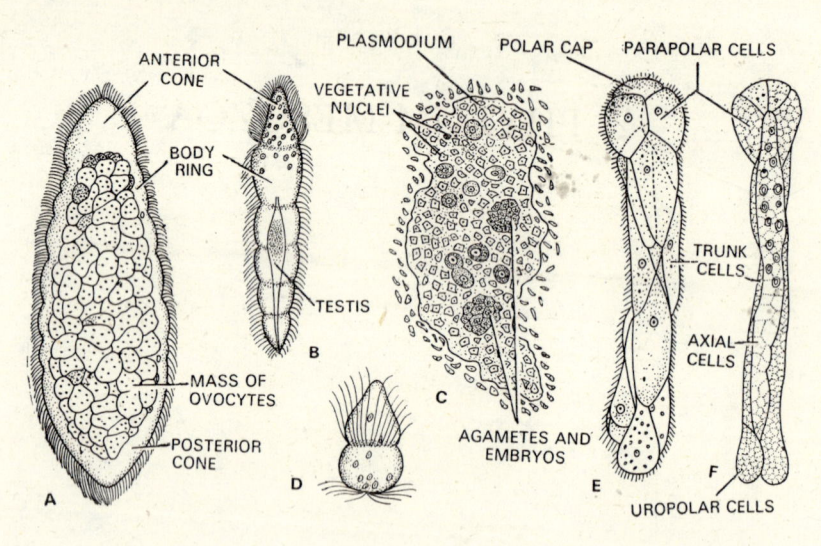

FIG. 2.1. A—*Rhopalura granosa*; B—*Rhopalura granosa*, male; C—Male plasmodium of *Rhopalura*; D—Embryo of same; E—Nematogen of *Pseudicyema truncatum*; and F—Optical section of same.

arranged in two tiers forming what is known as the **polar cap** or **calotte** (Fig. 2.1). The remaining 12-17 trunk cells are large highly vacuolated provided with long cilia and often containing inclusions. Sometimes the two anterior trunk cells just next to the polar cap are described as **parapolar** and are intermediate in size and appearance; whereas the last two trunk cells are known as uropolar cells and are rich in inclusions. The axial cell is elongate vermiform or cylindrical with vacuolated cytoplasm. This originally uninucleate cell soon undergoes division and forms reproductive bodies, which are nourished by a part of the cytoplasm that retains vegetative function. The young dicyemids swim about in the fluid of the host kidneys but the mature forms attach to the spongy tissues of the kidneys by the cilia of the polar cap and do not damage the tissue, it is presumed. They absorb nutritive substances present in the nephridial fluids.

Their life-history is supposed to be complicated but not clearly understood. According to those who believe that the life-cycle is complicated presenting alternation of sexual and asexual phase the ordinary form of parasite described above is called the **primary nematogen** (Gersch). The reproductive bodies contained in the axial cell are agametes which multiply, develop into new nematogens and finally leave the mother nematogen and wander about in the kidney. Fresh nematogen may produce another generation of nematogens. But when the host attains sexual maturity the parasite assumes a new form, the **rhombogen**, essentially similar to nematogen differing only in cell-inclusions and nature of offspring they produce. Within the rhombogens special clusters of cells, **infusorigens**, are produced from the surface of which bud off pseudoeggs that multiply to produce free-swimming ciliated larvae, which leave the host and escape into the sea-water. Their furhter history is unknown but it is presumed that they enter some intermediate host where the sexual phase of the life-cycle occurs. The mode of infection of the young cephalopods is also not understood.

## ORDER ORTHONECTIDA

The Orthonectida occur as parasites in Turbellaria, brittle-stars, annelids, nemerteans and clam. Unlike the dicyemids the asexual stage consists of a multinucleate amoeboid plasmodium (Fig. 2.1), located in the tissues and spaces of the host which are badly damaged. The plasmodia reproduce by simple fragmentation (**plasmotomy**) for sometime and finally produce male and female forms asexually. Usually a plasmodium produces forms of one sex only but those producing forms of both sexes of even hermaphrodites are also common. The sexual forms appear like dicyemid nematogens. The outer ciliated layer in this case appears ringed due to the presence of circular grooves at intervals. The males are smaller (about half the size) than the females, which also have genital openings behind the middle of their bodies. The sexual forms escape in sea-water where fertilization occurs by contact. The fertilized egg develops into ciliated larva (like the larva of dicyemids) within the body of the mother. The larva escapes into the sea-water reinfects new hosts of the same kind, wherein, the larva loses the somatic layer, the interior cells scatter, each one giving rise to a plasmodium. Majority of the species belong to the genus *Rhopalura*.

# 3 PHYLUM PORIFERA—SPONGES

The **Porifera** (Latin *porus*, pore+*ferre*, to bear) are lowly multicellular animals incapable of movement as they are attached to the substratum like a plant (Fig. 3.1). To the casual observer they show plant-like unresponsive behaviour. That is why for a long time they were variously described as animals, plants, both animal and plant, and even as non-living substances secreted by many animals that take shelter in the cavities of sponges. In fact, it was not until about one hundred years ago that the last sceptics were finally convinced of the true animal nature of sponges. Sponges are not regarded to be on the direct line of evolution, they appear to form an evolutionary dead end. This is often emphasised by classifying them in a separate sub-kingdom, the **Parozoa** (Gr. *para*, beside) which simply means that they form just a side issue. They differ on one hand from the rest of many celled animals (**Metazoa**), and on the other from acellular animals. In terms of numbers of living forms, the sponges (of which there are probably more than 4000 species) do not constitute a minor group. Ecologically they make up a measurable part of the biomass in relatively shallow waters in some areas of the seas. A few species have colonised certain fresh water habitats. The marine forms are found in shallow water, and up to depths of more than 5600 metres. What is commonly used as bath sponge is really the skeleton of sponges which were even used by the ancient Greeks for bathing and for scrubbing tables and floors. Today sponges have a wide variety of uses with the result that a full-fledged "sponge fishing" industry has grown. It produces over one thousand tons of sponges every year.

The sponges are of various shapes. They are flat crusts, vase-like, branched globular or goblet-like. Many are dark-coloured or grey, whereas, others are

# Phylum Porifera—Sponges

brilliant red, orange, yellow, blue, violet or black. The Porifera are multicellular organisms invariably sessile and aquatic. It is with the Porifera, therefore, that the description of cellular-layered animals, is begun, but the cellular grade here is at its lowest. Most of the diagnostic characteristics of the phylum are negative.

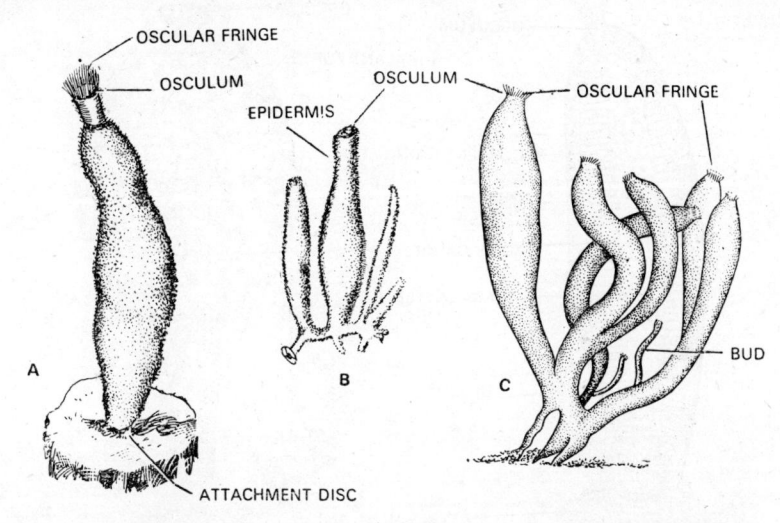

FIG. 3.1. External features of sponges. A—Simple sponge *Leucosolenia* single upright tube; B—A tuft of upright tubes; and C—Syconoid sponge *Grantia*.

The cells do not form perfect tissues. On the contrary, they look like mere loose aggregations of cells without definite form or symmetry. Definite organs or systems for the co-ordination of the body functions are lacking. The various body functions are performed by the activities of individual cells more or less independently. Some sponges having a vase-like shape are radially symmetrical, whereas, the majority grow irregularly in a plant-like manner forming flat, rounded or branching structures without any symmetry. The sponges are without a true gut. The only internal space is the so-called **paragaster,** often in the form of a series of cavities, pores, canals and chambers through which water is caused to flow. This water current which subserves all functions of feeding, respiration, excretion and reproduction of the sponge cells is created by the peculiar cells which line the paragaster. These are the **collared flagellate** cells or the **choanocytes.** The body wall has numerous pores through which water passes in **(incurrent openings)** and one or more larger openings through which water passes out **(excurrent openings).** Nearly all sponges possess an internal skeleton consisting of separate crystalline bodies, the **spicules,** or of organic fibres or of both. The spicules are generally calcareous and sometimes siliceous or horny. Sponges have power of regeneration. Asexual reproduction takes place by the formation of special cell-masses, the **gemmules**. Sexual reproduction is quite common and takes place by the union of typical eggs and sperms. A free swimming larva is produced.

Usually *Sycon gelatinosum* is described as a type of sponges, but it is not the simplest type of sponge, it is an intermediate form between the simplest and the most complicated forms. Thus, to give an idea of simplest type of sponge structure *Olynthus* is described first. *Olynthus* (Fig. 3.2) is vase-shaped, contracted at the base to form a short stalk, expanded at the extremity below to form an attachment disc. Its wall is very thin and the pores on the surface (**inhalant**

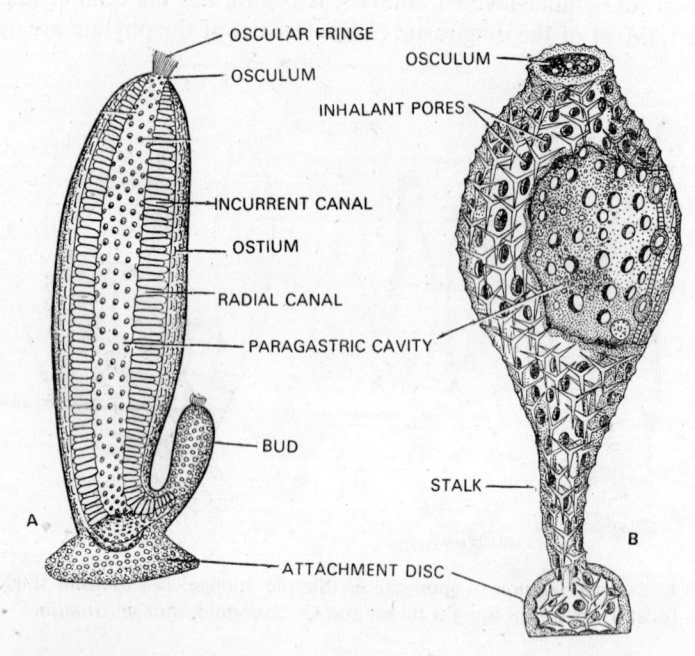

**FIG. 3.2.** Internal structure. A—Diagrammatic longitudinal section of Sycon; and B—*Olynthus* a larval form.

**pores**) directly open into the paragastric cavity. The structure of *Olynthus*, is, no doubt, simple but it is not a mature form. It is simply a passing larval phase of simple calcareous sponges such as *Clathrulina*. There is another calcareous sponge *Leucosolenia*, which has similar structure to that found in *Olynthus*. It has thin walls traversed by simple canals in the adult stage, therefore, here *Leucosolenia* has been described as type.

## PHYLUM PORIFERA

| | |
|---|---|
| **Class:** | CALCAREA OR CALCISPONGIAE |
| **Subclass:** | CALCARONEA |
| **Orders:** | Leucosolenida (Homocoela) |
| | Sycettida (Heterocoela) |
| **Subclass:** | CALCINEA |
| **Orders:** | Clathrinida |
| | Leucettida |
| | Pharetronida |
| **Class:** | HEXACTINELLIDA OR HYALOSPONGIAE |
| **Orders:** | Hexasterophora |
| | Amphidiscophora |
| **Class:** | DEMOSPONGIAE |
| **Subclass:** | TETRACTINOMORPHA |

# Phylum Porifera—Sponges

**Orders:**  Homosclerophorida
Choristida
Epipolasida
Hadromerida
Axinellida
Lithistida
**Subclass:** CERACTINOMORPHA
**Orders:** Halichondrida
Poecilosclerida
Haplosclerida
Dictyoceratida
Dendroceratida

### TYPE LEUCOSOLENIA

*Leucosolenia* (Fig. 3.3A), represents the simplest of sponges attached to sea shore rocks just below the low tide line. It consists of a tuft of slender vase-shaped upright tubes united at their bases by irregular horizontal tubes. Each vase-shaped sac has a large excurrent opening, the **osculum** (Figs. 3.1, 3.2) at the top and numerous small openings scattered all over the surface acting as incurrent pores (Fig. 3.3), and are known as the **ostia**. The osculum is surrounded by a fringe of larger spicules called **oscular** fringe (Figs. 3.1, 3.3). A closer examination reveals that the surface is not smooth, as it appears, but is made up of innumerable elevations of polygonal shape (**ectoderm**), which fit together like the tiles in a mosaic (Fig. 3.3). Minute spicules project from the outer surface.

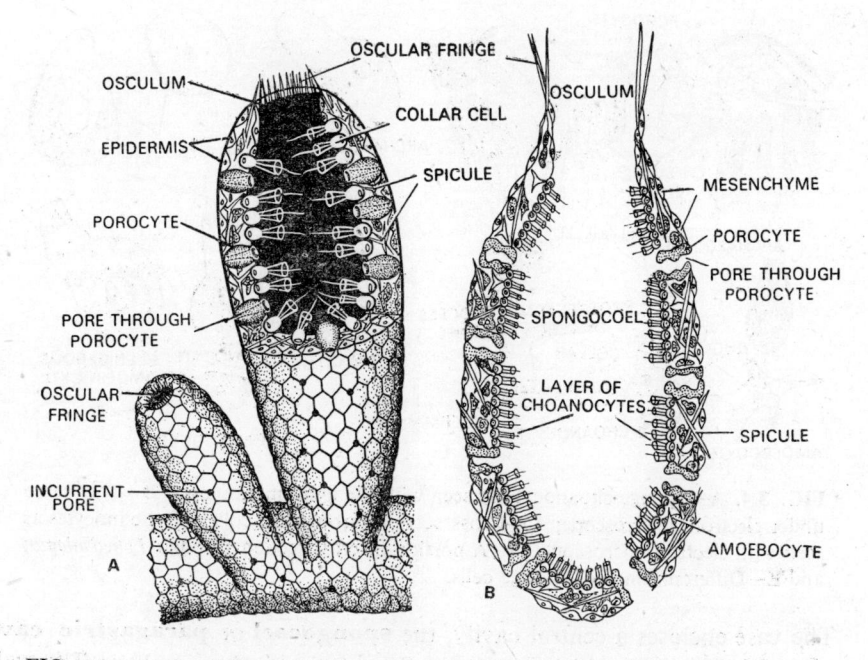

FIG. 3.3. A—*Leucosolenia* surface covered by flat hexagonal epidermal cells (*after* Dendy); and B—Diagrammatic longitudinal section of the same showing osculum, porocytes, spicules and collar cells.

*Sycon gelatinosum* also consists of a tuft (one to three inches long) of branching cylinders (Fig. 3.1C) of grey or light brown colour. All the cylinders are connected together at the base at which the sponge is attached to the substratum. To a layman these two sponges may appear similar but *Sycon* has got a more complicated canal-system and is not a simple form.

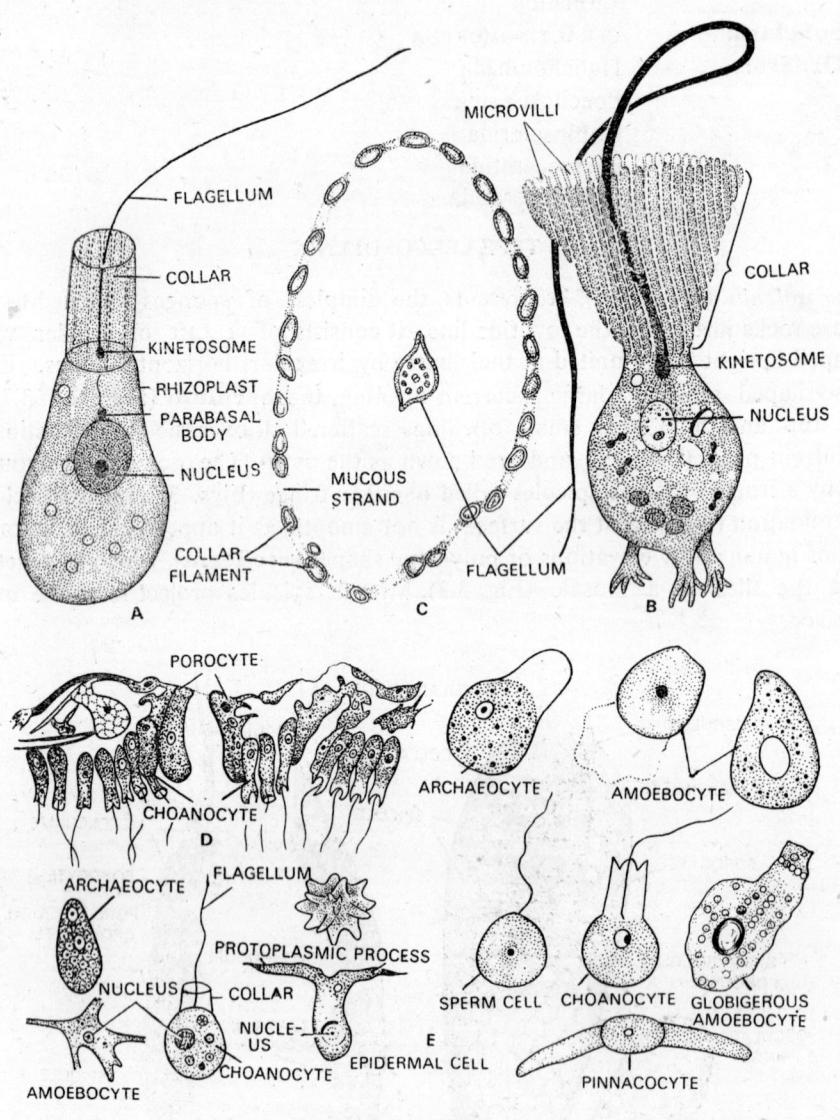

FIG. 3.4. A—Sponge choanocyte as seen with the light microscope; B—The same under electron miscroscope; C—Cross section through the collar of choanocytes as seen with electron microscope; D—A portion of cross section through *Leucosolenia*; and E—Different kinds of Sponge cells.

The vase encloses a central cavity, the **spongocoel** or **paragastric cavity** or **cloaca** (Fig. 3.2) which opens to the exterior at the osculum. The wall is pierced by numerous microscopic apertures, the **incurrent pores** or **ostia** which extend from the external surface to the spongocoel. Each pore runs

# Phylum Porifera—Sponges

through a tubular cell, the **porocyte** (Fig. 3.3), as an intracellular canal. The wall consists of an outer and an inner epithelium between which is located a mesenchyme. The outer epithelium is termed **epidermis** and consists of a single layer of thin flat cells. The inner epithelium, lining the spongocoel, is composed of a single layer of a special kind of cells, the **flagellated collar cells** or **choanocytes** (Figs. 3.3, 3.4). These cells are so called because its free end is encircled by a delicate collar of protoplasm (Fig. 3.4B) through which projects a flagellum. The beat of the flagella of the collar cell[1] creates a water current which passes through the incurrent pores into the spongocoel and out through the osculum. The choanocytes form a complete continuous lining of the spongocoel interrupted only by the inner ends of the porocyte (Fig. 3.3).

The inhalant pores and apopyles are surrounded by elongated contractile cells, often prolonged into narrow fibres (Fig. 3.4). They bring about the closure of the apertures and are considered to be of the nature of muscle fibres. The osculum is also surrounded by a similar set of fibres forming the **oscular sphincter.**

Between the covering cells and the layer of collar cells is a non-living jelly-like material containing living **mesenchyme cells.** Some mesenchyme cells secrete **calcareous spicules** that form the skeletal support of the wall and keep the vase erect. If a drop of acid is poured over the spicules they will dissolve with effervescence thereby confirming that they are made up of carbonate of lime. They are slender elongated microscopic rods pointed at both ends or are three-rayed. Other cells move about in amoeboid fashion and are known as the **amoebocytes.** These are the least specialised cells and can develop into any of the more specialised ones. They receive partly digested food particles from the cells, complete the digestion, and carry the digested food from one place to another. They probably also transport waste material to surface, from where it can be carried away by the outgoing current of water.

## Canal-system in Sponges

The most characteristic feature of the sponges is the presence of a system of canals through which water passes into the spongocoel. These canals traverse the wall of the vase in a simple or complicated manner. The evolution of small simple types of sponges into complex forms revolves chiefly around the problem of increasing the surface proportion to the volume. This problem has been solved by a simple folding of the body wall (Fig. 3.5A) which increases the surface and makes the canal system complicated. Three types of canal systems are found in sponges, **asconoid, syconoid** and **leuconoid**.

**Asconoid canal-system.** In the simplest-types of sponges the canals originate from certain pores or **ostia** (also called **incurrent** or **inhalant pores**) on the surface of the wall and lead directly to the spongocoel. Each canal is intracellular and runs through a tubular cell called **porocyte.** (Fig. 3.3). Such a simple of canals is described as **asconoid** or **ascon type**. It is found in *Leucosolenia*, described here, and in *Olynthus* (Fig. 3.2), the passing larval stage of

---

[1] The collar cells look and behave almost exactly like the collar flagellates. For this reason it has been suggested that sponges have evolved from the same group of ancestral Protozoa that gave rise to the living Choanoflagellates.

some sponges such as *Clathrulina*. The flagella produce water currents that pass through these incurrent pores into the spongocoel and carry food and oxy-

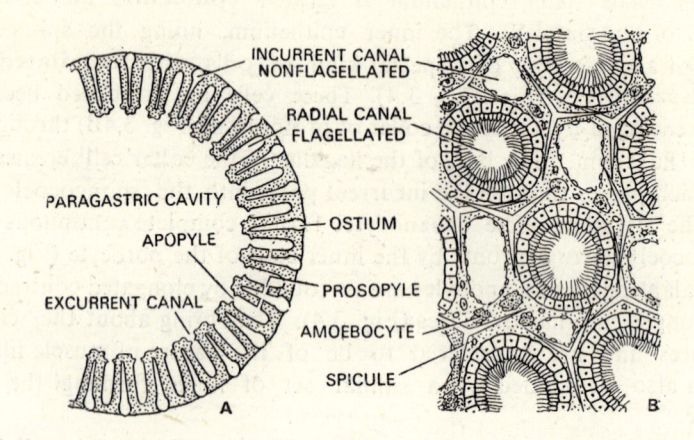

FIG. 3.5. A—A portion of the transverse section of *Sycon gelatinosum* through the wall of a cylinder; and B—Section through the wall of a cylinder taken at right angles to the long axis of the canal.

gen along. The currents pass out through the osculum carrying away metabolic waste products. Sponges with such a simple structure are described as asconoid. Such a type occurs only in a very few sponges, the vast majority show a more complicated construction.

**Syconid canal-system.** The first stage above the asconoid type is termed the **syconoid** type as represented in *Sycon*, taken as a type in many texts. The syconoid sponge can be easily derived (theoretically) from the asconoid condition. The walls of the asconoid sponge become folded up in such a way that a number of finger-like processes appear at regular intervals. Between two such projections a canal lined by ectoderm is formed. This is the **incurrent canal**. Internally the projections lodge the **radial canals** lined by choanocytes (Fig. 3.5), hence, also called the **flagellated canals**. Further the walls of the radial canals fuse in such a way that the incurrent canals appear to be opening to the exterior by minute apertures (between the tips of two radial canals) known as the **dermal pore**. The original ostia or the incurrent pores of the asconoid type are, in this type, situated within the incurrent canal and are termed the **prosopyles**. Thus, the thick wall of the syconoid sponge carries alternating incurrent and radial canals and the choanocyte layer no longer lines the whole interior but is limited only to the radial canals, which open into the spongocoel by means of minute openings known as the **apopyles**. Two stages of syconoid structure are met with. In the first state (*Sycon*) the external surface is made up of the blind outer ends of the radial canals the spaces between which serve as dermal ostia. In the second stage the epidermis and mesenchyme spread over the outer surface forming a thin or thick **cortex**, often containing special cortical spicules. The cortex is pierced by definite pores that lead into the narrowed incurrent canals directly or form large cortical spaces, the **subdermal spaces**, from which water is drawn into the flagellated chambers (Figs. 3.5 and 3.7).

**Leuconoid canal-system.** As a result of further outfolding of the choanocyte layer the third or **leuconoid** type of canal-system is produced. In this the radial canal of the syconoid structure becomes divided into a cluster of small

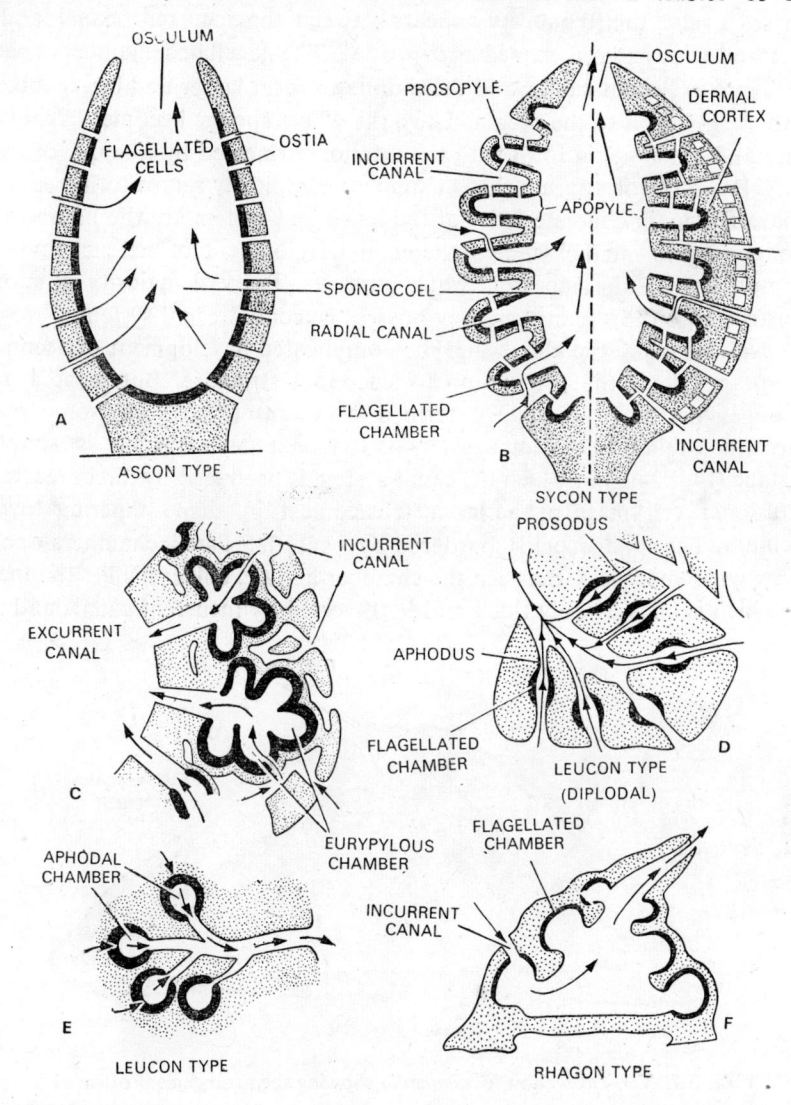

FIG. 3.6. Canal system in sponges. A—Simple ascon type of canals; B—Sycon-type (or syconoid) canal system; C, D and E—Leuconoid canal system with eurypylous chamber; diplodal chambers and aphodal chambers, and F—Rahagon type canal system.

rounded or oval flagellated chambers (Fig. 3.6D) by further foldings. Thus, the simple elongated flagellated chamber of the syconoid type is replaced by several small chambers to which only the choanocytes are limited. Mesenchyme increases and fills the space around the flagellated chambers. The outer surface is usually covered by an epidermal epithelium pierced by dermal pores and oscula. The incurrent passages leading from the dermal pores branch irregularly in the mesenchyme. The spongocoel is more or less obliterated. Sometimes the apopyles open directly into the excurrent channels. When it is so the system is called **eurypy-**

**lous** (Fig. 3.6C). Sometimes a narrow canal, the **aphodus**, intervenes between the chamber and the excurrent channel, such a system is called **aphodal**. In some cases a tube, the **prosodus** appears between the incurrent canal and the chamber and the system is named as **diplodal**. The flagellated chambers open by apopyles into excurrent canals and these unite to form larger and larger tubes of which the largest lead to the oscula. Thus, the whole sponge becomes irregular in structure and indefinite in form with the interior permeated by a maze of water channels. From this it appears that in sponges complexity serves to increase the efficiency. The main characteristics of the leuconoid system are the limitation of the choanocytes to small chambers, the great development of mesenchyme, and the complexity of the incurrent water passages. The vast majority of sponges are constructed on this plan, *i.e.*, they possess leuconoid canal-system.

It is evident from the above that the complicated development of ascon and sycon types results in the production of leuconoid structure. But most leuconoid sponges derive from a stage termed a **rhagon** (that is why some people have described this type of canal-system as rhagon type). Rhagon is simply a larval stage (Fig. 3.6F) in which the canal-system is produced by direct rearrangement of inner cell mass of the larva. It is conical in shape tapering towards the osculum. The spongocoel is bordered by oval flagellated chambers opening into it by wide apopyles. Between the chamber and the epidermis lies the mesenchyme which is considerably thick and is traversed by incurrent canals and sub-

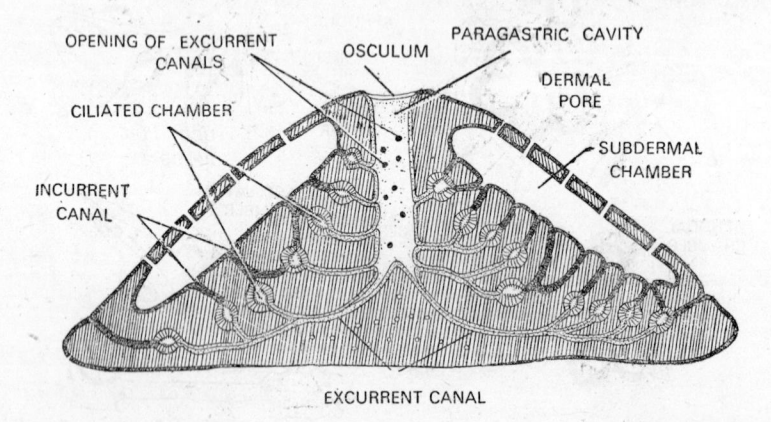

FIG. 3.7. Vertical section of *Spongilla* showing the arrangement of canal system.

dermal spaces. The leuconoid types develop directly from the rhagon by evaginations of the flagellated chambers and other subsequent developments.

The beat of the flagella of the choanocytes creates a pressure which draws water current through the dermal pore and the canal. This water carries along with it food particles, oxygen and sweeps away metabolic waste products. The function of the canal-system, therefore, is ingestion, respiration and excretion. The food particles brought by the current are ingested by the choanocytes and digested; oxygen brought by the current is exchanged by the cells and waste product is passed on to it, to be carried out.

In the simplest type there are lesser number of flagellated cells but as the canal-system becomes complicated, the number of flagellated cells increases and

*Phylum Porifera—Sponges* 163

the force to draw the current is also increased. The syconoid structure, therefore, is more efficient than the asconoid, and the leuconoid system still more. Increase in the total area of the choanocyte surface, thus, enhances the efficiency of three vital functions—ingestion, respiration and excretion.

## Histology

The cells of sponges are quite simple (Fig. 3.4). Except choanocytes all others are only slight modifications of simple amoeboid cells of mesenchyme. The surface epithelium (**epidermis**) in some sponges consists of large flat polygonal cells, the **pinacocytes**, each with a thickened central enlargement containing

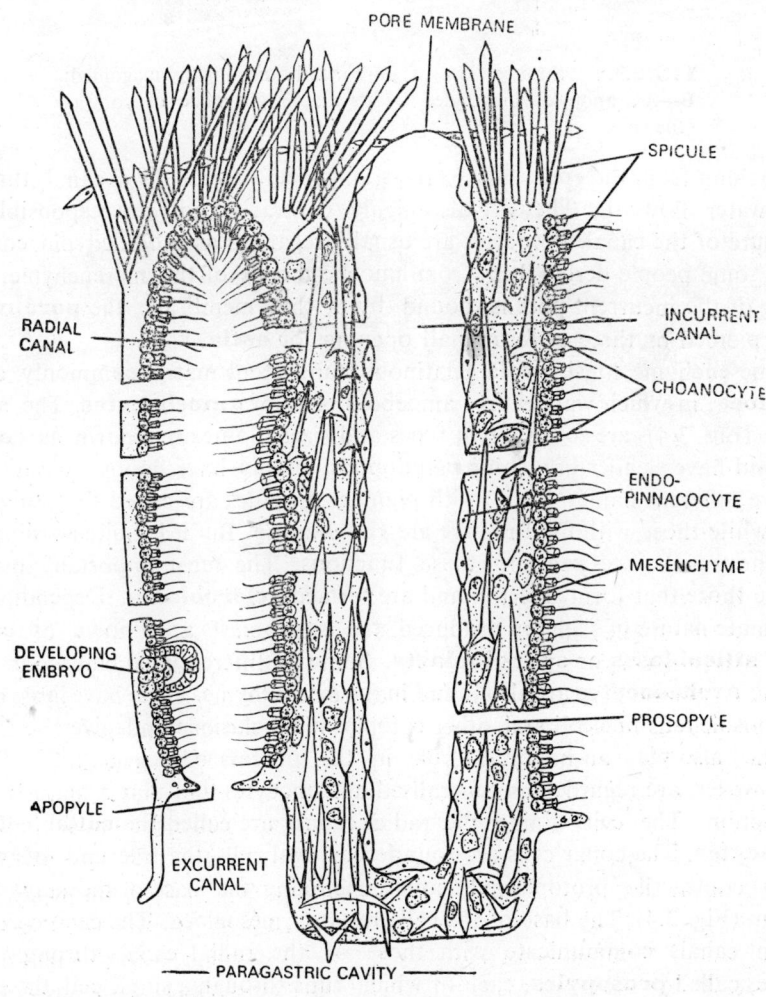

FIG. 3.8. A portion of cross section of *Sycon*.

the nucleus. From above these cells appear closely fitting flat polygonal cells (Fig. 3.3) and from side they present a central nucleated lump and thin margins continuous with those of adjacent pinacocytes. These cells also line the incurrent canals and spongocoel. They are highly contractile and are capable of reducing the surface greatly. In many sponges the epithelium is not cellular, but a continuous membrane containing scattered nuclei (syncytium) is known as **epithe-**

**lioid membrane.** In the simple asconoid sponges the epidermis is interrupted by large **pore-cells** or **porocytes** at frequent intervals (Fig. 3.3). They are tubular

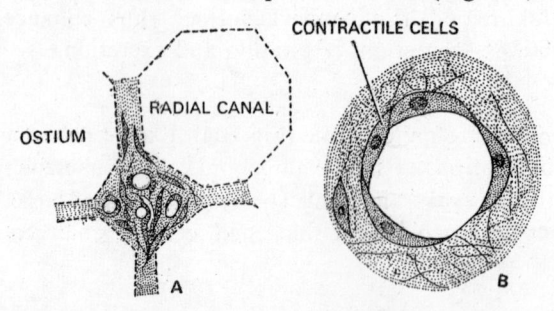

FIG. 3.9. A—Surface view of pore-membrane, highly magnified; B—An apopyle surrounded by its diaphragm made of contractile cells.

cells reaching from the epidermis to the spongocoel pierced by a canal through which water flows in. They are also highly contractile and are responsible for the closure of the canal. Porocytes are usually regarded as modified pinacocytes, though some people derive them from amoeboid cells of the mesenchyme. The opening of the incurrent pore is bound by a thin membrane, the **pore membrane**, pierced by three or four small openings or **ostia**.

The mesenchyme consists of a gelatinous transparent matrix commonly called **mesogloea**, in which wander free amoeboid cells or **amoebocytes**. The amoebocytes (Fig. 3.4) are of different types and sizes. Some are known as **collencytes** and have slender branching pseudopods, others have lobose pseudopods and have different names. Those with pigment granules are called the **chromocytes**, while those with food reserve are **thesocytes**. But it is quite possible that any of the amoebocytes can serve these functions. The most important amoebocytes are those that form skeleton and are termed **scleroblasts**. Depending on the ultimate nature of skeleton produced, the scleroblasts are known as **calcoblasts, silicoblasts** or **spongioblasts**. Some undifferentiated embryonic cells known as **archaeocytes** are also found in the mesenchyme. These have large nuclei with conspicuous nucleoli and other cytoplasmic inclusions and give rise to sex cells. They also play an important role in the process of regeneration. Those cells, however, are regarded as generalised amoebocytes that play a major role in reproduction. The cells lining the radial canals are called the **collar cells** or **choanocytes**. The collar cells are rounded or oval cells the free end of which bears a contractile protoplasmic collar encircling the base of the single long flagellum (Fig. 3.4). The base of the cells rests on mesogloea. The cavities of the incurrent canals communicate with those of the radial canals through small apertures called **prosopyles**, each of which runs through a single cell, the **porocyte**. Many years ago, careful light microscopy revealed that the choanocytes were also involved in the ingestion of food particles, the site of ingestion being invariably around the outside of the collar (Fig. 3.10). Until recently it was difficult to understand how these food particles got there. Electron microscopy has given a clear picture of ultrastructure of the collar, which seems to resemble nothing so much as a rather open palisade or picket-fence. It is evident from the study of the reconstruction that the water which is impelled by the

# Phylum Porifera—Sponges

choanocyte toward the centre of the paragaster must first pass *through* the tiny gaps in the microscopic fence. In living healthy choanocytes the "collar" or "fence" is withdrawn at intervals. Thus, the choanocyte not only creates the water current using its flagellum, but also filters particles from the water with its "collar" and then ingests the particles by amoebic action.

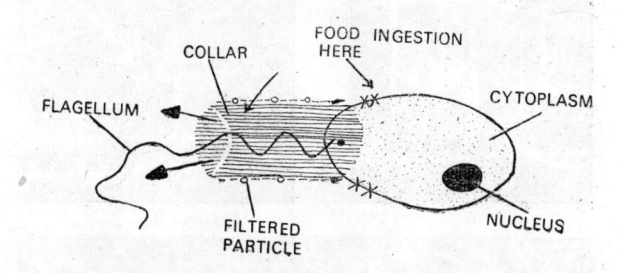

FIG. 3.10. An electron micrograph of a collar flagellate cell indicating how food particles are filtered out of water flowing through the body.

It has been mentioned earlier that there occur certain fusiform contractile cells around the osculum and form the **oscular sphincter** (Fig. 3.9). These contractile cells are called **myocytes**. The flagellated chambers are separated from the excurrent canals by thin **diaphragms**. Each diaphragm is pierced by a large central aperture, the **apopyle**, which is also surrounded by myocytes.

## Skeleton

The creation of positive pressures and currents of water on any scale requires that the cell layers of more complex sponges be supported against the deformation by skeletal tissues secreted by the mesenchyme. The skeleton consists of **spicules** or of **spongin fibres** (Fig. 3.11) or of a combination of both. The spicules are definite bodies of either calcium carbonate or hydrated silica. Basically the spicules are of two types: **megascleres** and **microscleres**. The megascleres are large spicules forming the main framework of the sponge. The microscleres are the smaller flesh spicules that occur throughout in the mesenchyme. The distinction is not absolute and does not hold for some groups. In its simplest form a spicule is like a simple small needle pointed at one or both ends. Such spicules are known as **monaxon** (axis is one). Monaxons (Fig. 3.11) with one broad end and one pointed are called **monactinal monaxons** or **styles**: whereas, monaxons in which the blunt end is knobbed like a pin-head are called **tylostyles**. Usually those pointed at both ends are termed **diactinal monaxons** or **diactines**. The diactines may be variously modified. Their ends may be curved inwards forming a C-shaped structure or may be variously tipped.

The spicules may be **triradiate** or **triactinal** (Fig. 3.11) in which three spines radiate from a point. This is the most common form of spicule of the calcareous sponges. Spicules with four rays radiating from one point, called **tetraxons** or **tetractines** or **quadriradiate** are also common. The four rays of the tetraxon may be more or less equal (**calthrops**), but generally one ray is elongated bearing a crown of three rays. Such a tetraxon is called **triaenes**. By the loss of rays they may become three- two- or even one-rayed. Sometimes three axes cross each other at right angles producing six rays extending at right angles from a central point. These six-rayed forms are known as the **triaxon** or **hexactinal** spicules with several rays radiating from a central point, **poly-**

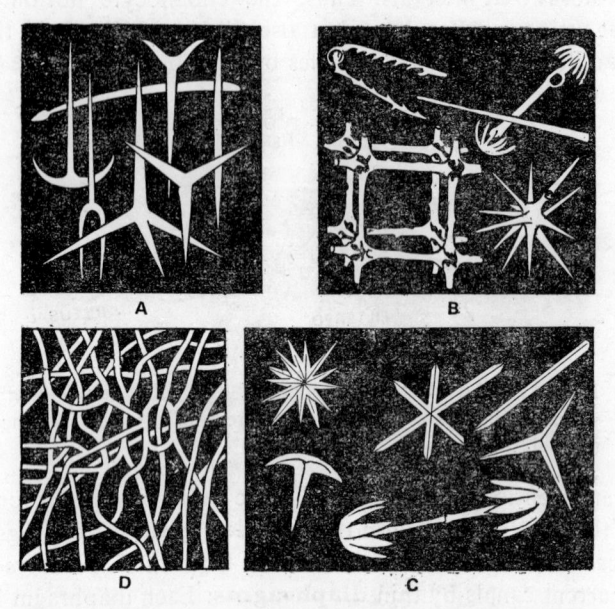

FIG. 3.11. Spicules of sponges. A and B—Calcareous spicules; C—Siliceous spicules; and D—Sponging fibres.

**axons** are also sometimes met with. In some cases rounded or star-shaped spicules are found. These are known as **spheres** or **sterrasters** (Fig. 3.11) respectively.

**Formation of spicules.** When a spicule is to be secreted scleroblast cell becomes binucleate (Fig. 3.12). Between the two nuclei a minute piece of cal-

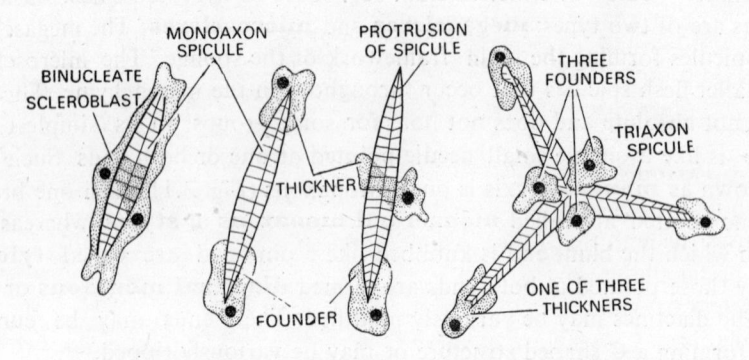

FIG. 3.12. Formation of spicules (monoaxons or triaxons).

cium carbonate is deposited. How this is secreted actually is not understood. As the spicule grows, the two nuclei draw apart until the mesenchyme cell separates into two cells. One of these cells (the **founder**) is situated at the inner end, the other (the **thickener**) at the outer end of the spicule (Fig. 3.12). The cell at the inner end moves inward establishing the shape and length of the spicule. The other adds to its thickness. When the spicule is completed both the cells wander into the mesogloea. Triradiate calcareous spicules are secreted by three scleroblasts that come together from three sides and each divides into two and secretes a minute spicule. Finally these three spicules meet (Fig. 3.12) forming a triradiate spicule. Each ray is formed in the same way as a monaxon, four-rayed spicules likewise are formed by four scleroblast cells. The calcium carbonate in all these cases is extracted from the sea-water. In water without calcium the sponges fail to secrete spicules. Development of siliceous spicule is not known well. They are formed, so far as known, within one scleroblast (better **silicoblast**). For large types several silicoblasts may co-operate. The hexactinal spicules of one order (Hexactinellida) arise in the centre of a multinucleate syncytial mass probably formed by divisions of a single silicoblast.

Spongin skeletons are made up of fibres of **spongin**, a protein secretion, that are either branched or form a network. Spongin fibres are also secreted by mesenchyme cells, termed **spongioblasts**. A number of spongioblasts is arranged in rows and the spongin rod secreted by each fuse with adjacent rods forming long fibres. The cells later become vacuolated and degenerate.

## Nutrition

The natural food of sponges consists of minute organisms and degenerating organic material. Besides these, the sponges are also known to digest dissolved nutrients in water. As the water current passes through the spongocoel the collar capture and ingest food organisms in the same way as certain flagellate protozoa do. The method is described on page 165. The collar cells digest food in food vacuoles either whole or in parts. Later the partly digested food particles are passed on to amoebocytes of the mesenchyme where digestion is completed. In other sponges no digestion occurs in the small choanocytes, but the food caught by them is immediately passed on to adjacent amoebocytes. Such a process exists in the simple sponge *Leucosolenia* and its other calcareous relatives (Calcarea). In others the food, as it passes through the incurrent canals, goes into the amoebocytes directly. Digestion is, therefore, entirely intracellular (as in Protozoa) occurring in food vacuoles which are acid first and alkaline later. Undigested particles are ejected and removed by the outgoing water. The digested food is stored into amoebocytes (then called **thesocytes**) as glycogen, fat and glycoprotein masses.

## Respiration and Excretion

Though Metazoa, sponges do not have special organs of respiration; oxygen diffuses through the surface of the cells within which energy is released. Sponges prefer places where water is rich with oxygen. If kept in foul water or water with less oxygen content the sponges undergo reduction in size and die. Similar results are obtained if the dermal pores are clogged. The oxygen consumption is dependent upon the rate of water current. Thus, if the oscula are closed it is reduced, but compensated when the oscula open. The upper surface below the osculum is reported to consume more oxygen than the basal half. This is an indication of a polar physiological difference.

The excretory products include ammonia and other complicated nitrogenous substances as has been particularly found out for siliceous sponges. Urea, uric acid and other common nitrogenous wastes found in higher animals have not been reported yet in the sponges. According to the opinion of some workers the amoebocytes containing excretory granules are discharged. Some sponges exude mucous or slime in large quantities and these are believed to carry excretory materials also. Some animals produce toxic substances which evoke irritation on the skin, when touched. Some such extracts of fresh-water sponges injected into mice prove fatal. But whether such products are excretory-materials or are secreted as a measure of defence is not clearly understood.

**Water current.** Under normal conditions all the apertures of the body remain widely open and a closer examination of a living sponge in nature will reveal that a current of water flows through the animal and out at the oscula. This current is created by the beat of the flagella of the collar cells in the flagellated chambers. The flagella of these cells, however, do not beat in co-ordination, therefore, it is difficult to understand how a continuous current, in one direction, is produced. Van Tright has suggested that the "flagellar movement consists of a spiral undulation passing from base to tip and creating a water current in the same direction. As the choanocytes in each flagellated chamber are grouped near the prosopyle, with their collars more or less pointed toward the apopyle, the water currents tend to flow from the flagellar tips to the apopyle. The mechanisms will obviously be more effective when the apopyles are larger than the prosopyles, as is usually the case in sponges, since the water will tend to seek the larger outlet. Small flagellated chambers are also more efficient than larger ones, since in the latter the flagellar currents will tend to be directed toward the centre of the chamber where some degree of stagnation will occur."

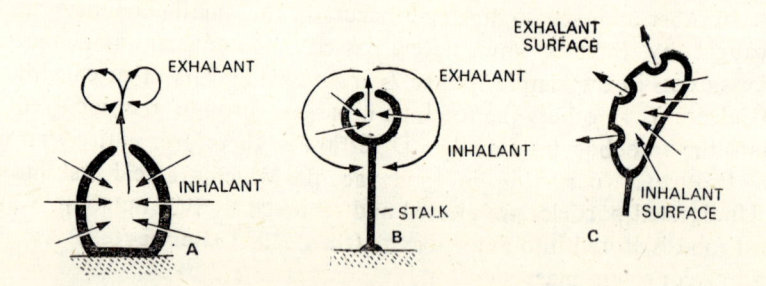

FIG. 3.13. Attachment of sponges to prevent repeated circulation of the same water.

It is evident from the above that in the ascon type the water current is slow, the leuconoid structure is the most efficient. As the current passes to the flagellated chambers it has to pass through narrow and more narrow channels thereby slowing down. This helps gaseous exchange which needs a little time. From the flagellated chamber the current enters larger canals where it moves at a more rapid speed, which further increases, when it emerges out of the osculum.

Many upright sponges show a series of adaptations tending to prevent repeated circulation of the water current, by isolation of the inhalant ostia from the exhalant oscula (Fig. 3.13). An ecological fact obviously related to this is that most encrusting forms of sponges will not live in completely still water (because the water gets polluted and is not changed).

The amount of water passed out of the sponge body has been worked out. It has been estimated that a specimen 10 cm high and 1 cm in diameter had 2,250,000 flagellated chambers and about 12.5 litres of water passed through it per day. A large specimen with millions of flagellated chambers will naturally force hundreds of litres of water to pass through it daily. So far as the rate of flow of water is concerned it has been calculated for a small leuconoid calcareous sponge (*Leucandra*) in which water from the osculum flows at a rate of 8.5 cc per second.

### Behaviour

No adult sponge is capable of locomotion. The majority have contractile powers and exhibit localised or generalised contraction because of the change of form of the pinacocytes and other cells. Some are devoid of contractility, only porocytes change shape. The body contracts if the individuals are disturbed or handled. The oscula close if the sponges are exposed to air, oxygen is lacking or if temperature is extreme. Addition of harmful chemicals to water has the same effect. Light is ineffective but touch induces closure in some forms. All the reactions, however, are slow. This is so because the sponges are devoid of nerve cells or other sensory structures and these reactions are directly in response to stimuli. Power of conductivity is also very slight, only developed at the osculum, the most sensitive part being the oscular rim.

### Colour and Odour

The sponges have disagreeable odour which is more marked in some species than in others. How the odour is produced is not understood, but its function is protective. Because of bad odour and projecting spicules they are seldom eaten by other animals.

The majority of sponges are of an inconspicuous flesh, drab or brownish coloration, but some are bright orange, yellow or red due to the presence of lipochrome pigment in some of the amoebocytes. Bright green forms are also common and blue, violet and black forms also occur. Some species are almost white. Sponges living in the muddy water are often nearly black because the cells of their body are gorged with minute particles of silt, which clears off in a few days if the sponge is kept in fresh water. Generally speaking, deep-water sponges are of dull appearance while those of the shallow water are often brightly coloured. Some fresh-water sponges are brownish green due to the presence of zoochlorellae.

**Variation**. Sponges are variable organisms and even a slight change in the environment of the fresh-water species often produces a considerable change in form and structure. Some species show such variation with the change of seasons and others without apparent cause. In most cases the changes apparently affect the breed for the period. The changes are so well marked that formerly subspecies or "varieties" were created on their basis.

### Regeneration

Sponges have great power of regeneration. Any part cut off is capable of producing a full sponge, but the process being slow, takes a long time. In some this power is still noteworthy. Even their individual cells, if separated, display regenerative behaviour. When certain sponges are squeezed through fine silk the cells are separated and come through either singly or in small groups, placed in suitable medium these cells show amoeboid movements. If they happen to come in contact with one another they stick together forming masses of cells. Finally these aggregates of cells (**reunion masses**) grow up into new sponges. Under adverse conditions many marine and fresh water sponges disintegrate leaving only rounded **reduction bodies**, each of which consists of a covering epidermis and an internal mass of amoebocytes. The reduction bodies develop into complete sponges on the return of favourable conditions.

**Asexual reproduction.** Asexual reproduction takes place by budding and by the formation of special reproductive units, the **gemmules**. A simple sponge

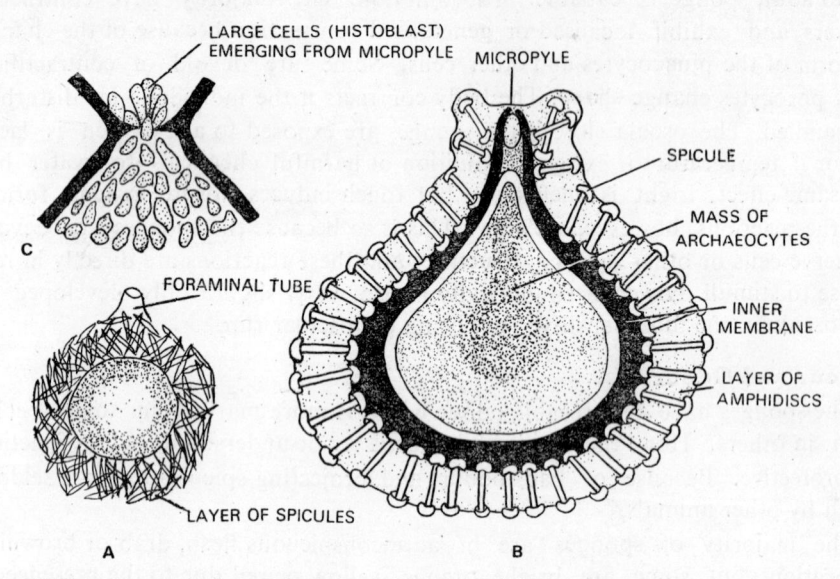

FIG. 3.14. Asexual reproduction. A—Gemmule of *Spongilla*; B—Gemmule enlarged to show mass of archaeocytes, layer of amphidiscs, etc.; and C—Histoblasts emerging from micropyle.

like *Leucosolenia* sprouts horizontal branches which grow over rocks and give rise to extensive colonies of upright vase shaped individuals. Often buds break off at their bases and grow into new individuals. Such buds are produced rather rarely. Almost all fresh-water sponges produce gemmules (Fig. 3.14), each of which consists of a mass of food-filled mesenchyme cells (amoebocytes) surrounded by a heavy protective coat strengthened by spicules. Some cells of the central mass form a layer of columnar cells around the food-filled mass of amoebocytes that ultimately secrete a thick inner membrane and later a thin outer membrane. Then some spicules, with knobs at both ends (Fig. 3.14) secreted by scleroblast cells, become arranged into the columnar layer rather radially between the inner and outer membrane. This layer is usually of great, relative thickness and is honey-combed by spaces, containing air and rendering the structure buoyant. For this reason this layer has often been called a "pneumatic coat" (Fig. 3.14). A completed gemmule has a small outlet, the **micropyle**, which is ordinarily closed. In some cases the mouth is surrounded by a tubular extension of the external membrane, the **forminal tubule** (Fig. 3.14A) rendering the whole structure flask-like in appearance. In India the gemmules are formed more frequently at the approach of hot weather and survive drying and carry the sponge over the period of drought. Under favourable conditions the sponge cells (**histoblasts**) emerge through the micropyle (Fig. 3.14C) and aggregate into small masses that grow into new sponges.

Gemmule formation has also been reported in some marine sponges. In these also an accumulation of amoebocytes becomes enclosed by a thin membrane of flat cells. The surface cells of the mass become columnar and flagellated except at the posterior pole and the gemmule then escapes as a flagellated larva. This swims about for a time, then attaches near the non flagellated pole, loses the flagella and develops into a young sponge by rearrangement and differentiation of the cells of the interior. It is quite remarkable that these larvae resemble those arising from fertilised eggs and develop like them. As the cells that form gemmule are derived from the same stock that produces sex cells, the gemmules have been regarded as products of parthenogenesis by some zoologists. Production of gemmule in the marine sponges and its ultimate behaviour further substantiates this view. The only apparent difficulty is that the gemmule is not formed by the multiplication of one cell but by the association of many.

**Sexual reproduction.** Ordinarily all sponges reproduce sexually by means of typical ova and spermatozoa. Some sponges are monoecious i.e., one individual produces both male and female sex cells. In others different individuals produce different sex cells, i.e., they are dioecious. The egg-mother cell or ovocyte is first noticed as an enlarged **amoebocyte (archaeocyte)** with a large nucleus and conspicuous nucleolus. It grows and acquires food stores by engulfing or re-fusing with other similar amoebocytes. It is called **oogonia** and passes into the cavities

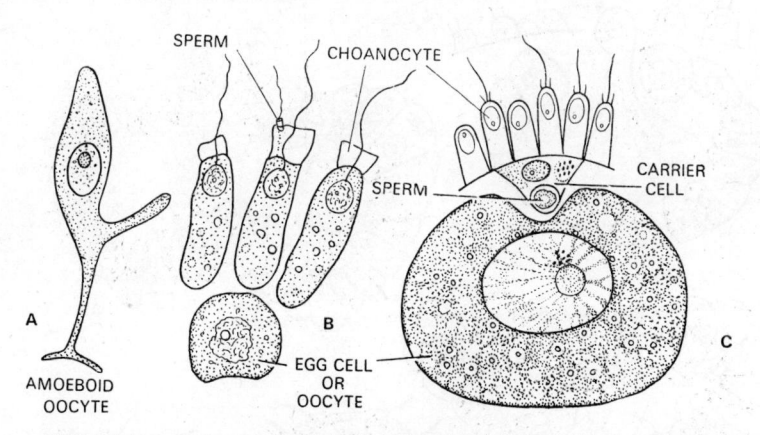

FIG. 3.15. Sexual reproduction. A—Amoeboid oocyte; B—Sperm entering a collar cell to be carried to the ovum; and C—Sperm inside carrier cell.

of the radial canals where it divides mitotically twice producing four ovocytes. These pass through the choanocyte layer into the mesenchyme, where they increase greatly in size, being nourished by special nurse cells or trophocytes (modified choanocytes). One such trophocyte attaches to the surface of each ovocyte and eventually fuses with it.

The sperm-mother cell or spermatogonium is described as an enlarged amoebocyte. It soon becomes enveloped by flattened cover cells derived by the division of the mother cells or consisting of other amoebocytes. The whole is called a **spermatocyst**. The enclosed sperm-mother cell divides two or three times into spermatocytes which develop into sperms. According to some authors (Gatenby) choanocytes develop into sperms which have typical forms, *i.e.*, rounded head and a vibratile tail, the lashing movement of which enables it to move through water. The sperms are carried to the female sponge with water current and the eggs are fertilised in *situ*. Sperm does not enter the egg directly but through a medium i.e., choanocyte (Fig. 3.15) which absorbs the sperm, loses its collar and flagellum and migrates to a position adjacent to the egg, thus acting as a **carrier** for the sperm.

The **fertilisation** in these sponges is peculiar. The spermatozoa enters a choanocyte adjacent to a ripe oocyte. The choanocyte loses its collar and flagellum, becomes amoeboid and plasters to the surface of the oocyte, which forms a conical depression to receive it. In the meantime the sperm loses its tail and the head becomes surrounded by a capsule. This capsule carrying the head enters the oocyte and the sperm carrying choanocyte moves away.

**Development.** It is only in *Sycon* and *Grantia* that the development has been

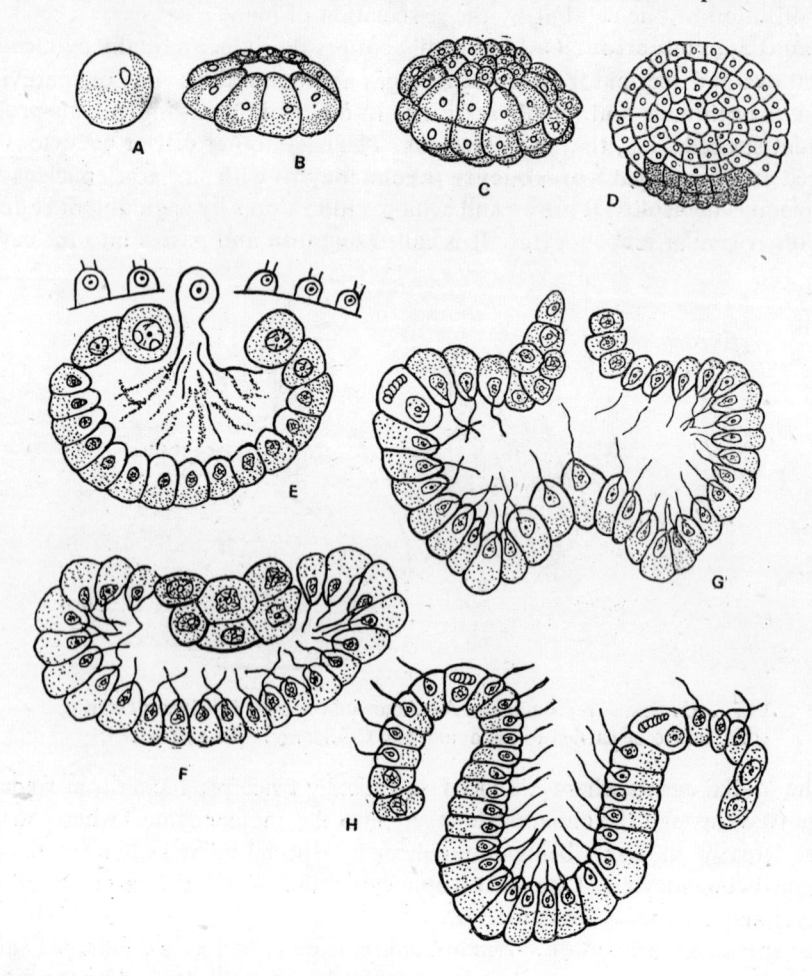

**FIG. 3.16.** Development of a calcareous sponge. **A**—Ovum; **B**—8-cell stage; **C**—48-cell stage; **D**—Hatching stage; **E**—Stomoblastula eugulfing collar cell; **F**—Blastula with flagella directed inwards; **G**—Inversion begins; and **H**—Inversion continues.

## Phylum Porifera—Sponges

adequately studied. The process of the formation of sex cells and fertilisation is as described above. The egg undergoes maturation after it is fertilised. The cleavage is **holoblastic** (Fig. 3.16). The first three cleavages are vertical resulting in the formation of a disc of eight pyramidal cells. A horizontal cleavage then produces a 16-cell embryo. It is a flattened disc-shaped body (Fig. 3.16 B) lying just beneath the maternal choanocyte layer. The position of the blastomere with respect to the maternal layer of choanocytes determines their fate. The tier of eight cells next to the maternal layer of choanocytes is the future epidermis and the tier of other eight cells represents the future choanocytes. The embryo, at this stage, is called the **blastula** (Fig. 3.16). A blastocoel appears in the interior. The eight small cells increase rapidly and elongate and each acquires a flagellum on its inner side facing the blastocoel. The few large cells remain undivided for sometime, become rounded and granular, and in their middle an opening appears that functions as a mouth and ingests adjacent maternal cells (Fig. 3.16 E). At this stage the embryo is called the **stomoblastula**. The embryo then turns inside out (like a stage in the colonial protozoan *Volvox*) through its mouth (**inversion**) bringing the flagellated cells outside (Fig. 3.17). The form produced after inversion is termed an **amphiblastula**, the typical larva of the Calcarea. It is at this stage that the larva becomes free and moves out. It consists of two types of cells—the greater part of the surface is formed of small, narrow flagellated cells that keep in front, while swimming, and the remainder, usually considered the posterior pole, consists of large, rounded, granular cells devoid of flagella. As the development proceeds beyond the blastula stage the embryo becomes enclosed in a **trophic membrane,** formed of maternal choanocytes, which supply the embryo with food. When completed the amphiblastula forces its way into the adjacent radial canal and is carried to the exterior. For some hours the amphiblastula swims freely, then the flagellated half is invaginated into or overgrown by the large granular cells. This process has been referred to as **gastrulation**. The larva then attaches to a solid object by the blastoporal end and begins to grow as a young sponge (*Olynthus* stage). Mesenchyme cells arise from both layers. Since the invaginated half seems to represent the animal pole of the egg, the germ layers of a sponge are not considered comparable with those of the metazoan embryos.

In some forms a typical gastrula is not formed and the invaginated flagellated cells form a solid mass in the interior. Such a larva is called **parenchymula** or **stereogastrula**. It attaches by the blastoporal pole and its interior mass hollows out to form the *Olynthus* stage. A parenchymula is formed in some species of *Leucosolenia* also but in a different way. The egg cleaves to form a **coeloblastula** composed entirely of narrow flagellated cells except at the posterior pole where there is a group of non-flagellate cells, the archaeocytes, the ancestors of all future archaeocytes of the sponge. These wander into the interior and fill it with a mass of cells. Some flagellated cells have also been reported to lose flagella and move with the archaeocytes to the interior. After free swimming of some hours the larva attaches by the anterior pole and flattens out. The cells of the interior migrate to the external surface and form the epidermis, and the mesenchyme and flagellated cells become the choanocytes in the interior.

**Affinities**. It was in the year 1765 that Ellis noticed the water currents and movements of the oscula of sponges and considered them to be animals. In spite of this sponges continued to be regarded as plants for a long time. Linnaeus, Lamarck and Cuvier placed them under Zoophytes, regarding them allied to an-

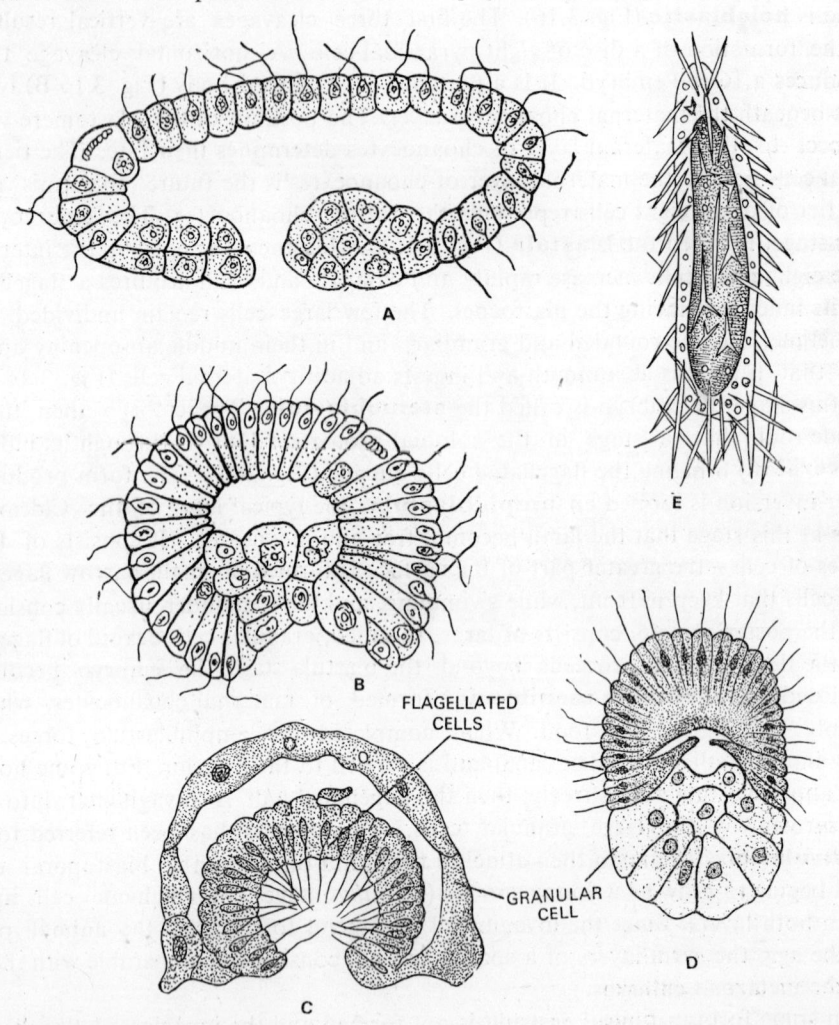

**FIG. 3.17.** A to C—Stages of inversion; D—Amphiblastula; and E—Later asconoid stage.

thozoan coelenterates, and through a major part of the nineteenth century the sponges were placed with the coelenterates. But now many important differences have been found out between these two groups. The most important of them is the absence of a true mouth in sponges. The study of embryology of sponges reveals that the osculum does not correspond with the mouth of the coelenterates. This point alone provides sufficient ground to separate the two groups, although there are many more points supporting it. They are: the presence of a inhalant apertures; the presence of peculiar collared gastric cells; the absence of stinging capsules; and the different way of formation of embryonic membranes. In sponges tissue formation is limited to the production of epithelia on free surface alone. The surface layer, which in other animals is generally differentiated at an early stage, remains undeveloped in sponges. Instead, differentiation proceeds in the mesenchyme. Epidermis and mesenchyme originate in sponges from the same embryonic cells and hence the mesenchyme may be considered theoretically to be an ectomesoderm. Besides, sponges have made no progress in the formation of the anterior end or head, a digestive sac is lacking and the mode of digestion is intracellular as in the Protozoa. From this it is evident that the sponges cannot be placed with the coelenterates.

The development of sponges presents peculiarities that are difficult to interpret. In most Calcarea and the simplest tetractinellids the coeloblastula invaginates into a typical gastrula; but it is the animal half that invaginates not the vegetal half as in other Metazoa. In other sponges a stereogastrula arises by ingression, but unlike other Metazoa the inner cell mass becomes the epidermis and the mesenchyme (not endoderm), while the surface epithelium (ectoderm of other Metazoa) transforms into choanocytes. Thus, the sponges cannot be placed in line with other Metazoa.

The intracellular digestion, the production of the skeleton in the protozoan fashion and some other such points take them close to the Protozoa. They cannot be placed with the Protozoa as there is an evidence of tissue formation. They resemble some colonial flagellated celled protozoans (proterosprongia) in having groups of flagellate cells and the flagellated sponge larva is probably reminiscent of a colonial flagellated ancestor. This view is greatly strengthened by the recent finding that the amphiblastula of syconoid Calcarea undergoes an inversion process identical with that occurring in the development of *Volvox*.

It seems sponges have evloved very little beyond the stage of colonial protozoan. Perhaps the reason for this is early adoption of a sessile mode of life and the inability of their cells to give up their protozoan habits. As the mouth and a digestive tract has not developed and the original flagellate cells have remained the main food getters and current producers, a unique anatomical type permeated with water channels has evolved. The flagellated surface cells were no longer needed for locomotory functions due to their assumption of sessile life, therefore, they have moved to the interior where they were more needed. Once the flagellated cells became located in the interior there followed the elaboration of the system of water channels, which is the chief structural feature of the sponges. Further evolutionary changes, in sponges, appear chiefly to be concerned with the perfecting of the water system and the development of a skeleton. Obviously the sponges seem to constitute a blind branch of the animal kingdom, not on the direct line of evolution, and, therefore, have rightly been called the **Parazoa**.

## CLASSIFICATION OF PHYLUM PORIFERA

It is difficult to classify Porifera because their characters are highly variable and the number of intermediate forms and collateral affinities is immense. The main basis of classification is skeleton (Fig. 3.18). On the whole there are three main types which have been assigned the status of classes for convenience.

I. CLASS CALCAREA OR CALCISPONGIAE. The Porifera with skeleton consisting solely of calcareous spicules which may be one- three- or four-rayed not distinguishable into megascleres and microscleres. The class includes asconoid, syconoid and leucocoid structures. The choanocytes are relatively larger. The osculum is always encircled by a fringe of upstanding spicules. Mostly under 15 cm in length. They are exclusively marine, found throughout the world. The class is divisible into two sub-classes.

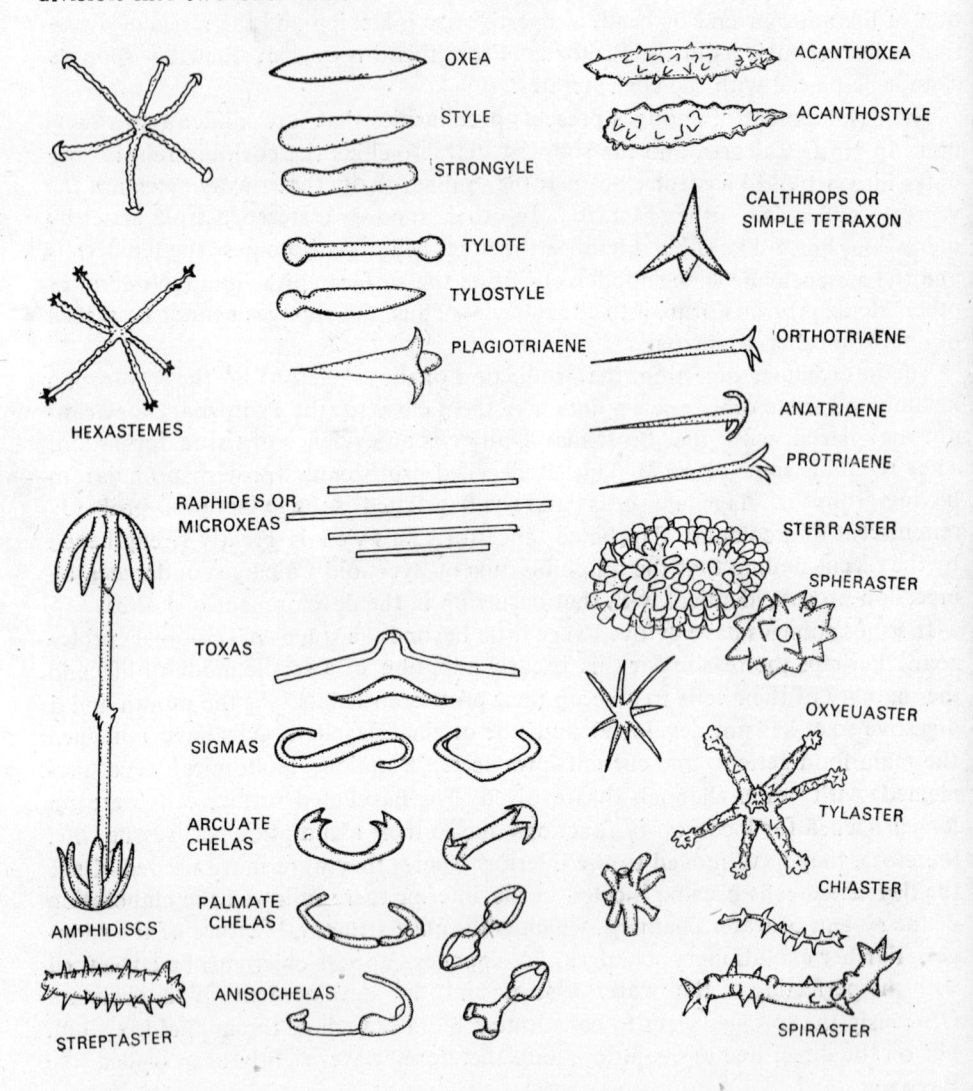

FIG. 3.18. Common spicule types of sponges used in their taxonomy.

(*i*) **Subclass Calcaronea**. Calcarea in which the larvae are amphiblastulae; the choanocyte nucleus is apical in position; the flagellum of each choanocyte arises directly from the nucleus; the triradiate spicules usually have one long ray.

1. **Order Leucosolenida** (=**Homocoela**). Calcarea in which the internal lining surface consists of flagellated collared cells throughout, the interior is not folded, the body wall is thin and canal system is asconoid.

Example: *Leucosolenia* (Fig. 3.3).

2. **Order Sycettida** (=**Heterocoela**). Calcarea in which the spongocoel is lined by flattened cells, the collared flagellated cells are restricted to chambers, body wall is thick, folded internally. Canal-system syconoid and leuconoid.

# Phylum Porifera—Sponges

Examples: *Sycon* (Fig. 3.1), *Grantia* (Fig. 3.1).

(*ii*) **Subclass Calcinea**. Calcarea in which the larva is **parenchymula** and the flagella of the choanocytes arise independently of choanocyte nuclei which occupy basal position in the cell (Fig. 3.10). Most species have triradiate spicules with equal rays. Leuconoid forms show no traces of the syconoid ancestry and have presumably evolved from reticulate asconoid forms.

1. **Order Clathrinida**. Calcinea possessing permanently asconoid forms lacking dermal membrane or cortex.

Examples: *Clathrina*.

2. **Order Leucettida**. Calcinea having syconoid or leuconoid forms with distinct dermal membrane and or cortex.

Examples: *Leucascus, Leucetta*.

3. **Order Pharetronida**. Calcinea with peculiar forms and with leuconoid canal system. Spicules quadriradiate, joined by calcareous cement or represented by rigid calcareous network (individual spicules not recognisable).

Examples: *Petrobiona, Minchinella*.

II. CLASS HEXACTINELLIDA OR HYALOSPONGIAE. The Porifera with purely siliceous skeleton composed of six-rayed spicules (hexactines). The internal structure consists of a network the **trabecular net**, made of thin strands forming open meshes. The canal-system is simple and choanocytes are small. They are without jelly and the osculum may be covered by a sieve plate of silica. Their skeleton is beautiful glass-like, hence they are commonly called the **glass-sponges**. They are mostly radially symmetrical, strongly individualised sponges of various shapes. They may be cylindrical or shaped like a vase, urn, funnel or other similar structures attached directly at the base or more usually by way of root-tufts of spicules (basal stalk). They are up to 92 centimetres in length, exclusively a marine group of deep-sponges occurring at depths of 90-5000 metres (approx). This class comprises two orders.

1. **Order Hexasterophora**. The Hexactinellida with hexasters are always present in the parenchyma. Generally, they are directly attached to hard objects. In some genera the skeleton is formed into a framework with irregular meshes.

*Euplectella*, popularly known a **Venus' flower basket** has a beautiful glassy skeleton, generally kept in museums. In life the skeleton is covered by dull grey flesh with many spicules, which has to be removed immediately on capturing the animal and the skeleton has to be bleached. *Euplectella*, the "beautifully-woven" sponge, is the most elegant of all glass-sponges. It is square-meshed in texture, yet every angle is rounded and ornamented by fine filaments which form a natural web and woof, soft, beautiful and fascinating. The glass filaments are not transparent. No artificer in the spun-glass handicraft can fashion such a marvellous frame produced in the living glass-sponge. In its living state it does not look so inviting as its skeleton looks. It occurs in mud of the sea-bed in the Eastern seas, a large number having been hunted out from seas around Borneo. The best known species *Euplectella aspergillum* is found off the Philippines. In life they are fixed to the sea-bed by large flinty filamentous anchors.

*Staurocalyptus* (Fig. 3.19A), another glass-sponge, has cup-shaped body usually without root spicules. A number of fine spicules projects from the surface. The filaments of these sponges are not transparent glass hairs, but threads of pure and lustrous white substance of which one of the constituents is flint.

2. **Order Amphidiscophora.** The Hexactinellida that are without hexasters, amphidiscs (monaxons with a crown of rays on both the ends) always present.

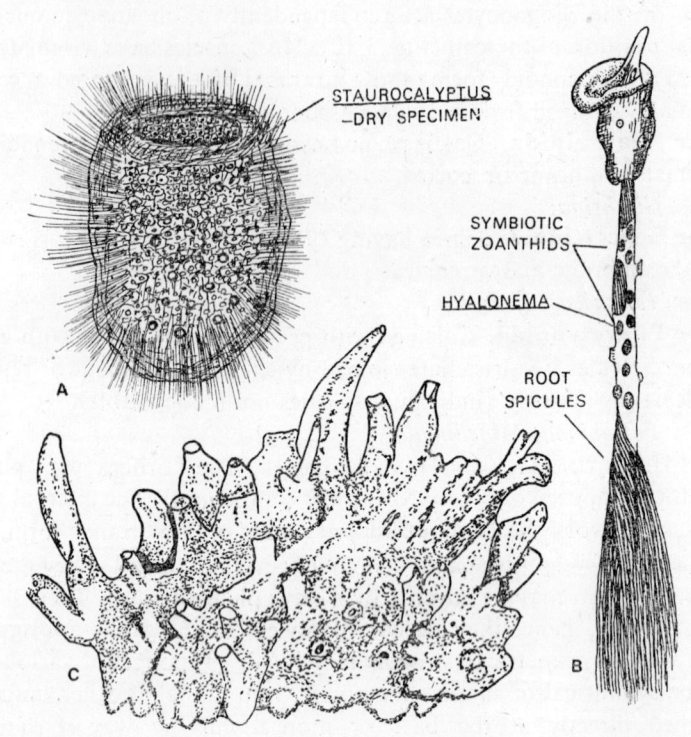

FIG. 3.19. A—*Staurocalyptus*, dry specimen; B—*Hyalonema*, dry specimen; and C—*Halichondria*.

*Hyalonema.* (Fig. 3.19B), the common glass-rope sponge having a rounded or avoid body with a single root tuft often spirally twisted (like a rope) forms a good example.

III. CLASS DEMOSPONGIAE. The Porifera that are either devoid of skeleton or with skeleton made up of spongin fibres alone; or in some cases having a combination of both siliceous spicules and spongin fibres. Those with only spongin fibres are known as horny sponges. The flagellated chambers are usually small and rounded having small choanocytes. The canal-system is usually complicated. The class is divisible into two sub-classes:

(*i*) **Subclass Tetractinomorpha.** Demospongiae having tetraxonid and monaxonid megascleres, usually lacking spongin fibres. The body is often radially constructed and cortical or axial development is advanced. The subclass has the following orders:

1. **Order Homosclerophorida.** Tetractinomorpha that are primitive and possess homogeneous leuconoid structure with little regional differentiation. Some forms lack skeleton.

Example: *Plakina, Oscarella* (without skeleton).

2. **Order Choristida.** Tetractinomorpha in which at least some of the megascleres are tetraxons, usually triaenes, microscleres are asters, the form of body is elaborate, spongin is typically present in the cortex.

Example: *Geodia* (Fig. 3.20)

# Phylum Porifera—Sponges

3. **Order Epipolasida.** Tetractinomorpha with radial construction and well-developed cortex; the megascleres are monaxons or styles and microscleres are asterose.

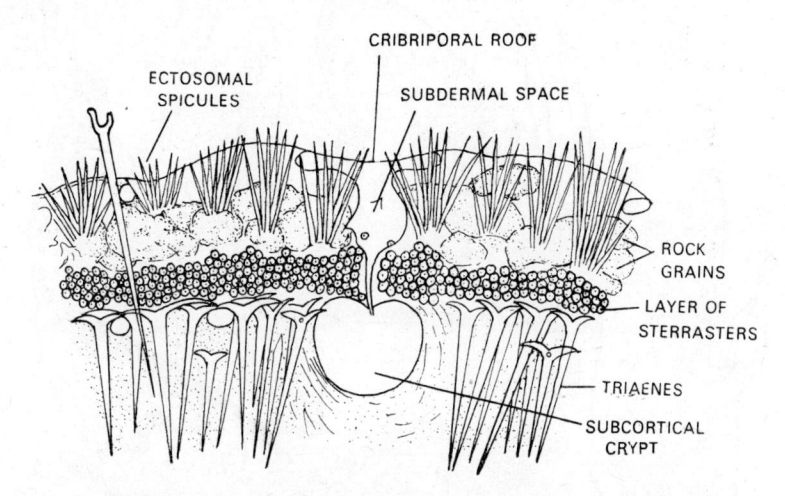

FIG. 3.20. Section at right angles to the surface of *Geodia* showing the differentiation into ectosome and endosome.

Examples: *Tethya* (Fig. 3.21), *Epipolasis*.

4. **Order Hadromerida.** Tetractinomorpha with typically radiate sponge body and a spicule cortex is present (histological differentiation not as well-marked as in the above two orders). Typical megascleres are monactinal tylostyles some forms adopt a boring habit.

Examples: *Cliona*, boring form (Fig. 3.18); *Polymastia*.

*Cliona* (Fig. 3.21), is a cosmopolitan genus which bores into the shells of molluscs (some species inhabit the interior of coral skeleton). Species of *Cliona* are often of a sulphur-yellow colour, others are green or purple. Some species bore into calcareous rocks. The larva settles on such objects, bores, occupies safe places in the interior and eventually permeates the entire objects emerging on the surface as a thin layer.

5. **Order Axinellida.** Tetractinomorpha, with monactinal or diactinal megascleres, spongin fibres abundant and encase spicules, microscleres are diverse.

Examples: *Axinella*, *Biemma*.

6. **Order Lithistida.** Tetractinomorpha including an artificial assemblage of all those forms in which the main skeleton is made of desmas—much-branched spicules with rays that are inextricably interwoven and often fused. They inhabit deep waters in temperate and tropical regions.

Examples: *Aciculites*

(*ii*) **Subclass Ceractinomorpha.** Demospongiae having incubated stereogastrula larvae, monaxonid megascleres, and sigmoid or chelate microscleres, asters are never present. Spongin is usually present but in extremely variable quantity and in most cases spongin fibres incorporate some spicules or detritus.

1. **Order Halichondrida.** Ceractinomorpha which have skeleton of monactinal or diactinal megascleres and lack microscleres.

Example: *Halichondria* (Fig. 3.21) forms with mostly flattened growth and looks like crumb of bread. It is, therefore, known as the "crumb-of-bread" sponge. It is a common British littoral form.

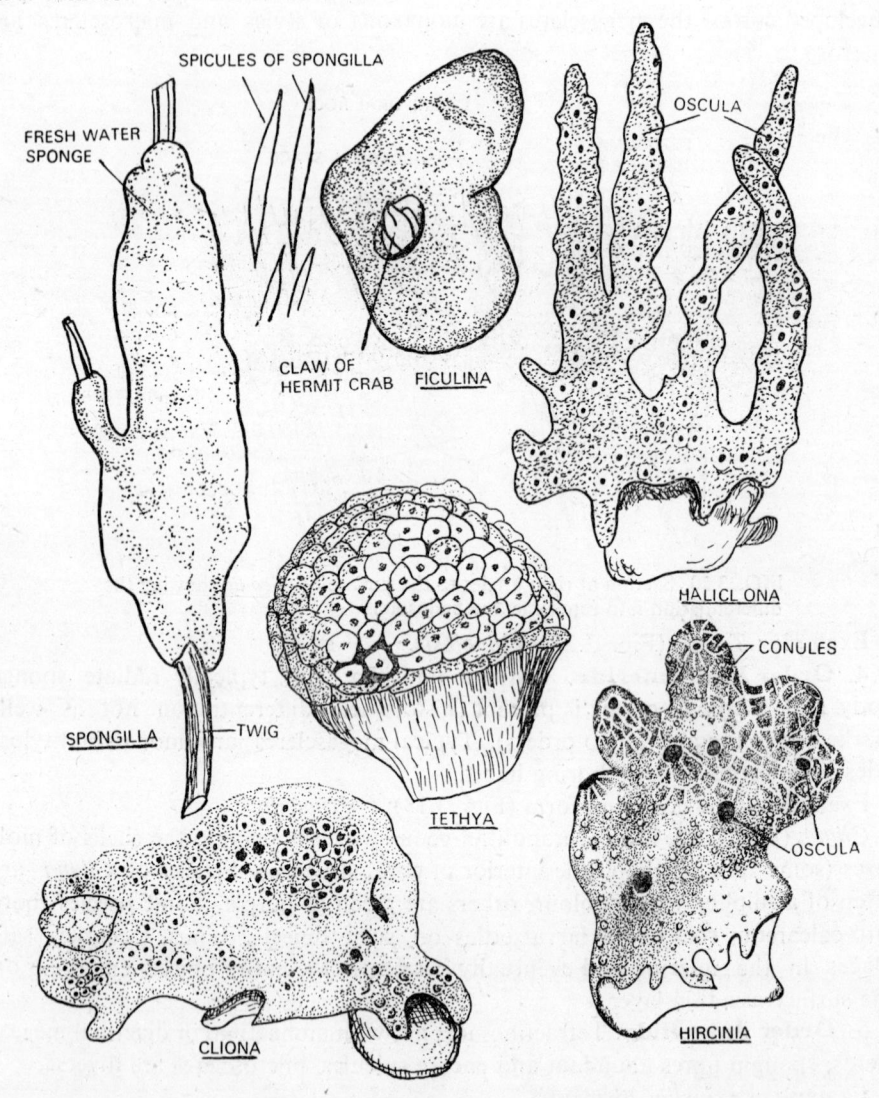

FIG. 3.21. Different types of sponges.

2. **Order Poecilosclerida** in which the skeleton always includes dermal and choanosomal megascleres.

Examples: *Clathria, Coelosphaera, Myxilla*.

3. **Order Haplosclerida**. Ceractinomorpha in which siliceous skeleton, when present, is made up of a single category of megascleres embedded in spongin fibres or joined in a network or reticulum by spongin cement.

Examples: *Haliclona, Spongilla* (Fig. 3.21).

*Spongilla* is a cosmopolitan sponge of the fresh-water family. The fresh-water sponge has no lasting skeleton. Many are brightly coloured during life, some are yellow, some are brown, others are green when exposed to sunlight because of zoochlorella (fresh-water algae) in the mesenchyme. Structurally also they differ from other sponges. Like other higher groups the flagellate cells are confined to certain enlargements of canals called **ciliated chambers**. The rest of

## Phylum Porifera—Sponges

the canals are lined by flattened cells. Internally the canals (the excurrent canals) open in the paragastric cavity and externally the incurrent canals open in characteristic **subdermal cavities** enclosed within thin **dermal membranes**, perforated by dermal pores. They grow as tufts of irregular masses up to the size of a person's first on sticks, stones or plants. Widely distributed all

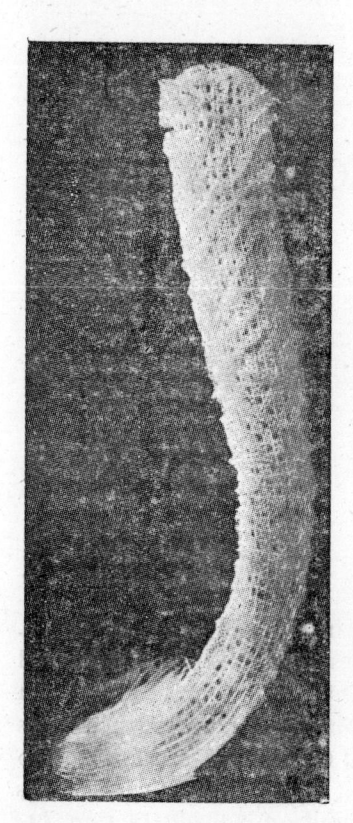

FIG. 3.22. A—Euplectella; and B—Staurocalyptus.

over the world, they occur in fresh-water streams, or ponds and lakes. *Spongilla lacustris* is common in sunlit running waters, whereas, *S. fragilis* avoids the light. Asexual reproduction takes place by the production of gemmules, although sexual reproduction producing typical larvae also occurs. The gemmules in this normally have cylindrical or subcylindrical spicules that are sharp or blunt or knobbed at the ends. The gemmules either lie free in the substance of the sponge or are attached to its support.

4. **Order Dictyoceratida.** Ceractinomorpha which lack spicules, the skeleton being entirely made up of a reticulation of spongin fibres which may enclose foreign particles such as sand, shell, spicules of other sponges and have a healthy dermis often reinforced by spongin.

Examples: *Dysidea, Spongia*.

5. **Order Dendroceratida.** Ceractinomorpha which have forms lacking spicules and in some cases also lacking a fibrous skeleton.

Example: *Aplysilla*.

# 4 THE METAZOA

The Metazoa are holozoic many-celled organisms which develop from embryos. The gametes never form within unicellular structures but rather are produced within multicellular sex organs or at least within surrounding somatic cells. To a beginner the Metazoa are animals whose bodies consist of numerous cells, as distinct from the Protozoa which are acellular. The cells are differentiated into many kinds for performing various functions, and at least part of their cells is arranged into layers. The Metazoa present higher grade of individualisation due to the formation of a nervous system which is usually centralised in one part of the body known as the anterior end or head. The occurrence of the embryonic development in the life-history is another characteristic of the Metazoa.

The Metazoa are believed to have evolved by colony formation, as was suggested by Haeckel who derived them from the hollow Volvox-like flagellate colony. Most writers have accepted this view, and thus picture the common ancestor of the Metazoa as a little hollow ball composed of a single layer of flagellated cells, swimming about with one pole directed forward and having an anteroposterior axis from this pole to the opposite one. Haeckel called this hypothetical organism by the name of the **blastea** which is generally believed to be reproduced in the blastula stage of the development of the Metazoa.

From this one-layered condition of the metazoan ancestor evolved the two-layered condition as envisaged by Haeckel in his hypothetical two-layered organism, the **gastrea**. The latter became the basis of his famous **gastrea theory** which asserts that the gastrea is a common ancestor of all the Metazoa and that it is reproduced by the gastrula stage during the development of the Metazoa.

There are several theories explaining the transformation of one-layered condition into two-layered condition, but all start from the assumption that the separation of locomotory and digestive functions would be advantageous. It is mostly accepted that the posterior cells of the blastea took on digestive functions, and thus Haeckel postulated that the two layered condition of the gastrea arose by the bending of the posterior half of the blastea into the anterior half producing a two-walled cup of which the inner endodermal sac formed the archenteron and its external opening the blastopore.

According to this view the original metazoan ancestors were two layered hollow sacs, with a hollow gastrovascular cavity and are represented by the Coelenterata. This view, that is generally held, hallowed by a long period of acceptance and teaching, regards the Coelenterata as the most primitive Metazoa. The sponges do not come into the picture, since, although multicellular, they are universally admitted to be a side line.

Another way in which the Metazoa might have been derived from the Protozoa is by the appearance of cell boundaries in a multinucleate syncytium, in other words, internal subdivision of the protozoan body by cellularisation. This view was not widely accepted because of lack of evidences. Recent work of Jovan Hadzi, a German, brings this view to the forefront. The researches that Hadzi has made into the structure and development of the lower animals for half a century, have led him to put forward a theory that multinuclear Protozoa evolved into Turbellaria Acoela, which would be the most primitive Metazoa. From them were descended the higher Turbellaria which gave rise to the main lines of invertebrate evolution, and in addition and independently to Coelenterate and to the Ctenophora.

This revolutionary view has been recently[1] critically analysed by G.R. de Beer who says that "Hadzi has been able to stimulate zoologists to re-examine one of the most fundamental of their beliefs, and at least to shake it." The conversion of the Protozoa into Metazoa by aggregation of single protozoan individuals (as suggested by the conversion of Volvox-like colony) has grave objections. This may have happened in sponges and "is the most likely explanation for the lack of co-ordination, integration and individuality" found in the sponges. The only way that answers this is cellularisation of a multinucleate protozoan such as a ciliate. There is no need of a hollow digestive cavity in ciliates, the same is true of the primitive Turbellaria (Acoela) which are imperfectly cellularised, for beneath the outermost layer or epidermis, in the more or less solid parenchyma of the body, there are irregularly shaped syncytial masses, some of which are concerned with digestion and others with the production of reproductive cells.

If this theory is to be acceptable at least three things need explanation: (*i*) conversion of the Protozoa into Turbellaria, (*ii*) derivation of the Coelenterata from Turbellaria, and (*iii*) the position and affinities of the Ctenophora.

Hadzi derives the Turbellaria from infusorian ciliate Protozoa, some of the multinucleate ones of which developed internal partition and formed the Turbellaria. Some small acoel Turbellaria, e.g., *Convoluta* are as small as the ciliates. The Turbellaria are hermaphrodite and perform copulation (derived from conjunction of ciliates). Their body is covered with cilia. The outer layer contains a

---

[1]*Evolution as a Process* (1945), George Allen and Unwin Ltd., London.

number of rod-like structures, capable of discharge, the **sagittocysts,** which are like the **trichocysts** of ciliates, on the one hand, and **nematocysts** of coelenterates on the other. These and other points such as small size, imperfect cellularisation, parenchyma, indistinct gonads and phagocytic digestive syncytium, make the case for derivation of Turbellaria from ciliate Protozoa attractive.

Hadzi has derived the Coelenterate form a rhabdocoelid form in which were already developed an ectodermal pharynx, a hollow gut cavity showing many lobes, and the tentacles containing hollow diverticula from the gut cavity. If a rhabdocoel adopted a sessile life and became affixed by its aboral end, its organisation possesses all that is needed for the construction of an Anthozoan. Some elements such as nerve centres, solenocytes and parenchyma would be expected to become reduced in response to the sessile habit. Although Rhabdocoela lack the sagittocysts of Acoela, Hadzi suggests that the Coelenterata inherited the nematocysts from prototypes such as the sagittocysts in their turbellarian ancestors. Hadzi has further drawn attention to the epithelio-muscular cells of the Coelenterata, which are not restricted to them alone, as believed hitherto, but believes that they have been derived from the epidermal cells of some Turbellaria which have muscle-fibres and which present a more primitive condition than that of the differentiated epithelio-muscular cells of Coelenterata.

Regarding the affinities of Ctenophora Hadzi suggested that they were evolved from polyclad Turbellaria by neotenous retention of the structural features of the larval form of the polyclad Turbellaria known as Muller's larva. This was also suggested by MacBride. The development of Ctenophora and polyclad Turbellaria is of a mosaic type, "there is no hollow blastula stage, gastrulation occurs by epiboly, the mouth, pharynx and gut cavity are formed in the same manner, aboral sense organs statoliths and paired tentacles are found, and there are great similarities in the gland cells of the epidermis" (de Beer). Based on the close affinity between the two groups MacBride interpreted that Muller's larva represented the pelagic, ctenophore-like adult ancestor of the Polycladida, and that Platyhelmia were evolved from the Ctenophora. Hadzi and de Beer prefer to regard the Ctenophora as descended from Polycladida by neotenous retention of many of the larval features of the ancestor. Some Ctenophora such as *Ctenoplana* and *Coeloplana*, in the adult stage show a spurious similarity with polyclad Turbellaria.

The history of germ-layers has also been convincingly answered by Hadzi's theory. In some ciliates there already exist different parts of the body, which may be called outer material, middle material and inner material, often specialised and differentiated into complex organelles. The cellularisation of these has naturally produced the ectoderm, mesoderm (called *mesonyle* by Hadzi as it does not primitively form a layer) and endoderm as illustrated in the Turbellaria Acoela.

On the whole the theory is quite attractive and convincingly brings the Turbellaria Acoela to the status of the primitive Metazoa, which gave rise to the main lines of invertebrate evolution. But as this is a revolutionary view and may well excite incredulity the same order of description has not been followed in the text at least for the present.

# 5 PHYLUM CNIDARIA (COELENTERATA)

The relatively familiar animals, which make up the phylum **Cnidaria**, including such forms as jelly fish, sea-anemones, corals, and hydroids are commonly called coelenterates. The characteristic feature of coelenterate organisation is that the body is made up of two cell-layers, which were named **ectoderm** and **endoderm** by Allman in 1853. Between these layers lies a form of connective tissue, the **mesogloea**, with a matrix containing mucopolysaccharides. Cells are not necessarily present in the mesogloea, and even when they are, their functions are in doubt. On the other hand, nineteenth-century references to this substance as a "structureless lamella" are certainly incorrect. If it were really a structureless lamella it would be difficult to account for the movements carried out by the coelenterate body. Because of the presence of these two layers these animals were formerly characterised as the **diploblastic metazoans**. In modern usage the term diploblastic has been abandoned because of the presence of cellular elements in the intermediate layer, the mesogloea.

The structurally significant aspect of the mesogloea is the extensive development in it of collagenous fibres that are systematically arranged to form a lattice. In the anemone *Calliactis parasitica*, for example, there are inner and outer sheets in which the fibres run mainly parallel with the epithelia and at 45° to the long axis of the body, while other fibres run across the thickness of the mesogloea from one layer of fibres to the other. Thus the system as a whole has the form of a three dimensional lattice. The source of these fibres is unknown, as also is the origin of their regular arrangement.

The cells making up intermediate layer in the coelenterates have always been derived as wandering amoeboid cells which have been detached individually from the outer epidermis and never originate as **mesoderm** (an initial epithelial mass of tissue derived from endoderm of the developing animal) of the higher phyla. With certain minor exceptions the coelenterates lack organs having their cells organised into tissues (tissue grade of organisation) which are specialised for various functions.

The coelenterates are most simply constructed successful group of animals. They are abundant in a wide variety of habitats in the sea, a few live in freshwater and none of them exists as terrestrial animal. Strictly on the basis of ecological assessment the Cnidaria are not really a major phylum, but in some regions of the tropics they are ecologically important making up a significant part of the biomass. On the other hand, they constitute an important group because of two conceptions, both involving uncertainties which have often been criticised, but both still largely defensible. This first is simply that the coelenterates are at least closely related to the stock from which the other metazoan groups were eventually derived. Obviously, it is not possible to prove that they are the most primitive of many-celled animals. The second conceptional basis for interest in the coelenterates is that their organisation involves a degree of structural and functional simplicity which is not found in other metazoan. This particular hypothesis has attracted two groups of biologists to them: developmental biologists interested in the problems of regeneration and differentiation, and neurophysiologists. Both groups of investigators have uncovered greater and greater degrees of complexity in recent years than were first thought to exist in these forms.

The nervous system and the skeletal mechanisms (which show clear functional homologies with the higher phyla) vary widely throughout the coelenterates. In coelenterate the nerve cells form nets immediately below the epithelia. The cells are usually multipolar, though in a few places in more highly organised cnidarians they may be bipolar, and are arranged in discontinuous two dimensional nets. The cells have synaptic contacts just as in neural organisation of higher many-celled animals. The difference with higher Metazoa is that the coelenterate synapses lack polarisation in transmitting action potential. Any branch or "fibre" of a cnidarian nreve cell can conduct in either direction. As regards nervous system each of the three main groups of coelenterates seems to have evolved its own method of coordinating its activities through conducting (or nervous) cells. It is almost impossible, on the present evidence to decide which type of nervous organisation was archetypic for the group. Similarly the skeletal mechanisms, or ways in which the animal maintains its shape in reaction to the environmental forces acting upon it, vary and form a poor subject for generalisation in the coelenterates.

It is evident that the stereotyped network of inter-relationships of functions so typical of the higher phyla does not need to exist at the simplest level of organisation found in coelenterates. Functional homology does exist. Coelenterates are predaceous carnivores of the tissue level of organisation, utilizing nematocysts as both a means of defense and of obtaining their prey. Nematocysts are only formed and used by cnidarians and by one species of ctenophore.

*Phylum Cnidaria (Coelenterata)*

Another general character of the coelenterates is the occurrence of polymorphism within many species. This may be temporal, with the species passing through a succession of different body forms in the course of its life cycle, or several different forms of the species may occur simultaneously in a colony. There is a considerable interdependence and division of labour between individuals (rather than between organs as in higher Metazoa). The **individuals** or **zooids**, are mainly of two types, the **polyp** and the **medusa**. The polyps (the hydroids) are the feeding zooids which have sac-like body, closed and attached at one end, while the upper end of the sac, just below the mouth, is drawn out to form circlets of tentacles. In colonial coelenterates the hydroids are joined to the main part of the colony by tubular prolongation from the lower end of the sac. The medusa is the free-swimming form specialised to carry reproductive organs. It has a gelatinous umbrella-shaped body, margined with tentacles and having the mouth on a central projection of the concave surface. The medusa can be easily derived from the schematic sac-like form described above. The sac flattens from end to end so that a somewhat bell-like structure is produced. At the same time there is a pronounced thickening of the mesogloea and the enteron becomes restricted to a small cavity into which opens the mouth, to the cavities of a number of radial canals, which join the central cavity, to a circular canal, and to the circular canal which runs around the margin of the bell.

Nematocysts constitute other diagnostic feature of the phylum. A nematocyst consists of a capsule having a coiled capillary tube, which when stimulated, discharges to the exterior by turning inside out and serves to paralyse the prey or to hold it by wrapping around bristles, etc., or is used for support, anchorage or adhesion. The coelenterates are commonly endowed with the power of forming skeleton either as exoskeleton, secreted on the external surface or as endoskeleton formed in the mesenchyme as separate pieces (**sclerites**) or as continuous mass. Budding is a common means of reproduction forming either free individuals or colonial zooids. Sexual reproduction usually produces an ovoidal, uniformly ciliated, larva known as the **planula**.

In certain older treatments the name **Coelenterata** was used as a phylum designation to include **Cnidaria** and **Ctenophora**. It seems better to separate these groups as direct phyletic entities. The Cnidaria closely approximate an idealised diploblastic condition and possess highly diagnostic intracelluar structures, the nematocysts, while the Ctenophora are triploblastic and do not posses nematocysts. The Cnidaria can be divided into three well-defined classes: the **Hydroza**, the **Scyphozoa** or true jelly-fish and the **Anthozoa** including the corals and sea-anemones. Their division is based upon the nature of the gastrovascular cavity or coelenteron. It is always a blind sac with only one opening (functionally both mouth and anus). This sac carries out the functions of both gut and circulatory system in higher animals. In the Hydrozoa it is usually a simple bag, but develops complicated branches forming canals and pockets in the Scyphozoa, while in the Anthozoa it is partially divided by ingrowths of a series of septa. The functional significance of both types of complication lies in the increased surface area of gastrodermis provided in both jellyfish and sea-anemones. The classification followed here presents the most convenient course:

## PHYLUM COELENTERATA

| | |
|---|---|
| **Class I:** | HYDROZOA |
| **Orders:** | Hydroidea |
| | Hydrocorallina |
| | Trachylina |
| | Siphonophora |
| | Graptolithina |
| **Class II:** | SCYPHOZOA (SCYPHOMEDUSAE) |
| **Orders:** | Stauromedusae |
| | Cubomedusae or Carybdeida |
| | Coronata |
| | Discomedusae (Seamostomeae) |
| | Rhizostomeae |
| **Class III:** | ANTHOZOA (ACTIONOZOA) |
| **Sub-Class:** | ALCYONARIA OR OCTOCORALLIA |
| **Orders:** | Stolonifera |
| | Telestacea |
| | Alcyonacea |
| | Coenothecalia |
| | Gorgonacea |
| | Pennatulacea |
| **Sub-Class:** | ZOANTHARIA OR HEXACORALLIA |
| **Orders:** | Actiniaria |
| | Madreporaria |
| | Zoanthideia |
| | Antipatharia |
| | Ceriantharia |
| | Rugosa or Tetracoralla |
| | Corallimorpharia |
| **Sub-Class:** | TABULATA |

### CLASS HYDROZOA

The **Hydrozoa** are polymorphic coelenterates with tetramerous or polymerous radial symmetry in which both polypoid and medusoid forms exist. Exclusively polypoid or exclusively medusoid forms are also found. Mesogloea is non-cellular and the gastrovascular cavity is without stomodaeum, septa or nematocyst bearing structures. The Hydrozoa are nearly always colonial typically with free or sessile medusoid phase, arising as a bud from the polyp colony. Some members of the group exist exclusively in the polypoid phases, for instance the hydras. The hydras are solitary fresh-water forms reproducing by budding and also sexually. Similarly there is another group (Trachylina) in which usually only medusoid stage is represented. These are mainly marine forms such as *Cunina* (Fig. 5.26C). The following two types, *Hydra* and *Obelia* present all the characters of polypoid and polymorphic forms.

*Phylum Cnidaria (Coelenterata)*

## TYPE HYDRA

*Hydra* lives in fresh-water ponds, lakes and streams and is found attached to submerged water plants rocks or other objects by its adhesive basal disc. The common little animal provides a typical example of a hydroid structure. The body is columnar in shape (Fig. 5.1) ordinarily of about 2 to 3 mm in size.

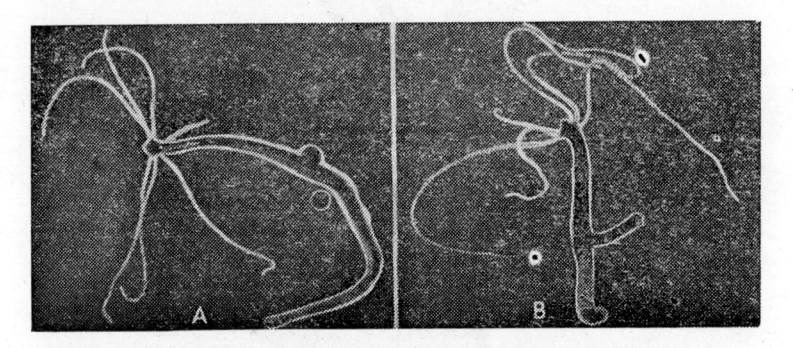

FIG. 5.1. Photographs of *Hydra littoralis*. A—Female with two eggs, one in the ovary and one extruded; and B—Hydra with bud.

Larger forms of 6 mm to one cm in size are also met with frequently. The closed lower end of the body is normally attached to the substratum or to water weeds by the basal disc. The mouth is situated at the apex of the oral cone or **hypostome** from the base of which arise a number of long hollow outgrowths, the **tentacles**. The number of tentacles is variable being 4-10.

### Body Wall

The body wall consists of two layers of cells the **ectoderm**, the outer layer, and the **endoderm,** the inner. These two layers are separated by a thin layer the **mesogloea** or **mesolamella**. The ectoderm is the outer layer of protective epithelium, the cells of which show considerable differentiation.

I. **Ectodeom cell types**. The apparently simple ectoderm consists of cells which are at least of seven different types: (*a*) **Musculo-epithelial cells,** (*b*) **interstitial cells,** (*c*) **nematoblasts,** (*d*) **nerve cells,** (*e*) **sensory (receptor) cells,** (*f*) **gland cells,** and (*g*) **germ cells.**

(*a*) The **musculo-epithelial cells** are cone-shaped cells with their broader bases towards the outer side forming the outer covering of the body. Being narrower towards the inner side they have interstices in the deeper parts of the ectoderm. Their inner ends are drawn out to form contractile processes, the **muscle-tails**, at opposite ends with a myoneme extending into them (Fig. 5.3C). These processes lie parallel to the long axis of the body, resemble muscles of higher animals and sometimes show cross striations, hence the name. They form the major part of the ectoderm.

The ultrastructure of the musculo-epithelial cell shows that the surface of the cell is covered with protective filamentous meshwork supposedly formed by liberation of mucous granules (Fig. 5.4). The plasma membrane is smooth with a few outward projections. There are large intracellular spaces inside the cells. The outline of the nucleus appears irregular and the nucleolus is central. The cytoplasm contains many free ribosomes and numerous typical mitochondria. Both rough and smooth-surfaced endoplasmic reticulum are present. Golgi apparatus lies parallel to the long axis of the cell.

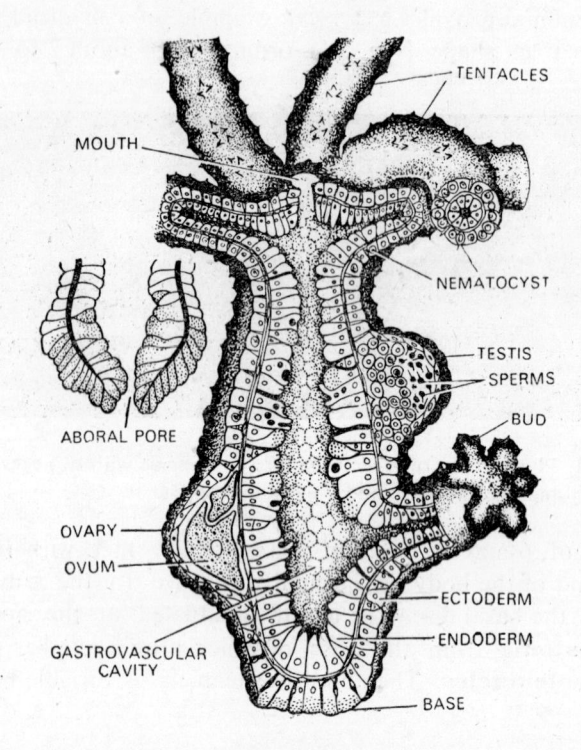

FIG. 5.2. Longitudinal section of *Hydra* (diagrammatic). Figure on the right shows a section of the base showing open aboral pore.

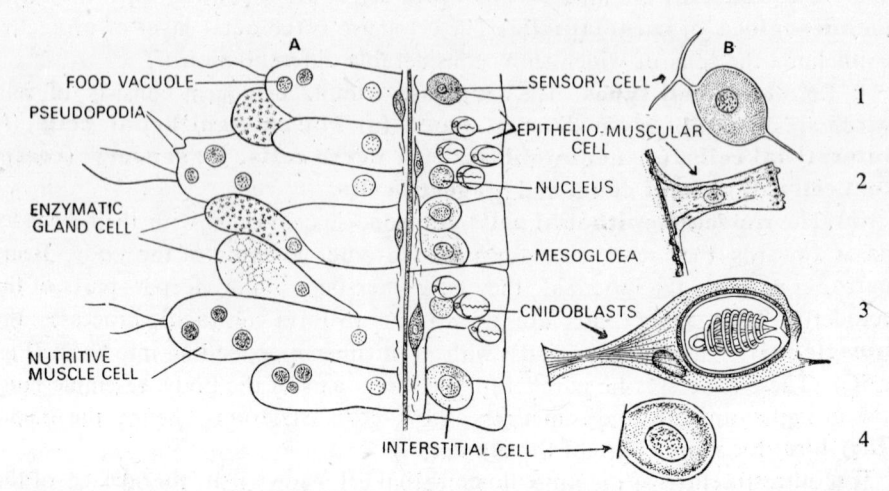

FIG. 5.3. *Hydra* structure as seen in the microscopic sections. **A**—A portion of cross section; **B**—Different types of cells: 1—Sensory cell; 2—Epithelio-muscular cell (highly magnified); 3—Cnidoblast; and 4—Interstitial cell.

The muscle processes at the base of the cell are filled with two types of myo-

## Phylum Cnidaria (Coelenterata)

filaments—numerous large myofilaments 50 Å in diameter, containing smaller myofilaments of about 200 Å diameter inside. The myofilaments terminate obliquely on the plasma membrane. Each myofilament is thickened at the point of

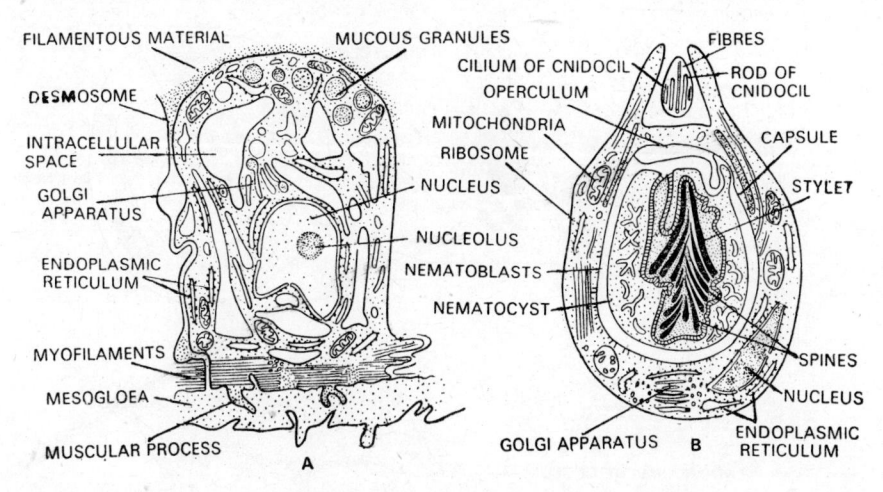

FIG. 5.4. A—Electron micrograph of a section of an epithelio-muscular cell of *Hydra*; and B—Electron micrograph of a cnidoblast containing a nematocyst of *Hydra* (both diagrammatic).

termination. Microtubules lying in the muscle process, it is believed, carry water or ions and may therefore, be involved in changes in electrical potential of muscle processes.

The ultrastructure of the musculoepithelial cell varies in different regions. The gland cells of the base are filled with mucous granules produced by Golgi apparatus. The endoplasmic reticulum is well developed. The basal region of the cells contains smaller mucous granules while apex has larger ones. The muscle cells of the peduncle are small with few intracellular spaces and filled with mucous granules and irregular masses of homogeneous material. The muscle cells of the tentacles are large, contain numerous cnidoblasts more apical mucous granules and more elaborate endoplasmic reticulum

(*b*) **The interstitial cells** are small rounded cells (Fig. 5.5) occupying the interstices between the musculo-epithelial cells. They are simple undifferentiated cells and are capable of giving rise to other types of ectodermal cells including the reproductive cells. The nucleus is central and contains scattered granules. The cytoplasm is filled with free-ribosomes, mitochondria and few smooth membrane-bound vesicles. The interstitial cells are busy replacing the mematoblasts that are being continually used up.

(*c*) The **nematoblasts** or **Cnidoblasts** are interstitial cells that have become extremely specialised forming a nematocyst in the interior. These cells are usually found in groups embedded in the superficial layer of the ectoderm. They are more plentiful in the tentacles. Each nematoblast secretes a **nematocyst** (Fig. 5.6), but how it is done is not clearly understood. The nematocyst consists of fluid-filled bladder or capsule of spherical oval or pyriform shape into which lies a hollow coiled thread, that is fastened to the narrower end of the capsule and is continuous with its wall. On one side of the nematocyst, on the outer surface, is a small protoplasmic projection, the **cnidocil** or **trigger**. When the cnidocil is properly stimulated the nematocyst is discharged. The

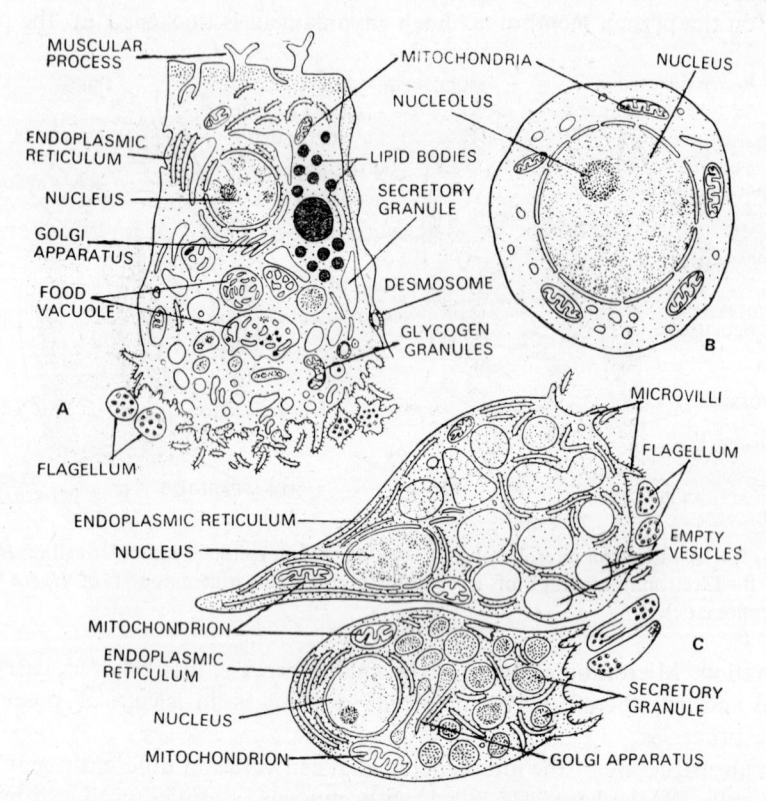

**FIG. 5.5.** Electron micrographs of *Hydra*. **A**—Digestive cell; **B**—Interstitial cell; and **C**—Gland cells (diagrammatic).

stimulation may be brought about by the touching of the cnidocil by food, prey or enemy or by the contact of chemicals released by these. Exact mode of stimulation is not understood. The undischarged thread has the appearance of a screw because of three left-hand helical spiral folds in its wall. On discharge the coiled thread is shot out and penetrates the tissues of the prey and discharges into them poisonous fluid which has powerful paralysing effect even on large animals like water-fleas. After the discharge the thread is a smooth, slightly tapered tube, the process of discharge apparently involving unfolding of these pleats.

In a high power section of a partially discharged thread where evenly-spaced equal-sized barbs are present the appearance is that given in Figure 5.7. This new evidence of microanatomy of nematocyst threads has made it easier to understand how these threads are able to penetrate the exoskeleton of the more highly organised animals such as crustaceans which provide a large part of the diet to small cnidaria. To understand how the nematocyst catches the prey one has not only to imagine a double tube everting toward one at a fantastically high velocity, but also the advancing edge of that tube (i.e. the fold between the tube already discharged and that undischarged contained within it) revolving at

## Phylum Cnidaria (Coelenterata)

enormous speed as the spiral pleats unfold, and at each turn a new ring of sharp barbs springing forward and then flicking backward like the blades of penknives.

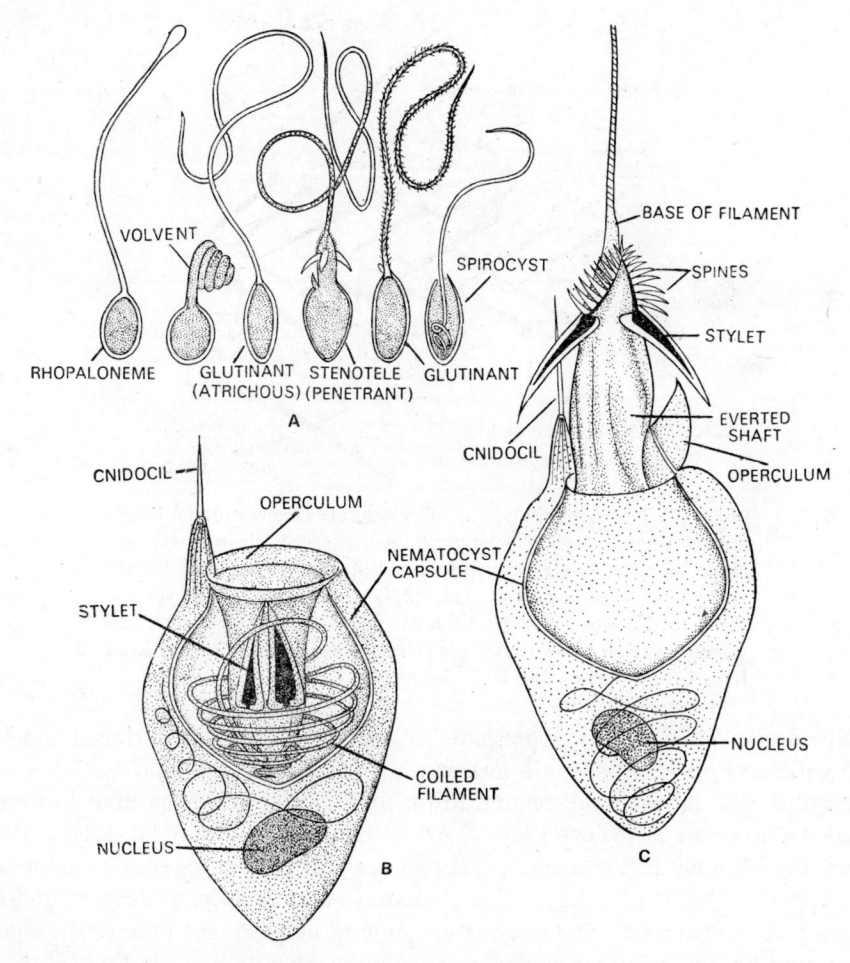

FIG. 5.6. Nematocysts of *Hydra*. A—Different types of nematocysts; B—Nematoblast with undischarged thread; and C—Discharged nematocyst.

In the development of a nematocyst, an interstitial cell produces in its cytoplasm a structure which has at first the appearance of a vacuole and which later develops into a minute capsule containing a thread that becomes surrounded by a fluid. Apparently the thread originates as an ingrowth from one side of the capsule. The nematocyst is not a cell but a capsule containing a hollow inverted thread and is produced within the cytoplasm of a modified interstitial cell, the cnidoblast. The cnidoblasts may migrate laterally from body to tentacles. The cnidoblasts are not under nervous control, they appear to be **independent effectors,** that is, structures responding directly to stimuli. The cnidoblasts and their contained nematocysts are not affected by mechanical stimuli, but suitable chemical stimuli are very effective in bringing about their discharge. These chemical substances diffuse into the water from the small organisms that are the food of the hydra.

As is shown in the diagrammatic representation of an electron micrograph (Fig. 5.4 B) of a cnidoblast of *Hydra* the cytoplasm shows usual ultrastructures.

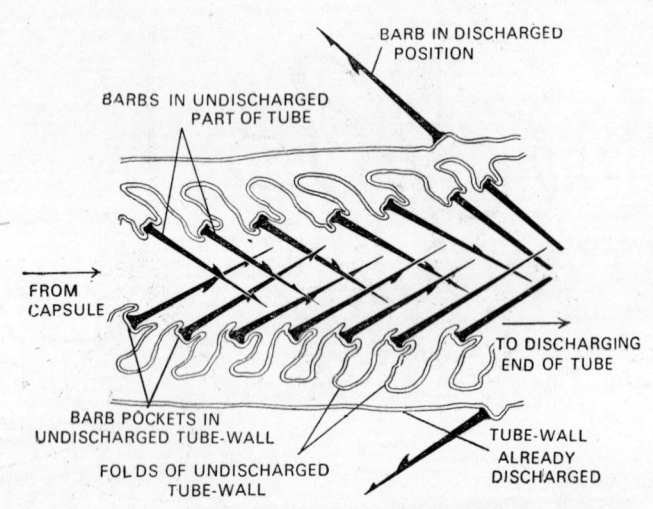

FIG. 5.7. The nematocysts of Coelenterat. Diagrammatic longitudinal section through the partially discharged tube of a *Holotrichous isorhiza*. Within the smooth-walled tube already discharged with its widely spaced and backwardly directed barbs is as yet undischarged tube with spiral folds in its wall and packed with forwardly directed barbs (based on electron microscope photographs from L.E.R. Picken and R. J. Skaer).

Endoplasmic reticulum is represented by rough and smooth surfaced lamellae and vesicles. Free ribosomes are scattered throughout the cytoplasm. The Golgi apparatus lies in the basal region. Mitochondria, lipid droplets and multivesicular bodies occur in the cytoplasm. A small bundle of myofilaments extends from the capsule of the nematocyst and lies in the basal region of cnidoblast. The capsule is made of collagen-like protein. Fine microtubules occur round the capsule of nematocyst. The nucleus lies on one side near the base of the cnidoblast, and has an indistinct nucleolus.

The cnidocil consists of a central core made of smaller fibres surrounded by large rods. The core is a modified cilium. Numerous dense granules are present on the periphery of the cnidocil both inside and outside (intra- and extracellularly).

The discharge of the nematocyst is brought about by both chemical and mechanical stimulation—each stimulation alone cannot bring about discharge. The nervous control is evident from the fact that acetylcholine, epinephrine and other neurohumoral amines produce discharge. The discharge is increased by mechanical stimulation in presence of enzyme substrate. The actual enzyme causing eversion is not known.

The nematocysts discharge by the eversion (turning inside out) of the tube to the exterior rather suddenly. The cause of the sudden eversion is not exactly known. Many theories have been put forward to explain this, but the most acceptable and probably the oldest suggests that the nematocyst discharges because of the increase of osmotic pressure inside the capsule. This increase of pressure

## Phylum Cnidaria (Coelenterata)

is brought about by the entry of water inside the capsule. With the help of coloured water it has been demonstrated that water enters in when the nematocyst discharges, further discharge takes place only in the fluid medium. Contraction of the nematoblast is also said to contribute to the discharge. Thus, the intake of water into the capsule plus the assistance, in some cases, of a nematoblast

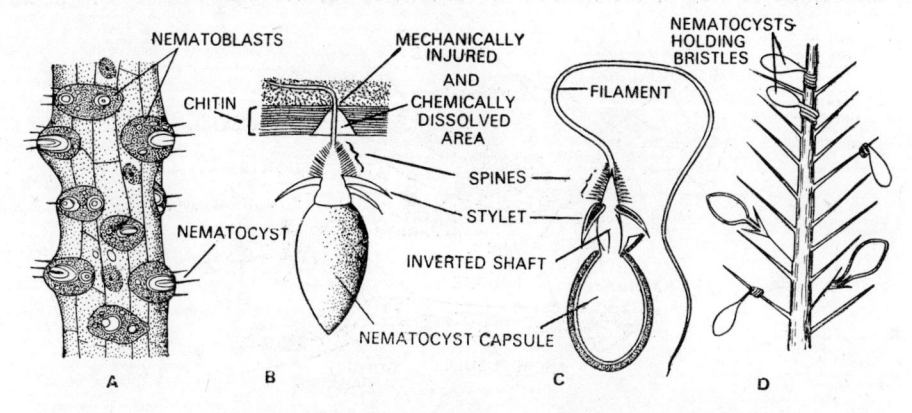

FIG. 5.8. A—Tentacle with nematocyst batteries; B—Penetrant nematocyst after penetrating the hard chitinous covering of the prey; C—Penetrant nematocyst; D—Several nematocysts have captured the bristle of prey.

contraction is the most satisfactory theory based on the available facts, that explains the discharge of a nematocyst.

There are three major types of nematocysts (Fig. 5.6) apparently to serve different functions. The largest in *Hydra* is the **penetrant** type. The tip of the thread in this case is open and it penetrates the tissues of the prey and releases poisonous fluids. The pear-shaped **volvent** types are closed at the tip and are used to hold the prey by binding about its bristles. The **glutinant** or **adhesive** type is used to fasten the tentacles to solid objects when the hydra is looping. The glutinant type presents two varieties in itself. The **streptoline glutinant** is oval, the thread of which bears minute spine-like structures and may coil upon discharge, whereas, the **stereoline glutinant** discharges a straight unarmed thread. A nematocyst can be discharged only once. Used ones are discarded and are replaced by new ones.

The poisons released by the discharge of nematocysts appear to be of several kinds, though they are probably all proteins, and mostly paralysing neurotoxins. Certain enzymes such as cholinesterase and phosphatases are also present in certain nematocysts. In the majority of cases it has paralysing action on the prey and is obviously toxic. In some cases the material evokes a burning sensation in the human skin, but not so among the Hydrozoa.

(*d*) The **nerve cells** are derived from the ectoderm but lie in close contact with mesogloea, which they actually penetrate and form a network on its endodermal as well as ectodermal surface. Each nerve cell has a cell-body containing the nucleus, and a number of branching processes, the finer ends of which form a **nerve net**. The oval nucleus is surrounded by a porous membrane. Nucleoli may or may not be present. Many free ribosomes are present, mitochondria may be scattered or clustered while the Golgi apparatus is quite prominent divided in two or three regions. The tubules are straight and neurites are longer than 10.0μ.

Some nerve fibres are connected to the receptor cells and others to the musculoepithelial cells and to the endodermal cells.

(e) The **sensory cells** or **receptors** are small columnar cells more plentiful in the ectoderm than in the endoderm. At its free surface each sensory cell bears

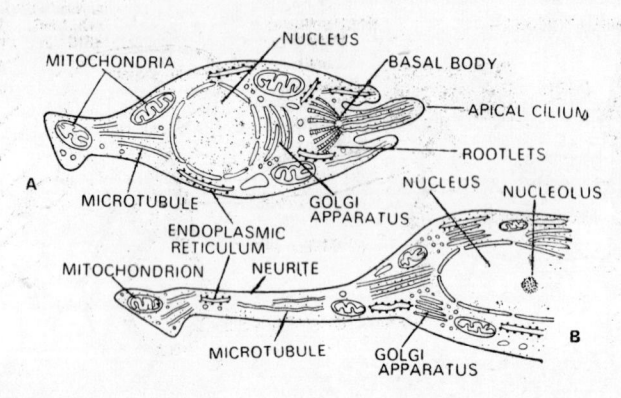

FIG. 5.9. Diagrammatic representations of electron micrographs of *Hydra*. A—Nerve cell; and B—Sensory cell.

a minute conical projection and the inner end is connected to a nerve fibre. The ultrastructure of a nerve cell is shown in the Fig. 5.9.

(f) The **reproductive cells** are produced by the repeated divisions of some of the interstitial cells in certain restricted regions of the body. These regions of active germ cell proliferation are called **gonads**, which may be differentiated into ovary or testis at a later stage.

II. **Endodermal cell types**. The endoderm of *Hydra* is primarily made up of nutritive muscular cells that form the main bulk of endoderm and gland cells. Interstitial cells and cnidoblasts also occur in the endoderm, though rarely. The distribution of sensory and nerve cells is similar to that of the ectoderm.

(a) **Nutritive muscular cells** are tall columnar cells lining the coelenteron. In the hypostome region the cytoplasm of the nutritive muscular cells is granular and homogeneous, but in the tentacles and basal disc these cells are highly vacuolated with little cytoplasmic content. They are highly developed in the column. The free surface of each nutritive muscular cell is thrown into delicate folds (microvilli) of varying lengths, forming a felt work of fibrillar or filamentous material. Each cell bears a pair of typical flagella. The cytoplasm of the cell is filled with mitochondria, glycogen granules, lipid droplets, vesicles and vacuoles. Endoplasmic reticulum (smooth and rough surfaced) is sparse. Golgi apparatus lies near the nucleus, free ribosomes are abundant. Nucleus is basal and contains a nucleolus. Muscular processes have myofilaments and act as circular muscles. When contracted the diameter of the body is reduced and consequently the length of the body is extended.

The shape of the nutritive muscle cell differs in different regions of the body. In the tentacles they are pyramidal and contain a large intercellular space in which occur lipid droplets and food material. Free microvilli on the surface and pinocytotic vesicles at the bases are also present. In the hypostome the cells are

## Phylum Cnidaria (Coelenterata)

of irregular form having numerous gland cells between their bases. In the peduncle the cells are small and cuboidal. Each cell has a large intracellular space containing lipid droplets and food materials. The nutritive muscular cells of the base are large, cuboidal containing large intracellular spaces in which microvilli, a few pinocytotic vesicles and lipid droplets are present. Nutritive muscular cells are of two types—**amoeboid** and **flagellated**. The broader end of amoeboid cells projecting in the coelenteron produces pseudopodia to engulf the particles of food. The cells with whip-like flagella cut the particles of the food into pieces.

Large endoderm cells of one species of hydra (*H. viridis*) lodge small spherical green algae called **zoochlorellae** in large numbers imparting green colour to the animal. They are symbiotic plant bodies as were noted in some radiolarians and are also found in many other coelenterates such as corals and sea anemones.

(*b*) **Gland cells** lie singly between the nutritive cells. They are present all through the endoderm excepting in the tentacles. They are numerous in the oral region but their number gradually reduces towards the aboral region. There are two types of gland cells. The first is smaller in size and oval in form. The end facing the coelenteron is broader and the other end narrows to a thread-like projection touching the mesogloea. The cytoplasm is basophilic and vacuolated. Nucleus is distinct and is situated at the narrow end of the cell. The second category of the gland cells exhibits less basophilic cytoplasm. Vacuoles of different sizes are dispersed within the cell-body. Nuclei are not distinctly visible. The ultrastructure shows mitochondria with closely packed cristae. The endoplasmic reticulum consists of rough-surfaced lamellae packed in groups. They occupy the basal region of the cell. The plasma membrane has few microvilli and flagella and is covered by a feltwork of fibrillar nature. Golgi apparytus occurs near the nucleus. The lamellae of Golgi apparatus are sometimes dilated and filled with material similar to secretory granules.

It has been observed that *Hydra* loses its old cells through the basal disc and tip of the tentacle. New cells are believed to be formed in a region, immediately beneath the hypostome. This subhypostomal region is called the "growth zone". The cells produced in this region migrate in two directions, towards the tentacles and the towards basal disc. This cell migration not only replaces the old cnidoblasts but also renews cells of the entire column. The body of *Hydra* thus never grows old. It is because of such a phenomenon of cell replacement that this animal is said to be "immortal."

**Mesogloea.** The gastrodermis and endodermis are separated by a jelly-like matrix containing cells or fibres called the **mesogloea** (compared to this the **mesolamella** is a thin acellular matrix, though at many places the words have been used interchangeably). The mesogloea in *Hydra* is $1.0\mu$ thick in which tiny filaments are embedded. The filaments run in various directions, they are parallel to the long axis or run obliquely to it. The mesogloea contains small granules of glycogen and is continuous with intercellular spaces between the epitheliomuscular and digestive cells.

### Movements

*Hydra* is a quiet animal which remains attached to the substratum by its basal disc. Some observers have seen it moving to favourable position by gliding along the substratum by amoeboid movement of the cells of the basal disc.

Occasionally the basal disc secretes a gas-bubble which enables it to rise in water and float on the surface (Fig. 5.10). Normally, however, the hydra moves

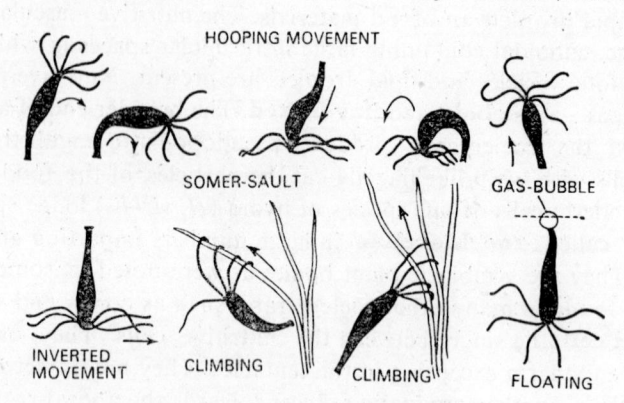

FIG. 5.10. Movement of *Hydra*.

by a **somersaulting** or **looping** type of movement. The erect body bends over, attaches the tentacles to the substratum by glutinant nematocysts, releases the basal disc, moves it to a new site in the direction of movement and then disengages the tentacles and again assumes an upright position. The looping process continues till the proper site is reached. Some hydras have been seen moving with the help of tentacles in an inverted position. An interesting mode of "climbing" has been reported in the case of the brown hydra, *Pelmatohydra oligactis*. It attaches its long tentacles to some objects, releases the basal disc and then contracts the tentacles. As compared to the brown and white hydras the green hydra is more active.

The body of hydras also shows movements. It can extend its body to a remarkable degree and may look like a long slender thread, almost transparent in consistency. On being disturbed it contracts immediately to the size of a small barrel. The contraction is brought about by the contractile action of the muscle-tails of the musculoepithelial cells, whereas, extension of the body is affected partly by the elasticity of the mesogloea and partly by the contraction of the muscle-tails of the endoderm cells which makes the body thinner. The tentacles are very mobile structures, they move in all directions. Their movements are apparently purposive. They also contract after the application of an unpleasant stimulus. The withdrawal of tentacles and contraction of the column are followed by a period of rest after which sets in a gradual extension of the tentacles and the column. Amidst abundant food material the extension is rather quicker.

## Feeding and Digestion

The food of *Hydra* consists of minute Crustacea, insect larvae and similar animals. It also damages trout nurseries by eating newly hatched ones. At times it swallows prey larger than itself. In feeding, the hydra remains attached with almost motionless tentacles trailing in the water. When a small crustacean or other animal contacts one of the tentacles in passing, the nematocysts explode and the neighbouring threads are shot out. Of these the penetrants pierce the body of the prey and inject a poisonous substance, contained in the capsule, that paralyses the prey. The volvents wrap about appendages and other parts where-

## Phylum Cnidaria (*Coelenterata*)

as the glutinants may fasten to its surface. Finally, the tentacles are wrapped round the prey and contract, drawing the prey towards the mouth, which opens widely to receive it. The victim is swallowed by means of muscular contractions of the body wall, aided by the slimy secretion from gland cells lining the inner side of the hypostome (Fig. 5.11). Localised food capture responses are well developed in all coelenterates which catch their food by means of tentacles around their mouth. In *Hydra* and sea-anemones the unstimulated tentacles will bend towards the side on which the prey is entangled. The number of tentacles which is thus involved will depend upon the stimulation caused by the food. If living food struggles violently more tentacles turn towards it and help the food capture.

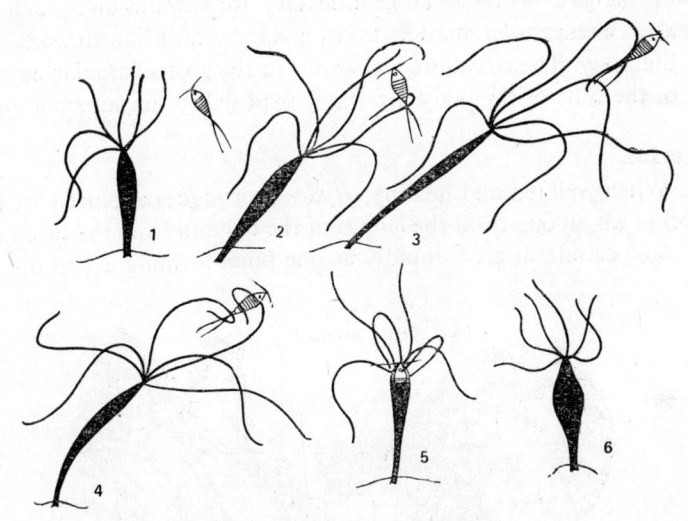

FIG. 5.11. *Hydra* capturing food.

**Digestion** takes place in the interior of the cavity. The ingested food is shifted by peristaltic contraction of the body to a position in the distal half of the enteron where the preliminary digestion takes place. As the food mass is never found at proximal levels, a physiological division of the enteron into gastric and intestinal regions is apparent, but there is no structural differentiation. The endoderm cells of the distal portion, however, contain relatively larger number of gland cells. If the food particles ingested by a hydra are examined during the earlier stages of digestion they are found disintegrating in the same manner as food in the stomach of any carnivorous animal. First the food ingested into the interior is killed by juice secreted by endoderm cells, containing enzymes chiefly of the protein and fat digesting types. The digestive fluid reduces the digestible part of the prey to a thick suspension containing many small fragments, which are then taken into food vacuoles by the pseudopodial activity of the nutritive endodermal cells. The digestion is completed within these cells and the indigestible remains are returned to the exterior. Thus, in the hydras the preliminary digestion takes place in the enteron outside the cells of the endoderm, *i.e.*, the process is **extracellular** (as in most higher animals) but the completion of the process is brought about within food vacuoles in the endoderm cells, *i.e.*, the process is **intracellular** like that of protozoans. It is clear that the process of digestion (intracellular and extracellular) is quite interesting in the hydras.

The digested food is passed by diffusion from cell to cell. Currents set up by muscular movement of the body and by the beating of the long flagella of the endoderm cells, circulate the food throughout the cavity of the body and of the hollow tentacles. The cavity serves two fold purpose of digestion and circulation, hence is called the **gastrovascular cavity**. The insoluble remnants of food particles left in the central cavity are eliminated with the water current through the mouth.

### Respiration and Excretion

No definite organs of respiration are present in *Hydra*. It is, however, concluded that both the gaseous interchange necessary for respiration as well as excretion of waste nitrogenous matter takes place by diffusion through the whole surface of the body. The circulation of water in the gastrovascular cavity ensures that most of the cells of the body are freely exposed to the surrounding water.

### Reproduction

Asexual. When well fed and healthy, *Hydra* reproduces asexually by **budding**. The buds occur about one-third the length of the column from the basal disc. Both endoderm and ectoderm grow rapidly at one point forming a protuberance into

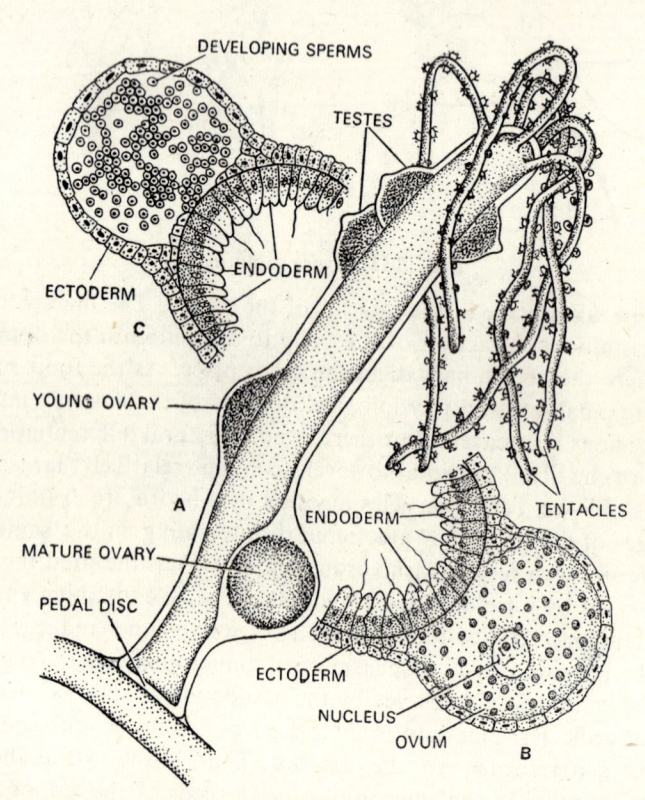

FIG. 5.12. Hermaphrodite species of *Hydra*. A—Showing the testis growing in the upper part of the column and the ovary on the lower; B—Section of ovary; and C—Section of testis

## Phylum Cnidaria (Coelenterata)

which extends a diverticulum of the gastrovascular cavity. This increases in size and soon develops tentacles at its outer end. In two or three days the bud looks like a little hydra, complete with the stalk, tentacles and mouth. At its base the enteron of the bud is continuous with that of the parent. Soon its attachment to the parent gets constricted, the enteric cavities of the two separate and the young individual detaches and takes up independent life.

### Regeneration

*Hydra* possesses a marked capacity for replacing injured parts and tentacles, etc. If it is cut into two parts each will build up its lost part growing into a full individual. Even quite small fragments are known to reform complete animals provided they are not below a certain minimum size, and that both ecto- and endoderm are present. *Hydra*, thus, provides a fairly good example of **regeneration**.

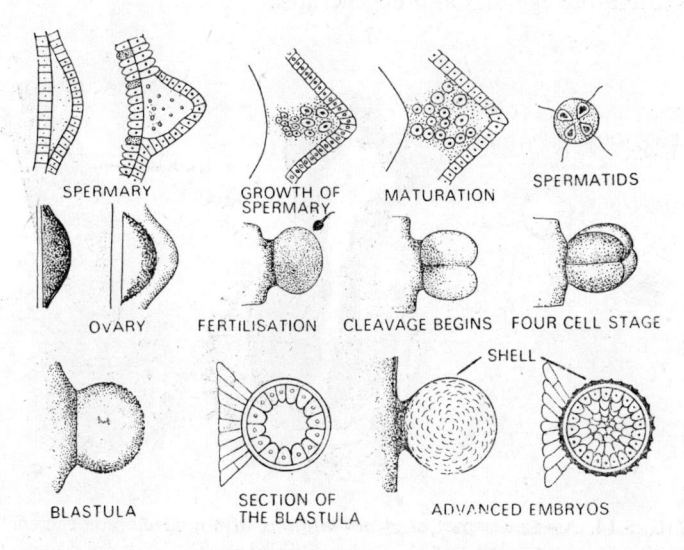

FIG. 5.13. Spermatogenesis, fertilization and development of the embryo in *Hydra*.

**Sexual**. At certain times of the year (generally autumn or winter) *Hydra* reproduces sexually. Both male and female sex cells occur in the same individual which is, therefore, described as **hermaphrodite**. In some other species the two sexes are separate and the male and female individuals can be distinguished. Even in the hermaphrodite species self-fertilisation is not possible, because as a rule, the testes become mature first and release the male cells, the **spermatozoa**, before the female cells, the **ova**, are ready to receive the sperms. The gonads are temporary structures formed by the active multiplication of some of the interstitial cells of the ectoderm which cause the body wall to bulge locally. It is difficult to distinguish a testis from ovary in the earlier stages because both consist of apparently similar masses of cells. This differs from that of the bud in that it contains no mesogloea and the endoderm cells do not take part in its formation (Fig. 5.12). The testes are commonly near the hypostome and the ovaries towards the basal disc, though this cannot be regarded as a definite rule.

During the earlier stages each gonad is filled with actively dividing **germ-mother cells**. Later these cells become transformed into sperm mother cells in the testes and are known as **spermatogonia**. These undergo a reduction division to form large numbers of tailed spermatozoa. The testes are simple conical protrusions often provided with a nipple for the exit of the sperms, but such nipples are lacking in the brown hydra in which the outer wall ruptures to allow the sperms to pass out. In the ovary, on the other hand, the interstitial cells fuse with, or engulf, each other and a single large food-filled **ovum** is formed after it undergoes a reduction division (Fig. 5.13).

**Fertilisation**. The mature egg breaks through the covering ectoderm and projects with its outer surface freely exposed to the water. The ectoderm withdraws forming a little cup or cushion on which the egg rests. Sperms discharged from the testes swim through the water and surround the egg. One of them fertilises the egg to form the zygote. If not fertilised within a short time after it is first exposed, the egg dies and degenerates.

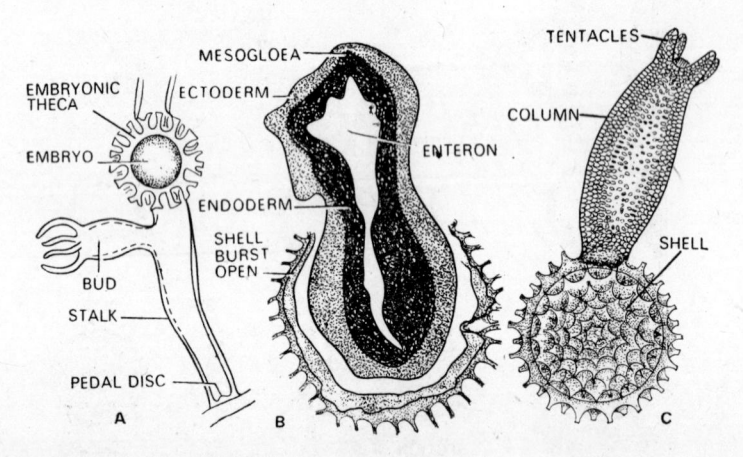

FIG. 5.14. A—Lower part of *Hydra littoralis* with a developing egg enclosed in the spiny theca; B—and C—Embryo hatching.

**Development**. Soon after fertilisation the zygote begins to divide (**cleavage**). First it forms two cells which promptly divide forming four cells (Fig. 5.13). As a result of continued division a ball of more or less equal-sized cells is formed. Internally a cavity appears in the cell-mass, which is known as a **blastula**, the cavity being termed **blastocoel** (Fig. 5.13). The blastula is thus one cell-layered hollow ball. The cells forming the single layer divide and migrate into the blastocoal from all directions gradually obliterating it. The process is known as **gastrulation**, whereas, the embryo is the **gastrula**. Meanwhile a new cavity, the enteron, appears in the mass of the central cells. As the gastrulation completes the outer layer of cells produces a heavy covering of horny envelope or **shell,** and the gastrula usually separates from the parent body and becomes attached, by sticky secretion, to the substratum. Under favourable circumstances the young hydra may hatch (Fig. 5.14 B) from the shell after a weak or so; or it may remain quiescent in the above state until next spring if the conditions are not favourable. When development is resumed, the outer layer forms the ectoderm, the inner cells form the endoderm, the embryo hollows out, tentacles develop and a mouth breaks through. Finally the shell ruptures and the young hydra hatches. At this stage the embryo is called larva by some people though essentially it is just like a young hydra.

**Behaviour**

Compared with the sponge *Hydra* shows more varied and complex behaviour. The hydras have a definite network of nerve cells (Fig 5.15) that is slightly more complicated around the mouth than elsewhere. Definite controlling centre of nerve cells or brain is not known and there are no definite pathways for the impulses which travel in any direction. Thus, an impulse picked up at any place in the network can spread from one nerve cell to another in all directions. A very strong stimulus applied to the tip of one tentacle may cause the whole animal to contract. Besides, the nerve impulses travel rather slowly here. Such a nervous system does not form an efficient mechanism, but it is evidently quite helpful to the animal basically due to its sedentary habits and limited activities. The stimuli are revived by the sensory cells from which the impulses are picked up by the nerve cells connected with the nerve net. They transmit the nervous impulses to muscle cells which contract or gland cells which secrete.

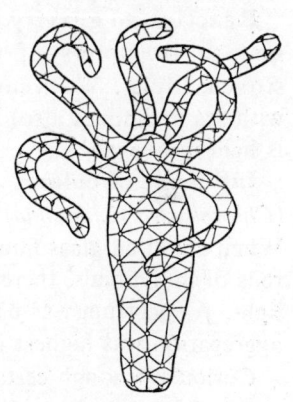

FIG. 5.15. Nerve net of *Hydra*.

**Reaction to mechanical stimuli (Thigmotropism).** The hydras are sensitive to touch. Slight touch causes them to turn the stimulated part away, whereas, a stronger stimulus causes them to contract completely and suddenly.

**Reaction to light (Phototropism).** The hydras are sensitive to light. They avoid very strong light as well as weak illumination and avoid unfavourable conditions by trial and error. This reaction to light is probably related to food getting because the animals on which the hydra feeds prefer well-lighted places. The green hydra prefers lights of greater intensity probably to expose the green alga to manufacture food easily.

**Reaction to temperature.** By nature hydras prefer cooler water, bright sunshine and a temperature betweed 90° and 95°F, seem to be ideal conditions for them. Hydras become scarce from March and are difficult to find in May, June and July. On lowering the temperature *Hydra* can live normally up to 54°F, they become sluggish with further decrease of temperature, contract their bodies and do not develop buds or gonads. When the temperature is lowered below 40°F, they become unhealthy. On further reduction of temperature, they show disintegration. At about 35°F, they die within a few hours. If the water containing them is heatad they become progressively unhealthy as the temperature rises above 70°F. At first they show great expansion and then suddenly contract. At 75°F disintegration sets in.

**Reaction to chemicals (Chemotropism).** The reaction of the hydras to injurious chemicals is negative. They try to avoid them. *Hydra* does not survive for more than a day in tap water due to the presence of large quantities of water purifiers like chlorine and alum.

It has been known for some years that the nematocysts of *Hydra* require chemical substances to stimulate their release. Only surface active materials are effective, but mechanical stimulation added to chemical produces a larger proportion of releases. It is now clear that the whole feeding reaction, at least in *Hydra littoralis*, shown as writhing of tentacles, tentacular contraction, and finally mouth opening, is initiated by **glutathione** released from the prey when it is pierced by the penetrant nematocysts. Thus, edible prey can be distinguished from inedible objects not containing glutathione.

**Reaction to gravity (Geotropism).** Ordinarily *Hydra* seems to be indifferent to the forces of gravity, the way it remains attached to the ground and stretched fully. When inverted it performs righting movements (as is the case with sea anemones also) but when *Hydra* moves by looping the righting reaction is held in abeyance.

**Influence of colour.** In *Hydra fusca* (*Pelmatohydra oligactis*) and *H. viridis* (*Chlorohydra viridissima*) there seems to be some evidence for colour preference. When coloured glass filters were placed against a tank containing some hundreds of individuals, there were gatherings on the walls over the area lit by blue light. A spectrum was used to study the reaction and it was found that the aggregation was highest in blue coloured areas.

Curiously enough certain unfavourable factors such as rich feeding, high temperature, and foul water throw the hydras in a lowered metabolic state known as "depression" under the influence of which gradual shortening of tentacles and column, beginning at the distal end, sets in. If the conditions remain unchanged this may result in ultimate disintegration, but the animal may recover with the improvement of conditions.

When *H. viridis* is at rest the tentacles and column, according to Wagner, exhibit rhythmical contractions in which those of the buds act in sympathy with those of the parent. These are believed to be due to some internal stimuli. In *H. vulgaris* no such movements have been observed. However, while waiting for the prey, it changes the direction of its tentacles about once in half an hour.

## TYPE OBELIA

*Obelia* is the nearest marine relative of the hydras that forms branching plant-like colonies attached to rocks, wharf pilings and the surface of the sea-weeds, etc. The colony arises by budding from a single hydra-like individual. The buds do not separate as in the hydras with the result that a tree-like colony is formed. The colony is polymorphic but the activities of the individuals are subordinate to the colony as a whole, therefore, they are sometimes referred to as subindividuals.

### Form and Structure

The colony of *Obelia* (Fig. 5.16) is small whitish in colour attached to the substratum by small tubes, the **stolons**, that form a branching root-like tangle called the **hydrorhiza** on the substratum. The stolons give off numerous upright buds that form polypoid colonies. The main stalk of the colony is known as the **hydrocaulus** and bears the zooids of the colony. The stolons and hydrocaulus consist of a living hollow tube, the **coenosarc**, which is surrounded by a

## Phylum Cnidaria (Coelenterata)

yellowish or brown chitinous tube, the **perisarc**. The coenosarc consists of an outer layer, the **ectoderm**, the **mesogloea** and the **endoderm**. The ectoderm

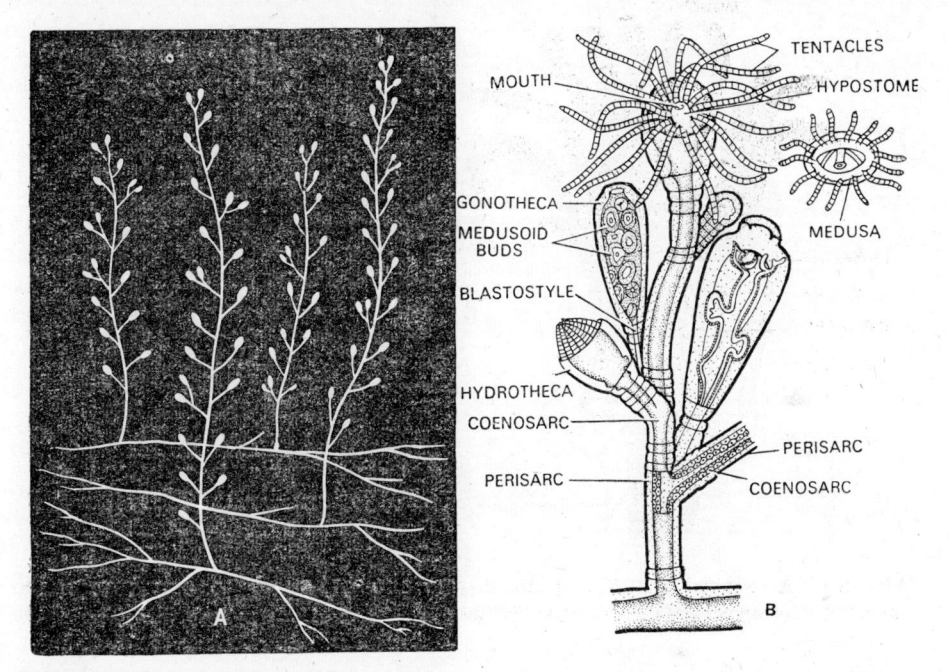

FIG. 5.16. A—*Obelia* colony attached to some object as it occurs in nature; and B—*Obelia* — a part of a colony and a medusa magnified; one reproductive polyp, gonotheca, and a part of the stalk has been drawn in section.

secretes the perisarc as a protective tube. The gastrovascular cavity enclosed by the coenosarc, is continuous throughout the colony. The zooids of the colony are of two types, **hydranths** and **blastostyles**, and are attached alternately on opposite sides of the hydrocaulus. Basically both the hydranths and blastostyles are modifications of the hydroid type of zooid and the difference in their structure is correlated with the functions they perform.

The **hydranths** (Fig. 5.17) are the feeding zooids of the colony and are of the hydroid type. The body is sac-like or vase-shaped connected by a hollow stalk at its lower end to the hydrocaulus. At the free end the hydranth bears an elongated cone-shaped structure, the **manubrium** or **hypostome** having a terminal mouth that is capable of great expansion. From around the base of the hypostome arises a circle of tentacles. The tentacles are long and tapering, **filiform**, strewn with nematocysts along the whole length. Unlike *Hydra* the tentacles here are solid and contain an **endodermal core** made of a single row of highly vacuolated, stiff, cylindrical cells. As mentioned above, the whole colony is covered with a protective chitinous secretion, the perisarc, which is absent in the hydras, and scanty in some other solitary forms (*Protohydra*). On stems the perisarc usually forms groups of rings or annular constrictions at definite points related to the branching, perhaps to lend flexibility. In *Obelia* the perisarc extends over the hydranth in the form of a cup, the **hydrotheca**, into which it can be withdrawn partly or wholly when disturbed. In some relatives of *Obelia* the perisarc does not extend over the hydranth (*Bougainvillia*). Between the hydrotheca and the stem the perisarc shows annulations. In the interior of each hydrotheca near its base is a ring-like shelf on which rests the base of the hydranth. The

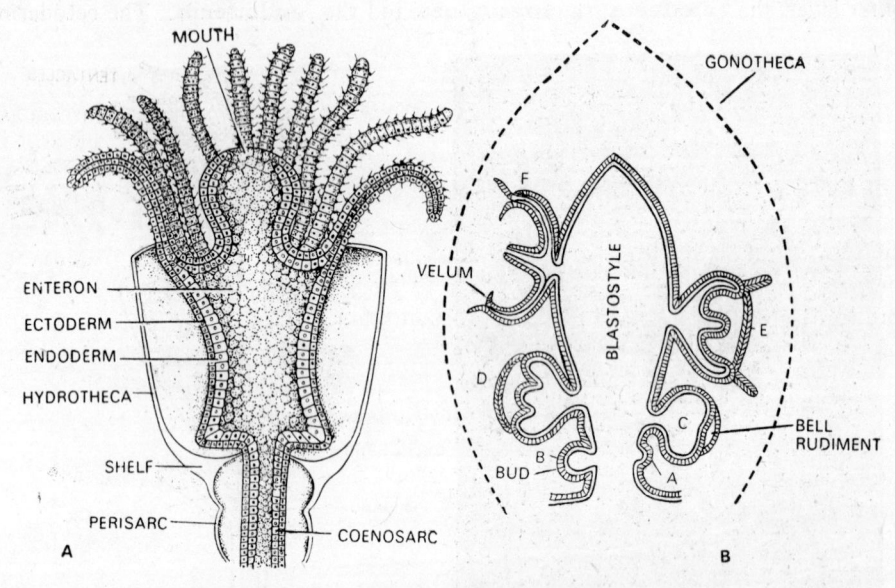

**FIG. 5.17.** A—Sagittal half of a hydranth showing the structure in detail; B—Diagram to show the development of medusa from blastostyle; C to E—Different stages in the formation of a medusa.

thin ectoderm consists chiefly of **musculoepithelial** cells. The **interstitial** cells are practically absent. **Namatocysts**, which are abundant in the tentacles being arranged in annular batteries, originate from the nematoblasts in the basal parts of the hydranths and in the coenosarc and migrate actively to their final positions. The endoderm cells are mostly furnished with flagella, that, according to some workers, may be absorbed and replaced by pseudopodia. On the whole, the endoderm cells are large and filled with granular inclusions. A **nerve-net** is present on both sides of the mesogloea.

The hydranth captures food in the same way as the hydra with the help of tentacles armed with nematocysts. The digestion is both extra- and intracellular. Partly the food is digested in the cavity of the polyp and the resulting fluid is circulated through the branches by the beating of the flagella of the endoderm, and by the muscular contraction of the polyp. Thus, food is distributed throughout the colony in a thorough co-operative fashion. The hydroids are chiefly carnivorous depending upon small animals including crustaceans, nematodes and worms, etc., for their food. Respiration and excretion take place through the entire surface as in the hydras.

**Budding**. The individuals increase their number by budding. In addition to the continuous budding on any vertical axis the horizontal stems also give rise to series of up right colonies by budding. Sexual reproduction does not occur in a polyp colony, but special reproductive zooids are formed which are specialised for asexual reproduction of a special type. These are the **blastostyles** or **gonozooids**.

The blastostyles are club-shaped projections bearing special **asexual buds** or **gonophores** on their sides. Basically a blastostyle is a modified hydranth the tentacles of which are reduced and eventually lost, the mouth is closed and the

## Phylum Cnidaria (Coelenterata)

gastrovascular cavity is reduced. The blastostyles are produced, when an erect stem has reached its full development, as buds in the axils of some of the lower hydranths. Each blastostyle is enclosed in a vase-like case of perisarc known as the **gonotheca,** the opening of which is closed by the tip of the blastostyle or by a lid. The gonotheca together with the blastostyle with its gonophore buds is called a **gonangium**. The lateral buds or gonophores develop into **medusea** (singular medusa) which when fully formed are set free and swim away from the parent colony.

**Medusa formation.** In the formation of a medusa first a bud appears as a simple diverticulum of the cavity of the blastostyle (Fig. 5.17B), soon the ectoderm splits off forming an outer layer that takes no part in further development but remains as a mantle covering. The gonophore (Fig. 5.17B), the inner layer, on the other hand, acquires a cavity which enlarges and assumes the shape of the subumbrellar cavity with the manubrium in its centre (Fig. 5.17B). The cavity, at this stage, is lined with ectoderm throughout, being closed externally by a layer of ectoderm which extends from one margin of the umbrella to the other. The layer, at last, breaks through and its remnant forms a circular shelf projecting inwards from the margin of the umbrella. (Fig. 5.17B). This shelf is known as the velum and is a permanent feature of most of the hydroid medusae but diminishes in *Obelia*, becoming almost vestigial. Later the mouth opens at the tip of the manubrium, tentacles develop as outgrowths at the margin and finally a constriction separates the medusa from the balastostyle. The little medusa thus separated escapes through the aperture at the tip.

**Medusa.** A medusa (Fig. 5.18) is a modified zooid which is solitary and free-swimming. A typical medusa consists of a bowl-shaped gelatinous disc or **umbrella** whose concave surface is described as the **subumbrellar** surface, and outer convex surface as the **exumbrella**. From the middle of the subumbrellar surface hangs a cylindrical or quadrangular projection, the **manubrium** at the apex of which is situated the **mouth**. The mouth leads into a small central cavity which is either rounded or quadrangular and is known as the **gastric cavity** or **stomach**. From the **gastric cavity** four

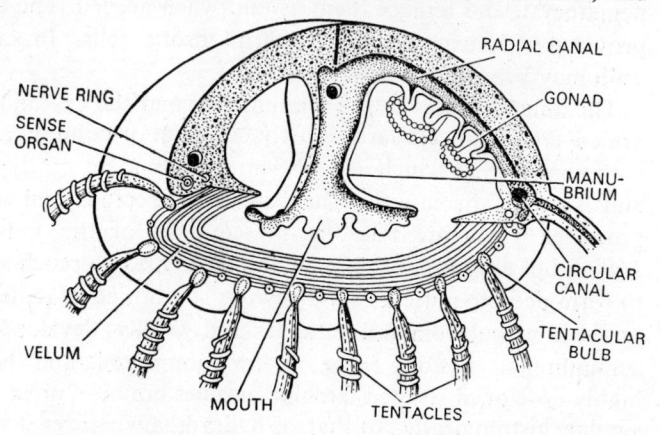

FIG. 5.18. General plan of the structure of medusa, a part of which has been removed to show internal structures (*from* Prasad and Grover).

gastrodermal canals radiate to the margins of the bell. These are the radial canals and mark out four principal radii known as **per radii**. The radial canals open into a **ring canal** (circular canal) running in the margin of the bell. Thus it is clear that the manubrium, stomach, radial canals and the ring canal constitute the gastrovascular system of the medusa. In most of the hydro-medusae the edge of the umbrella is produced inwards forming a very narrow fold or shelf known as the **velum**. The velum is rudimentary in the medusa of *Obelia*.

The margin of the umbrella is circular (Fig. 5.19) and bears numerous tentacles and sense organs. The tentacles have a solid core of endoderm and their number

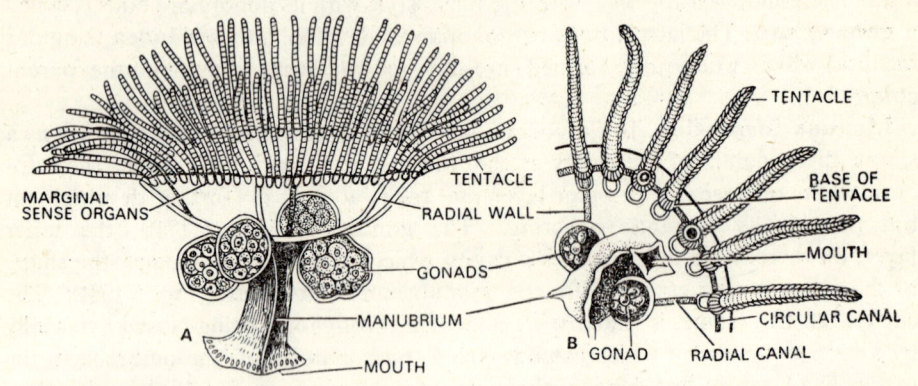

FIG. 5.19. Medusa of *Obelia*. A—Mature medusa with everted umbrella; B—A portion of the same presenting oral aspect.

increases with the age of medusa. The tentacles are given definite names depending upon their position on the margin. Those at the ends of per-radii are known as **per-radial** tentacles; those situated on the **inter-radii** (radii bisecting the angles between the perradii) are known as **inter-radial** tentacles; the **ad-radial** tentacles are situated at the radii drawn so as to bisect the eight angles between the per-radials and inter-radials. At the base each tentacle is swollen forming an enlargement or **tentacular bulb**, which is packed with interstitial cells that secrete nematocysts and replace them as and when needed. The tentacles are abundantly provided with nematocysts and with sensory cells. In some cases the tentacular bulb may bear an **ocellus** (Fig. 5.20).

The outer surfaces of the manubrium, and the ex- and subumbrellar surfaces are covered with ectoderm mostly cuboidal or columnar. It is liberally supplied with nematocysts which are particularly abundant on the margin of the umbrella and around the mouth. The bases of the ectodermal cells are drawn into long processes which are like the muscle tails of the musculo-epithelial cells of *Hydra*, but are more prominent and developed so much so that they may be said to form definite muscle cell. They are smooth contractile and longitudinally disposed in manubrium and tentacles and weakly developed radial fibres in the subumbrella. Besides these in the subumbrella and the velum (where present) highly developed striated circular muscles are also present. The endodermis is similar histologically to that of hydranths wherever it is digestive in nature. It lines the manubrium, the stomach and forms the walls of the radial and ring canals. In the aboral walls of the canals it is flattened. This flattened endodermis continues as a one-layered sheet the **endodermal lamella,** between the radial canals. The mesogloea is a thick gelatinous mass lying wholly between the endoderm and the exumbrellar epidermis. It is crossed by fibres of unknown origin.

The **nervous system** consists of a nerve-net (like that of *Hydra*) on each side of the mesogloea, but in the medusa, nerve cells and fibres are specially concentrated around the margin of the bell to form a double **circular nerve**

## Phylum Cnidaria (*Coelenterata*)

**ring**. The upper ring receives fibres from tentacles and sensory structures of the margin and the lower ring receives fibres from the **statocysts** and supplies the subumbrellar musculature.

The **sense organs** make their appearance for the first time in the invertebrate world in the medusa. Their appearance is associated with the active free swimming life. The sense organs, essentially structures for the perception of the factors of the external world, would be of little use to the sedentary colonial phase. The **statocysts** and **ocelli** ("eyes") are the chief sensory organs of the hydromedusae. The statocyst or **marginal vesicle** is an ectodermal pit or vesicle on the subumbrellar surface just inside the margin (at the velar base where velum is present) filled with fluid. This sac contains a movable round particle made of organic material and calcium carbonate called the **statolith**, secreted by a large cell, the **lithocyte** (Fig. 5.20 A). The wall of the sac bears a patch

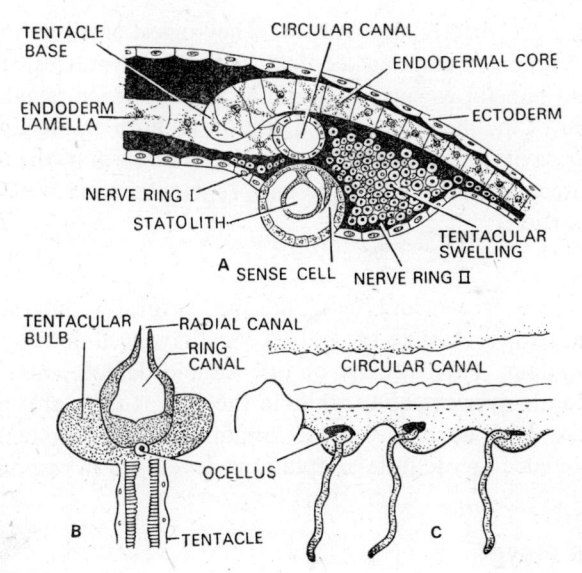

FIG. 5.20. A—Longitudinal section of the base of an adradial tentacle; B—Tentacle base of *Sarsia*; and C—Bell margin of a anthomedusan.

of sensory cells bearing fine processes which arch over the statolith. The statocysts are usually located between the bases of the tentacles and are considered to be the organs of equilibrium. As mentioned above they are connected to the lower nerve ring. When the medusa is swimming horizontally the statolith rests at the bottom of the statocyst, but when the medusa is inclined the statolith falls against the process of the sensory cells and stimulates them. A nerve impulse is initiated and transmitted to the nerve ring and by the contraction of the required ectodermal muscles, the medusa is again brought into the horizontal position. The statocysts are also believed to aid the swimming movement of the medusa.

The **ocelli**, when present, appear red, brown or black spots usually one on each tentacular bulb (Fig. 5.20 B, C), which may or may not persist as a tentacle. In the forms in which the tentacles occur in clusters (*Bougainvillia*) the single tentacular bulb of the cluster bears as many ocelli as there are tentacles. An **ocellus** consists of a group of pigment cells (probably epidermal) interspersed with sensory nerve cells in its simplest form, and presumably serves for clear and exact perception of light.

**Swimming movements.** The medusa swims by alternate closing and opening movements of the bell. The closure is brought about by contraction of ectodermal muscle-tails. As the contractile elements are best developed in the subumbrellar surface the wave of contraction spreads from there and ends at the upper surface, where it does not occur at all. Opening of the bell is brought about by the elastic mesogloea that regains its shape. The opening is also partly helped by contraction of the muscle-tails in the middle of the upper surface. Thus, the rhythmic alternating closing and opening movements are responsible for their locomotion but they can't withstand currents and winds and hence are easily swept away.

## Feeding

The medusae are strictly carnivorous. They ingest small animals of appropriate size that happen to come in contact with the tentacles. The crustaceans, nematodes and annelid worms, eggs and larvae and even small fish constitute common food of the medusae. The digestion is both extracellular (in the main gastrovascular cavity) and intracellular in the endoderm of the manubrium, stomach and tentacular bulbs, when present. Tentacular canals, when present, also ingest food particles.

## Adaptations

The medusae are free-swimming forms and, naturally they show many adaptations for such a life. Unlike the polyps they have well-developed sense organs and nervous system. Their organs of offence and defence, *i.e.*, the nematocysts are distributed all over the body while in the polyps they are present only on restricted areas. Well-developed sense organs, nervous system and organs of offence and defence are definite adaptations developed in response to free-swimming life.

## Medusa and Polyp

As compared to the polyp the medusoid form seems to be very different, but they are strictly homologous structure (Fig. 5.21), and one can be derived from the other. If a medusa is kept with manubrium and subumbrella upwards the comparison becomes very easy. The mouth and manubrium of the medusa are homologous with the mouth and hypostome of the polyp. The tentacles occupy comparable positions in both and the apex of the umbrella of the medusa corresponds with the base of the polyp. Suppose the tentacular region of the polyp is pulled out into a disc-like form and then bent in the form of a saucer with concavity to-

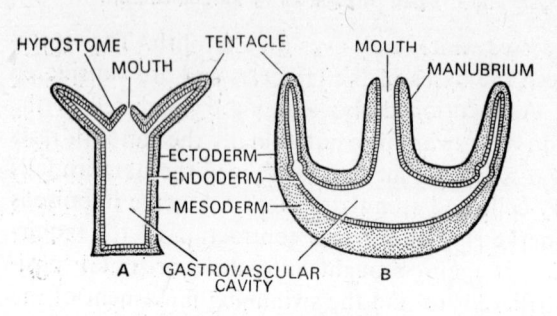

**FIG. 5.21.** Structure of a medusoid (B) form derived from a polypoid (A) form.

## Phylum Cnidaria (Coelenterata)

wards the manubrium, an umbrella-like structure will be formed. The wall of the umbrella will be double enclosing a narrow space *i.e.*, gastro-vascular cavity which is lined by endoderm. The endoderm grows slowly and the layers from all sides of the cavity meet forming a compact endoderm lamella and thus obliterating the gastro-vascular cavity, which persists at four meridional areas (radial canals) and a circular area close to the edge of the umbrella (circular canal).

### Reproductive Organs

Medusae are sexual individuals of the Hydrozoa. Nearly always they are dioecious (except the monoecious *Eleutheria*). The gonads are four in number, one in the middle of the length of each radial canal. They arise as ventral diverticula of the radial canals which push the mesogloea and the ectoderm before them and project from the subumbrellar surface. Their germ cells develop from various sources. Usually they develop from the interstitial cells located either in the ectoderm or endoderm. They are also known to develop from ordinary ectoderm and endoderm. The germ cells in the medusa of *Obelia* make their appearance quite early when the medusa itself is being formed on the blastostyle. They originate in the ectoderm of the manubrium, then migrate to the endoderm, eventually to the gonads. Before maturity they leave the endoderm and between the ectoderm of the subumbrellar surface and the mesogloea. The route of the sex cells from the place of their development to the gonad is more or less definite for the species. The gonads are simply accumulations of sex cells which when ripe escape into the sea by the bursting of ectodermal covering. The spermatozoa are of the ordinary flagellate type and are formed after typical spermatogenesis. The ovocytes form the egg in the same manner as in *Hydra*.

### Development

Fertilization takes place in the sea. The development is also simple but apparently without definite arrangement of blastomeres in the embryo. The segmentation ultimately results in the formation of a hollow blastula which becomes solid by the budding off of cells from the blastula wall (compare *Hydra*). Later the outer layer of cells becomes uniformly ciliated and forms the larva known as **planula** (Fig. 5.22). The outer layer of the planula is the ectoderm, the inner is the endoderm which splits forming a cavity, the enteron. The planula swims freely for a time, then settles down and attaches itself by one end to some submerged object. It then **elongates** develops a mouth, sprouts tentacles and thus changes into a hydroid zooid called **hydrula**. The hydrula later develops lateral buds and by a continuous process of budding forms a new colony. The *Obelia* colony is without sex organs and reproduces asexually by budding.

From the above description it is evident that the colonial hydroid phase in *Obelia* reproduces asexually by budding and certain of its buds (medusae) develop gonads, are released from the colony and reproduce sexually producing the asexual colony. Thus, there is an alternation of asexually reproducing colonial generation and sexually reproducing medusoid generation. The asexually and sexually reproducing generations here are very different from each other in form. This phenomenon of alternation of asexual and sexual generations has been called **alternation of generation** or **metagenesis**. But in this case both the generations are diploid. Strictly speaking alternation of generation indicates an alternation of **diploid** (asexual) and **haploid** (sexual) generations (as in certain plants).

Here both the generations are diploid (only gametes being haploid) and the so-called generation bearing gametes does not arise by the division of a single

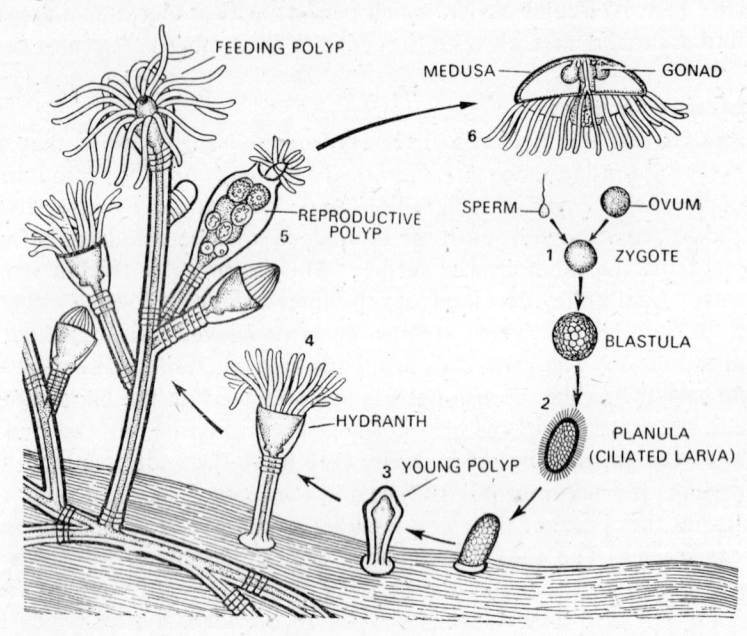

**FIG. 5.22.** Life-cycle of *Obelia*. 1—Zygote; 2—Planula; 3 and 4—Planula attaches to the substratum and develops into a polyp; 5—Polyp, which produces a full-fledged colony by budding; and 6—Medusa of *Obelia*.

cell. The medusa is merely a zooid modified for leading a free-swimming life to enable the sedentary colony to effect dispersal of gametes and prevent overcrowding of the species at a place. Thus, it can be easily concluded that the zooids are **polymorphic**; some are specially modified for feeding (hydranths), some for budding (blastostyles) and some for sexual reproduction and dispersal of gametes (medusae). It follows, therefore, that in *Obelia* there are several different types of individuals in a colony to perform different functions in life. This is called **polymorphism**.

**Polymorphism.** The phenomenon of polymorphism is very well illustrated by some of the coelenterates. *Hydra* is a simple individual polyp that performs all vital functions of life including reproduction. *Obelia*, on the other hand, is a colonial polyp showing polymorphism. The feeding hydranths of *Obelia* perform all functions of life except reproduction for which there are different zooids called Blastostyles, which produce medusae for dispersal. Polymorphism shows a little more advance in *Hydractinia*, which lives on shells of hermit crabs. It has, apart from separate feeding and reproductive zooids, a third form, the fighting polyp, which is slender without mouth and bears many spherical growths on which are situated large number of nematocysts.

Polymorphism attains a wide range in the complexity of colony structure in the members of the order Siphonophora, which form floating or swimming colonies of specialised individuals. The polypoid zooids are modified to form

## Phylum Cnidaria (Coelenterata)

three kinds of individuals, **gastrozooids, dactylozooids** and **gonozooids**. The gastrozooids (**siphons**) are the feeding zooids like the polyp but without usual tentacles. From near the base of each gastrozooid springs a long and contractile hollow tentacle that bears lateral contractile branches, the **tentilla**, each of which terminates in a complicated knob or coil of nematocysts. The dactylozooids (**feelers** or **palpons**) do not have mouth and their basal tentacle is not branched. The gonozooids may be like gastrozooids (as in *Velella* and *Porpita*) possessing even a mouth but are without a tentacle. In others the gonozooids form branched stalks (probably blastostyles) and bear grape-like clusters of gonophores. With such gonozooids are associated some tentacle-like dactylozooids known as **gonopalpons**.

The medusoid forms become modified into the **swimming bells** (Fig. 5.23), **bracts** and **pneumatophores**. The swimming bell (or **nectocalyx**) is a

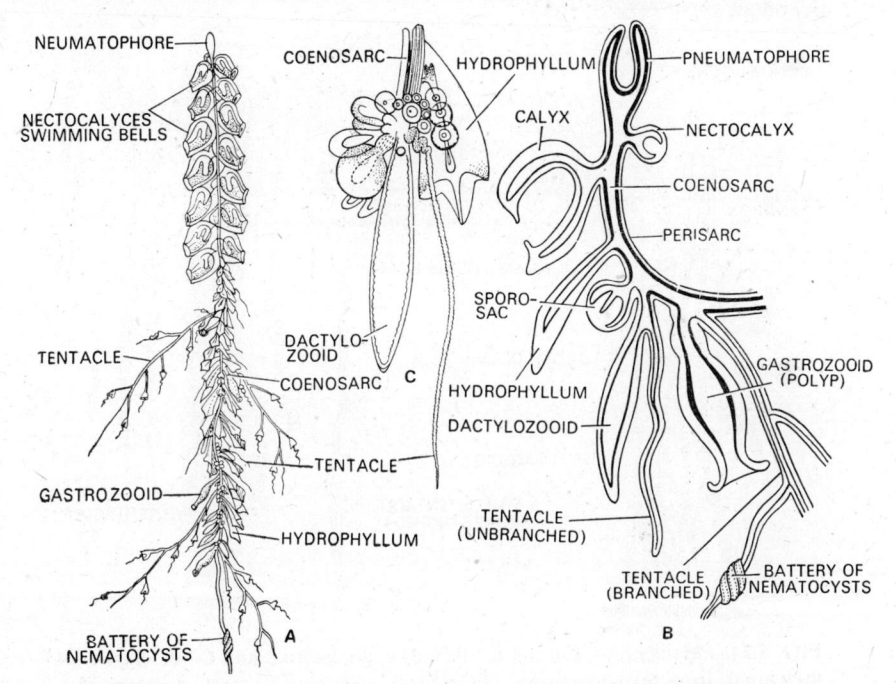

FIG. 5.23. A—*Halistemma*; and B—Siphonophore structure (diagrammatic).

medusoid bell often of various shapes with velum, four radial canals and a ring canal, but without manubrium, mouth, tentacles or sense organs. They are highly muscular hence form good swimming organs. The bracts, as they appear, do not look like medusae at all. They are protective in function and are thick gelatinous leaf-like medusoid forms containing a much reduced gastro-vascular cavity. The pneumatophore or float is an inverted medusan bell without meso-gloea. The exumbrellar surface is known as **pneumatocodon** and the umbrellar wall forms the air-sac or **pneumatosaccus**. Both the external wall and the wall of the air-sac have the usual two-layered structure. Between the two walls is the gastro-vascular space, that, in complicated floats, is divided by vertical partitions. The ectoderm of the air-sac secretes a chitinous layer that lines it internally. At the bottom of the air-sac there is an enlarged chamber, the funnel where the lining epithelium becomes glandular forming the **gas gland**, which secretes gas similar to air and fills the air-sac.

*Halistemma* (Fig. 5.23) is a siphonophore characterised by an apical float that is succeeded by a long stem bearing swimming bells closely pressed together in two alternating rows (Fig. 5.23A). This part of the colony is called the **nectosome**. Below the bells the remaining portion of the stem is called **siphonosome**. This long and tubular structure bears various types of individuals including the gastrozooids, dactylozooids, gonozooids, bracts (hydrophyllia) and tentacles, etc.

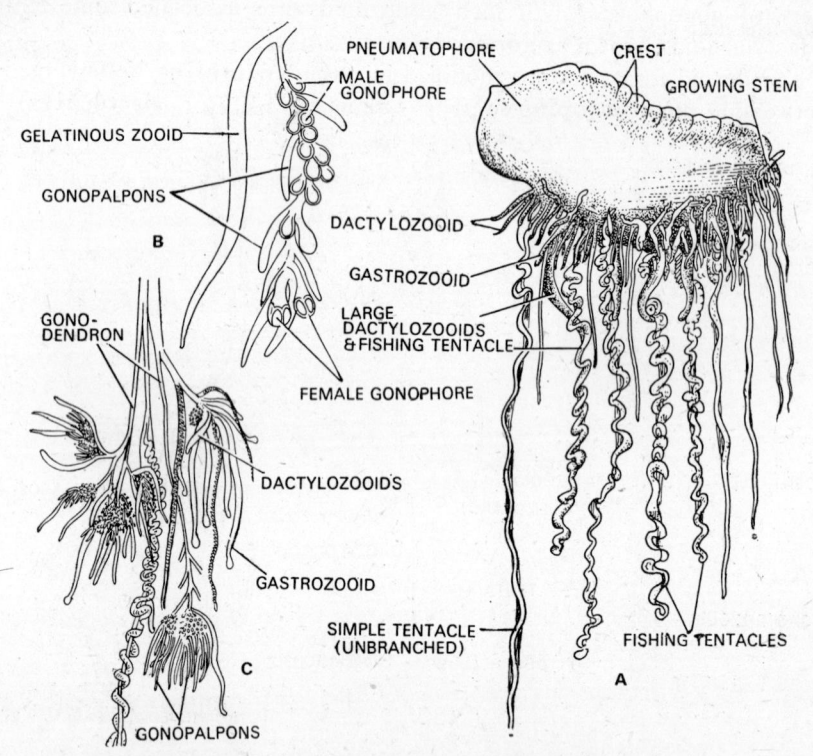

FIG. 5.24. *Physalia*. A—Young; B—Part of gonodendron; and C—A cluster of individuals from sexually mature *Physalia*.

*Physalia*, the Portuguese man-of-war, represents the other type of the group in which there is large float devoid of swimming bells or bracts. The shortened stem consists only of a budding coenosarc on the ventral surface of the large contractile crested float. From the underside of the float hang down various kinds of specialised zooids such as blastostyles (with attached medusae) dactylozooids, gastrozooids, and tangles of long tentacles that may reach a length of 60 feet and that are armed with specially large stinging capsules which can readily paralyse a large fish. The vivid blue floats provide beautiful and familiar sight on the surface of the warm seas, but it is not a welcome sight for swimmers, who know that the trailing tentacles an inflict serious and sometimes fatal injury on man (Fig. 5.24).

*Vellella* or "little sail" (Fig. 5.23, D) is a common siphonophore of warm seas. The stem is shortened to a flat coenosarc, which together with the float

## Phylum Cnidaria (Coelenterata)

forms a firm round disc containing many concentric air chambers and provided with an erect sail-like process on the upper surface. Around the margin is a single row of tentacle-like dactylozooids. In the centre of the underside is the feeding polyp, or gastrozooid. The gastrozooid is surrounded by numerous reproductive polyps which bud off free-swimming medusae.

*Porpita* is also similar to *Velella*, but is without sail (Fig. 5.28A, B). The colony consists of a disc-like body enclosing a chambered chitinoid shell containing air comparable with the float of *Physalia*. Around the margin of the disc are long dactylozooids. The under surface is beset with blastostyles provided with mouths and bearing medusae. The gastrozooid is in the centre as in *Velella*.

## CLASSIFICATION OF HYDROZOA

It is divided into five orders, *viz.*, Hydroida, Hydrocorallina, Trachylina, Siphonophora and Graptolithina.

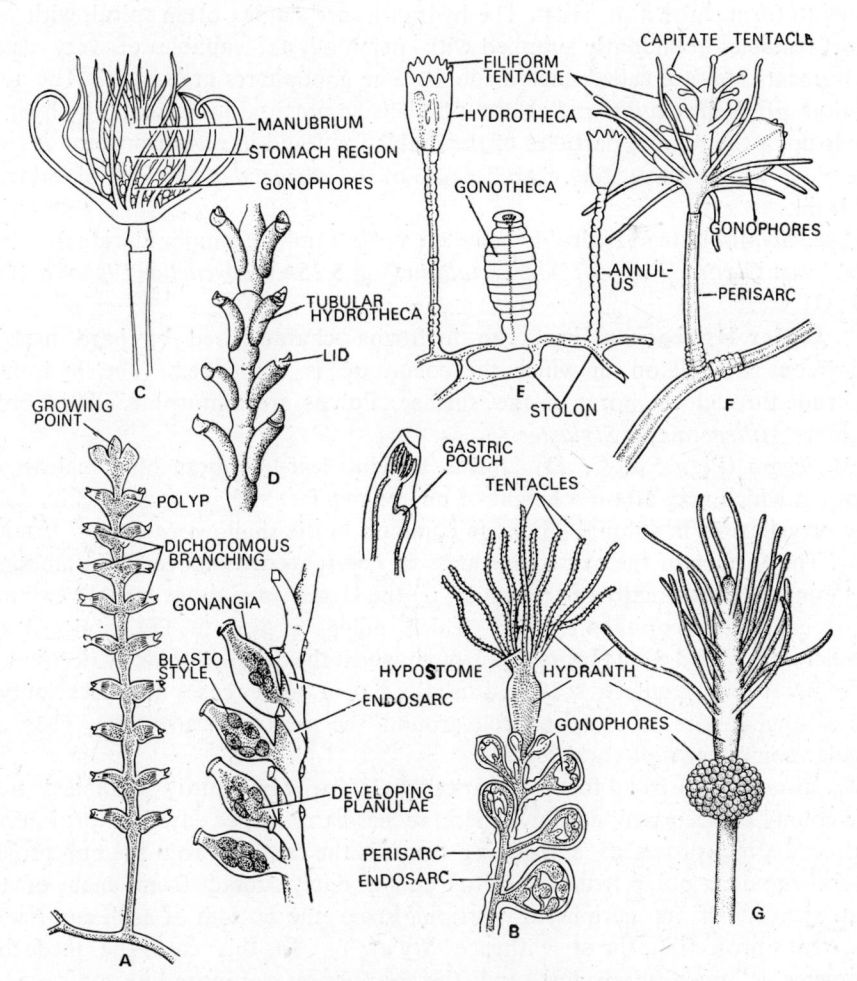

FIG. 5.25. Some calyptoblastean hydroids. A—*Sertularia*; a branch, a hydroid of *Sertularia* showing gastric pouch, (above B) a branch showing gonangia (between A and B); B—*Bougainvillia*; C—*Tubularia*; D—*Diphasia*; E—*Clytia*; F—*Pennaria*; and G—*Clava*.

1. **Order Hydroida** in which the polyp generation is well developed, solitary or colonial, and usually buds off small free-swimming medusa that bear ocelli and ectodermal statocysts. The hydroid types are either without theca (Athecata or Gymnoblastea) or are thecate (Calyptoblastea or Leptomedusae or Thecophora). To the Athecata belongs the family Hydridae including the fresh-water polyps or the hydras such as *Hydra*, *Pelmatohydra*, etc., without medusa stage. The foregoing account of the fresh-water polyp applies to these.

Other well-known athecates include *Tubularia* (Fig. 5.25) that forms loose racemose colonies of long-stemmed pink hydranths distinguished by oral and proximal circles of filiform tentacles and grape-like bunches of gonophores. *Bougainvillia* (Fig. 5.25B) is another type without hydrotheca in which bell-shaped medusae are usually produced, but not always set free. Gonads are located in the walls of the manubrium and have no statocysts. Marginal eye-spots are present. *Hydractinia* (Fig. 5.26A) is a curious genus in which the coenosarc is massive, very often found coating a shell, inhabited by Hermit-crabs, so as to form brownish crust. The hydranths are simple, often spiral, with very short tentacles abundantly supplied with nematocysts, capable of very active movements and are called dactylozooids. The gonophores are sessile. The association of *Hydractinia* and Hermit-crab is commensal in nature. The hydroid feeds upon the minute particles of the crab's food and in return protects it from enemies, which leave the crab because of the presence of the inedible stinging hydroid.

The calyptoblastean hydroids include *Obelia* (family Campanularidae), *Campanularia*, *Clytia* (Fig. 5.25E), *Sertularia* (Fig 5.25A), *Lovenella*, *Diphasia* (Fig. 5.25D), etc.

2. **Order Hydrocorallina.** The hydrozoa characterised by hard massive calcareous exoskeleton in which the coenosarc is embedded. The hydranths protrude through the pores on the surface. Polyps are dimorphic. This order includes *Millepora* and *Stylaster*.

*Millepora* (Fig. 5.26 C, D), forms upright leaf-like branching calcareous growths which may attain a height of one or two feet and are mostly white flesh-like or yellowish in colour. They are common in the shallow waters of tropical seas. The surface of the calcareous mass or **coenosteum** is covered with numerous minute pores usually of two types, (*i*) the larger ones about one or two mm apart, called **gastropores** through which polyps project in life, and (*ii*) the smaller ones called **dactylopores** through which the dactylozooids project in life, these pores are irregularly scattered usually but in some cases the dactylopores are arranged more or less in a circle around the central gastropore. They are regular components of coral reefs.

*Stylaster* (Fig. 5.26 F) forms remarkably elegant, abundantly branched tree-like colony of deep pink colour. The branches carry little cup-like projections with radiating processes converging towards the centre. From the cup project several zooids, a polyp from the centre and a dactylozooid from each of the compartment of its peripheral portion. From the bottom of each cup rises a calcareous projection, the style (hence *Stylaster*). In this case the medusoid character is much more lost and the gonophores are more like sporosacs or degraded reproductive zooids lodged in spherical chambers of the coral. They are found in warm tropical and sub-tropical seas from shallow to deep waters.

*Phylum Cnidaria (Coelenterata)*  217

Some extend into temperate zone.

3. **Order Trachylina**. The Hydrozoa in which usually only the medusoid stage is represented. The hydroid stage is either minute (Fig. 5.27D), or wanting

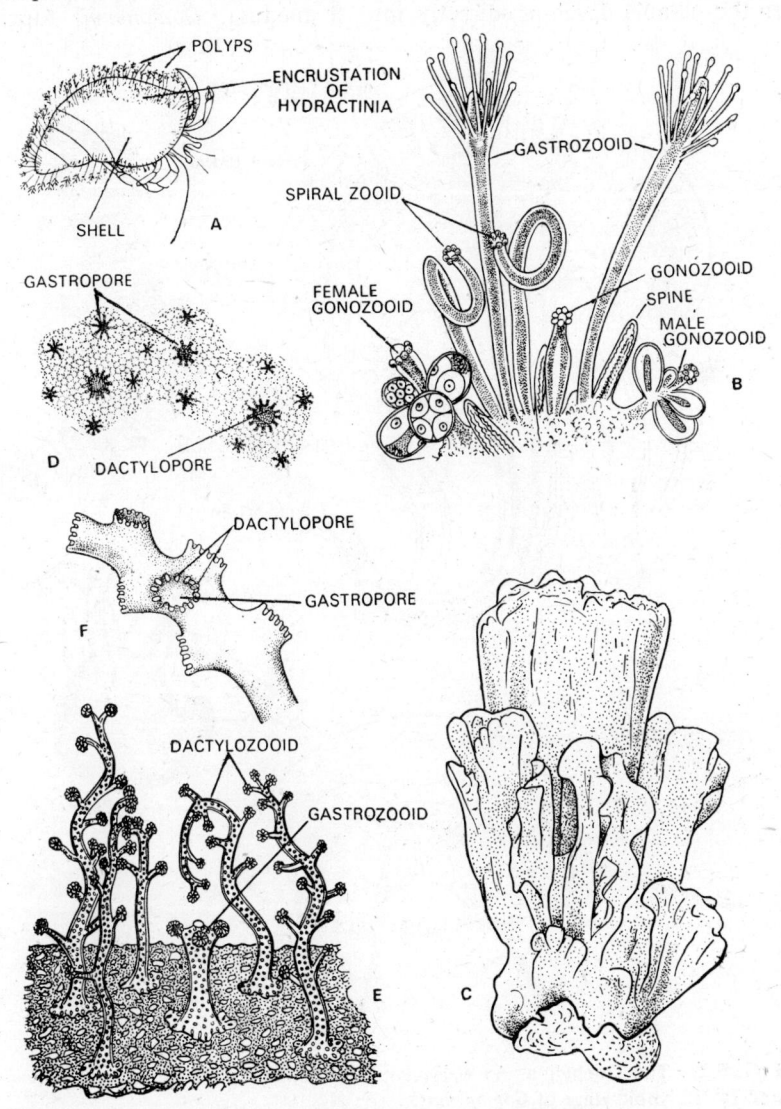

**FIG. 5.26.** A—*Hydractinia*, growing on a mollusc shell occupied by a hermit-crab; B—*Hydractinia* showing polyp types, from life (redrawn *after* Hyman); C—Piece of dry *Millepora*; D—The same magnified showing pores; E—Polyps of *Millepora*: and F—*Stylaster*, a piece of dried specimen magnified showing zooid pores.

being represented by the planula larva which directly metamorphoses into a medusa. Statocysts are present. The generative organs are located on the radial canals or on the floor of the gastric cavity. They are divided into two groups, the **Trachymedusae** and the **Narcomedusae**. The Trachymedusae have marginal tentacles and the gonads are situated on the radial canals, whereas, in the **Narcomedusae** *tentacles* originate from the exumbrella, the manubrium is lacking, mouth opens directly into the large stomach and the gonads are located in the floor of the gastric pockets.

*Geryonia* (Fig. 5.27A), *Rhopalonema* and *Liriope* represent Trachymedusae in which the planula develops directly into a medusa. *Gonionemus* furnishes an

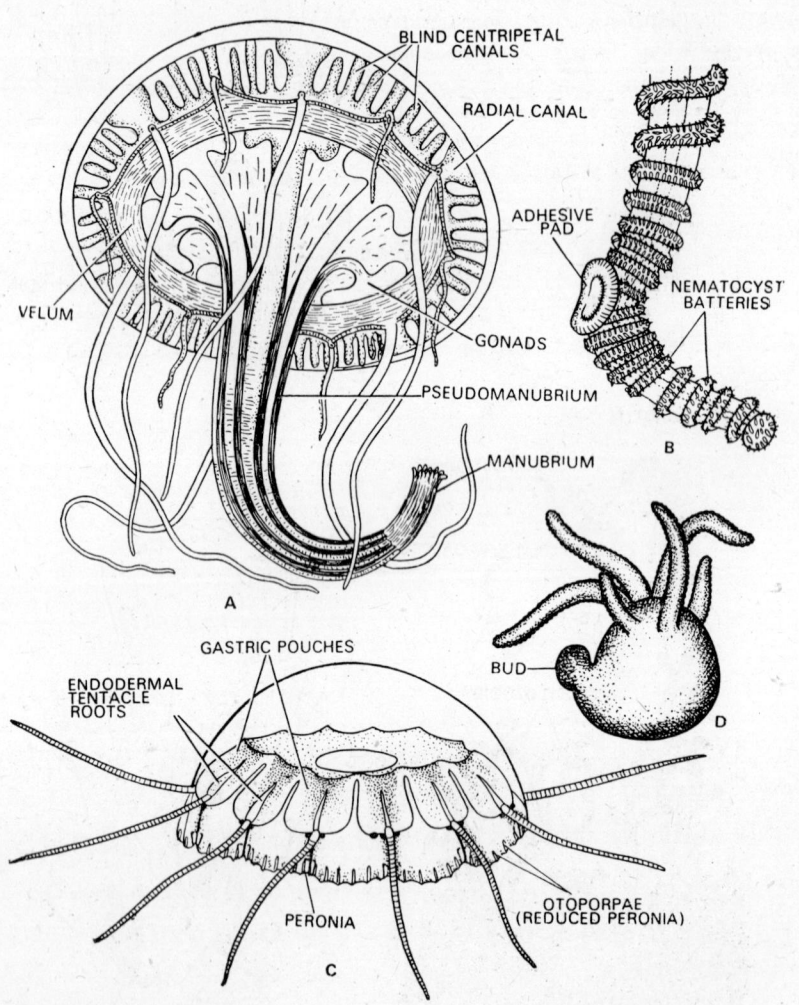

**FIG. 5.27.** The Trachylina. **A**—*Geryonia*; **B**—A tentacle of *Gonionemus*; **C**—*Cunina*; and **D**—Polypoid stage of *Gonionemus*.

example of a more complicated life-history. In this case the planula attaches and develops into a small solitary polyp (Fig. 5.26 D) from which medusae arise, budding from the sides. *Gonionemus* has four radial canals with ruffled gonads, one set of hollow tentacles with conspicuous tentacular bulbs. The tentacles bear besides nematocysts an adhesive pad behind the tip. It is a bottom living form that clings to sea-weeds with the help of its pads.

*Cunina* (Fig. 5.27 C), *Solmundella* and *Aeginopsis* represent the Narcomedusae and are characterised by firm, mostly flattened glassy bells with stiff solid tentacles emerging well above the bell margin from the ex-umbrella. The life-cycle of *Cunina* is quite complicated. The egg develops in the maternal tissue with the

*Phylum Cnidaria (Coelenterata)* 219

help of a nurse-cell (*phorocyte*), which is simply an enlarged blastomere, and, according to some, produces a reduced medusa that sheds its sex products and degenerates. The planulae from this generation take to parasitism. In this case they attach to the Trachymedusa *Geryonia* and develop parasitically into a flattened stolon from which definitive medusae are produced by budding.

4. **Order Siphonophora**. The Hydrozoa which form swimming or floating colonies of polymorphic individuals formed by the modification of polypoid and medusoid forms to subserve particular function. They present polymorphism of the highest degree. The individuals may be arranged in groups along a long central axis. (Fig. 5.23 A) or collected beneath a large float (**pneumatophore**)

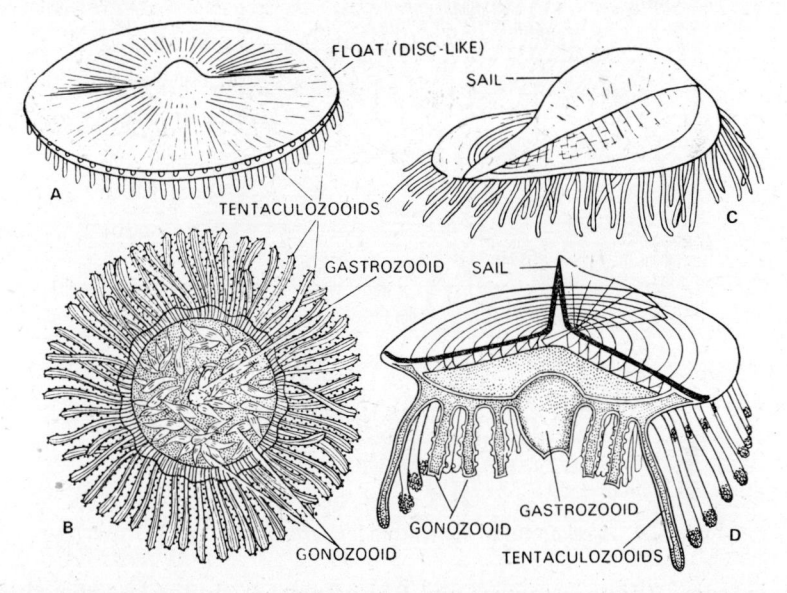

FIG. 5.28. **A**—*Porpita*, seen from above; **B**—*Porpita* seen from beneath; **C**—*Velella* seen from above; **D**—Vertical section of *Velella*.

(Fig. 5.23). The gonophores seldom develop into complete medusae and are seldom set free.

Examples: *Halistemma* (Fig. 5.23); *Physalia* (Fig. 5.24); *Porpita*; and *Velella* (Fig. 5.28); etc.

5. **Order Graptolithina**. This group includes extinct forms probably related to the Hydrozoa. If so, the polyp generation is dominant, the medusoid unfossilized or probably represented by the **prosicula**. The individuals budded off from one another and remained in contact with the parent; there was no definite coenosarc, and the perisarc was produced round the polyps as hydrotheca.

## CLASS SCYPHOZOA

The **Scyphozoa** (G. *skyphos*, cup+*zoon*, animal) present strongly developed tetramerous radial symmetry, the body can be divided into four structurally identical quadrants. Exceptions do occur and they are presented by such structures as tentacles that become numerous. The bell varies in shape in different orders, but is usually very gelatinous. The jelly occurs both on the ex- and subumbrellar surfaces and extends into the tentacles and oral arms. Velum is absent but in the Cubomedusae an analogous structure, the **velarium** is present. It is sub-umbrellar extension and is also seen in the genus *Aurelia* (Fig. 5.29). Nematocysts are borne by both the surfaces of the bell either scattered all over or in patches. The margins bear both tentacles and sensory bodies and are cut by notches into several **lappets**. The sense organs are lodged between two adjacent lappets, when these are present. The mouth is four-cornered and its angles are drawn out into four per-radial lobes, mostly short, but extended into long oral

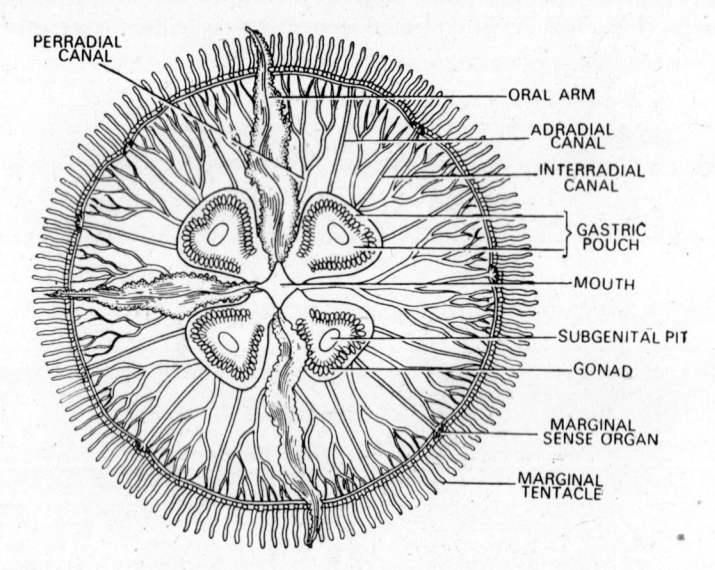

FIG. 5.29. *Aurelia* ventral view, from life (*from* Prasad and Grover).

arms in some (Semaeostomeae and Rhizostomeae). In the last named the oral arms are branched and their edges are fused obliterating the original mouth and forming thousands of suctorial mouths on their surfaces. Oral arms are well-provided with nematocysts. The manubrium present is not stomodaeal as it is lined by endoderm and not ectoderm. Four septa composed of endoderm and mesogloea project into the gastro-vascular cavity from ex-umbrella along the inter-radii and divide it into a central **stomach** and four per-radial, **gastric-pockets**. Such a division of stomach is not found in the hydromedusae. The muscular system is wholly ectodermal and limited to marginal and sub-umbrellar structures.

The nervous system is like that of the Hydromedusae, only the marginal nerve ring is absent. The marginal sense organs are represented by **rhopalioids** (in the Stauromedusae) or **rhopalia** (in other orders). Each rhopalium is located in a sensory pit roofed by a hood-like extension of the ex-umbrella and is essentially a **tentaculocyst**. It may be provided with statolith and in some cases with ocelli.

The sexes are separate. The sex cells develop in the endodermis which forms epithelium over them. The sex-cells are released in the gastro-vascular cavity and escape by mouth. The egg develops in the sea water or in folds of the oral arms and forms a **planula**, which in most scyphozoans attaches to substratum,

## Phylum Cnidaria (Coelenterata)

develops tentacles forming a polyp-like larva known as the **scyphistoma** larva, which gives rise to the medusoid forms by transverse fission or **strobilation**.

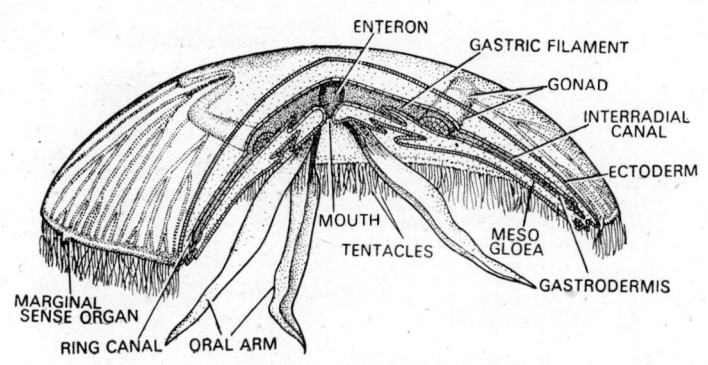

FIG. 5.30. *Aurelia*, median vertical section.

### TYPE AURELIA AURITA

*Aurelia aurita* is a common jelly-fish inhabiting coastal waters. It is often seen swimming singly or in groups, by rhythmic contractions of the bell. It is often found cast upon the sea-shore and is recognised by its gelatinous saucer-shaped umbrella, 3-4 inches in diameter and by four red or purple horse-shoe-shaped bodies (the gonads) in its centre (Fig. 5.29).

### External Characters

In general the structure is similar to that of the hydrozoan medusa described above. The exumbrellar surface is shallow convex above, and the subumbrellar surface is fringed with closely-spaced delicate marginal tentacles. The margin is indented with eight equally-spaced notches, each with a **sense organ** located between two small **marginal lappets**. The **mouth** is four-cornered and is situated on a short **manubrium** in the centre of the subumbrellar surface, between four, long frilled, tapering oral-arms (Fig. 5.30). The oral arms are provided with nematocysts. They are per-raidal in position. Along its subumbrellar surface each oral-arm bears a groove, two ridges of which are convoluted and are beset with many small tentacles. The velum (as in *Obelia* medusa) is lacking, in its place there is a narrow flap-like **velarium**.

Histologically the ex- and subumbrellar surfaces are covered by ectoderm and the stomach and canal-system is lined by endoderm, ciliated throughout. The mesogloea is gelatinous and forms the main mass of the umbrella. It is not structureless but is invaded by amoeboid cells drawn from endoderm and is traversed by branching fibres.

### Digestive System

It consists of the **mouth**, short **gullet** running through the manubrium and connecting the mouth to the spacious **stomach**, situated in the centre of the disc and produced into four large inter-radially situated **gastric pouches** containing slender tentacle-like **gastric filaments** or **digitelli** with nematocysts. Many radial **gastrovascular** canals arise from the gastric pouches and proceed towards the circumference of the disc opening ultimately into a ring-canal running in the margin of the umbrella. The canal that arises from the middle of the

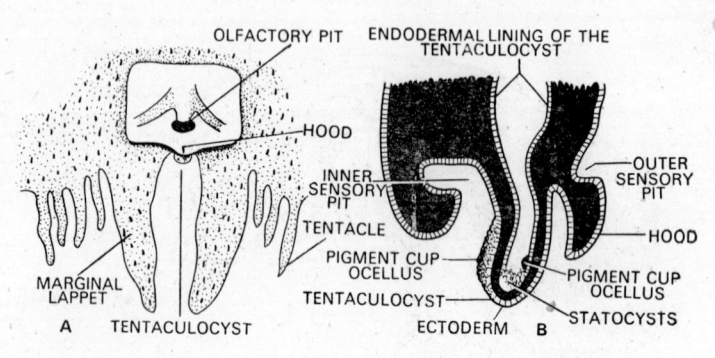

FIG. 5.31. A—marginal sense organ of *Aurelia*; and B—*Aurelia*, vertical section of marginal sense organ (tentaculocyst).

pouch is the **inter-radial canal** those, arising, on either side of this are called **ad-radial canals**. Between the two pouches lies another canal called **per-radial canal**. The inter-radial canals and per-radial canals are branched, whereas, the ad-radial canals are not. The stomach and the set of canals associated with it are ciliated throughout. The food of *Aurelia* mainly consists of small invertebrates, seized and paralysed by nematocysts on the oral arms. Nematocysts of the gastric filaments further aid quieting the prey.

## Muscular System

It is practically wholly ectodermal and limited to marginal and subumbrellar structures. Lying outside the mesogloea on the subumbrellar surface particularly around the edge, are highly developed striated circular muscles. The contraction of these muscles is responsible for the swimming movements of the bell. The musculature of the oral-arms and tentacles consists of smooth muscle fibres.

## Nervous System

It consists of a nerve-net like that of the Hydromedusae, but the marginal nerve ring is lacking in this case (except in the Cubomedusae). The nerve-net is best developed about the bell margin associated with the marginal sense organs. There are numerous long bipolar cells, forming a plexus between the epithelium and the muscle fibres on the subumbrellar side showing special concentrations into radial strands along the radii.

The **marginal sense** organs are lodged in the notches between the marginal lappets and are sometimes called **rhopalia** (rhopalioids in Stauromedusae) or **tentaculocysts**. Each rhopalium consists of a hollow club or **statocyst**, the interior of which is filled with a mass of polygonal cells especially towards its free end. Each of these cells contains a **statolith**. Besides this the rhopalium contains a hood-like process overhanging the statocysts, a pigmented eye-spot sensitive to light and two **sensory pits**, probably chemoreceptors having to do with food recognition. One of the pits lies on the subumbrellar side of the hood over the sensory niche and the other lies in the floor.

*Phylum Cnidaria (Coelenterata)* 223

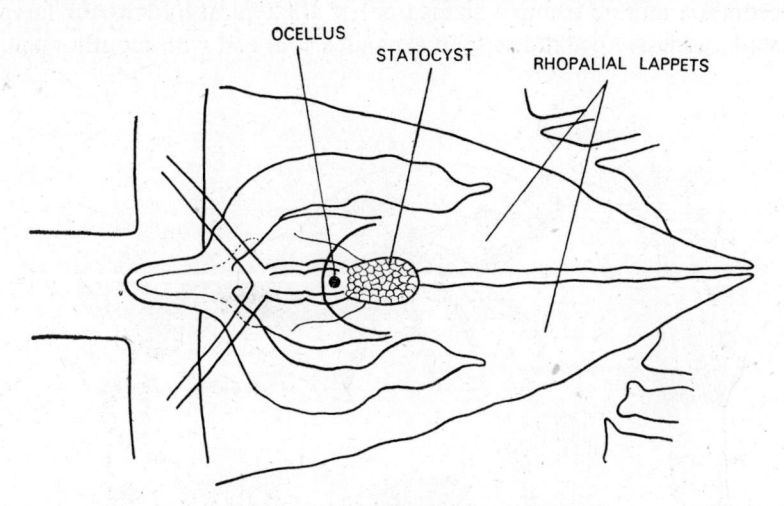

FIG. 5.32. Diagram of a rhopalium of *Aurelia*.

Although *Aurelia* looks like the medusa of *Obelia* yet these two differ in many ways. The differences are summarised below:

| *Aurelia* | *Obelia* (medusa) |
|---|---|
| 1. Marginal lappets around sense organs which are eight in number. | 1. Margin entire, marginal lappets absent. |
| 2. Velarium present. | 2. Ill-defined velum. |
| 3. Many radial canals, some are branched, others not. | 3. Four unbranched radial canals. |
| 4. Gonads horse-shoe-shaped. | 4. Gonads round situated on radial canals. |
| 5. Statocysts are present. | 5. No statocysts. |
| 6. An ocellus is present. | 6. No ocellus present. |

**Reproductive System**

The sexes are separate but the individuals of both sexes are similar. The sex-cells originate and ripen in the base of the endodermis and occur in the floor of the gastric pouches. They form four horse-shoe-shaped structures. The sex-cells rupture into the gastrovascular cavity and escape through the mouth. The sperms from the males enter into the enteron of a female to fertilise the eggs produced in its gonads.

**Development.** The zygotes emerge and get lodged in pockets formed by the frills of the oral-arms, where they develop into the ciliated **planula** stage (Fig. 5.33), after passing through usual **morula, blastula** and **gastrula** stages. The morula is a mass of cells. When this develops a cavity (**blastocoel**) in the interior it becomes the blastula. The blastocoel is filled with a fluid. The blastula invaginates on one side forming a double-layered gastrula (Fig. 5.33). This finally develops into two-layered ciliated planula. The planula escapes, swims for a little while and finally attaches itself to the bottom of the sea. Soon it loses cilia and becomes a minute trumpet-shaped polyp, the typical **hydratula** larva with basal stalk, adhesive pedal disc, and expanded oral end with mouth, manubrium

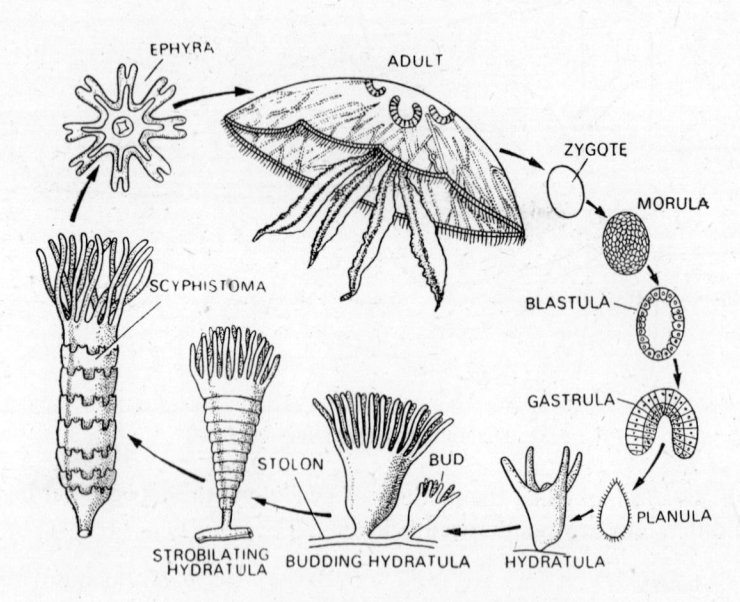

FIG. 5.33. Life-cycle of *Aurelia*.

and tentacles. The hydratula feeds, grows to about 12 mm long and remains unchanged through fall and winter. It may produce fresh hydratulae asexually by budding. In autumn and winter a type of transverse fission called **strobilation** sets in. At this stage the hydratula is called **scyphistoma**. In this process horizontal constrictions appear around the body and deepen with the result that the whole structure looks like a pile of minute saucers with fluted borders, because the edges of each are produced into eight double lobes. These flat eightlobed structures are known as **ephyrae** (singular **ephyra**). The ephyra (Fig. 5.34) when complete is constricted off by a muscular contraction and swims about as a small medusa. The next ephyra then develops as the first. Each is released after sometime. The ephyra ultimately develops into an adult jelly-fish. After a period of strobilation in the winter and spring they stop producing ephyrae and resume life as polyps feeding and budding off scyphistomae. Strobilation sets in again in the following winter. Thus, the scyphistomae live for several years.

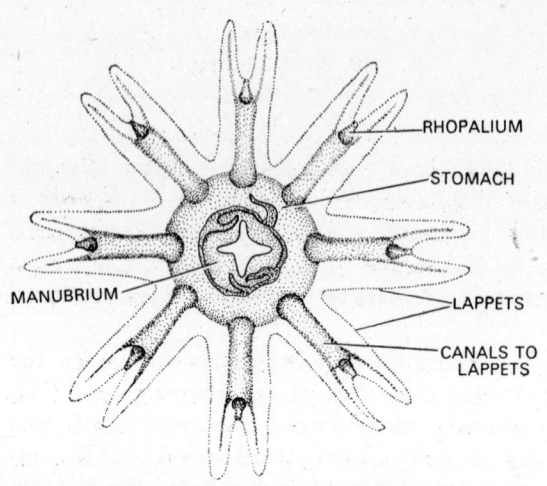

FIG. 5.34. An ephyra of *Aurelia*.

# Phylum Cnidaria (Coelenterata)

*Aurelia* reproduces sexually, whereas the hydratula and scyphistoma reproduce asexually. *Aurelia* is medusoid in nature and the hydratula and scyphistoma are polypoid. Therefore, it can be said that the medusoid phase alternates with polypoid phase. But *Aurelia* does not present a true case of alternation of generations because here the medusoid phase develops as a result of metamorphosis of an ephyra developed as one of the several segments of scyphistoma. The life-cycle simply presents a case of metamorphosis complicated by multiplication in the larval condition.

### Comparison of life-histories of *Aurelia* and *Obelia*

| *Aurelia* | *Obelia* |
|---|---|
| 1. Planula develops from the zygote directly. | 1. Planula develops from the zygote directly. |
| 2. Planula develops into hydratula and is able to reproduce by budding (as *Obelia* itself) | 2. Planula develops into hydrula. |
| 3. The Hydratula multiplies also by transverse fission (strobilation) and is then called scyphistoma. | 3. Hydrula develops lateral buds and forms the asexual colony. |
| 4. Scyphistoma releases ephyra as a result of strobilation. | 4. Not formed. |
| 5. Ephyra metamorphoses into medusa. | 5. Medusa directly formed asexually by budding. |

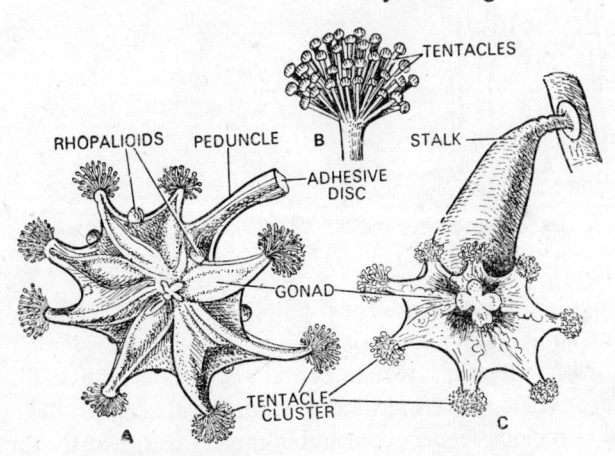

FIG. 5.35. The Stauromedusae. A—*Haliclystus*; B—Tentacle cluster of the same; and C—*Lucernaria*.

## CLASSIFICATION OF SCYPHOZOA

The Scyphozoa are all marine with more than two hundred species. They are divided into five orders—Stauromedusae or Lucernariida, Cubomedusae or Carybdeida, Coronatae, Semaeostomeae and Rhizostomeae.

1. **Order Stauromedusae.** The scyphozoa having a conical or vase-shaped umbrella. These have narrow stems or **peduncles** arising from the exumbrellar

surface by which they attach to external objects. Velum is absent and there are no tentaculocysts. The mouth is situated on a well-developed manubrium and opens into a spacious gastric cavity divided by four partitions, the inter-radial mesenteries, which in the genus *Haliclystus* extend and fuse in the centre of the stalk forming four longitudinal canals in it. They are not swimmers and keep hanging to sea-weeds etc. Some can detach and re-attach at will moving about like hydras.

Examples: *Haliclystus* (Fig. 5.35 A), *Lucernaria* (Fig. 5.35 C).

2. **Order Cubomedusae** or **Carybdeida**. The Scyphozoa having deep cup-shaped umbrella with four flattened sides, appearing square in transverse section. Tentaculocysts are four per-radial. The margin of the umbrella bears four tentacles inter-radial (in position), which spring from conspicuous gelatinous lobes (**pedalia**). The margin of the umbrella is produced into a horizontal shelf, the **velarium** acting as velum. Mouth is at the top of a short manubrium-

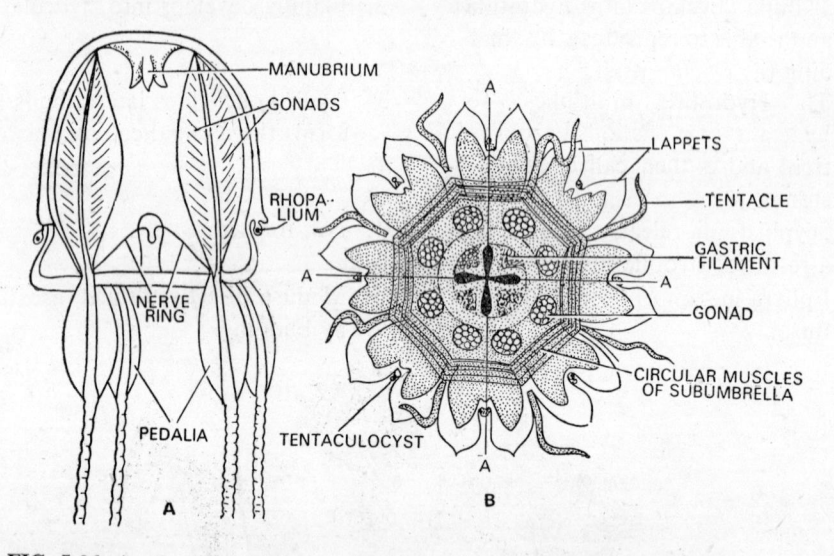

FIG. 5.36. A—*Carybdea*, entire; and B—*Nausithoe*, entire from the oral side. A—A per radii.

and is quadrangular. Gonads are four pairs of narrow plate-like organs. The nervous system, unlike other scyphozoans, comprises a marginal nerve ring that is connected with rhopalia, which contain statoliths and ocelli. They inhabit bays and shore-waters of tropical and subtropical regions. They are graceful swimmers and voracious feeders consuming mostly fish. Mostly they are between 2 and 4 cm though some reach a height of 10 to 25 cm. The members of this group are known as "sea-wasps" because they are known to "sting."

*Carybdea* (Fig. 5.36), *Tripedalia* and *Chiropsalmus* form typical examples of the group. Of these the last named is one of the most dangerous genera, whose sting can cause very serious illness and even death. For this reason Japanese and Philippine natives fear these very much and call them, "fire-medusa."

3. **Order Coronata**. The Scyphozoa having conical, dome-shaped or flattened scalloped bells, divided by a horizontal **coronary groove** (Fig. 5.37) into an apical cone and a marginal crown. Mouth is large, quadripartite, manubrium is wide

## Phylum Cnidaria (Coelenterata)

and gastric-filaments are numerous. Tentaculocysts are four to sixteen. The gonads are eight in number and ad-radial. They inhabit deeper waters (*Periphylla*) of the ocean, only some forms such as *Nausithoe* (Fig. 5.36) and *Linuche* are found near surface of warmer waters. This group includes a number of beautiful medusae.

4. **Order Discomedusae (Semaeostomeae).** The Scyphozoa having flat saucer-shaped umbrella that are the only ones seen in the temperate regions. Tentaculocysts are eight, four per-radial and four inter-radial. Tentacles are 3 to 7 or many in some. In the latter condition they may be distributed along the margin (*Aurelia*) or grouped in bunches (*Cyanea*). The mouth is four-sided borne on a short or conspicuous manubrium. The angles of the square mouth are produced into four long oral arms. A velarium is present. The gonads are coloured horseshoe-shaped patches on the floor of the large central part of the gastrovascular cavity. They are found in the coastal waters of all oceans mostly in warm and temperate regions. One genus *Cyanea* reaches even polar regions.

*Aurelia* (Fig. 3.28) is one of the commonest of the scyphozoan jellyfishes and occurs all over the world.

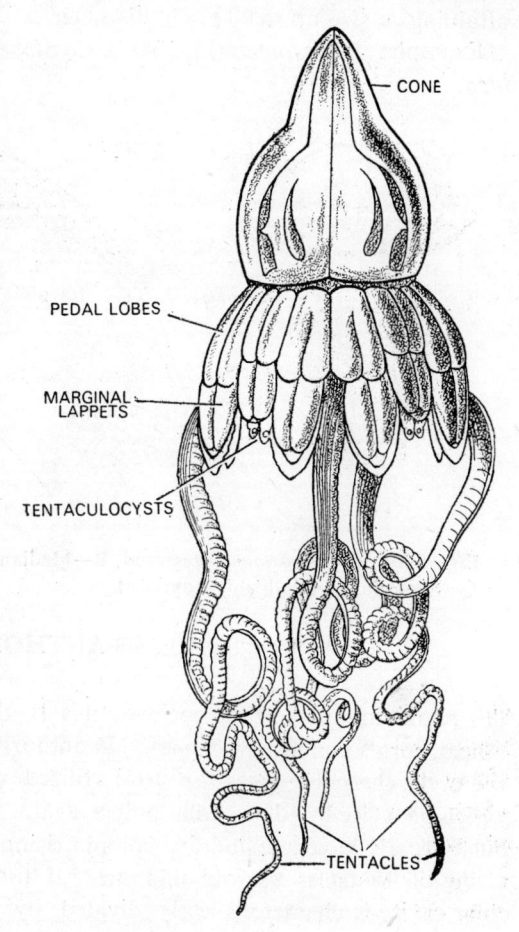

**FIG. 5.37.** *Periphylla quadrigata*, external view (*after* Haeckel).

It ranges in size between 8 cm and 30 cm across the bell. Exceptional individuals may reach a diameter of 60 cm. The sexes are separate and the eggs develop to a planula in the pockets formed by the frills of the oral arms. The planula develops into a typical trumpet-shaped scyphistoma larva that by strobilation produces typical medusae.

Examples: *Aurelia, Pelagia, Cyanea* and *Chrysaora* (hermaphrodite genus).

5. **Order Rhizostomeae.** The Scyphozoa in which the original four arms divide longitudinally forming eight arms which often attain great lengths. The oral-arms are variously lobed and folded. The young *Rhizostoma* has a single mouth but in the adult the mouth is obliterated by the growth of the oral-arms and replaced by numerous small "sucking mouths," which lie along the course of the closed-in grooves of the lips. These lips act even as organs of external digestion. Small food animals such as fish are enclosed by the lips, digested fluid absorbed through the "suctorial mouths," which are too small to admit solid particles. The marginal tentacles are absent. They inhabit the shallow waters in

the tropical and subtropical zones chiefly the Indo-Pacific region. *Rhizostoma* is found in temperate zone also. The members of the group become quite large attaining a size up to 80 cm in diameter.

Example: *Rhizostoma*, (Fig. 5.38), *Cassiopeia, Cephea, Mastigias* and *Cotylorhiza.*

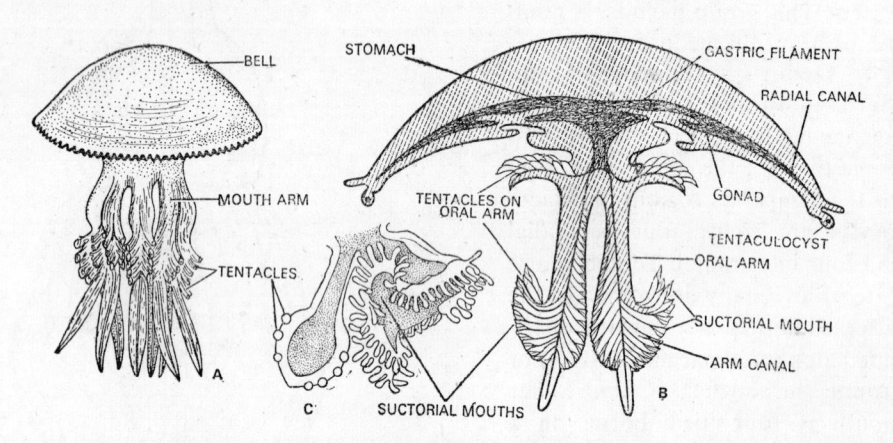

FIG. 5.38. A—*Rhizostoma*, preserved; B—Median vertical section of *Rhizostoma*; and C—Suctorial mouth, highly magnified.

## CLASS ANTHOZOA

The remaining class of the coelenterates is the **Anthozoa**, comprising, among others, corals and sea-anemones. No anthozoan has any medusoid stage in the life cycle; the polyp gives rise to fertilized eggs which develop into planulae, which give rise to adult sexual polyps again. They may be solitary or colonial, but typically have a cylindrical column rising from a basal disc (Fig. 5.39) with a ring of tentacles surrounding an oral disc at the other end. The gastrovascular cavity is characteristically divided by radial mesenteries, like the gastric ridges of the Scyphozoa, but more numerous and more complicated. The ectoderm is always tucked into a characteristic stomodaeum, and the mouth is typically slit-like or oval. In general the Anthozoa tend to exhibit bilateral (or at least biradial) symmetry, never tetraradial, as in the preceding groups. Anthozoan internal organisation is always bilateral, even in the sea-anemone which externally appears to show beautiful radial symmetry like that of a flower, and this primary bilateral organisation arises in embryonic development and persists into the adult. The gonads are endodermal and usually borne on the mesenteries (Fig. 5.39). The mesenteries are in-pushings of the endoderm up to the level of the stomodaeum but above this they fuse with it. The free edges of the mesenteries below the stomodaeum are usually thickened and form mesenterial filaments (which may be much folded to become considerably longer than the mesentery itself). At the end of the mouth slit there are (either one or two) ciliated gutters or siphonoglyphs, and the mesenteries associated with them differ from the others in their arrangement of muscle bands and are called directive mesenteries. The tentacles are hollow and contain extensions of the gastrovascular cavity. Myoepithelial cells are organised into well-developed muscles including the longitudinal bands on the mesenteries and circular sphincters around the disc and stomodaeum.

*Phylum Cnidaria (Coelenterata)*

The elaborate pattern of these well-developed muscle allows for quite complex behaviour, especially in sea-anemones. The mesogloea is always relatively thick

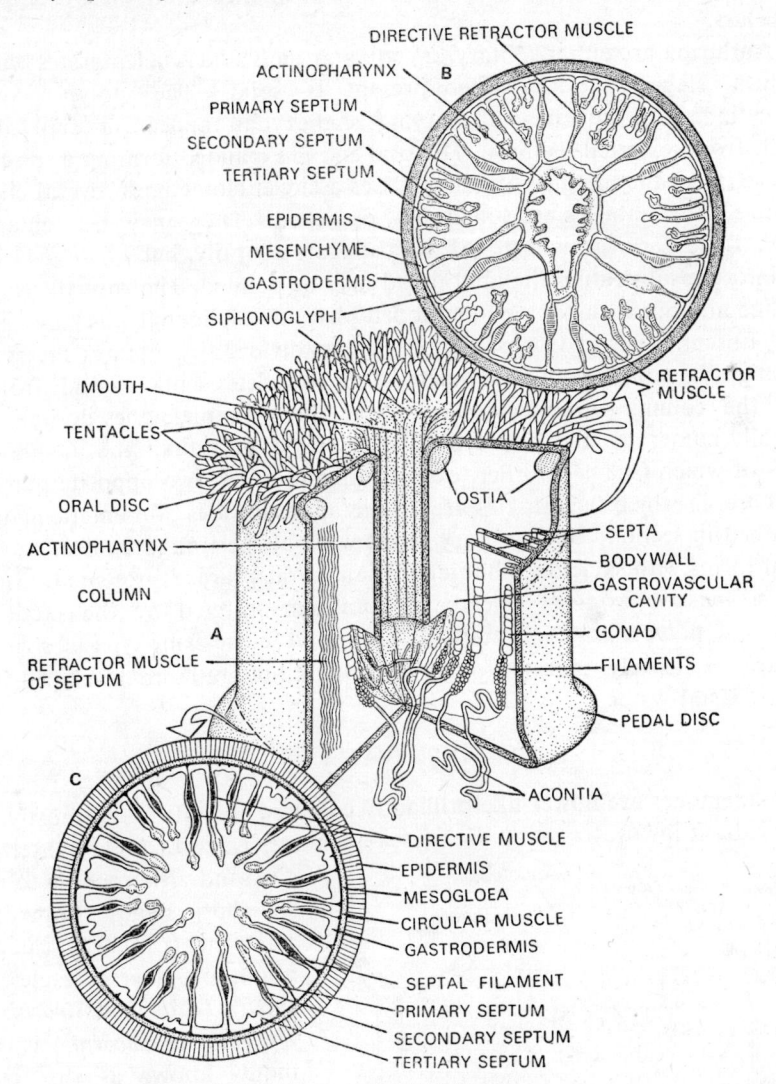

**FIG. 5.39.** A—Diagrammatic longitudinal section of an anemone, internal morphology; B—Cross section of *Metridium* through gullet; and C—Cross section below gullet.

and often supported with fibrous material. The nematocysts are usually relatively powerful and often localised into batteries. They reproduce sexually, as mentioned above, and asexually by budding and fragmentation.

The class Anthozoa is divided into two subclasses, the **Octocorallia** (or **Alcyonaria**) and the **Hexacorallia** (or **Zoantharia**). This primary division appears to reflect a natural difference in organisation and in relationships. The **Octocorallia** have their tentacles and mesenteries arranged in an eightfold symmetry and are always colonial. Their tentacles are pinnate and there is a single siphonoglyph. The **Hexacorallia**, on the other hand, have the tentacles and mesenteries often in multiples of six (but never in eights). The tentacles are always simple not divided or pinnate. Sea-anemones are typical examples of this subclass.

The Anthozoa are either solitary or colonial polypoid coelenterates without a medusa stage. Skeleton when present is solid. The polyp consists of a cylinder-like **column** of varying shapes and appears radially symmetrical from external features. The oral end is expanded forming a disc, the **oral disc**, from the margins of which arises a circlet (sometimes several circlets) of tentacle. The tentacles are sometimes branched. They may be solitary or colonial. In solitary forms a **pedal disc** appears aborally, but where it is intersected into the substratum the aboral end may be pointed. The **mouth** is oval or slit-like and opens into a well-developed pharynx. Siphonoglyphs may be two, one or absent. The gastrovascular cavity is divided by septa or mesenteries which might be complete or incomplete. The complete septa extend from the wall of the column to the pharynx, whereas, the incomplete ones do not. They are usually numerous and are typically arranged in pairs, the longitudinal muscles of which face each other, except in the case of two opposite pairs, the **directives**, in which the muscles are on the opposite sides. Mesenteric filaments are trilobed in section. Both solitary forms without skeleton (sea-anemones) and colonial forms with skeleton (madreporarian corals) are represented. In the corals, a massive external calcareous corallium secreted by the ectoderm is present. The polyps, however, may all be referred to the same type of structure. They are exclusively marine found everywhere but are more developed in the warmer coastal water.

## TYPE TEALIA

The sea-anemones are flower-like brilliantly coloured animals found mostly fixed to rocks about low water mark. Some sea-anemones (*Edwardsia*) are inserted in the sand. All are able to move by short stages. Some reef-anemones (*Cradactis*) can crawl about on their tentacles. The type *Tealia crassicornis* (or *Urticina crassicornis*) is commonly known as the "Dahlia Wartlet," and is found half-buried in sand and gravel. It is green or red in colour.

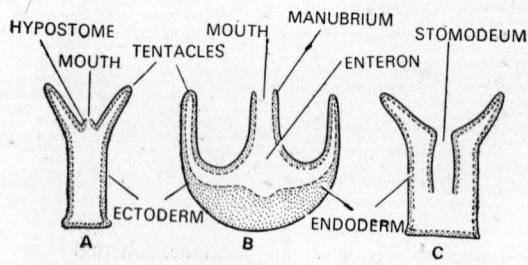

FIG. 5.40. A—Comparison of the hydroid polyp; B—The medusa (inverted); and C—The anthozoan polyp.

### Form of the Body

*Tealia* is column-like in shape and is broader than high, its diameter being about 8 cm. The mouth is a slit-like aperture situated in the middle of the **oral disc** from the edges of which spring out numerous short conical **tentacles** arranged in five circlets. The outermost whorl carries forty-eight tentacles, followed inwards by whorls containing twenty-four, twelve, six and six in succession. On the aboral side is the **basal disc** for attachment. The structure is like that of a hydroid polyp in this case, apart from the largeness of size, an invagination

## Phylum Cnidaria (Coelenterata)

of the oral disc occurs forming a **gullet (pharynx)** which hangs into the general cavity (Fig. 5.40 C). The gullet is a flat tube, along both sides of which is a smooth ciliated furrow, the **siphonoglyph**, in which water passes to the enteron.

**Mesenteries.** The internal cavity of the polypoid structure is subdivided into radial compartments by six complete septa extending from the wall to the gullet (Fig. 5.39). These complete septa are known as **primary mesenteries**. Between these are present incomplete septa. The septa that do not reach the gullet are known as **secondary** and **tertiary mesenteries** depending upon their size. Each septum has two **ostia** beneath the oral disc, through which the inter-mesenteric compartments communicate directly. The free inner margins of the septa are thick and convolute forming **mesenteric filaments** which continue below as thread-like **acontia** (singular **acontium**). The mesenteric filaments (Fig. 5.39) and acontia both bear gland cells and nematocysts. Acontia aid in subduing prey for which purpose they protrude through the mouth openings in the body wall known as **cinclides**.

**Body wall.** As in all coelenterates the body wall consists of outer ectoderm and inner endoderm separated by an extremely thick tough layer of mesogloea composed of cellular tissue. The septa are formed by the reduplications of endoderm, consisting of endoderm, on both the sides supported internally by

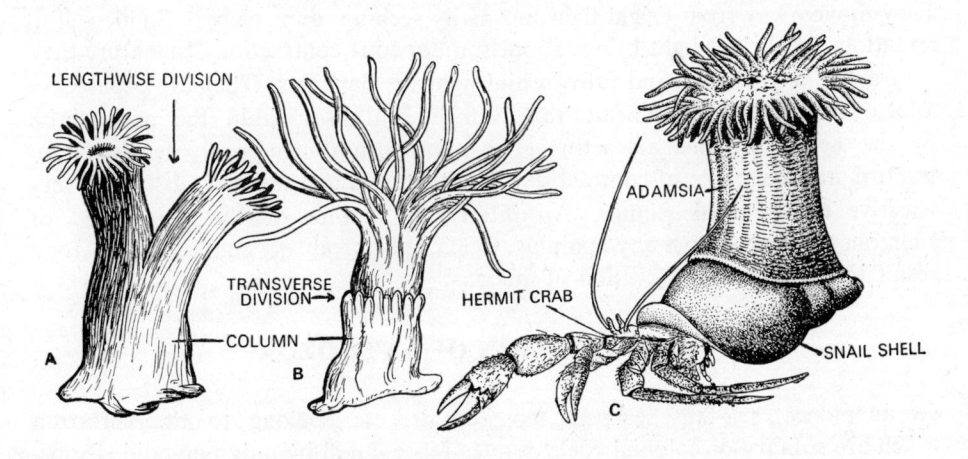

FIG. 5.41. Asexual reproduction takes place in one of the two ways represented above. A—*Metridium* dividing lengthwise; B—*Gonactinia prolifera* dividing across, tentacles for daughter fission are sprouting; and C—*Adamsia* on a snail shell occupied by the Hermit-crab presents a good example of commensalism.

plates of mesogloea. The septa, however, are provided with well-developed muscles, some of which run vertically along the septa and are termed the **retractor** muscles, others run across the septum and are called the **transverse** muscles, whereas, some pass obliquely across the lower and outer angles of the septa and are described as the **parietal** muscles. These muscles are responsible for the retractile and extensile movements of the body. The epidermis contains a nerve-net, and similar nerve-net occurs in the endoderm, which also extends into the walls of the septa, but localised sense organs are not present.

### Reproduction

The sexes are separate. The gonads develop along the margins of the septa. The sex cells are endodermal. The sperms are released in the enteron of the male and escape along water currents by way of mouth. They are carried to the enteron of the female by way of mouth and fertilize the eggs in the body. The fertilized egg develops into a planula, which leaves the parent, swims for sometime, attaches to some suitable spot and finally metamorphoses into an adult. The metamorphosis is complicated and often passes through different stages. Some anemones reproduce asexually by fission either longitudinal (*Sagartia*) or transverse (*Gonactinia*) (Fig. 5.41), and others fragment pieces off the base to form new individuals (*Epiactis*). In the latter case as the animal slides about pieces of the body are left behind either due to injury or due to poor coordination. These fragments regenerate into tiny anemones.

### Behaviour

The anemones are always solitary and sessile but they are never fixed immovably. Some glide slowly by means of muscular action, some "walk" upside down on the tips of their tentacles, some reach the surface film and float head downwards and some release their foothold, inflate and roll about in the waves.

Anemones generally move when disturbed or under unfavourable conditions. They move away from bright light, normally seeking dim places. Sudden illumination by bright light brings about spontaneous contraction. In nature they contract in day and expand fully in dim light or darkness. They are capable of a lot of contraction. Some can draw even the oral discs within the coelenteron by the action of their retractors. Those forms that have poorly developed retractors are incapable of contraction. The tentacles and the oral disc are very sensitive to chemical stimuli. An interesting feature of the behaviour of anemones is that almost any stimulus, touch, chemical or contact with food result in the abundant secretion of mucous.

## CLASSIFICATION OF ANTHOZOA

Sea anemones, sea-fans, sea-pens, stony corals, etc., belong to the Anthozoa which are solitary or colonial coelenterates represented by only polypoid forms. No medusa stage is known in any member of the class. Coelenteron is divided by mesenteries or vertical partitions and stomodaeum is present. Gonads are endodermal in origin and position. The class is divided into two subclasses **Octocorallia** or **Alcyonaria** and **Hexacorallia** or **Zoantharia**.

(*i*) **Subclass Octocorallia (Alcyonaria)**. The Anthozoa, with eight pinnate tentacles and eight mesenteries, has the form of a short cylinder terminating orally in flat disc. The eight tentacles arise from the oral disc, are stout at base, tapering towards the free end and bear short-pointed projections, the **pinnules** (hence **pinnate**) The tentacles are hollow, lodging extensions of the gastrovascular cavity and are motile. The oral disc carries the oval mouth in its centre that opens into the pharynx which is lined by smooth glandular epithelium having a ciliated groove, the **siphonoglyph**, on one side (generally described as

# Phylum Cnidaria (Coelenterata)

ventral). The gastrovascular cavity has eight **mesenteries** or **septa** which are more or less evenly-spaced and consist of endoderm and mesogloea. The free

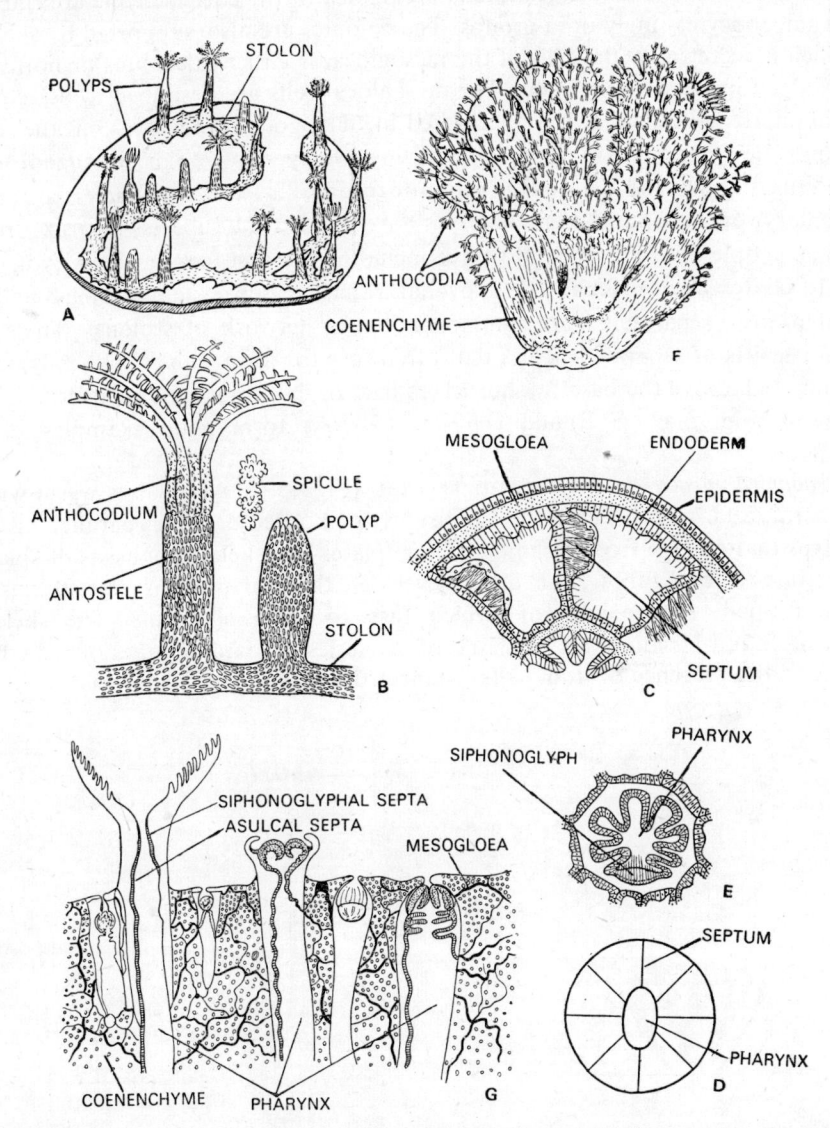

**FIG. 5.42.** A—A colony of *Clavularia*; B—Polyps of the same; C—Cross section through the wall of *Alcyonium* showing a septum; D—Diagrammatic section representing the septal arrangement; E—Cross section of the pharynx of *Alcyonium*; F—*Gersemia*, fleshy alcyonacean; and G—Vertical section through the coenenchyme of *Alcyonium*.

edges of the septa are thickened forming cord-like **septal** or **mesenterial filaments**, two of which are long and flagellated to create water current and the rest six are short and glandular having a digestive function. The Alcyonaria are exclusively colonial and their polyps are connected by endodermal tubes, the **solenia**, from which new polyps spring out. In most cases the solenia form a network that is embedded in mesogloea forming a fleshy mass covered by ectoderm. This often encloses the basal portion of the polyps, the projecting parts of which are termed **anthocodia**. The common flesh thus formed is known as **coenenchyme**. In the gelatinous mesogloea of the coenenchyme are embedded amoebocytes singly or in groups. The colonies are also supported by skeleton which is secreted by the cells of the mesogloea, is either calcareous or horny and is lodged in the coenenchyme. The mesogloeal cells are described to be ectodermal interstitial cells that have wandered in. The gonads develop on the septa. Some alcyonarians are dimorphic having two types of zooids, **autozooids**, as described above and the smaller **siphonozooids** in which tentacles are rudimentary or lacking. Recent workers have divided the subclass into six orders. Some people divide it only into three including the first four in the Alcyonacea.

1. **Order Stolonifera**. The Alcyonaria that lack coenenchymal mass and the polyps arise separately from common stolon or network of stolons. The skeleton consists of separate spicules that often fuse to form tubes. The polyps are connected above the base by short cross bars or by transverse platforms. *Clavularia* (Fig. 5.42 A, B) and *Tubipora* (5.43 A) form good examples of the order.

*Tubipora musica* or the organ-pipe coral (Fig. 5.43 A) inhabits warm waters or coral reefs. In this case the colony is composed of long parallel upright polyps that emerge from a common basal plate. The skeleton consists of spicules that fuse to form tubes. At other levels of the polyps transverse stolons are also formed, the spicules of which fuse to form platforms. The skeleton that is found in museums consists of a series of parallel tubes of red colour (due to the presence of iron salts) connected by transverse platforms.

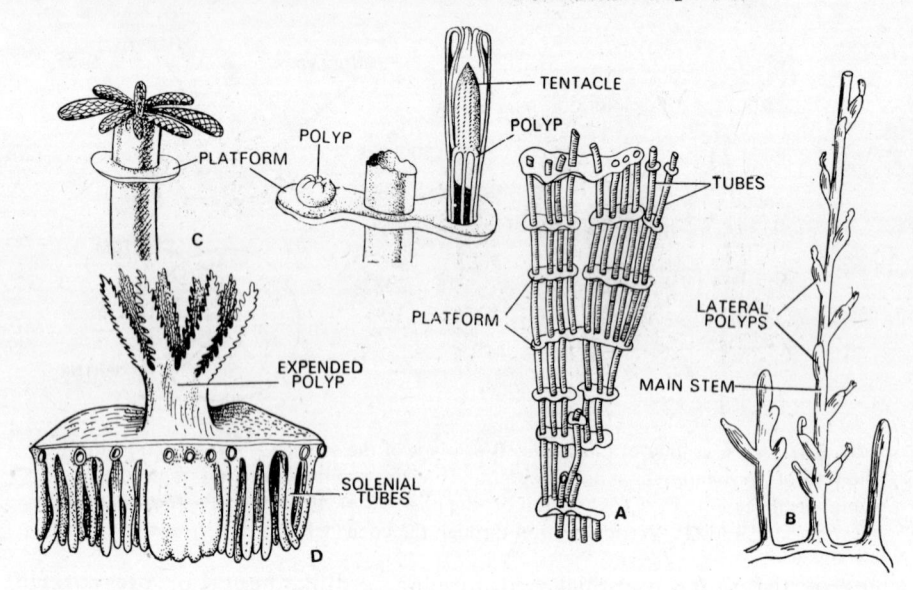

FIG. 5.43. Some Alcyonaria. **A**—*Tubipora* skeleton also showing polyps as they are found in life; **B**—*Telesto* preserved; **C**—*Heliopora* showing one polyp; and **D**—Many solenial tubes, the skeleton is removed (*after* Bourne).

2. **Order Telestacea**. The Alcyonaria with colonies of simple or branched stems that bear lateral buds and arise from a creeping stolon. The main stem is in fact formed by the elongation of a single polyp.

Examples: *Telesto* (Fig. 5.43 B), *Coelogorgia*.

3. **Order Alcyonacea.** This order includes soft-corals. The Alcyonaria in which the polyps are embedded in a fleshy mass (coenenchyme) from which the oral ends alone protrude. The polyps are located only on the free (distal) end, whereas, the proximal part is sterile being devoid of polyps. Skeleton is made up of separate calcareous spicules. They are found in warmer regions of the earth mostly in the Indo-Pacific Ocean.

*Alcyonium digitatum*, "Dead Man's fingers", is a typical colony of the type mentioned above. In life, the polyps project from the general surface of the colony and are delicate. They withdraw on slightest stimulus and the anthocodia are withdrawn in the coenenchyme. Isolated calcareous spicules form the skeletal system. The spicules are present in such a large number that they give a solidity to the colony and on death the spicules remain behind as a mass of calcium carbonate, or a series of connected tubes for the individual polyps.

4. **Order Coenothecalia.** The Alcyonaria in which the massive calcareous skeleton is composed of crystalline fibres and perforated by numerous closely-set cylindrical cavities closed below. This group is represented by the genus *Heliopora* (Fig. 5.43 D) or the coral of the Indo-Pacific region.

5. **Order Gorgonacea.** These are horny corals. The Alcyonaria usually of plant-like form including the sea-fans, sea feathers and sea-whips, etc., that are found in the tropical and subtropical shores. An axial skeleton of horn-like material, **gorgonin** is present. The polyps are dimorphic in some cases,

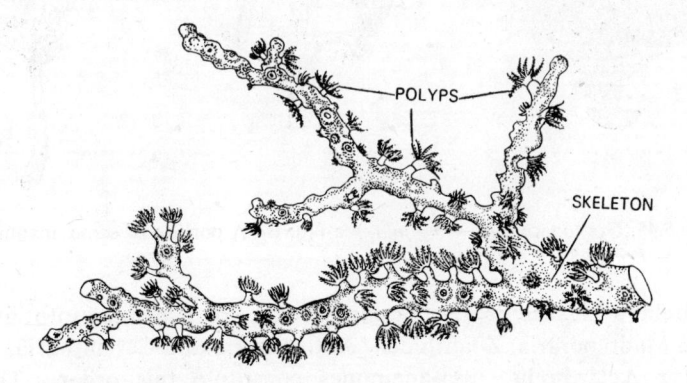

FIG. 5.44. A piece of a branch of red coral, *Corallium rubrum*.

particularly in the red coral, *Corallium*. The gorgonians are found in all seas at various depths. They are abundant in the Indo-Pacific Ocean.

*Corallium rubrum* (red coral) has an upright (Fig. 5.44) branched colony with a rigid axis clothed by the delicate tissue of coenosarc from which the short polyps arise. This is valuable as it is used in jewellery and is found at considerable depths in the Mediterranean and the seas of Japan.

*Gorgonia*, the sea-fan (Fig. 5.45A), another gorgonian, has upright branching colonies and forms another common example of the order.

6. **Order Pennatulacea.** The Alcyonaria in which the fleshy colony is composed of a single elongated axial polyp from which spring numerous dimorphic polyps laterally. In the representative genus *Pennatula* the axial

polyp grows to relatively enormous length (sometimes 3-4 metres), and contains a long horny axis (possibly endodermal). The proximal end of the stem, the **peduncle** or **stalk** is buried in the mud at the sea-bottoms and the distal end, the **rachis** bears zooids springing directly from the coenosarc or from flattened lateral branches. As the stem itself is regarded equivalent of a polyp all these zooids are called secondary polyps and belong to two types of individuals, the normal **anthozooids** which feed the colony and **siphonozooids** which maintain the circulation of water in the canals of the colony. The anthozooids in *Pennatula* are arranged in rows on either side forming regular lateral branches giving the colony a feather-like appearance. The siphonozooids are mainly formed on the back of the colony. They are found in warmer coastal waters mainly where bottoms are soft. They are coloured usually yellow, orange or red.

Examples: *Pennatula* (sea-pen, Fig. 5.45 C), *Renilla* (sea-pansy).

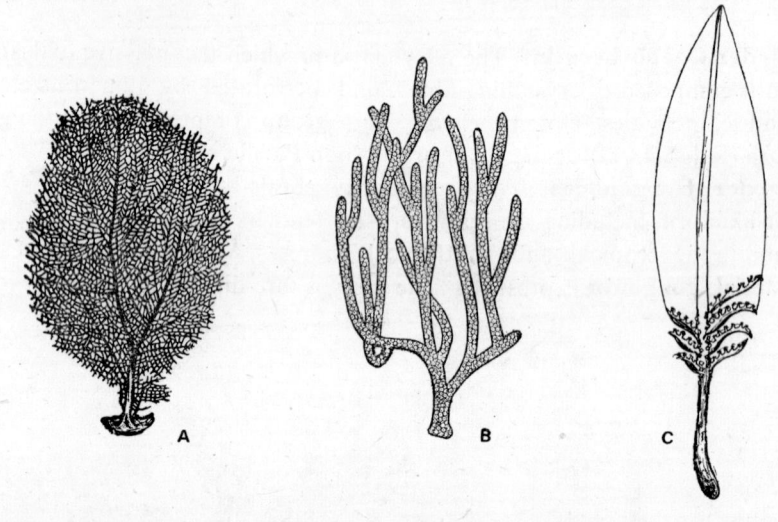

**FIG. 5.45.** Gorgonacea. A—*Gorgonia*, sea-fan; B—A portion of same magnified; and C—*Pennatula*, a sea-pen.

(*ii*) **Subclass Hexaborralia (Zooantharia).** It is divided into five orders: Actiniaria, Madreporaria, Zoanthidea, Antipatharia and Ceriantharia.

1. **Order Actiniaria.** Sea-anemones constitute this order. The Zoantharia which usually remain simple and form no skeleton; stomodaeum usually with siphonogyphs; septa paired, often in multiples of six; they are essentially solitary forms though some may be closely grouped. They are generally found attached to rocks or buried in sand or even on invertebrates.

*Edwardsia* is a typical anemone, in which the mesenteries present the most primitive condition. In this case there are eight mesenteries from which is derived the "hexactinian" type of other anemones. It lives partly buried in sand, enclosed in a tube formed of discharged stinging capsules. *Minyas* is a pelagic Actiniaria in which the aboral end is dilated into a sac containing air. This acts as a float and enables the animals to swim at the surface of the sea. Another genus *Adamsia* (Fig. 5.41C) furnishes a good example of commensalism. The individuals of the genus attach themselves to gastropod shells inhabited by

# Phylum Cnidaria (Coelenterata)

Hermit-crab, which is responsible for the movement of the anemone from place to place. The anemone gets varieties of abundant food supply and on the other hand protects the Hermit-crab from fishes that do not like to feed on tough and tasteless anemone with poisonous stinging capsules.

2. **Order Madreporaria.** These are true or stony corals. The Zooantharia that secrete compact calcareous exoskeleton, thus contribute to the formation of coral reefs and islands. The individuals otherwise are like the Actiniaria, without a siphonoglyph. Mostly colonial though solitary forms are also met with. The coral polyp is described as a skeleton forming anemone. It has no pedal disc, it occupies the skeleton-cup. The oral disc is typical with a circular or oval mouth and tentacles in cycles of six. The gullet is short but without siphonoglyph. The septa follow the hexamerous plan and bear filaments that are often highly convoluted. The lower part of the coelenteron becomes folded because of the skeletal ridges formed at the base.

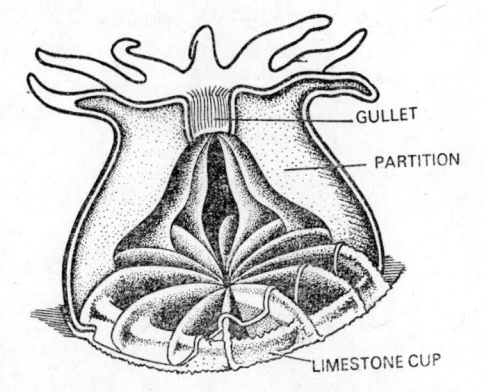

FIG. 5.46. Coral formation. Diagram of a vertical section through a coral showing formation of stony cup (*after* Pfurtscheller).

**Corals.** The skeleton of colony is known as **corallum** and that of each polyp is termed **corallite**. How actually the skeleton is formed is not understood, the most accepted view suggests that calcareous crystals are precipitated in a colloidal matrix that the ectoderm secretes towards its outer side. The stony corals particularly the reef-building ones require warm and shallow waters. They are thus limited to continental and island shores in tropical and subtropical zones. Small solitary cup corals grow in temperate coastal waters of many countries. Coral colonies exist in most diverse shapes and forms from the slender tree-like colonies of *Madrepora* to the massive rounded forms like Porites. A single planula forms each colony. First planula settles down and forms a polyp from which arise hundreds and thousands of polyps by division or gemmation. The skeleton is secreted later.

**Fossil corals.** The corals represent the coelenterates in the fossil record. There are more than 600 described fossil species that belong to three principal groups: the existing order Madreporaria; the wholly extinct order Rugosa and the subclass Tabulata. The rugose corals were largely solitary and horn-shaped, with a ridged surface (Fig. 5.47), and had conspicuous major sclerosepta. The tabulate corals were colonial and housed in external skeletal tubes connected together in a variety of ways. The base of the tube rested on a transverse platform, called a **tubula**, but sclerosepta were absent or poorly developed. The rugose and tabulate corals were inhabitants of the Palaeozoic seas where they contributed to the formation of great reefs, especially during the silurian period.

**Coral reef.** The coral reefs are principally built of stony corals but there are many other organisms that play a considerable role in their formation. Certain branching coralline algae impregnated with lime grow upon coral colonies and contribute a good bit to the formation of reefs. Foraminifera shells, much calcified alcyonarians, the gorgonians, *Millepora, Tubipora* and *Heliopora*, etc., form major part of such reefs The coral reefs are mainly of three types. **Fringing** reefs which grow in shallow water or border the coast closely (hardly a few

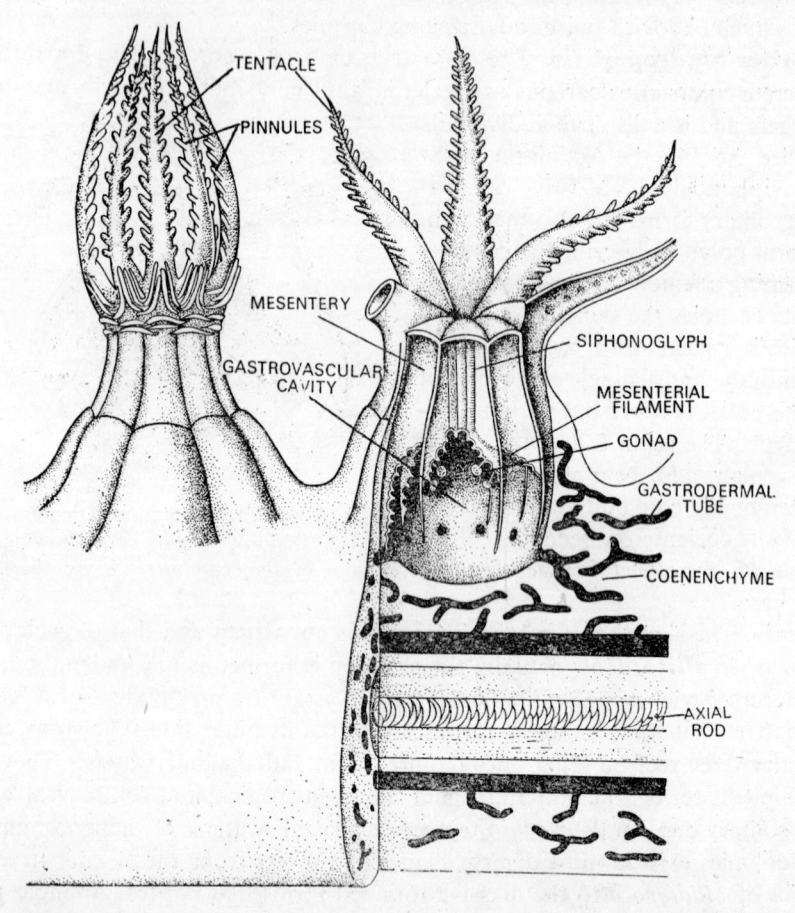

FIG. 5.47. Structure of a gorgonian coral of the anthozoan subclass Octocorallia (*after* R. C. Moore).

feet) or are separated from them at the most by a narrow stretch of water (about 400 metres) that can be waded across when the tide is out. **Barrier reef** (Fig. 5.48) is just like a fringing reef but the channel separating it from land is deep enough to accommodate large ships. This may be 15-300 metres deep and the width may range between 80 metres and 16,000 metres or more. The Great Barrier Reef of Australia, which is over 19,200 km long and in some cases 144 km from the shore, is of this type. Captain Cook sailed within this reef for 960 km without even suspecting its presence, until the channel narrowed and he was wrecked on the reef. An **atoll** (Fig. 5.48) is a more or less circular or horseshoe-shaped reef, not enclosing any island but a central lagoon which varies from a kilometre to 64 or 80 km across. Thousands of such atolls are found in the South Pacific. It must be borne in mind that the reefs are not continuous walls, but they are broken up into many reefs and islands by passages, the larger of which are valleys sunk below the sea.

*Phylum Cnidaria (Coelenterata)* 239

There are various views regarding reef formation. Of these only the following appears more acceptable. Darwin believed that the reefs begin as fringing reefs

**FIG. 5.48.** Coral reef formation. A—A fringing coral reef growing around an oceanic islet; B—A small barrier reef widely separated from the island; and C—Formation of an atoll enclosing a central lagoon.

on a sloping shore. Then the island subsides very slowly, so slowly that the reef grows upward at about the same rate, naturally the island becomes small, the channel between the reef and land widens and thus the fringing reef transforms into a barrier reef. Further subsidence of the land till it sinks completely out of sight results in the formation of an atoll. This is substantiated by the fact that all the known coral reefs were in regions where a sinking of the land was known to have taken place, or where there was evidence that it had probably occurred.

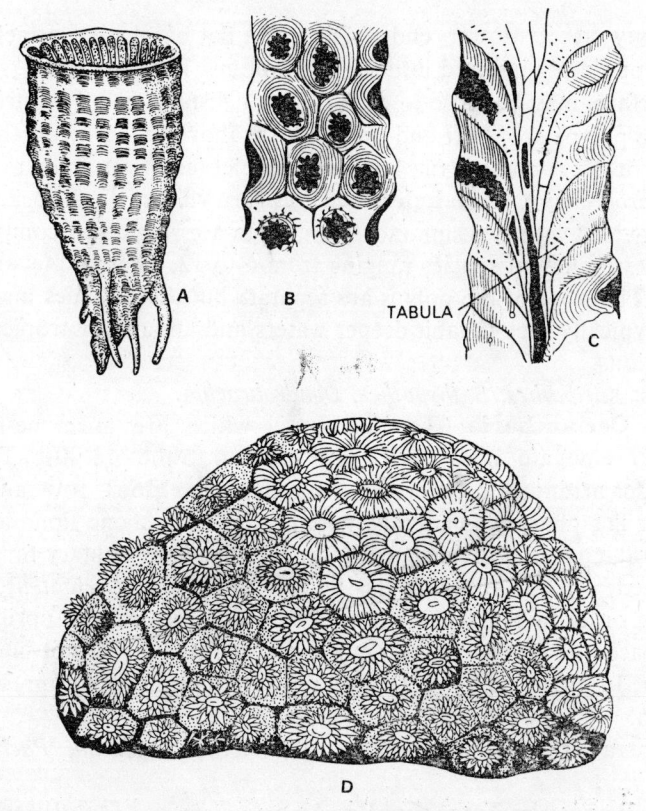

**FIG. 5.49.** A—A rugose coral (*after* Orbigny); B and C—End view and longitudinal section of a tabulate coral (*after* Rominger); and D—A living colony of *Astraea pallida*.

There are about 2500 living species of corals. The extinct species are about 5,000 far outnumbering the living ones. These animals, thus, reached their peak in the past and are now on the decline. The important genera include *Dendrophylia, Madrepora, Astraea, Fungia, Astrangia, Meandra, Isophyllia, Montipora,* etc.

3. **Order Zoanthidea.** The Zooantharia devoid of skeleton and pedal disc, and united by basal stolons (as in the Alcyonaria). Some solitary forms are stalked. The arrangement of septa differs from other living Anthozoa. Except directives, the septal pairs mostly consist of one complete and one incomplete septum (Fig. 5.49). The dorsal directives are incomplete, siphonoglyph is only one on the ventral side. In the zoanthids the sexes are separate but some genera are hermaphrodite. They are most abundant in warm waters though they are found in deeper waters ranging from the tropics to quite cold latitudes. Some solitary forms have their lower ends thrust into substratum. Most genera habitually grow on other animals such as sponges, hydroids, corals, gorgonians, etc.

Examples: *Epizoanthus, Parazoanthus; Zoanthus,* etc.

4. **Order Antipatharia.** The black or thorny coral comes under this order. The Zoantharia that form slender branching plant-like colonies, which like gorgonians consist of a skeletal axis covered by a thin coenenchyme bearing polyps.

The colony at its lower end consists of a flat basal plate attached to some firm object or simply inserted into the substratum. The polyps are scattered all over the surface, may be close to each other or distant or only restricted to one surface. The polyps are short and project little above the general coenenchyme. Tentacles are six, nonretractile borne in a circlet (eight retractile pinnate tentacles in *Dendrobrachia*). The **skeletal axis** is brown or black bearing thorns. Gullet is provided with two siphonoglyphs, septa are single and complete, usually 10, but vary in different species ranging from 6 to 12. These septa are arranged in couples. The sexes of the polyps are separate but the colonies may be hermaphrodite. Typically they inhabit deeper waters and abound in tropical and subtropical regions.

Examples: *Antipathes, Bathypathes, Dendrobrachia.*

5. **Order Ceriantharia.** The Zoantharia which are anemone-like in form with slender elongate body buried into sand up to the oral disc. Tentacles are simple slender arranged in two rows, an outer **marginal** row and an inner **labial** row. The pharynx is laterally flattened and bears one siphonoglyph. The round aboral end is provided with a pore. They are solitary forms and live in slime-lined vertical tubes in the bottom of the sea. *Cerianthus,* a representative of the order, is remarkable in as much as it regains an upright position when displaced from it. It becomes restless if its body is not in contact with some object. In direct sunlight it withdraws into its tube. It has great power of regeneration.

Other genera include *Arachnanthus, Arachnactis, Anactinia, Pachycerianthus,* etc.

6. **Order Rugosa** or **Tetracoralla.** An extinct order of mostly solitary corals possessing a system of major and minor sclerosepta, Cambrian to Permian.

Example: *Zaphrentis.*

7. **Order Corallimorpharia.** Tentacles radially arranged. Resemble true corals but lack skeleton.

Example: *Corynactis.*

Phylum Cnidaria (Coelenterata)

(*iii*) **Subclass Tabulata.** Extinct colonial anthozoans with heavy calcareous skeletal tubes containing horizontal platforms, or tabulae, on which the polyp rested. Sclerosepta absent or poorly developed.

Example: *Halysites*

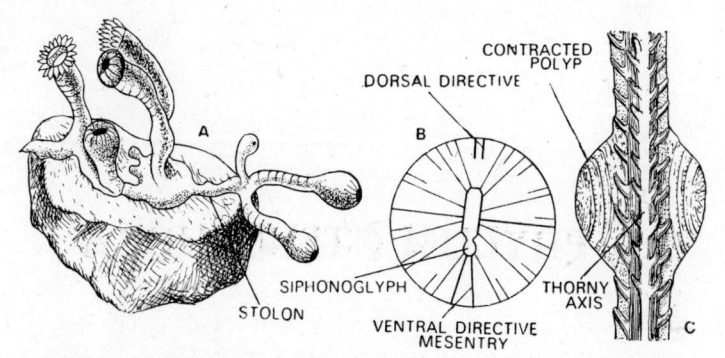

FIG. 5.50. The Zoanthidea. *Zoanthus sociatus.* A—A part of a colony; B—Diagrammatic cross section of a polyp; and C—A piece of *Antipathes* cleared and magnified to show thorny axis. Contracted polyps are also visible.

# 6 PHYLUM CTENOPHORA

**CLASS CTENOPHORA**

When alive, Ctenophora are among the most beautiful of marine organisms. The ctenophores are free-swimming marine animals with transparent gelatinous bodies showing radial symmetry based on an underlying bilateral symmetry. They are never colonial. They have no fixed stage except that they lack nematocysts. They bear comb-like ciliary plates (hence called "comb jellies") as characteristic locomotor structures. The Ctenophora (Gr. *Ktenos*, comb+*phoros*, bearing) present a biradial (radial+bilateral) symmetry and thus constitute the last group of the radiate phyla. Like coelenterates they have radial symmetry; with parts arranged on an oral-aboral axis. There is no internal space except the gastrovascular system which is branched. Mesogloea is gelatinous and definite organ-systems are lacking. The ctenophores principally differ from other coelenterates in the possession of the **ctenes** or **comb-plates**, mesenchymal or mesodermal muscles, an aboral sensory region and in the absence of nematocyst

The ctenophores are transparent gelatinous animals which float in the surface waters of the sea, mostly near shores. They are carried about by currents, being feeble swimmers. Commonly they are known as "sea goose-berries" or "sea-walnuts." They have eight equally spaced meridional rows of ciliary combs which help them in swimming. Each row consists of a succession of little plates formed of cilia fused together at their attached ends like the teeth of a comb. The rows are radially arranged on the globular body. The ctenes beat one after the other, from the aboral to oral pole. These meridionally arranged rows are termed **costae**. The effective beat of the ctenes in the costa is toward the aboral end to facilitate swimming of the ctenophores with mouth onward in the fashion opposite to medusae. The beating is metachronal and when the sense organ at the aboral pole is functioning properly, all eight costae beat in unison. Since the ctenes are iridescent, a flock of ctenophores swimming together in still sunlit water is an unforgettable sight. From opposite sides of the broader end arise two branched muscular tentacles which have no stinging capsules, but are covered with special adhesive cells which stick to and entangle the prey. The mouth is situated in the centre of the lower pole and leads into a branched gastrovascular cavity which extends through the jelly and eventually gives off eight branches, one below each row of combs. All ctenophores are hermaphrodite. Both ovaries and testes occur on the walls of each gastrovascular branch that runs below the rows of combs. The eggs and the sperms are shed through the mouth. The free-swimming larva is not of planula type but can be described as **cydippid larva** because of its resemblance to one of the primitive genera of ctenophores.

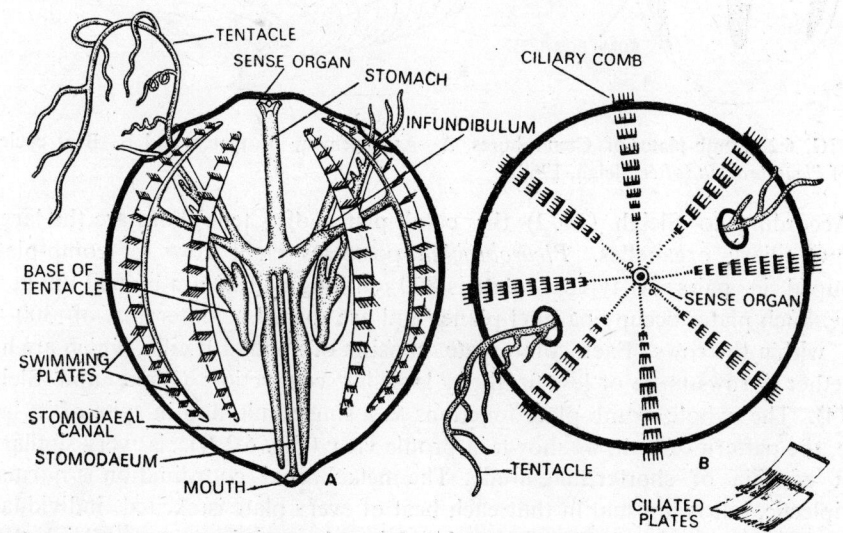

FIG. 6.1. The Ctenophora. A—*Pleurobrachia*, seen from side; B—The same seen from above; and C—Ciliated plates.

## PHYLUM CTENOPHORA

| | |
|---|---|
| **Class:** | CTENOPHORA |
| **Subclass:** | TENTACULATA |
| **Orders:** | Cydippida |
| | Lobata |
| | Cestida |
| | Platyctenea |
| **Subclass:** | NUDA |
| **Orders:** | Beroida |

### TYPE PLEUROBRACHIA

**Form of the Body**

The basic form of ctenophore is perhaps best illustrated in one of the sea-goose berries, *Pleurobrachia*, which inhabit the colder waters of both the Atlantic and the Pacific ocean. The body of *Pleurobrachia* (Fig. 6.1) is more or less spherical in form, highly transparent, and about 2 cm in diameter. The mouth is situated in the centre of the lower or **oral pole** and the **aboral pole** is marked by the presence of of a sense organ. The **oral-aboral** axis forms the main axis of the body. The **comb-plates** are eight equally spaced meridional rows, each of which consists of a slight ridge bearing a succession of small transverse combs formed of fused cilia. (Figs. 6.1 and 6.2).

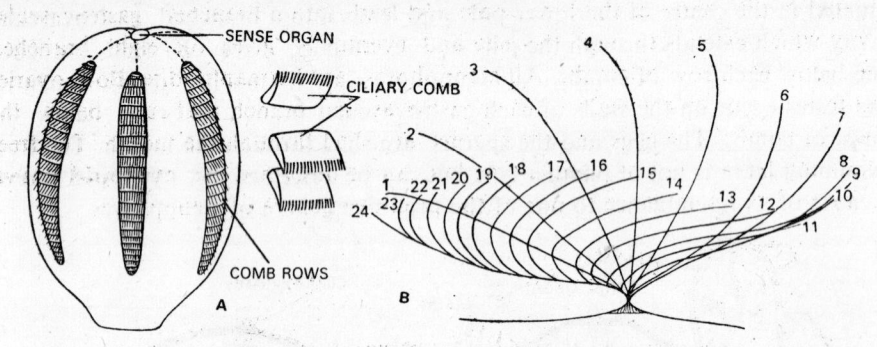

FIG. 6.2. Comb-plates of Ctenophores. A—Arrangement of plates; and B—Beat cycle of *Pleurobrachia* (after Sleigh, 1968).

According to Sleigh (1972) the comb-plates of Ctenophora are the largest known ciliary organelles. *Pleurobrachia* possesses eight rows of comb-plates, grouped in pairs. A typical plate is 800 $\mu$m long and 30 $\mu$m by 600 $\mu$m at the base. Such plates occupy parallel planes and are spaced at intervals of 300-400 $\mu$m within the rows. Each comb-plate consists of about $10^5$ cilia, which are held together in rows more or less firmly by lamellar connection (Blake and Sleigh, 1974). The whole comb-plate functions as a single unit, and in spite of its large size, the pattern of beat, as shown in profile view (Fig. 6.2 B), is very similar to that of cilia of shorter magnitude. The metachronal coordination is normally antiplectic, but is unusual in that each beat of every plate is excited individually, normally by a movement of the plate lying above it in the row (Sleigh, 1972, 1974). While effecting locomotion the combs are rapidly lifted in the direction of the aboral pole and then lowered slowly to their normal position. Those in each row beat after one another starting from the aboral to the oral end. All the eight rows beat in unison and the animal is propelled through water with the oral end in front. The rapid movements of the combs refract light and produce a constant play of changing colours. The ctenophores are known for the beauty of their daytime iridescence. The aboral end bears two long-branched tentacles, one on each side, which can be withdrawn into a blind pouch. The tentacles are solid (Fig. 6.3) containing muscle fibres and devoid of nematocysts. On its surface, however, are special adhesive cells known as **"lasso"** or **glue cells (colloblasts)** that secrete a sticky substance to entangle small animals which are then conveyed to the mouth. The tentacles are of no help in locomotion. They may be contracted or may be trailed out to about 15 cm. The position of the tentacles and some features of the digestive tract exhibit bilateral symmetry.

# Phylum Ctenophora

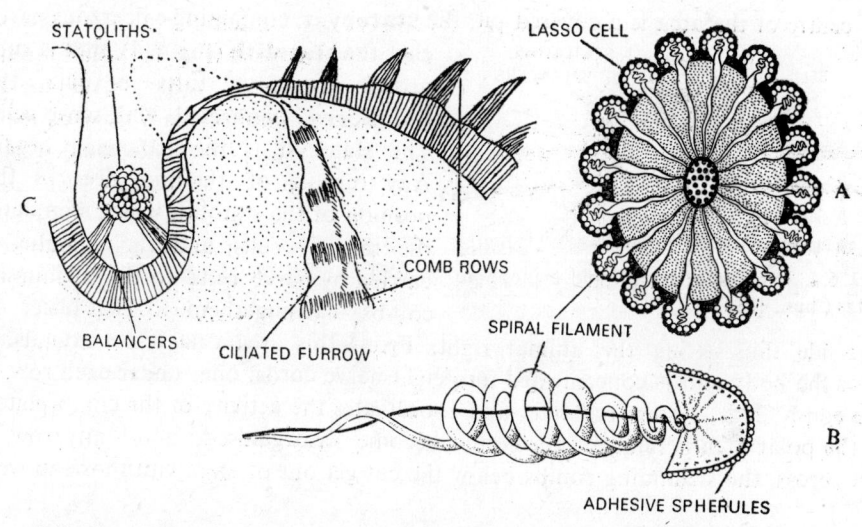

FIG. 6.3. A—Cross-section of a tentacle; B—Colloblast (*after* Koma); and C—Part of anterior end of a ctenophore showing statocyst, ciliated groove and Comb-plates.

## Digestive System

The **mouth** (Fig. 6.1) opens into a **gullet** or **pharynx** (stomodaeal) which opens into the **stomach**. From the stomach arise eight branches, extending one below each row of combs. The system is gastrovascular, that is, it serves the twin purpose of digestion and distribution of food. From the aboral side of the stomach an **infundibular canal** runs to the underside of the statocyst where it gives off four branches, the so called **excretory canals**, which run up to the aboral surface terminating in little sacs. Of these two are blind and two, the diagonally opposite ones, open to the exterior on opposite side of the sense organ by two pores, the **anal pores** (formerly known as "excretory pores"). All ctenophores are carnivorous, feed on small shrimps, fishes and larvae of small animals. All ctenophores use lasso-cells or colloblast cells to capture the food. The prey is captured by the attachment of very many colloblast cells whose filaments cannot be broken by the struggle of the prey. In tentaculate ctenophores the tentacles are extended fully while catching the prey and their curving and sweeping in water covers wider area. Digestion, which is extracellular, begins in the pharynx. Undigested food passes out through the mouth and the anal pores.

## Body Wall

**Ectoderm** forms the outer epithelium of the mouth, pharynx and tentacles, etc. **Endoderm** lines the gastrovascular cavity and is ciliated. **Mesogloea** is gelatinous and contains scattered muscle fibres, connective tissue cells and amoebocytes.

## Nervous System

At the aboral end a definite area of sensory cells and nerve cells is present. In the centre of the area is a covered pit, the **statocyst**, containing calcareous particles, the **statolith** (Fig. 6.4), that is supported on four tufts of cilia, the **balancers**, connected with sense cells. The statocyst is the balancing organ. Any turning of the body disturbs the position of the statolith which stimulates the sensory cells, and the stimulus is carried by nerve cells to the swimming combs, causing them to beat faster on one side, thus setting the animal right. From this area a nerve-net extends all over the body and is concentrated into eight nerve cords, one under each row of the comb. This system regulates and coordinates the activity of the comb-plates. If the polar area is removed the combs become disorganised, and if any row is cut across, the swimming combs below the cut get out of step with those above.

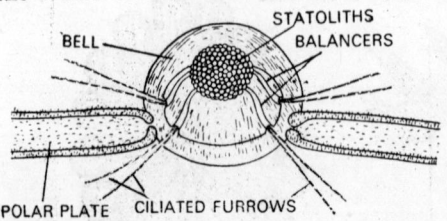

FIG 6.4. Statocyst of a cydippid ctenophore (*after* Chun, 1880).

### Reproductive System

Both the eggs and sperms are produced by the endoderm of the digestive canal underneath the comb plates. The individuals are monoecious. Sex cells are shed through the mouth and fertilization occurs in water. The larva transforms into the adult directly. As the larva resembles the primitive members of the group belonging to the order Cydippida, it is often referred to as **cydippid larva**.

### Affinities

The ctenophores resemble the coelenterates in many characters such as: (*i*) radial symmetry, (*ii*) arrangement of parts along oral-aboral axis, (*iii*) absence of coelom, (*iv*) gelatinous mesogloea, (*v*) branching gastrovascular canals, and (*vi*) general lack of organ-systems. But it is not possible to derive the ctenophores from any coelenterates. The ctenophores also possess some scyphozoan-anthozoan characteristics: (*a*) the stomodaeum, (*b*) the cellular mesogloea, (*c*) the four-lobed condition of the gastrovascular cavity of the larva, and (*d*) general tetramerous symmetry. The ctenophores have been directly linked with Hydrozoa by way of trachyline medusae (such as *Hydroctena*) having an aboral statocyst and two opposite tentacles in sheaths. The ctenophores, however, differ widely from the existing coelenterates in possessing (*i*) eight rows of comb-plates, (*ii*) mesenchymal or mesodermal muscles, (*iii*) higher organisation of the digestive system, (*iv*) an aboral sensory region, and (*v*) no nematocysts except in the ctenophore species *Euchlora rubra* (in which the true nematocysts were discovered in 1957). It seems that the ctenophores must have diverged very early from the trachyline stem form about the time it gave off the coelenterate stocks: trachyline-hydrozoan, scyphozoan and anthozoan lines.

Some authors consider that the ctenophores lead directly to certain lower Bilateria (the polyclad flatworms). The flattened Platyctenea which superficially suggest a polyclad have been considered the missing links between the coelenterates and the flatworms. This view is not considered to be correct now. Among the flatworms the Acoela are the most primitive and not the polyclad, hence the statement "ctenophores-platyctenea-polyclads" has become untenable. It is certain that the ctenophores present certain advanced structural features that appear to look forward to the Bilateria. These features include (*i*) the prominence of an apical nervous region, (*ii*) the mode of origin of musculature, (*iii*) the presence of gonoducts (sexual ducts), and (*iv*) the determinate type of cleavage.

In short, the ctenophores are regarded as a blind early off-shoot from the trachyline stem form that attained a considerable grade of differentiation without leading to any higher forms. They, however, appear to indicate the lines along which structural complication will proceed in the next grade (organ-system grade of construction) without themselves being directly in the line of ancestry of this grade. The stock which gave rise to ctenophores went on differentiating along the same path, but in another direction (that of bilaterality) and gave rise to higher forms.

## CLASSIFICATION OF CTENOPHORA

The Ctenophora are divided into two subclasses: I. Tentaculata with tentacles, and II. Nuda, without tentacles. Of these the Tentaculata is divided into four orders and Nuda consists of only one. The whole group comprises about 80 species only.

**I. Subclass Tentaculata.** These are Ctenophora in which tentacles are present.

**1. Order Cydippida.** The Tentaculata with simple rounded or oval form, the gastrovascular branches of which end blindly. Tentacles are two, branched and can be withdrawn in sheaths.

Examples: *Pleurobrachia* (Fig. 6.1 A), *Hormiphora*.

**2. Order Lobata.** The Tentaculata in which the body is laterally compressed and there are two large oral lobes. The tentacles are without sheaths. The oral ends of gastrovascular canal are anastomosed.

Examples: *Bolinopsis, Leucotheca* (=*Eucharis*).

**3. Order Cestida.** The Tentaculata in which the body is elongated and compressed in the sagittal plane imparting a ribbon like appearance to it. This order has only two genera *Cestus* (*Cestum*) and *Velamen* (Fig. 6.5).

*Cestus* (Fig. 6.5) is a transparent ribbon, some 1.5 metres long, and is found coiling and uncoiling as it curves its way through water. The long edges of the

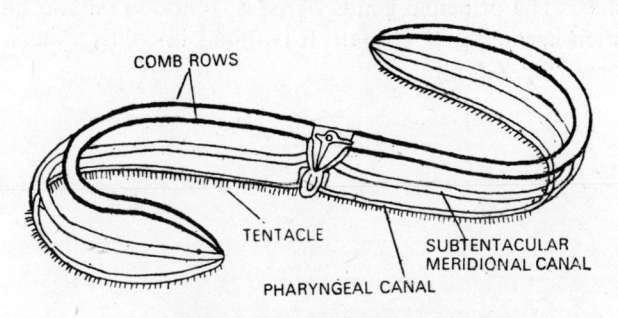

FIG. 6.5. *Velamen* (*after* Mayer).

ribbon are fringed with tentacles. The bases of the principal tentacles are enclosed in sheaths. Anastomoses of the meridional and stomodaeal vessels takes place. *Cestus* lives in warm seas but is sometimes carried north. Sun light playing on it reflects a thousand wonderful tints of silver and azure, pink and amethyst. Because of its elongated band-like shape and because of the beautiful colours that it reflects *Cestus* has been rightly called "Girdle of Venus."

4. **Order Platyctenea**. The Tentaculata in which the body is compressed along the oral-aboral axis forming flattened creeping forms. The two tentacles are with sheaths. Comb-plates may be present in larvae only.

Examples: *Coeloplana, Ctenoplana* (Fig. 6.6 A and B).

II. **Subclass Nuda**. These are Ctenophora which do not have tentacles.

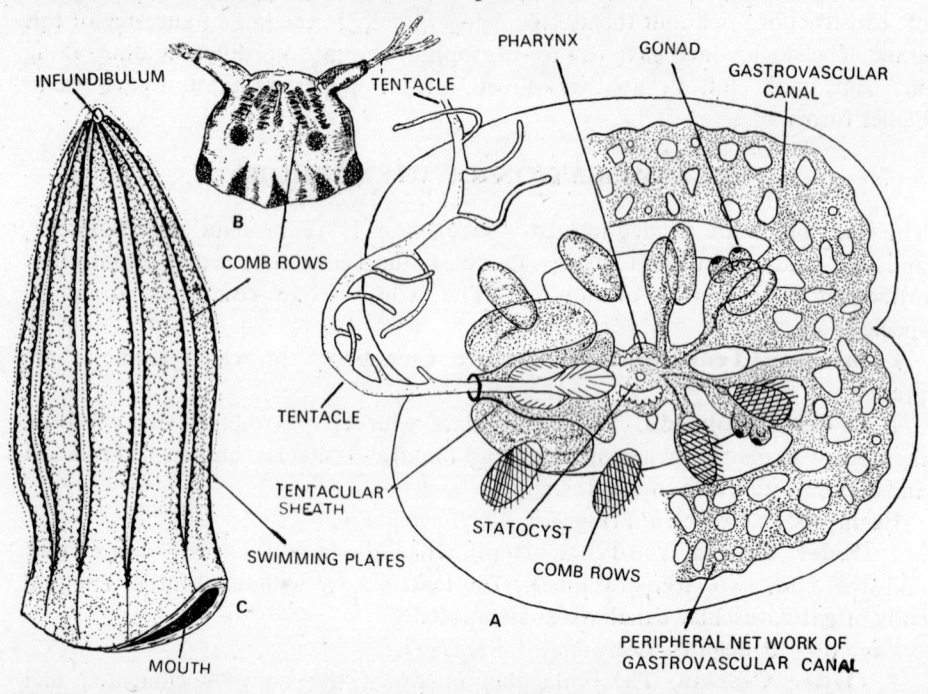

FIG. 6.6. A—*Ctenoplana*, seen from above; B—The same seen from side; and C—*Beroe*.

**Order Beroida**. The Nuda are of conical forms with a very wide mouth and pharynx. The meridional vessels are produced into a complex system of anastomosing branches. The principal genus *Beroe* is found in all seas and is of rose-pink colour, often becoming 20 cm tall. It is found in colder waters.

Example: *Beroe* (Fig. 6.6 C).

# 7 PHYLUM PLATYHELMINTHES

The remaining phyla of the animal kingdom are characterised by the presence of a regular three-layered condition, which appears quite early in development,

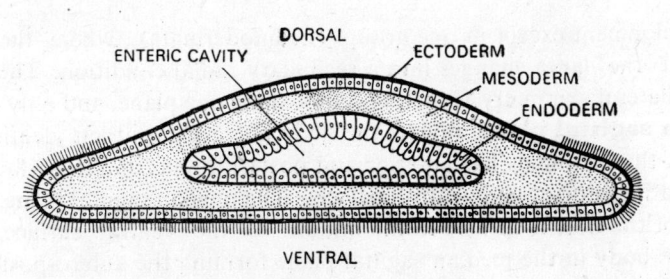

FIG. 7.1. Triploblastic condition. Diagrammatic cross-section of a planarian showing three-layered body plan.

hence known as the **Triploblastica**. A third layer of cells, the **mesoderm**, appears between the **ectoderm** and **endoderm** and opens up further possibilities of increase in size and complexity. Though a third layer is indicated in some coelenterates and also in the Ctenophora, yet a third layer with a definite fate is found only in the Triploblastica. In these animals three layers of cells, the primary germ layers, are present only in the embryo. As development proceeds the cells actively divide, become specialised and grouped together to form tissues which ultimately form organs. The flatworms can also be said to have reached the organ system level of complexity; in other words, they show a higher degree of interdependence of parts than do the coelenterates. Not only are their tissues specialised for various functions, but two or more types of tissue cells may be combined to form an organ of specific function.

It has been seen that the coelenterates were radially symmetrical animals but the Triploblastica are characterised by bilateral symmetry which appears quite

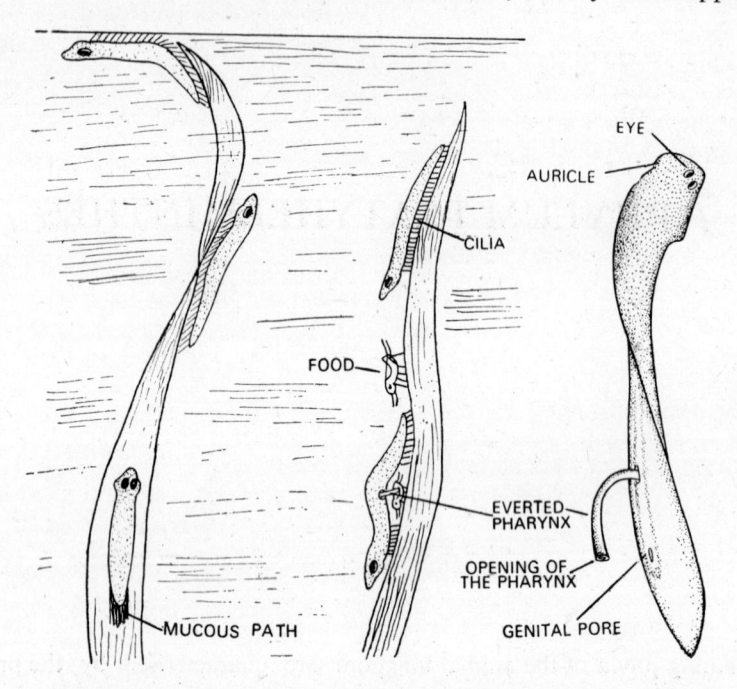

FIG. 7.2. A—Planarians as seen in nature; and B—*Planaria*, side view.

early in development except in one group (Echinodermata), where the bilateral condition of the larva changes into a secondary radial condition. The essential feature of bilateral symmetry is that a section along one plane, and only one plane, the **median sagittal plane**, divides the body into right and left identical halves. The body of these animals has anterior and posterior ends and its surfaces can be distinguished into dorsal and ventral surfaces. The median sagittal plane runs from the middle of the dorsal surface to the middle of the ventral surface, the main axis of the body in the median sagittal plane forming the antero-posterior axis. Since radial animals tend to lead a sessile or sedentary life, the assumption of bilateral symmetry appears to have been an important step forward. In contrast to the Radiata, the phyla in which bilateral symmetry occurs are known as the **Bilateria**.

There is no extensive cavity within the body except the gut, which is in the form of an endoderm-lined blind sac. The simple opening to the exterior must serve both as mouth and functional anus. In one group of small Turbellaria which may or may not be primitive, there is not even a differentiated gut cavity, and in one highly modified group of flatworms, the Cestoda or tapeworms, there is no gut at all, nutritive uptake being through the general body wall from the host's gut cavity. In no flatworm is there any other internal space, and certainly nothing which corresponds to the body cavities of more complex animals. Flat-

*Phylum Platyhelminthes*

worms are thus acoelomate. They are also the simplest animals (if one excludes the planula larvae of some coelenterates) to have a definite front end and rear end, upper-surface and under-surface, and therefore left and right sides, and true bilateral symmetry. This symmetry they share with all other more complex Metazoa excluding only adult echinoderms. It should be noted that having a definite front end does not imply having a head, although there are some beginnings of cephalisation (head formation) in Turbellaria with anterior concentrations of nerve and sense cells.

The Platyhelminthes are acoelomate Bilateria without a definitive anus. The enterior end of the body is differentiated into a head region and the body is flattened dorsoventrally. These dorsoventrally flattened worms include one group of free-living forms (class Turbellaria) found in fresh waters, in the sea, and in damp soil. Two classes are parasitic: class Trematoda of flukes (mainly parasites of higher animals), and class Cestoda, the tapeworms, which show replication of parts and are mainly parasites of the alimentary canal of higher animals. In general appearance they are worm-like varying from long flattened ribbon-like tapeworms to broad leaf-like forms of some polyclads. Most of the flatworms are of moderate size but some tapeworms are as long as 10 to 15 metres while others are microscopic. The forms are generally colourless or white or owe their colour to ingested material. Some freeliving forms are brown, grey or black, whereas, among the polyclads and planarians bright-coloured individuals are also met with. In the free-living forms the ventral surface is well-marked and bears the openings of the mouth and genital pores. This is the surface on which locomotion also occurs.

## PHYLUM PLATYHELMINTHES

**Class I:** Turbellaria
**Orders:** Acoela
Rhabdocoela
Alloeocoela
Tricladida
Polycladida

**Class II:** Trematoda
**Orders:** Monogenea (Heterocotylea)
Aspidobothria
Digenea (Malacocotylea)

**Class III:** Cestoda
**Subclass I:** Cestodaria (Monozoa)
**Orders:** Amphilinidea
Gyrocotylidea

**Subclass II:** Eucestoda (Merozoa)
**Orders:** Tetraphyllidea or Phyllobothrioidea
Lecanicephaloidea
Proteocephaloidea
Diphyllidea
Tetrarhynchoidea or Trypanorhyncha
Pseudophyllidea or Bothriocephaloidea
Cyclophyllidea or Taenioidea.

## CLASS TURBELLARIA

The Turbellaria are mostly free-living unsegmented flatworms that are clothed with a cellular epidermis which is usually ciliated, sometimes only in parts. They

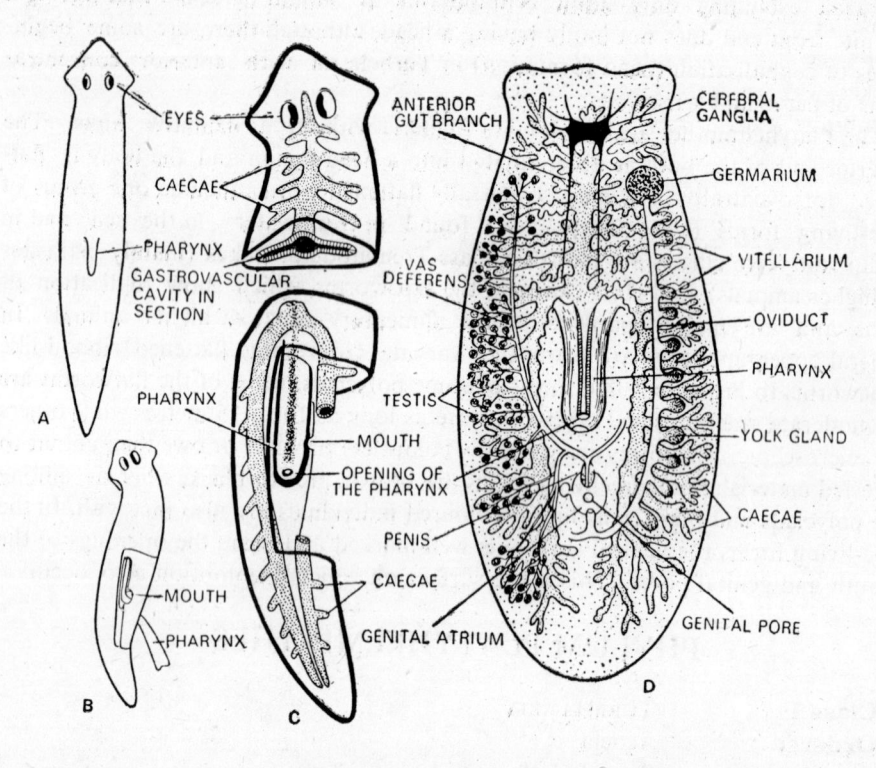

**FIG. 7.3.** Structure of a planarian. A—Ventral view; B—Dorsal view, a portion of body removed to show the structure of eversible pharynx and mouth; C—Digestive organs of planarian; and D—General anatomy of the same.

have simple life cycles. Some parasitic or commensal forms are also met with. They are mostly small animals of about 5 mm in length. Some forms are of microscopic dimensions, whereas, some land planarians may reach a length of 50 cm. Mostly they are of vermiform shape, but the larger forms are long and slender (*Bipalium*) or leaf-like (*Temnocephala*). The body is dorsoventrally flattened, with flat ventral surface and move with the help of vibratile cilia of the epidermis. When disturbed the animal hurries away by marked muscular waves which start at the anterior end and pass towards the posterior end thus setting up a regular series of ripples in the body wall that drive the body along. This type of movement is characteristic of the planarians and the muscular movement is particularly pronounced in larger free-living flatworms. The planarians trace out a wriggly path because the head is continually bending from side to side.

### Body Wall

The body wall (Fig. 7.4) of the Turbellaria is clothed with one-layered ciliated

# Phylum Platyhelminthes

epidermis consisting of cuboidal cells resting on an elastic basement membrane. The cells are quite large and contain large rounded nuclei towards their inner sides. The cytoplasm of the cells contains rod-shaped hyaline bodies, the **rhabdites**, (Fig. 7.4) the function of which is unknown. The epidermis has many deep-lying unicellular gland cells. They are located in the mesenchyme and are apparently insunk epidermal cells. These cells have a club-shaped body with a long neck which serves as a duct for the discharge of the secretion. The cells are packed with granules. The musculature forms a layer just beneath the basement membrane and is thus referred to as subepidermal musculature, which in its

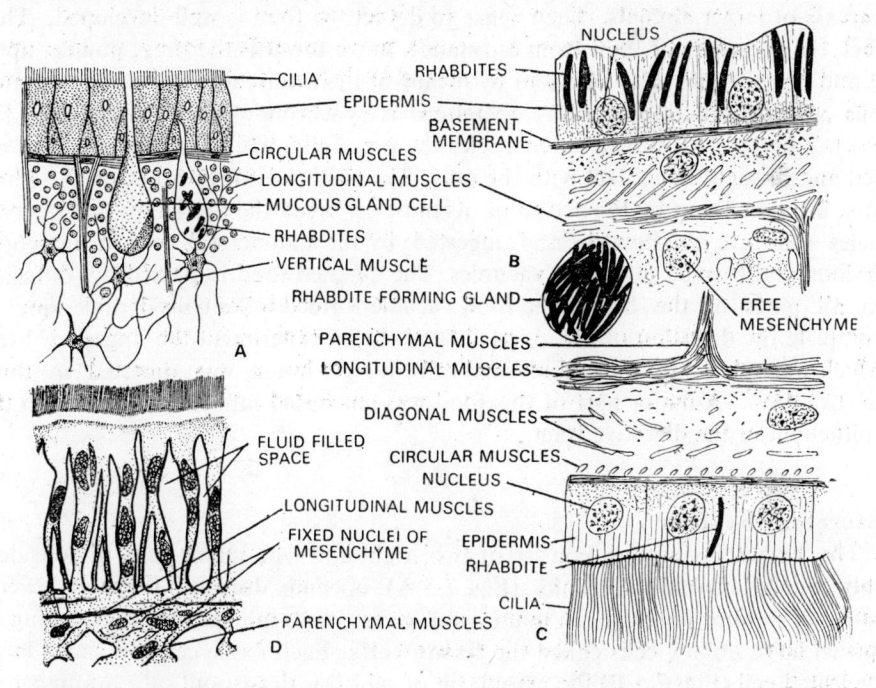

FIG. 7.4. Structure of the body wall. A—Transverse section of the body wall of a turbellarian; B—Longitudinal section through the dorsal epidermis of a freshwater planarian; C—The same through the ventral epidermis (note cilia and paucity of rhabdites); and D—Longitudinal section through the epidermis of an acoel.

simplest form consists of two layers—an outer layer of circular fibres and an inner layer of longitudinal fibres (Fig. 7.4). The interior of the body is packed with simple connective tissue or parenchyma. In some it consists of contiguous rounded cells. But in the planarian, as in most cases, it consists of a loose meshwork of cells lacking definite cell walls (syncytium), which produce new parts during regeneration. It is natural therefore that this layer cannot be regarded as a simple packing tissue as its component cells perform different functions and help in the assimilation and transport of food, and in the excretion of waste products.

## Digestive System

The mouth, situated near the middle of the ventral surface, opens into a cavity that contains a tubular muscular organ, the **pharynx**, which is attached only at its anterior end and consists of complex musculature and gland cells. Normally the pharynx is folded inside but is capable of eversion during feeding. The pharynx then becomes longer and is thrust out as a **proboscis** through the mouth for some distance. The pharynx opens into the intestine which, in the present case, consists of three main branches, one runs to the anterior end of the body and two are posterior and pass backwards one on either side of the pharynx. All these branches have numerous fairly regularly spaced side branches or caeca. The branches as well as the caeca are all blind. There is no anal aperture. The epithelium of the digestive tract consists of endoderm.

The food of a planarian consists of small living animals. They also feed on carcass of larger animals. Their sense to detect the food is well-developed. They feel the presence of food from a distance, move towards the prey, pounce upon it and press it against the bottom by means of their muscular bodies. Simultaneous with this action a slimy secretion is synchronously released from the worm's body. This substance entangles the prey fully, while the pharynx is everted and brought in contact with the prey. The suctorial movements of the pharynx, brought about by the action of its muscles, break the food into minute particles that are swallowed, and ingested by the endodermal cells in amoeboid fashion and formed into food vacuoles. The digested food is passed by diffusion to all parts of the body. The food vacuole formed takes considerable time to complete the digestion of the ingested food. In an experiment the ingested liver, which was taken into the endodermal cells in eight hours, was digested in three to five days. A major part of the food was converted into fat and stored in the epithelium of the digestive tract.

### Excretory System

The excretory organs consist of two, right and left, main, pairs of considerably coiled, longitudinal trunks (Fig. 7.5 A) opening dorsally through several pairs of minute pores. Each main trunk gives rise to minute branches ending in special large hollow cells called the **flame-cells**. Each flame cell is a single large nucleated cell (Fig. 7.5 B) the cytoplasm of which is drawn out into a number of of special prolongations. Internally the cytoplasm is hollowed out to form a central cavity continuous with that of the fine tubule. A bunch of flagella hangs down into the cavity of the cells and performs regular undulating movements. These movements bear a resemblance with the flickering flame of a candle, hence the name flame-cell. Bundles of vibratile cilia are found in the branches of the main tubule. Similar bundles of cilia, known as lateral flames, are also found in the main tubules. These help in keeping the fluid in circulation. In a number of turbellarians large cells with conspicuous nuclei and vacuolated cytoplasm are found in close contact with the tubes. These are called the **athrocytes** or **paranephrocytes** and it is probable that these collect excretory matter from the parenchyma and discharge into the lumen of the flame-cell. Since metabolic wastes seem to be excreted largely by way of the endodermal epithelium and mouth, the flame-cell system appears to function primarily for the regulation of water contents (osmo-regulation). They are, thus, also called the water-vessels.

**Gaseous interchange**. The gaseous interchange is carried out over the whole surface which is relatively increased by the flattening of the body. Free-living flatworms are fairly active animals and require continuous supply of

*Phylum Platyhelminthes* 255

oxygen which is ensured by quicker diffusion through increased surface of the body.

### Reproductive System

The reproductive organs are more complicated in the flatworms. The ovaries and testes are derived from the mesoderm and there is a system of tubules and chambers in which fertilisation occurs. For the transfer of sperms also some complications have taken place. The animals are hermaphrodite having both male and female sex organs in every individual, the self-fertilisation is avoided. Exchange of sperms takes place to effect cross-fertilisation. After the reproductive season the sexual organs degenerate and appear afresh at the onset of next sexual period.

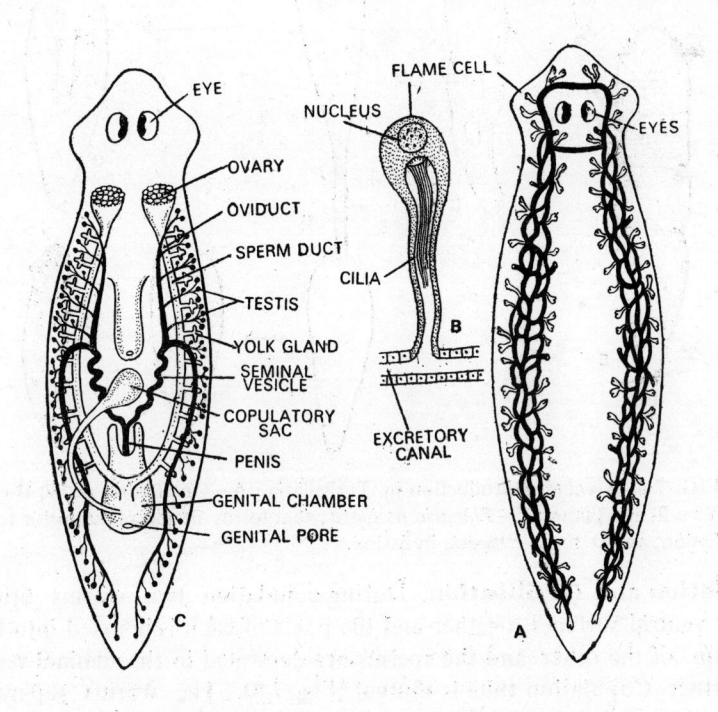

**FIG. 7.5.** A—Organs of excretion of a planarian; B—One flame-cell (enlarged); and C—Reproductive organs of *Planaria*.

**Male.** The male generative organs consist of **testes** of which there are many along the sides of the body (Fig. 7.5 C). From these arise delicate tubes, **vasa efferentia**, which unite to form a prominent **vas deferens** on each side. The vasa deferentia from two sides unite and form a common tube which is often distended to form a **seminal vesicle**. The tube opens into a protrusible copulatory organ, the **penis**, which helps in transference of sperms to another planaria. The penis projects into a chamber, the **genital chamber**, which opens to outside by the genital pore, situated behind the mouth on the ventral surface.

**Female.** The female generative organs (Fig. 7.5 C) consist of a pair of **ovaries** situated close behind the eyes. Each ovary is connected to a long oviduct running along the edge of the body parallel to the nerve cord. Into each oviduct open several **yolk** or **vitelline glands** consisting of clusters of yolk cells. The two oviducts unite to form a median vagina which opens into the genital atrium. To the vagina is attached a long stalked sac known as the **seminal receptacle** or **copulatory sac** (Fig. 7.5 C).

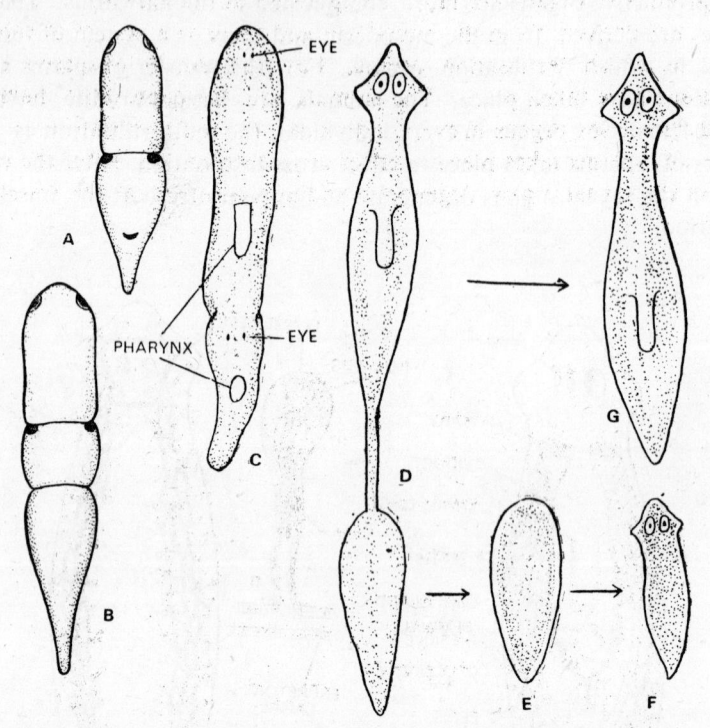

FIG. 7.6. Asexual reproduction in Turbellaria. A—One-fission plane; B—Two-fission planes; C—*Planaria fissipara*, that forms fission planes prior to fission; and D to G—*Dugesia* in fission.

**Copulation and fertilisation.** During copulation two worms bring their posterior ventral surfaces together and the penis of each is inserted into the genital atrium of the other and the sperms are deposited in the seminal receptacles of its partner. Copulation thus is mutual (Fig. 7.8). The worms separate after the deposition of the sperms. The sperms soon reach the ovaries via oviducts to fertilise the eggs as and when discharged. While passing down the oviducts the fertilised eggs become associated with the yolk cells that are ejected from the yolk glands. In the genital chamber the zygotes and the yolk cells are surrounded by a shell to form an egg capsule. Thus, it becomes apparent, that the zygotes (which do not contain yolk) are supplied with yolk cells. Each capsule contains about ten zygotes and thousands of yolk cells, on coming out of the genital pore the capsule often gets fastened to some objects in the water where development takes place. Development is direct. No larval stage is met with. The young one that hatches out is just like the adult (Fig. 7.8) but the reproductive organs are not formed.

**Asexual reproduction.** Some Turbellaria multiply asexually commonly by transverse fission, No preparatory morphological changes take place before

# Phylum Platyhelminthes

fission. When the animal has attained a certain size the worm constricts at a region behind the pharynx (Fig. 7.6). The animal keeps on gliding in this condition. Suddenly the posterior part may grip the bottom and hold on, while the anterior head piece struggles to move forward, and thus the fission region is pulled out to an elongated shape and finally ruptures. Both the pieces soon regenerate their lost parts. Asexual reproduction is restricted to some species only. Some species reproduce asexually in one season, and may develop sex organs in the others, but there are some species that reproduce exclusively asexually and rarely develop sex organs.

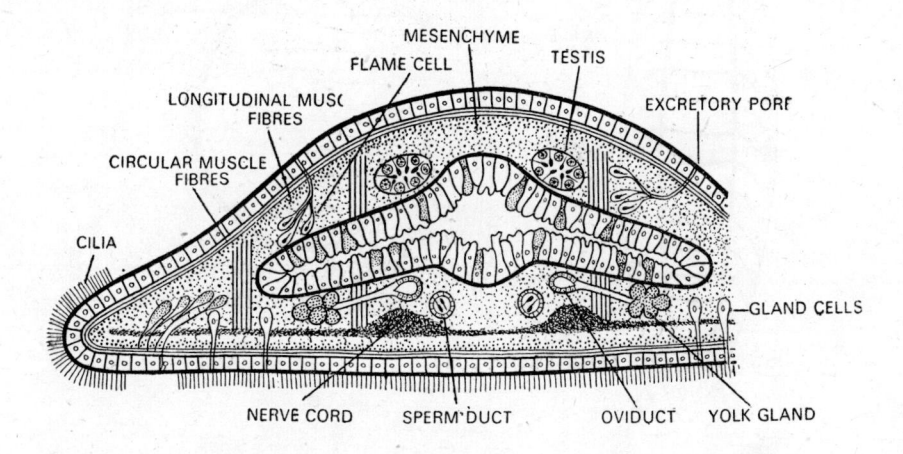

FIG. 7.7. Diagrammatic cross-section through a sexually mature *Planaria*.

## Nervous System

In the head region of a planarian the nervous tissue is concentrated into a bilobed mass called the **brain** (Fig. 7.8 A). This marks the first appearance of a centralised nervous system, the kind of which is found in all higher animals. From the brain arise two nerve cords, which are simple strand-like concentrations of nerve cells, and run backward through the parenchyma near ventral surface. The two cords are connected with each other by many cross-strands giving a ladder-like appearance to the system. Further the cords give off several branches to the sides of the body. The brain of a planarian has no control on the muscular co-ordination involved in locomotion. Thus planarian will keep on moving in a coordinated fashion even without its brain. The brain here receives stimuli from the sense organs and transmits them to the rest of the body. The nerve-net as was seen in *Hydra* does not disappear in the planarian, but persists in addition to the centralised nervous system. In some of the most primitive Turbellaria the nervous system is in a condition resembling that of the coelenterates, which, according to some authors, appears to be the original type of nervous system of the group.

**Sense organs**. The Turbellaria are richly supplied with sensory cells of various types. Some are simple tactile cells slender and elongated, and lie with their pointed ends projecting from the body surface. Some are specialised to receive chemical stimuli and are termed the **chemoreceptors**, which perceive and orient toward food juices at a distance. The chemoreceptors are mostly restricted to the head region and comprise ciliated pits, and grooves, etc. Some are similarly specialised to receive temperature stimuli, whereas, some are modified to

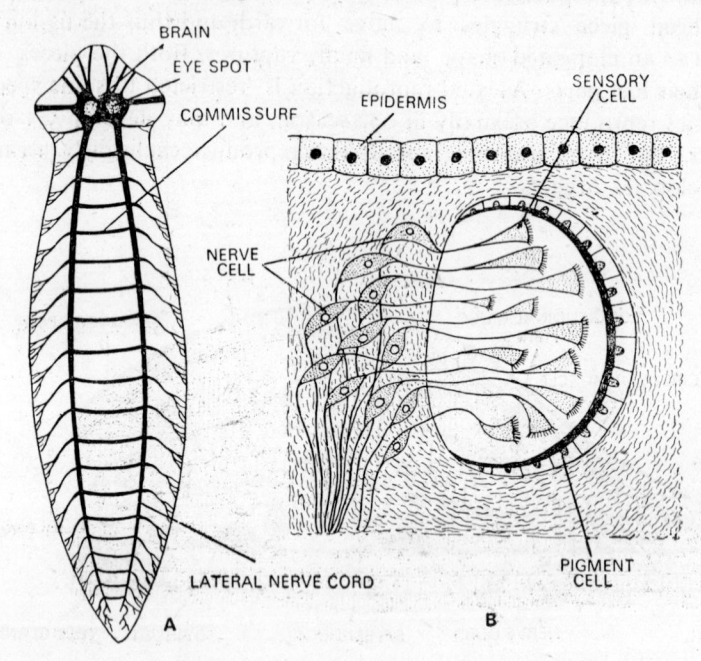

FIG. 7.8. A—Nervous system of a turbellarian; and B—Section through an eye of the same

act as current detectors. Some sensory cells are grouped in the head region forming special sense organs. The sensory lobes, auricles, on each side of the head are known to possess sensory cells sensitive to touch, water currents and probably also to food and other chemicals.

For the perception of light specialised sense organs, **eyes** (Fig 7.8B) or **ocelli**, are present, although the general surface of the body is also sensitive to light. Each eye consists of a bowl of black pigment filled with special photosensitive cells, or **retinal cells**, whose ends continue as nerves and enter the brain (Fig. 7.8A). The free ends of these bipolar cells projecting in the cavity of the pigment cup are marked by longitudinal striations forming the rod-border of uncertain function The pigment shades the sensory cells from light in all directions but one, and so enables the animal to respond to the direction of light. The ectoderm immediately above the eye is devoid of pigment and thus permits light to pass through to the sensory cells. In some more developed eyes this epithelium probably possesses some refractive power also. This evidently increases the efficiency of the eyes.

**Behaviour**

The behaviour of a planarian is more varied and shows much more rapid responses. The planarians avoid light (Fig. 7.10) and are generally found in dark places, under stones or leaves of aquatic plants. If placed in a dish exposed to light, they immediately move toward the darkest part of the dish. On applica-

## Phylum Platyhelminthes

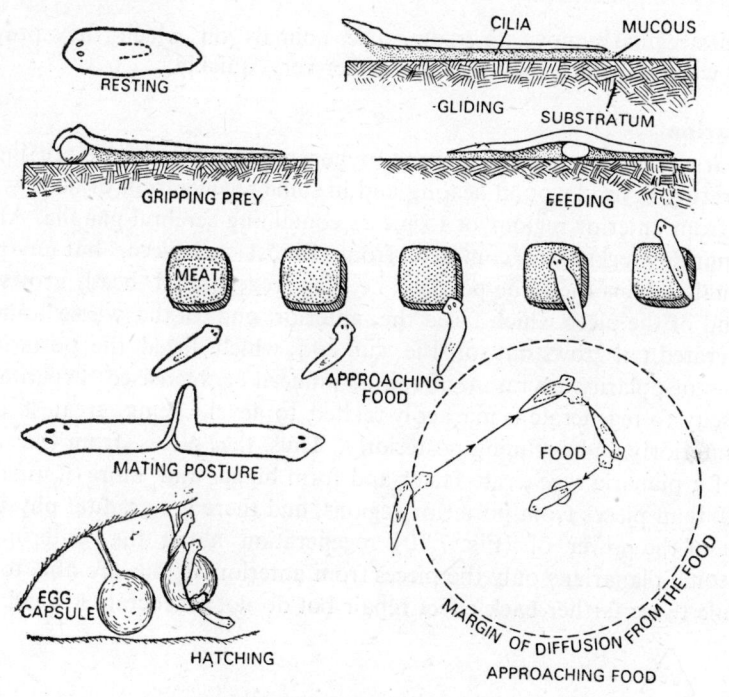

FIG. 7.9. Behaviour of a turbellarian.

tion of weak mechanical or chemical stimuli to the head region, either laterally or medially, the worm reacts positively by turning toward the stimuli, but strong stimuli result in a negative reaction. They quickly react to the presence of food

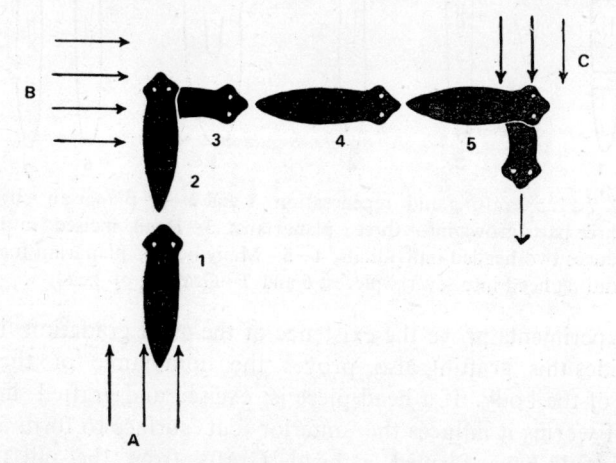

FIG. 7.10. Effect of light on a planarian. The planarian turns away from sources of light at *A, B, C*.

by moving directly towards it. That is why a piece of raw meat placed in a stream inhabited by planarians attracts hordes of them, which glide upstream toward the food, guided by the meat juices in the current of water (Fig. 7.9). The planarians are highly positive to contact and tend to keep the undersurface of the body in contact with other objects. They also react to water currents and some species regularly move upstream. The animals on which they prey often ripple the water to which they respond rather very quickly.

### Regeneration

Turbellarians possess a great power of regeneration. Only some are incapable of regenerating beyond wound healing and in some regeneration of head is limited to pieces from anterior regions or to pieces containing cerebral ganglia. Although some planarians regenerate complete from almost any piece, but any piece of such animals retains the same polarity, i.e., the regenerated head grows out of the cut end of the piece which faced the anterior end of the whole animal and the regenerated tail grows out of the cut end which faced the posterior end. Alterations of polarity are rare, but can be induced or increased experimentally. The capacity to regenerate is markedly related to level, being greatest or most normal anteriorly and declining posteriorly. Thus, the pieces from the anterior regions of a planaria regenerate faster and form bigger and more normal heads (Fig. 7.10) than pieces from posterior regions, and there is a gradual physiological gradation of the power of (Fig. 7.10) regeneration along the anteroposterior axis. In some planarians only the pieces from anterior regions are able to form a head, while those farther back effect repair but do not regenerate a head.

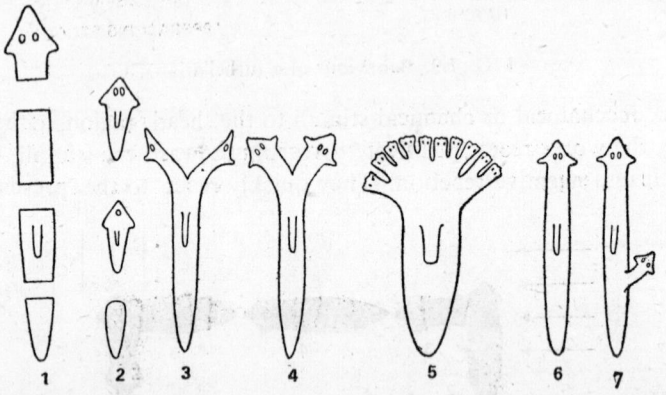

FIG. 7.11. Grafting and regeneration. **1 and 2**—A planarian cut into three parts grows into three planarians; **3**—Head incised in two; produces two-headed individuals; **4—5**—Many headed planarian formed by cutting head into several pieces; **6 and 7**—Grafting of head.

Grafting experiments prove the existence of the axial gradation in an effective manner. Besides this, grafting also proves the dominance of the head region over the rest of the body. If a head piece is excised and grafted in windows in other parts of worms it induces the anterior cut surface to form a tail, though ordinarily it would have formed a head. Grafts from the tail regions do not have these effects and are generally absorbed. The dominance of the head over the rest of the body is limited by distance. If the animal grows to a sufficient length its rear parts may get beyond the range of dominance of the head. This happens in asexual reproduction of the planarian when the rear part starts to act as if it were physiologically isolated and then finally constricts off as a separate animal. Experimentally, it has been shown that the cutting off the head induces separation of the bud in asexual reproduction.

# Phylum Platyhelminthes

Some other interesting features of planarian regeneration may be dealt with briefly. If the anterior end of the head is cut down the middle, and the two halves are prevented from growing together again into a single head, then each half will regenerate into a head forming a two-headed individual. Likewise, if the head is cut repeatedly and the cut edges are not allowed to grow together a multi-headed individual (Fig. 7.11) may be produced. Isolated heads cut off just behind the eyes also often regenerate a reversed head at the cut surface. Very short pieces especially from anterior regions may form heads at both cut surfaces. Similarly, in some species, short posterior pieces produce biaxial tails.

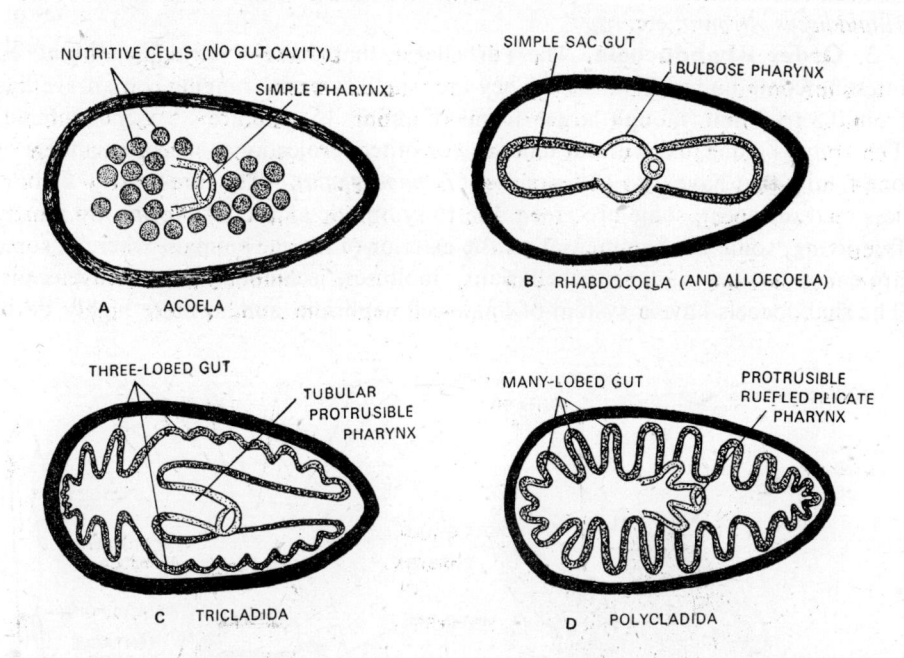

FIG. 7.12. Diagrams of gut and mouth structures in the five principal groups of free-living flatworms (A-D).

An oblique cut results in the formation of head at the most anterior point and a tail at the most posterior point.

## CLASSIFICATION OF TURBELLARIA

The Turbellaria are divided into five orders—**Acoela, Rhabdocoela, Alloeocoela, Tricladida** and **Polycladida**. This is the most conservative system of classification that is followed here, though some recent workers (Meixner, 1938) proposed alternative classification under which the rhabdocoels are split into three orders and the triclads are made a suborder of alloecoels.

1. **Order Acoela**. These are Turbellaria in which a digestive cavity is lacking, food is digested in the mass of the endoderm cells either scattered or aggregated into a central digestive zone in which temporary space appears around the digesting food particles. Mouth is present. They have no excretory system. In primitive forms the nervous system is diffused and epidermally located, but in most others the nervous system consists of a sub-muscular plexus with 3 to 6

pairs of longitudinal strands. The gonads are definitely delimited and the sex cells originate from the mesenchymal cells directly. Copulatory mechanism is well-developed. The Acoela are small free-living, exclusively marine forms, some live in the intestine of sea-urchins and sea-cucumbers. They live under stones, among algae and most commonly on the bottom extending from shallow to deep waters. They are generally flat, oval and translucent. One of the common forms is the genus *Convoluta* that inhabits sea-shore tide pools and is green in the adult condition due to the presence of symbiotic algae.

Other known genera include *Nemertoderma, Diopisthoporus, Ectocotyla, Haplodiscus Afronta*, etc.

2. **Order Rhabdocoela**. The Turbellaria that have a straight unbranched intestine come under this order. They are small worms ranging on an average from 0.3 to 3 mm, though larger forms of about 15 mm are also encountered. The body is elongated without tentacles or other projections except members of one family that have 2 to 12 tentacles (*Temnocephala*). They are mostly colourless or translucent, some are green due to symbiotic algae. The forms are mostly free-living, some are commensal on the exterior of marine animals, whereas, some are endoparasites in other turbellarians, molluscs, echinoderms or crustaceans. The rhabdocoels have a system of flame-cell nephridia and a more highly deve-

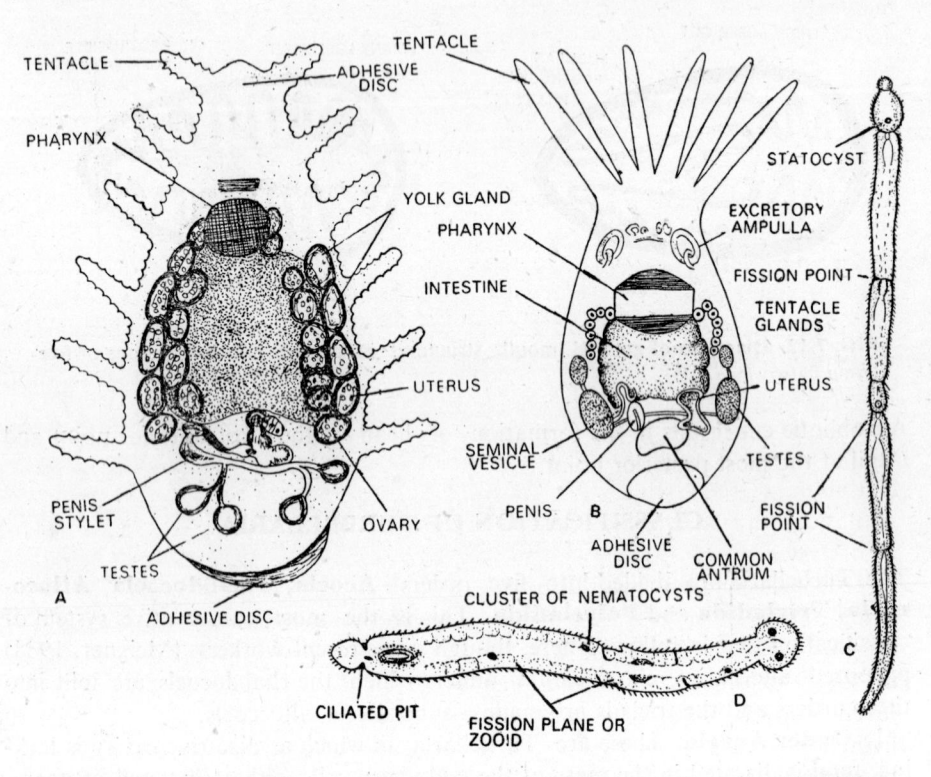

FIG. 7.13. Rhabdocoela. A—*Actinodactylella*; B—*Temnocephala* (*After* Haswell); C—*Catenula lemnae*; and D—*Microstomum* (*after* Hyman).

loped nervous system. The reproductive system consists in male, of well-developed testes, sometimes broken up into lobes or separated follicles, in female,

## Phylum Platyhelminthes

one ovary or a pair or ovaries, a pair of yolk glands and the copulatory complex. In some rhabdocoels asexual reproduction occurs as in the planarians. In *Microstomum* (Fig. 7.13 D) the parts fail to separate at once so that chains of eight to sixteen sub-individuals each with its own mouth, are often formed. This genus also possesses nematocysts, believed to be taken from hydras upon which it feeds. From the gastrovascular cavity the nematocysts pass through the mesenchyme to the ectoderm where they are lodged in some epithelial cells and used in defence.

The rhabdocoels are found all over the world in fresh standing and rarely flowing waters, in damp terrestrial habitats and littoral zones of ocean shores. *Microstomum* (Fig. 7.13 D), *Macrostomum, Mesostoma, Catenula* (7.13C), *Stenostomum, Temnocephala* (Fig. 7.13 B), *Actinodactylella* (Fig. 7.13 A) etc., are some important genera of the order.

3. **Order Alloeocoela.** These are Turbellaria with simple intestine which may be like a contoured sac or with diverticula. The paired protonephridia have two or three main branches on each side and a number of nephridiopores. The nervous system has three or four pairs of longitudinal cords transversely connected. They have usually many testes and one pair of ovaries. The alloeocoels present characters mostly found in the Tricladida, though some of its primitive members are like Acoela. Some are minute but most of them are larger than the acoels and rhabdocoels. They are predominantly marine and some occur in fresh water. They are found among vegetation especially on sandy bottoms.

Some of the known forms include *Prorhynchus* (Fig. 7.12 B), *Plagiostomum* (Fig. 7.12 C), *Hypotrichina, Otoplana,* etc.

4. **Order Tricladida.** The Turbellaria in which the intestine is three-branched and the pharynx is plicate belong to this order. They are commonly known as planarians and form familiar classroom types. They inhabit both fresh and salt water and humid terrestrial region. Some of the terrestrial forms are large and often brilliantly coloured. Mucous plays an important role in the life of terrestrial planarians. It forms a coat that protects them from drying. While moving they form a slime trail. Sometimes slime threads are formed with the help of which they can descend from leaves or branches or even drop from heights.

The genus, *Dugesia* (*Euplanaria*), is the typical planarian with dark colour, triangular head, projecting auricles, two eyes and is included as a type in many text-books. Other genera are *Planaria, Phagoctata, Polycelis, Bipalium* (land planarian) and *Bdelloura,* etc.

5. **Order Polycladida.** These are Turbellaria in which the gut has many diverticula which ramify all through the body and which are exclusively marine, living typically on the bottom. They are greatly flattened, thin and leaf-lke and in general, are the largest of Turbellaria. Some species are known to attain length up to 15 cm. There are some pelagic forms also, they are transparent or translucent but the bottom dwellers are either white, brown or gray. Some polyclads living among the coral reefs are brightly coloured. A pair of tentacles is present projecting from the dorsal surface near the anterior end or from the sides. The eyes are numerous and small. They are situated either as a pair of cerebral clusters in the brain region and a tentacular cluster near each tentacle or may be found scattered over the anterior end. Flame-cells occur. Ovaries are small,

numerous and testes scattered throughout the sides of the body. Yolk glands are lacking. Complicated set of organs to transfer sperms along with gelatinous secretion that forms adhesive thread containing a row of eggs. This thread is woven variously to form a gelatinous egg-mass. Development is not direct, it leads to the production of a larva known as Müller's larva (Fig. 7.15) with eight posteriorly directed lobes bearing especially long cilia. This larva has an apical sensory tuft of cilia and may have a caudal sensory tuft at the opposite end. In some species of *Stylochus*, the larva has only four lobes and this is termed Gotte's larva. This was long regarded as a stage in the development of Müller's larva but Kato has noted that it does not grow any further.

The larval lobes are absorbed after a few days, the sensory tufts disappear, additional eyes differentiate, and the larva transforms into a young polyclad as a result of flattening. Some authors drew a relationship between the eight ciliated lobes of the Müller's larva and the eight ciliated rows (comb-plates) of ctenophores, but this is very doubtful.

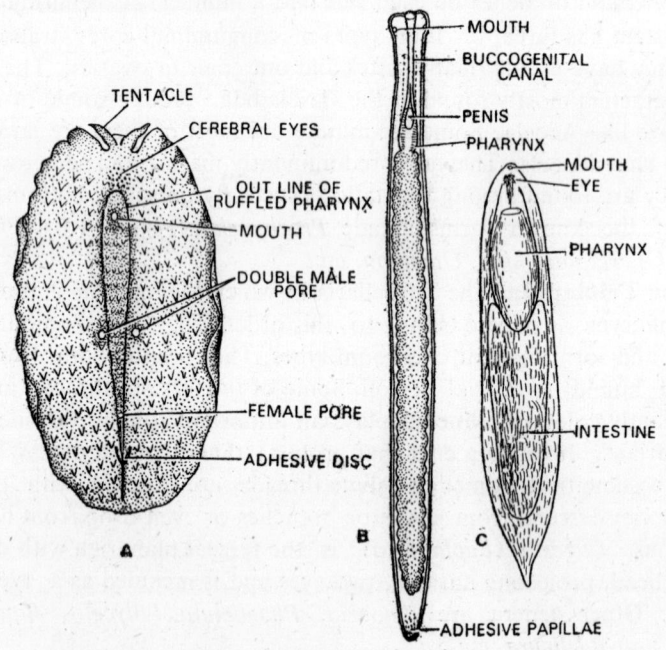

FIG. 7.14. A—*Thysanazoon*; B—*Prorhynchus stagnalis*; and C—*Plagiostomum* (*after* Hyman).

The genus *Stylochus* has prominent dorsal tentacles containing eyes and often feeds on oysters, so much so that it becomes harmful to the oyster industry, and is thus known as "oyster-leech." *Leptoplana*, another genus characterized by the lack of tentacles, does not have marginal eyes of which generally there are four clusters in the brain region. Another peculiar genus is *Thysanozoon* (Fig. 7.14 A) in which the dorsal surface is covered with papillae into which run caeca from the intestine. There is a small adhesive disc behind the genital pore and is probably of copulatory function. Other notable genera include *Notoplana*, *Pseudoceros* (like *Thysanozoon* but the papillae are without intestinal caeca), *Graffizoon*,

*Phylum Platyhelminthes*

*Stylostomum* and *Cycloporus*, etc.

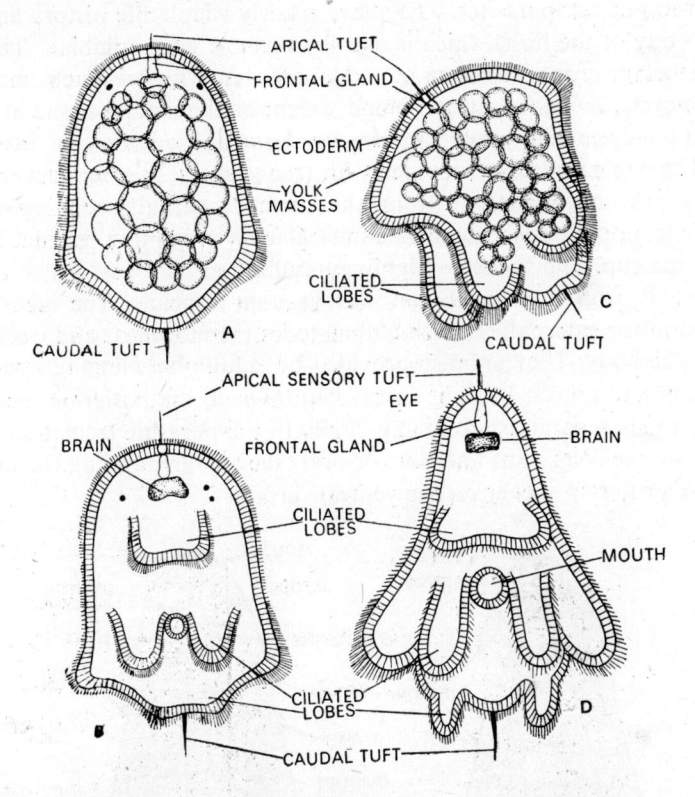

FIG. 7.15. Development of Müller's larva. A—Lobe formation begins; B—Later stage of lobe; C—Side view of B; and D—Completed Müller's larva, seen from above.

## CLASS TREMATODA

The Trematoda are parasitic flatworms resembling the free-living turbellarians, but the epidermis is absent. Instead nonciliated cuticle covers the whole body. They may be external parasites, **ectoparasites**, that live on or near the outer surfaces of their hosts and have simple life-histories; or they may be internal parasites, **endoparasites**, that live within the bodies, in the gut or in tissues, and whose life-cycle takes a complicated course, at least one stage being spent in another host. The Trematoda are flat in shape, hence are commonly known as **flukes**. They have one or more adhesive organs or **suckers** by which they attach themselves to their hosts. The suckers have a cavity hence the name Trematoda (Gr. *trema*, a hole). The trematodes are elongated oval in form, but the forms that live in blood of some animals tend to be slender. Some rounded disc-like forms are also known. They are usually small, but often minute to larger-sized forms (up to several centimetres long) are also found. They are usually colourless or slightly tinted, parts of the reproductive system may be dark or the intestine may be coloured by food.

The class Trematoda consists of parasitic flatworms called flukes, whose general body form is like that of the free-living Turbellaria. The group divides clearly into two subclasses, the first of which, the **Monogenea**, are relatively minor group of ectoparasites. They have a fairly simple life history and live outside the body of the host, which is usually a fish or an amphibian. The large and more important group **Digenea** are endoparasites with a much more complicated life-cycle, living in gut, the blood stream or the actual tissue of their hosts. A typical life-cycle will involve a **primary host** and one or two **intermediate hosts**. The Trematoda are provided with true suckers. A true sucker is a cup-shaped depression separated from the surface externally by a constriction and internally from the body tissues by a muscular layer running parallel to the concavity of the cup. The sucker is highly muscular and is provided with adhesive gland but it works primarily on the vacuum principle. The suckers are well-developed in the externally parasitic trematodes (Monogenea) and occur at both ends of the body. They are also provided by additional clinging structures such as the claws and hooks. In some cases (*Polystomun*) the posterior sucker may be divided into six separate suckers in a circle. In the parasitic trematodes there are usually two suckers, an anterior or oral sucker surrounding the mouth and a ventral or posterior sucker on the ventral surface.

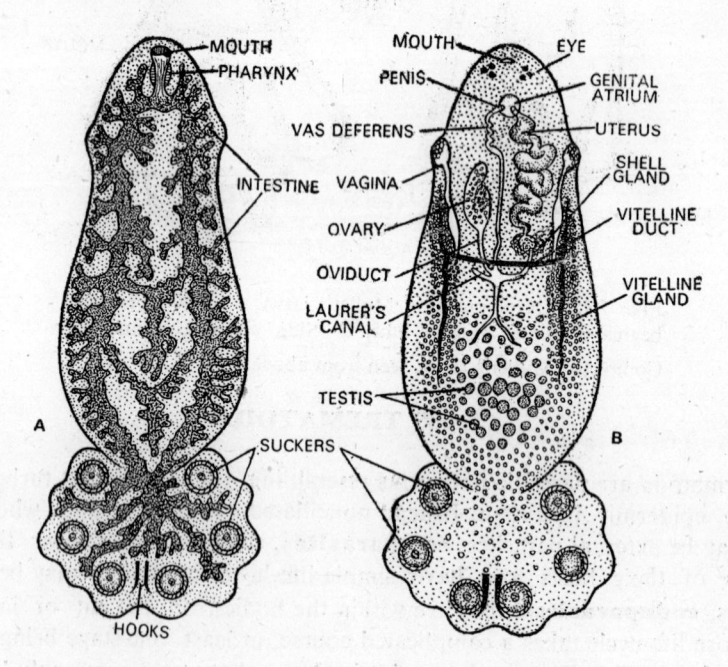

FIG. 7.16. Structure of monogenetic trematodes. A—Digestive organs of *Polystomum*; and B—Reproductive system of the same.

The mouth is anterior (Fig. 7.16) except in one family (Bucephalidae) in which it is in the middle of the ventral surface. The mouth leads through a muscular pharynx, with or without the intervention of an oesophagus, into a two-forked intestine (Fig. 7.16), the branches of which are known as **crura** or **caeca** that extend posteriorly as blind tubes. They may give off lateral branches in some cases that may eventually anastomose into a network. In some (monogenea and digenetic family Bucephalidae) the intestine is a simple sac as found in rhabdocoels. There is an excretory system of flame-cells and canals, each of which runs along the side and commonly recurves itself. In some the two canals are completely separate (Monogenea) and open separately near the mouth on the dorsal surface, but in others (Digenea) the two tubules unite to form a median oval bladder opening by a single pore, the nephridiopore, at the middle of the posterior end usually dorsally. In the trematodes each flame-cell is nucleated and is not attached directly to the main trunk, but to the recurved portions of the tubules or to the branches. Sometimes lateral flame-cells are also met with. In some (Digenea), the terminal flame-cells occur in clusters of two or four attached to a single collecting tubule by capillaries. The number and arrangement of flame-cells is used for the purpose of classifying the group.

The nervous system is usually well-developed. Like Turbellaria it consists of a pair of cerebral ganglia and a submuscular plexus running into longitudinal cords connected by transverse connectives. The trematodes have no special sense organs. Most of the external parasites alone have eyes, but in the endoparasites they are absent in the adults, being present in the larval stages only. The reproductive system presents a high degree of complexity seldom found even in higher animals.

## ECTOPARASITIC FLUKES (MONOGENEA)

The **Monogenea** includes trematodes that have a single host in the life-cycle. They are also distinguished from the digenetic trematodes in the possession of a large posterior adhesive organ called an **opisthaptor**. The opisthaptor is a complex muscular organ which attaches the parasite to the host and typically bears a number of suckers plus sclerotized pieces called **hooks**, **anchors** and **bars**. *Polystomum* is a good example of typical monogenetic trematodes. It is described below:

### TYPE POLYSTOMUM

**Form and Structure**

*Polystomum* (Fig. 7.16) is a common external parasitic fluke frequently found in the cloaca or bladder of the frog, and sometimes attached to the gills of tadpoles. In all essential features it resembles a turbellarian, but certain special characters have developed in response to the parasitic mode of life. The epidermis secretes a thick cuticle and the cells then sink inwards into the body and no longer form a continuous layer. Cilia are lacking. The posterior end of the body bears six suckers (Fig. 7.16), as well as a number of curved chitinous hooks so that the parasite is enabled to cling securely to its host. The mouth is anterior and opens into the pharynx which is represented by a suctorial bulb instead of an eversible pharynx like that of a turbellarian. This is correlated with the change in the mode of feeding. The parasite feeds by sucking juices from the tissue of its host and naturally a suctorial pharynx is needed. The pharynx is followed by a short oesophagus which leads into the intestine which consists of two main branches passing to the posterior end of the body and uniting in the region of the suckers. From the main branches arise several lateral caecae.

The excretory system consists of two lateral excretory canals, each begins anteriorly, runs backward to the posterior end and there bends upon itself running forward finally opening by a nephridiopore on the dorsal side near the mouth. Before opening the canal may be dilated to form a contractile bladder.

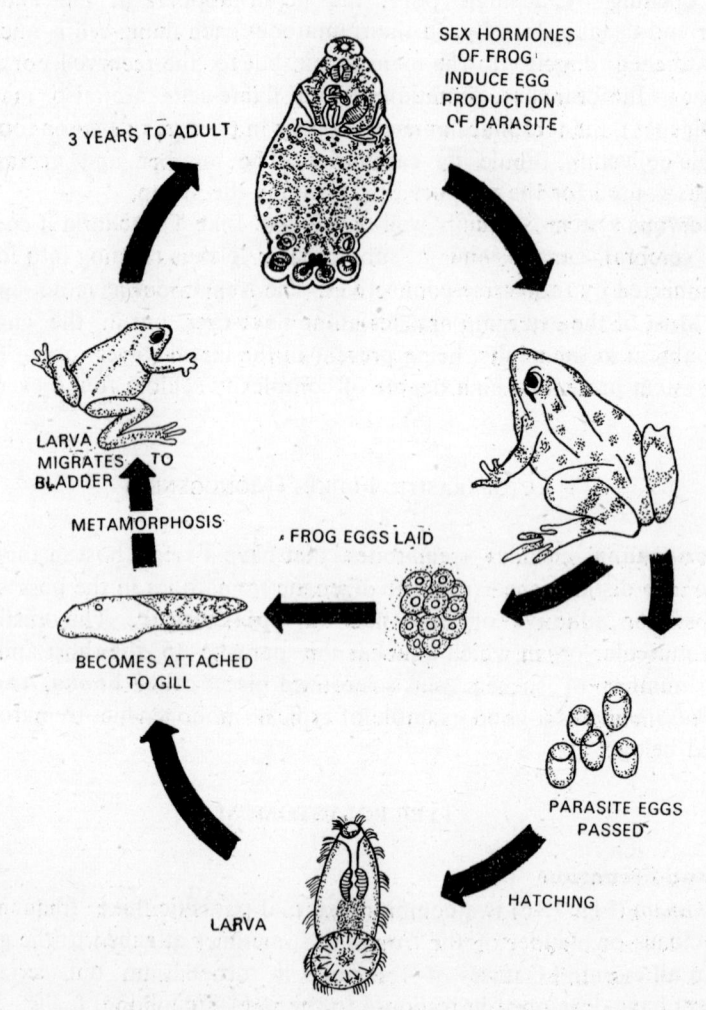

FIG. 7.17. Life-cycle of *Polystoma integerrimum*, a monogenetic trematode found in the bladder of frog.

The nervous system presents the turbellarian plan, special sense organs and eyes are generally absent, though may be present in some Monogenea.

The reproductive organs (Fig. 7.16 B) are more complicated than those of the Turbellaria though the fundamental plan is the same. The **testis** is a single follicular structure lying in the middle of the body rather towards the posterior side. From it arises the **vas deferens** that runs towards the anterior side to the copulatory organ before which it may widen to form a **spermiducal vesicle**. The copulatory system comprises a muscular **penis bulb** from which projects the **penis**, a muscular or fibrous papilla often armed with hooks. The male apparatus opens in an **atrium** common with the female uterus. The atrium opens by a common gonopore situated on the ventral surface. The female generative organs comprise enlongated **ovary**, a pair of follicular **vitelline glands**, lying in the lateral body regions, and the **oviduct** that widens into the **uterus**, which opens into the genital atrium. From the main vitelline duct of each side a transverse duct arises and joins another duct on the opposite side to form a common vitelline duct or **yolk reservior** from which a common yolk duct extends to the oviduct. Shortly after leaving the ovary the oviduct receives the common yolk duct, then receives a **genito-intestinal canal** from the right intestinal caecum and then widens to form small ootype, surrounded by numerous gland cells, **Mehilis's glands** (also called the **shell glands**). Beyond the ootype duct is widened out to form the uterus. Beside these structures there occur paired **copulation canals (vaginae)** which open separately and independently on the lateral surface towards the anterior end of the body. They receive the penis in copulation and may be widened below to form seminal recepatcle.

## Fertilization, Development and Life History

After leaving the ovary the egg is immediately fertilized, becomes surrounded by yolk cells and moves into the ootype where the shell is formed. The shell is formed mainly from the product of accompanying yolk cells and not from the secretions of the shell glands which probably lubricate the uterus, and make the passage of the capsule easy. The shell has an operculum at one end, often bears, at one or both ends, long filaments with which it attaches to the host. The fertilized egg develops into a small ciliated larva which has two eyespots and the rudiments of the adult organ-system. The larva swims about for some time and then attaches to the gills of a tadpole, a new host, after entering the branchial chamber (Fig. 7.17). The parasite then grows and metamorphoses into the adult form. The life-history in this case is simple involving only a single host, hence **Monogena**. From the branchial chamber the parasite moves into the pharynx and thence to the cloaca through the gut. It matures sexually in the cloaca. If the larva happens to attach itself to a very young tadpole it may attain sexual maturity whilst still on the external gills. These gill parasites always die before acquiring all their adult features and they never migrate to the cloaca of their host.

## ENDOPARASITIC FLUKES (DIGENEA)

The Digenea are all internal parasites, some of which are responsible for causing fatality in man and domestic animals. The life-history in each case is quite complicated needing at least two hosts, in some case three and four different hosts are involved, and there are typically four larval stages. The adult primarily occurs in the digestive tract and its appendages, such as lungs, liver, gall-bladder, and bile ducts, etc. of vertebrates. They may also be found in the kidneys and ureters, air-sacs of birds, parts of respiratory system, coelom, eye and other head cavities. One stage of their life-history is usually spent in an invertebrate host, almost invariably a mollusc. Their life-history is said to be digenetic one because of the fact that they need at least two hosts. This alternation of hosts is responsible for the naming of the group Digenea.

## TYPE FASCIOLA HEPATICA

*Fasciola hepatica* lives in the bile duct of a sheep as an internal parasite and therefore it is popularly called **liver fluke**. Typically they inhabit the bile duct but sometimes invade other organs and cause a disease known as **liver rot**. They may occur in cattle, other mammals and rarely in man. Heavy infestations even prove fatal. Sheep liver fluke thrives only on pastures with marshy areas where snails occur.

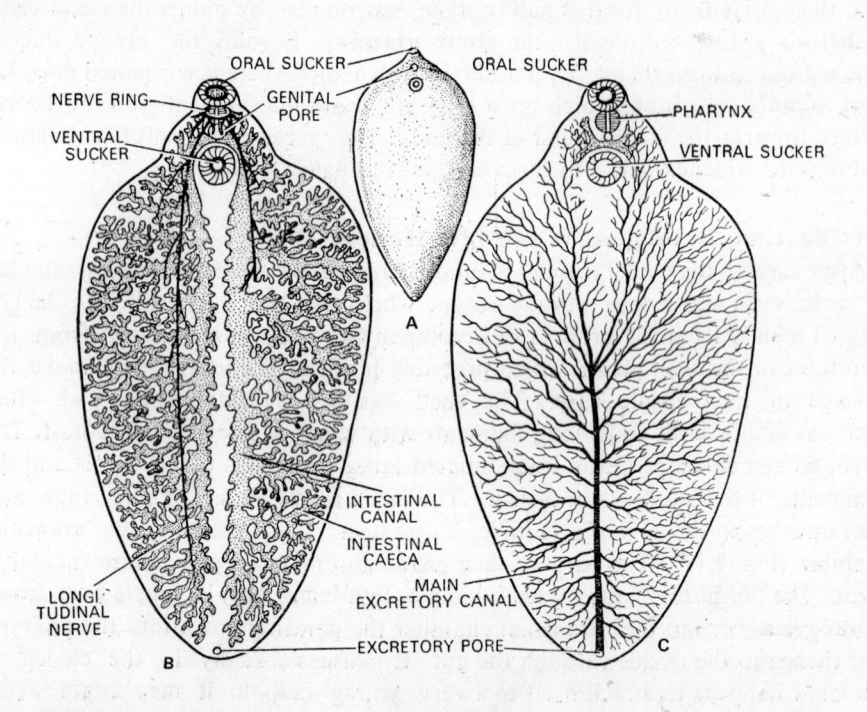

**FIG. 7.18.** *Fasciola hepatica*. A—Entire; B—Digestive system; and C—Excretory system.

### Form of Body

Sheep liver fluke is about 1-4 cm long and 2.5 cm broad. The anterior end is drawn out to form a conical **head lobe** (Fig. 7.18), at the apex of which is situated the oral anterior sucker, a cup-shaped structure with an oval opening, the **mouth**, in the middle. A short distance behind the head lobe is located another cup-shaped sucker, the **ventral sucker**. Between the two suckers lies the **genital aperture** placed slightly towards the posterior side. The excretory aperture is single and lies at the extreme posterior end of the body which is bluntly pointed.

### Body Wall

The body is covered by a thick layer of cuticle externally, the cilia are absent and the epidermis is a distinct cell layer (Fig. 7.19). The cuticle is very thick and homogeneous layer and bears numerous spine-like thickenings (Fig. 7.20) the **spinules** or cuticular scales. Below the cuticle, lies the muscular layer

## Phylum Platyhelminthes

of circular muscle fibres followed by a layer of longitudinal fibres. Scattered among the muscle fibres are found gland cells which open out on the surface of the

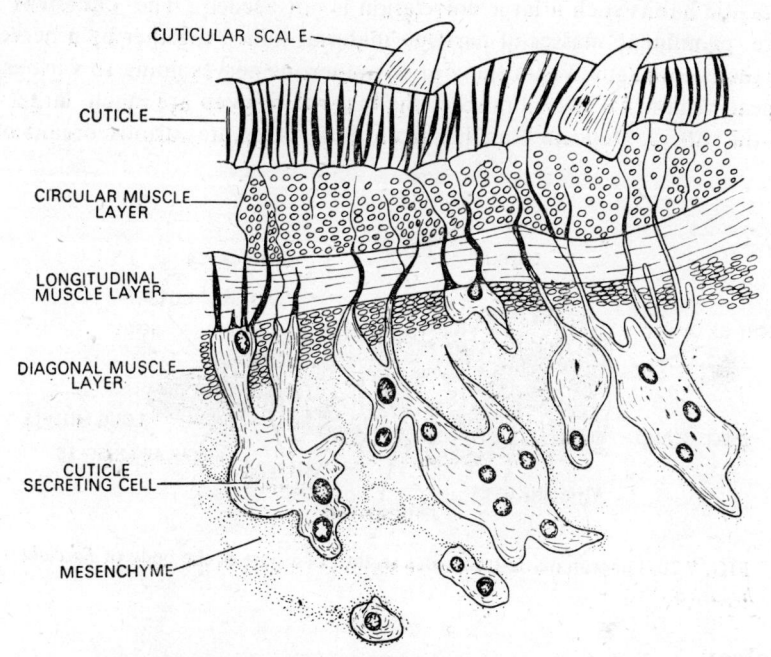

FIG. 7.19. Cross-section of the body wall of a trematode.

cuticle and secrete a nonmucous substance. The gland cells are scanty as compared to the Turbellaria. The ectodermal cells are elongated nucleated cells sunk into the parenchyma (Fig. 7.19). They are connected with the cuticle, which they secrete, by protoplasmic processes. Some writers regard them as special mesenchymal cells. The space between the organs is filled by parenchyma consisting of peculiar branching vacuolated cells.

### Alimentary Canal

The **mouth** leads into a short ovoid **pharynx** which is a suctorial structure well provided with muscles (Fig. 7.18 B). Following the pharynx is a short thin-walled tube, the **oesophagus,** that bifurcates almost immediately forming two main branches of the intestine, from each of which arise numerous branched **caecae** which run practically through every region of the body and between which are wedged other organs.

### Excretory System

As in the planarian, the **excretory system** consists of much branched tubes (Fig. 7.18 C) with characteristic flame-cells. The main excretory duct is one large vessel opening to the exterior at the posterior end of the body. In front it gives off four large trunks which branch and rebranch forming smaller capillary vessels which end in flame cells. The main trunk receives many large ducts, into which drain smaller tubules whose ultimate branches end in flame-cells.

## Nervous System

The **nervous system** is well developed in spite of the fact that due to sluggish and parasitic habits such a large correlation is not needed. The **cerebral ganglia** are prominent masses of nervous material joined together by a nerve ring around the oesophagus. From these arise numerous nerves going to various parts of the head region. Of these branches the **lateral nerves** are much larger than any of the others. Delicate branches from these innervate various organs of the body.

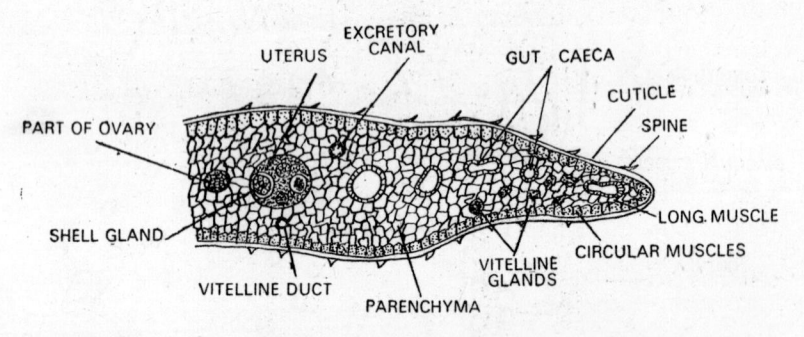

FIG. 7.20. Diagrammatic transverse section of a part of the body of *Fasciola hepatica*.

## Physiology

Much physiological work has not been done because of the difficulty in keeping the flukes alive outside the body of the host. Even in the tissue fluids of their hosts or in blood they cannot be kept for sufficient periods. Whatever is known is based upon fragmentary information available here and there. Flukes generally ingest food by mouth, the food comprises blood, tissue cells or food particles from the host intestine. The liver fluke feeds on the bile duct epithelium and the lung flukes of frogs feed exclusively on blood. Digestion is mainly **extracellular.** Some flukes release fluids in the host tissues that degenerate and dissolve into a nutritive fluid that is eventually ingested by the flukes. The nutrients may also be absorbed through the general body surface. Food is stored chiefly as glycogen which has been found in the parenchyma, musculature including the sucker and in ovaries. Fat droplets have also been reported mostly in the walls and lumen of excretory system.

The respiratory metabolism of the flukes is quite different. It lives in an environment whose oxygen content is very low so that it cannot respire in the same way as most animals do. It has been found that they respire anaerobically consuming glycogen and producing carbon dioxide and fatty acids (anoxybiotic respiration), although they do consume oxygen when this is available.

The physiology of excretion is not clearly understood. The bladder and main nephridial stems often contain some prominent pherules or granules probably of excretory nature. They are found more in the larval stages and their chemical nature is not known except that some are fat droplets. It is, therefore, presumed that the normal excretory products of the flukes are the fat droplets. Further, it has been experimentally shown that the flame-cells are of water-regulatory function, and the tubular part of the system secretes waste products.

*Phylum Platyhelminthes*

## Reproductive System

The individuals are hermaphrodite, the male gonads consist of a pair of much branched tubules, the **testes,** the walls of which produce spermatozoa and shed

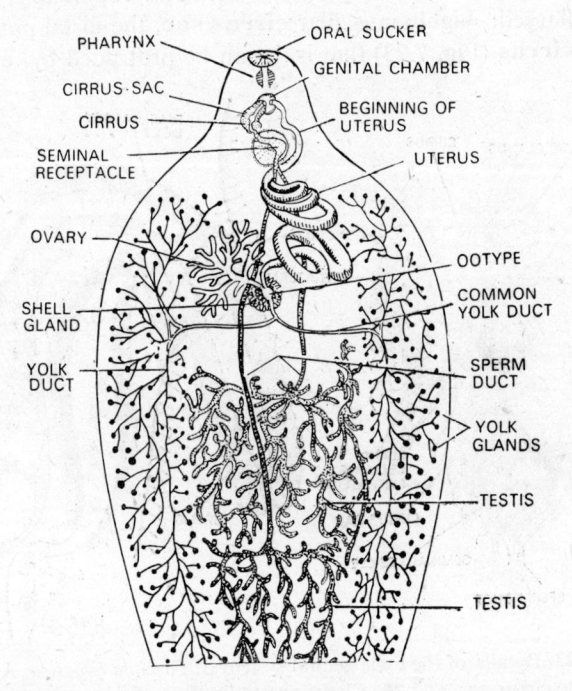

FIG. 7.21. Diagram of the reproductive system of *Fasciola hepatica*.

into the lumen. These occupy a good deal of space in the middle of the body, one lying anterior to the other. From each testis a sperm duct, the **vas deferens**

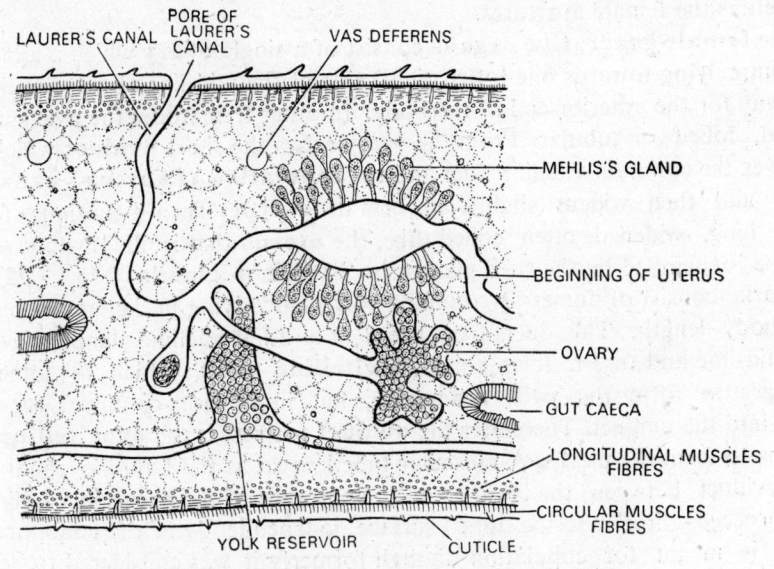

FIG. 7.22. The female reproductive organs of the Digenea.

extends forwards and fuses with its fellow forming a common **sperm duct**. This extends forwards as a wide coiled tube, the **vesicula seminalis**, lying in front of the ventral sucker, which is, in turn, followed by a narrow tube, the **ejaculatory duct**, opening into the copulatory apparatus. The copulatory apparatus consists of an enlarged, highly muscular **cirrus sac**, the distal portion of which is eversible as a **cirrus** (Fig. 7.23) that is, it can be protruded by evagination. The

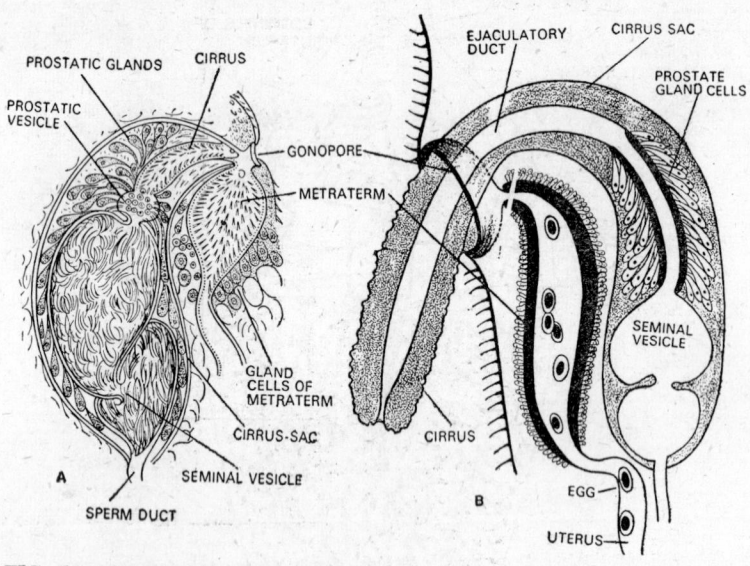

FIG. 7.23. Details of the male genital system of *Fasciola hepatica*. A—Cirrus inside the cirrus sac; and B—Cirrus protruding out of gonopore.

male pore is situated at the tip of the cirrus. In other flukes a true **penis papilla** is occasionally present. When the cirrus is withdrawn into the cirrus-sac a small space is left in front. This is **genital atrium**, a common chamber for the male as well as the female apertures.

The **female generative organs** consist of a single **ovary** which is a branched structure lying towards one (often right) side of the mid-line about one-third of the way for the anterior end of the body. In other trematodes the ovary may be round, lobed or tubular. From the ovary arises the short **oviduct** which soon receives the common vitelline duct and the **copulation (Laurer's) canal** (Fig. 7.22) and then widens slightly to form the **ootype**. Then it continues forward as a long, widened often coiled tube, the **uterus** that is filled with ova. The ootype is encircled by the shell glands or **Mehlis's glands** (Fig. 7.22). The vitellaria consist of numerous rounded yolk glands lying in the lateral region of the body length. The ducts from the right and left vitelline glands converge to the mid-line and fuse to form a median **vitelline duct**, which in some cases enlarges to form the **yolk reservoir**, but in all cases continues forward to open into the oviduct. Thereafter the **oviduct** (or rather the **ovovitelline duct**) widens to form the uterus. A muscular tube known as the Laurer's canal leaves the oviduct between the ovary and the entrance of the common vitelline duct and proceeds directly to the dorsal surface, independently of the gonopore. This canal is meant for copulation though formerly it was considered to serve as

*Phylum Platyhelminthes* 275

exit for superfluous sperm and yolk cells. Laurer's canal is absent in a number of flukes. In such cases the terminal part of the uterus receives the cirrus during

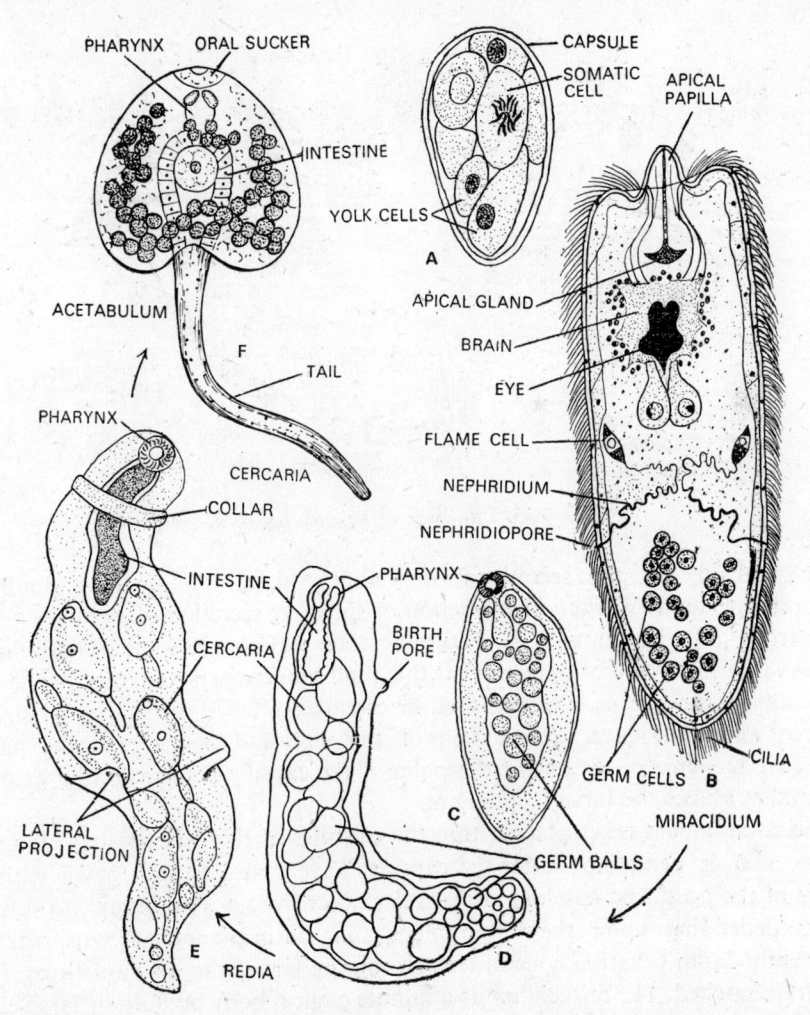

**FIG. 7.24.** Larval stage of *Fasciola hepatica*. **A**—Cleavage of fertilized egg; **B**—Miracidium; **C**—Sporocyst; **D**—Redia of amphistome; **E**—Redia of *Fasciola*; and **F**—Cercaria.

copulation, and for this purpose this part is strengthened by muscles and supplemented with spines and gland cells, etc. When the terminal part of the uterus is modified thus for copulation it is known as **metraterm**. The sperms enter the female tube directly by auto-copulation in which the cirrus enters the opening of the uterus in the genital atrium, or the sperms directly wander from the male into the female canal. This happens in the self-fertilization. In cross-fertilization the sperms enter by way of Laurer's canals or by way of metraterm where the former is absent. To enable this the copulation may be either one sided or reciprocal.

**Life-history**. Fertilization takes place probably in the lower part of the oviduct. As the fertilized eggs pass into the uterus they become surrounded by yolk cells, which, emit their shell forming droplets in the ootype. They form the characteristic shell of yellowish brown colour. Further hardening of the shell takes

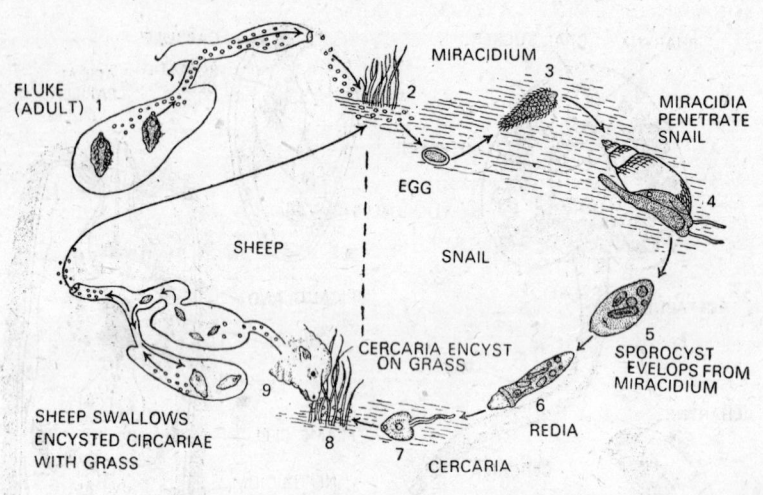

FIG. 7.25. Life-cycle of *Fasciola hepatica*.

place probably by the secretions of Mehlis's gland, though this is doubtful. According to many workers the function of the clear secretion of Mehlis's glands discharged into the ootype is very uncertain (Hyman). Some (Goldschmidt Nauss and others) are of opinion that this fluid lubricates the passage of the egg-shell along the uterus. The egg-shell or capsules differ in the shape and size in different animals. Hence, with the help of the size and shape even the parent flukes are recognized. At one of the poles of the capsule is formed a lid or operculum that allows the larva to pass out.

The development takes place within the capsules as they are passing along the uterus and is completed after the capsules have passed to the outside with the faeces of the host. The hatching of capsules takes from a few hours to several weeks depending upon the development going on in the interior. On reaching sufficiently damp situation a small free-swimming larva, the **miracidium** (Fig. 7.24B) is hatched. The miracidium is a minute conical body built up of large ciliated cells arranged in five rings. At the broader end is a small conical apical papilla and behind this is a pair of pigment spots or eye spots. A small blind gut and a pair of flame cells are present. The rest of the body is filled with loose mesodermal tissue. The miracidium (Fig. 7.25) swims freely in water or on moist vegetation and soon dies if it does not reach a species of common water snail, *Limnaea truncatula* or *Planorbis*, its secondary host.

Finding the snail the miracidium bores its way with the help of the papilla into the pulmonary sac, where it casts off the coat of ciliated cells, and grows rapidly into a different type of larva, the **sporocyst**. It is an elongated sac covered with thin cuticle beneath which is the usual muscle layer and the parenchyma. It contains germ cells or balls and propagatory cells apart from a rudimentary gut and few flame-cells. The germ balls may multiply by dividing up into daughter germ balls. In fact except the germ balls the sporocyst contains no other organ of any importance. The sporocyst absorbs nourishment from the

*Phylum Platyhelminthes*

host tissue and moves about a little, thus, often becomes highly destructive. The germ balls produce daughter sporocysts and under normal circumstances they pro-

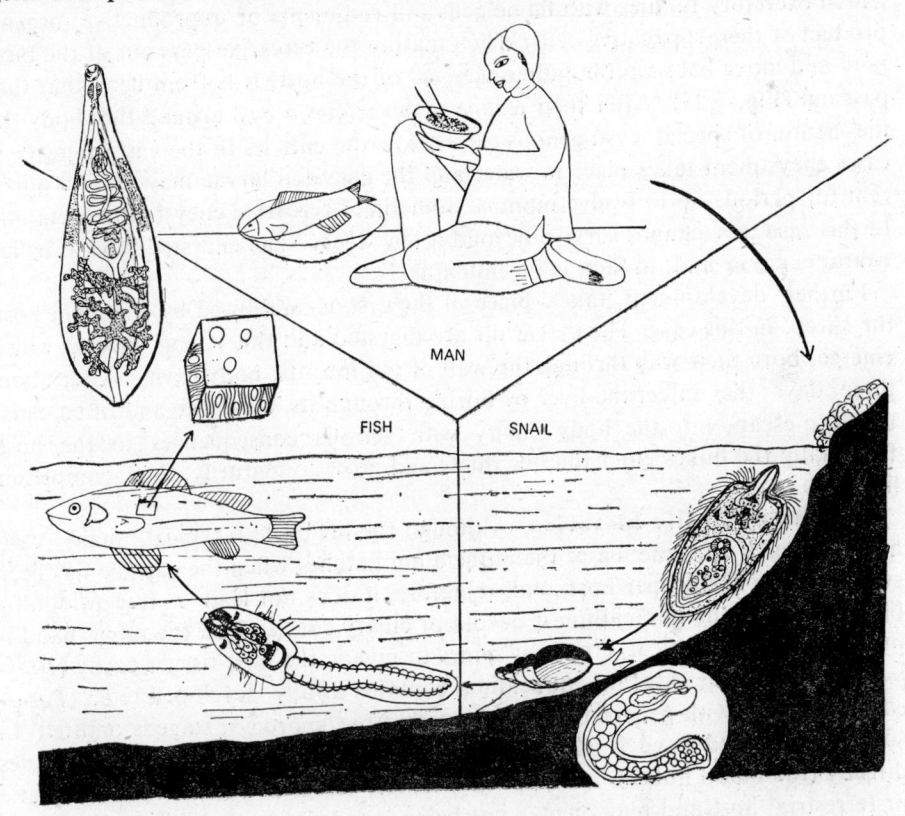

FIG. 7.26. Life-cycle of *Opisthorchis* (*Clonorchis*) *sinensis*, the Chinese liver fluke (*after* Faust).

duce the third type of larva, the **redia**, a stage (Figs. 7.24 and 7.25) often omitted in many flukes. Each redia is an elongate sac-like larva with a small mouth, a suctorial pharynx and a simple intestine. The posterior end of its body bears a pair of muscular projections, the **posterior processes** or lateral projections, whereas, at the anterior end the body wall is produced to form a muscular **collar**. Besides this near the anterior end there is a **birth pore**. The body wall consists of the outer cuticle and usual muscular and parenchymatous layers. The redia may have flame-cells. The **germ balls** lying in the cavity give rise to the secondary rediae, in the summer months only, or to the fourth type of larva the **cercariae** (singular cercaria). In any case the end product of the normal redia-generation is the cercaria (Figs. 7.24 and 7.25).

Before this happens the rediae pass out of the sporocyst and by means of the collar and posterior lobes of the body they slowly migrate to the digestive gland through the tissues of the snail. It is here that cencariae are formed inside the rediae. The cercaria is formed from a germ ball and has a higher grade of organization than any of the preceding larval stages. It is a small heart-shaped larva (about 0.5mm., long) bearing a tail (Fig. 7.25). The rudiments of most of the adult organs are present. There is a small oral sucker perforated by the mouth and a ventral sucker. There is a suctorial pharynx, a forked intestine, paired excretory tubules with flame-cells and rudiments of reproductive organs, product of the propagative cells. When mature the cercariae pass out of the birth pore and move back on the pulmonary sac of the host. It is from here that they pass out (Fig. 7.25). After their escape each secretes a cyst around the body by the action of special cystogenous cells below the cuticle. In the vast majority of cases encystment takes place in water and the encysted larvae may remain quiescent for periods up to twelve months. Sometimes cercariae encyst on vegetation. In this case they cannot survive beyond a few weeks. The encysted larva is called **metacercaria** and, in fact, is juvenile fluke.

Further development takes place if the cyst is swallowed by the main host, the sheep, in this case. The cyst walls are digested and the young flukes, which emerge, bore their way through the wall of gut into the body cavity. After about three days they infect the liver by boring through its substance and often cause blood to escape into the body cavity with serious consequences to the host. Eventually the flukes enter the bile ducts and grow to maturity, thus completing life-cycle.

**Variations in life-histories.** Although the life cycle is typical, many variations occur. In some cases the miracidium hatches when the capsule has been swallowed by the proper host, and, therefore, it does not lead a free-swimming life. Naturally, the miracidium is devoid of ciliary coat and on the other hand is provided with an apical crown of spines to invade the host-tissues easily (*Didymocystis*). In some cases the miracidium contains a fully developed redia (*Parorchis*) provided with penetrating glands so that the sporocyst stage is omitted. In some the encysted stage is skipped as the cercariae penetrate the host tissues directly (the blood flukes). In some cases tailless cercariae occur in either aquatic or terrestrial host and may or may not be encysted.

**Nature of digenetic life-history.** The production of the larval stages in the Digenea raises a vital issue regarding the nature of reproductive processes involved. Formerly the reproductive processes in the sporocysts and rediae were believed to be parthenogenetic. Thus, the propagative cells of these stages (germs balls) were regarded as eggs that developed parthenogenetically and some authors even described the maturation processes. This theory has now been given up and the reproductive process in the sporocysts and rediae are regarded to be an example of **polyembryony**, which means formation of more than one embryo per zygote, by fission, at some early stage of development. The rediae arise by the division, differentiation and subsequent growth of cells which are the direct lineal descendants of the propagative cells, set aside at the first division of the fertilized egg. There is, thus, from the very early stage of development of the zygote, a distinction between germinal and somatic cells and the germinal cells alone give rise to subsequent generations of the larvae by asexual multiplication. The digenetic life-cycle, therefore, is similar to that of the hydroid coelenterates comprising a period of asexual multiplication (during immature stages) followed by sexual reproduction in the mature stages, but it is better to regard the whole life-cycle as a continuous development involving asexual multiplication in larval stages.

*Phylum Platyhelminthes*

## Flukes and Man

A liver fluke *Clonorchis sinensis* parasitizes man and illustrates a life-cycle involving two intermediate hosts (Fig. 7.27). It is commonly found in China,

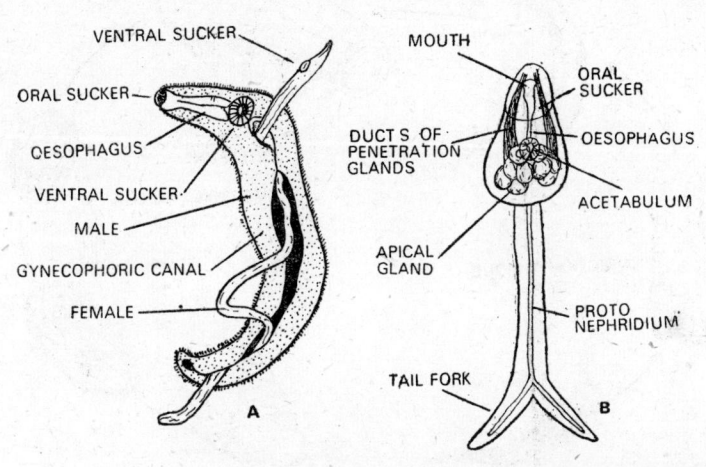

FIG. 7.27. The blood fluke. A—*Schistosoma haematobium*, adult; and B—Cercaria of *S. mansoni* (*after* Stunkard), a furcocercous cercaria.

Japan and Korea and is, therefore called the **Chinese liver fluke**. The adult is a little more than one centimetre long and has two suckers, oral and ventral. Its thick cuticle is highly resistant to digestive fluids. It is hermaphroditic, and capsules pass from the liver into the intestine and escape with the faeces. If the faeces get into water, as they commonly do, the capsules do not hatch into free-swimming **miracidia**, as in the liver-fluke, but are eaten by snails and hatched within the latter's digestive tract. The miracidum works its way through the wall of the digestive tract into the tissues of the snail and develops into a **sporocyst** which produces **redia**. The development of redia further increases the number of young larvae and eventually they produce **cercariae** which escape from the snail and swim about for some time. The cercariae encyst, not on grass like those of the sheep fluke, but in the muscles of a fish, which serve as the second intermediate host. They bore through the skin of the fish, lose their tails, and encyst. The fish secretes an additional capsule around the cyst produced by the parasite. They remain there till the fish is eaten by the final host—man. In the stomach the cysts are digested out of the flesh and the young flukes emerge. They make their way up to the bile duct and into the smaller bile passages of the liver, where they feed on blood. They may persist for many years, causing serious anaemia and disease of the liver by blocking the bile passages.

Another fluke commonly found in man is the blood-fluke (Fig. 7.27A), *Schistosoma japonicum*, an elongated and slender fluke that lives in the blood vessels of the intestine, clinging to the walls of the vessel by means of suckers and feeding on blood. The blood flukes are not hermaphroditic but the sexes are completely separate and they present sexual dimorphism. The male is generally broader than the female and its sides fold over to form a groove, **gynecophoric canal**, in which the slender and longer female is held (Fig. 7.27). The surface of the body is covered with minute spines or papillae. Both the suckers are present. A pharynx is lacking and the long oesophagus leads to forked intestine which fuses again posteriorly to form a single stem for a distance. There are

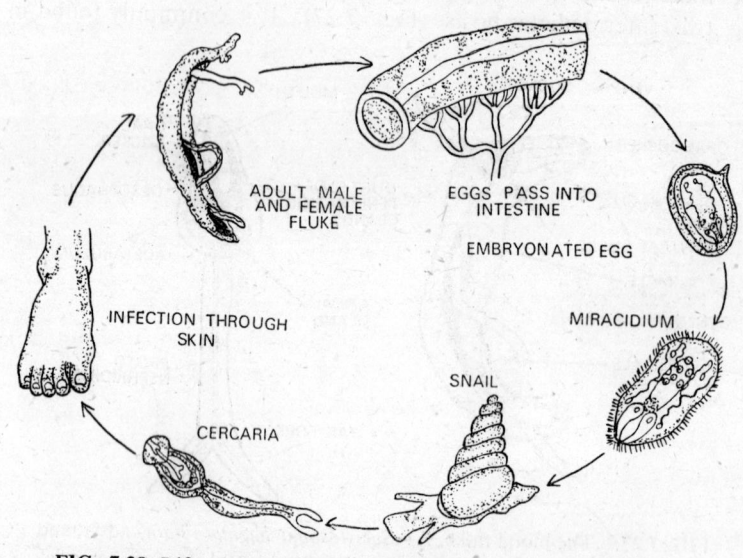

FIG. 7.28. Life-cycle of the blood fluke, *Schistosoma mansoni*.

several testes in male, and a long tubular ovary, voluminous yolk glands, long oviduct, ootype and straight uterus containing relatively few eggs, in the female.

The capsules are laid by female in the small blood vessels of the intestine wall, close to its lumen. In laying them the female leaves the gynecophoric canal of the male and pushes herself into the smallest possible vessels where she deposits the capsules. As the blood vessels of the host become congested with capsules the walls rupture, the intestinal epithelium breaks and the capsules are discharged into the cavity of the intestine. From there they are carried out with the faeces. The capsules hatch when they get access to water and free-swimming miracidia are produced. This is possible in countries where human faeces are used to fertilize the soil, and for that purpose are conserved in reservoirs on the banks of canals or irrigation ditches. The miracidium perishes after about twenty-four hours if it does not encounter suitable pulmonate snail. On getting the snail it penetrates and within it two generations of the sporocysts ensue and eventually a large number of cercariae are produced. They make their way out of the snail and infect the definitive host, man, directly through the skin while bathing or swimming, or through the mucous membranes of the mouth if taken with drinking water. They cannot survive in the stomach. They proceed to small blood vessels and are carried by the blood stream to the blood vessels of the intestine where the young flukes grow into adult forms, finally mating with another that entered at the same time or with one already established from a previous infection.

## Host-Parasite Relations

If the infection of trematodes is light usually no definite symptoms are produced, but heavy infections with either larval or adult forms are usually very harmful to the host. The general symptoms are loss of weight, retardation of

## Phylum Platyhelminthes

growth, emaciation and general unhealthy conditions, which in the extreme, cases may result in death. Flukes that live in such sites as the bile ducts, blood vessels, and lungs are generally more harmful than those inhabiting the digestive tracts. Heavy infection of snails with larval trematodes often results in partial or complete destruction of favoured sites of infection such as the gonads or digestive gland. The presence of the parasites evokes usual results such as inflammation of the place, aggregation of leucocytes at the site, phagocytosis, and production of connective tissue. The human liver flukes may persist for many years, causing serious anaemia and diseases of the liver due to blocking of the bile passages. The presence of blood flukes in man causes a disease, **schistosomiasis**, characterized by body pain, a rash and a cough in the early stages followed by severe dysentery and anaemia later on. Victims may live for many years, gradually become weak and emaciated and finally may die following exhaustion or succumb to other diseases due to already weakened condition and low vitality.

**Effect of parasitic life on the parasite.** In response to parasitic life the parasite has undergone many morphological changes with the loss of unnecessary organs and development of others. (*i*) The body is compressed dorsoventrally and the anterior end becomes conical so that the animal attaches itself to the wall of the bile duct while feeding. (*ii*) It is covered by cuticle secreted by cells beneath and cilia are absent in the adult condition. (*iii*) Adhesive organs have developed —there is usually a sucker around the mouth and one or several others on the ventral surface, hooks develop in some. (*iv*) The mouth is anterior and ∩ shaped digestive tract has smaller branches to distribute food. The food, consisting of body tissues or fluids, is sucked by the action of muscular pharynx. Anus is absent. (*v*) Organs of special sense such as the eyespots are lacking in the adults (except in ectoparasites). (*vi*) Definite respiratory organs are lacking and anoxybiotic respiration occurs in response to peculiar environment in which the parasite lives. (*vii*) Special development is seen in the organs of reproduction. As they cannot lay their eggs or larvae in suitable places, but must leave them to chance, large number of protected eggs are produced. (*viii*) Adaptation to parasitic life complicates the life-cycle, necessitating the acquisition of an intermediate host for shelter (primary) and for development and transmission (secondary) of the stages to the definitive host. To increase the chances of perpetuation of the race by further multiplication, some of the intermediate stages have become highly specialized (for instance the sporocyst). The eggs are covered with protective chitinous shell to prevent the action of digestive juice when they pass through the intestine of the host. The miracidium that hatches out of the egg is ciliated being freemoving, it has eyespots, a conical papilla accompanied by penetration glands for boring the body of the snail. In the body of the snail the miracidium loses cilia and eyespots and becomes sporocyst, which contains germ cells and is lined by cells that can produce germ cells. These germ cells produce rediae that force their way out of the sporocyst. In the interior of redia cells divide and form cercariae which have motile tails as they have to move inside the body of the snail. On passing out they lose their tails and encyst.

## Control

Control methods have evolved on the consideration of the life-history of the parasite. On this basis several possible methods of preventing the disease from spreading have been devised. The elimination of the adult stage is not very practicable, hence the parasite is generally attacked in its larval stages by the destruction of the secondary host, the mollusc. For the sheep liver fluke this may be done either by the introducing ducks to the pastures. This helps in reducing snail population being consumed by the ducks or by draining the pastures when the pond snails cannot survive. It is this last method that is usually effectively adopted. In the case of human liver fluke the control is relatively simple. It needs thorough cooking of the freshwater fish, its second intermediate host, to destroy the encysted cercariae. Yet in certain regions in the south of China, 75-100 per cent of the natives are infected probably because they like to eat raw fish or are forced to eat that way for economic reason *i.e.*, they cannot afford fuel required for cooking. The control of human blood fluke is still easier. It needs simple sanitary disposal of human faeces by modern sewage system or the same may be deposited in dry place. Where it is necessary to use the faeces as fertilizer it is better to educate the farmers to conserve it for a few weeks before utilising it in the fields, so that the young schistosomes within the capsules may die.

## CLASSIFICATION OF TREMATODA

The Trematoda are now divided into three orders: Monogenea, Aspidobothria and Digenea; of these the members of the order Aspidobothria have long been of

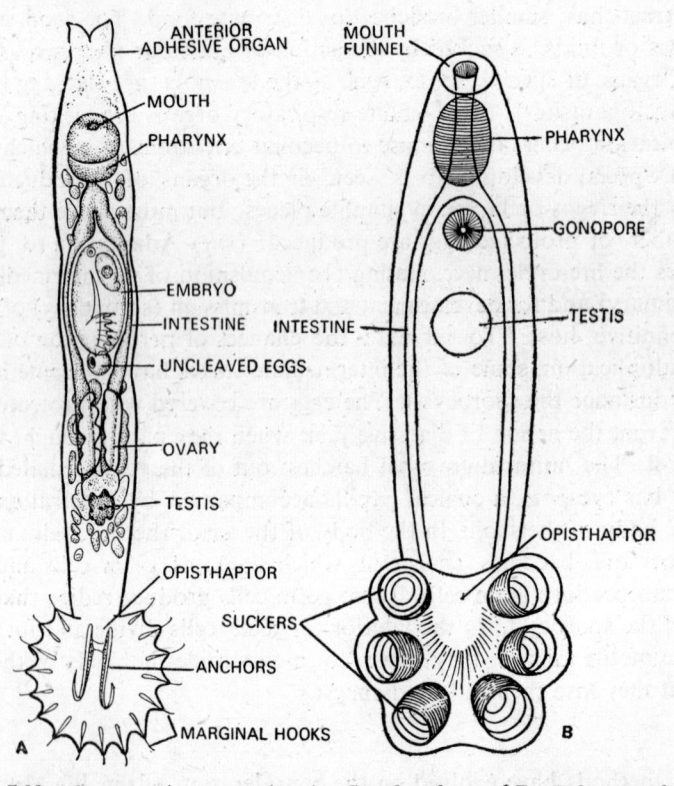

**FIG. 7.29.** Monogenetic trematodes. **A**—*Gyrodactylus*; and **B**—*Polystomoides*.

*Phylum Platyhelminthes* 283

uncertain status. Though separated quite early (1856, Burmeister) they were sometimes included in Monogenea and sometimes in Digenea but now they have been

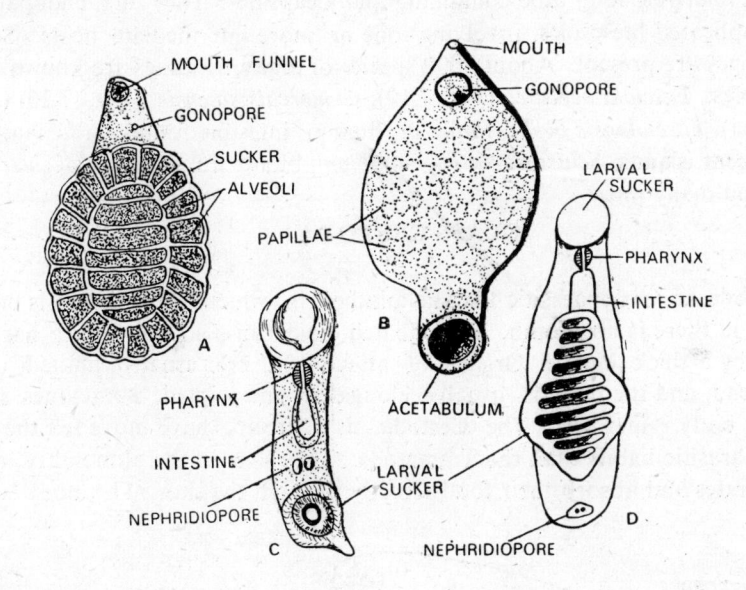

FIG. 7.30. A—*Cotylaspis* (*after* Osborn); B—*Homalogaster*; and C and D—Larval stages of *Aspidogaster*. (C—newly hatched, D—later stage).

finally separated under the above name.

1. **Order Monogenea (Heterocotylea)**. These are ectoparasitic Trematoda with relatively simple life-histories that do not require an intermediate host. Both hooks and suckers are present to enable the animal to attach itself to the host from outside. The excretory pores are paired and lie near the anterior end of the body.

Examples: *Gyrodactylus* (Fig. 7.29), common parasite on the gills of freshwater fishes; *Polystomum* (Fig. 7.17), *Monocotyle*, *Heterocotyle*, *Microbothrium*, etc.

2. **Order Aspidobothria**. These are Trematoda in which the oral sucker is wanting and the anterior end is without adhesive structures. The ventral sucker is very much enlarged and subdivided into compartments. There are no hooks present, nephridiopore is single and posterior. They are endoparasites but with simple life-cycles without an alternation of hosts. Mouth is terminal or subterminal with a large funnel-like opening, pharynx is small and the intestine is simple rounded or elongated sac. The reproductive system resembles that of the Digenea. The best known genus is *Aspidogaster*. *A. conchicola* is parasitic in the pericardial and renal cavities of the freshwater clams. Another genus *Cotylaspis* (Fig. 7.30) has three rows of depressions or alveoli on the ventral sucker and occurs in the intestine of turtles and also in the branchial chamber of clams. *Cotylogaster* is found in the intestine of fishes. Other notable genera include *Lophotaspis* and *Stichocotyle*.

3. **Order Digenea (Malacocotylea)**. These are Trematoda in which two suckers are present mostly, an anterior sucker that generally encircles the mouth

(hence oral sucker) and a ventral sucker (often called acetabulum). But either or both may be reduced. No hooks present. Nephridiopore is single; uterus is usually a relatively long tube containing many capsules. They are endoparasites with complicated life-cycles involving one or more intermediate hosts. Several larval stages are present. About 3000 species of digenetic flukes are known.

Examples: *Fasciola hepatica* (Fig. 7.19), *Clonorchis sinensis* (Fig. 7.26) (Japan and China), *Fasciolopsis buski*. Fluke of human intestine found in China, India and adjacent islands; *Schistosoma haematobium*, blood fluke, *Homalogaster* (Fig. 7.30B) and many more.

## CLASS CESTODA

The Cestoda are endoparasitic Platyhelminthes in which the enteron is entirely absent, and there is no mouth. The ciliated epidermis, in the adult, has been replaced by a thick cuticle. Organs of attachment are usually limited to the anterior end, and the body is usually elongated and jointed, sometimes simple fluke-like body is met with. The Cestoda, as a group, have more felt the influence of parasitic habits than the Trematoda. They live in the alimentary canals of vertebrates and absorb their food directly through the skin. The body is elon-

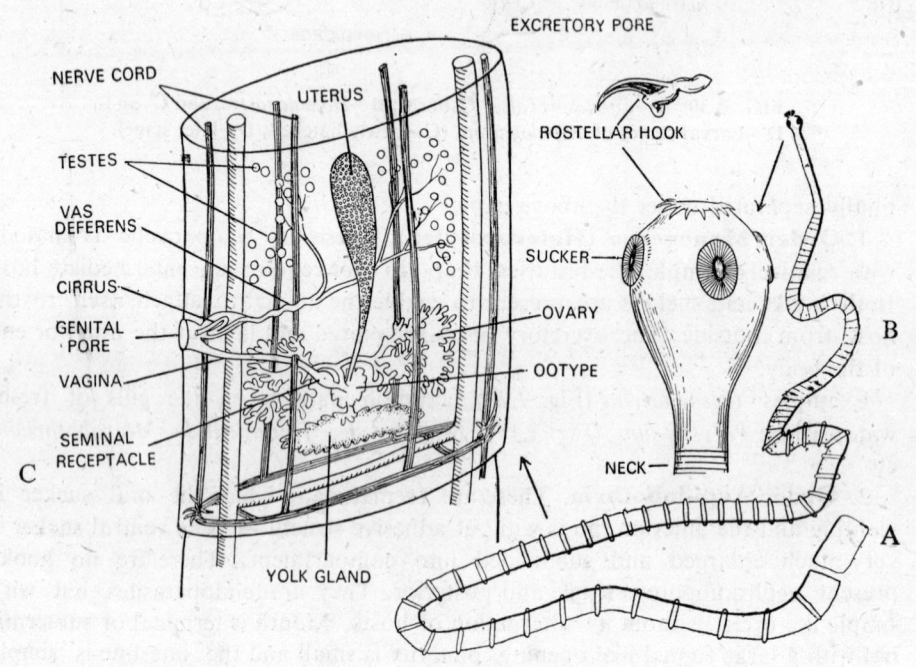

FIG. 7.31. Structure of a cestode. A—A typical tapeworm; B—Scolex of the same enlarged; and C—A mature proglottid enlarged (diagrammatic).

gated tape-like, hence they are commonly known as tapeworms. Typically the body is made up of a number of segments or proglottides, separated by transverse constrictions. A few tapeworms resemble flukes in general appearance and have a flattened undivided body (Fig. 7.38). Each proglottis of a segmented worm is, in many ways, comparable to a complete fluke-like worm in that it contains a hermaphrodite set of reproductive organs and portions of excretory and nervous systems which are common to all the segments (Fig. 7.31). The number of segments varies from 4 (*Echinococcus*) to 4000 and the length of the body may vary from 1 mm to 10 or 12 metres (40 feet). The body consists of knob-like head or **scolex** (Fig. 7.31 A) bearing organs of attachment, followed by a relatively short undivided **neck**, in which the proglottides proliferate. Following the neck is the main body or **strobila,** comprising a long chain of segments gradually increasing in dimensions towards the posterior side. The head is radially or biradially symmetrical and the neck and strobila are notably flattened as such it is difficult to define surfaces more so, when the external surfaces present almost identical appearance. To define the surfaces, therefore, internal features have been employed. The surface nearer the testes is defined **dorsal** and the one nearer the female system is **ventral**. When these characters cannot be recognised the term **surfacial** is conveniently used to indicate the surfaces.

The scolex has solid muscular structure. It has nephridial canals, and the central part of the nervous system. It bears organs of attachment which are represented by **hooks** and **suckers**. The suckers are of three kinds: (*i*) **bothria** consisting of a pair of elongated sucking grooves that have weak musculature (*Dibothriocephalus*); (*ii*) **bothridia** consisting of broad leaf-like structures with thin and very flexible and mobile margins (*Myzophyllobothrium*); they are four in number symmetrically placed and may be sessile or stalked; and (*iii*) **true suckers** or **acetabula** (Fig. 7.31) consisting of hemispherical depressions, four in number, sunk into the side of the scolex (*Taenia*). The genus *Spathebothrium* is without scolex or attachment organs.

The neck is slender unsegmented portion in which new proglottides proliferate. The strobila consists of segments as pointed out above. With regard to the segmentation of tapeworms there is a controversy. Some people suggest that the fundamental characteristic of segmentation, that is, the serial repetition of structural parts at regular intervals along the anteroposterior axis, is present as such they may be regarded as segmented. But the segmentation of tapeworms differs from that of annelids and arthropods in that the zone of proliferation is anterior (behind the head) so that the youngest segment is the anteriormost, whereas, in the annelids and arthropods the youngest segment is the most posterior one, excepting the terminal segment. Because of this difference in the manner of formation of segments the segmentation of the tapeworms is regarded different from that of other segmented animals. That is why most of the textbooks maintain the absence of segmentation in tapeworms.

Four types of tapeworms infect man. These include *Taenia solium, Taenia, saginata, Diphyllobothrium latum* and *Hymenolepis nana*. Intermediate host of *Taenia soli m* is the pig, of *Taenia saginata* is beef-cattle and for *Diphyllobothrium* is first the water flea (cyclops) and then fish. Man is infected by eating pork, beef or fish in raw or insufficiently cooked state. *Hymenolepis nana* has no intermediate host.

## TYPE TAENIA SOLIUM

The pork tapeworm, *Taenia solium*, is a common inhabitant of the human intestine, and maintains its place despite the constant flow of materials. *Taenia sagi-*

*nata,* the beef tapeworm, is another relative of the pork tapeworm that occupies the same place. The description below applies mainly to the pork tapeworm, but it also covers the salient features of the beef tapeworm.

**Form of Body**

*Taenia solium* is a giant among worms and may be two or four metres long, a quarter of an inch wide but only one sixteenth or so of an inch thick. The animal is attached to the wall of the intestine by the **head** or **scolex** which is not more than a pin's head in size.

The terminal part of the scolex bears a cone-shaped prominence, the **rostellum,** which is capable of protrusion and retraction to a slight degree. At the base of rostellum is a double row of pointed chitinous hooks **rostellar hooks** (Fig. 7.31), which are about twenty-eight in number. The scolex bears four cup-shaped **suckers,** the **acetabula,** projecting from the surfaces slightly behind the circlet of hooks. The head is followed by a slender **neck** which is capable of producing tiny segments at the rate of 7 or 8 segments per day. The neck is followed by the body or **strobila,** which has a segmented appearance as it is made up of about 850 proglottides.

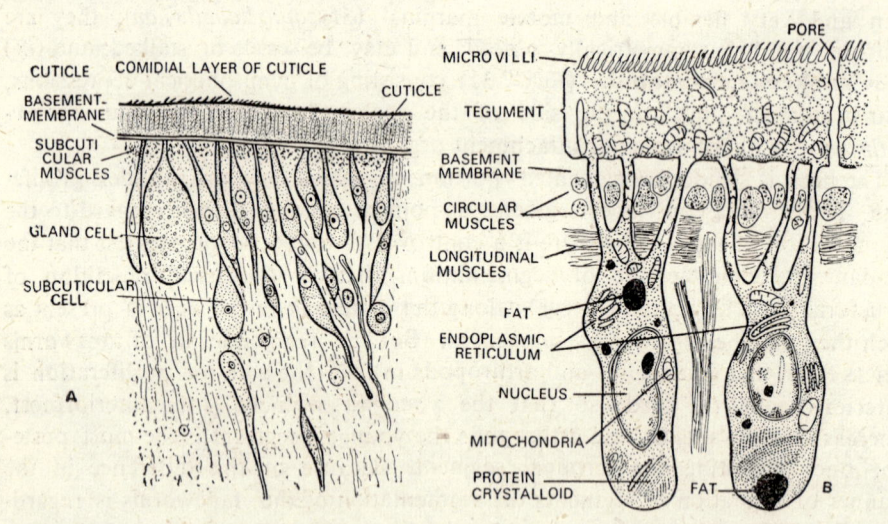

FIG. 7.32. Body wall of a cestode. A—Section through a taenoid showing subcuticular cells and structure of the cuticle; and B—Vertical section of the body wall (from electron microscopy, modified *after* Threadgold).

**Body Wall**

The body is covered externally by a protective **cuticle** (Fig. 7.32 A) that consists of protein impregnated with calcium carbonate and is several layers thick, perforated by fine canals at interval. Usually three layers of cuticle are met with: (*i*) the outer **comidial** layer which is fringe-like and may in some develop spines or scales, etc.; (*ii*) the thickest **central homogeneous layer**; and (*iii*) the **basement membrane.** It is secreted by the subcuticular cells which appear separately sunk within the parenchyma, which fills all space between internal parts. The muscular layer lies beneath the cuticle and consists of outer circular and inner longitudinal layers (Fig. 7.32 B). Beside this layer of muscles there are dorsoventral, transverse and longitudinal (rarely diagonal) muscle fibres also. They are known to constitute the **mesenchymal musculature** which is generally well developed. The subcuticle cells are found just below the longitudinal muscles sunk in the outer part of the parenchyma; the necks of these cells traverse the circular and longitudinal muscle layers. Gland cells are usually absent except in the scolex, where, in some cases, they may form apical glandular region. Such apical glands are always found in the tapeworm larvae. Figure 7.32 gives a vertical section of body wall as seen in electron micrograph.

### Digestive System

The tapeworm has no mouth and no trace of a digestive system, quite unlike the flukes which feed actively on the tissues of their host and perform their own digestion. The tapeworm lives in the intestine of its host where digested food is readily available and it diffuses through the wall to the interior, but how does this occur is not exactly known. As the scolex is deeply embedded in the intestinal wall it seems some fluid from host tissues is absorbed by it. Some authors (Chandler) believe that the worms obtain their sugar supply from the digested food of the host, but they get their proteins and vitamins from the living tissue of the host by secreting digestive enzymes which dissolve the cells. These results were obtained by studying a tapeworm (*Hymenolepis diminuta*) found in rats. The infected rats were subjected to different types of diets and the effects on the parasites were estimated. When the host is starved of carbohydrates, the worms become small and stunted. If the rats are deprived of protein, they lose weight, while their parasites remain healthy. The food reserve of tapeworms chiefly consists of glycogen and to a less extent various lipoid substances. Glycogen, a chief source of energy, is used in the process of respiration as described under Respiratory System.

### Nervous System

The nervous system is like that of a turbellarian or a fluke but less well developed. It generally consists of two longitudinal nerves running along the lateral margins of the body (Fig. 7.33). In some cases additional cords are also present. The main lateral longitudinal nerves continue in the scolex and these swell into a pair of ganglia transversely connected by a thick cross commissure. Besides being connected in the transverse axis they are also connected by a ring. The whole structure, thus, forms a brain complex from which nerves pass in front and sideways. The longitudinal nerves are connected by transverse commissures which in some cases are broken up into a network. The nervous system is located in the parenchyma just inside the mesenchymal muscle layer. Special sense organs are lacking, the body surface and organs of the scolex are richly supplied with free sensory nerve endings.

### Excretory System

The excretory system consists of two **longitudinal excretory** canals (Fig. 7.33) one on each side running parallel to the nerve cords. Of the two one is ventral and larger and other is dorsal and narrower. In the posterior part of each proglottis the ventral canals (seldom the dorsal canals) are connected by a transverse canal. Besides this the ventral canals run throughout the whole length of the body, whereas, the dorsal ones disappear in the more posterior part of the body where ripe proglottides are found. In the young worms in which any proglottides are not shed the excretory canals in the last segment terminate in a median pulsatile **excretory bladder,** the **caudal vesicle,** which opens to the exterior by a single pore. After this segment has been shed the bladder is lost and the broken ends of the excretory canals serve as pores. In the scolex the longitudinal canals break up into a network of delicate tubules through which both are connected. From the main trunks arise numerous branches which further divide into finer branches that eventually terminate in flame-cells.

**Respiratory System**

The tapeworm, like the flukes, live in an environment having very low oxygen content so that it cannot respire like other animals. The worm consumes oxygen whenever available in the surrounding. This is a true oxidative metabolism which is generally higher in the anterior part of the worm, less in mature and the least in ripe proglottides. But the main respiratory process of tapeworm is **anoxybiotic** in which glycogen is utilized as the source of energy and carbon dioxide and organic acids are released. This process continues irrespective of presence or absence of oxygen in the surrounding medium.

**Reproductive System**

The cestodes are hermaphroditic (except the taenioid genus *Dioecocestus*). The reproductive organs lie embedded in the parenchyma and develop progressively from anterior to posterior end of the strobila. As the male organs develop first the anterior segments contain only the male generative organs, further posteriorly placed segments consists of mature proglottides containing fully developed

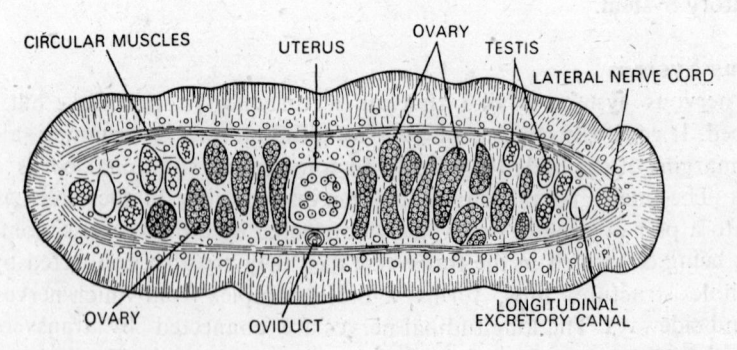

FIG. 7.33. Transverse section of a proglottid of a tapeworm (diagrammatic).

organs of both sexes and finally the rear part of the body comprises ripe or gravid proglottides occupied by the uterus full of capsules or embroys (Fig. 7.34).

The testes (Fig. 7.33) are numerous small rounded bodies lying scattered throughout the parenchyma more towards the dorsal side and connected by fine tubules with a single large convoluted **sperm duct** the end of which is modified to form a muscular body, the **sperm sac.** Inside the **cirrus sac** the sperm duct is made up of a proximal non-eversible portion, the **ejaculatory duct**, that is often coiled, and a distal eversible portion, the **cirrus** (Fig. 7.34), which is often armed with hooks and spines, etc. The ejaculatory duct is sometimes enlarged to

# Phylum Platyhelminthes

form a **seminal vesicle**. Likewise the sperm duct may also be enlarged before entering the cirrus sac forming the **external seminal vesicle**. The cirrus sac opens in the genital chamber which opens to the outside through a genital pore.

The ovary is a single bilobed structure, the posterior side of the proglottis (Fig. 7.34) approximated to the ventral surface. Both the lobes of the ovary are almost equal in size and connected by a narrow bridge. The **oviduct**, a short tubular structure springing from the bridge, is joined by the **vitelline duct**. From the point of junction a narrow tube, the **vagina**, runs obliquely forward to open at the **genital pore**. The inner portion of the vagina is slightly enlarged forming the seminal receptacle, that receives the sperms during copulation. From the base of the oviduct arises a blind sac and passes towards the anterior end. This is the **uterus**, a simple tube in younger proglottides, but becomes branched in more mature proglottides. It is interesting to note that in some primitive tapeworms this also opens to the exterior corresponding to the duct termed **vagnia** (or **right and left vaginae**) in the monogenetic trematodes. At the junction of the vagina, the uterus and vitelline duct the oviduct widens to form the **ootype** (Fig. 7.34), surrounded by **shell glands**. In the present case, however the shell glands are reduced very much forming a small mass surrounding the ootype. The **vitelline gland** in this case is a reduced median structure lying near and posterior to the ovary. The narrow vitelline duct arises from the gland and opens into the oviduct. As the uterus lacks an aperture and consequently the embryos are released by the disintegration of the proglottis.

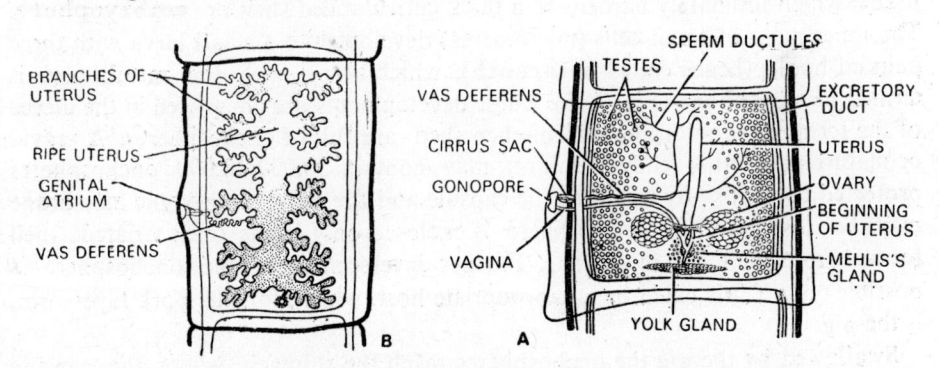

FIG. 7.34. A—A mature proglottid; and B—A ripe or gravid proglottid.

**Fertilization**. The common method of fertilization is probably self-fertilization by the eversion of the cirrus into the vagina of the same proglottis. Sometimes cross-fertilization can take place between the proglottides of different worms when two or more are in the same host. Sometimes one section of the worm (anterior) folds back and inseminates the more matured proglottides of the posterior section.

**Development**. During fertilization eggs become surrrounded by firm chitinous **shells** probably secreted by the shell gland, and then pass on to the uterus. In the uterus the development begins and the first division separates a large yolk filled **vitelline cell** (Fig. 7.35) from the embryonic cell. The embryonic cell undergoes further divisions forming a ball of cells, the **morula**, which consists of cells of different sizes and nature. There are two or three large cells **macro-**

**meres**, three or more **mesomeres** and a number of small cells, **micromeres**. The macromeres contain the yolk droplets and soon fuse to form a **syncytium**

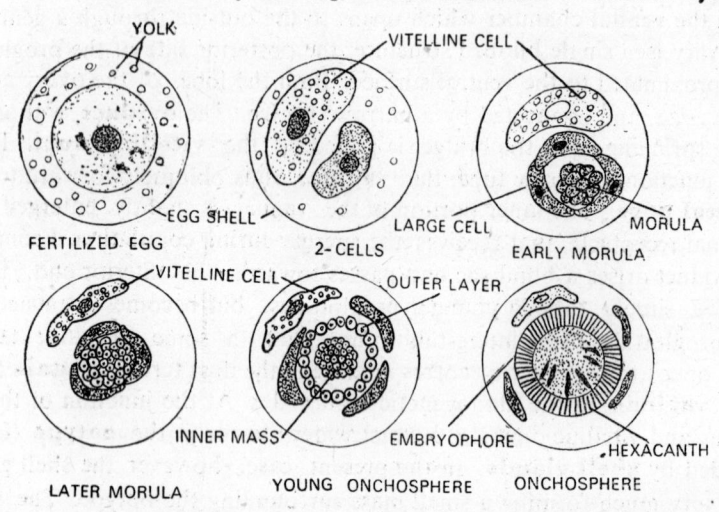

FIG. 7.35. Stages in the development of *Taenia solium*.

that surrounds the other blastomeres as the outer embryonic membrane of nutritive nature. After further cleavage the mesomeres form the inner embryonic membrane, which ultimately hardens to a thick cuticularized shell or **embryophore**. The inner mass of small cells (micromeres) develops into a small larva with three pairs of hooks (hence called **hexacanth**), which along with the membranes is termed **onchosphere** (Fig. 7.35). These developments are completed in the uterus of the terminal proglottides which when shed are full of onchospheres. A gravid or mature proglottis, ready to fall off, may contain 30,000-40,000 onchospheres protected by shells. By this time the capsule and the outer embryonic membrane may be lost and the onchosphere is enclosed only in the thick striated shell formed from the inner membrane. Further development of the onchosphere is possible only if it is eaten by an appropriate host, which, for the pork tapeworm, is the pig.

Swallowed by the pig the onchospheres reach the stomach where the capsule and the embryonic membrane are digested and the hexacanth is released. It bores its way rapidly through the walls of gut and enters the blood stream and on reaching some tissue it becomes encysted. Encystment takes place preferably in the muscles, frequently in the muscles of the tongue, though other organs such as liver, spleen, etc., may also be infected. The parasite here gradually develops into the second larval stage, the **bladder worm** (Fig. 7.37) or **cysticercus**. A central cavity appears in the onchosphere and enlarges until a fluid filled vesicle is formed. The wall of the vesicle thickens at one point and invaginates into the vesicle as a hollow knob (Fig. 7.36 C) that eventually develops into a scolex with its adhesive organs facing the cavity of invagination. This is the **cysticercus** larva or **bladderworm** with invaginated introverted scolex and a fluid filled vesicle devoid of tail. Such bladders occur in the musculature, liver and other organs of the host and may attain a considerably large size. The number of bladderworms varies with infection, in heavy infection there may be 3000 in one pound of pork

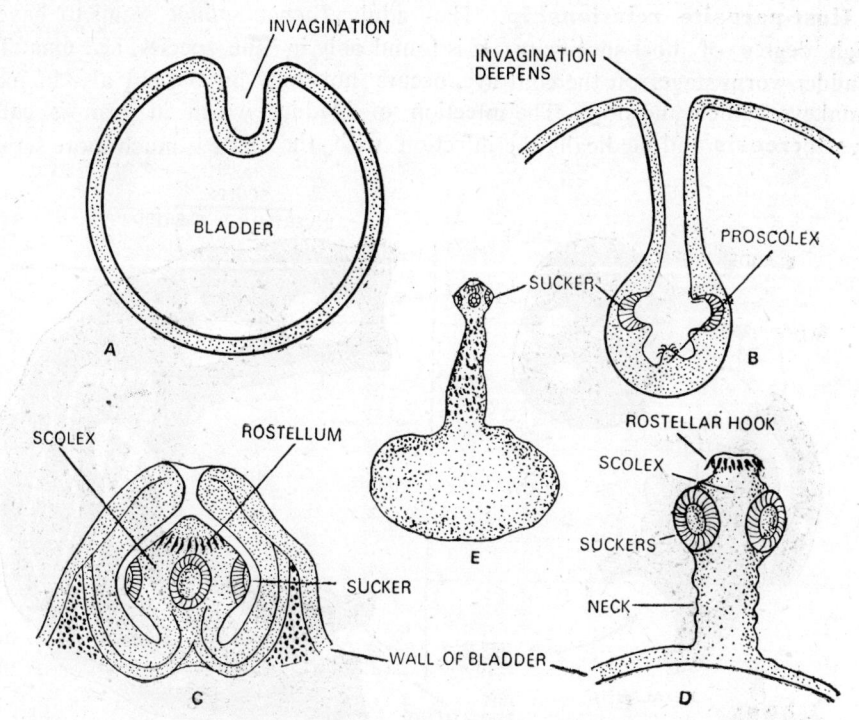

FIG. 7.36. Further stages in the development of *Taenia solium*.

(Brails food). They take four months to develop from the onchosphere, and they remain alive but quiescent for some years.

The bladderworm does not grow further till it reaches the gut of the primary host, man in this case. When eaten by a human being, they are immediately aroused to activity. On reaching the gut, with insuffciently cooked pork, the scolex emerges and attaches to the intestinal wall, the bladder is dissolved, and an area of proliferation is developed that begins to produce new proglottides.

**Asexual multiplication and regeneration.** Generally speaking the tapeworm cycle lacks asexual multiplication. But some notable cases of this type have been reported. Larvae of a rare human tapeworm *Sparganum porlifer* from Japan are slender, unsegmented worm-like individuals without scolex and reproduce asexually by giving off lateral buds. Likewise some taenioid larvae are known to produce additional cysticercoids by both internal and external buddings. External budding is seen in the bladder of *Cysticercus longicollis* of rats (larval stage of *Taenia crassiceps* of dogs). Numerous cysticerci are produced in the form of external buds that can detach and bud themselves in the same manner. In internal budding the inner surface of the bladder puts out buds into the cavity that develop into cysticercoids. This occurs in the larval stage of a taenioid tapeworm of birds called *Paricterotaenia nilotica*.

Power of regeneration has also been studied by various workers and it is concluded that the scolex and neck region of tapeworms can regenerate the stroblia.

Taking bothriocephaloid plerocercoides Iwata (1934) made various cuts, incisions and transections and then reimplanted the operated specimens into the appropriate host. He recorded a high regenerative capacity.

**Host-parasite relationship.** The adult *Taenia solium* seems to have a high degree of host specificity. It is found only in one species, i.e., man. The bladder-worm stage, on the contrary, occurs not only in pig but also in man, monkeys, camels and dogs. The infection of bladder worm in man is called **cysticercosis** and medically the infection with the larva is much more serious

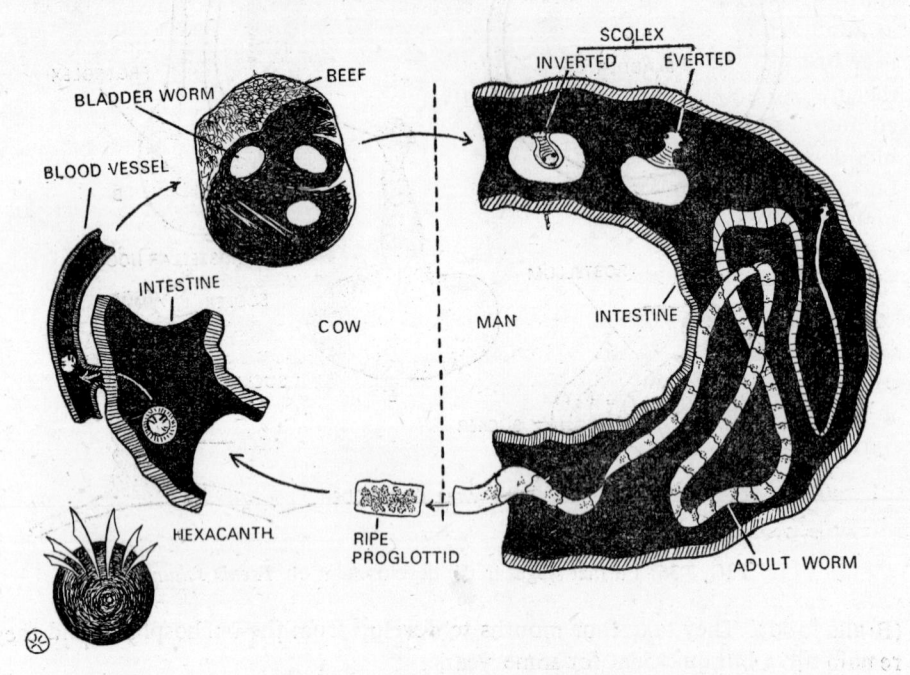

**FIG. 7.37.** Life-cycle of *Taenia solium*.

than with the adult as the larva may settle down in any part of the body and cannot be removed except surgically. It is very interesting to know that the bladder-worm is not treated as most foreign bodies are treated by host—it is not attacked or destroyed by phagocytes. Perhaps it has some kind of chemical defence.

De Waele has made interesting investigation of the susceptibility of various stages of the tapeworm to digestion. The adult worm, which is completely resistant to the alkaline juice of the intestine in which it lives, will be digested if exposed to both acidic and alkaline digestive juices successively, and the same is true of onchosphere. Now the environment in which the worm lives contains enzymes which are able to digest the proteins and fats of which the worm is made. But how the worms protect themselves against this hostile environment is not clearly understood. Some early experiments seemed to indicate that they secrete antienzymes, but new evidence is contradictory. Again both the onchosphere and the bladder-worm have to pass through the acid stomach and are then subjected to alkaline intestinal juices, but in each case the important parts of the parasites are protected against the stomach juice. The onchosphere is protected by its egg shell which is digested away in the intestine while the bladder-worm has the head of the future tapeworm safely tucked within the bladder, and the bladder, like the egg shell, is digested away in the intestine after it has protected the head from the gastric juice. It is evident from this that the relation between host and parasite is clearly one involving the most intricate biochemical adaptations.

Another interesting problem is what limits the growth and development of the larvae within the larval host. The larvae do not attain sexual maturity unless they are transferred to the definitive host of the adult. Certainly the substances or conditions essential for maturation are provided by the adult, but not the larval host. Symth, working on a tapeworm which lives as larva in sticklebacks and as adult in the duck, was able to keep the larvae alive for some months in peptone broth. If this was kept at room temperature the larvae did not develop further but if they were incubated at duck's temperature (40°C) they rapidly matured and produced eggs. In this case a change of temperature from that of a cold-blooded to warm-blooded host seems all that is necessary to induce maturation, but other factors control the process in worms which live in warm-blooded animals, both as larva and adult.

**Pathology.** Although the tapeworms are aggressive, grappling themselves to their host's gut wall with four strong muscular suckers and a circle of hooks, they are not medically very important in the adult condition, of course, they are feared very much. One or two tapeworms are not so harmful but hordes of them may cause physical obstruction. The taenias of man occur singly or occasionally in pairs. Ten beef tapeworms, *Taenia saginata*, have been reported in a man from Syria. The man has been known to carry the same tapeworms for 35 years. The life of the parasite is usually limited by the death of the host. It is wrong to believe that deleterious effects of tapeworms are produced because they rob the host of his digested food. In fact the damage to the host's intestinal walls caused by hooks and suckers is harmful. It may cause inflammations that may result in serious complications. Further, it has been found that the tapeworms also release toxic substances that are responsible for the development of severe symptons including pernicious anaemia. On the whole the worms prove inimical to children and weak patients. The effects are usually generalized and include nausea, abdominal pains, nervous disorders resembling epilepsy and anaemia accompanied by an increase in the number of eosinophil white corpuscles. Some people are hypersensitive to the worms and show various nutritional and nervous disturbances. The bladder-worm stages, on the other hand, are very harmful to the host. They may settle down in any part of the body and encyst. The cysts can only be removed surgically in some cases, that, too if present in very small number. Cysts may occur in the eye and may cause blindness. A cyst may be present in the brain giving rise to serious disorders. The bladder-worms die after five or six years and may become calcified. The bladder-worm of pork tapeworm is found not only in pig but also in man when they cause **cysticercosis**. The bladder-worm of the beef-tapeworm in man has been reported only twice. The larval tapeworms may stimulate tumours in the tissues. The liver tissue around the cysticercus undergoes inflammation and necrosis, and the final result is a malignant growth (**sarcoma**) composed of host connective tissue and endothelial linings of live capillaries.

**Parasitism and the parasite.** In order to fit into the parasitic way of life the parasite undergoes several modifications. The prominent features are given below:

(*i*) The body has become simple, dorsoventrally flattened and divided into numerous segments; (*ii*) The ciliated epidermis is replaced, in the adult, by thick cuticle, which is freely permeable to water. As the osmotic pressure of the parasite is generally lower than that of the surrounding host tissue or fluid, the worm can lose or gain water very readily if the osmotic pressure of the medium is varied; (*iii*) Special organs of adhesion including hooks and suckers have developed but these are restricted to the anterior end of the body; (*iv*) Mouth and digestive tracts are absent, the food is absorbed through the entire surface of the body; (*v*) Definite respiratory organs are lacking and the animal respires anaerobically (*see* page 288); (*vi*) Organs of special sense are absent; and (*vii*) The reproductive organs have become highly specialized and life-cycle complicated necessitating the acquisition of an intermediate host.

**Control**

The control of the infection of the pork or beef-tapeworms is easy. Efficient disposal of human excrement removes the parasite at the egg stage as the pigs will not be able to feed on faeces. Secondly, thorough cooking of pork or beef will destroy the cysts, if any. As some people prefer slighly under-cooked beef the beef tapeworm is now more common than other tapeworms. In, Abyssinia practically all the natives have beef-tapeworms, and their control has become a part of their life. They set aside a day every three months on which they take an infusion of **kousso** flowers, a powerful **vermicide**.

The worm is stubborn and resists attempts to get rid of it. Drugs will successively paralyze the main body so that it breaks up and is discharged from the host's gut with faeces, but the head and neck are dislodged with great difficulty and remain attached to grow the new body inch by inch. Several drugs are used as treatment but the most effective is **Acramil** which should be administered orally or by duodenal tube as directed by the physician.

## CLASSIFICATION OF CESTODA

Systematic accounts of the Cestoda (Latin *cestusm*, a girdle) are of a comparatively recent date, but have been subject to the usual disputes and disagreements. The class Cestoda[1] was formerly divided into two subclasses: **Monozoa** including all unsegmented tapeworms and **Merozoa** including the segmented worms. Now, however, it has been found out that some of the unsegmented tapeworms are more closely related to the segmented worms and should be placed with them; thus the class cannot be divided only on the basis of presence or absence of segments (proglottides). The class is now divided into two subclasses—**Cestodaria** including worms having a 10-hooked larvae, and **Eucestoda** including worms having a 6-hooked larvae. Further classification of Cestodaria does not present difficulties but that of Eucestoda is controversial. Four chief orders of Eucestoda are universally recognized. They are: **Cyclophyllidea (Taenioidea)**, **Pseudophyllidea**, **Trypanorhyncha** and **Tetraphyllidea**. These are differentiated in a general way on the characters of the scolex. The present work follows mainly the classification of Southwell (Fauna of British India, 1930)

---

[1]The original name of the class—Cestoidea—was proposed by Rudolphi in 1809. It is still used by some authors.

## Phylum Platyhelminthes

adding to it one new order including his genus *Echinobothrium* of uncertain systematic position recognized since his publication.

(*i*) **Subclass Cestodaria (Monozoa)**. Those are cestoda whose bodies do not form proglottides and whose larvae are 10-hooked. This is a small group of worms inhabiting the coelom and intestine of lower fishes. This is divided into two orders: Amphilinidea, Gyrocotylidea.

1. **Order Amphilinidea**. Body flattened, oval, usually elongate and tapelike ranging from a few to 30 to 40 cm. A definite scolex is lacking while a protrusible proboscis is present which is poorly developed in the genus *Amphilina* (Fig. 7.38). The cuticle is thin and delicate without cuticular spines and hooks. The excretory system takes the form of a peripheral network of vessels opening at the rear end by way of a small excretory bladder. Flame-cells are peculiar, each

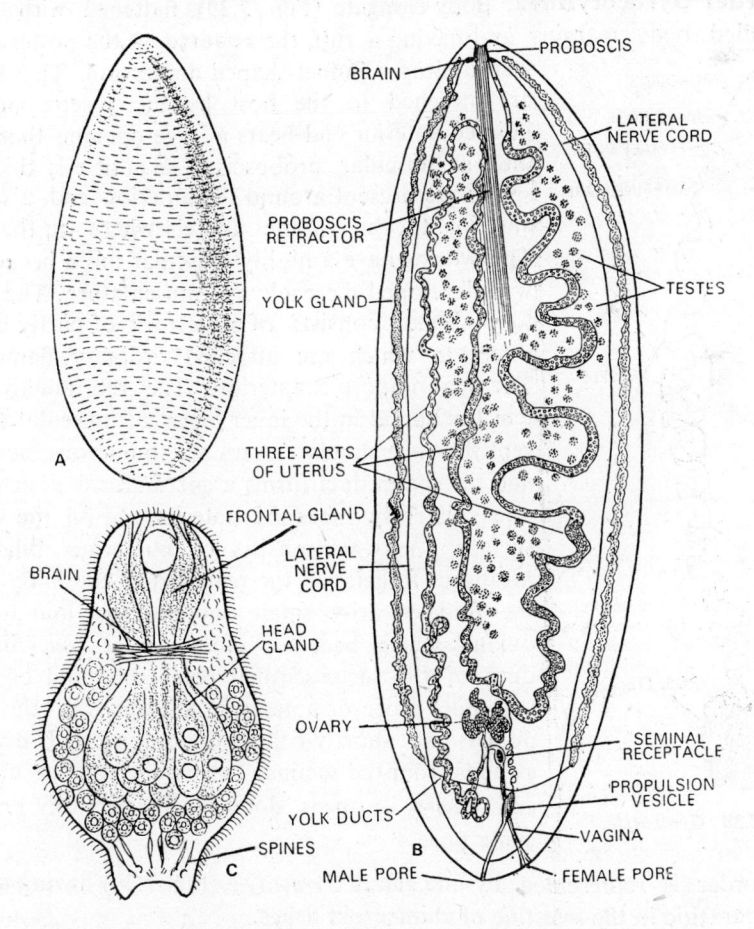

FIG. 7.38. *Amphilina foliacea*. A—Entire animal under low power; B—General anatomy of the cestode; and C—Lycophore larva newly emerged from the egg showing the conspicuous head gland (frontal gland).

cell enclosing a rosette-like cluster of 18-30 flame-bulbs. The numerous testes are strewn throughout the whole body (*Amphilina*) or in two stripes one along each lateral margin (*Gephyrolina*). The common sperm duct is provided with a muscular propulsion region and receives ducts from prostatic glands. It opens at the posterior end of the worm in some forms, through a penis papilla. Ovary is single or lobed in the rear part of the worm. The oviduct becomes the uterus, after receiving the vagina, and the common yolk duct, and traverses the body three times after the manner of three limbs of the letter N and finally opens at the anterior tip alongside the proboscis. Thus, it is apparent that the uterine and vaginal pores are situated at opposite ends of the body.

This order includes six genera of which the most common is *Amphilina* having strewn testes, single lateral vaginal opening and a seminal receptacle is wanting. It is a parasite in the coelom of the sturgeon, *Acipenser*. Other genera include, *Austramphilina* in the coelom of bony fishes and fresh-water turtles; *Gigantolina*, *Nesolecithus*, *Gephyrolina* and *Schizochoerus*.

2. **Order Gyrocotylidea.** Body elongate (Fig. 7.39), flattened with generally ruffled body margins and having a ruff, the **rosette**, at the posterior end surrounding a funnel-shaped depression. The worms are attached to the host by the rosette end. The pointed anterior end bears a large opening that leads into a muscular proboscis. The cuticle is thick and spines are present around the anterior end, also just anterior to the rosette and elsewhere on the body. The worms have a highly muscular build because of well-developed mesenchymal musculature. The excretory system consists of a network of partly ciliated vessels to which are attached ordinary flame-cells. The nephridipore is anterior. The reproductive system lies buried in the inner layer of musculature except the yolk glands. Testes are numerous scattered and the sperm duct forms a spermiducal vesicle that opens laterally near the anterior end on the ventral surface. The ovary consists of numerous bilaterally arranged follicles in the rear part of the body. From the follicles arise small ductules that join to form oviduct which becomes uterus, after receiving the duct of the large seminal receptacle and common yolk duct proceeds anteriorly and opens in the mid-dorsal line shortly behind the proboscis. The vagina extends from the seminal receptacle to the anterior end where it opens dorsally and laterally near the male pore.

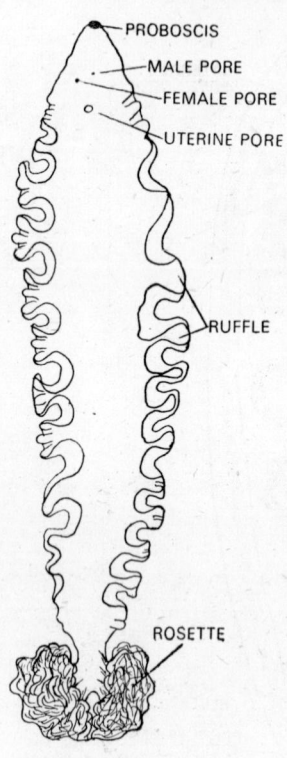

FIG. 7.39. *Gyrocotyle*, entire.

The order is represented by one genus *Gyrocotyle* (Fig. 7.39) having several species parasitic in the intestine of chimaeroid fishes.

(*ii*) **Subclass Eucestoda (Merozoa).** The Cestoda whose body is usually very long and ribbon-like divided into a few to many segments, but sometimes undivided. The anterior end of the body has a scolex furnished with organs of fixation such as hooks or suckers. They are generally provided with more than one set of reproductive organs and the larva is 6-hooked. Further subdivisions of Eucestoda depend upon the shape of the scolex.

# Phylum Platyhelminthes

1. **Order Tetraphyllidea** or **Phyllobothrioidea**. Segmented tapeworms with an armed or unarmed scolex bearing four bothridia often subdivided with septa. The bothridia may be sessile or pedunculated. One or more small accessory suckers may also be present. The head may also bear a terminal sucker; neck present or absent and the strobila definitely segmented. Segments are usually shed

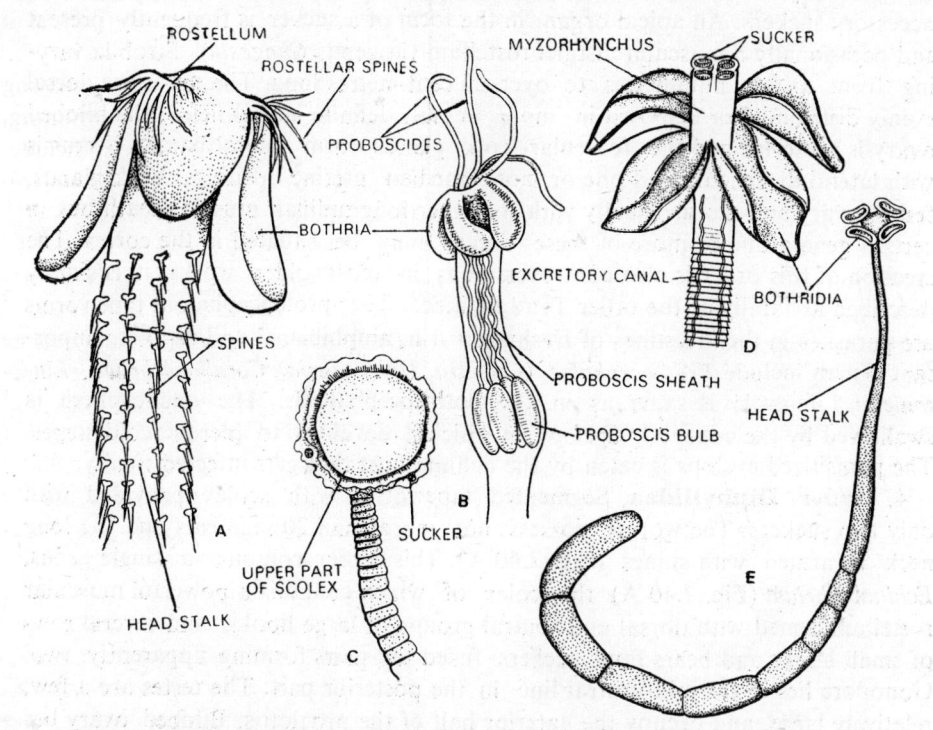

FIG. 7.40. Different kinds of scolices in cestodes. A—*Echinobothrium* (*after* Southwell); B—*Tetrarhynchus matheri* (*after* Southwell); C—*Lecanicephalum* (*after* Linton); D—*Myzophyllobothrium*; and E—*Anthobothrium*.

before they are ripe in which case they ripen and become gravid in the intestine of the host. The members of this group are parasitic in the intestine of elasmobranch fishes. They are mostly small in size not exceeding 10 cm in length, with not more than a few hundred proglottides. A typical onchosphere larva occurs in the capsules either before or after they have been laid in sea water. The onchosphere larva enters a copepod and develops into a larva plerocercoid in which condition it lives in the copepod till it is eaten by the primary host in which it develops into the adult. Important genera include *Phyllobothrium*, parasitic on elasmobranch fishes, reptiles and mammals, *Echinobothrium* (Fig. 7.40 A) with 4 grooved bothridia looking like the sucker of *Echeneis*, *Anthobothrium* (Fig. 7.40 E) with simple bothridia, *Myzophyllobothrium* (Fig. 7.40 D) and *Onchobothrium*, etc.

2. **Order Lecanicephaloidea**. Segmented tapeworms with scolex bearing four suckers like those of taenioids. Genital organs arranged as in the previous order. Since the reproductive organs of these are identically similar to those of tetraphyllid worms this group was included in that order (Tetraphyllidea). But the characters of scolex are very different as such they are considered as separate

order here. They are relatively small tapeworms inhabiting the intestine of elasmobranch fishes. Important genera include *Lecanicephalum* (Fig. 7.40 C), *Tylocephalum, Cephalobothrium, Polypocephalus,* etc.

3. **Order Proteocephaloidea.** Segmented tapeworms with scolex unarmed or armed with minute spines and with four sessile suckers devoid of areolae or accessory suckers. An apical organ in the form of a sucker is frequently present and occasionally a muscular distinct rostellum (in genus *Gangesia*). Strobila varying from a few millimetres to over 60 centimetres long. The testes are dorsal evenly distributed or disposed in more or less definite lateral fields, the bilobed ovary is posterior and the follicular yolk glands form lateral bands. Uterus is with lateral diverticula and one or more median uterine openings. Yolk glands, testes, ovary and uterus usually with the inner longitudinal muscle sheath but in certain genera one or more of these organs may be situated in the cortex. The creation of this order is also controversial as in most older works it is simply described as family of the order Tetraphyllidea. The proteocephaloid tapeworms are parasites in the intestines of freshwater fish, amphibians and reptiles. Important genera include *Proteocephalus, Gangesia, Ophiotaenia, Corallobothrium, Lintoniella.* Life-cycle is same as in the bothriocephaloids. The onchosphere is swallowed by the copepod, *Cyclops,* in which it develops to plerocercoid stages. The parasitized cyclops is eaten by the definite host that gets infected finally.

4. **Order Diphyllidea.** Segmented tapeworms with scolex provided with only two suckers. The worms possess not more than 20 segments and the long neck is armed with spines (Fig. 7.40 A). This order contains a single genus, *Echinobothrium* (Fig. 7.40 A) the scolex of which contains a powerful muscular rostellum armed with dorsal and ventral groups of large hooks, and several rows of small hooks and bears four suckers fused in pairs forming apparently two. Gonopore lies in the mid-ventral line in the posterior part. The testes are a few, relatively large, and occupy the anterior half of the proglottis. Bilobed ovary lies in the rear part of the proglottis. The follicular yolk-glands occur as lateral strands and the young uterus is a median tube that later expands to a sac without opening. The larva (a cysticercoid) is found in the prawn *Hippolyte* and the adult is found in the spiral intestine of selachians and other elasmobranchs.

5. **Order Tetrarhynchoidea** or **Trypanorhyncha** (Fig. 7.40 B). Segmented tapeworms with scolex provided with four suckers, each armed with a long spiniferous retractile process or proboscide. Small to moderate in size ranging from a few to 1000 millimetres; segmentation is complete. Single set of genitalia present in each segment except in the genus *Dibothriorhynchus* (=*Coenomorphus*) which has a double set of genitalia. The plan of reproductive system resembles that of tetraphyllids except that the vagina and sperm duct do not cross and the yolk glands usually encircle other reproductive organs. Adults parasitize elasmobranch fishes and occasionally teleosts. The genus *Haplobothrium* occurs in the ganoid fish *Amia calva.* Larvae occur in teleosts, reptiles and invertebrates.

Important genera include: *Tetrarhynchus,* (=*Tentacularia*) (Fig. 7.40), *Gymnorhynchus, Grillotia, Dibothriorhynchus, Aporhynchus, Haplobothrium,* etc. Life-cycle is not fully described.

6. **Order Pseudophyllidea** or **Dibothriocephaloidea.** Segmented or unsegmented tapeworms with a scolex provided with two to six shallow bothria or

*Phylum Platyhelminthes* 299

sometimes without adhesive organs. The bothria may assume various forms (tubular, flask-shaped, etc.) by the fusion of their edges or they may be replaced by a pseudoscolex, i.e., the scolex may degenerate and the anterior part of the strobila may function as scolex. Sometimes a terminal sucker may be present. The strobila is without segmentation or monozoic (*Caryophyllaeus*) (Fig. 7.41A)

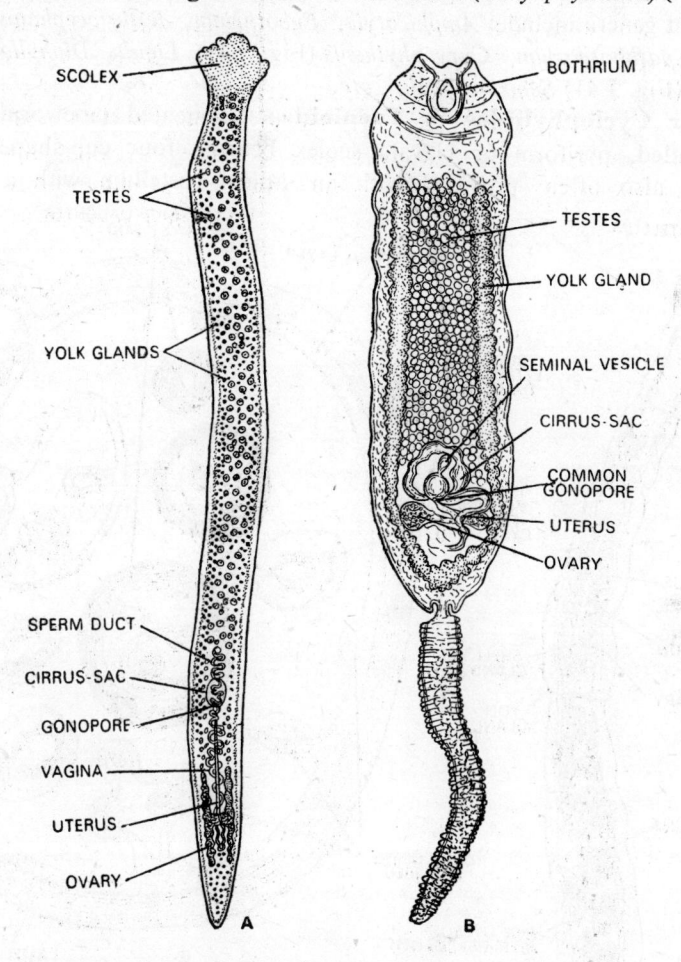

FIG. 7.41. A—*Caryophyllaeus mutabilis*; and B—*Archigetes*.

or segmented. In some cases the segmentation may be indistinct (*Triaenophorus*). One set of reproductive organs in each proglottis, but two sets are present in some (*Diplogonoporus*). Genital pores marginal or superficial (i.e., on the flat side). Yolk glands usually in the cortex, scattered, not condensed into a single gland. Testis situated either in the cortex or medulla. Uterus persistent usually a coiled tube, but may form a simple or lobulated sac in whole or in part. The uterine pore almost always superficial in ventral surface. The position of the vagina and cirrus sac is varied. Openings of the vagina and vas deferens may be close to that of the uterus, or on the opposite side or along the lateral margin. Eggs which may or may not be operculated are passed whilst the segments are still attached to the strobila. Onchosphere frequently with a ciliated covering (and called **coracidium**). Development, where known, results into a procercoid in the body cavity of Entomostraca succeeded by a plerocercoid in teleosts. The plerocercoid of *Caryophyllaeus* and *Archigetes* develops gonads paedogenetically so that there is no adult stage. Generally larva inhabits crustaceans and the adult is found in mammals, birds, reptiles and fishes.

Important genera include: *Amphicotyle, Eubothrium, Bothriocephalus, Cyathocephalus, Spathbothorium, Caryophyllaeus* (Fig. 7.41), *Ligula, Diphyllobothrium, Archigetes* (Fig. 7.41) *Schistocephalus*, etc.

7. **Order Cyclophyllidea** or **Taenioidea**. Segmented tapeworms with a small rounded, pyriform or clavate scolex bearing four cup-shaped suckers (acetabula) also often provided with an apical rostellum with a crown of

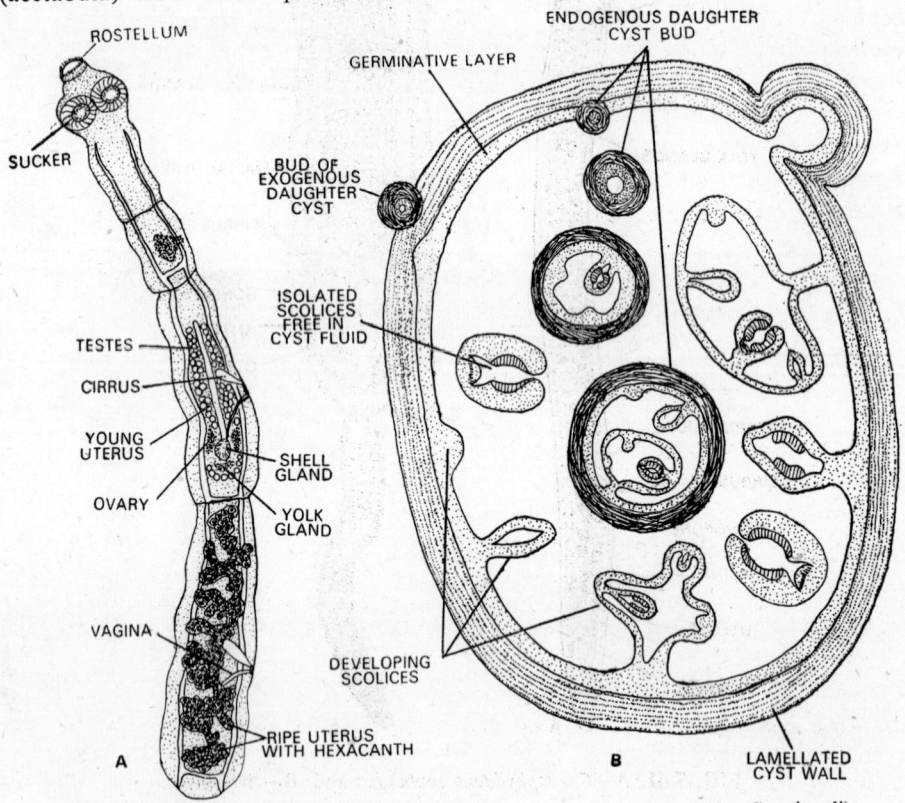

**FIG. 7.42.** *Echinococcus granulosus*. A—Entire worm from dog (*after* Southwell); and B—Scheme of hydatid cyst (*after* Hyman).

hooks. Small to moderate in length from 1-3 mm up to 10-30 mm, but some forms attain a length of 10 metres being made up of 1000 or more proglottides. The suckers are sometimes wanting (*Priapocephalus*) and sometimes even the rostellum is absent (*Taenia saginata*). The neck is short or long. The strobila is strongly flattened. Mostly hermaphrodite only genus *Dioecocestus* is dioecious. Reproductive system as that of *Taenia*. Members of this group are parasites of birds and mammals seldom occurring in reptiles.

Important genera include: *Tetrabothrium, Davainea, Anoplocephala, Prototaenia, Thysanosoma, Mesocestoides, Dilepis, Dipylidium, Nematotaenia, Hymenolepus, Taenia, Echinococcus* (Fig. 7.42), etc.

*Phylum Platyhelminthes*

## INTERSPECIFIC ASSOCIATIONS IN PLATYHELMINTHES

It is well known that the members of natural communities are linked by a system of relationship that operates in subtle ways to ensure efficient exploitation of the environment. Among these relationships there are certain inter-specific partnerships that are characteristic of particular species, and that are marked by their intimacy, by being obligatory for at least one of the partners. In such cases the relationship involves high degree of specialization. Thus, the life-cycles of many helminths, and for that matter, other parasites are not examples of isolated phenomena, but instances of specialized types of ecological relationships.

In the class Trematoda the Monogenea are characterized by being mostly ectoparasites with a complicated posterior attachment disc, and with a *direct* life-cycle involving no vector. The eggs give rise to a larva that metamorphoses on a new host. A few species parasitize amphibians and turtles, but as many as 95 per cent are found on fish, particularly on their gills. Of the fish-infesting forms the majority attack elasmobranchs, showing a very high degree of host-specificity. Apparently the Monogenea are ancient parasites that have remained confined to the lower vertebrates.

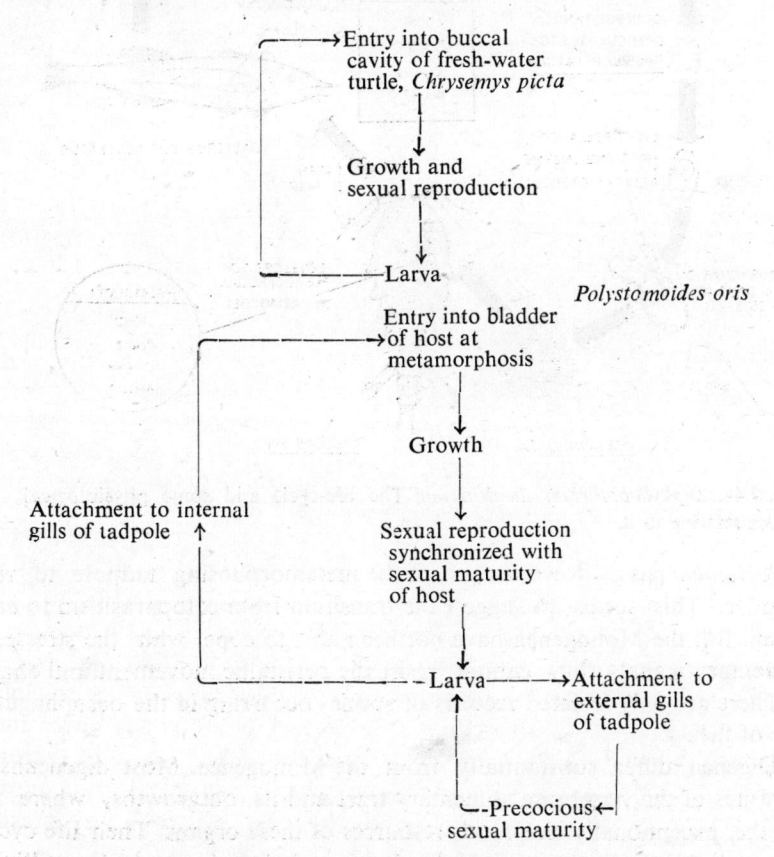

FIG. 7.43. Patterns of life-cycles of Monogenea. In *Polystoma intergerrimum* the normal cycle is shortened by the development of precocious sexual maturity (neoteny) if the larva attaches to external gills instead of internal ones.

302                                                                    LIFE OF INVERTEBRATES

*Polystomum integerrimum* is parasitic in the bladder of frogs. It is one of the species of Monogenea that has successfully exploited hosts other than fish. This they have done by moving into internal cavities such as the mouth, nostrils and bladder, never into the digestive regions of the alimentary tract although the

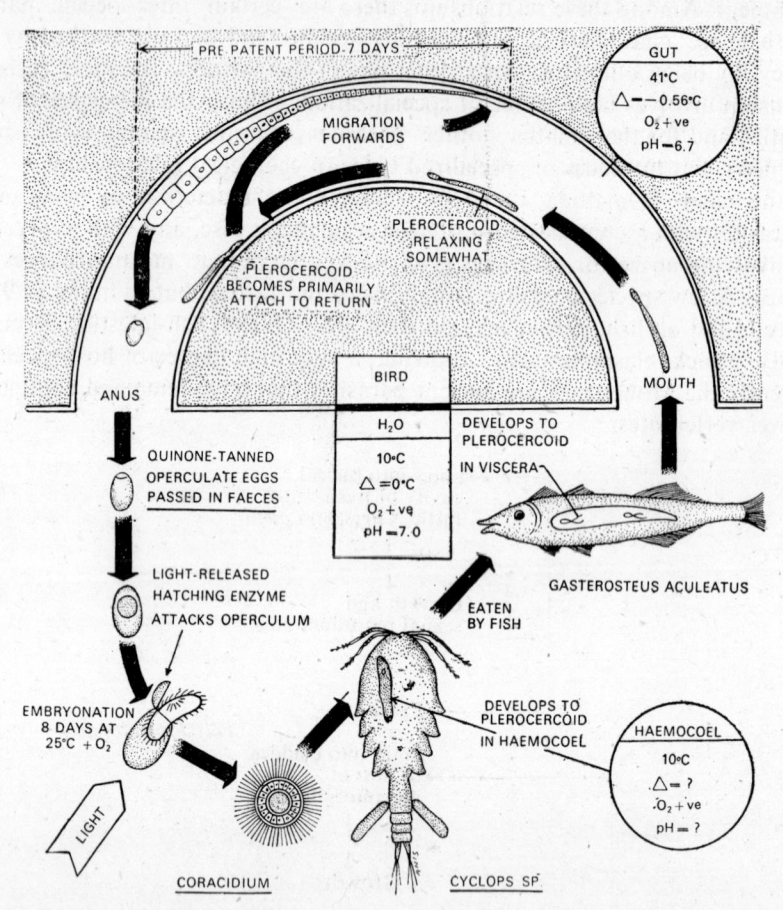

FIG. 7.44. *Diphyllobothrium dendriticum*: The life-cycle and some physiological factors relating to it.

young *Polystoma* passes down the gut of the metamorphosing tadpole to reach the bladder. This seems to suggest the transition from ectoparasitism to endoparasitism. But the Monogenea have not been able to cope with the stresses in the alimentary canal, they cannot resist the peristaltic movement and enzyme action. There are only isolated records of species occurring in the oesophagus or intestine of fish.

The Digenea differ substantially from the Monogenea. Most digeneans are endoparasites of the vertebrate alimentary tract and its outgrowths, where they exploit the exceptionally rich food resources of these organs. Their life cycle is complex and indirect having several developmental stages in which multiplication may occur. They involve more than one intermediate hosts, of which the first is invariably a mollusc, usually a gastropod, but rarely a lamellibranch or a scaphopod. This is so irrespective of whether the final host is marine, fresh-

## Phylum Platyhelminthes

water or terrestrial. When there is a second intermediate host this may be a member of almost any group of animals. It is very significant that the digeneans show high degree of specificity towards molluscan hosts, and a low degree towards their definitive vertebrate hosts. It is apparent from this that they were primarily associated with molluscs and began to parasitize vertebrates.

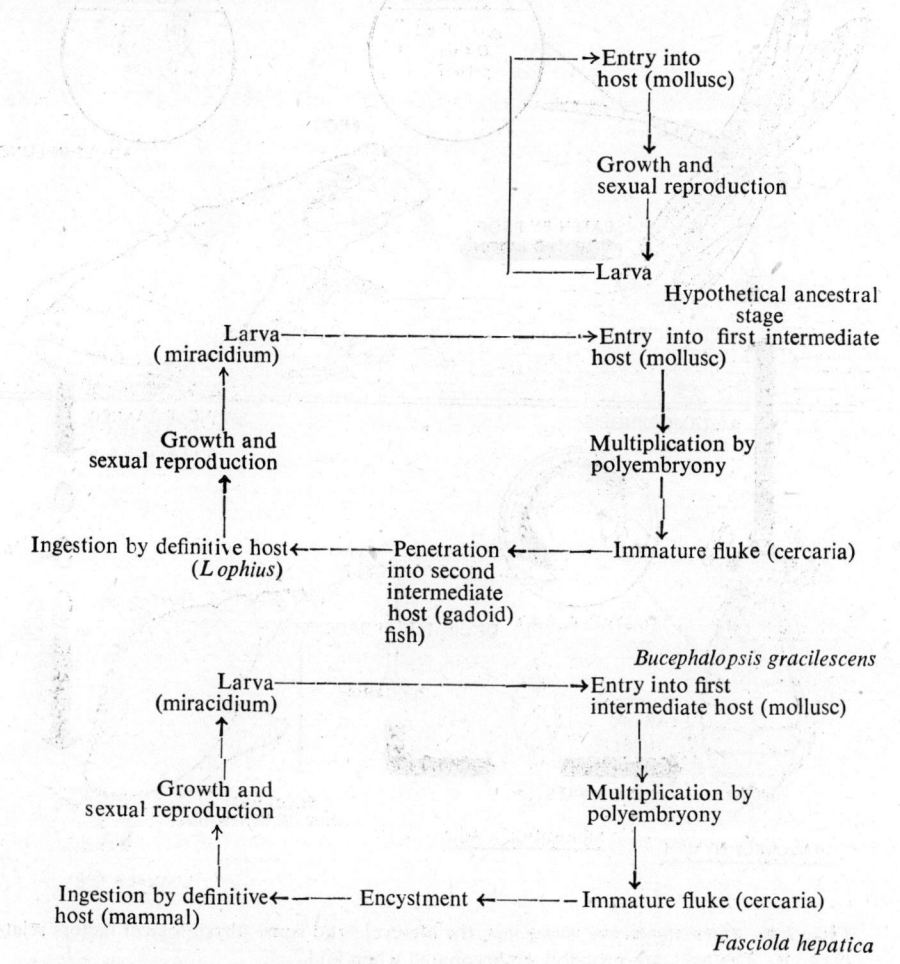

```
                              ┌──────→ Entry into
                              │        host (mollusc)
                              │             │
                              │             ↓
                              │        Growth and
                              │        sexual reproduction
                              │             │
                              │             ↓
                              └─────── Larva
                                            Hypothetical ancestral
                                                    stage
        Larva ─────────────────────→ Entry into first intermediate
      (miracidium)                     host (mollusc)
            ↑                               │
            │                               ↓
        Growth and                      Multiplication by
        sexual reproduction             polyembryony
            ↑                               │
            │                               ↓
   Ingestion by definitive host ←── Penetration ←──── Immature fluke (cercaria)
        (Lophius)                   into second
                                    intermediate
                                    host (gadoid)
                                    fish)
                                              Bucephalopsis gracilescens
        Larva ─────────────────────→ Entry into first
      (miracidium)                     intermediate host (mollusc)
            ↑                               │
            │                               ↓
        Growth and                      Multiplication by
        sexual reproduction             polyembryony
            ↑                               │
            │                               ↓
   Ingestion by definitive ←── Encystment ←──── Immature fluke (cercaria)
   host (mammal)
                                                   Fasciola hepatica
```

**FIG. 7.45.** Patterns of life-cycles in the Digenea. In the hypothetical ancestral stage there would have been only one host, a mollusc. Later this becomes the fresh intermediate host. A second intermediate host may (*Bucephalopsis*) or may not (*Fasciola*) be involved.

The entry of the parasite in the definitive host takes place in various ways. *Schistosoma mansoni*, one of the devastating blood flukes of man, is transferred from the molluscan vector to its primary host by the cercaria directly penetrating the skin of human beings that enter the water. *Fasciola hepatica* takes advantage of the amphibious habits of the snail *Lymnaea truncatula*, the motile cercariae encyst on vegetation as metacercariae and are eaten by sheep and cattle. In *Haematoloechus* species occur alternative routes into terrestrial vertebrates. It is parasitic in the lungs of amphibians arriving there by a life-cycle that passes first through molluscan hosts (*Lymnaea* sp.), and through the nymph of the dragon fly, in which metacercariae are formed. These are then ingested by the primary host when it eats the imago. (Fig. 7.46).

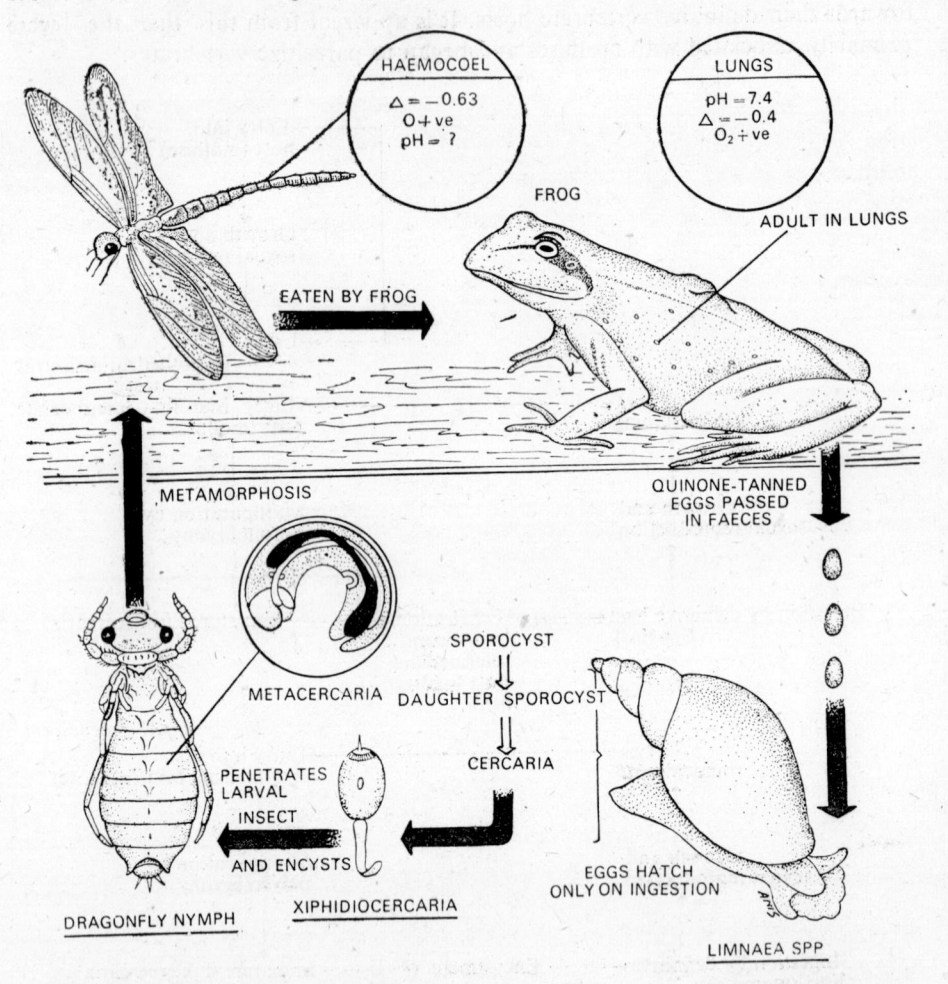

FIG. 7.46. *Haemıtoloechus variegatus*, the life-cycle and some physiological factors relating to it. The eggs are probably embryonated when laid.

The third major group of helminth parasites is the Cestoda. The organs of attachment of cestodes and trematodes differ, and there are important differences also in the reproductive system, particularly as regards the uterus in trematodes, this opens into a genital atrium with the male duct, but it is entirely independent of the atrium in the cestodes. Further cestodes are without alimentary canal and the eucestodes have a tape-like chain of proglottids. The parasites are limited to the alimentary canal of vertebrates. It is presumed that cestode parasitism, like that of Digenea, began with the ingestion of rhabdocoel-like organisms. It is, therefore, natural to conclude that in any case the first host must have been fish. This is supported by the fact that to this day more primitive tapeworms survive in the alimentary tract of fish, using other aquatic animals, as intermediate hosts.

*Phylum Platyhelminthes*

Following this the eucestodes clearly evolved in parallel with the vertebrates. Each vertebrate group has its own genera and species. Thus, the orders Tetraphyllidea and Trypanorhyncha are exclusive parasites of elasmobranchs, whereas the Pseudophyllidea chiefly parasitizes teleosts, birds and mammals. The cestodes of the last group have moved into land-living hosts by the fish-eating habit. *Diphyllobothrium dendriticum*, parasitic in gulls, provides an example of this type. Figure 7.44 gives the details of its life-cycle. A more complete adaptation to terrestrial hosts occurs among the Cyclophyllidea (Taenioidea), mainly parasites of birds and mammals, although found also in reptiles and amphibians. We have already seen the involvement of cattle in the distribution of the pork tapeworm.

The host specificity of the host parasite relationship depends upon physiological adaptations. The miracidium of the trematode *Opisthorchis felineus* is attracted to a prosobranch snail *Bithynia leachi* but not to the closely related *B.tentaculata*, although both occur in the same habitat. Little is known about the basis of these attractions. It is presumed that they involve chemoreception of some type of secretory products of the host. Thus, it is said that the cercariae of *Schistosoma mansoni* locate the skin of their human host by chance but their entrance depends upon the secretion of enzymes known as hyaluronidase, with which they are able to break down the epidermis so efficiently that they can enter the lymphatics within 20 minutes.

Once an association has been established between a parasite and a host its further evolution depends upon the reaction of both organisms. In certain instances it is known that the physiology of the parasite has become so closely integrated with that of the host that it depends upon the latter for the regulation of its reproduction. An elegant example of this is seen in *Polystoma* which enters on sexual maturation in spring, when the host frog will be preparing to enter water to effect its own reproduction. As a result the parasites' eggs are released when the frogs are in water.

With these relationships, as with all other aspects of invertebrate life, new and improved techniques of investigation continue to enlarge our appreciation of the elegance and precision of the adaptation by which animals have succeeded in coming to terms with their environment.

# 8 PHYLUM NEMERTINA

Closely related to the flatworms is the minor phylum of ribbon-worms, Nemertina, sometimes also called Rhynchocoela. Nemertines differ from flatworms in several ways, but resemble them in possessing a mesoderm without a general coelomic cavity and in the structure of their nervous and excretory systems. However, they have a tubular alimentary canal with both mouth and anus, a circulatory system, and an eversible proboscis which is enclosed in a special cavity (the rhynchocoel), lying just above the anterior part of the gut. The position of the Nemertina is difficult to settle. They were described with the Platyhelminthes formerly because of the character of the integument and nervous system, the presence of flame-cells and the lack of coelom and external segmentation, but they have now been separated and raised to the status of an independent phylum because of the presence of a complete digestive tract with an anus, and of a vascular system and the higher organization of the organs and tissues in general. They also show some characters found in the hemichordates and chordates. Among these are the nemertinean dorsal nerve resembling the vertebrate spinal cord, the lateral nerves placed as the lateral line nerves of fishes, the cephalic ganglia like the brain of vertebrates, the transverse nerves simulating the spinal nerves and the proboscis sheath suggesting the notochord.

There are about six hundred species of nemertines. The great number of these are marine scavengers living between tidemarks, under stone or sea-weeds or in burrows in muddy sand. A few are commensal, a few parasitic and a few have managed to invade freshwater (genus *Prostoma*) and tropical land habitats. *Malacobdella* lives in the mantle cavities of bivalve molluscs and *Carcinonemertes* lives among the gills or egg-masses of certain crabs. *Geonemertes* lives in damp soil. Their elongated bodies range in length from less than an inch to many feet, and in colour they are white or yellow but some are often red, orange or green with contrasting patterns of stripes and bars. The body may be cylindrical or more or less flattened. The body is covered with a ciliated coat. They are bilaterally symmetrical and the anterior end is more or less marked off as a head and bears numerous simple eyes and sensory cells. The most distinctive character is the **proboscis**, a long muscular tube which can be thrown out to catch prey. This is free in a cavity, the **rhynchocoel** situated above the digestive tract in the mid-dorsal line. The posterior end of the proboscis is attached to the wall of the rhynchocoel by a retractor muscle. The mouth is situated ventrally near the anterior end. In some cases a separate mouth is absent. The alimentary canal which opens by the proboscis pore is often distinguished into several regions and the intestine frequently bears paired lateral diverticula. The alimentary canal opens to the outside by an anus. There are many pairs of excretory tubules provided with terminal flame-cells. A circulatory system is also present. It consists of a pair of lateral vessels and many forms are provided with an additional unpaired dorsal vessel. The nervous system comprises a well-differentiated brain from which arises a pair of main lateral nerves. The gonads are simple sac-like structures opening directly to the exterior and complicated copulatory organs are mostly absent. The development in some cases includes a free swimming larva, the **pilidium**. The nemertines are **acoelomate**, that is all the space between the digestive tract and the body wall is filled with parenchyma and muscle fibres. They are the least known animals as they are not usually seen by visitors to the sea coast nor they have any economic or medical importance.

### TYPE TUBULANUS

*Tubulanus* presents typical structure of primitive nemertines which are the most complex acoelomates of one form or another. Although they are true acoelomates having solid constructions, the body wall of nemertines is much more highly organized than that of flatworms. In palaeonemertean the musculature of the body wall is two-layered or three-layered with the longitudinal layer located between two circular layers. All are littoral forms.

### Body Wall

The outer covering of the body consists of ciliated columnar epithelium which contains numerous gland cells. This layer is called the **epidermis**. The ciliated columnar cells are broader towards the outer side and narrower towards the inner side (Fig. 8.1). The gland cells are scattered irregularly between the narrower portions of the ciliated cells. Besides the unicellular gland many nemertines have clusters of gland cells known as **Packet glands** opening at the surface by a common duct. The packet glands may be included within the epidermis as in the plaeonemertines (*Tubulanus*, Fig. 8.1B) or they may sink into the subepidermal tissue as in the heteronemertines. Beneath the epidermis, in some nemertines, are found thick muscular layers, an external circular and internal longitudinal. In other nemertines (heteronemertines), on the other hand, a more complicated structure is met with. Below the epidermis is a thin **basement**

**membrane** (apparently a form of connective tissue) which presents a layer of homogeneous hyaline tissue, and this layer is followed by a thick layer of fibrous

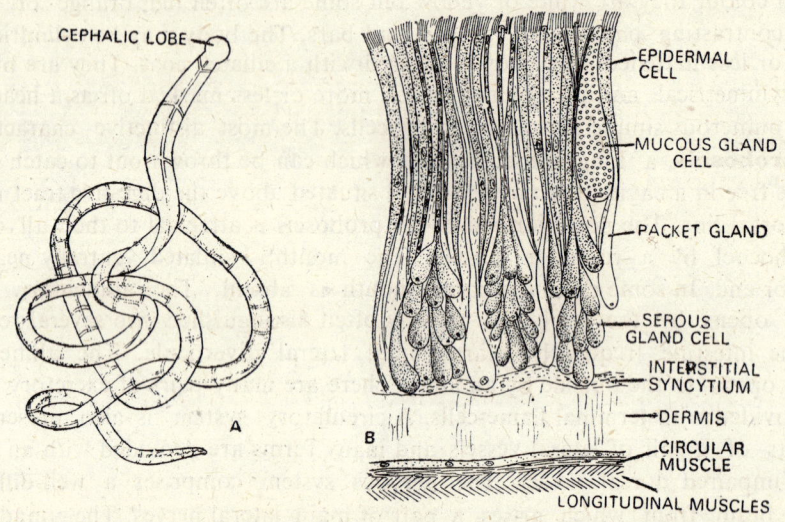

FIG. 8.1. A—*Tubulanus capistratus* (*after* Coe) with a distinct cephalic lobe; and B—Vertical section through the body wall of *Tubulanus*.

connective tissue containing some longitudinal muscle fibres also. Both these layers are called the **dermis** or **cutis**. Below the cutis lies the layer of body wall musculature consisting of smooth muscle fibres. The arrangement of its layers differs in different groups of nemertines. In some (palaeo-and hoplonemertines) there are usually two layers—an outer circular and an inner longitudinal layer—between which frequently occur two thin layers of diagonal fibres, but in many palaeonemertines there is an additional circular layer to the inside of the longi-

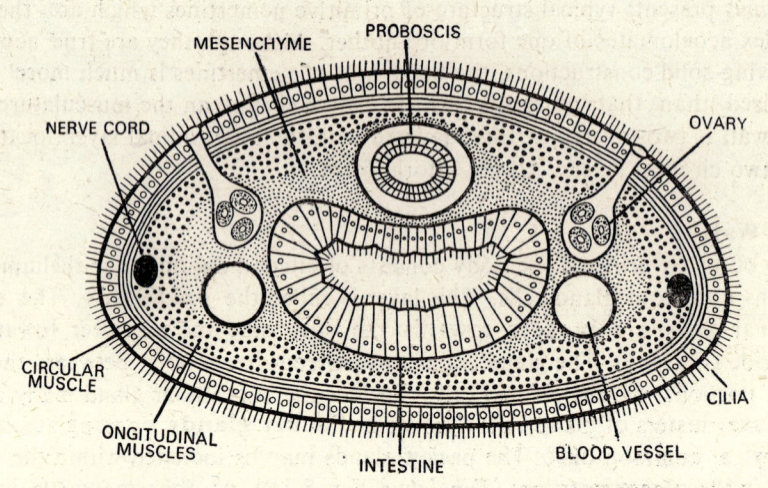

FIG. 8.2. Transverse section of the body of a palaeonemertinian.

tudinal layer. This may be a continuous layer or only restricted to the anterior region. In the heteronemertines the body wall musculature is three-layered consisting of an outer longitudinal, middle circular and an inner longitudinal layer.

*Phylum Nemertina* 309

Diagonal fibres may be present between the first two.

## Movement

The majority of the nemertines glide along the substratum in the manner of Turbellaria. It has been experimentally proved that the cilia are responsible for locomotion. The worm secretes a slimy track along which it moves rather slowly. The movements of cilia are directly controlled by the central nervous system through peripheral plexus. Some nemertines swim by the undulations of their flattened bodies. These movements are brought about by the musculature of the body which otherwise has nothing to do with locomotion, an exclusive function of cilia.

**Proboscis.** The **proboscis** (Fig. 8.3) is the most characteristic organ of the group. It is an elongated muscular tube lodged in the **proboscis sheath**. It is blind at its inner end. In length it may be two or more times the body length and is, therefore, more or less coiled within its sheath that may be shorter than

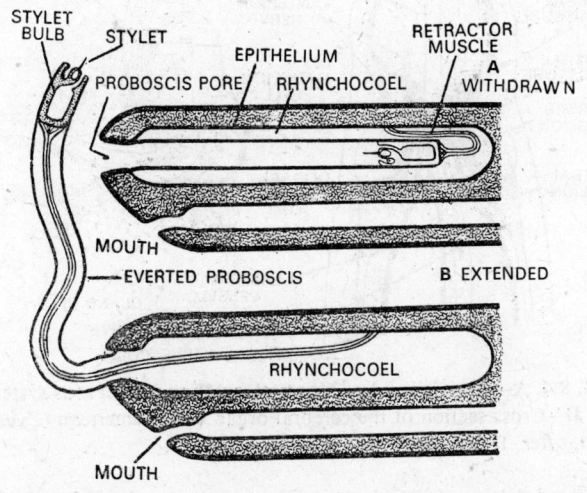

FIG. 8.3. Structure of the proboscis. A—Proboscis withdrawn; and B—Proboscis extended.

the body. The proboscis may be unarmed blind tube or may be armed with stylets at the tip. The stylets are formed of hardened glandular secretions. The proboscis lies free in a tubular cavity, the **rhynchocoel**. To the walls of the posterior end of the rhynchocoel is attached a **retractor muscle** (absent in some such as *Cerebratulus lacteus*). The wall of the rhynchocoel is termed **proboscis sheath.** The cavity is blind at both ends and contains a fluid in which float amoeboid cells. The opening through which the proboscis emerges out is known as the **proboscis pore**. Between the proboscis pore and the point where the proboscis begins there is a small tubular cavity, the **rhynchodaeum**. It is very short in some and absent in others (the Bdellomorpha).

Histologically the proboscis apparatus is similar to that of the body. The lining epithelium of the proboscis closely resembles the surface epidermis from which it is derived embryologically as an invagination. The muscle layers of the proboscis plus those of the proboscis sheath are similar to the body wall musculature. The proboscis is shot out with explosive force through muscular contractions exerting pressure in the fluid of the rhynchocoel. During this process the proboscis turns inside out (everts) and protrudes. Withdrawal of the proboscis is brought about by the retractor muscle. In cases where the retractor is absent the withdrawal is brought about by other muscles of the body. The proboscis wraps round the prey and entangles it with the aid of sticky mucous secretions. In those with the armed proboscis the stylet pierces the body of the prey and makes a wound into which is poured a poison secreted from glands in the proboscis.

**Nervous System**

The nervous system consists of a four-lobed brain, with two lobes, a dorsal and a ventral, on each side (Fig. 8.4). The two dorsal ganglia are connected by a **dorsal commissure**, passing above the **rhynchodaeum** and the ventral

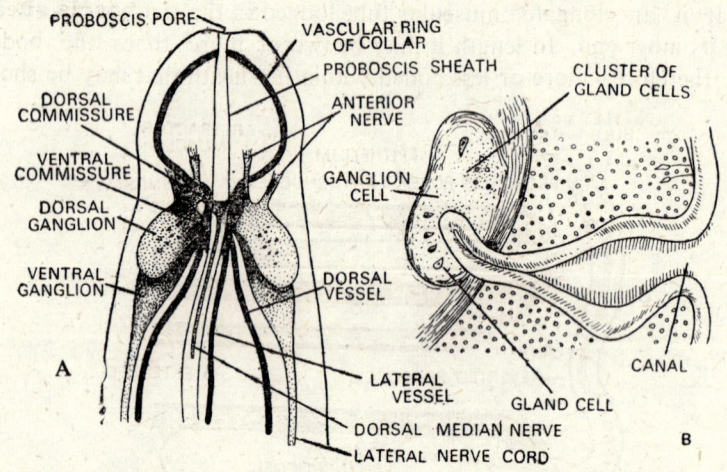

FIG. 8.4. A—Anterior end of a nemertean showing nervous system; and B—Cross section of the cerebral organ of a nemertean (*Zygeupolia, after* Thompson).

ganglia are connected by a ventral commissure beneath **rhynchodaeum**. This forms a nerve ring surrounding the proboscis apparatus. From the ventral ganglia arise two large lateral nerve cords that proceed posteriorly in a lateral or ventrolateral position. In addition to the lateral cords a number of minor nerves, some ganglionated, arises from the anterior faces of the cerebral ganglia. These are **cephalic nerves**. In some cases **middorsal nerve** arises from the dorsal commissure and runs along the length of the body. This nerve may present ganglionic swellings at regular intervals. The brain and the main cords are situated in the epidermis (*Carinina*) or in the dermis (*Tubulanus*) and most other palaeonemertines), in the body wall musculature (*Lineus*) or in the parenchyma internal to the body wall musculature (hoplonemertines).

**Sense organs**. Unlike the Platyhelminthes sense organs are found in the Nemertina. They comprise sensory nerve cells, sensory pits and eyes. The special organs of sense are mostly limited to the anterior part of the body but the sensory nerve cells are found scattered in the epidermis of anterior and posterior ends. They are slender cells each bearing a hair-like process on the outer side. They are probably tactile in function. Sensory pits are also found all over the body. Each such pit is a group of sensory nerve cells mixed with some support-

*Phylum Nemertina* 311

ing cells. In *Tubulanus* there is situated a special pit near the excretory pore, one on each side. These are known as the **lateral organs** and each consists of narrow epithelial cells below which lies nervous tissue. The whole structure is protrusible being provided by special muscles.

Most of the nemertines have eyes restricted to the anterior end of the body mostly to the region anterior to the brain. The number of eyes varies from two to six (*Prostoma*) while other have several up to 250 (*Amphiporus, Zygonemertes*). The eyes have the same structure as those of the platyhelminthes, consisting of pigment-cup ocelli of the inverted type and are nearly always found beneath the epidermis. In the hoplonemertine genus *Otolyphlonemertes* **statocysts (otolithic vesicles)** are also found. They are found on the ventral ganglia of the brain. There is a pair of the statocysts each consisting of a vesicle formed by non-ciliated epithelium the interior of which contains a spherical or dumbbell-shaped statolith.

The nemertines have a peculiar organ, the **cerebral organ**, closely associated with the dorsal ganglia of the brain. They are a pair of inverted epidermal canals whose inner end is surrounded by a mass of gland and nerve cells connected with the brain. They open directly on the body surface sometimes by way of grooves or pits. A water current is maintained in the canals and is intensified in the presence of food, it is, therefore, concluded to be of chemoreceptive nature. Becuase it is a glandular-nervous mass closely related to the blood, some authors have suggested an endocrine function of these organs.

**Digestive System**

The mouth is ventral and more or less, terminal, only in some it is posterior in position (*Procephalothrix*). The alimentary tract is a simple unmodified tube in most cases (*Tubulanus*) but in some **foregut** can be distinguished from the **midgut** or **intestine** from which arise numerous lateral diverticula (Fig. 8.5). The posterior portion of the foregut is also provided with a single or a pair of lateral **diverticula**. The intestine

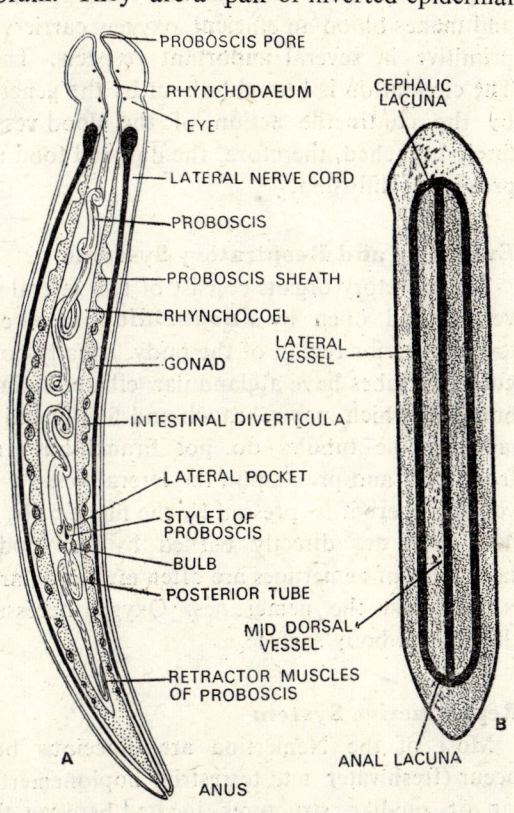

FIG. 8.5. A—Digestive organs of a freshwater namertean *Prostoma rubrum* (*after* Hyman); and B—Diagrammatic representation of circulatory system of a simple namertean.

opens to the exterior by the **anus**. The nemertines feed usually at night on living or dead animals mainly annelids, and also consume small molluscs, crustaceans and fish.

## Circulatory System

The Nemertines are the first animals that present a new organ system, the **circulatory system**, which takes over the circulatory function of the old gastrovascular cavity. This makes it possible for the intestine to become more efficient digestive apparatus, and leave the distribution of food, oxygen and other such substances to this new system. The circulatory system is of the closed type and consists of three longitudinal contractile muscular tubules, the **blood vessels** (Fig. 8.5B). These lie in the parenchyma one on each side of the intestine and one just above it. The lateral vessels communicate with each other both anteriorly and posteriorly by spaces lined by only delicate membranes known respectively as **cephalic lacuna** and **anal lacuna**. The longitudinal vessels give off other lateral branches that open into a system of lacunae in the tissues. The blood is usually colourless consisting of colourless fluid in which float oval nucleated blood corpuscles and some amoeboid lymphocytes here and there. In some cases the blood is red because the corpuscles contain haemoglobin, the pigment that combines readily with oxygen and makes blood an efficient oxygen carrier. The circulation of nemertines is primitive in several important respects. There is no pumping organ or heart. The circulation is brought about by the general movement of the body aided by the contractile action of the blood vessels. Also the blood vessels are not finely branched, therefore, the digested food must pass to the interior by slow process of diffusion.

## Excretory and Respiratory Systems

The excretory organs consist of two lateral tubes which lie close to the lateral vessels and open by a **nephridiopore** on each side. The tubules are restricted to the anterior region of the body, usually not extending back beyond the foregut. The tubes have a glandular ciliated lining and in most cases give off lateral branches which may rebranch and finally end in flame-cells. In most palaeonemertines the tubules do not branch but a number of flame-cells arise directly from them and press upon the lateral vessels. Sometimes the wall of the blood vessels, where it is pressed by the flame-cells, disappears with the result that the flame-cells are directly bathed by the blood but they do not open into it. The flame-cells of nemertines are often multicellular. No special respiratory mechanism is found in the nemertines. Oxygen necessary for respiration diffuses through the general body surface.

## Reproductive System

Most of the Nemertina are dioecious but some hermaphroditic forms also occur (freshwater and terrestrial hoplonemertines). The gonads are simple tubular or sac-like structures located between the diverticula of the intestine or in that region where diverticula are wanting. Each gonad opens on the dorsolateral surface of the body independently. There is no distinction between male and female except in *Nectonemertes mirabilis* in which the males have a pair of cirri not found in the females. Some forms, on maturity, display different coloration perhaps due to the colour of ripe-sex cells seen through the body wall.

The male and female may spawn without contact or the male may crawl over the female discharging the sperms. In some, however, two or more worms enclose in a common sheath of mucous within which the sex-cells are discharged. In some the sperms pass inside the ovaries and fertilize the eggs and the development may take place inside the ovaries (*Prostoma, Geonemertes, Lineus*). In others the eggs are fertilized when discharged to the exterior. The development is usually direct but in most of the heteronemertines the young are hatched as helmet-shaped, free swimming larvae, known as **pilidium** (Fig. 8.6).

The pilidium was formerly described as a species of a supposed independent genus. It has a ventral mouth (the blastopore forms it) but the gut is blind, and at the end opposite to the mouth is the apical sense organ topped by a tuff of long stiff flagella. A broad lobe grows out on each side of the mouth, on the edges of this extends the circum-oral ciliated band. This larva grows into the adult worm after an elaborate metamorphosis.

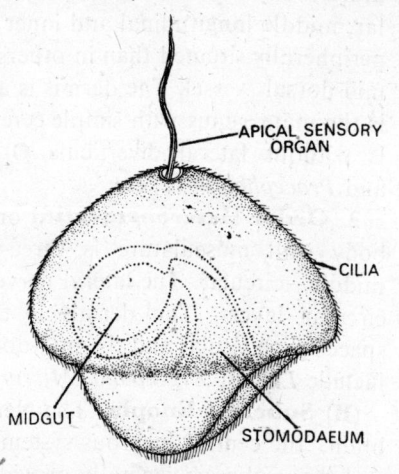

FIG. 8.6. Pilidium larva of Nemertenea (*after* Coe).

Asexual reproduction is unknown in the group but the power of regeneration is great. In some cases the body breaks up into pieces when touched and these pieces regnerate the lost parts.

**Phylogeny**

Phylogenetically, nemerteans appear to be an offshoot from free-living flatworms. Hymen (1951) believes that the proboscis of nemerteans is evidence of a link with certain rhabdocoel flatworms that possess a short, glandular anterior proboscis. Moreover, a solid mass of mesenchyme filling the space between the body wall and the intestine, strongly suggests a flatworm origin. Both flatworms and nemerteans possess protonephridia, a ciliated epidermis and similar nervous system and sense organs.

Despite their ancestry the nemerteans are more highly organized than flatworms and they display a number of changes that anticipate conditions found in high vertebrate groups. However they do not seem to be on direct line of protostome evolution leading to coelomates.

## CLASSIFICATION OF NEMERTINA

The Nemertina is divided into two subclasses each of which is divided into two orders. The two main divisions of the phylum have not been called classes because of the great similarity of structure existing throughout the group.

(A) **Subclass Anopla**. The Nemertina in which the mouth is situated posterior to the brain, the proboscis is unarmed and the nervous system is situated beneath the epidermis among the muscle layers of the body wall. This subclass is divided into two orders.

1. **Order Palaeonemertini**. The Anopla in which the body wall musculature is two-layered or three-layered. In the latter case the outer layer is circular, middle longitudinal and inner circular. The nervous system is relatively more peripherally situated than in others. The circulatory system is usually without a mid-dorsal vessel. The dermis is a thin gelatinous layer. *Tubulanus* (=*Carinella*) is the main genus with simple cerebral organs and without eyes, the intestine is without lateral diverticula. Other genera *Procarinina, Carinina, Cephalothrix* and *Procephalothrix*, etc.

2. **Order Heteronemertini** or **Heteronemertea**. The Anopla in which the body wall musculature is three-layered, the outer and inner—longitudinal, middle—circular. The lateral nerve cords lie between the outer longitudinal and circular layers. The dermis is thick and fibrous. The intestine has regularly spaced diverticula, and the mid-dorsal vessel is always present. Important genera include *Lineus, Zygeupolia, Micrura, Cerebratulus, Baseodiscus,* etc.

**(B) Subclass Enopla**. The Nemertina in which the mouth is anterior to the brain, the central nervous system internal to the muscle-layers of the body wall, and the proboscis generally provided with stylets. This subclass is again divided into two orders.

1. **Order Hoplonemertini** or **Hoplonemertea**.[1] The Enopla in which the proboscis is armed with one or more stylets. The intestine has a caecum besides lateral diverticula. The mid-dorsal blood vessel is present and is connected with the lateral vessels by transverse vessels. Important genera include: *Emplectonoma. Nemertopsis, Paranemertes, Zygonemertes, Tetrastemma, Prostoma, Gononemertes, Pelagonemertes,* etc.

2. **Order Bdellomorpha** or **Bdellonemertini**. The Enopla with an unarmed proboscis, the intestine is sinuous, without diverticula. This order includes the single genus *Malacobdella*, with three species, which live as ectocommensals in the mouth cavity of marine clams and one is found in the pulmonary sac of freshwater snail. For attachment the body is provided with an adhesive disc on the posterior end of the body. It feeds on plankton brought into the mantle cavity of the host, hence is commensal.

---

[1]Some authors prefer to call this order Metanemertini which is supposed to include the Bdellomorpha also.

# 9 PHYLUM ACANTHOCEPHALA

The flatworms and nemerteans, among the bilateral animals, are acoelomate i.e., the space between the digestive tract and the muscle layers of the body wall is filled with mesenchyme. All other bilateral animals have a fluid-filled space called a **body cavity** between the body wall and the internal organs. This cavity may be a **coelom** or **pseudocoelom**. Coelom is the main body cavity situated in the mesoderm and lined by **mesothelium**. Pseudocoelom is the **blastocoel** or the **primary body cavity** which persists in the adults as a spacious cavity. It is not lined by mesothelium. The vast majority of animals possesses a coelom as a body cavity, but a smaller number possesses a pseudocoel as a body cavity and are called pseudocoelomates or better **Pseudocoelomate Bilateria**. The pseudocoelomate phyla include the Acanthocephala, the Nemathelminthes (Aschelminthes and Entoprocta).

The **Acanthocephala** (Gr. *acantho*, spiny+*cephale*, head) are parasites of peculiar structure and function that live as adults in the intestine of vertebrates and as larvae in arthropods. They are slender vermiform parasites used to be considered as a class of nematodes. But it is difficult to reconcile their unique body plan with that of roundworms, they are now separated as an independent phylum. The most characteristic feature of the group is the organ of attachment consisting of a retractile **proboscis** forming the anterior end of the body (hence referred to as **"spiny-headed"**). Behind the proboscis is a short **neck** region and then the **body** proper, which is roughly cylindrical. By means of the spiny proboscis the worm clings to the intestinal lining of its host, absorbing nourishment through the delicate cuticle. There is no trace of a digestive tract. There are no circulatory or respiratory structures. Two branched nephridia with cilia are connected to a common posterior excretory duct. A nerve ganglion in the proboscis sends nerves to the proboscis and to the posterior part of the body.

The sexes are separate. The ova are fertilized in the body cavity and are shed in the faeces of the host generally advanced in development, if the host is an aquatic vertebrate. The eggs are probably eaten by a crustacean or an aquatic insect and in these animals the larvae develop. They get back into a vertebrate when the intermediate host is eaten by the final vertebrate host. The life-history for land vertebrates is similar, but it involves a land insect. There are about 300 known species found in various vertebrates, mainly fish, birds and mammals and occur throughout the world in the sea, on land or in freshwater.

### TYPE ECHINORHYNCHUS

## External Characters

The genus *Echinorhynchus* is a common parasite in the intestine of mammals, birds, reptiles, amphibians and fishes. The largest species *E. (Gigantorhynchus) gigas* (*Macracanthorhynchus hirudinaceus*) is the best known acanthocephalan, found the world over in pigs less often in dogs and other mammals. Once it has been reported in human being. It varies from 50 to 65 cm in length. Another species, *Gigantorhynchus echinodiscus* from South American ant-eaters, is 45 cm in length. Most species, however, are small ranging from 1.5 mm to 1 cm the majority falling under 25 mm.

The body is cylindrical (Fig. 9.1), though other species may be short plump,

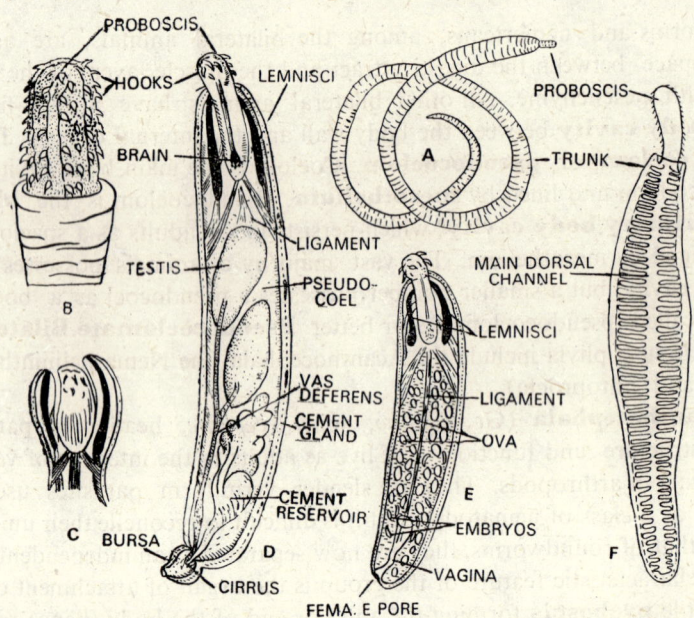

FIG. 9.1. Structure of spiny-headed worm. A—Entire; B—Proboscis enlarged; C—Proboscis withdrawn; D—Reproductive organs of male (*from* lynch); E—Reproductive organs of female (*from* Lynch); and F—Lacunar system (*After* Meyer).

fusiform, clavate or laterally flattened. Many forms are slightly curved and in one (*Hamanniella microcephala*) the body is spirally coiled. In all these forms it is difficult to distinguish dorsal from ventral surface from external criteria. When the body is curved the concave surface is ventral and when the proboscis is covered with hooks, of unequal sizes, the larger hooks are ventrally located, but in most cases dorsoventrality can be determined with the help of internal structure. For example, the cerebral ganglion is located in contact with what is considered to be the **ventral wall** of the proboscis receptacle. The acanthocephalan worms do not have a colour of their own. Frequently they appear red, orange, yellow or brown depending upon the absorption of food from the host.

In many acanthocephalans body cuticle is ringed or constricted at regular intervals and presents a more or less segmented appearance. In the genus under discussion the cuticle is only slightly ringed. The body is divided into a short slender forebody or **presoma** and the much longer stouter **trunk**. The presoma consists of the anterior proboscis that bears hooks and a short neck region without hooks. The trunk is cylindrical. In several genera the trunk is more or less noticeably differentiated into a broader **fore-trunk** and a slender **hind-trunk** (*Bolbosoma*). There is no trace of mouth, anus or excretory pore. The gonopore occurs at or near the posterior extremity.

The **body wall** is covered with a distinct cuticle of homogeneous structure. Beneath the cuticle lies the **epidermis** or **hypodermis**. The epidermis has very remarkable structure. It is a thick layer of fibrous syncytial construction (Fig. 9.2) comprising three fibrous strata: (*i*) **outer strata** of parallel radial fibres, only

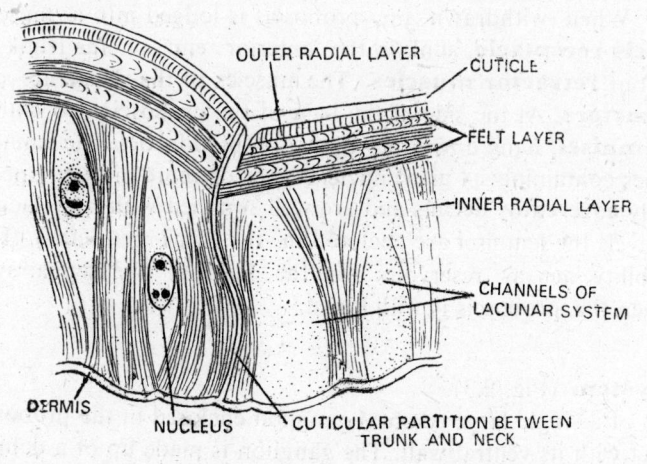

**FIG. 9.2.** Longitudinal section through the epidermis of a spiny-headed worm.

slightly thicker than the cuticle; (*ii*) **middle strata**, somewhat thicker feltwork of layers of fibres running in different directions; and (*iii*) an **inner strata** of radial fibres. The inner layer is the thickest and is regarded as the epidermis proper by many authors, the outer layers being considered to be parts of the cuticle. The nuclei of the epidermis are situated in this layer. They are quite large oval or ellipitical in shape. In the acanthocephalans the number of nuclei is approximately constant for each species. They are few in number and more or less fixed in position.

In the inner radial layer of epidermis is located a set of channels without definite walls forming what is known as the **lacunar system**. The lacunar system has more or less a definite pattern. It consists of two longitudinal vessels with regularly spaced transverse connections. The main channels, in this case, are medially located, i.e., they are dorsal and ventral (Fig. 9.1 D). The lacunar system is confined mainly to the epidermis. It does not communicate with the exterior or with any other body structure. It is filled with a granular fluid probably absorbed from the host. It is supposed to serve as a food-distributing system. The nutritive fluid in the channels moves with the movements of the body.

The epidermis is followed by a thin layer, the **dermis** (also called **basement** or **binding layer**). The material forming this layer also permeates the underlying musculature which is relatively thin and consists of two layers: outer **circular** and inner **longitudinal**. These muscles are also syncytial, forming a fibrous network and probably containing a constant number of nuclei. Each muscle fibre consists of a cytoplasmic and a fibrillar portion, the fibrils may run along one side of the cytoplasmic part or may encircle the latter. The pseudocoel is not lined by any distinct membrane.

**Proboscis apparatus**. The proboscis is cylindrical and is covered with many rows of recurved hooks of similar shape varying in size and arrangement (Fig. 9.1 B). The larger hooks have got their roots sunk into the proboscis wall. Some smaller ones are rootless and have been called spines by some authors. The exact chemical nature of hooks and spines is not known. They are probably of the same nature as the dermis from which they seem to arise. They are covered with cuticle. When withdrawn the proboscis is lodged into a muscular sheath, the **proboscis receptacle**, sunk in the anterior end of the trunk, and is provided with four **retractor muscles**. The muscles of the sheath are circular and act as **protractors**. At the side of the base of the proboscis two club-shaped organs, the **lemnisci**, hang down into the trunk pseudocoel. The lemnisci are long slender bodies containing six giant nuclei (fewer or more in other species). They are externally covered by dermis and are well supplied with the lacunar system. At certain levels the lemnisci are enclosed in the neck retractors. The lemnisci in all probability act as reservoirs for the fluid of the lacunar system of the presoma when the proboscis is withdrawn.

## Nervous System (Fig. 9.3)

It consists of a single large cerebral ganglion enclosed in the proboscis receptacle in contact with its ventral wall. The ganglion is made up of a definite number of ganglion cells (86 in the present case) surrounding a central fibrous mass. From this arise two single and three pairs of nerves. The single nerves are an **anterior median** and a **central anterior** nerve to the musculature and sensory papillae of the proboscis. The paired nerves are represented by a pair of **lateral anterior** nerves to the lateral protrusors; a pair of **lateral median** nerves to the receptacle wall, and a pair of **main lateral posterior** nerves innervating the posterior end of the animal. In the male an additional pair of **genital ganglia** is present near the base of the penis and connected with each other by a **ring commissure** (Fig. 9.3 C). These ganglia are connected with the main lateral posterior nerves. Branches from these ganglia supply the **reproductive** organs

# Phylum Acanthocephala

including the bursa. In the female branches from the main lateral posterior nerves supply the reproductive organs.

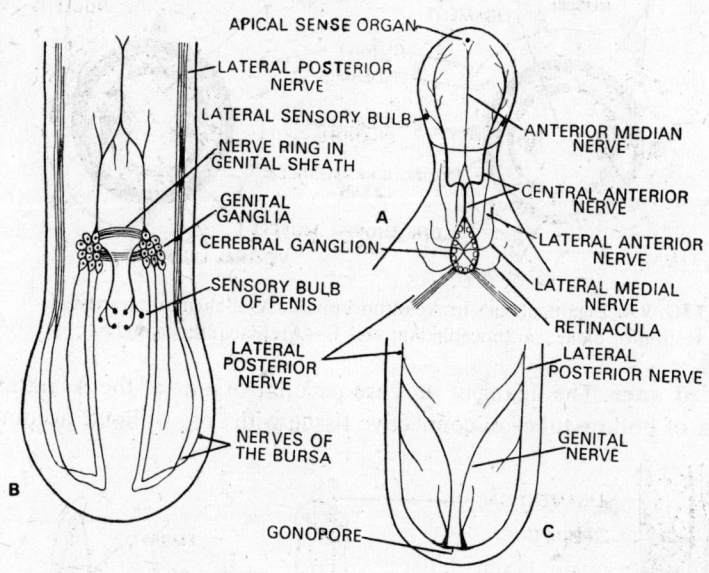

FIG. 9.3. Nervous system of *E. gigas* (*after* Brands). A—Nervous system on the anterior side; B—On the posterior side of a female; and; C—The same of a male.

**Sense organs**. Organs of special sense are not developed in correlation to endoparasitic life. The only structures that can be described under this heading include three structures in the proboscis and several in the male bursa and penis. All are tactile sense organs. The **proboscis sense organs** comprise a **terminal sense organ** in the centre of the tip and, in some genera, a pair of **lateral proboscis** sense organs, one on each side in the neck. The terminal sense organ consists of a small pit beneath which there is a fusiform nerve ending of a nerve fibre that makes a coil just below its termination. In the lateral proboscis sense organs several nerve fibres are coiled below the termination. In the male branches from the genital ganglia terminate in spherical tactile bulbs of which there are seven or eight around the rim of the bursa and several in the penis.

## Excretory Organs

The excretory organs consists of a pair of small bodies (**protonephridia**) situated at the posterior end near the genital aperture. Each protonephridium consists (Fig. 9.4) of a branching mass of flame bulbs attached to a common stem. The two stems unite to form a wide median dorsal channel which opens behind in the female into the unpaired portion of the oviduct and in the male in the ejaculatory duct or common sperm duct. Thus, the terminal canals of the reproductive system are, in fact, **urinogenital canals**. The female bulbs in each protonephridium range from 25 to 700 and are simple without nuclei (hence not cells); three nuclei occur in the wall of the chamber. The flame consists of a linear row of cilia. According to some at the end of each flame-bulb is a number of fine perforations through which its canal communicates with the pseudocoel.

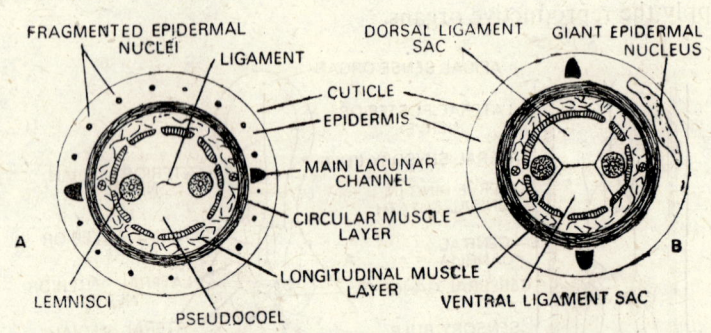

FIG. 9.4. Ligament sacs in Acanthocephala. A—Schematic transverse section of palaeacanthocephalan; and B—Archiacanthocephalan.

**Ligament sacs.** The ligament sacs are peculiar organs of the Acanthocephala and consist of hollow tubes of connective tissue with or without accompanying

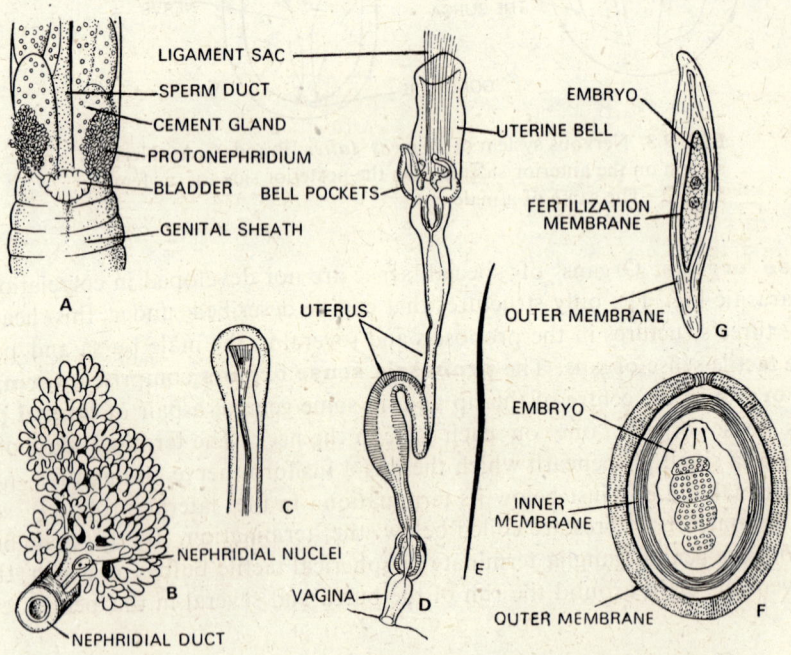

FIG. 9.5. Excretory and reproductive systems of *Acanthocephala*. A—Urinogenital organs of *Hamanniella* showing the position of the excretory organs (*after* Kilian); B—A protonephridium enlarged (*after* Kilian); C—A flame cell (*after* Meyer); D—Female reproductive system of *Bothosoma* (*after* Yamaguti); E—Sperm of an acanthocephalan; F—Shelled embryo of terrestrial acanthocephalan; and G—Shelled embryo of aquatic acanthocephalan.

muscle fibres. They run through the whole length of the body and enclose the reproductive organs. They are attached to the posterior end of the proboscis receptacle or the adjacent body wall anteriorly, and posteriorly they terminate on some part of the reproductive system. In the females there are two ligament sacs, a dorsal and a ventral, whose medial walls contact each other (Fig. 9.4).

Anteriorly the two sacs communicate with each other by a common opening. In the male there is only one sac, the ventral sac is lacking. The dorsal sac embraces the testes and the cement glands and posteriorly becomes continuous with the genital sheath.

Haffner (1942) has reported a nucleated strand, the **ligament strand**, between the two ligament sacs in the female and along the ventral face of the single ligament sac in the male. In both the gonads are attached to this strand, which is regarded as the endoderm or midgut.

**Pseudocoel**. The pseudocoel is the space devoid of any lining membrane between the body wall and the ligaments. It is small in forms with two ligament sacs but becomes quite large in those with one sac only. It also extends into the presoma between the muscle bands. The pseudocoel is filled with a clear fluid.

## Reproductive System

The sexes are separate and the female is larger than the male, The male has a pair of **testes** enclosed in the ligament sac lying one behind the other and attached to the ligament strand (Fig. 9.1A). From each testis arises a sperm duct (**vas deferens**) and runs posteriorly inside the enlargements, the **seminal vesicles**, for the storage of sperms. A cluster of unicellular gland cells, the **cement glands** (usually six or eight in number) open into the sperm duct shortly behind the more posterior testis. The common duct formed by the union of two sperm ducts, into which also open the cement glands, is known as the **ejaculatory duct**. The sperm ducts, the cement ducts and the protonephridial canals are all enclosed in a muscular tube, the **genital sheath**, continuous with the ligament sac. The ejaculatory duct opens into the **bursa** or bell-like copulatory organ and has at its opening a short conical protrusion, the **penis**. The bursa is a hemispherical cavity composed of inturned body wall and is eversible to the exterior. It grasps the rear end of the female during copulation. The sperms are long filament-like without definite heads (Fig. 9.4 B).

The female reproductive system is remarkably different from others. No persistent ovary occurs. On the contrary the original ovary breaks up into fragments called **ovarian balls** (Fig. 9.1 E) that float free in the dorsal ligament sac, which soon ruptures leaving the balls swimming freely in the pseudocoel where they are fertilized. The ducts are also very peculiar. Connected with the end of the ligament sac is a muscular **uterine bell**, a funnel-shaped organ that by peristaltic contractions, engulfs the developing eggs and conducts outwards (Fig. 9.4 A). The bell has a single (or a pair of) posterior ventral openings through which the immature eggs, which are spherical, pass back into the body cavity. The mature eggs, which are spindle-shaped and covered with a chitinous investment, make their way through a **uterine tube**, a double passage, to the **uterus**. The uterus is a long muscular tube and is followed by a short non-muscular **vagina** leading to the exterior. The nephridia lie alongside the bell and the common canal formed by their union opens into the beginning of the uterine tube. The ovarian balls comprise a central syncytium from which ovogonia separate and pass to the periphery where they develop further.

**Copulation and development**. The everted male bursa grasps the posterior end of the female, the penis is inserted into the vagina and the sperms are discharged into the uterus. Finally the secretion of the cement gland is discharged.

This plugs the pore preventing the escape of the sperms. The mature ova are elliptical surrounded by a membrane. Fertilization takes place in the body cavity and after which a membrane arises inside the original egg membrane. In the meantime the eggs have escaped into pseudocoel (or in some cases inside the dorsal ligament sac) where they undergo further development up to the larval stage provided with rostellum armed with hooks. Meanwhile a third membrane, the **shell**, appears between the two membranes (Fig. 9.4 C, D). This is hard in forms that have terrestrial intermediate hosts, softer in those with aquatic intermediate hosts. These are known as **ovic** or **acanthor larvae** and have three pairs of larval hooks, like those of the cestode hexacanth, on their anterior ends which ultimately becomes the anterior end of the worm. They are picked up by the uterine bell and conducted to the exterior. They can develop further only if ingested by the proper intermediate host which in this case is the grub of the June beetle or of the other scarabeaid beetles.

The acanthors thus pass out with the hosts' faeces. They are swallowed by the larvae of June beetle. On reaching the intestine of the intermediate host the shell splits along a line that appears earlier and the acanthor is released. It perforates the wall of the gut and either fixes to it or makes its way in the body cavity. While passing through the gut wall the acanthor loses its shape, becomes nearly spherical and the larval structures begin to degenerate. But it elongates in the haemocoel of the host. At this stage some authors call it **acanthella**. The endoderm that till now was a solid mass splits into an outer layer in contact with the ectoderm and a solid central axis, the inner nuclear **mass**, which gives rise to the reproductive organs, ligament sac musculature, proboscis and ganglion. The cavity formed by the splitting of the endoderm is the body-cavity or the pseudocoel. The ectoderm gives rise to the protoplasmic layer of the body wall, to the whole system of vessels and to the leminisci. The larval cuticle is thrown off and a new one is formed. This development in the intermediate host may take from six weeks to three months depending upon the species. The larva attains adult proportions and sexual maturity only after entering the definitive vertebrate host.

**Parasitism**

Undoubtedly the acanthocephalan worms are the most injurious of helminth parasites. The proboscis hooks are buried in the intestinal wall and thus they damage the tissue, especially in the mammals. At the damaged places, special nodules are formed and they persist even after the elimination of parasites. This causes mechanical injuries but apart from such injury these injured places are good for infection. Heavy infection is very harmful to the host. Schwartz (1972) found the pseudocoel fluid of *E. gigas* to contain a substance haemolytic to the red blood corpuscles of cattle and pig. As in other parasites the reproductive capacity of the acanthocephalan is very high. Kates (1944), in a gravid female of *Macr. hirudinaceus*, reported 10,000,000 shelled acanthors at one time without any impairment to their egg producing capacity. In this species the emission of shelled acanthor begins about 60-80 days after infection and continues at an average rate of 260,000 per day for a period of ten months.

**Affinities**. The relationship of the Acanthocephala is difficult to establish. The worms were noticed about the beginning of the 18th century but were not clearly distinguished from other intestinal worms until 1771 when Koelreuther proposed the name *Acanthocephalus* for one from a fish. In 1776 Zoega and O.F. Müller described one *Echinorhynchus* from a fish. Zeder (1803) called them hooked-worm (Haken-würmer) and Rudolphi (1809) changed this into Acanthocephala.

Cuvier included them with flatworms. Vogt was the first who distinguished flatworms from roundworms upon which Gegenbauer gave the name Nemathelminthes to the roundworms. The position of Acanthocephala, however, has remained uncertain. Some authors relate them with flatworms and some with roundworms. The nematode similarities of the acanthocephalan worms include: (*a*) the presence of pseudocoelom in both, though it develops differently in the acanthocephalan worms; (*b*) division of the body into presoma and trunk like that of the gordiacean larva; (*c*) armed proboscis as found in the echinoderids and gordicacean larva; (*d*) superficial segmentation; (*e*) cuticle chemically identical to that of the nematode worm; (*f*) pseudocoel is divided by partitions and tissues; (*g*) the presence of syncytial nucleated epidermis; (*h*) nervous system and flame-bulbs are similar; (*i*) close relationship between protonephridia and genital ducts as occurs in rotifers; and (*j*) the nuclear constancy. All these points link the acanthocephalan worms with the Nemathelminthes. They, however, differ from the Nematoda in (*i*) the presence of a proboscis; (*ii*) absence of a digestive tract; (*iii*) presence of circular muscles; (*iv*) presence of ciliated excretory organs; and (*v*) peculiarities of the reproductive organs.

The acanthocephalans are linked with flatworms because of the following points: (*i*) an armed proboscis which can be invaginated and withdrawn occurs in cestodes and the proboscides of Trypanorhyncha and Acanthocephala and is similar so far as the shape and arrangement of hooks is concerned; (*ii*) syncytial nucleated epithelium as occurs in the turbellarians; (*iii*) similar arrangement of the body wall musculature comprising circular and longitudinal muscles; (*iv*) the reproductive system of the acanthocephalan worms is more suggestive of flatworms than of Nemathelminthes; and (*v*) the embryology is like that of the cestodes, the main difference between the cestode hexacanth and the acanthor seems to be that the latter retains its epidermis while in the former it becomes one of the larval envelopes.

From the above no decisive conclusions can be drawn. Chitwood (1940) and Van Cleave (1941) favour platyhelminth affinity. But since they show no closer ties to other animals it is probably best to consider them as an independent phylum.

The Acanthocephala are divided into three orders: (*a*) Archiacanthocephala: proboscis hooks in concentric circles—*Gigomtorhynchus*; (*b*) Palaeacanthocephala: hooks not in circles, main lacunar channels lateral—*Echinohynchus*, and (*c*) Eoacanthocephala: nephridia wanting, host fishes—*Neoechinorhynchus*.

# 10 PHYLUM NEMATHELMINTHES (ASCHELMINTHES)

The nature of the body cavity studied in the phylum Acanthocephala was pseudocoelomate. Pseudocoel occupies a space between the mesoderm of the body wall and the endoderm of the gut. In this type of a body cavity there are no mesenteries suspending the internal organs and no muscle layer surrounding the gut. This condition has functional implication that in no pseudocoelomate animal does muscular peristalsis move food through the alimentary canal. The phylum including animals with such a body cavity has now been named Aschelminthes (*ascos*=space), but the name Namathelminthes is preferable because only one "major successful" group of animals built on pseudocoelomate plan is that of nematode worms, and because the name is hallowed by long usage.

The phylum Nemathelminthes (=Aschelminthes) is divided into five classes:

## PHYLUM NEMATHELMINTHES

**Classes:**
1. ROTIFERA
2. GASTROTRICHA
3. KINORHYNCHA OR ECHINODERA
4. NEMATODA
5. NEMATOMORPHA OR GORDIACEA

The first of these five classes is Rotifera having about 1,500 species living mainly in freshwater, although there are a few marine species also. They are mostly microscopic animals, having almost the same dimensions as the larger ciliate protozoans, and only a few species reach length in excess of 1 mm. The Gastrotricha is another minor group of about 150 species of minute animals living in fresh waters and sea. The third group, called Kinorhyncha, (Echinodera or Echinorhyncha) includes minute marine animals without cilia and with a regularly segmented covering of cuticle.

The Nematoda, as mentioned above, is the most successful class of the pseudocoelomate animals. The Nematomorpha is a group of the free-living adults the larvae of which are parasites of arthropods. The Entoprocta is another group of pseudocoelomate animals including about 60 species of small sedentary animals which superficially resemble hydroid coelenterates that have developed some mesodermal structures. They are probably more closely allied to the Nematomorpha.

These minor groups are often classified in various ways. Some modern text books use the superphylum Aschelminthes to include most of them (except the Acanthocephala) along with the nematodes. Old texts incorporated them in a superphylum Gephyrea which also included some obvious coelomate phyla such as the Sipunculoidea. According to others such composite groups would be far looser in their relationships than the major phyla, and therefore it is best to separate them as independent phyla, the minor pseudocoelomate phyla. In the present book they have been described as classes of the phylum Nemathelminthes (Aschelminthes). The relationship between the various classes of this group is not so close as that of the classes of other phyla but because the evidences of relationship are concrete and specific they cannot be ignored.

## CLASS ROTIFERA

The Rotifera or wheel animacules are minute aquatic animals ranging between 0.4 mm and 2 mm in size., .5 mm being the average size of the majority. The elongate body is divisible into an anterior head region bearing ciliary apparatus, the **corona,** often an elongate trunk and the posterior slender region called **tail** or **foot.** Apart from this simple type (Fig. 10.1) various other types of body forms are found. Some forms are sac-like (*Asplanchna*), some are broad-types (*Brachionus*) and some are slender (*Rotaria*). In one type (*Seison* Fig. 10.10) both the foot and the neck are elongated. Body is cylindrical in transverse sections, but lateral and ventral surfaces may be flattened or concave. Most of the animals are bilaterally symmetrical, but in some cases asymmetries are also found in external features, e.g., in the presence of toes. In some forms the body is ventrally turned and in some it is twisted.

The body is covered with cuticle that is yellowish in colour. It is often ringed simulating segmentation. In some cases the cuticle of the trunk region is specially thickened forming a hard encasement, the **lorica**, made up of one to several plates. The anterior part of the body is withdrawable within the lorica.

The anterior end of the body is not clearly delimited known as head, but it is called one only for the sake of convenience. It is typically broad and truncate, or slightly convex having an unciliated central region, the **apical field**, encircled by a ciliated zone, the **corona**. Apical field is commonly provided with several projections on some of which open certain ducts of glands that lie below.

Some projections are sensory and bear stiff bristles and hairs, etc. The corona

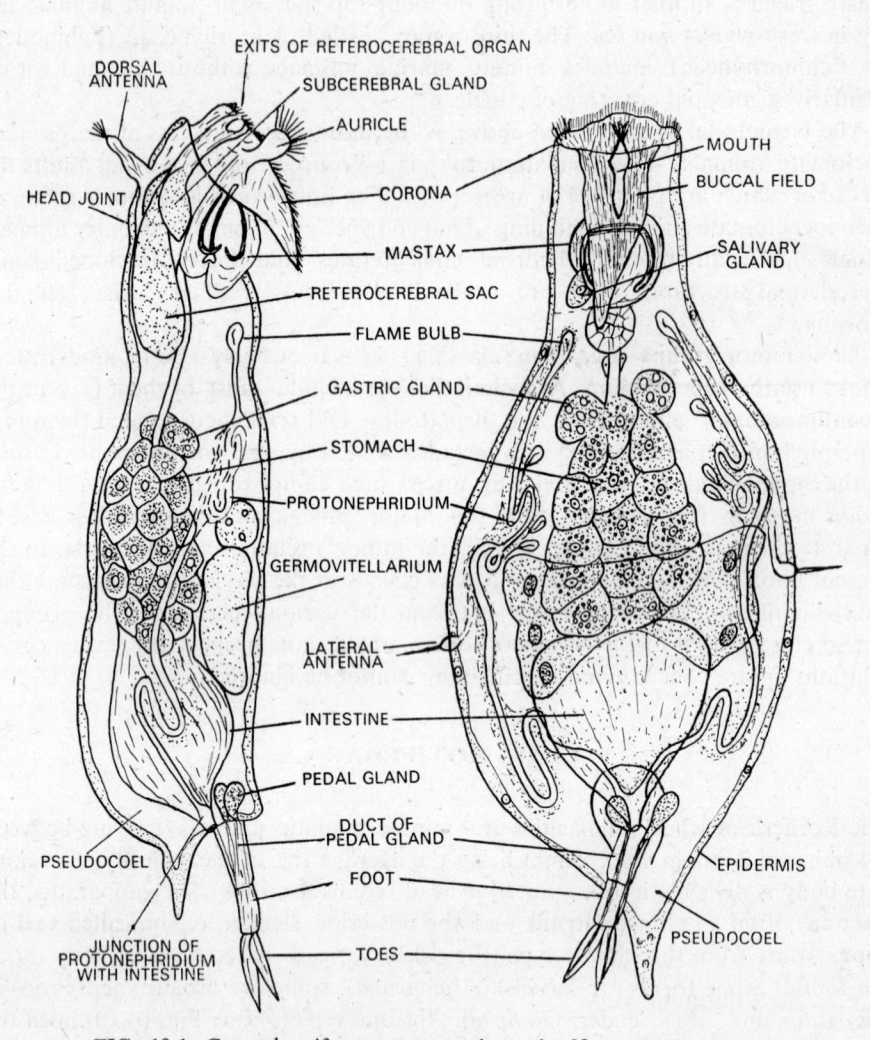

**FIG. 10.1.** General rotifer structure as shown by *Notommata copeus*.

may be entire or lobulated. In sessile forms (Flosculariacea and Collothecacea) the corona is drawn out into a number of lobes forming a funnel-shaped anterior end (Fig. 10.12). In one variety of rotifers (bdelloids) the corona is particularly subdivided into two retractile **trochal discs** (Fig. 10.13). The head may also bear a pair of prominent ciliated lateral retractile projections termed **auricles** (Fig. 10.2 C). In many forms there is present a mid-dorsal projection in the head region called the **rostrum** (Fig. 10.2 A) The tip of the rostrum is provided with cilia and sensory bristles that are protected by thin plates termed as the **rostral lamellae**. In bdelloids the rostrum is used in locomotion by looping method, as the anterior end.

Most of the rotifers are provided with eyes which appear as red flecks and occur singly or in pairs in the brain (Fig. 10.7). In others the eyes may be

# Phylum Nemathelminthes (Aschelminthes)

present in or near the corona, in the apical field or on the rostrum. The eyes may be elevated on papillae or sunk in depressions. The mouth is located in the corona in the mid-ventral line of the head, and often a protrusion of apical field serves as ventral lip. In the funnel-shaped stalked forms the mouth is situated at the bottom of the funnel-shaped anterior end.

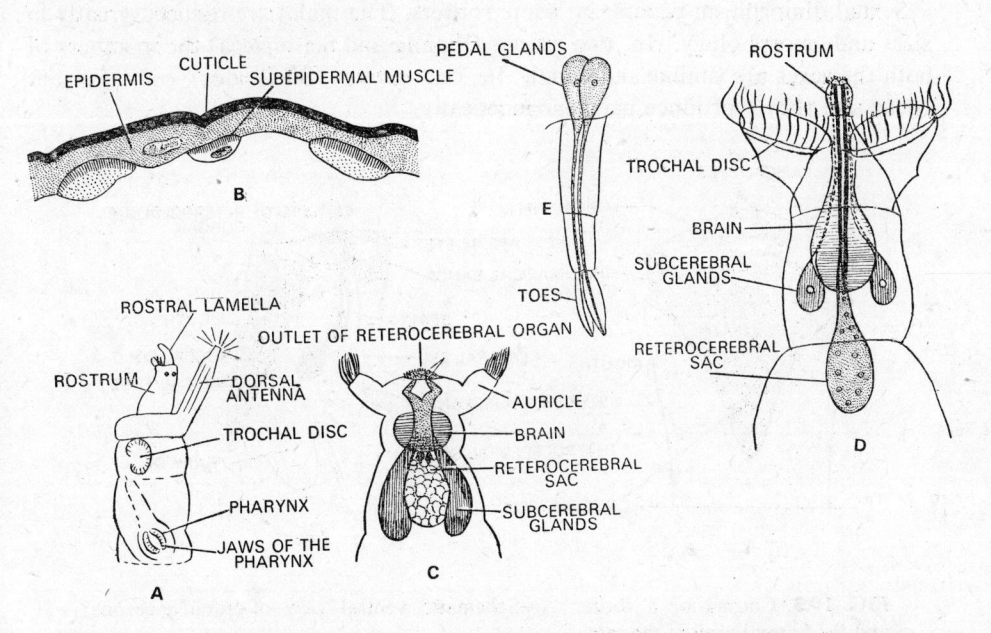

**FIG. 10.2.** Rotifer body wall and associated glands. A—Anterior end of a bdelloid rotifer; B—Cross section of the body wall; C—Reterocerebral organ of *Notommata*; D—Reterocerebral organ of a bdelloid; and E—Pedal glands of *Notommata*.

The trunk may be cylindrical or variously flattened and is often enclosed in a **lorica** which may be ornamented or spiny. The spines in some cases are very long, movable (Fig. 10.12) and help in producing skipping movements. In *Pedalia* such movable projections are very well-developed they are extensions of the body tipped with bristles (Fig. 10.13). The trunk is provided with certain other projections, viz., **lateral antennae** or **palps** and a **dorsal antenna**. The lateral antennae are situated on either side of the trunk (Fig. 10.1); slightly toward the anterior side in the sessile types and are wanting in many. The dorsal antenna (Fig. 10.1 A) is a finger-like projection usually single, situated in the mid-dorsal line of the anterior end. The **anus** is situated in the mid-dorsal line at or near the boundary of the trunk and foot. Posteriorly the body tapers gradually into the foot in many forms but in some loricate forms (those that have lorica) the foot is sharply set off from the stout trunk. The cuticle of the tail is ringed into a few to many joints. The foot serves as a clinging structure for creeping types and acts as a rudder for swimming types. In sessile forms the foot takes the shape of a long **stalk**. It usually bears one to four movable projections known as the **toes** used in holding the substratum while creeping. The toes may be short or conical (Fig. 10.1) or slender and spine-like (Fig. 10.13). At the tip of the toes open the ducts of pedal glands that are lodged in the interior of the foot.

The rotifers are usually transparent. They appear slightly yellowish due to the colour of the covering cuticle. Brown, red, or orange forms are also seen but their colour is localized in the digestive tract because of the coloured objects that they swallow.

Sexual dimorphism occurs in some rotifers. The males are reduced greatly in size and morphology. In two groups (Ploima and Seisonacea) the members of both the sexes are similar and equal. In one group (Bdelloidea) only females occur and they reproduce parthenogenetically.

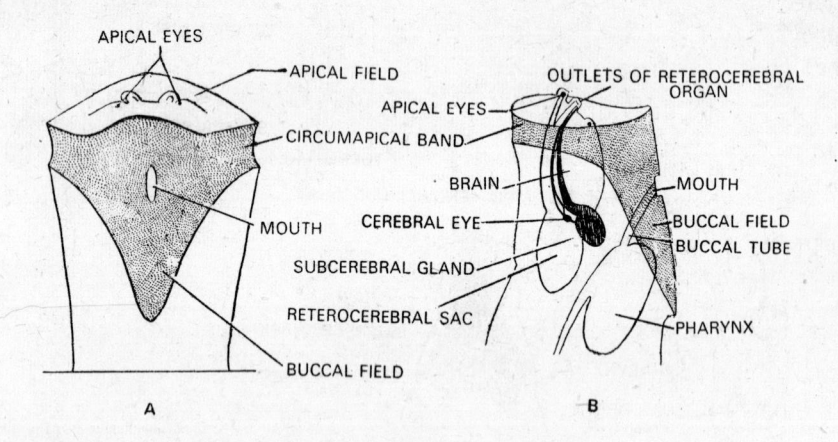

FIG. 10.3. Corona of a rotifer. A—Schematic ventral view of primitive corona; and B—Lateral view of the same.

**Occurrence**. The rotifers are common inhabitants of fresh water, some live in brackish water and a few in the ocean. Some inhabit land in damp sites, some in the axils of leaves of moss plants. They have adopted a variety of habitats. Some members are also known to be parasitic.

**Corona**. The corona or the "wheel organ" (Fig. 10.1, 10.3) is the most striking feature of the rotifers. It creates a current of water that brings along with it food particles and helps the animal in feeding. The metachronous beating of cilia produces illusion of a rotating wheel. The interpretation of the corona has been very controversial. As its correct interpretation is important from phylogenetic point of view, it engaged the attention of zoologists for a long time. Probably rotifers have originated from ventrally ciliated creeping forms. In such forms there was a ventral ciliated field around the mouth, used primarily for creeping. It is from such a ventral ciliated field that the corona has evolved. This is evident, seeing the ground plan of the corona in some primitive, rotifers, that comprises a large oval ventral field, the **buccal field**, evenly covered with short cilia, surrounding the mouth and a circumapical band extending from the buccal field to encircle the margin of the head, that remains naked and forms the apical field (Fig. 10.3). Such a condition occurs in many notommatids and is associated with creeping habits. It is from such a primitive condition that various modifications of the corona have taken place. The cilia of the corona frequently become complicated forming cirri, membranelles or styles, etc. A corona is absent in the

# Phylum Nemathelminthes (Aschelminthes)

adult females of the genera *Atrochus*, *Cupelopagis*, and *Acylus*, but is present in the normal form in the males and young females.

## Body Wall

The body wall consists of cuticle, epidermis and subepidermal muscles. The **cuticle** is secreted by the epidermis and probably consists of sclero-proteins. In creeping forms the cuticle is ringed, that facilitates body movements and helps telescoping of the segments. The cuticle is variously ornamented, the polygonal pattern being common. The **epidermis** consists of a thin syncytium containing scattered nuclei, that are bilaterally arranged and are constant in position and number for such species. In the region of the corona the epidermis presents numerous heaps of cytoplasm around the nuclei that project into the body cavity. Below the epidermis lie the subepidermal muscles (Fig. 10.2 B).

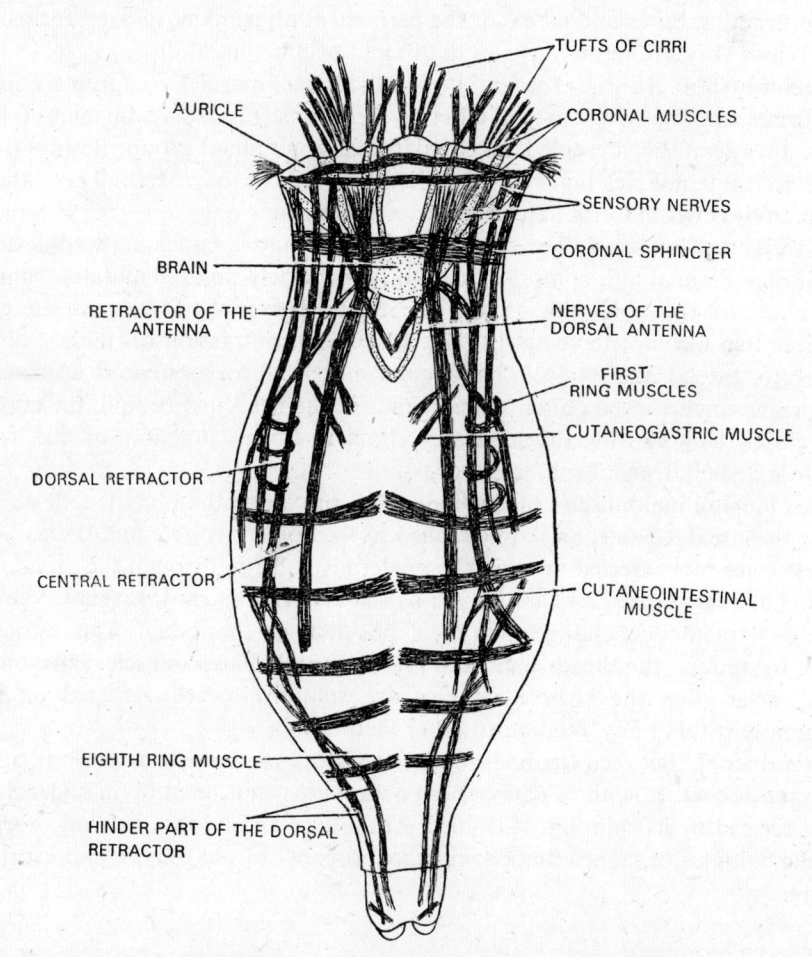

FIG. 10.4. Musculature of *Epiphanes* as seen from dorsal side.

**Epidermal glands.** Associated with the epidermis there are two sets of glands, the **retrocerebral organ** and the **pedal glands** (Fig. 10.2). The retrocerebral organ lies above and posterior to the brain. In its typical form the retrocerebral organ consists of a **median retrocerebral sac** and a pair of

**lateral subcerebral glands**. From the sac arises a duct, runs forward, forks and finally opens on the apical field on a single papilla; sometimes it opens on two papillae separately. Both the sac and glands are syncytial and secrete droplets that give them vacuolated appearance. Sometimes both the sac and the glands are filled with strongly diffractive granules and sometimes red pigment grains also. It is supposed to have produced adhesive secretions originally but now it is in the process of degeneration. In some cases the sac is reduced or even lost, whereas, in others it is prominent and the glands are reduced. In fact the sac and glands vary much in relative and absolute sizes in different rotifers (Fig. 10.2).

The pedal glands are unicellular glands situated in the foot (Fig. 10.2). From each gland a duct arises and runs through the entire length of the foot, finally opening on the tips of the toes. They secrete adhesive material used in creeping, in constructing tubes and cases, or for permanent attachment in fixed forms. In some cases they are numerous (up to 30) and of different kinds.

**Musculature**. In the rotifers the sub-epidermal muscles comprise a number of muscles running irregularly in different directions (Fig. 10.4). In some typical cases they can be identified as circular and longitudinal group. Besides these, there are some muscles that run from the body wall to the viscera. These are the **cutaneovisceral** muscles. Some muscles which run only on the viscera are called **visceral** muscles. The circularly disposed muscle bands are well-developed in some Ploima, and comprise three to seven widely spaced muscles running just close to the epidermis on the under side. These may form complete rings but are often incomplete ventrally (Fig. 10.4). The contraction of these muscles serves to extend the body. The circular muscles form a **coronal sphincter**, comprising one to seven, often united bands in the head just behind the corona. This closes the neck after the corona is retracted. At the junction of the trunk and foot a similar pedal sphincter occurs.

The longitudinal muscles of the body wall run along the body length directly under the circular bands and are attached to the epidermis here and there. Some muscles are not inserted at all in the epidermis and run through the pseudocoel and act as **retractors**, and are known as the **central, dorsal, lateral, ventral** pairs of retractors depending upon their position in the body. The retractors serve to retract the head and foot into the trunk. These muscles have one or more nuclei. On the whole muscles are smooth or cross-striated or both apparently without any relation to other factors.

**Pseudocoel**. Between the body wall and viscera there exists a spacious cavity, the **pseudocoel**. It is not a true coelom because it is not lined by mesoderm and is not formed by its splitting. It is filled with a fluid and a loose network formed by [the union of branched amoeboid cells, probably of phagocytic and excretory nature.

### Digestive System

The **mouth** (Fig. 10.3) opens into a ciliated buccal tube formed by an invagination of the apical field. Therefore, the histological structure of the buccal tube is similar to that of the apical field. A buccal tube occurs only in those forms that gather microscopic food by coronal currents. It is not found in raptorial forms in which the mouth directly opens into the pharynx.

*Phylum Nemathelminthes (Aschelminthes)* 331

The **pharynx** or **mastax** (Fig. 10.5) of rotifers is a peculiar structure and is characteristic of the group. It is a muscular organs of various forms (rounded,

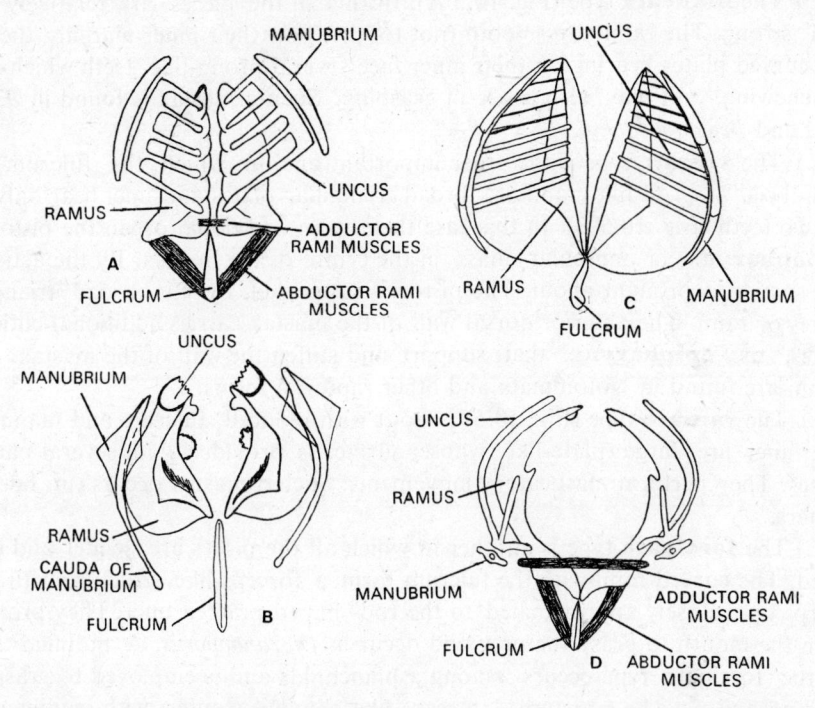

**FIG. 10.5.** Mastax of rotifers. A—Malleate trophi; B—Virgate trophi; C—Ramate trophi; and D—Incudate trophi.

elongated or trilobed) and contains a complicated masticatory apparatus in the interior. This apparatus is made up of hard cuticularized pieces called **trophi**. The trophi consist of seven main pieces, a median **fulcrum** and paired **rami**, **unci** and **manubria** (Fig. 10.5). The paired pieces are arranged sideways. The right and left pieces often differ in shape and size since they are constructed to interlock when used. They are arranged in two sets, the fulcrum and rami go together and are collectively termed **incus**, whereas, the unci and manubria are called the **malleus**.

fulcrum and rami } incus      unci and manubria } malleus

The fulcrum is the median usually thin posterior plate-like piece lying in the sagittal plane of the body. The rami extend forward from the fulcrum (Fig. 10.5) and are usually thicker triangular pieces with the apex directed in front. The unci are pieces that lie along the anterior tips of the rami with their long axes transverse to those of the rami. The manubria are elongated pieces attached anteriorly to the outer ends of the unci and extending backward somewhat parallel to the rami. Their pointed ends are called **caudae** (Fig. 10.5) and are directed posteriorly.

Several different types of trophi are found in different rotifers. Their structure is correlated with the mode of feeding of the animal. The most primitive among these is:

(*a*) The **malleate** type (Fig. 10.5 A). In this all the pieces are relatively stout and strong. The rami are smooth (not toothed) on their inner margins; the unci are curved plates bearing on their inner face several prong-like teeth which serve for chewing, but may also assist in grasping. Such a mastax is found in *Epiphanes* and *Brachionus*, etc.

(*b*) The **virgate** type is the other important one, in which the fulcrum and manubria are rod-like, rami are broad triangular plates and unci bear only one or two teeth (Fig. 10.5 B). In this case there is an additional organ the piston or **hypopharynx**, a muscular mass in the centre of the mastax. By the action of this suction is brought about. The piston is supported by the broad triangular plates or rami. The anterior dorsal wall of the mastax carries additional cuticular plates, the **epipharynx**, that support and stiffen the wall of the mastax. Such trophi are found in Notommata and other raptorial forms.

(*c*) The **ramate** type (Fig. 10.5) is stout with reduced fulcrum and manubria. The unci are large plate-like whose surface is provided with several parallel ridges. They perform masticatory movements. Such a mastax occurs in bdelloid rotifers.

(*d*) The **forcipate** type is another in which all the pieces are slender and elongated. The curved rami with the fulcrum form a forceps-like structure with their sharp tips closely approximated to the rod-shaped pointed unci. They protrude from the mouth to grasp the prey and occur in *Dicranophorus*. A molified from of the forcipate type occurs among asplanchnids and is employed to grasp the food organisms. The structure is forceps-like, though stouter with rudimentary manubria. Such a mastax is known as the **incudate** type.

The lumen of the mastax is lined by cuticle and is ciliated only in the current feeding forms. Epithelium forming the wall is wholly syncytial or partly syncytial and partly cellular. It is this epithelium that secretes the trophi. The muscles of the mastax are simply fibrillar extension of its cytoplasm.

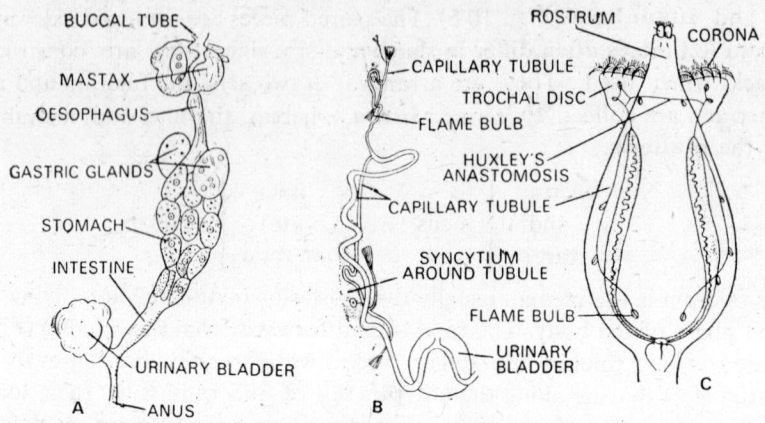

FIG. 10.6. Internal strucure of rotifers. A—Digestive system; B—Excretory system in *Rotaria*; and C—Excretory system of *Euchlanis*.

Associated with the mastax wall are found the salivary glands (absent in many). They are 2-7 in number and are uninucleate or syncytial masses with granular or vacuolated cytoplasm. Usually there is a ventral pair of such glands. The ducts from these glands open anterior to the trophi and their secretion is probably digestive or it may help ingestion of food.

The **oesophagus** follows the mastax (Fig. 10.6). Its length varies in different types. Internally it may be smooth or ciliated. Its wall is syncytial and its opening into the stomach is guarded by sphincter. No glands are present in the oesophagus. The **stomach** (Fig. 10.6) is an enlarged thick-walled sac (U-shaped tube in some) the wall of which consists of a definite number (30-45) of large granular usually ciliated cells. These cells are filled with inclusions. In Bdelloidea the stomach wall is syncytial and the lumen is a narrow tube inside the syncytium. The stomach is provided with a muscular layer. Both circular and longitudinal fibres form a net over its external surface. Usually a pair of **gastric glands** (Fig. 10.6) occur at the junction of the oesophagus and stomach. Each gland is syncytial having a constant number of nuclei and opens directly into the stomach by minute pores. These glands are known to secrete minute droplets of enzymatic material. In some cases (Collothecacea) the gastric glands are situated at the posterior end of the stomach.

The stomach narrows down into the **intestine**, there being no apparent distinction between the two. In such cases the whole structure is called the **stomach-intestine**. In others a constriction occurs between the two (Fig. 10.6). The wall of the intestine is syncytial and is externally covered by muscles that cover the stomach.

The last part of the intestine is usually termed the **cloaca** because it receives the protonephridial ducts and the oviducts. Histologically the cloaca is similar to the intestine. Sometimes it is constricted off forming a sac-like chamber and assumes the functions of a urinary bladder. The cloaca opens to the outside at the **cloacal aperture** or **anus**.

## Excretory System

Typical protonephridial tubules provided with **flame-bulbs**[1] constitute the excretory system of the rotifers. The main tubules run lengthwise on each side of the body opening posteriorly into a common urinary bladder. Each tubule is often forked into an anterior and a posterior (Fig. 10.6) branch. Into each branch open ciliated capillary ducts from flame-bulbs. Anteriorly the main tubules are often connected by transverse anastomosis, (Fig. 10.6 C), called **Huxley's anastomosis**, into which also open some flame-bulbs.

Usually there are two to eight flame-bulbs on each side. They open into main excretory channels through ciliated capillaries. The flame-bulbs vary in shape from a tubular to a flattened triangular form and each contains a membranelle of fused cilia, that keeps on moving constantly. Sometimes several delicate protoplasmic strands arise from the thickened cap-like end of a flame-bulb and connect it to the body wall. Some bulbs, on the other hand, bear external flagella. Posteriorly the tubules open into the cloaca separately or by a common stem (Bdelloidea) that finally opens to the outside.

---

[1] They are called "flame-bulbs" and not "flame-cells' because they are parts of a nephridial syncytium and do not have separate nuclei of their own. They are mere channels in a nephridial syncytium often provided with scattered nuclei.

## Nervous System

The nervous system is more or less like that of a turbellarian flatworm. The **brain** is a bilobed mass (Fig. 10.7), which may appear rounded triangular or

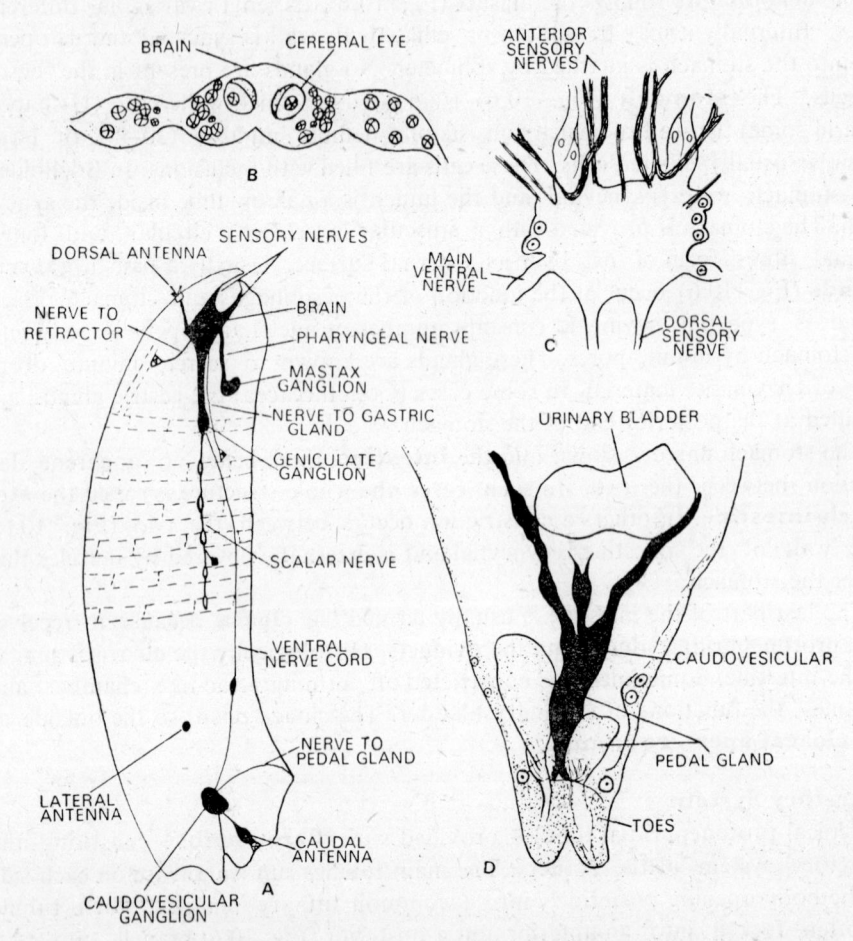

FIG. 10.7. Nervous System of a rotifer. A—Nervous system of *Lindia* (*after* Dehl); B—Section through the brain of *Synchaeta* showing cerebral eyes; C—Brain of *Epiphanes* (*after* Martini); and D—Caudovasicular complex of *Epiphanes* (*after* Martini).

quadrangular body, lying dorsal to the mastax. It is composed of a central fibrous mass surrounded by a cortical layer of ganglion cells of four types differing in size and nuclear details. A number of sensory nerves (paired) connects the brain with various sensory organs of the head, viz., the eyes, sensory bristles, dorsal antenna, and rostrum, etc. The brain also sends motor nerves to the anterior parts of various muscles and to the salivary glands.

The complicated masticatory structure, the mastax, is supplied by a pair of **pharyngeal nerves** that also innervate the mastax musculature. In the mastax a loose unpaired mass of nervous tissue lies in the midventral wall. This is the mastax ganglion and was formerly thought to correspond with the subenteric ganglion of annelid. From the mastax ganglion arises a **visceral nerve** and passes into the digestive tract on each side.

The main ventral nerve cords are two and are ganglionated. They spring from the sides of the brain and proceed backward into the foot in a lateroventral position. They bear an anterior ganglion near the brain and further posteriorly another ganglion, the **geniculate ganglion** (Fig. 10.7) from which arise a number of nerves. These include: (*i*) the **lateral sensory nerve** to the lateral antenna on each side, (*ii*) the **scalar nerve**, a longitudinal nerve to the ring muscle-bands of the trunk, having a ganglion cell for each muscle-band; (*iii*) a nerve to the **coronal sphincter**, (*iv*) minor branches to main longitudinal retractors (that may or may not arise from the geniculate ganglion). All these nerves may sometimes arise directly from the ventral cord.

Posteriorly the ventral cords terminate in ganglia supplying the urinary bladder and the foot by the vesicular and pedal or caudal ganglion respectively. Sometimes these ganglia are fused in one mass, the **caudo-vesicular** ganglion (Fig. 10.7 A and D).

**Sensory structures.** The rotifer body is richly supplied with sensory structures particularly on the anterior end of the body. These include the **sensory styles** which are single stiff bristles situated near the inner edge of the circumapical band. They may be dorsolateral, lateral or ventrolateral according to their positions. The sensory styles of the apical field are the apical styles, whereas, those near the mouth are the oral styles. Below each there is a nerve cell that conveys sensory fibres to the brain. They are tactile in nature. **Ciliated pits** are other sensory structures situated on the apical field. They are paired and are simple ciliated depressions probably chemoreceptors. Conical finger-like palps tipped with sensory hairs are other sensory structures of the apical field.

The **ocelli** or **red pigment spots** are common in Ploima, Bdelloidea and free-stages of the sessile orders. The ocelli are sometimes double (usually single) embedded in the dorsal or ventral surface of the brain (Fig. 10.7 B). Each consists of a single cell resembling a brain cell in the simplest forms. The lateral and apical eyes are epidermal cushions with one or more nuclei. The red pigment is located inside the optic cell or the epidermal cushion. Sometimes a lens-like structure may be present but the same is not cuticular.

**Reproductive Organs**

The sexes are separate and in the majority of forms the males are smaller than the females and appear only in the breeding season. The males may be as much as one-tenth of the females. In appearance they resemble females but display reduction in internal structures. The female gonad consists of a syncytial **ovary** (Fig. 10.8 A) in the majority with a syncytial **vitellarium**, both surrounded by a common membrane (hence called **germovitellarium**) that continues to the cloaca as a simple tubular oviduct. Germovitellaria and oviducts are paired in the bdelloids; whereas, the Seisonacea have paired ovaries without vitellaria. The male gonad comprises a single large sac-like **testis** (Fig. 10.8 B), which has a ciliated spermduct opening at the gonopore (the cloaca is lacking in males). Into the spermduct open a pair (sometimes more) of **prostate** glands. The posterior end of the spermduct is eversible, acting as **cirrus**. The cirrus is lined with hardened cuticle. In some cases the spermduct bears a cuticular tube that is protrusible and acts as a penis.

**Copulation.** Copulation by the insertion of the copulatory organ into the cloaca occurs only in a few rotifers (Fig. 10.8). In the majority, as a rule, the sperms are injected through the body wall into the pseudocoel. This has been described as **hypodermic impregnation**.

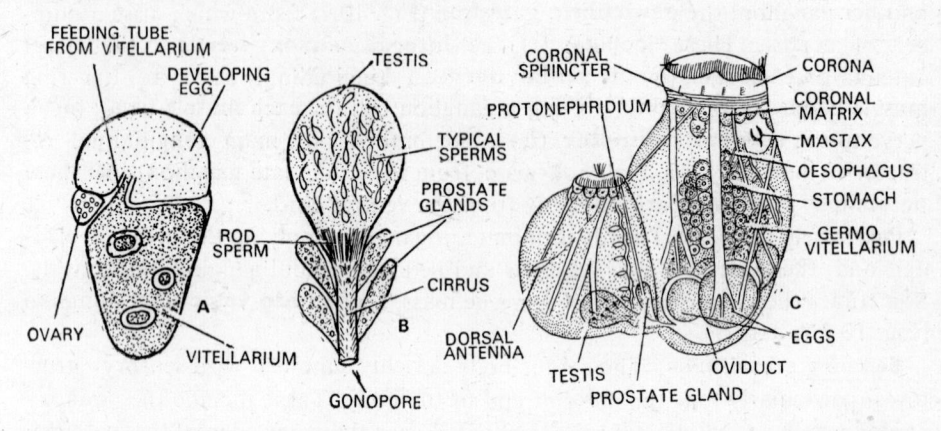

FIG. 10.8. Reproductive organs of a rotifer. A—Germovitellarium of *Synchaeta* (*after* Peters); B—Male system of *Rhinoglena* of (*after* Wesenberg-Lund); and C—*Asplanchna* in copulation.

The egg, as it matures, constricts off from the syncytial ovary by a delicate membrane that comes to lie in contact with the vitellarium. Later the edge ruptures, and the edges at the rupture turn back into the egg forming a **feeding tube**, through which the nourishment passes into the developing egg. Eggs are laid on the substratum or are stuck to the body of the female (Fig. 10.9 D) or to other animals. Each female can lay only a fixed number of eggs. This number corresponds with the number of nuclei present at the time of birth. The eggs are usually oval enclosed in an evident shell within which there are one or two thinner membranes closely adhering to the egg. Outside the shell there may be an additional thin membrane or a gelatinous layer acting as a float. The eggs that develop without fertilisation are different in structure, have thin shells and develop parthenogenetically sometimes they are called **amiotic eggs**.

The sperms are of two types: (*i*) typical sperms with large rounded or oval heads and a tail provided with an undulating membrane; and (*ii*) typical sperms in the form of rod-shaped bodies (Fig. 10.9 A).

**Fertilization**. The sperm penetrates the egg while it is still immature. The egg then grows larger, and receives more nourishment from the vitellarium. Later it changes in colour, shell becomes hairy and thick and often ornamented, spiny, warty or ridged.

**Development**. The development of a typical rotifer has not been studied up to now. Here an account of the development of *Asplanchna* has been given after Nachtway (1925). The egg divides into four blastomeres A, B, C, D; of these D is the largest (Fig. 10.9) and cleaves more rapidly. Further divisions produce a ten-cell embryo consisting of two tiers of cells. Those formed by the division of D are four in a row, the lowest of which is again the largest forming the vegetal pole. This cell is the primordial sex-cell, therefore, as the development proceeds it passes into the interior and becomes surrounded by other cells. Formerly this cell was wrongly called the **endoderm** cell, but in *Asplanchna*

# Phylum Nemathelminthes (Aschelminthes)

there is no endoderm formation or gastrulation.

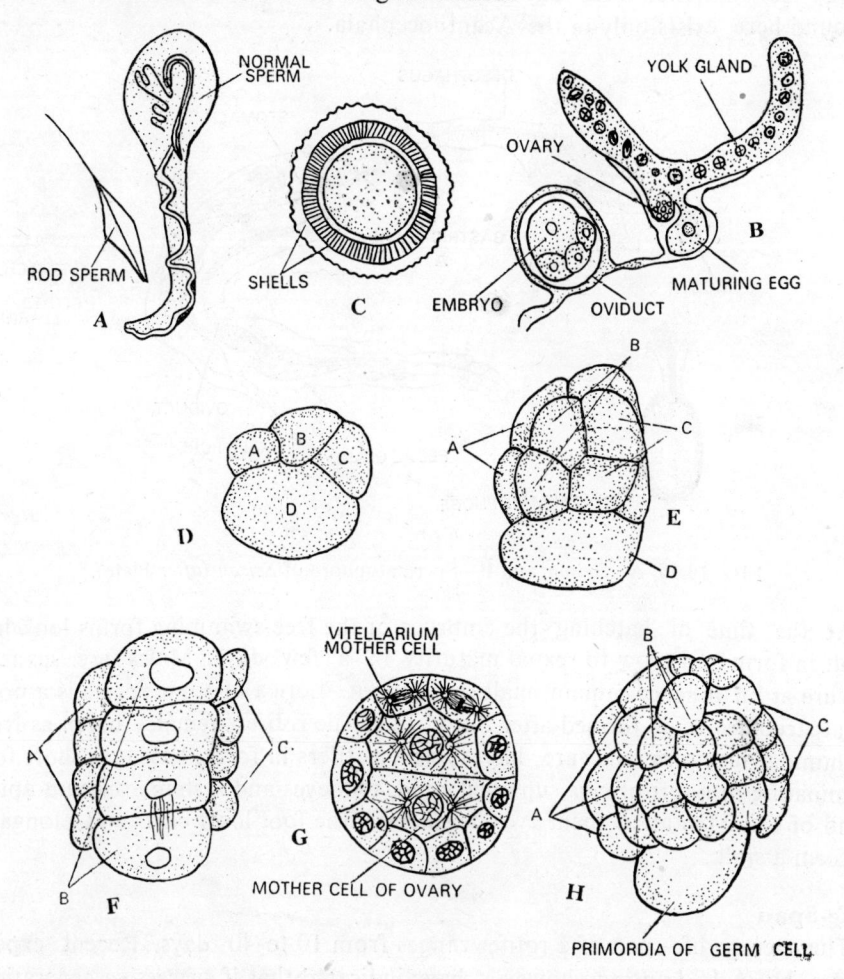

**FIG. 10.9.** Reproduction and embryology. **A**—Sperm of *Sinantherina* (*after* Hamburger); **B**—Female system of *Asplanchna*; **C**—Fertilized dormant egg of *Asplanchna*; and **D** to **H**—Development of *Asplanchna*. [**D**—4-cell stage; **E**—10-cell stage; **F**—16-cell stage (B-quadrant in centre flanked by A- and C-quadrants); **G**—35-cell stage; and **H**—Section through the embryo after invagination] (**B**—**G** *after* Tannreuther).

The primary germ cell is distinguished from other blastomeres by being full of a cloud of granules or **ectosomes**. It first gives off two minute cells and then divides into two equal cells. The posterior one of these receives the ectosomes and becomes ovary at a later stage. The anterior one forms a syncytial vitellarium. All other organs come from the surface layer of blastomeres that continue to divide. The vegetal pole becomes the anterior end and the animal pole (determined by the polar body) is the posterior end. D-quadrant cells are ventral A-, B- and C-quadrant cells constitute the dorsal and lateral regions.

As is evident from the above the early cleavage pattern suggests the **determinate spiral** type of development, but the later stages present **bilateral** symmetry. Due to these and many other peculiarities the development of rotifers cannot be compared with that of any other invertebrate. The bilateral pattern, as found here, exists only in the Acanthocephala.

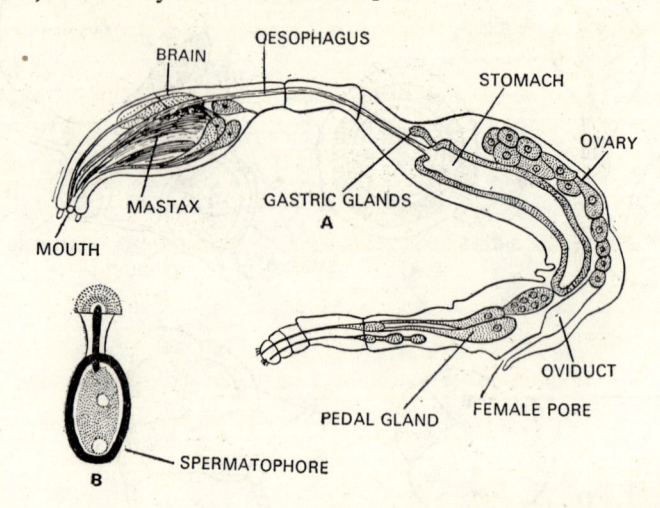

FIG. 10.10. A—*Seison*; and B—Spermatophore of *Seison* (*after* Plate).

At the time of hatching the embryos of the free-swimming forms look like adult in form and grow to sexual maturity in a few days. Males are sexually mature at birth hence remain small throughout. Lorica of some forms is a postnatal structure being formed after birth. In sessile rotifers females hatch as free-swimming juveniles and are like typical rotifers in form, having a short foot terminating in a tuft of cilia, unlobed head, two eyes and a simple circum-apical band of cilia. On attachment eyes degenerate, the foot loses cilia and elongates to form a stalk.

**Life-Span**

The average life-span of a rotifer ranges from 10 to 40 days. Recent experiments by A.I. Lansing, however, have indicated that if successive generations of rotifers are grown using only the eggs of young animals, the life-span may be considerably lengthened. If, on the contrary, successive generations are reared using the eggs of old animals only, the life-span will be shortened. This can be seen quickly by referring to the following experiment. Rotifers were selected from a colony which had an average life-span of 4-20 days, and successive generations were grown from parents of different ages. In one group, the first generation and succeeding generations were derived from eggs of young parents that were only 4 days old when the eggs were laid. In another group, all generations were grown from eggs of middle-aged parents who were 11 days old. In the third group, all generations were grown from eggs of 17-day old senile animals. It will be observed that the 17-day group died out (did not reproduce) by the third generation. The 11-day group was wiped out by the fourth generation. The 4-day group not only survived but also the life-span was lengthened in three of the four generations. Rotifers, therefore, appear to be another group of organisms which will enable us to acquire additional information concerning the problems of aging.

*Phylum Nemathelminthes (Aschelminthes)*

**Affinities.** The rotifers have been linked with all invertebrates especially with the arthropods and annelids. The arthropod relationship is based upon the resemblance of *Pedalia* (Fig. 10.13 B) with the arthropodan types. *Pedalia* has movable bristle-bearing arms apparently like those of the appendages of a crustacean larva. This view was later abandoned in favour of Hatschek's **trochophore theory,** because *Pedalia* is a highly modified rotifer, and to prove affinity the relationships should be studied between primitive forms.

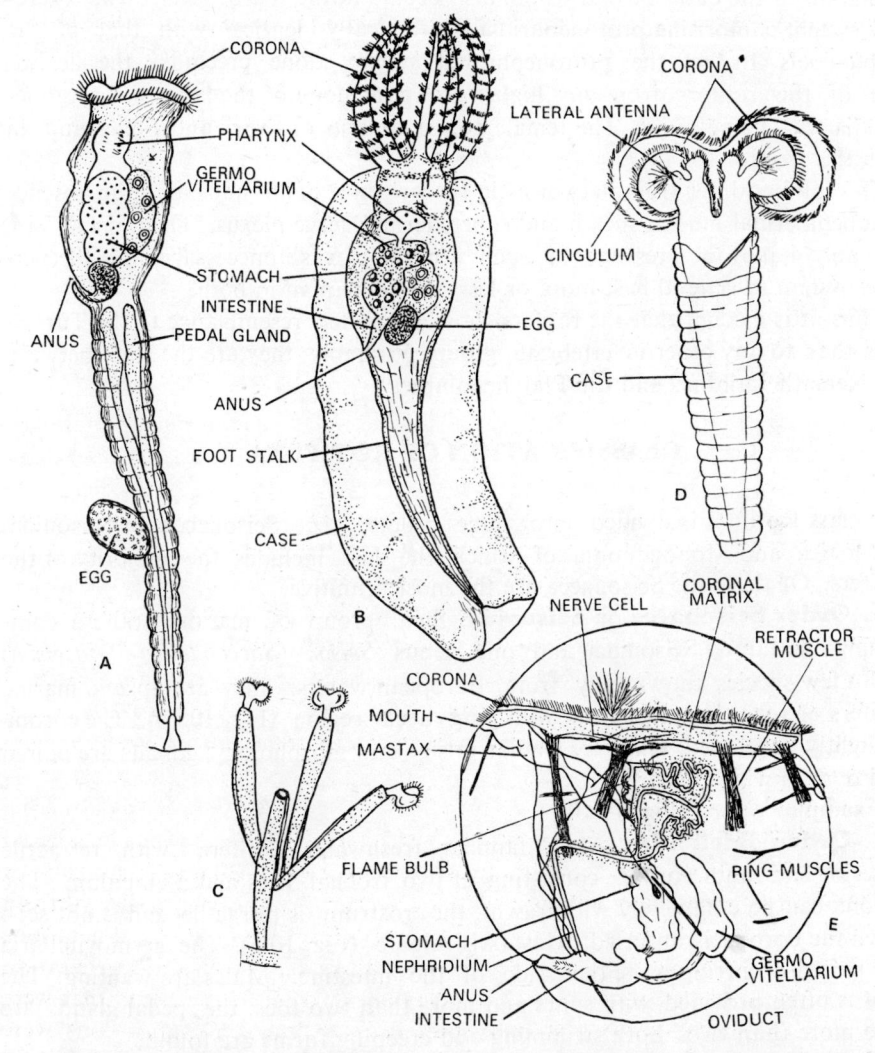

FIG. 10.11. Different types of rotifers. A—*Ptygura* (*after* Weber); B—*Stephanoceros* (*after* Weber); C—An aggregation of *Limnias*; D—*Limnias* (*after* Hyman); and E—*Trochosphaera*, female (*after* Valkanov).

According to the **Annelida theory** the rotifers are compared with the trochophore larva of the annelids and are considered as annelids that have remained in the larval stage. *Trochosphaera*, a rotifer, is the basis of this relationship. In structure *Trochosphaera* (Fig. 10.11) has many things in common with the trochophore larva-ciliary girdles, bent intestine, and excretory organs resemble topographically in both the types. Again *Trochosphaera* is not a primitive but a highly modified rotifer and cannot be regarded as a basis of relationship.

The embryology of rotifers (whatever has been studied) shows that the rotifers are lowly organized animals and not derived from retrogression of higher forms. The anatomy points to their origin from some lowly organized creeping bilateral form such as a primitive flatworm. It has been pointed out earlier that the rotifer corona has originated from ventral ciliation as found in the Turbellaria. The formation of the cuticularized trophi also occurs in the Turbellaria. The excretory system comprising protonephridia is practically identical with that of the rhabdocoels. In fact the protonephridial system alone precludes the derivation of the rotifers from any higher group, as none of the latter has protonephridia with flame-bulbs. The female gonad consists of ovary and vitellarium in both the groups.

The rotifers differ from flatworms in the presence of an anus and the absence of subepidermal muscle sheath and subepidermal nerve plexus. Their small size probably makes the presence of such a nerve plexus unnecessary. But the nervous system in general has, more or less, the same plan in both.

Thus, it is evident that the rotifers show a greater resemblance to the Turbellaria than to any other invertebrate group. Probably they are the link between the Nemathelminthes and the Platyhelminthes.

## CLASSIFICATION OF ROTIFERA

The class Rotifera is divided into three orders, viz., Seisonacea or Seisonida, Bdelloidea and Monogononta, of which the last includes the majority of the rotifers. Of these the Seisonacea are the most primitive.

1. **Order Seisonacea** or **Seisonida**. Small group of marine Rotifera comprising one family Seisonidae and one genus *Seison* (*Saccobdella*; *Paraseison*) and a few species known only from European waters. They are epizoic marine rotifers of very elongated form with long neck region (Fig. 10.10). The corona is slightly developed, sexes are similar in size and morphology, gonads are paired and ovaries with vitellaria.

Example: *Seison* (Fig. 10.10).

2. **Order Bdelloidea**. Most familiar freshwater Rotifera with retractile anterior end and corona consisting of two trochal discs and a cingulum. The corona can be completely withdrawn; the rostrum is retractile and is not seen when the corona is unfurled. Mastax is ramate (Fig. 10.5). The germovitellaria are paired occurring on either side of the intestine. Males are wanting. The foot is often provided with spurs and more than two toes, the pedal glands are also more than two. Both swimming and creeping forms are found.

Examples: *Philodina*; *Habrotrocha*; *Taphrocampa*; *Rotaria* (Fig. 10.11), etc.

3. **Order Monogononta**. Swimming or sessile Rotifera with one germovitellarium; males more or less reduced with one testis. The order is divided into three important suborders.

(*i*) Suborder **Ploima**. This suborder includes the majority of rotifers with more normal body shape and corona. Three important superfamilies of this suborder are known. (*a*) **Notommatoidea** including a large number of roti-

# Phylum Nemathelminthes (Aschelminthes)

fers of normal elongate form (Fig. 10.11) without a lorica, with mostly a virgate mastax, (*b*) **Brachionoidea** includes mostly stout Ploima often dorsoven-

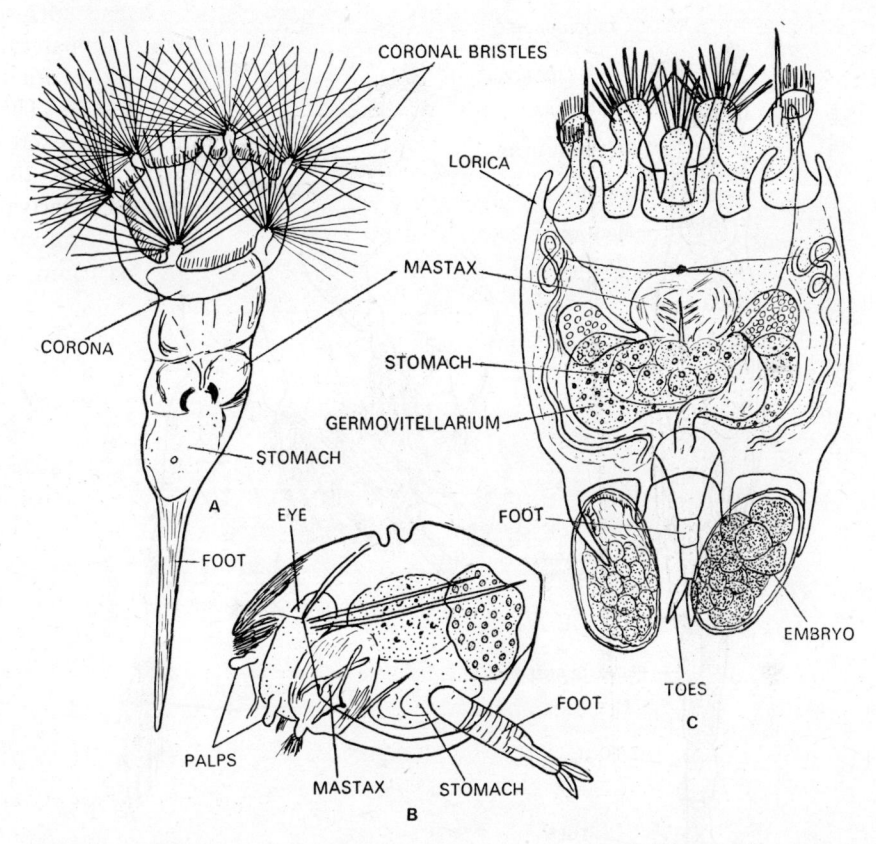

FIG. 10.12. Different types of rotifers (contd.). A—*Collotheca*, a sessile rotifer with seven-lobed corona; B—*Platyias*, a brachionoid rotifer; and C—*Ploesoma*, a ploimate pelagic rotifer.

trally flattened having heavy lorica and malleate trophi. *Brachionus* is one of the common forms of this superfamily. (*c*) **Asplanchnoidea** including pelagic rotifers with delicate sac-like bodies and corona reduced to a simple circumapical circlet. Foot absent or ventrally displaced. *Asplanchnopus* is a form with foot but without intestine or anus; *Asplanchna* without intestine, foot or anus, and *Harringia* is with anus and a small foot.

(*ii*) Suborder **Flosculariacea**. Sessile or free-swimming rotifers, with often a circular or lobed corona provided with trochal and cingular circlets. Trophi modified ramate, foot without toes and more than two pedal glands. Males greatly reduced.

Examples: *Testudinella* (free swimming form), *Ptygura* (Fig. 10.11 A), *Limnias* (Fig. 10.11 D) and *Conochilus*.

(*iii*) Suborder **Collothecaca**. Mostly sessile rotifers characterized by the expanded funnel-shaped anterior end, central mouth; devoid of definite ciliary circlet often provided with motionless bristles; males greatly reduced, foot without toes.

Example: *Collotheca* (Fig. 10.12 E)

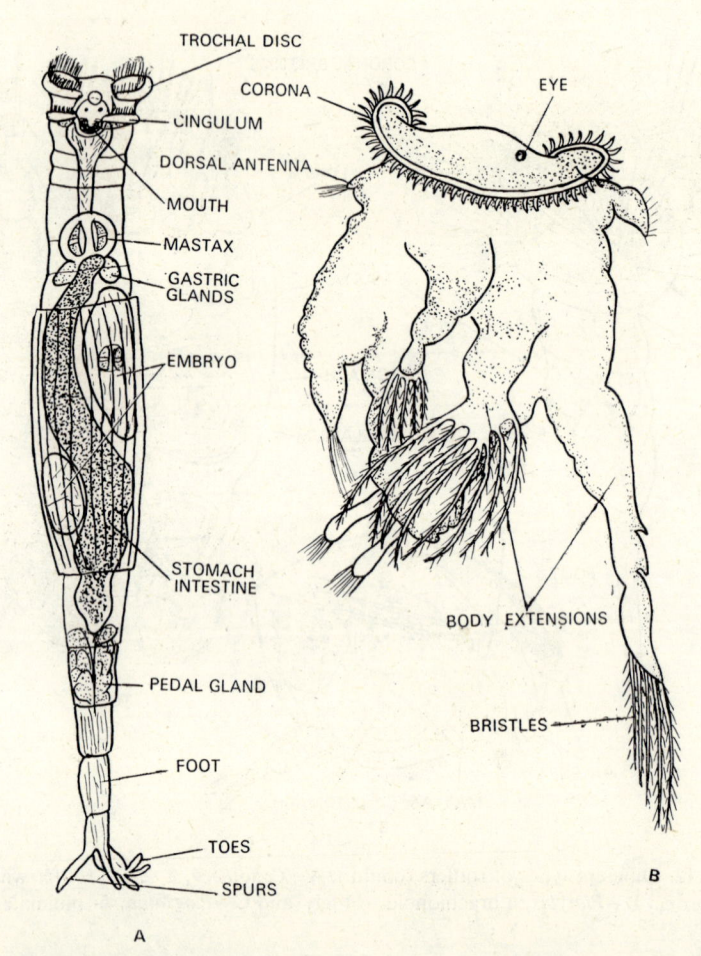

**FIG. 10.13.** Different types of rotifers (contd.). **A**—*Rotaria*, a bJelloid rotifer; and **B**—*Pedalia*, with movable arms tipped with setose bristles.

## CLASS GASTROTRICHA

The Gastrotricha (Gr. *gaster*, belly + *trichos*, hair) are free-living aquatic Nemathelminthes without corona ciliation being restricted to limited areas. Unsegmented cuticle, often provided with spines, plates, scales, is furnished with many adhesive tubes. Pharynx is like that of nematodes and excretory organs, if present, are a pair of protonephridia with one flame-bulb each. The Gastrotricha are microscopic animals found in fresh and salt water. They were seen even by early microscopists. O.F. Müller (1786) has figured them though included with "Infusoria" (protozoans). They vary in size from 0.1 to 1 5 mm. In shape they are elongate ventrally flattened animals and glide on ventral cilia like rhabdocoels. In the common freshwater form, *Chaetonotus*, the anterior part is marked out into a rounded head borne on a slightly constricted neck (Fig.

*Phylum Nemathelminthes (Aschelminthes)* 343

10.14). The moderately elongated trunk terminates in a fork posteriorly. In another gastrotrich, *Macrodasys*, the anterior end is not delimited as a head and

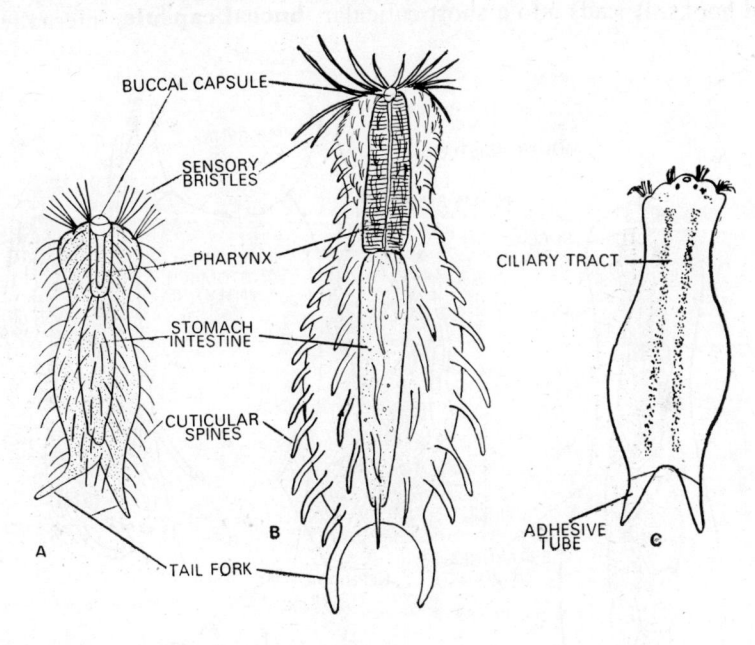

**FIG. 10.14.** A—*Chaetonotus*, a gastrotrich in natural habitat; B—*Chaetonotus*, enlarged; and C—Ciliary tracts of the same.

the posterior end is drawn out into a long slender point. Usually the dorsal surface is convex. The gastrotrichs are colourless and transparent. Sometimes ingested food may lend them colour. Cilia occur on the head lobe and also on the ventral side, which may be entirely ciliated (*Macrodasys*) or show only two longitudinal bands of cilia (*Chaetonotus*).

The surface of the body consists of a thin cuticle that forms scales covered over by spines, giving the familiar bristly appearance. The **adhesive tubes** are two projecting, cylindrical, movable tubes, each supplied by an adhesive gland cell containing 1-3 nuclei forming the **tail-fork**. The sticky substance is used to glue the animalcule to objects. In other gastrotrichs there may be more adhesive tubes which are variously situated.

**Body Wall**

The body wall consists of cuticle below which lies a thin syncytial epidermis interrupted by the epidermal glands that supply the adhesive tubes. There is no subepidermal muscle-sheath as in the rotifers; but there are delicate circular fibres without nuclei lying just next to the epidermis. The muscles of the movable and adhesive tubes arise from these fibres. These muscle fibres may take transverse direction. Longitudinal muscles are also present that run along the ventrolateral margins. Between the body wall and the viscera occurs a narrow space probably of the nature of a pseudocoel.

## Digestive System

The **mouth** is terminal, slightly ventrally placed and is encircled by numerous curved hooks. It leads into a short cuticular **buccal capsule**, internally bearing

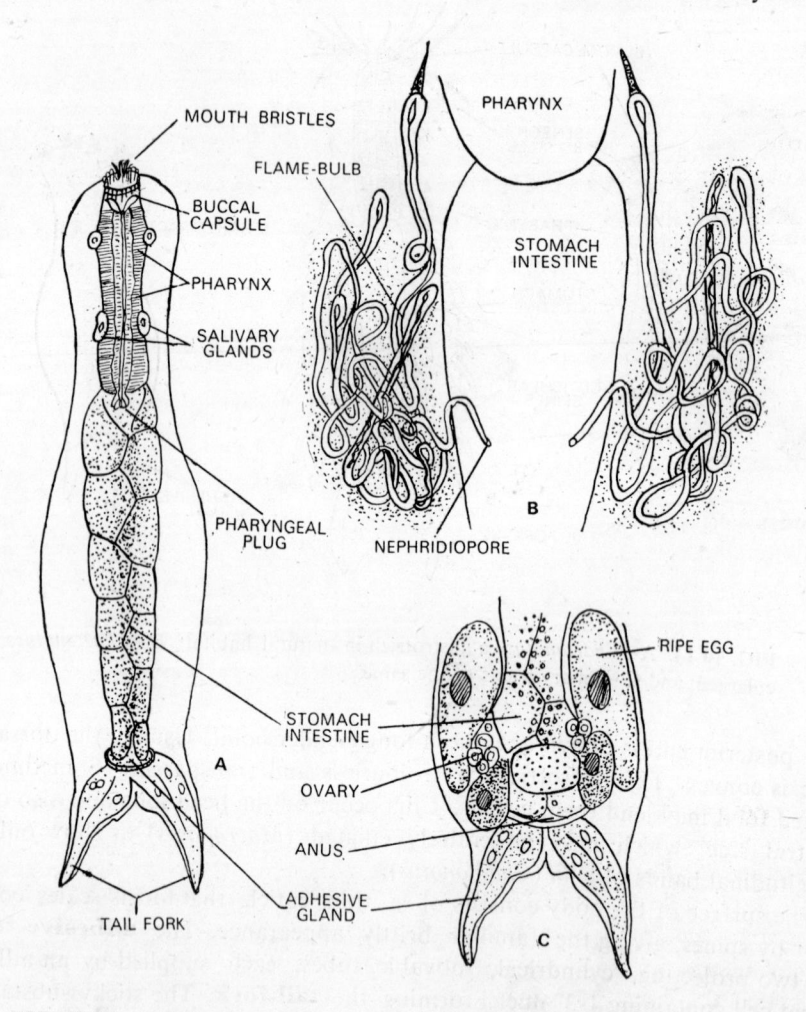

**FIG. 10.15.** A—Digestive system of *Chaetonotus*; B—Protonephridia of *Chaetonotus*; and C—Reproductive system of *Chaetonotus* (all *after* Zelinka).

longitudinal ridges (and sometimes projecting teeth). The buccal capsule opens into the **pharynx**, without a mastax, and is of elongated tubular shape occupying one-sixth to one-third of the body length. Like a nematode pharynx, it presents one or more bulbous enlargements. The lumen of the pharynx is three-angled as in the nematodes. The pharynx leads into the **midgut** or **stomach-intestine,** a simple straight tube without external glands. The anterior portion of this tube is wider, presumably **stomach**, and the posterior portion is narrow (**intestine**) but the two portions are not separated externally or internally. The **anus** is situated between the bases of the tail-fork (Fig. 10.15).

The food of gastrotrichs comprises bacteria, protozoans, diatoms and other minute organisms. Nothing is known about their digestion, apparently it takes place in the midgut lumen. They keep on moving about seeking food rather continuously, but some chaetonotids are said to attach by adhesive tubes of the tailfork and collect food in the same manner as the rotifers do.

**Excretory System**

On either side of the middle part of the digestive tract is situated a single-non-nucleated flame-bulb having long flame (Fig. 10.15B). A much coiled tube leads from each flame-bulb and opens separately on the ventral side about the middle of the body. In many other forms there are no protonephridia.

**Nervous System**

The brain is large consisting of two masses, on either side of the anterior part of the pharynx, connected by a broad dorsal commissure. A pair of lateral nerves arise from the brain and extend to the body length. From the brain also arise delicate fibres that connect it to the sensory cilia on the head lobe. Similar sensory or tactile hairs, situated all over the body, are connected to the lateral cords or the brain. Sensory pits are also present on the head lobe just behind the most posterior ciliary tuft.

**Reproductive System**

Only females occur in the genus *Chaetonotus* and allied freshwater forms, male system has degenerated, and they reproduce parthenogenetically. The marine forms are hermaphrodites. The simple ovary fills much of the body cavity, each of the proportionately large eggs is encased in a tough shell bearing hooks that fasten to material in the water (Fig. 10.15). There is no larval stage.

## CLASSIFICATION OF GASTROTRICHA

About 1500 species of the Gastrotricha are known. The class is divided into two orders—one comprising freshwater forms and the other marine.

1. **Order Chaetonotoidea.** Mostly freshwater Gastrotricha (a few species are marine) in which the adhesive tubes are limited to the posterior end of the body and are two or four in number. A pair of protonephridia is present and each having only one flame-bulb. Only females occur and reproduce parthenogenetically.

Examples: *Chaetonotus, Dasydytes, Neodasys,* and *Xenotrichula,* the last two are marine with male reproductive organs.

2. **Order Macrodasyoidea.** Marine Gastroticha in which a number of adhesive tubes is present along the body. They are without protonephridia and are hermaphrodites.

Examples: *Cephalodasys, Lepidodasys, Macrodasys* and *Urodasys* etc.

**Affinities.** The Gastrotricha are usually linked with the Rotifera and placed under Trochelminthes. The similarities between the two groups include (*i*) the presence of cilia; (*ii*) the structure of integument and musculature; and (*iii*) the presence of protenephridia with flame-bulbs. The dissimilarities comprise the specialization of Gastrotricha, the structure of the digestive tract, and reproductive system. The Gastrotricha present many similarities with the Nematoda. They are (*i*) identical digestive tract; (*ii*) cuticular spines and bristles; (*iii*) adhesive tubes; and (*iv*) the lateral sense organs of the head, etc. Thus, it appears that Gastrotricha are more closely related to the Nematoda and both together are related to the Rotifera and have probably originated from some common ciliated ancestor, perhaps the Turbellaria.

## CLASS KINORHYNCHA

The **Kinorhyncha** (Gr. *Kineo,* movement+*rhynchos,* snout) are also minute animals that were known to the early microscopists. They are less than 1 mm in length and have a segmented body made up of 13 or 14 joints or **zonites**. The segmentation is superficial like that of the Acanthocephala or Nematoda. The body surface is without cilia and consists of a thick cuticle covered by various spines and other similar cuticular-specializations. The elongated body is divisible into a spherical head connected by a neck to the jointed trunk. The head is retractile and can be withdrawn into the second and third zonite. The head bears central terminal mouth and 5-7 circlets of posteriorly directed spines.

### Body Wall

The body wall comprises the cuticle below which lies a more or less syncytial epidermis, that forms longitudinal chords similar to those of nematodes. The

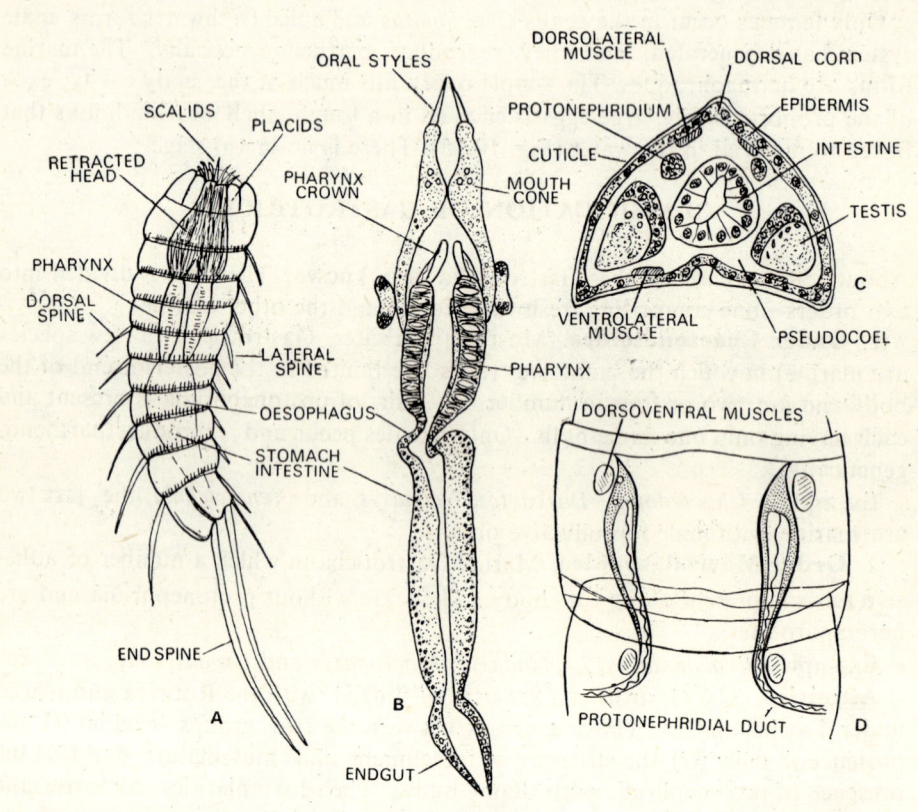

FIG. 10.16. A—*Echinoderella*; B—Digestive tract of *Pycnophyes*; C—Section through intestinal region of *Pycnophyes* showing general str cture; and D—Protonephridia of *Pycnophyes* (*after* Zelinka).

*Phylum Nemathelminthes (Aschelminthes)* 347

muscular system is like that of gastrotrichs or rotifers and is arranged in relation with the body segments. Both transverse ring muscles and longitudinal muscles are present. The muscle fibres are inserted on the cuticle directly, all are nucleated and cross-striated except the ring-muscles of the head and neck (Fig. 10.16 C).

### Digestive System

It is like that of gastrotrichs and nematodes. The **mouth** opens into a conical **buccal cavity** lined by a syncytial epithelium. The buccal cavity leads into the **pharynx**, a muscular fusiform body, the lumen of which is three-rayed in some and rounded in others. The anterior end of the pharynx possesses a cuticular ring bearing teeth-like structures that project into the buccal cavity. The pharynx is externally surrounded by 10 longitudinally disposed muscle bands. The remaining alimentary canal is composed of a slender tubular **oesophagus, stomach-intestine** and the **hindgut.** The stomach-intestine is without cuticular lining or gland cells, and is covered externally by loose net of circular and longitudinal muscle fibres. At the junction of the oesophagus and stomach are found two or more pancreatic glands, also called **salivary glands**. The hindgut follows the stomach-intestine and leads to the terminal anus, situated in the last zonite (Fig. 10.16).

The pseudocoel is spacious between the body wall and the alimentary canal and is filled with a fluid containing numerous active amoebocytes.

### Excretion

It is brought about by a pair of protonephridia, which lie in the tenth zonite, on either side of the gut. Each consists of a flame-bulb with a long flagellum followed by a short tube that opens on the side of the eleventh zonite at the nephridiopore (Fig. 10.16).

### Nervous System

The nervous system lies in close contact with the epidermis. The brain encircles the base of the mouth-cone forming a circumenteric ring. From the nerve ring springs a ventral ganglionated cord that runs in the mid-ventral line just in contact with the epidermis. The organs of special sense of the kinorhynchs include the eyes, in *Echinoderes*, and the sensory bristles. Each eye is a cup-shaped mass of red pigment enclosing a lens-like body.

### Reproductive System

The sexes are separate though the males and females are exactly alike. The gonads are a pair of sac-like syncytial bodies opening separately on the thirteenth zonite. At the anterior end of each gonad is found an apical cell that gives rise to the other cells of the gonad. It is during development of the sex-cells that the changes are apparent. The ova differentiate from the syncytial mass of nuclei, that is the ovary, and absorb nutritive material. The sperms arise similarly and attain usual shape with a broad elongated head and a short tail. The laid eggs are unknown hence nothing is known about embryology. Sexual reproduction appears to occur throughout the year. The eggs hatch into larvae different from the parents and undergo a metamorphosis before attaining the adult form.

Examples of the class are *Echinoderes, Echinoderella, Centroderes*. There are about 100 species and all are marine, the known species have been collected chiefly on the European coasts.

**Affinities**. Ever since their discovery by Dujardin in 1801 the Kinorhyncha have been linked with various groups, viz., copepods, sipunculids, acanthocephalans, rotifers and nematodes because of some anatomical likeness or the other. They have been even linked with the arthropoda or considered as connecting forms between annelids and arthropods, because of their apparent segmentation, molting and general similarity to an insect larva. Their segmentation, however, is superficial like that of nematodes; and differs fundamentally from the annelid and arthropod segmentation. Absence of a coelom, absence of a definite muscle layer in the body and intestinal wall, presence of protonephridia and the association of the nervous system with the epidermis show a lower grade of organization. All these are found in the Nemathelminthes. Strong affinity to nematodes is seen in the anatomy of the digestive tract, the division of the epidermis and musculature into longitudinal chords, the circumenteric form of the brain, the presence of a single midventral nerve cord, the molting of the cuticle in the larval stages and the copulatory spicules in the male. The pharynx is also similar in these groups. Thus, it is apparent that the Kinorhyncha appear to be close relatives of Nematoda.

## CLASS NEMATODA

We are familiar with the minute pinworms that inhabit the alimentary canal of children. We are also familiar with large "earthworm-like" ascarids living in the alimentary canal of man, and we are well-acquainted with the filaria worm causing elephantasis or roundworms or nematodes. But all the nematodes are not parasitic; there are very many that reside in many diverse kinds of environment in all latitudes. They occur in soil, in both freshwater and marine conditions, in hot mineral springs, and even in the icy polar regions. Some of them attack plants, causing heavy damage to crops, others associate with insects which they often kill. It is very interesting to note that all species, whether free-living or parasitic, have the same basic structure; the body is filiform and elongate and is invested by a longitudinal muscle layer and bounded externally by a cuticle. The mouth is terminal and the anus is at the opposite end of the body. Food material, which may be largely liquid, is pumped straight into the intestine by means of a muscular pharynx. The sexes are usually separate but some forms are hermaphroditic or parthenogenetic.

The free-living nematodes are usually microbivorous, living on decomposing organic material or on bacteria and other micro-organisms. The great diversity of their habitats and their capacity to adapt to widely different temperature conditions suggests how they have been able to exploit the many ecological niches offered by a living organism and become successful parasites. The nematodes are clearly preadapted to parasitism, since larval development is the same in the free-living and parasitic forms.

## Phylum Nemathelminthes (Aschelminthes)

They can easily be distinguished by their characteristic thrashing movements. The free-living forms are generally white or yellowish or transparent. In fact the nematodes on the whole lack coloration, the yellowish or pinkish tint is seen due to the coloration of the cuticle. Free-living forms are usually small in size below 1 mm in length, among these the freshwater and terrestrial forms are hardly within a couple of millimetres but the marine forms rise upto 50 mm. The parasitic nematodes attain considerable size. Some (*Dracunculus medinensis*, the guinea worm) may attain a metre or more. In these cases the females are generally longer than males. The fecundity of nematodes is enormous. The common roundworm of man (*Ascaris*), according to Caullery lays annually sixty-four millions of eggs equal to 1,700 times the weight of its own body.

### TYPE ASCARIS

*Ascaris lumbricoides* is a common parasite of the human intestine. About fifty different species of roundworms have been found in man, but only about a dozen of these are common. Some are harmless and do not even make their presence

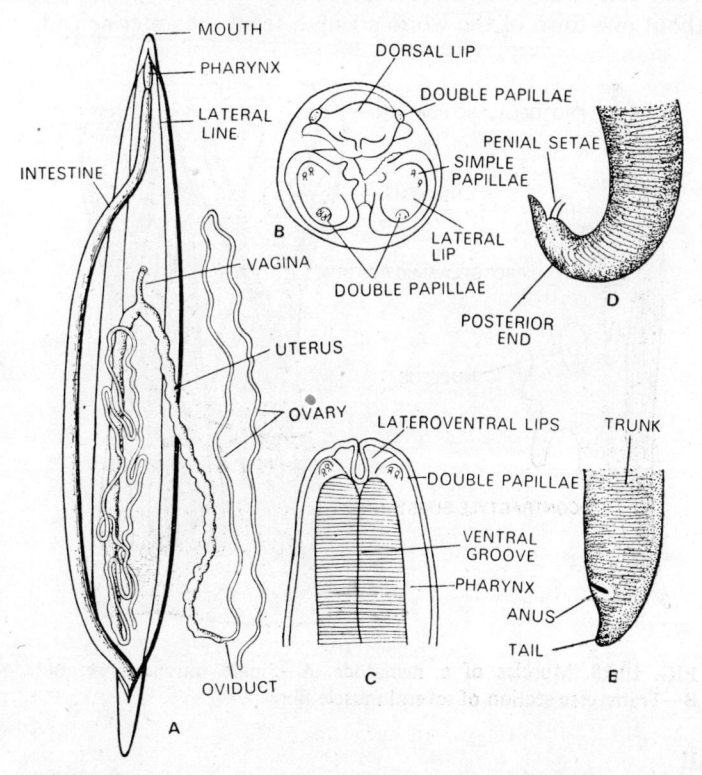

FIG. 10.17. Structure of a nematode. A—*Ascaris* dissected to show alimentary canal and reproductive organs; B—Face view of *Ascaris* showing lips and papillae; C—Anterior end ventral view; D—Tail end of male *Ascaris*; and E—Tail end of female *Ascaris*.

known, while others cause mild to very serious diseases. *Ascaris* lives in the intestine in the adult stage and is relatively harmless. Sometimes the ascarids occur in large numbers and block the intestine completely resulting in even death of the host.

**Form of Body**

The body is elongated cylindrical (Fig. 10.1 ), pointed at both ends. The body surfaces are marked by narrow white lines. The dorsal surface has a dorsal longitudinal line, the ventral surface has a ventral longitudinal line, whereas, the sides have relatively deeper lateral longitudinal lines. The **mouth** is at the anterior tip of the body and is bounded by three **lips** (Fig. 10.17B), one **median dorsal** and two **ventrolateral**. Each lip retains three labial papillae, one single and two double, other head sense organs are absent. The **excretory pore** is a minute aperture on the ventral side about two millimetres from the anterior end. The **anus** is also ventrally situated and is a transverse opening near the posterior end of the body. In the males it serves as a reproductive aperture and, therefore, a pair of bristle-like chitinous **penial setae** project from it (Fig. 10.17 D). The body of the male, on the whole, is smaller and the posterior end is curved ventralwards. The body of the female is large and the genital pore lies in the mid-line about one-third of the worm's length from the anterior end.

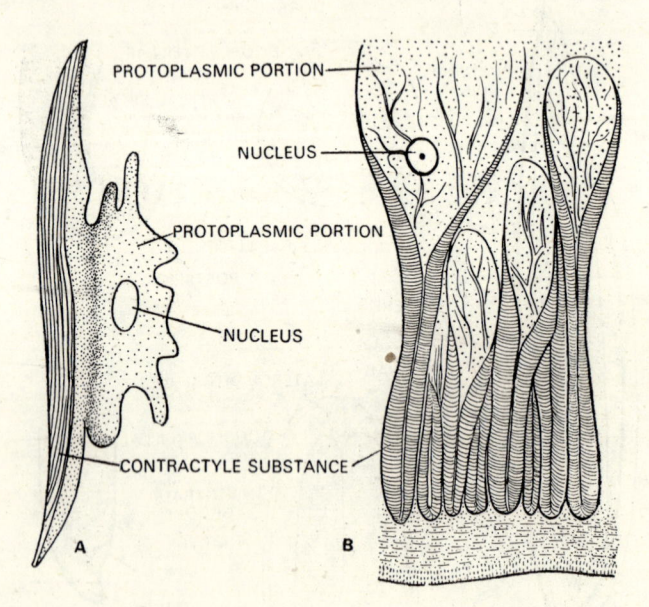

**FIG. 10.18.** Muscles of a nematode. A—Single muscle fibre; and B—Transverse section of several muscle fibres.

**Body Wall**

The body is covered with a thick tough **cuticle** (Fig. 10.19), often several layers deep. It is of albuminoid composition secreted by the underlying layer. The cuticle is transversely ringed imparting a segmented appearance to the animal from outside. Beneath the cuticle lies the epidermis or **hypodermis** also known as **syncytial ectoderm** because of its nature. It is a soft, granular, nucleated protoplasmic layer without cell limits, the nuclei alone indicate its cellular nature (syncytium). It bulges into the pseudocoel at four places forming four longitudinal ridges (Fig. 10.26), the **longitudinal chords**, at **mid-dorsal, mid-ventral** and **lateral** positions. Of these the lateral chords are more prominent and show on the surface as faint pale lines. In most nematodes the chords are better developed anteriorly and tend to diminish posteriorly, the lateral chords persisting from end to end. The nuclei of the epidermis are usually situated in the chords, thus the epidermis between the chords is devoid of nuclei. Beneath the epidermis lies the well-developed muscular layer consisting exclusively of longitudinal fibres having peculiar structure. The muscular layer is divided into four longitudinal bands by the ectodermal thickenings. Thus, there are two **dorsolateral** and **ventrolateral quadrants** of the muscular layer. Each muscle cell, apparently derived from a single cell, consists of a **fibrillar zone** and a **protoplasmic** zone or **medullary substance**. The fibrillar zone (Fig. 10.18) situated next to the epidermis consists of longitudinal bands or ribbons of homogeneous contractile substance separated by non-contractile material. The medullary substance consists of a clear, sometimes granular, protoplasm containing the nucleus, projecting into the body cavity often with prolonged processes.

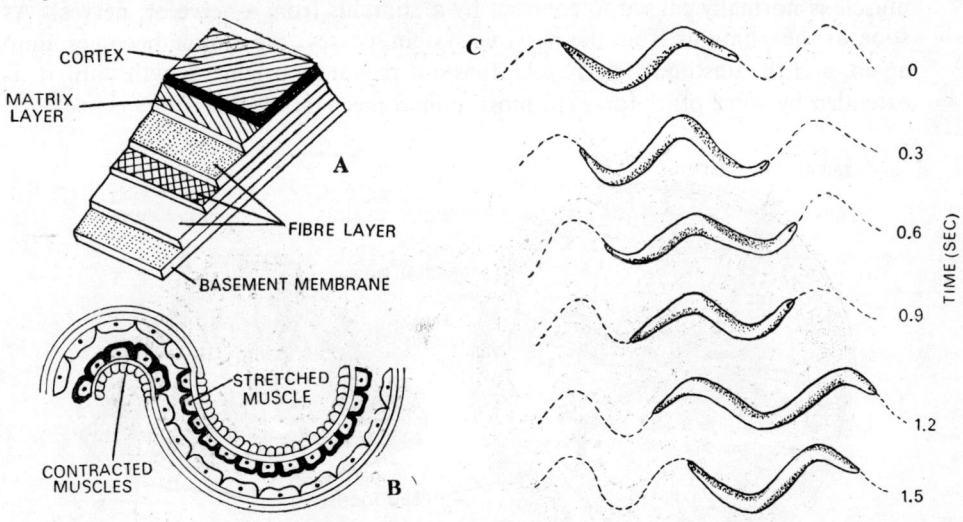

FIG. 10.19. Structure of cuticle and movement in nematodes. A—layers of cuticle; B—State of contraction of the body muscles of a nematode showing a static wave (animal shows no progression) and C—Locomotion in a nematode. In the diagram the nematode is moving in a thin water film. Wave formation and water distribution are indicated in black. The posterior edge of each wave pushes against the water which exerts an equal and opposite thrust on the nematode thus causing the animal to move forward.

**Pseudocoel**. The **body cavity** or **pseudocoel** is the space between the body-wall and the viscera. This cavity is not lined by a mesodermal epithelium, nor is crossed by mesenteries supporting the viscera. The pseudocoel is filled with fluid and usually contains fibrous tissue and fixed cells or nuclei. Distinct cells arranged more or less in rows may also occur usually along one or more longitudinal chords. These cells are termed **pseudocoelocytes** and are connective tissue cells reported to be phagocytic by some and to have an oxidative function by others. A small puncture in the body of living *Ascaris* will produce a sharp jet of pseudocoelomic fluid, demonstrating the high internal turgor pressure (e.g. 16 to 225 mm Hg) under which the fluid is held. To withstand such pressures the body wall and the cuticle are considerably thickened. The fluid-filled muscular tube thus forms a highly developed **hydrostatic skeleton** which maintains a body shape and is also involved in movement.

**Locomotion**

Most nematodes, including *Ascaris*, live in fairly viscous fluids, e.g. mud, plant or animal tissues, through which they move by undulating the body in dorso-ventral waves. These are caused by differential contractions of the longitudinal muscles of the body which are stretched by the hydrostatic pressure of the body fluid in conjunction with the cuticle. At this stage it seems necessary to discuss the importance of pressure of the body fluids (**hydrostatics**) among some lower invertebrates.

To understand the mechanism of locomotion clearly some relatively simple facts are to be accepted. A muscle can exert force only by contracting. A muscle is normally caused to contract by a stimulus from a nerve or nerves. As soon as the stimulus from the nervous system ceases, the muscle becomes limp again, and the unstimulated muscle does not regain its original length until it is extended by some other force. In most animal mechanism this force is provided

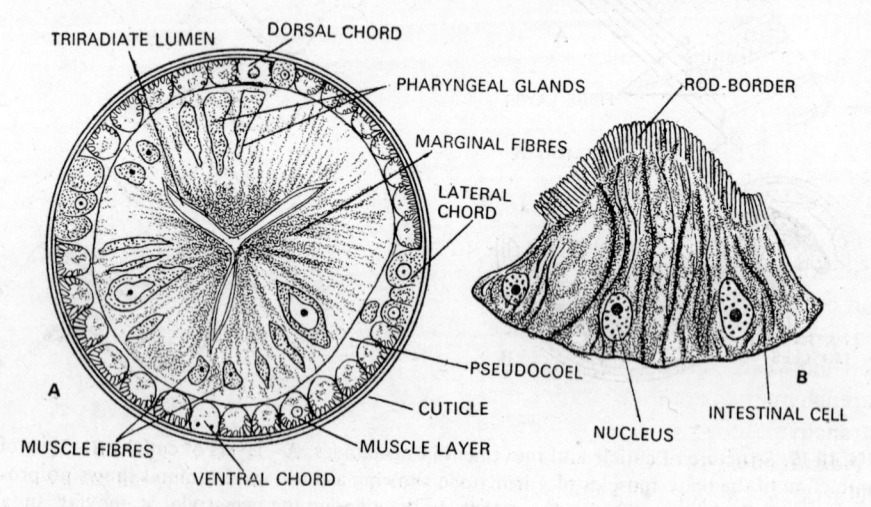

FIG. 10.20. A—Transverse section of *Ascaris* passing through pharynx. B—Transverse section of the wall of intestine of *Ascaris* showing rod border.

by the contraction of another muscle or set of muscles which is said to be **antagonistic** to the first. An antagonistic pair of muscles of course acts upon skeletal elements which transmit the contracting force of one to cause the stretching of the other (unstimulated). In the absence of rigid skeleton the pressure of body fluid plays the part of skeleton. It is, therefore, called **hydrostatic** or **hydraulic skeleton** which causes the stretching of any flaccid muscle. It is clear that the antagonistic (or opposite) relationship to the muscle in lower invertebrates is provided by the pressure of the body fluids. The need is met by the ready availability of water. Two properties of water (which it shares with other fluids) are important in this connection: incompressibility, and the capacity for transmitting pressure changes equally, in all directions. Additionally water has low viscosity with the result that it can easily be deformed. Water, therefore, provides the physical basis of the **hydrostatic skeleton.** The functioning of such a system depends upon the musculature being so arranged that it surrounds an enclosed volume of fluid. In these circumstances the contraction of any one part of the muscular system sets up a pressure in the fluid which is then transmitted in all directions to the rest of the body. The presence of the hydrostatic skeleton enables the two sets of muscles to act antagonistically to each other, bringing about changes in shape and making locomotion possible.

**Coelenterates.** Modern workers have reported the existence of hydrostatic skeleton in coelenterates and flatworms. In the mesogloea of many coelenterates (e.g. sea-anemone *Metridium*) there is a lattice-work of fibres (muscular tissue of the body) which is connected with cell bodies by protoplasmic strands. The interstices of these strands contain fluid, which, it is thought, may have hydrostatic properties that facilitate the movements of cell-bodies during contraction and protect them from excessive local strain. It also aids in the diffusion of metabolites. The coelenteron of the column of *Metridium* contains **coelenteric fluid**. When the coelenteron is closed (by way of the closure of mouth) with fluid enclosed it is able to build up in it a level of pressure that makes possible the translation of muscular contraction into movement. The relationship of coelenteric fluid with the surrounding layers of ectodermal and endodermal muscle fibres constitutes a simple form of **hydrostatic skeleton**.

**Flatworms.** In the flatworm develops a regular mesoderm, an extensive mass of cells which separates the ectoderm from the endoderm. The mesoderm has been called the third germ layer which produces true **mesodermal parenchyma**. The structure of this is somewhat obscure although it is commonly regarded as a syncytium with interstices which are filled with fluid. Whatever its structure its presence greatly aids the morphological and functional differentiation of the organs as a new feature of the group. It relieves the ectoderm and endoderm of some of their primitive and generalized functions, gives them space in which to extend, and provides a measure of transport and communication between them by diffusion and by the passage of fluid. The mesodermal parenchyma is used as deformable **hydrostatic skeleton**. The possibilities of a parenchymatous hydrostatic skeleton are considerable in this group of animals.

**Nematodes.** In *Ascaris* the collagen fibres of the inner layers of the cuticle are arranged to form a spiral lattice-work around the body with the fibres at an angle of about 75° to the longitudinal axis of the body. Any changes in the shape of the body will tend to distort the angle between the fibres and so alter the body volume.

The muscle cells of the body wall are attached to the basal layer of the **cuticle (hypodermis)**, and the remaining portion of each cell projects inward. The muscle cell layer is one-cell thick and the fibres run longitudinally. Four rigid projections of the hypodermis called cords divide the longitudinal musculature into four functional fields. The lateral cords are larger than the dorsal and ventral ones and are seen externally as the lateral lines. Unlike ordinary striated muscles which act between two points of attachment, the muscles of *Ascaris* may contract in groups, i.e., fields, which produce a local shortening (Fig. 10.21), because the volume of internal fluid is constant and incompressible and the in-

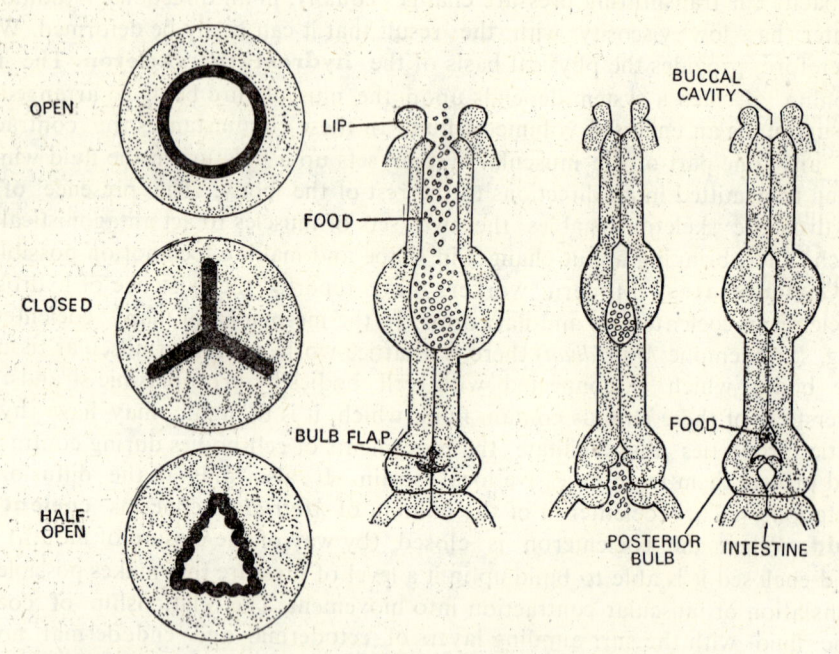

FIG. 10.21. A—Diagrammatic cross sections of the pharynx showing three positions—closed, half open and open; and B—Diagrams of the pharynx during food ingestion.

ternal pressure when increased causes extension of muscle cells in another region of the body. There are no circular muscles, and the high and changeable hydrostatic pressure antagonizes the longitudinal muscles and provides the resorting force. Through this system the dorsal and ventral muscles act as antagonist, producing sinusoidal waves along the length of the body and ultimately causing movement of the whole worm (Fig. 10.21).

Most nematodes lie on their sides, the dorsolateral undulations thus being in the horizontal plane. For progression to occur, the medium in which the worm lies must be relatively viscous. Because relative viscosity increases with decrease in size smaller forms can swim in less absolutely viscous media than larger forms.

## Digestive System

The **mouth** is terminal and is bounded by lips carrying papillae (Fig. 10.17) acting as sense organs. The mouth leads into a narrow **buccal cavity** which opens into the **pharynx** which is usually called the oesophagus. The pharynx acts as a suctorial tube internally lined by cuticle, posteriorly it is enlarged to form a muscular bulb-like portion. The lumen of the pharynx is triradiate (Fig. 10.21). The three narrow grooves apparently divide the wall of the pharynx into three sectors—one dorsal and two ventrolateral corresponding to the three lips in position. In life the pharynx exhibits rhythmic pumping movements (Fig. 10.21) and is bounded externally from the pseudocoel by a thin membrane. The

# Phylum Nemathelminthes (Aschelminthes)

wall of the pharynx consists of a syncytial epidermis which is traversed by muscle fibres and glandular tissue (Fig. 10.20), hence is not very distinct. Numerous radial fibres in clusters extend from the cuticle to the bounding membrane. Typically the pharyngeal wall has three-highly branched **pharyngeal glands**, one dorsal and two ventrolateral. The ducts of these gland open into the lumen of the pharynx. The pharynx leads into the intestine which is a thin-walled straight tube formed of one-layered cellular epithelium and bounded by thin membrane externally. Its walls are cellular and non-muscular. The intestinal epithelium consists of large columnar cells packed with inclusions and provided with **rod-border** (Fig. 10.20), that are probably cilia that have become fixed. The terminal portion of the alimentary tract is known as the **rectum** (Fig. 10.17) that opens to the exterior at the **anus**. It is a short flattened tube having a special investment of muscle fibres which make it contractile. It is lined

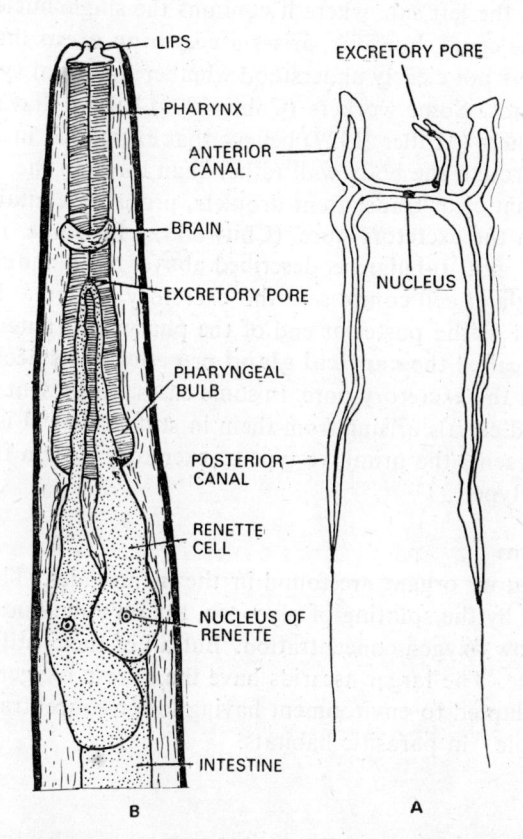

FIG. 10.22. Excretory organs of *Ascaris*. **A**—Excretory tubules forming a typical H-system; and **B**—Renette system (two-celled renette) of *Rhabdias*.

with cuticle being proctodaeal in nature. In many parasitic nematodes the rectum is provided with large unicellular rectal glands. The rectum in the female opens independently at the anus, which is a transverse slit operated by muscle. In the males the rectum serves as a part of gonoducts. The sperm-duct enters the ventral wall of the rectum which serves as a cloaca and is provided with a

**spicule-pouch** that lodges the **copulatory spicules** (Fig. 10.27). The food of the adult nematodes living in the alimentary tracts consists primarily of the intestinal contents. Some people believe that the ascaroids ingest blood, to some extent at least. Food is stored as glycogen and fats.

### Excretory System

The excretory system is peculiar in nematodes. It is devoid of flame-cells or cilia or any current producing mechanism. It consists of two **longitudinal canals** (Fig. 10.22) enclosed in the lateral chords. In the anterior part of the body the two canals are connected by a **transverse canal**. As the transverse canal lies in the anterior part of the body the posterior limbs of the canals are longer than the anterior ones. In *Ascaris* limbs are relatively reduced and the transverse canal and the adjacent parts of the lateral canals form a network which is distinct on the left side where it contains the single nucleus of the system. From transverse canal, however, arises a **common stem** that leads to the **excretory pore**. It is not clearly understood whether the canal system is excretory in function or not. Some workers (Cobb, 1890) believe that urea is ejected by the canals, but others (Müller, 1929) believe that excretion in ascarids probably takes place through the body wall rather than through the canal system. But it is certain that clear transparent droplets, probably containing excretory material, exude from the excretory pore (Chitwood). In some nematodes the excretory system is not tubular, as described above, but glandular. In typical cases a single large gland cell comprises the excretory system. It is ventrally situated in the region of the posterior end of the pharynx or anterior part of the intestine. The cell is called the **cervical gland** or **renette**. Its neck may be short or long and opens at the excretory pore. In some cases are present two gland cells which have elongated canals arising from them in still others. It is now believed that the renette presents the primitive type of excretory system from which has evolved the tubular type.

### Respiratory System

No definite respiratory organs are found in the nematodes. They respire by **anaerobic method** by the splitting of glycogen, the only method possible in the environments with low oxygen concentration. But all of them utilize free oxygen if and when available. The larger ascarids have the lowest oxygen requirements and are thus best adapted to environment having lower concentration of oxygen as is always available in parasitic habitats.

### Nervous System

The nervous system is buried entirely in the tissues of subcuticular layer. It mainly consists of a ring, the **circumenteric ring** (Fig. 10.23), encircling the pharynx and connected with a number of ganglia from which longitudinal cords are given off both anteriorly and posteriorly. The main ganglia include the paired **lateral** and **ventral ganglia** besides which are also present a small **dorsal ganglion**, a pair of **subdorsal ganglia**, a pair of **sublateral ganglia** behind the laterals and a pair of **postventrals** behind the ventral. The lateral ganglia correspond to the cerebral ganglia of other animals. The various ganglia are close to the ring and are connected with it or each other by short

## Phylum Nemathelminthes (Aschelminthes)

commissures. Anteriorly there are six longitudinal nerves (Fig. 10.23) supplying the sense organs of the head. There are eight posterior cords, a **median ven-**

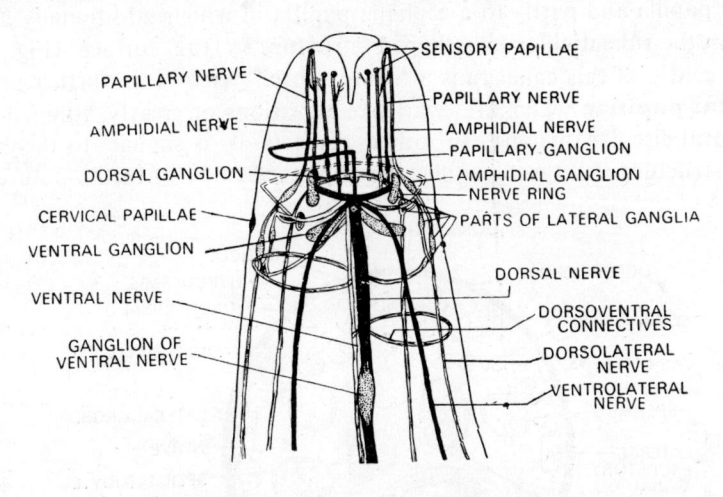

FIG. 10.23. Nervous system of *Ascaris*.

**tral** (largest), a **median dorsal** (next in size), two **laterals** and four **sublaterals**. These longitudinal nerves are connected with each other by asymmetrically placed **transverse connectives** at intervals. Another characteristic of *Ascaris* (and of other nematodes) is that the nerve cells are constant in number, location, shape and course of their fibres. Thus, each ganglion contains a fixed number of cells.

**Sense organs.** Parasitic nematodes lack sense organs like sensory bristles and papillae which the free-living nematodes are abundantly provided with. The

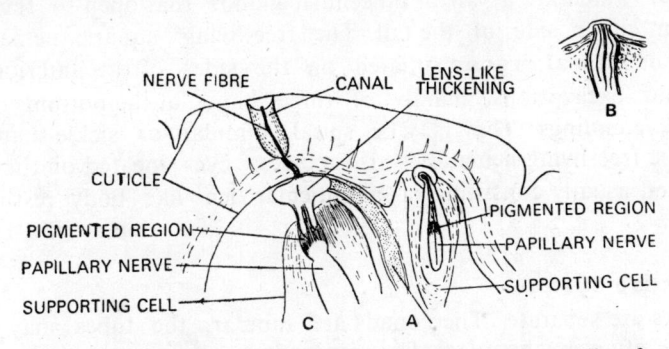

FIG. 10.24. Papillae of Nematoda. A—Outer labial papilla of *Ascaris*; B—Genital papilla of *Ascaris*; and C—Cephalic papilla of the same.

only sense organs found in *Ascaris* are the **labial** and **cephalic papillae** borne by the lips. These papillae appear as circular grooves on the surface perhaps because the cuticle covering the sensory structure is thinned out. The main structure of each sense organ (Fig. 10.24) consists of a stout nerve fibre that ascends towards the surface, narrows and terminates beneath the cuticle in a bulb-like ending. Shortly before it reaches the undersurface of the cuticle each fibre presents a region that strains intensely and each nerve fibre is accompanied by two cells, an **inner supporting cell** that surrounds the fibre, and an **outer supporting cell** surrounding the inner one. This description applies fully to a labial papilla and partly to a cephalic papilla in which additionally a canal passes from the subcuticular lens-shaped structure to the surface (Fig. 10.24). Along the centre of this canal runs a nerve fibre. *Ascaris* is further provided with **genital papillae** which are wart-like elevations of greatly thinned cuticle with a central circular opening. In other respects it is similar to the cephalic papilla in structure. All these papillae are supposed to be of tactile nature.

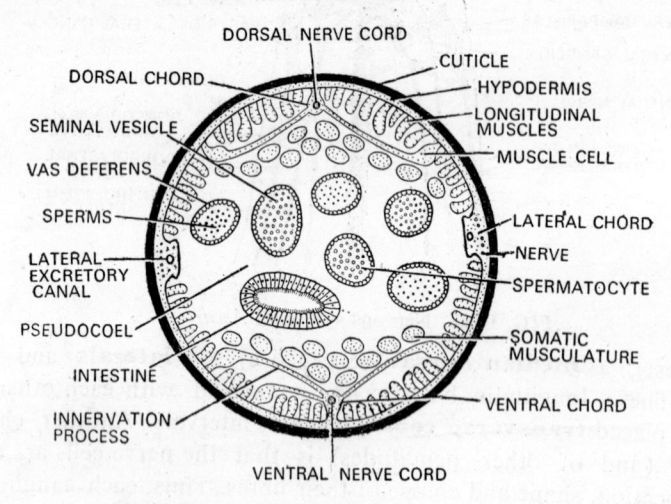

FIG. 10.25. Transverse section of a male *Ascaris*.

Some parasitic nematodes have characteristic glandulo-sensory organs, the **phasmids**, which are a pair of unicellular glands that open to the outside by a pore on either side of the tail. The free-living aquatic nematodes have **amphids** or **lateral organs** situated on the sides of the anterior end. They are cuticular excavations, mainly of three shapes, at the bottom of which are located nerve endings. They may be spiral, circular or sickle-shaped. Besides all these the free-living nematodes also possess eyes located on the sides of the pharynx and usually consisting of a cuticular lens-like body resting on a pigment-cup.

**Reproductive System**

The sexes are separate. The gonads are tubular, the tubes may be single or double and often much coiled. The generative products are developed at the upper end of the tube.

The male generative system (Fig. 10.28A) consists of an unpaired coiled tube occupying a considerable portion of the pseudocoel. The germinal zone of the tube lies at the blind free end and is known as the **testis** as it produces the male germ cells or sperms which are amoeboid in this case. The **sperm-duct** (**vas deferens**) is single continuous with the surface epithelium of the testis. Posteriorly the sperm duct is widened and somewhat muscularized to form the seminal vesicle. Finally, the terminal part is more heavily muscularized forming the **ejaculatory duct** (Fig. 10.28A), which continues to the posterior side and

## Phylum Nemathelminthes (Aschelminthes)

meets the rectum. All these divisions of the testis are continuous, and a sphincter may occur at its junction with the rectum. For copulation each male is provided with a pair of **copulatory spicules** that are lodged in a pair of muscular pouches, the **spicule pouches** (Fig. 10.27). The spicule pouches are formed as dorsal evaginations of the cloacal chamber and they themselves secrete the spicules. In some nematodes the walls of the spicule sacs are provided with special cuticular structures to assist the spicule properly. The dorsal wall of the spicule pouch is provided with the **gubernaculum** (Fig. 10.28), and the ventral wall (in Strongyloidea) is provided with another piece, the **telamon** (Fig. 10.28). Both are formed as a result of sclerotization of the walls and protect the walls from getting ruptured during copulation, and also direct the spicules to function properly.

The **female generative** organs (Fig. 10.17) are represented by paired coiled tubes of the type found in the male, occupying the pseudocoel. The free ends of the tubes produce ova and as such are the **ovaries**. The epithelial covering of the ovary continues behind as the **oviduct** which is slender at first. Soon it widens into a broad tube, the **uterus**, lined with flat or cuboidal cells and provided with muscles. The beginning of the uterus functions as the seminal receptacle that stores sperms, and where fertilization occurs. The seminal receptacle has distinctive epithelium provided with tall processes that almost block the lumen and apparently serve to phagocytize excess of sperms. The uterus stores eggs. As the eggs pass down they are fertilized and develop shell, and to some extent, undergo embryonic development. Distally the two uteri unite and form a common tube lined with cuticle. This

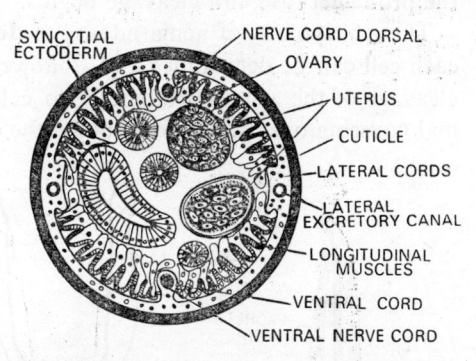

FIG. 10.26. Transverse section of female *Ascaris*.

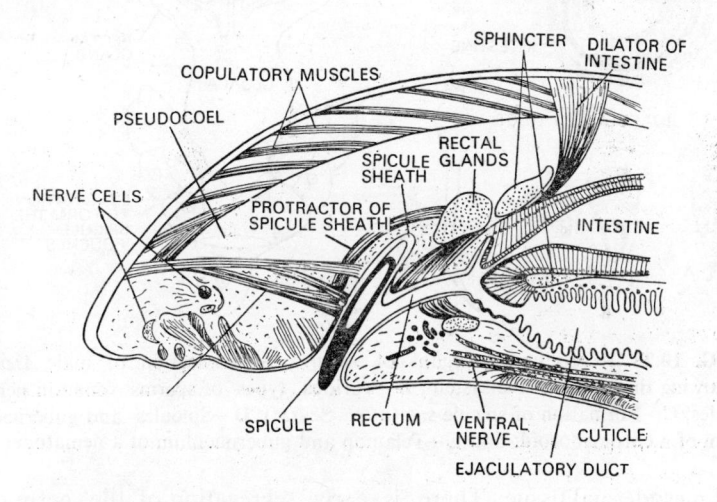

FIG. 10.27. Longitudinal section of the posterior end of male *Ascaris*.

muscularized tube is the vagina that opens to the exterior at the female **gonopore** or **vulva**, which is a mid-ventral transverse slit with bulging lips, situated about one-third of the worm's length from the anterior end.

**Fertilization.** After sperms are transferred from the male to the female, they travel up the female reproductive tract and either fertilize the mature ova or are stored in the seminal receptacle of the female for subsequent fertilization. As soon as a sperm has entered an egg cell, a **fertilization membrane** appears around the egg and a shell begins to form. A lipoidal **vitelline membrane** is produced on the inner side of the shell, and as the eggs passes down the uterus, protein is added on the outside of the shell. The egg of most nematodes are covered by three distinct membranes: (*i*) the external **protein coat** from the uterine wall, (*ii*) a true egg shell formed by the egg itself, and (*iii*) the **vitelline membrane**. When the membranes are being formed meiotic divisions occur, the pronuclei fuse and cleavage begins.

The embryology of nematodes is of **determinate type**, that is, the fate of each cell can be determined and followed from the very beginning. The first cleavage of the zygote produces two cells, one of which forms the ectodermal covering over most of the body, the other forms the stomach, pharynx, intes-

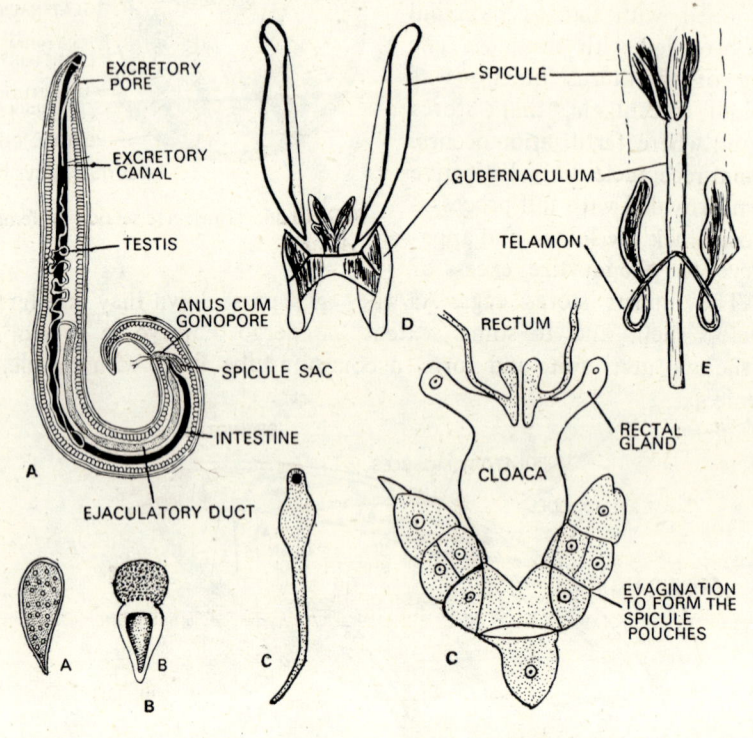

**FIG. 10.28.** Reproductive organs of *Ascaris*. A—Dissection of male *Ascaris* showing the reproductive system; B—Various types of sperms found in nematodes; C—Formation of spicule sac (*after* Seurat); D—Spicules and gubernaculum of a chromadoroid; and E—Telamon and gubernaculum of a nematode.

tine and mesodermal tissue. There is early segregation of the germ cell. The genital primordium is formed by the latter cell at the second division. Progeny of the second cell accumulates in the ventral side and gastrulation by epiboly occurs so that the blastocoel persists to become the pseudocoel of the adult. Following gastrulation, the former spherical, cell-cluster (morula) elongates and assumes a tadpole-like form. By the end of the week the first-stage juvenile, a recognizable nematode develops. This moves within the egg with alternating period of activity and rest. It is elongate with a rounded anterior end and a tapering tail. It is covered with a smooth cuticle with a **flange** or **ala** running along the whole length on each side. At first the first stage juvenile is not visible because large number of granules are scattered throughout the body, but these are gradually eppleted especially from the pharyngeal region and after the first moult the structure becomes clear.

*Ascaris* deposits eggs that have not undergone cleavage and the eggs have a proteinaceous outer coat. The cleavage and development continue. Having completed their development the juvenile worms hatch. Hatching is a strenuous operation. These juveniles are generally fully developed except for the gonads and body size. A significant feature of post-embryonic development is the occurrence of moults.

**Moulting.** Free-living nematodes are easily cultured from the egg stage, using suitable media. They develop through five stages separated by four moults, the fifth stage transforming into the adult. Schematically this can be shown as follows:

$$\text{Egg } L_1 + M_1 \rightarrow L_2 + M_2 \rightarrow L_3 + M_3 \rightarrow L_4 + M_4 \rightarrow L_5 = \text{adult}$$

where $L_{1-5}$ are the larval stages and $M_{1-4}$ the moults.

Examination of cultured worms shows that the different larval stages have different biological characters. For instance, stage, $L_3$ is still sheathed in the uncast cuticle from moult $M_2$ and is extremely resistant to external conditions, especially desiccation. If the cultures were to dry up only the third-stage larvae would survive and these would be able to exist in a desiccated condition for several months or even years in a state of **cryptobiosis**. The term cryptobiosis is used to mean the ability of organisms to survive long periods of desiccation and become active again in the presence of water. Consequently moistening the dry and inert larvae with water would allow them to become active again and to complete their development. In natural condition this specialized larval stage aids survival and even dissemination of nematodes.

**Host.** Host is an organism on which another lives as a parasite. There may be a **definitive host**, an **intermediate host** and a **paratenic host**. The term definitive host is used for the host in which the parasite becomes adult and lays its eggs. The intermediate host contains the infestive larva and must be eaten by the definitive host or used as a means of introducing the larvae in the final host. The intermediate host is, therefore, indispensable to the completion of the life cycle. There is also a category of host known as the paratenic host in which the infestive larvae accumulate without being able to undergo further development. This occurs when the intermediate host is eaten by a host which is not definitive host and in which the life cycle cannot continue. The paratenic host is not necessary to the fulfilment of the cycle but it can help to disseminate the parasite, and under certain ecological conditions could favour completion of the life-cycle.

## LIFE-CYCLES IN NEMATODES

Three different types of life-cycles are known among nematodes. A life cycle which involves no intermediate host is termed **monoxenous**, while the one

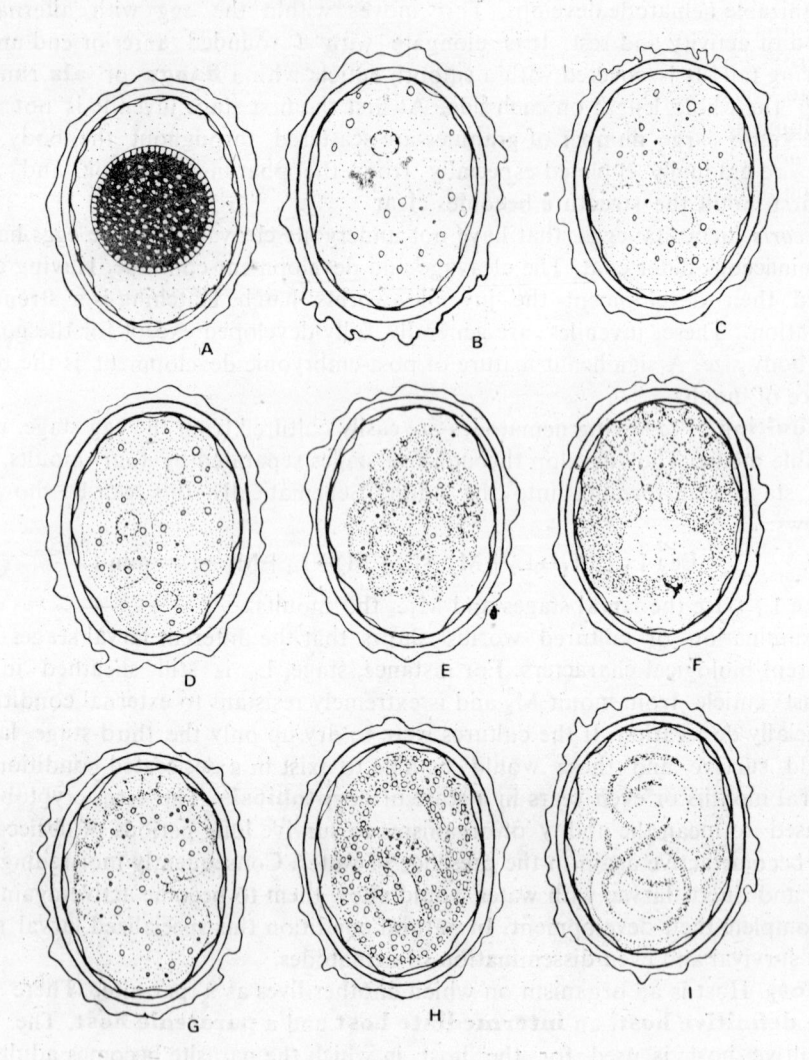

FIG. 10.29. Stages in the development of *Ascaris suum*. A—Fertilized egg; B—Ready for cleavage; C—First cleavage; D—4-celled embryo; E—Morula F—Blastula (segmentation cavity appears); and G to I—Embryo formation.

involving an intermediate host is termed **heteroxenous**. There is also a condition where the definitive host can also serve as an intermediate host, as occurs in *Trichinella*. This kind of life history is known as **autoheteroxenous**.

### 1. Monoxenous Cycles

Monoxenous cycles may be with (*a*) a free-living larval stage, or (*b*) without it.

(*a*) In the first category come those primitive life-cycles in which a parasitic generation and a generation free-living in the soil are produced successively. In *Rhabdias bufonis*, for instance, the parasitic generation lives in the lungs of frogs and toads as a protandrous, hermaphrodite female. Eggs laid with faeces hatch in humid soil. They develop into the third stage larvae that develop in two ways depending upon the circumstances. The third-stage larvae may invade a new host or may develop into free-living adult male and female worms. This is heterogeneous generation which lays eggs that produce infestive third-stage larvae. Then infestive larvae penetrate via the skin of the host and are carried to the lungs in the blood stream. The same kind of life-cycle occurs in *Strongyloides stercoralis*, an intestinal parasite of man.

The parasite with no heterogeneous generations may have life-cycles of two types: (*i*) with no larval migration in the host; and (*ii*) with larval migration in the host. The example of the first type is *Trichostrongylus* spp. of ruminants that lives in the intestine from where the eggs are evacuated. These hatch on damp ground and the third-stage larvae migrate to blades of grass where they remain until eaten by the host. They enter the intestine directly.

In the second type fall the hookworms of man and carnivores. They lay eggs in the gut. These hatch in the soil to produce third stage larvae which penetrate the skin and are carried to the lungs in the circulatory system. They spend little time in the lungs and pierce the alveolar walls, pass up the trachea and are swallowed, to complete their development in the gut. They have to migrate in the host body to reach the definitive place for their development.

(*b*) Monoxenous cycles with no free-living larval stages may also be of two types: (*i*) with no larval migration in the host; and (*ii*) with larval migration in the host. The example of the first type is found in the oxyurids of the genus *Oswaldocruzia* inhabiting the intestine of amphibians, the larvae moult twice in the egg so that this (egg) contains a third-stage larva which can directly infest a new host. In the case of the human oxyurid, *Enterobius vermicularis*, the female leaves the intestine and lays eggs on the peri-anal skin. The irritation excites scratching and the eggs with infective juveniles reach the mouth through finger-tips.

The example of the second type is provided by the ascarids of man, horse and pigs. The eggs are laid in the intestine and reach the ground in a comparatively undeveloped stage. They, therefore, have to remain in the soil until they become mature and contain a second-stage larva. When the egg is swallowed, the second-stage larva is released and bores through the wall of the intestine, to be taken to the lungs in the blood stream. The third-stage larva develops in the alveoli and from here migrates via the bronchi into the trachea and back down the gut where maturity is reached.

## 2. Heteroxenous Cycles

These are of three types: (*i*) heteroxenous cycles with a single intermediate host and a free-living larval stage; (*ii*) heteroxenous cycles with a single intermediate host and no free-living larval stage; and (*iii*) heteroxenous cycles with two intermediate hosts and a free-living larval stage.

(*i*) *Habronema muscae* lives in the stomach of horses, the eggs hatch in the intestine, and the first-stage larvae pass out in the dung. The larvae are eaten by the maggots of houseflies and develop into infestive third-stage larvae which accumulate in the labium of the intermediate host as soon as the adult fly leaves the pupa. The fly is attracted to the lips of horses by their humidity and as soon as it settles there the infective larvae actively penetrate and traverse down the animal's stomach. In tropical regions the fly vector is often attracted to uncovered sores which the nematode larvae penetrate. The worms do not develop here but cause irritation, termed **cutaneous habronemiasis**, which complicates the injury. Here there is no larval migration in the definitive host.

The protostrongyles are largely lung parasites of ruminants. Eggs hatch in the alveoli and are passed out via the intestine. First-stage larvae enter the feet of various molluscs in which the third-stage larvae are formed. The host acquires the infestive larvae by eating infested snails. From the host intestine the larvae migrate to the lungs.

(ii) *Dracunculus medinensis* (the Guinea worm) is a well-known parasite of man living under the skin usually of the leg. The female worm ruptures the covering of the skin and emits hundreds of first-stage larvae which develop into third-stage larvae in the body of a copepod. Man is infested by drinking water containing copepods and the larvae leave the intestine and spend some time in the mesentery. Infestation with the adult worms becomes apparent only after about twelve months.

*Portocaecum ensicaudatum* inhabits the intestine of various birds. Eggs are evacuated on to the ground and the first-stage larva moults within the egg. This with the enclosed second-stage larva is carried to the intestine of an earthworm where the larva hatches and migrates to the blood vessels, and develops into infective third-stage larva. When the bird feeds on the earthworm, the larva moults in the intestine and develops into adult nematode.

The filaria worms of man inhabit the connective tissue of different regions according to the species in question. They are all viviparous and, apart from one exception (*Onchocerca*), the first-stage larvae, called microfilariae, occur in the blood. Their mass appearance in the blood follows a periodicity appropriate to the particular species of filaria worm. The microfilariae are ingested by haematophagus insects (flies or mosquitoes), and the larvae moult within the bodies of these vectors. The infective larvae become concentrated in the labium of fly vectors and when these bite a new host the labium ruptures and the larvae are released on the skin entering the body rapidly and migrating to the preferred site in which the adult filaria worms occur.

Filariasis, caused by the *Culex* mosquito, begins with a slow, inflammatory swelling of the leg, scrotum or arm. If left unattended, the swelling can grow to elephantine proportions. Over eight million Indians bear the mark of this disease, which is prevalent throughout the country, except in the northwest and in some eastern states. The National Filariasis Control Programme started in 1955, tackled the disease by initiating three steps: mass administration of diethyl carbamazine (DEC), an anti-parasitic drug; spraying of dieldrin, a common insecticide in endemic areas; and attacking the mosquito larvae with the help of mosquito larvicidal oil (MLO) at weekly intervals. But a reassessment in 1962 showed that DEC had severe side-effects and moreover many mosquitoes had become resistant to dieldrin. Since then only the third measure had been employed and filariasis continues to pose a serious threat to the people.

*Phylum Nemathelminthes (Aschelminthes)* 365

(*iii*) *Heterogenous cycles* with two intermediate hosts and a free-living larval stage. No larval migration in the definitive host: *Gnathostoma spinigerum* is a parasite of the intestine of a cat where it often produces tumours. The eggs, laid into water, enclose a second-stage larva which hatches and has to be eaten by a copepod. Curiously enough however, it does not develop to the third-stage larva here. This stage is reached in a fish which acquires the infestation by eating copepods.

### 3. The Autoheteroxenous Cycle with No Free-Living Larval Stage

(*i*) **Larval migration in the host.** This is very specialized kind of life-cycle, in which the definitive host inevitably becomes the intermediate host. It is found only in *Trichinella spiralis*, a parasite of man, and a large number of domestic and wild mammals. The adult worms occur in the intestine where the viviparous females burrow into the mucosa and here produce first-stage larvae, which are carried to the heart and then circulated by blood throughout the body. But they can continue their development only in striated muscle where they reach the infective stage and cause the host tissues to react and enclose the larvae in a cyst. Within this cyst the larvae may remain infective for several years. A new host becomes infected by consuming the flesh of the parasitized host. Transmission of *Trichinella* must originally have been favoured by cannibalism as occurs, normally, among rodents. Man gains the infestation by eating contaminated pork and pigs, acquire *Trichinella* by eating pork scraps or occasionally infested rats. Numerous other mammals can act as reservoir hosts in all latitudes, among them being the polar bear in the Arctic and the bush pig in Africa. In temperate regions rats are the main reservoir hosts.

### PLANT PARASITIC NEMATODES

Several hundred species of nematodes living in the soil become parasites of the plants, thus causing a variety of plant diseases. Plant parasitic nematodes are

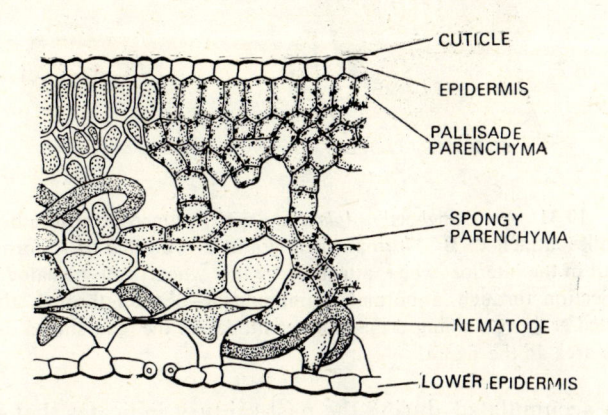

FIG.10.30. Transverse section through the leaf of a plant showing nematodes. They enter the tissue through the stomata and feed on the sap inside.

thousand times larger than bacteria, they are just a little too small to be easily seen with the naked eye, even when separated from the soil. The length of the full grown plant parasitic nematode may be less than one sixty-fourth of an inch and seldom exceeds one eighth of an inch.

The life-history of plant parasitic nematodes is simple enough. Eggs may be deposited in the soil or in the plant on which the female feeds. In the eggs the immature forms develop and eventually hatch. If plants on which they can feed are available, they may begin to feed immediately developing through several distinct stages. After the last moult the nematode becomes sexually mature and able to reproduce.

Most of the forms that have been closely studied have a minimum length of life-cycle, from egg to egg-laying female, of several days to several weeks. The maximum time may be much longer as sexual maturity is not reached until the nematode begins to feed on living plant. The species found in cold climate can easily live in winter and are not killed by freezing of the soil. Some species on drying enter a dormant state in which they can remain alive for years and from which they can revive in short time if moistened. The most remarkable of these are parasites of wheat, rye, and other grains and grasses. The wheat nematode has been revived after dry storage for 28 years. A species that parasitizes rye has been revived after 39 years.

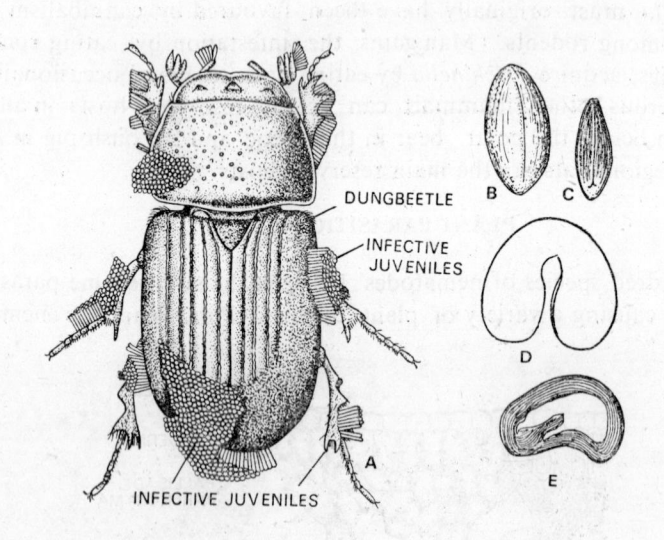

FIG. 10.31. A—Dungbeetle *Aphodius* with numerous ensheathed juveniles attached; B—Normal wheat grain; C—*Anguina tritici* forms a gall in the interior when attached by the wheat gall nematode; D—Section through a normal grain; and E—Section through the infected grain containing a mass of nematodes in the gall shown by white area in the figure.

Information accumulated during the past century indicates that all crops and ornamental plants grown in the world can be attacked by plant parasitic nematodes. That does not imply that any species of plant parasitic nematode can attack any kind of plant. All plant parasitic nematodes are more or less specialized, attacking some plants freely, leaving others totally unharmed. The species range from highly specialized (those that attack only a few kinds of plants) to phytophagous (those that attack different kinds of plants). The reasons, therefore, are not known. The damage to the plants assaulted by nematodes is due primarily to their feeding on plant tissue. All the important plant parasitic nematode species have a special feeding organ known as stylet or spear.

Some members of the order Rhabditoidea (family Tylenchidae) include most important plant parasites (phytoparasites). As these have a buccal capsule armed with a spear they pierce plant tissues and cells and suck juices. The pharynx in these has a median muscular bulb. Three genera of this type are very important. They are *Tylenchus* (*Anguillulina*), *Ditylenchus*, and *Anguina*. Of these *Tylenchus* is associated with fungi, moss and liverworts. *Ditylenchus dipsaci*, the stem and bulb eelworm attacks a number of plants including rye, oats, and other grasses and numerous such bulbs as lilies, onions narcissi and gladiolus, etc. *Anguina tritici,* wheat-gall eelworm, converts wheat grains into galls containing dormant young ones that migrate to the soil, as the galls soften and decay, and attack new plants. They enter the developing inflorescence, produce galls within which they mature, breed, and produce young ones (Fig. 10.32 B, D).

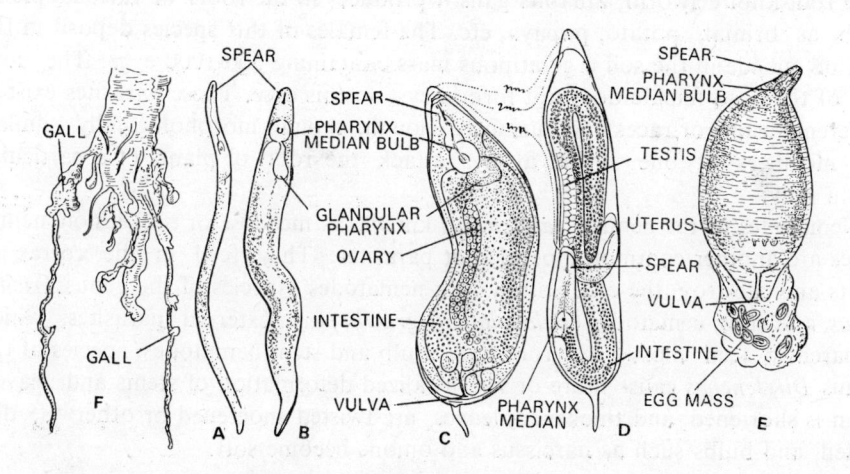

FIG. 10.32. Root-knot eelworm *Meloidogyne marioni*. A-C—Second, third and fourth stage females; D—Fourth stage male inside third cuticle; E—Mature female; and F—Root infected with eelworm showing galls.

Other phytoparasitic Rhabditoidea are *Rotylenchus, Aphelenchus, Meloidogyne* and *Heterodera. Rotylenchus similis* is found in tropical and sub-tropical countries and attacks roots of coffee, tea, banana, pepper, sweet-potato, etc. *Aphelenchus arenae* occurs on the roots of a number of plants and is often associated with fungus infection. Probably the easiest of the nematode disease to recognize is root knot caused by various species of the genus *Meloidogyne* (formerly grouped under the name *Heterodera marioni*). They form galls (knots) on roots of great variety of plants including trees and shrubs. Therefore, they are called **root-knot** worms. The genus also shows sexual dimorphism. Young ones after one month penetrate the host plant by piercing and sucking the cell contents. On entering the tissue they usually produce galls within which they develop to sexual maturity. In the beginning they are tender filiform but with successive moults, usually, three, the female becomes more and more plump finally assuming a pyriform shape (Fig. 10.32). They remain within the galls, the outer tissues of which often disintegrate at the top through which they often protrude. The males develop in the same way, after three moults, but remain vermiform in appearance. They leave the parent gall, wander through the soil to find and fertilize females on other plants. Ultimately they die. The females also degenerate after producing 200-500 eggs leaving their cuticle as a protective cyst for the eggs. The cysts become free in the soil as the infected part of the host plant disintegrates and may remain there inactive for a long time, even withstanding adverse conditions. Within them development occurs and second stage juveniles are produced. They escape into the soil and attack fresh plants or reinfect the uninfected parts of the same plant. They may live for months without feeding on and attacking fresh plants. *Heterodera schachtii*, the sugarbeet eelworm, attacks mainly sugarbeets and also a variety of other plants of the families Chaenopodiaceae and Cruciferae. *H. rostochiensis*, the golden potato eelworm, attacks the plant throughout the world and is highly injurious. The root-knot eelworm, inhabits galls it produces in the roots of various plants such as brinjal, potato, papaya, etc. The females of this species deposit in the gall or outside in the soil a gelatinous mass containing 500-1000 eggs. The cuticle of the dead female does not form a cyst in this case. These parasites exist in different strains or races and curiously enough, though morphologically similar, the members of one strain always attack the roots of plants of a particular strain only.

Nematodes of the genus *Pratylenchus* known as meadow or root lesion nematodes are another common type of root parasites. They feed in the cortex of roots and destroy the cell. Stubby root nematodes (species of the genus *Trichodorus* and sting nematodes (*Belonolaimus gracilis*) are external parasites, which apparently feed mostly on root-tips. Bulb and stem nematodes, species of the genus *Ditylenchus* cause more or less localized deformation of stems and leaves. Stem is shortened and thickened, leaves are twisted shortened or otherwise distorted, and bulbs such as narcissus and onions become soft.

### NEMATODE PARASITES OF ANIMALS

Many nematodes are parasitic on other animals and some cause many diseases in higher animals including man. Many, however, parasitize and destroy insects. The number of nematodes so far reported from insects probably exceeds 1000 species. Most of the reported hosts are in the orders Lepidoptera (approximately 300); Coleoptera, Orthoptera, and Diptera, having at least 100 species each as hosts to nematodes.

The young or juvenile stage of mermithids (order Mermithoidea) are parasitic chiefly in insects (Fig. 10.33) and other invertebrates such as spiders, crustaceans and snails. The hosts become infected either by eating mermithid eggs or sometimes the juvenile worms directly penetrate. The parasites live in the haemocoel and kill the host or at least render it incapable of undergoing metamorphosis and reproduction. Juvenile mermithids are usually provided with stylets for

*Phylum Nemathelminthes (Aschelminthes)* 369

boring, generally lacking in the adults.

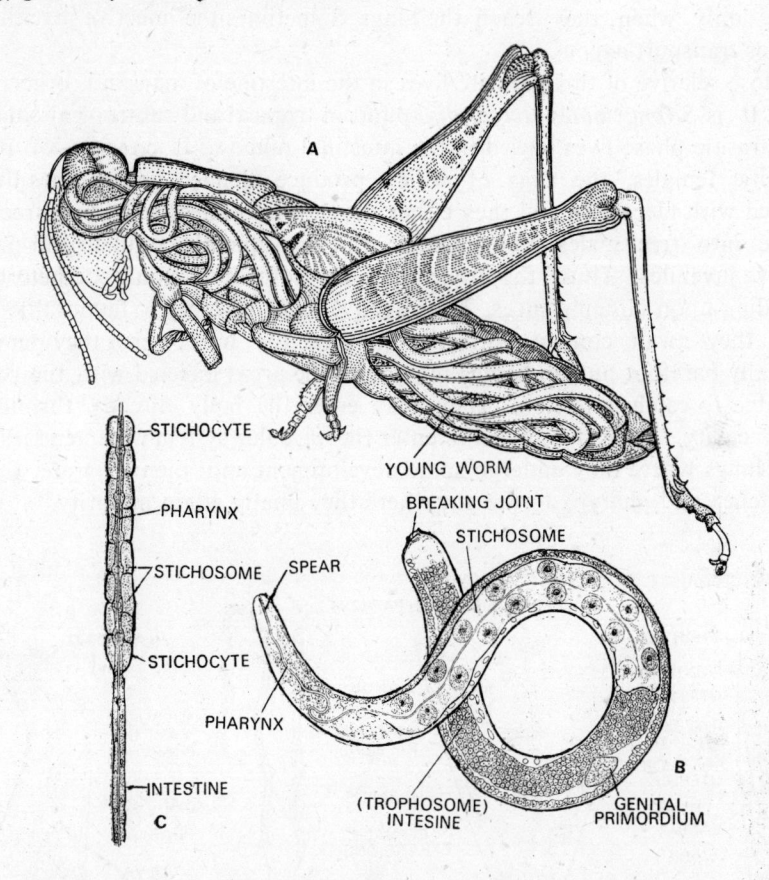

FIG. 10.33. Nematodes of Mermithoidea are parasitic on insects. A—A grasshopper infected with young ones of *Agamermis decaudata*; B—Juvenile *Agamermis* after 6 days in the host (*after* Christie); and C—Stichosome arrangement in a mermithid.

It may be of interest to note that many insect-destroying species are these days being used in biological control and destruction of harmful insects.

## Lung Nematode of Frog

*Rhabdias bufonis* is the common lung nematode of the frog. It has a complicated life-cycle. The parasite that lives in the lungs is a protandrous hermaphrodite with the structure of the female. In the early phases the gonad produces sperms that are stored in the seminal receptacle. The developing eggs pass into the buccal cavity of the host and hence they are swallowed into the digestive tract where they hatch into young ones which accumulate in the cloaca and finally move out with the faeces. In the soil they mature into free males and females. The females after impregnation produce elongated slender young ones within their uteri (viviparously). The young ones develop within their mother and then escape as ensheathed juveniles which are the infective stages and must penetrate into frogs to escape death. They work their way through the sink, enter the lymph channels and from there reach various other organs. They mature only when they reach the lungs. Sometimes the infective juveniles use snails as transport agents.

A close relative of this parasite lives in the intestine of man and other mammals. It is *Strongyloides stercoralis* found in tropical and subtropical countries. The parasitic phase lives buried in the intestinal mucosa. It consists of parthenogenetic females, the eggs of which produce slender young ones that are excreted with faeces. In soil they either develop into infective stages directly or mature into free males and females who subsequently copulate and produce infective juveniles. Thus, the cycle may be both homogonic or heterogonic depending upon circumstances. The infective juveniles creep to favourable places where they await contact with the skin of the host which they penetrate. Naturally barefoot human beings moving in the areas infected with the parasite are liable to catch infection. They may enter the body directly through the mouth cavity. Then the juveniles enter the vascular system which carries them to the lungs where they undergo some development and then migrate through the trachea and pharynx to the gut where they finally attain maturity.

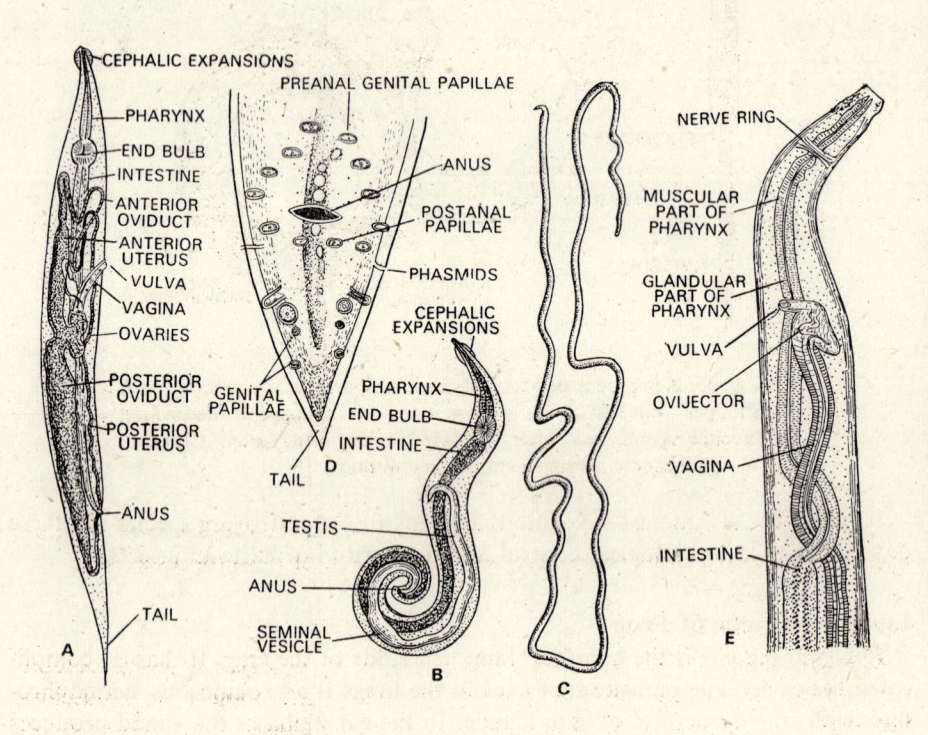

FIG. 10.34. Some known animal parasites. A—Female *Enterobius vermicularis*; B—Male of the same (*after* Leucart); C—*Dracunculus medinensis*, female; D—*D. medinensis* posterior end; and E—*Wuchereria bancrofti*, female.

As a rule the parasite does not cause very serious ill-effects though it lives quite happily buried in the intestine. Formerly they were believed to be the chief agents of dysentery and diarrhoea, but now these cases are known to be due to other agents, strongyloids are more or less innocent bystanders. They are, however, reported to cause certain nervous symptoms. In many cases where a diseased condition is brought about by other causes, the strongyloids increase in number and make the condition worse.

The Spiruroid worms have complicated life-cycle involving an intermediate arthropod host or sometimes two intermediate hosts. The eggs that are laid contain a developing young one which hatches only if swallowed by the intermediate host. In the body of the intermediate host the eggs hatch and the young ones pierce the haemocoel and other tissues, they finally become encapsulated by the host's tissues and further develop only when the infected intermediate host is devoured by the final host. The main genus *Rictularia* (Fig. 10.36 F), has two ventral longitudinal rows of comb-like spines, the mouth opening is surrounded by small teeth and the buccal capsule also possesses teeth and spines. This lives in the intestines of mammals. The genera *Thelazia* and *Oxyspirura* are known as eye-worms. They parasitize eyes in the domestic animals such as horses, camels and pigs, etc., and occasionally man. They live on the surface of the eye in the inner corner and may move to other parts of the eyes. *Thelazia callipaeda* is the Chinese human eye-worm (Fig. 10.39 D). In this case the egg membranes swell to form a float so that the young ones may not drown in the conjunctival fluid. *Pneumonema* is another genus that lives in the lungs of reptiles.

## PINWORMS OF CHILDREN

The most common human parasite is the pinworm of children *Enterobius* (*Oxyuris*) *vermicularis* (Fig. 10.34). This parasite occurs almost universally in children

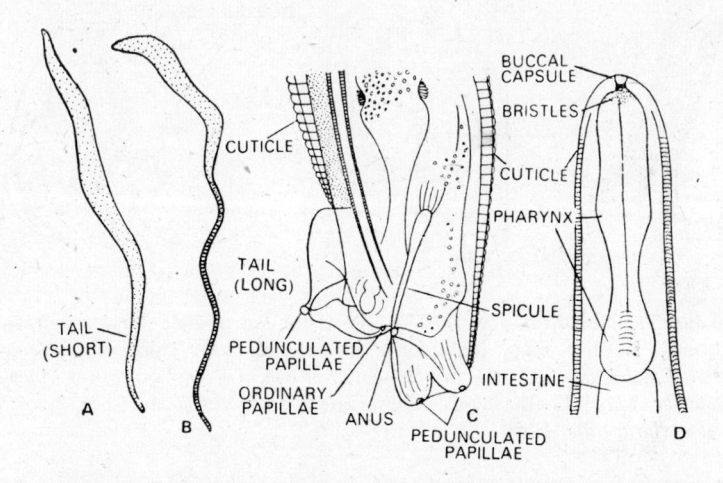

FIG. 10.35. *Oxyuris equi.* **A** and **B**—Short and long tailed females; **C**—Posterior end of male, side view; and **D**—Anterior end of a female, dorsal view.

at one time or another, in temperate as well as tropical countries. The parasites inhabit the terminal part of the colon and have the diameter of a pin. The males are only about half as large and have their posterior end curved ventrally. The mature females filled with eggs creep out of the anus, especially in the evening or at night, causing intense itching. As they come in contact with air they are stimulated to lay eggs which are laid in the moist grooves between the buttocks. From the scratching and rubbing the fingers and finger-nails become infected with eggs that may be transferred to the mouth directly or indirectly causing reinfection. They may be transferred from one person to another by unclean hands. Infection may also occur by eating raw vegetables and other food material which are polluted by the eggs. The eggs of pinworms may also be scattered by flies which visit infected faeces.

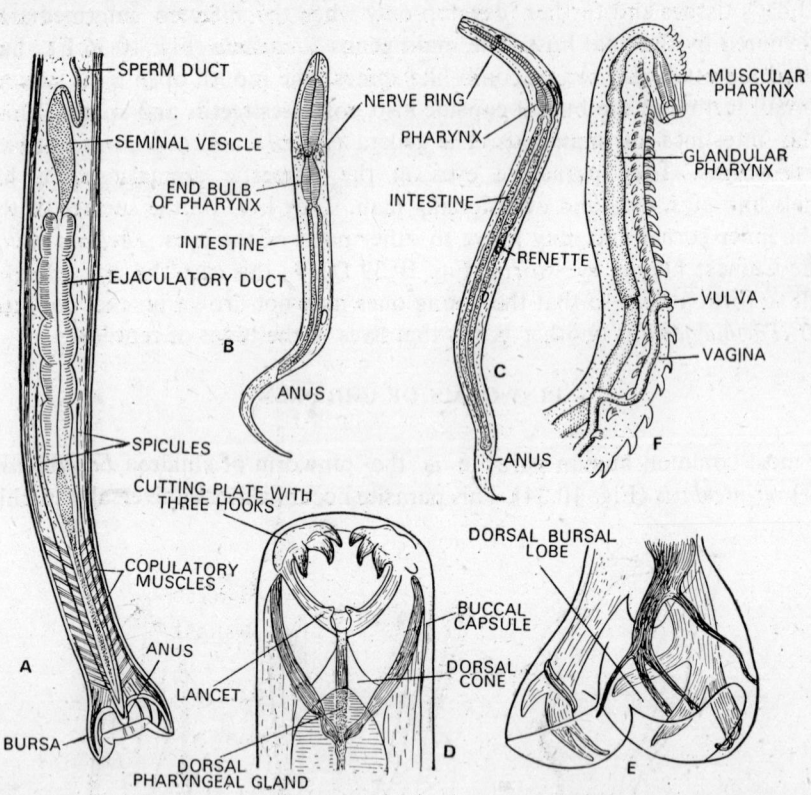

**FIG. 10.36.** The human hookworms. **A**—Posterior end of male *Ancylostoma duodenale* (*after* Looss); **B**—first stage rhabdiform juvenile; and **C**—Third-stage strongyliform juvenile of *A. duodenale* (*after* Looss); **D**—Anterior end of a dog-hookworm *Ancylostoma caninum* (*after* Lane); **E**—Bursa of *Bunostomum trigonocephlaum* (sheep hook worm); and **F**—*Rictularia* (*after* Hall).

The presence of pinworms does not inconvenience the children as they do not suck blood, and seldom cause wounds in the alimentary tract, though they sometimes produce reflex nervous symptoms probably by secretions of toxins. They may also interfere with the normal movement of bowels. Pinworms are believed to be the original cause of lesions in the appendix causing appendicitis. The intense itching which they produce by creeping in the vicinity of the anus is the most disagreeable effect of their presence.

The pinworm belongs to the order Oxyuroidea. Members of the order are found in other mammals and birds.

# Phylum Nemathelminthes (Aschelminthes)

Of these *Oxyuris equi* is a parasite occupying the colon and caecum of the horse. The females have varying tail lengths and those with very long tails were sometimes regarded as separate subspecies. Ripe females protrude from the anus and deposit their eggs around the anal region fastening them by a sticky secretion. The eggs develop on exposure to air. The advanced young ones fall off and are ingested by horses while grazing (Fig. 10.35).

*Heterakis* is another important genus of the order Oxyuroidea. The males have well-developed caudal alae and they are parasitic chiefly in birds and a few species occur in mammals also. The most cosmopolitan species is *Heterakis gallinae*, that lives in the caeca of domestic poultry. The life-cycle is simple as in other oxyuroids.

**Ascarid worms.** The worms belonging to the order Ascaroidea are gut parasites of vertebrate animals. The genus *Ascaris*, which probably includes the oldest known parasite of man, formerly had nearly one thousand species, but is now restricted to only a few members parasitic in mammals. Of these the most familiar is *Ascaris lumbricoides*. It is so long that it could hardly be overlooked, when passed from the body, even by the prehistoric people. The adults assume a size of 22-25 cm and are as thick as lead pencil.

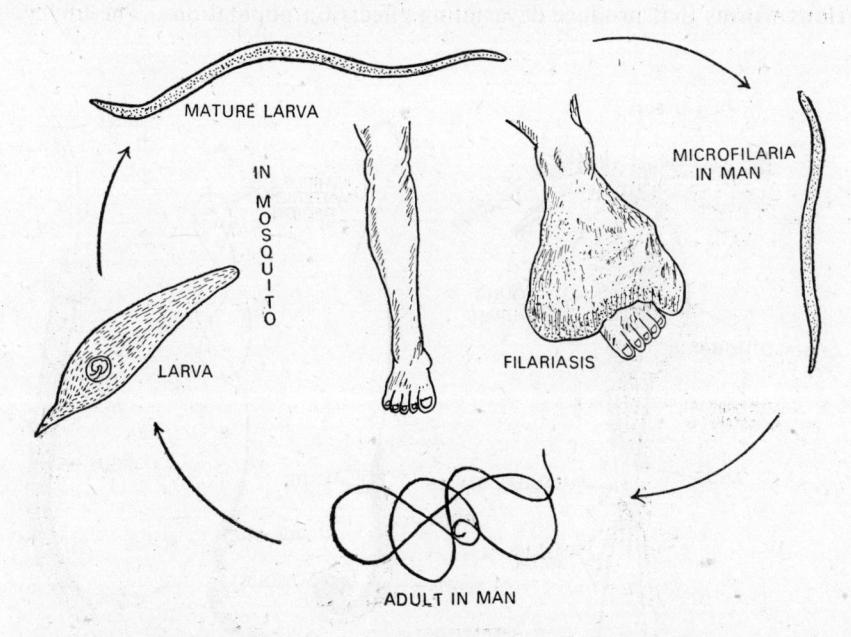

FIG. 10.37. Life-cycle of *Wuchereria bancrofti* that causes elephantiasis.

The worm is found in groups of five or six in the upper part of the small intestine, and may easily pass into the stomach and be vomited therefrom. It is widely distributed and is common in children, especially in tropical countries. The eggs are ejected with the faeces of the host. They are oval have a thick and warty, but transparent shell, and when freshly passed are unsegmented; and are fully developed in 10 days to a month. Under adverse conditions they may remain dormant for years. They are extraordinarily resistant to extremes of temperatures or moisture, even to chemical disinfectants or to acids and alkalis.

They enter the host along with vegetables or unfiltered water and hatch in small intestine. The larvae penetrate the gut wall and are transported by blood to liver or lungs. They may provoke mild bronchitis. They pass into the bronchial tubes of the lungs, to the trachea, to mouth and are carried back to intestine where they settle and grow.

They affect the general health by absorbing nutritive elements present in the diet, and sometimes mechanically by causing an intestinal obstruction. The production of toxic and allergen-like substances by these worms is indicated by the eosinophilia. Adult roundworms can penetrate the gallbladder and bile ducts. The larvae in the lung produce infiltrations which can be seen by X-ray. The ubiquitous roundworm is the most frequent intestinal nematode parasitizing man. Children are especially affected.

Other ascarid genera include *Neoascaris* with a short glandular region at the rear end of the pharynx, parasitic in bovines; *Parascaris* parasitic in horses; *Toxocara* and *Toxascaris* parasitic in carnivores; *Ascaridia* having rimmed preanal sucker is parasitic in birds.

**Human hookworms.** The most known worms are the human hookworms *Ancylostoma duodenale* (Fig. 10.36) and *Neacter americanus*. They are the most injurious worms that produce devastating effects on populations. The life-cycles

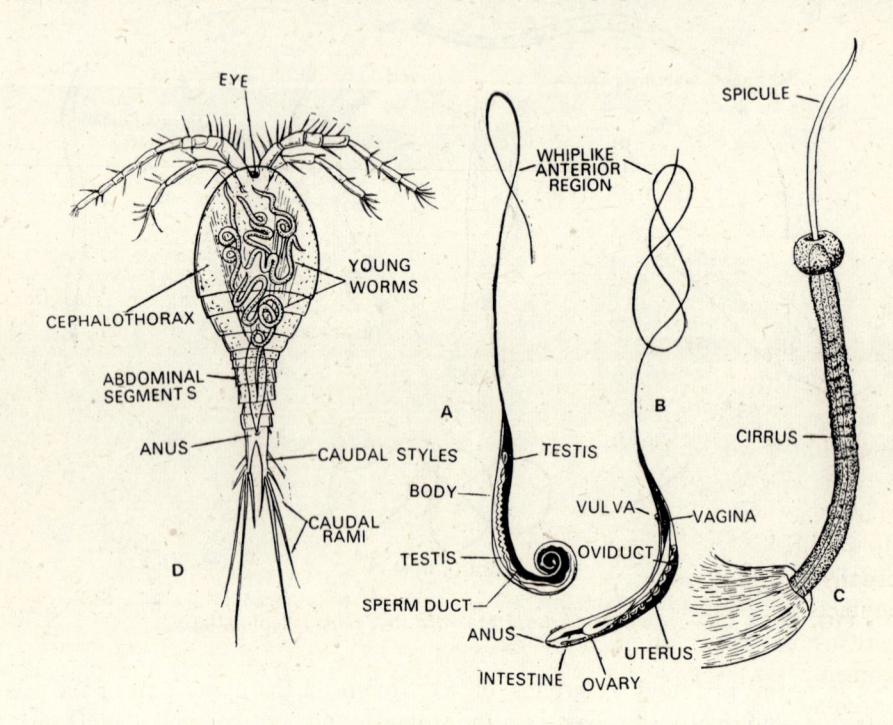

FIG. 10.38. A—*Trichuris ovis* female; B—Male of the same; C—Cirrus and everted spicule of the same; and D—Cyclops carrying young worms.

in both the cases are similar. The adult *Neacter americanus*, for instance, contains some plates in the mouth cavity by which it grasps the intestinal wall and sucks in blood and tissues while it holds the host. The eggs pass out in the faeces and develop in the soil under warm and moist conditions, where they hatch into slender worms feeding mainly on bacteria. They moult twice and reach the infective stage in which they remain ensheathed in the cuticle of the second moult. In moist conditions they move to suitable places where they may get opportunity to penetrate into a suitable host. On getting an opportunity they penetrate the human skin, leaving their sheaths behind, and enter the lymphatic system, whence they are carried to the lungs. From the lungs they move to the pharynx and are swallowed back to the digestive tract. If the infection takes place by way of mouth they directly mature omitting the passage through lungs. The infection occurs in habitual barefoot walkers in localities where the soil is likely to contain human faeces. *Ancylostoma duodenale* is found in western Europe, mainly in mines, and in tropical and subtropical areas of Asia, (including India), Africa and various Pacific Islands and South America. This is known to cause disability among labourers working in mines throughout Western Europe. *Neacter americanus* is known as "the American killer" and is found in tropical and subtropical countries including the south eastern United States. Hookworm disease is generally confined to rural parts with adequate moisture and temperature necessary for the development of the juveniles in the soil. The symptoms of the disease are widely known anaemia, laziness, and general lack of physical and mental energy. These conditions lead to retardation of physical and mental development so that an infected child of 15 years of age may appear to be only ten years old. Negroes harbour these worms but are not so susceptible to their harmful efforts.

Related hookworms of economic importance include *Gaigeria pachyscelis* in sheep and goats; *Uncinaria stenocephala* in dogs, wolves and foxes; *Bunostomum* (=*Monodontus*) *trigonocephalum* and *phlebotomum* in cattle, etc.

## THE GUINEA WORM

The common guinea worm or *Dracunculus medinensis* is a well-known human parasite that inhabits the deeper subcutaneous tissues of man in tropical countries especially India, Arabia, Egypt and adjacent areas of Africa. The females attain a length of one or more than one metre and when mature contain the uteri stuffed with young, anus, vulva and intistine degenerate. The male worm is rare and is about four centimetres. It lives beneath the skin and copulation takes place when both sexes are young. When mature, the female approaches the surface and causes blisters due to poisonous secretions. Such blisters appear usually on the arms or legs and on such parts of the body that frequently make contacts with cold water. The blister breaks exposing a shallow depression with a hole in its centre. When the ulcers suddenly plunge in cold water (as by women while washing clothes or man wading through the water for cutting crops or collecting material) the female worm contracts convulsively, a loop of ruptured uterus is forced through a cuticular rupture near the anterior end of the body, and cloud of milky fluid containing large number of tiny larvae is ejected from the hole. They are small worms about 0.6 mm in length. They swim about till they establish contact with a *Cyclops* in which they develop. The larvae are introduced into the human system when he drinks unfiltered water containing the *Cyclops* (Fig. 10.38).

The guinea worm infects a large number of people in particular places during certain part of the year. On the formation of blisters some symptoms like

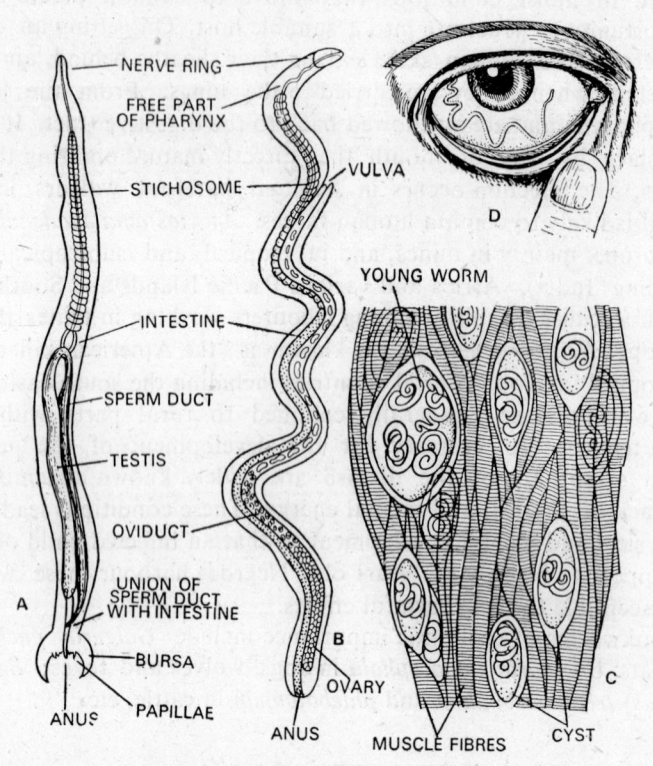

FIG. 10.39. A—*Trichinella spiralis*, the trichina worm, male; B—Female; C—Encysted young within the muscles of host; and D—Eyeworm in the cornea.

vomiting, dizziness and that of diarrohoea appear. Native medicinemen extract the worms by slowly and painfully winding it out on a stick. This often results in new infection followed by loss of limb or by death. The method will be successful if done by a doctor. Control is simple. If infected persons keep out of water and if drinking water is properly filtered the infection may be checked.

*Dracunculus* has also been reported from such mammals as foxes, raccoons, muskrats, weasels and dogs, etc. The other two genera of the order include *Philometra* and *Micropleura Philometra* is found in the coelom and tissues of fish, and *Micropleura* occurs in crocodilians

## FILARIA WORM

*Wuchereria* (*Filaria*) *bancrofti* is a human parasite of tropical and sub-tropical countries belonging to the order Filarioidea. It parasitizes human lymphatic system. The adult worms look like coiled white strings as they lie in the lymph glands or ducts of an infected person. The female is 8-10 cm long and the male is about one centimetre. The accumulation of living and dead block the lymphatic system resulting in various pathological conditions, of which the most

*Phylum Nemathelminthes (Aschelminthes)* 377

spectacular one is an immense swelling of the legs and other parts termed **elephantiasis**. The female lays small larvae called **microfilariae** which get into the blood vessels and do not develop further unless sucked by a mosquito of appropriate species. Within the mosquito's stomach the microfilariae lose their sheaths, penetrate the walls of the stomach, migrate to the thoracic muscles where they develop into long slender forms that moult twice to form the infective stages, and migrate to the labium of mosquito. When the insect bites another person, the worms creep out into the skin of the victim, pierce near the bite and finally develop the symptoms of the disease. The worms are quite common in the eastern districts of Uttar Pradesh particularly Gorakhpur.

The genus *Onchocerca* is represented by several species in horses, cattle and related mammals and by *O. volvulus* in man. The adult worms cause nodules of

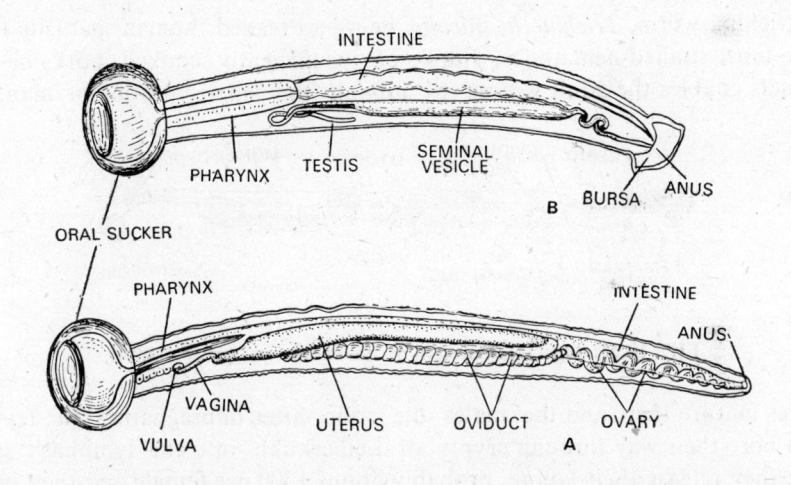

FIG. 10.40. A—*Sobollphyme*, female; and B—Male of the same, occur in the digestive tract of foxes and domestic cat.

connective tissue to be formed beneath the skin (**onchocerciasis**) and the microfilariae occur in the dermis of adjacent regions. *Onchocerca* of man is transmitted by black flies of the genus *Simulium* which have aquatic larvae inhabiting swift water as of mountain streams. While the fly is biting the human host microfilariae seem to be attracted to the site by its salivary secretions, for a greater number of microfilariae are subsequently found in the fly than can be accounted for by their distribution in the skin at the place of the bite. Adult filariae develop in the regions of the body where infective microfilariae were inoculated by the vector. Microfilariae infesting the eye cause blindness which is quite common in certain regions of Africa.

The black flies carry a number of other filariae, e.g., the cattle filaria *Onchocerca gutturosa* (Steward, 1937), the avian filaria *Splendidofilaria fallisensis* (Anderson, 1954) and *Mansonella ozzardi* (Batista *et al.*, 1960) which occurs in South America in humans. Voelker and Garms (1977) have applied the collective term **agamofilariae** to the immature worms in intermediate hosts on the basis of which generic determination is not possible. These authors have divided the development of filaria worms into three stages: (*i*) Microfilariae in the final host; (ii) agamofilariae in the intermediate host and final host; and (iii) adult filaria in the final host.

In other filarioids also similar cycle is met with. *Acanthocheilonema perstans* inhabits the coelomic membranes and perirenal tissues of man in Africa and South America. The life-cycle of this parasite is similar to that of *Wuchereria*, the intermediate hosts are nocturnal biting midges (Chironomidae). Other notable genera of the order include *Filaria* found in subcutaneous tissues of carnivores and rodents; *Diplotriaena* in the connective tissues of birds; *Mansonella* in the coelomic membranes of man; and *Setaria* in the abdominal cavity of mammals.

## THE TRICHINA WORM

The trichina worm, *Trichinella spiralis*, being a dreaded human parasite is one of the most studied nematodes. Intake of insufficiently cooked porks or pork products enables the adult worms to inhabit the small intestine of man. The

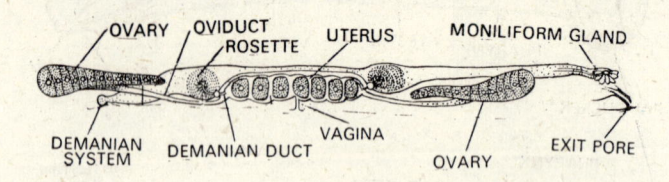

FIG. 10.41. Demanian system of *Adoncholaimus* (*after* Cobb).

females mature there and the males die soon after impregnating the females, which bore their way through crypts of Leiberkühn into the lymphatic spaces where they release their young, probably about 1500 per female and then perish. The larvae immigrate by way of the lymph and blood vessels throughout the body eventually entering every organ and tissue. They develop to the infective stages only in voluntary musculature. In the final stage they coil up into quiescent condition. Around them is formed a cyst by the host. The cyst finally become calcified. The young larvae in this condition degenerate unless eaten by suitable hosts where they undergo final development and maturity Recent investigations indicate that pigs usually catch infection feeding on garbage containing pork scraps.

The adult worm is directly harmless, the injury, however, is caused by the migrating larvae. When such a large number of larvae simultaneously pierce through almost all the tissues of the body, there naturally develop a **excruciating muscular pain**, muscular disturbance and weakness. Fever, anaemia and swellings of the various parts of the body are the consequences. It is at this stage that **trichinosis**, as the disease is called, is fatal. If the victim survives the period, the larvae become encysted and the symptoms subside, though some muscles may be permanently damaged. In less infected cases there may be no symptoms and are likely to be diagnosed as intestinal trouble. Serious cases are often diagnosed as typhoid fever. In America about 20 per cent of the population suffer from trichinosis at some time or the other.

# Phylum Nemathelminthes (Aschelminthes)

Among other notable genera are *Trichuris*, known as the whip-worm because of its whip-like body, parasitic in the intestine of man and other mammals; *Capillaria* parasitic in birds and some mammals; *Trichosomoides crassicauda*, parasitic in the urinary bladder and other parts of the urinary tracts of domestic rat; and *Cystopsis* parasitic in the skin of sturgeons.

## KIDNEY WORM

*Dioctophyme renale* (Fig. 10.42) or the kidney worm is the species found in the abdominal cavity and kidney of mammals especially the dog. It generally

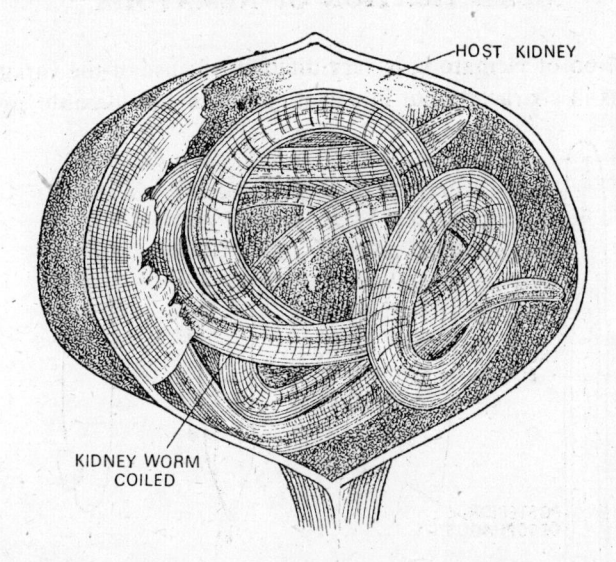

FIG. 10.42. Kidney worm. *Dioctophyme renale* in the kidney of its host.

lives in the abdominal cavity but invades the kidney also, therefore, it is known as the kidney worm. Curiously enough the worm has been reported from the right kidney alone. It destroys the kidney tissue completely so that the infected kidney consists merely of a connective tissue capsule containing a coiled mass of worms. The female is blood-red in colour and about one metre long. The males are much smaller, about 45 cm. The eggs pass out along with urine and develop into infective stages in about six months. Some epizoic oligochaetes serve as their first intermediate hosts. Young worm penetrates the digestive tract of the oligochaete and occupies coelom and its various organs. Finally, it encysts and develops only if it is devoured by fish along with its first intermediate host. Within the fish the young one leaves the cyst and migrates to the mesenteries where it again encysts but continues growing to infective stage. It reaches its definitive host that feeds upon fish. The entire cycle is very slow and takes two years to complete.

Other notable genera of the order include *Eustrongylides*, and *Hystrichis*. Both these are found in the proventriculus of various aquatic birds. *Soboliphyme* (Fig. 10.40) is found in the digestive tract of foxes and the domestic cat in Siberia and also occurs in other mammals in America. The genus is small up to 20 mm and possesses a large oral sucker.

### Number of Nematodes

Long ago Cobb pointed out that every vertebrate animal is infested with at least one and usually with more than one kind of nematode. Thus, those parasitic on vertebrates alone are estimated to be about 100,000 as there are about 40,000 species of vertebrates. Add to these those that are parasitic on invertebrates and plants. On a rough estimate the total comes to about 500,000 species of nematodes in the world. A single acre of cultivated soil may contain hundreds of millions of nematodes but they are rare, if ever seen.

## CLASSIFICATION OF NEMATODA

The classification of Nematoda is very difficult because of the varieties of forms and variations in characters but in recent years a considerable agreement and

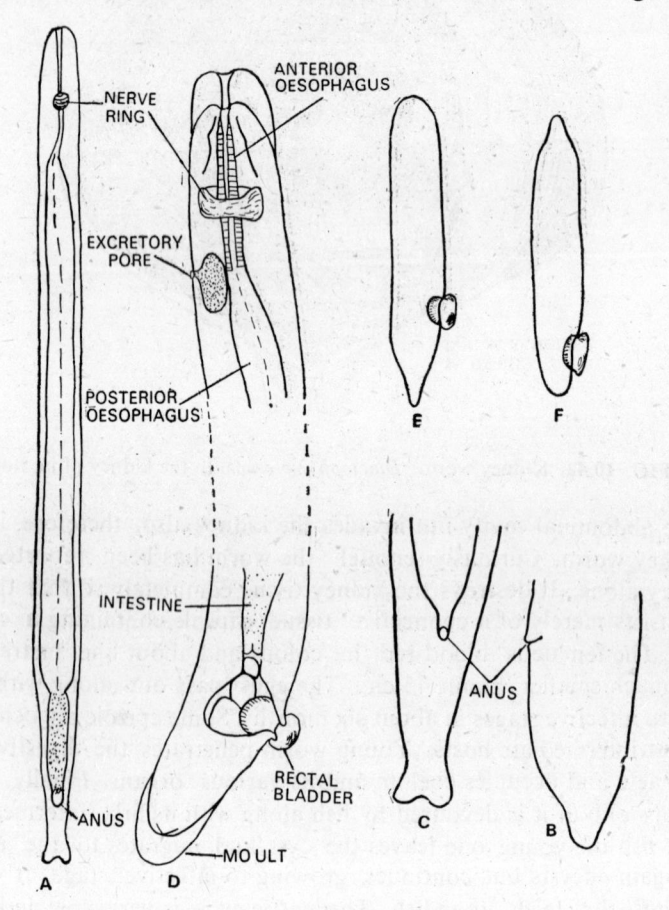

FIG. 10.43. Young ones of *Onchocerca* obtained from black fly. A—Ventral view; B—Lateral rear end; C—Ventral rear end; D—Third larval stage; E and F—Other larval stages.

stability has been reached chiefly through the study of free-living forms. All the known nematodes are divided into seventeen orders of which seven are mostly free-living. The free-living forms have many common characters. Most of them are relatively small in size (except mermithids), they have conspicuous sensory bristles or papillae or both on the anterior end in definite circles, they have relatively large amphids and possess caudal adhesive glands.

**1. Order Enoploidea.** Free-living chiefly marine Nematoda with smooth cuticle and typically with cyathiform amphids. The head is provided with six labial papillae and 10 to 12 sensory bristles arranged in one circle of ten or two circlets of six each. Additional sense organs in the form of a pair of cephalic slits are present. Excretion takes place through a one-celled renette. Three caudal glands are present. The female system is usually didelphic (has two ovaries) with reflexed end and the males may bear few genital papillae. The females of the marine Enoploidea possess a peculiar system of tubes known as the demanian system (Fig. 10.39) after its discoverer, de Man (1886). It is an inimately associated system of tubes and glands.

Examples: The non-marine forms include genera *Triphyla*, *Trilobus*, etc., while the marine enoploids include: *Anticoma*, *Endoplus*, *Pelagonema*, *Metancholaimus*, *Adocholaimas*, etc.

**2. Order Dorylaimoidea.** Common freshwater Nematoda that have smooth cuticle and are without bristles. They are in many respects like the Enoploidea of which they were formerly regarded as a family. Anterior end has two circlets of papillae of six and ten each. The buccal cavity is furnished with a protrusible spear, odontostyle, a hollow tube with an oblique terminal aperture through which food (plant or animal juice) is sucked in. Rear part of the pharynx is enlarged. Amphids are sickle-shaped.

Example: *Dorylaimus* with about 200 species.

**3. Order Memithoidea.** Smooth filiform Nematoda that are parasitic in invertebrates, chiefly insects, in juvenile stages. Adults are usually free-living inhabiting soil or water. The head sense organs are reduced to papillae and are 16 in number. The amphids may be large sickle-shaped or may be small pore-like. A buccal capsule is wanting. The pharynx proceeds directly from the mouth opening and is also peculiar (Fig. 10.33). It is slender and long apparently embedded in a longitudinal row of cells, the **stichosome**, the cells of which (the **stichocytes**) are pharyngeal glands opening into the pharynx. Further the pharynx is blind, it does not communicate with the intestine although it usually adheres to it. The intestine is also peculiar. It consists of two or more rows of greatly enlarged cells packed with food reserves. The intestine is also blind and serves to nourish the adults which probably do not feed. The ovaries in the females are two (didelphic) oriented in the opposite directions. The males are smaller than the females and have two testes (diorchic).

Examples: *Paramermis contorta* parasitic in the larvae of the midge *Chironomous Agamermis decaudata* and *Mermis subnigrescens* parasitic in grasshoppers (Fig. 10.33), *Mermis nigrescens* parasitic in earwigs; and *Allomermis myrmecophila* parasitic in ants.

**4. Order Chromadoroidea.** Aquatic (mostly marine) Nematoda with usually ringed cuticle heavily ornamented with bristles, knobs and other markings. The buccal capsule is armed with teeth and the pharynx is provided with posterior bulb, amphids are spiral. The reproductive system is double with reflexed gonads. Males provided with copulatory aids such as genital papillae. Lips six and papillae of bristles ten arranged in a circlet.

Examples: *Chromadora, Richtersia, Paracyatholaimus, Spilophorella*, etc.

5. **Order Araeolaimoidea**. The Nematoda with smooth cuticle sometimes provided with bristles. Individuals are characterized by the presence of four conspicuous cephalic bristles clearly marked off from the labial circlets which consist of papillae or reduced bristles; amphids, spiral or loop-like. This form is mostly found in freshwater or soil. Representative genera include *Plectus* with a long tubular buccal capsule and *Wilsonema* with strange cuticular expansions on the head.

Examples: *Odontophora, Araeolaimus, Triphyloides, Camaeolaimus.*

6. **Order Monhysteroidea**. Terrestrial and aquatic (mostly marine) Nematoda with smooth or slightly ringed cuticle often beset with scattered bristles. Ovaries double or single and males with preanal papillae. Anterior end with four, six or eight bristles or multiples thereof. Amphids circular.

Examples: *Cylindrolaimus* with long cylindrical buccal capsule, *Siphonolaimus*, marine form with a buccal stylet, *Theristus* and *Monhystera* with eyes, and *Sterineria* notable for long body bristles.

7. **Order Desmoscolecoidas**. Marine Nematoda with short, plump bodies enclosed by heavily ringed cuticle covered with bristles. Head is armoured and set off from the body. Amphids are hemispherical. Buccal capsule is not differentiated.

Examples: *Desmoscolex*, the chief genus has a body provided with coarse annulations and sparse bristles, thus resembling an insect larva in appearance. In the genus *Tricoma* all rings of the body are alike.

8. **Order Rhabditoidea** or **Anguilluloidea**. Free-living or parasitic Nematoda with ringed or smooth cuticle. The worms of this order are small sized with the head sense organs in the form of papillae. Amphids are like small pockets. Phasmids are present and caudal glands are lacking. The pharynx has generally two bulbs, one without valves internally (pseudobulb) and the other valculated. Both renette cell and tubular canals are present for excretion in different members of the group, but the system is usually asymmetrical. Female reproductive system is double and the ovaries are reflexed, males possess caudal alae that form a bursa for copulation which is aided by genital papillae. Spicules are also present, they are equal and similar. The dorsal wall of the spicule pouch forms the gubernaculum, that directs the spicules towards the anus and prevents the wall of the pouch from being pierced by the spicules. The life-cycle is direct but sometimes the infective stage, which is a sheathed juvenile, is transported by an invertebrate usually an insect. It may be carried internally or externally. Sometimes the development is completed within an intermediate host. These nematodes are usually terrestrial, dwelling preferably in sites rich in organic decay (decaying plants, decomposing animal carcasses, dung, manure, etc.). Some rhabditoids are free-living inhabiting aquatic habitats.

Example: *Rhabditis.*

Genus *Rhabditis* has numerous species, many free-living others with semi-parasitic habits. Many that live in animal droppings utilize coprophagous insects for transport to places with new food supply. In some cases the transport is obligatory. Thus the third stage juvenile of *Rhabditis coarctata* cannot develop further unless transported to fresh dung by dung-beetles to whose body they adhere in ensheathed stage. They will not adhere to dead beetles. On arrival at fresh places they develop to sexual maturity. Some species employ different insects that feed on dung or rotten plant material as transporting agents. In some species the stage requiring transport is distinct morphologically. Some forms utilize other invertebrates for transport. *R. maupasi*, for instance, uses earthworms. Active juveniles inhabit the nephridia.

9. **Order Rhabdiasoidea.** The rhabditoid Nematoda with complicated life-cycle involving vertebrate hosts (hence given the status of a new order). They are mostly smooth worms without definitive pharyngeal bulb. The life-cycle includes both parasitic and free-living stages like the Rhabditoidea, but in this case the parasitic phase is represented by either a protandrous hermaphrodite (the same tubular gonad first produces sperms which are stored and then eggs), or a parthenogenetic female. The cycle is of two types—direct or homogonic and indirect or heterogonic. In the direct life-cycle the eggs from the parasite give rise to free-living young which directly develop into parasitic forms. In the indirect life-cycle the egg may develop into free-living males, the offsprings of which take to parasitic life. In the direct cycles males are absent. Some of these (family Rhabdiasidae) are parasitic in the lungs of amphibians and reptiles and others (family Strongyloididae) are parasitic in the intestines of mammals.

Examples: *Rhabdias bufonis, Strongyloides stercoralis*.

10. **Order Oxyuroidea.** Small or medium-sized Nematoda of fusiform shape with a long slender tail in the female (males are also tailed sometimes). They have three or six lips, buccal capsule is small and the pharynx is provided with a posterior bulb, usually valvulated. Excretory system is tubular. Female generative organs are single or double. The members of this order are obligatory zooparasites chiefly of vertebrates with simple life-cycles involving only one host.

Examples: *Oxyuris equi, Enterobius (=Oxyuris) vermicularis*.

11. **Order Ascaroidea.** Stout Nematoda that are obligatory parasites of the intestines of vertebrates. Mouth is surrounded by three prominent lips, pharynx is without a true posterior bulb. Excretory system is tubular, with reduced anterior limbs. Males without caudal alae but with two equal or nearly equal copulatory spicules. The life-cycle in known cases is direct without intermediate host but the juveniles after hatching in the host intestine enter in the lungs on a migratory phase.

Example: *Ascaris* (Fig. 10.1).

12. **Order Strongyloidea.** The Nematoda in which the wide caudal alae meet posterior to the tail and form a copulatory bursa, provided with muscle rays (in others the bursa is a simple cuticular expansion without muscle rays). Typically the bursa, in this group, consists of two large lateral lobes, each supported by six muscle rows and a median lobe supported by one muscle ray that may be branched and subdivided. Definite prominent lips are absent, instead some have well-developed teeth, pharynx is simple without bulbs or other specialization, and there are no caeca of the digestive tract. The excretory system is tubular with which are connected two renette cells. The ovaries are double and opposite with highly developed muscular ovijectors. Members of this order are obligatory parasites of the digestive tract of vertebrates, especially mammals. They are among the most harmful parasites of man and domestic animals. The life-cycle is direct, usually including two free-living juvenile stages.

Examples: *Ancylostoma duodenali, Neacter americanus* (Fig. 10.36).

13. **Order Spiruroidea**. Mostly slender Nematoda of moderate size having two unilobed or trilobed lips (called pseudolabia by Chitwood). The head sense organs consist of six reduced papillae forming an inner circlet and an outer circlet of four double or eight single papillae. Buccal capsule is cuticularized, sometimes toothed, pharynx consisting of anterior muscular and posterior glandular portion. Body shows various cuticular ornamentations especially restricted to the anterior end in some cases. Tubular excretory system lacks anterior limbs of the lateral canals forming a U-system. Female pore is in the middle of the body and an ovijector is always present. The caudal alae are beset with papillae. The spicules are two, usually unequal in size, dissimilar in shape. Members of this order are obligatory parasites in the digestive tract, respiratory system, nasal cavities, sinuses and eyes of vertebrates.

Examples: *Rictularia* (Fig. 10.36), *Thelazia, Oxyspirura*.

14. **Order Dracunculoidea**. Filiform Nematoda without definite lips in which the buccal capsule is not cuticularized. The head sense organs consists of an inner circlet of well developed labial papillae and an outer circlet of four double or eight separate papillae. Pharynx is divided into an anterior muscular portion and a broad glandular portion posteriorly. Vulva in the females is near the middle of the body, usually not functional. The female reproductive system is double, uteri are opposite and the development is viviparous. Males are provided with equal filiform spicules. The members of this order are parasitic in the connective tissue of vertebrates or their coelom and its membranes. The life-cycle is complicated involving an intermediate host, usually a copepod.

Examples: *Dracunculus medinensis* (Fig. 10.34C).

15. **Order Filarioidea**. Filiform Nematoda of moderate to large size in which lips and buccal capsule are lacking, and the pharynx consists of an anterior muscular and posterior glandular portion. Males are considerably smaller than the females. The vulva in females is placed anteriorly and remains functional. The copulatory spicules are dissimilar and unequal in males. The members of this order are parasites of the circulatory system, coelomic cavities and muscular and connective tissues of the vertebrates and require a blood-sucking insect (intermediate host) for the transmission of the young.

Example: *Wuchereria (Filaria) bancrofti* (Fig. 10.37).

16. **Order Trichuroidea** or **Trichinelloidea**. The Nematoda whose body is filiform anteriorly. Mouth is without lips and the pharynx slender provided with a stichosome. Males are without copulatory apparatus or are provided with an eversible spicule sheath, hence a cirrus. The female reproductive system is single, and mostly species are oviparous laying unsegmented eggs. Usually the members of this order are parasitic in the digestive tracts of birds and mammals, but may be found in other vertebrates and other places. The life-cycle is simple without any intermediate host.

Example: *Trichinella spiralis*. (Fig. 10.39)

17. **Order Dioctophymoidea**. The Nematoda in which the bursa is muscular bell-like without rays. Lips are not present and the mouth is surrounded by one or two circlets of papillae and a sucker in some members only. Pharynx is elongated without bulb-like enlargement. The female is oviparous and possesses single reproductive organ. Males have only one copulatory spicule. Members of

this order are parasitic in birds and mammals and probably acquiring fish as intermediate hosts.

Example: *Dioctophyme renale* (Fig. 10.42).

## CLASS NEMATOMORPHA OR GORDIACEA

The **Nematomorpha** are extremely long and slender unsegmented worms with a uniformly cylindrical body. Due to their extremely slender nature a myth became prevalent during the fourteenth and fifteenth centuries that they are transformed horse hairs. They were thus called **hair worms** or **living horse** hairs. Their body length varies from a few millimetres to one metre. The anterior end is bluntly rounded (Fig. 10.44), whereas the posterior end is swollen and coiled. The males are usually shorter than the females. They may be yellowish or buff to dark brown in colour, but the anterior tip, called **calotte**, is usually white. The mouth is situated terminally in the calotte and the anus is also terminal or sometimes ventral (Fig. 10.44). There are probably less than 250 species placed in genera such as *Gordius* or *Paragordius* living in freshwaters and in one marine planktonic genus *Nectonema*.

### Body Wall

It comprises the **cuticle, epidermis** and musculature. The cuticle is rather thick and is rough as seen from outside. In section the cuticle seems to be made up of two layers, an outer thin **homogeneous** and an inner **fibrous layer**. The epidermis is one-layered epithelium of cuboidal cells (Fig 10.44). Musculature consists of longitudinal fibres only. The muscles are histologically similar to those of nematodes.

The **pseudocoel** is spacious cavity occupying the space between the body wall and the digestive tract. This is filled in the adult by mesenchyme cells, that are so arranged as to form partitions looking like mesenteries.

### Digestive Tract

It is degenerate in all nematomorphs both in adult and juvenile stages. Apparently they do not ingest food. A normal mouth is absent and the anterior part of the digestive tube is perhaps pharynx. Toward its posterior end the intestine receives the genital ducts and thus forms a **cloaca**. No other organs are present.

### Nervous System

The nervous system comprises a **nerve ring** surrounding the oesophagus and is connected to a single mid-ventral **nerve cord**. All these are closely related to the epidermis. There are two minute **eyes** and many **sensory bristles** forming the organs of special sense.

### Reproduction

The sexes are separate, each with two gonads in the body cavity. In males each gonad continues posteriorly and enters cloaca separately. No copulatory structures are present. In some species (*Gordius, Paragordius*) the cloaca bears spines probably to help during copulation. The sperms are simple elongate bodies. In the female the gonads are like the testes (Fig. 10.44) in the early stages of development but later on numerous diverticula arise from its body. The eggs ripen in these diverticula, re-enter the main ovarian tube that narrows posteriorly and opens separately in a common chamber or antrum that is glandular internally. The antrum opens into the cloaca. The eggs are laid in strings which can be easily collected. The strings may be sometime as long as 2-3 metres).

The eggs hatch into small larvae that swim about in water and enter aquatic insects, often May fly larvae. It may also infect other hosts that frequent the banks of streams. Such hosts include grasshoppers, crickets, cockroaches and beetles, etc. Development in the host occupies several weeks during which period the

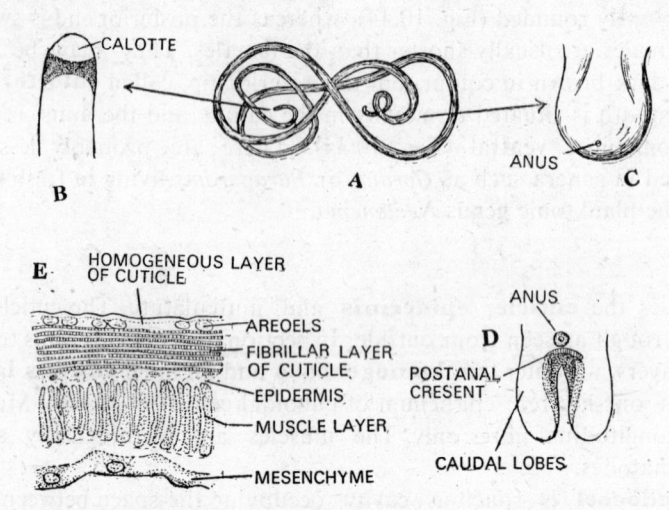

FIG. 10.44. Nematomorpha. **A**—A gordian worm (*after* Janda); **B** – Anterior end of *Gordius*; **C**—Posterior end of *Gordius*; **D**—Posterior end of male *Gordius*; and **E**—Section through the body wall of *Paragordius*.

juvenile worms attain the adult characters. They then leave the host when the latter is near water. The worms then lead free life in the water. It is now certain that only one host is involved in the parasitic phase of the life-cycle or nematomorphs.

## CLASSIFICATION OF NEMATOMORPHA

The Nematomorpha is divided into two orders.

1. **Order Gordioidea.** Terrestrial and freshwater Nematomorpha with parasitic stage in terrestrial or aquatic arthropods. The cuticle is without bristles; subcuticular muscle sheath interrupted only ventrally and the pseudocoel is much reduced by mesenchymal tissue.

Examples: *Gordius, Paragordius, Gordionus, Chordodes, Parachordodes*, etc.

2. **Order Nectonematoidea.** Marine Nematomorpha with parasitic stage in Crustacea. They possess a double row of natatory bristles and both dorsal and ventral chords. Pseudocoel is not reduced. This order consists of single genus *Nectonema*, a marine pelagic nematomorph.

## Phylum Nemathelminthes (Aschelminthes)

**Affinities.** The general structure of the Nematomorpha is pseudocoelomate type. They resemble nematode worms in following characters: (*i*) structure of

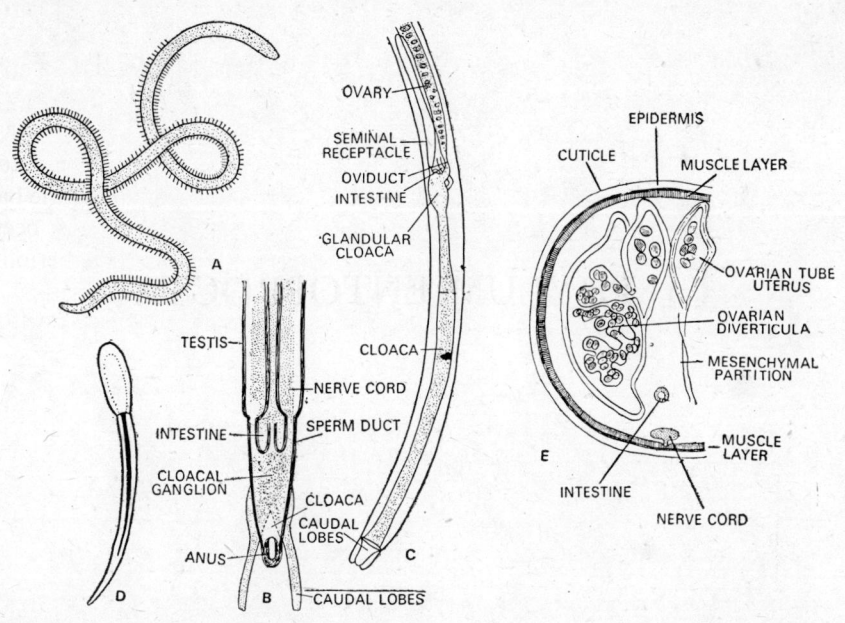

**FIG. 10.45.** A—*Nectonema agile* (*after* Fewkes); B—Male systems of *Paragordius* (schematic); C—Female system of *Paragordius* (schematic); D—Nematomorph sperm; and E—Transverse section of a female showing main ovarian tube and its diverticula.

the cuticle; (*ii*) presence of the epidermal chords; (*iii*) similar subepidermal musculature; (*iv*) the occurrence of mesenchyme in the pseudocoel; (*v*) nervous system intimately associated with the epidermis; (*vi*) the utilization of the rear part of the gonads as sex ducts; and (*vii*) the moulting of the larva as in the nematodes and kinorhynchs. These points furnish sufficient justification to include the Nematomorpha as a class of Nemathelminthes (Aschelminthes).

# 11 PHYLUM ENTOPROCTA

The **Entoprocta** or **Calyssozoa** are small solitary or colonial sessile animals which look superficially like hydroid coelenterates that have developed some mesodermal structures. They are found attached to sea weeds or other submerged objects. They are pseudocoelomate Bilateria having a distal circlet or ciliated tentacles, encircling both the mouth and anus. There are about sixty species of small animals all of which are marine except for one genus. They are all ciliary filter feeders. The characteristic feature of the Entoprocta (Gr. *endon*, within+ *proktos*, anus) is that the anus lies within the circlet of tentacles which also encloses the mouth. They are small, almost microscopic animals below five millimetres in length growing singly or in colonies; attached to objects or to other animals and having the general appearance of the hydroid polyps. In the entoprocts, however, the tentacles are ciliated. Genus *Pedicellina* is typical representative of the phylum. Other equally known genera include *Loxosoma* and *Urnatella*. *Urnatella* is a beautiful form found only in freshwater, living on the underside of stones, in running waters or in lakes.

### TYPE PEDICELLINA

A colony of *Pedicellina* was described by Pallas in 1774, but the generic name was given to it by Sars in 1835. *Pedicellina* (Fig. 11.1 B) is characterized by a crown of tentacles, the cup-shaped body mass or calyx, the stalk and the basal attachment of the stalks called stolons.

**Calyx**. The **calyx** is slightly laterally compressed or cup-shaped structure, oriented slightly obliquely on the stalk towards the mouth. The free surface bet-

*Phylum Entoprocta* 389

ween the tentacular crown is **ventral** (as shown by the study of embryology) and the convex surface of the cup is **dorsal**. Both the **mouth** and **anus** open on the ventral surface at opposite ends of the sagittal axis. Posterior to the mouth, in the same plane, is situated the **nephridiopore**, behind which lies the **gonopore** (Fig. 11.3). The concave surface between the mouth and anus is termed the **vestibule** (or atrium) and the part of this that lies between the gonopore and anus serves as a brood chamber, or **embryophore**, for the developing eggs. The dorsal surface of the calyx is smooth (covered with spines in some and by a cuticular dorsal shield (in *Chitaspis* and *Loxosomatoids*).

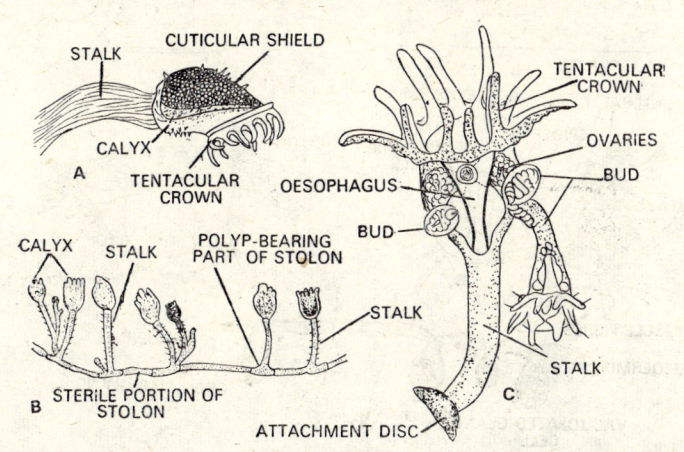

**FIG. 11.1.** Typical entoprocts. A—*Loxosomatoides* (*after* Annandale); B—Colony of *Pedicellina* (*after* Ehlers); and C—*Loxosoma* with growing buds.

**Tentacular crown.** The tentacles are evenly spaced simple outgrowths of the rim of the cup-shaped calyx. The number of tentacles varies from 8 to 30 in different species and are usually held in an incurved attitude. Along the inner side of the tentacular bases runs a ciliated vestibular groove. The tentacles are also ciliated on their inner surfaces and their cilliation runs into that of the vestibular groove. The older tentacles occur on either side of the mouth and the youngest on each side of the anus. This is so because the tentacles first appear near the mouth and then keep on growing towards the anus.

**Stalk.** The stalk of *Pedicellina* is a simple cylindrical outgrowth of the calyx and is partially separated from the calyx internally by an incomplete septum formed by the infolding of the body wall. The stalk may be smooth or spiny and in some other entoprocts (*Urnatella*, *Arthropodaria*) it may be beaded in appearance. In solitary forms the stalk is flattened out to form an attachment disc (11.1 C).

**Body Wall**

The outer surface is covered with **cuticle** (Fig. 11.2), the tentacles and vestibule being without it. It varies in thickness in different regions and may be several layers thick in the stolons. From the cuticle, in some species, arise cuticular spines, which in some cases, *Myosoma* (Fig. 11.2) take the form of adhesive tubes with glandular cells at their bases to secrete adhesive fluid. Below the cuticle lies a layer of epidermis comprising a single layer of mostly cuboidal cells. The tentacles are simply epidermal tubes (Fig. 11.2 B) enclosing loose mesenchyme. The epidermal cells of the vestibular tube and the inner surfaces of tentacles are longer and ciliated (Fig. 11.2 B). In some (*Loxosoma*) the wall of

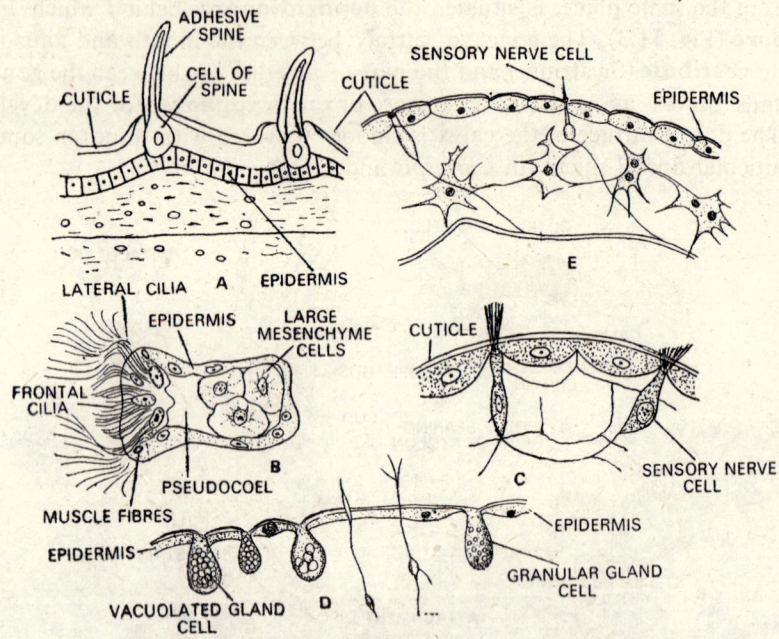

**FIG. 11.2.** Structure of entoprocts. A—Body wall of *Myosoma* with adhesive spines; B—Section through the tentacle of *Pedicellina*; C—Epidermis of tentacle of *Pedicellina*; and D—Epidermis of calyx of *Loxosoma* with sensory nerve cells and mesenchyme.

the calyx is glandular bearing opaque gland cells and transparent gland cells filled with vacuoles. Next to the epidermis occurs the body wall musculature consisting of longitudinal fibres. It is sparser in the calyx than in other parts of the body and is responsible for the movement of the tentacles or stalks, etc.

**Pseudocoel.** The spaces within the tentacles, stolons, and between digestive tract and the body wall form the **pseudocoel**. It is filled with a gelatinous material containing mesenchyme cells. The pseudocoel of the tentacles is filled with large rounded cells, imparting rigidity to them (the tentacles). The mesenchyme of the calyx consists of stellate wandering amoeboid cells and somewhat fixed cells with long processes. The pseudocoel of the stalk is separated from that of the calyx by a plug of cells, through which (probably) material can pass (Fig. 11.3).

### Digestive System

The digestive tract is U-shaped and occupies the greater part of the interior of the calyx (Fig. 11.3). The **mouth** is situated at the anterior end of the ventral surface and is transversely elongated ciliated aperture lying in the vestibular groove. The upper lip is not much indicated but the **lower (posterior) lip** is often present. The mouth leads into a funnel shaped **buccal cavity** that narrows into the tubular **oesophagus**, which opens into an enlarged sac-like **stomach** (Fig. 11.3). The stomach is the most conspicuous organ of the entoproct. From the stomach the narrowed **intestine** proceeds ventrally and is usually separated by a constriction from the terminal **rectum**. The rectum opens at the posterior end of the vestibule. The opening or the **anus** is situated on a projecting eminence, the **anal cone**.

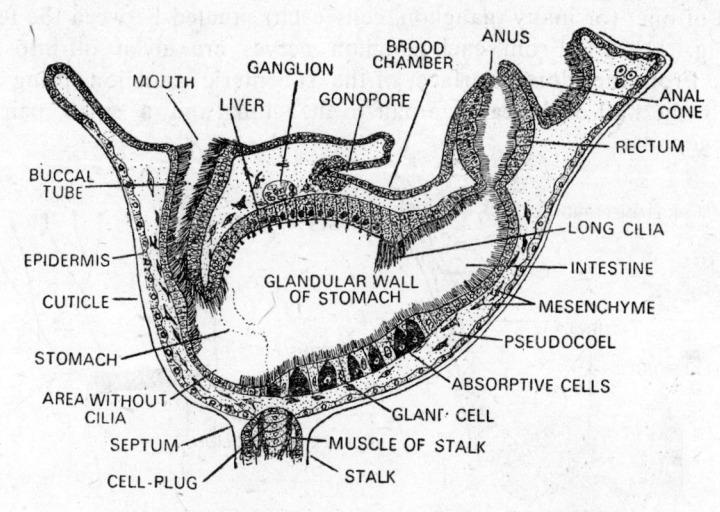

**FIG. 11.3.** Digestive system of *Pedicellina* (*after* Becker).

A single-layered epithelium of varying thickness lines the alimentary tract throughout. The cells are cuboidal on the outer side of the buccal cavity and columnar on the inner side. They are ciliated and carry fat and protein spherules as they are storehouses for food. At the beginning of the stomach some cells have long cilia that help to rotate and drive food particles. The **floor** of the stomach (ventral wall) has tall columnar cells filled with brown inclusions (Fig. 11.3) hence the whole area is called "**liver**," it, however, does not function as a liver. The inclusions are probably of excretory nature. The **roof** (dorsal wall) of the anterior part of the stomach has ciliated cuboidal cells. The posterior part, on the other hand, has glandular cells of two kinds, dark granular cells containing enzymatic granules and paler cells that appear to be of absorptive nature. The intestine is lined by ciliated cuboidal cells of decreasing height towards the rectum, in which the epithelium consists of larger cells.

## Nephridial System

The Entoprocta have a single pair of flame-bulbs in each calyx, situated ventral to the stomach between the oesophagus and subenteric ganglion. The flame-bulb (Fig. 11.4 A) is followed by a number of enlarged amoeboid cells, the **athrocytes** (Fig. 11.4 A) through which passes an intracellular canal or duct. The ducts from both the sides converge and unite and open by a single **nephridiopore** in the sagittal plane shortly posterior to the mouth.

## Nervous System

The central nervous system consists of one main ganglionic mass located ventral to the stomach and dorsal to the vestibular wall. It apparently represents a subenteric ganglion, the cerebral ganglion (found in association with the apical plate) is lost during metamorphosis. The ganglion is of rectangular to bilobular shape (Fig. 11.4 B) and consists of ganglion cells surrounding nerve fibres in the centre. From the ventral surface of the ganglion three pairs of nerves proceed to supply the crown of tentacles. Their branches terminate in large swellings of one (or many ganglion cells each) situated between the tentacular bases (Fig. 11.4 B). From each ganglion nerves are given off into adjacent tentacles. From the dorsal surface of the subenteric ganglion spring a pair of nerves to the wall of the calyx, a pair to the stalk and a small pair to the adjacent gonads.

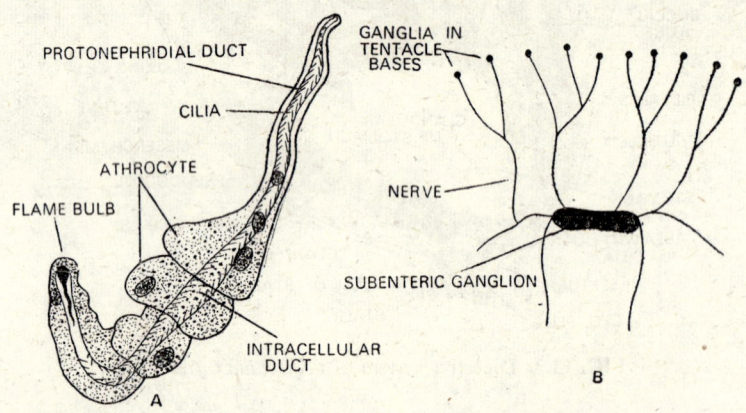

FIG. 11.4. A—Nephridium of an entoproct; and B—Brain of an entoproct.

### Sense Organs

Numerous sensory bristles occur on the outer surface of the tentacles and along the margins of the calyx. Each bristle (Fig. 11.2 C, D) arises from a nerve cell situated beneath the epidermis. Sensory pits are also found in the stalks and stolons of *Pedicellina*.

In Loxosomatids, either throughout life, or in larval stages only, there occurs a pair of sense organs on the sides of the calyx near its oral ends. Each consists of a tuft of bristles connected below with a ganglion cell. This pair of sense organs bears a remarkable resemblance to the antennae of rotifers and is of phylogenetic significance.

### Reproduction

**Asexual reproduction.** The Entoprocta multiply extensively by asexual budding. In some entoprocts buds are produced by the calyx (*Loxosomn*, Fig. 11.1 C), but in *Pedicellina* buds are produced by stolons and stalks, never by the calyces, and as the buds do not separate colony formation results. The bud begins as an epidermal proliferation (Fig. 11.5 A), which cuts off internally as an epithelial vesicle. This soon constricts into two vesicles (Fig. 11.5 C) of which the outer one becomes the free surface of the calyx and the tentacular crown and gives rise to the ganglion from its inner wall; while the inner part of the vesicle forms the digestive tract. The muscle cells, gonads and other mesodermal structures arise from the mesenchyme of the parent. The bud gradually elongates and a constriction separates the stalk from calyx. The entire animal originates from the ectoderm and mesoderm without any participation of the endoderm.

*Phylum Entoprocta*

**Regeneration.** The entoprocts possess good power of regeneration. Under adverse conditions the colonies shed their calyces, but the stalk and stolons remain alive for a long time and as soon as favourable conditions return they regenerate new calyces.

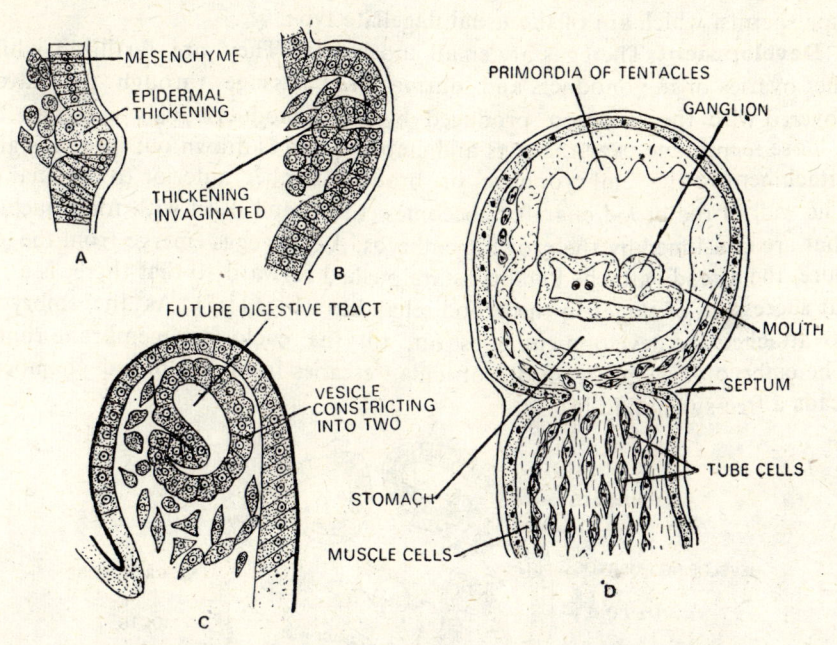

FIG. 11.5. Formation of a bud in *Pedicellina* (*after* Seelinger).

**Sexual reproduction.** It is evident from the available information that some entoprocts are hermaphroditic and others dioecious. Many species of *Pedicellina* are dioecious but one common species, *P. cernua*, is peculiar—sometimes it is apparently hermaphroditic and sometimes dioecious.

The **gonads** (Fig. 11.6) are a single pair of sac-like bodies located ventral to the liver region of the stomach either anterior or posterior to the ganglion. In hermaphrodites a pair of testes occurs posterior to the pair of ovaries. From each gonad a short duct proceeds medially and unites with its fellow to open

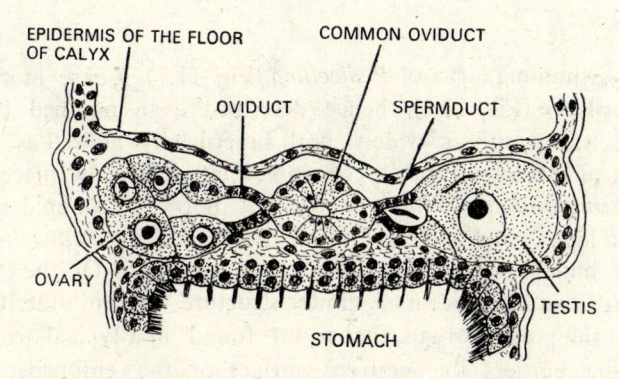

FIG. 11.6. Transverse section of *Pedicellina* showing reproductive organs.

on the ventral surface of the calyx by a common **gonopore**. In hermaphrodites the sperm duct unites (Fig. 11.6) with the oviduct of that side before the formation of the common duct. The gonoducts are glandular. The sperm duct, in males, may present an enlargement, the **seminal vesicle**, for the storage of ripe sperms, which are of the usual flagellate type.

**Development.** The eggs are small and yolky. They are fertilized while in the ovaries or in gonoducts and during their passage through the latter get covered with the secretion produced by the gonoducts. This secretion forms a loose membrane over the eggs and embryos and is drawn out into a stalk for attachment to the embryophore or brood chamber anterior to the anal cone. The wall of the brood chamber becomes thick and filled with food inclusions that are consumed by the growing embryos. As new eggs emerge from the gonopore, the already attached embryos are pushed forward so that there is a regular succession of stages in the brood chamber (Fig. 11.7). As the embryos are so attached the development goes on, till the enclosing membrane ruptures. The embryo develops a girdle of cilia, escapes from the brood chamber and leads a free-swimming life.

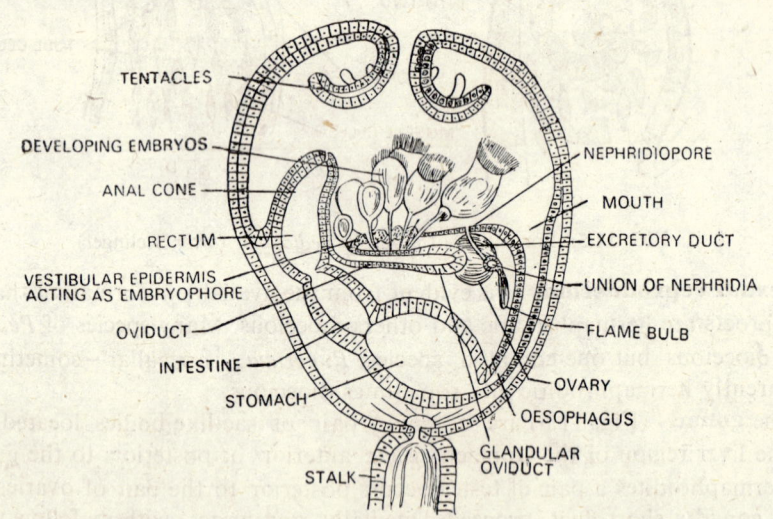

**FIG. 11.7.** Sagittal section of the calyx of *Pedicellina* with embryos in the brood chamber.

The free-swimming larva of *Pedicellina* (Fig. 11.7) looks, more or less, like the **trochophore** (Fig. 12.2), hence described as a modified trochophore in most texts. But it differs widely both superficially as well as structurally. At the summit of the dorsal surface or somewhat anteriorly placed there occurs the **apical organ** comprising a ciliary tuft borne by an epidermal sac below which lies a large ganglion (Fig 11.8). This organ of the *Pedicellina* larva corresponds to a similar structure in the trochophore larva. On the anterior surface of the *Pedicellina* larva occurs a similar structure with an underlying ganglionic mass called the preoral organ, that is not found in a typical trochophore. The ciliary girdle borders the **ventral** surface of the entoproct larva whereas the corresponding girdle surrounds the **equator** of the trochophore. The digestive tract of the *Pedicellina* larva is **U-shaped**, the anus being at the posterior edge of the ventral surface. The digestive tract of the trochophore is **L-shaped** with the anus opposite the apical plate. The **vestibular invagination** between the mouth and anus, supplied by three paired clusters of gland cells, is peculiar to the *Pedicellina* larva and is not found in the trochophore. The entoproct larva possesses protonephridia of the **flame-bulb type** whereas the trochophore has **solenocytic** protonephridia. The trochophore larva possesses mesodermal bands which are lacking in the entoproct larva. The entoproct larva is very **motile** and changeable. The apical and preoral organs can be protruded and retracted, the ciliary girdle can be expanded and contracted and the areas bearing the mouth and anus can be conspicuously exerted or withdrawn. From this it is apparent that the entoproct larva is not a modified trochophore. Whatever similarity exists between the two is due to identical mode of life.

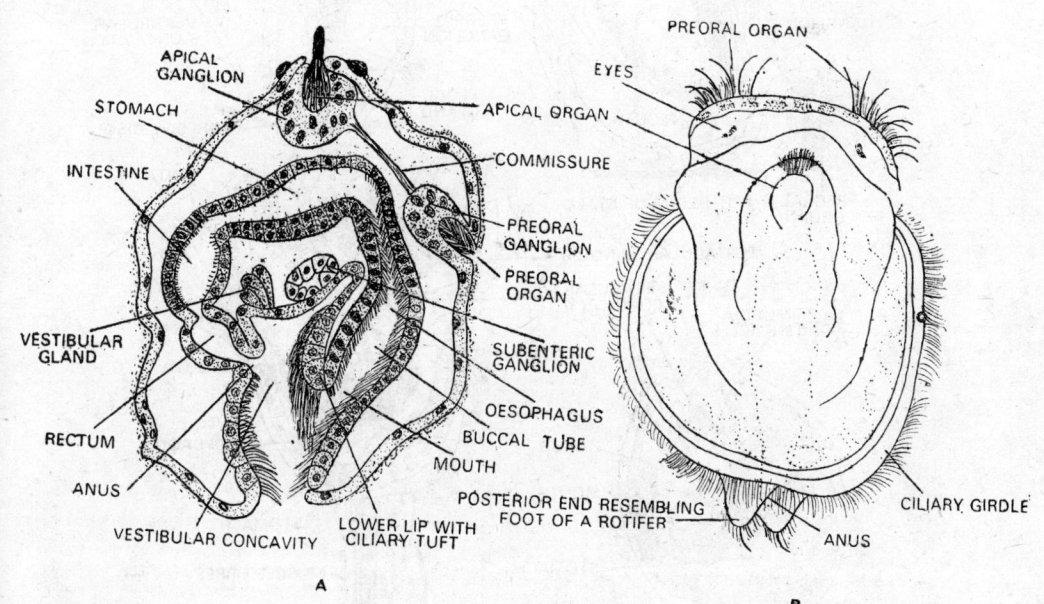

FIG. 11.8. Entoproct larval forms. A—Larva of *Pedicellina*; and B—Larva of *Loxosoma*, dorsal view.

**Metamorphosis.** After a free-swimming life for a short period the entoproct larva attaches to a suitable object by the **ciliary rim** (Fig. 11.9 A) and undergoes a process of **metamorphosis**. The ciliary rim loses its cilia, gradually contracts, eventually fusing to a one-layered epithelium that ultimately forms the attachment disc. The epithelium of this region may become glandular and secrete sticky material for attachment. In Pedicellina only this layer represents the pedal gland which, in others, develops as a prominent epidermal invagination. The closure of the ciliary girdle cuts off the vestibular cavity from the exterior (Fig. 11.9 B), and mesenchyme cells accumulate between the vestibule and the attachment disc. The mesenchyme cells multiply and pack the cavity, and as a result of their further multiplication this region of the larva elongates forming a definitive stalk. The apical and preoral organs degenerate. The vestibular cavity and digestive tract rotate by 180 degrees with the result that the vestibular cavity comes to lie opposite the stalk (Fig. 11.9 C), and tentacles sprout at the vestibular margin. This is how the original ventral surface of the larva becomes the free surface of a stalked individual. But for the sense organs and ciliary girdle of the larva all the larval organs persist in the adult.

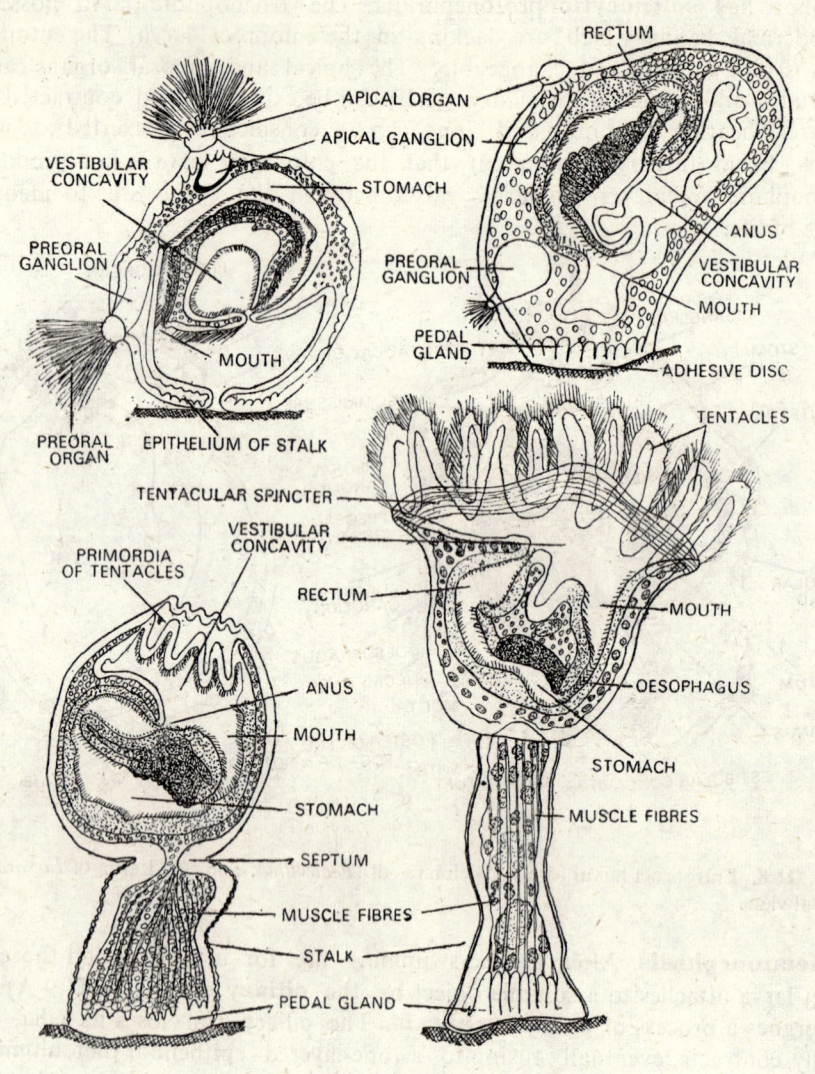

**FIG. 11.9.** Metamorphosis of *Pedicellina*.

**Affinities.** The Entoprocta were formerly regarded to be related to the Ectoprocta and included as a class under the phylum Bryozoa, as a subclass when the Bryozoa were placed as a class under phylum Molluscoidea (Bassler, 1922). It was in 1870 that Nitsche separated the Entoprocta, including *Pedicellina*, *Urnatella* and *Loxosoma* and Ectoprocta including other Bryozoans. Hatschek, on evidences obtained from the study of embryology of *Pedicellina*, raised the Entoprocta to the status of a phylum and suggested them to have relationships with the rotifers. A.H. Clark recognized their non-coelomate nature. Thus, now the Entoprocta are considered much lower in organization than the Ectoprocta that are distinctly coelomate. They were linked with Ectoprocta because of some common features including the tentacular crown of ciliated tentacles, and looped digestive tract, but such features are common in sessile animals, hence cannot be regarded of phylogenetic importance. In other anatomical features the two groups do not agree much. On the basis of the resemblance between the larvae of the two groups a relationship can be emphasized. But in their further development the two larvae differ altogether. Thus it is presumed that the similarities between the larvae are because of the pelagic habits of both.

The entoprocts were further related with the annelid-molluscan types because of the trochophore similarity, but it has been seen (page 394) that the entoproct larva departs much from the typical trochophore and that the coelom formation takes place in the two in a different manner, the body cavity of the Entoprocts is of a pseudocoelomate type.

Among the pseudocoelomates the rotifers come quite close to the Entoprocta. The loxosomatid resembles a collothecacean rotifer considerably: (*i*) Both have a trumpet-shaped body with the free surface bordered by ciliated or bristle-bearing projections that are simple extensions of the body wall; (*ii*) The stalk in both is a post-embryonic development and is provided with pedal glands at least temporarily; (*iii*) The mouth lies within the tentacular crown, the anus is not included within the crown in rotifers. It lies, however, quite near the tentacular crown and is clearly getting nearer and nearer to the mouth. Besides this, the digestive tracts are similar in many ways in both (mastax, found in the rotifers, is in the process of degeneration in the collothecacean rotifers); (*iv*) The excretory organs are similar in both the groups comprising protonephridia of the flame-bulb type; (*v*) A pair of eyes is present in both loxosomatid and collothecacean young ones; (*vi*) The ciliary rim of the ventral surface of the entoproct larva is probably a remnant of an originally ciliated ventral surface, as is the rotifer corona; (*vii*) The preoral organ of the entoproct larva, that persists in the loxosomatid adults, is homologous with the lateral antennae of the rotifers. In the loxosomatid adults they are situated toward the free end of the calyx. In the collothecaceans too they have migrated to an anterior position. Thus, it can be easily concluded that the Entoprocta are pseudocoelomate animals with greater affinities with the rotifers.

# 12 PHYLUM ANNELIDA

The phylum Annelida includes the **segmented worms** that are provided with a true **coelom**. Hitherto we have studied animals that were either devoid of coelom or were pseudocoelomate i.e., the body cavity was the original blastocoel which became spacious and occupied space between the body wall and alimentary canal. Like the development of mesoderm and internal musculature the appearance of a cavity known as **coelom** within the mesoderm is a definitely an advancement in the life of invertebrates.

### Coelom

The coelom is a cavity that arises within the endomesoderm, and that is covered on its outer surface by the **somatic mesoderm** and on the inner surface by the **splanchnic mesoderm**. It contains a fluid, and is lined by an epithelium, the **coelomic epithelium** or **peritoneum**. Coelom decidedly provides many advantages to the evolving animals. One obvious advantage derived from the coelomic cavity is that it provides a space around the alimentary canal, which is now surrounded by the peritoneum, and in many animals, suspended in the cavity by a double fold of peritoneum called a **mesentery**. This, together with the development of a visceral musculature from the associated mesoderm, permits free movement of the digestive tract, an advantage conferred also upon the heart and other mobile organs. This freedom improves efficiency in the handling of the food material.

The coelomic fluid serves as a medium of transport. The excretory organs, whether nephridia or coelomoducts, open into the body cavity and extract dis-

*Phylum Annelida*

solved nitrogenous waste from it. The coelomic fluid also plays an important part in osmoregulation.

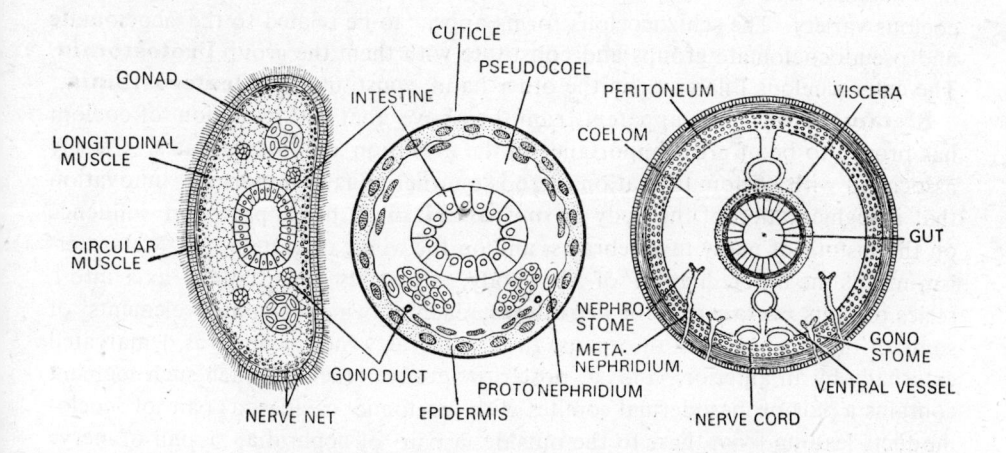

**FIG. 12.1.** Comparison between A—Acoelomate; B—Pseudocoelomate; and C—Coelomate, grade of animal types in diagrammatic cross sections.

The coelom is an important factor in reproduction for the germ cells mature in it, either in the cavity as a whole or in restricted part of it, and are eventually discharged through the coelomoducts. These ducts were probably genital in their initial function, and only later became concerned with excretion and osmoregulation.

The most important function of the coelom is that its contained fluid provides the structural basis for a more highly organized hydrostatic skeleton. In this respect it probably played a major part in ensuring the successful survival of the lower coelomates during the period when calcium and phosphorus metabolism had not developed to a point at which rigid and jointed skeletons could be constructed. The movement of annelid worms, echinoderms and many of the smaller groups are based upon hydrostatic principles and they show a greatly increased flexibility and speed of response. This is partly, of course, because of the more highly differentiated structure of the body as a whole, and to a much greater elaboration of the nervous system. Development of the coelom is a major factor in the improvement of locomotion in these animals. As a matter of fact this function of the coelom formation is the most important single factor in determining its appearance.

Embryologically speaking the coelom arises by two general methods, the **schizocoelous** and the **enterocoelous**. In the former the coelom originates as a space in the mesoderm, which splits into two layers, a **somatic layer** next to the skin, and a **splanchnic layer** around the endoderm. The space between these two is the coelom. In the enterocoelous type the coelom appears as pouches from the embryonic gut that become cut off and lie in the blastocoel. Their cavity is the coelom, their walls the mesoderm. These sacs expand until they touch the body wall and the gut wall and the end result is, therefore, the same as by the schizocoelous method. There is a third way of coelom formation also, in which the mesenchyme so arranges as to enclose a space, the **coelom**. This method is the **mesenchymal** method and is said to occur only in *Phoronis*. But for all purposes of study *Phoronis* is included with the schizocoelous variety. The schizocoelous forms appear to be related to the acoelomate and pseudocoelomate groups and constitute with them the group **Protostomia**. The enterocoelous Bilateria, on the other hand, constitute the **Deuterostomia**.

**Metamerism.** It is apparent from the above that the evolution of coelom has proved to be of great importance in the evolution of Invertebrates. Closely associated with coelom formation there is another equally important innovation that is segmentation of the body or **metamerism**. It has a profound influence on the history of many invertebrates, and on the whole of vertebrates. Metarmerism means the differentiation of the body, along its longitudinal axis into a series of units **metameres** or **segments**, each of which contains elements of some of the chief systems of organs. In invertebrates each segment is demarcated externally by an anterior, and a posterior groove. Ideally, each such segment contains a pair of mesodermal somites with coelomic cavities, a pair of coelomoducts leading from these to the outside, a pair of nephridia, a pair of nerve ganglia borne on the paired ventral nerve cord, and often a pair of appendages. It follows, therefore, that the essence of metamerism is the serial succession of segments, each containing unit subdivision of several organ system.

The significance of metamerism is undoubtedly to be found in the modes of locomotion of metameric animals, and especially of the more primitive ones. In tapeworm the repetition of the body structure is a parasitic adaptation and not a locomotor specialization. Annelid worms form that primitive group of invertebrates which throws light on the importance of segmentation. Annelids are essentially creeping and burrowing worms, and it is reasonable to accept their present day modes of life as being those that influenced the origin and early evolution of metamerism.

The earthworm possesses a plan of structure that conforms very closely in some respects to the idealized type of hydrostatic skeleton. The body wall is composed of continuous layers of longitudinal muscles (on the outside) and circular muscles (on the inside), these being placed so that they can exert pressure upon the coelomic fluid they surround. The longitudinal muscle is often said to be segmented in conformity with the external annulation, but this is not strictly correct. The longitudinal fibres are long enough to extend through two or three segments, so that in this respect the segments are linked in small groups. Each segment, however, has its own complement of segmental nerves so that precise and localized control of muscular contraction can certainly be exercised by the central nervous system. The subdivision of the coelomic cavity by transverse septa makes an essential contribution to the localization of muscular activity. The hydrostatic skeleton of the earthworms differs in this important way from that of the unsegmented animals. In the earthworm the septa restrict the pressure changes and tend to limit them to particular regions of the body. They can do this better because they are well provided with intrinsic muscle fibres which enable them to resist the stress. The septa of earthworms are perforate for continuity of coelomic fluid, but in actively moving worm the foramina are closed by sphincter muscles. Thus, the pressure in the fluid of one segment is substantially isolated from adjacent segments. The segmentation of the muscles and their nerve supply facilitate the passage of waves of contraction down the body. A glance at the details of locomotion of earthworm given later will confirm this.

The identification of the phylum Annelida (Lat. *annulus*, little ring) is: triploblastic, coelomate animals with metameric segmentation and tubular gut running from mouth to anus. The implication of the first character was discussed earlier (Page 398) and that of the other two has been described above. The functional significance of a tubular gut running from a subterminal mouth to a terminal anus through a long, wormlike body is of course that it allows for one-way traffic of food material past a series of specialized tissues for trituration, secretion, digestion and absorption. In other words, annelid worms have a functional alimentary layout similar to that in the gut of vertebrates, and, like them are capable of forming a dis-assembly-line for food materials.

In fact all the essential parts of the body are metamerically repeated. This fact is the essence of metamerism. All the body segments must be identical in an ideally segmented animal (homonomous condition). No such animal exists since one or more segments are fused to form head or anal segments, and these segments differ from typical body segments. Some of the annelid worms (the polychaete annelids) approach the ideal condition. Thus, it is evident that in most animals the segmentation has undergone local alterations through loss or fusion of segments (heteronomous segmentation). The perivisceral coelom communicates with the exterior through metamerically placed nephridia. The gonads arise from the coelomic epithelium and the reproductive elements pass out through a series of paired ducts, the **coelomoducts**. The nervous system consists of a pair of cerebral ganglia, a pair of peripharyngeal connectives and a double ventral nerve cord typically presenting a double ganglion in each segment.

Development is direct in some annelids (Oligocheata, Hirudinea) or indirect (Polychaeta, Archiannelida), with an intervening larval stage represented by the free-swimming trochophore larva (Fig. 12.2). The annelids are from minute to large in size. Some of the smallest Oligochaeta (*Chaetogaster*) are under 1 mm long, but the giant earthworms (*Rhinodrilus fafneri* of Ecuador and *Megascolides australis* of Australia) grow over two metres in length and 2.5 cm in diameter. Most earthworms are only a few centimetres long. The smallest Polychaeta are minute, but some are large—*Nereis brandi* of Californian coast attains a length of 1.5 metres and *Eunice gigantea* is about three metres long. The leeches are small ranging from 5 to 200 mm.

The phylum Annelida is divided into three major and three very minor classes. The most numerous and the most diverse group are the bristle-worms, the class **Polychaeta**, including more than 5,500 species, almost entirely marine. Among the bristleworms, the Errantia group is probably the least specialized of all annelids, and includes forms like the ragworm *Nereis* which shows a long series of similar segments, each bearing a pair of parapodia with setae which are used in swimming, walking and burrowing. The next largest group is the class **Oligochaeta** numbering more than 3,000 species of mainly land and freshwater forms. Earthworms and freshwater Oligochaetes have fewer and smaller setae on the segments, never have flap-like parapodia and usually have a reduced head and sense-organs if compared to the polychaete pattern. A third group is the leeches, class **Hirudinea,** with a more modified and much more uniform pattern of body involving attachment suckers and a modified mouth and gut The leeches are found in freshwater, in the sea and on land of tropics. There are about 300 known species. In leeches setae are absent, fixed points for locomotion are provided by suckers. In both Oligochaeta and Hirudinea the majority of species are hermaphrodite and some form of copulation results in cross fertilization, usually internally. There are three other minor groups of annelid worms each numbering about 50 species of little ecological importance. These are **Archiannelida,** small marine annelida probably related to the Polychaeta, but some are without parapodia and setae, the **Echiuroidea** comprising burrowing, marine forms showing little or no traces of segmentation, the **Sipunculoidea**, without segmentation, and the **Priapulida**, a small group of cylindrical worms living in mud.

## CLASS POLYCHAETA

The **Polychaeta** (Gk. *polus,* many+*chaite,* hair) are marine worms, with only a few exceptions, in which the setae are numerous and are borne upon special pro-

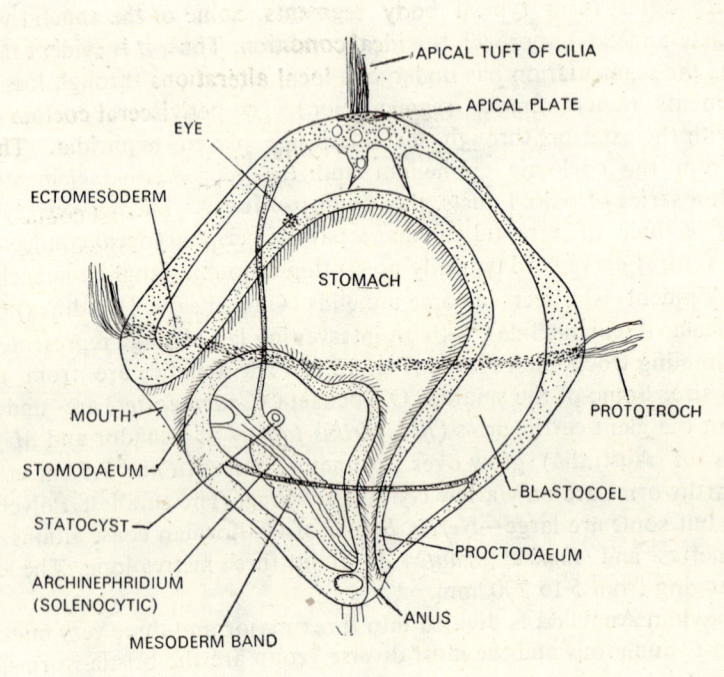

FIG. 12.2. Trochophore larva of an annelid (*after* Shearer).

cesses of the body wall known as **parapodia**. The head consists of two parts, the **prostomium** and the **peristomium**. The coelom is extensive and is incompletely divided into a series of chambers by muscular partitions extending from the body wall to the alimentary tract which is provided with an **eversible buccal region** and a **protrusible** (not eversible) **pharynx**. A vascular system is present in the majority and organs of respiration in the shape of gills or branchiae are usually developed. The excretory organs comprise segmentally arranged

*Phylum Annelida*

pairs of tubes, the **nephridia**. The nervous system is well-developed comprising a dorsal brain formed by double ganglia connected to a ventral nerve cord consisting of a double chain of ganglia extending throughout the body. As a rule there are no special generative ducts and the generative products released in the coelom pass out by any or some of the nephridia. Polychaetes are mostly dioecious, and spawn their eggs and sperms into the sea so that fertilization is external and early development involves ciliated planktonic larva, the **trochophore** (Fig. 12.2).

### TYPE NEREIS

*Nereis*, the rag-worm, is one of the less specialized bristleworms, which forms a convenient type for study. Several species of these are common in the inter-tidal zone on the sea-shore under stones and among seaweeds, almost in all parts of the world. In size the various species range from 5 to 30 cm. They vary considerably in colour though they invariably have richly tinted iridescent hues. *Nereis* (Fig. 12.3) forms most deadly baits for the salt water angler, no fish being able to resist it.

## External Characters

The body of *Nereis* is long and slender. The most noticeable feature is the segmentation of the body which is not merely external but involves nearly all the

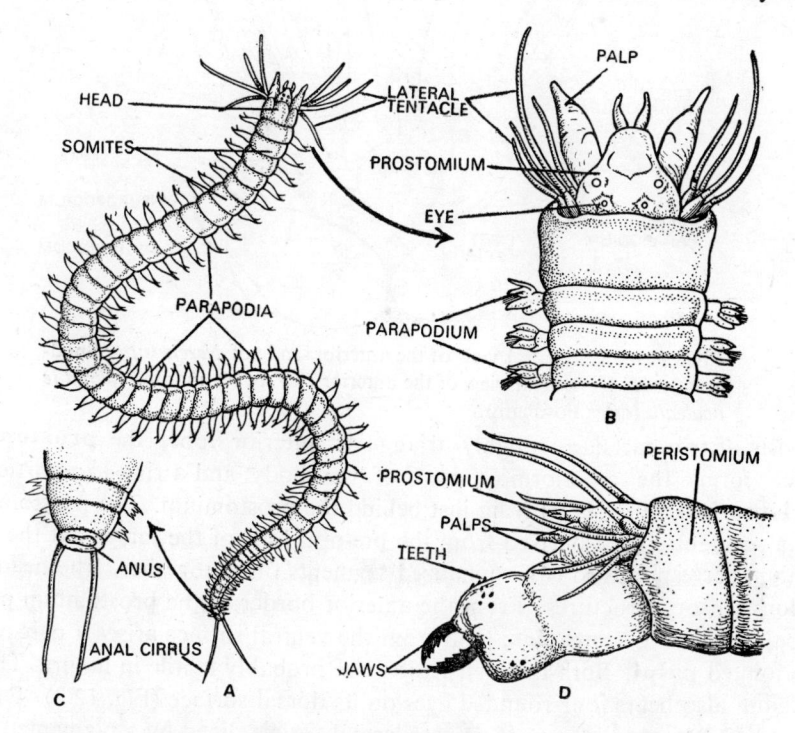

FIG. 12.3. External features of *Nereis*. A—*Nereis* entire; B—Anterior part of *Nereis*; C—Anal end of the same; and D—Everted pharynx.

internal structures as will be seen later. Except for the **head** and the last segment, the "**tail**," all the segments of the body are externally alike. A series of ring-like grooves mark off about eighty segments besides those of the head. The head, at the anterior end, is a distinct structure bearing tentacles and eyes. It

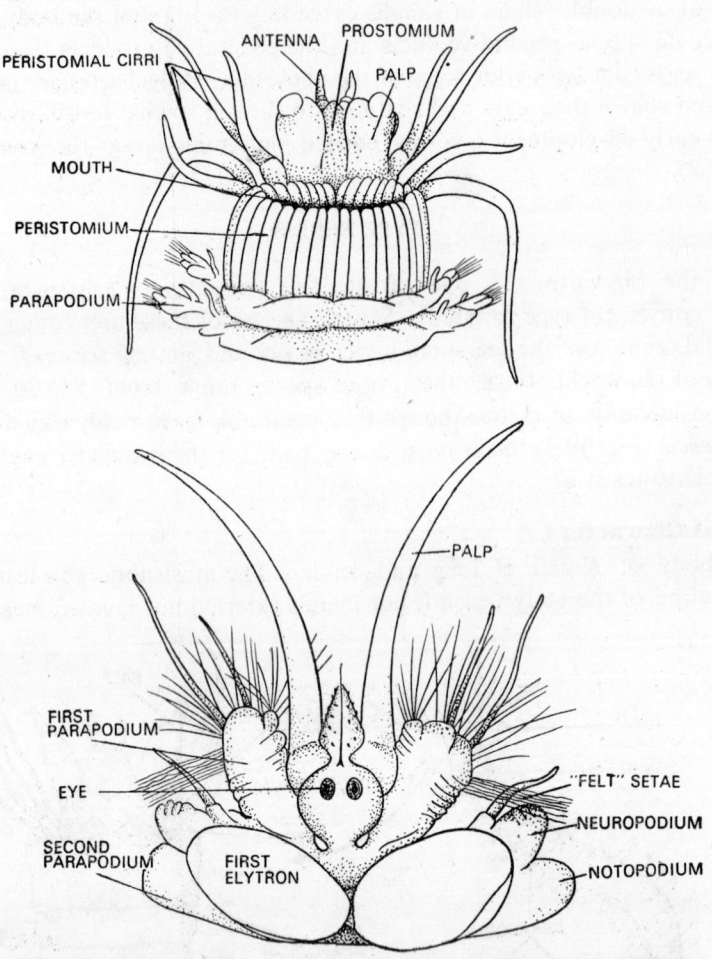

FIG. 12.4. A—Ventral view of the anterior end of *Nereis* (after Snodgrass); and B—Dorsal view of the anterior end of sea-mouse, *Aphrodite aculeata* (after Fordham).

consists of two parts: a roughly triangular anterior lobe, the **prostomium**, which forms the anteriormost part of the body; and a ring-like portion, the **peristomium** (Fig. 12.3) lying just behind the prostomium. The prostomium is not a segment as it is derived from the pre-oral lobe of the larva, but the peristomium corresponds to two cephalized segments fused together. The head bears various sensory structures. From the anterior border of the prostomium projects a pair of short **tentacles**, and from the ventral surface arises a pair of stout two jointed **palpi**. Both these structures are probably tactile in nature. The prostomium also bears four rounded eyes on its dorsal surface (Fig. 12.3). Each eye (Fig.12.12 B) is made up of an ectodermal vesicle, lined by a pigmented layer, containing a cuticular lens. The visual elements which form the walls of the vesicle are modified ectodermal cells and are continuous with the epidermis at the edges. On each side of the prostomium there is a ciliated pit, the **nuchal organ**, the function of which is doubtful.

## Phylum Annelida

The peristomium bears four pairs of **cirri** (Fig. 12.4) which arise from its anterior part, four on each side. These cirri are homologous with the dorsal (**notopodial**) and ventral (**neuropodial**) cirri borne by the parapodia (see below) of more posterior segments. Each body segment bears only two cirri on each side; as the peristomium bears four cirri on each side it is concluded that two segments have fused to form the peristomium during cephalization. On the ventral surface of the peristomium is a transverse aperture, the mouth.

Each body segment is usually convex dorsally, the ventral surface is usually flat, but may even become concave when certain of the body muscles contract. Each segment carries a pair of projecting appendages **parapodia** (side feet) consisting of flattened fleshy projections of the body wall (Fig. 12.4) from which protrude bundles of horny bristles or **setae** (**chaetae**). In structure each parapodium

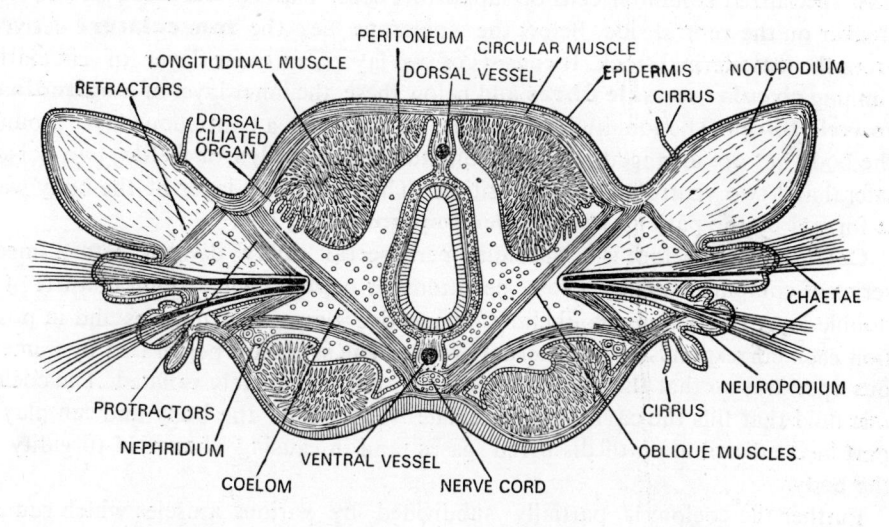

FIG. 12.5. Transverse section of a segment of *Nereis* showing parapodia.

is **biramous** consisting of an upper part, the **notopodium** (Fig. 12.4) and a lower part, the **neuropodium**. Each of these is bilobed bearing a bundle of setae, and from each arises a slender filament, the **cirrus**. The **dorsal cirrus** (Fig. 12.5) is a short cylindrical tentacle-like appendage borne by the notopodium, whereas, the one borne by the neuropodium is known as the **ventral cirrus**. The setae are grouped into two bundles, one inserted in the notopodium and the other in the neuropodium and in each bundle two main types of setae are found. Each of the bundles of setae is lodged in a sac formed by invagination of the epidermis, the **setigerous (chaetigerous) sac**, and is capable of being protruded or retracted and turned in various directions by strands of muscular fibres in the interior of the parapodium. In the interior of each bundle there is a stout straight dark coloured seta, the **aciculum**. The ordinary setae are fine chitinous rods. Each consists of a **shaft** articulating with a long **blade** at the free end. Structurally there are two types of setae: (*a*) with a longish shaft and a long, slender, more or less, straight blade; and (*b*) a short shaft with a short-hooked blade. The setae originate from a single cell, the **formative cell**, lying at the base of the sac. The slender setae assist in locomotion by acting as minute paddles, whilst the aciculum has a skeletal function.

The anal segment is without parapodia, bearing a pair of filamentous processes, the **anal cirri**. The anal segment represents the posterior part of the larva and does not arise by a process of subdivision or segmentation like true body segments, it is therefore, frequently termed the **pygidium**. Towards its posterior extremity the anal segment carries a rounded opening, the **anus**.

## BodyWall

The **cuticle** (Fig. 12.4) forms the thin outermost chitinous covering of the body. It is perforated by the openings of the epidermal glands. Next to the cuticle lies the epidermis, an ectodermal contribution comprising a single layer of cells of which some are glandular, some are sensory, whilst others are more or less specialized columnar cells or supporting cells. The **epidermis** is relatively thicker on the ventral side. Below the epidermis lies the **musculature** derived from the mesodermal stock. It consists of two layers, an outer layer of circularly running **circular muscle fibres** and below these, the inner layer of **longitudinal muscle fibres**. The longitudinal muscles do not form a continuous layer around the body but are arranged into four longitudinal **bundles** or **fields**, two dorsolateral and two ventrolateral in position. The innermost layer of the body wall is formed by the parietal layer of **peritoneum** (Fig. 12.5).

**Coelom**. The coelom is a spacious perivisceral cavity divided into a linear series of compartments by cross partitions or **septa**. Each septum consists of a double layer of coelomic epithelium, connective tissue and muscles and in position each corresponds to an external groove. The septa are perforated by numerous apertures so that the compartments are not completely isolated. The coelomic fluid that fills the cavity can circulate throughout the body and can play a part in the distribution of dissolved gases and in causing change in turgidity of the body.

Further the coelom is partially subdivided by various muscles which run an oblique course from the dorsal to the ventral body wall. The most obvious **oblique muscles** (Fig. 12.5) take origin at the sides of the nerve cord and are inserted in the parapodia. The exact anatomical relationships of these and the parapodial muscles (the **retractors** and the **protractors**) are not fully known, but perhaps they (oblique muscles) control the movements of parapodia. This is evident from the fact that they arise at the sides of the nerve cord, and are inserted in the bases of parapodia. In every segment there are two pairs of oblique muscles; one pair is inserted in the anterior and the other in the posterior face of the bases of the parapodia. Each oblique muscle appears in reality to be composed of two bundles of muscle fibres—one running to the dorsal part of the base of parapodium, the other to its ventral part. It has been suggested that one function of these muscles is the retraction of the whole parapodium. Protrusion of the parapodia is probably very largely due to coelomic fluid being forced into their cavities by the contraction of the body wall muscles, while controlled flexures of the parapodia in the vertical as well as in an anterior-posterior direction could be effected by differential contractions of the various oblique muscle bundles against the pressure of the coelomic fluid.

The movements of the various lobes of the parapodia and setae inserted in them are controlled by the **parapodial muscles** (Fig. 12.5). The protractors of

# Phylum Annelida

the acicula are the largest groups of parapodial muscles. They take origin from the circular muscles of the body wall around the base of each parapodium and

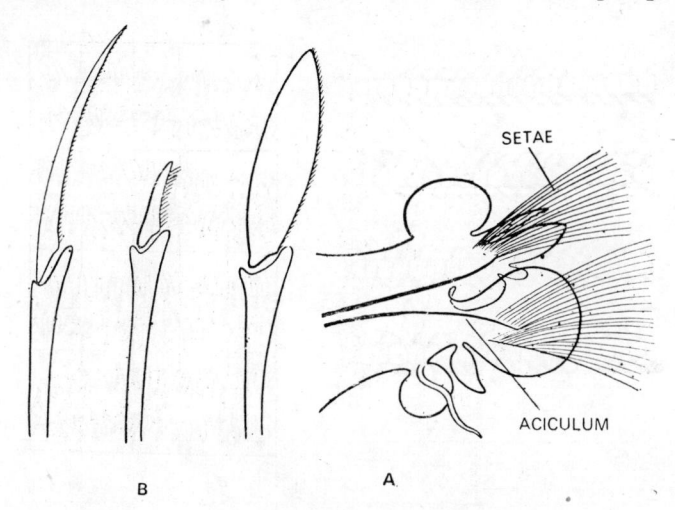

FIG. 12.6. A—A parapodium of a heteronereid; B—Structure of chaetae; C—From ordinary parapodia; and D—Oarlike chaeta of a heteronereid.

are inserted in the setigerous sacs of the acicula. These are radiating muscles, the simultaneous contraction of which causes the protrusion of the acicula and central lobes of the parapodium. The protractors are thus regarded skeletal structures. The retractors of the acicula are other parapodial muscles which withdraw the acicula. Thus, the varied movements of the parapodia and associated parts are controlled by complex musculature, together with the action of the coelomic fluid.

## Locomotion

Two structural elements are especially important in the locomotion of *Nereis*: the longitudinal muscles of the body wall, and the parapodia. The segmentation of the muscles and of their nerve supply facilitates the passage of waves of contraction down the body. The longitudinal muscles in *Nereis* do not form a continuous layer. Instead they are broken up into two pairs of blocks, one pair dorsal and the other ventral. Because of this the muscles of the two sides of a segment can be in opposite phases, one contracted and the other relaxed so that the passage of waves of contraction along the body can throw it into lateral undulations (Fig. 12.7). These are the characteristic feature of the locomotion of *Nereis*, and contrast markedly with the peristaltic waves of the earthworm. The waves of contraction in *Nereis* pass forwards. In this animal the circular muscle layer is relatively weak, more particularly because it is interrupted laterally where muscles derived from the circular layer run into the parapodia. Some support is probably given by fibres that are present in it, and a contribution is also made by the oblique muscles that run upwards and outwards from the region of the ventral nerve cord. Despite this supplementary strengthening the body is clearly ill-adapted for the peristaltic movements that are so well provided for in earthworms. It is the lateral undulations that are important in locomotions of *Nereis*. Assisted by the parapodia the lateral undulations displace the

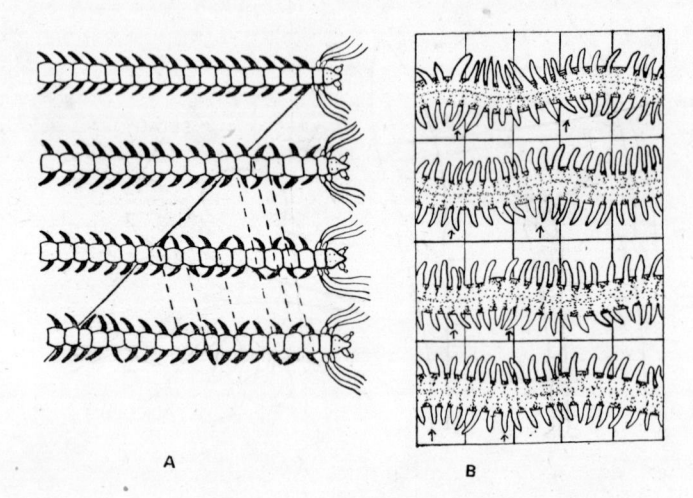

FIG. 12.7. Figures showing movements and locomotion in *Nereis*. A—Cycle of activity of parapodia during slow crawling. Broken lines connect successive parapodia which are about to execute a power stroke, the solid line marks the spread of excitation backward along the body; and B—Inclination of parapodia during undulatory swimming movements; arrows indicate successive positions of two parapodia.

body forward. This is the type of movement that anticipates the locomotory mechanisms of arthropods.

The parapodium in *Nereis* is clearly quite versatile in its mode of action. *Nereis* and other polychaetes remain dependent upon the hydrostatic skeleton to provide firm basis for muscular action. The properties of musculature and parapodia of *Nereis* interact to provide for several different types of movement. One of these is a slow creeping, which depends almost entirely upon the use of the parapodia as a series of levers. Initially a parapodium moves forwards, with its tip lifted from the ground and with the aciculum withdrawn, this phase of the movement constituting the **preparatory stroke**. At the end of this stroke the parapodium makes contact with the substratum, the aciculum is protruded, and the oblique muscles contract so that the body is pulled forwards. The parapodium comes to be directed backwards during this phase which is now called **power stroke**. The movements of parapodia are so integrated that the two members of any one segmental pair alternate with each other in phase. Moreover, the movement of any one parapodium begins slightly after that of the one next behind it. The actions are thus seen as waves of movements that travel forwards over the length of the body.

Slow creeping readily passes into rapid creeping. This involves a similar rhythmic pattern of waves, but with the difference that the longitudinal muscles of the body wall are now of primary importance. They contract serially in parallel with the movements of the parapodia. The oblique muscles of the latter are now probably of much less importance than in slow creeping, for rapid creeping is mainly effected by the longitudinal muscles pulling against points of friction established between parapodia and substratum. The contractions of the longitudinal muscle throw the body into lateral sinusoidal waves that pass anteriorly along its length. In the free-swimming phase the pattern of movement is essentially the same as that of the rapid creeping. Sinusoidal waves of the body interact with parapodial movement, but there is a marked increase in the length of the waves and also in their amplitudes and frequency. Here again the worm is propelled forwards by the body waves that also pass forwards (situation that presents a striking contrast with fish, which swim forwards through the agency of waves that pass backwards). This difference is again a result of the presence of parapodia. These move backwards during their active stroke and so by creating a backward flow of water, give a forward thrust to the body.

## Alimentary Canal

The **mouth** leads into the **buccal cavity**, which is a wide chamber lined by the ectodermal epithelium continuous with the epidermis of the head (stomodaeal in nature). Posteriorly, the buccal cavity opens into the **pharynx** (Fig. 12.8) which is also lined by an ectodermal epithelium like the buccal cavity. The cuticle lining the buccal cavity and the pharynx is thickened at various places

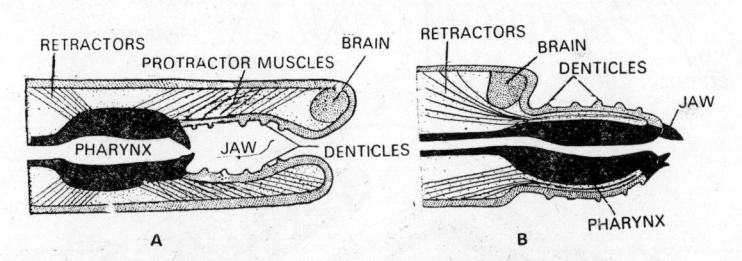

FIG. 12.8. Diagrammatic representation of the pharyngeal apparatus of a predatory polychaete. A—Retracted; and B—Protruded.

forming small dark brown chitinous denticles also known as **cuticular teeth** and **pharyngeal teeth** (in the pharynx). A pair of the pharyngeal teeth, near the posterior end of the pharynx, is notably enlarged and is termed the **"jaws"** (Fig. 12.8). The whole of the buccal and pharyngeal region is eversible. When needed the buccal cavity becomes turned inside out, while the pharynx is thrust forward so that the "jaws" are exposed and come to lie in front of the head. Eversion of the buccal cavity is brought about chiefly by the pressure of the coelomic fluid forced into the buccal region by the contraction of the muscles of the body wall, and the forward movement of the pharynx is due to the contraction of the protractor muscles of the pharynx. These muscles comprise a number of bands or sheets of muscles running backward and inward from the wall of the prostomium to the wall of the buccal cavity. The folding in of the buccal cavity and retraction of the pharynx is brought about by the action of retractor muscles aided by a relaxation of muscles of the wall, which also bring about a redistribution of the coelomic fluid.

The part of the alimentary tract (Fig. 12.9) lying posterior to the pharynx is known as the **mesenteron** and internally it is lined by endoderm in contrast to the bucco-pharyngeal region. The mesenteron is divisible into two main subdivisions, the anterior **oesophagus** followed by the **intestine** (Fig. 12.9) The oesophagus, a narrow tube, runs through five segments and opens into the intestine. Into the oesophagus open a pair of laterally placed caecae or pouches known as the **oesophageal glands** because they secrete digestive juices. The intestine is a straight tube, constricted in each segment, and opening into the **rectum** posteriorly. The rectum is the last part of the gut which opens posteriorly by the **anus**. It is lined by ectoderm continuous with that of the outer surface and as such constitutes the **proctodaeum**.

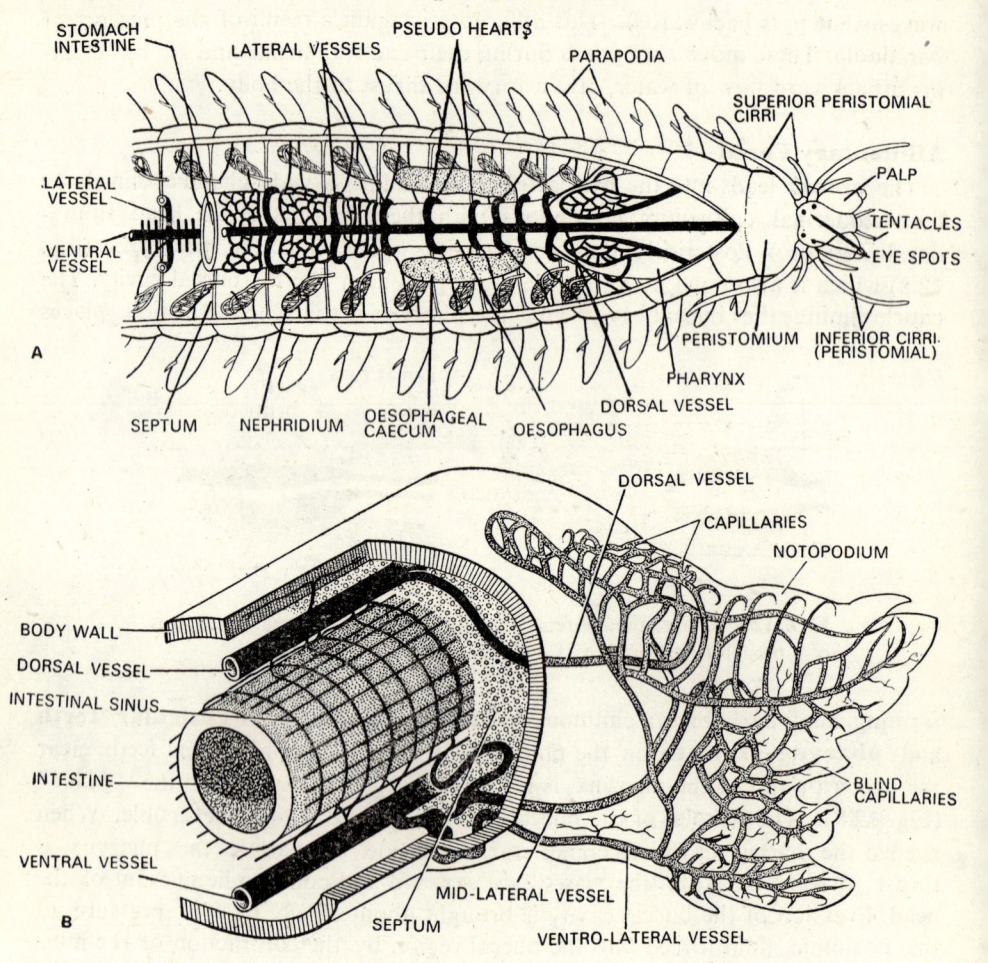

FIG. 12.9. A—Dissection of *Nereis* seen from dorsal side; and B—Blood vascular system of a segment.

The food of *Nereis* consists of small molluscs, small crustacea and other animals which it grasps by means of the everted jaws and swallows by withdrawing the pharynx. The digestive juices are poured by the oesophageal glands and also by the lining epithelium of the mesenteron, part of which is also responsible for the absorption of digested food. The food moves from one part of the gut to the other by the contraction of muscles in the wall of the digestive

# Phylum Annelida

tube. The contraction of these muscles produces a succession of rhythmic waves of constriction, a type of muscular activity called **peristalsis**, which pushes the food along, independently of the movements of the whole body. This provision of muscles in the gut wall outside the digestive epithelium presents a definite advance over the condition found in the nemerteans.

**Filter feeding.** In some errant polychaetes filter feeding occurs. Under some circumstances *Nereis diversicolor* (Fig. 12.10) produces body secretion within its burrow. This takes the form of a net. Water is pumped through this net so that it can be used for a simple form of filter feeding. Particles collected in the secretion, as though in a bag, are swallowed from time to time. It can be visualized that the further elaboration of some such mechanism might have been aided by the presence on the head of tentacles and palps, which are used by errant worms for sensory purposes and to assist in the manipulation of food. In sedentary worms the structures have given rise to complex and beautiful system of tentacular outgrowths often called **gills**, or **branchial crown** because at one time regarded as primarily respiratory in function. No doubt they are respiratory but they also provide mechanisms for collection and sorting of food particles; they are aided in this by the production of mucous which is distributed over tracts of ciliated epithelium (Figs. 12.1, 12.17 and 12.18).

## Circulatory System

The circulatory system of *Nereis* is well developed, compared with it the nemertean circulatory system appears a crude apparatus. In essentials it consists of a distributing system of vessels which communicate with a collecting system by means of networks of capillaries, and by means of certain contractile vessels, the blood is maintained in constant circulation. The main collecting vessel is the **dorsal vessel** which runs above the digestive tract and the main distribut-

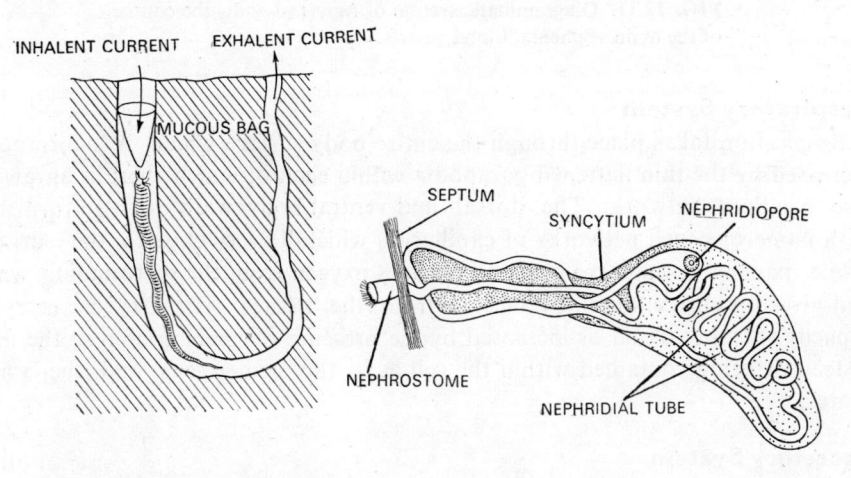

FIG. 12.10. A—Filter feeding in its burrow by *Nereis diversicolor*; and B—Nephridium of *Nereis*.

ting vessel is the **ventral vessel** just below the gut. They are connected with each other by a series of segmental **commissural vessels** (Fig. 12.9). There are two of these in every segment, but they do not pass direct from the ventral to the dorsal vessel, but first supply the body wall and parapodia where they break into an extensive capillary network. The gut also receives its blood supply from **segmental vessels** arising from the ventral vessel and after circulating through a capillary network in its wall the blood is returned to the dorsal vessel by two pairs of vessels in every segment. Most of the larger vessels are contractile. In the dorsal vessel rhythmic waves of muscular contraction run forward from behind driving the blood anteriorly. In the ventral vessel the blood flows posteriorly.

The circulating fluid is blood in which is dissolved a red pigment called haemoglobin and in which blood corpuscles float The pigment is concerned with the conveyance of oxygen from the respiratory surfaces to the tissues. Besides respiratory gases blood also distributes dissolved food substances and nitrogenous wastes, etc. These diffuse out through the thin walls of the capillaries whose extensive ramifications ensure that the substances are delivered almost to every cell and do not have to move long distances by the slow process of diffusion.

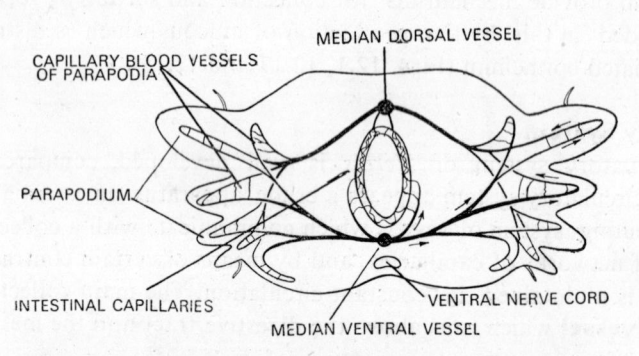

**FIG. 12.11.** Diagrammatic section of *Nereis* to show the course of the main segmental blood vessels.

### Respiratory System

Respiration takes place through the entire body surface which is enormously increased by the thin flattened parapodia within each of which there is an extensive capillary network. The dorsal and ventral body walls are also furnished with numerous such networks of capillaries, which lie very close to the surface. While passing through them, blood receives oxygen from the surrounding water and gives up carbon dioxide collected from the tissues. The oxygen carrying capacity of the blood is increased by the presence of haemoglobin in the fluid instead of being contained within the cell as in the nemerteans and the vertebrates.

### Excretory System

The excretory system consists of a series of segmentally arranged organs, the **nephridia** (singular: nephridium) (Fig. 12.10B). Essentially each nephridium consists of a coiled tube which opens at one end by a ciliated funnel, **nephrostome**, into the coelom and at the other end by a pore, the **nephridiopore**, to the exterior. In *Nereis* the coils are embedded within an oval syncytial mass of pro-

*Phylum Annelida* 413

toplasm. The nephridiopore lies at the base of the parapodium. The internal end runs forward, passes through the septum, and opens into the coelomic chamber just anterior to that in which lie the main body of the nephridium and the nephridiopore. Waste products, extracted from the blood which passes through the excretory organ and also from the coelomic fluid, are sent to the exterior by means of cilia lining the excretory tube.

**Nervous System**

The nervous system of *Nereis* conforms to a plan which is very common in invertebrate animals. It comprises a **central nervous system** (Fig. 12.12A) and a **peripheral nervous system**. The central nervous system consists of a mass of nervous tissues concentrated dorsally to the pharynx in the prostomium hence known as **suprapharyngeal ganglia ("brain")**. The brain is connected by a **nerve collar (circumpharyngeal connectives)** to a **ventral ganglionated nerve cord** situated in the mid-ventral line and running throughout the ventral nerve cord is represented by two parallel cords joined together by segmental cross connections, but the dual nature of the two cords is not evident ordinarily in *Nereis* as they have fused and become invested in a common connective tissue sheath. The peripheral nervous system consists of nerves arising from the brain and innervating the eyes, the cirri, etc. and the **segmental nerves** arising from the ventral nerve cord which are connected with a well-defined nerve plexus lying just below epidermis.

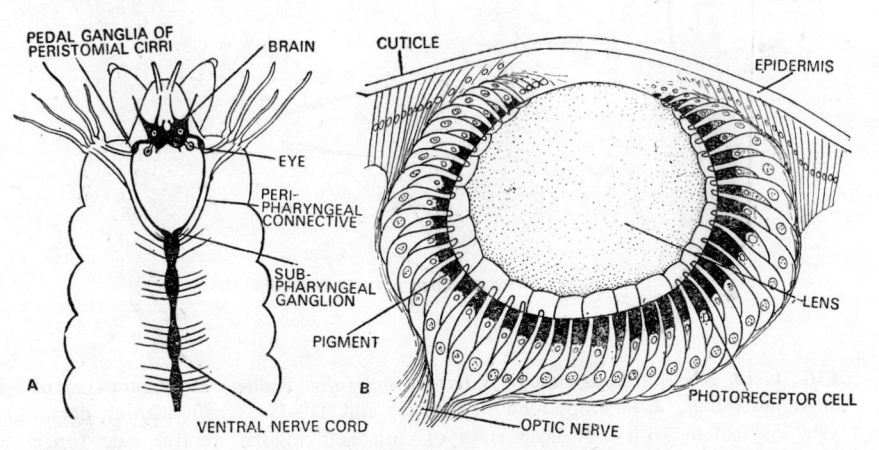

FIG. 12.12. A—Nervous system of *Nereis*; and B—Vertical section of the eye of the same.

In *Nereis* the tendency towards centralization of the nervous system is carried farther than that found in the flatworms and others. The brain, like other parts of the central nervous system, consists of nerve cells and nerve fibres. The major part of these is concerned with receiving and relaying sensory impulses from the organs of sense situated in the head. In the brain of *Nereis* three main centres can be recognized, an anterior, a middle and a posterior. A pair of short nerves from the anterior centre supplies the prostomial palps. The middle region sends a stout nerve to each of the four eyes and a pair of short nerves to the prostomial tentacles. The posterior centre receives nerve fibres from the sensory cells lining the nuchal organs. In all these regions the nerve cells surround a centrally placed mass of nerve fibres. The nerves to the peristomial cirri arise from the ganglia on the nerve collar. The two halves of the collar unite in the ventral part of segment three where they join the **sub-oesophageal ganglia** formed by the fusion of two anterior ganglia of the ventral nerve cord. A pair of nerves from the sub oesophageal ganglia innervates the posterior pair of peristomial cirri and another pair innervates the body wall and parapodia of the segment. The ventral nerve cord is also made up of the nerve cells and nerve fibres. In each segment it enlarges into a segmental ganglion, which gives off nerves to the muscles of the body wall and parapodia.

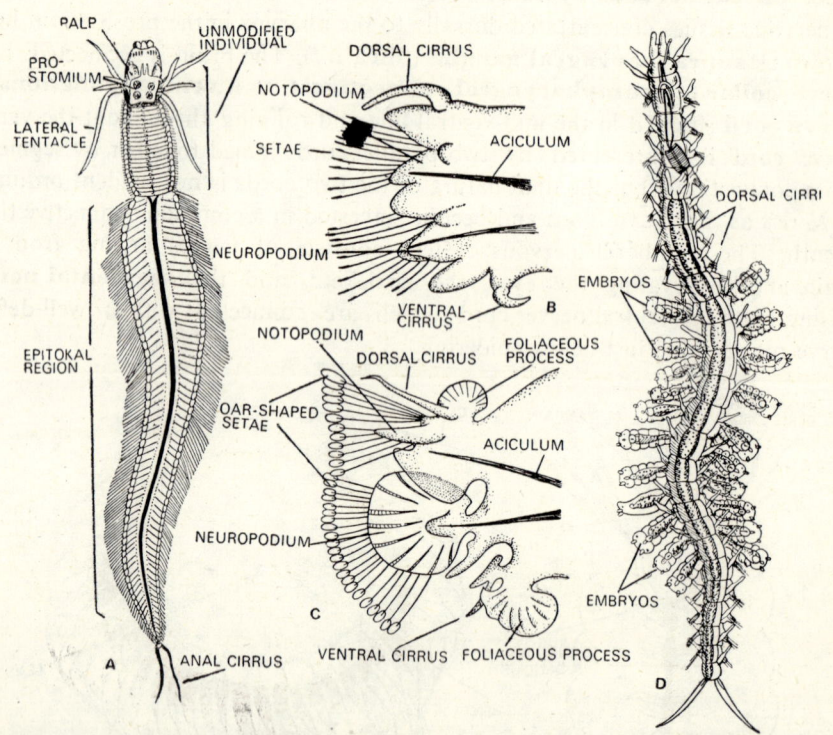

FIG. 12.13. A—Epitokous male of *Nereis irrorate* (*after* Rullier); B—Parapodium of atoke individual; C—Parapodium of epitoke; and D—*Pionosyllis elegans*, dorsal view of female with advanced embryos attached to the ventral side (*after* Pierantoni).

The primitive brain, as that of a planarian, served chiefly as a sensory relay, a centre for receiving stimuli from the sense organs and then sending impulses down the nerve cord. This is also true of *Nereis*, for, if the brain is removed, the animal can still move in a coordinated way; in fact, it moves about more than normal. If it meets some obstacle, it does not withdraw and go off in a new direction, but persists to make unsuccessful attempts to move forward. This is an unadaptive kind of behaviour and shows that in the normal *Nereis* the brain has an important function, that of inhibition of movement in response to certain stimuli. Such a function of the brain was lacking in the flatworms. The main function of the central nervous system and the peripheral nerves is the integration (unification) of the responses of the separate effector organs. The animal is, thus, able to react immediately to any stimulus, no matter where it is received, by suitable movements of the whole body.

## Reproductive System

The sexes are separate. The gonads are temporary structures developing by the proliferation of coelomic epithelial cells. They cannot be detected as distinct organs particularly after the gametes have been shed. The coelomoducts are also absent and are represented by the so-called dorsal ciliated organs, segmental patches of ciliated epithelium on the peritoneum, just below the dorsal column of the longitudinal muscles. They resemble closely the coelomoducts of other polychaete worms and the condition in *Nereis* has probably been secondarily arrived by the failure of the funnels to acquire duct to the exterior.

The **testes** in *Nereis dumerilli* are one pair of proliferating masses between the 19th and 25th segment of the body. In other species there are numerous testes. During breeding season these masses undergo active cell division forming smaller cells which develop into tailed spermatozoa while floating in the coelomic fluid. The **ovaries** are similar masses of cells of coelomic epithelium overlying the principal blood vessels of the body. From these bud off oogonia that undergo further development in the coelom. The development and proliferation of the gametes is accompanied by changes in the structure of the posterior half of the body in which the gametes are especially abundant. The anterior 15 to 20 segments are not greatly modified. In this part the parapodia become larger and acquire flattened foliaceous outgrowths, while the setae are thrown off and replaced by new setae of the flattened form with a fan-like arrangement. Besides this, the eyes become enlarged, and dorsal cirri altered. There is also a change in the segmental musculature.

The reproductive individual, thus formed, has been termed *Heteronereis* or **epitoke**. (The nonsexual form is **atoke**.) The process of the formation of the reproductive individual is called **epitoky**. Apart from nereids such reproductive phenomenon has also been found in syllids and eunicids. Instead of creeping about on the floor of the sea they start swimming actively in the surface waters. The changed parapodia aid in swimming.

In *Eunice viridis*, the Samoan palaloworm, the anterior part of the worm is unmodified, and the epitokal region consists of a very long chain of narrower but longer posterior segments, each with an eyespot on the ventral side (Fig. 12.13). Epitokous individuals arise from an atokous form either by direct transformation or by budding. Transformation is characteristic of the nereids and the eunicids, while asexual budding of epitokes is common in the syllids, in which epitokes usually bud off at the caudal end as a single body or a chain or cluster of individuals (Fig. 12.13).

All epitokous polychaetes swim to the surface in large numbers. This is called swarming. Some nereids and syllids are known to perform nuptial dance in which males and females swim rapidly in small circles. The females attract the males, by producing a substance called **fertiliun** (different from fertiliun produced by animal egg), and stimulate shedding of the sperms which in turn stimulate shedding of the eggs. To ensure fertilization swarming in the males and females coincides and lasts for a relatively brief period. Naturally the swarming is marked by distinct periodicity often coinciding with the lunar periods. The lunar periodicity as displayed by the palaloworm is the most striking example of such a phenomenon. After the sex cells are discharged the worms die.

**Development.** The coelomoducts are absent, therefore, the eggs are released by the rupture of the body wall. The eggs so released are enclosed in a thick transparent gelatinous envelope. A closer study reveals two membranes within the gelatinous envelope. Of these the outer is very thin and delicate and the inner is thick and distinctly striated radially hence called zona radiata. The protoplasm of the egg is filled with yolk-spherules and a number of oil-drops. The eggs are fertilized in the surface layers of the sea. After fertilization the yolk-spherules move away from what is destined to become the upper pole of the egg, leaving a granular polar area. The zona radiata disappears, the egg for a time becomes irregular in shape and then again spherical. Cleavage begins after two small polar bodies are extruded from the upper pole.

The cleavage is of the spiral type and resembles that of a Polyclad egg up to a fairly advanced stage. As a result of first two cleavages four cells are formed of which one is larger than the rest. These are the **macromeres**. From these

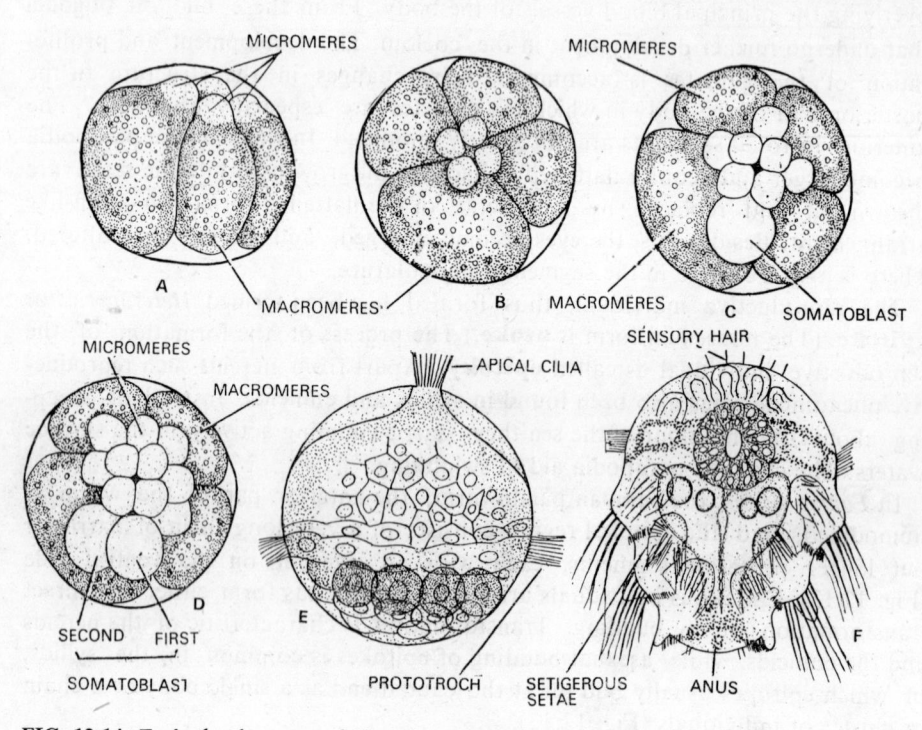

FIG. 12.14. Early development of *Nereis*. **A**—Four-cell stage; **B**—Micromeres and megameres separate; **C** and **D**—Formation of somatoblast cells; **E**—Formation of prototroch and apical cilia; and **F**—Early trochophore.

first four small cells or **micromeres** are separated off, then again four cells and finally four cells are separated off for the third time. At this stage there are four macromeres and twelve micromeres, of which one is somewhat larger than the other and is called the **first somatoblast** (Fig. 12.14 C). Soon after this

*Phylum Annelida*

the same macromere, that gave rise to the first somatoblast, gives rise to the **second somatoblast** (Fig. 12.14 D).

The micromeres of this give rise to the ectoderm, destined to give rise to the epidermis and all its derivatives, to the cerebral ganglion and nerve cord to the oesophagus and rectum. The macromeres constitute the endoderm, destined to form the inner epithelium of the alimentary canal. The second somastoblast gives rise to the entire mesoderm of the annelid and also gives a few small cells to the endoderm of the intestine. The cells of this stage represent the primary germ-layers.

The micromeres multiply and encroach upon the macromeres and the descendants of the somatoblasts, and finally cover them except at one place at the lower pole. This is the **blastopore**, which also closes after some time. Soon after this an apical plate with apical tuft of cilia is formed in the middle of future head-end. In close relation to the apical plate lies the rudiment of the cerebral ganglion and a pair of pigment spots (larval eyes). The **prototroch**, a thickened ciliated ridge, develops around the body of the larva. The rudiment of the **mouth** and **oesophagus** appears as small depression, the **stomodaeum**, just behind the prototroch in the middle of the future ventral surface; the **anus** appears later, but its position is indicated early by a pigmented area, at the point where the blastopore existed. The product of the first and second somatoblasts form a plate of small cells, the **ventral plate**, extending on the ventral surface behind the prototroch and the mouth, the deeper cells of this plate form a pair of mesoderm-bands (muscle-plate) that give rise to body wall musculature, the musculature of the alimentary canal is also formed from certain cells of the

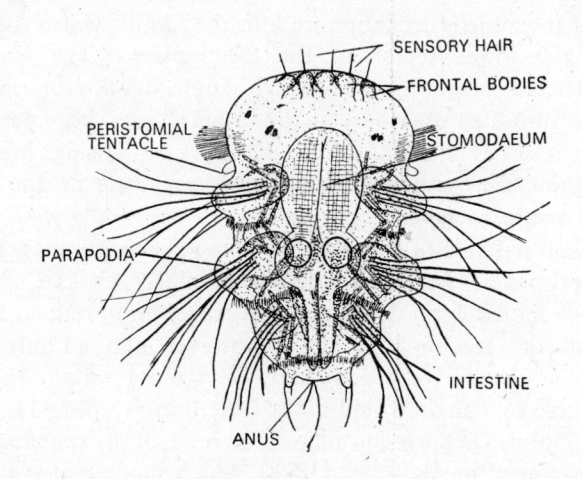

FIG. 12.15. Larva with three segments (*after* B.B. Wilson).

same stock. A superficial thickening of the ectoderm along the middle of the ventral plate forms the ventral nerve cord. By this time the larval nephridia also appear from a pair of micromeres separated from the rest at an early stage. First they are situated at the upper end but later they migrate below the prototroch. The larva hatches at this stage.

The ciliated larva is called **nectochaetes** (Fig. 12.15) from which the new adult develops after a gradual metamorphosis in which most of the larval characters are lost. The body grows in length and additional segments with their setigerous sacs appear. Finally the tentacles appear, the parapodia develop with their cirri and permanent setae (which replace those that are formed first), the full number of segments are formed and the internal organs are completely formed to give rise to the adult.

The larvae which hatch from the eggs of some polychaetes are called trochophores. The **trochophore** larva is somewhat biconical or oval structure with a protruding **equator**. Its external surface consists of a one-layered epithelium (ectoderm) thickened at the apical pole into a sensory plate, the **apical plate**, which bears a tuft of cilia. Around the equator there is a ciliated band termed the **prototroch**, which serves as the chief organ of locomotion and also directs a food bearing current towards the mouth which lies just below (Fig. 12.2). In some a second equatorial girdle, the **metatroch**, passing below the mouth is present, and also a ciliated circlet, the **paratroch**, around the anus. The larva has a complete ciliated digestive tract with oesophagus, large bulbous stomach, and a short intestine that opens through an anus at the lower pole. There is a large cavity, the **blastocoel**, between the digestive tract and the ectoderm. There may be present a **ganglionic mass** under the apical plate with a number of longitudinal nerves radiating from it, and there may be one to several nerve rings. The larval excretory organ contains a **flame-cell** and an excretory tube which opens to the exterior near the anus.

The typical trochophore (Fig. 12.2) is present in only a few annelids. When the young *Nereis* hatches from the egg membrane it has already passed through the trochophore stage and has three segments and bristles (Fig. 12.15). The development of the annelid trochophore into the adult worm begins with the elongation of the lower region of the trochophore. The elongated region becomes constricted into segments which soon develop bristles. The ciliated bands disappear and the upper part of the trochophore becomes the head. The young worm then settles at the bottom, takes up a burrowing life and continues to grow throughout life by the addition of new segments in the hinder region just in front of the last segment.

The trochophore larva is of considerable importance because the same type occurs in several phyla. Apparently diverse animals such as a segmented worm and a snail have similar early developmental stages and produce similar trochophores. Beyond the trochophore stage, however, marked differences begin to appear, and the adults are very dissimilar. The close relationship thought to exist between these two phyla would never have been suspected but for the study of their trochophores. On such considerations is based the **trochophore theory**, elaborately developed by Hatschek (1878), which suggests that the trochophore is the larva of an ancestral form, the **trochozoon**, which was the common ancestor of most, if not all, the bilateral phyla; and which, among living forms, most nearly resembled a rotifer. It is to be noted, however, that the trochophore is regarded as recapitulating the larva of the ancestor, not the ancestor itself.

The polychaetes also reproduce asexually by budding (Fig. 12.13). This power is closely associated with the power of regeneration. In many cases it has been

*Phylum Annelida* 419

observed that if the body is cut in two parts the anterior part will produce a new hind end, and the posterior part a new head.

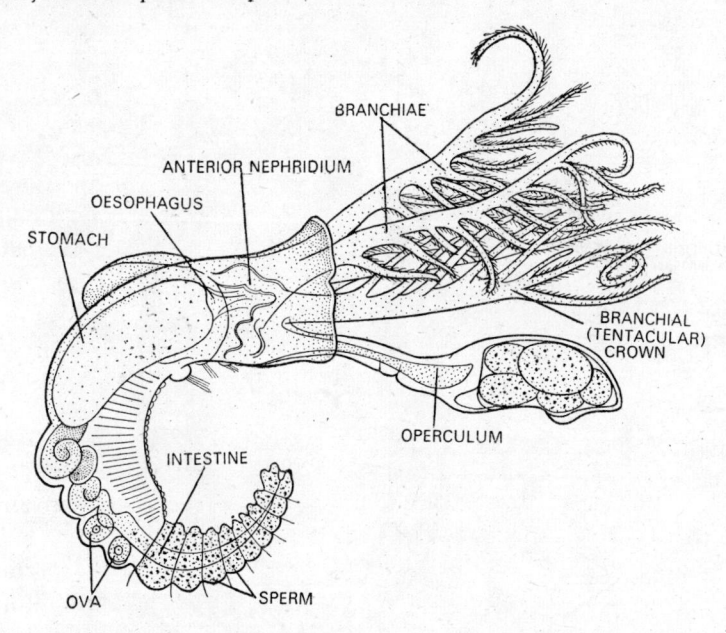

FIG. 12.16. *Spirorbis laevis*, a hermaphrodite tubicolous polychaete taken out of its tube.

## Autotomy and Regeneration

In worms asexual reproduction is more complex. It usually involves fragmentation (breaking up) of the body and subsequent regeneration of the missing part. Thus in worms there is a close relation between regeneration and asexual reproduction. In some cases damaged parts are replaced while in others there are specialized acts of fission that produce individuals closely related with sexual reproduction.

Some worms shed any part of the body damaged by predators. This is **autotomy**. This part is regenerated later on. This ability occurs in tube dwelling polychaete worm *Chaetopterus* (Fig. 12.17). If the anterior end of this worm is pulled out by a predator, a contraction of the circular muscle occurs between segments 12 and 13 and the body breaks into two pieces at that point. This is autotomy and it is followed by regeneration. A complete worm can be replaced from a single isolated segment. Both anterior and posterior parts will grow from the same segment provided that this segment is one of the anterior 14 segments. Posterior part of the body regenerates from any body segment.

If *Autolytus* is placed in dilute sea water it breaks into pieces by very strong contractions of the longitudinal musculature. The breaking occurs along certain septa the positions of which are indicated by white transverse lines. The anterior and posterior pieces of the worm may regenerate respectively the missing tail and the head. Sometimes regeneration of head or tail begins before the pieces are actually separated. In *Autolytus edwardsi* a new head forms in a segment situated within 16-22 segments from the hind end. More heads then appear in front of it so that a chain of individuals or zooids is formed. As many as 29 such heads have been observed in *Myrianida*. All remain attached to the parent stock for a time.

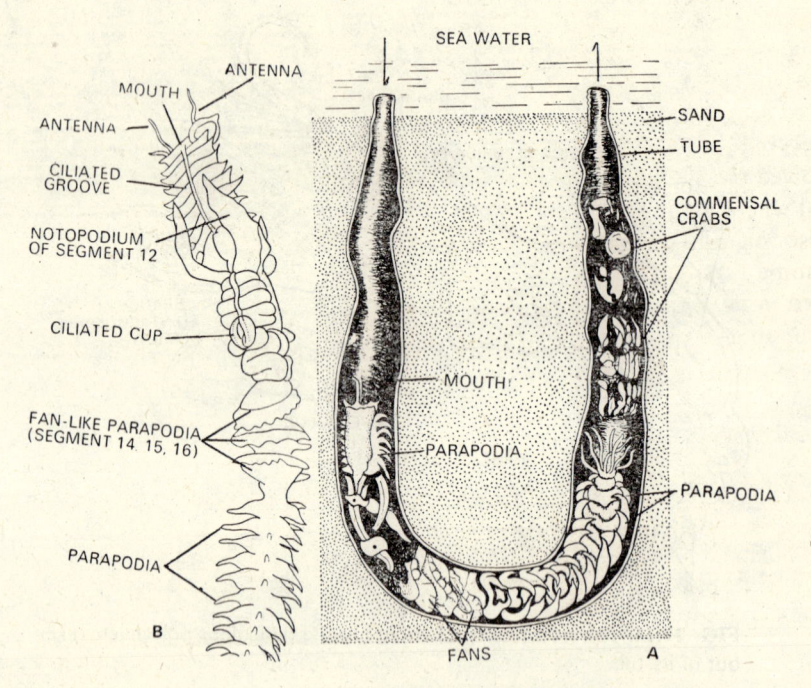

FIG. 12.17. **A**—*Chaetopterus* in its tube; it is a specialised worm dwelling in a secreted tube on the sea bottom. The "fans," modified parapodia, draw water through the tube; and **B**—The worm taken out of its tube.

In *Autolytus pictus* the new head is always said to appear on the anterior half of the fourteenth segment. In many cases segment 14 seems to have inherent power for producing head. In *Syllis ramosa*, which is commensal within a siliceous sponge, lateral buds grow out tail first from various segments. Some of these buds then produce secondary lateral buds before they separate from the parent.

## CLASSIFICATION OF POLYCHAETA

Annelida with chitinous setae developed in setigerous sacs. Parapodia are highly developed and in most cases tentacles are present. Cirri and branchiae are present on the segments of the body. Sexes are separate. Gonads are simple and metamerically repeated. A free-swimming trochophore is usually present. The class is divided into two subclasses:

I. **Subclass Errantia**. The Polychaeta in which the pharynx is usually protrusible and armed with chitinous jaws. Except in head and anal region the body segments are all alike, with parapodia provided with cirri and equally developed throughout. They are predominantly free-swimming, often pelagic. Some forms are found in free or attached tubes, often they are predatory. Some genera are known for asexual budding. It is divided into the following families:

**Family Aphroditidae**. Dorsal surface with elytra.

*Phylum Annelida* 421

Examples: *Aphrodite* (Fig. 12.20), *Hermione, Laetomonice.*

*Aphrodite* (Fig. 12.20) is a short broad form which burrows in mud and which covers its back with a blanket made from interwoven setal threads formed from the notopodium. Between this blanket and the back is a space into which water is drawn by a pumping action of the dorsal body wall being filtered through the matted setae. In this there are special plate-like modifications of the dorsal cirri, the **elytra**, around which circulates the water from which they possibly obtain dissolved oxygen. *Aphrodite* has remarkable segmental caeca of the alimentary canal where the digestion of food particles takes place. *Aphrodite* is called sea-mouse on account of its furry coat and its habit of creeping on the sea floor. Its home is in fairly deep water, where it hides amongst the stones often half-buried in mud. If seen in shallow waters its hairy covering converts the sunlight falling on it into a veritable rainbow of colour. Why such glorious colours should be bestowed upon a creature of such sluggish habits, and one which lives buried in mud for a greater part of its life, is not easy to understand. Probably the colours have some connections with its natural enemies. The larger spine-like

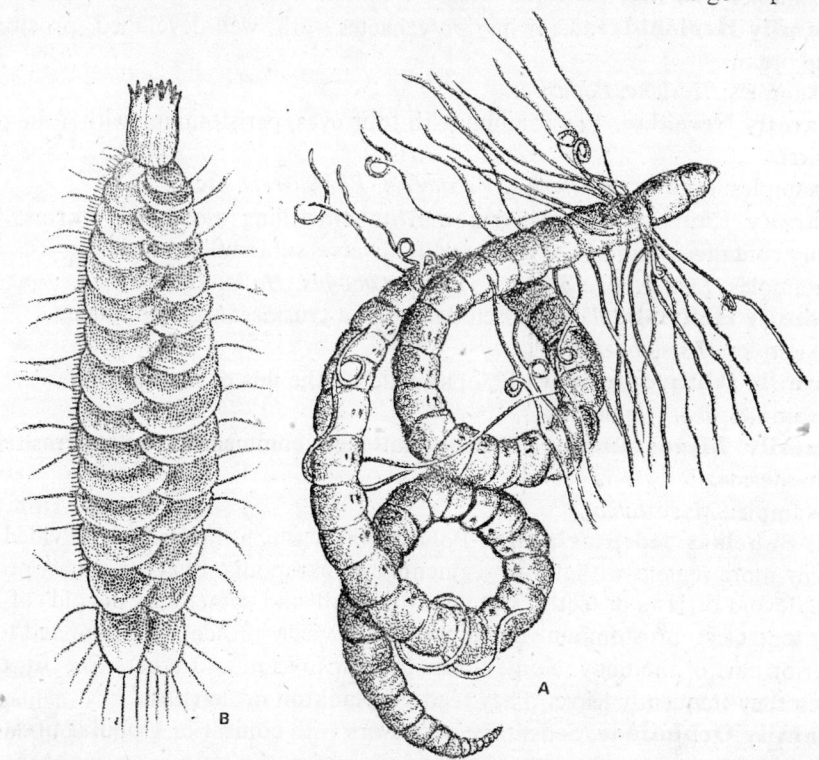

**FIG. 12.18.** A—*Cirratulus cirratus* (*after* McIntosh), a polychaete with long thread-like dorsal cirri functioning as gills; and **B**—*Polynoe.*

bristles of blackish colours are retractile and when the animal gets annoyed they are erected transforming the animal into a miniature porcupine.

**Family Polynoidae.** Dorsal surface with elytra, the scale-worms.
Example: *Polynoe* (Fig. 12.18 B).

**Family Phyllodocidae.** Uniramous parapodia with flattened leaf-like dorsal cirri, crawling polychaetes.

Examples: *Phyllodoce* (Fig. 12.77), *Notophyllum*.

**Family Alciopidae.** Pelagic worms with transparent bodies and two large eyes.

Example: *Alciopa*.

**Family Tomopteridae.** Pelagic worms with membranous pinnules in place of setae.

Example: *Tomopteris*.

**Family Glyceridae.** Errant burrowing worms with a long proboscis armed with at least four jaws.

Example: *Glycera*.

**Family Nephthyidae.** Rapid crawling worms with well-developed prostomial sense organs and gills.

Example: *Nephthys*.

**Family Syllidae.** Small crawling worms with long delicate bodies and uniramous parapodia.

Example: *Syllis*.

**Family Hesionidae.** Crawling polychaetes with well-developed prostomial sense organ.

Examples: *Hesione, Podarke*.

**Family Nereidae.** Prostomium with four eyes, peristomium with four pairs of cirri.

Examples: ragworms—*Nereis, Perinereis, Platynereis*, etc.

**Family Eunicidae.** Free-living and tube-dwelling elongated worms. This family contains a number of ecologically diverse subfamilies.

Examples: *Eunice, Marphysa, Lysidice, Onuphis, Halla, Hyalinoecia*.

**Family Histriobdellidae.** Ectoparasitic in crustacean gill-chambers.

Example: *Histriobdella*.

**Family Ichthyotomidae.** Ectoparasitic on the fins of marine eels.

Example: *Ichthyotomus*.

**Family Myzostomidae.** Greatly flattened commensals and parasites of echinoderms.

Example: *Myzostoma*.

II. **Subclass Sedentaria.** The Polychaeta in which the body is divided into two or more regions with unlike segments and parapodia. Pharynx is non-protrusible devoid of jaws or teeth. The head is small and sometimes devoid of eyes and tentacles; prostomium indistinct. Gills, when present, are confined to the anterior part of the body. Some forms burrow in sand and others live in tubes which they frequently leave. They feed on plankton or detritus.

**Family Orbiniidae.** Sedentary burrowers with conical or globular prostomium without any appendages.

Examples: *Orbinia, Scoloplos*.

**Family Spionidae.** Prostomium with two long palps, tube-dwellers.

Examples: *Spio, Nerine, Polydora*.

**Family Chaetopteridae.** Tube-dwellers in parchment like tubes.

Example: *Chaetopterus* (Fig. 12.17).

**Family Sabellariidae.** Tube-dwellers in sandy tubes.

Example: *Sabellaria*.

**Family Capitellidae.** Burrowers in sand and mud with conical prostomium

and long body.
Examples: *Capitella, Dasybranchus.*
**Family Arenicolidae.** Sedentary burrowers without head appendages.
Examples: *Arenicola* (Fig. 12.19), *Abarenicola*. *Arenicola marina* (Fig. 12.19) is a burrowing polychaete, whose body is divided into three regions: (*i*) the anterio

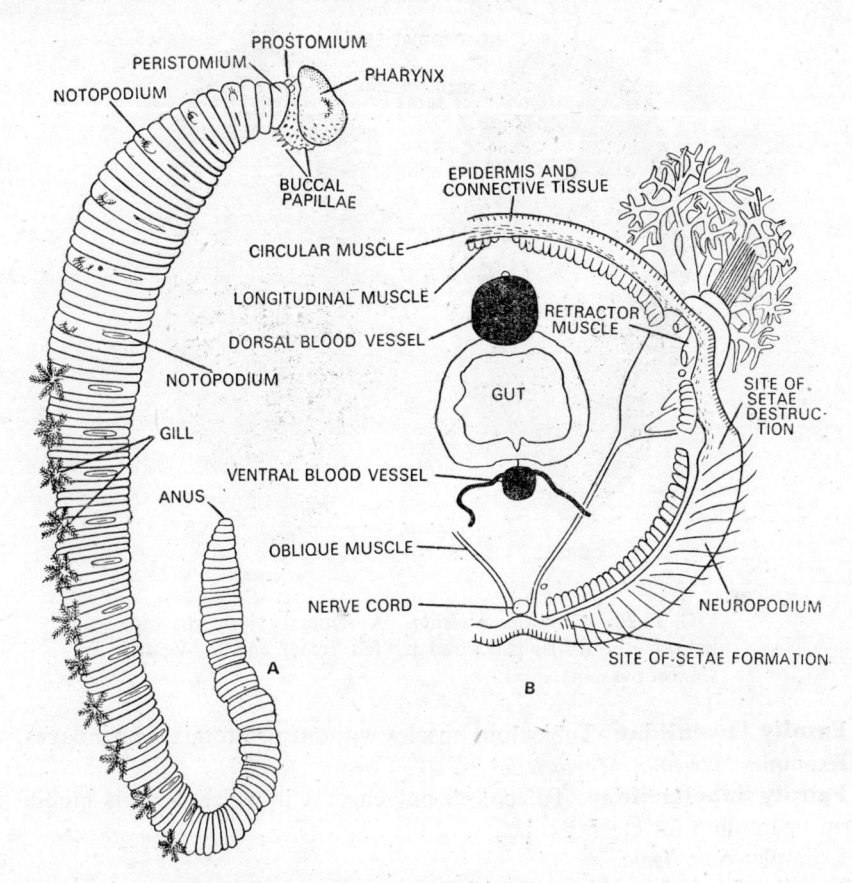

FIG. 12.19. A—*Arenicola*, lateral view (*after* Brown); and B—T.S. of a setigerous segment of *Arenicola marina* (modified *from* Wells).

region consisting of seven segments of which the peristomium is devoid of setae but each of the other six segments has a notopodium with capillary setae, and a neuropodial ridge with setae resembling crotchets; (*ii*) the middle region follows the above and the segments of this region are provided with gills; and (*iii*) the posterior region of the body has segments devoid of parapodia and setae. The prostomium is much reduced without any appendages and there is an eversible pharynx covered with minute papillae, which is the organ of locomotion through the sand as well as of feeding. The body cavity is spacious, not encroached by longitudinal musculature and without vertical paritions or septa.

**Family Opheliidae.** Burrowers in sand and mud.
Examples: *Ophelia, Thoracophelia.*
**Family Maldanidae.** Tubicolous polychaetes with small prostomium fused

to peristomium and without head appendages. The bamboo worms.
Examples: *Maldane, Clymenella.*
**Family Cirratulidae.** Segments bear long thread-like gills.
Examples: *Cirratulus, Ctenodrilus.*

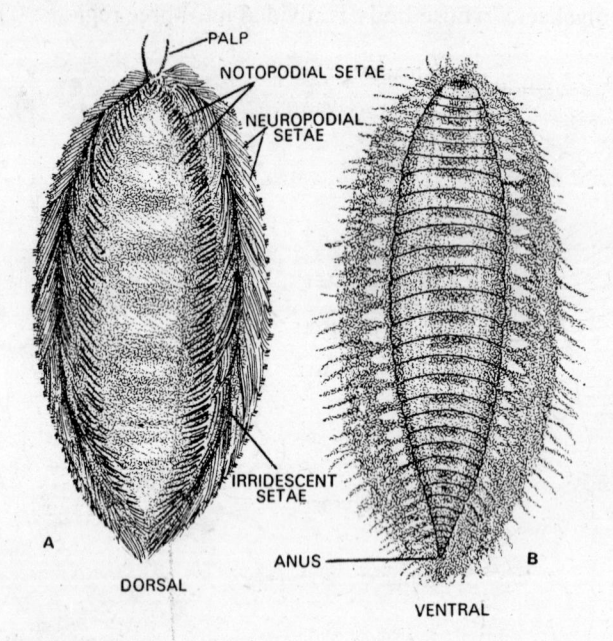

FIG. 12.20. *Aphrodita aculeata.* A—Dorsal view, the wide uniform dorsal strip is covered by felt setae; and B—Ventral view of the same.

**Family Oweniidae.** Tubicolous species without prostomial appendages.
Examples: *Owenia, Myriochele.*
**Family Sabellariidae.** Tubicolous polychaetes in which head is modified to form operculum for closing tube.
Example: *Sabellaria.*
**Family Amphictenidae.** Tubicolous polychaetes making conical tubes.
Examples: *Pectinaria, Cistenides.*
**Family Ampharetidae.** Tubicolous polychaetes with retractile buccal tentacles.
Example: *Amphicteis.*
**Family Terebellidae.** Tubicolous or sedentary burrowers in which peristomium is with numerous, long filiform tentacles.
Examples: *Terebella* (Fig. 12.75 B), *Amphitrite.*
**Family Sabellidae.** Fanworms in non-calcareous tubes.
Examples: *Sabella, Myxicola,* etc.
**Family Serpulidae.** Fanworms in calcareous tubes.
Examples: *Serpula, Pomatoceros, Spirorbis* (Fig. 12.16).

## CLASS OLIGOCHAETA

The **Oligochaeta** (L. *oligos,* few+*chaete,* spine) comprises some small, and for

## Phylum Annelida

the most part, freshwater worms, and all the worms are known as earthworms. The oligochaetes differ from the polychaetes in several important ways. There are no parapodia and the setae emerge from pits in the body wall. The setae are, as a rule, few in number. The head consists of a prostomium, sometimes much elongated (*Nais lacustris*), and is generally separated from the peristomium or oral segment by a distinct groove. The peristomium is always without setae. There are no cephalic appendages either on pro- or peristomium. The segmentation of the body is well-marked externally by external grooves, and internally the coelomic septa are always present (Fig. 12.21). Because the oligochaetes and

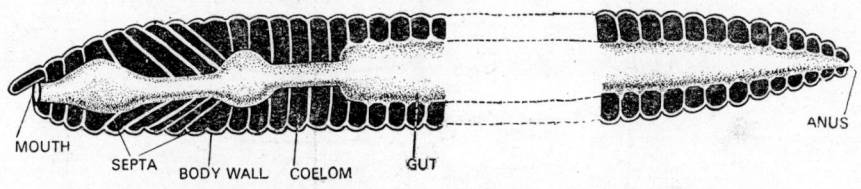

**FIG. 12.21.** General body construction of an earthworm

the polychaetes both are segmented and bear setae which are moved by muscles and serve in locomotion they are grouped under Chaetopoda ("bristle footed"). The gonads are localized thickenings of the coelomic epithelium and special generative ducts are present to convey their products to the exterior. Both male and female sex organs are present in the same individual (hermaphroditic). There is always present a glandular development of the ectoderm, generally having a marked annular form called the clitellum or girdle, which secretes the cocoon in which the eggs are laid. The oligochaetes are derived from the polychaetes which are more primitive. The oligochaetes present great constancy in the arrangement and structure of many important organs such as the body form, the ventral nervous system, etc , but they differ remarkably so far as the excretory and generative organs are concerned.

### TYPE PHERETIMA

The earthworms and their allies (Oligochaetes) are, in many respects, rather specialized annelids, and although many species of earthworms are found in the Indian soil, one of them, *Pheretima posthuma*, is usually selected for detailed description because its anatomy and habits have been fully studied. The earthworm differs from *Nereis* in its adaptation to a subterranean life. As in other burrowing animals the body is stream-lined and has no prominent sense organs on the head or any projecting appedages on the body which would interfere with easy passage through soil. For the same reason the eyes are also lacking. The lack of prominent sense organs on the head does not mean that the earthworm is insensitive to stimuli, but only that there is no concentration of sensory cells into highly specialized organs at the anterior end.

### Form of Body

*Pheretima posthuma* has a long narrow body (Fig. 12.22), nearly circular in cross section, divided into a number of **segments** or **metameres** by distinct transverse grooves. Mature worms measure 150 mm in length and 3 to 5 mm in thickness which is, more or less, uniform throughout, the thickest part of the body being a little behind the anterior end. The body is divided into 100 to 120 segments by a series of distinct transverse grooves. Some of the anterior segments

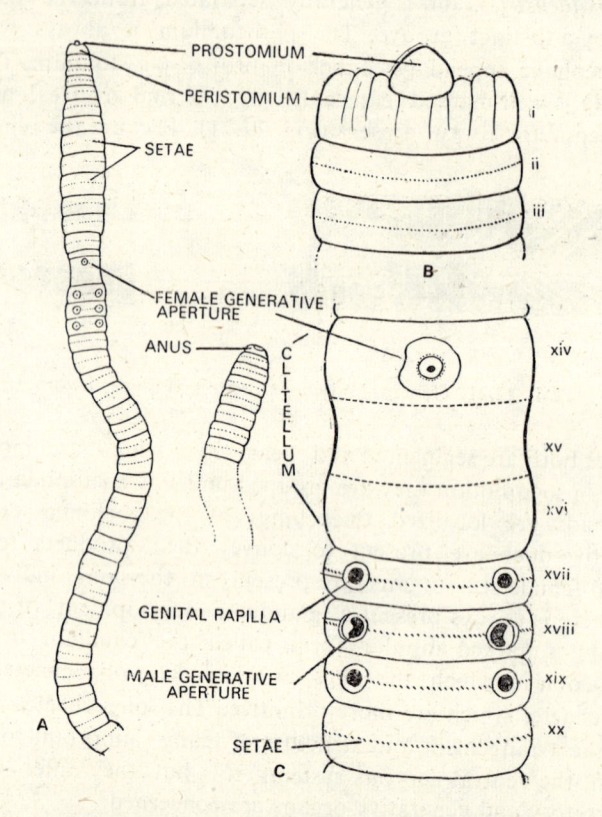

FIG. 12.22. *Pheretima*, external characters. A—Ventral view of the body; B—Dorsal view of anterior three segments; and C—Genital segments of the same, highly magnified.

are subdivided into two or three annuli by secondary annulations which are just **superficial furrows** as compared to the deeper **intersegmental grooves**. The worms are of rich brown colour, the dorsal surface being darker than the ventral. Further the dorsal surface presents a dark median line, the **dorsal blood vessel**, running from end to end and the ventral surface is distinguished by the presence of **generative apertures** and **papillae**, etc.

The **mouth** is a crescentic opening situated at the anterior end. A fleshy lobe, the **prostomium** (Fig. 12.22), overhangs the mouth, which is pushed rather ventralward. The first segment bearing the prostomium and the mouth is the **buccal segment** or **peristomium**. The prostomium is considered to be only a process of the buccal segment and not a true-segment itself. Situated towards the anterior end of the body is a circular band of glandular tissue called the **clitellum** (Fig. 12.22) or **cingulum**. To be exact this extends over 14th, 15th and 16th segments of the body counting from anterior end. The clitellum is

## Phylum Annelida

developed only in mature worm and forms an important landmark in the exterternal features of the worm by dividing its body into a **pre-clitellar, clitellar** and **post-clitellar** (Fig. 12.22) regions. The last segment of the body bearing the slit-like opening, the **anus,** is known as the **anal segment.**

Besides the mouth and anus there are other apertures on the surface of the body. There is a series of minute openings, the **dorsal pores,** situated along the mid-dorsal line in the intersegmental grooves beginning from the furrow between 12th and 13th segments but absent in the last segment. Scattered all over the body surface, behind the first two segments, are numerous minute pores (seen in sections only), the openings of the integumentary nephridia. On the ventral surface of the 18th segment are two openings of the male generative organs, one on either side. These are the **male genital openings.** On the same surface, and in line with the male genital openings, are two pairs of papillae, the **genital papillae,** one pair on the 17th and one on the 19th segment. The **oviducal aperture** or the female genital opening is a median aperture situated in the middle of the 14th segment in a saucer-shaped depression. Further, there are four pairs of small elliptical openings, the **spermathecal apertures,** situated ventrolaterally in the intersegmental grooves 5/6, 6/7, 7/8 and 8/9. These are the openings (Fig. 12.23) of the **spermathecae,** the accessory female organs.

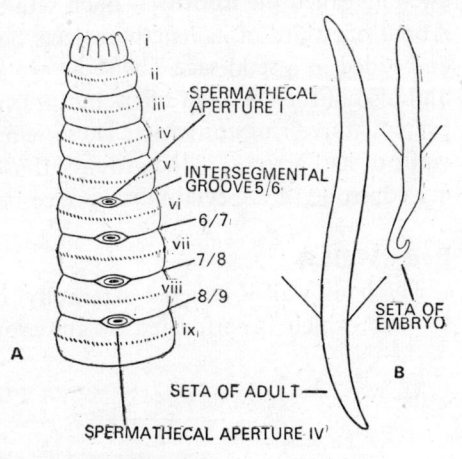

FIG. 12.23. *Pheretima.* A—Lateral view of the first nine segments; and B—Setae, highly magnified.

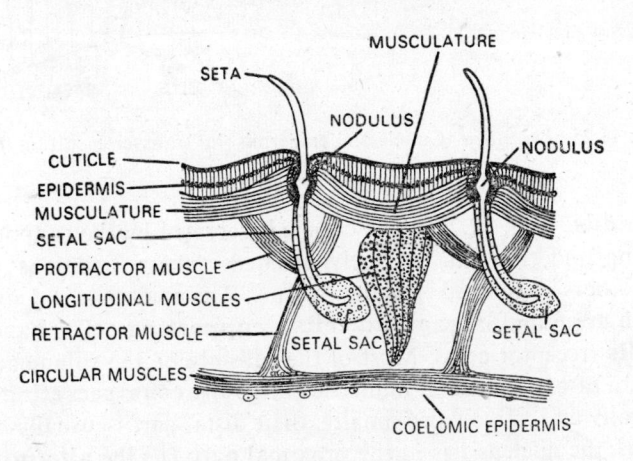

FIG. 12.24. *Pheretima.* Transverse section of the body wall through setal ring (*after* Bahl).

**Setae.** The setae in the earthworms are not numerous or so strongly developed as those of the polychaetes. The setae usually begin on the second segment and are arranged in the middle line of each segment in a ring. Such an arrangement is known as **perichaetine** arrangement. In the English earthworm, *Lnmbricus*, the setae are arranged in two pairs on each side of each segment (the **lumbricine** arrangement). Each **seta** is a chitinoid structure of a faint yellow colour and shaped like an elongated "S" (Fig. 12.23), the distal end of which is pointed and the proximal end is more or less blunt. In its middle there is a swelling called the **nodulus**. Each seta is about 26 mm long and .032 mm thick. About one-third of its length projects beyond the body wall, the remaining being embedded in a setal sac. The setae are absent in the first and the last segment and also in the clitellum of a mature worm. The clitellum of a young worm presents three rings of setae. These setae help the worm to secure a hold on the walls of its burrows or the ground, thus helping in locomotion. Their movements are controlled by special muscles (see below).

## Body Wall

The body wall is covered externally by a thin non-cellular membrane, the **cuticle**, which is perforated by numerous pores through which open the epi-

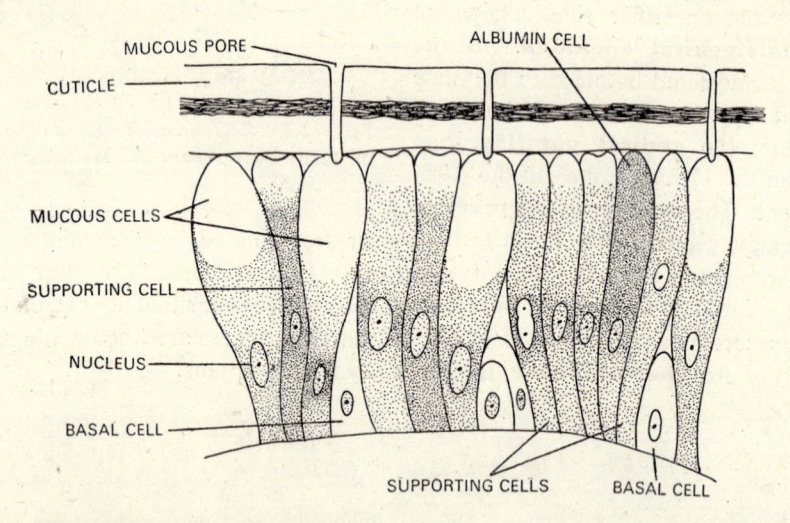

FIG. 12.25. *Pheretima*. Cuticle and epidermis in transverse section (*after* Bahl).

**dermal glands** (Fig. 12.25). The cuticle is secreted by the supporting cells of the underlying epidermis and is finely striated. These striations produce the iridescent colours displayed by the worms. The epidermis is a single layer of cells in which are found large **gland cells, supporting cells, basal cells** and **sensory cells** (receptor cells). Most of the cells are of the columnar type. The gland cells are of two kinds: (*i*) numerous large **mucous secreting** ovoid cells, rounded distally and narrow proximally; their distal part is usually full of mucous, whereas, the nucleus lies in the proximal part; (*ii*) the **albuminous gland** cells are few and far between with uniformly distributed, secretory granules and the nucleus lying towards the basal end. The supporting cells are tall and narrow with oval nucleus in the middle of each cell. They form the bulk of epidermis. The basal cells are conical or rounded with distinct nuclei lying in the spaces between the basal ends of supporting and glandular cells. The **sensory cells** whose function is to receive stimuli (hence also called **receptor cells**) are found in groups. Their nature and functions will be dealt with later. The musculature lies below the epidermis and consists of two layers, an outer layer of **circular muscle fibres** and an inner layer of **longitudinal muscle fibres**. The circular muscles form a thin continuous sheet of muscles, which in some is lacking in the region of intersegmental grooves. The longitudinal muscles run in long parallel bundles, each of which is separated from the other by a thin connective tissue layer. In the region of the setae, the sacs penetrate the whole thickness of the body wall and divide the bundles of longitudinal muscles completely from one another. Each setal sac is simply an invagination of epidermis that secretes a seta and lodges it. At the base of each are inserted two sets of muscles—the **protractors** and **retractors** which control the movement of setae.

The innermost layer of the body wall consists of a thin membrane of a single layer of flat cells. This is the **parietal layer** of the coelomic epithelium. The integument not only protects the internal organs but secretes mucous over its surface that keeps the latter slimy, clean and does not allow foreign organisms to settle on it. Mucous helps in plastering the walls of the burrows. Further the integument acts as a respiratory organ, lodges receptor organs that keep the worm in touch with its surrounding and lodges the locomotor organs.

## Locomotion

The locomotion of earthworms is interesting since it provides a clear-cut and relatively simple example of the results of coordinated movements in an animal with a body built on metameric plan. When an earthworm starts to crawl, the first few segments become thinner but longer. This is due to a contraction of the circular muscle fibres and a relaxation of the longitudinal muscles in that region, elongation being directly caused by an influx of the coelomic fluid, which is driven away from the contracting regions into more anteriorly placed segments. The setae of these segments protrude and get a grip of the substratum. The thinning and extension spreads to more posterior segments, in fact a wave of extension is produced which spreads backwards, throughout the body. The longitudinal muscles of the anterior segments then contract so that the body here becomes shorter and fatter and the hinder regions of the body are pulled forwards. The wave of shortening, thus caused, also spreads backwards. The whole body is thus traversed by a wave of extension immediately followed by a wave of contraction (Fig. 12.26). The anterior segments contract and pull the next few segments forwards. These in turn pull the next few segments and so on. These alternating series of waves of extension and contraction aided by the leverage afforded by the setae bring about the locomotion of the worm. Coordination of the waves of muscular contraction is maintained in two distinct ways: by reflexes passed on from one segment to the next by mechanical stimulation; and by impulses passing along the ventral nerve cord. The contractions of both circular and longitudinal muscles are also controlled by reflexes passing through the ventral nerve cord from one segment to the next behind it. Experimentally it has been shown that either method of coordination will singly succeed in promoting orderly locomotion, but normally both methods reinforce each other. Further, it is interesting to note that it is only when the worm is in contact with the substratum (or if when suspended it is stretched by its own weight or other means) that the locomotory waves of contraction pass along the body. This means that the tactile receptors in the skin require to be stimulated to initiate the contraction waves which when first started are self-propagating.

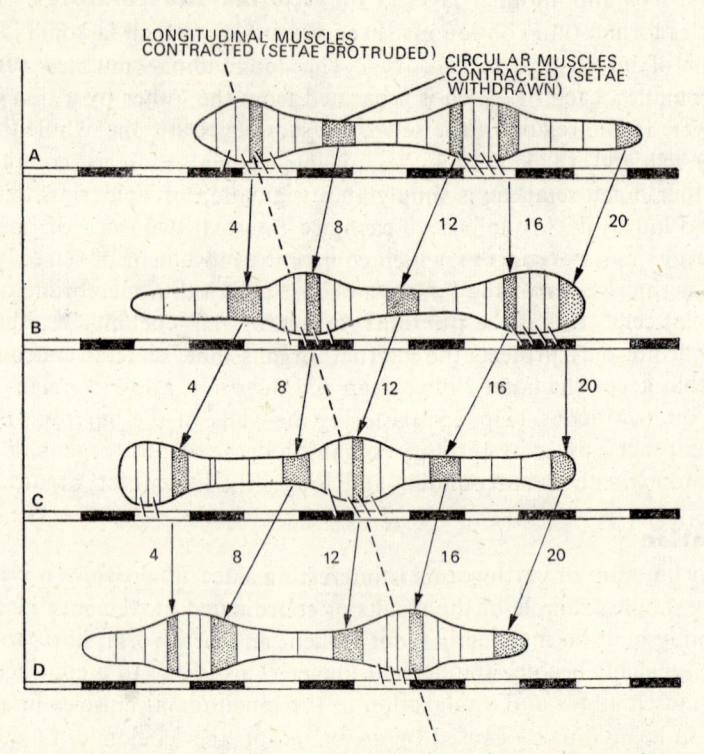

FIG. 12.26. Locomotion in a twenty-segment earthworm. Every fourth segment is marked and linked by the numbered lines, and the broken line links a region of contact.

In order to illustrate the sequence of locomotion in earthworms, Gray and Lissman (1938) figured a hypothetical earthworm consisting of only twenty segments (Fig. 12.26). In those segments in which the longitudinal muscles are contracted, the segment as a whole is at its thickest and shortest with setae protruded and in contact with the ground. Since the volume of individual segments remains constant around each unit of coelomic hydrostatic skeleton, those segments in which the circular muscles are contracted are at their most elongate and thinnest, have their setae withdrawn and are not in contact with the ground. Neurally controlled waves of muscle contraction pass anteriorly along the worm. This means that segments behind the region of contact enlarge and get in contact with the ground, and taken away anteriorly. As a result the region of contact (that is the region where the longitudinal muscles are contracted and the setae protruded) moves back slowly in relation to the ground. However, the body of earthworm moves forward more rapidly in relation to the region of contact so that there is net forward movement. It must be clearly understood that the bulges in an earthworm move **backward** during **forward** locomotion of the worm as a whole. This can be easily verified by any student by watching a living worm on which a few segments have been marked to allow their identification (Fig. 12.26). The main coordination of this type of locomotion, which also serves for normal burrowing of earthworms through soft soils, is carried out by the relatively short fibres called internuncial neurons in the ventral nerve cord. Among other elements involved, there are stretch receptors in the longitudinal muscles of each segment which are stimulated by the contraction of the circular muscles and these are connected through the short internuncial neurons to the motor neurons controlling the muscles of a more posterior segment.

### Body Cavity

The body cavity is coelomic bounded on all sides by a layer of peritoneum (coelomic epithelium), which when it covers the intestine, etc., is known as the

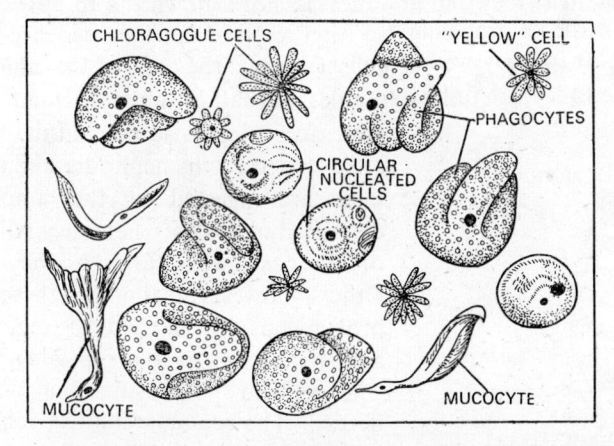

FIG. 12.27. *Pheretima*. Coelomic corpuscles (*after* Bahl).

splanchnic layer and when it lines the body wall is known as the parietal layer. The spacious coelom is filled with milky coelomic fluid in which are suspended colourless corpuscles which may be of several types. The largest and the most numerous of these corpuscles are the nucleated **amoeboid cells** with several

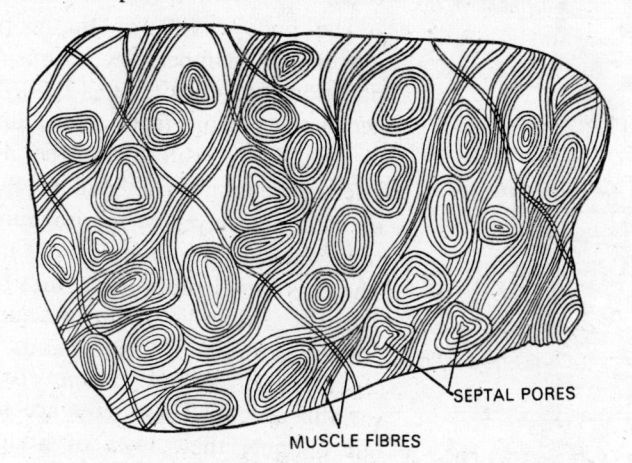

FIG. 12.28. *Pheretima*, a portion of intersegmental septum which is perforated (*after* Bahl).

membranous folds on the surface and a deep concavity on one side. They always contain large number of ingested granules and are known as the **phagocytes** (Fig. 12.27). Smaller than the phagocytes and forming ten per cent of the coelomic corpuscles are **circular nucleated cells** which resemble blood corpuscles in shape, have a clear protoplasm, and present characteristic markings on the surface. Then there are elongated cells, the **mucocytes**, having a fan-like process at one end and a narrow nucleated body at the other. The **yellow cells** are small but are almost as numerous as the phagocytes. They are easily recognized by their intense yellow colour in iodine solution and by their peculiar vesicular bulgings. These **yellow chloragogue cells** are found covering the gut. They are believed to be excretory in nature and the yellow-green granules contained within are of waste matter. But the evidence in this direction is contradictory. The true nature of the granules is not known up to now. Some people suggested that the chloragogue is a food reserve, but the evidence for this too is conflicting. It is fairly well established that the ends of the chloragogue cells break off when they are full of granules and fall into the coelom. They disintegrate according to some, and the small particles are excreted by the nephridia and the larger particles are ingested by the amoebocytes and deposited in places where they will do no harm or are extruded through the body-surface. Others, however, claim that these cells or their broken ends are carried to regions of growth and repair or to the reproductive organs or wherever an extra supply of nourishment is needed. The coelomic fluid is slightly alkaline in reaction, is ejected through its dorsal pores, and kills myriads of bacteria of the soil that would otherwise grow on its body. Along with it are also removed excretory products. It keeps the integument moist and helps respiration.

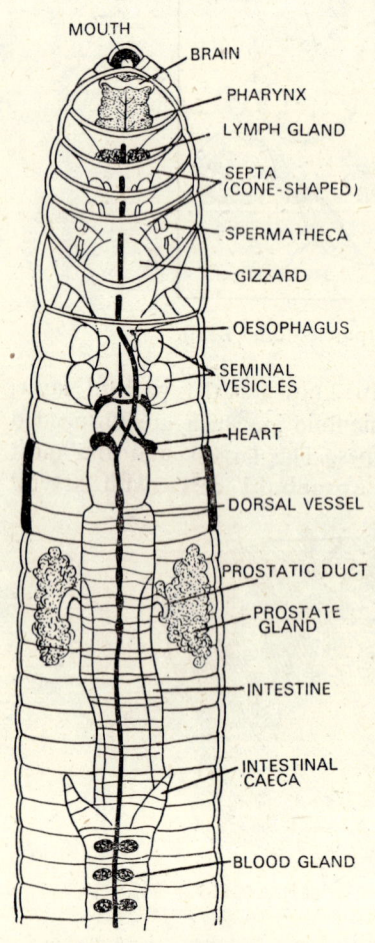

FIG. 12.29. A dissection of the anterior region of the worm showing the general anatomy (*after* Bahl).

The body cavity of the worm is divided into many chambers by means of transverse partitions or **septa** (Fig. 12.29). Septa stretch across the coelom from the alimentary canal to the body wall along the line of each intersegmental groove. Each septum is formed of a double layer of peritoneum and numerous interlocking bundles of muscle fibres. There is no septum in the first four segments. The first septum lies between segment 4th and 5th and is thin and membranous. The next five septa are thick, and septum 9/10 is almost invariably absent. These septa are not transverse but oblique, their lines of attachment on the alimentary canal lie considerably behind the lines of attachment on the body wall. Behind

*Phylum Annelida* 433

the 11th segment there is a regular series of thin membranous partitions separating successive coelomic compartments from one another. All the septa up to septum 13/14 form complete partitions and have no perforations on them, but beginning from septum 14/15 all the septa are perforated (Fig. 12.28). Each half may have as many as 68 apertures. Each aperture is oval or circular, 1.6 mm in diameter, surrounded by a thick sphincter made up of unstriped muscle fibres. The perforated septa are characteristic of the genus *Pheretima*. The coelomic fluid flows backwards and forwards through these apertures though there is no regular circulation. By closing the sphincters the coelomic fluid is restricted in a particular segment which is made turgid and stiff. This turgidity acts as hydrostatic skeleton and helps in securing a firm leverage of the setae on the substratum thus aiding locomotion.

**Alimentary Canal**

The alimentary canal although not uniform in diameter or in the nature of its walls takes the form of a straight tube from the anterior opening, the **mouth**,

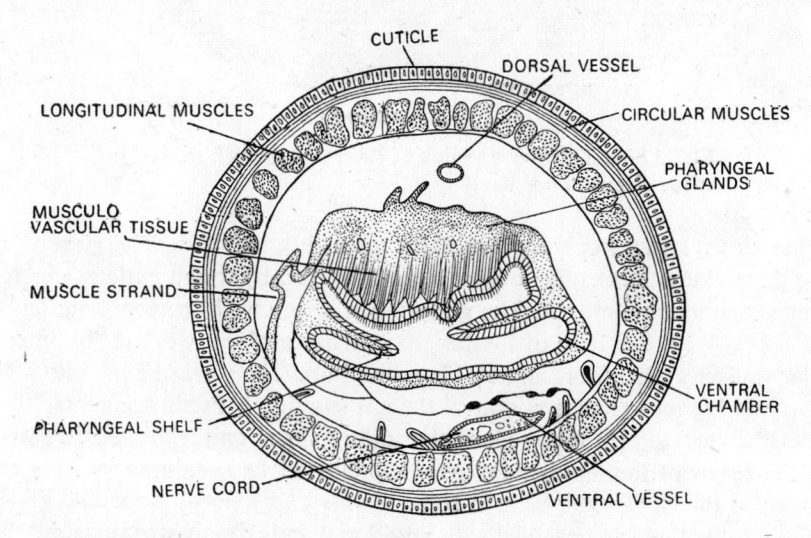

FIG. 12.30. Transverse section of earthworm passing through the pharynx (*after* Bahl).

to the terminal anus. The mouth which is overhung by the **prostomium** opens into a short thin-walled **buccal chamber** (Fig. 12.29) which extends up to the middle of the third segment. Its walls are slightly folded. The buccal chamber is protrusible and as such acts as an organ of ingestion of food. The buccal chamber leads into pear-shaped structure, the **pharynx**, externally marked off from the former by a groove. The dorsal surface of this highly vascular structure is irregular and rough being lobulated. Its lumen is compressed dorsoventrally because of the pressure of the **pharyngeal mass** or **bulb** (Fig. 12.29). The sides are pushed inwards forming a narrow horizontal shelf on either side (Fig. 12.30). The shelves divide the cavity into a dorsal or **salivary chamber** and a ventral or **conducting chamber** (Fig. 12.30). The pharyngeal bulb consists of **ciliated pharyngeal epithelium**, a thick mass of **musculo-vascular tissue** and an aggregate of **pharyngeal gland cells**. They produce mucous and digestive enzymes that are lodged in the salivary chamber before use.

The pharynx opens into the oesophagus which runs from fourth to eighth segment of the body. It is a simple, long narrow tube that swells up forming a

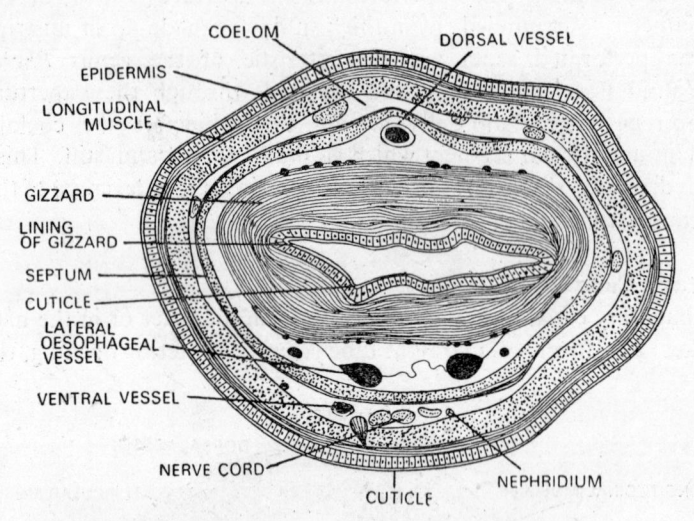

FIG. 12.31. Transverse section of *Pheretima* through the region of the gizzard (*after* Bahl).

prominent oval structure, the **gizzard**, in the eighth segment. The gizzard (Fig. 12.31) is a hard muscular organ internally lined by tough cuticle, secreted by columnar epithelium lining it. The gizzard has a very well-developed musculature (Fig. 12.31). With the help of the tough internal surface worked by powerful circular muscles the food material is ground up. The gizzard runs into a short, narrow and highly vascular region of the gut running between segments 9th and 14th. This part is also richly glandular and its inner epithelium is thrown into folds. The cells of the epithelium are non-ciliated and columnar. This region is known as the stomach, which in some worms (*Eutyphoeus* and *Lumbricus*) is profusely folded and is provided with additional **calciferous glands**.

In the 14th segment the gut widens forming a thin-walled large chamber running right up to the end. This is the intestine. Part of the intestine between 15th and 26th segment is folded internally and is highly vascular. In the 26th segment arise two conical outgrowths, the **intestinal caecae**, that run up to 22nd or 20th segment. Behind the 26th segment the intestine is provided with a median dorsal, internal fold, the typhlosole (Fig. 12.32) which runs up to the last but 23 or 25 segments. The intestine thus has a pre-typhlosolar region and a typhlosolar region. In *Pheretima* the typhlosole is poorly developed. It is just a light fold enclosing the typhlosolar vessel (Fig. 12.32). The post-typhlosolar region is the rectum which opens at the anus.

Histological study of the gut reveals that it is made up of three layers of cells. The outermost covering is formed by **peritoneal epithelium**, which consists of tall and narrow cells. In the intestine and stomach some of these cells are full of yellow refractive granules, hence called "**yellow**" or **chloragogen cells**.

## Phylum Annelida

These cells, according to one view, are excretory and take up waste material from blood capillaries and store it as yellow granules. When full they drop off into the coelom and are got rid off through nephridia or dorsal pores. According to another view they act as storehouse of reserve nutriment. **Musculature** forms the second layer. Ordinarily it consists of an outer layer of **longitudinal muscle fibres** and an inner layer of **circular fibres**. Both these layers are well developed in oesophagus but in the gizzard the main bulk of musculature is represented by circular muscle fibres only (Fig. 12.31). The entire thickness assists the trituration of food. In the intestine both the layers of musculature are feebly developed. All the muscles are non-striped. Internal epithelum in the buccal region consists of tall columnar cells covered with thin cuticle and devoid of cilia. In the pharyngeal region the columnar epithelium is ciliated on the roof and devoid of cilia on the floor. Oesophagus has tall columnar cells which become short in the gizzard, where they are covered by thick **internal cuticle.** The inner epithelium of the intestine has **absorptive** and **glandular cells.** The absorptive cells are long and narrow, broad at free ends, whereas, the glandular cells have vacuolated appearance.

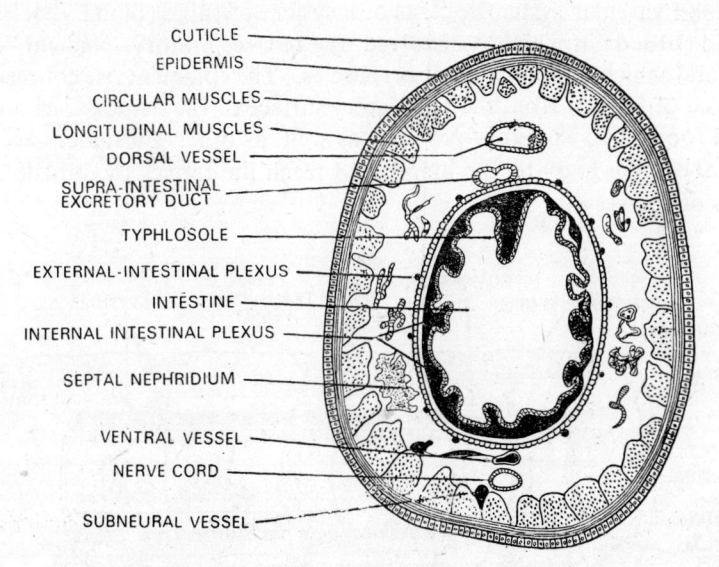

FIG. 12.32. Transverse section of *Pheretima* through the region of intestine (*after* Bahl).

Earthworm obtains nourishment from organic material of the soil. The proportion of such material being less in the soil the worms have to swallow a lot of earth, that passes through the gut. The digestive fluid poured in the pharynx contains mucin and some proteolytic ferments. Mucin lubricates the food and the ferment starts digestion of proteins. The food is ground up in the gizzard, its final digestion takes place in the intestine, which pours out a number of ferments. Intestinal digestive fluid of the earthworm corresponds with pancreatic juice of higher animals. The food is absorbed by the absorptive cells and is passed into the blood-stream for distribution.

## Respiratory System

The gaseous interchange necessary for respiration takes place mainly through the skin which is richly supplied with blood capillaries which sometimes extend between the cells of the epidermal layer. The gaseous interchange is possible only if the skin is kept moist by the moist earth, by the secretion of the epidermal mucous gland cells and also by coelomic fluid passing out through the dorsal pores. Oxygen in solution passes through the cuticle and epidermis into the blood where it enters into chemical combination with haemoglobin, dissolved in the plasma, to form oxyhaemoglobin, which gives up oxygen once again to tissues which have a low oxygen tension. In no instance, however, does the blood come into direct contact with the tissue cells so that the oxygen must be handed on through the agency of tissue fluids, and it is probable that the coelomic fluid plays a part in the distribution of dissolved gases to the body. Carbon dioxide is removed from the tissue by the blood and carried in solution in the plasma to the skin where it is got rid off.

## Blood Vascular System

The blood vascular system consists of a system of tubes (blood vessels) containing a fluid (**blood**) in which is dissolved the red respiratory pigment (**haemoglobin**) and which contains blood corpuscles. The pigment is concerned with conveyance of oxygen from the respiratory surface to the tissues. Carbon dioxide, dissolved foodstuffs, excretory products as well as other substances are also distributed about the body by the blood, and reach the tissues by diffusion through the walls of the vessels.

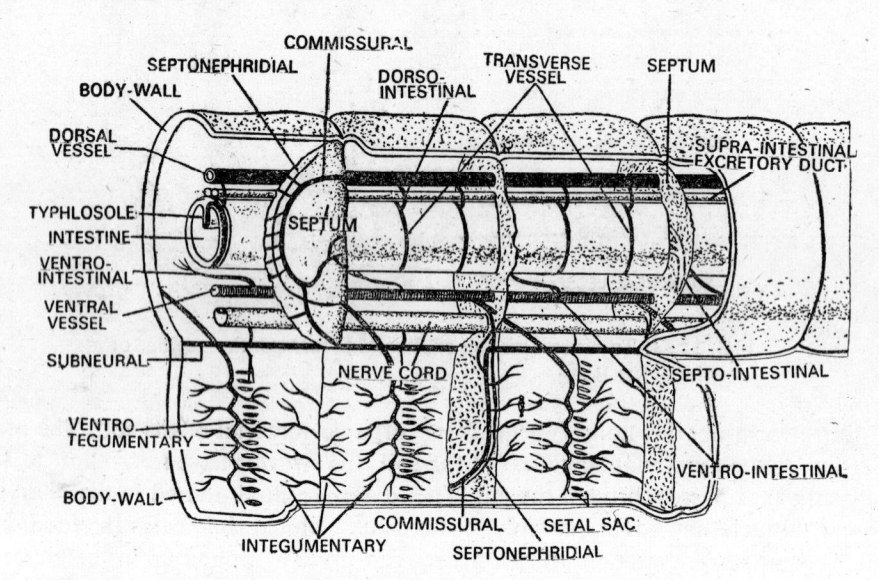

FIG. 12.33. Circulatory system of *Pheretima*. Arrangement of blood vessels in the segments of the intestinal region (*after* Bahl).

The arrangement of the vessels is very complicated, but the general plan is that a distributing set of vessels communicates with a collecting system of longitudinal vessels through networks of capillaries and circularly running vessels in each

## Phylum Annelida

segment; and the blood is kept in constant circulation by means of certain contractile vessels.

The arrangement of vessels in the intestinal region, i.e., behind the 13th segment is different from that of the first thirteen segments. For convenience the arrangement of vessels in the intestinal region will be discussed first. The main collecting vessel of this region is a large longitudinally running **dorsal vessel** (Fig. 12.33), which for the greater part of its length, is in close contact with the gut. It has thick muscular walls and contracts rhythmically driving blood forward. To prevent back flow of the blood the vessel is provided with valves in front of each septum. In each segment the dorsal vessel receives one **commissural vessel** (Fig. 12.34) on each side. Each commissural runs along the posterior face of a septum and is ventrally connected with the **subneural vessel** and during its course receives blood capillaries from nephridia and body wall. A pair of **dorsointestinal vessel**, one on each side of the segment, arises from the intestine and opens into the dorsal vessel. Their openings are guarded by valves to prevent the back-flow of the blood.

The main distributing vessel of the body is the **ventral vessel** (Fig. 12.33) that runs from end to end, suspended in a mesentery just below the gut. The vessel is not provided with valves. In each segment it gives off a pair of **ventro-**

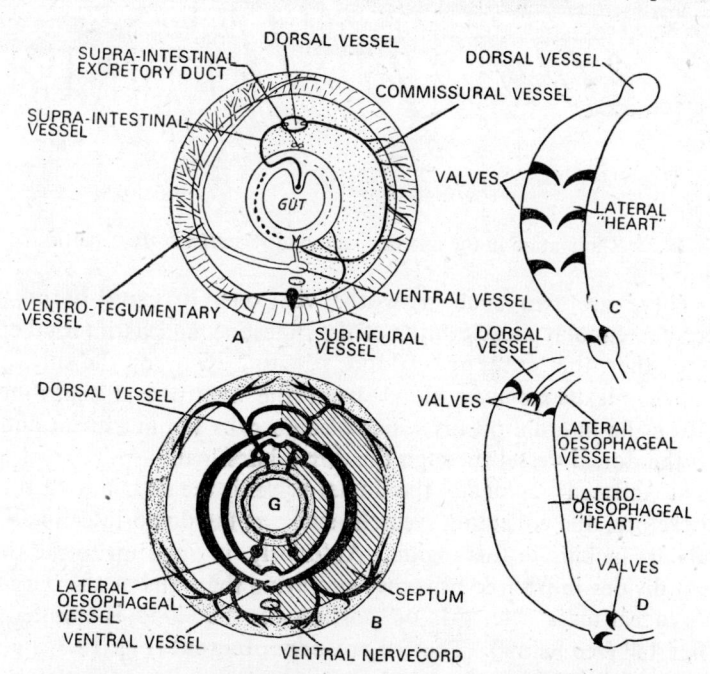

FIG. 12.34. Circulatory system of *Pheretima*. A—Transverse section of a segment of intestinal region showing the arrangement of blood vessels; B—Transverse section of anterior segment showing arrangement of blood vessels; C—Lateral "heart"; and D—Latero-oesophageal "heart" (all *after* Bahl).

**tegumentary** vessels (Fig. 12.34), each of which runs for a short distance, along the anterior face of the septum, then pierces it (septum) and enters the succeeding segment. It runs along the body wall to the mid-dorsal line, supplying the body wall and the integumentary nephridia. Before the ventro-tegumentary pierces the septum it gives off a delicate branch, the **septonephridial**, to the septal nephridia. The intestine is supplied by a small median vessel, the **ventro-intestinal vessel** (Fig. 12.34) that arises from the dorsal surface of the ventral vessel.

The subneural vessel (Fig. 12.35) is another important longitudinal vessel runing below the nerve cord in the mid-ventral line. It is a slender vessel that collects blood from the ventral region of the body wall and the ventral nerve cord and sends a major portion of it to the dorsal vessel through the commissurals. A small branch, the **septointestinal** from the commissural, takes part of this blood to the intestine. In the wall of the intestine there are networks of blood

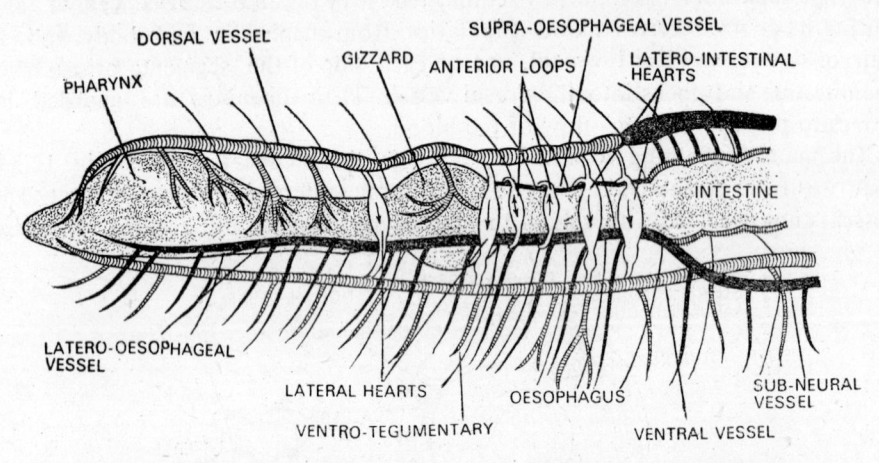

FIG. 12.35. Circulation in the anterior region of *Pheretima* (*after* Bahl).

capillaries. There are two such networks, viz., one, **internal blood plexus**, lying between the circular muscle layer of the intestine and its internal epithelial lining and the other, the **external blood plexus**, lying on the surface of the gut. The external plexus receives blood from the ventro- and septo-intestinals and passes it on the internal plexus which also absorbs the nutriment and passes the blood to the dorsal vessel through the dorso-intestinal.

In the first thirteen segments the dorsal vessel is the distributing trunk and naturally, therefore, the collecting vessels such as the dorso-intestinals and the commissurals are lacking in this region. Anteriorly it runs up to the third segment where it divides into three branches supplying the walls of the buccal chamber and pharyngeal mass, etc. Major portion of the blood is emptied into the so-called "hearts" (see below). The **supra-oesophageal** (Fig. 12.35) is a single small longitudinal vessel lying dorsal to the oesophagus between the ninth and thirteenth segment. This is a collecting vessel for the stomach and gizzard regions. In the 10th and 11th segments it is connected with the lateral oesophageals by large vessels called the **anterior loops**, whereas, in the 12th and 13th it is connected with the **lateral hearts**.

The **lateral oesophageals** (Fig. 12.35) are fairly large vessels on the ventro-lateral aspect of the oesophagus. They arise by the bifurcation of the subneural vessel in the 14th segment. In each segment they receive delicate branches from the gizzard and stomach. They collect blood from this region and send it on to the supra-oesophageals through the anterior loops. The ventral vessel runs up to segment two. The ventro-tegumentaries of this region distribute blood to the body wall, septa and the nephridia of the same segment (not of the hinder segment like the intestinal region). All the organs of this region such as the spermathecae, testis sacs, seminal vesicles, ovaries, oviducts, etc. receive blood from the ventro-tegumentaries. The ventro-intestinals are absent in this region.

In the anterior region there are four pairs of pulsatile "**hearts**" (Fig. 12.35) located in 7th, 9th, 12th and 13th segments. The posterior two pairs communicate dorsally both with the dorsal vessel and supra-oesophageal, hence called the latero-oesophageal "hearts," others are simply known as the lateral "hearts." Each "heart" has thick muscular wall enclosing spacious cavities (Fig. 12.34). At the ends they are connected with the blood vessels and have valves that direct the flow of the blood. The anterior loops that lie in the 10th and 11th segments and connect the supra-oesophageals with the lateral-oesophageals are thin walled, non-pulastile vessels devoid of valves. The wall of the stomach has a series of circular vessels, the ring vessels, about 12 vessels per segment opening into the supra-oesophageals. They are characteristic vessels of the stomach situated within the muscular coat. Through these the blood flows upward from the lateral-oesophageals to the supra-oesophageal.

The dorsal blood vessel collects blood behind the 13th segment and drives it forwards by posteroanterior peristalsis aided by valves. No blood leaves the

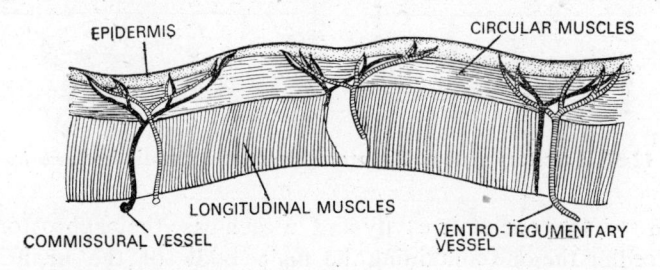

FIG. 12.36. Parallelism between arterial and venous capillaries in the body wall of *Pheretima* (after Bahl).

vessel in this region. In the anterior thirteen segments it pumps out and distributes all the blood collected behind. The ventral vessel is the main distributing channel. Posterior to the "hearts" the blood flows backwards and supplies the body wall and the viscera through the ventro-tegumentaries and the ventro-intestinals. The subneural collects blood from the region of the body wall and sends it to the dorsal vessel. Intestine, in this region, has a double supply through the ventro-intestinal (from ventral vessel) and the septo-intestinals (from the commissural vessel). In the anterior thirteen segments the dorsal vessel pumps major portion of its blood into the ventral vessel through the lateral-oesophageal and lateral "heart" and the remaining blood is distributed to the pharynx the pharyngeal nephridia, the oesophagus and the gizzard through paired branches. The ventral vessel in this region distributes blood, the flow of which is towards the anterior side. As the gut receives blood directly from the dorsal vessel there are no ventro-intestinals in this region. The lateral-oesophageals collect blood from the peripheral structures and pour it into the supra-oesophageals through the anterior loops. The ventral vessel receives all its blood from the dorsal and supra-oesophageal vessels.

**Excretory System**

The organs of nitrogenous excretion consist of segmentally arranged tubules called **nephridia**, which are present in all segments except the first two. Each nephridium opens from the coelom by a ciliated funnel, the **nephrostome**, into a tube which loops about in a complicated manner before opening to the exterior by the **nephridiopore**. In the simplest case (*Lumbricus*, *Perionyx*) there is a pair of large nephridia in each segment (Fig. 12.37) lying in the ventral and

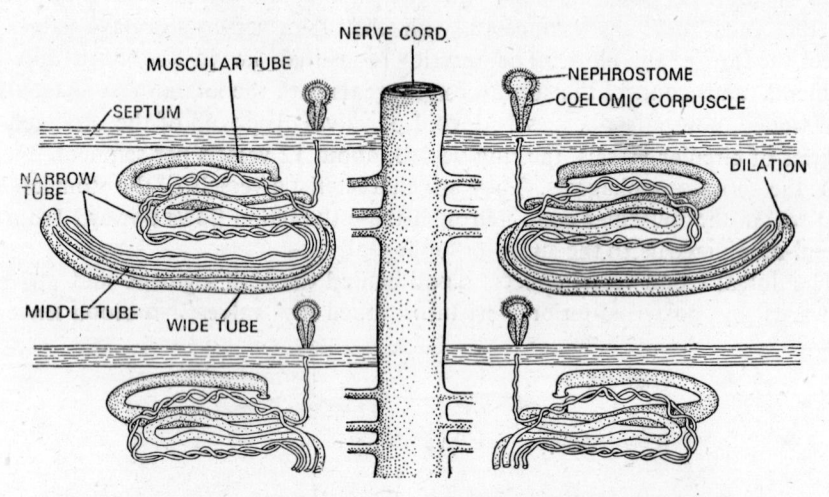

FIG. 12.37. Excretory system. The arrangement of nephridia in *Lumbricus*.

lateral region of the coelomic cavity and in each case the nephrostome lies in segment preceding the one containing the main body of the nephridium. In *Pheretima*[1] the nephridia are numerous and are quite small and the major part of the excretory fluid is passed on to the alimentary tract through an elaborate system of ducts and canals. There are three types of nephridia in *Pheretima* named according to their positions in the body. Those arranged in relation to the septa are called the **septal nephridia**, those attached to the inner surface of the body wall are known as the **integumentary nephridia**, and those found in the region of the pharynx are the **pharyngeal nephridia**.

The septal nephridia (Fig. 12.38) are attached to each intersegmental septum behind the 15th segment. Each septum bears four rows of nephridia, two on its anterior face and two on posterior. On an average each septum has 40 to 50 nephridia attached to its anterior face and the same number on its posterior face. Thus each coelomic compartment has 80 to 100 nephridia. Each nephridium consists of a funnel or **nephridiostome** (**nephrostome**), followed by a short narrow tube, the **neck**, the body of the nephridium consisting of a short

---

[1] In other worms all the fluid passes out through the nephridiopores.

*Phylum Annelida*

**straight lobe** (Fig. 12.38) and a long spirally **twisted loop**, and the **terminal nephridial duct** which varies in length in different cases. The funnel is a rounded structure having a hood-like ciliated upper lip and a comparatively small

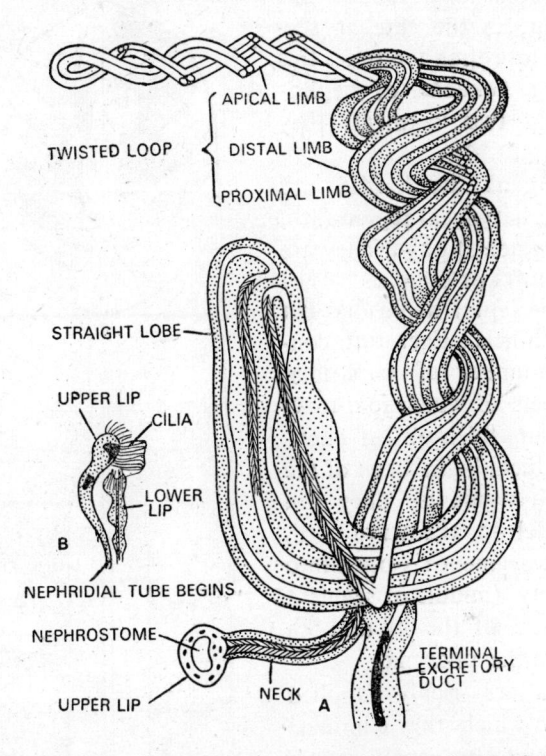

**FIG. 12.38.** *Pheretima.* A septal nephridium (*after* Bahl).

lower lip separated by an elliptical opening the mouth of the funnel. The neck is a narrow ciliated tube leading from the funnel to the body of the nephridium, which consists of several loops of the tubule distinguishable into two parts. The straight lobe is a shorter part almost half the size of the twisted loop. In the straight lobe there are four tubules running more or less parallel to one another. The twisted loop is longer than the straight lobe and has three parallel tubules within. It is again distinguishable into (*i*) a **proximal limb,** and (*ii*) a spirally twisted **distal limb**. The proximal limb receives the funnel on one hand and gives off the terminal nephridial duct on the other. The terminal ducts of the nephridia attached to the septum, on one side, open into **septal excretory canals** that run on each septum parallel to the commissural vessels of each side. The septal excretory canals in their turn open into a pair of **supra-intestinal excretory** ducts, which lie dorsal to the gut beneath the dorsal vessel and extend from fifteenth to the last segment. These ducts open through several narrow ductules, each located near each intersegmental septum. The excretory fluid collected by each septal excretory canal is poured into the supra-intestinal excretory duct of its side, from where the fluid passes on to the intestine to be excreted through the gut. Such an arrangement of excretory system is known as the **enteronephric nephridial system** as the excretory products are discharged into the gut (Fig. 12.40).

The **integumentary nephridia** are small nephridia projecting in the body cavity. Each nephridium is closed internally, has no funnel and opens independently to the exterior by the nephridiopore. They are usually 200 to 250 in each segment, in the clitellar segment they are more than ten times this number forming **forests** of nephridia. In the first two segments they are absent (Fig. 12.39).

The **pharyngeal nephridia** are confined to the pharyngeal region (4th, 5th and 6th segments) only. They are just like paired tufts or bunches. They are large like the septal nephridia but lack coelomic funnels. Each tuft consists of a large number of nephridia, the terminal ducts of which join together and form long thick-walled ducts. A pair of such ducts lie in each of the 4th, 5th and 6th segments. They run forwards parallel to the nerve cord, and those of the segment six open into the buccal cavity (middle of segment two), while those of the 4th and 5th segments open in the pharynx.

The nephridia are richly supplied with blood, from which they eliminate excess of water and nitrogenous waste. Excretory material from the coelomic fluid is also picked up and got rid of by septal nephridia. It has been, further,

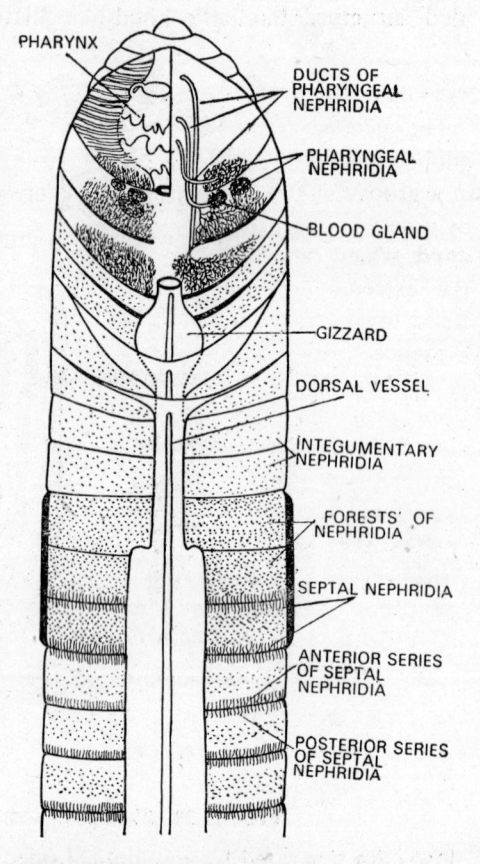

FIG. 12.39. General plan of nephridial system in *Pheretima* (*after* Bahl).

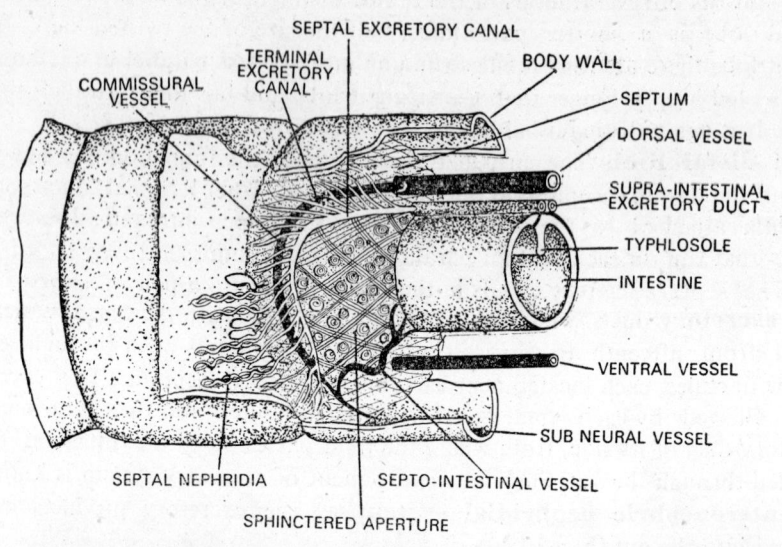

FIG. 12.40. *Pheretima*. Diagrammatic representation of the enteronephric system (*after* Bahl).

*Phylum Annelida*

suggested that the **enteronephric** type of nephridial system has an additional advantage of conserving water which would ordinarily be lost to the worm. This is more useful to the earthworms living in a dry climatic region.

## Nervous System

The nervous system is of the same general type that is found in *Nereis*. A bilobed ganglionic mass, the **cerebral ganglia** ("**brain**"), lies in segment three in a groove separating the buccal chamber from the pharynx, and is connected by a nerve collar, the **circumpharyngeal commissure**, to the **ventral nerve cord**, which runs backwards in the mid-ventral line, just inside the coelom, to the extreme posterior end. It is actually made up of two longitudinal cords fused together. In the middle of each segment the cord is swollen due to the presence of two fused ganglia (Fig. 12.41). At the points of swelling three pairs of minute nerves are given off. These nerves contain both afferent and efferent fibres and are distributed to the body wall and the viscera. The afferent fibres start from a receptor organ in the epidermis and terminate in the ventral nerve cord in a tuft of fine branches. Forming a synapse with these branches arises the efferent nerve that innervates the muscles. This circuit forms, in a way, a **reflex arc**. The nervous impulses are brought by the afferent fibres to the ventral nerve cord whence they are reflected back as motor impulses. Often a third neurone, the **adjustor neurone**, is interpolated between sensory neurone and the motor neurone to establish connection between the two elements.

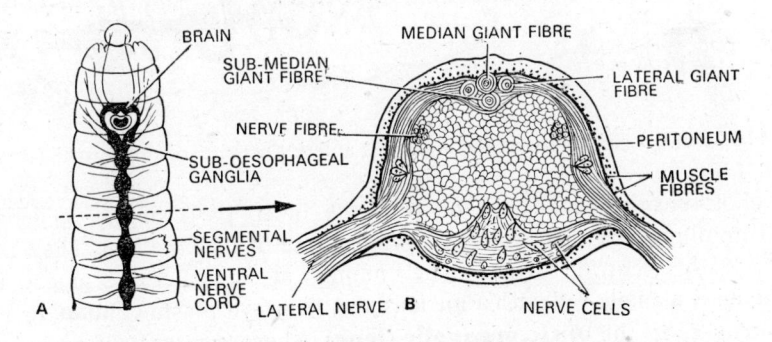

FIG. 12.41. A—Anterior part of the nervous system of earthworms; and B—Transverse section through nerve ganglia.

**Sense organs**. As in other Metazoa, the organs concerned with the reception of stimuli from the outside world are situated on or near outside of the body. These are called sense organs or the receptor organs (Fig. 12.42), and are quite simple in structure in the earthworms, consisting of either single or of small groups of specialized ectodermal cells. From the base of each cell a sensory (afferent) fibre continues into the ventral nerve cord without interruption and synapses with a fibre from an adjustor neurone. Three types of receptors (Fig. 12.43) have been described for *Pheretima*—the **epidermal receptors, buccal receptors** and **photoreceptors**. The epidermal organs are found in the epidermis and consist of a group of cells each of which terminates in a small hair-like process. They cause little elevations of the cuticle and are found all over the body but are numerous on the sides of the body and on the ventral surface and less on the dorsal surface. The buccal receptors, as the name indicates, are confined to the buccal chamber and are more numerous. In structure they are like the epidermal receptors but they have a broad distal end and their sensory hairs are much better developed. The nuclei of the cells of these receptors lie deeper in the cells than the nuclei of the surrounding cells of the buccal epithelium. The epidermal receptors respond to tactile stimuli and in all probability, also to chemical stimuli and changes in temperature.

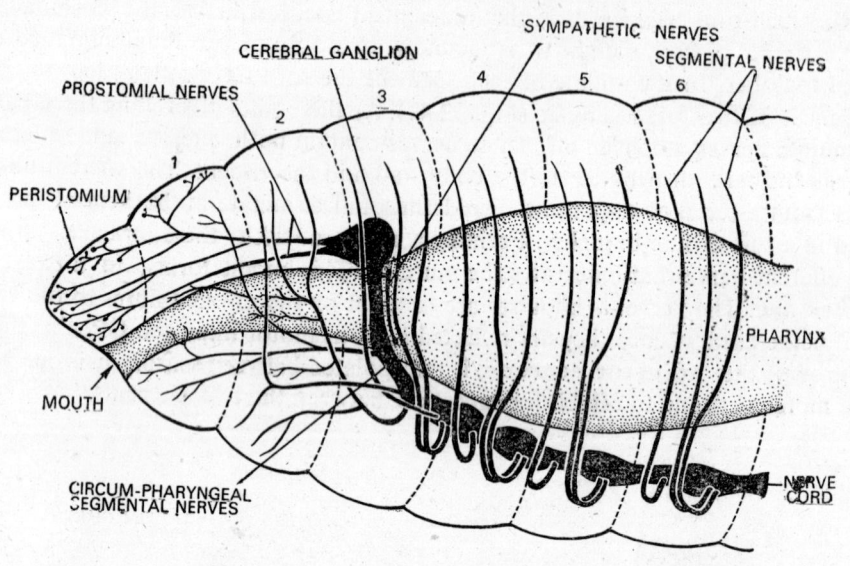

FIG. 12.42. Lateral view of the brain and anterior nerve cord of the earthworm.

The photosensitive organs are found in the inner parts of the epidermis of the prostomium and all segments except those of the clitellum. They are also absent on the ventral surface. Their number in the posterior segments is less. Each organ is a single cell with a nucleus and clear cytoplasm containing a small transparent rod, the **optic organelle (lens)**, which focuses light on the neurofibrillae which ramify through the cell. By means of these organs the worms can detect changes in the intensity of light. They avoid intense light and withdraw to poorly-lit areas.

If touched the worm contracts at once and disappears into its burrow. If an attempt is made to pick up the worm it will try to escape by violent reaction to the grasp. A similar reaction occurs if a bright light is shown on to the front end. These reactions are possible because of the presence of special giant nerve fibres in the nerve cord. They are thicker than others and are very long, as long as the nerve cord itself. Each fibre has been shown to be a chain of short compound axons. They conduct impulses very rapidly from one end of the worm to other and make the instantaneous reaction possible.

## Reproductive System

The earthworm is **monoecious** or **hermaphroditic**. The reproductive organs are restricted to a few anterior segments. In this earthworm differs from the less specialized annelids, the Polychaetes. Though hermaphroditic, the

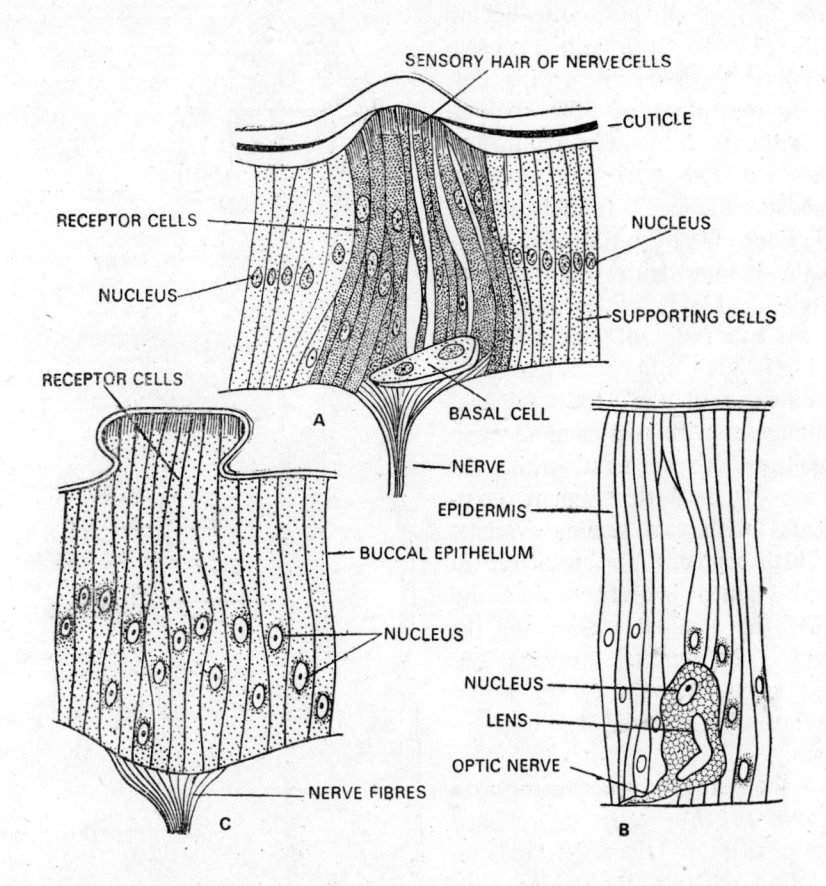

FIG. 12.43. Sense organs of an earthworm. A—Epidermal receptor; B—Photo-receptor; and C—Buccal receptor.

earthworms are not self-fertilizing, but pair together (copulate) and later deposit their eggs in cocoons. Thus, the reproductive processes are determined by two conditions; the ensuring of cross-fertilization and the deposition of the cocoon. The complicated reproductive apparatus of earthworms consists of three complemental systems that work in harmony to ensure these ends. The parts of the reproductive apparatus are the **gonads** and **gonoducts** (essential reproductive organs) together with other regions of the body wall specialized to play a part in the reproductive activities. The gonads (as in all other coelomates) are produced by the proliferation of the coelomic epithelium and the gametes are liberated into the coelom wherefrom they pass to the exterior by way of coelomoducts which, thus, function as gonoducts.

There are two pairs of testes (Fig. 12.44) lying in the 10th and 11th segments. Each testis is a minute white body which lies close to the middle line and is made up of 4-8 digitate processes arising from a narrow base. They are very small and can be seen only under a strong lens. The **testes** are shut off from the general body cavity by partitions developed from the coelomic epithelium forming **testis sacs** (Fig. 12.44). Thus there are two testis-sacs lying ventrally beneath the oesophagus, one behind the other, in the 10th and 11th segments. Each testis-sac is a wide flattened compartment of the coelom, which appears bilobed in front, and encloses a pair of testis and a pair of ciliated spermiducal funnels (Fig. 12.44). Each side of a testis-sac communicates behind with the **seminal vesicle** (Fig. 12.45) of its own side. There are two pairs of seminal vesicles lying in 11th and 12th segments. The testis-sac of the 10th segment communicates with two seminal vesicles of the 11th segment, while the testis-sac of the 11th segment communicates with two seminal vesicles of the 12th segment. The testis sac of the 11th segment is quite spacious and encloses the seminal vesicles of the segment. The seminal vesicles are formed as outgrowths of the septa, hence are also referred to as septal pouches.

From the testes the **spermatogonia** or **sperm-mother** cells are released in the cavities of the testis-sac from where they pass into the seminal vesicles where they develop into spermatozoa. The mature sperms move back to the testis sacs and pass out through the **spermiducal funnels**. Each spermiducal funnel leads into a **spermiduct** or **vas deferens**. There are two slender vasa deferentia on each side that run backwards on either side of the nerve cord and open to the exterior in segment 18th. Before opening they are joined in their course by the prostatic ducts enclosed in common sheaths, and open independently at the base of a pit in the 18th segment.

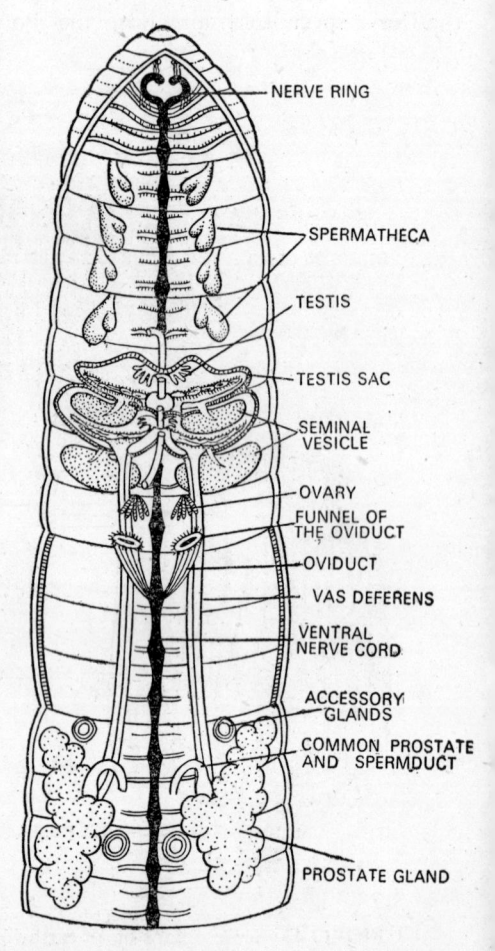

FIG. 12.44. Reproductive system of *Pheretima* (after Bahl).

The **prostate glands** (Fig. 12.44) are irregular white solid structures lying between 16th and 21st segments. Each consists of glandular as well as non-glandular portions. The glandular portion comprises large racimose glands which secrete profusely. Non-glandular portion has small ductules that finally form the common prostatic duct which is curved tube that is joined by the two vasa deferentia during its course immediately on emergence from the gland. Three ducts are encased in common muscular sheath to form a thick muscular duct. The three tubes inside remain separate and mixing of the spermatic fluid with the prostatic fluid is not possible within the body. The function of the

*Phylum Annelida*

prostatic fluid is not known.

The essential organs of the female system consist of a pair of **ovaries** (Fig. 12.44) and **oviducts**, whereas, the **spermathecae** are the **accessory organs**.

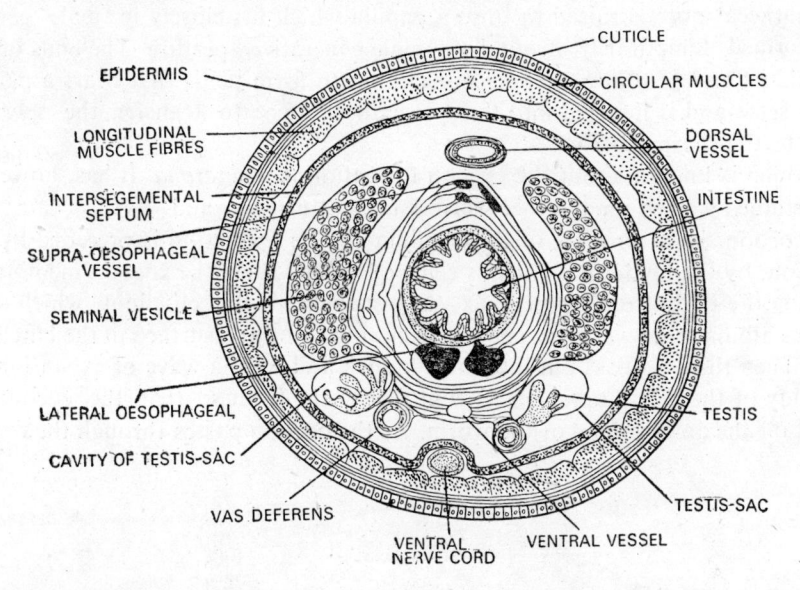

FIG. 12.45. Transverse section of earthworm through testes and seminal vesicles (diagrammatic).

The ovaries hang in the 13th segment, being attached to septum 12/13, one on either side of the nerve cord. Each is a minute whitish mass in which the ova, in various stages of development, are arranged in a linear series. The oviducts are two short conical tubes, each with a large oviducal funnel at its free end. Each oviducal funnel is more or less saucer-shaped structure with folded and ciliated margins, and lies immediately behind the ovary. The oviducts run behind, perforate the septum 13/14 and converge, meeting finally in the body wall beneath the nerve cord to form a short common duct that opens the oviducal aperture or the **female generative opening** on the 14th segment. Internally the oviducts are ciliated. There are four pairs of spermathecae (Fig. 12.44) situated in the 6th, 7th, 8th and 9th segments. They are flask-shaped structures, the body of which is known as the **ampulla**. From the narrow neck of each arises a simple elongated **diverticulum** which in *Pheretima* stores the spermatozoa. In other worms the sperms are stored in the ampulla.

**Copulation**. Two worms meet and become apposed to each other in head to tail position so that the male generative apertures of each worm lie opposite the spermathecal pores of the other (Fig. 12.46). The areas of the male generative apertures are raised into papillae which are inserted successively into the various pairs of spermathecal pores from behind forwards,

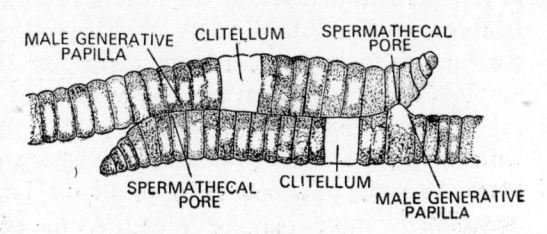

FIG. 12.46. Copulation in earthworms.

thus filling the spermathecae with spermatozoa and prostatic fluid. After the completion of this process they separate. In others the method of copulation may differ in minor detail. In *Eutyphoeus waltoni* the area surrounding each spermathecal pore is raised to form a papilla which fits closely in male genital cup formed temporarily around the male generative aperture. The ends of the spermiduct and prostatic duct are everted out to form penis that bears a pair of penial setae and is inserted into the spermathecal pore to transfer the spermatozoa to the spermatheca.

Nothing is known about the cocoon formation in *Pheretima*. It has, however, been studied well in many other worms such as *Eisenia* and *Rhynchelmis*, etc. The cocoon is a product of the clitellum and its formation is preceded by the secretion, by the epidermal mucous cells, of a slime tube, the cocoon membrane, covering the body from segment six to just behind the clitellum, which also secretes albumen between the cocoon membrane, and the surface of the clitellum itself. Then the eggs pass out into the albumen and soon a wave of expansion of the body of the worm takes place from rear to the front so that the cocoon is forced off the anterior end of the worm. As the cocoon passes through the region

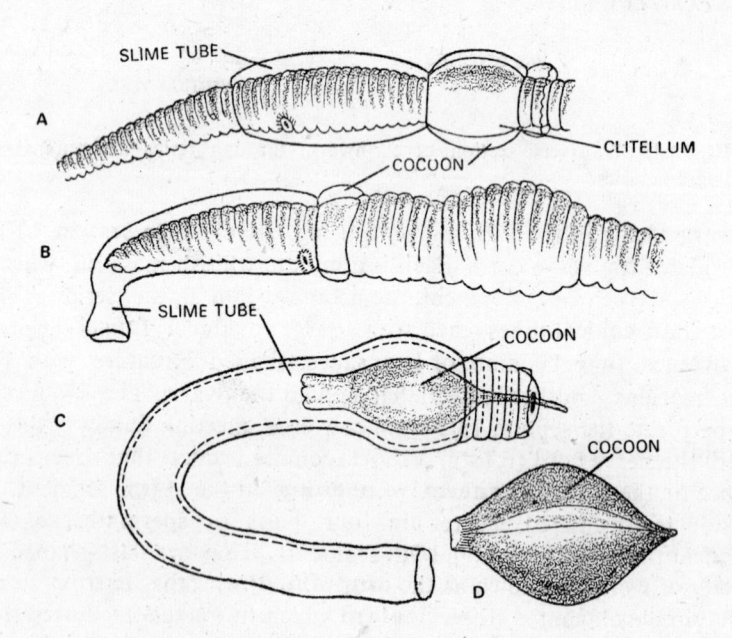

FIG. 12.47. Cocoon formation in earthworm.

of the spermathecae sperms are also passed out into it. As soon as the cocoon is thrown off its both ends close up owing to the elasticity of its walls. The cocoon thus formed is laid in the earth. The slime tube dries up and the cocoon becomes progressively darker. The cocoons of *Pheretima* are more or less spherical in form and of light yellow colour. Within, the entire space is filled with a thickish albumen in which the eggs are fertilized and development takes place. As a rule only one embryo attains final stage of development though occasionally more may be found. As the eggs develop in cocoons, it may be mentioned, that here the trochophore stage (Fig. 12.2) found in the polychaete

*Phylum Annelida*

worms is suppressed. Deposition of cocoon begins about 24 hours after copulation and may continue at least for a year without further pairing. The seminal fluid from one copulation is sufficient for several months of viable cocoon production but under natural conditions copulation occurs at every opportunity.

It has been observed that cocoons may be laid by individuals which have never copulated. This has been reported in case of many species but not known so far in *Pheretima*. Such cases were believed to be due to self-fertilization that might have taken place in some way. Recently Muldal has shown that in most such cases parthenogenesis occurs. This may be obligatory or facultative.

The cocoons of *Pheretima* are laid during and after monsoon when there is plenty of moisture in the soil and the temperature is suitable for the development of the embryo. In gardens and other moist places, on the other hand, the cocoons may be found in April, May and June.

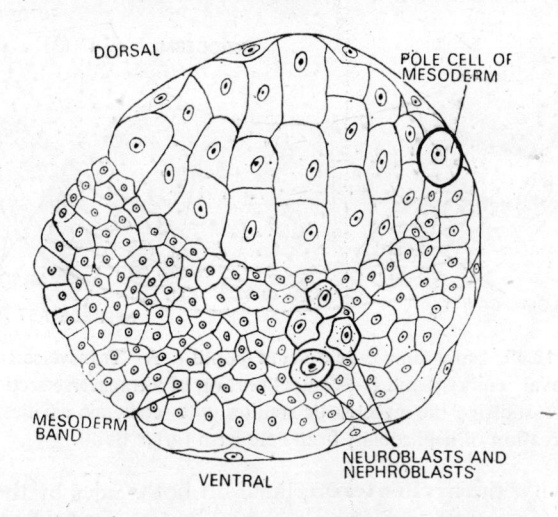

FIG. 12.48. Embryonic development in earthworm. Ventro-lateral view of a young embryo.

**Development.** The rounded ovum of earthworm contains very small quantity of yolk. Cleavage is unequal but holoblastic. Two kinds of cells are produced after sometime. The small ones are the **micromeres** and the larger ones are the **megameres**. A segmentation cavity appears in the interior forming a **blastula**, whose walls are composed of a single layer of cells, but the ones at the lower end are larger than those above. Soon after the segmentation cavity is formed, a certain cell, the **pole-cell of mesoderm** (**mesoblast** or **stomoblast** cell) is cut off. It divides forming two pole-cells which project into the cavity of blastula near the equator and remain inactive for a time while the other cells of the blastula continue to multiply. But soon the pole cells become active, multiply and form two rows of small cells which are the mesoblastic bands of the embryo (Fig. 12.48).

Now the blastula elongates and flattens forming a structure with the small cells above and large columnar cells below, while the mesoblast cells lie on the equatorial plane. **Gastrulation** begins at this stage by invagination. The lower surface made up of large cells is pushed inwards. Meanwhile the micromeres multiply and keep on growing over the megameres. The invagination produces a groove, the **blastopore**, on the lower surface. The blastopore leads into the

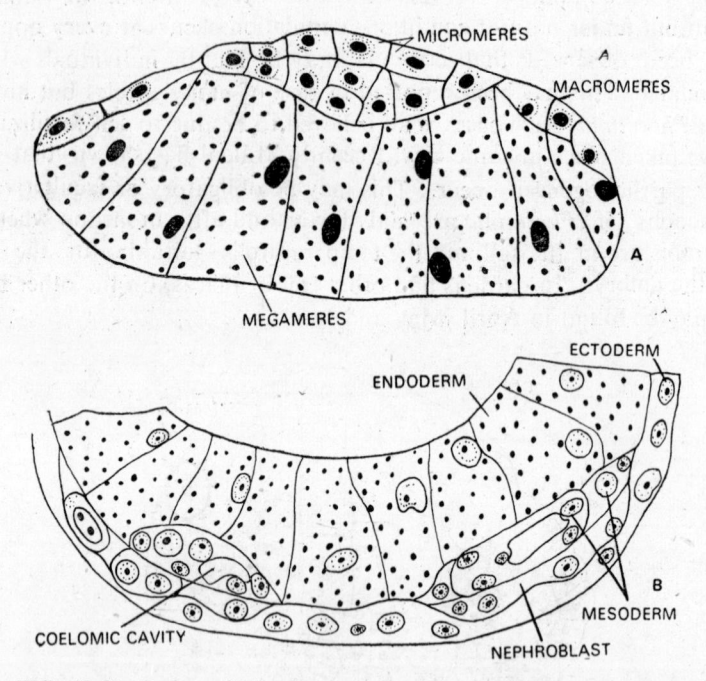

FIG. 12.49. Embryonic development (contd.). A—Transverse section of the oval embryo; and B—Ventral portion of a transverse section of an embryo showing the mesoblastic somites with coelomic cavities and the demarcation of nephroblast from ectoderm (*after* Bahl).

invaginated cavity, the **archenteron**, lined on both sides by the products of the megameres. The result of gastrulation is the formation of three primary germ-layers. The outer surface made up of micromeres forms the **ectoderm**, the products of the megameres lining the enteron constitute the **endoderm**, while the pole-cells and their product fill in the space between the two and form the **mesoderm** (Fig. 12.48).

The blastoporal groove begins to close from behind forwards till only a small opening is left at the anterior end. This persists as the mouth and leads into a small stomodaeum lined by ectoderm. The stomodaeum leads into the archenteron. The embryo now consumes yolk contained in the cocoon. The anus is formed in a very advanced embryo as an ectodermal invagination fusing with the enteron in the last segment of the worm.

The cells of the mesoblastic band multiply and ultimately form a number of mesodermal blocks, the **mesodermal somite**, in the anteroposterior direction. First they are solid but soon become hollow forming the beginning of the coelom. The lateral somites with their cavities grow dorsally and ventrally ultimately fusing with each other, with the result that a continuous space is formed in the mesoderm. The layer of mesoderm in contact with the body wall is the somatopleure and the inner layer is the splanchnopleure. The somatopleure forms the body-wall musculature and coelomic epithelium internal to it while

*Phylum Annelida*

the splanchnopleure forms the gut musculature and the coelomic epithelium surrounding it. The partitions between the successive mesoblastic somites persist as transverse septa.

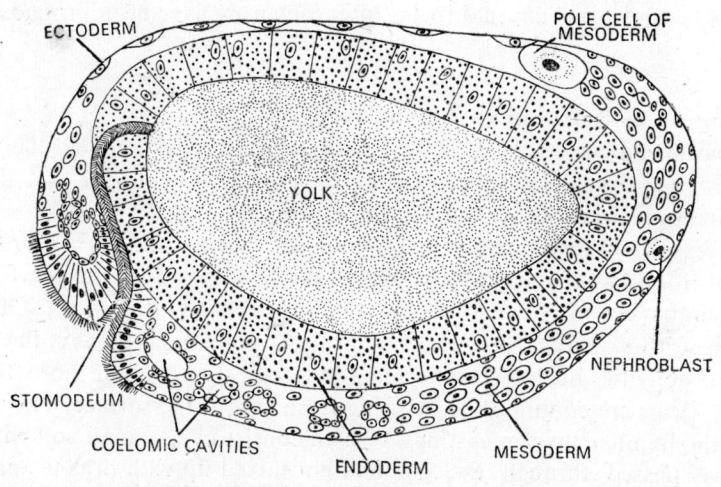

**FIG. 12.50.** Embryology (contd.). A longitudinal section of the embryo showing the formation of the stomodaeum and the coelomic spaces (*after* Bahl).

Meanwhile ectoderm gives rise to certain 3-4 rows of cells ending in large **teloblast cell**. The teloblast cell and the row of cells in front of them sink inward and come to lie between ectoderm and mesoderm. Of these that row of cells which lies on either side of the midventral line forms the nerve cord, hence the cells forming it are known as the **neuroblast** cells. The next two rows of cells (**nephroblast**) form the nephridia and setal sacs. All kinds of the nephridia are produced by these cells but they differentiate later.

The ectoderm thus gives rise to the epidermis, the nervous and excretory systems as well as the stomodaeum and the proctodaeum. The mesoderm produces the musculature, the coelomic epithelium and the blood vessels, and the endoderm produces the lining of the gut and its appendages and glands, etc.

Further development of the worm takes place by a continuous production of new body segments. First new somites are produced by budding from the mesomere, the internal segmentation is ultimately followed by an external segmentation of the embryo. When the required number of segments is formed the development is completed.

**Regeneration**

Earthworms possess the ability to regenerate lost parts, but this ability is limited. A new head may be regenerated after the removal of up to fifteen anterior segments, but the capacity to regenerate decreases rapidly behind the 9th segment. Regeneration is prevented by removal of the cerbloral ganglia. The posterior segments are more easily regenerated. If the cut is made near the tail end, a new anterior end will not regenerate on the tail-half, but instead another tail will develop from the cut surface. This produces an animal with two tails and no head, and death from starvation results. When any portion of the tail region is cut off, the lost parts readily regenerate. The number of segments regenerated is usually less than the number removed. Numerous grafting experiments have also been performed on earthworms. Almost any part of an individual grafted to the cut surface of another (if properly located) will fuse to it and grow. Numerous unusual forms of earthworms have been produced in this way.

### Earthworms in Relation to Agriculture

Earthworms have a strong claim to be ranked as beneficial animals being better cultivators than the most efficient machine devised by man. In the long past they have made a great portion of our soil most valuable and fertile, and even now they are ceaselessly renewing and improving it. Their continual burrowing in search of food results in the loosening of the earth particles and the formation of innumerable channels that open the way alike for air, rain drops and plant roots. They thus make the soil soft by bruising the soil particles in the gizzard-mills. The activities of the worms drain and aerate the soil. But more than this the earthworms are continuously bringing fresh soil to the surface. This takes the form of the familiar "worm castings" which consist of the finest soil particles as they have passed through the gut and got mixed up with organic matter and nitrogenous materials. As the wet worm casts dry up the wind scatters the fine powder over the surface of the soil and a new layer of tilled earth is in the process of formation. The importance of their humble labour is actually sublime. Darwin showed that there are, on an average, over 53,000 worms in an acre of garden soil, that ten tons of soil per acre pass annually through the bodies of the inhabitants (earthworms) and that they bring up mould from below at the rate of 7.5 cm of thickness in 15 years.

## CLASSIFICATION OF OLIGOCHAETA

CLASS OLIGOCHAETA. There are annelida with segmentation conspicuous both externally and internally. Parapodia are lacking, a distinctive head is not formed; setae usually few per somite; seldom with gills, individuals are hermaphroditic, gonads few and anterior. Development takes place within cocoons secreted by the clitellum. No larval stage is met with as the development is direct. They are chiefly inhabitants of freshwater and moist soil. They are divided into several families which are classified into two groups according to habitat—earthworms living in soil and aquatic oligochaetes living in water. This grouping, however, cannot be made a basis of classification. The classification here followed is that given in many new textbooks.

1. **Order Plesiopora**. Male gonopores located in the segments immediately following that containing the testes. Spermathecae, when present, in the region of the genital segment. Mostly aquatic.

**Family Aeolosomatidae**. Small worms with relatively few and poorly defined segments.

Example: *Aelosoma*.

**Family Naididae**. Minute freshwater species reproducing mostly by asexual means.

Examples: *Nais, Stylaria* and the carnivorous *Chaetogaster*.

**Family Tubificidae.** Aquatic oligochaetes constructing tubes in mud.

Examples: *Tubifex, Branchiura.*

**Family Phreodrilidae.** Small family of freshwater, marine and commensal species related to tubificids.

2. **Order Plesiopora prosotheca.** Male gonopores in the segment following the testes. Spermathecae considerably in front of the genital segments.

**Family Enchytraeidae.** Aquatic and terrestrial worms often white. An important component of the soil fauna.

Examples: *Enchytraeus, Lumbricillus.*

3. **Order Prosopora.** Male gonopores located on the same segment as the one containing the testes or on segment containing the second pair of testes (when two pairs present). Aquatic.

**Family Lumbriculidae.** Small to large Oligochaetes that look similar to earthworm.

Example: *Lumbriculus.*

**Family Branchiobdellidae.** Small leech-like parasites on the gills of crayfish.

4. **Order Opisthopora.** Male gonopores located usually some distance behind segments bearing testes.

**Family Haplotaxidae.** Freshwater and semi-terrestrial oligochaetes transitional between aquatic species and earthworm.

**Family Alluridae.**

Example: *Allurodes.*

**Family Moniligastridae.** A small family of terrestrial worms occurring in South-east Asia and India. Includes some giant species.

**Family Glossoscolecidae.** Earthworms.

Examples: *Pontoscolex, Alma.*

**Family Lumbricidae.** Earthworms.

Examples: *Lumbricus, Eisenia, Allolobophora.*

**Family Megascolecidae.** Earthworms. Nearly half the known species of Oligochaetes belong to this family.

Examples: *Megascolex, Pheretima, Megascolides, Microscolex.*

**Family Eudrilidae.** African earthworms.

## CLASS HIRUDINEA

The **Hirudinea** (Lat. *hirudo*, a leech) form a relatively small but well-defined group of segmented worms which are generally adapted to an ectoparasitic mode of life. The class contains over 300 species of marine, freshwater, and terrestrial worms commonly known as leeches. Although they are all considered to be blood suckers, a large number of them are not parasitic. As a group the leeches are the most specialized annelids. Undoubtedly they have evolved from the oligochaetes and display many features of the latter, but most of the distinguishing characteristics of the class have no counterparts in the other two annelid groups.

The body is elongated, vermiform, and consists of a limited and definite number of segments, each of which is externally marked with a number of transverse rings, the **annuli** or **secondary rings**, which are short and may be more or less indistinct. These rings do not correspond with the internal segments, which may be indicated by incomplete partitions. The number of annuli per segment differs in different animals. They may be three (*Branchellion*), six (*Pontobdella*), five (*Hirudo*) and twelve (*Piscicola*), corresponding to one true segments. Towards the two ends of the body the segments are reduced in size

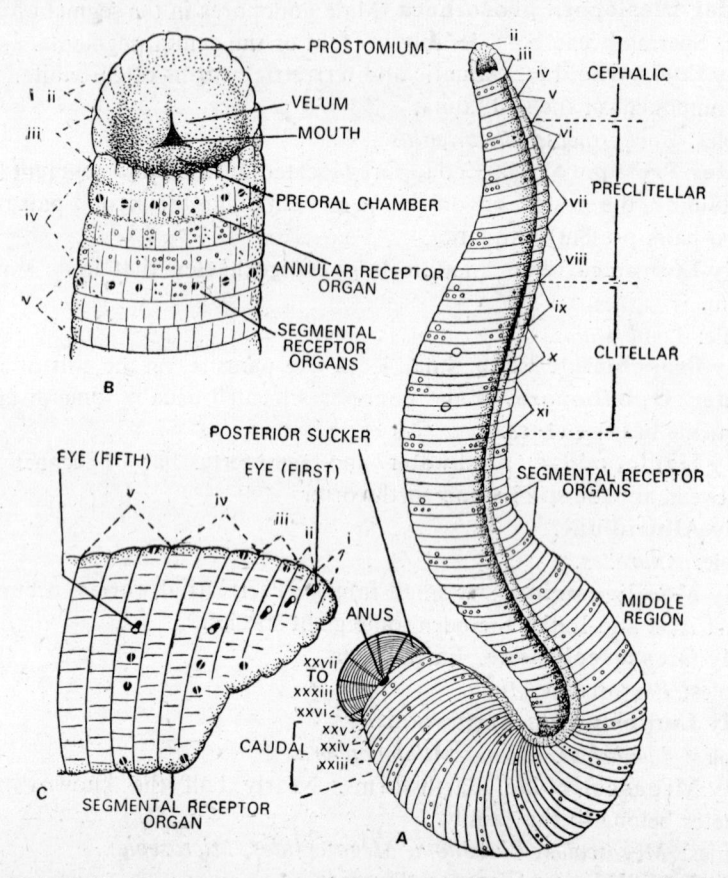

FIG. 12.51. *Hirudinaria*, external features. A—The entire worm; B—The ventral view; and C—The lateral view of the cephalic region (*after* Bhatia).

and other respects, the number of annuli is also reduced. There is always a **posterior sucker** formed by the fusion of six or seven true segments; and generally an **anterior sucker** at the anterior end of the body. Both the suckers are ventral. The **anus** opens dorsal to the posterior sucker, and the mouth is on the ventral surface in the anterior sucker. The animal moves in a looping manner by means of its suckers. There are no parapodia or setae (except in **Acanthobdella**), or head appendages, but in one genus (*Branchellion*) there are leaf-like branchial processes on the middle part of the body. The segmental senseorgans or sensory papillae are arranged on the annuli in transverse rows. On the head some of these organs are modified into eyes. The skin is often highly coloured owing to the presence of pigment. The clitellum consists of glandular development of the skin of the 9th, 10th and 11th segments.

In the typical leeches the coelom is difficult to be traced completely in the adult, but it is undoubtedly present in the embryo and is of usual annelidan

*Phylum Annelida*

form having a series of paired cavities in the mesoblastic somites. In the adult these give rise to the sinus system; and indirectly to other structures such as the tubes of the botryoidal tissue, gonads and nephridia, etc. A vascular system is present. The nervous system is of the typical annelidan plan. The excretory organs are segmentally disposed nephridia. The reproductive organs are complicated. The openings of the male and female organs are placed one behind the other, the male being anterior, in the midventral line of the anterior region of the body. The reproductive organs are hollow structures, and are continuous with their ducts. The eggs are usually laid in cocoons either on stones or plants or in damp earth. The cocoon is the product of clitellar secretions. Asexual reproduction is unknown in the group. Most members of the group live in freshwater and have a blood sucking habit, a few are marine and prey upon fishes and other animals in the sea. Some live in the moist soil and suck blood from the leg of a passing man or beast. A few carnivorous forms are also known and they devour earthworms, insect larvae, tadpoles and even small fishes.

## TYPE HIRUDINARIA

*Hirudinaria*, the Indian cattle-leech, is found all over India. It is found in freshwater ponds, tanks or reservoirs. Like most ectoparasitic leeches it leads a life sucking the blood of vertebrates with the help of suckers. The leeches have to swim about to locate victims, as they are less modified than the parasitic flukes or tapeworms, and have eyes. Ordinarily the leeches attain a length of 10-15 cm but a mature specimen may be 30-35 cm long. In contracted condition the body appears almost cylindrical, but in a state of extension it is dorsoventrally flattened. The animal is beautifully coloured. The dorsal surface is dark green generally, in younger specimens it may be olive green, mottled with dark stripes and spots throughout, whereas, the ventral surface is orange-yellow or orange-red. The sides often show distinct stripes, orange or yellow. The lateral stripes are bounded below by broad black or dusky submarginal stripes and dorsally by a series of segmentally arranged black or dusky spots. Colour often presents individual variations. The markings on the surface of the body form a characteristic pattern.

Under natural conditions the body is soft and flaccid and the skin always remains moist and slimy due to profuse secretion of mucous from the skin. If a large number of leeches are kept in small quantity of water, the slime turns the water into a viscid fluid which becomes uninhabitable for the leeches themselves.

The external segmentation of the body does not correspond with the internal segmentation. Externally there appear more than hundred annuli or rings separated by well-marked grooves, though there are only 33 true body segments. Each true body segment, thus, includes five annuli. The integument of each annulus is divided into several roughly rectangular areas by short longitudinal wrinkles. Each area bears elevated papillae in which are lodged receptor organs.

Both the **anterior** and **posterior suckers** (Fig. 12.51) are recognizable in this type. The anterior sucker, also called the **cephalic sucker**, is formed by the fusion of some anterior body segments with the **prostomium**. Situated at the ventral surface of the anterior end, it is oval in outline with a cup-like depression in the centre. Its anterior border is free and mobile and is bent over the mouth forming a sort of **upper lip**. Its function is to aid locomotion and

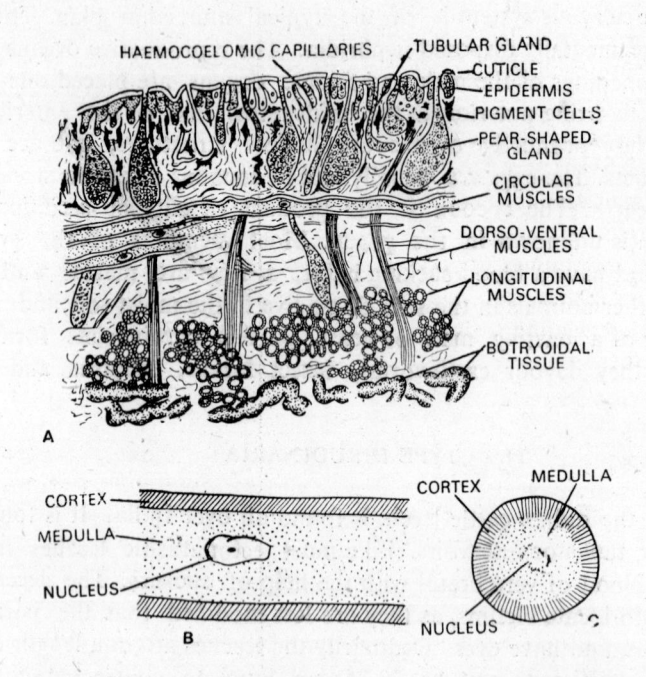

**FIG. 12.52.** A—Cross-section of the body wall of leech (*after* Bhatia); B—Longitudinal section of a part of a muscle fibre; and C—The same in cross-section.

prehension. The **posterior sucker** or **anal sucker** is a highly muscular circular disc at the posterior end of the body directed ventrally. It is formed by the complete fusion of seven body segments. This is indicated by seven sets of segmental receptors, each arranged in a semicircle. The sucker acts as a powerful organ of adhesion.

The mouth is situated in a cup-shaped depression, the **preoral chamber**, of the anterior sucker. It is a triradiate aperture. The **anus** is a minute opening situated dorsally at the root of the posterior sucker (Fig. 12.51). Besides these openings, there are seventeen pairs of minute apertures, the **nephridiopores**, situated on the ventral surface of the body beginning from the 6th somite. Each pair of openings lies on the last annulus of each somite, the last being on the 22nd. The generative apertures are median and ventral. The male generative aperture is situated in the groove following the second annulus of the 10th somite on a swollen area encircled by a furrow. Sometimes a filamentous penis is seen protruding through the aperture. The female generative aperture is a little behind the male pore in the groove between the second and third annuli of the 11th somite. The latter is small and less prominent (Fig. 12.51).

The **head region** (or **cephalic region**) is made up of the first five segments, the **ocular segments**, of the body in which the nephridiopores are absent (Fig. 12.51). Corresponding to the prostomium of other annelids in this case also there

*Phylum Annelida*

is a small preocular lobe, the **prostomium**. The five segments following this bear five pairs of eyes (hence called ocular segments). The segment just behind the prostomium bears the first pair of eyes. The second segment bears the second pair and so on. The first two segments are not marked by grooves, in other words, they are unannulated. The third segment is biannulated and the 4th and 5th segments are each triannulate. In each case the first division of the segment bears the corresponding pair of eyes. The eyes appear as black dots on the dorsal surface. The upper lip is formed by the first three segments with the prostomium (Fig. 12.51).

No permanent clitellum exists in *Hirudinaria*, but during breeding season the walls of complete 9th, 10th and 11th segments become glandular and usually develop different colours though the separate identity of the segments is retained. On this basis these segments are referred to constitute the **clitellar region** and the three preceding segments (6th, 7th and 8th) from the **preclitellar region**. Of these the 6th segment is triannulate and two are completely quinquannulate. Likewise segments 12th to 22nd are known to form the **middle region** of the body. This is the longest part of the body and all the segments carry nephridiopores. This region is followed by the **caudal region** which consists of four incomplete segments (23rd to 26th). Of these the 23rd is triannulate but the remaining three are biannulate. They are without nephridiopres. The last three segments of this region form the stalk of the posterior sucker which consists of seven segments.

## Body Wall

The **cuticle** forms the outermost covering of the body wall (Fig. 12.52), and is followed by the **epidermis**, the **dermis**, the **musculature** and the **botryoidal** tissue. The cuticle is a transparent colourless membrane of variable thickness forming a firm but moderately elastic outer covering. It is perforated by the external openings of the unicellular epidermal glands. It is secreted by the underlying layer of epidermis and is cast off in the form of thin transparent shreds. Normally a well-fed leech casts off its cuticle once in every two or three days, but if it is kept in dirty water the cuticle is cast off more frequently.

The **epidermis** consists of a single layer of ectoderm cells underlying the cuticle. The cells are broader towards the outer side and narrower towards the inner side. The narrow inner side carries the nucleus. The blood capillaries penetrate between the cells of the epidermis forming a vascular membrane. The epidermis also shows various types of unicellular glands, which become especially enlarged and sink down in the deeper layers. These glands are scattered all over the body. They are pear-shaped and also produce slimy mucous which covers the entire body. They are often referred to as slime glands. Likewise the gland cells situated in suckers are known as the **sucker glands**. They are always found in groups. In the prostomium the **prostomal glands** are located in a series of distinct groups. The clitellar glands are found in the clitellar region. Associated with the epidermis are also found groups of modified epidermal cells, the receptors, which show a high grade of differentiation.

The **dermis** is the next layer inward. It varies in thickness in the different regions of the body. It consists of a network of connective tissue fibres in the meshes of which are found various forms of connective tissue cells, pigment and fat cells, scattered muscle fibres and branches of haemocoelomic channels. Pear-shaped and tubular continuations of the ectodermal glands also extend into this layer.

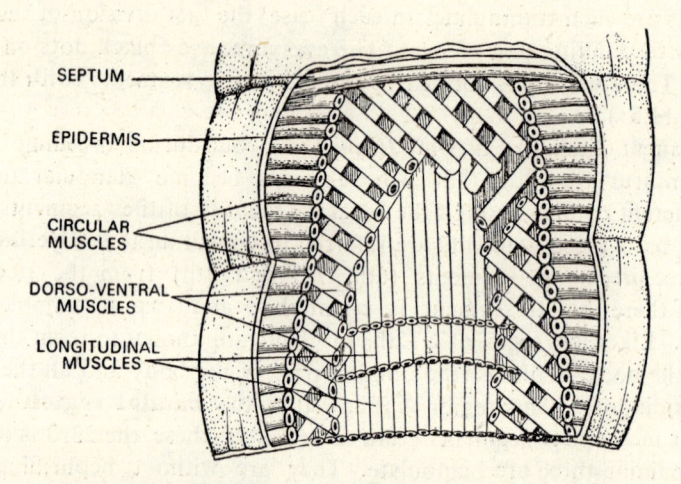

FIG. 12.53. Diagrammatic representation of the body wall musculature of leech.

The **musculature** consists of two layers: an outer thin layer of **circular muscle fibres** (Fig. 12.53) just below the dermis, and an inner layer of **longitudinal muscle fibres** consisting of several strata. The fibres of the inner portion of the circular muscle-fibre layer are arranged in oblique manner, so much so that the layer of oblique cells has been referred to as a separate layer by some authors. The longitudinal muscles form the thickest part of the musculature. Besides these layers of muscles there are some isolated bundles of muscles that run either obliquely or in straight lines from the dorsal to the ventral surface. They run across the circulars and longitudinals and their ends spread out and terminate beneath the epidermis. Some of these are centrally placed, running by the sides of the alimentary canal and are known as the **dorso-ventrals**, whereas, those lying near the flanks of the leech are termed **vertical muscles**. The elaborate musculature helps the animal in performing necessary movements needed for locomotion.

The **muscle fibres** have a characteristic structure consisting of elongated fusiform cells arranged either singly or in bundles. Each cell, moreover, consists of a cortex of striated contractile substance and a medulla of unmodified protoplasm containing a large oval nucleus. In transverse section a muscle fibre appears tubular being made up of a radially striated ring enclosing a central mass of granular cytoplasm (Fig. 12.52).

The **botryoidal tissue** lies beneath the layer of the longitudinal muscles. It consists of masses of mesenchymatous tissue, the cells of which are arranged from end to end and contain intracellular capillaries filled with red fluid, in life. The walls of these cells are pigmented, being loaded with dark brown pigment.

## Locomotion

On a substratum leeches move by performing looping movements like those of a caterpillar. First the animal fixes the posterior sucker (Fig. 12.54) to the substratum. The adhesion of the sucker is helped by the secretions of the sucker-glands. The fixation of the posterior sucker initiates the contraction of the circular muscles and relaxation of the longitudinal muscles with the result that the body elongates. The elongated body finally bends its anterior edge towards the substratum and fixes the anterior sucker. The adhesion of the anterior sucker initiates the relaxation of circular muscles and contraction of the longitudinal muscles with the result that the posterior sucker is let off and is drawn close to the anterior, where it is fixed forming a loop. As soon as the posterior sucker is fixed the anterior one is released and the body elongates, and thus the animal moves in the desired direction by alternating changes in the length of its body. Leeches also perform swimming movements. The dorsoventrally flattened body assumes the form of a flattened band which moves in water by successive undulating movements.

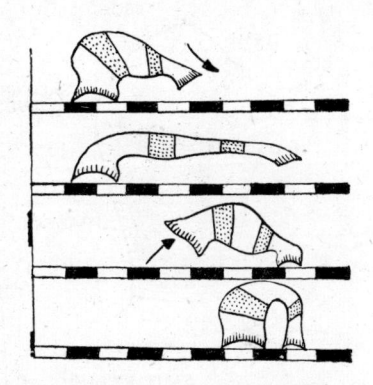

FIG. 12.54. Locomotion of leech.

## Digestive System

The digestive system comprises the alimentary canal (Fig. 12.55) which is a straight tube extending from the mouth to the anus. On the ventral side of the anterior sucker there is a cup-shaped depression, the **pre-oral chamber**, at the base of which is the triradiate opening, the **mouth**. The pre-oral chamber helps in sucking blood, as such, it is very well developed in blood-sucking leeches. Between the mouth and the buccal cavity there is a more or less complete partition, the **velum**. The buccal cavity is a short chamber behind the velum. The wall of the chamber is marked out by the presence of three muscular cushion-like thickenings, the **jaws** (Fig. 12.55), of which two are ventrolateral and one mediodorsal corresponding to the slits of the triradiate mouth. When a leech bites the jaws are protruded through the mouth into the pre-oral chamber, the three lips of the velum contract to allow the protrusion. At the summit of each jaw there are rows of minute teeth, the **denticles**, borne on a ridge formed by special thickening of the cuticle. The denticles numbering 103-128 in the median jaw and 85-115 in the paired jaws form a sort of semi-circular ridge. On each side of the jaw there are numerous button-shaped protuberances, the **salivary papillae**, each of which bears a number of minute openings of the salivary glands.

The **buccal cavity** (Fig. 12.55) leads into an oval muscular sac, the **pharynx**, which is externally surrounded by large masses of unicellular pyriform glands, the **salivary glands**. The ductules of these glands run forwards in groups or bundles and enter the three jaws. The salivary glands secrete a substance which prevents the coagulation of blood facilitating the act of sucking. The pharynx leads into the oesophagus, a short narrow tube opening into the crop. The **crop (mid-gut)** is the largest portion of the alimentary tract occupying almost two-third of the visceral space between segments 9th and 18th. It has ten pairs of **lateral caecae**.

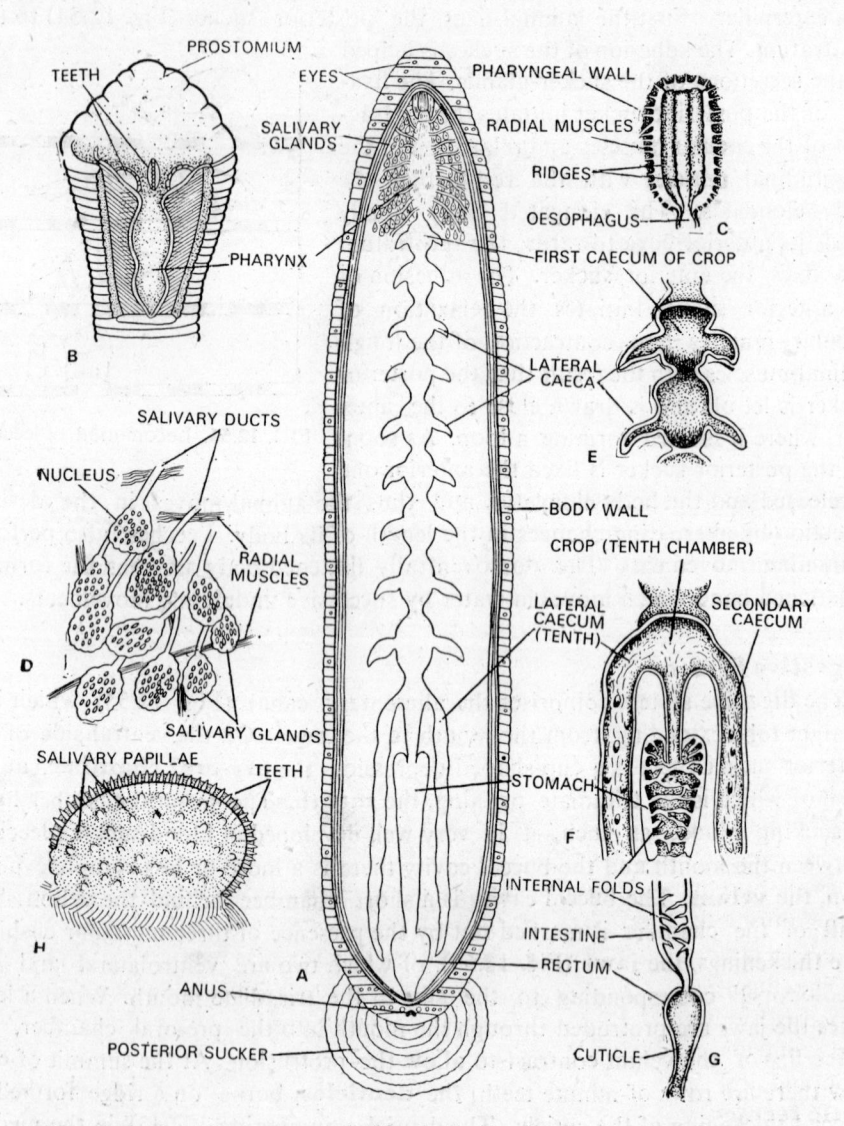

**FIG. 12.55.** A—Alimentary canal of leech; B—Longitudinal section of the anterior end of leech; C—Longitudinal section of pharynx; D—Salivary glands of leech; E—Crop with caeca (dissected); F—Tenth pair of lateral caecae and stomach (dissected); G—Intestine and rectum (dissected); and H—Lateral view of jaw (all *after* Bhatia),

The anterior caecae are usually small, but the succeeding ones go on increasing in size, the last pair being the largest reaching the 12th segment. They are simply elongated blind sacs lying one on either side of the intestine. The **stomach** is the next part of the gut lying in the 19th segment. It is a small, more or less heart-shaped thin-walled chamber, the walls of which are internally folded. It leads into a straight narrow tube, the **intestine**, from which it is not clearly demarcated externally. It extends from the 20th to the 22nd segment. Internally its walls are folded forming villi-like structures to increase the absorptive surface

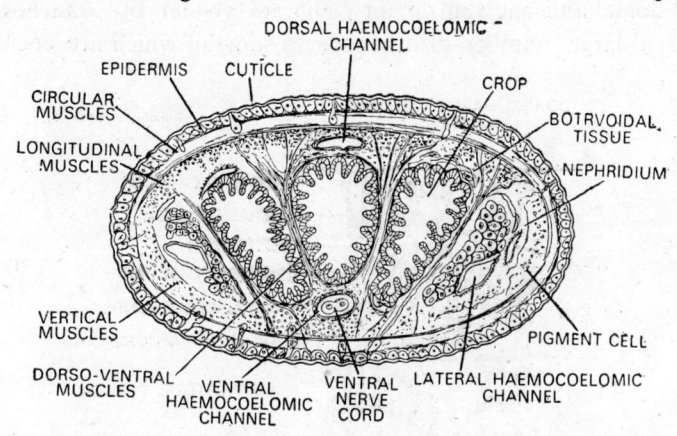

FIG. 12.56. Semidiagrammatic transverse section of the leech passing through the crop and diverticula.

of the intestine. The intestine leads into the **rectum**, which is a thin-walled tube without any inner folds. It opens to the exterior through the anus (Fig. 12.55).

**Feeding**. In feeding a leech attaches itself to its prey by the posterior sucker, applies the anterior sucker to the skin and makes a wound with the help of little jaws inside the mouth. The incision is triradiate and it lays open the cutaneous vessels from which the blood is sucked in with the help of the pre-oral chamber, the buccal cavity and the pharynx that form highly efficient apparatus for the purpose. A leech can suck blood several times its own weight in a single meal. When the digestive tract is filled with blood it drops off. It then prefers darker places and remains hidden till this meal exhausts. It remains turgid while digesting the meal. Large blood meals are few and far between, but the lateral pouches of the digestive tract hold enough blood to last for months. The salivary glands of leeches produce a substance, **hirudin,** which prevents the coagulation of the blood during their feeding time. For this reason a wound made by a leech continues to bleed for a long time after it has detached itself.

The digestion of a full meal takes a long time, as mentioned above. The blood is **haemolyzed**, that is, the blood corupscles burst and their haemoglobin contents dissolve. The stomach and the intestines are the regions of digestion and absorption. Some authors have even reported the presence of **proteolytic ferments** in the system. The absorption of the digested blood takes place slowly in the intestine. The faecal matter is discharged through the anus.

## Coelom

The **coelom** is reduced in leeches due to the growth of botryoidal tissue, which brings about the adhesion of the alimentary canal to the integument. It is spacious only in **Acanthobdella** and is divided by septa as in the oligochaete worms. In all other leeches it is much broken up and though retains its relations with some of the organs like the ventral nerve cord and the principal blood vessels, etc., it is largely without a perivisceral character. It has the form of longitudinal canals or sinuses connected by a complicated system of intercommunicating spaces. In the leech morphology this system is called the **sinus system**. In *Hirudinaria* there are four haemocoelomic longitudinal canals, two lateral, one dorsal and one ventral, an elaborate system of branches of these canals and a large number of other spaces some of which are coelomic while

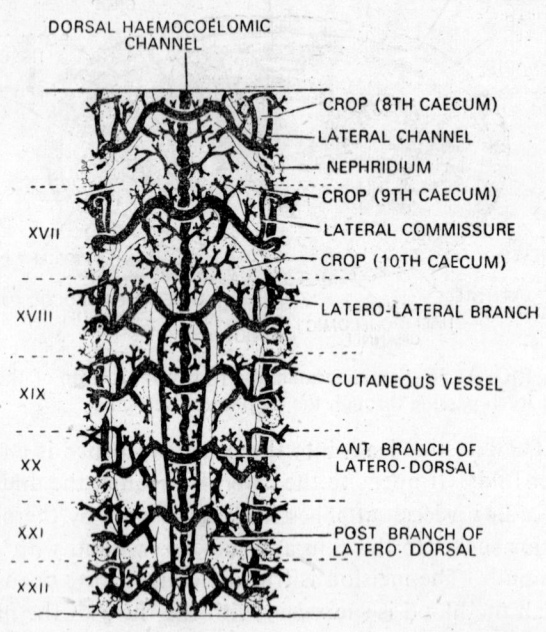

FIG. 12.57. The haemocoelomic system of the middle region of the leech (*after* Bhatia).

the nature of others is doubtful. All the canals are thin-walled, their walls consisting of connective tissue and coelomic epithelium; but the walls of the lateral canals have become secondarily muscular. All the four canals communicate with one another. Through the coelomic canals circulates the coelomic fluid in which is dissolved the respiratory pigment, **haemoglobin**, and which contains numerous colourless amoeboid corpuscles floating in it. Because of the presence of the respiratory pigment the fluid is called the **haemocoelomic fluid** and the canals are called the **haemocoelomic canals**. In *Hirudinaria*, thus, the coelomic canals are filled with blood-like fluid and true blood-channels are lacking. The same is the true of *Hirudo*.

Leydig, on the other hand, held that the sinus system and the vascular systems were distinct. He was supported by various workers. But now the majority recognizes that the two systems are distinct, but they are continuous through their finer branches. Goodrich proved that in *Hirudo* the contractile and non-contractile vessels were continuous with each other and that the coelomic fluid circulates through them. To the fluid he gave the name **haemolymph** and thus called the system as **haemolymph system**. The same is true of *Hirudinaria*.

The true coelom containing fluid without the respiratory pigment is restricted, in *Hirudinaria*, to the testis-sac and ovisacs. In the early stages the oviducts open in spaces surrounding the ovaries, but soon these spaces close to form ovisacs

# Phylum Annelida

before the formation of the respiratory pigment The testis-sacs also close in the same manner.

**Longitudinal canal.** The **dorsal haemocoelomic canal** runs below the body wall in the mid-dorsal line along a slightly sinuous course. It is attached to the gut from the anterior to the posterior end of the body. The fluid flows in the canal from behind forwards. From the dorsal canal arise, in each segment, two pairs of **dorsolaterals**, each of which runs outwards and finally breaks up in a **capillary plexus** in the dorsal and dorsolateral region of the body wall,

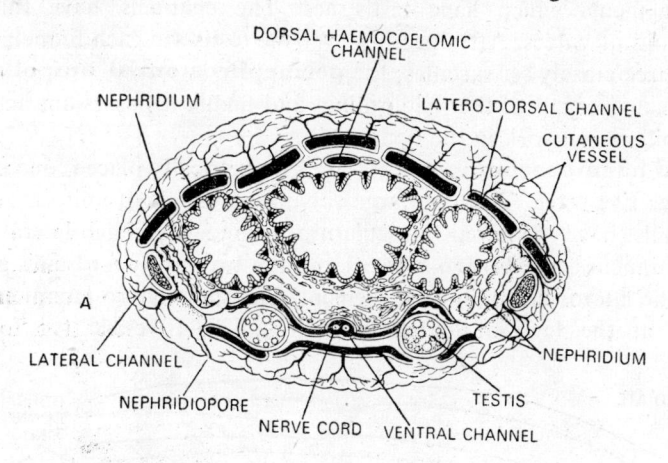

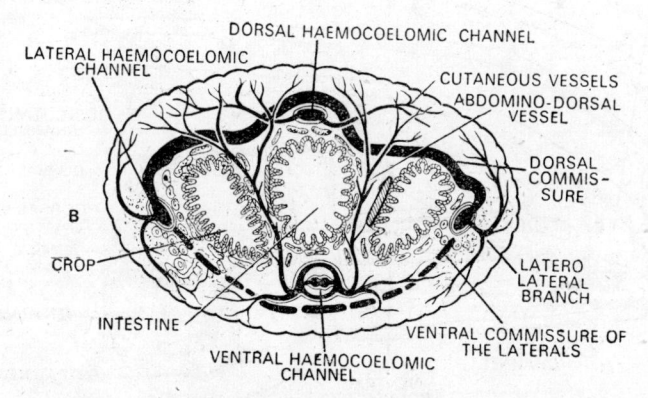

FIG. 12.58. A—Diagrammatic cross-section of the leech through the region of the nephridial branch; and B—Diagrammatic cross-section of the leech passing through the region of the latero-dorsal haemocoelomic channel.

and a number of **dorsointestinals** running to the alimentary canal. In the 6th segment the dorsal canal breaks up into a capillary network supplying the first five segments. Posteriorly in the 22nd segment it bifurcates, the branches passing ventrally around the rectum enter the posterior dilation of the ventral canal with which a direct communication is established here (Fig. 12.57). The dorsal canal and its branches are without valves and muscular walls.

The **ventral haemocoelomic canal** runs below the alimentary canal and, like the dorsal, extends from the anterior to the posterior end of the body. It is wider and lodges the cerebral ganglia, the nerve cord, the peripharyngeal nerve ring and the sub-pharyngeal ganglia. The course of blood in this vessel is from anterior to the posterior end. From each segment two pairs of branches arise from the ventral canal, the first pair arises at the level of each nerve ganglion and each branch divides into two, the **ventral** and the **abdominodorsal branches**. The ventral branch forms a **capillary network** in ventrolateral side of the body wall while the latter passes upwards and forms a **dorsolateral cutaneous plexus**. The second pair consists of the nephridial branches supplying the testis-sacs and nephridia. These are found only in the eleven segments which have testis sacs. The channels have thin walls and are devoid of valves. After passing over the testis-sac each branch enlarges into two or three closely set saccules, the **perinephrostomial ampullae**, containing the ciliated organ of the nephridium and finally it passes out and breaks to form a capillary network in the body wall.

The **lateral haemocoelomic canals** are symmetrically placed, one on either side of the digestive tract. They are large valvulated vessels of uniform diameter and their walls have acquired musculature secondarily. Each lateral channel receives two branches, the **latero-lateral** and the **latero-dorsal,** and gives off one branch, the lateroventral, in each segment. The short latero-lateral enters the lateral canal at the level of the base of each nephridial vesicle. It is formed by

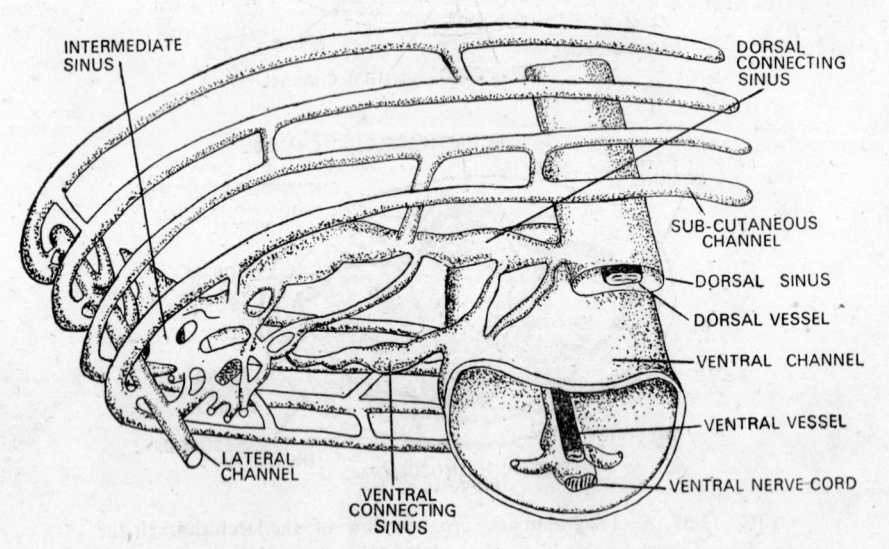

**FIG. 12.59.** Reconstruction of part of the coelomic sinus system of *Glossiphonia complanata* (*after* Grasse).

the union of a branch from the skin with a few branches from the nephridium. The latero-dorsal enters the lateral canal at the level of the main lobe of the nephridium. Each is formed by the union of some branches from the superficial cutaneous capillaries of the dorsal and dorsolateral regions, and a number of branches from the wall of the gut. In each segment the two latero-dorsals of the opposite side are joined by a **transverse loop,** the dorsal commissure of the lateral canal, above the dorsal canal. There are seventeen such loops in segments 6th to 12th. There is another commissure in each segment connecting the latero-

# Phylum Annelida

lateral and latero-dorsal branches of each side. This is known as the **lateral commissure**. The latero-lateral and the latero-dorsal are collecting vessels and pour the haemocoelomic fluid in the lateral canal. Their openings into the lateral canals are provided with valvular arrangements.

Three sets of capillary networks are distinguished in *Hirudinaria*. The first is an inner deeper set of **botryoidal capillaries,** which receive their supply from the branches of the ventral canal and cover the viscera by a fine capillary network. The second is an **intermediate** set of capillaries arising from the lateroventral branches of the lateral canals and the third is the **superficial cutaneous capillary** network which extends into the epidermis and its branches also enter the latero-dorsal branches.

**Course of circulation.** The fluids flow in a definite direction. The dorsal and ventral canals are distributing vessels, while the lateral canals are collecting as well as distributing vessels. All the four canals communicate with one another at the posterior end (Fig. 12.60). The dorsal canal distributes the fluid in the region

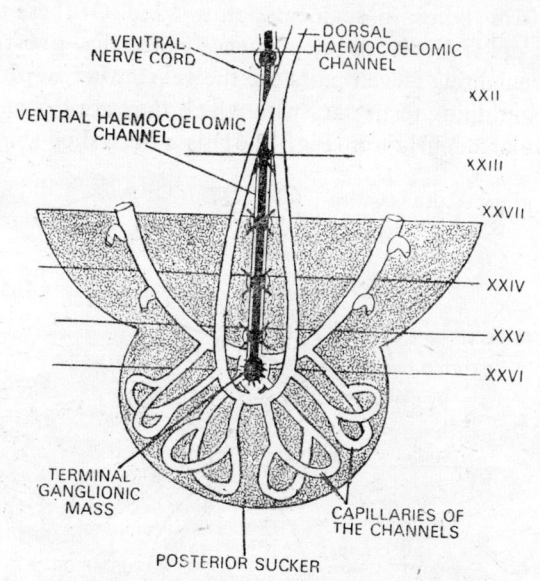

FIG. 12.60. Diagrammatic sketch showing the union of longitudinal haemocoelomic channels. Horizontal lines indicate body segments XXII to XXVI.

of the dorsal, and dorsolateral parts of the body wall and also to the entire alimentary canal. The fluid from these parts is collected by the branches of the latero-dorsals and carried to the lateral canals. The stomach and the intestine are drained by two latero-intestinals and a ventral intestinal, all these pour their contents in the dorsal commissure of the laterals. The ventral canal distributes the fluid to the ventral, the ventro-lateral, lateral median dorsal regions of the body wall and parts of nephridia. From all these parts the fluid is collected by the latero-dorsal and latero-lateral branches of the lateral canals. The lateral canals on their turn supply the nephridia, the genital organs, the gut and the ventral body wall through the lateroventrals. The fluid from these parts is returned to the lateral canal by the latero-lateral and the latero-dorsal branches. Thus, the lateral canals are simultaneously distributing as well as collecting canals.

**Respiratory System**

Like the earthworm the leeches are also devoid of special respiratory organs. The skin which has a rich supply of the haemocoelomic fluid acts as the respiratory surface for the exchange of gases. The epidermis acts as the permeable membrane through which the oxygen dissolved in water passes in and the carbon dioxide from the haemocoelomic fluid diffuses out. For this purpose the skin is always kept moist and a copious secretion of mucous prevents it from drying. Some leeches such as *Branchellion* (Fig. 12.67) possess lateral outgrowths of the body wall that act as respiratory surfaces. They are, thus, known as gills.

**Excretory System**

The **excretory organs** of the leech comprise seventeen pairs of segmentally arranged **nephridia** lying in segments 6th to 22nd. Of these the first six pairs are located in the pre-testicular segments hence called the **pre-testicular nephridia**, and the remaining eleven pairs are the **testicular nephridia** as they lie in the segments containing testis-sacs with which they are intimately associated.

A typical testicular nephridium (Fig. 12.61) is a horseshoe-shaped structure of

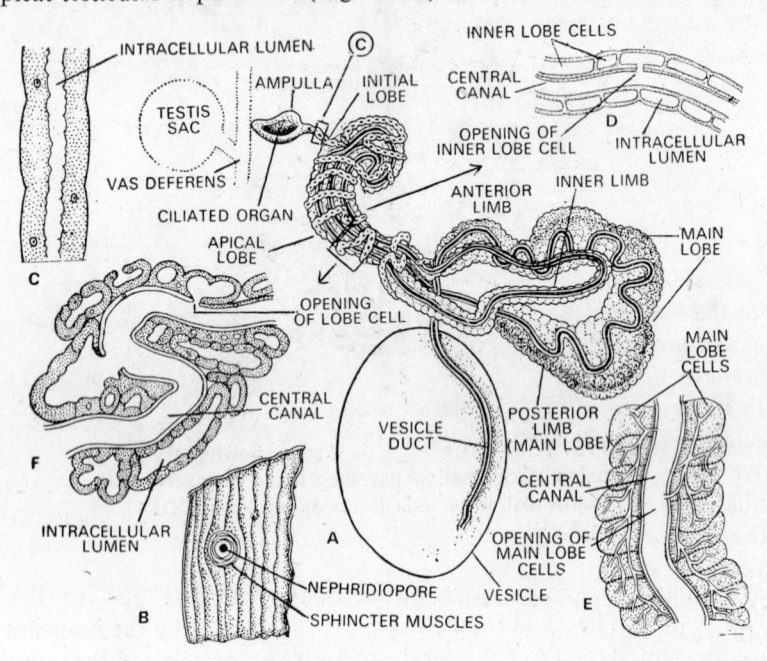

FIG. 12.61. Excretory organs of the leech. A—Testicular nephridium; B—Ventral wall of the vesicle as seen from inner side to show the sphincters surrounding nephridiopore. C—F—Longitudinal sections of different parts of the nephridium; (C—Initial lobe; D—Inner lobe; E—Main lobe; F—Apical lobe).

which one arm is prolonged into the **testis-sac** and the other enters the terminal vesicle of the nephridium. The horseshoe proper is known as the **main lobe** that occupies the latero-ventral position. It consists of two unequal limbs, an **anterior** (the longer of the two) and a **posterior**. From the ventral end of the anterior limb arises a narrow duct, the **vesicle duct**, passes backwards alongside the spacious terminal vesicle into which it ultimately opens. The vesicle is a very large oval sac lying behind the rest of the nephridium and closely applied against the ventro-lateral surface of the body wall, and opens to the exterior through a rounded aperture, the **nephridiopore**. The posterior limb passes backwards and forms a stout lobe, the **apical lobe**, lying in an anteroposterior position beneath the gut. Besides these, there are two other lobes, the **inner** and **initial** lobes The inner lobe lies all along in the inner concavity of the main lobe and also runs forward along the outer border of the apical lobe for about half its length. It joins the main lobe at its posterior extremity. The initial lobe, extremely long and slender, is closely twined around the apical lobe. It joins the main lobe at its posterior extremity. At its anterior extremity it runs as a slender cord of cells towards the testis-sac, where it ends abruptly by the side of the perinephrostomial ampullae which lodge the ciliated organ within. The ciliated organ is a compound structure corresponding to the funnel or **nephridiostome** of a typical annelid nephridium and each consists of (*a*) a central reservoir, a **perforated chamber,** appearing more or less spongy, with an outer wall formed of a single layer of cells and a central mass of connective tissue cells, the seat of the manufacture of the coelomic corpuscles; (*b*) the numerous **ciliated funnels** each fitting into a pore of the reservoir. They cover the reservoir from outside. Each funnel is like an earlobe, with about one-fourth of its margin incomplete along one side. The distal end is broad and ciliated, while the proximal end or "neck" is narrow and fits into a pore of the reservoir. In the adult *Hirudinaria* the ciliated organ has no connection whatever with the body of the nephridium although there is a distinct connection between the two in the embryo. It loses excretory function in the adult and is only associated with the haemocoelomic system, for which it manufactures coelomic corpuscles and drives them into the haemocoelomic stream. It lies a little towards the inner side of the testis-sac and not on it.

The pre-testicular nephridia are first six pairs lying, in the pre-testicular segments. In these segments there are no testis-sacs, hence haemocoelomic ampullae and ciliated organs are absent. The initial lobes end loosely in the general connective tissue on either side of the ventral nerve cord.

The ciliated organ, as mentioned above, is not connected with the nephridium proper (Fig. 12.62) in the adult, and has become completely subservient to the haemocoelomic system. The nephridium proper, on the other hand, is truly excretory in function. It serves to eliminate excess of water and nitrogenous waste with which the terminal vesicles are invariably found filled. Some workers have credited botryoidal tissue also with excretory function.

## Nervous System

The nervous system satisfies the usual pattern of Annelida. The **cerebral ganglia** and the **ventral nerve cord** are enclosed within the ventral haemocoelomic canal. The nervous system is divisible into three parts: (*i*) the central nervous system; (*ii*) the peripheral nervous system; and (*iii*) the sympathetic nervous system.

468   LIFE OF INVERTEBRATES

The **central nervous system** consists of a pair of cerebral ganglia situated in the 5th segment lying on the roof of the anterior part of the pharynx immediately behind the median dorsal jaw and connected to a ganglionated ventral nerve cord by a pair of peripharyngeal connectives. Both the cerebral ganglia are closely connected forming, more or less, a compact mass. They supply nerve branches to various structures of the anterior region. The peripharyngeal connectives closely surround the anterior end of the pharynx forming a sort of collar

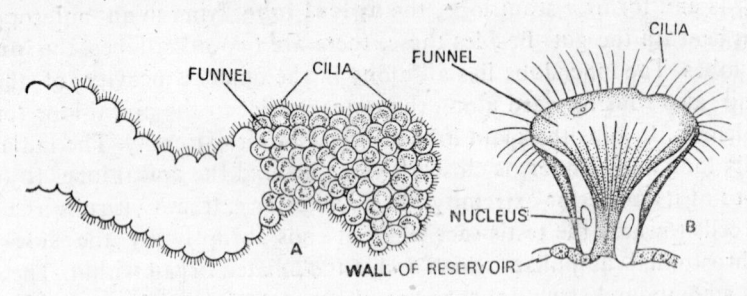

**FIG. 12.62.** A—A portion of the ciliated organ of the nephridium; and B—A separate funnel (magnified).

around it. Ventrally the peripharyngeal connectives unite to form the **subpharyngeal ganglionic mass** which lies below the pharynx exactly opposite to the cerebral ganglia (Fig. 12.63). It is a composite structure of more or less trian-

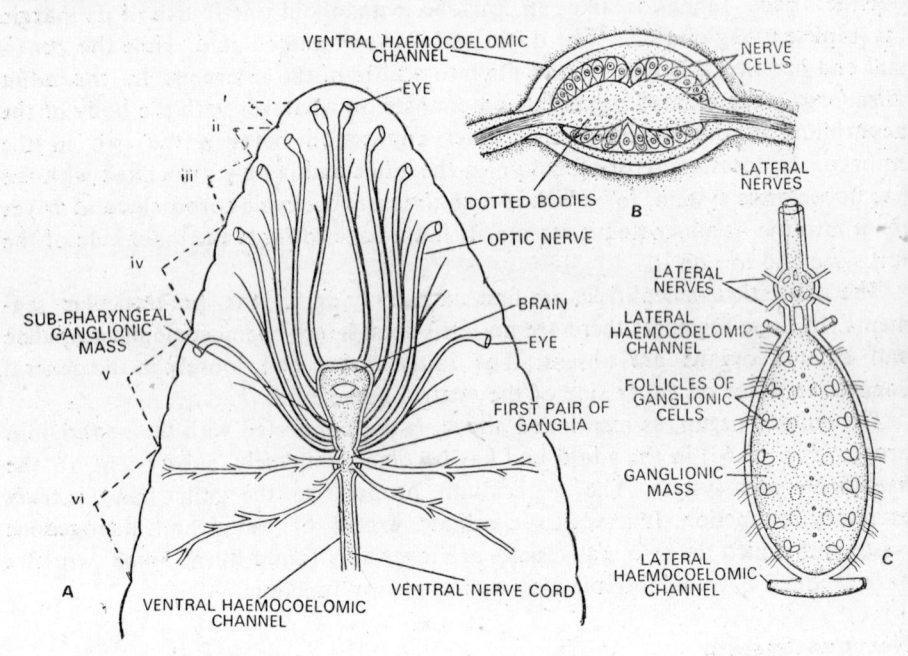

**FIG. 12.63.** Nervous system of the eech. A—A dissection of the brain and anterior nerves; B—Cross-section through a segmental ganglion; and C—Enlarged view of the posterior ganglionic mass.

gular shape, apex of the triangle pointing backwards formed by the fusion of four pairs of ganglia. From its apex arises the ventral nerve cord and runs along the mid-ventral line. It is a chain of closely apposed nerve cords segmentally swollen because of the fusion of ganglia. There are twenty-one such ganglia, each lying within the first annulus of its own segment. The nerve cord terminates in a large ovoid **terminal ganglionic mass** formed by the fusion of seven pairs of ganglia belonging to segments that form the posterior sucker.

The peripheral nervous system consists of several pairs of nerves which arise from the central nervous system and innervate the different parts of the body. Each paired ganglion gives rise to two pairs of nerves, (a) the **anterior laterals** which are stout nerves arising from the anterior part of the ganglion, and (b) the **posterior laterals** arising just posterior to the anterior lateral. From the anterior end of each cerebral ganglion arises a nerve that supplies the eye of the first pair of its own side and also sends a branch to the prostomium and the roof of the buccal chamber. The second, third, fourth and fifth pair of eyes are supplied by four pairs of stout nerves arising from the subpharyngeal ganglionic mass. From these ganglia originate several other nerves that supply the roof of the buccal cavity, the muscles of the body wall and also the segmental receptors of the anterior five segments. The posterior sucker and segmental receptors of that area are supplied by nerves arising from the terminal ganglionic mass.

The sympathetic nervous system consists of an extensive nerve plexus lying between muscle-layers of the body, beneath the epidermis, and one on the wall

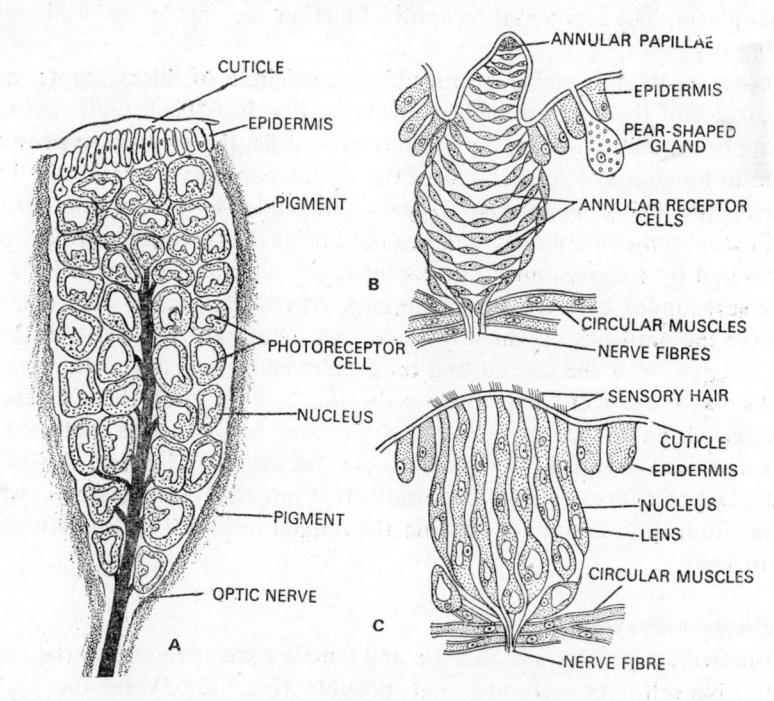

FIG. 12.64. Sense organs of the leech. A—Vertical section of the eye; B—Annular receptor; and C—Dorsal segmental receptor.

of the alimentary canal. It is connected with cell-follicles on both sides of the nerve collar and with irregularly distributed multipolar ganglionic cells over the entire alimentary nerve plexus.

**Sense organs.** The sense organs (Fig. 12.64) of the leech include four types of receptors: the free **nerve endings,** the **annular receptors,** the **segmental receptors,** and the **eyes**. The free nerve endings are found all over the body between the epidermal cells. The ganglion cells of these nerve endings lie beneath the epidermis. The annular receptors are 36 in number—18 on the dorsal and 18 on the ventral surface. Each receptor is formed of a group of flattened overlapping cells and projects out of the integument. They are known as annular receptors as they are arranged in a line across the middle of each annulus. They are tactile in nature. The segmental receptors are small whitish areas elevated on elliptical papillae. They are segmentally located and occur on the first annulus of each segment. Each segment of the body bears four pairs of three receptors dorsally and three ventrally. These are named **median, inner lateral, outer lateral** and **marginals** according to their position on either side of the mid-dorsal line, dorsally, and median, lateral and marginal on either side of the midventral line. The **ventral medians** are more widely separated from each other than the **dorsal medians**. Each receptor consists of two types of cells—ordinary tactile cells or **tangoreceptors** and the light sensitive cells or **photoreceptors**. The tactile cells are tall, slender cells separated from one another by clear spaces and provided with hair-like processes at their free ends. The light perceiving cells show a crescentic hyaline substance, the **optic organelle** in their cytoplasm. The segmental receptors function as tactile as well as light perceiving organs.

The **eyes** are ten in number arranged in a semicircle of black spots on the dorsal surface of the anterior sucker. Each eye (Fig. 12.64) is cylindrical in shape consisting of a number of large, clear refractive cells, the **photoreceptor cells,** arranged in longitudinal rows. Each of these cells contains a crescentic hyaline substance, the **optic organelle (lens),** surrounded by a very thin peripheral layer of cytoplasm containing a small round nucleus. The outer surface of the eye is covered by a convex and transparent epidermal cap, whereas, all the other cells are surrounded by a thick pigment cup. At its deep basal end enters the optic nerve the branches of which run into each photoreceptor cell. All the eyes are not of equal size, the second and third pairs being larger than the remaining ones. As each eye can receive light only from one direction the eyes are differently directed. Some face forward (Ist pair), some forward and outward (2nd pair), some face upward (3rd pair), some backward and outward (4th pair), while the last pair faces entirely backward. It is not clearly understood whether the eyes from any image or enable the animal to locate the direction of the source of light.

**Reproductive System**

Each individual possesses both male and female generative organs (i.e., hermaphrodite), but self-impregnation is not possible (Fig. 12.65). In the male the **testis-sacs** are eleven (sometimes 12 or 13) pairs of rounded sac-like structures, a pair lying in each of the segments from 12th to 22nd, sometimes 23rd and 24th. Each pair lies in the middle of the segment, one on each side of the ventral nerve cord. The spermatogonia are produced by the wall of the sac and released in the coelomic fluid, contained in the sac itself, where they develop into spermatozoa. From the posterior border of each sac arises a short sinuous duct, the **vas**

**efferens**, runs outward and meets the vas deferens of its side. The **vasa deferentia** are a pair of slender longitudinal ducts lying on either side of the ventral nerve cord between segments 11th and 22nd, and sometimes 23rd and 24th. The vasa deferentia are enclosed within reduced coelomic spaces which are packed with amoeboid cells similar to those seen in the haemocoelomic fluid. The coelomic space can be seen only if good histological slides are examined. After emerging from the coelomic covering in the 11th segment the vasa deferentia run forward up to the posterior end of the 10th segment, and here they bend inwards, forming a closely compact convoluted mass, the **epididymis** or **sperm-vesicle**. From the inner and anterior end of each epididymis arises a short and narrow tube, the **ejaculatory duct** which runs into the **atrium**, a pyriform sac-like structure extending into the 9th and 10th segments. The atrium consists of two parts: (*a*) the anterior base-like **prostate** consisting of a large number of unicellular glands covering thick muscular wall; and (*b*) a backwardly directed narrow and elongated chamber the **penis-sac**, containing a filamentous coiled eversible tube, the penis, often seen protruding through the male genital aperture.

In the female the **ovaries** (Fig. 12.65) are not visible from outside because they are contained within a pair of coelomic sacs, the **ovisacs**, situated in the

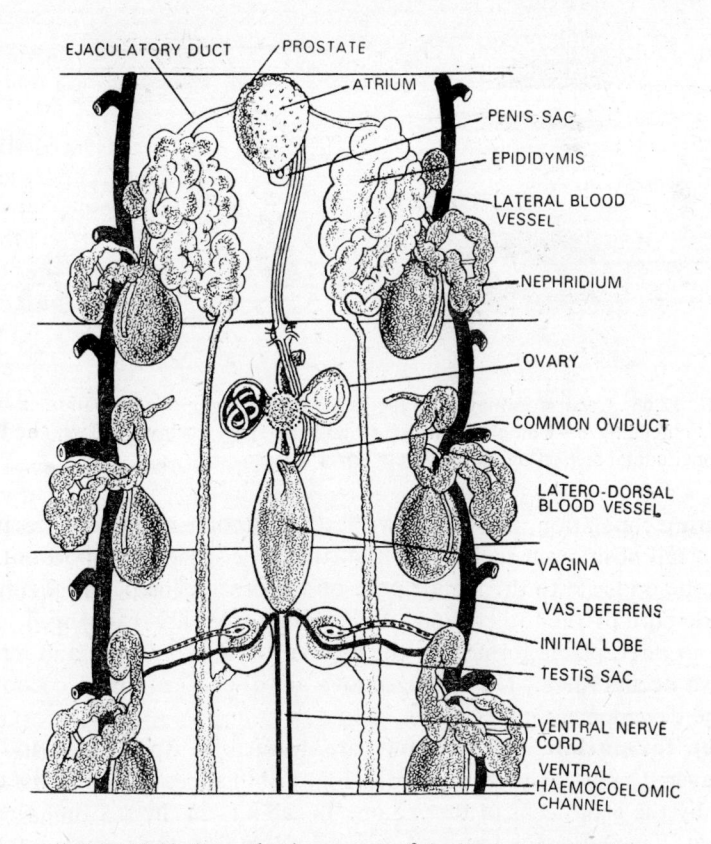

FIG. 12.65. Reproductive organs of the leech (*after* Bhatia).

11th segment. Each ovary is a delicate filamentous cord formed of nucleated protoplasm with club-shaped terminations. The filamentous structure is coiled and remains floating in the coelomic fluid within the ovisacs. From the base of each ovisac arises a short and slender tube, the **oviduct**. The right oviduct passes beneath the nerve cord and meets the left one in the 11th segment forming a common oviduct. Their place of union is covered by a thick layer of unicellular **albumen glands**. Each gland is filled with secretory materials towards its base and contains a small nucleus on one side. The ductules of the glands are narrow and open into the lumen of the common oviduct. The posterior part of the common oviduct, not covered by the albumen glands, becomes folded and opens into the **vagina**, large pear-shaped muscular bag, lying in the posterior part of the 11th segment. During breeding season it becomes larger and folds inwards. The vagina narrows to a duct that terminates at the female generative aperture.

**Fertilization.** The spermatozoa released within the testis-sacs are stored in the epididymis. Thence they pass into the posterior chamber, where they are glued together by the prostatic secretion to form packets of sperms, the **spermatophores**, that are transferred into the vagina of the female through the

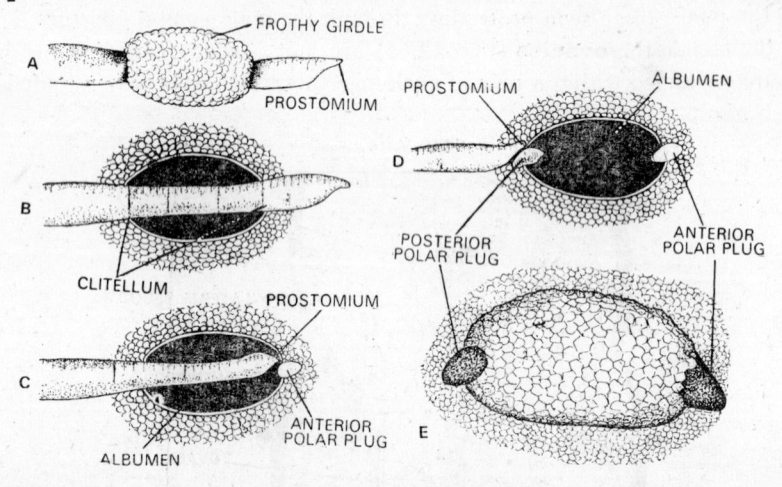

FIG. 12.66. Cocoon formation in the leech. A—Frothy girdle formation around the clitellum; B—Complete girdle; C and D—The leech withdraws; and E—Longitudinal section of the cocoon.

penis during copulation, for which two leeches become apposed to each other in a head to tail position in such a way that the male generative pore of the one comes to lie opposite to the female pore of the other. The penis of one enters the female pore of the other and the spermatozoa are exchanged. They are received in a receptacle formed by the vagina for the purpose and fertilization of the ova occurs there. The fertilized ova are discharged into a cocoon, within which the development takes place.

**Cocoon formation.** The cocoons are formed in April, May and June. A several layered snow-white frothy girdle (Fig. 12.66) appears around the clitellum, secreted by the gland cells of the region. In about an hour's time the girdle is completed. Thereafter the fertilized ova emerge out of the female pore and the leech withdraws its anterior end by rhythmic movements. As it withdraws the secretion of the prostomial glands forms the **plug**; first the **anterior plug** is formed and then the **posterior** one. The process of cocoon formation is quite slow extending over a period of about six hours. On exposure to air the cocoon hardens. It is yellow or amber-coloured, chitinous, barrel-shaped structure about 25-33 mm in length, and 12-15 mm in diameter having a distinct conical projection, the **polar plug**, at each side. It is filled with dark jelly-like albumen, hence appears swollen. The cocoon contains 1-24 embryos. For hatching the polar plugs drop off and young leeches emerge out. The whole process is completed in a fortnight.

The number of eggs in a cocoon varies, but is never large. The eggs of Gnathobdellidae are small and the young are hatched early and float as larvae in the albumen upon which they feed. In *Hirudinaria* the eggs are larger and hatch at a later stage. *Clepsine* presents a different case. Its cocoons are attached to stones, over these it broods till the young hatch. The young then attach themselves to the ventral side of the mother, and are carried about by her, living on albumen secreted by her.

## CLASSIFICATION OF HIRUDINEA

CLASS HIRUDINEA. Mostly ectoparasitic annelida with highly modified body. Fixed number of segments, broken up into annuli; without parapodia or setae.

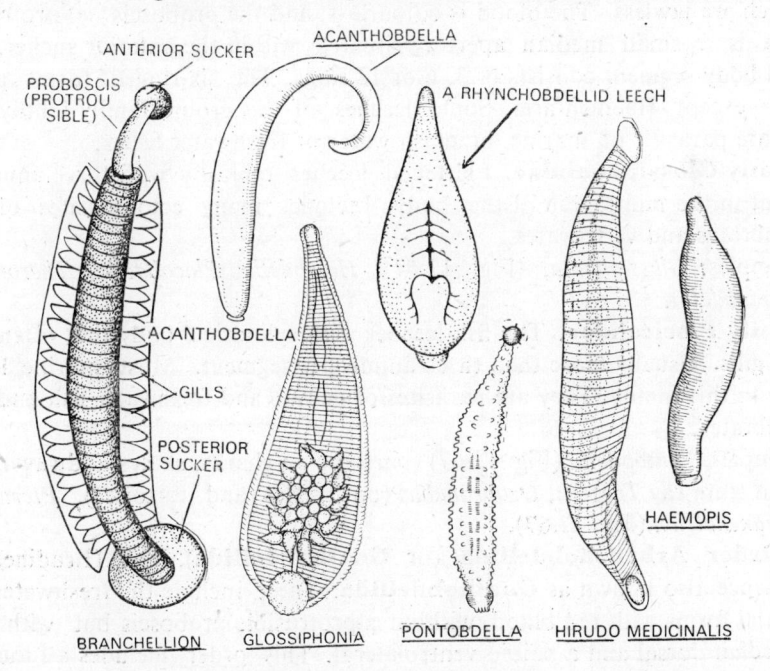

**FIG. 12.67.** Different types of leeches (from various sources).

Several segments fused to form suckers at the anterior and posterior ends of the body. Body cavity packed by the growth of the mesenchymatous tissue, coelom being reduced to longitudinal tubular spaces (sinuses). Hermaphrodite eggs are laid in cocoons. The Hirudinea are closely related to the Oligochaeta, so much so, that Michaelsen considers that the two may be grouped together under class Clitellata. Here the classification of the Hirudinea follows that of Harding and

Moore as given in the Fauna of British India. The class is divided into three orders:

1. **Order Acanthobdellida.** The Hirudinea in which the body is divided into thirty segments. The anterior sucker is lacking, the posterior sucker is composed of four segments and is well-developed. The members of this order are the only Hirudinea that possess setae. There are two double rows of setae in the first five segments. The setae are embedded in sacs provided with retractor muscles. The coelomic body cavity is a continuous perivisceral space, interrupted by segmental septa. Nephridia open on the surface between the segments. Vascular system consists of a dorsal and ventral vessel. Nervous system has twenty ventral ganglia of which the first and last are composite. The reproductive system is constituted by a pair of tubular ovaries and testes (really seminal vesicles) running through several segments. The testes are filled with developing spermatozoa and their epithelial wall is continuous with that of the perivisceral coelom. The specialized hirudinean characters are only partly developed in this order which, therefore, forms a link between the Oligochaeta and Hirudinea. It includes a single genus *Acanthobdella* which is parasitic on the fins of the Siberian salmon. No representative of this order has been reported from India.

2. **Order Rhynchobdellida.** Marine and freshwater Hirudinea the members of which are jawless. The blood is colourless, and the proboscis is protrusible. Mouth is a small median aperture situated within the anterior sucker. Each typical body segment consists of 3, 6 or 12 rings. The six-ringed forms are all marine except Haementaria. Some leeches of this group (family Ichthyobdellidae) are parasitic on marine, brackish water or freshwater fishes.

**Family Glossiphoniidae.** Flattened leeches typically with three annuli per segment in the mid-region of the body. Includes many ectoparasites of both invertebrates and vertebrates.

Examples: *Glossiphonia*, (Fig. 12.67), *Helobdella*, *Placobdella*, *Theromyzon*, *Marsupiobdella*.

**Family Piscicolidae.** The fish-leeches with cylindrical body that often bears lateral gills. Usually more than three annuli per segment. Most marine leeches belong to this family. They are parasites of marine and freshwater fish and some invertebrates.

Examples: *Pontobdella* (Fig. 12.67) (parasitic on sharks, skates and rays), *Piscicola* (on sting ray *Trygon*); *Ozobranchus* (on turtles and tortoises). *Pterobdella* and *Branchellion* (Fig. 12.67).

3. **Order Arhynchobdellida** (or **Gnathobdellida**). The Hirudinea with jaws, hence also known as **Gnathobdellida**. These include the freshwater and terrestrial forms with red blood, without a protrusible proboscis but with jaws, one median dorsal and a paired ventrolateral. This order includes all the large leeches which are fully adapted to a blood-sucking habit.

**Family Hirudinae.** It comprises chiefly aquatic blood sucking leeches.

Examples: *Hirudo* (sanguivorous and attacks man, buffalo and other domestic wild animals), *Hirudinaria*, *Macrobdella*, *Philobdella*, *Haemopis* (Fig. 12.67).

**Family Haemadipsidae.** Terrestrial tropical leeches attacking chiefly warm-blood vertebrates.

Examples: *Phytobdella*, *Haemadipsa*.

*Phylum Annelida*

**Family Erpobdellidae.**[1] Predaceous leeches feeding on small oligochaetes, planarians and insect larvae.

Examples: *Erpobdella, Herpobdella, Foraminobdella,* an amphibious burrowing leech.

## CLASS ARCHIANNELIDA

The class Archiannelida includes small marine annelids with simplified structure. This is a minor group of worms some features of which resemble those of larval Polychaeta, they may thus be either primitive or degenerate. They are elongated and thread-like (*Polygordius, Protodrilus*) and are entirely without setae or parapodia. The head has two tentacles and two ciliated pits (Fig. 12.68). Externally the segmentation is slightly marked by either faint grooves (*Polygordius*) or by ciliated rings (*Protodrilus*). Internally it is marked by coelomic septa. A peculiar feature is the absence of circular muscles from the body wall. The cerebral ganglia are in the preoral lobe and the ventral cords, devoid of ganglionic swellings, lie in the epidermis. The cords may be separate (*Protodrilus*) or

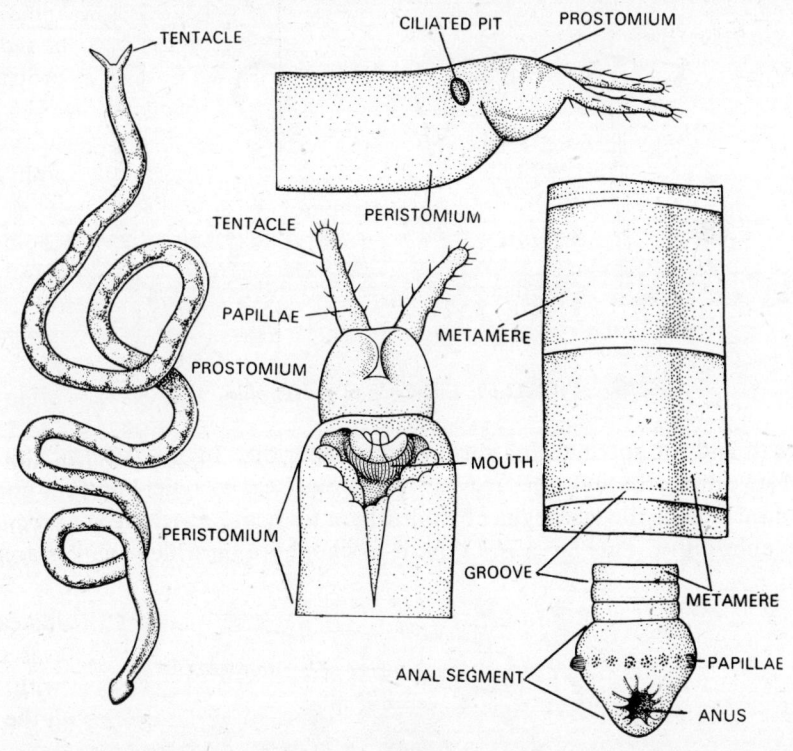

FIG. 12.68. *Polygordius,* external features. A—Entire worm; B—Side view of anterior end; C—Ventral view of anterior end; D—Segments of the trunk; and E—Anal segment.

fused (*Polygordius*). In some the buccal region is also eversible (*Polygordius*) and is followed by a small oesophagus. There is a simple nephridium in each segment

---

[1]Some authors place this family in a separate order Pharyngobdellida.

opening internally in the preceding segment. Supporting the gut in the body cavity there is a dorsal and ventral mesentery. The vascular system consists of a dorsal vessel in the dorsal mesentery. These are linked by connecting vessels. The blood is red, yellow or green or colourless. The reproductive organs develop from the coelomic epithelium and discharge their products in the body cavity. In *Polygordius* (Fig. 12.68) the ova probably escape by the rupture of the

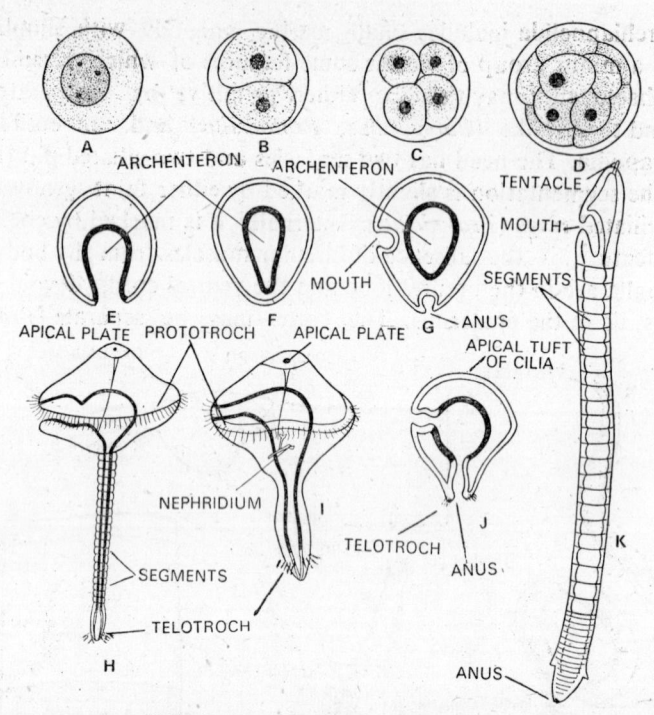

FIG. 12.69. Life-cycle of *Polygordius*.

body wall and the spermatozoa through the nephridia. In *Protodrilus* the eggs pass backwards through the meshes of the body cavity reticulum to a pore on the ventral side of the last segment from where they are discharged. *Protodrilus* is hermaphroditic. The larva is a typical trochophore on which somites are pro-

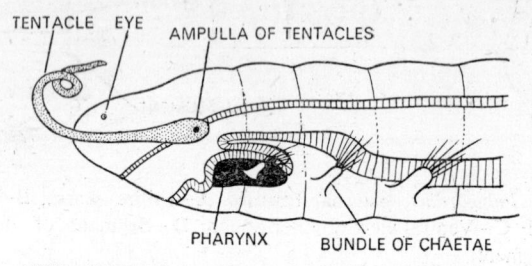

FIG. 12.70. *Saccocirrus*, anterior end, side view.

duced posteriorly during metamorphosis. The cilia of the larval form are retained by the adult. These are regarded as simplified annelids not primitive.

*Saccocirrus* (Fig. 12.70) is another genus of this class in which the segments are numerous, and each of them, except the first, carries two dorso-lateral bundles of simple setae. There are no parapodia, but the lower parts of the bundles of setae are enveloped in a cutaneous sheath, which can be protruded or retracted into the body. The hind end of the body is provided with two characteristically marked appendages on either side of the anus, by which the animal fixes to the stones on which it creeps. The circular muscles are less developed than the longitudinal. The sexes are separate and the gonads are found in all segments behind the 14th. There are paired patches of coelomic epithelium on the posterior faces of the septa which produce and discharge the germ cells in the body cavity where they mature. The genital apparatus is complicated, the females are provided with spermathecae and males with a pair of penes in each segment behind the oesophagus.

*Dinophilus*, another well-known form, is a small marine worm of about 2 mm length in which the body is made up of five segments, each encircled by two bands of cilia, that serve for locomotion. No setae are present.

## CLASS ECHIUROIDEA

The Echiuroidea are Annelida which show little or no traces of segmentation in the adult, the body cavity (coelomic cavity) is spacious, with a well-marked preoral lobe and a single pair of ventral setae. They were formerly united with Sipunculoidea in the class Gephyrea. They have now been separated as distinct classes of Annelida, though some workers prefer to give them independent phylum status with affinities to the Annelida. The echiuroids comprise about sixty species of peculiar worms of grey, reddish, or yellowish colour that inhabit the mud and sand between rocks under shallow coastal waters of all warm and temperate seas. The largest known species, *Urechis caupo* of California, grows to a length of 450 mm but others are smaller, down to 25 mm or less. They live in wide V-shaped burrows dug out by themselves. The body is more or less cylindrical which is anteriorly drawn out into a long, highly contractile pre-oral proboscis, with a ciliated groove on its ventral surface leading to the mouth. There are almost always two-hooked setae on the ventral surface close to the anterior end. In *Echiurus* there is an additional double row of setae, the **anal setae** round the hind end. Openings of the anterior nephridia are situated a short distance behind the hooked setae. These vary from four pairs to a single in number. The skin is covered with numerous small papillae arranged in rings. The body wall is highly muscular. The anus is posterior and terminal. The digestive tract is complete beginning at the mouth. The pharynx is muscular, intestine long and much coiled, rectum with two long vesicles and terminal anus. The circulatory system comprises dorsal and ventral vessels connected anteriorly, but lateral vessels are absent. Coelom is large, undivided but crossed by muscle strands that support the alimentary tract. Nephridia are one to three pairs anteriorly or one large, all serve as gonoducts. The nervous system consists of a ventral nerve cord lying entirely within the body wall. In front, this cord divides into two, which pass round the oesophagus and extend up to the anterior end of the proboscis, where they join one another. There are no ganglionic swellings. There are no special sense organs. The sexes are separate, dimorphic forms occur in some (*Bonellia*) and the gonad is single median and posterior. The larva is trochophore.

*Echiurus* (Fig. 12.71) is the most known form found almost everywhere. It lives in burrows and feeds by extending the mucous-covered proboscis above the mud and trapping plankton and debris. If disturbed the proboscis is cast off and then regenerated. *Urechis caupo* lives permanently in a tunnel dug by its setae and flushed with water expelled from the rectum. This worm has a peculiar feeding mechanism. Behind the setae there is a band of epidermal cells that secrete a mucous cylinder (with pores only $0.004\,\mu$ in size) attaching the fore part of the body to the burrow. The movements of the body will draw water through the burrow. As the water passes through the mucous net microorganisms are sieved and later the mucous cylinder with the entangled microorganisms is swallowed. Another genus, *Bonellia*, is known for sexual dimor-

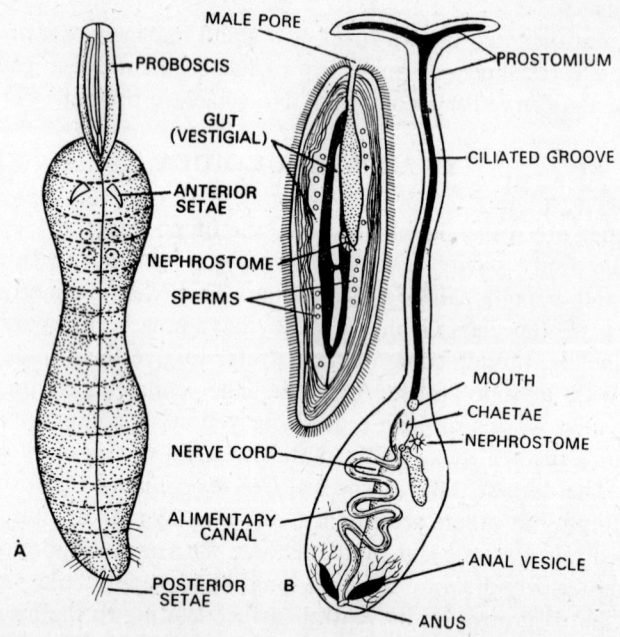

**FIG. 12.71.** A—*Echiurus*, entire; and B—*Bonellia* (*after* Spengler), male (small ciliated) and female with long proboscis.

phism. They live in deserted burrows on the sea bottom. The female has a small green ovoid body and a long proboscis, bifurcated at the end, capable of enormous elongation and is extremely mobile. A single segmental organ (brown tube) acts as uterus. The males are reduced to small ciliated organisms (Fig. 12.71), like a turbellarian, and is without proboscis, mouth or anus. Very early in life the male enters the female's gut and finally occupies the segmental organ as a "parasite." Larvae of *Bonellia* carry the potentialities of both sexes. If they develop independently they become females. If they come in contact with the body of the adult female they develop into males (probably through the action of some specific secretion).

## CLASS SIPUNCULOIDEA

The Sipunculoidea are Annelida with a spacious coelom, anterior dorsal anus

*Phylum Annelida*     479

and one pair of nephridia. They are elongated worms of yellowish or greyish colour living in sand or mud at the seashore or sheltering in empty shells about rocks. They also bore in coral rocks. The anterior part of the body is such that can be retracted within the body, such a part is called the **introvert**. The mouth is placed at the anterior end of the introvert and is in relation with a row of ciliated hollow tentacles which may be arranged in a circle. Preoral lobe is lacking. The anus is situated on the dorsal surface about one-third of the length of the body from the anterior end. The body is devoid of setae, but the introvert is often covered by rows of backwardly directed horny hooks or by small scale-like papillae which overlap one another. There is no trace of segmentation, but the skin is often thrown into ridges both transverse and longitudinal. The body wall is highly muscular and consists of the outermost roughened cuticle, epidermis, dermis with glands and sense organs, and three layers of muscles—a layer of circular muscles, a thin layer of oblique muscles, and a thick layer of longitudinal muscles—which are often arranged in bundles forming lattice-like pattern. On the interior there is a layer of flat coelomic epithelium, the cells of which bear isolated cilia. The alimentary canal is a thin-walled tube beginning at the mouth, extending to the posterior end and spiralled back on itself, the anus is dorsal near the base of the introvert. Coelom is spacious

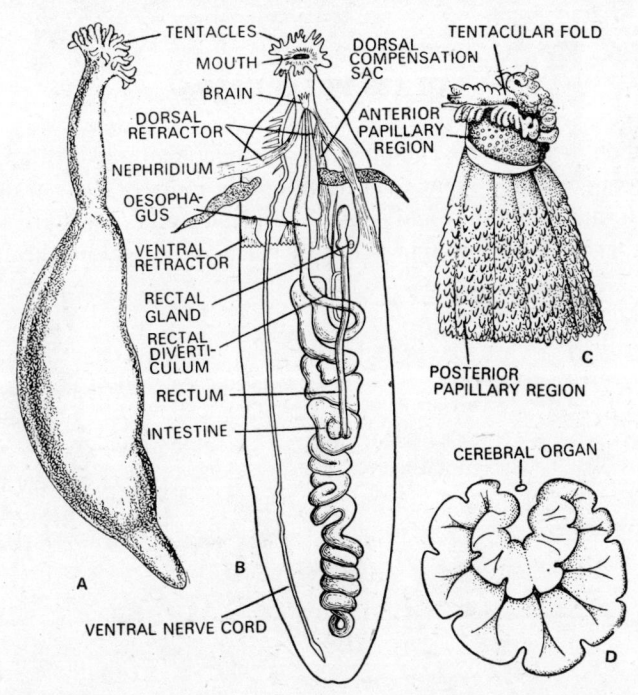

FIG. 12.72. A—*Sipunculus*, entire; B—Internal anatomy of the same; C—Anterior end, magnified; and D—Tentacular fold.

undivided but traversed by many strands of connective tissue and muscles attached to the gut. It is filled with the coelomic fluid in which float several types of corpuscles. The circulatory system consists of dorsal and ventral vessels with ring sinus below tentacles. The nephridia are 2 or 1, often called brown tubes, opening externally on the ventral surface in the anterior region near anus. The nervous system consists of a dorsal bilobed ganglion near the tentacles. This is connected by two connectives to a single ventral unsegmented nerve cord that runs through the body. From this arise many lateral nerves. In the mid-dorsal line, outside the tentacular circlet, there is an ectodermal pit which reaches as far as the supraoesophageal ganglion. It is called the cerebral organ and is lined by ciliated cells. The sexes are separate but alike. No permanent gonads are formed. The gonads are formed by simple proliferations of the coelomic epithelium, at the point of attachment of the retractor muscle to the body wall. The germ cells are released in the body cavity where they mature. Fertilization is external, the germ cells escape through the nephridia. The larva is usually similar to a trochophore without any trace of somites.

*Sipunculus* (Fig. 12.72) is a common genus of this class. Like all sipunculids it is also marine and capable of but little movement. The tentacles are extended, while feeding on the sea bottom and the beating cilia entrap micro-organisms in mucous which is swallowed along with mud or sand. When disturbed, the introvert is drawn entirely into the anterior part of the body. Other common genera include *Dendrostoma* and *Phascolosoma*. Another genus *Phascolion* is found at the depth of a thousand fathoms from the West Indies to the Arctic. It lives in a snail shell full of sand to form a tube which is carried by the worm wherever it moves.

## CLASS PRIAPULIDA

The Priapulida are marine cylindrical worm-like creatures with an introversible **presoma**, a superficially segmented trunk and a, more or less straight digestive tube with terminal mouth and anus. The urinogenital pores are separate from the anus and excretion is brought about by solenocytic protonephridia. They are

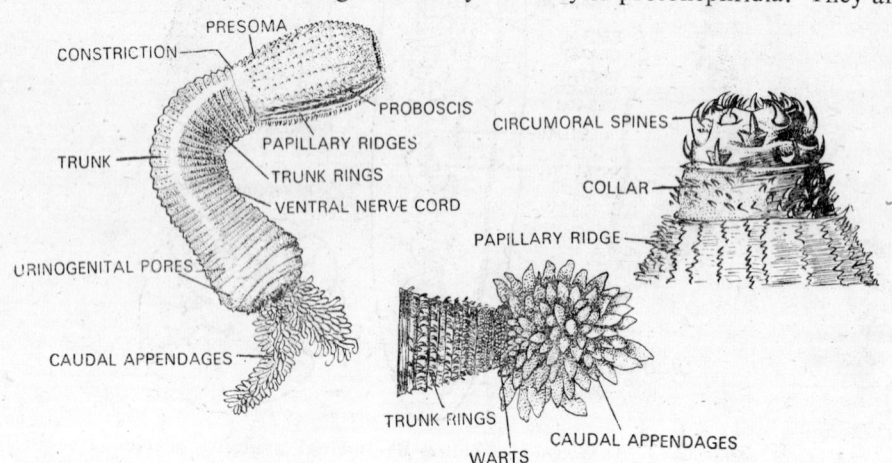

FIG. 12.73. A—*Priapulus*, entire; B—Anterior end with everted circumoral spiny areas; and C—Rear end of *P. caudatus*.

yellow or brown creatures up to eight centimetres in length. They live in crevices of rocks or between sessile animals or in empty mollusc shells, but can burrow and leave piles of castings on the bottom. The body (Fig. 12.73 A) is divisible into an anterior shorter region, the presoma, a longer trunk and one or two warty appendages at the posterior end (in genus *Priapulus* only). The presoma is somewhat plump and occupies about one-third of the body length (*Priapulus*) it may be less in others (*Halicryptus*). The mouth is terminal and is surrounded by a number of strong spines (Fig. 12.73B) numbering 5-7 in *Priapulus*. The spiny circum-oral region is ordinarily invaginated into the interior in which condition the spines point backward, but the region is everted when needed. In *Priapulus* a prominent circular band or collar is found posterior to the spine-bearing area. The main part of the body is the proboscis, a somewhat bulbous region lying posterior to the collar. It is invaginable but usually remains exposed. The trunk has 30-40 annuli in *Priapulus* and is covered with small spines and papillae. The posterior end of the trunk has three openings, the anus and two urogenital pores and is provided with one caudal appendage in *Pr. caudatus* (Fig. 12.73 C) and two in *Pr. bicaudatus* (Fig. 12.73). The caudal appendages are hollow stems thickly beset with hollow oval vesicles and are of unknown function. Some people called them respiratory structure but the same has been disproved (Lang, 1948).

**Body Wall**

The body wall consists of outer cuticle followed by epidermis below which lies the musculature comprising circular and longitudinal muscle-layers and finally there is the lining membrane. The papillae, spines and warts of the body surface are either sensory or glandular structures. The papillae are conical structures the interior of which is packed with long, slender epidermal cells.

**Body cavity.** Between the body wall and the digestive tract exists a cavity the true nature of which has not yet been found out. It is lined by structureless membranes without nuclei. The space is filled with a fluid in which float numerous round cells.

**Alimentary Canal**

The mouth opens into a muscular pharynx, lined by epidermis and a cuticle covered with spines, special retractor muscles connect the anterior end of the pharynx with the body wall and bring about the invagination of the proboscis. The pharynx opens into the midgut or intestine into which it invaginates. The midgut is a straight tube and is constricted from the short end-gut or rectum which opens at the anus. They feed upon marine algae and bottom detritus. When hungry they plough the mud on all directions, apparently seeking food, otherwise they lie quiescent in a vertical position with a widely opened mouth on a level with the surface (Lang).

**Nervous System**

The nervous system consists of a circumenteric ring surrounding anterior end of the pharynx and a mid-ventral cord lying in the epidermis. The whole system is strikingly similar to that of the kinorhynchs.

**Reproductive System**

The sexes are separate. A pair of urinogenital organs lies in the body cavity, one on either side of the intestine. Each organ consists of a urogenital duct bearing the gonad on the side next to the mesentery and clusters of solenocytes on the other side. Each gonad is fairly compact made up of a network of numerous tubules. The excretory part of the urogenital system consists of several large clusters of solenocytes opening into the common duct by narrow collecting canals.

Nothing is known about the embryology and development of priapulids.

**Affinities**

Ever since its discovery by Linnaeus *Priapulus* has presented difficulties to establish relationship of the worm with others as their characteristics indicate no immediate relationships to any living phyla. The concept of the phylum Gephyrea including the priapulids, sipunculoids, and echiuroids gave a relief and the same is maintained till now. Hyman (1951) has suggested the inclusion of the group with Aschelminthes, but as nothing is known about the way of the formation of its body cavity the relationship cannot be held as valid, particularly because the body cavity in priapulids is lined by structureless membrane of doubtful nature. It seems, therefore, better to keep them under the annelids with which they have been associated in the past. Those who do not approve this relationship can very well erect a separate phylum for priapulids with affinities with sipunculids and echiuroids.

## GENERAL ORGANIZATION

**Metamerism.** The Annelida include the metamerically segmented worms. Some of them have cylindrical body (round in cross-sections), while the majority are dorsoventrally flattened. The perivisceral body cavity is coelomic. They resemble the Arthropoda in many respects, but differ from them in having pervisceral divisions of the coelom. They also resemble the Mollusca in the arrangement of the central nervous system and in the widespread occurrence of the trochophore larva, but they differ from the Mollusca in the fact that the body is segmented. The segmentation of the Annelida is exhibited not only by the body but by a number of organs of the body. This segmentation develops from the segmentation of the mesoblastic somites of the embryo. Ideally constructed homonomous segmentation does not occur, since one or more segments are fused to form head or anal segments and these segments differ from the typical body segments. In other words the segmentation is heteronomous. The body consists of the prostomium, the trunk (soma) and the terminal segment, the pygidium. The prostomium comprises a preoral lobe, the prestomium and a postoral portion, the peristomium. Sometimes the prestomium is called the first segment, but in the enumeration of segments the peristomium is counted as the first. Sometimes the prestomium is quite small and inconspicuous. In many it is very large and elongated into a proboscis like organ (*Nais lacustris*, the Echiuroidea). In some cases it bears special sensory organs, tentacles and palps. In the Polychaetes sometimes certain number of body segments are fused to form a secondary composite head.

The significance of metamerism has been discussed in the beginning of this chapter. The problem of the origin of metamerism is quite obscure. It may be relevant that the lower groups of animals show a marked tendency for repeating certain organ systems. Some platyhelminths, clearly unsegmented, have multiple gonads alternating with pouches of the alimentary canal. This is also found in nemertines. In the echinoderm-chordate line gill-slits, gonads and alimentary pouches are repeated in the hemichordates. The advantage of such repetition probably lies in the resultant increase in surface area of the organ concerned. It is of particular significance in acoelomate animals in which the absence of a coelom and often of a vascular system limits the efficiency of transport of metabolites. It does not throw any light in the origin of metamerism. The role of metamerism in locomotion has been discussed in sections dealing with locomotion in polychaetes, oligochaetes and hirudinians. In annelids (as also in arthropods) locomotion depends upon segmentation of the musculature, coelom and nervous system. The subdivision of musculature and coelom in annelids is not only important but is intimately linked with locomotion. Therefore, locomotion seems to have influenced the origin of metamerism in these animals.

**Body Wall**

The Annelida possess a dermomuscular body wall that is to say, muscular tissue enters largely into the composition of the integument. The result is that the body is extremely contractile. They all possess chitinous spines, the setae, which are secreted by the ectoderm and are embedded in pits of skin. The setae are very conspicuous in the class Chaetopoda, less so in the classes Hirudinea, Echiuroidea and Archiannelida. The epidermis of the body wall is a single layer of cells including gland cells. Externally the cuticle covers it except the ciliated and glandular parts.

**Nervous System**

The central nervous system in annelids is much more compact than in the Platyhelminthes. It consists of two ventral nerve cords. These two nerve tracts are generally closely approximated mainly in the mid-ventral line over the greater part of their course, but in some, (e.g., Polychaeta-Sedentaria) they are widely separated. In the Echiuroidea they are fused ventrally to form a single ventral cord. But in all forms, even in those in which there is a complete union between them posteriorly, they separate from one another in front, and embracing the anterior part of the alimentary canal, become continuous with one another on the dorsal side at the front end of the body. Where the two nerve tracts meet dorsally a single or bilobed swelling called the cerebral ganglion (or ganglia) or brain is formed. The nerves that enter the brain are the sensory nerves of the anterior end of the body in which important sense organs are located. The ventral nerve cord is almost always swollen at segmental intervals into ganglia, which generally give off nerves to adjacent organs. The first swelling on the ventral cord is called the suboesophageal ganglion and the subsequent swellings the ventral ganglia. The ganglionated condition is due to the fact that there are more nerve cells and more nerve fibres in the ganglia, not due to the exclusive presence of nerve cells at any one point. As a matter of fact nerve cells are found along the whole length of the central organ. The central nervous system is generally separated from the ectoderm and placed within the muscular layer. In the Archiannelida and in one or two Chaetopoda it lies in the ectoderm. In Amphinomidae two additional ganglionated nerve cords are present, one on either side of the ventral nerve cord with which the additional nerve cords are connected by loop-like commissures. The ganglia of the additional nerve cords send nerves to the parapodia and are termed podial ganglia. The nervous system with four nerve cords is called tetraneural nervous system while the one with double nerve cord is the dineural nervous system. As most larvae possess four pairs of longitudinal nerves the tetraneural condition is believed to be primitive. It gives rise to the dineural type by the disappearance of one pair.

The brain is made up of two ganglia, a principal ganglion and a somatic ganglion. The principal ganglion develops from the apical plate of the trochophore larva while the somatic ganglion becomes associated with it as a secondary step. In some worms, on the other hand, only a single ganglion (principal ganglion) is present. The nerves from the principal ganglion supply the sensory organs such as the eyes, tentacles, palps and nuchal organ, while the somatic ganglion gives rise to the sympathetic or stomatogastric nervous system. In some forms (*Arenicola*) a stomatogastric or visceral system is given off from the somatic part of the cerebral ganglion or from the circumpharygeal commissures or from both. In some Polychaeta—errantia—the brain is further differentiated into corpora pedunculata and palp glomeruli, an enlargement that gives nerves to the palps. This condition is often used to establish the origin of Arthropoda from Polychaete-like ancestors, for in Onychophora and Arthropoda also similarly differentiated brain exists.

In the nereid worms bipolar sensory cells are well-developed and are numerous on the parapodia (especially on the cirri) and on the ventral body wall. Information from these receptors is conveyed into the central nervous system by afferent fibres running in the segmental nerves, four pairs of which arise from each of the segmental ganglia. Of these nerves first and fourth carry fibres from receptors on most of the body surface, second from the parapodia and third from the mechanoreceptors (Proprioceptors) of the dorsal and ventral longitudinal muscles.

There is a remarkable disparity between the number of receptor cells in each segment and the number of fibres in each segmental nerve. The number of afferent fibres in each of the four segmental nerves is only 3-4, 6-8, 2 and 6 respectively, yet there are probably not less than 1000 sensory cells per square millimetre of the body surface alone. This disparity can be explained by the presence of a nerve plexus which lies close below the basement membrane of the epidermis and which comprises the nerve of multipolar association cells. The motor fibres of the segmental nerves are even fewer than the sensory fibres, the four nerves of each segment having respectively 1, 3, 1-2 and 4. These fibres which arise from the cell body in the segmental ganglia, are more numerous towards the peripheral end of the nerves. Because of this arrangement two conclusions have been suggested: the great preponderance of peripheral fibres in both the sensory and motor pathways may provide for short circuiting of the passage of the nerve impulses, so that these do not have to traverse the central nervous system. Thus responses to stimulation can be mediated by local reflexes at the periphery without involving the central nervous system at all. Secondly, the limited connections between the central nervous system and the periphery through the segmental nervous system will permit some central integration. Since so few fibres pass to the central nervous system it seems likely that it can receive broad patterns of information rather than precise details.

The mode of action of the metameric nervous system of annelids is conveniently illustrated in locomotion. This involves the interaction of a muscular body wall with a hydrostatic skeleton. It is apparent that the segments can act independently, yet their individual activities are integrated into the unified behaviour of the whole individual. This is seen when a polychaete worm begins to move. The first movement forwards is effected by a stepping action of parapodia situated in about 4-segmented intervals from each other. This action begins at the front end of the body, and spreads rapidly backwards. It is then linked by the movement of the parapodia of the intervening segments, so that the whole of the body is involved in the peristaltic cycle.

The eyes, tentacles and cirri, nuchal organs (or ciliated pits) and statocysts are the various types of the sense organs of the annelids. The eyes are absent in the Oligochaeta and in some tube-dwelling Polychaeta, as well as in some freeliving forms of that order. Usually the eyes are confined to the prostomium, but in others they extend to many of the segments of the body. *Leptochone*, for instance, has a pair of eyes on each segment, in *Fabricia* there is a pair on the anal segment, in many species of *Sabella* and all the species of *Dasychone* there are eyes on the branchial filaments. In the leeches the number is subject to considerable variations. There are five pairs in *Hirudo* and *Hirudinaria*. In some the eyes may be present on the posterior sucker, or may be absent altogether. The structure of

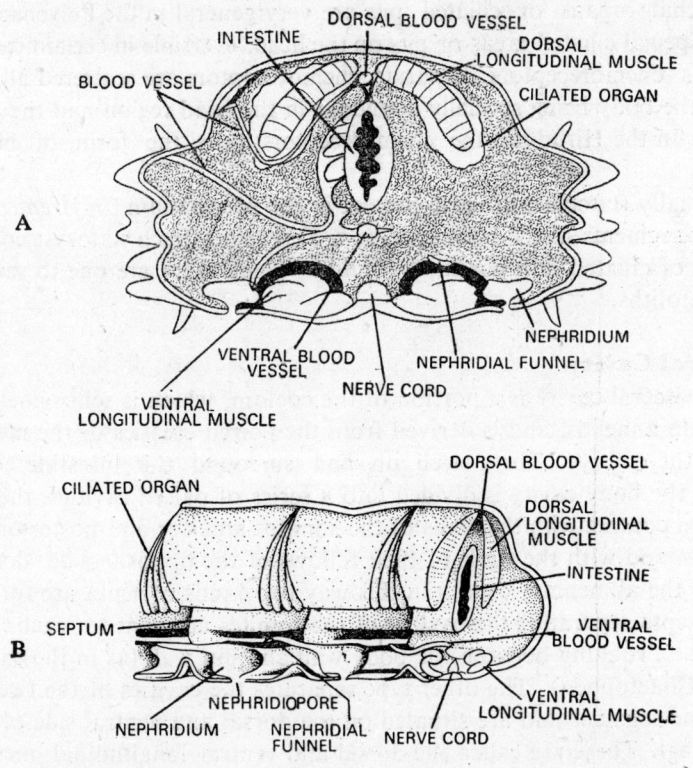

FIG. 12.75. A—Transverse section of *Nereis*; and B—Stereogram with the ectoderm of one side removed to show segmental structures of annelids (*after* Goodrich).

the eye, as a rule, is very simple, but in some forms the structure reaches quite a high grade of development. The eyes on the branchial filaments of many tube-forming Polychaeta consist each of a group of retinal cells having its own lens-like body and quite independent of others (not unlike a compound eye). In others the eye is a spherical capsule (*Nereis*) with a wall composed of a single layer of cells, which are elongated on the inner side that is the side turned towards the brain, while on the outer side they are flattened. The outer part of the wall is thin and transparent (cornea). Sometimes it unites with the epidermis. When the two layers remain as distinct transparent layers, they form the outer and inner cornea. In many cases a thickening of the surface cuticle over the cornea forms a cuticular lens. The cells of inner portion of the wall of the capsule form the elements of the retina. Some pelagic polychaetes have eyes with lens and have power of accommodation. The structure of the eye of a leech is peculiar (see page 469). Each eye is cylindrical, its outer layer is formed of black pigment tissue surrounding a layer of large clear refractive cells, which occupy the greater part of the organ. A nerve enters at one side, and is continued upto the axis of the cylinder by a row of sensory cells.

The tentacles, palps and cirri are regarded as tactile organs. They also carry taste organs which also occur on the general surface of the head and in the pharynx. The taste organs exist in the form of sensory bristle or may lie in hollow cuticular bristles (Arthropoda).

The nuchal organs or ciliated pits are very general in the Polychaeta. They consist of special ciliated areas or pits on the head, eversible in certain cases. They function as chemoreceptors. Separate chemoreceptors are scattered all over the surface of the body being specially abundant in the head region and the posterior most zone. In the Hirudinea the chemoreceptors take the form of beaker or buds.

Exceptionally statocysts are also present in the prestomium (of *Arenicola*) or in some other segments of the body (*Fabricia, Terebella*). Each statocyst consists of a capsule of ciliated cells containing fluid in which there are one to several calcareous statoliths.

## Perivisceral Cavity

The perivisceral cavity is a portion of the coelom, which is schizocoelous (see page 399) in annelids, and is derived from the paired cavities of the mesoblastic somites of the embryo. These swell up and surround the intestine. Thus, to begin with, the body cavity is divided into a series of paired cavities, the walls of which are in contact with the walls of the somites anterior and posterior to them in the series, and with the walls of their fellows of the opposite side dorsal and ventral to the alimentary canal. In this way two kinds of septa are formed, the transverse septa separating the cavities of the somites adjacent to each other in the series, i.e., running between the body wall and gut wall (as in the earthworm and other Chaetopoda). The other type separates the cavities of the two somites of the same segment and are situated on the dorsal and ventral side of the alimentary canal. These are called the dorsal and ventral longitudinal mesenteries. They also run between the body wall and the gut in a longitudinal direction. These two types of coelomic partitions coexist in the Archiannelida and possibly in one or two Chaetopoda, but as a general rule the dorsal and ventral mesenteries break down in the adult, so that the two sides of the body cavity become continuous with each other on the dorsal and ventral sides of the alimentary canal. The transverse septa occur in the adult Oligochaeta and some Polychaeta, but they also break down partially or wholly so that the perivisceral cavity becomes a continuous space from end to end of the animal (Echiuroidea, some Polychaeta). The coelomic wall in contact with the gut is the visceral layer and that in contact with the body wall is the parietal layer. The coelomic epithelium is usually non-ciliated but is ciliated in *Aphrodite*, *Glycera* and some others. The movement of the cilia brings about an active circulation of the coelomic or perivisceral fluid in the coelom. The characteristic botryoidal tissue in Hirudinea is developed in the peritoneal cells, and assumes such greater proportion that it obliterates the coelomic cavity, which is reduced to the spaces of the testicular and ovarian sacs and to the lacunae and canals which become part of the blood-vascular system. *Acanthobdella* is the only leech with a spacious body cavity divided by septa as in the oligochaete worms.

The coelomic fluid, which fills the coelom, in any annelid is usually under a slight positive pressure. This amounts in a quiescent lugworm (*Arenicola*) to a positive pressure of 14 cm of water; in a totally anesthetized lugworm to about three centimetres of water, while in an actively burrowing lugworm to about 30 cm of water.

Positive pressures of more than 90 cm of water can occur in some annelids, and have been recorded in leeches which are standing rigidly upright on their posterior suckers. In this rather special case it is probable that most of the muscles of the body wall were simultaneously in a state of slight contraction. Normally, however, contraction of one set of muscle in an annelid means stretching of another.

The coelomic fluid provides the structural basis for a hydrostatic skeleton, more highly organised than that of coelenterates and platyhelminths. The movements of annelid worms are based upon hydrostatic principle in essentially the same way as are the movements of the coelenterates but they (annelids) show greatly increased flexibility and speed of response.

**Alimentary Canal**

The alimentary canal is a nearly straight tube running from mouth to the anus. The mouth leads into the buccal cavity which is followed by a muscular pharynx. Both the buccal cavity and pharynx are of stomodaeal nature being formed in the embryo by an invagination of the ectoderm. The muscular pharynx is absent in some of the Sedentaria. In others it is frequently protrusible to a great or less extent and is furnished with papillae, jaws and teeth of chitinous material (Fig. 12.55). Feeding devices of some annelids are illustrated in Fig. 12.78. With the eversion of the pharynx the jaws come into play. A gizzard with thick walls may follow upon this protrusible pharynx, and is sometimes preceded by an oesophagus which may be dilated behind into a crop. The intestine is nearly always more or less deeply constricted intersegmentally and in *Aphrodite* it gives rise to a pair of branched caeca in most of the segments with the exception of one or two of the most anterior and one or two of the most posterior segments. In *Hesionidae* a pair of caeca which opens into the anterior part of the intestine frequently contains gas, and probably has a hydrostatic function. In some of the terrestrial Oligochaeta (earthworms) a fold of the intestinal wall, the typhlosole, projects into the lumen. In most annelids the intestine is straight, it is somewhat coiled in *Sternaspis*, and others. The posterior part of the alimentary canal is proctodaeal, it is formed by ectodermal inpushing. The anus is usually terminal in position sometimes slightly dorsally placed. In response to the blood sucking habit of the leech its alimentary canal becomes highly modified. The mouth is surrounded by three jaws, one median dorsal and two ventrolateral. It produces the characteristic triradiate bite in the skin of the animal upon which it preys. The pharynx is muscular and is surrounded by unicellular salivary glands. The crop is the most conspicuous organ of the alimentary system of the leech. It is a huge thin-walled tube extending from 8th to the 18th segment and produced into eleven pairs of lateral pouches (Fig. 12.55). The number of lateral pouches

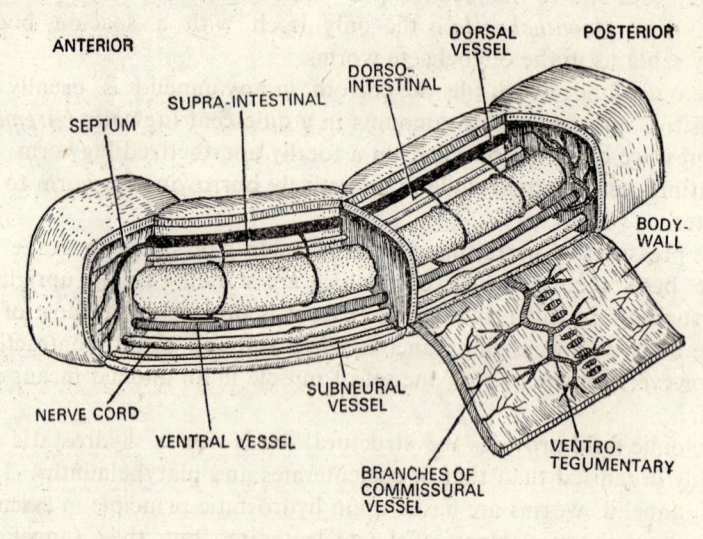

FIG. 12.76. A model of two segments of an earthworm from the intestinal region showing organs of each segment. Position of a nephridium of a primitive worm is shown with respect to the septum in the middle. For the position of nephridia in *Pheretima* see Fig. 12.40.

to the crop varies in different leeches being just one pair in the horse leech while absent in *Herpobdella*. In the Rhynchobdelida there is a distinct slender gullet between the pharynx and the crop. When the proboscis is retracted it is thrown into coils. The crop opens into the stomach, the digestive portion of the canal, which leads into a narrow intestine. The intestine opens into a dilated rectum that terminates at the anus.

**Filter feeding.** In many polychaetes elaborate filter-feeding has developed. Like many creeping and burrowing animals many polychaetes produce over their body surface a mucous secretion which protects the surfaces and forms temporary linings to burrows. This also helps in filter feeding as found in *Nereis diversicolor* (Fig. 12.9). Terebellid worms (Fig. 12.77) which live in permanent tubes in mud are deposit feeders. They obtain detritus by extending long ciliated tentacles from their head over to the surface of the substratum. The food particles are trapped in mucous and swept along ciliated groove (on the tentacle) into the mouth. The lugworm (*Arenicola*) obtains detritus by swallowing the mud in which it is contained. *Sabella*, on the other hand, is a real filter feeder. It extracts its food suspended in water (suspension feeder) from water current which are created by coordinated cilia. These are set upon the branchial crown (Fig. 12.77 D).

*Pomatoceros*, that is common on rocky shores and lives in calcareous tubes, is an excellent filter feeder. So is *Chaetopterus* which lives in a U-Shaped tube of parchment-like material. In this animal there is no branchial crown. Instead water is drawn through the tube by the beating of three pairs of fans that are presumably derived from the parapodia of related forms.

In most of the annelids digestion is extracellular. The gut is differentiated in many parts and enzymes are poured on the food to complete digestion. In *Arenicola marina*, on the other hand, digestion is completed in wandering amoebocytes that take up from the alimentary epithelium particles that its cells have ingested.

## Vascular System

A vascular system is nearly always present. Some Chaetopoda (family Glyceridae, Errantia) are entirely devoid of blood vessels, but in such cases the perivisceral fluid assumes some of the functions of the blood, and contains numerous red corpuscles with haemoglobin. In the majority of the Chaetopoda there is a highly developed vascular system. Very commonly the blood is bright red in colour with haemoglobin dissolved in the plasma. In some cases the blood is colourless. In *Serpula* and its allies the blood is bright green, owing to the presence of chlorocruorin, a green respiratory pigment, which functions in the same way as haemoglobin. Chemically it is closely related to haemoglobin and occurs in four families of Polychaeta (the Ampharetidae, Chlorhaemidae, Sabellidae and Serpulidae).

The principal blood vessels are the dorsal and ventral longitudinal vessels which are connected together by metamerically arranged transverse branches. In some annelids (Sedentaria) the dorsal vessel is not present in the greater part of the length of the body, its space is taken by a peri-intestinal sinus or a plexus of vessels lying in the wall of the alimentary canal. As a result of this arrangement a short thick-walled dorsal vessel is left. The circulation of blood is effected by the peristaltic contractions of the dorsal vessel or peri-intestinal sinus or plexus. Sometimes the peri-intestinal gives rise to a "heart" which helps in maintaining the circulation. In the earthworm and some others (the Sedantaria) specially dilated lateral vessels exist. These are pulsatile and their contractions bring about circulation.

The blood system of the Hirudinea is complex. In *Hirudinaria*, for instance, there are four haemocoelomic longitudinal canals, two lateral, one dorsal and one ventral, an elaborate system of branches of these canals and a large number of other spaces, some of which are coelomic, while the nature of others is doubtful. All the canals are thin-walled, but the walls of the lateral canals have become secondarily muscular. All the four canals communicate with one another. Through these canals circulates the coelomic fluid in which is dissolved the respiratory pigment, haemoglobin, and which contains numerous colourless amoeboid corpuscles floating in it. Because of the presence of the respiratory pigment the fluid is called the haemocoelomic fluid and the canals are called haemocoelomic canals. The haemocoelomic canals of the leech do not correspond developmentally to the true vessels of the Chaetopoda, which are formed independently of the coelom.

## Excretory System

In the Annelida the organs of excretion are a series of segmentally arranged tubes, the nephridia. As a rule a pair of nephridia occurs in each body segment, only a few anterior and a few posterior segments are without them. In its simplest form the nephridium is a primary ectodermal internally ciliated curved tube opening on the exterior by a laterally placed pore at the one extremity and at the other ending in a ciliated funnel or nephrostome, which opens into the coelom either of the same segment as that on which the external aperture is situated (most Polychaeta), or of the segment in front (all or most Oligochaeta, some Polychaeta). The nephridia, as a matter of fact, establish connection between the coelom and the exterior. They pick up waste products from the coelomic fluid and throw them out. In many cases the cells lining the tube can separate out waste matters and are loaded with granules and concretions.

In many Polychaeta the ciliated coelomic funnel is absent and the nephridial tube ends blindly. The inner end of the tube bears a peculiarly modified cell or solenocyte, which may be single or in a group. Each solenocyte is a rounded cell lying in the coelom and connected with the nephridium by a long slender tubular

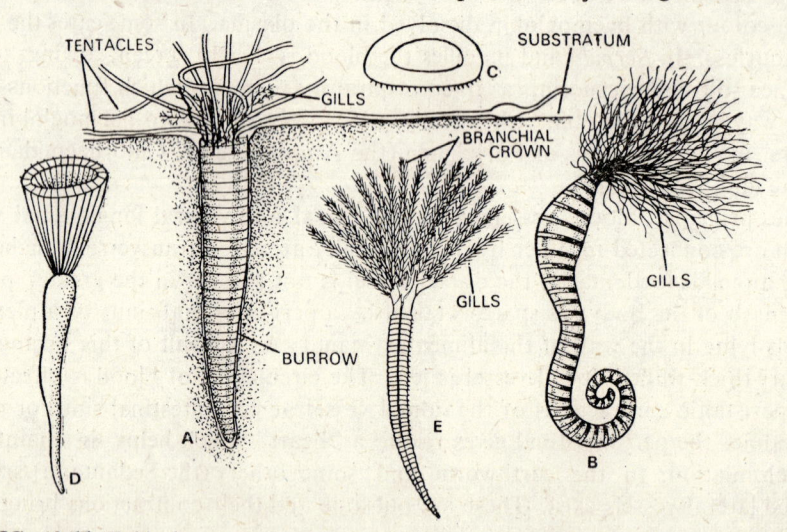

FIG. 12.77. Tubicolous polychaetes showing gills. A—*Amphitrite* in its burrow with tentacles outstretched on substratum (*after* Dales); B—*Terebella*; C—Cross section through tentacle of *Terebella lapidaria* creeping over substratum; D—*Sabella*; and E—*Myxicola*.

process. A single extremely long vibratile flagellum is attached to the rounded end of the cell and it may extend for some distance in the interior of the nephridium. The solenocytes resemble the flame-cells of the Platyhelminthes. The nephridium with solenocytes is called a protonephridium. In contradistinction to the protonephridium the nephridium with a ciliated coelomic funnel is called metanephridium.

The fusion of the nephridium and coelomoducts takes place in a number of ways. In the first place the coelomoduct becomes united with the nephridia (Fig. 12.79) as in Phyllodocidae and Goniadidae (protonephromixium). In Nephthyidae the coelomoduct becomes reduced to a ciliated organ. In *Dasybranchus*, on the other hand, there occurs a combination of nephridia with nephrostomes and separate coelomoducts (Fig. 12.80). Most usually, however, the segmental organs are formed by the union of nephridia with nephrostomes and coelomoducts. Finally there occurs the condition in which there are nephridia with nephrostomes and the coelomoducts are reduced to ciliated organs (as in *Nereis*, etc.)

A series of pairs of ciliated funnels, the coelomoducts, also occur in these animals as a set of segmentally repeated structures. The coelomoducts are genital ducts derived from the coelomic epithelium. In *Nereis*, the coelomoducts do not open externally and are represented by the dorsal ciliated organ. In many Polychaeta they do not remain independent, but fuse partially or completely with the

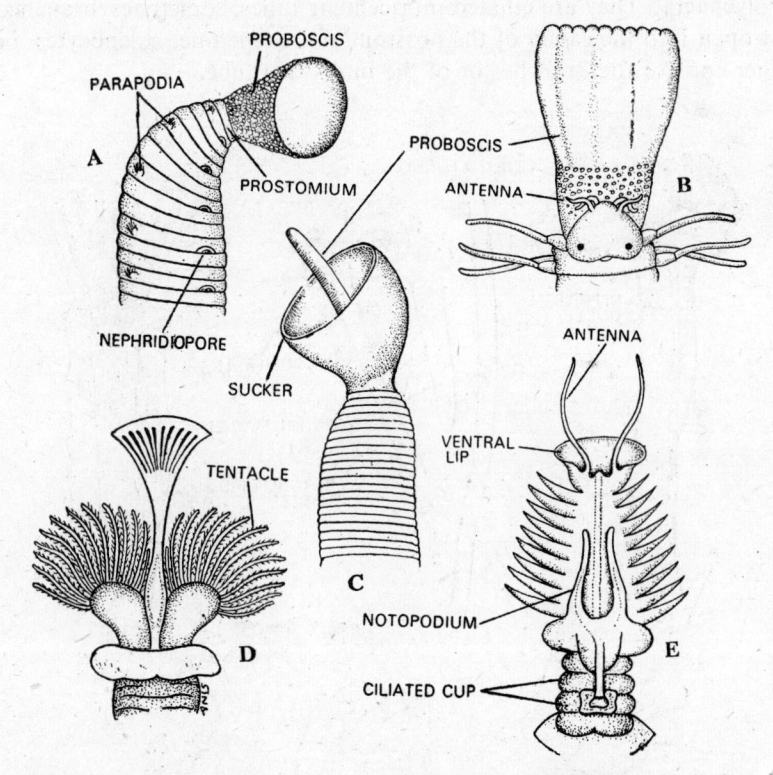

FIG. 12.78. Feeding devices in some annelids. A—*Arenicola*; B—*Phyllodoce*; C—*Piscicola*; D—*Serpulla*; and E—*Chaetopterus* (from various sources).

nephridia and the functions of excretory organs and reproductive ducts become combined in one set of the segmental organs (described as nephromixium by some authors). In *Serpula* and its allies there is a single pair of large nephridia in the anterior region of the body, with smaller pairs in the posterior segments. The large nephridia have an excretory function, while the smaller ones act as genital ducts.

In tubedwelling worms there is a division of labour so far as the nephridia are concerned. The anterior ones are specialized for excretion, and the posterior ones for the escape of ova and sperm. In Serpulids, for instance, there are three pairs of long nephridia on the anterior side of the body and posteriorly there are small nephridia with which the coelomoducts have also become fused (nephromixia) for the discharge of sex cells.

In the Oligochaeta the nephridia are usually simple elongated and coiled tubes, one or more pairs in each segment. In some the nephridia of the most anterior segments open into the mouth or pharynx (*Pheretima*) and all the nephridia of the posterior region of the body communicate with a pair of longitudinal canals which opens posteriorly into a median vesicle communicating with the rectum (*Allolobophora antipae*) or communicate with intestine in each segment (*Pheretima*).

In the larval chaetopod provisional or embryonic nephridia are present. These have been found to occur in the head (prostomium) of many larval Oligochaeta and Polychaeta. They are ciliated intracellular tubes, sometimes branched, and do not open into the cavity of the prostomium. Sometimes solenocytes occur at the inner ends of the branches or of the undivided tube.

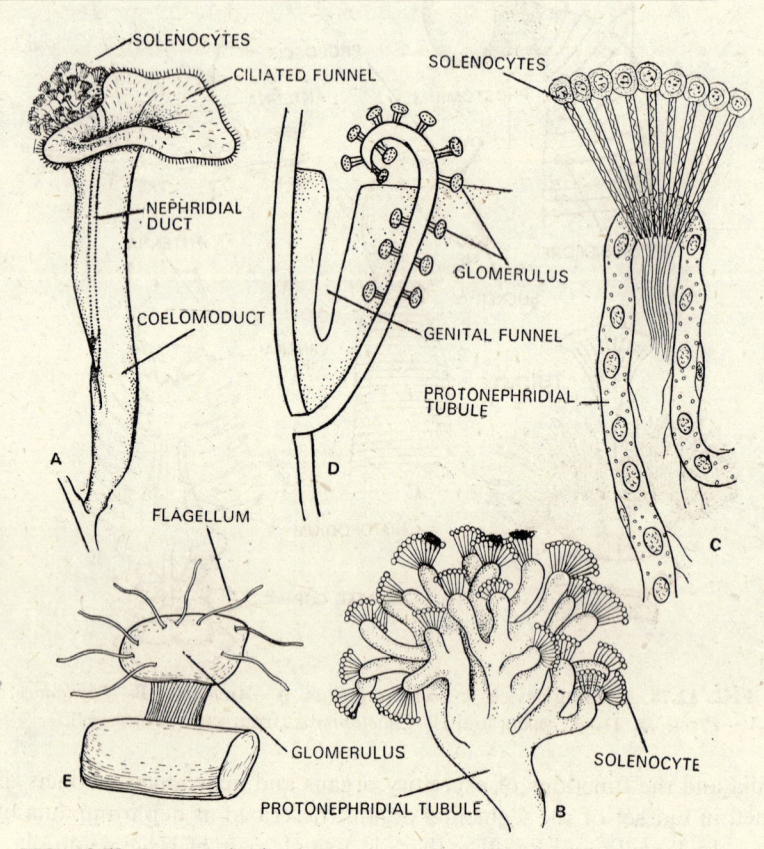

FIG. 12.79. Protonephridium and coelomoducts in polychaete worms. A—*Phyllodoce* sp.; B—Branched end of protonephridium; C—Solenocyte of one protonephridial branch; D—Nephridium of *Vanadis*; and E—Nephridium of *Asterope* (all *after* Goodrich).

The nephridia of the Hirudinea are formed on the same general type as those of *Hirudinaria* (pages 466-467), but differ in the structure of the ciliated funnels. In *Pontobdella*, on the other hand, a very interesting modification of the nephri-

# Phylum Annelida

dial system occurs. Instead of distinct nephridia, there is found on the ventral surface of the body a very complex network, which on each side of each segment sends off a short branch terminating in a ciliated funnel, and a similar branch which opens externally. In *Branchellion* also a similar modification is found.

## Respiratory System

In general respiration takes place through the surface of the skin, but in many cases special respiratory organs such as branchiae or gills are developed. In many of the Polychaeta the branchiae are borne on the dorsal surface of more or fewer of the segments. Sometimes they occur in all or nearly all the segments, sometimes they are present only in the middle region of the body, while in most of the tubedwelling forms they occur only at the anterior end. The branchiae are of different forms, they may be filiform, pinnate, compressed and leaf-like and sometimes branched in a tree-like manner. The surface of the branchiae is usually ciliated, and they have a rich blood supply. In *Terebella* the branchiae are situated on the dorsal surfaces of some of the anterior segments. In *Serpula* and its allies they form two incomplete lateral circlets of elongated appendages

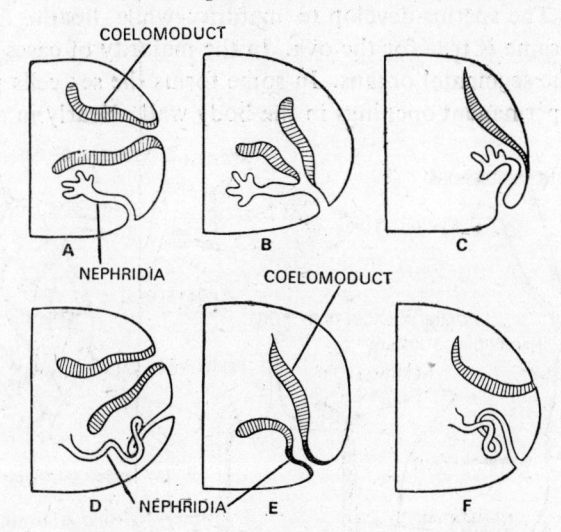

FIG. 12.80. The relation between excretory and reproductive ducts in the annelids (*after* Goodrich). A—Hypothetical stage with closed nephridia and separate coelomoducts; B—Coelomoducts have become united with the nephridia (*e.g.* in Phyllodocidae and Goniadidae); C—Coelomoduct becomes reduced to a ciliated organ (Nephthyidae); D—Nephridia with nephrostomes and separate coelomoducts (*Dasybranchus*); E—Formation of 'segmental organs' by the union of nephridia with nephrostomes and coelomoducts (found in most worms); and F—Nephridia with nephrostome and coelomoducts reduced to ciliated organs (*Nereis*) (*after* Goodrich).

situated at the anterior end of the body. One of them is enlarged to form an operculum often armed with calcareous plates and spines. Simple or branched metamerically arranged branchiae are also present in the aquatic Oligochaeta.

Sometimes they are retractile and are usually present on the segments of the posterior region. In *Branchellion*[1] differentiated respiratory organs or gills are present in the form of plicate lateral outgrowths of the segments.

**Luminescence.** The annelids are known for their power to produce light. Various Polynoids and *Chaetopterus*, etc. are the animals that become brilliantly luminous in the dark. In *Polyophthalmus* the light is produced by definite luminous organs which are eye-like in appearance.

## Reproductive System

Most of the Polychaeta are unisexual (only a very few cases of hermaphrodite forms are recorded), while the Oligochaeta are mostly hermaphrodite. The gonads in the polychaetes are masses of cells formed by active proliferation of the coelomic epithelium in certain positions. These organs become conspicuous in the breeding season and occur in the great majority of segments. In certain cases they are confined to a certain region of the body. The place of the gonad within the segment is variable, sometimes they are situated in the sides in the bases of the parapodia, whereas, in others they surround one of the principal blood vessels. The sperms develop to maturity while floating in the coelomic fluid, and the same is true for the ova. In the majority of cases the sex cells pass out through the segmental organs. In some forms the sex cells pass out through temporary or permanent openings in the body wall. Nearly in all cases fertiliza-

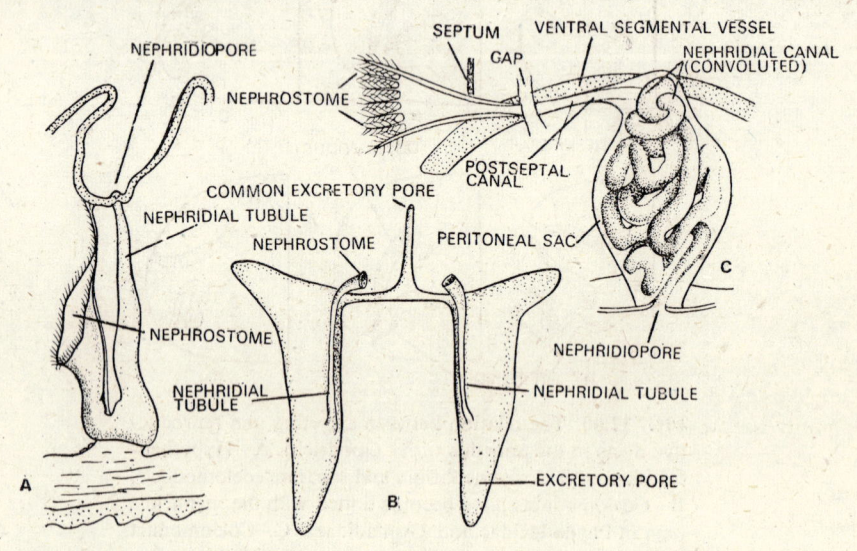

FIG. 12.81. Metanephridia of three polychaetes. A—*Nerine* (*after* Goodrich); B—Serpulid fanworm, *Pomatoceros* (*after* Thomas); and C—*Nereis vexillosa*.

tion takes place inside. The Oligochaetes present a different plan of reproductive system. It is restricted to a limited region of the body. The testes may be a single pair (aquatic forms) or two pairs (earthworms). They are small and frequently become reduced to mere vestiges in the adult animal. The sperm-mother

---

[1] A leech belonging to the order Rhynchobdellida parasitic on the electric rays (*Torpedo* and *Hypnos*) and on one of the Australasian skates (*Raja nasuta*).

cells from the testes reach the seminal vesicles and develop into sperms. The seminal vesicles are large sacs and vary in number and arrangement in different genera. In some cases the seminal vesicles coalesce to form median sperm reservoirs. There are two or four ciliated funnels depending upon the number of testes and they are situated within the sperm reservoirs. There are two or four ducts on each side. In *Pheretima* the four ducts remain separate. They open at the male generative aperture. In many earthworms special penial setae are found near the aperture of the vas deferens. Ovaries are always two. The ova become mature in the ovary. Oviducts are also two and open into the coelom by funnel-shaped structure. The seminal receptacles (spermathecae) are also present.

In many forms the phenomenon of swarming is very common. Normally a worm keeps on crawling or burrowing on the sea bottom, but when sexually mature it rises suddenly to the surface, swims vigorously and eventually discharges the genital products. Soon after it swims back to the bottom as vigorously as it came. In most nereids this occurs irregularly through the summer months, but in at least two forms (*Leodice viridis*, the "palalo" of the Southern Pacific, and *Leodice furcata* of the West Indian reefs) the phenomenon has aquired strict periodicity. In the first case the palalo breaks off the posterior half of its body as the day of the last quarter of the October-November moon dawns. These fragments rise to the surface in large numbers so much so that the water writhes with worms and is milky with eggs and sperms discharged. The remaining end regenerates the broken piece in a year. In the West Indian species the phenomenon takes place in the third quarter of the June-July moon. Swarming insures fertilization of largest number of eggs.

The gonads are usually located in the posterior part of the body which is detached as a free-swimming unit which may develop a head but never jaws and pharynx. It cannot feed but can live for sometime. In majority of forms a single bud is produced, but in *Autolytus* a proliferating region is established at the end of the original body. From this a chain of sexual individuals is budded off. The most posterior bud is the oldest. The worm drags the chain of sexual individuals as it moves and one by one the buds detach and lead short independent existence.

The leech is hermaphrodite. The testes usually have the segmental arrangement as found in *Hirudinaria* (page 470). The number of testes, however, varies from five to twelve pairs. In *Herpobdella*, the testes are numerous and are not segmentally arranged. In the order Rhynchobdellida the muscular penis is absent, its place is taken by an eversible sac. As a rule, there is an elongated hollow ovary producing ova from its epithelial lining. The method of copulation in *Glossiphonia*, a freshwater Rhynchobdellid, is quite interesting. The method has been described as hypodermic impregnation. One leech deposits one or more spermatophores on any part of the body of another, often on the back. Under the influence of the spermatophore the skin becomes permeable to the spermatic fluid which streams through the skin into the coelomic spaces, where it reaches the ovaries and effects fertilization.

**Eggs and development**. The eggs are sometimes laid in cocoons (Hirudinea and Oligochaeta). In the oligochaetes they are either buried in earth (the earthworms) or attached to water plants. The cocoon contains several eggs together plus a quantity of albuminous fluid. In others the eggs are laid in water. In those cases where eggs are laid in cocoon the development is direct (as there is no larval stage), and takes place at the expense of the enclosed albuminous matter. In the other case the young undergo only a small part of their development in the egg-membranes, and are hatched out as larvae at an early stage, to undergo the rest of their development in the free state. The larval form of the annelids is the trochophore or its modification. The larva of *Polygordius* which is known as the larva of Loven is a typical trochophore and so are the larvae of the Chaetopoda and Echiuroidea. Although in the Oligochaeta no free larval stage similar to the trochophore is found, the stage intervening between the completion of the gastrula and the commencement of the segmentation of the mesoderm band corresponds to the trochophore in essential respects, and in some forms even a feebly developed circlet of cilia comparable to the prototroch is recognizable. In some such cases a pair of head-nephridia is also met with.

The fertilization in Chaetopoda takes place outside the body and so also the development. In some cases development takes place within ciliated segmental

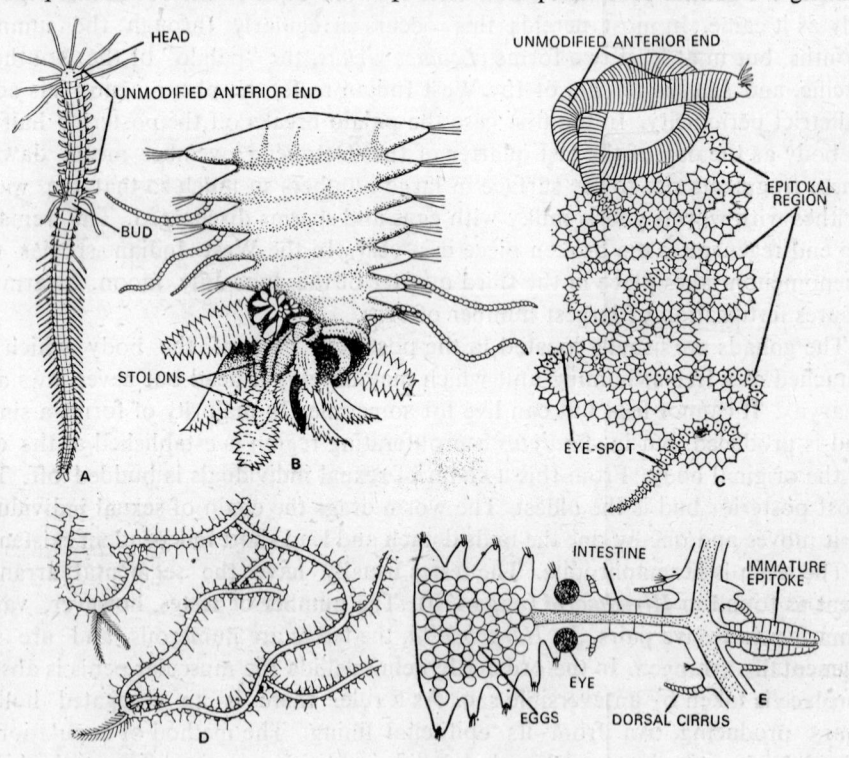

FIG. 12.82. A—*Autolytus cornutus*, budding (*after* Agassiz); B—Posterior end of the syllid, *Trypanosyllis* with a cluster of stolons (buds); C—*Eunice viridis*, the Samoan palalo worm (*after* Woodworth); D—Portion of *Syllis ramosa* (*after* Mcintosh); and E—*Syllis ramosus* showing epitokes developing from sides of atokes (*after* McIntosh).

organ. In most of the Polychaeta during fertilization the sperms come in contact with the ova when both have become discharged, and the development of the embryos goes on while they are floating freely in the sea. Many interesting places of development have been recorded for other worms. In certain Errantia a brood pouch develops on the ventral surface in which the fertilized ova undergo

## Phylum Annelida

development. In *Polynoe* the fertilized ova adhere in masses to the dorsal surface covered by the elytra. In the Sedentaria they develop in the cavity of the operculum.

The cleavage of the ovum in the Polychaeta is of the spiral determinate type. In jelly fishes and echinoderms the individual blastomeres may be separated up to the 16-cell stage and each will produce a complete but proportionately smaller embryo. The cleavage in which each blastomere, if separated, can produce a separate and complete individual is called the **indeterminate** cleavage. If the blastomeres (cell groups) of the eggs of ctenophores, some molluscs (*Dentalium*) and annelids are separated each will produce only that part of the embryo which it would form in an undisturbed egg. This is the **determinate cleavage** and indicates organization of material in the egg prior to cleavage. In such a cleavage the fate of each cell can be determined and followed from the very beginning (cellular predestination) and there is a very early segregation of the germ cell line.

The spiral cleavage occurs rhythmically, affecting the whole or greater part of the blastomeres at the same time. The first two divisions are equal as a result of which four cells A, B, C, and D are produced. They lie in the same plane. Each cell cleaves further and gives rise to one of the quadrants of the embryo. Thus four quadrants are formed. Of these D quadrant tends to be larger than the others and forms the dorsal surface of the embryo, B forms the ventral surface and A and C lateral. Further three divisions, i.e., third, fourth and fifth, are unequal and at right angles to the first two. The result is that three quartets of micromeres are separated from the macromeres (as A, B, C and D are then termed). The region on which lie the micromeres forms the animal pole and the one with macromeres is the vegetative pole. The micromeres are not directly over the macromeres from which they are formed but in one quartet they are

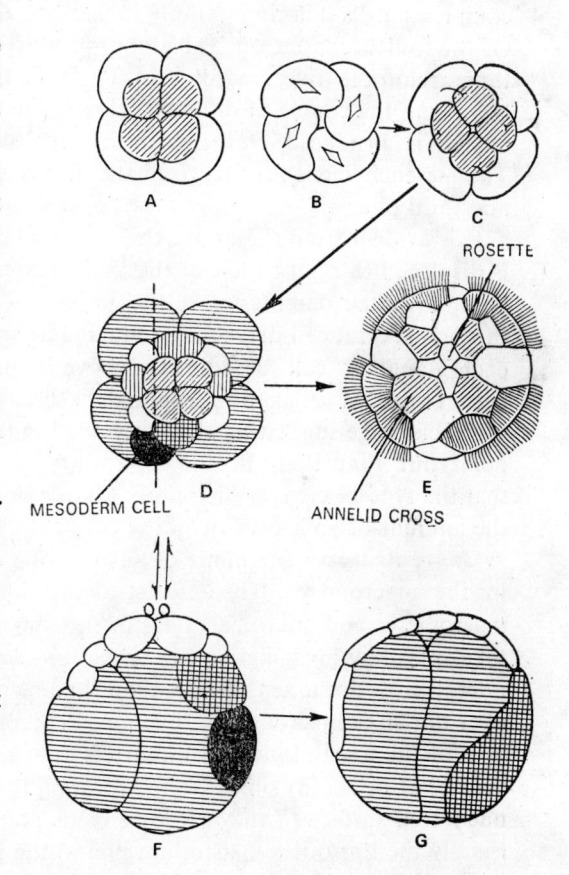

FIG. 12.83. A—8-cell stage radial cleavage; B, C—8-cell stage spiral cleavage; D—Embryo of *Nereis* showing macromeres (stippled, the first three quartets of micromeres, the mesoblast cell (black), in the fourth quartet; E—Later stage seen from the animal pole showing rosette, the annelid cross, and the prototroch cells (horizontal shading); F—Vertical section of D along dotted line; and G—Vertical section of the embryos showing gastrulation (*after* Borradaile, Potts, etc.).

displaced to the right and in the other to the left of the embryonic radius and the next will be displaced to the right again. It is for this reason that the cleavage is called of spiral type. At a later period it is replaced by cleavage in which there is no alternation with the result that the embryo becomes bilaterally symmetrical. The cells of the first three quartets give rise to the ectoderm of the larva and of the adult.

At the sixth division a fourth quartet is separated from the macromeres. This comprises cells differing notably in size and density from those of the first three. Of the fourth quartet one cell alone produces the mesoderm, while the other three reinforce the macromeres to form the endoderm. The mesoderm in the larva is in the course of differentiation alone and a larval mesoderm or mesenchyme is produced from which the musculature of the trochophore is derived. The mesenchyme is formed from the inward projections of cells of the second and third quartets.

It is evident from the above that in spiral cleavage the second quartet of cells is displaced to right or left of the first quartet. Consequently the daughter cells lie above their parents but in the grooves between them. This pattern continues through several cell divisions producing a spiralling effect. In radial cleavage each daughter cell lies directly above its parent cell (not displaced on grooves between parent cells). Some animals, such as annelid worms and molluscs, have spiral cleavage suggesting that they are fundamentally related. Some biologists point out that there are non-spiral groups even within these phyla and believe that the type of cleavage is purely a mechanical problem that is dependent on the amount of yolk present in the cell.

**Gastrulation.** The mode of gastrulation depends upon the amount of yolk in the macromeres. The amount of yolk in *Polygordius* is very little and so the macromeres and micromeres are of the same size. The gastrulation in such forms takes place by invagination. In *Arenicola*, *Nereis* and nearly all Polychaeta the micromeres are much smaller than the macromeres and they multiply and grow over the large macromeres, thus an "epibolic" gastrulation takes place. The cells of the fourth (and fifth) quartets approach each other from two sides. The mesoblast cells (4d) sink inwards forming the blastocoel, and the blastopore (the uncovered surface of the macromeres) is gradually reduced and becomes slit-like. Finally the lips of the blastopore join in the middle leaving an anterior opening and another posterior opening. The anterior opening forms the mouth, while the posterior one closes and in its neighbourhood the anus is formed by an invagination. The blastopore is on the ventral surface of the larva. By now the macromeres withdraw further forming the archenteron. The cells of the fourth and fifth quartets (4a, 4b, 4c, 5a, 5c, 5d) take part in its formation. The cells of the second quartet (2d) form the ventral plate including ventral nerve cord. With a little further cell division the gastrula changes to the trochophore The micromeres of the first quartet differentiate into: (*i*) the apical rosette of four cells that form the apical plate; (*ii*) the cells of the so-called annelid cross, which form the cerebral ganglia, the prototroch; and (*iii*) the intermediate girdle cells. The body of the trochophore behind the prototroch is formed by the proliferation of a single cell in the second quartet (2d), the **somatoblast**. This forms a plate which spreads round the sides. Other cells of the second quartet (2a, 2b, 2c) form the stomodaeum and complete the alimentary canal. The larval gut opens

## Phylum Annelida

at the mouth in the equatorial region and consists of an ectodermal oesophagus (stomodaeal) opening into an endodermal stomach that leads into an ectodermal hindgut (proctodaeal) opening out at the anus. The space between the ectoderm and gut is the original blastocoel and is occupied by mesenchyme, larval muscle, nerves and two larval nephridia (Fig. 12.2).

### Phylogeny

It is well-known that the Annelida are ancestral forms of Mollusca, Arthropoda and Echinodermata, but the ancestry of the Annelida themselves is not fully understood. It is not supported by strong palaeontological evidences. There are two notable views about their ancestry, one links them through the trochophore larva and is called the trochophore hypothesis and the other derives them from the turbellarians and is called the turbellarian hypothesis.

**Trochophore hypothesis.** According to this hypothesis the annelids have been derived from some trochophore-like ancestor, and in the absence of such a definite form, a hypothetical animal, *Trochozoon*, has been placed in that position. It is imagined that this form resembled the larva of *Polygordius*, and this form is believed to have been the ancestor of all those animals that have a trochophore larva. On the basis of this assumption the annelids come quite close to the Rotifera through a rotifer genus *Trochosphaera*. In structure *Trochosphaera* (Fig. 10.11 E) has many things in common with the trochophore larva—ciliary girdle, bent intstine and excretory organs resemble topographically in both the types. But *Trochosphaera* is a highly specialized (not primitive) form and cannot be regarded as a basis of this relationship.

Some authors derive the *Trochozoon* from a ctenophore-like ancestor. They compare the cross-arms-like germinal matter of the annelid embryo with the ctenophore swimming plates, and the statocyst, with the apical plate of trochopore. This view is not tenable, for the development plan in these two groups is very different. The annelid embryonal germ and that of the ctenophore should show full radial symmetry if the trochophore hypothesis is to be taken as correct assumption, but this is not so.

**Turbellarian hypothesis.** This hypothesis derives the annelids from turbellaria-like forms. This view is amply supported by the phenomenon of spiral cleavage that has been extensively used as a basis for establishing phylogenetic relationship. The groups of animals in which this type has been generally recognized (Turbellaria, Platyhelminthes, Nemertinea, Sipunculoidea, Echiuroidea, Annelida, Mollusca) are considered more or less related to one another. Similarities in embryology, including cleavage, fate of the blastopore and the origin of the mesoderm are among the most widely used characters as the basis of phylogeny. The development pattern among the Turbellaria and Annelida bring them close together, and, apart from that, the larval forms of acoel or rhabdocoel turbellarians are similar to those of annelids. The plan of nervous system, the protonephridia, ecto- and endomesoderm and general form of the body of the larval forms resemble greatly. The brain of the Acoela resembles the apical plate of the trochophore. All these points favour the turbellarian hypothesis of the origin of the annelids. Further, it has been suggested that the metameric animal may have arisen from an unsegmented stock in forming chains of individuals by asexual fission, and with the products of fission remaining united and acquiring both structural and physiological unity. Another theory presumes the segmental division of muscles, nerves, coelom, nephridia, etc., within a single individual as seen in the formation of somites by both larvae and adults of some animals.

# 13 PHYLUM ECTOPROCTA (BRYOZOA)

Among the coelomates there are certain groups of animals that look moss-like in their appearance. For this reason Ehrenberg (1831) named the group as Bryozoa. Older Bryozoa included Entoprocta and Ectoprocta. But now Entoprocta has been found to be a pseudocoelomate and has been placed with Aschelminthes, while Ectoprocta, being coelomate has been retained with Bryozoa. Another term Polyzoa was applied to these animals by J.B. Thompson (1830) but this was not so precisely defined and a long controversy arose concerning the two rival terms. The fact that the animals of this group possess a lopophore (see below) sets them apart from other Eucoelomata. In many text-books the three groups (Ectoprocta, Phoronidea and Brachiopoda) are united under the name Molluscoidea or Tentaculata. Sometimes the Brachiopoda are separated as a phylum and only the ectoprocts and phoronids are retained under the name Tentaculata. Here all the three have been raised to the status of independent phyla and placed along with the schizocoelous Eucoelomata although the coelom of brachiopods arises by the enterocoelous method. The name Bryozoa has been given up and Ectoprocta has been used for the phylum as almost all the known Bryozoa, both fossil and recent, belong to this subdivision.

The **Ectoprocta** (Gr. *ektos*, without + *proktos*, anus) are characterized by the position of the anus without (outside) the row of tentacles surrounding the mouth (unlike the Entoprocta). The Ectoprocta colonies are best known from the leaf-like papery structures and "sea-mats." Some calcareous ectoprocts are coral-like in appearance, hence they were popularly called "Corallines." Some colonies are shrub-like, and hang from blades of sea-weeds or grow out from crevices of rocks. others form flat encrusting growths on sea-weeds and rocks. The ordinary individuals in a colony of Ectoprocta, at the first instance, appear to resemble hydroid polyps in their general shape and size and circlet of tentacles. Closer examination, however, reveals that they are triploblastic animals. In the Ectoprocta each individual consists of two different parts: (*i*) the **zooecium** or **body wall** and (*ii*) the **polypide** consisting of the alimentary canal, the tentacles and tentacular sheath (Fig. 13.1), into which the tentacles are withdrawn when contracted. The cuticle is thickened and forms a protective covering for the support of the colony and is often called the cuticula.

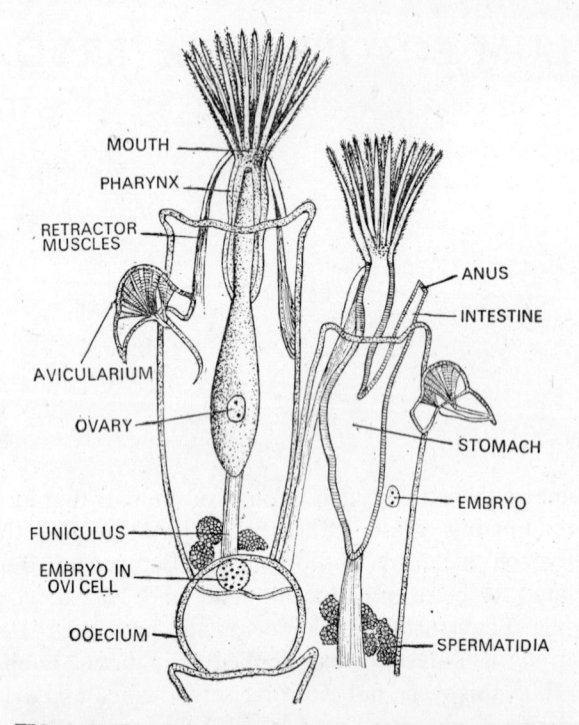

FIG. 13.1. *Bugula avicularia*, general anatomy (*after* Parker and Haswell).

The zooecia of *Bugula avicularia* are closely united together and arranged so that they appear to form stems of a dichotomously branched colony. They are approximately cylindrical in shape but relatively broader distally, proximally. Each is four or five times longer than broad. The distal end bears a wide crescentic aperture—the mouth of the zooecium. On either side of the mouth is a short spine. The anterior end of the body bears a circlet of usually fourteen long slender filiform tentacles. The tentacles are situated on a circular ridge or lopophore surrounding the mouth. Along with the lopophore a flexible part of the body wall is also invaginated and this is called the **tentacular sheath**. The lopophore is either circular, as in the above case (*Bugula*), or is horseshoe-shaped as in the **Phylactolaemata**, and the tentacles are set along its edges.

The whole of the anterior part of the individual can be involuted like the finger of the glove, *i.e.*, it is an introvert. In times of necessity the whole portion

## Phylum Ectoprocta (Bryozoa)

is withdrawn. The tentacles are visible when the introvert is everted. Special musculature is developed to control the movement of the introvert. The extru-

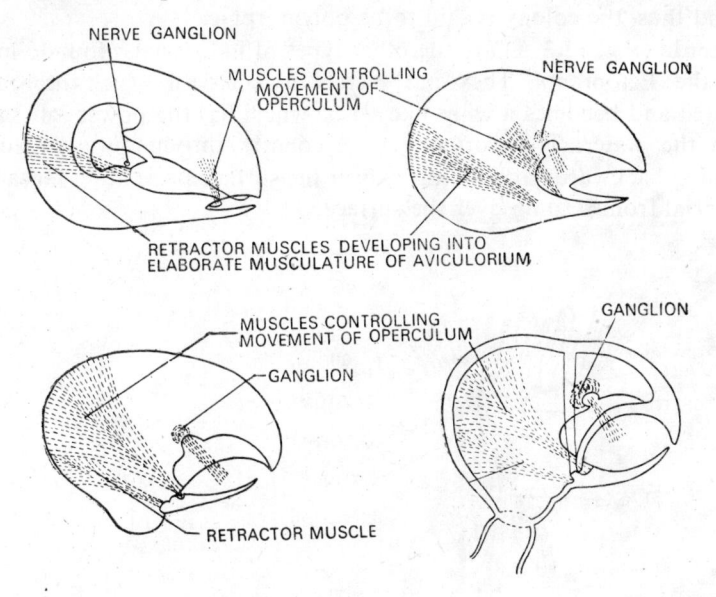

FIG. 13.2. *Bugula.* Transformation of a polypide into an avicularium (*after* Delage and Herouard).

sion of the polypide depends on musculature. In some ectoprocts (Cheilostomata) there is an additional structure, the compensation sac, that brings about extrusion of the polypide. In these, the zooecium emerges only if an equal volume of water is introduced to compensate for the extrusion. The **compensation sac** or **compensatrix** is placed beneath the zooecium and communicates with the aperture. At the moment of extrusion of the polypide, the muscles attached to the compensation sac contract, thus, enlarging the sac, and the operculum opens. Water then enters the sac pushing out the polypide.

The tentacles bear cilia profusely, only the outer surface is devoid of cilia. The cilia vibrate actively in such a way as to drive current of water and with it food particles towards the mouth. The tentacles are tactile and act as organs of respiration.

**Polymorphism.** Each individual bears a remarkable appendage, the **avicularium**, supported on a very short stalk (Figs. 13.1 and 13.2). It looks like a bird's head and is not found in zooecia at the extremities. The avicularium possesses a movable **mandible** which is homologous with the operculum of an unmodified polyp, and is provided with powerful muscles. The avicularia suddenly snap their jaws and catch small roving animals which touch them particularly the larvae of encrusting animals. Probably they are defensive in function. In most primitive cases an avicularium is found in the same position in the colony as an ordinary zooecium and may even possess a functional polypide. Such examples are provided by *Flustra* and *Cellaria*. Further evolution led to displacement of the avicularia so that they became appendages of other zooecia situated near the external orifice (Fig. 13.1). Close examination of the living individual shows the avicularia to be in constant motion, turning from side to side, and the lower part moves like the lower jaw of the bird opening widely and "snapping" food material. The whole avicularium is regarded as an individual and thus, the colony is said to be polymorphic.

The **vibracula** (Fig. 13.3 A) are the other types of individuals found in the colonies of the Ectoprocta. They are merely avicularia in which the beak is much elongated and becomes a whip-like set or which has the power of sweeping through the water. They often act in concert throughout a part of the colony sweeping backwards and forwards over the surface preventing larvae and noxious material from settling over the surface.

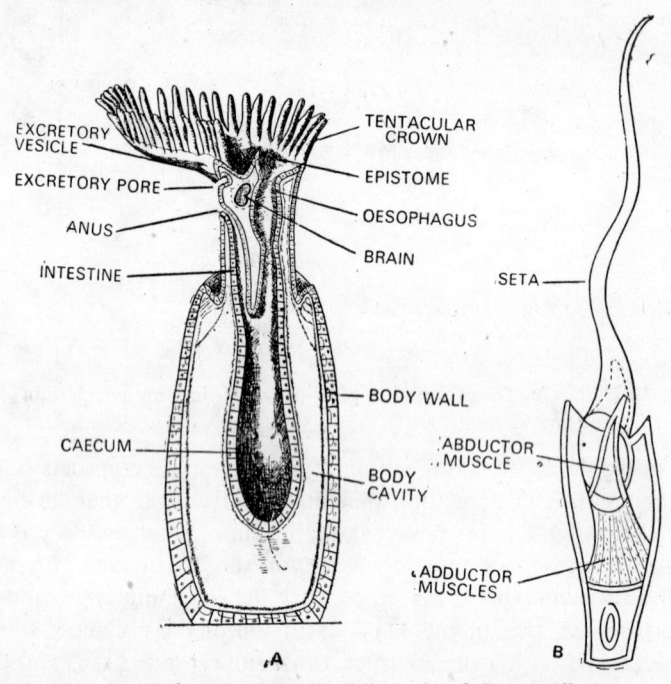

FIG. 13.3. A—Phylactolaemata; B—Vibracula of *Serupocellaria*.

The **ovicells** are other structures found in some **Cyclostomata**. They are zooecia with vestigeal polypides. They, therefore, represent another form of individual in the colony.

**Body Wall**

The body wall consists of the outermost layer of **cuticle**. Beneath the cuticle is the **epidermis**, composed of a single layer of large flattened cells firmly united together by their edges. Then there is a **muscle layer** consisting of two layers, one outer circular and an inner longitudinal. Finally, there is a layer of irregular cellular tissue or **parenchyma**. The peritoneal cells are ciliated in the Phylactolaemata.

**Body cavity**. The body cavity, a true coelom, is extensive and filled with colourless corpuscles floating in it. It is lined by definite coelomic epithelium which is ciliated in parts in some forms, and thin parenchymatous in others. In *Bugula* it is of the latter type. Internally it is lined by a visceral layer of the same tissue. Across the cavity pass numerous strands of spindle-shaped cells. Connecting the aboral end of the alimentary canal with the corresponding wall

# Phylum Ectoprocta (Bryozo a)

of the zooecium is a large double strand, the **funiculus**. The coelom is not continuous, but is divided into two incomplete chambers by a transverse partition. The anterior chamber, that also continues within the tentacles, is known as the **ring canal** or the **circular canal**. Besides the funiculus, in some cases, there is another mesenteric band situated anteriorly near the mouth suspending the alimentary canal.

**Musculature.** The ectoprocts possess an efficient contractile mechanism controlled by musculature. The most important feature is the parietal system of muscles. By their contraction the internal pressure is raised and the polypide is protruded. Of these the most important ones are the special **parieto-vaginal** muscles passing from the introvert to the body wall. The retractor muscle which runs from the lopophore to the opposite end of the zooecium has an opposite action. A pair of special muscles passes from the body wall to the stomach and act as retractors of the alimentary canal.

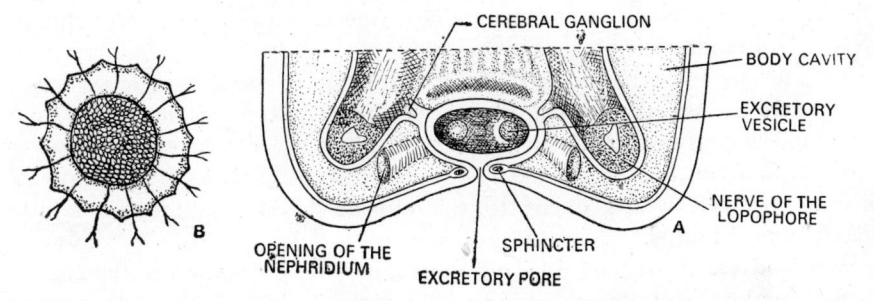

FIG. 13.4. A—Excretory organs of Phylactolaemata in transverse section (*after* Delage and Herouard); B—Statoblast.

## Digestive Organs

The digestive organs (Fig. 13.1) lie freely in the coelom supported by a mesodermal network (mentioned above). The **mouth** is situated at the centre of the circular (or horseshoe-shaped) lopophore. In some freshwater forms, a movable epiglottis-like process, the **epistome**, is found on the dorsal side of the mouth. The alimentary canal is bent on itself and consists of an elongated ciliated oesophagus, often dilated at its upper end forming a **muscular pharynx**, which lies just behind the bases of the tentacles. From the pharynx a somewhat narrow tube leads to the stomach from which it is marked out by a constriction (Fig. 13.1). From the stomach arises a long conical outgrowth, the **caecum**, directed towards the aboral end of the zooecium. The caecum is attached to the wall of the zooecium by the funiculus. A narrow intestine arises from the oral end of the stomach, quite close to the entrance of the oesophagus and runs upwards, almost parallel to the oesophagus. The intestine opens by the anus near the mouth *outside* tentacular crown. The anus is capable of being distended to a considerable size.

The side on which the anus is situated is the dorsal side. In the Phylactolaemata the epistome and the concavity of the horseshoe-shaped lopophore, are dorsal. In the Ctenostomata there is an additional thick-walled chamber, the **gizzard**, having chitinous teeth internally, between the oesoghagus and the stomach.

The entire alimentary canal is lined by ciliated epithelium, only small portion of the stomach is devoid of cilia. The cells are arranged in a single layer and vary in length in different regions being the longest in the pharynx. In some of the Cheilostomata, it is stated, that the cells of the oesophagus bear numerous striated muscle processes.

**Nervous System**

The nervous system is represented by a small round **ganglion** situated between the mouth and the anus (placed on the oesophagus). It lies in the ring canal of lopophore and gives off small nerves to various parts of the body. In the Phylactolaemata the ganglion is placed in the concavity of the lopophore and is attached to the oesophagus by a delicate circumoesophageal ring and sends off numerous branches to the tentacles and oesophagus. According to Fr. Müller in *Serialaria* there is a so-called colonial nervous system which connects the individual zooids of a colony and enables them to coordinate their activities. Claparede describes the same in other species. Special sense organs are absent. No trace of vascular system exists.

There is no definite excretory system. The function of collecting nitrogenous waste matters is assigned to the leucocytes or the corpuscles of the coelomic fluid and to the cells of the funicular tissue. In Phylactolaemata there exists a pair of **nephridia** for excretion. The two nephridia (Fig. 13.4) open into an **excretory vesicle**, which opens to the outside by a minute pore, a little distance above the anus.

**Brown bodies**. A remarkable phenomenon is the formation of the brown body, a normal process of periodical renewal of the individuals. Such a process is a normal feature in most Ectoprocta. Tentacles, alimentary canal and nervous system, rather the whole polypide degenerates and forms a brown compact mass contained in the zooecium. A new polypide is formed from the persistent body wall of such a degenerated polypide as an internal bud. The brown pigmented mass often comes to lie in the new stomach and is evacuated through the anus. Formerly, this process was believed to be of excretory nature, in the absence of specific excretory organs. It was explained that the polypide degenerated due to the accumulation of excreta which is ultimately evacuated through the stomach.

**Reproduction**

Asexual reproduction may be effected by the so-called **statoblasts**, which are found only in the Phylactolaemata. They arise from masses of cells which appear on the funiculus towards the end of summer. They usually possess a lens-like biconvex form (Fig. 13.4 B) and are enclosed by two watch-glass shaped, hard chitinous plates the edges of which are usually enclosed by a flat ring formed of cells containing air (acting as float) and sometimes provided with crown of spines. The Phylactolaemata generally die down in the winter and the statoblasts are left out for the perpetuation of species. They germinate in spring and give rise to new colonies.

In the fresh water ctenosomes *Paludicella* and *Victorella* the colonies also die down in winter, but before this certain external buds develop. The growth of these buds is arrested. They develop after winter into zooecia. Such buds are called **hibernacula**. In some cases parts separated off from the colony are able

# Phylum Ectoprocta (Bryozoa)

to develop into new colonies as in *Cristatella* and *Lophopus*. This is so as the animals possess good power of regeneration. Budding is a common method of colony formation and is similar to that of the entoprocts (Fig. 13.5).

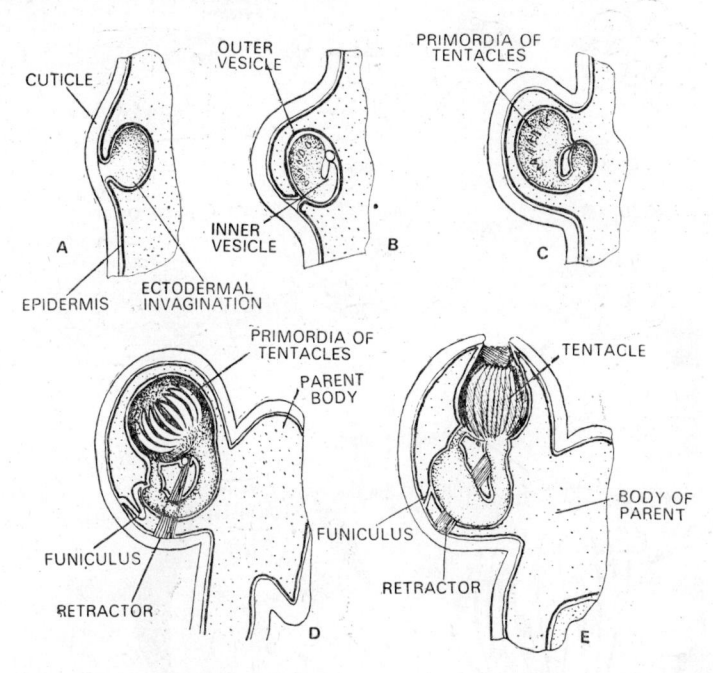

**FIG. 13.5.** Budding in Ectoprocta. A—Ectodermal invagination; B—Formation of inner and outer vesicles; C—Primordia of tentacles; D and E—Formation of polypide.

**Sexual reproduction.** The zooides are usually hermaphrodite, **ovary** and **testes** are found in the same individual. They arise from specially modified cells of the parenchyma of the body wall or from the cells of the funiculus. The testes are developed either on the upper parts of the funiculus or near the point of attachment of the latter to the body wall. From the funicular tissues arise spherical masses of cells, the **spermatidia**, from which sperms with long motile tails develop. They are then released in the body cavity where they swim freely. Definite sperm ducts are absent (Fig. 13.1). Whether the sperms pass out to the exterior is itself doubted.

The **ovary** is formed by the rapid multiplication of the cells of the parietal layer of the parenchyma about the middle of the zooecium. It consists of a small number of cells of which only one becomes a mature ovum. Certain smaller cells of the ovary form an enclosing follicle. The mature ova are reported to be fertilized in the coelom. The fertilized ovum passes into the interior of a rounded outgrowth of the zooecium, the **ooecium**, lined with parenchyma in which it undergoes development. In other species (i.e., Ctenostomata) the development takes place in the tentacle sheath or in special pouches as in Cheilostomata. The eggs or larvae escape either by the intertentacular canal, or by the rupture of the body wall of the parent or by the degeneration of the polypide. In *Alcyonidium duplex* the sexual zooecia produce two polypides to complete sexual process; the first of these produces spermatozoa and then undergoes disruption into a brown body, whereupon is formed a second polypide which produces ova.

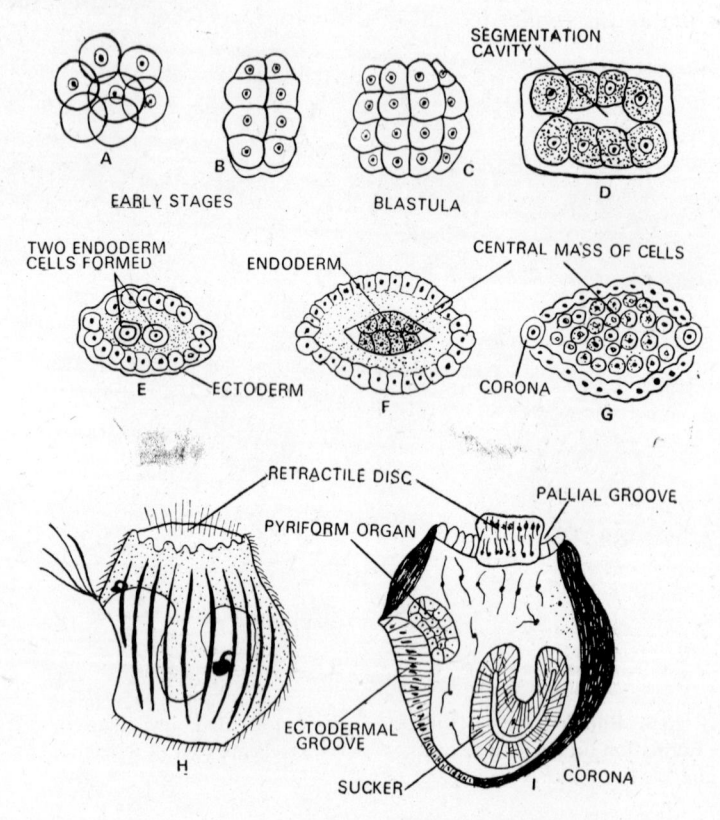

FIG. 13.6. Segmentation and development of *Bugula*.

**Embryology.** In *Bugula* the cleavage is complete and regular forming a biconvex **blastula** (Fig. 13.7). In the segmentation cavity of the blastula four cells are produced. These are the **primitive endoderm** cells and divide and fill the entire cavity in some time, when they represent both the **endo-** and **mesoderm**. In the **endomesoderm**, as it may be called, appear small cavities that unite and finally form the primitive coelom. Around the equator of the embryo a broad ring-shaped thickening, the **corona,** is formed and becomes ciliated later on. On the oral side on the corona develops a circular **pallial** groove (Fig. 13.7). Another important larval structure, the **sucker,** appears on the future oral side of the larva. This is formed by simple invagination of the ectoderm. In the beginning it is sac-like but finally it becomes beaker-shaped. Another ectodermal depression appears in the region of the corona on the oral side. This is the ectodermal groove. On the aboral side appears a retractile disc provided with motionless sensory cilia. In close relation to the ectodermal groove is formed a mass of cells, the **pyriform organ.** There is no alimentary canal in the larva (Fig. 13.7).

**Metamorphosis.** After an interval of free-swimming life the sucker of the larva everts and fixes to some object. The aboral side of the larva becomes greatly extended, the pallial groove disappears and the corona curves downwards (towards the oral side). At this stage the larva looks like an umbrella (Fig. 13.7).

## Phylum Ectoprocta (Bryozoa)

The edges of the umbrella bend further downwards and fuse with the broad plate formed by the sucker and enclose a circular cavity, the **vestibule**, the walls of which are lined by coronal cells. At a later stage the wall breaks and the vestibular cavity merges with the internal cavity of the larva. All the larval structures except the basal plate of the sucker and the retractile disc disappear. The basal plate gives rise to the basal part of the wall of the primary zooecium and from the retractile disc arise both the ectodermal and endodermal structures of the primary zooid. Most of the internal organs from the first brown body after which arises a new polypide.

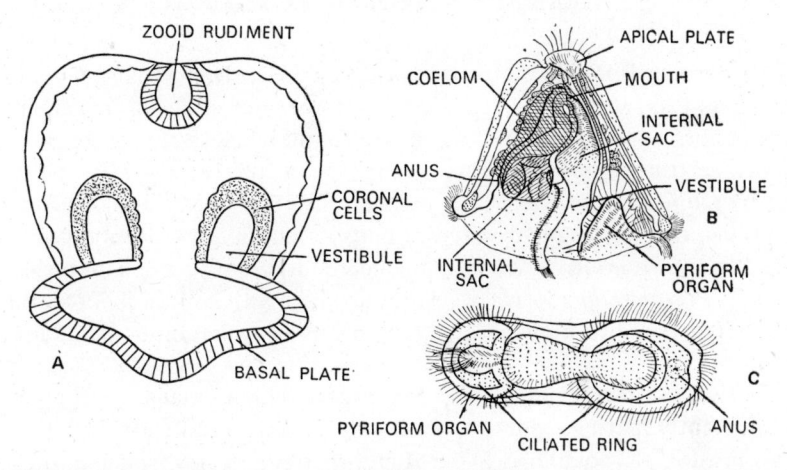

FIG. 13.7. A—Umbrella stage in the metamorphosis of *Bugula*; B—Structure of cyphonautes larva; and C—Front view of the larva.

This is the account of a form in which the early stages are passed through under the protection of the parent and thus the larva is without functional alimentary canal. In the forms in which the development takes place outside the body a free-swimming larva with a functional alimentary canal is produced. Such a larva is called the **cyphonautes**.

The **cyphonautes** (Fig 13.7) has the form of a laterally compressed bell with a circle of cilia round the base. There are two shell plates placed right and left on the aboral surface. The valves meet in front and behind but gape ventrally. At the aboral apex there is an opening between the two valves. Through this opening projects the ciliated disc or the **apical organ**. The ciliated ring (corresponding to the prototroch) can also be seen projecting from between the valves. The base of the bells is the ventral surface; it has two openings both enclosed by a special lobe of the ciliated ring. The larger and posterior of these leads into a depression called the **vestibule**; the smaller and the anterior into an ectodermal depression called the **pyriform organ** of unknown function. The vestibule is divided into two parts: (*i*) the **anterior chamber** at the base of which is the mouth, and (*ii*) a **posterior chamber** into which opens the rectum. The alimentary canal of the larva consists of the **pharynx, oesophagus, stomach** and **rectum**. In front of the anus, at the floor of the vestibule, is an invagination of thickened ectoderm, the **internal sac,** by which attachment is effected prior to metamorphosis.

On attachment the alimentary canal degenerates and is in no way concerned in the formation of the digestive tube of the adult. Other larval organs including the pyriform organ also disappear and the larva becomes reduced to a layer of epithelium surrounding a central mass of cells and broken down larval organs. The attached organism is the primary zooecium. Its organs degenerate forming the first brown body which gives rise to an inner layer of ectoderm and an outer mesoderm. The ectoderm gives rise to the tentacles and tentacle sheath, the ganglion and the alimentary canal of the new polypide.

## CLASSIFICATION OF ECTOPROCTA

The Ectoprocta is divided into two classes: Phylactolaemata and Gymnolaemata.

I. CLASS PHYLACTOLAEMATA. These are the Ectopraocts in which the lopophore is horseshoe-shaped and which are found exclusively in freshwater. They produce gelatinous housings bearing zooids on the surface. The mouth is overhung by a lip (epistome) and they have a special mode of surviving unfavourable conditions, *i.e.*, by means of internal buds or statoblasts. Important species include *Plumatella princeps*, zooecium calcareous, colony much branched erect or creeping; *Pectinatella* zooecia a gelatinous mass with zooids exposed on surface; *Cristatella mucedo* forming an elongate creeping gelationous mass up to 25 cm long in ponds.

II. CLASS GYMNOLAEMATA. The Ectoprocta with a circular lopophore and are predominantly marine. Mouth is not provided with an epistome. This class includes most of the fossil and living Ectoprocta, and is divided into five orders: (*i*) **Ctenostomata** with chitinous or gelatinous zooecia provided with tooth-like process to close aperture; Ordovician to recent. Examples: *Alcyonidium*, *Serialaria* (marine), *paludicella* (freshwater). (*ii*) **Cyclostomata** with calcareous zooecium without operculum, embryos in enlarged ovicells; Ordovician to recent. Examples: *Crisia*, *Idmonea*, *Tubulipora*. (*iii*) **Cheilostomata** with chitinous or calcareous zooecium of various shapes provided with an operculum. Jurassic to recent. Examples: *Bugula*, *Menipea*, *Membranipora*. (*iv*) **Treptostomata** with calcareous zooecium of two parts, some colonies are of diameter of 60 cm. Ordovician to Permian. (*v*) **Cryptostomata** with short zooecium; Ordovician to Permian.

### Affinities

The phylogenetic relationships of the Ectoprocta are uncertain. Formerly they were linked with the Entoprocta on the basis of certain superficial resemblances and the similarities of the larvae. The larvae of both are similar in structure but they behave differently in two groups during metamorphosis; and it must be remembered that pelagic larvae present certain similarities. The development of the two has proved the difference, the Entoprocta are pseudocoelomate and as such much lower in organization than the coelomate Ectoprocta.

With all this information it seems most consistent to seek for the affinities of the Ectoprocta among the Phoronidea and Brachiopoda as their possession of a lopophore sets them apart from the Eucoelomata. It has been seen that the Ectoprocta are coelomate animals with a true coelom and usually well-developed endomesoderm, excretory organs, when present, are simple with or without nephrostomes and with anus. These characters along with the presence of a circular or cresentic or spirally coiled lopophore, bearing ciliated tentacles and looped intestine links them with the Phoronidea and Brachiopoda. But as the objections against this view are weighty it is advisable to raise them to the status of an independent phylum with affinities with the Phoronidea and Brachiopoda. (Also see affinities of Phoronidea and Brachiopoda).

# 14 PHYLUM PHORONIDEA

Phoronids comprise a small group of tube-dwelling, lopophorate, coelomate worms found in marine coastal waters in tropical and temperate regions. They are sessile animals occurring from about low-water down to, in certain species, about 50 metres. They are non-colonial and live in tubes secreted by themselves. They are slender worm-like creatures of various sizes that inhabit the bottom under shallow seas. About 15 or more species are known; of these a majority are included in the genus *Phoronis* and are from 1 to 1.5 mm long but one species, *Phoronopsis californica*, grows to 30-37.5 cm and its tube may be 45 cm long. The adult is cylindrical, lives in leathery tube, to which particles of foreign matter, such as sand grains, and sponge spicules are often found adhering. The animal appears to be only loosely attached to this tube and can protrude the anterior part of the body from the opening of the tube. Some are scattered singly and others are so densely placed as to cover the sand for nearly an acre. Often a number of individuals may be found curiously associated together, their tubes twisted round each other (Fig. 14.1 B) without any connections between them. The reason of this association is not understood. The tube is secreted by cutaneous glands in the anterior region of the animal. In some cases the straight cylindrical tubes lodging the animals are embedded vertically just below the surface of the substratum in mud and sand. The bodies and tentacles of different species are red, orange, or green, and when numerous, they may give brilliant colour to the sea floor.

# Phylum Phoronidea

## Lopophore

At the anterior (oral) end each individual bears a horseshoe-shaped **lopophore** (Figs. 14.1, 14.2), often spirally coiled, at the ends (*P. australis*) bearing

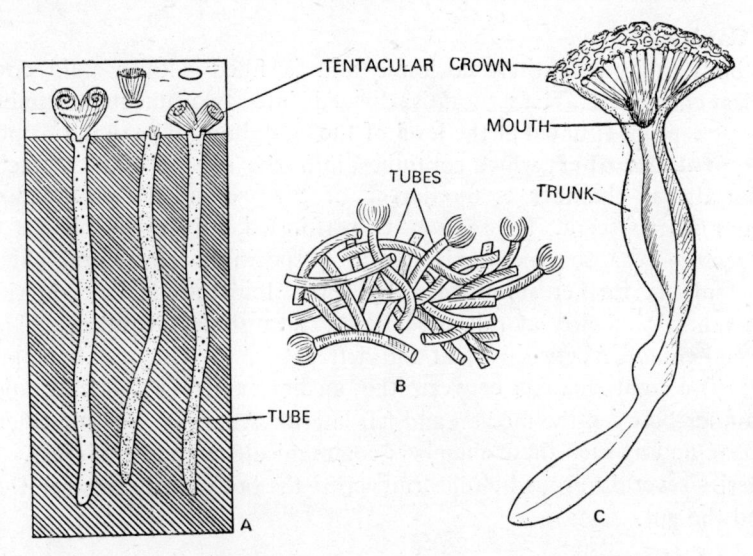

FIG. 14.1. *Phoronis*. A—Individuals buried in sand; B—A group of individuals in their tubes; and C—One individual taken out of tube.

ciliated tentacles (Fig. 14.1 A) that help in catching food. The convacity of the lopophore is dorsal. The tentacles are hollow and arranged in two rows (Fig. 14.2), of which the one on the inner side of the lopophore is incomplete in the middle line. Between the two rows of tentacles the mouth is medially situated in a groove. Between the mouth and the inner tentacles of the lopophore (i.e., dorsal to the mouth) there occurs a laterally extended flap of the body wall, the epistome. The anus is dorsal and lies outside the lopophore at the summit of a median longitudinal ridge. On each side of the anus (Fig. 14.2) is the external opening of the nephridia. Developmentally speaking the area between the mouth and anus marks the dorsal surface and the whole of the aboral extension of the body is ventral. The lopophore has a great power of regeneration, any part, if torn off, regenerates.

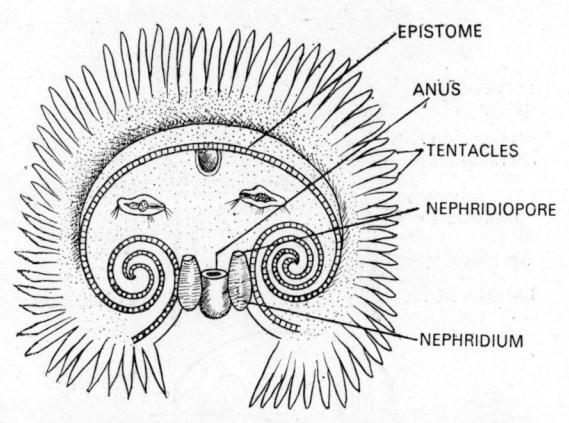

FIG. 14.2. Lopophore of *Phoronis* seen from above.

## Body Wall

The body wall (Fig. 14.4) consists of the cuticle, an ectodermal epidermis, followed by a **basement membrane** below which occurs the body wall musculature comprising two layers, **circular** and **longitudinal**. The last layer internally is the **parietal** layer of coelomic epithelium. The tentacles possess a ring of skeletal tissue of mesoblastic origin within the epidermis.

### Body Cavity

The body cavity is entirely coelomic and is filled with a fluid containing colourless corpuscles. The coelom is divided into two unequal chambers by a transverse septum situated at the level of the lopophore. The anterior portion is smaller, **oral chamber**, which continues into the epistome and tentacles. The posterior aboral chamber is the larger of the two and lodges the alimentary canal and other viscera. The septum is perforated by the oesophagus but not by the rectum because the anus is aboral in position. The aboral chamber of the coelom is further sub-divided by three longitudinal mesenteries, one median ventral and two lateral mesenteries, into three chambers (Fig. 14.4): (*i*) a rectal chamber between right and left lateral mesenteries, containing the rectum; (*ii*) a right chamber between the median and right lateral; and (*iii*) a left chamber between the median and left lateral. Aborally the mesenteries are incomplete and as such these chambers communicate with one another. Besides mesenteries several fibrous bundles run across the body cavity between the body wall and the gut.

### Digestive Organs

The alimentary tract is V-shaped (Fig. 14.3) and occupies the major portion of the aboral extension of the body wall. It consists of the **oesophagus,** the **stomach**, the **intestine** (sometimes called the second stomach) and the **rectum,** which opens at the anus. The inner lining is ciliated except the terminal part and the wall of the stomach is glandular. Phoronids feed by extending the mucous-coated tentacles to lay on the bottom and take minute organisms from the water or in the surface debris.

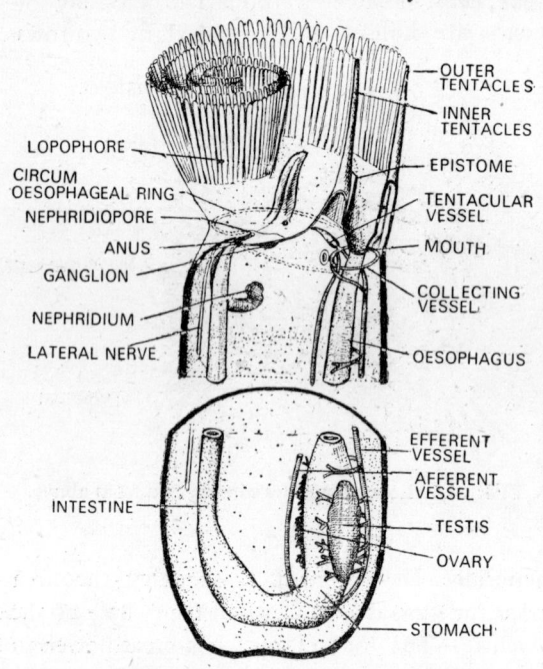

**FIG. 14.3.** Sagittal section of *Phoronis* showing general anatomy.

### Excretory System

Excretion is brought about by two nephridia (Fig. 14.3), each of which opens externally on either side of the anus. Each nephridium passes towards the aboral side by crossing the septum and opens into the body cavity by two openings, the smaller in the lateral chamber of the body wall and the larger into the rectal chamber. The nephridia also function as generative ducts.

## Vascular System

*Phoronis* has two longitudinal (Fig. 14.3) vascular trunks, one lying in the rectal chamber between the two limbs of the alimentary canal and the other on the left side of the oesophagus. Aborally the vessels are continuous and anteriorly they pass through the septum and give rise to blood vessels to the tentacles and other parts lying in the area. These vessels are contractile to maintain the circulation of the fluid that fills them. The blood consists of a colourless plasma containing nucleated blood corpuscles which possess haemoglobin. No definite respiratory organs are found in the phoronids.

## Nervous System

The nervous system lies below the epidermis in the body wall. It consists of a circumoral ring round the mouth. On the dorsal side the ring is slightly en-

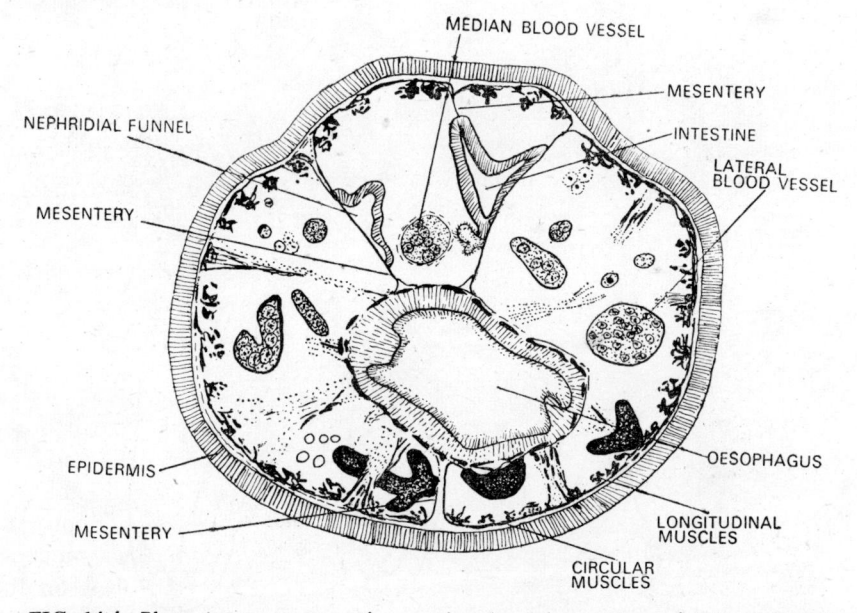

FIG. 14.4. *Phoronis*, transverse section passing through anterior end (*after* Benham).

larged between the mouth and the anus forming a dorsal supraoesophageal ganglion. From the ring nerves are given off to the tentacles and nephridia, etc. Two large fibres from the ring run along the lateral mesenteries.

There are no organs of special sense. Two ciliated pits are, however, found on the dorsal side of the base of the inner series of the tentacles. These organs have been regarded sensory by some authors and glandular by others.

## Reproductive System

*Phoronis* is hermaphrodite. The gonads lie on the left side of the stomach in the aboral region of the body cavity (Fig. 14.3). The **ovaries** lie on one side of the oesophageal blood vessel and the **testes** on the other. They are formed by the special proliferations of the coelomic epithelium. The reproductive cells are released in the body cavity and pass out through the nephridia, and reach the spaces enclosed by the tentacles and it is here that fertilization is effected. Some authors have suggested that the fertilization takes place in the body cavity. Development takes place while the embryo remains attached to the tentacles.

**Development.** The cleavage is complete and slightly unequal cells are produced. At the four-cell stage two darker, larger blastomeres, the **endoderm**, can be distinguished from the two smaller clearer cells, the **ectoderm**. A **blastula** is formed with clearer ectoderm cells on one side, and a **gastrula** is formed after invagination; the **blastopore**, which assumes a slit-like form, closes up behind, but remains open in front as the mouth. The anus is formed as a pit at the hind end of the closed-up portion of the blastopore. The **mesoderm** arises partly as cells budded off from the endoderm on each side and partly as a pair of diverticula from the anal pit. A large preoral lobe is formed, and the anus becomes surrounded by circlet of cilia (Fig. 14.5) and the part bearing it

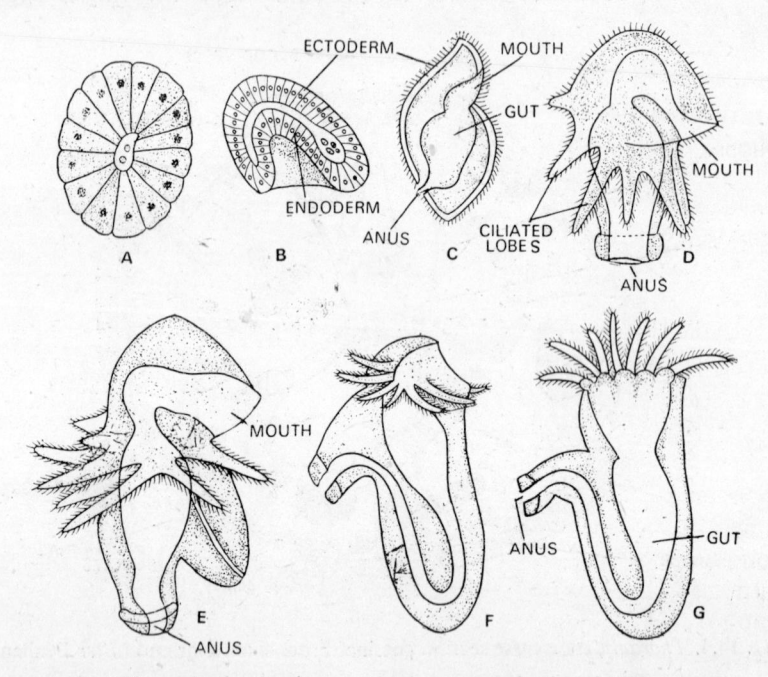

FIG. 14.5. Stages in the development and metamorphosis of *Phoronis*. A—Blastula; B—Gastrula; C—Early embryo; D—Embryo with ciliated lobes; and E to G—Larva everts to form adult.

becomes elevated into a conspicuous process. Behind the mouth there is a circlet of cilia and from this region grow out rudiments of larval tentacles (Fig. 14.5 D). The larva, on the whole, has a gelatinous transparent body and is known as the **actinotrocha**.

The preoral lobe is covered with cilia and has an apical thickening of ectoderm forming the **larval ganglion**, and in some species **eyespots** are also present. The cilia on the margin of the preoral lobe are longer than the rest and form a **velar ring**. Excretory organs are like those of the trochophore larva (Fig. 12.2), have no coelomic apertures but are provided with solenocytes. They are said to change into nephridia of the adult during metamorphosis. Due to the presence of the velar ring of cilia and the apical plate the actinotrocha is regarded as a modified trochophore by many authors. The actinotrocha possesses a coelom in the form of two pairs of mesoblastic sacs which constitute the body cavity of larva and also give rise to that of adult. The trochophore larvae of molluscs and annelids do not possess this.

**Metamorphosis**. After a period of free life the larva sinks to the bottom and undergoes metamorphosis. There is an invagination on the ventral surface between the mouth and anus of the larva. This portion becomes evaginated and forms a tubular projection, the visceral sac, at right angles to the ventral surface. This projection forms the body of the adult. By this time the rudiments of adult tentacles also appear. The alimentary canal now passes into the visceral sac forming a U-shaped loop. Simultaneously with this the anal papilla becomes reduced, more and more, eventually taking the adult positions, with the extraordinary development of the visceral sac. The larval tentacles and the preoral lobe along with its ganglion are now absorbed and the lopophore appears. The adult tentacles become fully formed and the attachment is effected by the aboral apex of the new body.

**Distribution**. The Phoronidea are widely distributed occurring in all the chief geographical regions. Some species occur at low depths but some are found up to moderate depths of about thirty fathoms. In some cases the phoronid tubes lie in the excavations in stones or in shells of molluscs and some species have even been reported from the bodies of sea-anemones.

## Affinities

The Phylogenetic affinities of Phoronidea have long been a subject for speculation and difference of opinion. The group has been linked at one time or another, with the annelids, gephyreans, ascidians, brachiopods and ectoproct bryozoans. Masterman (1900) attempted to relate the phoronids to the Hemichordata in the group "Trimetamera." It is due to his influence that some old textbooks include the phoronids in the phylum Chordata. Its protrusible crown of tentacles, tubicolous life and the position of the anus on the dorsal side, not far from the mouth, do bring it quite close to *Cephalodiscus*; but observations of Masterman have not been substantiated by subsequent work and the relationship between the phoronids and hemichordates has thus been discredited.

Lately, the phenomenon of spiral cleavage has been used extensively as a basis for establishing phylogenetic relationships.[1] The groups of animals in which this type has generally been recognized (Turbrellaria,) Platyhelminthes, Nemertina, Sipunculoidea, Echiuroidea, Annelida, Mollusca) are considered as, more or less, related to one another, although many factors are concerned to establish this point of view.

---

[1] Joan C. Rattenbury, *Journ. of Morph.*, Vol. 95, 1954.

# 15 PHYLUM BRACHIOPODA

The **Brachiopoda** (L. *brachio*, arm: Gr. *pod*, foot) are another group of lopophore-bearing unsegmented coelomate animals found in the sea, and like the phoronids are also sessile, benthic and non-colonial. Unlike the phoronids they secrete a bivalved shell, which caused an unnecessary confusion with the phylum Mollusca, but the shells of the brachiopods are dorsal and ventral in position and not lateral as in the molluscs. They have inhabited the sea-bottom since animal life was first common in Cambrian time but are not very widely spread. Numerous species are known from ecological times some 456 genera are known from Paleozoic rocks, and 177 from the Mesozoic. They were important fauna of the invertebrate world of that time. About 70 genera covering some 225 living species are their sole representatives of the modern time. Modern genus *Lingula* (Fig. 15.1 F) is almost identical with that of the Ordovician time, found at least 5000,000,000 years ago and has the "honour" of being the oldest known animal genus. Most of the living forms occur in shallow waters, a few occur in deep waters. The animals are called "lamp-shell" because of the resemblance of a valve to an old Roman oil lamp. The larva of Brachiopoda is free-swimming and may be considered as modified trochophore.

**Shells**

The body of a Brachiopoda (Fig. 15.1) is enclosed within a bivalve shell, but the two halves are not right and left as in the Pelecypoda, but are dorsal and ventral. The dorsal shell is, as a rule, the smaller of the two and usually lies towards the lower side of the animal in natural state (Fig. 15.2). The gape

# Phylum Brachiopoda

of the brachiopoda shell is anterior and the hinge is posterior. The posterior end of the body is extended into a stout muscular stalk or peduncle by which the animal is attached usually to rocks and stones. The **peduncle** is absent in some cases (*Crania*) in which the animal is attached by the ventral valve. The posterior end of the shell narrows, and the ventral valve projects behind the dorsal and may be produced into short **beak** or **funnel** through the aperture out of which the stalk protrudes.

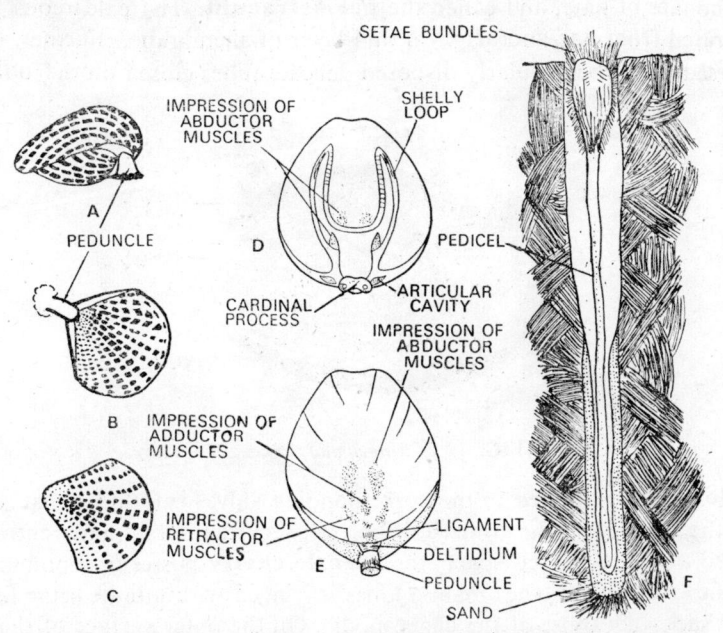

**FIG. 15.1.** *Magellania*, the brachiopod. A—Entire animal seen from side; B—The same seen from above; C—Dorsal valve of the same; D—Interior of the dorsal valve; E—Interior of ventral valve; and F—The inarticulate *Lingula* in feeding position within burrow.

The **ventral valve**, as already stated, is produced posteriorly into a beak terminating in a foramen for the peduncle. Distally the shell does not complete the margin of the valve (Fig. 15.1) but a separate double plate, the **deltidium**, closes it. Anterior to the beak is the curved hinge-line along which the valve articulates with its fellow. Still anterior, the inner surface of the shell is produced into a pair of large **articular teeth** or **dent articulaire**. Posteriorly the inner surface of the valve shows a number of shallow depressions which are impressions of **superior** and **inferior adductor** muscles, **superior** and **inferior abductor** muscles and **adjustor** muscles meant for the attachment of shells (Fig. 15.1).

The **dorsal valve** has no beak (Fig. 15.1) but at the hinge its posterior edge is produced into a strong **cardinal process** or **apophysis**. During actual articulation the cardinal process fits into the space between the articular teeth of the ventral valve. In the middle the inner surface of the shell presents a **median ridge** or **septum** which continues posteriorly up to the cardinal process. With the base of the septum are attached the two arms of a peculiar **calcareous ribbon**, the **shelly loop**, which projects freely into the cavity enclosed between the two valves (Fig. 15.1). The inner surface, besides this, presents impression of the muscles for attachment.

The valves externally present a series of concentric rings the line of growth. In many forms the shell is perforated by minute pores which, however, are closed, externally by the outer layer of the shell. They lodge the tubular prolongations of the mantle. The shell is made up of **prismatic** rods of calcium carbonate placed obliquely to the external surface made up of chitinous substance mixed with carbonate of lime, and called the **periostracum**. The calcareous spicules are separated from one another by a thin layer of membrane. Further, the shell is traversed by perpendicularly disposed delicate tubes closed on the outer side.

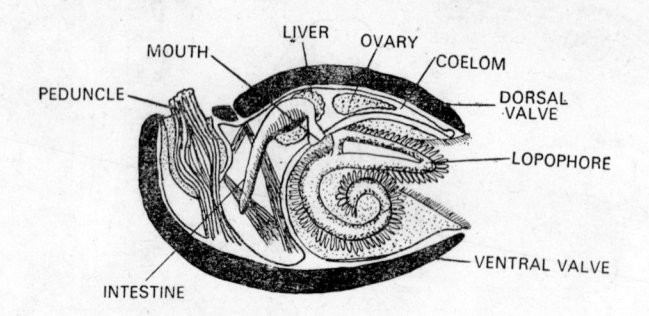

FIG. 15.2. *Magellania*, internal anatomy.

**Mantle.** Closely applied to the corresponding valves is found a flat continuation of the body wall, the **mantle lobe**. Between the dorsal and ventral lobes is enclosed a wedge-shaped space, the **mantle cavity** posteriorly bound by the body. The margin of the mantle lobes is fringed with minute setae lodged in muscular sacs (like those of the chaetopods). On the outer surface of the mantle arise delicate hollow processes which extend into the tubules of the shell. Each of the mantle lobes is formed by the reduplication of the body wall and encloses a prolongation of the coelom (Fig. 15.2).

**Lopophore**

The body of the animal lies at the posterior end of the cavity occupying about one-third of the space enclosed by them. The dorsal and ventral surfaces are in contact with the two valves, whereas, the anterior surface faces the gape. The most conspicuous structures of the body are the two spirally coiled tentacular arms or the horseshoe-shaped **lopophore** (Fig. 15.2). The two arms of the lopophore curve inward and coil to fit into the mantle cavity. The middle of the concave edge, which is dorsal in position, is produced into a spirally coiled offshoot and lies coiled towards the dorsal side between the two arms. The lopophore is hollow internally containing a spacious cavity or sinus. Its two main arms also receive prolongations of the coelom into which project the **digestive glands**. A ciliated food groove runs along each arm of the lopophore. The outer boundary of this groove is formed by a row of long ciliated tentacles, whereas, on the inner side the groove is bounded by a wavy **ridge** or **lip**. The tentacles are long and may protrude from the gape. The cilia on the groove and those on the mantle surfaces produce two ingoing currents of water that carry food particles and also maintains a steady supply of oxygen.

## Body Wall

The body wall comprises three layers: (*i*) an outer layer of **epidermis** consisting of a single layer of cells; (*ii*) next layer of **connective tissue** which is more or less cartilaginous at many places; and (*iii*) a ciliated **coelomic epithelium** lining the body cavity.

**Body cavity.** The body cavity or the perivisceral cavity is a true coelom and extends into the folds of the mantle. It is divided into a right and a left half by a dorsoventral mesentery. Transverse mesenteries also occur. The coelom communicates with the exterior by one pair of nephridia. It is lined by ciliated coelomic epithelium and is filled with fluid into which float colourless corpuscles.

**Musculature.** The valves are attached to each other by well-developed muscles (Fig. 15.1 and not by elastic ligaments. There are two large **adductor** muscles arising from the dorsal walls, one **superior** and the other **inferior**. As these two muscles pass ventral-ward they unite and thus have only a single insertion on the ventral valve. These muscles close the shell. A small pair of **abductor** or **divaricator** muscles arises from the ventral valve and are inserted into the cardinal process, which they depress. The function of these muscles is to open the shell. Besides these, there are two pairs of **adjustor** muscles, one arising from the ventral, and the other from the dorsal valve and inserted into the peduncle. Their function is to adjust the position of the animal as a whole by turning in various directions.

## Digestive System

The **mouth** is a narrow cresentic aperture situated in the middle of the lopophore in a transversely directed **buccal groove**, which, in the simplest case has a somewhat circular course (Fig. 15.2). The mouth is on the posterior side but the anterior part of the groove is bent backwards (Fig. 15.2). In other forms the buccal groove does not form a complete circle, but is incomplete anteriorly and the two ends so formed are coiled into a spiral which is variously disposed in different forms (Fig. 15.2). The mouth leads into a ciliated alimentary canal, which is V-shaped comprising a sac-like **stomach** and **intestine** a straight tube which extends from the stomach downwards and backwards towards the ventral side and ends blindly in some forms (*e.g. Magellania=Waldhesmia*). In other forms (*e.g.* in *Lingula* and *Crania*) the intestine ends in an **anus**. Into the stomach opens the digestive glands composed of branching tubes.

## Excretory Organs

The renal organs are **nephridia**. Usually there is a pair of nephridia, but in *Rhynchonella* there are two pairs, one dorsal and one ventral. In the body cavity each nephridium opens by a funnel-shaped **nephrostome** and externally on either side of the mouth near the base of the arm. These tubes also function as gonoducts.

## Nervous System

The well-developed nervous system comprises a **supraoesophageal ganglion** in front of the mouth and a larger **suboesophageal ganglion** behind the mouth both connected by **circumoesophageal connective**. The suboesophageal gives off nerves to the dorsal mantle lobe, the arms and abductor muscles and to two small ganglia which supply the ventral mantle lobe and the muscle of the peduncle. All the ganglia and commissure are in immediate contact with the ectoderm. Special organs of sense have not been described up to now.

### Circulatory System

The circulatory system is simple and consists of a longitudinal vessel in the dorsal mesentery. One region of this is distinguished as a contractile vesicle, the **heart**, placed on the dorsal side of the stomach. The heart gives off a vessel which passes forward along the oesophagus and supplies the tentacular region and the oesophagus, etc. There are two pairs of small lateral vessels running to the generative organs. The blood is colourless.

### Generative Organs

The sexes are usually separate, some forms however, may be hermaphroditic. The generative organs develop from the coelomic epithelium near the intestine. The gonads appear as thick yellow bands or ridges which project into the body cavity and extend into the lacunae of the mantle where they may be considerably branched. The ripe sex cells are released into the body cavity whence they pass to the exterior through the nephridia. In some genera (*Thecidium, Cistella, Argiope*) the eggs undergo the first stages of development in brood pouches placed near the openings of the nephridia. The brood pouch may be median and ventral (*Thecidium*) or paired (*Argiope*).

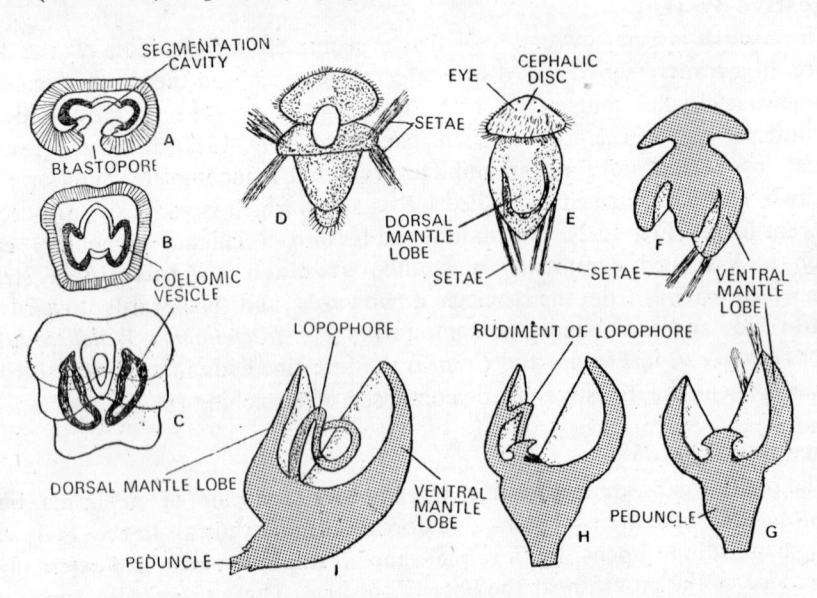

FIG. 15.3. Stages in the development of a brachiopod, *Argyrotheca*. A to C—Formation of the coelomic vesicles from segmentation cavity; D and E—Later stages in development; and F to I—Metamorphosis of the same.

**Development.** Much work on the development has yet to be done. The egg is small and the segmentation is complete resulting in the formation of a ciliated **blastula**. **Gastrula** is usually formed by invagination. The blastopore narrows and gradually closes. In the interior two diverticula arise from the archenteron and these are gradually constricted off forming **coelomic sac** (Fig. 15.3 B.C.). An ectodermal thickening in front forms the **apical plate** from which is derived the supraoesophageal ganglion. The embryo then becomes constricted by an annular furrow, the **mantle groove**, into two parts. The anterior of these soon divides in a similar manner so that three segments are formed. Of these the anterior is the **head region**, the middle one is the **body** or **mantle** region, the third **foot** or **peduncle** region. The enteron is confined to the two anterior segments and does not extend to the caudal segment.

As the development proceeds the head region develops into an umbrella-like disc which develops cilia and four eyespots. The second segment or mantle lobe grows back projecting over the caudal region and develops four groups of provisional setae, that are replaced by a second set of setae in the adult. In this condition the embryo leaves the brood-pouch and begins its free-swimming life, which is of a short duration. During larval life the enteron remains entirely closed. After some time it comes to rest and fixes itself by the peduncular segment and soon the mantle folds turn forwards and envelop the first segment.

The larva now assumes the adult condition. Valves of the shell are secreted by the mantle folds, the provisional setae are replaced by the permanent ones, the umbrella-like head region decreases in size and perhaps forms the lip which is at first confined to the part immediately behind the mouth. The lopophore first appears on the inner surface of the dorsal mantle lobe, but gradually extends and surrounds the mouth. At its earlier stages it is circular (but becomes horseshoe-shaped by sending out paired extensions. The caudal region of the larva becomes the peduncle of the adult.

## Affinities

There are four main opinions regarding the natural affinities of Brachiopoda:

(1) **The Mollusca theory**. At first they were brought nearer to acephalous molluscs with which their bivalve shell gives a striking external resemblance. The position of the valves, the peduncle, the arms and all the interior organizations are essentially different in two groups. The development has nothing common in either. The molluscan affinities have, therefore, nothing more than a historical interest.

(2) **The Chaetognatha theory**. The second opinion links them with the Chaetognatha (see later) based on the resemblance in the formation of the coelom and mesoderm which is enterocoelic in both. This resemblance is indeed very real, but nowadays the embryogenic processes, specially those of the first stages are supposed to be regulated by certain factors and laws common to several types and thus do not have any phylogenetic significance. Even if phylogenetic importance may be given to the development it will be seen that in the present case only the appearance of mesoderm has some similarity between the two and the further development between the two is quite different and so is the organization of the adult.

(3) **The Annelida theory**. Comparatively more admissible appears their relationship with annelids. The larva is believed to be of trochophore type. This is suggested considering that the mouth in the larva is formed on the ventral side of the umbrella-like head segment which is the preoral lobe and which soon dis-

appears or is reduced to the epistome (lip of the buccal groove). Many people are opposed to this view. Further, the presence of setae secreted in cutaneous sacs, indication of segmentation plan of the nervous system and relations of the coelom in the adult also suggest annelidan affinities. In the light of this view the Brachiopoda can be regarded as annelids with three segments, marked in the embryo by the annular constrictions of the integument. Thus, the Brachiopoda may be considered with some probabilities of reason like the segmented annelids (*Cephalobranchs* or *Oligonereis*) which might have become fixed and transformed their tubes into a pair of calcareous scales (valves) adhering to the epidermis. But the segmentation of the body is only superficial and there is no segmentation either of the body cavity, or of the genital organs, or of the nervous system, or in segmental organs, etc. Hence, this view had to be given up.

4. **The Phoronis theory**. The opinion which tends to prevail links them with *Phoronis* and ectoprocts. The arms with their tentacles are comparable to the lopophore and the lip to the epistome of *Phoronis*. Further, the possible presence of three mesoblastic segments, the relations of the nephridia and the coelom, and the mode of attachment by caudal region can also be simulated. About the caudal region it is said that it is regarded as equivalent to the evaginated foot of *Phoronis*. Thus, the suggestion that the Brachiopoda are allied to Phoronidea appears attractive. It is strengthened by the arguments that (*i*) in both the preoral lobe shrivels up or disappears leaving at most the epistome as its indication; and (*ii*) in both, the dorsal surface between the mouth and anus is extra-ordinarily shortened.

But even these principal arguments are open to serious objections. On the whole the increased knowledge of its anatomy and development suggests that the status of independent phylum may be assigned to the group with special affinities to the Annelida by (*a*) the form of their central nervous system; (*b*) their setae; (*c*) the presence of the well-developed perivisceral coelom; and (*d*) the traces of an imperfect segmentation.

## CLASSIFICATION OF BRACHIOPODA

The phylum Brachiopoda is divided into two classes **Inarticulata** or **Ecardines** and **Articulata** or **Testicardines**, on the basis of the nature of their shells and the presence or absence of anus.

I. CLASS INARTICULATA or ECARDINES. These are the Brachiopoda having two valves of the shell almost alike and of horny texture and without hinge, being held together by muscles. Shelly loop is absent and anus is present. Among the recent forms this class includes *Crania* in which the ventral valve is flat, fastened to the rock and the dorsal valve is conical. *Lingula* found in the Pacific and Indian Oceans lives between the tide lines and burrows in mud with long contractile peduncle. This includes many fossil forms.

II. CLASS ARTICULATA or TESTICARDINES. This comprises Brachiopoda having two dissimilar valves formed of calcareous material; a hinge is usually present and usually with a beak for the peduncle and shelly loop for the lopophore. Anus is absent. Besides many fossil forms, this group includes most of the living genera such as *Rhynchotremata, Terebratulina, Magellania* (=*Waldhesmia*).

# 16 PHYLUM ARTHROPODA

The Arthropoda (Gr. *arthros*, joint+*podos*, foot) are jointed limbed animals in which the segments of the body are variously fused to form the head, thorax and abdomen or the like. The exoskeleton, containing chitin, is moulted entire at intervals. Cilia are entirely absent. The Arthropoda contains animals that have attained the greatest biological success. This group has the largest numbers of species and of individuals, occupying the widest stretches of territory and the greatest variety of habitats; consuming the largest amount and kinds of food, and most capable of defending themselves against their enemies. More species of arthropods have been described than of all other kinds of animals put together. More than 800,000 species have been described and many of them are enormously abundant as individuals. This includes approximately 80 per cent of all known animal species. The group includes prawns, lobsters, shrimps, crabs, water-fleas, barnacles (Class **Crustacea**), the insects (Class **Insecta**), the centipedes (Class **Chilopoda**), the millipedes (Class **Diplopoda**), the spiders, scorpions, ticks, and their allies (Class **Arachnida**) and other less familiar and fossil forms. *Peripatus*, representing a division of Arthropoda which have preserved features that are clearly intermediate in structure between two modern phyla (**Annelida** and **Arthropoda**) is now included in this phylum under class **Onychophora**.

The Arthropoda are bilaterally symmetrical animals in which the outer protective layer becomes highly intensified. In the annelids the body is invested by a cuticle secreted by the epidermis, but in the arthropods, although still secreted by the epidermis, the outer covering assumes much greater proportion and forms the exoskeleton. The exoskeleton is made up of chitin and takes various forms in different groups. This protective covering enables the arthropods to live successfully in every conceivable environment. Due to the natural rigidity the exoskeleton is sloughed off from time to time. This phenomenon is **ecdysis**. Due to the formation of chitinous exoskeleton cutaneous breathing, as in annelids and other lower animals, is not possible. Therefore, in aquatic forms **gills** are formed and in terrestrial forms a system of air-tubes, the **tracheae**, has been developed by means of which oxygen of the air is conveyed directly to the

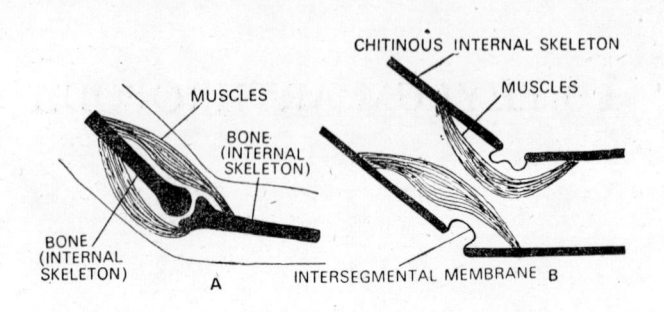

FIG. 16.1. Diagram illustrating the structure of a jointed appendage (*arthros* + *podos*) and its comparison with the endoskeleton of vertebrates.

tissues. Among other internal modifications due to the formation of a tough exoskeleton is the reduction of the coelom that is replaced by **haemocoel**. The body shows metameric segmentation. Some anterior segments undergo cephalization forming a distinct head. Metamerism is clear in the embryonic development of all arthropods and is a conspicuous feature of many adults, especially the more primitive species. In the arachnids and crustaceans metamerism has become reduced. In some forms like mites and crabs it has almost disappeared. Loss of metamerism has occurred in three ways. Segments have become lost, segments have become fused together, and segmental structures, such as appendages, have become structurally and functionally differentiated from their counterparts on other segments. Different structures having the same segmental origin are said to be serially homologous. Thus, the second antennae of a crab are **serially homologous** to the **chelipeds** (claws), for both have evolved from originally similar segmental appendages. The limbs are jointed. Proctodaeum and stomodaeum are lined by cuticle continuous with the exoskeleton. A peculiar feature of the Arthropoda is the complete absence of cilia except in *Peripatus*. True nephridia are absent. Muscles are striated, usually complex, capable of rapid action.

### Arthropoda and Annelida

The Arthropoda have much in common with the Annelida and must be regarded as derived from the same stock as **Polychaeta** in that phylum. These are the two invertebrate phyla with conspicuous segmentation. Both show this in the body muscles and nervous system. The Arthropoda differ from the Annelida in: (*i*) the general lack of intersegmental septa, (*ii*) reduction of the coelom, (*iii*) concentration of the excretory organs and the gonads, (*iv*) presence of chitinous exoskeleton, (*v*) presence of jointed appendages, (*vi*) presence of compound eyes, (*vii*) absence of cilia, and (*viii*) separation of sexes.

## General Characters

The cuticular skeleton basically consists of two layers: an outer **epicuticle** and inner **procuticle**. The epicuticle is a thin layer composed of proteins and lipids, but lacks **chitin**. **Procuticle**, on the other hand, is a much thicker layer which contains chitin as the principal constituent (Richards 1951). It is separated into two layers, an outer **exocuticle** and an inner **endocuticle**. Chemically chitin is an acetate of the polysaccharide that contains glycosamine as its most important constituent. The procuticle may be impregnated with mineral salts in some arthropods. For instance, in the Crustacea the procuticle is impregnated with

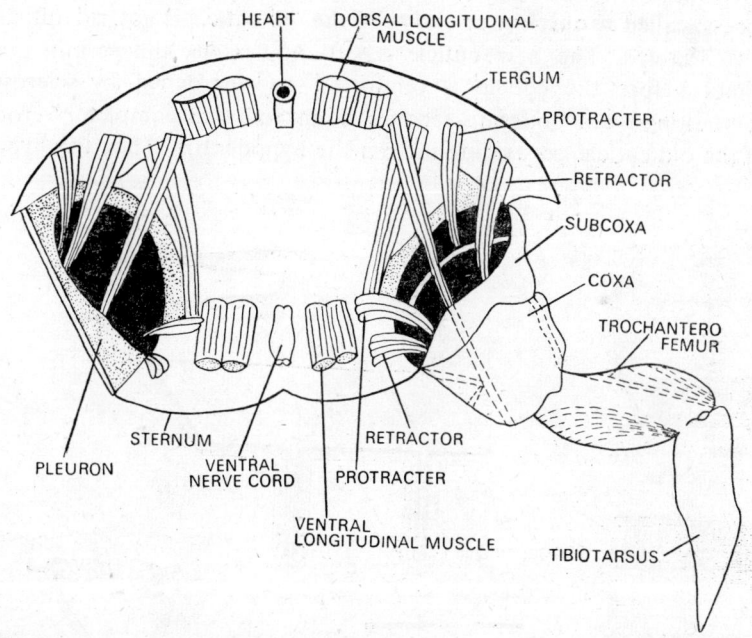

FIG. 16.2. A transverse section of an arthropod body segment: showing appendicular articulation and various segmental muscles.

calcium carbonate and calcium phosphate. Where it is thin the culicle is permeable to gases or to water. The arthropodan cuticle also extends in the derivatives of stomodaeum and proctodaeum. Thus, both the foregut and the hindgut are lined by chitinous cuticle only the midgut is without it. The tracheal tubes of insects, chilopods, diplopods, the booklungs of scorpions and spiders, and parts of the reproductive systems of some groups also have cuticular lining.

Often body colouration originates from the deposition of brown, yellow, orange and red melanin pigments within the cuticle. Sometimes the colour is produced by subcuticular chromatophores or is caused by blood and tissue pigments, which are visible through a thin transparent cuticle. Iridescent greens, purples and other colours result from fine striations of the epicuticle which produce colour effect by refraction of light.

The problem of movement that a continuous external skeletal covering would have created has been eliminated by the division of the cuticle into separate plates. Basically there are four such plates in each body segment, a dorsal plate, the **tergum**, a ventral plate, the **sternum**, and two lateral plates the **pleura**. The plates of the adjoining segments are connected by means of an articular membrane, a region in which the cuticle is very thin and flexible. Sometimes secondary fusion or subdivision of segments may also occur.

The cuticular covering of the appendages is also divided into tube-like segments connected to one another by articular membrane, thus creating a joint at each junction. In some arthropods the special articular condyles and sockets appear at the joints.

In order to meet the requirement of growth in spite of the presence of an external skeleton a new device of periodical shedding of cuticle has developed. During this process, called **moulting** or **ecdysis**, the old cuticle is got rid off and new cuticle is secreted. The new cuticle is soft and flexible, and permits growth of the animal before the cuticle is reinforced and hardened by sclerotization. During moulting the hypodermis secretes an enzyme that completely erodes the base of the old cuticle, separating it from the hypodermis. Then the hypodermis

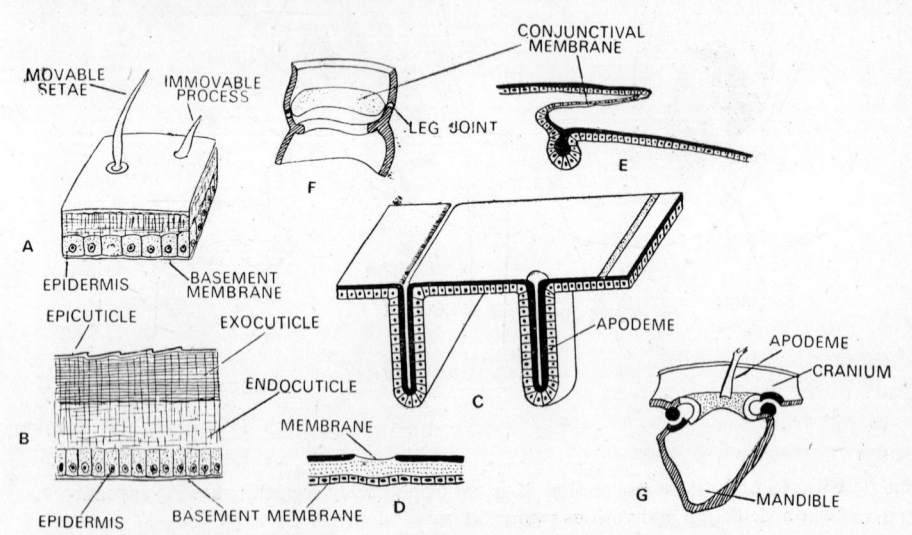

FIG. 16.3. A—Three dimensional section of an insect integument; B—Vertical section of arthropod integument; C—Apodemes; D—A simple membranous suture; E—A conjunctival membrane; F—A dicondylic leg joint; and G—The typical extrinsic dicondylic articulation of the mandible with the cranium.

secretes a new epicuticle, which is impervious to the previously secreted enzymes. After this the hypodermis secretes new procuticle. Once the new cuticle is formed the old one is cast off. The process of shedding is simple. There is a longitudinal rupturing of the cuticle on the dorsal and lateral sides of the body and the adult then pulls itself out of the old cuticle. The new cuticle is soft and unprotective. Usually the animal goes into hiding until the new cuticle hardens.

The nature of body musculature has greatly changed in the arthropoda, because of change in the method of movement. In annelids the somatic musculature takes the form of two cylinders, the outer comprises circular fibres, while the inner cylinder consists of only longitudinal fibres. These muscular cylinders have disappeared in the arthropods and distinct striated muscles have formed. These are attached to the inner surface of the skeletal system. The muscles are attached to the procuticle by special fibres inserted on the inner surface. Flexion and extension between plates are effected by the contraction of these muscles, with muscles and cuticle acting together as a lever system. This presents an example of co-functioning of muscular system and skeletal system to bring about locomotion, as found in vertebrates. In the arthropods the muscles are attached to the inner surface of the external skeleton, while in vertebrates to the outer surface of the internal skeleton. The jointed appendages of the arthropods act as paddles in the aquatic species or as legs in the terrestrial forms.

S.M. Manton (1951) has made extensive studies on the arthropodan locomotion on land. The locomotor appendages of arthropods tend to be more slender, longer and located more ventrally. The body usually sags between the limbs. In the cycle of movement of a particular leg, the effective step or stroke, during which the tip of the leg is in contact with the substratum, is closer to the body than is the recovery stroke, when the leg is lifted and swung forward. Speed depends upon the length of the stride which itself depends on the length of the legs. In centipedes, the arthropods with a large number of legs, the fields of movements of individual legs overlap those of other legs. Thus, the difference in proximity of the legs to the body during the effective and recovery strokes prevents mechanical interference. Field overlap tends to be considerably reduced in those arthropods utilizing five, four or three pairs of legs. Stride length is greater and the problems of mechanical interference are further decreased because of difference in leg length and the relative placement of the leg tip.

The arthropodan gait involves a wave of leg movements, in which the posterior leg is put down just before or a little after the anterior leg is lifted. The movements of the legs on opposite sides commonly alternate with one another, sometimes they may coincide. Alternate leg movement tends to induce body undulation, because one limb of a given pair is moving through its effective stroke while its opposite is making a recovery stroke. This tendency (of body undulation) is counteracted by increased body rigidity, such as the fused leg-bearing segments that form the thorax and cephalothorax of insects, some crustaceans, and arachnids.

Arthropodan skeletal muscles are typically innervated by two types of motor neurons—fast fibres, which effect rapid contractions and slow fibres. In addition to these there are some inhibitory fibres in crustaceans. Each neuron usually innervates numerous muscle cells, and the total number of axons supplying a given muscle is relatively small. The nature of contraction in such a multiple innervating system is dependent upon the type of neuron used, as well as the interaction of different types of neurons. For example, two different extensor muscles are innervated by the same motor fibre in a crayfish claw, but the two muscles function independently because each possesses separate inhibitory fibres.

For the attachment of muscles internal skeleton (endoskeleton) has developed. An infolding of the procuticle produces inner projections or **apodemes**, on which the muscles are inserted. The nature of endoskeleton varies in different arthropods.

The well-developed metameric coelom as found in annelids does not exist in the arthropods. Instead there occur spaces or sinuses filled with blood hence called **haemocoel**. The coelom is restricted to the gonads (gonocoel) and its associated ducts, and in some arthropods to space in the excretory organs.

The circulatory system of arthropods is open. There is a dorsal heart (compare dorsal vessel of annelids) which varies in length and position in different arthropodan groups. The heart consists essentially of one or more chambers with muscular walls arranged in a linear fashion and perforated by pairs of lateral openings called **ostia**. Blood flows from the surrounding sinus to the

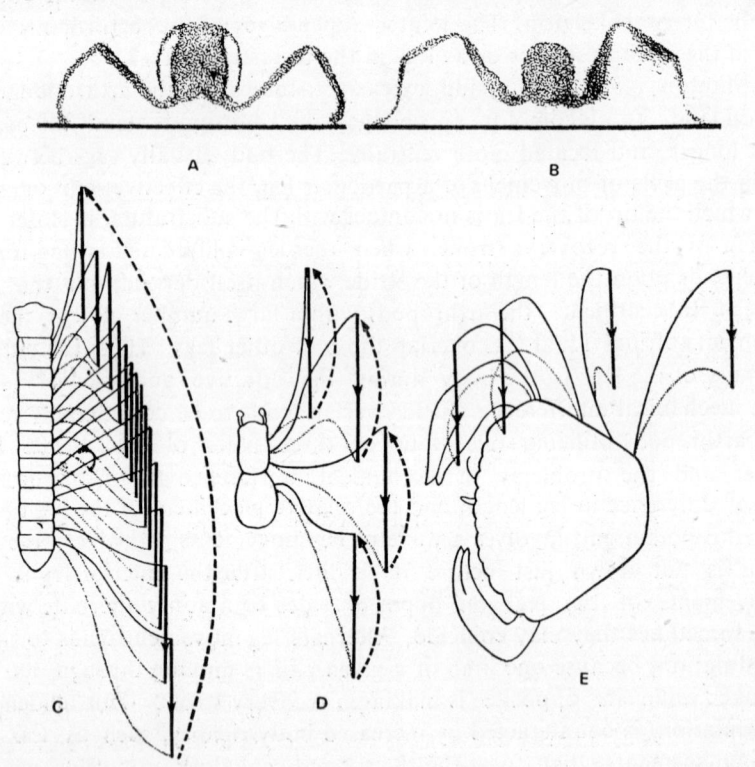

FIG. 16.4. Locomotion in arthropod. A—Stance of a crayfish; B—Spider; C to E—Leg movement in the gait of different arthropods. The effective stroke and the stride length shown by heavy bars, therefore, each leg is drawn twice, at the beginning and at the end of effective stroke. The recovery stroke is shown by dotted lines; C—Strokes of *Scutigera*; D—Strokes of spider; and E—Strokes of a crab.

heart through the ostia. The membrane that separates the chamber containing the heart from the general body cavity is called **pericardium,** which does not derive from the coelom as in the vertebrates, but instead is a part of the haemocoel. After entering the heart, blood is pumped out to the body tissues through vessels frequently called arteries, and is eventually dumped into sinuses

# Phylum Arthropoda

in which it bathes the tissues directly. The blood then returns by various routes to the pericardial sinus.

The blood contains several types of **amoebocytes** and the respiratory pigments such as haemoglobin and haemocyanin are sometimes found dissolved in plasma.

The nervous system has developed in the same pattern as that found in the annelids, but shows higher degree of cephalization. The brain is larger and is correlated with well-developed sense organs (eyes and antennae). The behavioural patterns of arthropods are complex. The brain of arthropodans shows

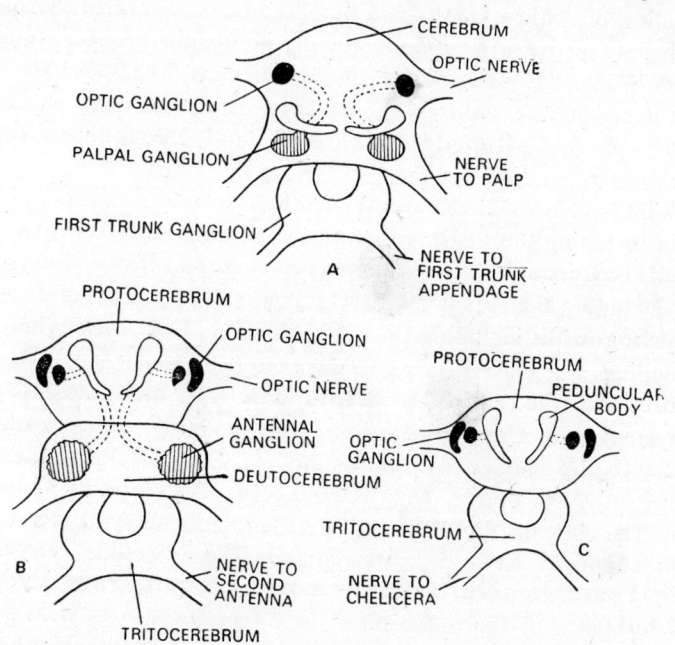

FIG. 16.5. Comparison between an annelid and arthropod brain. A—Polychaete annelid; B—Mandibulate arthropod; and C—Chelicerate arthropod.

three subdivisions, an anterior **protocerebrum**, median **deutocerebrum** and posterior **tritocerebrum**. The protocerebrum has one to three pairs of optic centres (**neuropiles**) which send nerves to the eyes. The optic and other neuropiles of the protocerebrum integrate photoreception and movement and are probably the centres for the initiation of complex behaviour.

The deutocerebrum sends nerves to the antennae (first antennae in Crustaceans) and contains their association centres. Arthropods that have no antennae (scorpions, spiders, mites) are without deutocerebrum.

The tritocerebrum sends nerves to the labrum, the digestive tract (stomatogastric nervous system), the chelicerae of chelicerates, and the second antennae of crustaceans. The commissure of the tritocerebrum is located behind the forgut (postoral).

There is a considerable controversy regarding the evolutionary origin of the arthropodan head and brain.

Majority of the arthropods are dioecious. Many employ modified appendages during copulation. Fertilization in terrestrial forms is always internal, but may be external in aquatic species. The eggs of most arthropods are extremely rich in yolk and are centrolecithal. Some arthropods display traces of spiral cleavage, but most have a modified type of cleavage that is associated with centrolecithal eggs and is called superficial cleavage.

## CLASSIFICATION OF ARTHROPODA

The classification of Arthropoda is in a flux because of divergent views concerning the phylogeny of the group. Consequently no definitive system for its classification exists. Some zoologists have divided it into five subphyla including **Branchiata** (crustaceans, trilobites), **Insecta** or **Tracheata** (insects), **Chelicerata** (spiders, etc.), **Chilopoda** (centipedes), and **Progoneata** (millipedes, etc.), with several appendices for forms of uncertain status.

Some authors have divided the group into three subphyla: Subphylum **Trilobitomorpha** including the fossil trilobites, subphylum **Chelicerata** including classes (i) **Merostomata** (living horseshoe crabs and fossil Eurypterids; (ii) **Arachnida** including the scorpions, harvestmen, spiders, ticks and mites; and (iii) class **Pycnogonida** including the sea-spiders; and the subphylum **Mandibulata** including six classes, viz., **Crustacea** (the crustaceans), **Insecta** (the insects), **Chilopoda** (the centipedes), **Diplopoda** (the millipedes), **Symphyla** (the symphylans) and **Pauropoda** (the pauropodans). Some authors divide Arthropoda into 12 classes with no subphyla. Some feel that there are four major groups of Arthropoda, viz., **Crustacea, Myriapoda, Insecta** and **Arachnida**. The classification followed here presents a synthesis of modern views. Seven subphyla have been recognised. These include **Onychophora, Tardigrada, Pentastomida, Trilobitomorpha, Chelicerata, Pycnogonida** and **Mandibulata**. Of these the first three represent forms with doubtful or only superficial relationship with other Arthropoda. Trilobitomorpha includes a number of extinct classes. The subphylum Chelicerata (classes Merostomata and Arachnida) is distinguished by the absence of antennae and the presence of a pair of chelate appendages. The subphylum Pycnogonida include the sea spiders. The Mandibulata includes the Crustacea. Some authors keep Crustacea as a subphylum and include myriapods and Insecta in one subphylum Antennata, because both the Myriapoda and Insecta lack chelicerae and have one or two pairs of antennae and a pair of mandibles. Of these the most ancient is Onychophora which is being described first. The sequence of description of classes is mainly for the sake of convenience.

## PHYLUM ARTHROPODA

**Subphyla:**  **ONYCHOPHORA** (the peripatus)
**TARDIGRADA** (the tardigrades)
**PENTASTOMIDA** (the pentastomides)
**TRILOBITOMORPHA** (the fossil trilobites)
**CHELICERATA**

*Phylum Arthropoda*

| | |
|---|---|
| **Class:** | MEROSTOMATA |
| **Subclasses** | XIPHOSURA (horseshoe crabs) |
| | EURYPTERIDA (fossil Eurypterids) |
| **Class:** | ARACHNIDA (the largest of Chelicerata) |
| **Orders:** | Scorpionida (scorpions) |
| | Pseudoscorpionida (false scorpions) |
| | Solifugae (sunspiders or wind scorpions) |
| | Papligarda (papligardes) |
| | Uropygi (whip-scorpions) |
| | Amblypygi (amblypygids) |
| | Araneae (Spiders) |
| **Suborders:** | Orthognatha (mygalomorphs) |
| | Labidognatha (araenomorphs) |
| **Orders:** | Ricinulei (ricinulids) |
| | Opiliones (harvestmen) |
| | Acarina (mites and ticks) |
| **Class:** | PYCNOGONIDA (sea spiders) |
| **Subphylum:** | **MANDIBULATA** |
| **Class:** | CRUSTACEA (crustaceans) |
| **Subclasses:** | CEPHALOCARIDA |
| | BRANCHIOPODA |
| **Orders:** | Anostraca |
| | Notostraca |
| | Choncostraca |
| | Cladocera |
| **Subclass:** | OSTRACODA |
| **Orders:** | Myodocopa |
| | Cladocopa |
| | Podocopa |
| | Platycopa |
| **Subclasses:** | MYSTACOCARIDA |
| | COPEPODA |
| | BRANCHIURA |
| | CIRRIPEDA |
| **Orders:** | Thoracica |
| | Acrothoracica |
| | Apoda |
| | Rhizocephala |
| | Ascothoracica |
| **Subclass:** | MALACOSTRACA |
| **Classes:** | DIPLOPODA (millipedes) |
| | CHILOPODA (centipedes) |
| | SYMPHYLA (symphylans) |
| | PAUROPODA (pauropodans) |
| | INSECTA (insects) |
| **Subclass:** | OLIGOENTOMATA |
| **Order:** | Collembola |

| | |
|---|---|
| **Subclass:** | MYRIENTOMATA |
| **Order:** | Protura |
| **Subclass:** | APTERYGOTA |
| **Orders:** | Thysanura |
| | Aptera or Diplura |
| **Subclass:** | PTERYGOTA |
| **Division:** | Exopterygota (Hemimetabola) |
| **Orders:** | Orthoptera |
| | Grylloblattodea |
| | Phasmida |
| | Dictyoptera |
| **Suborder:** | Blattaria |
| **Orders:** | Dermaptera |
| | Plecoptera |
| | Isoptera |
| | Zoraptera |
| | Embioptera |
| | Psocoptera |
| | Mallophaga |
| | Siphunculata |
| | Ephemeroptera |
| | Odonata |
| **Suborders:** | Anisoptera |
| | Zygoptera |
| **Orders:** | Thysanoptera |
| | Hemiptera |
| **Suborders:** | Heteroptera |
| | Homoptera |
| **Division:** | Endopterygota |
| **Order:** | Neuroptera |
| **Suborders:** | Megaloptera |
| | Palnipennia |
| **Orders:** | Mecoptera |
| | Trichoptera |
| | Lepidoptera |
| **Suborders:** | Zeugloptera |
| | Monotrysia |
| | Ditrysia |
| **Order:** | Coleoptera |
| **Suborders:** | Adephaga |
| | Archostemmata |
| | Polyphaga |
| **Orders:** | Strepsiptera |
| | Hymenoptera |
| **Suborders:** | Symphyta |
| | Apocrita |
| **Order:** | Diptera |

| | |
|---|---|
| **Suborders:** | Nematocera |
| | Brachycera |
| | Cyclorrhapha |
| **Order:** | Aphaniptera |

## I SUBPHYLUM ONYCHOPHORA

The onychophorans have been described as the "missing link" between annelids and arthropods, because this small group of animals does have many interesting similarities to both annelids and arthropods. There are only about 65 species of existing onychophorans. All species were originally included in a single genus, *Peripatus*. Now the species are referred to 12 genera in two families. Members of this group are commonly called peripatus. The Onychophora (Gr. *onychus*, claw + *phorus*, bearing) which means "claw bearing" and refers to the curved claws on the feet, is the most ancient group, and does not appear to have changed greatly since the Cambrian period, from which the only certain fossil specimen was discovered. They are elongate, unsegmented animals with many cross rows of fine papillae and 15 to 43 pairs of stumpy legs with claws. Head not distinct, eyes present, and antennae short. Sex openings mid-ventral before anus. The onychophores show affinities with the true worms. They are terrestrial animals with a thin soft cuticle. The head consists of only three segments of which the preoral one bears preantennae and two postoral segments bear jaws and oral papillae, etc. The body segments are alike, each bearing a

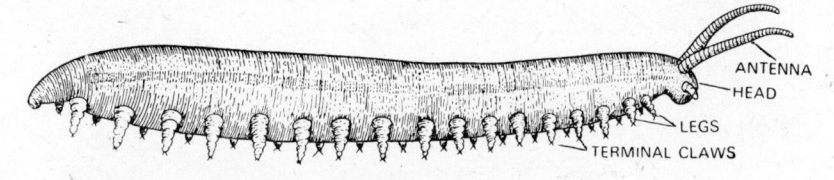

FIG. 16.6. External features of *Peripatus*.

pair of appendages. The legs of onychophores bear arthropod-like claws. The animals breathe through trachea, the openings of which (**stigmata**) are scattered all over the body. The body cavity is haemocoel, there are paired lateral ostia opening into the pericardium. The reproductive organs are furnished with cilia. The existence of an animal with structures peculiar to two different phyla is the result of the process of organic evolution. But such a case creates difficulties in classification. Some such groups of dual nature have been raised to the status of independent phyla. The same has been done for the Onychophora by some. Others, however, think that the onychophores are definitely annelids (with soft cuticle, segmental nephridia and ciliated gonoducts, etc.) and that their arthropod-like characters have arisen independently. Still there are some zoologists who think that the arthropodan characters of these animals (including the jaws derived from appendages, the body cavity being haemocoel, the dorsal "heart" with ostia, the coelom reduced to the cavities of the nephridia and reproductive ducts; the tracheate respiration and the general structure of the reproductive organs, etc.) are more emphasized and as such place them with the Arthropoda as a separate subphylum. *Peripatus*, the animal belonging to this group, is regarded as a connecting link, that has survived the pressure of time and exists as an intermediate stage between the two phyla. The same view has been adopted here.

The known onychophorans are confined to tropical regions (the East Indies, the Himalayas, the Congo, the West Indies and northern South America) or to the south temperate regions (Australia, New Zealand and South Africa). No

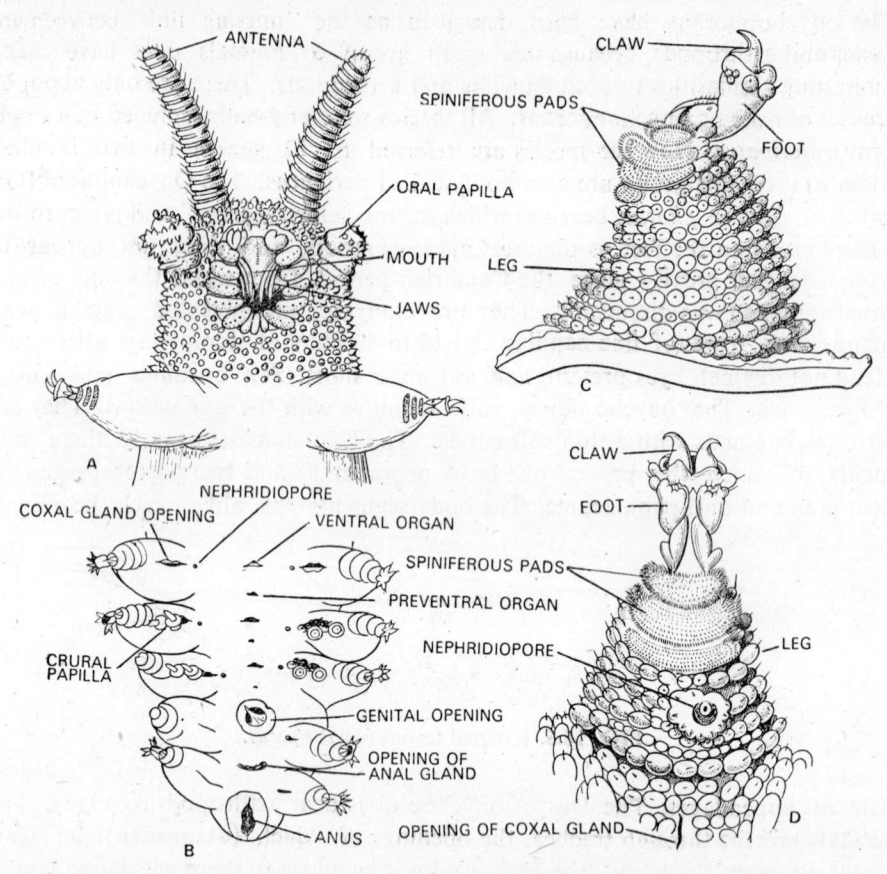

**FIG. 16.7.** A—Ventral view of the anterior end of *Peripatus* (*after* Balfour); B—Ventral view of posterior extremity of the same (*after* Bouvier); C—Enlarged diagram of leg of *Peripatus*, lateral view; and D—Enlarged diagram of the same, front view.

species have been reported north of the Tropic of Cancer. Most onychophorans live in humid habitats, such as the tropical rain forests, beneath logs, stones and leaves or along the banks of streams. During low temperature such as in winters they become inactive and remain in hiding.

## TYPE PERIPATUS

*Peripatus* is a rare animal found in moist places under rocks, in crevices, under stones and logs, in the tropical forests of Africa, Australia, Asia, South and Central America, West Indies, Malaya Peninsula, Sumatra and a few other

## Phylum Arthropoda

regions. Its occurrence only in local regions in such widely separated parts of the world suggests that it was probably a more successful and widespread form in the past but is now gradually disappearing. About 65 species are known and they all are shy creatures living in the dark avoiding the day. Because of its light avoiding nature it is rarely seen.

*Peripatus* looks much like a caterpillar (Fig. 16.6), 1.4 cm to 15 cm long with soft velvety skin and many pairs of stumpy legs, each ending in two claws (Fig. 16.7 B). It shows no external segmentation, the only features of segmentation are several (14-43) pairs of legs, a pair for each internal segment of the body, depending upon species and sex. The stumpy legs are not jointed. The anterior end lacks cephalization, no distinct head is formed. It, however, bears two short **antennae** and a pair of simple annelid-like eyes dorsally. The **mouth** is ventral and is surrounded by a fleshy fold. It is further furnished by two small horny jaws used to tear or grind food. On either side of the mouth are situated two blunt processes, the **oral papillae**, on which open a pair of glands that gives off

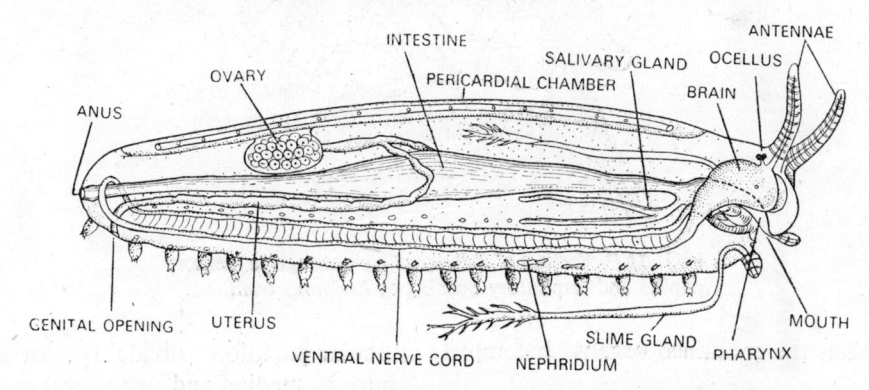

**FIG. 16.8.** Internal organs of *Peripatus*, seen from side (*after* Balfour).

slimy secretions which are discharged when the animal is attacked. Each of the head appendages (antennae, oral papillae, jaws) occurs on one of the three segments that compose the head. The three-segmented head of *Peripatus* is thought to indicate a condition midway between that of annelids and arthropods (that have six segmented head). The skin is thin and lightly chitinized with many transverse rings of fine papillae, each with a spine. At the base of each leg towards the inner side is a **nephridial opening**. The anus opens at the bluntly conical posterior end. The **genital opening** is also singly ventral pore, a little anterior to the anus. The ventral surface is reddish in colour and dorsal darker, presenting an elaborate pattern produced by minute mottling of various colours such as green, red and brown, etc. Onychoporans, as a whole, are coloured blue, green, orange, or black and the papillae and scales give the body surface a velvety and iridescent appearance.

### Body Wall

The outer covering is a thin cuticle like that of annelids. It is ridged and covered with microscopic tubercles, small and large, which give it a velvety texture unknown in other animals. The **epidermis** is the next layer that also secretes the cuticle on its outer surface. Beneath this lies the musculature consisting of a thin but complex series of muscle layers comprising **circular, longitudinal** and **transverse muscles**. A layer of epithelium lines the body cavity and

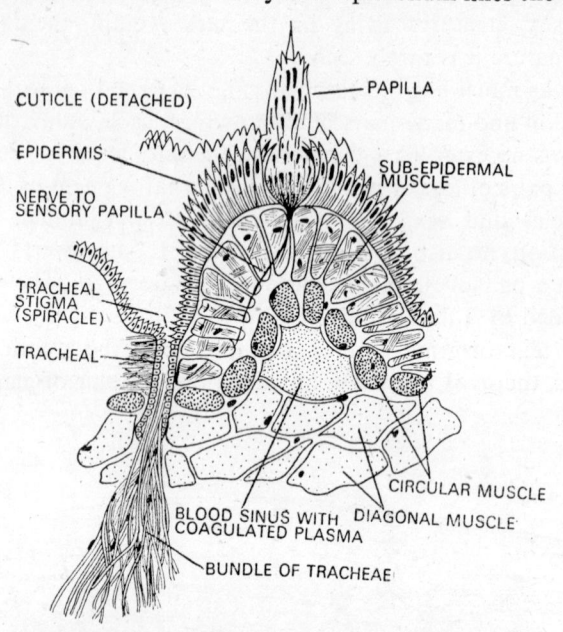

FIG. 16.9. Section of skin passing through a sensory papilla and respiratory opening of *Peripatus trinitatus*.

invests the contained organs. Incomplete muscular partitions divide the cavity into a median and two lateral compartments, besides the pericardium which lodges the heart. The prolongations of the lateral compartments extend into legs. Developmental studies show that the body cavity is not a true coelom but haemocoel.

### Digestive Organs

The internal anatomy is a mixture of annelid-like and arthropod-like structures. The digestive tract is simple and not particularly distinctive. The buccal cavity is formed secondarily by the union of a ring of papillae and folds surrounding the true mouth into a circular lip. On its roof it bears a slight prominence, the **tongue,** and a row of small **spine** or **teeth.** The buccal cavity also encloses the base of the jaws. It is

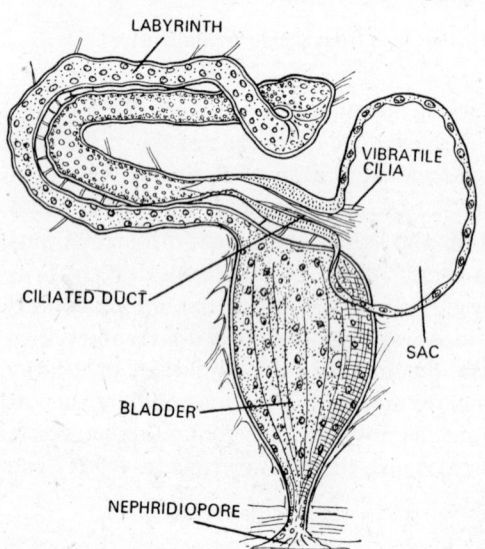

FIG. 16.10. Nephridium of *Peripatus*, highly magnified (*after* Gaffron).

# Phylum Arthropoda

followed by a thick-walled **pharynx**, in turn, followed by the **oesophagus**. This leads into a thin walled tube extending nearly to the posterior end of the body and known as **stomach intestine** or **mesenteron**. The **rectum** is a narrow terminal part of the intestine that opens at the anus. There are two long, narrow, tubular **salivary glands** that open into a small diverticulum leading backwards from the buccal cavity. These are supposed to be modified nephridial tubes (Fig. 16.10).

## Circulatory System

The circulatory system consists of a single mid-dorsal tube running through the entire length of the body. It is enclosed in a **pericardial sinus** by a longitudinal partition, the **pericardium**. The heart communicates with the pericardial cavity through segmentally placed pairs of valvular **ostia**, one pair of which lies opposite each pair of legs.

## Respiratory System

The respiratory system consists of delicate unbranched **tracheal tubes**, which are lined by a thin chitinous layer exhibiting fine transverse striations. They extend throughout the body and open on the external surface, in little depressions of the integument, the **tracheal pits**, through openings called **stigmata**, which

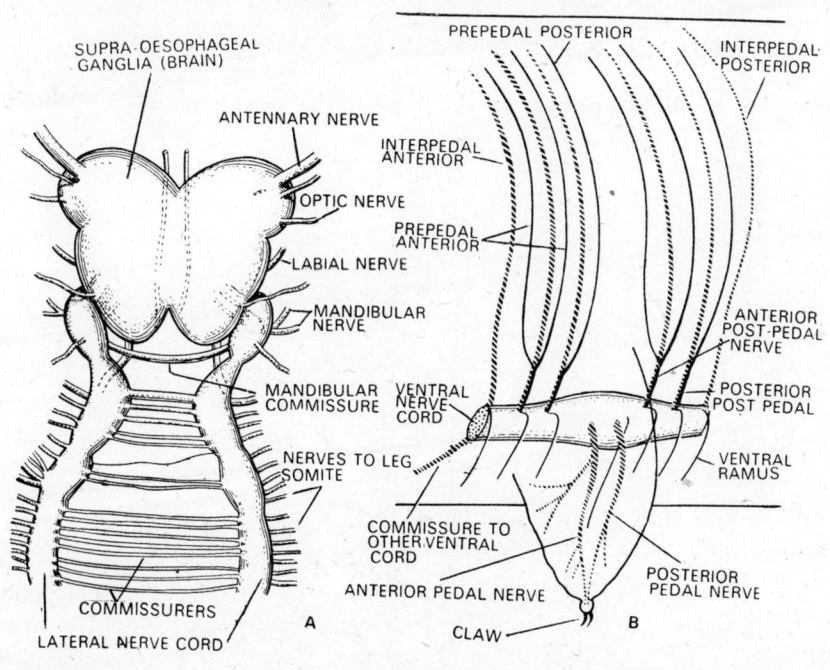

**FIG. 16.11.** A—Brain, highly magnified (*after* Balfour); and B—Segmental nerves, lateral view.

are scattered throughout the surface. Although such structures occur nowhere else in the animal kingdom except in terrestrial arthropods, they are considered to have arisen independently in the two groups. The tracheal tubes are lined with ciliated epithelium. The number of stigmata varies in different species.

### Excretory System

The most annelid-like character is the excretory system. This consists of segmentally arranged pairs of coiled tubes, the **nephridia**, which open by external

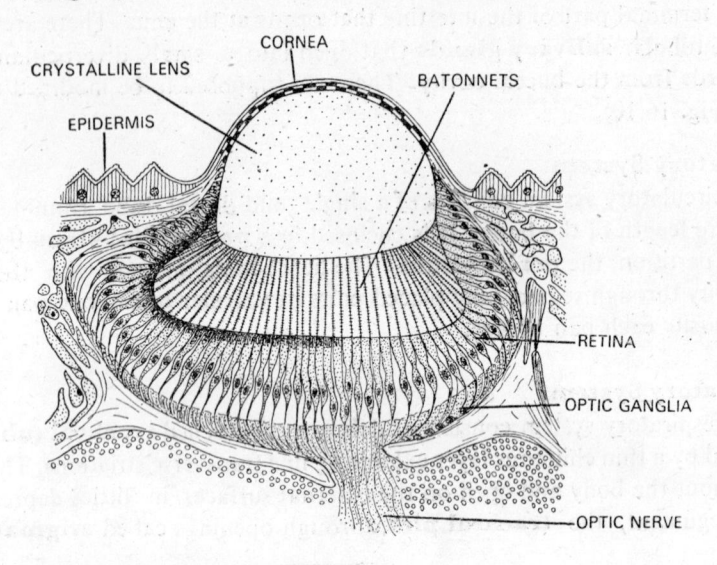

FIG. 16.12. Sagittal section of the eye of *Peripatus* (*after* Dakin).

pores, the **nephridiopores** at the bases of the legs. Each has a coelomic sac

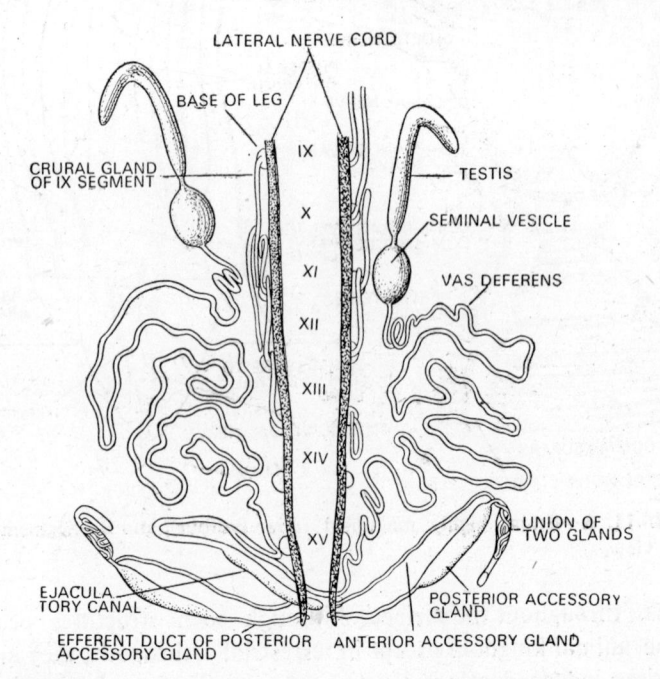

FIG. 16.13. Male genital system of *Opisthopatus cinctipes* (*after* Purcell).

# Phylum Arthropoda

closed at its inner end, a ciliated funnel and duct, a slightly expanded bladder opening at the nephridiopore (Fig. 16.10). The wastes are collected from the coelomic sac presumably entering there by diffusion from the large internal blood space. Cilia in the tube drive the excretory products out.

In some species there exist **coxal** or **crural glands** lying in the lateral compartments of the body cavity. Their ducts open on the lower surface of the legs (Fig. 16.7), just outside the apertures of the excretory ducts. They are absent in the female except in *P. capensis*. Their number and arrangement differ in the males of various species. A pair of large **slime glands** (Fig. 16.8) lies in the body cavity in similar position and opens at the tips of the oral papillae. These are supposed to be modified coxal glands. When the animal is irritated their secretion is discharged often in the form of a number of fine viscid threads probably as a measure of defence.

## Nervous System

The nervous system is primitive. It consists of a pair of oval ganglia, the brain in the head, from which arise two widely separated ventral nerve cords

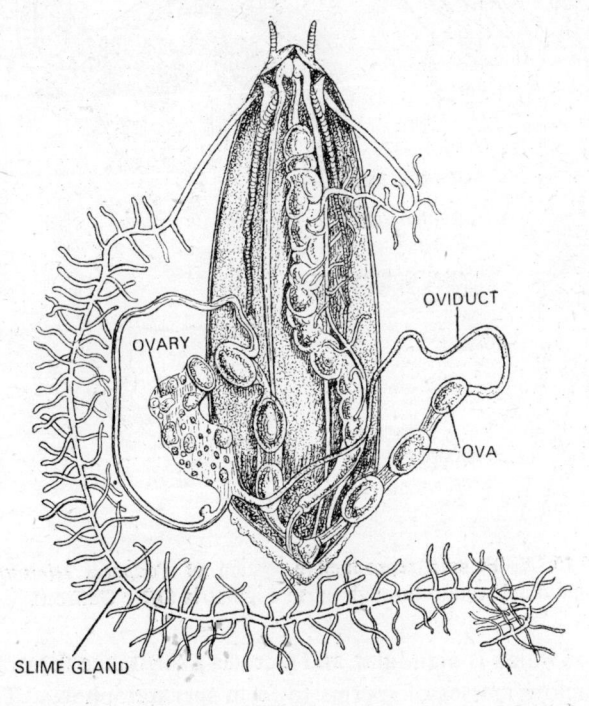

FIG. 16.14. Female generative organs of *Symperipatus oviparus*.

which run parallel to one another to the posterior end of the body, where they join together behind the anal aperture (Fig. 16.8). The cords show slight thickenings in each segment opposite each pair of limbs. Several transverse nerves connect the two cords imparting them a ladder-like appearance. From the brain nerves go to the antennae and other anterior structures. From the cords arise secondary nerves innervating the legs and other parts of the body. The eyes (Fig. 16.12) resembles those of annelids.

## Reproductive System

The sexes are separate and the reproductive organs are ciliated as in annelids. The gonads are tubular and paired, opening with a single aperture. The male generative organs consist of paired **testes**, each with a narrow **vas efferens**, opening like a funnel-like structure into a **seminal vesicle**. This is followed, on each side, by a long, narrow coiled **vas deferens**. The vasa deferentia from the two sides unite to form a common median tube, the **ejaculatory duct**, the

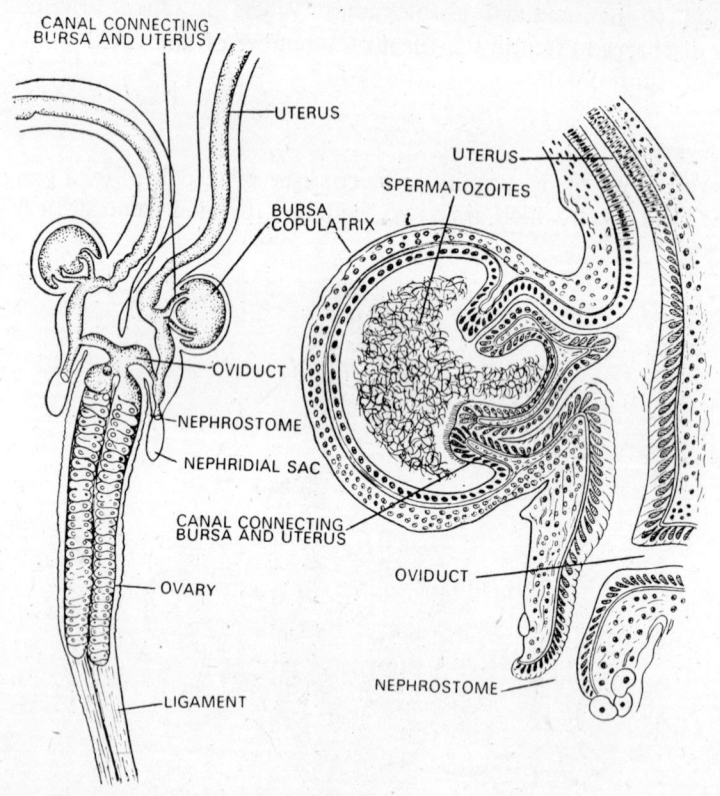

FIG. 16.15. A—Female reproductive system of *Peripatus trinitatis*; and B—Longitudinal section through bursa copulatrix (*after* Gaffron).

proximal part of which is glandular and secretes a substance forming complicated cases which enclose masses of sperms to form spermatophores. The ejaculatory duct opens to the exterior on the ventral surface between the legs of the last pair or behind them.

In the female the ovaries are two tubular structures followed by **oviducts** which join to form a **common atrium** (termed vagina by many) at the genital opening situated at the same place as in the males (Fig. 16.14). Most species of Onychophora are viviparous, and part of each oviduct is modified as a uterus where the embryos develop. The uteri join to form the common vagina. In the oviparous form the female pore is situated at the summit of a long cylindrical

# Phylum Arthropoda

process, the **ovipositor**. In some viviparous forms a placenta-like **trophoblast** develops through which the embryo draws nourishment from the uterine wall. In most forms fertilization is internal and the eggs develop within the females and the young are fully developed at birth. A large female may produce 30 to 40 young ones per year.

## Relationship

*Peripatus* is, in some respects, a connecting link between the Annelida and Arthropoda, but the reason for its inclusion in the latter phylum is not evident from superficial examination. Perhaps it is to be derived from polychaete which gave up marine habitat and became terrestrial. Parapodia are not present as

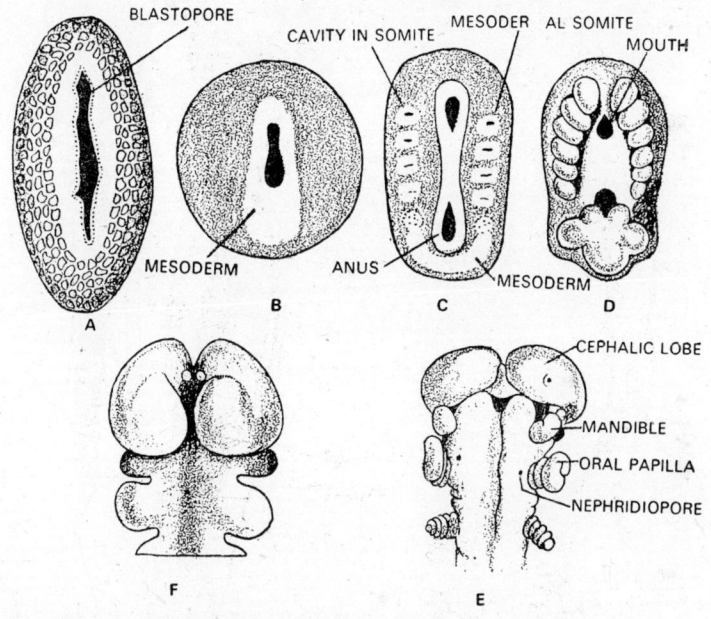

FIG. 16.16. A—Gastrula of *Eoperipatus weldoni* (*after* Evans); B to D—Evolution of blastopore of *Peripatus capensis* (*after* Sedgewick); and E to F—Embryos of *Peripatus trinitatis* (*after* Kennel).

swimming organs but have become modified for locomotion on land without having acquired the joint arthropod character. The integument is soft, no chitinous exoskeleton having developed, and the excretory organs take the form of metamerically repeated nephridia. Arthropodan features are the possession of tracheae, salivary glands and the terminal claws in the appendages. The presence of jaws of an appendicular nature, the paired ostia of the heart, the pericardium, the haemocoelic body cavity and reduced coelom are further important characters linking them with the Arthropoda.

## II. SUBPHYLUM PENTASTOMIDA

The **Pentastomida** are soft worm-like unsegmented animals with short cephalothorax and elongate ringed abdomen. Beside the mouth, there occur two pairs of retractile hooks. There are no circulatory, respiratory or excretory organs. Sexes are separate. The individuals are parasitic in vertebrates. The larva of *Linguatula serrata* is 5 mm long and lives in lung, liver etc., of rabbit, horse, goat. The adult female is up to 13 mm long and male up to 20 mm long and lives in the nasal cavities of fox, wolf and dog, occasionally in horse and goat but rarely in man. The eggs of *L. serrata* are sneezed out by the definitive host, and adhere to grass. When this grass is ingested by the intermediate host (a rabbit or hare) the egg-case is digested and the larvae released. The larvae

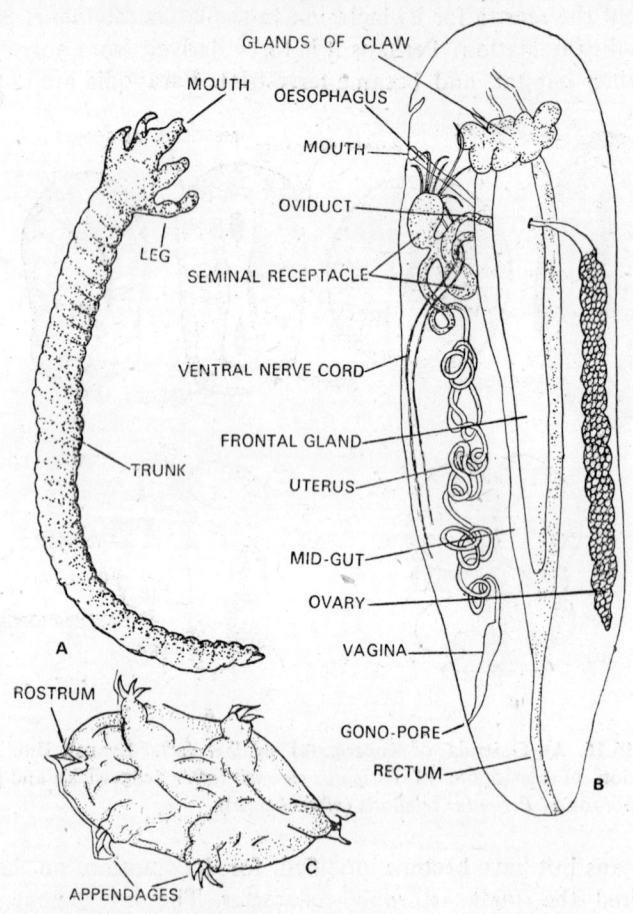

FIG. 16.17. A—*Cephalobaena tetrapoda* from the lung of a snake (*after* Heymons); B—Internal structure of a female *Waddycephalus*; and C—Larva of *Porocephalus clavatus*.

bore through the wall of the intestine into blood vessels and are carried to organs such as liver. Here the larvae form cysts and go through a series of molts, increasing in size and gradually changing to a form similar to the adult, but covered with circlets of minute teeth (Fig. 16.17). If the rabbit or hare is eaten by a dog or wolf, the larvae are released, and pass up the oesophagus to the pharynx and finally into the nasal cavities.

*Cephalobaena tetrapoda* has two pairs of rudimentary limbs, bearing claws apically, at the anterior end of the body. The alimentary canal is straight and simple. The nervous system is reduced. All muscles are striated. Sense organs are restricted to papillae on the surface of the body. Larvae show obvious arthropodan affinities. Among other genera *Reighardia* occurs in gulls and terns; *Sibekia* in crocodiles; *Kiricephalus* in water snakes; some others are found in rattlesnakes, boas and pythons (Fig. 16.17).

## III. SUBPHYLUM TARDIGRADA

The Tardigrada (waterbears and bear-animalcules) are small cylindrical unsegmented or vaguely segmented animals, not divisible into regions, rounded at both ends, about 1 mm long. Each possesses four pairs of stumpy unjointed legs (Fig.16.18) ending in two or more claws, the last pair being at the hind end of the body. The anterior end is provided with a retractile snout and teeth, mouth with stylet, no circulatory or respiratory organs are present. The mouth is surrounded by papillae, the buccal cavity contains a pair of horny, sometimes partly calcified teeth. Ducts from salivary glands open into the cavity of the

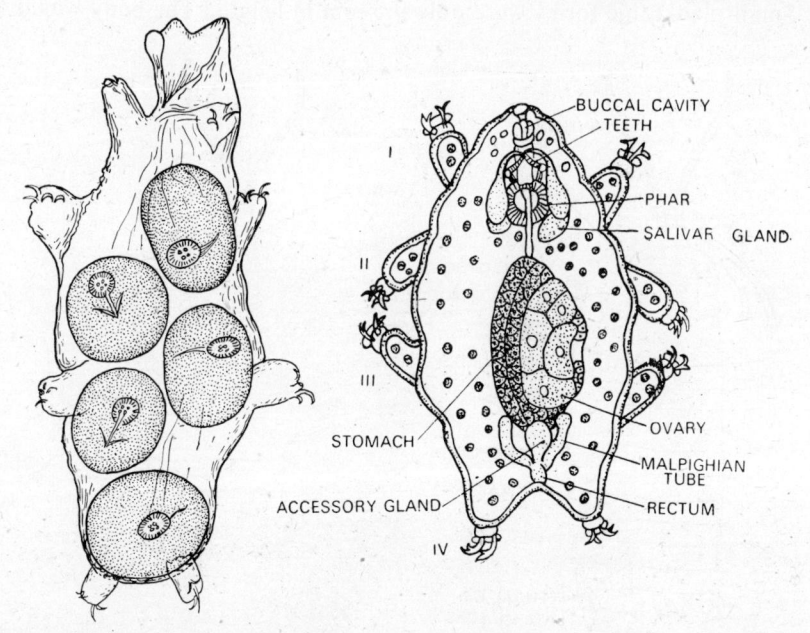

FIG. 16.18. A—*Macrobiotis hufelandi*; and B—Eggs left in the moult of *Hypsibius scabropygus*.

mouth. Muscular pharynx is followed by a narrow oesophagus which opens into an extensive stomach. The anus is subterminal situated in front of the last pair of limbs. For excretion peculiar rectal glands occur. The muscles are non-striated. There is a brain and a ventral nerve cord with four ganglia. The only sense organs present are two eyespots situated at the anterior end. Sexes are separate with only one saccular gonad each. Some have larval stages. More than 280 species have been recorded. *Echiniscus* is about 0.2 mm long, lives in moss on damp roofs, but can resist desiccation. *Macrobiotus* is 0.7 mm long and occurs in freshwater. Marine and terrestrial forms are also known.

## IV. SUBPHYLUM TRILOBITOMORPHA

The subphylum **Trilobitomorpha** is one of the most primitive groups of arthropods, that according to many, represents the starting point in discussing the arthropodan classes. The subphylum includes many classes of which the Trilobita is by far the most known. Others are **Pseudocrustacea, Marrellomorpha** and **Arthropleura**. All are extinct, known from their fossils. Trilobite fossils are well-preserved.

Trilobites are an extinct group of marine arthropods which were abundant and widely distributed in the seas of the Paleozoic era. They reached peak of their distribution and abundance during the Cambrian and Ordovician periods and disappeared at the end of the Paleozoic era. Over 3,900 species have been described from the fossils obtained so far.

The trilobite body was depressed, more or less oval and was usually about 3 cm long. Some were larger (about 10 cm long), while some were about 76 cm long. Small planktonic forms were only 0.5 mm in length. The body was divided

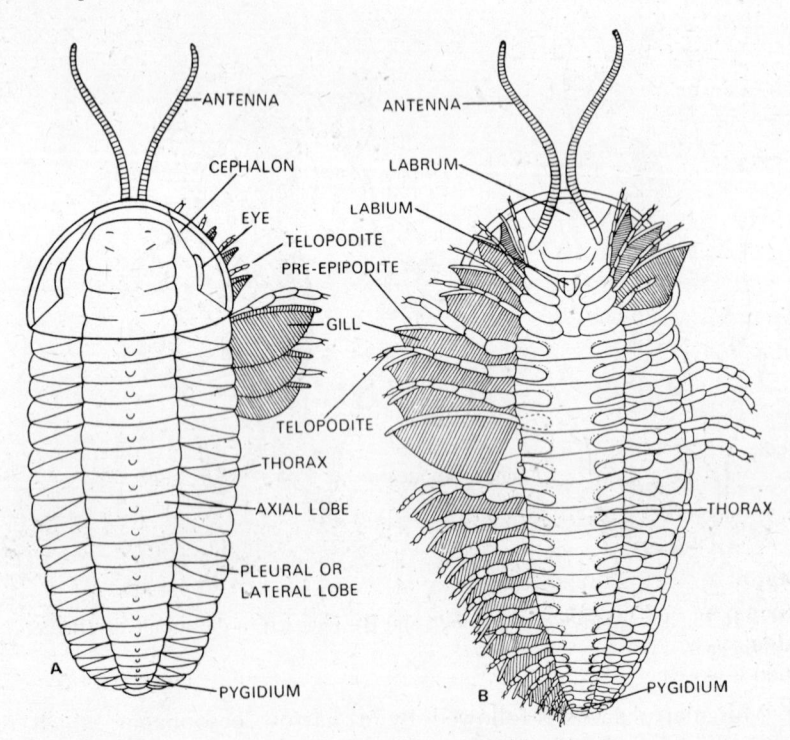

**FIG. 16.19.** The Ordovician trilobite, *Triarthrus eatoni*. A—Dorsal view; and B—Ventral view (*after* Walcott and Raymond).

into three, more or less equal parts—a solid anterior **head cephalon** or **prosoma**, an intermediate **thorax** or **trunk** and **abdomen** or **pygidium** (Fig. 16.19). Each of these body divisions was in turn divided in three regions by a pair of furrows running from the anterior to the posterior end, forming a median axial lobe flanked, on each side, by a lateral lobe. The name Trilobita refers to this division of the dorsal body surface into three parts. In some cases the thorax and abdomen formed the **opisthosoma**. The entire exoskeleton is calcareous. The dorsal cuticle of the exoskeleton was much heavier and thicker than the ventral, which carried the appendages. For this reason in some cases only the dorsal exoskeleton formed the fossil.

## Head

The head was covered by a **cephalic shield** or a carapace, the elevated median portion of which, the **glabella**, was usually divided by transverse groove indicating the number of segments fused to form it. Three to five segments form the head. **Cheek** or **genae** is the name given to the pleural parts. The last cephalic segment is called **occipital** or **neck-ring**. A furrow, the **neck-furrow**, separates the head from thorax. The furrows are more prominent in primitive forms than in comparatively recent types. The carapace continued ventrally into a **sub-frontal** plate, to the posterior end of which was attached a large **labrum** or **hypostome**. The posterior angles of the carapace were drawn out into spines. The mouth was located in the middle of the underside of the cephalon beneath the labrum.

## Thorax and Pygidium

The thorax was made up of 2-29 movably articulated segments, which were commonly bilobed, consisting of a median region or axis, and of lateral pleura often produced backwards and downwards. The pygidium was made up of a variable number of segments fused together. Since the thorax was mobile the trilobites, in many cases, were able to roll themselves up like wood-lice (Fig. 16.19).

## Appendages

Excepting the last or anal segment all body segments bore a pair of appendages. On each side of the labrum was a long sensory antenna (**pre-oral antenna**), homologous to the first pair of antennae of the crustaceans and the antennae of insects. The remaining head segments had four pairs of peg-like biramous appendages. The appendages of the thorax and abdomen were all biramous identical to the appendages of the head mentioned above. Each biramous appendage consisted of an inner walking leg (**telopodite**) and an outer gill bearing branch (**pre-epipodite**) (Fig. 16.20). The appendages articulated with the pleural membrane through a small **precoxal** segment that also bore on its outer side the pre-epipodite, which bore a fringe of filaments extending posteriorly. The fringe of one appendage overlapped the fringe of the next These filaments were probably **gills**. A large **coxa** carrying the walking leg (telopodite) articulated with the undersurface of the **precoxa** (Fig. 16.20). Considerable differences of opinion exist as to the exact homologies of the trilobite leg with those of other Arthropoda.

Most of the trilobites were bottom dwellers and crawled over sand and mud using the walking legs. This type of existence is proved by the flattened body and dorsal eyes. Some were perhaps adapted for burrowing as is evidenced by a shovel-shaped or plough-shaped cephalon. The bottom dwelling forms were probably scavengers or consumed mud or silt and then digested the organic materials from it like many annelids. Some people feel that the branchial filaments could have sifted food material from the surrounding water and then passed on to the mouth. The body of some trilobites was narrow and eyes were located on the sides of the head. They apparently led a swimming existence.

Fossil material has yielded a good understanding of the larval stages. Entire larval series has been preserved for some species of the genera *Sao* and *Olenus*. The trilobites passed through three larval periods, each with a number of instars.

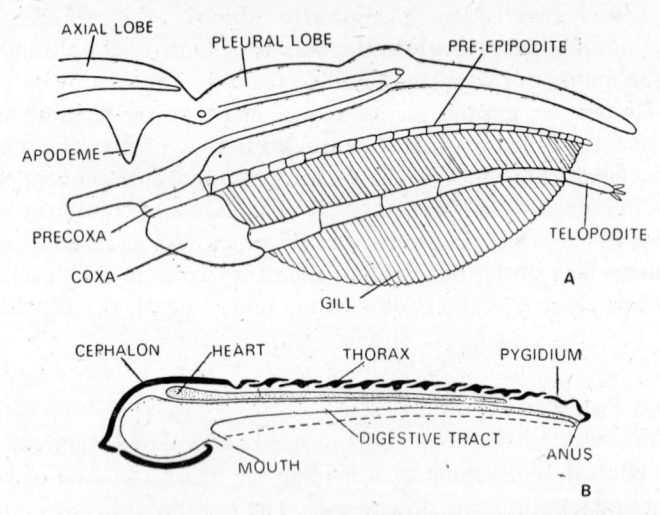

FIG. 16.20. A—Trilobite showing a trunk appendage, transverse section through left half (*after* Walcott and Raymond); and B—Trilobite, sagittal section (*after* Moore, Lalicker and Fischer).

The first larval form was **protaspis**. It was a small planktonic form measuring from 0.5 mm to 1 mm in length and the body had a single dorsal carapace consisting of the acron and four postoral segments. Segments were added subsequently and the **meraspis** larva was produced. In this larva the pygidium was located behind the cephalon. During subsequent moults thoracic segments were added between the cephalon and pygidium and the final larval stage called **holaspid (holaspis** larva) was formed. Although still small this larva displayed the general adult structure.

## V. SUBPHYLUM CHELICERATA

The Chelicerata are arthropods derived from an early offshoot close to that of the trilobites. The body of chelicerates is divided into two parts: an anterior **prosoma (cephalothorax)** and a posterior **opisthosoma (abdomen)**. There are no antennae. The first preoral appendages are feeding structures called **chelicerae** (from which the name Chelicerata has been derived). Originally the chelicerae are postoral in position but during development they move anterior. The chelicerae have been regarded as being homologous with the first antennae of Crustacea and antennae of insects, but evidence, now available, indicates that the

*Phylum Arthropoda* 549

chelicerae are homologous with the second antennae of Crustacea. The first postoral appendages are called **pedipalps** and are modified to perform various functions in different classes. The prosoma consists of six to eight postoral segments. The genital opening is on segment eight (usually second opisthosomatic somite). The chelicerates respire by book-gills, book-lungs, tracheae or skin.

The Chelicerata is divided into two classes, the **Merostomata** and **Arachnida**.

## CLASS MEROSTOMATA

The **Merostomata** are aquatic chelicerates in which the prosoma is provided with chelicerae and five similar pairs of appendages, the opisthosoma is divided into a **mesosoma** and **metasoma**. The mesosoma has five or six pairs of

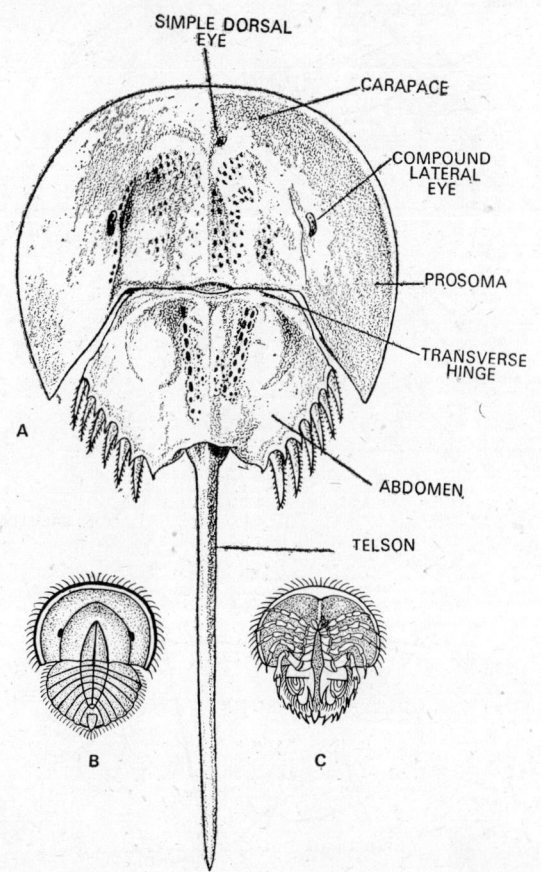

FIG. 16.21. A—King-crab (*Limulus polyphemus*), dorsal view; B—Lateral view; and C—Ventral view of young forms.

lamelliform appendages of which the first form a genital operculum, while the others are biramous and bear gills. The metasoma is without appendages and with a telson. The class is divided into two subclasses; **Xiphosura** and **Eurypterida**.

## Subclass Xiphosura

The Xiphosura are marine Merostomata represented by only three genera and five species living today. The fossil record of the subclass extends back to the Ordovician period. One of the known species is the horseshoe-crab or king-crab, *Limulus polyphemus,* common to the north western Atlantic coast and Gulf of Mexico. All other members of this group are found along Asian coast from Japan and south Korea through the East Indies and Philippines. Many are collected in the Bay of Bengal off the coast of Puri. Opisthosoma is hexagonal,

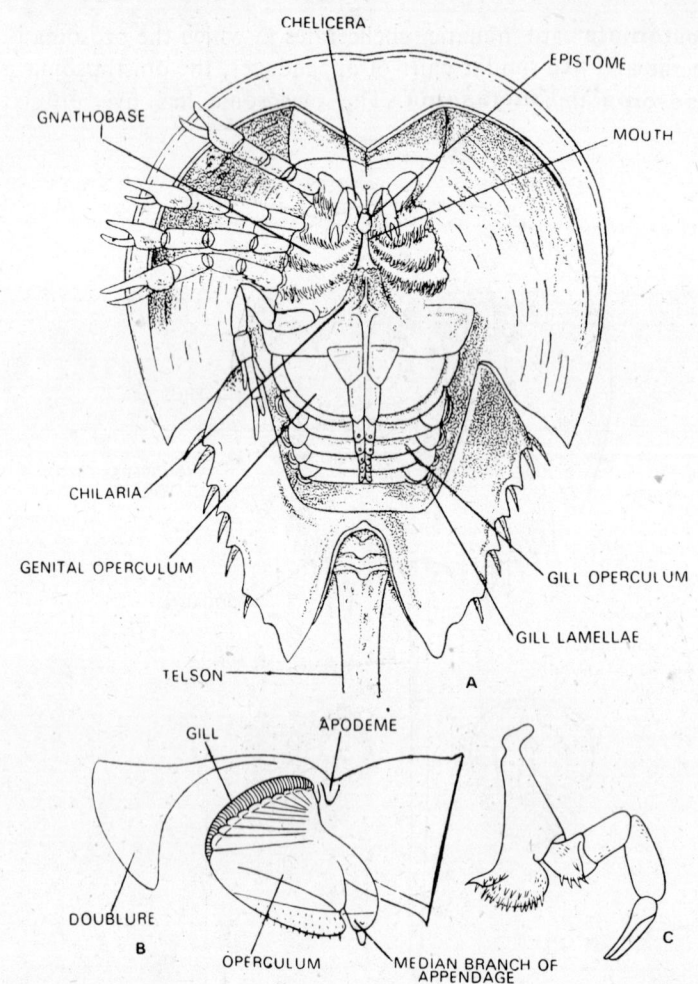

FIG. 16.22. A—King-crab, ventral view with legs of one side removed; B—Posterior surface of a gill operculum of *Limulus* in position beneath the body; and C—A walking leg.

broadly joined to prosoma, with six mesosomatic somites, a vestigial metasoma and a long spine-like telson. Chelicerae of three segments, small, chelate (pincer-like). Pedipalpi not similar to walking legs, all with six segments and gnatho-bases, lying round the central elongated mouth. Posterior legs with exopodite

# Phylum Arthropoda

and tibial apophyses forming a scoop. Chilaria behind the mouth. Mesosomatic appendages plate-like, the first genital operculum, the rest bearing gill-books, on the exopodite.

The king-crab, *Limulus*, is the only representative of this order having five living species. These animals have not changed much in form from their earliest fossil relatives and retain the archaic characters They are found in shallow water along sandy or muddy shores and can swim well but spend most of the time burrowing in the sand or mud for the worms and molluscs on which they feed.

The body of a king-crab consists of a **prosoma** having an unsegmented **carapace** and **opisthosoma** (abdomen) having a dorsal shield with serrated edges. The prosoma and opisthosoma are joined by a transverse hinge passing through segment eight whose tergite is split, one part being fused with the prosomatic carapace while the other forms the anterior lateral lobes of the opisthosomatic carapace.

The carapace of the prosoma is broad semicircular in outline with steeply sloping sides enclosing the space in which the appendages lie. Laterally its edges

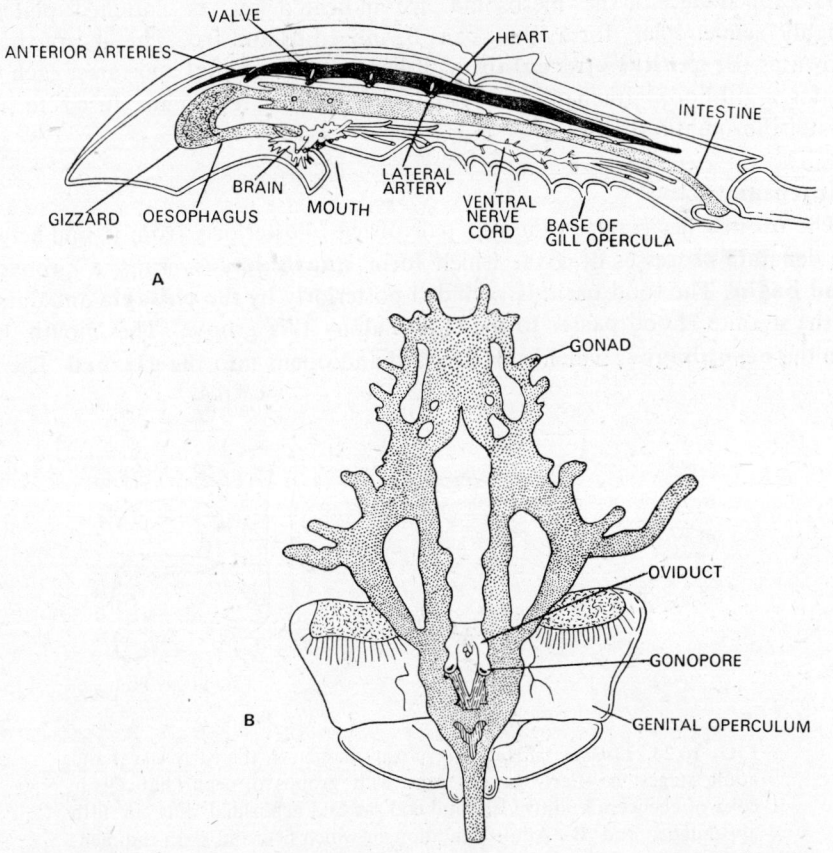

FIG. 16.23. A—Sagittal section through *Limulus* (*after* Patten and Redenbaugh); and B—Female reproductive system of *Limulus* (*after* Owen).

are bent over to form a **doublure.** It bears two median and two lateral eyes. The exoskeleton is of tough smoothly polished chitin. The prosoma is furnished with seven spines, one lying behind the median eyes, three in the middle row and three on the posterior margin of the prosoma which is a three-sided re-entrant into which fits the opisthosoma. The median eyes of *Limulus* are simple ocelli each with a single lens; the lateral eyes are compound. The opisthosoma is a broad hexagon without any trace of segmentation and with three spines along its middle line. Its posterolateral margins carry six small spines and terminate in a much larger one. Between these, six movable spines are articulated. The posterior margin is a three-sided concavity. It carries the **telson** articulated from its central edge. The appendages lie on the underside wholly within the cavity formed by the sloping sides of the prosoma. The **chelicerae** are small chelate organs of three segments, the **chelae** being smooth and finely pointed. The first legs of the male are modified as clasping organs. The second pair of appendages, which in all other living Merostomata, are known as **pedipalpi**, is like other ambulatory appendages of which there are four additional pairs (total five). These increase gradually in size from first to the fifth. All are composed of six segments, but a mark across the tibia of some perhaps indicates a **patella**, and restores the number of segments to seven—a characteristic of the arachnids.

The appendages of the mesosoma are indicated by six flattened plates of roughly semicircular form that can be moved to and fro. The first of these is known as the **genital operculum** as it lies over the genital apertures, each containing a gill-book attached to it. The tergites of this region are fused to form the opisthosomatic shield.

### Alimentary Canal

The **mouth** lies between the first pair of legs. Posteriorly from it, and between the denatate processes of coxae which form **gnathobases**, runs a groove or **food basin**. The food basin is bounded posteriorly by the **chilaria** and dorsally by the sternite. Food passes to the mouth along this groove. The mouth leads into the **oesophagus**, which runs forward and opens into the **gizzard**. The wall

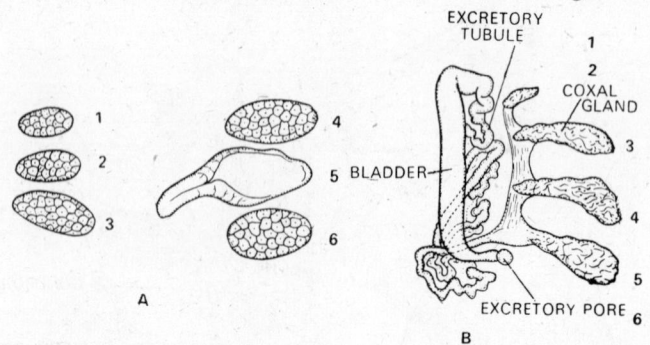

FIG. 16.24. *Limulus polyphemus* coxal glands in the early larval and adult stages. A—Early larval stage with groups of nephridial cells in coxa of chelicera and five legs and end sac and nephridial duct in fifth appendage; and B—Adult condition in which first and sixth rudimentary glands have disappeared.

of the oesophagus is folded (8 folds), while grooves in the gizzard are armed with chitinous teeth. Solid food is ingested and crushed in the gizzard. Indigestible

Phylum Arthropoda

portions are then ejected through the mouth, whereas fluid passes through a conical valve into the **midgut (stomach)**. The midgut has a pair of **hepatic caecae**, which terminate in ramifying glandular diverticula that occupy space in the cephalothorax and abdomen. Digestion is intracellular. The hind intestine is followed by chitinous rectum which opens at the anus on the last segment of the base of the telson. There is no salivary gland.

**Excretory Organs**

Excretion takes place through **coxal glands,** which in the larval stage consist of **saccules** derived from coelomic sacs, one in each segment one to six (Fig.16.24). During development the first and sixth degenerate, while those of the second, third, fourth and fifth segments lead by a short collecting tube into a common labyrinth which is much coiled before opening at the exit duct at the base of the fifth appendage.

**Respiratory Organs**

Gills are lamellate and borne an appendages of the ninth to thirteenth segments. They are protected by the lateral extension of the abdominal carapace.

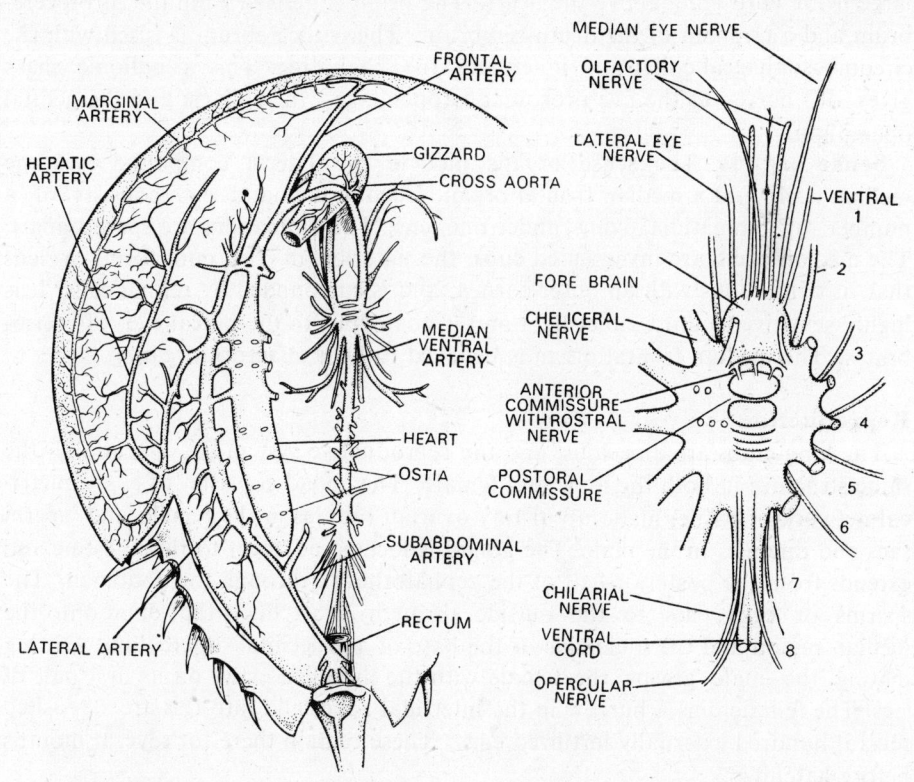

FIG. 16.25. A—Arterial system of *Limulus polyphemus*; and B—Brain of *Limulus polyphemus*.

A continuous current of water is maintained by the movement of the legs. The respiratory pigment is haemocyanin.

## Circulatory System

The circulatory system of *Limulus* is the most developed among arthropods. The **heart** (Fig. 16.23) is dorsal and is located in the prosoma. It is provided with eight pairs of **ostia**. From the heart arise one **anterior** and two **lateral arteries**. The anterior artery divides into three, one **median frontal** and two **cross arteries**. The cross arteries supply the intestine and lead into a ventral arterial system giving off five pairs of branchial arteries. The latter arteries pass along the sides of the heart and unite as the posterior abdominal artery behind the heart. The lateral artery gives off branches to the liver (**hepatic artery**) and to the marginal parts (**marginal artery**). A venous system comprising two longitudinal **ventral sinuses** and five pairs of **branchio-pericardial** veins leading into the ostia of the heart, also exists in *Limulus*.

## Nervous System

The **central nervous system** consists of (*i*) the **brain** (Fig. 16.23) with separate pairs of nerves to the median and lateral eyes and to the olfactory organ; (*ii*) the **circumoesophageal collar**; (*iii*) a **ventral abdominal** nerve chain of five **branchial ganglia,** and three **post-branchial** ganglia; and (*iv*) a large nerve cord lying above the heart. The brain is derived from the protocerebrum and a remnant of the deuto-cerebrum. The tritocerebrum is fused with the circumoesophageal collar and innervates the chelicerae. This ganglionic mass gives off nerves to the five prosomatic appendages, the chilaria and the genital operculum.

**Sense organs**. The sense organs include the lateral compound eyes, the median ocelli and a median frontal organ. Each compound eye consists of a number of ommatidia lying under one lens. The ommatidia are spaced apart. The median ocelli are invaginated cups, the interior of the cup forms a lens that is continuous with an outer cornea, and is surrounded by retinal cells. It is highly sensitive to ultra-violet light and is analogous to the pineal eye of vertebrates. The median frontal organ is believed to be a olfactory organ.

## Reproduction

The king-crabs are dioecious, and the reproductive system has essentially the same structure in both the male and female. The gonad consists of a symmetrical network of tissue, made up of tiny ovarian tubules in the female and sperm sacs and ductules in the male. The gonad is located adjacent to the intestine and extends from the posterior half of the cephalothorax through the abdomen. The sperms or eggs pass to the outside through short ducts that open onto the median region and the underside of the base of the genital operculum. During mating the male grasps the female with the modified claws on its first pair of legs. The female digs a burrow in the intertidal zone, and into this are deposited several hundred externally fertilized eggs. These remain there for several months before hatching.

**Cleavage** is total. A solid **gastrula** is formed containing two mesodermal (germinal) centres, one anterior and the other posterior in relation to the future embryo, which looks like a trilobite at the time of hatching, and is therefore, called a **trilobite larva**. This larva swims about and burrows in sand. The cephalothorax of the king-crab is very similar to the cephalon of the trilobite.

# Phylum Arthropoda

The telson is small and does not project beyond the abdomen. After successive moults the telson increases in size and the proper number of book-lungs appear and the young crab assumes the form of the adult. Sexual maturity is attained in three years.

**Distribution.** The subclass is marine but members may also occur in fresh and brackish estuarine waters. *Limulus* occurs along coasts of North America, while *Tachypleus* and *Carcinoscorpius* occur in South-east Asia. Members of the subclass burrow and forage for molluscs and worms in sand and mud with a head-on shovelling action of the carapace. The living xiphosurans have not changed from their ancient (Silurian) relatives.

## Subclass Eurypterida

The **Eurypterida** (or **Gigantostraca**) are Merostomata that occurred from Ordovician to Permian times. They are all extinct now. The carapace is not expanded and the abdomen of 12 somites is narrowed behind. They were a dominant group in seas of Paleozoic times before the cephalopods. In appear-

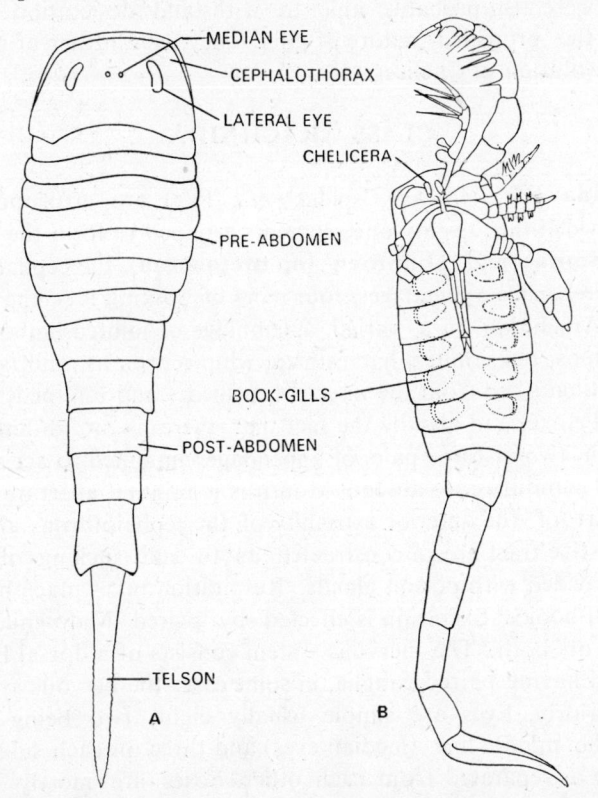

FIG. 16.26. A—*Eurypterus remipes* (dorsal view); and B—Another of an eurypterid (*Mixopterus*) showing appendages of one side, ventral view.

ance they were like scorpions to some extent attaining a length of about three metres. The Eurypterida were aquatic and probably swam or crawled in the bottom of saline and brackish waters. One species of genus *Pterygotus* was almost three metres long.

The prosomal appendages and telson of eurypterids were similar to those of king-crabs, only the cephalothorax was small and lacked posterior lateral extensions. Lateral and median eyes were present. The abdomen of eurypterids, on the other hand, was made up of separate segments, seven segments bearing appendages forming **preabdomen (mesosoma)** and five narrower segments without appendages forming **post-abdomen (metasoma)**. The **telson** was attached to the last abdominal segment. The appendages of the cephalothorax included a pair of small **chelicerae,** four pairs of **walking legs** and one pair of large elongated oar-like appendage. Often the fourth pair of appendages was elongated, and paddle-like. The abdominal appendages consisted of six pairs of **gills,** the first pair also formed the **operculum.**

Fossil larval stages indicate that the larvae had less number of segments than the adults and passed through a trilobite stage similar to that of *Limulus*.

Eurypterids inhabited brackish, fresh and occasionally marine waters. They probably fed on large invertebrates and also vertebrates. They were benthic and fossorial, but were also probably able to withstand desiccation, a conclusion drawn from the protected nature of gills. They were rather of new ecological phase in the evolution of Chelicerata.

## CLASS ARACHNIDA

The **Arachnida** (Gr. *Arachne*, spider+*oid*, like) are arthropods in which the body is divided into twenty-one segments grouped to form the **cephalothorax (or prosoma)** and **abdomen (or metasoma)**; the cephalothorax bears two pairs of prehensile appendages, four pairs of walking legs; the antennae are absent. The Arachnida is a varied assemblage of jointed-limbed animals like spiders, scorpions, ticks, mites, harvestmen, whip-scorpions and scorpions, etc. They are distinguished from the insects, centipedes and millipedes by the entire absence of antennae, and also by the fact that there is no differentiated head furnished with two or three pairs of appendages modified to act solely as jaws. Antennae and mandibles are absent. Mouth is a minute aperture placed near the lower part of the anterior extremity of the cephalothroax and the mouth-parts and digestive tract are so constructed as to suit sucking of food juices. Some are provided with poison glands. Respiration takes place by book-lungs, tracheae or gill-books. Excretion is effected by paired Malpighian tubules or coxal glands or both. The nervous system consists of a dorsal brain and ventral nerve cord having paired ganglia, in some cases the nervous system is concentrated anteriorly. Eyes are simple usually eight, two being situated close together in the middle line (median eyes) and three on each side (lateral eyes) set in a cluster or separated from each other. Sexes are mostly separate, sex opening single on second abdominal somite. Fertilization is usually internal. The arachnids are viviparous or oviparous. The young hatch in a form substantially resembling their parents, and except in the case of the Acari (mites and ticks) growth is not accompanied by metamorphosis.

The arachnids are chiefly terrestrial and solitary, either free-living and predaceous or parasitic. They compose the largest class, and from a human stand-

Phylum Arthropoda

point, the most important of the chelicerate classes. Of these the scorpion alone have a great nuisance value to farmers and cattle. Although not usually fatal to man the venom of scorpions kills invertebrates and disturbs farm animals. Some species, however, have venom that can kill man. Venom of these forms is neurotoxic, and causes convulsions, paralysis of the respiratory muscles, or cardiac failure in fatal cases. Great majority of them are numerous in warm dry regions than elsewhere. The spiders and some other arachnids have special glands that produce silk for different purposes. A few spiders and scorpions have bites or stings that may cause serious illness to man or may prove fatal. Several kinds of ticks are intermediate hosts and some parasitize man and animals, produce injury, sickness and death. The arachnids are of ancient origin, some stocks having been reported from Cambrian times. Most spiders are under 25 mm long. The smallest spider is **Microlinypheus** under 1.0 mm and longest is **Theraphosa lebondii** which attains a length of nine centimetres.

## TYPE SCORPION

Scorpions are familiar arachnids of hilly regions of India. They inhabit warm countries almost all over the world. The largest varieties are found in Africa

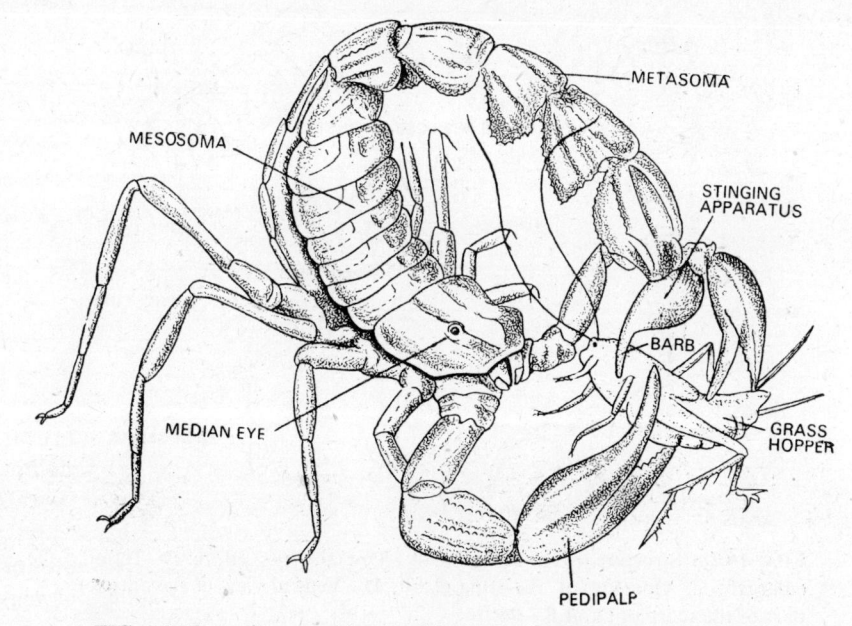

FIG. 16.27. A giant scorpion catching a grasshopper and paralysing it.

and America; none, however, occurs in New Zealand or on Oceanic Islands or in the extreme south of South America unless artificially introduced. They live in crevices and holes, under stones, fallen tree-trunks, the bark of dead trees and in deep burrows; they dig in soil during the day, and being nocturnal in habits come out at night and move in search of prey. They are strictly carnivorous, feeding mostly on spiders and insects. They catch their prey by means of pincer claws, sting to death (Fig 16.27) with the caudal spine and finally suck their juices. It was estimated in 1949 that there are 700 species of 70 genera of these animals. They vary in size. The smallest scorpion *Microbuthus pusillus* is 13 mm long while the largest species *Pandinus imperator* sometimes reaches a length of 18 cm; the largest Indian form, *Palamnaeus swammerdami* is 138 mm long. The description given below applies to all scorpions though mainly it centres round the local species *Palamnaeus bengalensis*.

## External Features

The body (Fig. 16.28) of the scorpion is long and narrow consisting of an anterior broader portion, including the cephalothorax or **prosoma** and the anterior region of the abdomen the **mesosoma**, followed by a narrow terminal part, the posterior abdomen or **metasoma** which is habitually turned upward over the back, in the living state. The prosoma is the small anterior portion dorsally covered by an unsegmented tergum the **shield** or **carapace** of, more or less, square shape. The anterior border of the carapace is notched forming a right and a **left frontal** lobe. The cephlothorax is formed by the fusion of the anterior

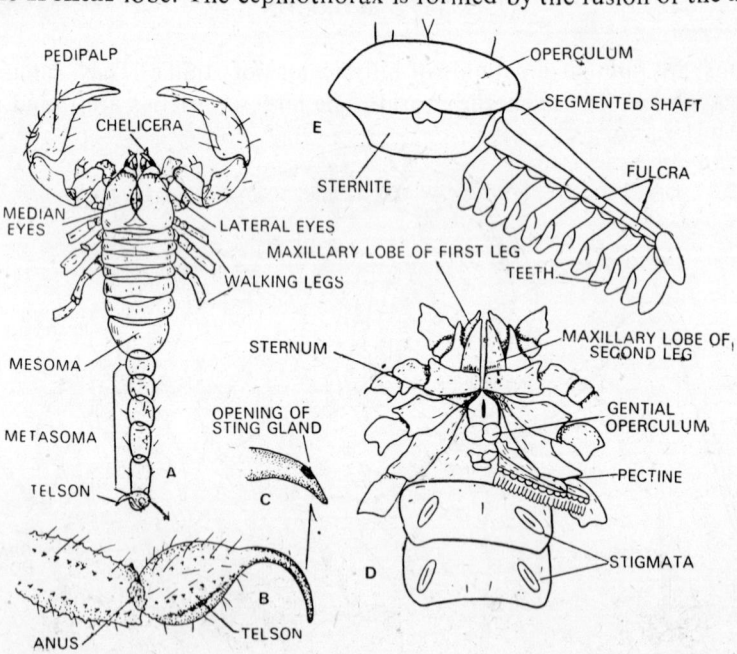

**FIG. 16.28.** Structure of a scorpion. A—Dorsal view, entire; B—Telson, enlarged; C—Opening of the sting gland; D—Ventral view of the anterior part of the scorpion; and E—Pectine.

presegmental region with originally seven segments, of which the first (precheliceral segment) exists only in the embryonic period. In the interior of the cephalothorax there is a cartilaginous plate, the **endosternite**, forming the internal skeleton and providing means of attachment for muscles. The carapace bears on the outer surface a pair of median eye about its middle and two or three pairs of small eyes on the anterolateral margin. Between the eye-groups a triangular area is formed, this is called the frontal area or anteocular triangle. Besides these the carapace carries many ridges or carinae which are of great systematic importance. A very small transverse opening, the **mouth,** is situated at the anterior end of the cephalothorax on its ventral side. The **labrum,** a small fleshy lobe, overhangs the mouth (Fig. 16.28).

The **abdomen** (Fig. 16.28) is the part following the cephalothorax. Anteriorly it is as broad as the **pre-abdomen** or **mesosoma** and consists of seven segments. As usual, each segment consists of a dorsal sclerite, the **tergum,** and a ventral sclerite, the **sternum.** Both of these are connected to each other by soft integument at the sides. Originally there are eight segments forming this portion, but the original first, the pregenital segment, becomes indistinct during embryonic life. The posterior part of the abdomen, the **postabdomen** or **metasoma,** consists of five flexibly jointed segments which are narrow, tubular, enclosed in a complete investing ring of chitinous material. The whole structure is often wrongly called the tail. Articulating with the last segment of the post-abdomen, occurs the **telson** or **caudal spine** (or **sting**) which is triangular in shape, swollen at the base and pointed at the apex. The basal swollen part of the telson (Fig. 16.28B) is often called the **vesicle** and lodges the poison glands and the spine-like portion is the **aculeus.** The ducts of the poison glands open a little behind the apex on the upper side (Fig. 16.28). The adult scorpion, thus, has 18 postoral segments in all but if the seventh suppressed segment is also counted the total becomes to 19. This is the maximum number of segments possessed by any arachnid.

The apertures on the ventral surface of the abdomen include the **stigmata,** which are a pair of oblique slit-like opening (Fig. 16.28) on the sides of the relatively broader sterna of the third, fourth, fifth and sixth segments of the pre-abdomen. They open into the pulmonary sacs (see later). The **anus** is situated ventrally in the last segment of the post-abdomen (Fig. 16.28) immediately in

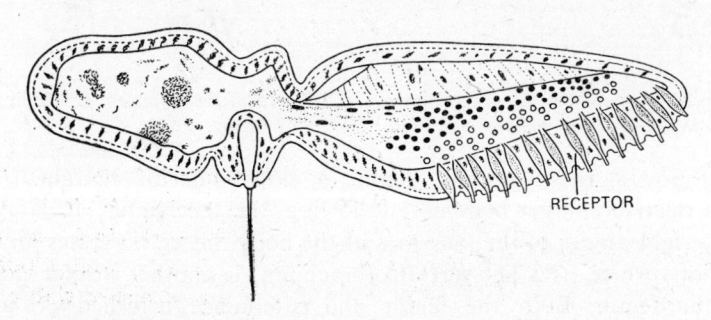

**FIG. 16.29.** Transverse section through the pectine of a scorpion.

front of the sting. It is located on a rounded protuberance guarded by four chitinous lips or **labia.** The opening itself is four-rayed. The openings of the **genital** ducts lie at the base of a soft rounded median lobe on the narrow sternum of the first segment cleft into two portions and known as the **genital operculum.**

## Appendages

The cephalothorax bears six pairs of segmental appendages including the chelicerae, the pedipalps and the walking legs. The first pair of these is preoral in position and prehensile in nature. It is called the **chelicera** (mandible of

Pocock) which is three-jointed and chelate (bears a tong-like process at the free end). The basal joint is small and is almost concealed under the carapace, the second is large and swollen furnished in front with a strong process, the **immovable finger**, armed with teeth on the upper and lower surfaces, and the third, the **movable finger**, is small working in horizontal plane, and furnished with teeth along its biting edge. Internally it possesses both an opening and a closing muscle arising in the hand. These two form the characteristic **chela** (Fig. 16. 30)

The second pair of appendages, the **pedipalpi** (chela of Pocock[1]) are the well known, powerful, six-jointed nippers of the scorpions, postoral in position. The following are the joints of pedipalpi: (*i*) The **coxa** is the basal joint that connects the appendage with the cephalothorax. Towards the mouth each coxa is drawn out into a membranous process, the **gnathobase**, which acts as jaw and bites against that of the other. (*ii*) The **trochanter** is an irregular-shaped

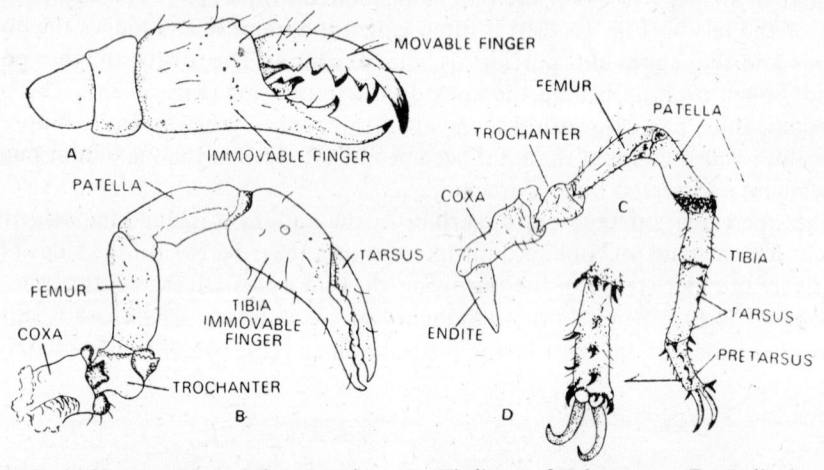

**FIG. 16.30.** Appendages of a scorpion. A—Chelicera of *Palamnaeus*; B—Pedipalp; and C—Walking leg of *Palamnaeus*.

podomere following the coxa and having spinous anterior margin. (*iii*) The **femur** is a relatively larger podomere following the trochanter. It is regarded to be set at right angles to the long axis of the body, hence possesses an anterior and posterior surface. (*iv*) The **patella** (brachium) is another stouter podomere following the femur. Both the femur and patella are furnished with granular crests. During resting the patella lies in a line parallel to the long axis of the body, thus, it has an outer and inner surface. The **hand** or **chela** is the terminal podomere, variable in size and form, being generally narrower and longer in females while broader and smaller in males. It consists of two portions. (*v*) The **under-hand** or **tibia** is the large flattish area articulating with patella. Its upper side is often evenly convex and the lower surface is flat and horizontal. Both the surfaces have prominent ridge on them. The ridge on the under surface runs up to the tip of the pointed end, the **immovable finger**, and as such is sometimes known as **finger keel**. The ridge of the lower surface is called the **keel** or the underhand. (*vi*) The other part of the chela is the movable finger

[1] Fauna of British India, *Arachnida*, 1900.

or tarsus which is a small pointed structure movably articulated with the distal narrow part of the underhand. This along with the immovable finger forms the pincer. The biting surface or the movable finger, as that of the immovable finger also, is provided with sharp teeth, that are often arranged in distinct longitudinal and oblique series. Internally it is furnished with powerful muscles.

The remaining four pairs of appendages are the walking legs. The four legs on each side are practically all alike. Each leg consists of seven[2] podomeres (Fig. 16.30): the **coxa, trochanter, femur, patella, tibia, tarsus** and the terminal **pretarsus** which bears curved and pointed horny claws. The coxae of both first and second pairs are furnished with forwardly directed **maxillary processes** (gnathobases or endites) lying immediately below the mouth. The coxae of the second pair meet each other in the middle-line. The coxae of the third and fourth pair are welded together, they do not meet across middle line of the body but are separated by the sternum of the cephalothorax which is triangular or pentagonal in shape. This is called the **metasternite** that buts against the coxae of the second pair anteriorly (Fig. 16.28D). In between the tibia and the tarsus is a structure called the **tibial spur**, and the tarsus bear the **pedal spur,** while the pretarsus is furnished on its lower side with hairs or spines or both. Each pretarsus bears at its apex a pair of larger claws called the **superior claws** and smaller **inferior claws** located ventral to the superior ones. The inferior claws are generally worn out. Sometimes the upper side of the pretarsus is produced, at its free end, into a process, the **claw lobe**.

**Abdominal appendages**. The abdomen of the scorpion does not possess typical appendages, but its first two adult segments bear rudiments of appendages or their modified forms. The first of these is the **genital operculum** of the narrow sternum of the first segment. On the sternum of the second abdominal somite are movably attached a pair of comb-like appendages called **pectines** (Figs. 16.28, 16.29). They vary in form in different genera of scorpions. Each of these consists of a **handle** or **shaft** made up of three pieces, the proximal part is the longest, and the middle one the shortest (Fig. 16.28). The shaft bears a number of narrow processes, resembling the teeth of a comb, called the **pectinal teeth** (Fig. 16.29). The number of teeth is

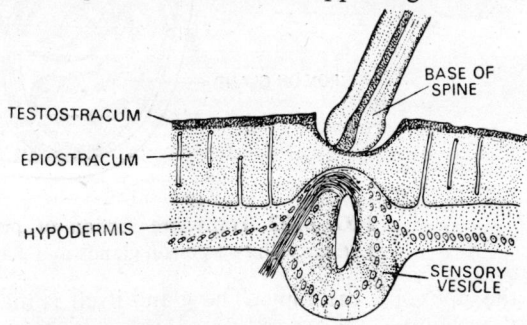

FIG. 16.31. Vertical section of the integument of scorpion (*after* Pavlovsky).

different in different species and varies from four to over thirty. The pectines are peculiar appendages quite characteristic of the scorpions. Their specific function is not known, although sexual and other roles have been suggested for them because the genital opening lies immediately in front of them. They are unquestionablly sensory organs, since the teeth bear numerous innervated sensilla.

---

[2]Snodgrass maintains that each leg has eight podomeres, but since the two parts in the tarsal region are not connected by muscles they are to be regarded as sub-segments of tarsus or tarsomeres.

## Body Wall

The body wall of a scorpion consists of an outer layer of chitinized **cuticle** secreted by **hypodermis** lying below. The hypodermis (Fig. 16.31) consists of a single layer of columnar cells filled with pigment granules of various types. Below it lies the basement membrane. The chitinous layer is brilliant in colour outwardly and consists of three distinct layers: (*i*) a superficial cuticle, **testostracum**, (*ii*) the middle layer, **epiostracum**, and (*iii*) the deep lamellous layer, **hypostracum**. This layer is traversed by canals which open outside on the surface. In these openings are lodged the ducts of the secretory gland which are located in the hypodermis. They themselves are broader and are covered with circular thickenings. Besides, there are some narrower canals with smooth inner lining that do not pierce through the cuticle. From the chitin arise several processes of different types. Of these the **dermatidia** are small outgrowths formed by the chitin alone and found on chelae, etc. There are some other similar derivatives involving the integuments in their formations. Such sculptural derivatives are called **coelodermatidia** and include such structures as **cristae, tarsal claws, spines, spurs, hairs,** etc.

The **dermal glands** include the **alveolar glands** in the chela of the pedipalpi, in the tibia and in the legs, and the **poison glands**. The poison glands are two sac-like glands clothed with a muscular membrane (Fig. 16.32) situated in the ampullar cavity. From the glands arise fine efferent ducts opening near

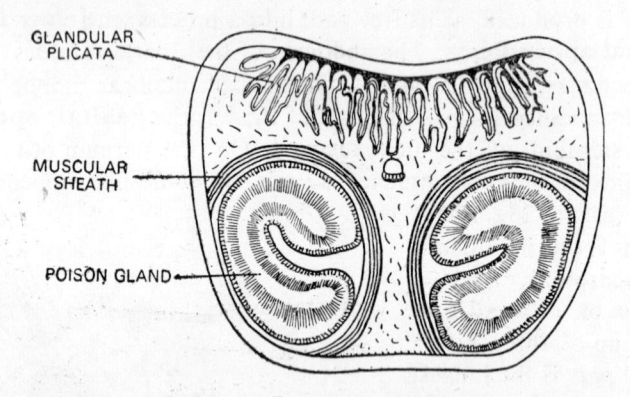

**FIG. 16.32.** Transverse section of poison vesicles of *Bothriurus vittatus* with two poison glands and glandula plicata.

the apex of the sting. The gland itself is made up of tall cylindrical epithelium between the cells of which are inserted flat supporting cells. Outside these is the basilar membrane and a connective tissue membrane. Internally, wall of the gland is smooth as in primitive embryonic gland or folded as in complex gland.

**Glandula plicata.** The males of some species of the genus *Bothriurus* (*B. vittatus*) are provided, on the dorsal surface of the poison vesicle, with a scutelliform depression, corresponding to which a special organ is situated underneath the integument. The chitin of this place is relatively thicker. Underneath, the hypodermis forms a special organ consisting of numerous longitudinal folds (Fig. 16.32) of the cubical epitheliam. These folds are longer in the midddle and lower at the circumference and are glandular. As they are found only in males they are supposed to play some role in sexual reproduction.

# Phylum Arthropoda

**Endosternite.** Inside the body at the junction of the prosoma and mesosoma there is an internal skeleton known as the **endosternite**. It is more or less a triangular piece which has holes for the aorta and the digestive tube (Fig. 16.33). Roughly it divides the body cavity. Its edges are produced into **anterior, posterior** and **lateral** processes, for the attachment of muscles which control the movement of various appendages. The endosternite, though cartilaginous in nature, differs from the vertebrate cartilage histologically. Another similar skeletal piece lies ventral to the nerve cord in the segment which bears the pectines.

**Musculature.** One pair of the muscles lies in the cephalothorax attached to it by one end and by the other to the endosternite above the nervous system. There are eight pairs of **dorso-ventral muscles** in the pre-abdomen on the right and left of midgut, running from the dorsal to the ventral surface of the body. The first pair is not easily seen as it lies closely applied to the posterior surface of the diaphragm (the partition that separates the thorax from abdomen). Dorsally these muscles are attached to the anterior end of the first abdominal tergite on the outside of the epicardiac ligament. Ventrally they are attached like the cephalothoracic pairs, to the endosternite. The dorsal attachments are to the outside of the epicardiac ligament. The ventral attachments are slightly farther apart than the dorsal, one pair for each abdominal sternite, the second pair lying at the sides of the genital opening, and third at the sides of the basal plate of the comb.

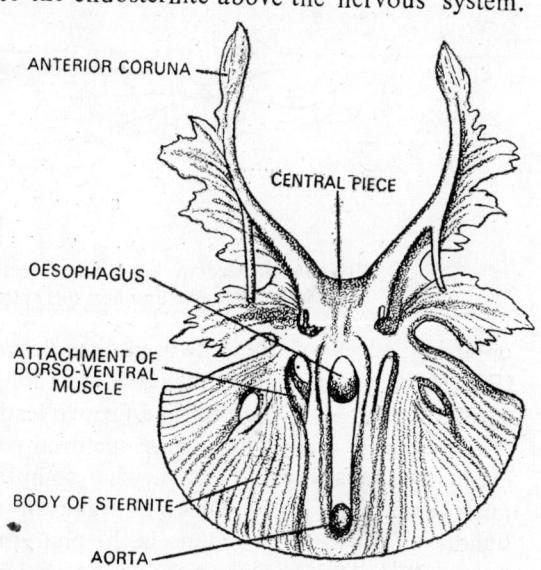

FIG. 16.33. Endosternite of *Lurus dufoureius* (after Pocock).

In some arachnids (spiders) there exists the power of self-mutilation (autotomy), and also autospasy that is the casting off of limbs when pulled by some outside agent. This is possible because the muscles at the joints become weak. On the other hand, in scorpions and kingcrabs (Xiphosura) the musculature is evenly arranged throughout and the legs and chitin is well provided with longitudinal fibres at all inter-articular membranes. Thus, in these animals there is no autotomy or autospasy.

## Digestive Organs

The **mouth** of the scorpion is concealed within a large open **preoral cavity** between the broad, soft inner surfaces of the coxae of pedipalp. Dorsally the chelicerae overhang the cavity and ventrally it is closed by a wide underlip formed of the closely approximated endites of the coxae of first and second legs. The mouth is seen on removing the chelicerae and pedipalps. It is a minute transverse aperture placed near the lower part of the anterior extremity of the cephalothorax. It is guarded by **labrum** above, and usually by the anterior sternal plate below, which acts as the lower lip or **labium**.

The labrum projects forward from between the bases of the pedipalps. Extended anteriorly from below the mouth is a basin-like lower lip, the upper part of which is formed of the concave dorsal surfaces of the first coxal endites. The

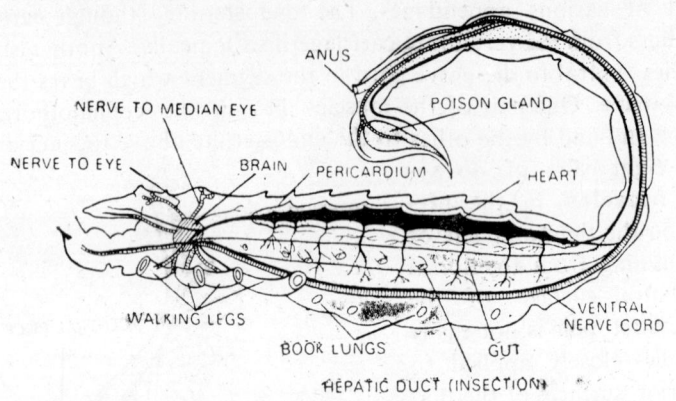

FIG. 16.34. Median longitudinal section of scorpion showing circulatory, digestive and nervous systems.

opposing edges of these two endites leave between them a median groove (Fig. 16.36) which is closed below by the long rigid supporting endites of the second leg coxae. The gutter-like groove leads directly to the mouth (Fig.16.35).

The mouth leads into a large suctorial pear-shaped bulb, the **pharynx**, with elastic chitinous walls. It is somewhat compressed laterally and rounded at its inner end. This is the sucking organ of the animal. A number of radiating bundles of muscle fibres connects the pharynx with the walls of the cephalothorax (Fig. 16.36). Due to the action of these muscles the pharynx can be dilated greatly. This increases the suction force and enables the animal to suck

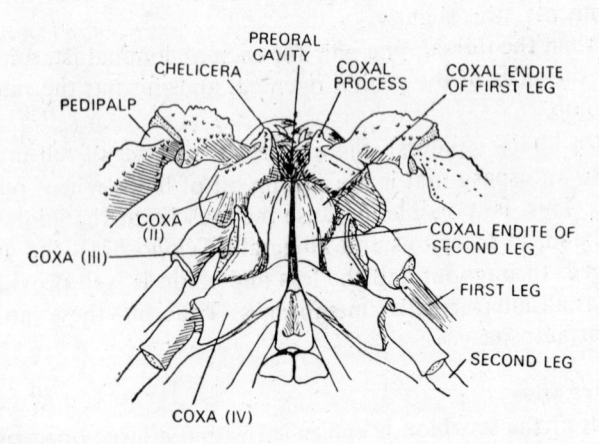

FIG. 16.35. Anterior part of the ventral surface of *Palamnaeus*.

juices efficiently. A short tube, the **oesophagus**, succeeds the pharynx. It emerges from the lower end of the pharynx. Into the posterior part of the oesophagus, which, in some cases is dilated, open the ducts of the salivary glands. All

## Phylum Arthropoda

these are stomodaeal parts and hence often referred to as the fore-gut. The next part of the alimentary canal is the **mid-gut** or **mesenteron**. It is long and narrow, straight tube with **glandular** (endodermal) interior and lies buried in the substance of the liver. Several narrow ducts (often five or six pairs) from the liver open into the mid-gut throughout its course. The liver is a large glandular body that occupies the dorsal part of the mesosomatic cavity. Though this bulky mass is called "liver" its hepatic function is doubtful.[3]

The liver and diverticula of the stomach form the major part of the alimentary system of the scorpion and occupy most of the space in the mesosoma. The mid-gut opens into the **hind-gut**, the terminal portion of the alimentary canal. The beginning of the hind-gut is marked by the presence of one or two pairs of delicate tubes, the **Malpighian tubules**, which act as organs of renal excretion. These originate from the mid-gut and, as such, are lined by endoderm, while in insects and most other arachnids they are lined with proctodaeal lining (of ectodermal origin). As the mouth of the scorpion is small, large pieces of food cannot be taken in. The scorpion has only a sucking apparatus for the ingestion of food and feeds only on liquids extracted from the prey mechanically

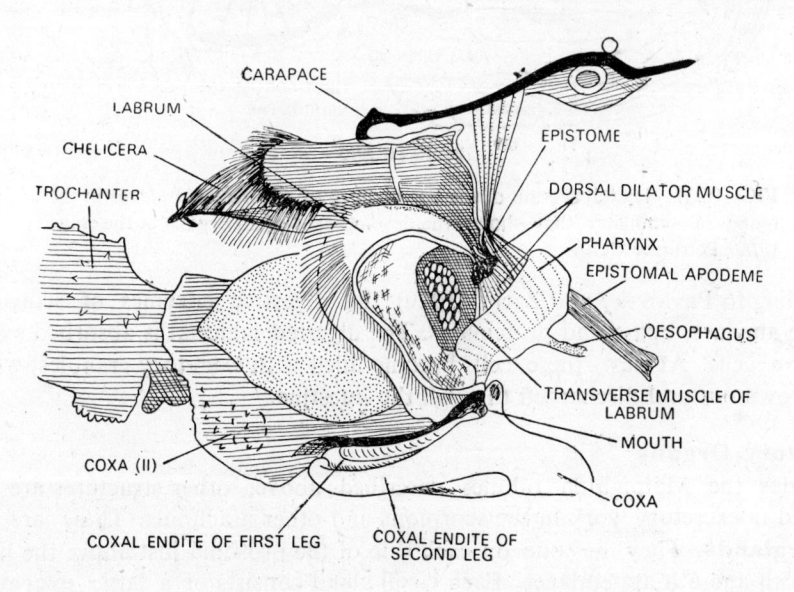

FIG. 16.36. Sagittal section of the scorpion showing mouth and pharynx, etc.

or by extraoral digestion. The prey (such as the spiders, myriapods or insects) is seized by the chela of the pedipalps, which in large species are able to crush hard-shelled beetles but if the victim is not killed by crushing, it is subdued by the abdominal sting. From the pedipalps the food is passed on to the chelicerae, one of which holds it while the other rips open the body and pulls out the vis-

---

[3]Snodgrass (*Arthropod Anatomy*, 1952) has given a different description of the digestive organs. From the sides of the axial tube of the stomach are given off six pairs of diverticula that expand into large sacs with soft infolded walls, all of which are closely packed along the sides of the body and bound together by a covering tunic of connective tissue. The first pair of diverticula is in the prosoma, the others arise in the five segments of the mesosoma.

cera. The extracted material collected in the preoral cavity is thoroughly cut by chelicerae, then reduced to a pulp by digestive juices discharged upon it, probably from the stomach, and finally it is sucked in liquid form by the pharynx.

This material undergoes preliminary digestion in the stomach. The epithelium of the stomach sacs includes secretory and digestive cells. The secretory cells produce enzymes that are poured in the cavity to bring about the digestion.

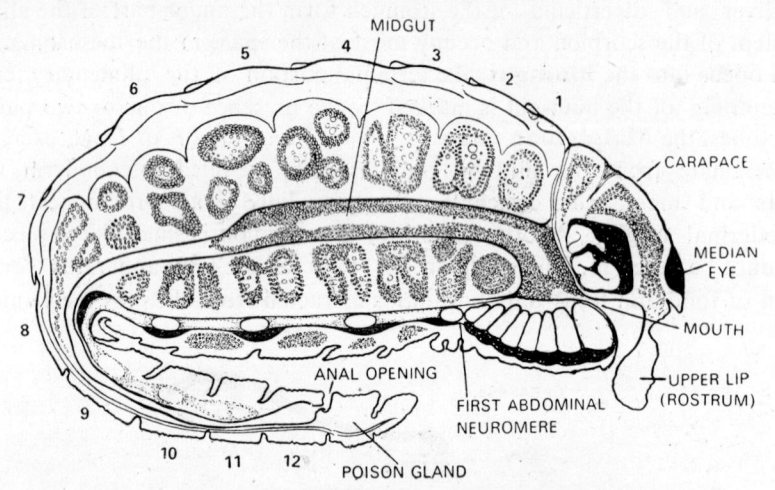

FIG. 16.37. Sagittal section of the late embryo of *Centrurus insulanus* in the plane of symmetry that shows the development of all organs at the stage (*after* Petrunkevitch).

According to Pavlovsky and Zarin (1920) the digestive enzymes of scorpions include amylase, lipase and proteinases. The digested material is absorbed by the digestive cells. At last these cells become filled with excretory granules, which are thrown out and discharged through the intestine.

## Excretory Organs

Besides the Malpighian tubules described above, other structures are also engaged in excretory work in the scorpions and other arachnids. These are the **coxal glands**. They lie one on each side of the prosoma just above the bases of the 5th and 6th appendages. Each coxal gland consists of a large **excretory saccule** (Fig. 16.38 A) internally lined with cubical or flattened epithelium lying outside the endosternite against the coxae of the 3rd leg in segment 5th and 6th. The saccule is followed by a convoluted tube, the **labyrinth**, that forms many coils around the former, lying in the fifth segment of the prosoma. The tubule opens to the exterior by small orifice on the base of the third walking leg. Before opening the tubule enlarges to form a reservoir. It is clear from embryological evidences that the coxal glands are derived from the coelomoducts. Many coelomoducts are represented in the young scorpions but only one persists in the adult. The coxal glands are regarded as nephridia homologous with the large nephridia of segments 6 and 7 of *Peripatus*.[4] Urate crystals are found in the

[4] Savory, *The Arachnida*, 1935.

saccule. Carmine injected in the body cavity is soon picked up by the coxal glands (Buxton).

## Circulatory System

The **heart** is an elongated muscular tube lying just beneath the terga in a groove in the liver mass, entirely in the preabdomen, extending from the diaphragm, which separates the preabdomen from the cephalothorax, (*i.e.* from the

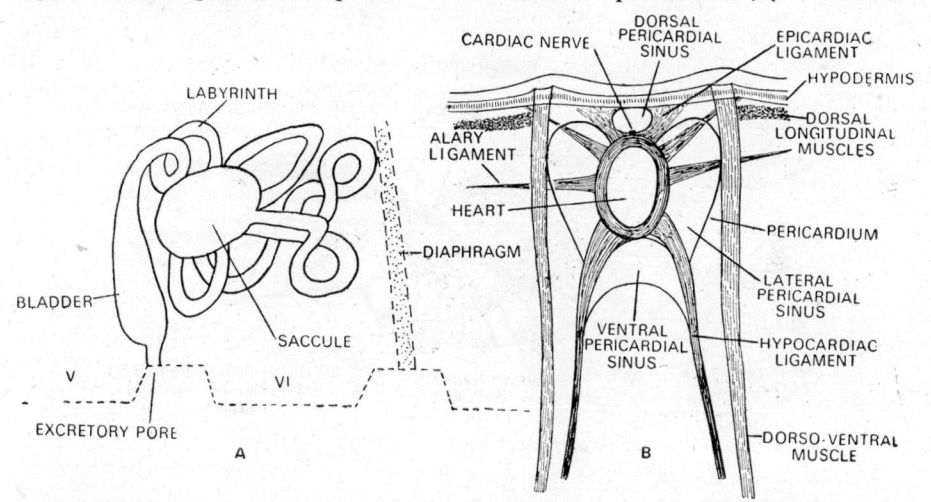

**FIG. 16.38.** A—Coxal gland of scorpion; and B—Transverse section through the heart of the scorpion, *Centrurus insulanus* (*after* Petrunkevitch).

7th to the 13th segment). Its anterior and posterior limits are clearly defined by valves. The heart has seven pairs of ostia, the presence of which gives the heart the appearance of an eight-chambered organ. In fact neither valves nor any constrictions exist to make it chambered. The heart is suspended by means of several ligaments. All told there are eight metamerically arranged groups of ligaments (Fig. 16.38B). Each group, except the first, is composed of four pairs. The shortest of these are the dorsal **epicardiac** ligaments. The second pair is called the **pteripyles**. Third pair is the **alary** ligaments (not muscles) directed almost at right angles to the long axis of the heart. The fourth pair is the **hypocardiac** ligaments, by far the strongest and longest, easily mistaken for muscles. Since there are no muscles for the dilation of the heart, diastole is accomplished through the elasticity of the heart ligaments. During systole the heart has to overcome the resistance of the ligaments, that is why the muscularis of the heart is very well developed.

The **pericardium** is a thin membrane running horizontally across the heamocoel. The pericardial cavity is divided into four regions owing to the presence of the ligaments. The spaces, thus formed, are often described as sinuses, of which the **lateral sinuses** are the largest, next in size is the **ventral sinus**, and the **dorsal sinus**, almost round in shape, is the smallest of the four. The cardiac nerve extends from one end of the heart to the other, partly embedded in a groove in the wall of the heart in the dorsal mid-line. The anterior and posterior aortic valves consist of muscle fibres—the anterior having a single layer of transverse muscular fibres and the posterior having a single layer of circular muscle fibres.

The heart continues anteriorly as the **anterior aorta** for a considerable distance. In the region of the third neuromere of the sub-oesophageal ganglionic mass it bifurcates and the two branches thus formed embrace the oesophagus on both the sides and meet below forming the arch which is short and connects the aorta with the right and left thoracic sinus. Ventrally the arch gives

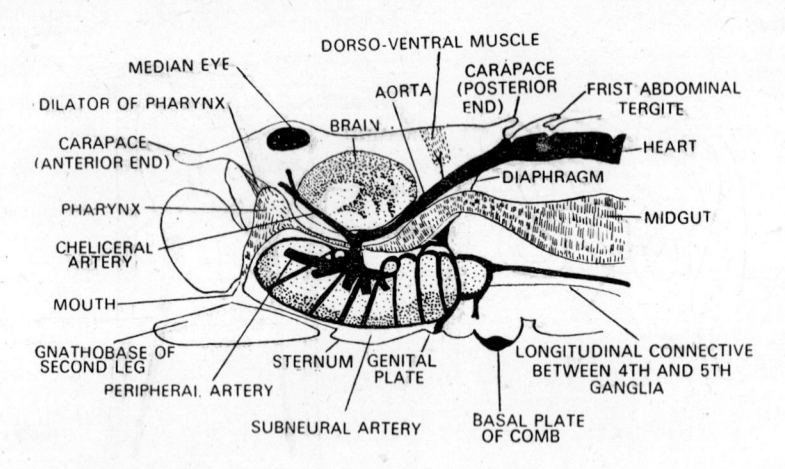

FIG. 16.39. The circulatory system of a young scorpion, *Centrurus insulanus* (*after* Petrunkevitch).

rise to a common blood vessel running over the nerve cord. This is the **supraneural artery**. Anteriorly it runs to the very end of the ganglionic mass, turning downward in its course and now continuing backward in the mid-ventral line below the ganglionic mass as **subneural artery** which ends behind the suboesophageal ganglionic mass. Both the supraneural and the subneurals are connected by nine **vertical arteries** (Fig. 16.39).

From the aortic arch arises a pair of large **cephalic arteries**, each of which gives off several branches, the most important of which is the **ophthalmic** artery. Beyond the ophthalmic artery the main vessel is termed **cheliceral artery** and supplies the chelicera. Each thoracic sinus gives rise to small and thin vessels connecting the sinus with the supraneural artery, and to four large vessels to the appendages. Of these the first is the largest and splits almost immediately into two branches, the first is the **pedal artery** and the second the longer and stouter **pedipalpal artery**. The continuation of the heart on the posterior side is the **posterior aorta**. It runs straight up to the tip of the postabdomen, supplying various organs including the intestine by various delicate branches.

The **ventral sinus** is a large space on the ventral side that collects blood from various parts of the body and sends it to the respiratory organs or **booklungs** for aeration. Between the roof of the ventral sinus and the floor of the pericardial sinus there is a special arrangement of muscles, which when contract enlarge the ventral sinus drawing venous blood into it and when relax the blood

## Phylum Arthropoda

is pumped to the pericardial sinus. Besides these, a series of **segmental vessels** or **veins** occur conducting blood from the book-lung to the pericardial sinus whence it enters the heart through ostia.

The microscopical structure of the heart reveals that its wall is composed of three layers, (*i*) the outer, the **adventia**, consisting of connective tissue; (*ii*) the middle **muscularis** composed of either spiral muscles or of symmetrically arranged semi-circular muscle fibres, which meet in the middorsal and midventral line; and (*iii*) the third layer or **intima**, whenever present, is a very thin transparent membrane.

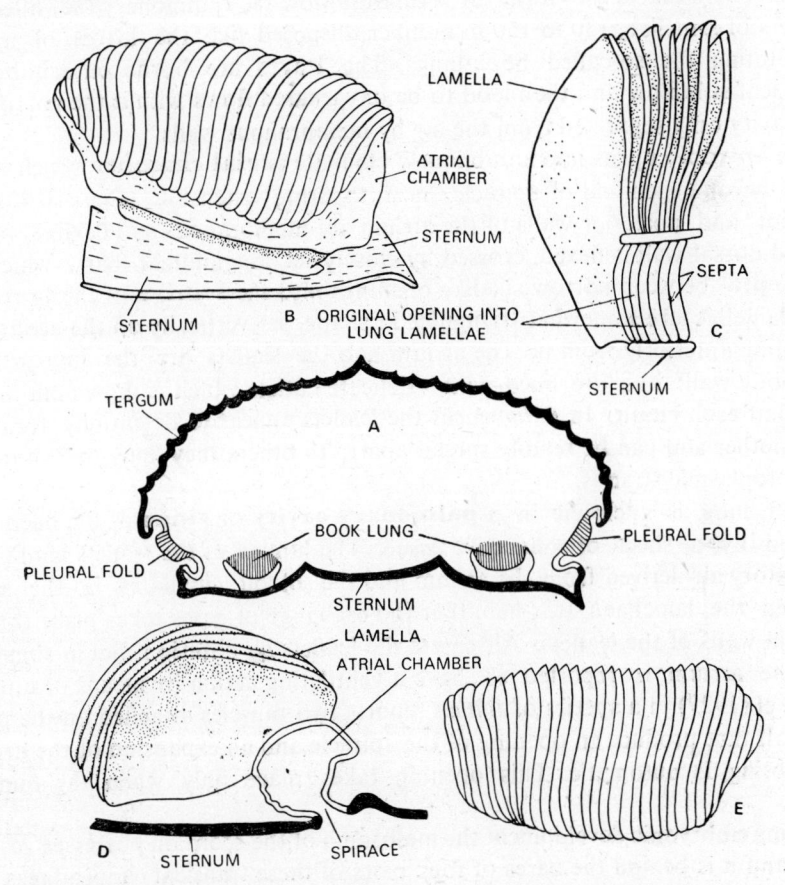

**FIG. 16.40.** Structure of the book-lung of scorpion. A—Cross section of the body of scorpion showing the position of book-lung; B—Dorsoposterior view of the right book-lung; C—A few lamellae in natural position; D—Ventral section of a book-lung; and E—Book-lung, dorsal view.

The blood system of scorpion is usually more developed than in Arthropoda in general. The blood contains **haemocyanin**, a blue coloured respiratory proteid. For this reason the freshly taken out blood of the scorpion is indigo in colour.

**Lymphatic glands.**[5] The scorpions are provided with a system of lymphatic

[6]E.N. Pavlovsky, "Studies on Scorpions," *Q.J.M.S.*, Vol. **68**, 1924.

glands in connection with the supraneural artery. The glands may be represented by a single nodule-like structure (*Buthus eupeus*), or a number of nodulose glands lying close to each other (*B. australis*) or a continuous band (*Lychas mucronotus*). The cavity of the gland is continuous with the cavity of the supraneural blood vessels. The glands are phagocytic in nature.

### Respiratory System

Respiration in scorpions takes place by means of **book-lungs**. There are four pairs of these structures opening at the four pairs of slit-like stigmata described above. Each takes the form of a small hollow sac (pulmonary sac) filled with clusters of lamellae, 130 to 150 in number, disposed like the leaves of a book constituting the so-called book-lungs. The leaves are borne on definite axis. Each leaf is hollow and the blood to be oxygenated flows within the narrow slit-like cavity only separated from the air by membranous walls.

The **spiracle** opens into an obliquely elongate **atrial chamber** which is produced beyond each end of spiracle in a tapering extension (Fig. 16.40). The anterior and posterior walls of the atrium are membranous and flexible, but the arched dorsal wall is flexed, crossed by closely-set longitudinal bars, which are the **septa** between narrow slit-like openings into the **lumina** of the **lamellae**. The lamellae are somewhat triangular in shape, set vertically on the atrium and extending anteriorly from it. The atrium and the leaflets are the ingrowths of the body wall, and are lined with a delicate cuticle which is drawn out and renewed at each moult. In *Palamnaeus* the leaflets appear to be entirely free from one another and can be readily spread apart. In others they may be connected by protoplasmic strands.

Each lung is enclosed in a **pulmonary cavity** or **sinus** of the haemocoel covered by a sheet of connective tissue. The lumina of the leaflets contain the respiratory air derived from the atrium and the blood circulates in the spaces between the lamellae. It is here that the exchange of gases takes place through the thin walls of the leaflets. Air enters the leaflets by diffusion, but in some species the atrium is reported to have a ventilating action by means of muscles. Fraenkel (1929) in a species of *Buthus* reports two muscles attached on the posterior wall that produce an opening of the spiracle and an expansion of the aritum, the closing is automatic. This opening takes place only when the animal is active.

During embryonic development the mesosoma of the scorpion possesses vestigial limbs and it is behind the bases of four pairs of these transient appendages that the book-lungs appear. Similar book-lungs are found in other arachnids such as the Pedipalpi, Araneae and Palpigradi. *Limulus* or the kingcrab has other homologous structures called gill-books for breathing. As it breathes dissolved oxyegn these organs consist of clusters of branchial lamellae or plates resembling the leaves of a book, hence the name. There are five pairs of gill-books in *Limulus*.

### Nervous System

The **brain** or **supraoesophageal ganglion** is a bilobed structure from which arise numerous delicate nerves which innervate the various anterior appendages and other structures including the median and marginal eyes (Fig. 16.28). From the brain arise **circumoesophageal connectives** which form a collar around

*Phylum Arthropoda* 571

the oesophagus and connect the dorsal brain with the ventral **suboesophageal ganglion** which lies below the oesophagus and, in fact, forms the beginning of the ventral nerve cord, and gives off nerves supplying the first six appendages, the pectines and operculum. The ventral nerve cord runs up to the fourth segment of the post-abdomen. Except the first ganglion of the ventral nerve cord, that lies in the fifth segment of the adult preabdomen or prosoma, one ganglion lies in each segment (Fig. 16.34). The last ganglion supplies the area around it and the post-anal sting.

The organs of special sense include the eyes and pectines, which are comb-like structures situated on the second segment of the prosoma (pre-abdomen) and are probably tactile and also olfactory organs. The structure of the lateral eyes is similar to that of the ocelli of insects, the median eyes are different. In the

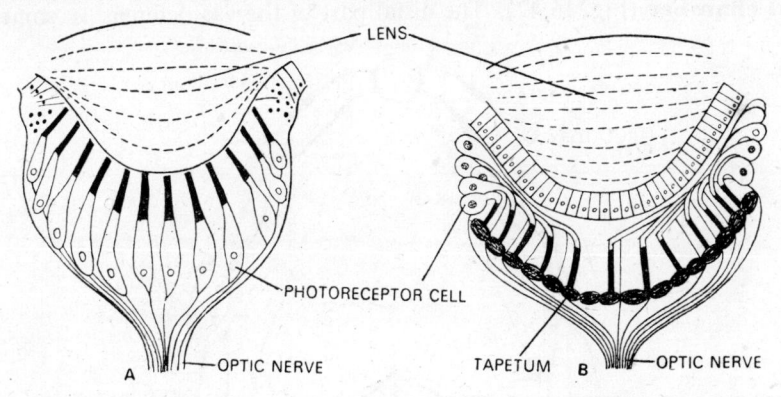

**FIG. 16.41.** A—Vertical section of the median eye of scorpion; and B—Vertical section of the lateral eye.

median eyes (Fig. 16.41A) the retinal cells are arranged in groups, as in the compound eye, but each possesses only a single cuticular lens, a character common in the simple eyes. The scorpions also have some perception of sound. This is indirectly indicated by the presence of stridulating organs, in some species, in the shape of ridges on the coxae of the pedipalps, across which file-like surface can be drawn to produce sound.

### Reproductive Organs

The accounts of the reproductive organs of the scorpions available in the literature are neither precise nor complete. There are many points that still need explanation. Here the best efforts have been made to provide as complete description of *Palamnaeus*, the commonest scorpion of Uttar Pradesh, as possible. The sexes are separate but it is difficult to distinguish the male from the female on the basis of external characters only. The mature female, however, possesses a fatter abdomen than that of the male.

In the **male** the testes are two (Fig. 16.42) lodged between the inferior lobules of the hepatopancreatic gland or liver, in the region of the last five mesosomatic segments. Each testis consists of a pair of longitudinal tubules, connected by four cross branches, in such a way that three quadrilateral meshes are formed. The tubules are extremely delicate, whitish semi-transparent structures of narrow and uniform calibre throughout the length. Each is circular in section. The wall of the testis is made up of flattened **germinal epithelium** resting upon the **basement membrane**. Below the basement membrane lies an extremely thin layer of musculature. The germinal epithelium gives rise to a large number of **spermatogonia** which form loose masses of small round cells with densely chromatic nuclei. There are several such masses each having spermatozoon in the different stages of development, viz. **spermatids, spermatocytes** or **spermatozoa**. Each group of similar cells is surrounded by a thin connective tissue wall in the form a **cyst**. All cysts lie closely adjacent to each other. When mature the spermatozoa are released in the lumen of the testis from which they pass on through the vas deferens to the seminal reservoirs.

The **vasa deferentia** arise from the outer angles of the foremost mesh of the testes. Each vas deferens runs anteriorly and opens into the inner side of the **genital chamber** (Fig. 16.42). The distal part of the vas deferens is somewhat

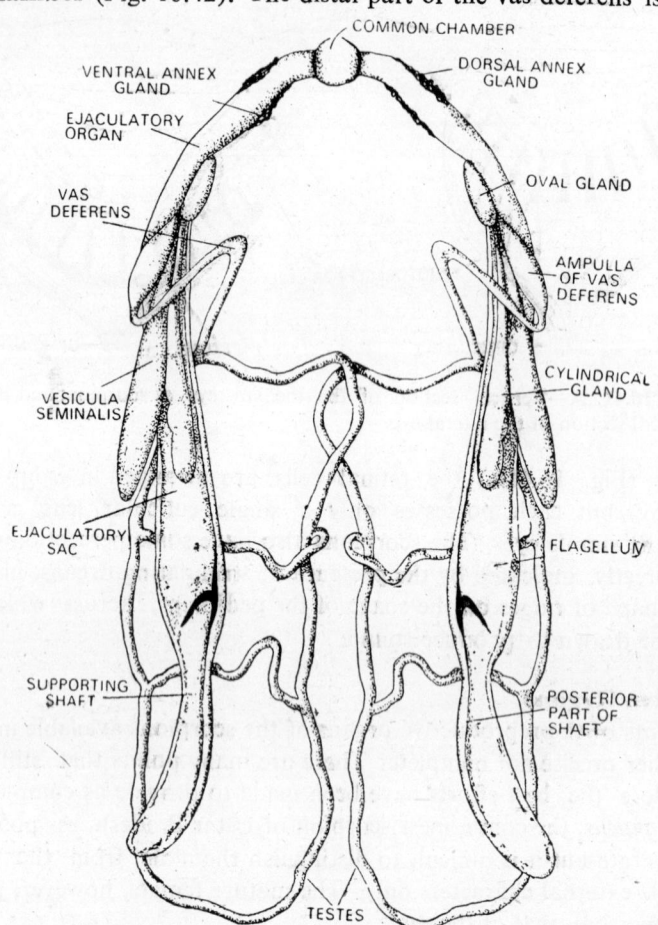

**FIG. 16.42.** Reproductive system of the male *Palamnaeus*.

dilated is some cases to form the **terminal ampulla**. The terminal portion of the vas deferens receives the openings of the accessory organs including the vesicula seminalis, the cylindrical and the oval glands. The **vesicula seminalis** is an elongated nearly club-shaped organ of slightly yellowish colour. It opens at

*Phylum Arthropoda* 573

the terminal portion of the dilated ampulla. It is a thin-walled tube usually filled with spermatozoa. The **cylindrical gland** is an elongated transparent whitish sac, shorter than the seminal vesicle. It opens into the dilated ampulla of the vas deferens at a point just beyond the attachment of the seminal vesicle. Its wall is folded and consists of cylindrical glandular epithelium, besides other usual

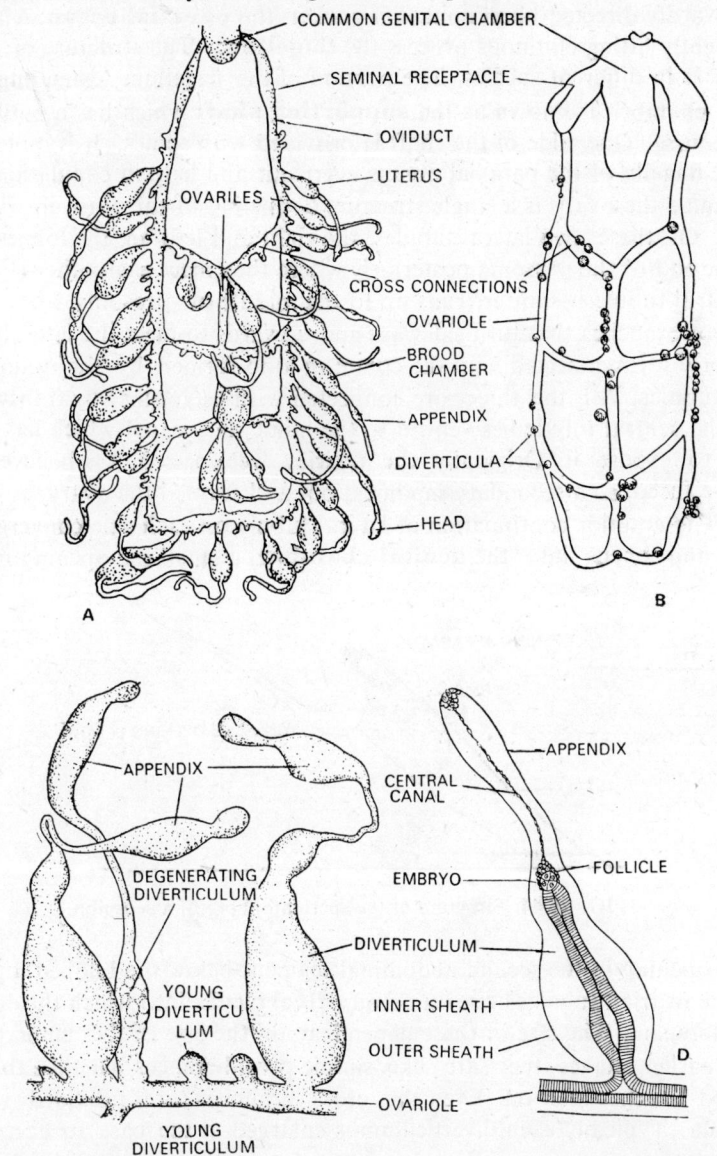

FIG. 16.43. Reproductive organs of female. A—*Palamnaeus*; and B—*Buthus*; C—Enlarged figure of diverticula; and D—Longitudinal section of a diverticulum.

structures and musculature. The **oval gland** is transparent, whitish nearly oval in form. It lies towards the inner side of the genital chamber into which it opens.

The **genital chamber**[6] is a prominent muscular tube about 15 mm long and 1-2 mm broad lying buried in the liver mass one on each side. The two genital chambers unite and form a comparatively short **common chamber** that opens to the outside by a narrow genital opening situated on the second segment, covered by genital operculum. The base of each genital chamber is provided with a backwardly directed blind epithelial pocket, the **paraxial organ**, within which lies a tightly fitting chitinous process the **flagellum**. The structure of the flagellum varies in different species. The portion of the flagellum extending into the genital chamber is known as the **supporting shaft** which has a gutter traversing its course. One side of the shaft is provided with spines. It is quite probable that the flagella of the paraxial organs evert out and help in copulation.

In female the ovary is a single structure consisting of three (4 in *Parabuthus*) tubules. Of these two lateral tubules, the right and left, are the longest and run up to the end of the prosoma posteriorly where they meet each other (Fig. 16.43). The central tubule is shorter, runs up to the middle of prosoma and bifurcates into two branches that turn sideways and upwards to meet the lateral branches. Posteriorly it is connected with the common loop formed by the union of two lateral tubules. All the three are connected with each other by transverse branches. The central tubule lies ventral to the alimentary canal which has to be removed to expose it. Originally the ovaries were paired, it is believed, but the two have fused, as a secondary modification, and formed the ovary as described above. The anterior continuation of lateral tubule on each side converges to the middle and opens into the **genital chamber**, a small sac opening in front by

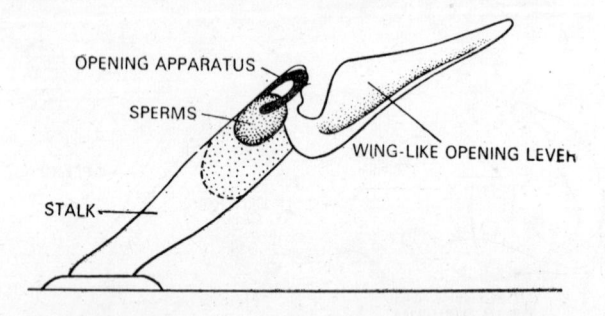

FIG. 16.44. Structure of the spermatophore of a scorpion.

a single opening in the second abdominal segment below the bilobed operculum. From the ovarian tubules arise numerous **diverticula** into which the eggs undergo development. The size of these depends upon the size of the embryo within. In the earlier stages they are like simple protuberances but take the form of elongated pouches later on. Even the cross branches are provided with such diverticula. Typically each diverticulum is enlarged at the base to accommodate the developing embryo and is drawn out into a narrow appendix distally, the tip of the appendix is enlarged forming the head (Fig. 16.43) which contains a cluster of absorbing cells. These cells rest against the maternal digestive caeca, from which nutritive material passes through the tubule to the embryo at the base. In *Buthus* these diverticula remain rounded throughout.

---

[6]Dufour (1856) called this organ in *Scorpio occitanus* as the penis. Pavlovsky (1917) called it *paraxial organs* as it is lateral in position. Abd-el-Wahab (1957) called it the *ejaculatory organ*.

*Phylum Arthropoda* 575

**Spermatophore.** The sperm is transferred to the female by means of a **spermatophore** produced by the male who attaches it to the subsratum at a convenient place. The spermatophore has a stalk for attachment to the substratum (Fig. 16.44). The free end carries sperms in a pocket and also an opening apparatus connected with a wing-like opening **lever**. As the lever is pressed the opening apparatus is released and the sperms are triggered into the female orifice. The male himself directs the female to the spermatophore.

**Courtship.** The Arachnida on the whole are known for courtship which in the case of the scorpion has been briefly described by Fabre. Male and female

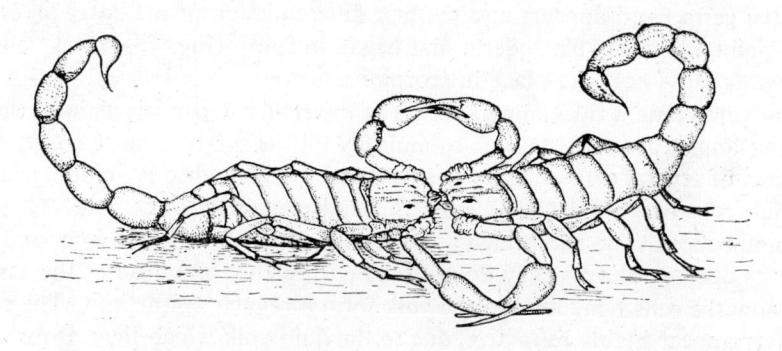

FIG. 16.45. Male scorpion pulling the female over the spermatophore planted earlier, seen as black line between 2nd and 3rd leg of left scorpion.

face each other, extend each abdomen high into the air, and move about in circles. The male then seizes the female with his pedipalps, and together they walk backward and forward. This behaviour may last hours or even days. Eventually the male deposits a spermatophore that is attached to the ground. A wing-like lever extends from the spermatophore. The male then manoeuvres the female

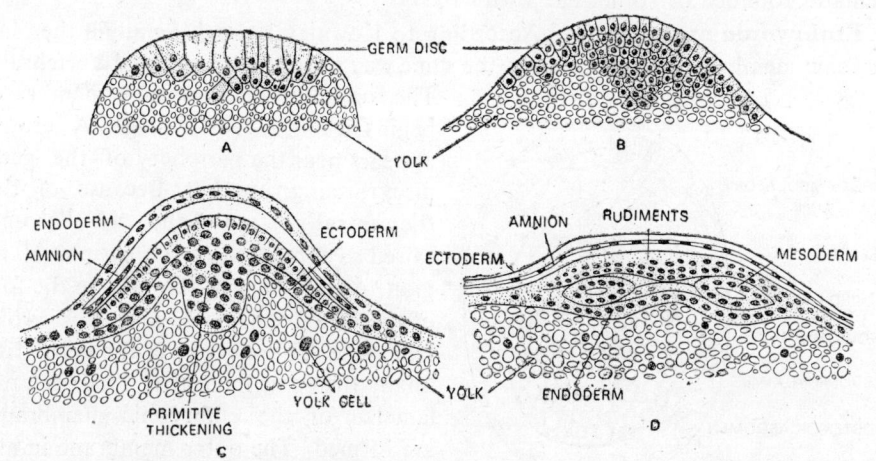

FIG. 16.46. Cleavage and earlier stages of development of scorpion.

so that her genital area is over the spermatophore. Pressure on the spermatophore lever releases the sperm mass, which is then taken up into the female orifice. The fertilization is internal.

The scorpions are viviparous. The oval or spherical egg which is rich in yolk and is surrounded by a thin membrane lies in a follicle that arises as outgrowth of the walls of the ovarian tube. Fertilization takes place either in the ovarian follicle or when the egg has left the follicle and passed into the ovarian tubes where they undergo further development.

**Cleavage**. The cleavage of the yolky-egg is discoidal (and that of yolk-poor eggs is equal and bilateral). At one pole which is directed from the follicle towards the ovarian tube a number of blastomeres form a small unilaminar cap on the yolk, this is the germ disc or the **blastoderm**. It spreads gradually from the point, advancing very slowly over the yolk. Long before it has grown round the yolk, the germ band appears and the first differentiation of the latter takes place at the point where the blastoderm first began to form (Fig. 16.46). A cleavage of the yolk does not take place in scorpions.

After some time a thickening appears at the centre and projects into the yolk as a swelling. The cells continue to multiply till the blastoderm has attained a thickness of several cells. The inner surface of the germ disc is quite irregular, for single cells become detached from this and migrate into the yolk. These give rise to amoeboid yolk-cells which are distributed throughout the yolk and bring about its disintegration. They do not take part in the formation of the embryo.

By now the cells lying next to the yolk form a regular epithelium. The cells of this layer appear highly refractive, due to the fluid yolk which they absorb. This is the **endoderm** layer (Fig. 16.46). The mass of cells which, after the differentiation of the endoderm, remains between it and the outer layer or ectoderm corresponds to the **mesoderm**. At first this is an irregular mass, extending over the whole region of the germ-disc, but at a later stage it takes the form of two symmetrically arranged bands situated near the middle line. These two bands, which fuse with one another posteriorly, become divided up later into the primitive mesodermal segments, each of which contains a cavity. The germ-disc remains rounded or somewhat oval.

**Embryonic membranes**. According to Kowalevsky and Schulgin the embryonic membranes are formed in the same way as in the Insecta and Vertebrata. The formation of embryonic membranes begins at a very early stage. A groove appears near the periphery of the germ disc, running round it. Because of this the central portion of the disc becomes raised as compared to the peripheral region. At the edge of the groove the fold of the ectoderm rises, and this grows from the periphery over the germ disc, finally fusing at its centre. Thus, the two lamellae of the embryonic membranes are formed. The outer membrane immediately below the egg-integument is the **serosa** and the inner the **amnion**.

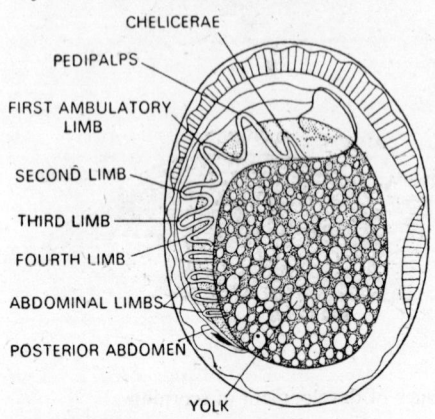

FIG. 16.47. Advanced embryo of scorpion showing rudiments of appendages.

During the formation of the germ layers and the development of the embryonic membranes the germ disc changes

## Phylum Arthropoda

its shape. It becomes broader at the anterior end and narrower at the posterior. In the middle appears a groove-like depression which soon disappears. The cephalic region becomes marked off from the primary trunk region by a transverse furrow near the anterior end of the germ disc, and about the same time, or very soon after, a few transverse furrows appear further back. These are the rudiments of body segments. Further separation of the segments continues for some time (Fig. 16.48).

Following this still smaller segments appear, one for chelicerae, a large segment for pedipalps, another for limbs. By the time ten segments appear the limbs become apparent (Fig. 16.48). They arise as outgrowths of the segments on each

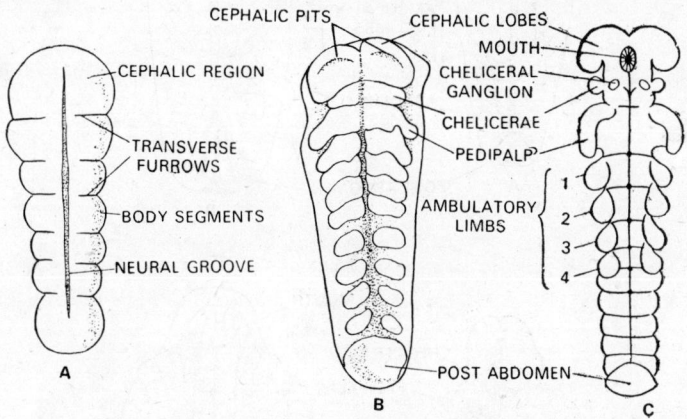

FIG. 16.48. Further stages in the development of the scorpion.

side of the middle line and are hollow and truncated. The development of limbs takes place from before backwards, but the chelicerae are remarkably late in development. When the pedipalps are already large the chelicerae are no more than mere prominences. The chelicerae as well as the pedipalps are without doubt post-oral in position, for the mouth first appears quite anteriorly between the cephalic lobes. The upper lip (rostrum) appears in front of the mouth a little later. Following the rudiments of chelicerae and pedipalps appear four pairs of appendages (those of the thorax) and finally another series of six pairs of abdominal limbs. In front of these is a small limbless segment. The first pair is specially small and soon degenerates, it is represented in the adult by a slight prominence covering the genital aperture (genital operculum). The second pair of abdominal limbs gives rise to the large combs. The four posterior pairs degenerate, but help in the formation of the lung-sacs. All the limbs are merely truncated in the beginning, but soon they take normal adult form.

**Organogeny.** The ventral longitudinal commissures arise early in the form of two thickened bands situated at the sides of a median groove, and extend the whole length of the body. Soon the band becomes segmented to form ganglia in different segments. The supraoesophageal ganglion arises in close relation to the invaginations on the cephalic lobes. Along with the brain the median eyes are also formed. The lateral eyes arise simply from two long pigmented areas that appear on the sides of the anterior part of the embryo. The lung-sacs arise as depressions on the posterior sides of the last four abdominal limbs. These depressions are at first shallow but they grow deeper and spread forward. In the last embryonic stages the lung-sacs assume the adult form. The alimentary canal develops differently in different parts of the body. The endoderm forms a small portion in the middle. In the post-abdomen the endoderm appears as a solid mass of cells which forms a tube which is neither thick nor long. The cells of

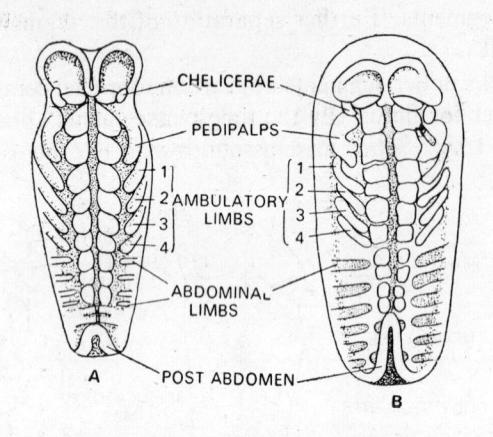

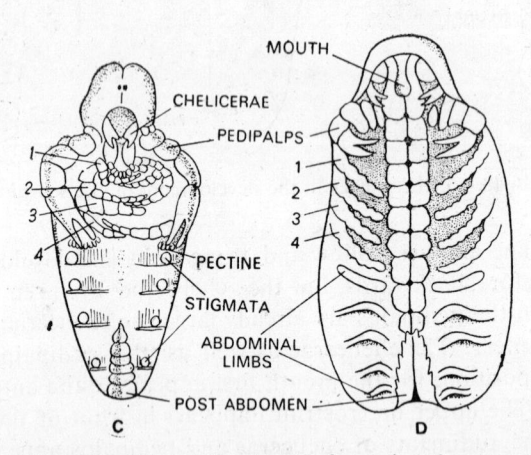

FIG. 16.49. Advanced embryos of scorpion showing differentiation of appendages.

this tube and those surrounding the endoderm form the provisional intestinal epithelium. The hepatic caecae arise as outgrowths of the provisional epithelium. The stomodaeum arises as invagination between the cephalic lobes and proctodaeum appears later on the posterior side. The coelom develops in the mesoderm which also forms the blood vascular system. The coxal glands appear as simple tubes, each running forward from the base of third ambulatory limb and opening by a funnel into the coelom. It later becomes coiled, and forms a glandular mass. The development of the genital organs is not fully known.

## TYPE SPIDER

The spider is the most familiar example of Arachnida. More than 2,600 species of spiders are known and this probably represents only a fraction of the actual number. Also, spider populations are very large; an acre of undisturbed, grassy meadow in Great Britain contained 2,26,500 spiders (Bristowe, 1938). They live in various habitats, from the sea level to the highest mountains, from the seashore and freshwater swamps to the driest of deserts, among rocks, in forests, in or on soil and in and about buildings. They are free-living, solitary and predaceous creatures that feed mainly on insects. Ordinarily the spiders entrap their food in webs, but the hunting spider, *Lycosa*, wanders for food and runs it down, and *Salticus* jumps on it. Individuals live only about a year. The enemies of spiders include birds, lizards and certain wasps. Spiders are unusual because of two characteristics, namely silk production and the modification of the pedipalps in the male to form copulatory organ. The silk produced is used for a great variety of purposes.

**External Features**

Spiders range in size from tiny species less than 0.50 mm in length to large tarantulas with a body length of 9 cm. When at rest the fang fits into a groove at the end of the large basal piece. A spider's body consists of two parts, a **prosoma** or **cephalothorax** and an **abdomen**. There is a waist, but no neck. The slender waist or **peduncle** connects the cephalothorax and abdomen. The cephalothorax possesses a dorsal carapace bearing eight eyes anteriorly and

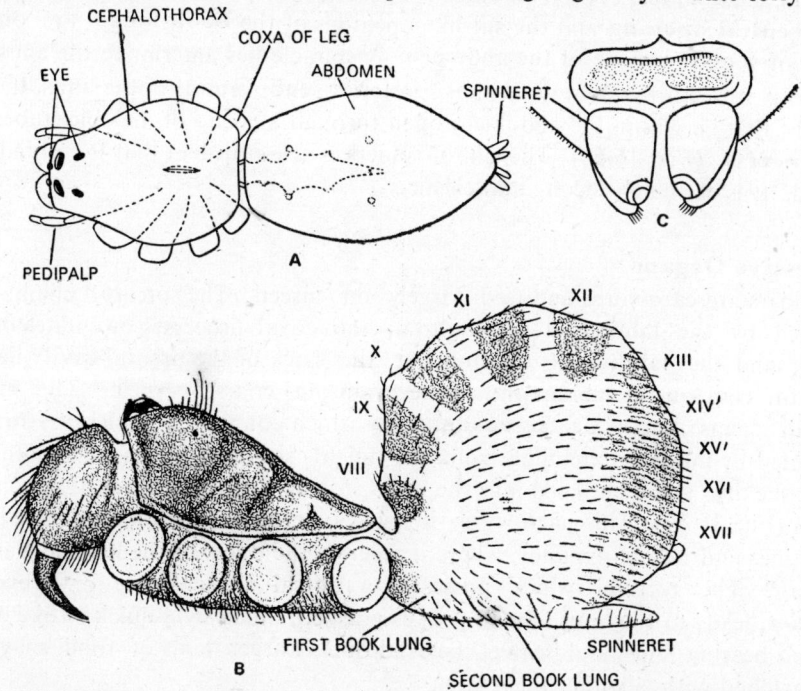

FIG. 16.50. A—A spider, dorsal view; B—The primitive Asian mygalomorph spider, *Liphistius malayanus* with legs removed (both *after* Millot); and C—Spinneret.

bears six pairs of appendages. The first is the chelicera or jaws, containing poison gland (Fig. 16.50) and terminating in a claw-like fang with a duct on its tip. Spiders are unique in having chelicerae provided with poison glands, located within the basal segments of the chelicerae and usually extending backward into the head. When the animal bites the fangs are raised out of the groove and rammed into the prey. Mostly the poison of spiders is not toxic to man, but some species such as black widows (*Latrodectus*) have venomous bites. The venom is neurotoxic and, therefore, the bite is severe and very painful. Pain in abdomen and legs, high cerebrospinal fluid pressure, nausea, muscular spasms and respiratory paralysis are various symptoms accompanying a bite. Small children usually die due to respiratory paralysis. The venom of *Lycosa raptoria*, and Brazilian wolf spider is haemolytic rather than neurotoxic and produces necrosis or decomposition of tissue that gradually spreads out from the bite, which is slow to heal. The second is the six-jointed **pedipalpi**, which is leg-like in female, but shorter and terminating in a bulbous enlargement in males, to act as a container for the transfer of sperm. Its enlarged bases are used to squeeze and chew the food.

The remaining four pairs are the eight **walking legs** typical of spiders. Each consists of seven joints, viz., **coxa**, **trochanter**, **femur**, **patella**, **tibia**, **metatarsus** and **tarsus**. In some spiders the tip of the legs bears pad of hairs, **scopula**, by which it clings to the wall or similar surface. The eyes are situated on the anterior dorsal surface of the cephalothorax; there are, in most common spiders, eight simple eyes. The **mouth** opening is situated between the maxillae. The **genital opening** and the slit-like openings of the **book-lungs** are on the anterior ventral surface of the abdomen. A spiracle lies anterior to the anus and connects to short tracheae. At its posterior end, ventral to the anus, the silk glands, that lie within the abdomen, open through a group of flexible tubes, the **spinnerets** (Fig. 16.50). The silk of spiders is scleroprotein that is emitted as a liquid. It is extremely strong and elastic.

### Digestive Organs

Spiders are carnivores and feed largely on insects. The preoral chamber is formed by the labium at the bottom, the coxal processes or endites on the sides, and the labrum at the top. At the back of the preoral cavity lies the mouth. Hair on the labium and endite strain out coarse particles. The minute mouth opens into a slender **oesophagus** which connects a **sucking stomach** operated by muscles extending from its dorsal surface to the cephalothorax. The sucking stomach leads into the main stomach with five pairs of pouches of caeca (Fig. 16.51). This narrows as it passes into the abdomen forming the intestines and here it expands, where it receives the ducts of the hepatic gland or "liver." The rectum, which possesses a dorsal **cloacal sac** or **stercoral pocket**, leads to the anus. Small prey is grasped, killed by a quick stroke of the poison-bearing fangs and "eaten" (sucked out). Larger items of food may first be fastened with silk and killed later.

### Circulatory System

The circulatory system is like that of an insect. The dorsal tubular **heart** lies in a pericardial cavity from which it receives blood through paired **ostia**. Blood is pumped out through vessels which empty into body sinuses. The large **ventral sinus** communicates with the sinuses of the book-lungs where the

# Phylum Arthropoda

blood goes for aeration and returns by "pulmonary veins" to the pericardial cavity to re-enter the heart. The blood is colourless containing amoeboid corpuscles and dissolved haemocyanin as a respiratory pigment.

## Respiratory System

The respiratory organs include the **book-lungs** which are peculiar to arachnids The book-lungs consist of folded lamellae or leaf-like horizontal plates, 15 to 20 in number, inside the respiratory chamber, each lamella enclosing a blood sinus.

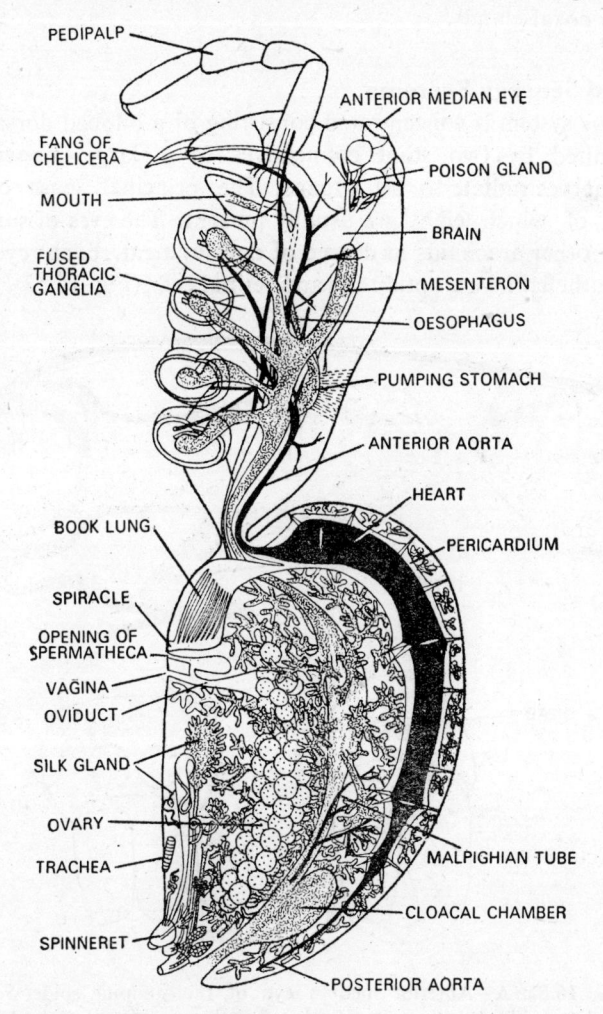

FIG. 16.51. Internal anatomy of a two lunged spider (*after* Comstock).

Air entering the external slits on the abdomen circulates through the lamellae where exchange of gases takes place. Tracheae may supplement these, or replace them. When present the tracheae are like those of insects.

## Excretory System

It comprises a pair of **Malpighian tubules** that empty into the cloacal sac.

These are similar in function to those of insects, but not structurally homologous, having a different embryonic origin. Excretion is further helped by a pair or two of **coxal glands**,[7] on the floor of the cephalothorax, that empty by ducts between the legs. Coxal glands are not as well-developed in spiders as in other arachnids. Groups considered primitive have two pairs of coxal glands opening on to the coxae of the first and third pairs of walking legs. In other spiders the anterior pair of glands persists and displays various stages of regression. Some people believe that silk glands and poison glands have perhaps evolved from coxal glands.

**Nervous and Sensory Systems**

The nervous system is concentrated consisting of a bilobed dorsal ganglion or "**brain**" united by two stout connectives to a large **ventral ganglion** from where nerves radiate to all organs. The principal sense organs are the simple eyes, of which eight are usually present. The eyes of some spiders surpass those of other arachnids in degree of development. Each eye has a chitinous lens, epithelial layer, optic rods and retinal cells (Fig. 16.52). Vision in some

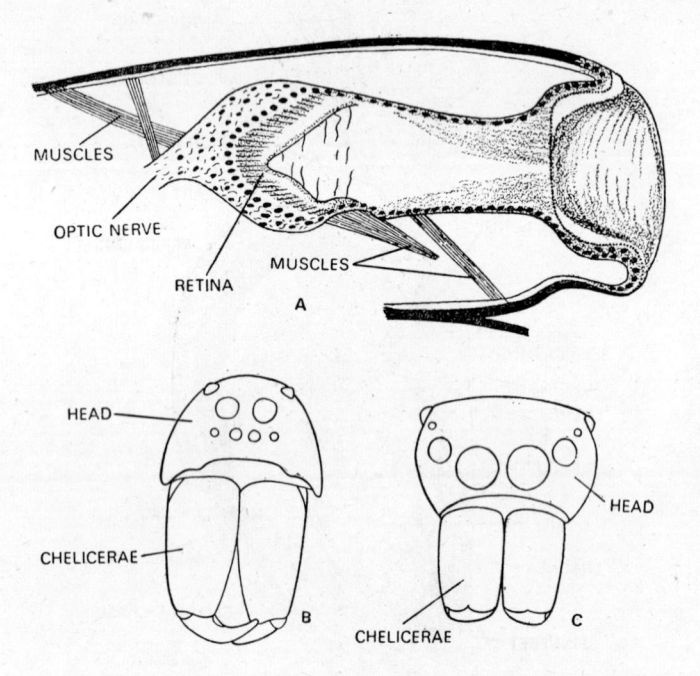

FIG. 16.52. A—Anterior median eye of the jumping spider *Salticus scenicus*, sagittal section; B—Head and chelicerae of wolf spider, *Lycosa aspersa*; and C—Head and chelicerae of the jumping spider, *Icius elegans* showing eye pattern.

spiders is keener than in others. Arrangement of eyes is characteristic of certain families. Sensitive tactile hair cover pedipalpi and other parts of the body. The response to sound is uncertain; they have, however, been seen moving in conso-

---

[7] Considered homologous with the green glands of the crustaceans.

## Phylum Arthropoda

nance with the vibrations of a tuning fork. Some spiders have definite sound-producing mechanism. Apparently well-developed sense of smell resides in minute "lyriform organs" found on the body as well as appendages.

The spiders have eyes that are very beautiful objects. Situated on the top or front of the head region (Fig. 16.52 B, C), there is to be seen a group of usually six or eight eyes shining like little gems. The exact dispositon of these varies greatly in different kinds of spiders, but in all cases they have a structure essentially like that of the ocelli of insects. An interesting point in connection with this type of eye is that the lens is a part of the outer skin and is cast off at the time of moulting. The animal thus becomes temporarily blind. In the spiders, it is reported, that all the eyes do not moult simultaneously. Sometimes all the eyes are not of the same appearance, some being black and the rest pearly white. Usually the anterior median eyes or direct eyes are black while the others (or indirect eyes) are provided with a tapetum which reflects light rays (hence pearly white).

In the wolf-spiders and jumping spiders the eyes have shifted to the anterior half of the carapace so that they can have a wider perception. The spiders are neither sharp-sighted nor long-sighted, relying more on a highly developed sense of touch in their dealing with the outer world. They do not form sharp images as the number of receptors in their eyes is less. The hunting spiders have a larger number of receptors while the sedentary web-builders have a very small number of receptors. Only the jumping spiders make a sharp image. In jumping spiders the median eyes are situated at a great depth, that is, there is a tube between the lens and the retina (Fig. 16.52A). This perhaps compensates the absence of any power of accommodation. Some spiders, like the funnel-web spider *Agelena* and the wolf-spider *Arctosa*, utilize the sun and the polarized sky light to orient themselves to their surroundings. *Arctosa* possesses an internal clock which corrects for the changing angle during the course of the day.

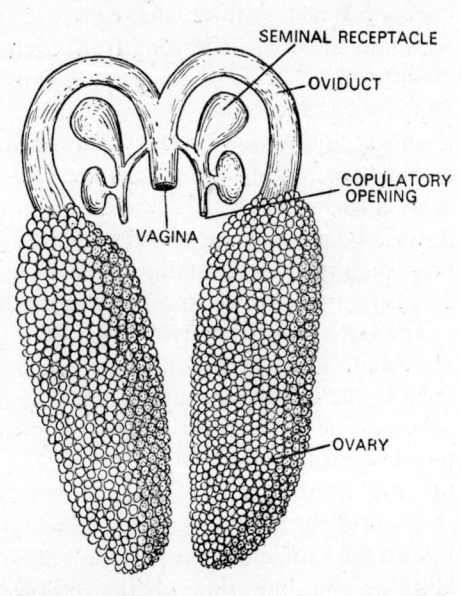

FIG. 16.53. Reproductive system of a female spider (*after* Millot).

The sense of touch is equally important in the life of spiders. The sensory hairs are the most important organs of sense in the sedentary spiders. Isolated or grouped slit sense organs are located over the entire surface of the body. In the web-building spiders the lyriform organs in the joint between the tarsus and the metatarsus enable the spider to discriminate vibrations transmitted through the silk strands of the web. The spiders can also find out the size of the entrapped prey when it is an insect that produces burring vibrations. The spiders themselves may produce vibration signals. Spiders also possess foregut chemoreceptors for certain pungent insects are rejected as food after being captured and killed. If chemically treated (say by quinine) food is given the spiders show this response. The chemoreceptors of taste are located in bands on each side of the central region of the pharynx.

### Reproduction and Development

The sexes are separate. The gonads are contained in the ventral part of the abdomen, opening to the outside near the anterior end of the abdomen on the ventral surface. The male has two testes below the intestine that join by coiled **vasa efferentia** to a single seminal vesicle leading to the genital opening. In the female the two **ovaries** (Fig. 16.53) are large and hollow, each with an **oviduct** joining the single **vagina**, into which open two lateral **seminal receptacles**.

Each ovary consists of an elongated sac, the epithelium of which produces **oocytes**. When enlarged, during **oogenesis**, the ova bulge outward giving the appearance of a bunch of grapes to the ovary. When fully formed the ovaries occupy about two-thirds or more of the abdomen. Some epithelial cells do not form eggs, but secrete adhesive material used to agglutinate the eggs in the cocoon. When mature the eggs are released into the cavity of the ovary and pass into the oviduct leading from each ovary. Each oviduct extends forward, downward towards the mid-line where it meets its fellow from the opposite side and forms a common tube (uterus). This tube extends ventrally and opens into a short chitinous vagina which opens at the middle of the epigastric furrow. Seminal receptacles and glands are also associated with the vagina and uterus. In most spiders these have separate opening to the outside and connected internally to the vagina. These external openings are for the reception of the male copulatory organs during mating. The copulatory opening is situated in front of the epigastric furrow on a special sclerotized plate, the **epigynum**.

The testis consists of two large tubes lying ventrally along each side of the abdomen. The sperm ducts are convoluted tubes which extend downward and open to the outside through a single genital pore in the middle of the epigastric furrow. The copulatory organs are located at the ends of the pedipalps. The tarsal segment of pedipalps has become modified to form a remarkable structure for the transmission of sperms (Fig. 16.54). Each palp consists of a bulb-like reservoir from which extends an ejaculatory duct. This leads to a penis-like projection the **embolus**. At rest the bulb and embolus fit into a concavity, the alveolus, on one side of the tarsal segment. The bulb and embolus become engorged with blood during mating, and project out of the alveolus. The embolus is then inserted into one of the female pores and the seminal fluid is passed on to the seminal receptacle. Fertilization occurs at the time of egg laying.

The structure of the male palp is simple in primitive spiders but becomes much more complicated in many families (Fig. 16.54). Many accessory structures become associated with it. These include the conductor, bulbal apophyses and other structures (Fig. 16.54). During mating the palps are dipped into the globule until all the semen is taken up into the reservoir. The male then seeks the female for mating.

In copulation the sperm cells of the male are transferred to the female by the modified palps. Fertilization is internal. The eggs are laid in silk cocoons, which, in some cases, are carried about by the females until the young **spiders**

**hatch.** Female *Lycosa* carries the young on her abdomen for some days after they hatch. The black widow, *Latrodectus mactans*, lays 25 to over 900 eggs in a cocoon and produces one to nine such cocoons in a season. Her eggs hatch

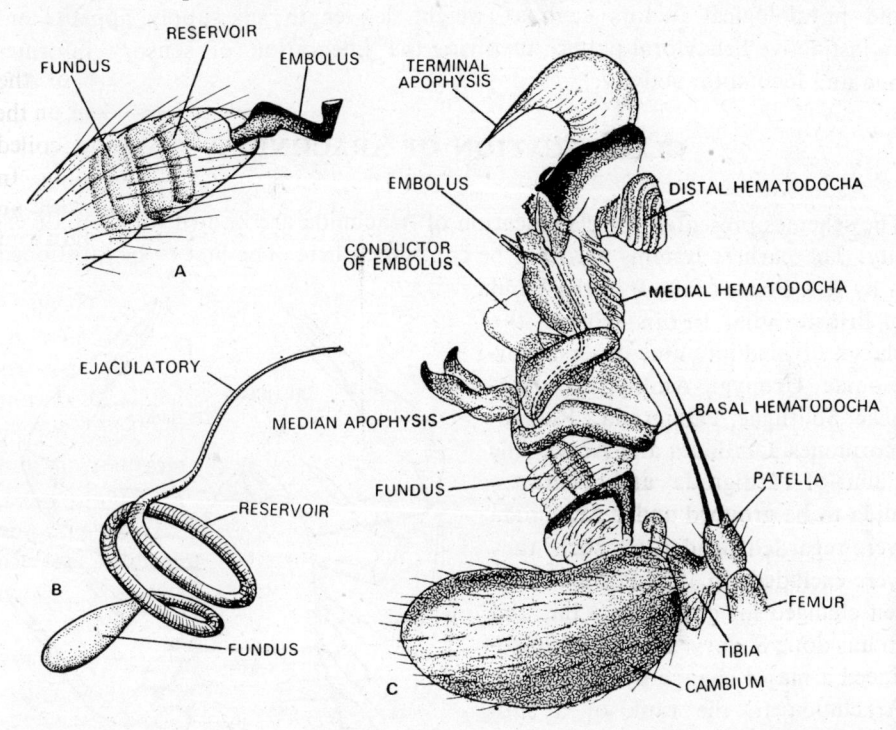

**FIG. 16.54. A**—Simple palp of *Filistata hibernalis* (*after* Comstock); **B**—Receptaculum seminis, the semen containing part of the male; and **C**—Complex palp of orb weaver *Aranea frondosa* in expanded state (*after* Comstock).

in 10-14 days and the young remain in the sac for two to six weeks after the first moult. Five moults are needed by the males to attain maturity and the females need seven to eight weeks.

**Spider silk.** The spider silk is a secretion of special abdominal glands and its chemical nature is that of a complex albuminoid protein. Its most striking physical property is its great tensile strength and its elasticity. A thread of .01cm in diameter has been found to carry a weight of more than 80 grams before breaking and to have stretched by 20 per cent of its original length. The silk web of some species is so strong that it is used by natives of some tropical regions for catching fish. Hardening of the silk results not from exposure to air but from the actual drawing out process itself. A single thread is composed of several fibres, each drawn out from liquid silk supplied at a separate duct opening. Most spiders produce more than one type of silk from a corresponding number of different silk glands. Spider makes many uses of silk which, unlike the silk of insect larvae, is used throughout life. It provides the gossamer threads on which the young "ballooning" spiders migrate; webs are made of silk and captured insects are bound in silk fetters. Wanderers leave a silk dragline behind them. Sedentary spiders live in a silk home or in a silk-lined burrow, and eggs are laid in silk cocoons. Snares, shelters, nests for hibernation or mating are spun with various kinds of silks by different species. The construction of a web by a spider is a remarkable feat. It depends upon morphological and physiological factors such as weight, leg length, silk supply, appetite and an instinctive behavioral pattern involving the integration of sensory information and locomotor activity.

## CLASSIFICATION OF ARACHNIDA

The schemes presented for classification of Arachnida are known for their diversity. The earliest systems need not be considered here. The first to be mentioned is by R.I. Pocock (1900) in the Fauna of British India. In this scheme the class is divided into nine orders: Scorpionida, Uropygi, Amblypygi, Araneae, Solifugae, Palpigrada, Pseudoscorpiones, Opiliones and Acari. The claims of Tardigrada and Pentastomida to be grouped under Arachnida were regarded slender, as such they were excluded. In 1929 Pocock[8] himself changed his groupings a bit and so has done Savory. Borradaile introduced a major change by promoting Arachnida to the rank of a subphylum. The modern trend is to treat arachnids as a class of the subphylum Chelicerata. The Arachnida themselves have been divided into ten orders which appear to be a heterogeneous group. There is a marked tendency of loss of metamerism in different groups. In scorpions, for example, the metamerism shows slight reduction, and reduction of metamerism is maximum in ticks and mites. Other groups grade between these extremes. The follwoing are the orders of Arachnida all having representatives:

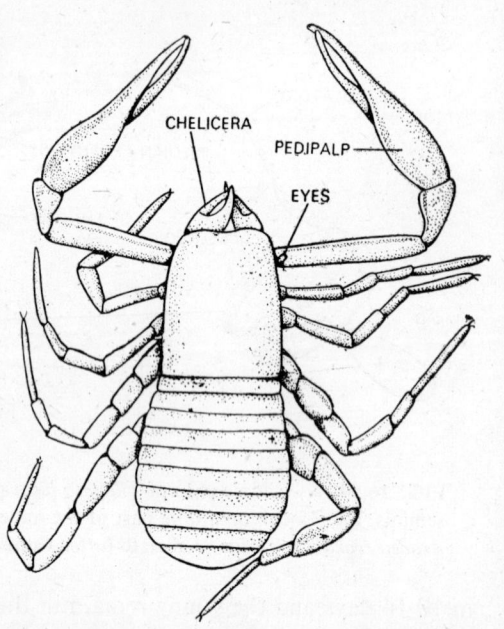

FIG. 16.55. Dorsal view of a pseudoscorpion (*after* Hoff).

1. Order Scorpionida (the scorpions)
2. Order Pseudoscorpionida (the false scorpions)
3. Order Solifugae (the sun spiders or wind scorpions)
4. Order Palpigrada (the palpigrades)
5. Order Uropygi (the whip-scorpions)
6. Order Amblypygi (the amblypygids)
7. Order Araneae (the spiders)
8. Order Ricinulei (the ricinulids)
9. Order Opiliones (Harvestmen)
10. Order Acarina (mites and ticks)

[8]*Encyclopaedia Britannica*, XIVth edition, 1929.

*Phylum Arthropoda*

1. **Order Scorpionida.** Arachnida in which the prosoma is uniformly covered with a cephalic shield and bears two median and six to ten lateral eyes, and the opisthosoma is divisible into mesosoma and metasoma each of six somites. Telson takes the form of a poison sting. Chelicerae of three segments, small, chelate. Pedipalpi of six segments, large, powerful, chelate. Legs of seven segments, tarsi with three claws. Sternum triangular or pentagonal. Second mesosomatic somite bears a pair of characteristic pectines. Respiration by four pairs of book-lungs, on third to sixth opisthosomatic somites. Includes 600 species of scorpions from Silurian to Recent. The hairy scorpion, *Hardurus hirsutus*, which is about 10 cm long and *Vejovis spinigerus* about the same size are well-known American species.

Examples: *Centrurus, Buthus, Hemibuthus, Lychas, Stenochirus, Scorpiops, Hromurus, Palamnaeus,* etc.

2. **Order Pseudoscorpionida (Chelonethi).** False scorpions. Arachnida in which the prosoma is covered above by a carapace, quadrate or triangular in shape, and bears not more than two pairs of lateral eyes. Opisthosoma or twelve visible segments, each provided with a distinct dorsal and ventral sclerite. No pedicel; no telson. Chelicerae of two segments, small, chelate. Pedipalpi of six segments, large, chelate provided with tactile setae and containing poison glands.

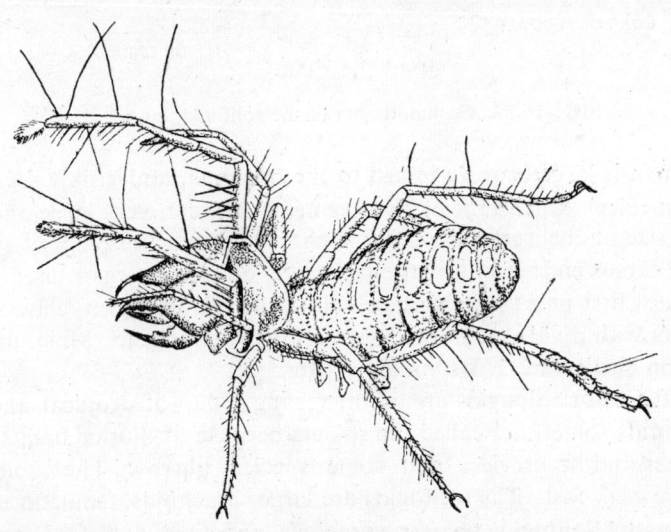

**FIG. 16.56.** North African solifugid, *Galeodes arabs* (*after* Millot and Vachon).

Sternum usually absent. Legs composed of five to seven segments, all tarsi with two claws. Forelegs of males occasionally modified to assist fertilization. Found under moss or stones, in trees or about buildings. About 1,000 species are known. Pseudoscorpions are tiny arachnids rarely longer than 8 mm. They live in leaf mould, in soil, beneath bark and stones, in moss and similar type of vegetation, and in nests of some mammals. A few species inhabit caves, and some species of several genera are common inhabitants of algae and drift in the intertidal zone.

Pseudoscorpions (Fig. 16.55) superficially resemble the true scorpions, but differ in several important respects other than size. The body is composed of eighteen segments, six fused prosomal segments and twelve segments forming the abdomen. The abdomen contains a pregenital segment that is lacking in scorpions. The prosoma is covered by a single rectangular carapace. One or two eyes are located at each anterior lateral corner, or the eyes may be absent. Ventrally the prosoma is entirely occupied by the coxae of the walking legs and pedipalps. From Tertiary to Recent.

Examples: *Chelifer cancroides* is the common house or book-scorpion 3.5 mm long found about buildings. *Carypus*, a Californian genus is about 7.5 mm long.

3. **Order Solifugae.** Arachnida in which the prosoma is more fully segmented than that of any arachnid. The unique feature exhibited by these two prosomal divisions is that they can articulate with one another in the non-burrowing species. The opisthosoma consists of ten somites, each with a dorsal and ventral sclerite.

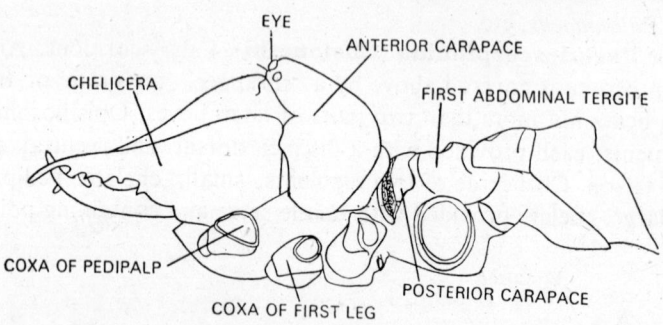

FIG. 16.57. Cephalothorax of the solifugid, lateral view.

The abdomen is large broadly joined to the prosoma, and visibly segmental. No telson; no pedicel. Chelicerae of two segments, chelate, very large and powerful, enormous size of chelicerae is a striking characteristic. Pedipalpi of six segments tactile, the tarsus ending in a suctional sac or adhesive organ used to capture prey. Legs of first pair tactile with one claw, the rest with two claws; third and fourth pairs with divided femora. Respiration by tracheae. Male usually with flagellum on chelicerae. No spinning organs.

The solifugids or solpugids are an interesting group of tropical and semitropical arachnids sometimes called sun spiders because of diurnal habits. They hide under stones and in crevices and some species burrow. The non-burrowing forms move very fast. The solifugids are large arachnids, sometimes reaching 7 cm in length. Solifugids possess voracious appetites and feed on all types of small animals including vertebrates. Found in warm dry regions. About 570 species known. Carboniferous to Recent.

Examples: *Eremobates*, *Galeodes*, North African solifugid (Fig. 16.56).

4. **Order Palpigrada (Palpigradi).** Micro-whip scorpions. Arachnida of small size, (not exceeding 3 mm in length) in which the prosoma is divided into two principal plates (anterior carapace and posterior carapace). Eyes absent. Telson in the form of a long jointed flagellum. Chelicerae of three joints, well-developed chelate. Pedipalpi undifferentiated made of six joints, leg-like, with small tarsal claws and used as a pair of walking legs. Legs of 12, 7, 7 and 8 segments respectively, tarsi with two claws. Mouth on a protecting rostrum. No gnathobases on any coxae. The abdomen (opisthosoma) is broadly joined to the prosoma and consists of 11 segments. The terminal segment bears the anal open-

Phylum Arthropoda

ing and a hairy flagellum of about 15 segments. Respiration cutaneous or by three pairs of book-lungs on fourth, fifth and sixth opisthosomatic somites. Live under stones and in soil or caves. Twenty species have been described from tropical and the warmer temperate regions.

Examples: *Koenenia* (Fig. 16.58), *Prokoenenia*.

5. **Order Uropygi (Pedipalpi)**. Whip-scorpions. A small order of arachnids having about 100 species found throughout tropical and semitropical parts of the world. They are small about 2 mm in size. Some species may be 65 mm long e.g., *Mastigoproctus giganteus* (Fig. 16.59). They are nocturnal and during the day hide beneath leaves, stones, logs and other debris. The prosoma is covered with uniform carapace or has the posterior somite free, and is jointed by

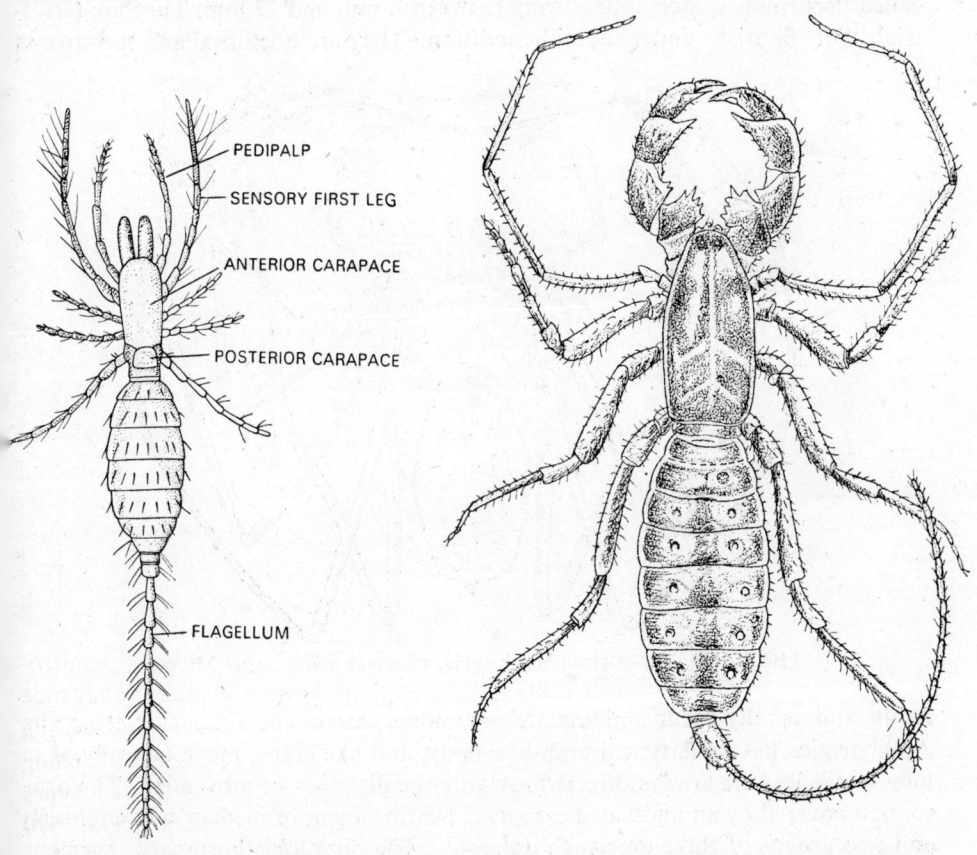

FIG. 16.58. *Koenenia*; a palpigrada (*after* Kraepelin and Hansen).

FIG. 16.59. American whip-scorpion, *Mastigoproctus giganteus*.

a pedicel to the segmented opisthosoma (abdomen). Telson of few or many segments present. Chelicerae of two segments, unchelate, without poison gland. Pedipalpi normally large chelate weapons, with trochanterial processes and tibial apophyses. Sternum of three segments, elongated. First pair of legs with two claws. Respiration by book-lungs. Sexual dimorphism slight. Odoriferous glands (anal glands) occur at the extremity of opisthosoma (abdomen). When irritated, the animal elevates the end of the abdomen and discharges a stream of fluid secreted from these glands. The secretion apparently contains formic or acetic acid used for defence. Because of the repellent odour of this fluid these animals are called wine-garrons. During feeding the prey is seized and torn apart by the pedipalps and then passed to the chelicerae, which complete maceration of the tissues. *Thelyphonus indicus* is a common whip-scorpion (Fig. 16.59) of India. Found in damp places beneath stones or pieces of wood in termites' nests or other crevices or shelter where they are protected from light. Some species excavate burrows for protecting themselves and their young. The young are hatched from the eggs, which the female carries about enswathed in a sticky membrane and attached to the abdomen.

Examples: *Thelyphonus, Uroeportus, Labochirus, Hypoctonus,* etc.

6. **Order Amblypygi (Phrynichida).** The arachnids that are commonly called "scorpion spiders" measuring between 8 mm and 37 mm. They are terrestrial, but flourish under humid condition. They are nocturnal and secretive in

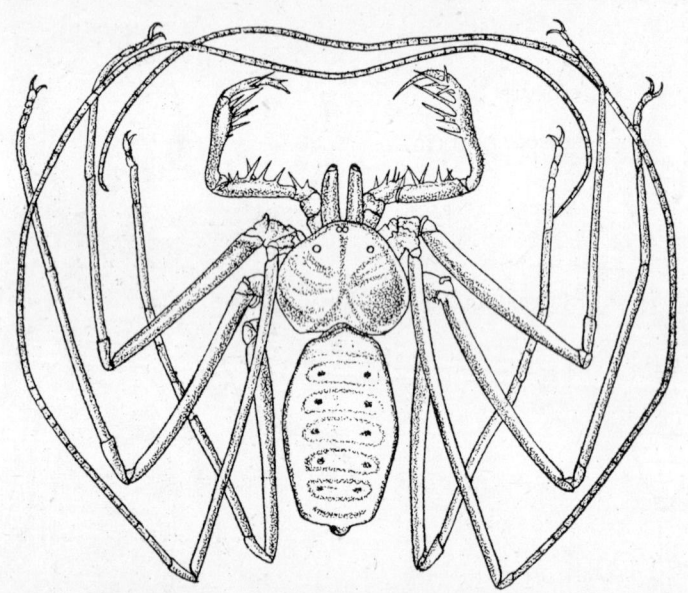

FIG. 16.60. The African amblypygid, *Charinus milloti* (*after* Millot).

habits and usually occur underneath logs, stones, leaves and similar objects. The Amblypygids have a flattened crab-like body, and like crabs, move laterally. The long tactile legs are always directed towards the direction of movement. The prosoma is covered by an undivided carapace, bearing a pair of median eyes anteriorly and two groups of three eyes each laterally. The first opisthosomatic segment forms a pedicel. The opisthosoma has twelve segments, each with a tergite and sternite. The chelicerae are moderate in size and similar to uropygid chelicerae (two jointed, non-chelate, and the second joint is fang-like, as in spiders). The pedipalps are heavy and raptorial. The first pair of legs is modified as sensory tactile appendages. They are extremely long and whip-like and are made up of many joints. The last three pairs of legs are used for locomotion. The abdomen has 12 visible segments. The first and the last segments are narrow. A terminal flagellum is absent.

They feed on insects. The prey is captured and killed with the pedipalps, and then passed to the chelicerae, where juices are sucked into the digestive system

*Phylum Arthropoda* 591

as in a typical arachnid. Occur in tropical and semitropical conditions. There are approximately 60 species.

Example: The African amblypygid *Charinus milloti* (Fig. 16.60).

7. **Order Araneae (Araneida)**. This order consists of spiders. Arachnida in which the prosoma is protected by a uniform shield without much traces of segmentation, and bears no more than eight eyes and is joined by a pedicel to the opisthosoma, which is usually unsegmented, bearing not more than four,

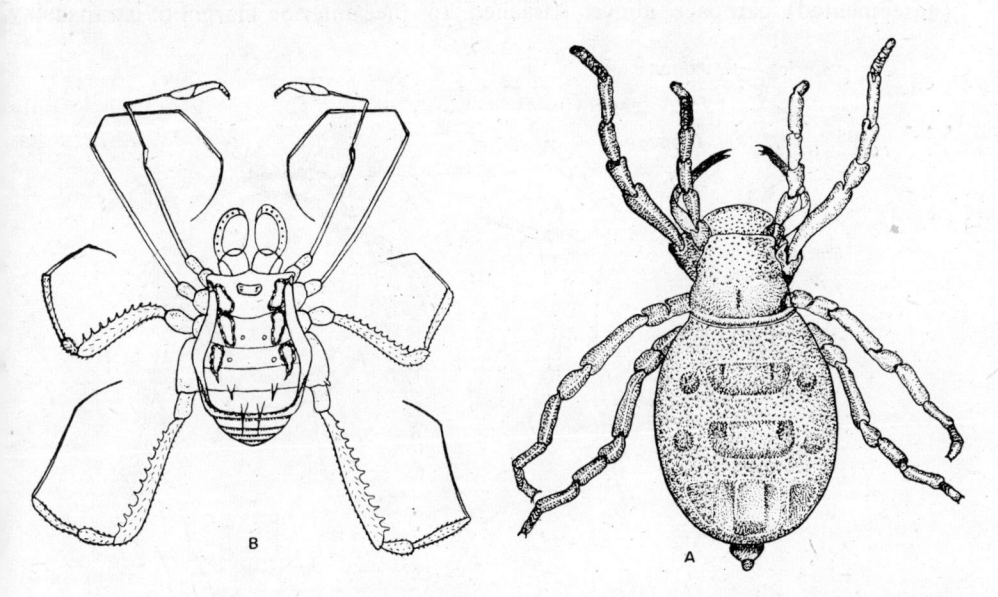

**FIG. 16.61.** A—*Ricinoides*, a ricinuleid (*after* Millot); and B—The opiliones, *Rhaucoides reviti*.

usually three pairs of spinnerets. No telson. Chelicerae of two segments, moderately large and unchelate, containing a poison gland. Pedipalpi of six segments, leg-like, tactile in function. Sternum present, usually ovoid. Legs of seven segments, tarsi with two or three claws. Respiration by book-lungs or tracheae or normally both. Pedipalp of male modified as intromittent organ. Oviparous, eggs commonly in cocoons. Chiefly terrestrial. About 20,000 species have been studied. Carboniferous to Recent. Spiders are cosmopolitan, occurring in all oceanic islands.

Some common genera include the following:

*Pachylomerus*, trap-door spiders (Ctenzidae), make nests in holes dug in the ground, closed, hinged door of earth and silk.

*Eurypelma*, bird spiders (Theraphosidae), lives on ground or trees, bite painful but not dangerous to man. Indian genera of this family include *Phlogiodes*, *Chilobrachys*, etc.

*Agelena naevia*, grass spider (Agelenidae), web flat with funnel at side, in grass and in houses. *Hippasa* is an Indian genus that makes flat webs on grass, etc.

*Lycosa*, wolf spider (Lycosidae), is a large hunting spider of 25 mm long, active, hunts prey, lives in silk-lined underground holes but does not spin webs; female carries egg cocoon and young.

*Miranda aurantia*, orange garden spider, *Epeira marmorea* (Argiopidae) with nests of leaves by the side of web are known as orb spiders.

The crab-spiders (Thomisidae) are short, wide with crab-like appearance and no web. *Misumena*, bright or white coloured, live in flowers, mimic their colour; *Philodromus*, mottled, *Xysticus*, lives under bark or leaves.

Jumping spiders (Attidae) are spiders without webs that run and jump in all directions.

Examples: *Salticus, Phidippus,* etc.

8. **Order Ricinulei (Podogona)**. Arachnida in which the prosoma bears single (unsegmented) carapace above. Attached to the anterior margin of carapace

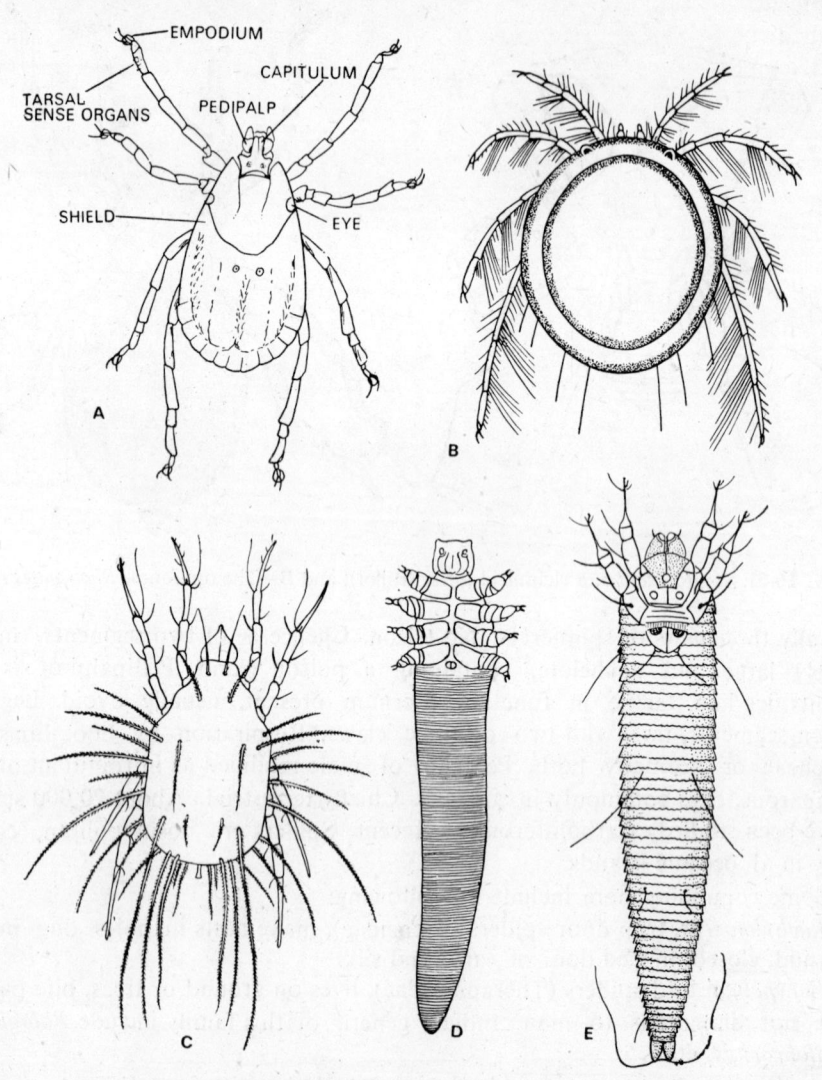

**FIG. 16.62.** Acarina. A—Tick, *Dermacentor variablilis* (*after* Snodgrass); B—Water mite, *Mideopsis orbicularis* (*after* Soar and Williamson); C—*Oudemansium domesticum*, dorsal view; D—*Eriophyes*, ventral view (*after* Nalepa); and E—*Demodex folliculorum*, ventral view.

there is a characteristic wide, oval, slightly convex plate (cucullus). It can be raised and lowered. When lowered the hood covers the mouth and chelicerae.

Eyes absent. Opisthosoma (abdomen) of nine somites, united to the prosoma by a pedicel which is normally hidden. No telson. Chelicerae, of two segments, small chelate. Pedipalpi of six segments, chelate, their coxae fused in the middle line. Legs of seven, eleven or twelve segments without spines, tarsi with two smooth claws. Respiration by tracheae with apertures on prosoma. Metatarsus and tarsus of third leg modified as sexual organs. They have been found only in two regions of the tropics, the central west coast of Africa and the Amazon basin. Carboniferous to Recent. Little is known about egg-laying or early development. A six-legged "larva" similar to those in mites and ticks has been described. Only a single family with a small number of genera.

Examples: *Cryptostemma, Cryptocellus* are the two known genera of which thirteen species are known and only 32 (18 *Cryptostemma*, 14 *Cryptocellus*) specimens have been collected.

9. **Order Opiliones (Phalangida)**. This order contains familiar long-legged arachnids known as "daddy long legs" or harvestmen. These animals live in temperate and tropical climates and all prefer humid habitats. Abundant on the forest floor, on tree trunks and fallen logs in humus and in caves. Many species are nocturnal, some diurnal. The uniform prosoma bears two eyes, usually on prominent tubercles (ocularium) No pedicel. No telson. Opisthosoma (abdomen) with ten tergites and nine sternites, more or less united. Chelicerae of three segments, small, chelate. Pedipalpi of six segments, usually small, with or without a terminal claw. Legs of seven segments, generally long, the second or fourth pair the longest. Tarsi many jointed with one, two or three claws. Respiration by tracheae, a pair of spiracles on the second sternite. A pair of odoriferous glands in the prosoma. The openings of the glands are located along the anterior lateral margin of the carapace. The secretions of these glands have been described as having an acrid odour similar to nitric acid, rancid nuts or walnut husks. Sexes are much alike. Occur in fields, woods and buildings, 2,200 species are known. Carboniferous to Recent.

Examples: *Phalangium, Liobunum, Gonyleptes, Purcellia*, etc.

10. **Order Acarina (Acari)**. Mites and ticks. Arachnida in which the uniform prosoma may bear simple eyes or not. No pedicel. No telson. Opisthosoma with segmentation almost completely visible. Chelicerae and pedipalpi usually small and associated with mouthparts, which are modified for biting, sawing, piercing and sucking. Legs of seven segments, tarsi usually with two claws and protarsus of varying form. Respiration by tracheae or cutaneous. Generally a larval stage with six legs, and other stages with eight legs. Parthenogenesis not unknown. They are free-living or parasitic and worldwide in distribution. About 6,000 species are known. Oligocene to Recent.

From human economic point of view the order Acarina is the most important of arachnid orders. Many species are parasitic on man, his domesticated animals, and his crop, while others are destructive to food and other products. Mites occur in polar regions, in deserts and even in hot springs. Terrestrial species are extremely abundant, particularly in moss, fallen leaves, humus soil, rotten wood and detritus. The number of individuals is enormous. A small sample of leaf-mould from a forest floor may contain hundreds of individuals belonging to numerous species. Taxonomy and biology of mites is still not well-known. Because of their economic importance a separate field of study called **Acrology** has developed on the study of mites. All members of this group are not parasitic, some are free-living. The Acarina are morphologically diverse so much so that the order is considered to be polyphyletic. Ticks attack all groups of terrestrial vertebrates. In man they are responsible for the transmission of Rocky Mountain spotted fever, tularemia, and other pathogens. There are parasitic mites that spend their entire life-cycle attached to the host, as for example, the feather mites of birds, the scab producing fur mites of mammals, and the "itchmite." The human itchmite (*Sarcoptes scabei*) in the case of scabies or seven year itch tunnels into the epidermis. The female is about 0.50 mm and the male is less than 0.25 mm in length. Irritation is caused by mite's secretion. The female deposits eggs in the tunnels for a period of two months after which she dies. As many as 25 eggs are deposited every two or three days. The eggs hatch in several days. Larvae follow the similar existence as the adult. The infection thus can be endless, and is transmitted to another host by contact with infected areas of the skin.

Examples: *Tetranichus*, the red spider mite, sucks plant juices; *Trombicula alfreddugesi* (*irritans*) the larvae of which puncture and irritate skin of labourers;

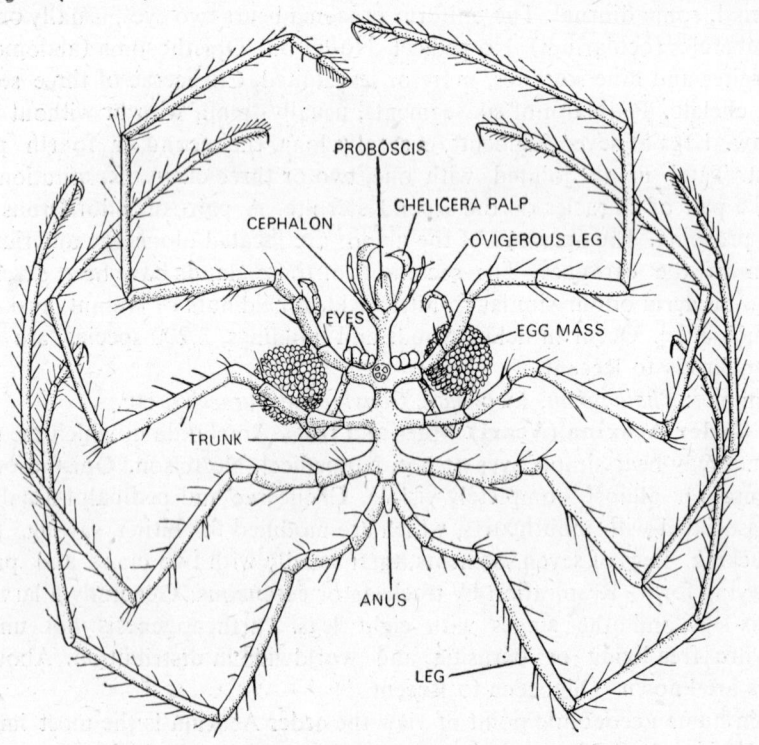

**FIG. 16.63.** *Nymphon rubrum*, a sea-spider (*after* Sars).

*Hydrachna* (water mite); *Argas persicus* (fowl tick); *Orinithodorus megnini* (ear tick), *O. moubata* carries African tick fever; *Psoroptes communis* (scale mite of sheep); *Eriophyes* (blister mite) damages leaves, buds and fruits; *Demodex folliculorum* lives in hair follicle and sebaceous glands of man and domestic mammals and many more.

The Order Araneae is subdivided into two suborders: (*i*) **Orthognatha (my-**

**galomorphs**) that includes the "tarantulas"; (*ii*) **Labidognatha (araneomorphs**) that includes the remaining spiders. The tarantulas are rather large spiders, possess two pairs of book-lungs, two pairs of coxal glands and the fang of the chelicera articulates in the same plane as the long axis of the body. The araneomorphs are divisible roughly in two groups depending upon ecological conditions, web-builders and cursorial forms. Web-builders are sedentary and catch the prey with silk webbing. Following are important families of this subgroup:

**Pholcidae.** Small and long legged, resembling phalangids (Daddy long legs). Members spin small webs of tangled threads in sheltered corners.

**Theridiidae.** A large family of comb-footed spiders (because of the presence of a series of serrated spines on the fourth tarsus) which build tangled web. Examples: *Latrodectus*, the black widow, and *Achaearanea tepidariorum*, the common house spider.

**Araneidae.** A family of orb-weaving spider. Members of this family build circular webs, large brightly coloured species.

Example: *Argiope aurantia*.

**Linyphiidae.** The sheet web-spiders, webs are horizontal silken sheets or bowls usually in vegetation.

**Agelenidae.** Funnel web-spiders. The web is built in dense vegetation or in crevices of logs or rocks and is easily visible specially in grass covered with dew.

The hunting-spiders do not use silk to catch prey which is captured directly by pouncing upon it. Third or middle claw is absent in this group. Important families are given below:

**Lycosidae.** These are rapid moving, hairy wolf-spiders with brown and black colours. Nocturnal in habit. *Lycosa* is the common example.

**Pisauridae.** These are called fish-spiders and have longer legs and live around the edges of the ponds, lakes and streams.

**Thomisidae.** Called the crab-spiders because of crab like movements and hunt on vegetation.

**Salticidae.** The jumping-spiders that are heavy bodied and capable of jumping short distances. Often brightly coloured and possess the best eyesight. Temperate and tropical regions.

## V. SUBPHYLUM MANDIBULATA

The subphylum **Mandibulata** includes such successful classes of Arthropoda as the **Crustacea** and **Insecta** besides the **Pauropoda, Diplopoda, Chilopoda** and **Symphyla**. Considering abundance, temporal persistence and diversity the Chelicerata, as represented by the class Arachnida, is unquestionably a successful arthropod subphylum; but the subphylum **Mandibulata** surpasses the Chelicerata by way of great success of its two classes **Crustacea** and **Insecta**. The Crustacea have attained success in acquatic environment, while the Insecta have attained a greater success in the terrestrial environment. The Crustaceans also occur on land just as the insects occur in aquatic situations. In the older classification the less successful terrestrial classes Pauropoda, Diplopoda, Chilopoda and Symphyla were grouped together as the **Myriapoda**.

The body of the Mandibulata is usually divided into three parts: an anterior head, a middle **thorax** and a posterior **abdomen**. In the Myriapods the body consists of a head followed by numerous body segments. The head is made up of six body segments, of which the first is without appendages. The second head segment bears **antennae** homologous with the **chelicerae** of the Chelicerata. In the Crustacea a **second pair** of **antennae** is present on the third head segment. The fourth head segment bears a pair of **mandibles** in all classes, a pair of **maxillae** is borne by the fifth segment while the sixth segment bears the **second pair** of **maxillae**. The thorax is not always distinct (cephalothorax in Crustacea), but where defined it is formed of three (the Insecta) or more segments. Respiration is **cutaneous, tracheal** or by means of various sorts of thin-walled body outgrowths. The members of the subphylum inhabit freshwater, brackish water, sea-water or land, the insects are predominantly aerial.

Those who believe in the polyphyletic origin of the major group of Arthropoda do not recognize the Mandibulata as a meaningful evolutionary grouping, and regard the similarities between crustaceans and Myriapods—insects as the results of convergence. Those who believe in the monophyletic origin of the Arthropoda have opposing viewpoint, but both schools agree that insects and myriapods are related. The subphylum Mandibulata is being retained in the present description more as a convenience, although morphologically speaking, the crustaceans and myriapods—insects—do have many features in common.

## CLASS CRUSTACEA

The **Crustacea** (L. *crusta*, a hard shell), the first group of the mandibulate classes, includes some of the most familiar arthropods such as prawns, crayfishes, lobsters, shrimps, water-fleas, barnacles, crabs and their relatives. There are about 26,000 species of Crustaceans. Myriads of tiny crustaceans that live in the sea, lakes and ponds of the world occupy a basic position in aquatic food chains. The Crustacea is the only large class of arthropods that is primarily aquatic.

Some, like barnacles, are sessile or parasitic, and there are still others that are commensal or parasitic on various aquatic animals from hydroids to whales. In response to parasitic life some are so much modified that it becomes difficult to recognize them as adults. The first five segments are fused to form the head, while the others usually form the thorax and the abdomen. Many or fewer of the thoracic segments may fuse with the head to form a cephalothorax. The head bears two pairs of antennae, one pair of lateral mandibles, for chewing, two pairs of maxillae, a median eye which frequently disappears in the adult, and a pair of compound eyes which is sometimes elevated on jointed stalks. The thorax consists of two to six segments distinct or variously fused. The abdomen is segmented and often bears limb-like appendages with a telson at the end. Often a carapace is present over the head and part of thorax, as a dorsal shield or as two lateral valves. Appendages are variously modified to subserve such diverse functions as locomotion, respiration, and food collection, etc. With the exception of the antennules the appendages are typically biramous, consisting of a stem or **protopodite** bearing two branches the **endopodite** and **exopodite**. Respiration takes place by the gills, rarely the body surface is also used for the purpose. The excretory organs are peculiarly modified coelomoducts, which may

*Phylum Arthropoda*

take the form of either **shell-glands** opening on the second maxillae or of **green (antennary) glands** opening on the antennae. Sexes are separate (except in Cirripedia and some parasitic Isopoda); openings of the generative organs are paired; eggs are often carried by the female. Parthenogenesis also occurs in some Brachiopoda and Cladocera.

Prawns, lobsters and crayfishes are mostly swimmers, although they belong to the order Decapoda, which owes its name to the fact that the hinder five pairs of thoracic limbs in these animals are typical walking legs. The true crabs, on the other hand, are real walkers although some possess flattened legs with which they can swim. The hermit crabs and coconut crabs, which also come under this group, are forms intermediate between prawns, lobsters, crayfishes and the true crabs.

### TYPE PALAEMON

The prawn *Palaemon* is found all over India inhabiting freshwater streams, rivers lakes and ponds, etc. The large river prawns found everywhere within the tropics belong to the genus *Palaemon*. Several species of the genus are found outside India also. The size is variable. Some are reported to be about one metre long from the tip of the telson to the tip of the extended legs. The large species are found in the south as well as in Bengal. Those found in Madhya Pradesh and

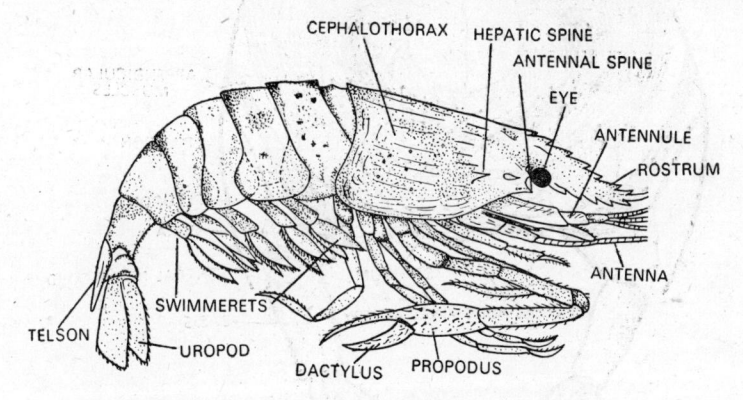

**FIG. 16.64.** External features of *Palaemon*, entire, lateral view.

Uttar Pradesh are generally 25-30 cm in length, rarely longer. These animals prefer slow-moving, clean water and are more at home at the bottom. At night they go in search of food and are often noticed among weeds in the crevices on the banks. They feed upon algae, moss and other weeds and also upon some insects. Debris from the bottom of the water are also swallowed with food. Fresh animals are pale-yellow or white in colour. Colour of different species is variable. In some the body is pale-blue or greenish with brown or orange-red patches and bands of different patterns.

### Form of the Body

The body is elongated and divisible into two main regions, the **cephalothorax** and **abdomen**. The cephalothorax is a rigid region formed by the fusion of the first thirteen segments of the body. It is, more or less, uniformly cylindrical in shape and unsegmented externally. The segmentation of the region is indicated by the presence of thirteen pairs of appendages on the ventral side. In some (*Astacus*) the head and thorax regions are externally demarcated by the **cervical groove**, but the same is not present in the prawn. The abdomen has preserved its original segmentation and consists of six movable **segments** and a terminal conical structure, the **telson** or **tail-plate**. The abdomen is rounded dorsally and flattened laterally and its hinder end is sometimes flexed inwards. The jointed-appendages are articulated to the ventrolateral margins of the body (Fig. 16.64).

The entire body including the appendages is covered by **cuticle** that forms the stout **exoskeleton** of the animal. The cuticle is secreted by the underlying epidermis and is not soft as in the annelid worm. Here it is **chitinized** and is made particularly hard by a deposition of calcium salts. Chitin is an organic substance chemically somewhat similar to silk or to spongin of the sponge skeleton.

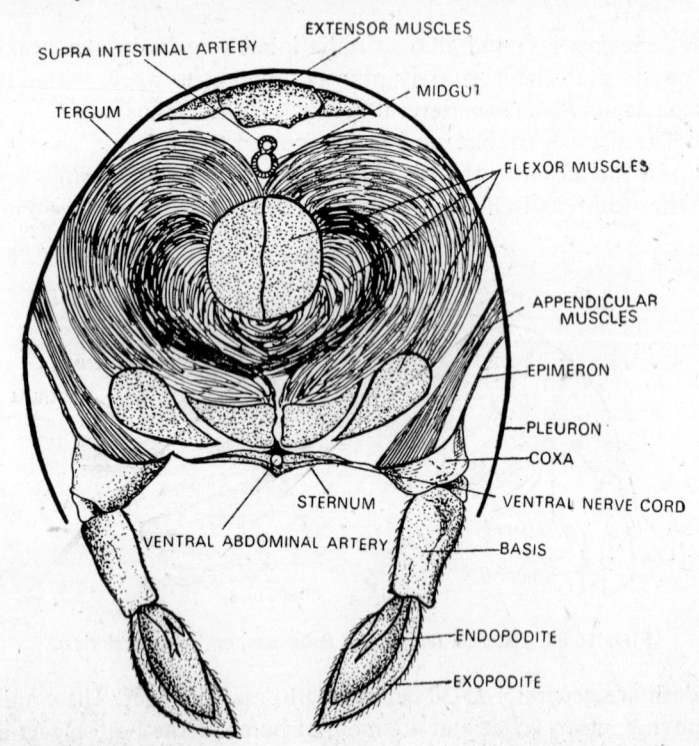

FIG. 16.65. Transverse section through the abdomen of the prawn showing sclerites and musculature.

It is unaffected by ordinary acids and alkalies, though soluble in sodic or potassic hypochlorite. The chitinization is not uniformly continuous all over the body but is localized forming hardened pieces of exoskeleton called **sclerites**, between which lie thin flexible memberanes known as the **arthrodial membranes**. These allow the movements of one sclerite on the other. In a typical segment as that of the abdomen of the prawn the dorsal sclerite is broad or convex and is known as the **tergum**, the ventral sclerite that forms the floor is narrow and transverse and is known as the **sternum**. On either side the tergum extends downwards forming side flaps, each called **pleuron** (Fig. 16.65). At each end of the sternum, close to its junction with the tergum, arises a jointed appendage called a **swimmeret**. Connecting the appendage of each side with the pleuron there is a small chitinous plate, the **epimeron**, really a part of the pleuron. The arthroidial membranes between the terga are known as the **intertergal**, whereas, those between the sterna are known as the **intersternal membranes**.

The joints between the adjacent abdominal segments are of the peg-and-socket type, a rounded peg fitting into a corresponding socket on the next segment. Such a joint allows the movement of one segment upon the other in a vertical plane. Besides this, each segment articulates with its adjacent segment by means of a pair of hinge-joints, one on each side. When extended all the segments of the abdomen lie in line with the cephalothorax, but it bends downwards when flexed.

All the abdominal segments are alike and agree with the plan of structure detailed above which is especially applicable to that of third segment. They, however, differ from each other in minor details. The pleura of the second segment is larger and overlaps the pleurae of both the first and third segments. The posterior half of the intertergal membrane connecting the cephalothorax with the tergum of the first abdominal segment is calcified and rigid. The hinge joints between the third and fourth abdominal segments are absent. The pleurae of the sixth segment are very small, each being triangular in shape, ending in a backwardly

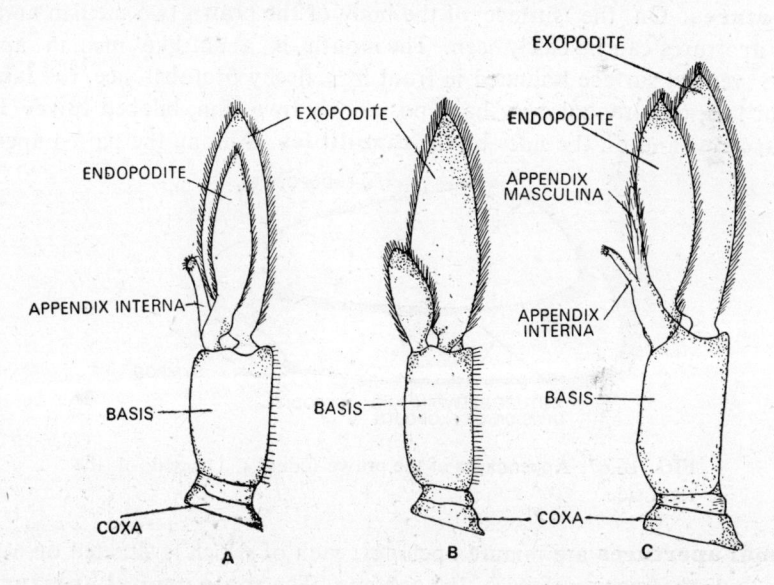

FIG. 16.66. Appendages of the prawn. A—First; B—Second; and C—Third abdominal appendage of the male prawn.

directed sharp point. The **telson** constitutes the last abdominal region. It is horizontally flattened structure attached by its base to the sixth abdominal segment. Its tip is sharply pointed and bears two spines and number of setae (Fig. 16.64).

The sclerites of the cephalothoracic region have all fused together to form a

**dorsal shield**. The dorsal shield is formed by the fusion of the **dorsal plate**, forming the exoskeleton of the head region and the **carapace**, the exoskeleton of the thoracic region. Between the dorsal plate and the carapace there is a cervical groove present in some (*Astacus*), but in *Palaemon* it is a continuous compound structure. The dorsal shield is further modified to form a **branchial chamber** on each side of the body to house the gills. From the dorsolateral margin of the dorsal shield projects, on each side, an extensive complex plate, the **branchiostegite** or **gill-cover**, which encloses the cavity of the branchial chamber between it and the lateral wall of the thorax. Extending forward from the anterior end of the dorsal shield is an elongated and a laterally compressed median outgrowth, the **rostrum** (Fig. 16.64), the upper and lower ends of which are serrated and beset with setae. About the middle of its length it is slightly bent and is turned upwards. Situated at the base of the rostrum are the paired **eyes** mounted on movable stalks, lodged in the **orbital notches**, on the anterior margins of the dorsal plate. Projecting forwards from the surface of the dorsal plate, just behind each eye stalk, there is a short spine known as the **antennal spine**. A similar though shorter spine, the **hepatic spine**, is situated a little posterior to the antennal spine. The sterna of the cephalothoracic segments have also fused together to form the floor the cephalothoracic box. The sternal plates are broad and well-developed posteriorly, but anteriorly they remain narrow.

**Apertures.** On the surface of the body of the prawn two median and three paired apertures can be easily seen. The mouth is a slit-like median aperture on the ventral surface bounded in front by a fleshy protuberance, the **labrum**, more or less, quadrangular in shape posteriorly by a thin, bilobed **lower lip** or **metastoma** and on the sides by the **mandibles**. Among the paired apertures,

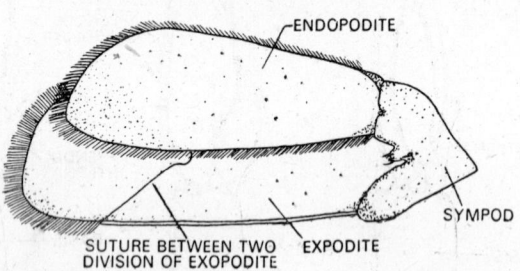

FIG. 16.67. Appendages of the prawn (contd.). Uropod of the right side.

the **renal apertures** are minute openings, each of which is situated on a raised papilla on the inner surface of the antenna. The **male genital apertures** are situated on papillae on the inner surfaces of the bases of the last pair of walking legs. The **female genital apertures** are similarly placed on the inner side of the bases of the third pair of walking legs. The third pair of apertures are the minute **openings of the statocysts**, each of which lies hidden beneath a small fold of integument in a deep depression on the basal segment of each antennule. The **anus**, another median opening, is situated on the ventral surface of the telson, close to its anterior border. It is a longitudinal slit with tumid lips surmounting a small elevation.

# Phylum Arthropoda

## Appendages

The appendages of the prawn, as also those of other Crustacea, are derived from the **biramous type**. The biramous limb consists of two "**rami**" or branches which arise from a basal portion articulating with the body. The name **protopodite** has been given to the basal portion, whereas the rami are called the **endopodite** (inner ramous) nearer the median line and the **exopodite** (outer ramous). The protopodite consists of two podomeres or joints, the **coxopodite** (proximal) and the **basipodite** (distal). From the protopodite also arises a thin-walled extension of the integument, the **epipodite** (probably having respiratory function). The fundamental type of appendage undergoes various modifications, in accordance with the purposes it fulfils in the life of the animal.

The prawn has nineteen pairs of appendages (Fig. 16.64) of which there are five pairs of head appendages, eight pairs of thoracic appendages and six pairs of abdominal appendages. The least modified ones are those of the abdomen and are known as the **pleopods** or **swimmerets**.

In a typical swimmeret (the third abdominal appendage) the basal piece or protopodite consists of two podomeres, a short ring-like segment devoid of setae (Fig. 16.66) known as the **coxa** and the **basis** which is cylindrical, bears setae on its surface and carries two rami (the exopodite and the endopodite) of the limb. The exopodite is a leaf-like structure having thickly set marginal setae and is larger of the two. The endopodite is also a leaf-like structure and gives off

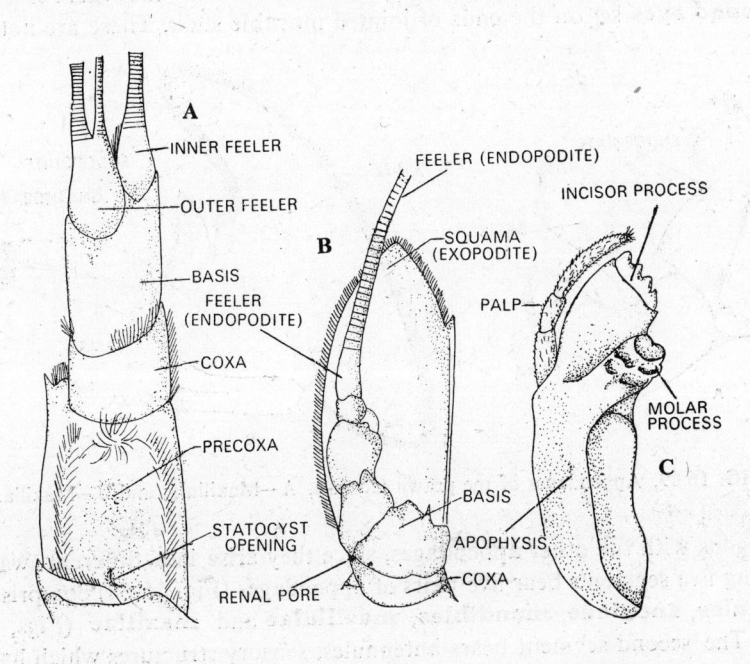

FIG. 16.68. Appendages of the prawn (contd.). A—Antennule, left; B—Ventral view of left antenna; and C—Mandible.

a special process, the **appendix interna**, which is a short, slender and slightly curved rod closely applied against the inner margin of the endopodite. In the first abdominal appendage the endopodite is reduced and the appendix interna is absent. The second abdominal appendage in the male prawn has an additional rod-shaped process the **appendix masculina** arising from the base of the appendix interna (Fig. 16.66 C) and lying on its inner side. It is covered over with long thorn-like setae. All the abdominal appendages are like the one described above, only the sixth differs. They are called the **uropods** and form broad flat plates, lying one on each side of the telson. The protopodite is represented by one triangular piece formed by the fusion of the coxa and basis and is often termed a **sympod** (Fig. 16.67). The endopodite is an oval and oar-shaped plate whose margin is fringed with setae. The exopodite is a similar plate but is larger than the endopodite and is divided into two unequal parts by a transverse suture. Besides this the outer border of the exopodite is devoid of setae. The uropods along with the telson form a fan-shaped structure called the **tail-fin** which helps the animal in locomotion (Fig. 16.64).

In the female prawn the second, third, fourth and fifth abdominal appendages serve to carry eggs in the breeding season. The appendix interna of each of these appendages becomes interlocked with its fellow of the opposite side, thus forming a series of ridges on the ventral side of the abdomen which support the eggs.

**Cephalic appendages.** The head of the prawn is fused with the thorax, but its component segments have been determined from careful studies of embryology of appendages and other structures, especially the nervous system. As in all arthropods, there are six segments of the head. On the first there is a pair of **compound eyes** set on the ends of jointed movable stalk. These are not serially

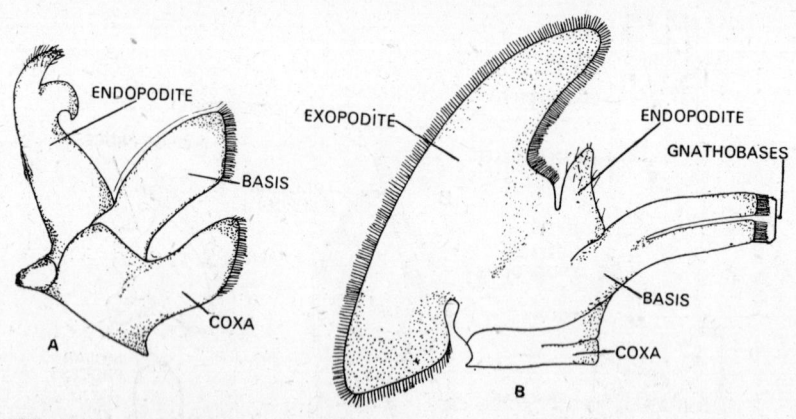

FIG. 16.69. Appendages of the prawn (contd.). A—Maxillula; and B—Maxilla.

homologous with the other appendages, since they arise in a different way. The remaining five segments bear five pairs of appendages (Fig. 16.68) comprising the **antennules, antennae, mandibles, maxillulae** and **maxillae** (Figs. 16.64, 16.69). The second segment bears antennules, sensory structures which have two filaments. The antennules are pre-oral appendages situated at the anterior end of the body immediately below the bases of the eye-stalks. Its protopodite consists of three podomeres (not two) called **precoxa, coxa** and **basis**. The basal or proximal podomere, precoxa, is very large, hollowed out dorsally. The cavity of the precoxa lodges the **statocyst**, whose opening is situated near its proximal end. From the outer margin of precoxa arises a spiny lobe the **stylocerite**. The

## Phylum Arthropoda

coxa, the middle podomere, is short and cylindrical beset with setae on its inner margin. The basis or the distal podomere is longer than the second and bears two flagelliform, many jointed feelers. Each feeler shows a swelling at its origin, then becomes slender. Of these the outer feeler is divided into two unequal branches. While antennules themselves are serially homologous with the rest of the appendages, their two feelers are not. It is wrong to regard the two feelers as the two rami of the biramous limb. Early developmental studies have shown that the feelers remain single until long after the other appendages have become two branched (biramous) and even when the larva emerges from the egg, the inner feeler is represented by only a small bud from the base of what finally becomes the outer feeler. Thus, the two feelers do not correspond with the endopodite and exopodite of a typical limb, but, perhaps they represent only the endopodite subdivided.

The antennae, located on the third segment, are also pre-oral appendages each with one long feeler, situated immediately below and behind the antennules. The protopodite has typically two podomeres, the coxa and the basis. Each one is swollen at the base as it lodges the excretory organs within. The coxa is small in size and bears the renal aperture on its inner margin. The basis bears two rami, a long narrow flagelliform **feeler** and a broad leaf-like process, the

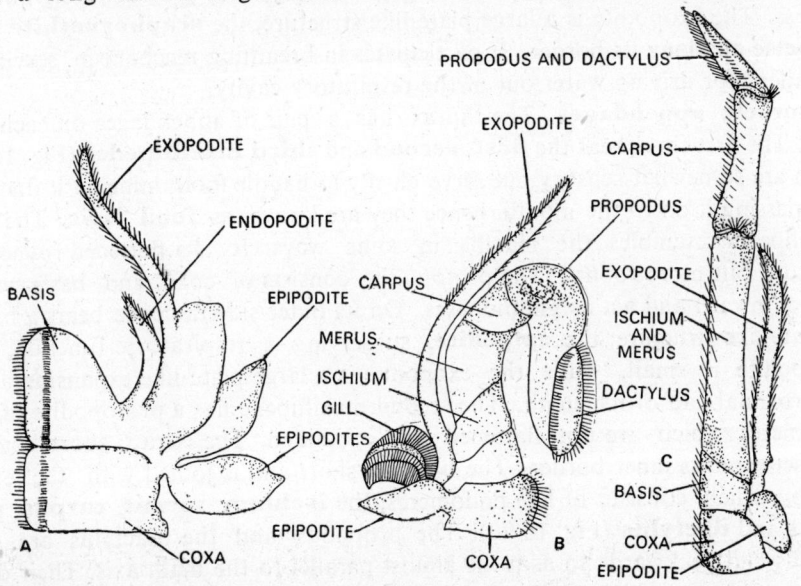

**FIG. 16.70.** Appendages of the prawn (contd.). A—First maxillipede; B—Second maxillipede; and C—Third maxillipede.

**squama**. Of these the feeler, which corresponds to the endopodite, consists of three large podomeres and a very large number of short joint-legs (Fig. 16.60). The leaf-like squama (exopodite), that acts as a balancer in swimming, is fringed with setae along its inner border. Its outer border is smooth terminating in a small spine in front.

The fourth head segment bears toothed jaws or **mandibles** that are short, stout masticatory appendages post-oral in position, lying one on each side of the mouth. The protopodite has a massive coxa which is densely calcified forming powerful jaws (Fig. 16.68). Proximal portion of the coxa is spoon-shaped, and called **apophysis**. Ths solid distal portion of the coxa is called the head and comprises two portions: a stout process, the **molar process**, bearing six dental plates at its end and lying at right angles to the apophysis; and a flat plate-like **incisor process** terminating in three closely set teeth. Each mandible bears a three-pointed **mandibular palp** that arises from the outer border of the head. The basal segment of the palp represents the basis of the protopodite, the outer two segments represent the endopodite and the exopodite is absent. On the fifth and sixth segments are the maxillulae and the maxillae, which pass food on to the mouth. The **maxillulae** (Fig. 16.69) are delicate flattened structures, each made up of three podomeres, lying close behind the labium or lower lip. The protopodite is represented by two larger plates, coxa and basis, projecting inwards as jaws or **gnathobases** which are covered with stiff spines. The endopodite forms distally a curved bifurcated process arising from the outer side of the basis. The exopodite is absent. The **maxillae** (Fig. 16.69) are larger than the maxillulae though still flattened and delicate. The inwardly directed protopodite consists of two lobes, coxa and basis, each again being subdivided into two. The basis is larger, projects inwards as forked gnathobase having stiff pointed setae at its inner end. The endopodite is represented by a narrow forwardly directed process. The exopodite is a large plate-like structure, the **scaphognathite** bearing setae all along its border. It participates in breathing mechanism, serving as a "bailer" for driving water out of the respiratory cavity.

**Thoracic appendages.** The thorax has a pair of appendages on each segment. The first three bear the **first, second** and **third maxillipedes** (Fig. 16.70) which are somewhat sensory but serve chiefly to handle food, mincing it first and then passing it on to the mouth, hence they are known as **food jaws**. The first maxillipede resembles the maxilla in some ways. It is a flattened foliaceous structure but more robust. The protopodite consists of coxa and basis which project inward and act as gnathobases. On its outer side the coxa bears a bilobed plate-like structure, the **epipodite**, subserving a respiratory function. The endopodite is small, while the exopodite is a large plate-like expansion along the proximal half of its length. The second maxillipede has a protopodite of two podomeres which are not flattened and expanded. The coxa is short, covered with setae on its inner border. The **basis** is short and is joined with the endopodite, which consists of five podomeres, the **ischium, merus, carpus, propodus** and **dactylus** (Fig. 16.70). The propodus and the dactylus are bent inwards and backwards so as to lie almost parallel to the main axis. Their outer edges are beset with tough setae as such they form a cutting edge. The exopodite is long, slender and unjointed bent over the endopodite. The third maxillipede is the largest of the series and is shaped much like the second, only the endopodite is still larger and exopodite smaller, giving it a distinctly leg-like appearance (Fig. 16.27).

**Peraeopods.** The next five pairs of the thoracic legs are the **peraeopods** or walking legs (Fig. 16.71). They do not have exopodites and epipodites, and the podomeres are arranged along a single axis reperesenting only the protopodite and endopodite. A typical peraeopod has seven ring-shaped, and a triangular basis with which articulate five podomeres of the endopodite. These are, as in the maxillipede, the **ischium, merus, carpus, propodus** and **dactylus**. Each

*Phylum Arthropoda* 605

podomere is elongated and cylindrical in shape and the terminal podomere, the **dactylus**, is clawed. They articulate with one another by means of hinge-joints. The bend between the merus and carpus is called **knee**. The fourth pair of walking legs conforms to this description.

The first and second pairs of walking legs, on the other hand, bear **chelae** or pincers for seizing objects and hence are known as the **chelipedes** or **chelate** legs (also known as the pinching legs or pincer claws, etc.). Each chelate leg (Fig. 16.71) consists of protopodite (two podomeres) and endopodite (five podomeres), the last two podomeres of which are modified to act as pincers. The propodus becomes enormously enlarged and the dactylus becomes freely movable. The two biting edges of the chela are furnished with a number of teeth, while their whole surface is beset with setae. The second pair of chelate legs is considerably larger. The ischium bears a deep oblique groove on its anterior surface. This is

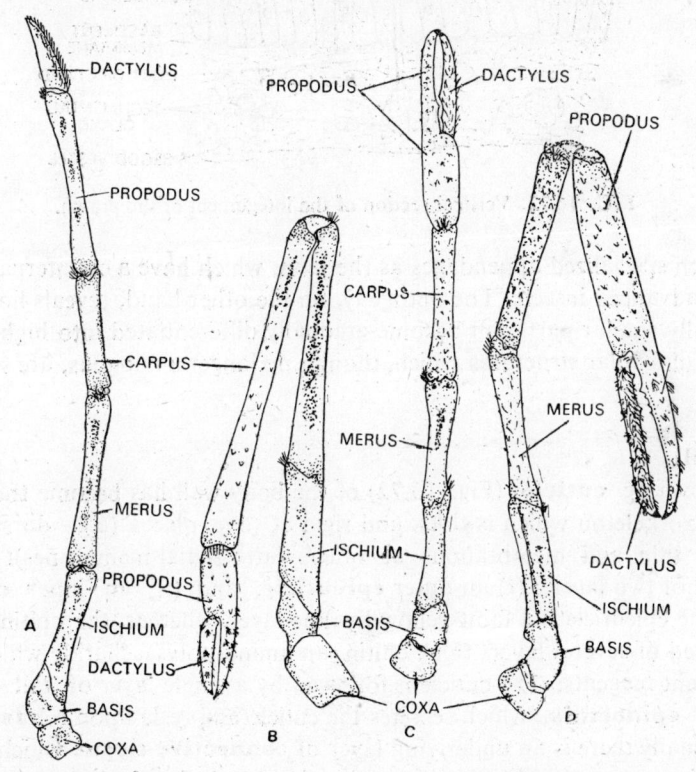

**FIG. 16.71.** Walking appendages of the prawn. A—Fourth walking leg; B—First chelate leg; C—Second chelate leg of female; and D—The same in male.

an organ of offence and defence and is very large in the male. The third and fifth pair of walking legs are similar to the fourth described above. But in the female prawn the coxa of the third pair of walking legs bears the crescentic genital aperture on its inner side; and likewise in the male the fifth pair bears the slit-like male genital aperture on the inner side of the arthroidial membrane connecting the legs to the thorax. All five pairs of legs have a gill attached to their bases. While walking the legs move the gills and stir up water in the respiratory cavity under the carapace. The last pair of walking legs is used also for cleaning the abdominal appendages.

The study of the appendages has been emphasized for a number of reasons. They furnish a striking example of specialization among appendages of different segments. While the flattened biramous swimmerets are not very different from the appendages of the hypothetical ancestral arthropod, on the same animal are

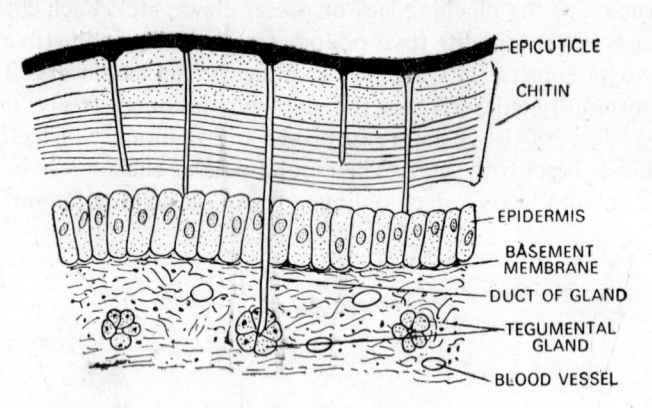

FIG. 16.72. Vertical section of the integument of the prawn.

found such specialized appendages as the jaws, which have a counterpart even in the most advanced insects. The ontogeny, on the other hand, reveals how a series of originally similar parts can become gradually differentiated into highly specialized and dissimilar structures which, though no longer analogous, are still homologous.

**Body Wall**

The covering **cuticle** (Fig. 16.72) of the body wall has become transformed into the exoskeleton which is thick and rigid at some places (e.g. dorsal shield, etc.) and thin and membranous at others (arthroidial membranes). Cuticle is composed of two layers: (*i*) an outer **epicuticle**, and (*ii*) an inner **chitinous layer**. The epicuticle is a faint yellow hyaline layer, whereas, the chitinous layer is composed of several layers thick chitin, an amino-polysaccharide which resists most solvent reagents. The cuticle is followed by a single layer of tall columnar cells, the **epidermis**, which secretes the cuticle and rests upon a **basal membrane**. Finally there is an underlying layer of **connective tissue** which supports the body muscles. Within the connective tissue layer itself are found clusters of secreting cells, the **tegumental cuticular glands**. From the gland cells arise capillary canals which join together in the centre of the gland and open into a long narrow **cuticular duct** leading to the exterior. The spines and setae of various sizes and shapes are simply outgrowths of the exoskeleton and have the same structure.

**Musculature**

The bulk of the body of the prawn consists mainly of muscles that form a good article of food. The muscles are enclosed within the exoskeleton and are mostly of the striated type. These muscles contract very rapidly and are, therefore, well suited for moving the body and appendages. The unstriated muscles are found in organs such as the digestive tube and blood vessels, which undergo slow rhythmic contractions. There are two sets of muscles in the body of the prawn; the **appendicular muscles** that are segmentally arranged and are used in moving the appendages; and **abdominal muscles** for the straightening and flexure of the abdomen. The muscles are arranged in opposing series. By the contraction of one set the limb is flexed and such muscles are called **flexors**. The contraction of the opposing set straightens the limb so that it returns to its original position. Such muscles are termed **extensors**. In the abdomen too the flexor and extensor muscles are found. In the terminal sclerites of a chelate limb, the flexor muscles become **adductors**, closing the pincers, and the extensors, **abductors**. In many jointed limbs it is noteworthy that the muscles which move one sclerite of a limb lie within the sclerite proximal to it. In many cases the articulation between the sclerites is such as to permit movement in one plane only, acting like a hinge; but where the limb is made up of several sclerites the plane of possible movement of the different sclerites is not always the same and the limb as a whole can perform complicated actions in several directions.

The extensors of the abdominal muscles include a pair of **dorsal extensors** lying in the mid-dorsal line; four pairs of **intertergal abdominal muscles**, connecting the side walls of two adjacent abdominal terga; a pair of **intertergal** muscles between the abdomen and thorax, connecting the thorax with the first abdominal segment; and two pairs of the **extensors** of the **telson**, a pair on each side, for straightening it.

The flexors of the abdominal muscles include the following six sets; five pairs of the **main flexors**, one pair in each of the first five abdominal segments (absent in the sixth segment); four pairs of accessory **flexor muscles** in the first four abdominal segment (absent in the last two); a pair of the **flexors** between the thorax and first abdominal segment; two pairs of the **lateral thoraco-abdominal muscles**; two pairs of the **flexor muscles** of the telson; and a pair of **ventral superficial flexor muscles** lying in each of the intersternal places, one on each side of the mid-ventral line.

All the appendages have their own set of muscles, the size of which is correlated with the amount of work the muscle is called upon to perform. Thus, the antennary and mandibular muscles, the muscles of the second chelate legs and of the uropods are large and powerful, while those of the pleopods, the maxillae, the maxillulae and the maxillipedes are relatively weaker. In each podomere there are typically four muscles, two extensors and two flexors.

## Endophragmal Skeleton

In the thorax and abdomen a number of ingrowths of the exoskeleton occur to provide attachments for the powerful muscles of the regions. These ingrowths are known as the **apodemes** and in the prawn they are present only in cephalothorax where they are connected together to form a regular framework that is known as the **endophragmal skeleton** (Fig. 16.73). This consists of a series of ingrowths from the lateral walls, the **endopleurites**, and upgrowths from the sterna, the **endosternites**. For the attachment of the appendicular muscles a rigid vertical double wall is formed lying between the bases of two consecutive appendages. Elements from two different segments take part in its formation: an **endopleurite posterior**, a projection from the anterolateral end of the epimeron belonging to one segment meets a similar projection the **endosternite posterior** from the sternal element of the same segment. To this is attached another bar formed by the **endopleurite anterior**, from the posterolateral end of the epimeron belonging to the preceding segment with a similar

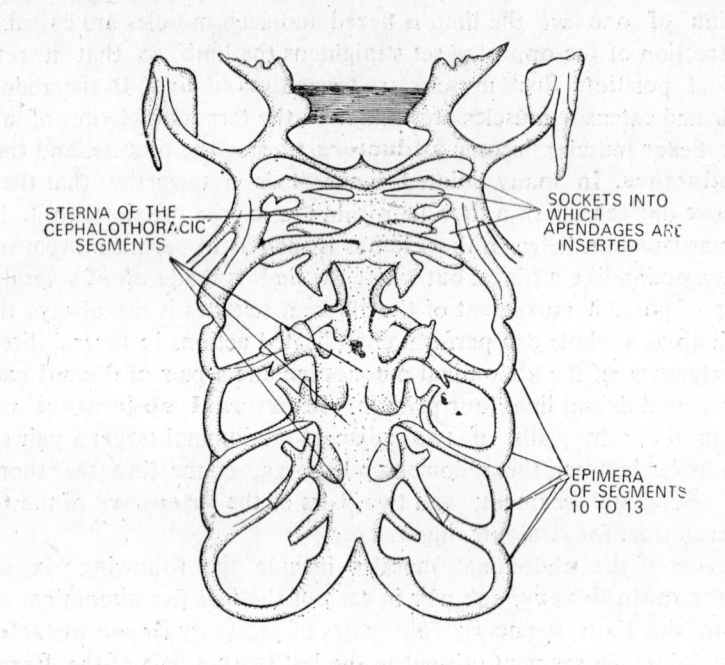

FIG. 16.73. The sterna and endophragmal-skeleton of the prawn, dorsal view.

projection, the **endosternite anterior**, from the sternal element of this very segment. From the inner ends of the adjacent endosternites of an apodeme, a vertical plate projects upwards and forks at its free end to form a Y-shaped structure for the attachment of the flexor muscles of the abdomen. A typical apodeme with complete number of endopleurites, endosternites and Y-shaped structures occurs only between eleventh and twelfth and thirteenth segments. In other segments the apodeme is incomplete; between the last thoracic and the first abdominal segment the endosternite is single and complete Y-shaped outgrowth is lacking; while in the anterior segments the Y-shaped outgrowths become less and less prominent. In the mandibular region there is an H-shaped structure, the **cephalic apodeme**, for the insertion of the muscles of the branchiostegites, small gastric muscles and the muscles of the mandibles.

**Respiratory System**

The prawn breathes with the help of the **gills** or **branchiae** lodged in the gill chamber (Fig. 16.74), each of which is enclosed between the branchiostegite externally and the wall of the thorax internally. Inner lining of the carapace and three pairs of the epipodites also aid respiration. There are eight pairs of

*Phylum Arthropoda* 609

gills but only seven are visible because one lies hidden beneath the dorsal part of the second gill.

Depending upon their position and relation to the thoracic appendages there are three kinds of gills. The **podobranch** is the gill attached to the basal podomere or coxopodite of an appendage; the **arthrobranch** is the gill attached to the arthroidial membrane between the body (thorax) and the appendages; and the **pleurobranch** is the gill attached to the lateral wall of the segment to which the appendage belongs. Originally the gills were outgrowths of the basal podomeres (precoxa and coxa) of the appendages. As the animal evolved, to provide more efficiency, the precoxa became absorbed into the flank of the body

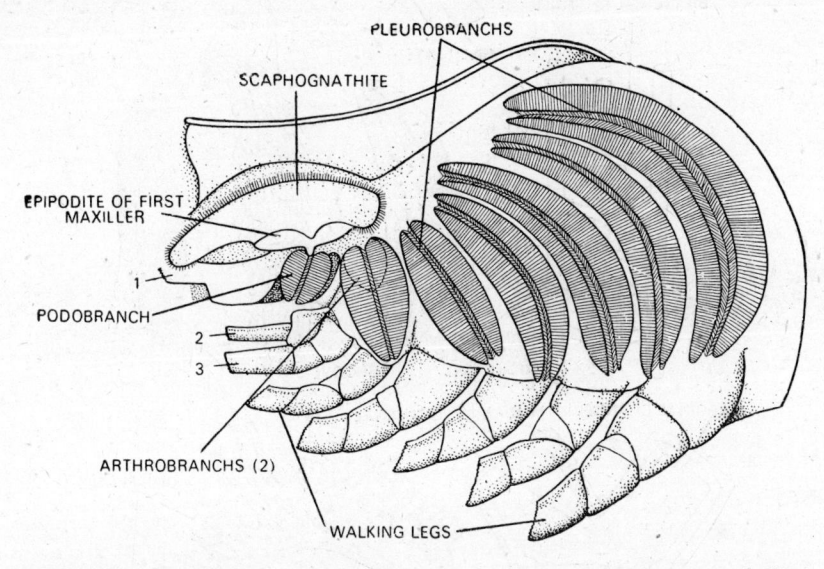

**FIG. 16.74.** Gill-chamber of the prawn exposed to show the arrangement of gills.

with the result that the gills originally attached to the podomere became attached either to the side of the thorax (pleurobranchs) or to the arthroidial membrane between the body and the limb (arthrobranch). Only podobranchs have retained their original position. With the improved respiratory mechanism the branchiostegites extended over the gills and formed the gill chamber.

The arrangement of the gills with respect to the thoracic limbs is as follows: (Fig. 16.74). The first maxillipede bears only a bilobed epipodite on the outer side of its coxa and no gill. The second maxillipede bears a podobranch as well as small leaf-like epipodite on its coxa. The third maxillipede bears two arthrobranchs and a small epipodite on its coxa. The five walking legs bear pleurobranchs borne on the lateral wall of the thorax. Thus, there are three pairs of epipodites (on the three maxillipedes), one pair of podobranch (on 2nd maxillipede), two arthrobranchs (on 3rd maxillipede), and five pairs of pleurobranchs (on five walking legs), eleven pairs in all.

The gills in the prawn are of the type describable as the **phyllobranch** (Fig. 16.75), that is, the gill filaments are thin and plate-like, often called **gill-plates**, and are arranged in two rows on a narrow axis, the **base**, like leaves of a book. Each gill, therefore, consists of a long narrow base which bears two rows of rhomboidal gill-plates which lie at right angles to the long axis. There is a deep groove between the two rows on the gill-plates. The groove opens in the gill-chamber at both the dorsal and ventral ends of the gill. The gill-plates are larger in the middle and become smaller in size towards the two ends. In the crayfish *Astacus* the gills are of the **trichobranch** type (Fig. 16.75). In this the gill filaments project round a central axis like the bristles of a bottle brush.

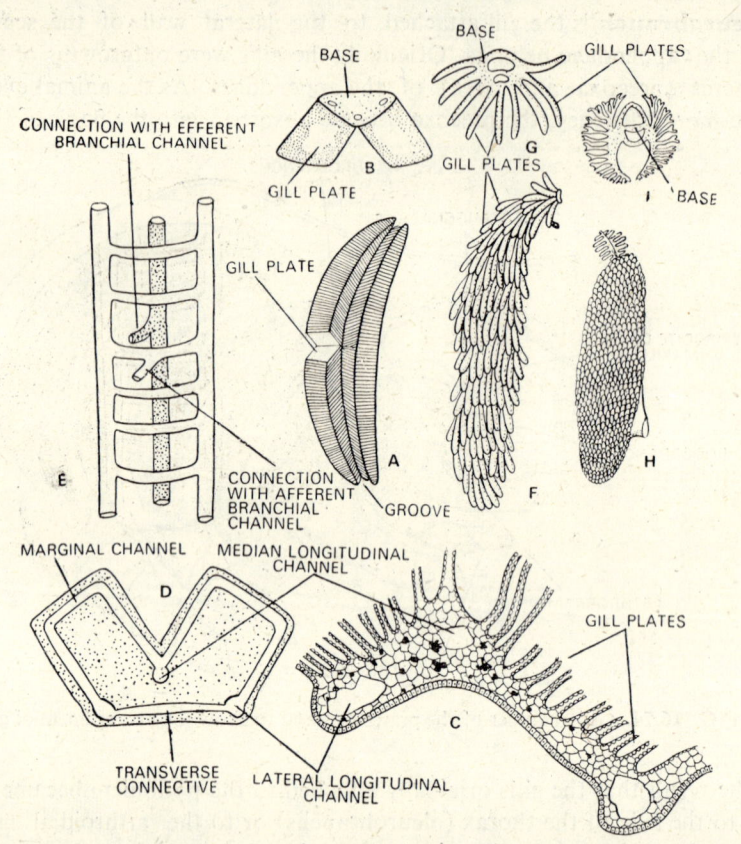

FIG. 16.75. The structure of gills and blood channels in a gill of *Palaemon*. A—Phyllobranch of *Palaemon*; B—Cross section of the same; C—Oblique cross section of the gill of *Palaemon*; D—Blood channels in a gill in transverse section; E—In vertical section; F—Trichobranch of *Astacus*; G—Cross section of the same; H—Dendrobranch of *Penaeus* (after Calman); and I—Cross section of the same.

Besides these, there are several other types of gills either intermediate in form between the above two or derived from any of them. A third type, called the **dendrobranch**, for instance, is a modification of the phyllobranch in which the gill-plates are further broken up into smaller filaments (e.g. in *Penaeus*). No two types of gills are found together in the same animal.

Histological examination of the gill reveals that each gill-plate is made up of a double layer of cuticle enclosing a single layer of cells within. There are two types of cells in this layer; the **pigmented cells** that alternate with the **transparent cells**. The axis or base is more or less triangular in cross-section and comprises

*Phylum Arthropoda*

a central core of connective tissue, surrounded by a layer of epidermis, itself bounded externally by a thin layer of cuticle. Within the axis run three longitudinal canals, two **lateral longitudinal channels** and one **median longitudinal channel** (Fig. 16.75). Throughout their course the lateral longitudinal channels are connected by several small transverse connectives making it ladder-like in appearance. The transverse connective lying immediately against the root of the gill is larger and receives blood for aeration from the **afferent branchial channel** from the body. The lateral channels are also connected with the median channel through **marginal channels** that arise from the lateral channels at the bases of the gill-plates, run along their margin and open into the median channel. The blood that enters the transverse connective through afferent branchial blood vessel passes to the median channel through the lateral and marginal channels of the gill plates. The gaseous interchange between the blood and the water takes place by diffusion during the circulation of the blood through the filaments. The aerated blood is returned by the **efferent branchial channel** to the pericardial sinus and the heart.

To ensure proper respiration, a regular supply of oxygenated water is to be maintained. Thus a continuous current is set in and maintained by the incessant paddle-like movement of the scaphognathites of the maxillae. The movements of the scaphognathite create a suction force that draws in water, which enters the chamber along the posterior and ventral margins of the branchiostegite, passes over the gills, and thence is expelled from the front end beneath the head. The

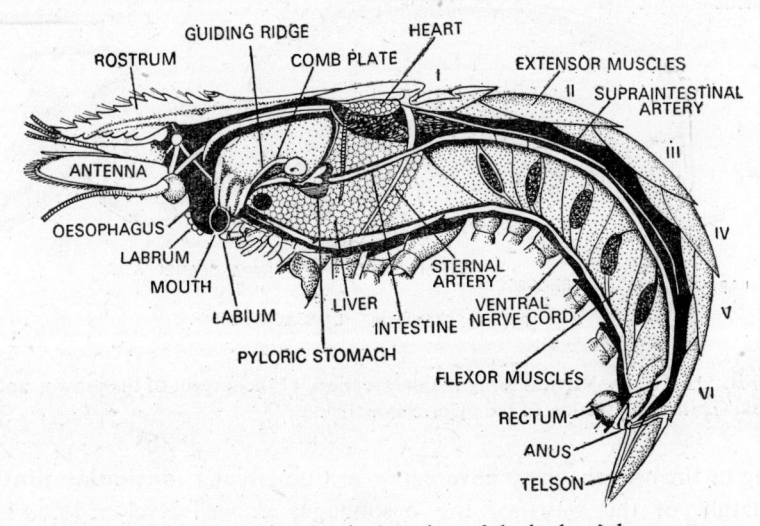

FIG. 16.76. Median longitudinal section of the body of the prawn.

epipodites of the maxillipedes supplement the vibratory movements of the scaphognathite. Thus, the gill chamber is constantly supplied with a current of freshwater. In small crustaceans like *Cyclops* and *Daphnia* respiration takes place through the entire surface of the body, but in the prawn it has become restricted to a particular region of the body, the gill chamber and other structures associated with it have become especially modified to ensure the efficient performance of the respiratory activities.

## Digestive System

The alimentary canal of the prawn (as in all coelomates) consists of three parts (Fig. 16.76): the **stomodaeum** or the **fore-gut** leading from the mouth, the **mesenteron** or the **mid-gut** in the middle having endodermal lining, and the **proctodaeum** or **hind-gut** opening out at the anus. The stomodaeum begins at the mouth, which is a large elliptical aperture on the ventral side of the head region bounded in front by the labrum, on the sides by the incisor processes of the mandible and behind by the labium which is deeply cleft in the middle forming two flattened plate-like lobes, the **paragnatha**. The mouth leads into the buccal cavity which is a short, anteroposteriorly compressed tube with thick chitinous lining thrown into irregular folds. The molar processes of the mandibles form the sides of this cavity and crush food within. The buccal cavity leads into the oesophagus (Fig. 16.76).

The **oesophagus** is a short though stout tube running upwards from the buccal cavity. The inner wall of the oesophagus is thrown into four prominent longitudinal folds of which the anterior one is short, and the lateral and posterior folds are longer and more prominent. The internal cuticle is covered with a large number of small bristles. The opening of the oesophagus in the stomach is guarded by valve-like structures formed by prominent projections of the upper ends of the posterior and lateral folds. The anterior wall of the

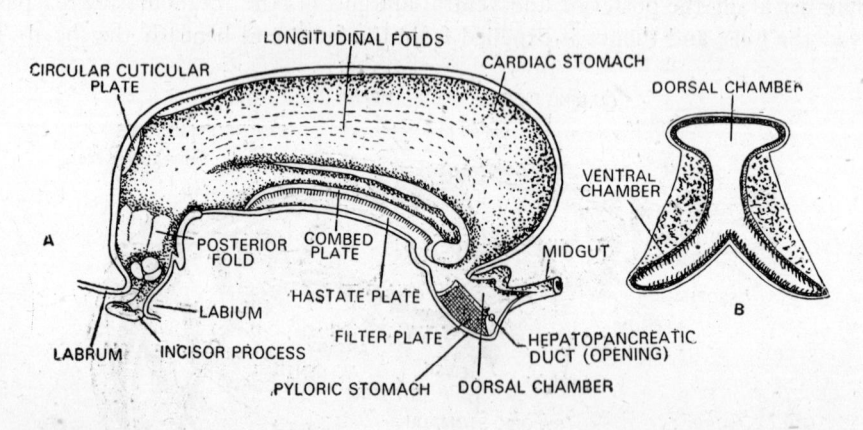

FIG. 16.77. A—Vertical longitudinal section of the foregut of the prawn; and B—Transverse section of the pyloric stomach.

opening of the oesophagus is covered by a thin circular **cuticular plate**. The musculature of the walls of the oesophagus is well-developed and its inner lining resembles the outer integument in structure, save for the fact that here epithelial cells are much more elongated and the tegumental glands are more numerous.

The oesophagus opens into the **stomach**, a spacious chamber occupying the greater part of the cephalothorax (Fig. 16.76). It is divided into two parts, a large bag-like **cardiac stomach** in front and a much smaller **pyloric stomach** behind. Between the cardiac and pyloric stomach there are many valves. Internally the chitinous lining is thick, produced into a very large number of low longitudinal folds and covered over by a velvety growth of small bristles. The cardiac stomach presents irregularly thickened cuticular lining. At certain places the thickening is heavy and the cuticle becomes even calcified and at others it is soft. This irregular thickening of the cuticle results in the formation of certain plate-like structures or ridges. On the anterior part of the roof of the cardiac stomach, just behind the opening of the oesophagus, there is a circular cuticular plate known as the **lanceolate plate**. The floor of the cardiac stomach presents a large triangular cuticular plate, the **hastate plate** (so called due to its resemblance with the head of a spear). Its upper part is covered with a thick growth of delicate setae. Posteriorly it forms a small triangular depression beset with setae, acting as the anterior valve of the cardio-pyloric aperture. The central portion of the plate is softer. From below, the sides of the hastate plate are supported by a pair of cuticular **supporting rods**. Running parallel to the lateral borders of the hastate plates are long cuticular ridges, the **lateral grooves**, which towards their inner sides are fringed with equally spaced setae forming comb-like structure, the **combed plate**, covering the groove and overlapping the outer margin of the hastate plate. The fine bristles of the comb keep moving to and fro over the outer margin of the

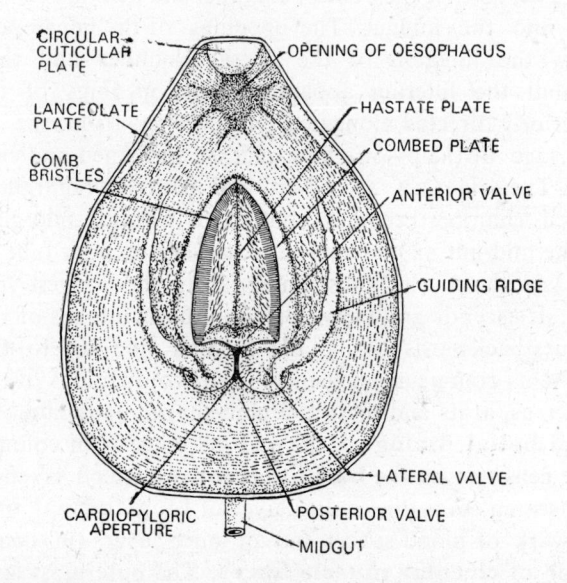

**FIG. 16.78.** Inner view of the cardiac stomach of *Palaemon carcinus* (*after* Patwardhan).

hastate plate. The floor of the lateral groove is like an open drain-pipe and is called the **groove plate**. The **lateral longitudinal** folds are two prominent longitudinal folds of deep blue colour in life situated on the either side of the comb-plates (Fig. 16.78). They are highest towards the posterior side and low in front. They guide the contents of the cardiac stomach towards the cardio-pyloric aperture, that is why sometimes they are referred to as the "**guiding ridges**" (Fig. 16.78).

In Some other Decapoda the chitinous lining is thick and specially hardened to form a complex mechanism of "ossicles" supporting three large, and highly calcified teeth. This mechanism, called the **gastric mill**, is absent in the prawn in which the food is crushed by the molar processes of the mandibles.

The inner cuticle of the stomach is thickened in such a way that it breaks the food and filters it. The cardiopyloric aperture is guarded anteriorly by an **anterior valve** formed by the posterior triangular area of the hastate plate, posteriorly by a **posterior valve** which is just a semilunar fold of the posterior wall of the cardiac stomach, whereas on the sides are situated large flap-like **lateral valves**. Due to these valves and folds the opening appears X-shaped. It is narrow and becomes sieve-like because of the presence of numerous setae on the folds. The side walls of the pyloric stomach form very prominent lateral folds that come close to each other leaving only a slit-like aperture in between. These folds divide the cavity of the stomach imperfectly into two chambers, the **dorsal** and **ventral chambers**. The floor of the ventral chamber has a pair of cuticular plates of rectangular outline. The side walls of the ventral chamber, as well as the surface of the two plates, all are covered with closely set bristles which form an efficient filter or strainer which prevents the passage of insufficiently crushed food into the mesenteron. The dorsal chamber leads behind into the narrow **mid-gut**, so that the ventral chamber lies at a lower level than that of the opening into the midgut. The openings of the hepatopancreatic ducts are situated below the junction of the dorsal chamber and the mesenteron immediately behind the filtering apparatus. The openings of these ducts are guarded by posteriorly directed elongated setae arising from the posterior end of the median ridge of the pyloric filter-plate, a thickened and calcified area of the lining cuticle. The setae act as valves and prevent regurgitation of food. Before the dorsal chamber becomes continuous with the mid-gut it gives off a short caecum. The mid-gut or mesenteron is a very slender tube which follows the stomach and is without chitinous lining. It is the longest portion of the alimentary canal. It ascends gradually between the two lobes of the hepatopancreas and continues backwards along the median line up to the sixth abdominal segment. At its commencement there are two openings of the hepatopancreatic ducts, whereas, at its end there is a club-shaped swelling of the hind-gut. The internal **epithelial lining** of the mid-gut consists of columnar cells with large nuclei. The cells rest on the **basal membrane** which is enclosed by the musculature consisting of a thick outer layer of **longitudinal muscle fibres**, containing a network of blood spaces and an inner layer (in contact with the basal membrane) of **circular muscle fibres**. The outermost layer of the wall of the mid-gut consists of connective tissue. The epithelial lining is thrown into longitudinal folds reducing the lumen.

The **hind-gut** is the shortest portion of the gut beginning where the mid-gut ends and terminating at the anus. Anteriorly it is swollen and forms a thick muscular sac which gradually narrows down and becomes tubular. The innermost lining is cuticular in nature. Next to this is an epithelial layer consisting of columnar cells. In the hind-gut the usual muscle layers are thin but there is an additional thick layer of longitudinal muscle fibres between the basement membrane and the circular layer of muscle fibres. The outer wall consists of connective tissues. The inner lining is thrown into longitudinal folds which are well-developed anteriorly. They become smaller in the narrow tubular posterior portion of the hind-gut that finally opens at the anus, with a longitudinal slit-like

# Phylum Arthropoda

sphinctered opening situated on a raised papilla.

The **hepatopancreas** is a large organ of orange-red colour in the living state. It occupies a major portion of the cephalothorax (Figs. 16.76, 16.80) and consists of two separate lobes. The hepatopancreas secretes digestive enzymes that are capable of digesting carbohydrates, proteins and fats. In fact it combines within itself the functions of pancreas, liver and intestine. It is also called the "liver" by some authors. The hepatopancreas consists of numerous branching tubules closely set and firmly held by connective tissue. These tubules, in fact, arise from the hepatic caecae that develop in the earlier stages. The canals of these tubules unite together forming larger canals which ultimately fuse forming two large hepatopancreatic ducts that open into the pyloric stomach, one on each side. Histologically each tubule consists of an outer coat, the **tunica propria**, composed of connective tissue with a network of circular and longitudinal muscle fibres. This is followed by a structureless **basement membrane**. Internal to the

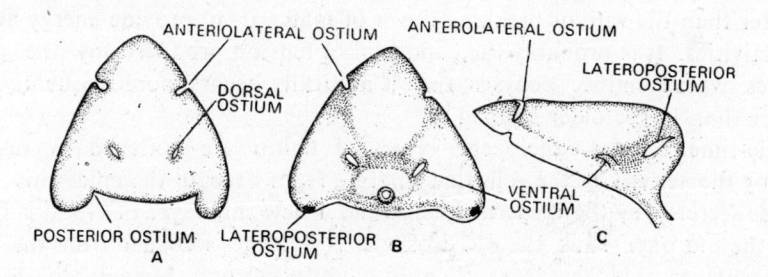

**FIG. 16.79.** The heart of *Palaemon*. A—Dorsal view; B—Ventral view; and C—Lateral view (*after* Patwardhan).

basement membrane lies the epithelial cell-layer consisting of tall cells of many varieties. Internally these cells form a distinct inner border. The cells of the epithelium are of the following types: (*a*) the **columnar** or **glandular** cells containing granular cytoplasm and large rounded nuclei; (*b*) the **hepatic cells** similar in shape but filled with fat-globules; (*c*) the **ferment cells** which are irregularly distributed and are broader than others (breadth slightly less than the length); they are filled with granular cyoplasm; and (*d*) the **basal** or **replacing cells** that lie close to the basement membrane on the outer side of the tall cells. The physiology of the hepatopancreas of *Palaemon* has not yet been worked out but as it has been shown in other animals, probably it produces digestive ferments and is responsible for the digestion of carbohydrates, proteins and fats. The secretion is poured in the pyloric stomach, whence it passes on to the cardiac stomach. Here it mixes with the food and dissolves it.

**Feeding.** The food of the prawn consists mainly of moss, algae and other weeds, etc., although the animal swallows the debris of the bottom, sand grains and even small insects and other small animals inhabiting the locality. The food is transferred to the mouth by the action of gnathobases of the maxillipedes, maxillae and maxillulae. In the buccal cavity the food is cut into pieces by the incisor processes of the mandibles. A wave of contraction passing along the walls of the oesophagus transfers the masticated food to the cardiac stomach which itself exerts suction force. From the cardiac stomach the liquefied food or small particles of food are conducted to the pyloric stomach by the lateral grooves. In the cardiac chamber, as well as in the pyloric chamber the food receives the secretions of the hepatopancreas which help in digestion. In the pyloric stomach the food is filtered through the pyoric filtering apparatus. The filtrate passes back into the hepatopancreas which absorbs the food, in this case; the refuse of the undigested or indigestible food is passed to the dorsal chamber of the pyloric stomach from where is passes back to the mid-gut, where the remaining digestible material is absorbed and the residue passes out through the anus.

**Growth and Ecdysis**

Like all arthropods, in the prawn actual increase in size takes place during the periods immediately following the sloughing off (ecdysis) of the rigid exoskeleton, and the formation of the new one. It must not be imagined, however, that growth only occurs at this time. Growth is much more than mere increase in size and will continue so long as the rate of absorption of "building material" is greater than the rate of breaking down of materials to provide energy for the vital activities. It is probably the condition of tension produced by the growth activities which induce ecdysis and it naturally occurs more frequently in the younger than in the older animal.

At the time of ecdysis the deeper layers of chitin are softened or dissolved either by the activity of the cells that migrate from beneath the epidermis or by enzymes secreted by the epidermal cells, and a new thin layer of chitin is formed below the old one. Thus, the old exoskeleton becomes loosened from the underlying new layer. The new layer of chitin is quite soft and remains so until the old armour has been sloughed off. Just at this time the new layer of chitin, unprotected by the epicuticle, is permeable to water which it absorbs resulting in an increase in the size of the animal. By the time this is done the tegumental glands become active and secrete the epicuticle which hardens on exposure and becomes impermeable to water.

Exactly how the animal gets rid of its old exoskeleton is not known for *Palaemon*. It has, however, been studied for the crayfish and other arthropods. The crayfish retreats into its burrow and for a time violently agitates its limbs, often rubbing them together and jerks its abdomen apparently in an endeavour to loosen the old armour. Then, it bends the body at the junction of the cephalothorax and abdomen, causing the exoskeleton to crack across. From this the anterior part of the body is withdrawn; then the appendages and eye stalks, etc., are slowly pulled out, and finally the abdomen is released and by a violent spring the animal becomes completely free. The removal of the limb is facilitated by an extensive split down the length of each, though occasionally a limb may break off as a result of some disturbance in the process of freeing.

**Autotomy.** Occasionally a limb may break off during ecdysis, as mentioned above, but under stress of circumstances the limb is cast off deliberately. This phenomenon is known as **autotomy**. This usually occurs when a limb is held by an enemy so that the animal cannot extricate itself. The tissues between the basal joint and the more distal part of the limb break off allowing the fracture and contract preventing loss of blood.

## Blood Vascular System

The blood of the prawn is a clear fluid in which colourless leucocytes float. The respiratory pigment in this case is **haemocyanin**, the base of which is copper (the base of haemoglobin is iron). When oxidized the blood is bright blue, whereas, deoxidized blood is colourless. The blood vascular system is open in the prawn as is the case of Arthropoda on the whole. The body cavity is a **haemocoel** and the blood vessels with definite walls open into it directly or into its extensions, the **blood lacunae**.

The **heart** (Fig. 16.80) is a triangular structure with its broader base facing posteriorly and the apex in front, situated in the dorsal region of the posterior part of the thorax, enclosed in a spacious pericardial sinus just beneath the dorsal wall of the thorax. A **horizontal pericardial septum**, lying just above the hepatopancreas and the reproductive organs, forms the floor of the sinus. A median longitudinal strand, the **cardiopyloric strand**, extending from the apex

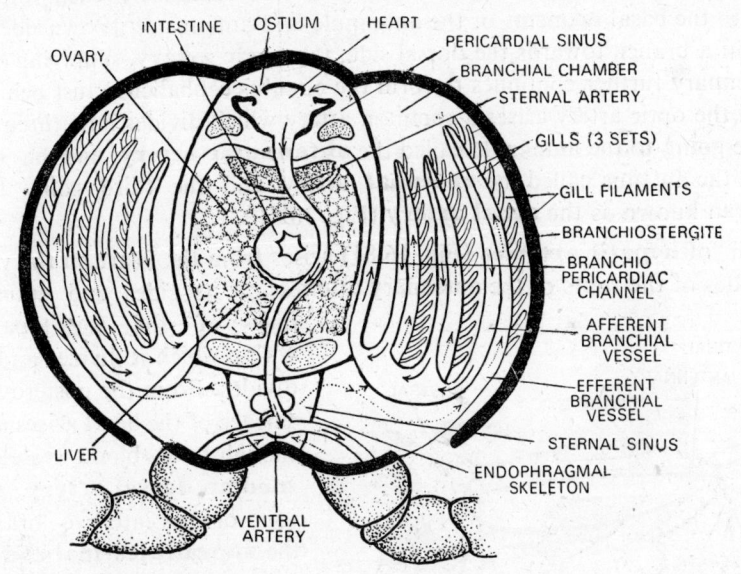

FIG. 16.80. Diagrammatic transverse section through the thorax of lobster showing circulation (modified *after* Huxley and Plateau).

of the heart to the dorsal wall of the pyloric stomach, and two lateral strands, stretching from the lateral angle of the heart to the body wall, support the heart in the pericardium. In the living prawn the heart can be seen beating regularly. The blood from the pericardial sinus enters the heart through five pairs of **apertures** or **ostia** that pierce its muscular walls. The ostia are guarded by valves allowing the blood to enter in but not to come out. The first pair of ostia lies a little behind the middle of the ventral surface one on each side of the median line. The second pair is situated almost opposite to the first on the ventral side, third pair lies on the posterior border of the heart, the fourth behind the apex, one on either side, and the fifth on the two lateral angles of the heart. The cavity of the heart is not continuous but is traversed by a large number of interlacing muscle fibres. Histologically the heart looks like a thick spongy meshwork of muscle fibres, the interstices of which represent the cavity of the heart.

The **arteries** are thick walled blood vessels through which the heart pumps out its blood received from the pericardial sinus. Three arteries arise from the apex of the heart. The **median ophthalmic artery** (Fig. 16.81) lies in the mid-dorsal line just beneath the carapace. It is a slender blood vessel that runs anteriorly up to the root of oesophagus and joins with the loop formed by the **antennary arteries**. Two **antennary arteries** arise from the apex of the heart from the outer side of the median ophthalmic artery. After its origin each antennary artery runs slightly obliquely towards the anterior side, and just behind the eyes anastomoses with its fellow from the other side and forms a loop, the **circulus cephalicus**, which gives rise to a **rostral artery** on each side. During its course each antennary artery gives off many branches. Along the outer side of the mandibular muscle it gives off three branches: (*i*) the **pericardial** to the pericardial sinus; (*ii*) the **gastric branch** to the cardiac stomach; and (*iii*) the **mandibular branch** to the mandibular muscles. Anteriorly each antennary runs up to the basal segments of the antennule and antenna of its own side where it gives off a branch towards the dorsal side, the **optic artery**, supplying the eye. The antennary further continues to form the circulus cephalicus. Just behind the origin of the optic artery arises a common artery which divides into three branches, one going to the antennule called the **antennular artery**, and the second going to the antenna called the **antennal** artery and the third going to the renal organ known as the **renal artery** (Fig. 16.81).

A pair of **hepatic arteries** (Fig. 16.81) arises from the heart on the ventrolateral sides of the roots of the antennary arteries of its sides and enters the mass of the hepatopancreas within which it divides and subdivides. From the posteroventral surface of the heart arises a short though stout blood vessel, the **median dorsal artery**, which soon divides into two branches: the **supra-intestinal** and **sternal arteries**. The supra-intestinal artery runs straight behind, dorsal to the intestine, and ends just over the hind-gut by forking into two branches, each of which lies on one side of the hind gut. It gives rise to many small branches supplying the

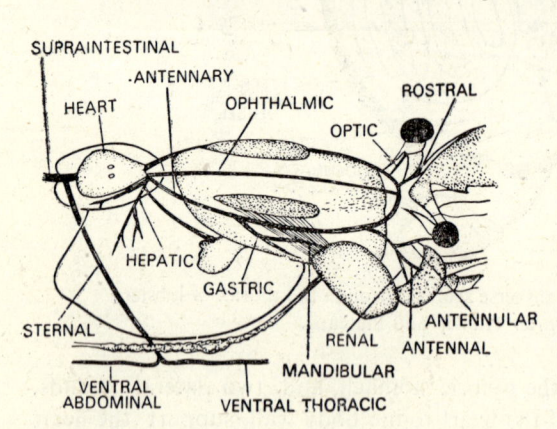

FIG. 16.81. Circulatory system in *Palaemon* (*after* Patwardhan).

intestine. The sternal artery is a large blood vessel that runs obliquely downwards, passing either from the left or right of the mid-gut. After crossing the ventral nerve cord it bifurcates into two branches, one goes to the anterior side lying below the nerve cord, and is called the **ventral thoracic** (Fig. 16.81) and the other runs behind, below the nerve cord, and is called the **ventral abdominal**. The ventral thoracic is larger of the two, runs up to the oral aperture and supplies the maxillae, the maxillulae, the maxillipedes and the first three pairs of walking legs. These blood vessels branch extensively forming

## Phylum Arthropoda

minute hair-like branches that open freely into the **blood spaces** or **sinuses** of the haemocoel.

Of the blood sinuses there are two **ventral sinuses** situated lengthwise beneath the hepatopancreas in the thorax. extending back into the abdomen for a short distance. The two sinuses are not separate but communicate with each other at several places. Ventrally they are supported by the sterna. The ventral sinuses receive blood from all parts of the body and send it to the pericardial sinus after aeration. For aeration the blood is sent to the gills through six pairs of **afferent branchial channels** (Fig. 16.82). These are definite channels lying on the inner sides of the wall of the thorax. Of these the first channel which is relatively small and ill-defined takes blood to the two arthrobranchs and also to the podobranch of the second maxillipede. The other five conduct

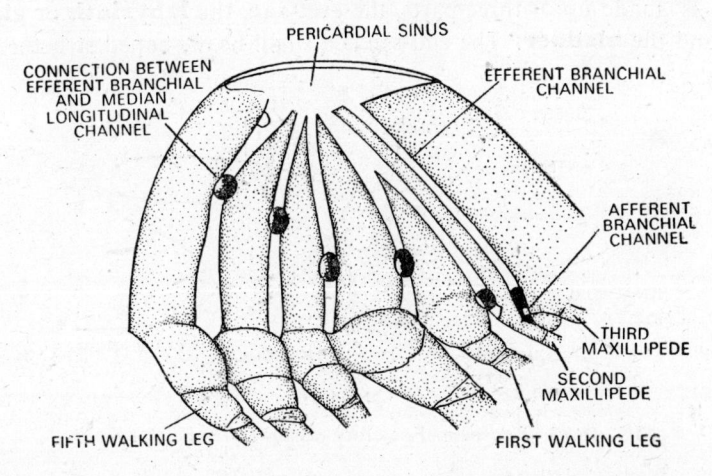

**FIG. 16.82.** *Palaemon.* Pericardial sinus and efferent and afferent channels.

blood to the pleurobranchs. In the gills the blood circulates through the **lateral longitudinal, marginal** and **median longitudinal channels** and is aerated. The aerated blood is returned to the pericardial sinus through six pairs of **efferent branchial channels** or the **branchio-cardiac canals,** which emerge out of the gill-chamber at the roots of the gills at places just dorsal to the openings of the afferent branchial channels. Thus, the two sets of channels pass across the thorax one above the other and their two openings are situated at the root of each gill. Both these openings are surrounded by a common chitinous wall. The six efferent channels are distinct canals which receive all the aerated blood from the gills and pass it on to the heart.

**Course of circulation.** The blood from different parts of the body is emptied into the ventral sinuses through various channels. From the ventral sinuses the blood passes into the gills, being conducted by the afferent branchial channels. After being oxygenated in the gills the blood is returned to the pericardial sinus through six pairs of efferent branchial channels. From the pericardial sinus the blood enters the heart through ostia and from there it is pumped to various parts of the body.

## Excretory System

The excretory organs of the prawn are a pair of **antennary glands** (Fig. 16.83) also called the **green glands**. In the Crustacea two pairs of glands, the **antennary** and the **maxillary glands**, form the excretory organs. In the majority of the crustaceans the antennary glands form the larval kidneys and the maxillary glands are found in the adults. In the Decapoda (to which the prawn belongs), on the contrary, the maxillary glands act as kidneys in the larval stages. Both the types of excretory organs do not occur simultaneously, as a rule, one succeeds the other, except in some freshwater Ostracoda, where both the glands are present together.

The antennary glands in the prawn lie within the coxal podomere of the antennae (Fig. 16.83). Each gland is an opaque white structure of the size of pea-seed and is made up of three parts, the **end-sac**, the **labyrinth** or **glandular plexus**, and the **bladder**: The end-sac is a small bean-shaped structure lying in

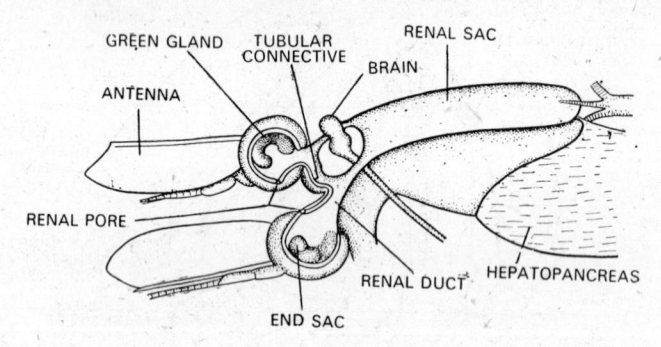

FIG. 16.83. *Palaemon*. Excretory organs (*after* Patwardhan).

the anterior part of the gland internal to the labyrinth and contains a large blood lacuna within its cavity. The inner epithelium of the end-sac is folded forming a number of radial septa which projects into the cavity of the end-sac. Its outer wall consists of connective tissue. The labyrinth or glandular plexus is relatively larger in size and consists of a mass of highly convoluted and branching excretory tubules which open by one opening into the end-sac and by many openings into the bladder. The lumen of each tubule is small and is lined by a single layer of excretory epithelial cells. The bladder is a thin-walled sac lying on the inner side of the gland proper and internally forms an excretory duct or ureter that opens to the outside on the inner side of the coxal podomere of the antenna by the **excretory** or **renal aperture**, which is a rounded opening situated on a papilla, immediately in front of the labrum.

Each antennary gland communicates behind with an elongated **renal** or **nephro-peritoneal** sac through a narrow **lateral duct**. The renal sac is a large thin-walled elongated sac lying in the cephalothorax just above the cardiac stomach. Anteriorly it communicates with two lateral ducts. Each lateral duct runs along the oesophagus and crosses the oesophageal nerve collar to enter the bladder of its side. Immediately after crossing the nerve collar, it is joined to its fellow of the other side by a transverse connective. Posteriorly the renal sac ends blindly and is in contact with the gonads. The wall of the renal sac is composed of a single layer of flattened excretory epithelium.

# Phylum Arthropoda

Uric acid and other nitrogenous compounds have been detected in the various parts of the excretory organs. The end-sac secretes ammoniacal compounds. They are thus believed to carry on excretion. Some authors believe that ecdysis is a special device to get rid of waste products of metabolism. The thick chitinous layer of the integument is a non-living nitrogenous product secreted by the ectoderm. This is cast off every now and then during ecdysis.

## Nervous System

Essentially the central nervous system consists of a pair of **supraoesophageal ganglia** or **brain** (Fig. 16.84) connected to a ventral nerve chain of paired ganglia by a pair of circumoesophageal connectives. There is also a visceral or **sympathetic nervous system** in connection with the oesophagus and stomach. Typically there would be a pair of ganglia for each body segment, but fusion of ganglia has taken place so frequently that the number of visible ganglia may not be the same as the number of segments. In the front of the head region, beneath the base of the rostrum, anterior to the junction of the cardiac stomach with the oesophagus are found the **supraoesophageal ganglia** (the brain), embedded in a mass of fat. Seen from the dorsal side it is bilobed in appearance and represents an extreme case of concentration and complete fusion of three pairs of ganglia indicated

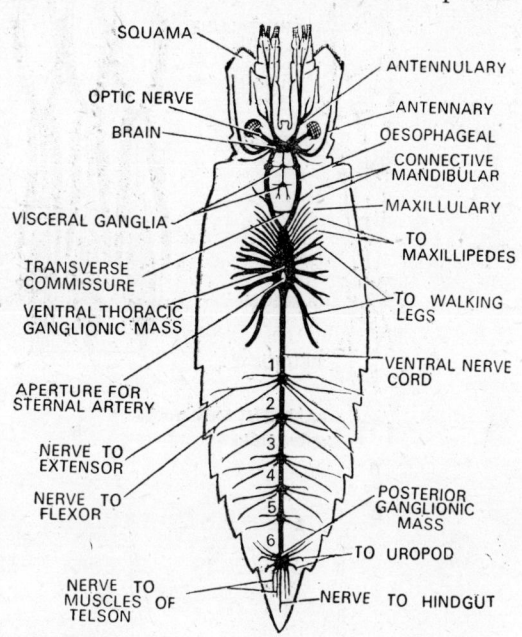

FIG. 16.84. Dissection of the nervous system of the prawn.

only by the three pairs of nerves that arise from it and supply the eyes, antennules and antennae, and consequently known as the **optic, antennulary** and **antennary nerves** (Fig. 16.84). From the supraoesophageal ganglia pass the **circumoesophageal connectives**, a pair of stout nerve cords arising from the hinder end of the brain and running downward on either side of the oesophagus, to the **sub-oesophageal ganglia** lying in the floor of the cephalothorax forming the anterior part of the large **thoracic ganglionic mass**. Both the connectives are joined by a transverse connective before meeting into the suboesophageal ganglia. Then follows the **ventral nerve cord**, a ganglionated chain of double cords having as many ganglia as the postoral segments of the body. But the gangila (Fig. 16.84) belonging to the region of the cephalothorax have fused, like the body segments, to form one ganglionic mass, the **thoracic ganglionic mass**, while those of the abdomen remain free. The thoracic ganglionic mass is a complex mass formed by the fusion of as many as eleven pairs of ganglia. It is oval in shape marked out into three regions, the anterior, the middle and the posterior. The anterior region (sometimes referred to as the sub-oesophageal ganglia) gives rise to six pairs of nerves: three pairs, the **mandibular**, the **maxillulary** and the **maxillary** supplying respectively the muscles of the mandibles, the maxillulae and the maxillae, and three pairs innervating the maxillipedes. The middle portion gives rise to three pairs of nerves innervating the first three pairs of walking legs; whereas, the posterior part gives rise to two pairs of nerves innervating the fourth and fifth pairs of walking legs.

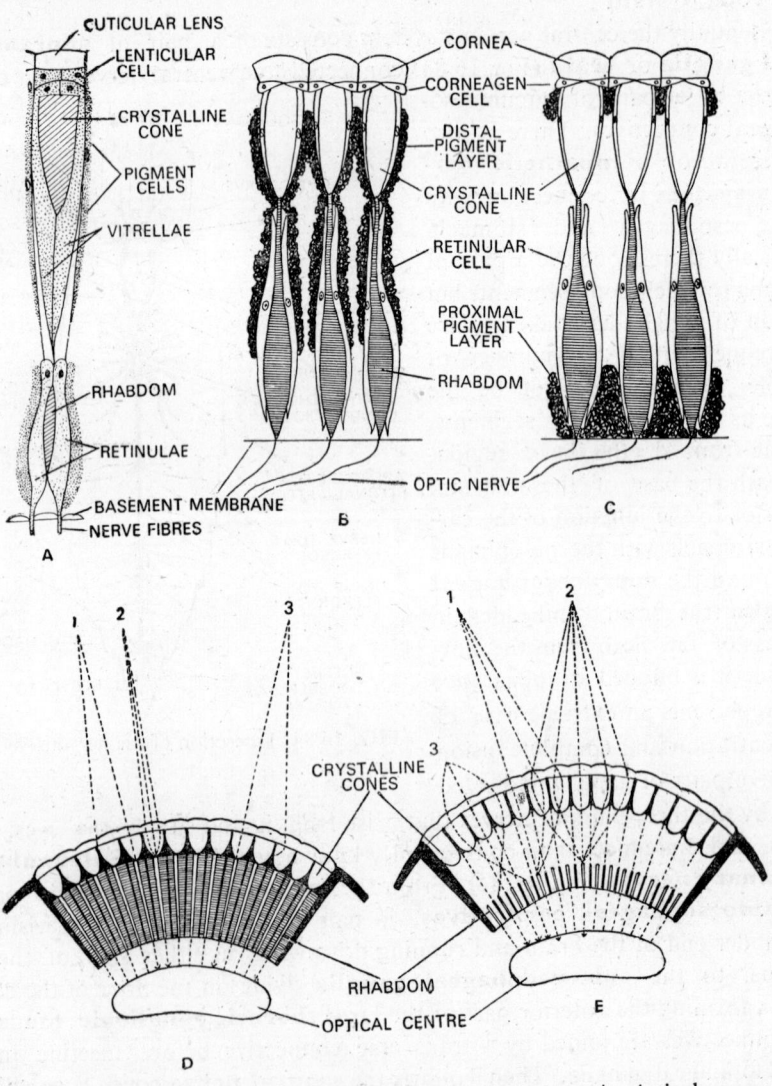

**FIG. 16.85.** Structure of the compound eye of a crustacean. A—A single ommatidium; B—Pigment cells isolate ommatidia; C—Pigment cells, extracted; D—Superposition image formation; and E—Apposition image formation.

In the abdomen there is a series of six **abdominal ganglia** connected by longitudinal commissures, the double character of which is not obvious. The first five ganglia are double and in each segment gives rise to three pairs of nerves, the **pedal** nerves, innervating the appendages, a pair of nerves to the extensor muscles, and a pair of nerves innervating the flexor muscles of the succeeding segment. The sixth ganglion is a relatively larger mass, because with it are fused a number of post-abdominal ganglia. From this arise several nerves of which two pairs supply the flexor muscles of the segment, two pairs control the muscles of the telson, two pairs innervating the uropods, and a single median nerve innevates the hind-gut.

The **visceral nervous system** (Fig. 16.84) consists of a small nerve that arises from the brain and connects with it two small **visceral** or **oesophageal ganglia** lying one behind the other on the roof of the cardiac stomach. Of these the anterior visceral ganglion is connected by two delicate connectives with the **commissural ganglia** on the circumoesophageal connectives.

Within the central nervous system the bodies of the **nerve cells** (neurons) are confined to the ganglia and the commissures, enclose the process from these cells by which intercommunication between them is established. This is a point of distinction from the condition in the annelids where although cells are more numerous in the ganglionic swellings, they are also found along the whole length of the nerve cord. The mechanism of nervous integration in the Crustacea naturally shows an advance over that of the annelids. The pattern of their behaviour is at a much higher level, indicating a greater degree of coordination. The arrangement of superficial receptors, intermediary and effector neurons is similar to that of annelids but their interrelations are more complex. There are also present simple receptors sending their fibres into the central nervous system and other receptors with afferent neurons in close relation with them. Besides, the intermediary neurons extend their influence further up and down the nerve cord than in annelids. Reflex arcs involving more than one or two body segments are, therefore, possible and coordinated movement of limbs remote from one another can take place. It has been experimentally shown that injury to the supra-oesophageal and suboesophageal ganglia results in a greater disturbance in behaviour than in annelids. This proves that a greater measure of control has been located in these nervous masses. Then there are giant fibres that arise from special cells in the brain and run throughout the whole length of the nerve cord. These fibres help the transmission of impulses along the whole length of the body, producing a violent convulsive movement, such as the rapid flexing of the abdomen, by which dangers may be avoided.

**Sense organs**. The sense organs of the prawn include the eyes, the statocyst, the tactile organs and the olfactory setae. The **eyes** (Fig. 16.84) are paired borne on short two-jointed movable stalks situated between the bases of the antennule and the rostrum. Each eye is a composite structure being made up of numerous visual elements, the **ommatidia**, that collectively form a compound eye. The outer surface of the eye shows a large number of roughly squarish **facets** (Fig. 16.85) each corresponding to a single ommatidium. Each ommatidium is a complete apparatus for the reception of light rays. Externally there is the outermost transparent cuticular covering, the **corneal facet** (cuticular lens) secreted by a pair of flat epidermal cells, the **corneagen cells** (lenticular cells) lying below the corneal facet. Beneath the corneagen cells are four tall cells, the vitrellae, the inner borders of which become retractive and form a **crystalline cone**. These parts are sometimes described as **dioptrical region** of the eye. Beneath the vitrellae is another group of seven elongated visual cells, the **retinulae**. They surround a more or less spindle-shaped elongated body, the **rhabdom**. The rhabdom is made up of seven **rhabdomeres** each of which is formed by the inner border of retinular cell becoming refractive. The retinular cells and the rhabdom together form the **receptor** or **retinal** part of the ommatidium. From the retinulae nerve fibres pass back to link up with the mass of nervous material forming the **optic ganglion** in the eye stalk. Surrounding each ommatidium is a pigment sheath of dark colour that separates each ommatidium from the other. The sheath in this case forms two portions, (a) the **retinal sheath** surrounding the rhabdom and retinular cells; and (b) the **iris sheath** surrounding the crystalline cone and vitrellae. Of these the retinal pigments penetrate the **basal basement** or **membrane**, a thin membrane of more or less cuboid cells, upon which rest the inner ends of the rhabdom and the retinal cells. The optic nerve fibres also pierce the basement membrane to enter the ommatidium.

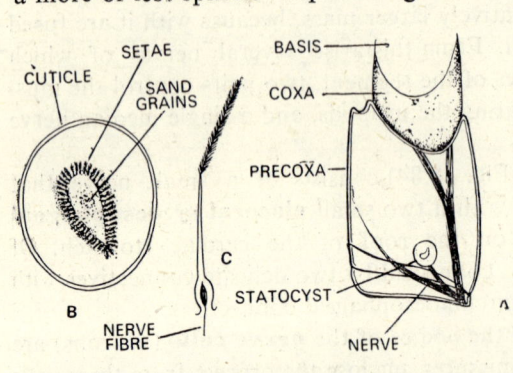

FIG. 16.86. Structure of statocyst. A—Antennule dissected to show the position of statocyst; B—Section of the statocyst; and C—A sensory hair.

The details of the working of the compound eye, such as the one described above, are too complex to be entered into fully here, but it may be mentioned that each ommatidium is capable of producing an image. When the ommatidia are optically isolated by the extended pigment sheaths a composite image is formed made up of a large number of separate "pieces" each of which is contributed by a single ommatidium (Fig. 16.85), for the image-forming cells are stimulated by only light rays paralled or almost parallel to its longitudinal axis. In this eye much light is wasted, because it enters the ommatidium at an angle it is absorbed by the pigment. Such an image that is formed of several components, each received through a distinctly separate visual element placed in juxtaposition is known as the **"apposition image"** and the vision is known as the **mosaic vision**. In this case the sharpness of the image depends on the number of ommatidia involved and the degree of their isolation from one another.

When, however the pigment is retracted, all the ommatidia work in unison and a complete image is formed. Such an image is called a "superposition image." All the crystalline cones here act together as a single refractive body and the whole of the retinular portion as a continuous retina. The vision in this case is not distinct. The animal is able to have some sort of an image of its surrounding objects, especially of their movements. The extension and retraction of the pigment are influenced by differences in the intensity of illumination. In weak light, as at night the pigment is retracted and a "superposition image" is formed. In some insects such as the fire flies and some moths the eyes are permanently set in this condition and are well-adapted to dim light but they are day blind. In bright light as during the day time the pigment is extended and an "apposition" image is formed. Again if the eyes are permanently set in this condition, as in butterflies, the animals can only see in bright day light and are night blind.

In the prawn probably both the types of images are formed depending upon circumstances. In bright light the pigment-sheaths are so arranged as to completely isolate each ommatidium from its neighbour and an "apposition" image is formed; whereas, in dim light a "superposition" image is formed.

**Gravity receptors.** Presssure acts in all directions on an animal whether it is in air or in water, but gravity acts only in one direction. Therefore, detectors for the pull of gravity can be also used to deduce the position of an animal with regard to gravity, in other words, whether it is upright or not. All these organs contain a mass of some substance which is denser than the surrounding tissues and will move when the animal moves its body axis out of the vertical. The mass is usually calcium carbonate, either resting loosely as granules in a chamber lined with sensitive cells which will respond to the pressure of the granules on themselves. Such an organ of the prawn is called the **statocyst**.

The **statocysts** (Fig. 16.86) are a pair of cuticular structures, each situated in the large cavity of the precoxal segment of each antennule. Each is a subspherical sac of cuticle attached to the lower surface of the concave roof and opens to outside at its base (Fig. 16.86). The essential part of the statocyst consists of a more or less oval ring formed by numerous delicate **receptor setae** connected to the statocystic nerve branches. Normally the space between the setae is occupied by sand particles which function as **statoliths**. Any change in the position of a prawn displaces the sand grains which press against the setae and stimulate them to convey the change of position to the brain. The statocysts are organs of balance with the help of which the animal knows its position and maintains equilibrium. They can, thus, be said to be organs for perceiving the direction of the force of gravity. Formerly they were believed to be auditory in function, but the same has not been proved so far. The statocyst is believed to have originated from a simple receptor setae, the area of integument below which sank down to form a pit that eventually transformed into a sac.

Experiments on the prawn *Palaemon xipnias*, on the other hand, show that the presence of the statocysts of the prawn is not essential for equilibrium, and the position sense of this animal depends upon interplay of impulses from the statocysts, of a dorsal light response and of the sense of touch in the tarsi. When a normal prawn is turned on its side, the legs which are uppermost make kicking movements which will right it again. During righting its upper antenna turns dorsalwards, the lower one downwards, its upper eye turns downwards and the lower one upwards. When it attains the right position the symmetry of the position of antennae, eyes and legs is restored. When turned upside down it makes no attempt to right itself though as soon as it begins to tilt away from this position it makes compensatory movements and may turn either clockwise or anti-clockwise. When its statocysts are removed it still reacts in the same way. If a prawn is tilted over with a piece of board held against its legs, it remains standing on the board on a symmetrical position, though its eyes may bend towards light. After the removal of both statocysts the light falling on the dorsal side of the body takes control. This has been described as **dorsal light response** of the animal. Even with one eye only the prawn will still regulate its position when it is tipped sideways. Such an operated animal has a new position of equilibrium with its dorsoventral axis at 45° to the vertical, the intact eye uppermost.

**Tactile organs.** Besides the many-jointed feelers or the antennae there are some **plumose setae** believed to be of tactile nature. These are hollow cuticular outgrowths internally supported by muscle fibres and supplied by nerve fibres. Each consists of two portions, (i) the hollow base bearing, (ii) a distal tapering blade having linear rows of tiny barbs. Such setae are found specially covering the flattened portions of the pleopods and other parts of the body.

The tactile senses play important roles in the life of the animal. Contact between the legs and bottom is maintained by these and as soon as this contact is disturbed, as by tilting, there is kicking movement following righting of the position. The swimming activity of the prawn *Palaeomonetes varians* ceases as soon as its legs make contact with the ground.

**Olfactory setae.** The olfactory setae are situated on a small, middle feeler between the two elongated feelers of the antennules. Each has a two-jointed shaft proximally attached to the integument by means of a flexible membrane. The distal segment is bluntly rounded at the free end and covered with a thin membrane. A single nerve fibre from the olfactory branch of the antennulary nerve innervates each seta.

**Reproductive System**

The sexes are separate in the prawn. There are a number of external features with the help of which the male can be distinguished from the female. One is size. Female is generally smaller than the male of the same age. In the males the bases of the thoracic legs are more closely approximated; the second pair of chelate leg is longer and more profusely covered with setae and spines, and the endopodite of the second pleopod carries an additional stylet, the **appendix masculina**, besides the appendix interna. The genital apertures in the males are paired, each situated on the arthroidial membrane above the coxa on the inner side of the last pair of walking legs and is covered over by a tongue-like process. In the female the epimera of the abdominal segments are larger and form deep recesses for carrying eggs; and the genital apertures are paired, each situated on a raised papilla on the inner side of the coxa of the third pair of walking legs.

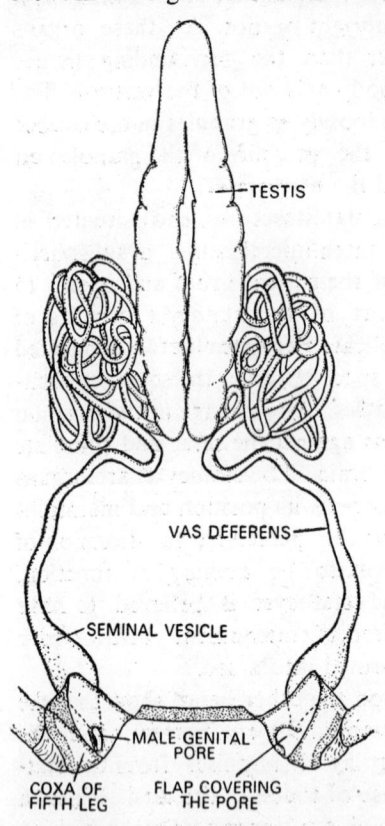

FIG. 16.87. Male reproductive organs of *Palaemon*, as seen from above (*after* Patwardhan).

In the male the **testes** (Fig. 16.87) are paired elongated structures lying beneath the pericardial sinus. The testes in a mature prawn extend forward up to the renal sac. Anteriorly both the testes fuse together to form a common lobe. Each testis consists of a large number of seminiferous tubules held together by connective tissue. Each tubule has a thin wall enclosing a large cavity lined by a single layered epithelium, cells of which proliferate and give rise to **spermatids** that separate from the wall and ultimately develop into **spermatozoa**. From the posterior end of each testis arises a vas deferens, which runs for a short distance within the substance of the testis before emerging out. On emerging each forms a coiled mass beyond which it descends as a straighter muscullar tube opening in the coxa of the last pair of walking legs. Before actually opening to outside each vas deferens swells up forming the **vesicula seminalis**, in which the spermatozoa are stored in the form of white compact bodies called the **spermatophores**. The spermatozoa are very small each consisting of

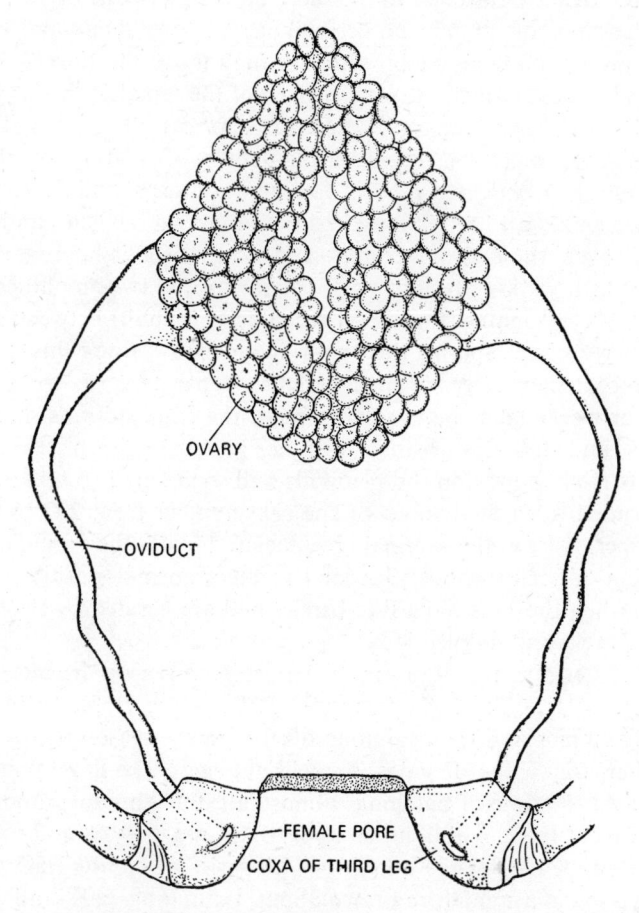

**FIG. 16.88.** Female reproductive organs of *Palaemon*, as seen from above (*after* Patwardhan).

hemispherical mass of cytoplasm drawn out at one end into a tail-like process. The nucleus is dark crescentic mass lying near the root of the elongated process.

In the female the **ovaries** (Fig. 16.88) are paired structures occupying relatively same position in the body as the testes in the male. Each ovary is a saccular organ whose shape varies with age and time of the year. Ordinarily the ovaries are compact sickle-shaped structures closely approximating with each other at their ends, leaving a space in the middle. They become larger in the breeding season and extend behind up to the first abdominal segment, and anteriorly upto the renal sac. Each ovary is composed of radially disposed strings of ova in the various stages of development, the immature ova lying towards the centre and maturer ones towards the outer side. Each ovary is enclosed within a membranous capsule which is continuous with the wall of the **oviducts**. The oviducts are short thin-walled tubes but broader than the vasa deferentia, and pass vertically downwards to their openings on the inner side of coxopodites of the third walking legs. The mature ova are rounded structures with large nuclei lying on the periphery of the ovary and full of yolk granules.

When laid the ova are fastened to the setae on the pleopods of the female by the sticky secretion of glands occurring both on appendages and on the body segments themselves. Immediately after laying they are fertilized, the male depositing spermatophores on the ventral surface of the female's body just before oviposition. Exactly how the male deposits spermatophores is not known for the prawn but Andrews (1904) has given the details of mating in the case of American crayfish, *Cambarus affinis*. The male grasps and inverts a female, stands over her, seizes all her walking legs with his two chelae, and flexes his telson tightly over the end of her abdomen, so that she is held motionless. He uses one of his fifth walking legs to press the tips of the two modified pleopods on his somite XIV against the sperm receptacles (annuli) between somites XII and XIII on her thorax. Sperms then pass in mucous along his pleopods to lodge in her receptacles, after which the animals separate.

Some days or weeks later, before oviposition, the female cleans her abdomen and pleopods and lies upside down with her abdomen sharply flexed. A slimy secretion emits from glands on the pleopods and soon small (2 mm in diameter) eggs are extruded from the oviducts. The eggs number from 200 to 400 and are fertilized by sperms from the seminal receptacles. Finally they are glued together by the secretion of her pleopods. She later resumes normal position and moves into a shelter when the eggs hang like berries and are aerated by the slow movements of the prawn and its pleopods.

It is easy to determine the duration of the breeding season because the females carry eggs attached to their pleopods. *Palaemon malcolmsonii* breeds in the latter half of summer and the beginning of the rainy season (May, June and July). At other times the females are without eggs. The development is direct (*P. lamarrei*). At the time of hatching almost all the thoracic and abdominal appendages, except the last abdominal appendage, are present and the five pairs of abdominal appendages are uniramous. A series of moults leads to the adult form. Each young is a miniature prawn about 4 mm long pale and translucent, and is often attached to the cuticle moulted within the shell before hatching. In crayfish they remain attached to the female until the second stage, stay near her few days more, and then become independent. The adult form is reached after a series of moults. The period taken for hatching directly depends upon temperature. In cold waters of Europe, *Astacus fluviatilis* produces eggs in October or November that do not hatch until the following June.

**Development**. The cleavage in the case of *Palaemon* is of the purely superficial type (Fig. 16.89). The nucleus of the fertilized egg, lying in the centre, divides regularly into two, four, eight, etc., nuclei surrounded by radiate masses

## Phylum Arthropoda

of protoplasm (yolk-pyramids), but areas of the separate cells do not become marked off by furrows cutting right through the egg. More and more cleavage nuclei are produced and migrate toward the surface forming a regular **blastoderm**, not all over the surface simultaneously as in *Astacus*, *Branchipus* and the free-living copepods, but the blastoderm appears first on the ventral side of the

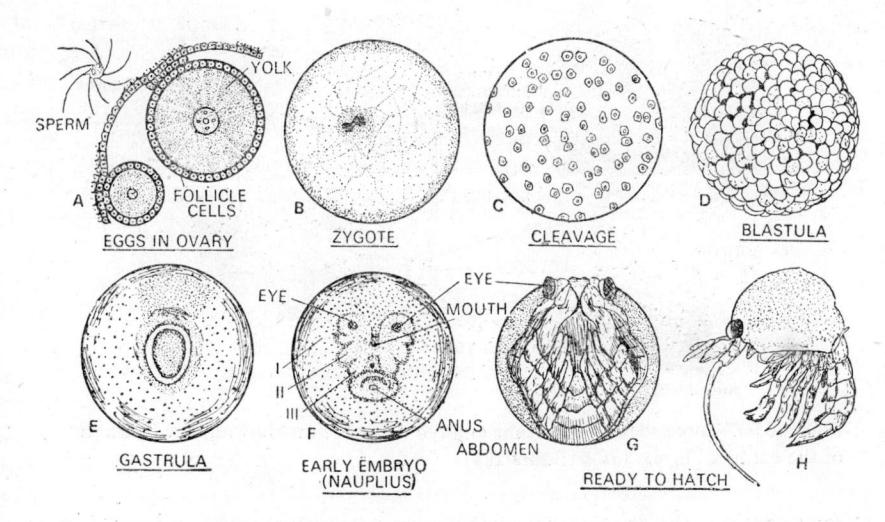

FIG. 16.89. Stages in the development of *Astacus*.

egg. Its formation begins at one point on the surface and proceeds gradually from this point, which always represents the future ventral side of the egg. In *Palaemon* and other Decapoda this point denotes the most posterior end of the ventral side, the spot at which later the gastrula invagination appears.

**Gastrulation**. In *Palaemon* (according to Bobretzky) a small invagination (Fig. 16.90) develops at a time when the blastoderm is not completely formed, *i.e.*, when the blastoderm cells have not separated from their yolk-pyramids over the whole circumference of the egg. This invagination marks the first step towards gastrulation and a small archenteron is formed with its opening, the **blastopore**, at the surface. After the blastopore closes the cells of the endoderm vesicle lose their epithelial character (Fig. 16.90). From the lateral walls of the endoderm vesicle arise elements which form the mesoderm at a later stage.

The endoderm formation, as reported by Bobretzky, is also interesting in *Palaemon*. The cells arising from the floor of the endoderm vesicle pass into the yolk, traversing it like wandering cells, and multiply within the yolk. Each of these endoderm cells swallows the surrounding food yolk in an amoeba-like manner, with the result that the yolk mass breaks into smaller spheres. In the later stages the nuclei, each with a certain amount of protoplasm, rise to the surface of the food yolk forming an epithelium which represents wall of the mid-gut containing food yolk within it. Some endoderm cells still remaining in the food-yolk are absorbed along with yolk.

When the mesoderm appears for the first time it consists of a mass of cells near the points of origin. These cells multiply rapidly and spread out apparently in an irregular manner between the ectoderm and the food-yolk. The distribution of the mesodermal elements into paired mesoderm bands is perceptible only in a few stages.

Gastrulation in *Astacus* takes place in a different way. By the time the blastoderm formation is completed, the rudiments of the embryo become visible on the ventral side as thickenings of the blastoderm. There are five distinct thickenings,

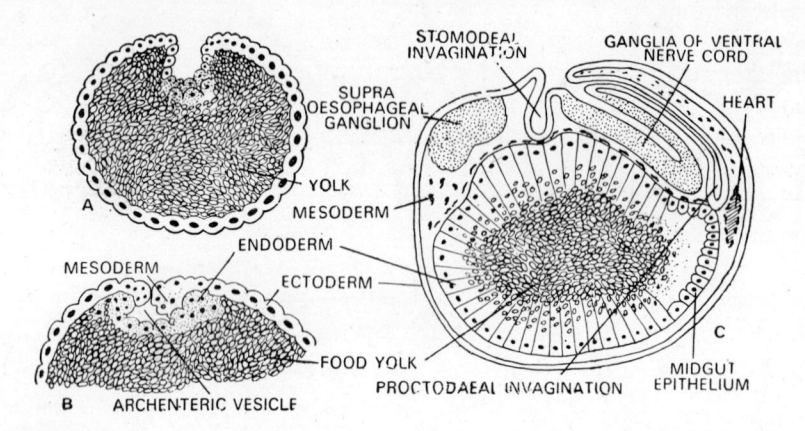

FIG. 16.90. Three sections from the embryo of *Palaemon* showing the formation of the germinal layers (*after* Bobretzky).

two representing the rudiments of the eyes, two thoraco-abdominal rudiments, and an unpaired thickening, the **endoderm disc**, behind these. After invagination this forms the archenteron.

The beginning of gastrulation is indicated by crescentic furrow at the anterior edge of the **endoderm disc**. This furrow is joined by a similar posterior furrow. The inpushing increases gradually and forms the **archenteron** opening at the **blastopore**. The archenteron is lined by endoderm cells.

Even before the process of invagination begins, active proliferation of cells takes place at the anterior edge of the **endoderm disc**. These cells multiply quickly and pass below the blastoderm (Fig. 16.91) forming the **mesoderm**. It is apparent that in *Astacus* the mesoderm originates from a definite point on the

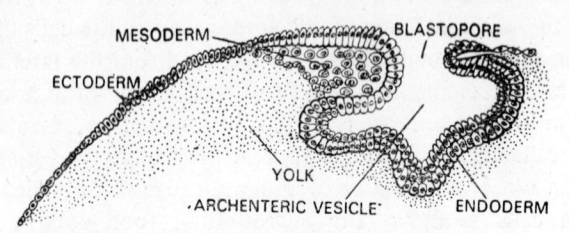

FIG. 16.91. Median longitudinal section through the gastrula of *Astacus fluviatilis* showing mesoderm formation (*after* Reichenback).

anterior margin of the blastopore, where the ectoderm passes into the endoderm.

In the case of *Palaemon* the mid-gut is formed by the interpenetration of the food yolk by the cells of the gastrula. In *Astacus* the gut is formed differently through the filtration of the food-yolk. The endoderm cells absorb the food-yolk which lies in the primary coelom. After this secondary yolk pyramids appear. Then the nuclei of the endoderm cells shift to the surface of the yolk and there they form the epithelium of the mid-gut. The characteristic of this method is that the archenteric cavity persists throughout development and then the lumen of endoderm vesicle passes into the lumen of the future mid-gut.

As the mid-gut develops in this way the fore- and hind-gut arise by ectodermal invaginations forming the stomodaeum and proctodaeum. These two approach the mid-gut with which ultimately they fuse, but this may be completed after hatching.

From the ectoderm arises the **external integument**, and internal **skeleton** by its infolding and invagination. The whole of the nervous system arises as an ectodermal thickening. The rudiments seen after the blastoderm formation develop into various parts and form a **nauplius larva**, which is never free-living in *Palaemon* and many other Decapoda. The next stage (**protozoaea stage**) also merges in the general embryonic development, which takes place in the egg. The embryo leaves the egg in an advanced **zoaea** stage, a stage that in many ways anticipates the **mysis** stage. Thus, it is evident that the metamorphosis in this case is distinctly abbreviated and the larva in the course of time change into the adult.

**Larval forms.** In some prawns (*Penaeus, Lucifer*), like many other crustaceans, the development is accompanied by metamorphosis. There are several larval stages, the younger of which are quite unlike the parent animals. However diverese the form of the adult may be there always occurs a larva with certain constant characters. The first and the most common of these is the **nauplius larva** (Fig. 16.92). It is oval in shape with an unsegmented body having median frontal eye and typically three pairs of appendages (the antennules, antennae and mandibles of the adult). A pair of long furcal setae projects on either side of the hind end of the body. The nauplius is a free swimming form that moves rapidly

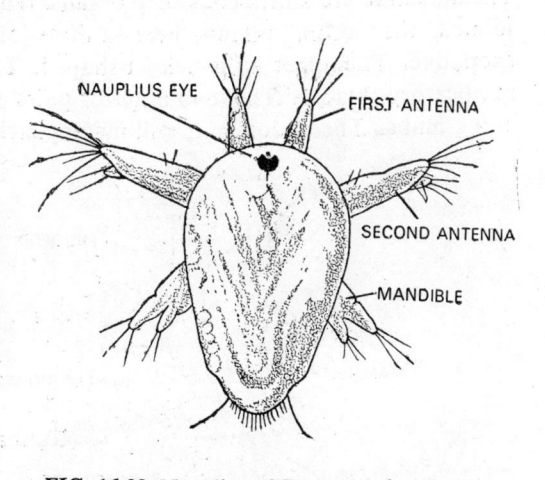

FIG. 16.92. Nauplius of *Penaeus* (*after* Green).

through the water by the strokes of all three pairs of limbs that are unjointed and are provided with swimming stage. In some (*Penaeus*) the young hatch out as nauplius larvae, but in others (e.g, *Palaemon*) it is apparently lacking. In such cases, it has been suggested, that a distinct nauplius stage, followed by the shedding of a larval cuticle, is passed through in the egg and the actual larval life begins at a later stage.

With successive moults the **metanauplius** and **protozoea** stages appear (Fig. 16.93). In the metanauplius a fourth pair of appendages (first maxillae of the adult) appears behind the three original pairs. Later, rudiments of four fresh pairs of appendages are formed. The body consists of the oval cephalothorax and of an elongated terminal portion which gives rise to the last thoracic segment and to the abdominal segments by progressive segmentation. It terminates in two truncated processes or **furcal processes**, forming a caudal fork provided with setae. The next stage is the **protozoea**, the third larval form. The limbs, which are rudimentary in the metanauplius stage are fully developed here,

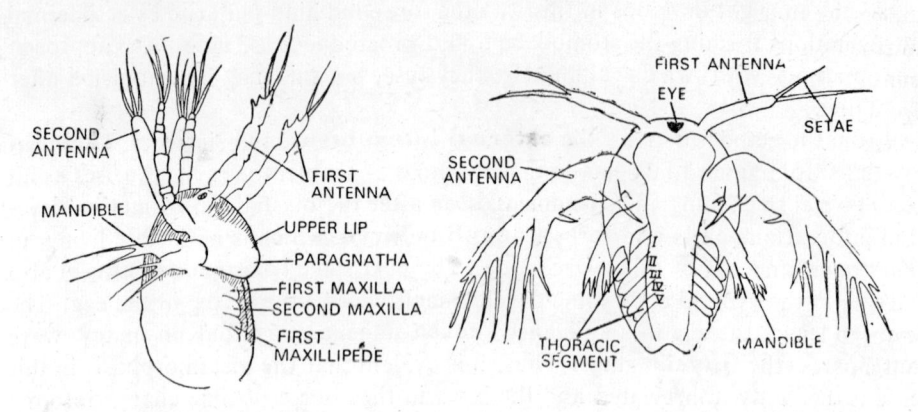

FIG. 16.93. A—Metanauplius of *Lucifer* just hatched; and B—Metanauplius of *Apus* just hatched.

The antennae are still locomotory organs. The first antenna (antennule) is four jointed, the second antenna bears a three-jointed endopodite and a four-jointed exopodite. The upper lip is helmet-shaped. The mandibles are palpless, toothed masticatory blades. The two anterior pairs of maxillipedes are biramous swimming limbs. The abdomen is still incompletely segmented and the paired eyes

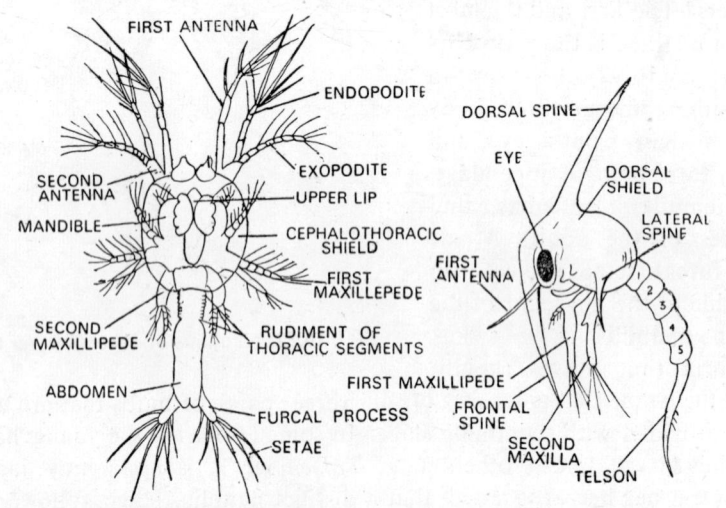

FIG. 16.94. A—Protozoea of *Penaeus* (*after* F. Müller); and B—Zoea larva of *Thia polita* (*after* Claus).

have begun to appear as lateral outgrowths of the head. In some cases (*Sergestes*) the larva that hatches is in the protozoea stage (Fig. 16.94).

Then follows the **zoea stage** (Fig. 16.94B) with a distinct cephalothorax and

## Phylum Arthropoda

distinctly segmented abdomen. The appendages of the head and the first and second (in some cases third in addition) thoracic segments are well-developed while the abdomen remains limbless. The movable stalked eyes have developed. A frontal spine is found between the eyes; a dorsal spine arising from the centre of the dorsal shield and a pair of lateral spines are also present. The movable abdomen is ventrally flexed and consists of five free segments while the sixth is united with the telson. Seven pairs of limbs are present in this stage. The antennule (first antenna) is simple, unjointed process, bearing a few setae at the tip. The second antenna grows out into long spinous process bearing other processes also which will develop into endopodite and exopodite. The mandible is exclusively a masticatory blade. The maxillae already show the structure typical of the Decapoda. The protopodite of the first maxilla has two masticatory blades provided with setae and a two-jointed palp (endopodite). The second maxilla has protopodite, endopodite and exopodite. The two anterior maxillipedes have developed as biramous swimming limbs. The eyes are large and compound and the abdomen has a forked telson. In some cases (marine *Palaemonetes varians*) the young are hatched as zoeae. The zoea moults into the **mysis** (schizopod) larva, so called because the larva resembles an adult *Mysis*. It has thirteen pairs of appendages on the cephalothorax and those of the thorax bear exopodites that help in swimming. A further moult results in the formation of the adult with nineteen pairs of appendages. The mysis larva (Fig. 16.95) is at the end of the metamorphosis of the prawn (*Penaeus*), while it initiates that of the lobster. In the lobster (*Panulirus*) the newly hatched larva is a strangely modified mysis-form called a glass-crab or *Phyllosoma*.

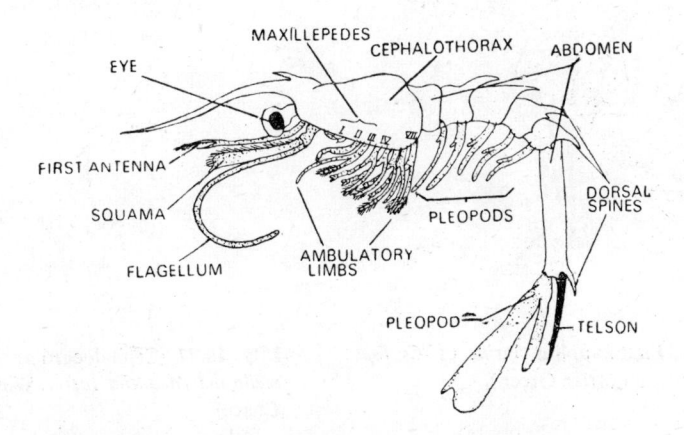

FIG. 16.95. Mysis stage of *Penaeus*, lateral view (*after* Claus).

In a crab the young hatches out in the zoea stage, but the larva has a helmet-like carapace with long dorsal and anterior spines and sessile eyes. The thorax bears two pairs of biramous swimming legs (maxillipedes) and the slender mobile abdomen lacks swimmerets. This larva moults into the **megalopa larva** (Fig. 16.96). In this case the cephalothorax is well developed (crab-like) but lacks spines, the eyes are large and stalked, the thorax has five pairs of "walking" legs, and there are functional swimmerets on the abdomen. The larva swims on the surface waters but later on sinks to the bottom and moults into typical crab form.

In the Cirripedia and Cypridae (Ostracoda) the nauplius that hatches usually has a delicate dorsal shield and after several moults, the larva changes into a **cypris** stage. It is enclosed in a bivalve shell and possesses a pair of compound eyes in addition to the median eye. The anterior antennae are four jointed and bear a characteristic disc at that end of the second segment on which opens the duct of the cement gland. The posterior antennae of nauplius have disappeared. There are six pairs of biramous and setose thoracic appendages, the short abdomen ending in a caudal fork. The cypris stage lasts from four days to ten or twelve weeks in different species. The larva then settles to the bottom, hunts a place of attachment to which it adheres by the antennules, aided by the secretion from the cement gland, and metamorphoses into the adult.

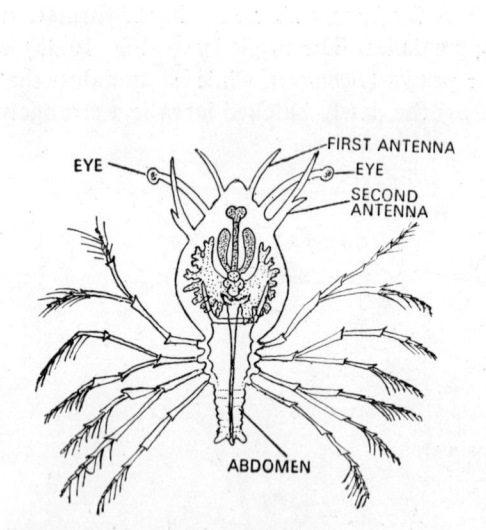

FIG. 16.96. First nauplius larva of *Cyclops fuscus*, a copepod (*after* Green).

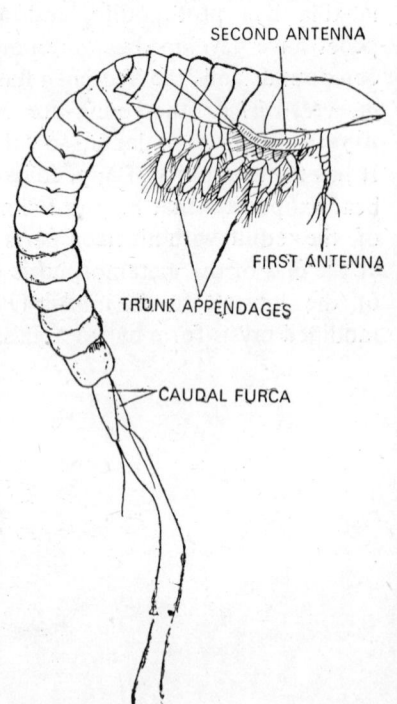

FIG. 16.97. Cephalocardian *Hutchinsoniella macracantha* (*after* Waterman and Chace).

## CLASSIFICATION OF CRUSTACEA

Crustacea essentially comprises acquatic Arthropoda including the prawns, lobsters, shrimps, crabs, cray-fishes, etc. that respire by gills or by the general surface of the body. The exoskeleton is massive. Antennae are two pairs borne by second and third segment; and three pairs of head appendages act as jaws while some of the anterior thoracic appendages also act as mouthparts.

I. **Subclass Cephalocarida**. The most recently discovered and perhaps the most primitive group having only four known species belonging to three genera.

## Phylum Arthropoda

The group includes tiny (less than 3 mm) shrimp-like crustaceans with a horseshoe-shaped head followed by an elongated trunk of 19 segments, of which only the first nine bear appendages. All trunk appendages are similar and resemble second pair of maxillae. The appendage appears triramous as the basal section of each bears a large flattened pseudepipodite. Both pairs of antennae short, eyes absent. First specimens collected from the bottom sand and mud of Long Island, Sound, and described in 1955. Also reported from the West Indies and Coasts of California and Japan.

Example: *Hutchinsoniella*.

II. **Subclass Branchiopoda.** The fairy-shimps, brine-shrimps, etc., are free-living Crustacea with a varying number of body segments provided with similar appendages usually foliaceous fringed with hairs. The coxa is provided with flattened epipodite which serves as a gill (Branchiopoda=gill-feet). The cephalic

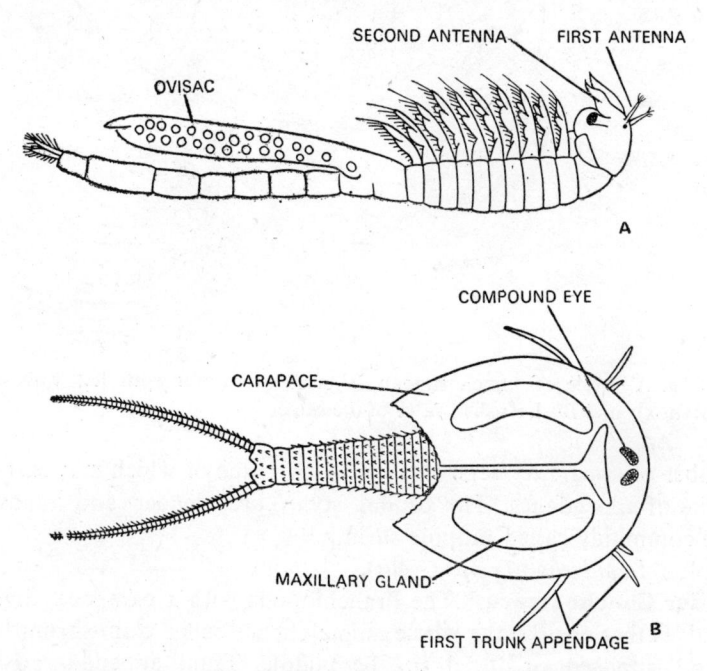

FIG. 16.98. A—Fairy shrimp *Branchinecta*, an anostracan, lateral view (*after* Calman); and B—Tadpole shrimp *Triops*, a notostracan, dorsal view (*after* Pennak).

carapace is shield-like or bivalve when present, sometimes it is absent. Paired compound eyes are usually present. Mostly found in freshwater. About 800 species are known (from Cambrian to Recent). The Branchiopoda has four distinct groups.

1. **Order Anostraca.** The Branchiopoda without carapace (=Anostraca), stalked eyes and with fairly well-sized antennae, which are not biramous. The fairy-shrimps, as they are called, have an elongate trunk of 20 or more segments of which the anterior 11 to 19 bear appendages which are all alike. The caudal styles are jointed, flat and subcylindrical.

Examples: *Branchipus* (Fig. 16.98), *Branchinecta*, *Artemia*.

2. **Order Notostraca.** The Branchiopoda with a large dorsal shield-shaped carapace, covering the head and anterior part of trunk. The compound eyes are sessile and close together. The antennae are much reduced. Trunk appendages numerous (30 to 70 pairs) of which the first two pairs differ from the rest.

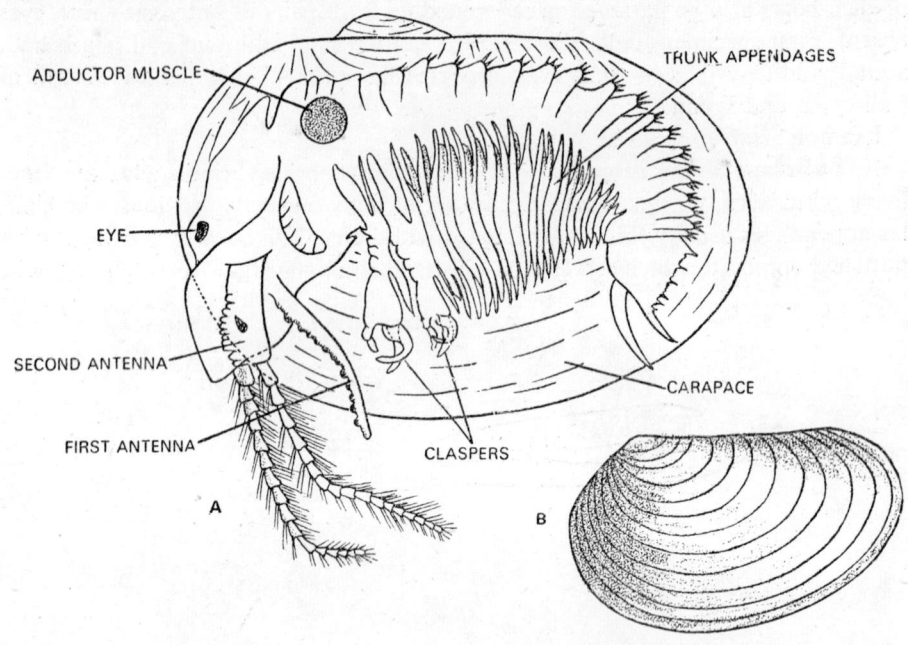

**FIG. 16.99.** A—Male of conchostracan, *Cyzicus mexicanus* with left valve removed (*after* Mattox); and B—Left shell valve of the same.

The number of apparent "segments" is 25-44, some of which may carry as many as 10 pairs of appendages. The caudal styles are slender and many jointed. They are commonly called tadpole shrimp.

Examples: *Lepidurus, Triops* (*Apus*).

3. **Order Conchostraca.**[1] The Branchiopoda with a carapace divided into lateral halves that enclose the whole animal, hence called clam-shrimps. Eyes are sessile and apposed or fused in the middle. Trunk appendages 10-32 pairs. Caudal styles are a pair of curved claws.

Examples: *Lemnetis, Cyzicus* (Fig. 16.99) (*Estheria*).

4. **Order Cladocera.** The Branchiopoda with a bivalve carapace having only 4-6 pairs of trunk appendages. The biramous antennae are used for swimming. Common name is water-flea.

Examples: *Daphnia* (Fig. 16.100), *Polyphemus*.

III. **Subclass Ostracoda.** Small crustacea called clam or seed-shrimps with a bivalved carapace and with or without compound eyes. The valves are rounded or elliptical and the outer wall of each valve is impregnated with calcium car-

---

[1]According to some authors Diplostraca is the third order containing laterally compressed branchiopods which are of two types, hence classified as two suborders, Conchostraca (clam-shrimps) and Cladocera (water-flea). Here Conchostraca and Cladocera have both been placed as orders.

# Phylum Arthropoda

bonate. Dorsally there is hinge-line formed by a non-calcified strip of cuticle. The body may be indistinctly segmented or unsegmented bearing four pairs of trunk appendages. The antennules and antennae, which are generally biramous,

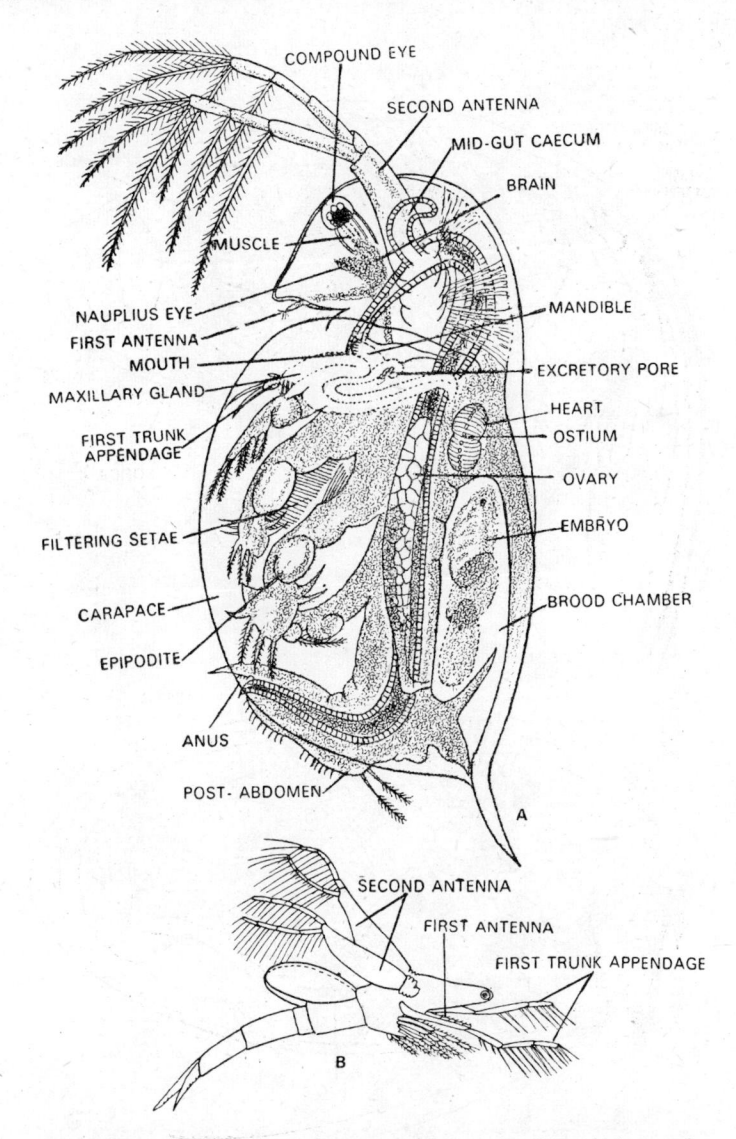

FIG. 16.100. A—Female cladoceran *Daphnia pulex*, lateral view (*after* Sebestyen); and B—Female of *Leptodora*, an aberrant pelagic cladoceran.

are used for swimming. The genital aperture is present on the 7th body segment. Fresh and saltwater forms occur, mostly living on or near bottom. 200 species are known (from Ordovician to Recent). The Ostracoda are divided into four orders.

1. **Order Mydocopa**. Carapace notched anteriorly: 2nd antenna biramous, two pairs of trunk appendages are present.

Examples: *Conchoecia*, found in northern oceans; *Cypridina* (Fig. 16.101).

2. **Order Cladocopa.** Marine Ostracoda in which the trunk is without appendages. Second antennae biramous.

Example: *Polycope*.

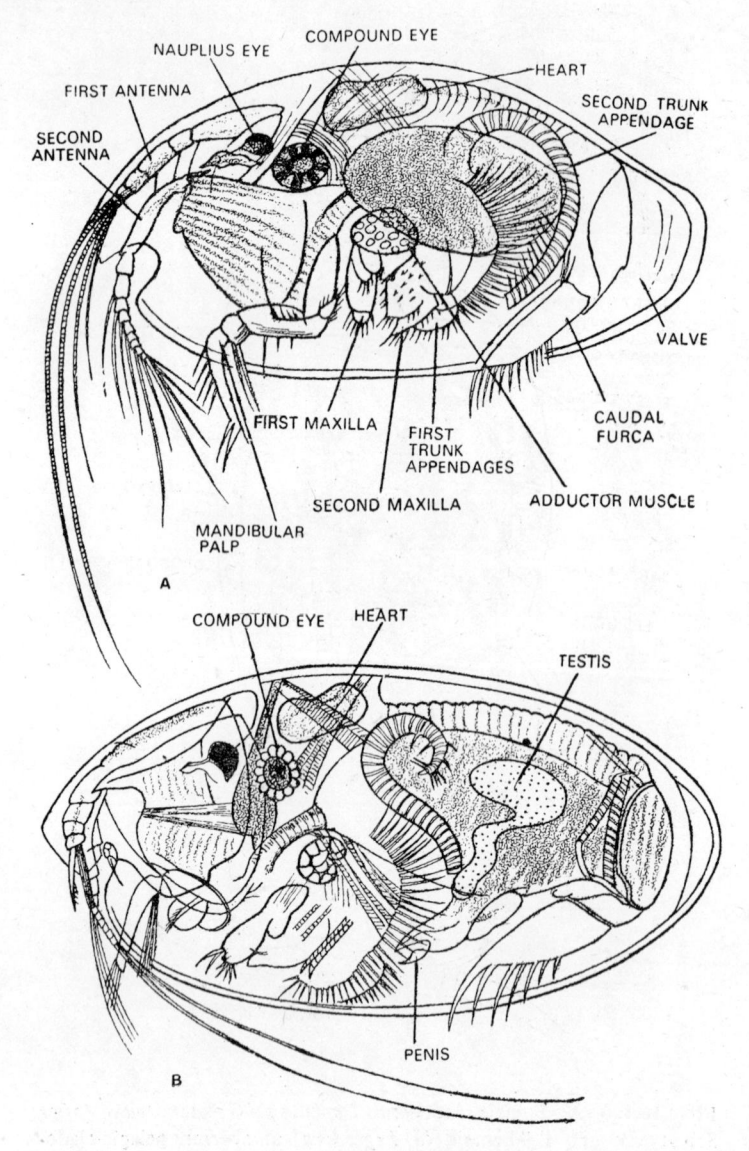

**FIG. 16.101.** A—Female marine myodocopid ostracod *Cypridina*, with left valve removed, lateral view; and **B**—A male *Cypridina*, lateral view (*after* Claus).

3. **Order Podocopa.** Carapace unnotched; second antenna uniramous. Two pairs of trunk appendages are present. Large order. Marine as well as freshwater forms.

Examples: *Eucypris, Darwinula*, both in freshwater; *Entocythere* on the gills of crayfish; *Cytheresis*, marine and *Cypris*.

*Phylum Arthropoda*  639

4. **Order Platycopa**. Marine Ostracoda in which the second antennae are biramous and there is only one pair of trunk appendages.

Example: *Cytherella*, the only genus.

**IV. Subclass Mystacocarida**. Microscopic crustaceans with an elongated body divided into head, five-segmented thorax and six-segmented abdomen. Head appendages well-developed, thorax has one pair of maxillipedes and abdomen has no appendages except the caudal styles. Compound eyes absent, only

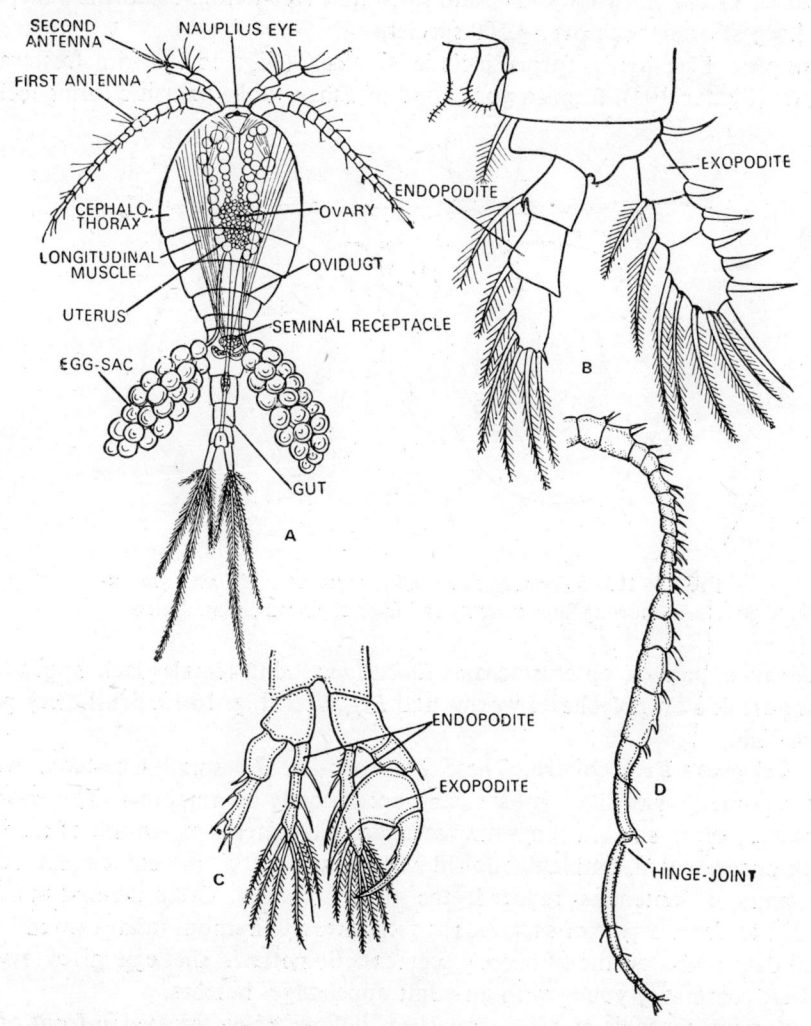

**FIG. 16.102.** A—*Macrocyclops albidus*, (dorsal view); B—Right second trunk appendage of cyclops (typically biramous); (*after* Sars); C—Last pair of thoracic appendages of male, in which exopodite is modified as a claw for clasping female abdomen; and D—The clasping first antenna of a male *Centropages* (*after* Sars).

one nauplius eye is present. Sexes separate. A metanuaplius is the earliest known larval form.

Example: *Derocheilocaris* is the only known genus.

**V. Subclass Copepoda**. It consists of water-flea, fish-lice, etc. The Crustacea without compound eyes or carapace which lead a free or parasitic life. Typically six pairs of appendages of which the first and the sixth are uniramous and the rest are biramous, none of the appendages is on the abdomen. Antennae and antennules are well developed; some of these characters may become lost. Mandible has a uni- or biramous palp, or none. The genital aperture is on a special genital segment preceding the abdomen. Sometimes, the eggs after leaving the oviduct remain attached to the female in oviscas, typically they hatch as nauplius larva. Occur in freshwater, and saltwater, free-living, commensal or parasitic. Largest order with over 4,500 species.

Examples: Free-living forms include *Cyclops* (Fig. 16.102) in freshwater, *Calanus* (Fig. 16.103), forms a good food of fishes, while parasitic forms include

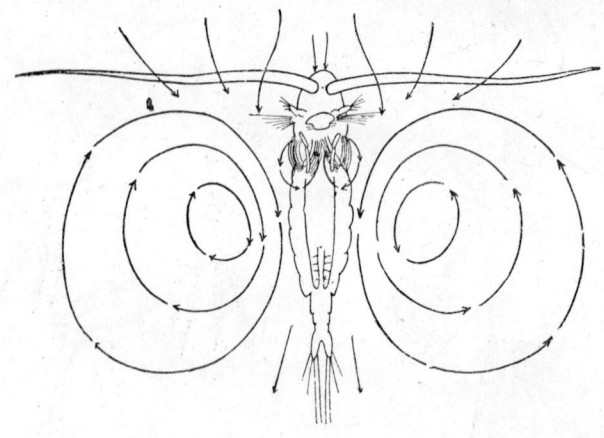

FIG. 16.103. Swimming and feeding currents in *Calanus finmarchicus* showing filter currents and filter apparatus, ventral view.

*Choniostoma*, parasitic on crustaceans; *Rhisorhina*, adult females lack appendages and is parasitic on polychaete worms and *Ergasilus* (Fig. 16.105) (fish-lice) parasitic on fish.

VI. **Subclass Branchiura**. These are carp-lice or small Crustacea which are temporarily parasitic on fishes and occasionally on amphibia. They possess compound eyes, suctorial mouth, carapace-like lateral expansion of the head and an unsegmented, limbless, bilobed abdomen, with a minute caudal furca. Both pairs of antennae reduced, the first is clawed. Often the first maxillae modified to form a pair of suckers, the second are uniramous and clawed. The genital duct opens on the fifth body segment. Sometimes the egg gives rise to nauplius, commonly young with all adult appendages hatches.

In *Argulus* there is a large sheathed hollow spine located in front of the mouth cone. It is used for puncturing the skin of the host and is supplied by a gland that is, perhaps, poisonous. The second maxillae are heavy, uniramous and terminate in claws. Maxillipedes absent. Four thoracic appendages are large and biramous with swimming setae (branchiurans can detach and swim from one host to another). The eggs are deposited at the bottom. In some species of *Argulus* a nauplius larva hatches from the egg, but usually the young have all of the adult appendages when they hatch.

Example: *Argulus* (Fig. 16.106).

*Phylum Arthropoda* 641

VII. **Subclass Cirripedia.** The barnacles come under this group They are hermaphroditic Crustacea which are sedentary, either sessile or stalked, in the adult condition. Some are parasitic. The carapace is folded to enclose the body. This fold is also known as mantle and secretes calcareous plates supporting the former. Typically there are six pairs of biramous cirriform body appendages. The

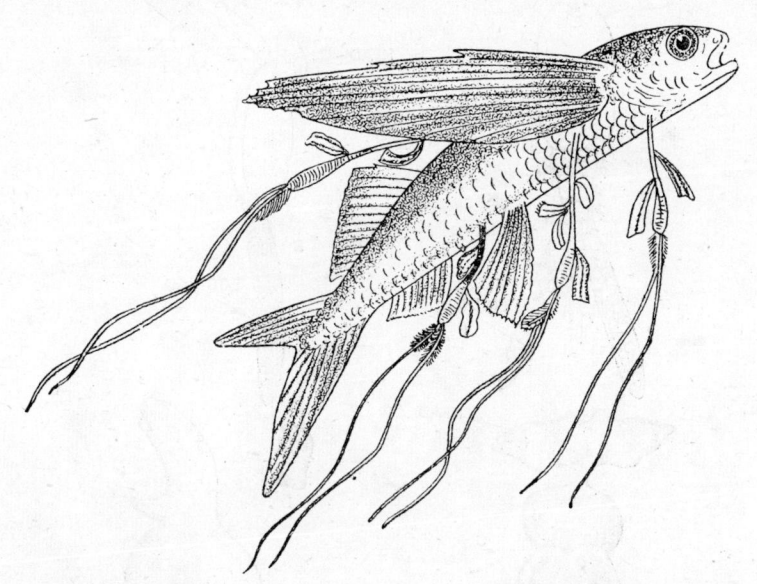

FIG. 16.104. Parasitic copepods, *Penella exocoeti* on a flying fish. The copepods are in turn carrying the barnacle *Conchoderma virgatum*, striped body (*after* Schmitt).

abdomen is limbless and rudimentary. The adult is without compound eyes. Larvae free-swimming, marine, about 800 species are known (from Ordovician to Recent). Up to 1830 the members of this subclass were placed with Mollusca. The discovery of larval stages in 1830 revealed their true relationship and they are now with Crustacea.

1. **Order Thoracica.** Non-parasitic Cirripedia having an alimentary canal, six pairs of biramous thoracic appendages, mantle usually covered with calcareous plates.

Examples: (*i*) *Lepas* (Fig. 16.106 B), the goose barnacle, is found all over the world, attached to floating objects in the sea, by a long stalk, or peduncle (which is the foremost part of the head and bears rudiments of antennae). The rest of the body is called capitulum and is enclosed in a carapace or mantle strengthened by five calcified plates—a median dorsal, carina, two lateral proximal, the scuta and two lateral and distal, the terga. The body within the mantle cavity is turned over its back with the appendages upwards.

(*ii*) *Balanus* (Fig. 16.107), the common acorn-barnacle, is not stalked. They occur in large numbers on rocks between tidemarks in all parts of the world. On the capitulum of *Balanus* are present a number of additional plates. The attached undersurface of the barnacle is called the **basis** and is either membranous or calcareous. This is the preoral region of the animal and contains the cement glands. Certain plates located on the capitular base of many stalked barnacles are present in the sessile barnacles as mural plates which form the carina and rostrum, a vertical wall ringing the animal. The top of the animal is

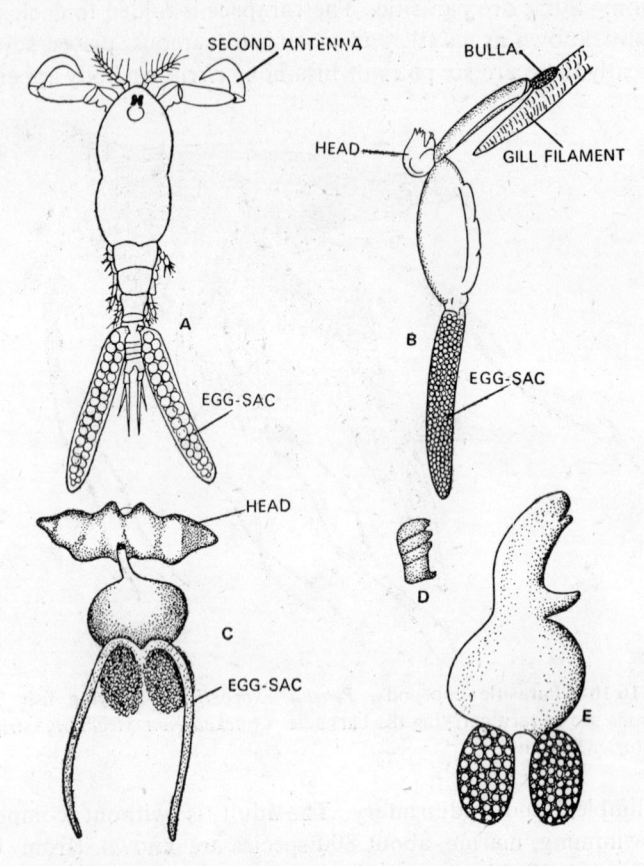

FIG. 16.105. Parasitic copepods. A—*Ergasilus versicolor*, parasitic on gills of freshwater fish, only adult female is parasitic hooking to fish with clasping antennae; B—*Salmincola salmonea* mature female attached to gill of European salmon; C—*Lesteira* head is embedded in the skin of fish, remainder of the body hangs free; D—Male; and E—Female of *Brachiella obesa* lives on gills of red gurnard.

covered by an operculum, formed by the paired movable **terga** and **scuta**. The plates composing the walls (Fig. 16.107) overlap one another and may be simply held together by living tissue by interlocking teeth or actually may be fused to some extent. The figure of the vertical section of *Balanus* (Fig. 16.107) gives an overall picture of the internal anatomy of the animal.

2. **Order Acrothoracica**. Minute cirripedia which bore into the shells of molluscs and corals. The sexes are separate and the throacic limbs are fewer than six pairs.

Examples: *Aclicippe, Trypetesa*.

3. **Order Apoda**. Cirripedia which are hermaphroditic and devoid of mantle, thoracic limbs or anus. The body appears to be like that of a maggot. They are parasitic.

*Phylum Arthropoda* 643

Example: *Proteolepas*, the only representative, that lives in the mantle cavity of the stalked barnacle.

4. **Order Rhizocephala**. Parasitic Cirripedia having no alimentary canal at any time and no appendages or segmentation in the adult. They parasitize decapod crustaceans exclusively.

Example: *Sacculina* is parasitic on crabs and is the best example of the group. It looks like a big tumour on the abdomen of the host, giving off root-like processes which extend through the body of the host (Fig. 16.109).

In *Sacculina carcini* the nauplius resembles that of other cirripedes but lacks both mouth and intestine. It is free-living and does not feed until it moults into cypris larva. There are two kinds of cypris, males and females. The female attaches to the base of a bristle on a crab by means of their antennae, lose locomotory apparatus and accessory muscles, leaving only a mass of cells which

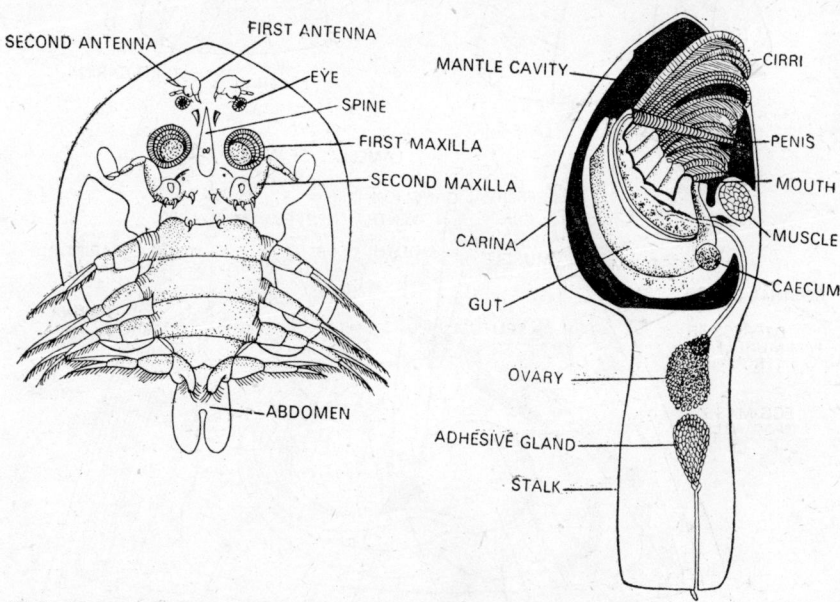

FIG. 16.106. *Argulus foliaceus*, a branchiuran fish parasite (*after* Wagler); and B—*Lepas*, a stalked barnacle (*after* Broch).

forms a hollow stylet. The rest of the cypris then degenerates leaving only what is now called **kentrogon larva** attached to the crab. The stylet pierces into the body wall of the crab as far as body cavity, and the cell mass then passes through this into the body of the host so that the larval parasite is actually injected into the host (crab). This larval mass now proceeds to grow differentiating into two main parts, a complex root system, the **internal sacculina**, which gradually extends throughout the body of the crab even into the limbs and serves to absorb the nourishment. And second part is a rounded outgrowth which appears after the root system forming the tube body of the adult parasite or the **external sacculina**. As it grows it pushes through the ventral body wall of the abdomen of the crab and emerges as a broad sac enclosing the female reproductive organs. The enclosed brood chamber communicates with the exterior by a small opening through which the larval cypris male, which is neoteinic, enter in order to fertilize the female. *Sacculina carcini* on the crab *Carcinus maenas* reproduces throughout the year but has a peak nauplius production between August and December.

5. **Order Ascothoracica.** Parasitic Cirripedia having six pairs of thoracic appendages. Appendages modified for piercing and mouthparts for sucking. They are often embedded in the tissues of the hosts.

Examples: *Petrarca* (parasitizes Hexacorallia) and *Laura* (living in *Gerardia*).

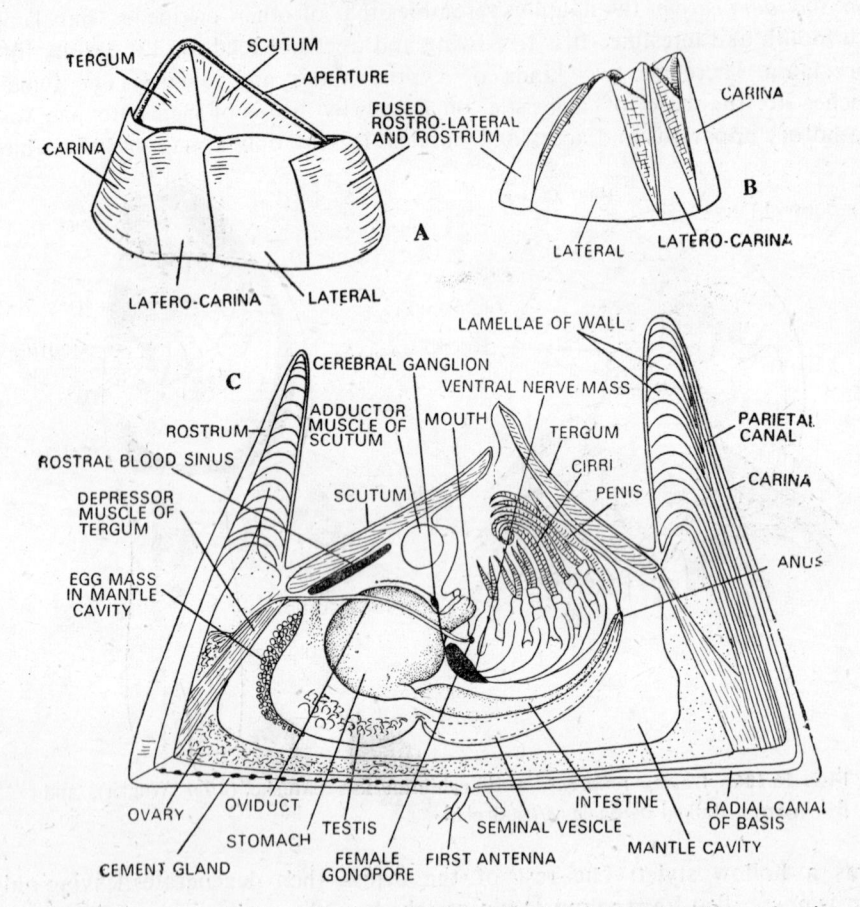

FIG. 16.107. A—*Balanus* sessile barnacle (vertical section); B—*Balanus* showing number and position of shell plates; C—*Balanus*, natural view, tergum and scutum hidden by lateral plates.

VIII. **Subclass Malacostraca.** It comprises the prawns, crabs, lobsters and cray-fishes, etc. The Crustacea in which the eye is typically stalked and a carapace covers the thorax which is formed by eight segments. The abdomen consists of only six segments bearing appendages. Caudal styles are absent and the tail is formed into fan-like structure by the uropods and telson. This subclass contains almost three quarters of all the known species of crustaceans, as well as all the larger forms.

The malacostraçans are of two types, those that have seven abdominal seg-

## Phylum Arthropoda

ments and those that have six. Thus, the group has been divided in two series, the Leptostraca and Eumalacostraca.

**A. Series Leptostraca.** Marine Malacostraca with 21 body segments of which seven form the abdomen. Large carapace encloses much of the trunk. Thoracic appendages usually foliaceous, eye stalked and telson has a pair of caudal styles.

**1. Order Phyllocarida.** Extinct order of Leptostracans reported from the Silurian period. The carapace was notched behind, and was either uni- or bivalved.

Example: *Nahecaris*, from the Lower Devonian (400 million years ago).

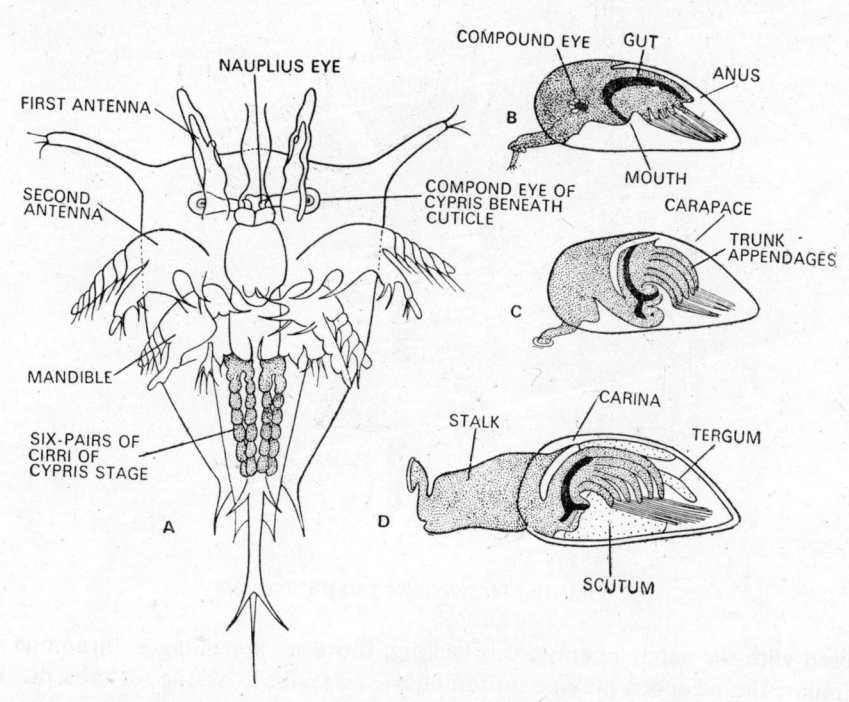

FIG. 16.108. A—*Nauplius* larva of *Balanus*; B to D—Metamorphosis of cypris larva of *Lepas*.

**2. Order Nebaliacea.** Extinct order of Leptostracans in which the carapace is folded, not fused to any thoracic segment and has an adductor muscle. The eggs hatch as post larvae (called **manca stage** by some.)

Example: *Nebalia* (Fig. 16.110) is a typical genus.

**B. Series Eumalacostraca.** Marine, freshwater or terrestrial Malacostraca with twenty body segments of which six form abdomen. Carapace, when present, never bivalved; thoracic appendages leg-like; eyes sessile or stalked, caudal styles absent.

**Superorder Syncarida** (also called Division Syncarida by some). Eumalacostraca without carapace and with the first thoracic segment either fused to the head, marked off from the head groove or free. There is little difference in form between thoracic and abdominal segments. Typically most thoracic appendages have exopodites, abdominal appendages (pleopods) may be uni- or biramous or absent. Commonly the uropods and telson from a tail fan. Eyes stalked or sessile, sometimes absent. Eggs shed into the water where they develop directly without any larval stages.

1. **Order Palaeocaridacea.** Extinct order of the Syncarida with eight free thoracic segments, stalked and biramous pleopods.

Example: *Palaeocaris*.

2. **Order Anaspidacea.** Syncarida in which the first thoracic segment is

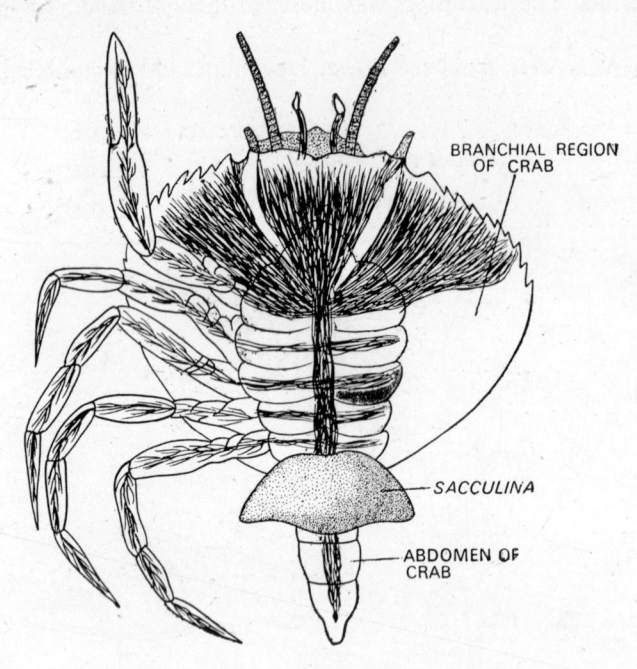

FIG. 16.109. *Sacculina* parasitic on crab.

fused with the head, a carapace is lacking, thoracic appendages biramous and similar, the pleopods may be uniramous; eyes stalked, sessile or absent. They swim and crawl over the bottom or unaquatic vegetation.

Examples: *Anaspides*, the Tasmanian genus that is reported to leap out of water sometime; *Paramaspides*, *Micraspides*, etc.

3. **Order Bathynellacea.** Syncarids with eight free thoracic segments, no eyes and reduced or no pleopods. The members of this order are smallest malacostracans.

Examples: *Bathynella* (blind European cave inhabitant) is a typical example, *Thermobathynella*.

**Superorder Hoplocarida** (or **Division Hoplocarida**). Marine Eumalacostraca called mantis-shrimps including only one order **Stomatopoda**.

**Order Stomatopoda.** Carapace is shallow, rather shield-like, fused with the first two thoracic segments only. Anteriorly the carapace terminates short of the front of the head, so that the eyes and the first antennae are attached beneath a small free rostrum. Third and fourth throacic segments vestigial, but the last four are well-developed and unfused. The surface of the carapace, the free thoracic and abdominal tergites and broad telson are usually ornamental with

*Phylum Arthropoda* 647

keel-like ridges and spines. Eyes large and stalked. First antennae large and triramous and with a fringed sac that is as long as or longer than endopodite. First five thoracic appendages uniramous and subchelate, second pair enormously developed for a raptorial feeding. Last three thoracic appendages slender and not chelate.

Examples: *Squilla* (Fig. 16.111), *Pseudosquilla*.

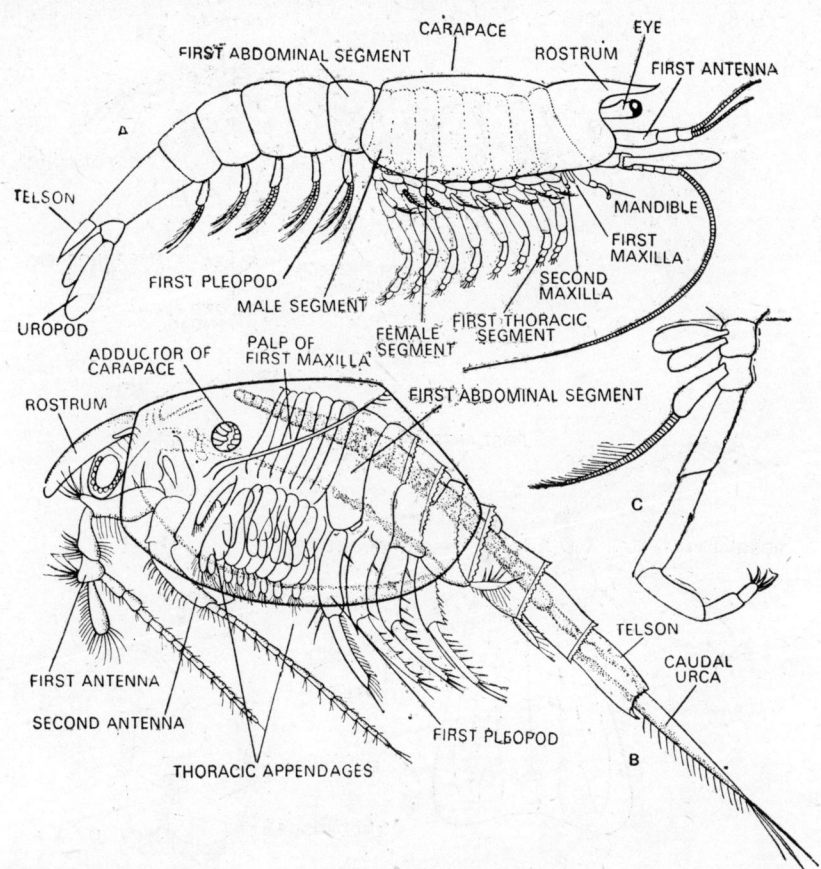

FIG. 16.110. A generalised malacostracan: A—Lateral view of body; B—Female *Nebalia bipes*; and C—Thoracic appendage (*after* Calman).

**Superorder (=Division) Peracarida**. Eumalacostraca in which a carapace is present or absent; first thoracic segment always fused with head, last four thoracic segments free, even when a carapace is present. Nauplius eye never persists in the adult. Eyes stalked or sessile. A ventral brood-pouch or marsupium present in the female. Marsupium formed by large plate-like processes (oostegites) on certain thoracic coxae, thoracic sternites form the roof. Development direct, eggs being incubated in the brood-pouch.

1. **Order Thermosbaenacea**. Small peracaridans with a small carapace fused to the first thoracic segment. Females have no oostegites but they develop brood-pouch dorsally from the enlarged carapace during breeding. Antennule is biramous. Live in fresh or brackish water.

Examples: *Thermosbaena, Monodella* (only two genera known).

2. **Order Spelaeogriphacea.** Blind peracaridans with a short carapaces fused with the first thoracic segment. Females possessing oostegites on the first five pairs of thoracic legs have exopodites except for the last pair, the first three pairs are natatory and the last three respiratory.

Example: *Spelaegriphus lepidops* (only one species).

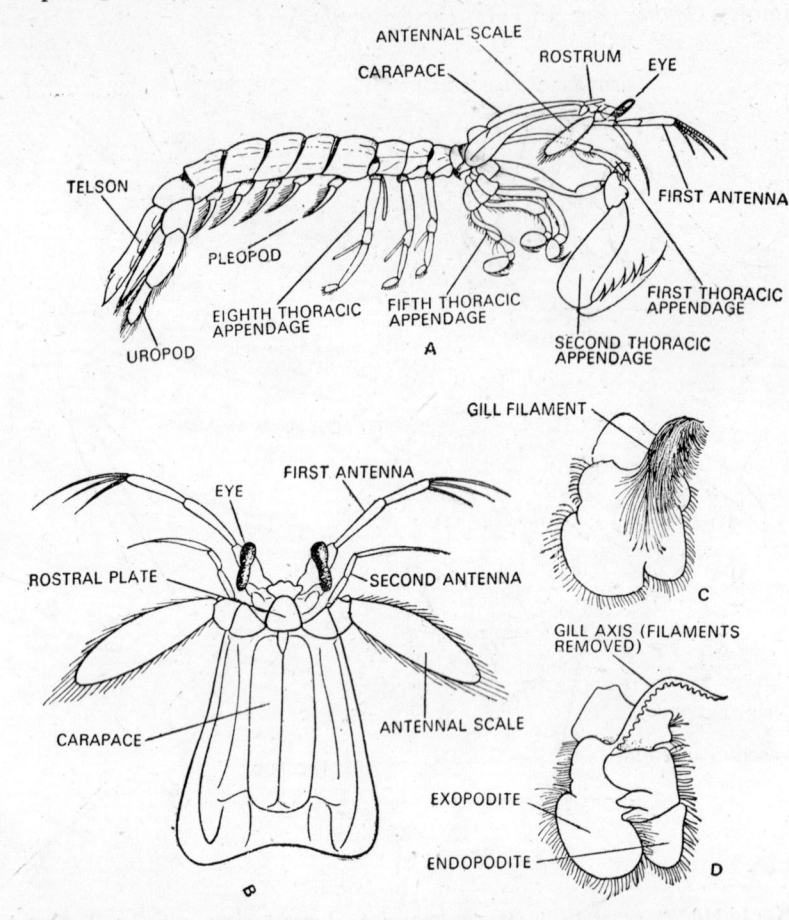

FIG. 16.111. Stomatopoda. **A**—*Squilla mantis*; **B**—Head and carapace of *Squilla*, dorsal view; **C**—Second pleopod and gill of *Squilla*; and **D**—Second pleopod and gill filaments removed (*after* Calman).

3. **Order Mysidacea.** Mainly pelagic, marine peracaridans in which a carapace is present and covers the thorax. The eyes are stalked when present. The exopodite of the antenna is scale-like and a tail-fan is well formed.

Example: *Mysis*.

4. **Order Cumacea.** The peracaridans in which the carapace is small and covers only three or four thoracic segments and forms laterally a branchial chamber and anteriorly a "rostrum." Other segments (four or five) are absent. The eyes sessile, when present. Abdomen slender and the uropods do not form a tail-fan. Breeding females have ventral brood-pouch.

Example: *Diastylis* (16.112).

*Phylum Arthropoda* 649

5. **Order Tanaidacea.** The peracarida in which the carapace is very small and covers only two segments of the thorax with which it fuses. Laterally the

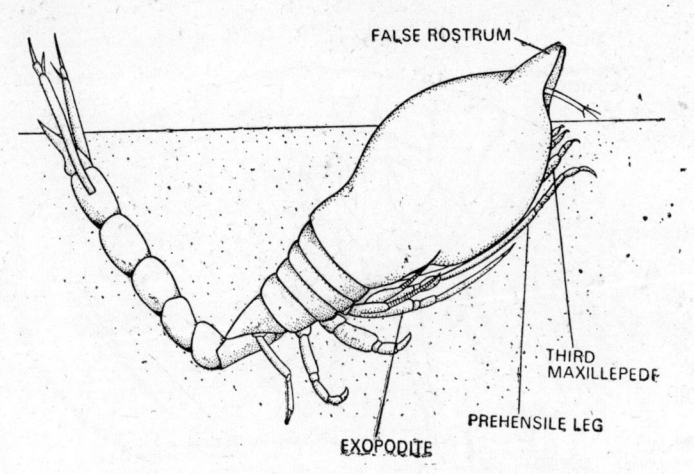

FIG. 16.112. *Diastylis*, a cumacean buried in sand.

carapace forms a branchial chamber. Breeding females have a ventral brood pouch. Antennules are, sometimes, biramous. Eye stalked, stalk short and immovable. Small marine forms that live in burrows or tubes.

Examples: *Apseudes, Tanais*.

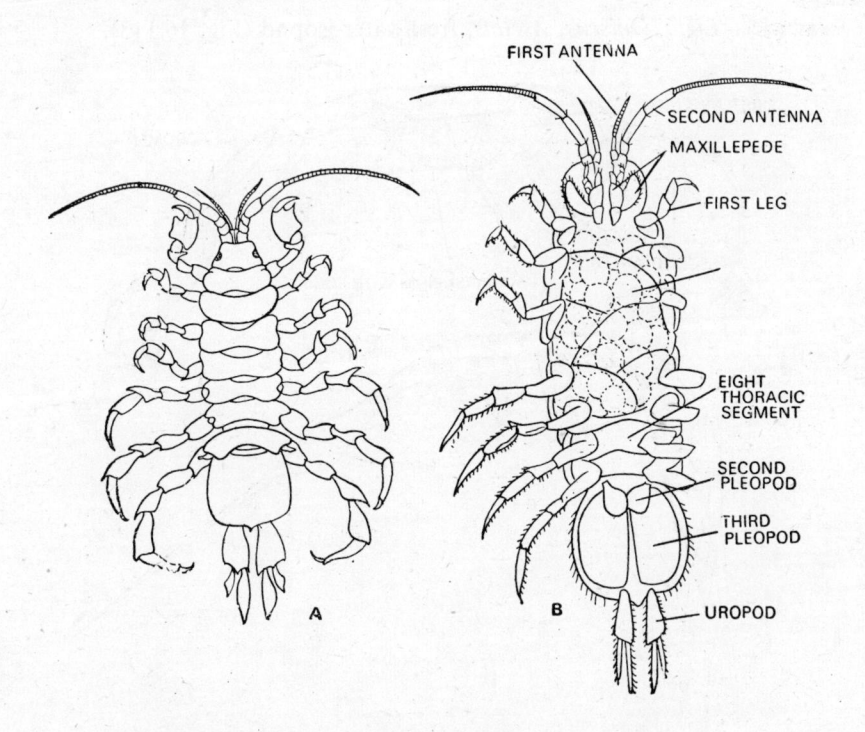

FIG. 16.113. Freshwater isopod. A—*Asellus*, dorsal view; and B—The same ventral view.

**6. Order Isopoda.** The Peracarida without a carapace and with sessile eyes. The body is usually dorsoventrally compressed. The thoracic appendages are

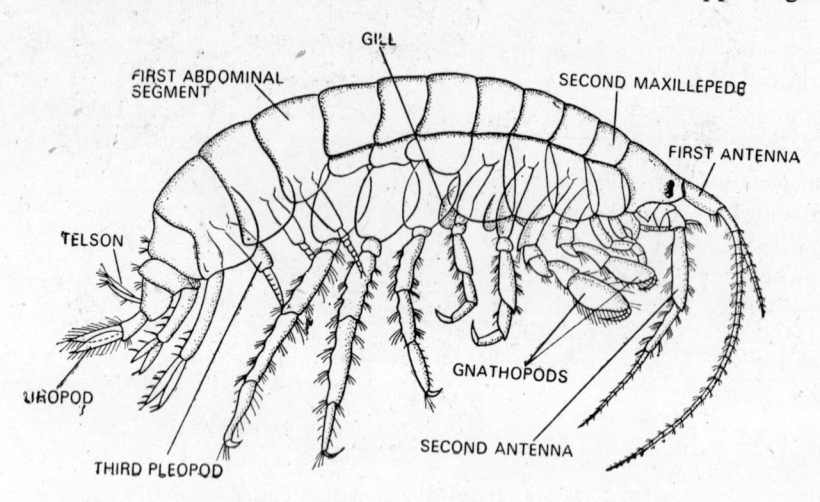

FIG. 16.114. Male of amphipod *Gammarus*.

devoid of exopodite (no gills). Second and third pairs of thoracic appendages not prehensile, abdominal appendages usually flattened, modified for respiration, antennules nearly always uniramous. They form a large group exhibiting much variety.

Examples: *Ligia, Oniscus, Asellus,* freshwater isopod (Fig. 16.113).

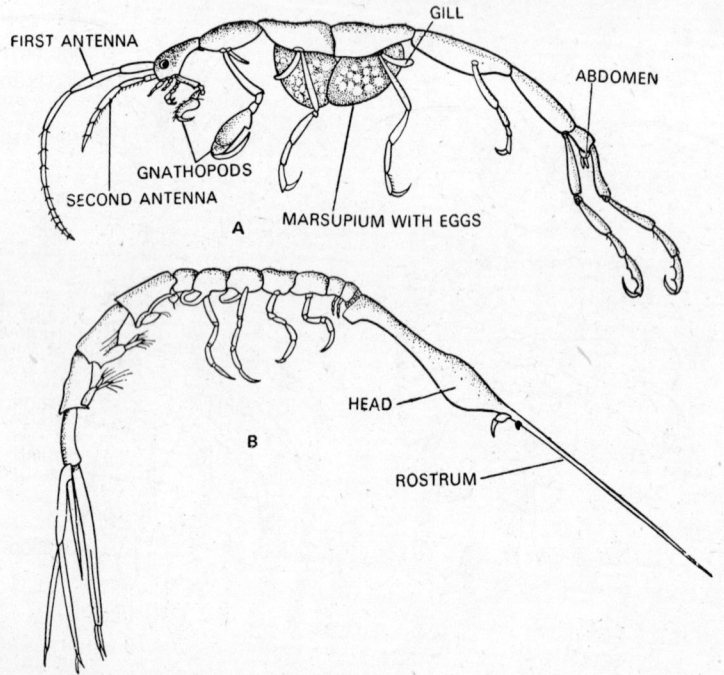

FIG. 16.115. A—*Phthisica*, a caprellid amphipod; and B—*Rhabdosoma*, an amphipod with elongated head.

*Phylum Arthropoda* 651

7. **Order Amphipoda**. Freshwater and marine Peracarida in which the carapace is absent and the body is laterally compressed and abdomen is elongated. Breeding females have a ventral brood-pouch formed from oostegites. Thoracic appendages have no exopodites, but basally some have lamellar gills. The first pair of thoracic appendages is modified as maxillipedes, the second is prehensile. Abdominal appendages, when present, consist of three anterior pairs (pleopods) and three posterior pairs (uropods). Eyes sessile or on short immovable stalks. Marine, freshwater and terrestrial.

Examples: *Gammarus* (Fig. 16.114), *Phthisica* (Fig. 16.115), *Rhabdosoma* (Fig. 16.115).

**Superorder** (=**Division**) **Eucarida**. Eumalacostraca in which the carapace fuses with all thoracic segments to form a cephalothorax. No oostegites or

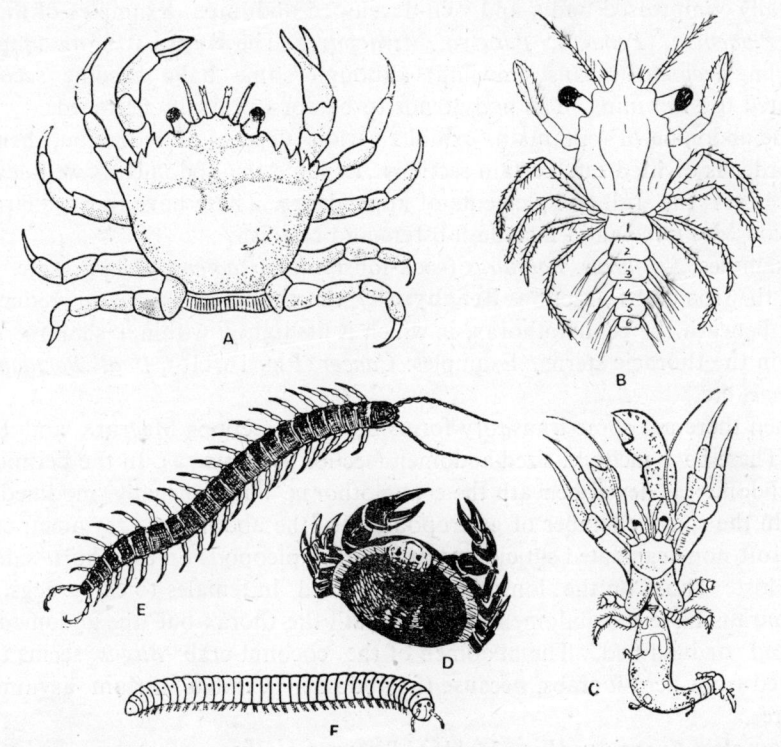

FIG. 16.116. A—Crab *Carcinus*; B—Megalopa of the crab *Carcinus*; C—The hermit crab *Eupagurus* out of shell, dorsal view; D—Ordinary crab found in freshwater; E—A centipede; and F—A millipede (*Julus*).

brood-pouch. Eyes on movable stalks. Development indirect producing a shrimp-like form. This superorder is composed of two orders, the Euphausiacea and the Decapoda.

1. **Order Euphausiacea**. The Eucarida in which the thorax is enclosed by the carapace and none of the anterior thoracic appendages is modified into maxillipedes. There is a single series of gills borne by the exopodites of the thoracic appendages. The larva is nauplius. No statocyst. Order is small.

Examples: *Euphausia*, *Nyctiphanes*.

2. **Order Decapoda**. The Eucarida in which the carapace is well-developed

and covers the thorax completely. The exopodite (scaphognathite) of the maxilla is large. Of the thoracic appendages three pairs are modified as maxillipedes and five pairs as walking legs. Usually more than one series of gills are present. Proximal joint of the antennule generally carries a statocyst. This order contains the most highly organized crustaceans such as the prawns, lobsters, crabs, hermit crabs, coconut-crabs, etc. It is the largest order of crustaceans having 8,500 described decapods (about one third of the known species of crustaceans). Most of the decapods are marine, but the crayfish, some shrimps and a number of crabs have invaded freshwater. Crabs are also amphibious and terrestrial.

Decapods are grouped into two suborders, the **Natantia**, including the shrimps, and the **Reptantia** containing the lobsters, crayfish and crabs. The **Natantia** are generally adapted for swimming (*Natant* means swimming) with laterally compressed body, and well-developed abdomen. Examples of this order are *Palaemon, Penaeus, Paratya, Atyaephyra*. The **Reptantia** are adapted for crawling (*reptant* means crawling) although some have become secondarily adapted to swimming. The body tends to be dorsoventrally flattened.

The abdomen of reptantians exhibits various degrees of reduction, hence the suborder is divided into certain sections. In some the abdomen is well-developed and bears the full complements of appendages. These have thus a "large tail" (section **Macrura**) and include lobsters and crayfish.

Examples: *Scyllarus, Palinurus* (rock-lobsters), *Astaciens, Homarus* etc.

In the true crabs (section **Brachyura**) the abdomen is greatly reduced and fixed beneath the cephalothorax, in which it fits tightly within a shallow depression in the thoracic sterna. Examples: *Cancer* (Fig. 16.116), *Uca, Portunus, Parthenope*, etc.

Then there are some transitory forms between sections Macrura and Brachyura. These have medium-sized abdomen (section **Anomura**). In the hermit crabs, the abdomen is flexed beneath the cephalothorax, but is usually modified to fit within the spiral chamber of gastropod shells, the abdomen is asymmetrical with thin soft non-segmented cuticle and at least the pleopods on the short side have been lost. Those on the long side are retained in females to carry eggs. Other anomurans have the abdomen flexed beneath the thorax but the abdomen is not reduced or modified. The abdomen of the coconut-crab *Birgus* seems to have evolved from hermit crabs, because the abdomen exhibits certain asymmetrical features.

Examples: *Eupagurus* (Fig. 16.116), *Birgus, Galathea*.

## CLASS DIPLOPODA—MILLIPEDES

The **Diplopoda** or **millipedes** (thousand-legged) are tracheate Arthropoda with a head bearing two pairs of jaws (mandibles and maxillae), a thorax consisting of four segments and a cylindrical body consisting of numerous, for the most part, double segments (Fig. 16.116F). The body wall includes the deposit of lime salts. The length varies from 2 mm (*Polyxenus*) to 200 mm (*Rhinocricus, Spirostreptus*). The head has six segments with the same appendages as in the centipedes except that the first maxillae, which appear in the embryo, do not persist in the adult stage. The maxillae are applied together to form a **gnathochilarium** (Fig. 16.117). The antennae are short and clubshaped and consist usually of

# Phylum Arthropoda

seven segments of which the last is concealed in the penultimate piece. The eyes superficially resemble compound eyes, but each is only an aggregation of many simple eyes set closely together. The thorax is short, of four single somites each with one pair of legs. In *Polydesmus* only three segments of the thorax have one pair of legs each and one being without leg. The long abdomen consists of from 20 to over 100 double somites, each containing two pairs of spiracles, ostia and

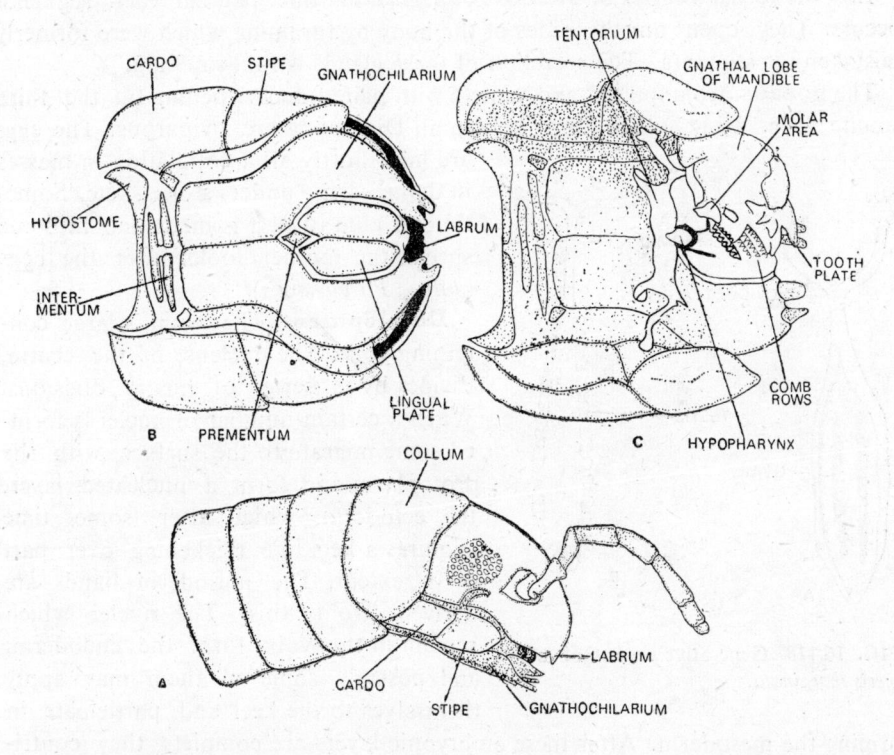

FIG. 16.117. Head and mouthparts of *Julus*. A—Lateral view; B—Ventral view; and C—View from the dorsal side after removing the roof of the head capsule and clearing away the soft parts.

nerve ganglia and bearing two pairs of seven-jointed legs. The parts of the legs are exactly the same as in centipedes. All the legs are apparently similar in size and shape. The genital openings are paired and placed, both in the male and female, between 2nd and 3rd pairs of legs. In the male on either side of the openings are copulatory structures or penes.

## Digestive Organs

The alimentary tract consists of a narrow oesophagus (stomodaeal) into which open two salivary glands: a mid-gut, a wide tube with short liver diverticula; and a hind-gut which receives two or four Malpighian tubules at its anterior end. Most millipedes are herbivorous.

**Nervous system** is like that of the centipedes (Fig. 16.119). The organs of special senses are represented by the simple aggregated eyes by the olfactory hairs on the antenna and the organ of Tomosvary as in the centipedes.

The **vascular system** consists of a dorsal tubular heart surrounded by the perforated pericardium. The tracheae are unbranched in most diplopods but in the Oniscomorpha they are branched. There are two kinds of tracheae—some are larger with spiral fibres and others smaller without spiral fibres. There are two pairs of stigmata in each double segment. The excretory organs are represented by two or four malpighian tubes which open in the gut.

In some forms a series of **odoriferous glands**, one pair in each segment, occurs. They open on the sides of the body by foramina which were formerly mistaken for stigmata. The secretion of these glands is foul-smelling.

The **gonads** are unpaired and ventral with paired ducts opening on the third somite (Fig. 16.117). So far as known all Diplopoda are oviparous. The eggs are laid shortly after copulation in masses in damp earth, under stones, etc. Sometimes a kind of nest is made and in some species the mother looks after the eggs (*Julus, Polydesmus*).

**Development.** The ovum is large containing yolk. The nucleus, in the centre, divides by a series of binary divisions. When a certain number of nuclei is formed some migrate to the surface with the protoplasm and form a nucleated layer, the ectoderm, which after some time acquires a keel-like thickening over part of its extent. The mesoderm bands are derived from this. The nuclei which remain in the yolk form the endoderm; and possibly some of them may apply themselves to the keel and participate in forming the mesoderm. After these embryonic layers are complete they contribute to the formation of various parts of the body. An embryonic cuticular envelope is formed by the ectoderm at an early stage and a second one later on. In some cases (*Julus, Polydesmus*) the embryo remains in this envelope for a short time after hatching and a kind of resting pupal stage is passed through superficially, the embryo at this stage resembles an apterous insect with three pairs of legs attached to the thorax.

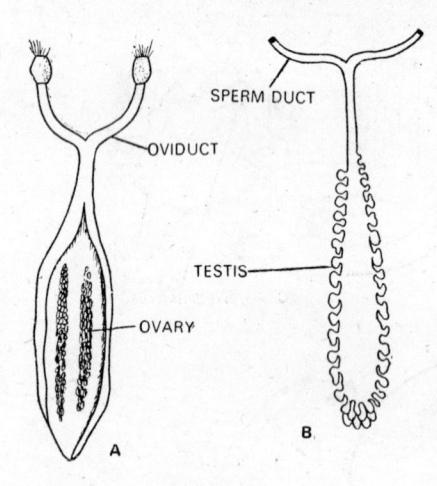

FIG. 16.118. Generative organs of *Glomeris marginata*.

The class is divided into two orders: (*i*) **Pselaphognatha** which includes small diplopods with soft chitin without calcareous deposits, with 10 to 12 body segments carrying 13 pairs of legs. *Polyzenus* is a small member of this order that destroys the vine-louse (*Phylloxera*). (*ii*) **Chilongnatha** is the other order which includes diplopods with chitin hardened by the deposits of calcareous salts, the body segments 11-13 (in the tribe **Oniscomorpha**) or 10 to 108 (in **Helminthomorpha**) bear two pairs of legs each, and the thorax bears 3 or 4 pairs of legs. *Julus, Glomeris, Polydesmus, Cryptodesmus, Brachydesmus, Polyzonium,* etc. are some of the examples.

*Phylum Arthropoda*

## CLASS CHILOPODA—CENTIPEDES

The **Chilopoda** or **centipedes** are terrestrial tracheate Arthropoda in which the head is clearly demarcated, the mandible is without palp, the genital opening is posterior and the body segments, bearing a pair of legs each, are all alike; the appendages of the first body segment bear poison claws and the last segment is legless.

The Chilopoda have an elongate dorsoventrally flattened body and the legs are borne ventrally. The head is distinct bearing four pairs of appendages, a pair of long antennae with 12 or more joints, a pair of mandibles and two pairs of maxillae. The head is covered by a cephalic plate, the **cephalite**. The body

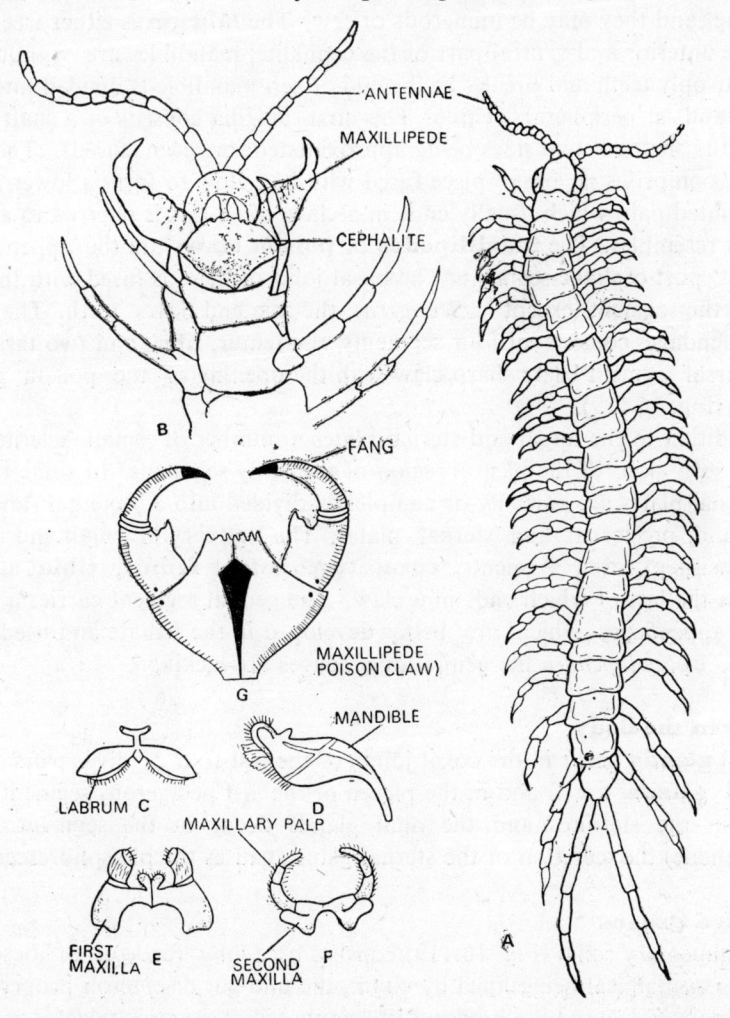

FIG. 16.119. A—A centipede, entire; B—Its head enlarged; mouthparts; C—Labrum; D—Mandible; E—First maxilla; F—Second maxilla; and G—Poison claw.

segments vary from 15 to 137 somites in different species. The first somite bears a pair of four-jointed poison claws and on each of the other somites, except the last two there is a pair of small 7-jointed walking legs. All the segments are alike. The last apodal segment bears the anus and the last but one bears the median generative opening. Some lay eggs and others are viviparous. The young resemble the adults in form. The centipedes are inhabitants of warm countries, hiding by day under stones or logs and running swiftly about at night to prey on earthworms and insects. The tropical species are 15-20 cm long. The house centipede, *Scutigera*, has fifteen pairs of extremely long fragile legs. It is very agile though small and eats insects that are harmless to man.

### Mouthparts

The **head** (Fig. 16.119) bears ocelli placed on the sides of the head behind the antennae and they may be numerous or few. The **labrum** is either free or fused with the anterior and ventral part of the cephalite, mandibles are without palps, and bear only teeth and bristles at the end. Each mandible is divided into a basal **cardo** and a peripheral **stipes**. The first maxilla consists of a shaft and two blades, the shafts of two sides being approximated (or even fused). The second maxilla comprises a basal piece fused with its fellow to form a lower lip and a three-jointed palp which usually ends in a claw, often this is referred to as labium which it resembles. The **maxillipedes** or **poison claws** are the appendages of the first post-cephalic segment. The basal joint of these is fused with the sternal plate of the segment (except in *Scutigera*); the free end bears teeth. The rest of the appendage consists of four segments, the femur, tibia, and two tarsalia, the distal tarsal segment bears sharp claw with the opening of the poison gland at its apex (Fig. 16.119).

In addition to the tergal and sternal plates a number of small sclerites occur in the soft skin of the pleural region of the body segments. In some the tergal and sternal plates are partially or completely divided into a pretergal and tergal plates and presternal and sternal plates. The legs vary in length and normally each consists of seven segments, **coxa**, **trochanter**, **femur**, **tibia** and three **tarsalia** the last of which ends in a claw. The genital segment carries a pair of genital appendages which are better developed in the female and used for oviposition. In *Scolopendra* the genital appendages are lacking.

### Glands in the Body

**Coxal glands** occur in the coxal joints of the last four or five pairs of legs; **pleural glands** are found in the pleura of the last pedigerous segment, sternal glands in the sternites and the anal glands occur in the segment. In some (Geophilidae) the secretion of the sternal glands causes the phosphorescence.

### Digestive Organs

The alimentary canal (Fig. 16.119) consists of a long fore-gut or oesophagus which is stomodaeal being lined by chitin, the mid-gut or enteron proper is without chitinous lining and the hind-gut or rectum which is proctodacal in nature, is lined with chitin and receives two long Malpighian tubes at its anterior end. The recrum shows S-shaped curvature in *Scolopendra* and is straight in *Lithobius*. The salivary glands open in connection with the mouth. In *Scolopendra* the two anterior pairs are located in the head and open into the buccal cavity. There are two larger posterior pairs extending up to the seven body segments in some cases and opening into the mouth in front of the second maxillae. The food of these consists of earthworms, slugs and insects. The prey is killed quickly by poison and is chewed with the help of the mandibles.

**Nervous System**

It consists of a bilobed cerebral ganglion in the head connected to the ventral ganglionated nerve chain by the circumoesophageal connectives through a suboesophageal ganglion. The sense organs include sensory hair and spines of

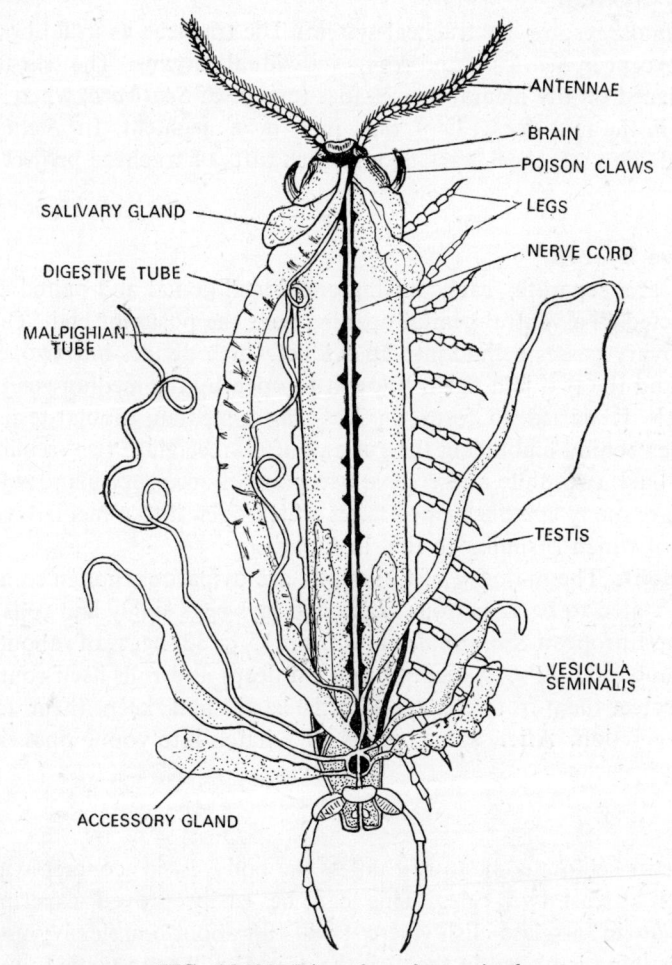

FIG. 16.120. Dissection of a centipede.

various kinds found on the body generally especially on the antennae and jaws. In *Scutigera* maxillary pits constitute additional sense organs. Each consists of a pit, beset with setae and supplied by a special nerve, situated on the inner side of the basal portion of the first maxilla. Near the base of the antenna there occurs an organ of Tomosvary which consists of a transparent projection of chitin covered with fine sensory hair. This is believed to be auditory in function. The eyes are simple ocelli. In *Scutigera* there are compound eyes, resembling those of the insects and crustaceans.

## Circulatory System

The tubular heart extends along the whole length of the body, surrounded by the pericardium, with a pair of ostia and of lateral arteries in each somite, and in front it is continued as an artery to the cephalic organs. The blood is colourless.

The fat body occupies the pericardial as well as the perivisceral cavities. It consists of aggregations of fat cells.

## Respiratory System

These animals respire by tracheal system. The tracheae as usual branch and anastomose except in *Scutigera* and reach individual organs. The slit-like stigmata are placed on the pleural membranes (except in *Scutigera*, where they are seven placed in the mid-dorsal line), one pair in a segment. In *Scutigera* the stigmata lead into tracheal sacs from which tufts of tracheae project into the pericardium.

## Reproductive System

The sexes are separate, each having one dorsal gonad and paired accessory glands connected to a ventral genital opening near the posterior end. The single and median ovary passes behind into an oviduct which divides into two, embraces the rectum and the two join again below it to open by the median genital opening. The testis is varied in form. In *Lithobius* a median, tubular testis occurs. Its duct divides behind embracing the gut and after receiving two seminal vesicles, further back two unite and receive two pairs of accessory glands. In *Scolopendra* there are many spindle-shaped testes which open into a median vas deferens the fate of which is similar to that in *Lithobius*.

**Development.** The majority of Chilopoda are oviparous though some scolopendras are stated to be viviparous. *Lithobius* lays eggs singly and rolls them in the earth. The European *Scolopendra* lays from 15 to 33 eggs of about 3 mm length, in June and July, in earth (3 to 8 cm deep) and rolls itself round them. Thus, they protect them from contact with the earth and keep them moist by their body secretion. After some weeks they hatch into young ones similar to the adults in appearance.

## Behaviour

The centipedes seldom come to rest unless the body is in contact with some solid object on at least two sides. This can be easily proved experimentally. Place *Lithobius* in a glass dish where it will run about ceaselessly; but if some narrow glass tubing is placed in the dish the animal will soon come to rest in the tubing, which affords a maximum of contact with the surfaces of the animal. This proves that the centipedes are positive in reaction to contact. The very fact that they prefer to live under cover and also hunt in darkness, probably guided by sense of touch in their movements, indicates that they avoid light, *i.e.*, they are negative to light.

The class is divided into two orders: (*i*) **Pleurostigma** with tracheal openings on the pleural areas, the number of sterna never exceeds that of the terga. These include more or less cosmopolitan genera *Geophilus*; *Mecistocephalus, Himantarium, Cryptops* and *Scolopendra*, etc. (2) **Notostigma** with tracheal openings

Phylum Arthropoda 659

in the dorsal middle line and with compound eyes. The number of sternites exceeds that of tergites. *Scutigera* is its example.

## CLASS SYMPHYLA

The **Symphyla** are small arthropods with 12 pairs of legs and a head possessing antennae, mandibles, maxillulae, maxillae and labium; the gonads open on the fourth postcephalic segment and there is a single pair of spiracles which is situated on the head. The name Symphyla signifies that the animals combine characters of insects and myriapods (Ryder). They are myriapodan in their number of legs and body segments, but their jaws and legs resemble those of the insects. The Symphyla, represented by *Scolopendrella* and *Scutigerella*, are small animals less than 3 mm in length. The head is distinct and the dorsoventrally

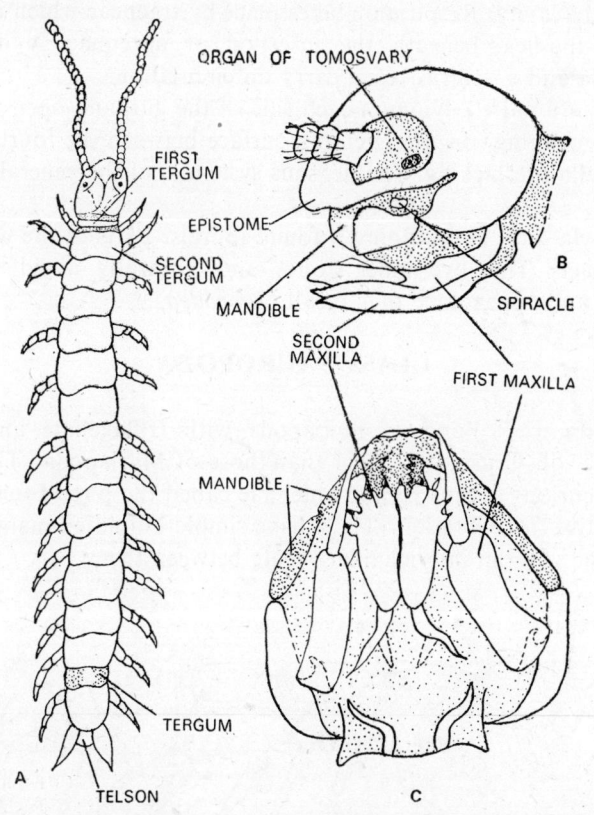

FIG. 16.121. Symphyla. A—*Scutigerella immaculata*, dorsal view; B—Head of *Hanseniella*, lateral view; and C—Head of *Scutigerella*, ventral view.

flattened body consists of 12 leg-bearing and 13 legless segments. The head is similar to that of the insects in constitution. It bears a pair of long many-jointed antennae and three pairs of jaws; a pair of 2-jointed mandibles; a pair of maxillae bearing short palps, and a labium consisting of a median basal part carrying two pairs of small lobes. Hypopharynx is also present and it articulates with a maxillula on each side (similar structures appear in Collembola and

Thysanura). An upper lip is also present. A pair of ocelli represents the eyes. The legs of the first pair are placed on the segment following the head and are 4-jointed and the remaining 11 pairs are 5-jointed, they are widely spaced. Close to the coxae of the legs of the last ten pairs and on their inner sides there arise movable processes called **parapodia** and just on the inner side of each of these a protrusible sac (Fig. 16.121). The parapodia probably serve as sense organs. Some authors (Haase) suggest that the ventral sacs are homologous with the coxal glands and function as respiratory organs. The last body segment bears conical cerci at the end of which open ducts of spinning glands, the threads secreted by which are used for attaching the egg.

The alimentary canal consists of pharynx, oesophagus, stomach, intestine and rectum. Two or three Malpighian tubes open into the intestine. Into the mouth opens a pair of salivary glands. Well-developed fat body occupies the great part of the body cavity. Respiration takes place by tracheae which open by a pair of stigmata on the head beneath the insertion of antennae. A dorsal tubular heart with ostia and a ventral vessel carry on circulation.

The gonads are paired, lying on each side of the gut, and open to the exterior by a median opening on the ventral surface between the fourth pair of legs. External genitalia are lacking. The nervous system is of the general arthropodan type.

The Symphyla are active animals found in most parts of the world and prey upon small insects. They live under stones and naturally avoid light. In their general habits and appearance they recall *Campodea*.

## CLASS PAUROPODA

The **Pauropoda** are a kind of myriapods with triflagellate antennae which resemble those of Crustacea rather than those of Myriapoda. They possess no eyes but have curious organs behind antennae called the **pseudoculi**. These are possibly sensitive to vibration. They have simple unjointed mandibles and one pair of maxillae with an intermaxillary plate between them.

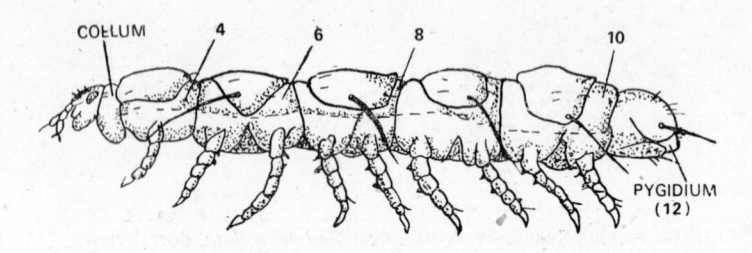

FIG. 16.122. The pauropod *Pauropus silvaticus*, lateral view (*after* Tiegs).

The body is rather narrower in front. The first trunk segment is the **collum**. This is followed by a further ten segments and a telson. Nine of those post-collum segments bear a pair of legs while the tenth is legless. The posterior legs are the longest. In *Decapauropus* there is an additional segment and this is podous. In *Brachypauropus* tergites cover each podous segment and the terminal apodous segment, but in *Pauropus* and *Eurypauropus* the first ten segments share five tergites. This is due to the loss of alternate tergites and perhaps foreshadows the condition in diplopods.

*Pauropus* was discovered by Sir John Lubbock who described it in 1866 as a small, white, bustling, intelligent little creature about one-twenty-fifth of an inch (0.5 to 2 mm) in length. It is without elaborate circulatory and respiratory organs (has no heart or tracheal system). The ovary is dorsal and the testis ventral. Pauropods are found under the bark of dead wood and in the soil. They were first found in the kitchen garden. *Pauropus* at first looks most like a chilopod, but differs from it in the form of antennae, in the absence of poison claws and in the form of mouthparts.

# 17 CLASS INSECTA

The study of insects is one of the most highly fascinating subjects among the biological sciences. Insects are the most abundant form of animal life on earth. About 80 per cent of the world's known animals are insects. They are now found in every conceivable location and inhabit every continent including Antarctica where many collembola or spring-tails were found at 1825 metres above sea-level on Mount Gran. Nearly a million distinct species have been described and catalogued, and more than 6,000 new species are being added each year. Some scientists believe that the total number of insect species may eventually reach several million.

The insects are an ancient tribe. They were already old when the dinosaurs roamed the earth. Thus, they were living on the earth several million years before man. The insects were flying 50 million years before the reptiles and birds took to this mode. The body of insect is covered by exoskeleton made of chitin which is resistant to acids and alkalies, but is slowly dissolved by water. Insects have a capacity to produce a large number of young ones. Their muscle power is enormous and respiratory powers are extraordinary.

The insects are considered helpful, harmful or neutral to man. People pay more attention to their harmful nature and forget their beneficial role. Insects may be helpful to man by producing, directly or indirectly, materials of economic value, such as silk, honey, beeswax, shellac, cochineal dye, cantharidin, and galls used in making tannic acid, permanent inks and dyes, by aiding in the production of fruits, vegetables, flowers and seeds because of pollinating activity, by serving as food for fish, birds and other wildlife, by destroying injurious insects either as predators or as parasites, by serving as research specimens in the study of toxicology, physiology, genetics and related fields, by acting as scavengers, attacking and destroying dead plants and animals, by destroying noxious plants, by their medicinal values, particularly honey-bee venom for arthritis and by serving as objects in art, ornamentation, and in other aesthetic ways.

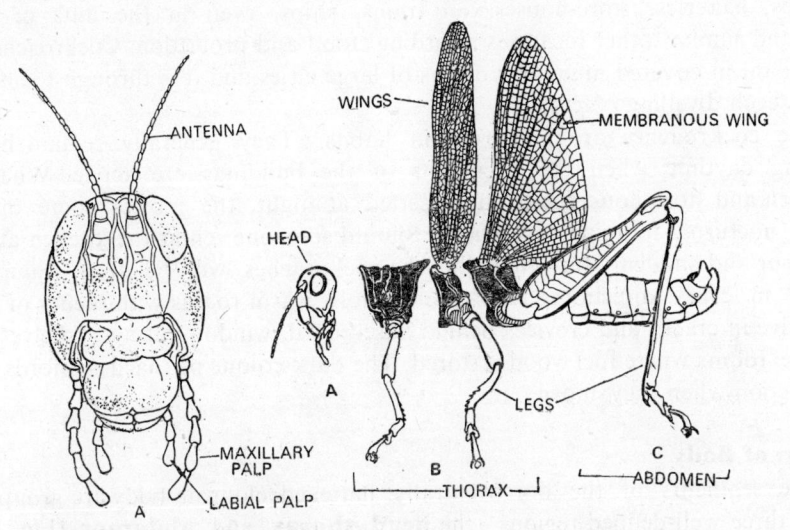

FIG. 17.1. Body of insect is incised into 3 parts. A—Head (front view); B—Thorax black with wings and legs; and C—Abdomen.

Insects may be equally harmful to man and cause great economic loss by damaging or destroying agricultural crops and other valuable plants; by aiding in the spread and growth of bacteria, fungi protozoa, helminths, and viruses that produce diseases which sometimes prove fatal in man, domestic animals and plants, by annoying man or animals or lowering the value of stored foods, other products and possessions. Only one per cent of all insect species is considered harmful to man but this harmful group alone causes losses averaging from 5 to 15 per cent of the annual agricultural production.

It is of advantage to begin the study of this fascinating group by understanding the morphology, anatomy and biology of some large insects as the grasshoppers or cockroaches. In the following pages a preliminary account of the study of cockroach is given.

## TYPE PERIPLANETA

The **cockroaches** (Fig. 17.2) are familiar animals common in houses, where sometimes they do considerable damage. Originally native of tropical Asia, cockroaches are now found almost all over the world. The oriental cockroach or **black beetle**, *Blatta orientalis*, is the oldest and the most important domesticated species. With this are found three other species, *Periplaneta americana*, other cosmopolitan species, the Australian cockroach, known as *Periplaneta australasiae*, and the brown cockroach as *Periplaneta brunnea*. The most common species in India are *Blatta orientalis*, in which the wings and wing covers of the females are rudimentary and in the male do not reach up to the end of the body, and *Periplaneta americana*, in which both sexes are winged and the wings and wing covers are longer than the body.

### Habitat and Habits

The cockroaches are found in houses, hotels, slaughter-houses, garbage dumps, bakeries, store-houses, on trains, ships, even in the huts of savages, and similar other locations affording food and protection. Cockroaches are abundant in covered municipal drains of large cities and it is through them that they reach dwelling places.

The cockroaches are nocturnal in habits. They generally remain hidden during daytime when the occupants of the buildings are active. When the kitchen and store-houses, etc., are deserted at night, the insects come out on their nocturnal foraging expeditions. Should someone enter the kitchen at such an hour and suddenly turn the light, the cockroaches will be seen scampering away in every direction. They prefer moist warm rooms with plenty of food, and live in cracks and crevices behind base-board, window-casings, shelves and in the rooms where fuel wood is stored. The dark colour of roaches affords them protection when they move out.

### Form of Body

The segments of the dorsoventrally flattened elongate body are grouped to form three well-defined regions—the **head**, **thorax** and **abdomen** (Fig. 17.2). The head is connected to the thorax by a slender **neck** or **cervicum**. The head represents six segments though all external indications of segmentation have

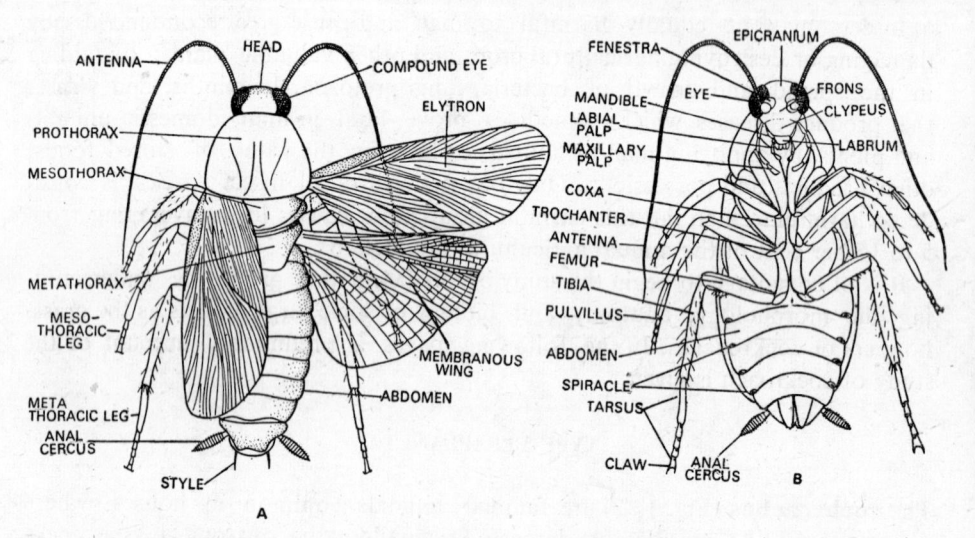

FIG. 17.2. External features of cockroach. A—Dorsal view; and B—Ventral view.

been lost, the thorax, three, the pro-, meso- and metathorax and the abdomen eleven, although only ten are recognizable and all of these are not visible externally.

**Skeleton**. The whole of the skeleton is covered with chitinous cuticle which is not hardened or reinforced with calcareous deposits as in the crustaceans. The chitinization is not continuous but localized forming sclerotized regions or

**sclerites.** The dorsal sclerite is known as the **tergum**, the ventral the **sternum** and the lateral ones are termed as the **pleura**. In the region of the thorax each tergal sclerite is known as **notum**. Thinner flexible **articular membranes** separate individual sclerites of each segment and lie between the borders of contiguous segments. The segments may show complete or partial fusion in some cases. In the head, for instance, several segments have fused, and the outlines of the sclerites are merely indicated by sutures. The complex **head capsule** thus formed consists of the **frons, clypeus,** and **labrum** in front, the vertex, composed of two **epicranial plates** above, and the **genae** at the sides (Fig. 17.3).

It is evident from the above that the whole body is encased in hard cuticular covering which is known to form the **exoskeleton.** As was seen in the prawn the exoskeleton is carried inwards in certain places to form apodemes (hardened process) for the attachment of muscles. This arrangement of apodemes constitutes the **endoskeleton.** In the cockroach the endoskeleton is not so well-developed as in prawn. In the head capsule the endoskeleton is known as the **tentorium.** It consists of **anterior** and **posterior arms** whose inner ends amalgamate to form the **body** of the **tentorium.** In *Blatta* additional slender dorsal arms may be present. The tentorium strengthens the head capsule, gives attachment to muscles and supports the brain and oesophagus. In the thorax and abdomen there are **dorsal apodemes, lateral apodemes** and **sternal (ventral) apodemes.**

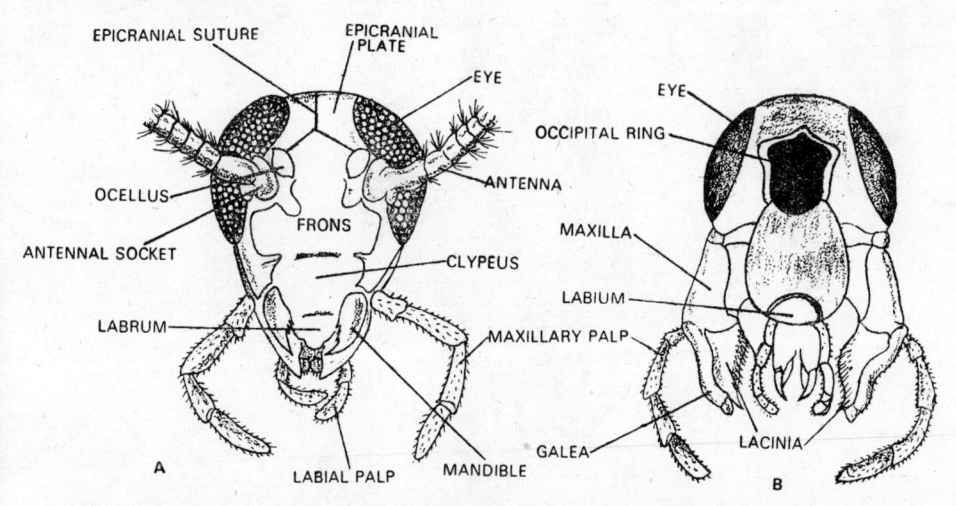

FIG. 17.3. Structure of the head of cockroach. A—Front view; and B—Hinder view.

**Head.** The head (Fig. 17.3) is held almost at right angles to the axis of the body so that the morphologically dorsal surface becomes anterior. Seen from the front it appears pearshaped and is dorsoventrally flattened, having an evenly rounded upper part, while the sides narrow down towards the mouth. Two strong chitinous plates, the **epicranial plates,** cover the whole of the dorsal and posterior side of the head. These plates are fused in the middle at the epicranial sutures. Between the arms of the epicranial suture lies an unpaired sclerite, the **frons**. Immediately below the frons lies a median sclerite, the **clypeus**, forming the lower part of the face. The two sclerites are not distinct owing to the obliteration of the **clypeo-frontal** suture. The side of the head is covered by a pair of vertical plates, the **genae** ("**cheeks**") lying below the eyes. The head bears a pair of very long filiform antenna and the mouthparts. The compound eyes are a pair of large reniform black patches on the sides of the head. On a careful examination each eye appears to be made of a large number of hexagonal areas or facets.

**Thorax.** The thorax consists of three segments, the **pro-, meso-** and **metathorax**. The tergum of the prothorax (the **pronotum**) is larger than those of the meso- and metathorax (**meso-** and **metanotum**). Each segment bears ventrally a pair of legs and dorsally the meso- and metathorax bear the wings.

**Abdomen.** The abdomen is dorsoventrally flattened and has seven visible segments, the remaining ones being telescoped within the hinder end. None of the anterior segments bears appendages, but the terminal ones have appendages that are modified in connection with the reproductive system. The last (eleventh) segment bears two external appendages known as the **anal cerci**. These are many-jointed projections from the hinder end of the body.

### Head Appendages

The **antennae** or **feelers** are *freely* mobile appendages articulated to the head just in front of the eyes. Each feeler consists of a tolerably stout basal

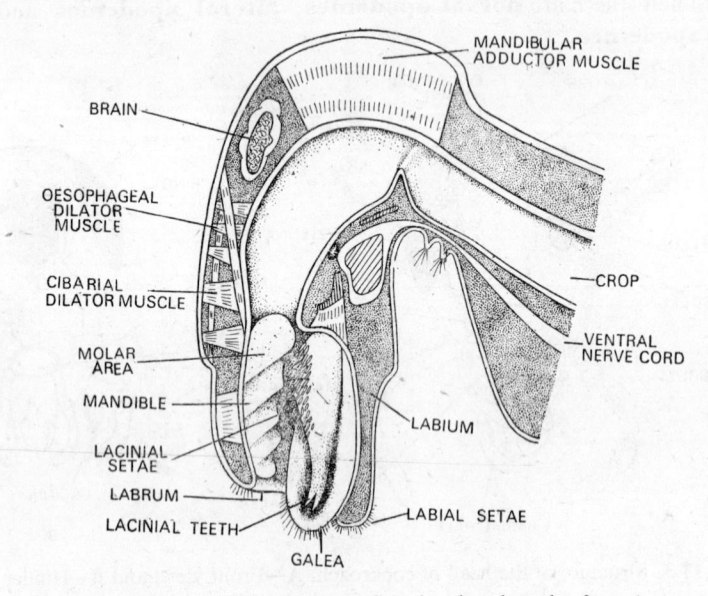

FIG. 17.4. Median vertical section of the head and neck of cockroach showing attachment of mouthparts and musculature of the head.

joint, the **scape**, set in a depression with flexible walls, the **antennal socket**, situated in the clypeus. The scape is followed by a segment called **pedicel**, following which lies the **flagellum** made up of a large number of small segments. The antennae often show marked differences in the two sexes. In the cockroach the males have antennae longer than the body, and in the females they are shorter. The movements of the antennae are controlled by means of **extrinsic muscles** (Fig. 17.4) usually arising from the tentorium and inserted into

## Class Insecta

the scape. The pedicel is also movably articulated with the scape due to **intrinsic muscles** between them, the remaining segments are immovably articulated. Functionally antennae are sensory appendages bearing olfactory (smell) and tactile (touch) receptors. Internal to the base of each antenna is a rounded white space representing a highly primitive **ocellus**, a simple type of eye found in most insects.

### Mouthparts

The mouthparts of the cockroach consist of an upper lip or **labrum** (Fig. 17.3, 17.4), the **labium** or lower lip, an anterior pair of jaws or **mandibles**, and a posterior (lower) pair of jaws known as **maxillae**. Arising from the floor of the mouth is a median lobe or hypopharynx.

The **labrum** is a broad oblong plate movably articulated to the lower edge of the clypeus projecting over the upper part of the mouth and covering the mandibles. Behind the genae and articulating with the sides of the epicranium are a pair of stout tooth-bearing structures, the **mandibles**, that move in a transverse plane, *i.e.*, from side to side. Essentially the mandibles are the biting apparatus, hence bear teeth. The **maxillae** lie beneath the mandibles and are a more flexible pair of jaws. Each maxilla consists of a basal portion of two sclerites, the proximal one hinder to the head is the **cardo** and the distal one is **stipes**. Both these are bent at an angle to each other. From the stipes arises a five-joined palp, the **maxillary palp**, on the outer side. It is like a miniature antenna and is sensory in function. At the base of the palp is a small sclerite, the **palpifer**. On the inner side the stipes bears a bipartite structure consisting of an outer **galea** and an inner **lacinia**. The galea is hooded in shape, while

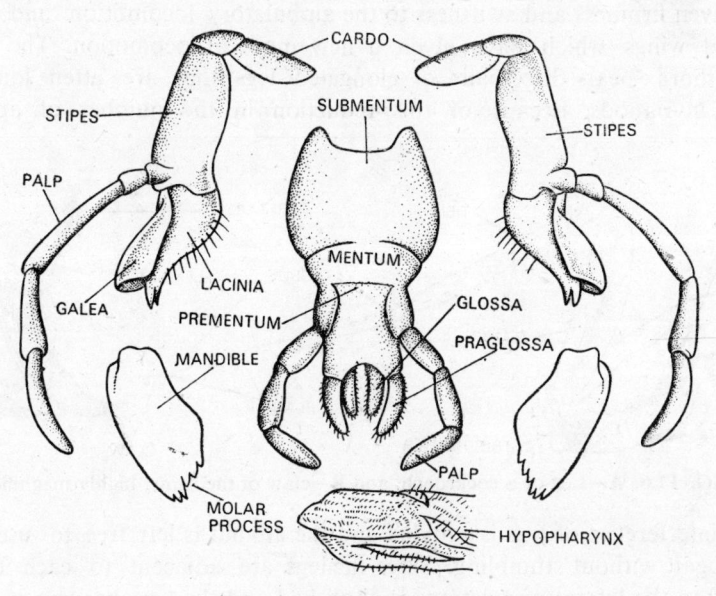

FIG. 17.5. Mouthparts of cockroach.

the lacinia terminates in a claw-like projection and is furnished with stiff bristles along its inner margin. The maxillae are used in holding and masticating the food.

The labium (Fig. 17.5) is apparently a median structure but is really derived from the fusion of a pair of maxilla like appendages. For this reason this is often referred to as the **second pair** of **maxillae** as distinguished from the **first pair** of **maxillae** described above. The broad basal portion of the labium is the **submentum** (Fig. 17.5). Its proximal border articulates with head capsule. The next smaller sclerite is known as the **mentum** that bears an obviously paired structure, each half of which is composed of an outer palp, the **labial palp**, and inner bipartite portion exactly similar to the distal region of the maxilla. The labial palp is three-jointed here. The bipartite inner portion consists of an inner **glossa** corresponding to the lacinia and an outer **paraglossa**, that corresponds to the galea. Paraglossae and glossae collectively form the **ligula**.

The **hypopharynx** or **lingua** (Fig. 17.5) is a median tongue-like process lying between the maxillae and above the labium. It is a cylindrical fleshy projection of the posterior wall of the buccal cavity and bears the aperture of the common salivary duct. The anterior wall of the entrance to the alimentary canal forms another structure, the **epipharynx**, lying on the inner side of the labrum.

### Thoracic Appendages

Typically the thorax bears two pairs of **wings** dorsally, and three pairs of **legs** ventrally. With the acquisition of wings and legs a correlated specialization occurs in the make-up of the thorax, which exists in its typical form only in wingless insects (Apterygota) or in many larvae. The **prothorax** in the cockroach is well-developed being large and shield-shaped, whereas, the **meso-** and **metathorax** are more or less intimately fused. The evolution of insect thorax is of special importance in the history of arthropodan locomotor mechanisms. It has given firmness and swiftness to the ambulatory locomotion and has also produced wings which have given a new mode of locomotion. The three-segmented thorax bears three pairs of elongated legs that are often longer than those of myriapods. Because of this reduction in the number of appendages

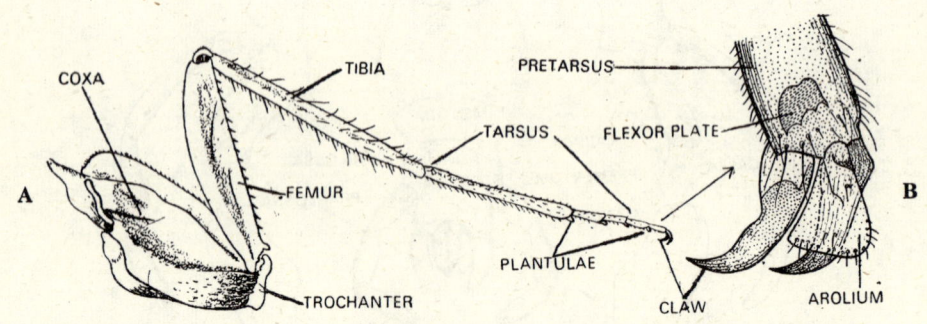

FIG. 17.6. A—Leg of a cockroach; and B—claw of the same, highly magnified.

mutual interference of legs is avoided and the animal is left free to use a flexibility of gait without stumbling. Since the legs are adjacent to each other the tendency to the lateral undulations is abolished and the legs become more effective. The localization of these three segments immediately behind the head aids in the support and manipulation of that region.

**Wings**. Wings (Fig. 17.2) are two-layered membranous expansions of integument arising as lateral folds from the region between the tergum and pleuron. In

## Class Insecta

the early stages the wings show the same layers as the integument. *i.e.* cuticle, epidermis and basement membrane. In the completed organ the original epidermal layers are greatly reduced in dimensions or lost. During development the upper and lower epidermal layers meet and fuse, except along certain linear channels, through which run narrow, sclerotized tubes known as **veins** or **nervures**. The veins form a supporting network, the complete system of which is termed venation. Developmentally the veins are tubular prolongations of the haemocoel and contain wing-nerves, tracheae and blood. In the cockroach the fore-wings have become leathery in texture and are known as **tegmina**. In the beetles the whole of each fore-wing becomes horny forming an **elytron**. The posterior (metathoracic) wings are larger, flexible and much folded when at rest. These are the wings that are used for actual flight. The movements of each wing during flight are effected by two sets of muscles, the **indirect** muscles and the **direct** muscles Fig. 17.7). The direct muscles are so called because they are inserted directly on the sclerites at the base of the wing. The indirect muscles are not inserted on the wings. One group of indirect muscles runs dorsally and longitudinally between the mesothorax and metathorax, the other group runs dorsoventrally between the tergum and sternum.

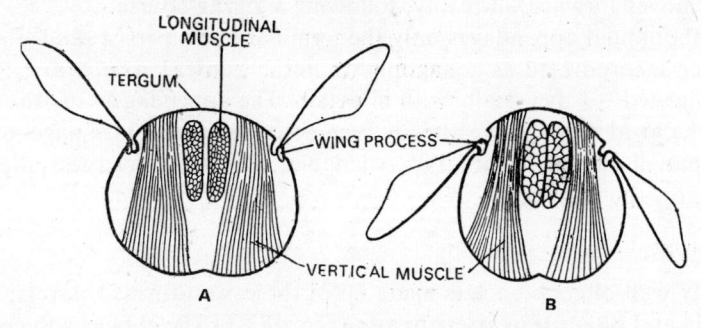

FIG. 17.7. Wings and wing muscles of an insect.

The cockroaches rarely fly. During flight the tough and leathery forewings are held at right angles to the body and do not beat. They probably act as stiff planes. Only the membranous hind wings beat in a complicated manner supporting the body as well as driving it forward. In the primitive insects (like the dragonflies, cockroaches and mantises) power is applied to the wings through the direct muscles, each of the four wings of a dragonfly possessing elevator and depressor muscles. These are called **lamellar muscles** because the protofibrils of the muscle cells are grouped together to form sheets or lamellae with sarcoplasm and mitochondria lying between them. Little is known about the physiology of lamellar muscles, but it may be primitive type of flight muscle. Its important characteristic is that each contraction of the muscle is the result of stimulation by a single nerve impulse. In this respect the muscle behaves like the general body musculature of insects. This type of flight mechanism has been called **synchronous type**, because the stroke of the wing is synchronous with the stimulation of their musculature by the nervous system. In higher insects the flight is **asynchronous** (see General Organization).

**Legs.** Attached to the sternum of each segment of the thorax is a pair of legs. Each consists of a stout, flattened basal portion, the **coxa** (Fig. 17.6), that connects the leg to the body. It is followed by the **trochanter**, a small intermediate segment freely movable over coxa, but fixed to the **femur** that follows it. The femur is the largest part of the leg, very well-developed due to strong underlying muscles. The femur is succeeded by **tibia**, a long slender part beset with a large number of stout spines, the **tibial spurs**. The distal portion of the leg is known as the **tarsus** and is made up of five similar segments. The last one bears two claws. The basal segments are called **metatarsus** while the claw-bearing sclerite is often called the **pretarsus**. The tarsal sclerites bear many fine hair, while at their lower edges occur soft, adhesive pads, the **plantulae**. Beneath the claw is a pad-like cushion, the **pulvillus**.

The cockroach is a swift runner. Strong legs, plantulae, claws and arolium form efficient gripping devices and help the insect in running to perfection. During walking an insect moves its legs in two series so that the fore- and hind-legs of one side and the middle legs of the opposite sides are carried forward. Thus, the body is momentarily supported on a tripod formed by the remaining three legs. The fore-legs act as a tractor having fixed its claws, it pulls the body forward, the middle leg supports and lifts the body on its own side while the hind-leg pushes the body forward and turns it. As a result of these movements the insect moves forward, in reality, following a zigzag course.

Of the abdominal appendages only the terminal ones persist and these have mostly been incorporated as gonapophyses in the **genital armature**, which is too complicated to be dealt with in detail. The appendages of the eleventh segment, the **anal cerci**, have already been described earlier (*see* page 666). The last segment, however, bears two triangular chitinous sclerites, the **podical plates**, below the anal cerci.

### Body Wall

The body wall of cockroach is made up of three well-formed layers: **cuticle**, **epidermis** and **basement membrane** (Fig. 17.8). The cuticle is the outermost covering and is two-layered, outer **primary cuticle** and inner **secondary cuticle**. The primary cuticle is without chitin but carries pigment and spines. Inner cuticle, on the other hand, is thicker. The epidermis is made up of a single

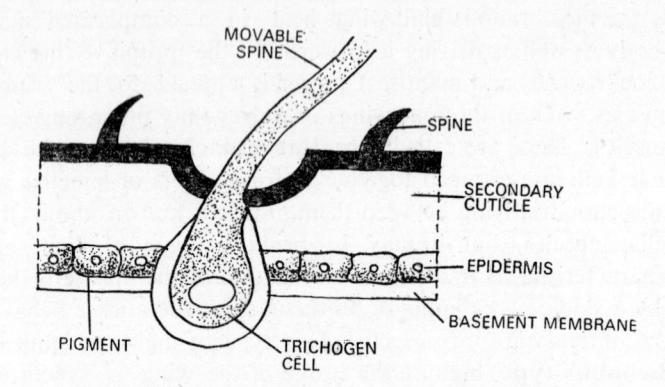

FIG. 17.8. Section of the integument of an insect.

layer of columnar cells, which are able to secrete the cuticle. Some cells of the cuticle produce movable bristles. The epidermal cells that form the cuticle are

*Class Insecta*

known as **trichogen cells**. The basement membrane lies below the epidermis. It is extremely thin and without any structure.

### Digestive System

The digestive system comprises a long tube, the **alimentary canal** occupying a larger part of the centre of the body. Morphologically the alimentary canal consists of (*i*) a long **fore-gut** or **stomodaeum**, (*ii*) a small **mid-gut** or **mesenteron**, and (*iii*) a long **hind-gut** or **proctodaeum**. Both the stomodaeum and proctodaeum originate as invaginations of the ectoderm anteriorly and posteriorly. The inpushing begins after chitin formation, therefore, both

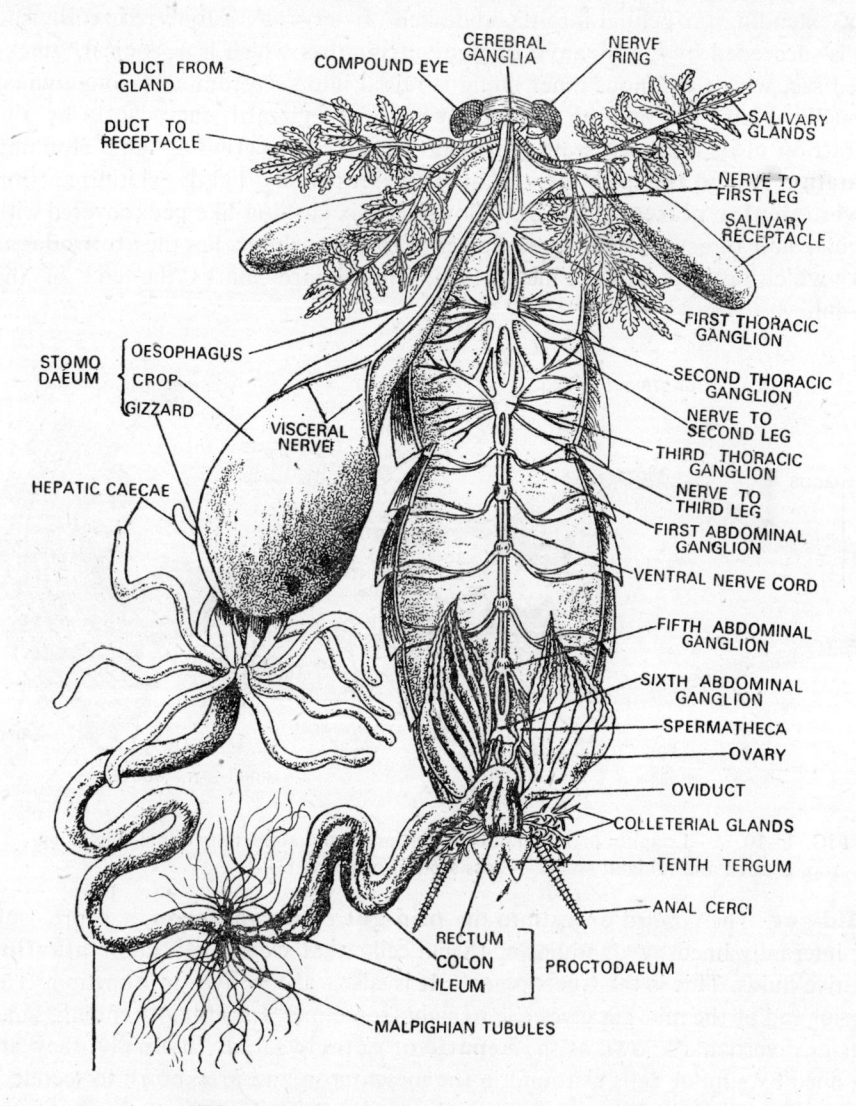

FIG. 17.9. Digestive system of female cockroach.

these have chitinous inner lining. These morphologically different parts have different functions to perform, as such they are separated by means of circular valves, the **stomodaeal** and **proctodaeal** valves.

**Fore-gut.** The **fore-gut** (Fig. 17.9) begins with the rather indeterminate **mouth cavity**, bounded by the epipharynx and labrum anteriorly, by the mandibles at the sides and by the hypopharynx and labium posteriorly. At the base of this ill-defined buccal cavity is a small opening leading into the **pharynx**. The pharynx is very well-developed in sucking insects. In the cockroach it is tubular and passes straight into the **oesophagus**, with which it is not delimited clearly. The oesophagus is a simple, narrow and laterally compressed tube passing from the hinder part of the head into the fore part of the thorax. In the thorax the oesophagus widens into a thin-walled, spacious pyriform sac, the **crop**, extending far behind into the abdomen. It serves as a food-reservoir. The crop is succeeded by the **gizzard** or **proventriculus** which is a conical, thick-walled sac, whose chitinous inner lining is raised into six prominent longitudinal folds (Fig. 17.10) forming the so-called **teeth** of the gizzard, they work by the contraction of its muscular walls, forming an excellent grinding and straining apparatus. Behind the teeth there is another set of six lightly chitinized thin **proventricular plates** which are followed by six cushion-like pads covered with cuticular hair directed backwards. Posterior to the cushions lies the **stomodaeal valve** which projects into the mesenteron. The gizzard marks the end of the fore-gut.

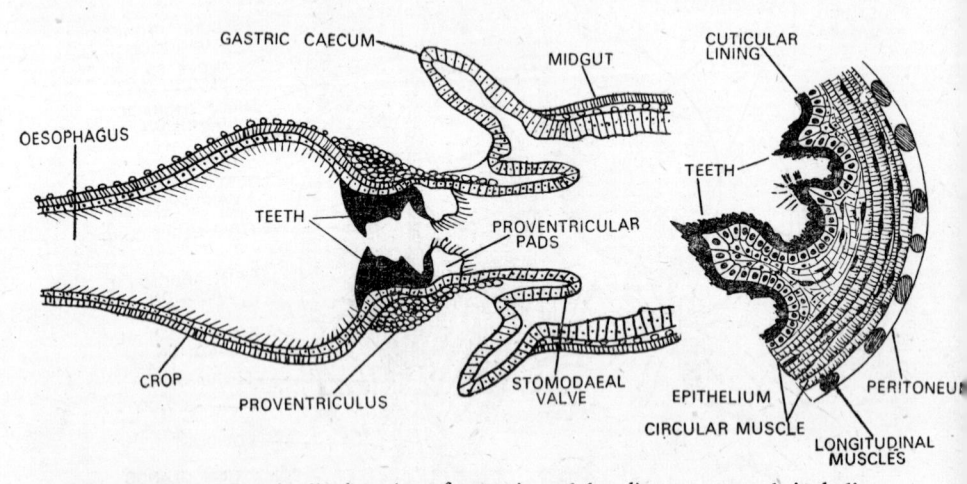

FIG. 17.10. A—Longitudinal section of a portion of the alimentary canal including crop, gizzard and midgut; and B—Cross section of gizzard.

**Mid-gut.** The gizzard opens into the **mid-gut** or **mesenteron**, a short, soft tube internally lined by glandular epithelial cells, that secrete neutral or **alkaline** digestive fluids. This is the true stomach. It is also absorptive in function. The anterior end of the mid-gut gives rise to eight (sometimes less) blindly ending, small tubes or diverticula known as the **hepatic** or **enteric** caeca. Internally they are also lined by similar cells as found in the mesenteron and are known to secrete a digestive fluid that dissolves proteins and starches. The posterior limit of the mesenteron is marked by the presence of a large number of extremely fine tubules, the **malpighian tubules** (Fig. 17.9), which really arise from the end of the mid-gut. Though the Malpighian tubules open into the alimentary canal yet they

## Class Insecta

are not digestive in function. On the contrary they are excretory in nature and shall be dealt with later.

**Hind-gut.** The **hind-gut** (Fig. 17.9) consists of three parts, (*i*) the proximal tubular part known as the **small intestine** or **ileum,** followed by (*ii*) the **large intestine** or **colon**, and finally terminating in (*iii*) the **rectum**. The chitinous lining of the hind-gut is often folded and sometimes provided with setae. The internal surface of the rectum is thrown into six prominent longitudinal folds (Fig. 17.10) terminating at the **anus**.

**Salivary glands.** Associated with the anterior part of the fore-gut are found a pair of **salivary glands** (Fig. 17.11) the main parts of which lie in the thorax. Each gland consists of a lobulated bipartite glandular portion and a thin-walled sac-like reservoir or the **salivary receptacle**. From the glandular portion of each side arises a duct which joins its fellow of the opposite side to form a common duct. Similarly, the ducts from the receptacles unite forming a single duct. The **common ducts** then join to give rise to the **efferent salivary duct** which opens into the mouth cavity on the hypopharynx (Fig. 17.11).

**Physiology of digestion.** As already mentioned the food of the cockroach comprises any kind of organic material, animal or plant. The food is seized with

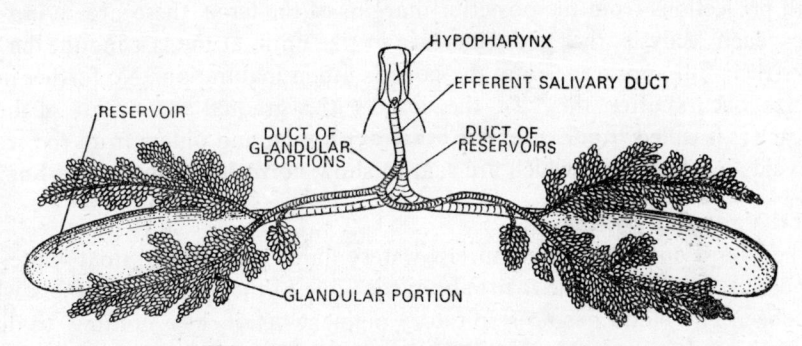

FIG. 17.11. Salivary glands of cockroach.

the help of legs, antennae and mouthparts, cut by mandibles assisted by the maxillae, and then pushed into the mouth cavity with the help of the labium. As these masticatory processes are going on, salivary secretion is poured out of the aperture on the hypopharynx into the food, that is, thus, moistened with saliva as it moves down to the crop. In the crop the food is retained for some time. The salivary secretion of insects usually contains such digestive enzymes, as **amylase, invertase, maltase** and **lactase** with the help of which starch, sugar, maltose and lactose are digested to some extent. In the crop some digestive enzymes are regurgitated from the mid-gut. The gizzard mostly acts as a filter mechanism allowing only food particles of suitable size to pass into the mesenteron, some bigger particles are also broken into smaller ones by the muscular contraction of its walls. Into the mesenteron the food is subjected more completely to the enzymatic secretions of the enteric caeca and mid-gut epithelium, thus, completing the digestive processes. The digestive juice produced by the gastric caeca and the anterior part of the mesenteron contains **amylase, invertase, lactase, maltase, tryptase** and **peptase**. Here carbohydrates, proteins and fats are digested. The food content of the mesenteron is separated from its epithelium by a thin, chitinous **peritrophic membrane** secreted by the gizzard. It forms a cylindrical sheath around the food and is permeable to both digestive juices and digested food. It protects delicate lining of the midgut. The digested food is absorbed in the mesenteron and also in the caeca. Absorbed food material, carbohydrates, proteins, as well as fats, are stored into the diffuse **fat-body**. Finally as the remainder of the food passes through the intestine and rectum, especially the latter, water is extracted from it and the faecal matter is got rid of through the anus.

### Growth and Ecdysis

As the cockroach feeds and grows it becomes necessary for it to cast off its exoskeleton and form a new one. The new covering is formed by the reactivated epidermis (hypodermis) before the old is sloughed off, the increase in size takes place before the new covering is hardened. This process of changing the exoskeleton is called **ecdysis**. When hatched the young cockroach is a small edition of the parent, but is devoid of wings and the gonads are immature. During the earlier stages it grows quickly and, thus, changes its exoskeleton several times. At the later ecdyses, on the dorsal surface of the meso- and metathorax, appear small projections from the posterior margins of the terga, these are **wing-pads**. After each ecdysis the pads increase in size until, at the last moult, the wings make their full appearance and the gonads attain maturation. No further increase in size occurs after this. In the insects this gradual assumption of the adult characters is called **incomplete metamorphosis**, and differs from the sequence of events in other cases which are said to show **complete metamorphosis**.

### Blood Vascular System

The **blood** consists of a colourless watery fluid into which float white blood corpuscles known as the **amoebocytes** due to their striking likeness with the amoebae. The blood has no respiratory pigment as it has nothing to do with respiration. It serves to distribute the absorbed nutriment from the alimentary canal, and carries waste nitrogenous materials to the Malpighian tubules through which they are passed out of the body.

For this reason the circulatory system is greatly modified. It consists of a dorsal median contractile vessel, the **heart** (Fig. 17.12), running through the whole length of the thorax and abdomen. It is divided into thirteen segmentally arranged chambers—three thoracic and ten abdominal. Externally these chambers are marked out by deep constrictions and internally by narrow passages guarded by valves. The arrangement of valves permits only forward flow of blood. Posteriorly the cardiac tube is closed. Anteriorly, it is continued into the head as a slender tube, the **aorta** (Fig. 17.12), which runs forward on the dorsal surface of the oesophagus and ends in front of the peripharyngeal nerve ring in a tunnel-shaped orifice.

The heart is enclosed in a space, the **pericardial sinus** by thin membranous **pericardial septum**, stretched across the body cavity immediately beneath the terga of the thorax and abdomen. The pericardium is perforated by a number of small openings through which the pericardial cavity communicates with the **perivisceral cavity** or the **haemocoel**. The pericardial sinus is also in communication with the heart, each chamber of which has two openings or **ostia** for this

## Class Insecta

purpose. The ostia are guarded by valves allowing the blood from the sinus into chambers of the heart but not in the reverse direction.

Presence of four pairs of segmental vessels (Fig. 17.12) has recently been reported in cockroach with the help of a live heart preparation from a nymphal

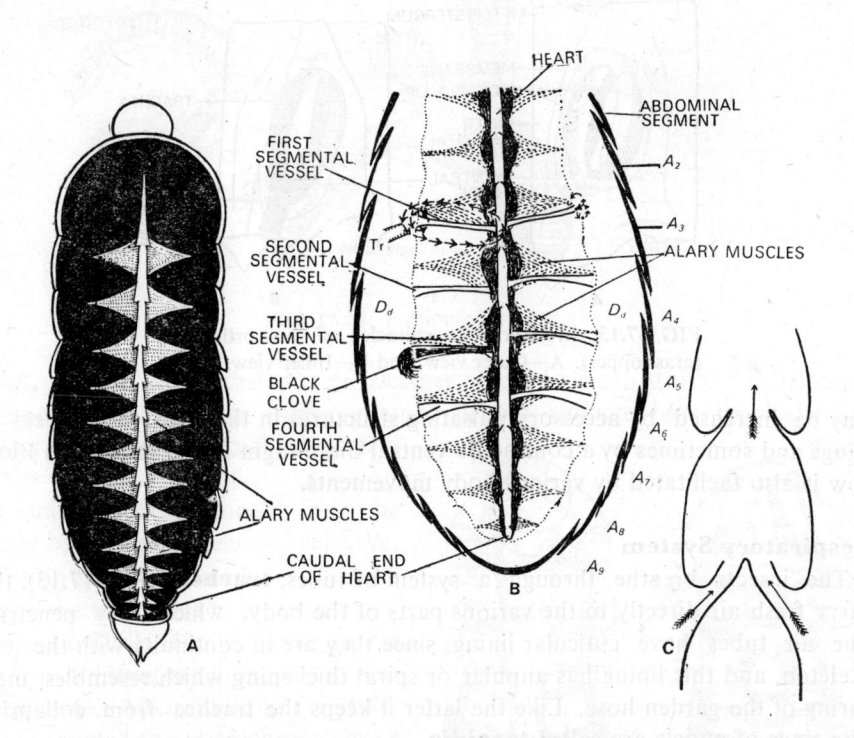

FIG. 17.12. Circulatory system of cockroach. A—Dissection of the heart and alary muscles; B—Diagram of the abdominal portion of the live heart preparation from a nymphal female showing segmental vessels; and C–A portion of the heart showing the ostia and course of circulation.

female. Particles of India ink suspended in saline solution, dropped upon the distal ends of vessels, did not go into vessels although much of it reached the pericardial cells surrounding the heart. Ink dropped upon the caudal end of the heart passed into the heart and the segmental vessel and was ejected from its distal end as a black cloud.

**Mechanism of circulation.** The perticardial wall is provided with a series of triangular, paired **alary muscles.** Each muscle originates from the underside of the **terga** and is inserted in the pericardial wall. The rhythmic contraction of these muscles forces the blood through the ostia into the heart. The chambers of the heart contract successively from behind forwards the wave of contraction succeeding one rapidly. The blood is thus driven forwards till it is poured out into the body cavity through the aorta (reversal of blood flow occurs in a few groups). In the body cavity the blood bathes all the visceral organs and then enters the pericardial sinus through the apertures in the pericardium, thus, completing the circulation. The circulatory system of cockroach has been des-

or by a ventilating mass flow of air down a pressure gradient or by a combination of both. Ventilating pressure gradients result from body movements largely abdominal, which bring about compression of the air sacs and of certain elastic

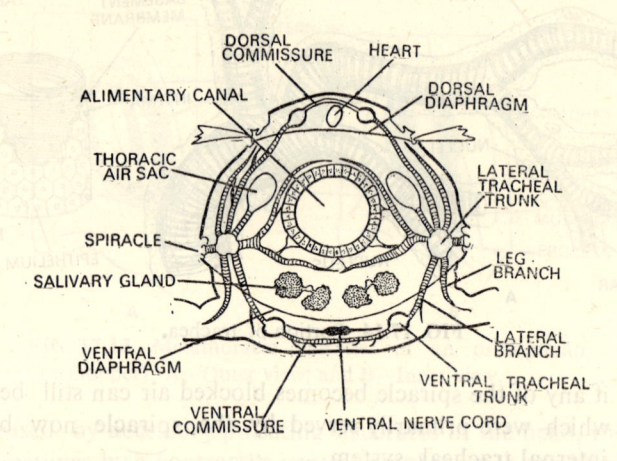

FIG. 17.16. Transverse section of a grasshopper passing through spiracles showing tracheation of a segment.

tracheae. Ventilation is facilitated by the sequence in which certain spiracles are opened and closed. Diffusion along a concentration gradient can supply enough oxygen for small insects but larger forms (weighing 1 gram or more) or those that are highly active require some degree of ventilation. Tracheoles are permeable

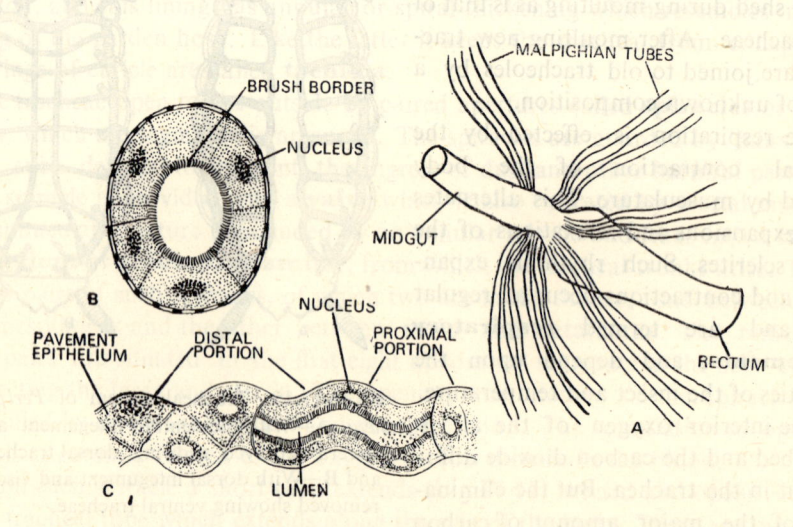

FIG 17.17. A—Malpighian tubules, entire; B—Cross section of the same; and C—Longitudinal section of a Malpighian tubule at the junction of distal and proximal portions.

to liquids and in most insects their tips are filled with fluid. This fluid is believed to be involved in the final transport of oxygen. The fluid pressure has been known to rise and fall depending on osmotic pressure in the surrounding tissue. At the tissue tracheole level gases are exchanged by diffusion across a concentration gradient. Some carbon dioxide is probably released from the tissue directly into the haemolymph and diffuses through the integument. It is noteworthy that no oxygen carrier is known in insect blood except in the aquatic larvae of certain midges (family Chironomidae) whose blood contains haemoglobin in the plasma.

From this it is evident that the insects possess the best developed type of respiration, whereby fresh air reaches the tissues of individual organs. With an ample food supply, this makes a rapid oxidation of the tissues possible, and undoubtedly is one of the chief reasons for the wonderful muscular activity, working power and endurance of insects.

### Excretory Organs and Fat Body

The main excretory organs are the **Malpighian tubes,** opening into the mid-gut directly far in front of the commencement of the hind-gut. The malpighian tubules of insects arise from the mid-gut while those of arachnids develop from the hind-gut. The histological structure of the tubule wall resembles that of the mid-gut. The lumen is lined by large cuboidal cells with brush border. The outer layer of the tubule wall, which is in contact with the haemolymph, is composed of elastic connective tissue and muscle fibres. The tubules are capable of peristalsis and can undergo some movement within the haemocoel. The number of the tubes in the cockroach is about sixty to seventy arranged in six groups, in others it varies from 100 to 2 0. In *Periplaneta* 60 malpighian tubules have total area of some 132 sq.mm. or 412 sq. mm. per gramme of insect.

Besides the malpighian tubes the excretory functions are performed by some cells and the fat body. These cells are called the **nephrocytes**, and occur as groups of special cells found in localized regions of the insect body. In most insects they occur in two principal groups: (*i*) the **dorsal** or **pericardial nephrocytes**, and (*ii*) the **ventral nephrocytes**. The dorsal nephrocytes are commonly termed the **pericardial cells** and occur as two chains of often binucleate cells arranged in a linear series, one on either side of the heart in the pericardial sinus. These cells have been regarded as accessory excretory organs, but it is uncertain whether their chief function is in this capacity. Nephrocytes have common property of absorbing ammonia carmine injected into the blood and retaining its precipitate in their cytoplasm. In fact they absorb unwanted colloid particles from the blood.

The **fat body** is present in all insects and is derived from the walls of the embryonic coelomic cavities. It consists of irregular masses or lobes of rounded or polyhedral cells which are usually vacuolated and contain inclusions of various kinds. Alternatively it may form loose discontinuous lining in the haemocoel with a parietal layer against the body wall and a splanchnic layer around the gut and organs. Since the fat body lies in the haemocoel it is immersed in blood which also circulates through the inter-spaces of this tissue. The main function of the fat body is the separation and the storage of reserve materials, such as fat, proteins and glycogen from the blood. In the young insect the cells of fat body are rounded, with a homogeneous cytoplasm free from vacuoles or inclusions. As the animal grows these cells increase in size, become vacuolated and their boundaries are then hard to see, some nuclear changes also take place. Ultimately two kinds of cells may be distinguished, the **trophocytes** and the **urate** cells. The trophocytes are storage cells and are typically vacuolated. In a well-fed animal they become enlarged because they store the reserve food material

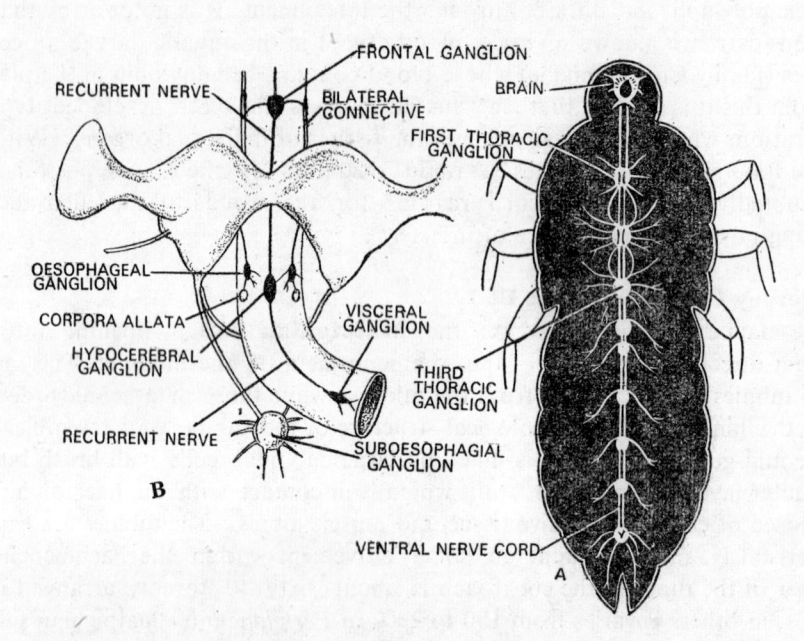

FIG. 17.18. Nervous system of cockroach. A—Nervous system; and B—Brain enlarged showing visceral nervous system.

which may be used during starvation, when the cells become smaller in size. Normally the reserve is consumed at certain periods only, i.e., during the changes from larva to pupa, during hibernation, and to some extent at each ecdysis. Many adults, particularly those that do not feed, rely on nutritive fats, proteins and glycogens stored in the fat during immaturity. Some observers feel that certain histological relationship between trophocytes and blood cells exists. Young trophocytes are almost indistinguishable from young blood cells. According to some the trophocytes are capable of phagocytosis and even that they may sometimes break away from the fat body and circulate in the blood stream as phagocytes. Conversely, blood cells filled with ingested nutritive material have been reported as incorporating themselves into fat body.

Urate cells differ mainly by the fact that they contain deposits of uric acid or urates, concretions which accumulate during life. The fat body, therefore, also performs an excretory function. In the cockroach the urate concretions are deposited throughout life. The cells continue to increase in size. In Collembola, an order lacking malpighian tubules, urate cells provide a major part of the solution to the excretory problem by affording more or less a permanent storage. In some insects without malpighian tubules the excretory function is performed by tubular cephalic glands.

## Nervous System

The nervous system is the primary mechanism of coordination and control in the body. This system is supplemented and complemented by internal secretions as in vertebrates. As in all animals receptors are placed on the body surface of the insect in such a way that activity or change in the environment can be detected. In complexity receptors vary from simple tactile setae to the multicellular highly organized eyes. Each receptor is associated with a dendron. The external stimulus, light heat, sound, chemical activity, touch, etc., is translated into an impulse which travels through the sensory neuron to a ganglion. Here the stimulus is identified and interpreted. If the external event required some response, an impulse then travels from the ganglion to an effector, generally a muscle or set of muscles, though sometimes a gland by way of motor neuron. The result is muscular or glandular activity. The nervous system comprises the central nervous system, the visceral nervous system and the peripheral sensory nervous system.

**Central nervous system** (Fig. 17.18). It includes a pair of **cerebral ganglia** fused together, forming the **brain** lying in the head just below the epicranial plates. The brain is connected to the **suboesophageal ganglia**, that lie under the oesophagus and within the head capsule, by the **circumoesophageal connectives** that pass around the oesophagus. The supraoesophageal mass has three parts—the **protocerebrum (forebrain)** derived from the closely fused ganglia of the first (preantennal) segment; the **deutocerebrum (midbrain)** derived from the paired ganglia of the second (antennal) segment, and **tritocerebrum (hindbrain)**, derived from the ganglia of the third (intercalary) segment. From the two tritocerebral halves the main nerve cords pass down and backwards on each side of the oesophagus to the suboesphageal ganglia formed from the ganglia of the mandibular, maxillary and lateral segments. The suboesophageal ganglia give rise to a pair of longitudinal commissures that pass through the neck to the prothorax and are connected to a pair of **prothoracic ganglia**. Then follow the meso- and metathoracic ganglia lying in the meso- and metathorax respectively and connected by paired connectives. Similar double ganglionated nerve chain continues in the abdomen as well. There are six pairs of **abdominal ganglia**, five in the first five abdominal segments and the last one some distance behind. The last pair of ganglia is larger than others and represents a number of fused ganglia.

The **optic** and **ocellar** nerves arise from optic lobe which is derived from the central mass. The antennary and the accessory antennal nerve (incorporated into the antennal nerve) arise from deutocerebrum. The laterofrontal nerves, each with two branches, arise from the hindbrain. The suboesophageal mass is always completely fused, hence it is not clear from adult structure just how many pairs of ganglia really contribute to it. Paired trunks supply the mandibles, maxillae and labium, smaller nerves go to hypopharynx, the salivary glands and some of the cervical muscles. Most of these nerves contain both sensory and motor fibres.

The brain is the seat of "will" whatever an insect may have. It gives off nerves to the eyes, antennae, palpi and other sensory organs of the head. The suboesophageal ganglia control the movements of mouthparts and also some other body movements. The thoracic and abdominal ganglia give off nerves to all parts of their segments, the nervous requirements of which they fulfil. From the last abdominal segment arise a number of nerves supplying the terminal abdominal segments.

**Visceral nervous system.** The principal component of the visceral nervous system is the **stomatogastric** or **oesophageal** system. It comprises a median

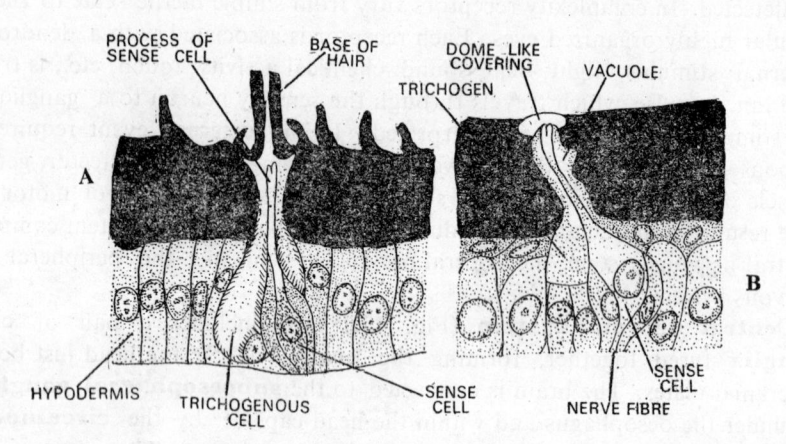

**FIG. 17.19.** A—Simple tactile hair from the cercus of *Gryllus campestris* (cuticle black); and B—Campaniform sensillum from cercus of *Blatta orientalis*.

**frontal ganglion** lying just in front of the brain (Fig. 17.18). This ganglion is connected to the brain by **bilateral connectives**, and gives off a **median nerve**, the **recurrent nerve**, that runs back to another ganglion lying on the upper side of the crop. This is the **visceral** or **ventricular** ganglion. Besides, there are paired **oesophageal** ganglia, just behind the brain. These are jointed to the brain by connectives and are also connected to a median **hypocerebral ganglion**. Closely associated with the oesophageal ganglia are the **corpora allata**, a pair of small ductless glandular bodies, that are universally present among insects and their larvae, and secrete a hormone into the blood that initiates moulting.

The peripheral sensory nervous system is composed of a fine network of axons and sensory cells lying beneath the integument. The nerve cells have branched distal process that ends in the epidermis itself. This system appears to be homologous with the nerve-net of lower invertebrates.

## Sense Organs

The sense organs of the cockroach differ somewhat from those of the prawn and are adapted for receiving stimuli from the air and the land environment where this insect lives. Basically sensory perception is achieved by means of structures termed **receptors** or **sensilla**. These take various forms and are situated at the peripheral ending of the sensory nerves. These receptors may be scattered in distribution, as for instance, the tactile receptors, or they may be aggregated as in the eyes and tympanal organs. In their least modified form receptors closely resemble ordinary body hairs and only differ in having a connection with the nervous system. Different types of receptors are evidently modified sense hairs. It is not possible here to classify all of the prevailing types of receptors. So far as the cockroach is concerned they include: (*i*) **tactile hairs** on various parts of body such as the antennae, mouthparts, palps, and cerci,

# Class Insecta

and distal leg segments; (*ii*) **olfactory organs** on the antennae; (*iii*) **organs of taste** on the palps and other mouthparts; (*iv*) the **ocelli** which are sensitive to

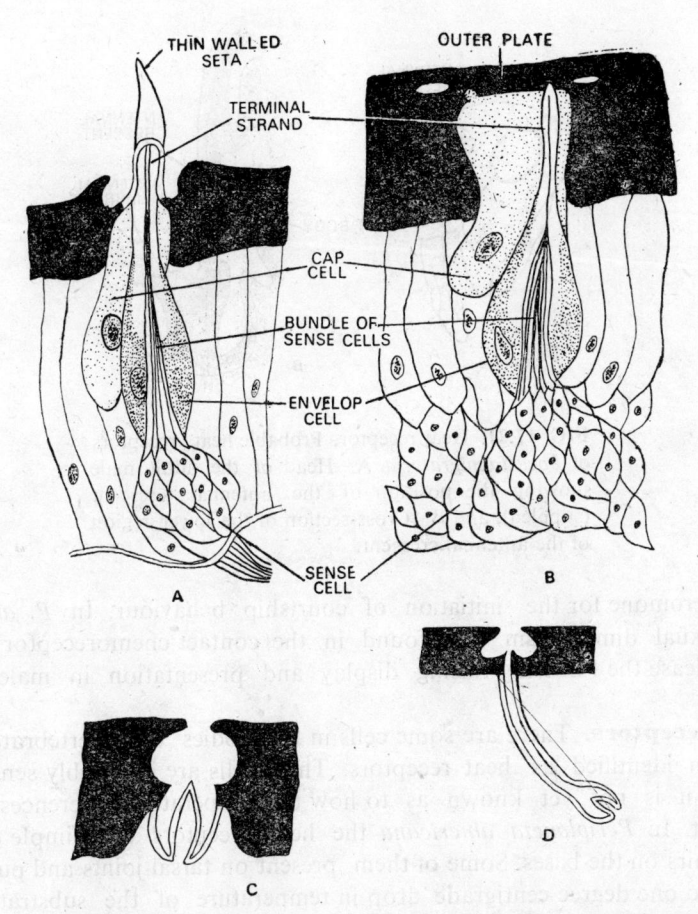

**FIG. 17.20.** Two types of chemoreceptors. **A**—Thin-walled chemoreceptor seta; **B**—Placoid sensilla of *Apis* antenna; **C**—Cuticular parts of coeloconic sensillum; and **D**—Ampullaceous sensillum.

light and shade and may form crude images at close range; and (*v*) the **compound eyes**. Some of the receptor hairs present on the tarsal joints and pulvilli of *Periplaneta americana* are sensitive (below 13°C) to 1°C drop in temperature of the substratum with which the feet of the cockroach are in contact, but they seem to be less sensitive to rise of temperature. The function of the ocelli is not clearly understood, but is probably not the same in all insects.

Sexual dimorphism in the distribution of antennal sense organs is common among adults of the genus *Periplaneta*. In three out of the four strains of *Periplaneta americana* examined, adult males had more contact chemoreceptors than females. In the fourth strain of *P. americana* and in *P. australasiae, P. brunnea, P. fuliginosa* and *P. japonica* no statistically supportable sexual dimorphism of contact chemoreceptors was found. However, in all strains and species of *Periplaneta* examined, sexual dimorphism was found in the total number and/or density of olfactory sensilla. Male adults had nearly twice as many olfactory sensilla as female. These observations are consistent with the behavioural observations that males within the genus *Periplaneta* rely on the reception of an air-

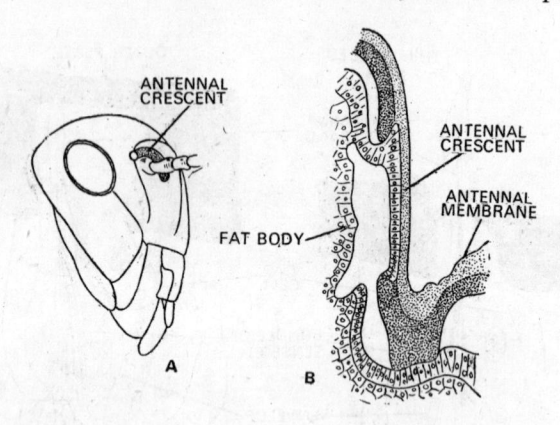

FIG. 17.21. Heat receptor. Probable heat receptors of *Locusta migratoria*. A—Head of the adult male showing the position of the antennal crescent (stippled); and B—Cross-section of the lower region of the antennal crescent.

borne pheromone for the initiation of courtship behaviour. In *P. americana*, where sexual dimorphism was found in the contact chemoreceptors, contact stimuli release the full wing raising display and presentation in males during courtship.

**Heat receptors.** There are some cells in the bodies of invertebrates which have been identified as heat receptors. These cells are invariably sensory cells although it is not yet known as to how the temperature differences produce their effect. In *Periplaneta americana* the heat receptors are simple cells with sensory hairs on the bases. Some of them, present on tarsal joints and pulvilli, are sensitive to one degree centigrade drop in temperature of the substratum with which the feet of the cockroaches are in contact. These cells seem to be less sensitive to rise of temperature being stimulated by 50°C increase.

**Position sense organs (Pro-prioceptors).** The existence of the pro-prioceptors among the arthropods is no longer in doubt, for muscle receptor organs and other pro-prioceptive devices have been described in various decapod crustacea and in insects. In the insects on the whole the nerve-endings are connected with the cuticle in such a way that they are stimulated by the pressure changes in air acting as pro-prioceptors. These nerve-endings may lie in the general body surface or in softer intersegmental membranes. Stretched across the internal spaces of the exoskeleton occur the **chordotonal organs**, which can detect alterations in position. Similar chordotonal organs are also found in the base of the wings and in the halteres. In the leg of *Periplaneta* certain tension receptors are also situated. They react to compression and to stretching strains. Some of them are campaniform, others are elongated and still others are rounded. Some of them enable the animal to perceive the variation of the surface upon which they stand or move. Flexion of joints of legs of *Periplaneta* can be detected by tactile sensory hairs on the inner sides of the joint.

## Class Insecta

**Humidity sensitivity.** Studies of humidity sensitivity have been carried out on many insects. Many of them lose water easily or relatively easily and, therefore, reactions which keep them in moist atmosphere are advantageous. The physiological state of animal influences the humidity. Cockroaches (*Blatta orientalis*) are basically hydro-negative (repelled by water) but after desiccation they react positively (attracted by water). Cockroaches (*Blatta orientalis*) when fully fed and kept in moist air are not attracted to water (hydro-negative), but when kept in dry air and prevented from drinking and eating they get attracted to water (hydro-positive). This shows that it is internal physiological state of the animal which determines whether they are attracted or repelled by water.

**Ocelli.** Structurally an ocellus consists of a biconvex **lens**, beneath which is a transparent **corneagen layer**, which overlies the sensory elements or retinulae. Between the retinulae and around the margin of the lens there are usually pigment cells. The most obvious difference between an ocellus and a compound eye is the presence of single corneal lens in the ocellus, whereas the latter has many.

If the eyes of cockroaches are painted in order to blind them they will still choose a dark place rather than a light one. This is so because isolated light sensitive cells (photoreceptors) are found scattered all over the surface. These cells are probably the cause of the generalized dermal light sense found in many insects.

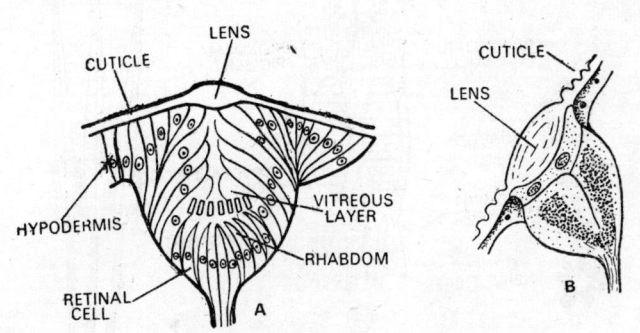

FIG. 17.22. A—Section through the dorsal ocellus of a young *Dytiscus* larva; and B—Lateral ocellus of an insect.

**The compound eyes.** The compound eyes (Fig. 17.23) are large and occupy a considerable portion of the sides of the head capsule. They are composed of **ommatidia** similar in structure to those already described for the prawn. There are about 2,000 ommatidia in each eye of the cockroach. In the insects in general there are two principal types of ommatidia in the compound eyes. In most diurnal insects the ommatidia are relatively short and each forms its own image, so that the picture obtained by the insect is probably a sort of mosaic. Such ommatidia are optically isolated from each other. In most nocturnal insects the ommatidia are more elongated, the pigment does not completely separate them and the image on a single rhabdom is formed by several ommatidia. This is the superposition image and is less distinct but brighter. Both the types of ommatidia may be present in the same comound eye as, probably, is the case in the cockroach. In some cases the ommatidia have pigment that is capable of migration. It can completely isolate an ommatidium from others or shorten in such a way that light may pass from one ommatidium to others forming a superposition image.

It is likely that an insect does not get as clear an image of objects as the vertebrates, and its sense of differentiation is not very well-developed. It is usual-

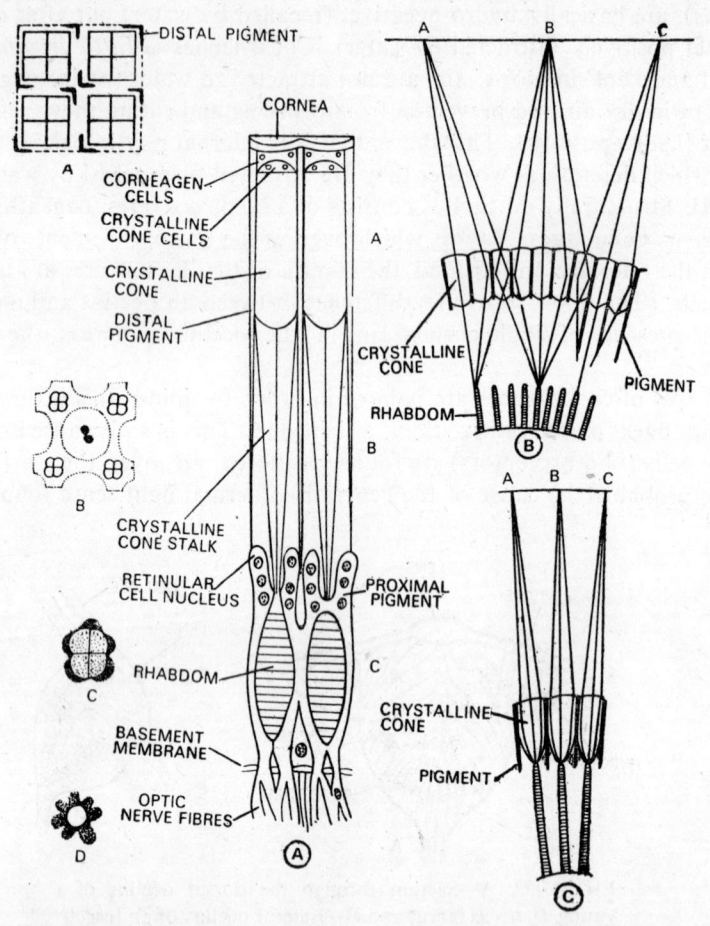

FIG. 17.23. Structure of a compound eye and image formation A—Two ommatidia of the compound eye, longitudinal view; Sections through regions a, b, c, shown on sides; B—Compound eye adapted for superposition image; and C—The same adapted for apposition image.

ly very sensitive to motion. The insect's visible spectrum is different from that of ours. Insects see further into the ultraviolet than we do, in some cases wavelengths as short as 0.257 microns (the limit in man is about 0.400 microns), and not see as far into the red (the limit is about 0.690 microns for insects, and 0.800 microns for man).

**Auditory organs**. Insects hear by means of delicate tactile hairs sensitive to sound waves and **tympanal organs**. Some insects, such as the male mosquitoes, detect sound by means of the antennal hairs; many other insects detect sound by means of their body hairs. Nothing is known about hearing in the cockroach.

## Reproductive System

The sexes are separate in the insects. The female reproductive organs consist of a **pair of ovaries** (Fig. 17.25) each of which consists of eight ovarian tubules

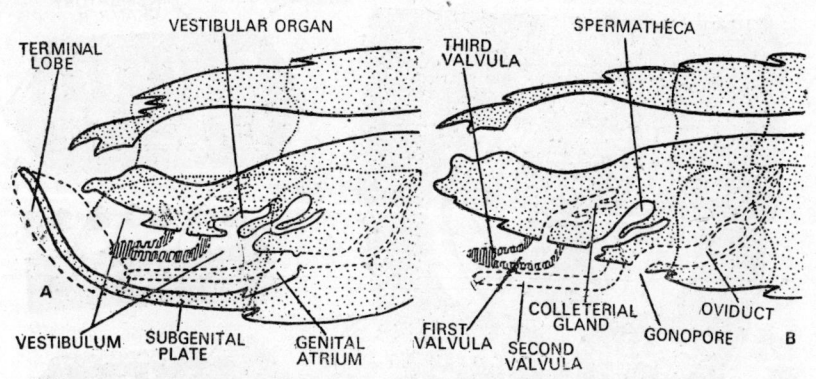

**FIG. 17.24.** Diagrams of the posterior segments of the insect abdomen cut in sagittal section. A—Fundamental arrangement of typical posterior abdominal segment; and B—The same as modified in *Periplaneta americana* (*after* Brunnet).

or **ovarioles** in which the eggs are elaborated. Each tubule presents a number of swellings, which contain developing eggs one behind the other in a single chain, the oldest oocyte being situated nearest the union with the oviduct. The wall of an ovariole is a delicate transparent membrane. The anterior end of each ovariole tapers to form a thread-like **terminal filament**. The flaments of the ovarioles of one ovary combine to form a common thread that ultimately gets lost in the fat body. Posteriorly all the ovarian tubules join to form a short rather than wider tube, the **oviduct**. The two oviducts combine to form a median **uterus** and the latter continues posteriorly with a somewhat wider passage or **vagina**. The two regions have no external distinction, but the vagina differs morphologically in that it is an invagination of the body wall. The vagina opens by a slit-like aperture, the **vulva**, into a **genital pouch** or **chamber** bounded by the terminal sclerites of the abdomen. Into the genital chamber opens a pair of branched glands, the **colleterial glands**, the secretion of which forms the ootheca (see later). Into the genital chamber also opens a pair of **spermathecae** of which only one is fully developed and functional in the present case.

The genital chamber (Fig. 17.24) is formed as a result of the elongation of the sternum of the seventh abdominal segment so much so that the sterna of the rest of the segments eighth, ninth and tenth are telescoped inwards enclosing the spacious chamber. The anterior portion of the genital chamber into which open the gonoducts is known as the **genitalatrium**, posterior to which there is a large cavity known as the **vestibulum**. The ventral wall of the vestibulum represents the intersegmental membrane between the seventh and eighth sterna. The dorsal wall of the vestibulum is formed by the ventral parts of the ninth and tenth segments. The ninth sternum is absent, the roof of the vestibulum in this region is formed from the ventral walls of the second and third valvulae (parts of the copulatory apparatus), of which there are three. The openings of the colleterial glands occur in the part of the roof formed from the second

**valvulae.** Behind the opening of the colleterial glands there is another opening of an invagination called the **vestibular organ.**[1] The opening of the sperma-

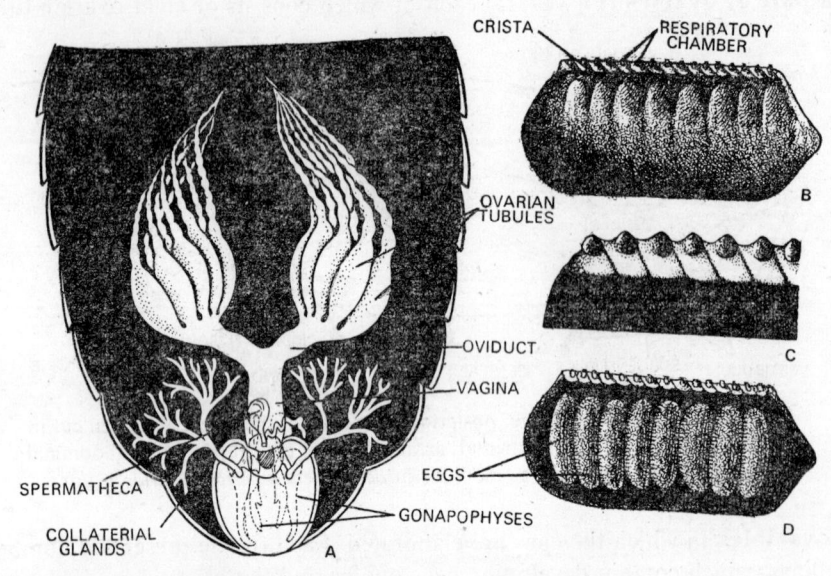

FIG. 17.25. A—Reproductive organs of a female cockroach; and B to D—Structure of egg-case of the same.

thecae is situated just behind the opening of the vestibular organ. To compensate, the terga of the segments eight and nine are reduced in anteroposterior length and telescoped under the seventh.

In the male the testes (Fig. 17.26) are paired, each being made up of a number of **follicles** or **vesicles,** and lie beneath the fifth and sixth abdominal terga, on either side of the body. The testes discharge their contents into two narrow tubular spermducts or **vasa deferentia**. The two **vasa deferentia** unite together to form a muscular **ejaculatory duct** which opens out on a median rod-like structure, a part of the complicated genital armature. Into the vasa deferentia and ejaculatory duct open the accessory glands. Where the vasa deferentia join the ejaculatory duct their ends become enlarged forming **ampullae** or **vesicles**, from which arises a large number of blindly ending diverticula. In the adult male the diverticula are filled with spermatozoa and are of glistening white colour. The whole structure is called the **mushroom** or **utricular gland**. The diverticula of mushroom gland are of three types: the long **peripheral tubules**, short **central tubules** and short **bladder-like** outgrowths, six to seven on each side. The bladder-like outgrowths form the seminal vesicles for storing sperms. In the oriental cockroach there are in all 350 to 450 diverticula. Based on their staining reactions Jurecka (1950) distinguished six types of tubes in males of *Blatta orientalis*; their period of most active secretion occurs for several hours following metamorphosis into the adult. This is followed by a resting period, secretion begins only after copulation when the secretion and spermatozoa are ejected and moulded into a spermatophore.

[1] Brunet, *Q.J.M.S.* Vol. 92, Part 2 (1951).

## Class Insecta

On the lower side of the ejaculatory duct opens the median **conglobate** or **phallic gland**. The exact function of this gland is unknown. At one time it was

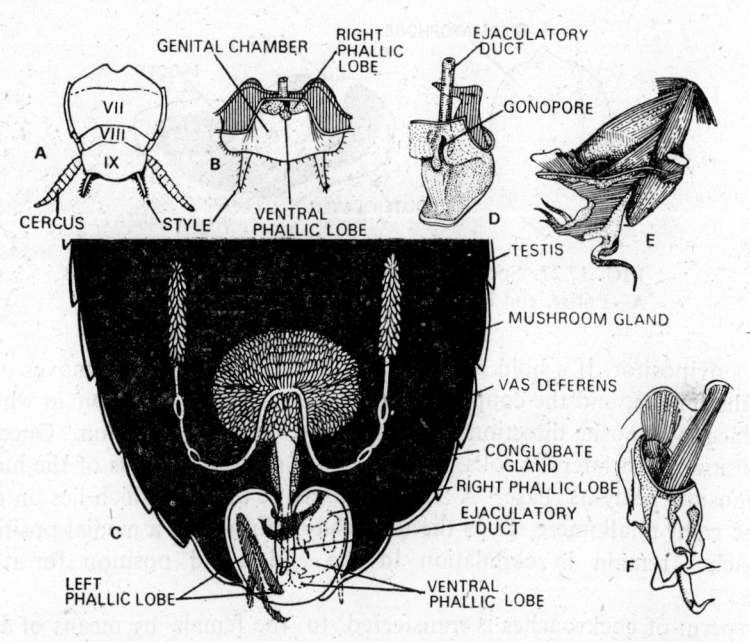

FIG. 17.26. Reproductive organs of male cockroach; A—Posterior part of the abdomen of male *Blatta orientalis*; B—The same in young nymph; C—Male reproductive organs of *Periplaneta americana* alongwith genitalia; D—Ventral phallic lobe (phallomere); E—Right phallic lobe; and F—Left phallic lobe, adult *Blatta*.

believed to produce volatile alkaline secretion with a strong mousy odour believed to be offensive to its enemies. Now it is believed that its secretion forms the outermost layer of the spermatophore.

Arranged around the male gonopore lie the three lobes forming the asymmetrical male genitalia. They are named according to their position. They are **right**, **left** and **ventral phallomeres**. The left phallomere is a complicated structure and has several parts arising from a common base (Fig. 17.26). On the extreme left it has a long slender arm ending in a curved hook (**titillator**). Next to it lies the hammer-headed **pseudopenis**. The hook-shaped **asperate lobe** lies next to pseudopenis. The duct of the phallic gland passes through the left phallomere and opens between the pseudopenis and asperate lobe. The ventral phallomere is a simple broad plate and carries the gonopore.

**Copulation**. A week after the last moult the insects become ready for copulation, which takes place during night in the months of March to September. As the female feeds on male's glandular secretion, he pushes his abdomen backward so that the female is directly above the male. This is the position just prior to copulation. Thus, the male carries the female on his back in the beginning of copulation with their heads in the same direction and they later assume end to-end position. This method of copulation has been observed in many species of cockroaches including *Blatella*, *Blatta*, and *Periplaneta*.

As the male cockroach pushes backward under the female, he extends his hooked left phallomere. This appendage clasps a large sclerite located near the

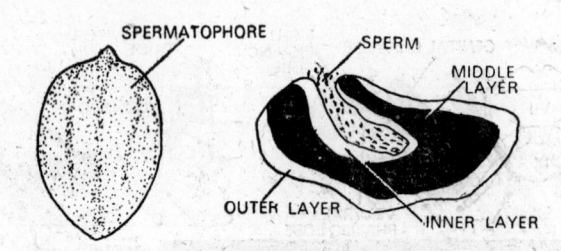

FIG. 17.27. Spermatophore of the oriental cockroach. A—Entire; and B—Section through the spermatophore.

female's ovipositor. If a hold is secured on the sclerite, the male moves out from under the female, and the couple assumes the opposed position in which their heads face in opposite directions. This is the copulating position. Once in the final position, two lateral hooks lying on either side of the anus of the male hold the ovipositor near its base. A small crescentic sclerite, which lies on one side near the right phallomere, grips the ovipositor firmly in a medial position. The cockroaches remain in copulation in the end-to-end position for at least 30 minutes.

The sperm of cockroaches is transferred to the female by means of a capsule or **spermatophore** formed by the secretions of the male accessory sex glands. The secretions from the various accessory glands pour into the pouch of the ejaculatory duct and form the spermatophore. The spermatophore begins to form on the male as soon as copulating pair is securely hooked together. The three layers of spermatophore (Fig. 17.27) are formed from three protein secretions. The innermost layer of the spermatophore is formed by the secretion of peripheral tubules of mushroom gland, middle layer by the secretion of the walls of ejaculatory duct, while the outermost layer is formed by the secretion of the phallic gland, during copulation. The spermatophore is attached to the spermathecal papilla. At one point in the formation of the spermatophore, sperms flow from each seminal vesicle into a milky middle layer within the spermatophore. Each of the two sperm masses forms a separate sac.

Following formation, the completed spermatophore descends the ejaculatory duct and is pressed by the male's endophallus against three sclerites lying on the left hand side of the spermathecal groove in the female, which serves for holding the spermatophore. The tip of the spermatophore, which contains the openings of the sperm sac, is inserted into the spermathecal groove. During 12 to 24 hours time the sperms migrate to the spermathecae. The empty spermatophore dries and shrinks and is eventually dropped by the female.

**Formation of ootheca.** The secretion from the colleterial glands flow out over the inner surface of the vestibulum or oothecal chamber. The vestibulum is closed posteriorly by the apical lobes of the seventh sternum. As the forming ootheca presses against these lobes a characteristic pattern (Fig. 17.25D) is imparted to the distal end of the ootheca. After a number of eggs has entered the vestibulum the distal end of the ootheca emerges beyond the end of the abdomen.

This projecting portion is of opaque white colour, it then becomes transparent, pink and reddish chestnut. It continues to darken after it is laid. As the eggs move posteriorly, the valvulae of the ovipositor move them into the oothecal

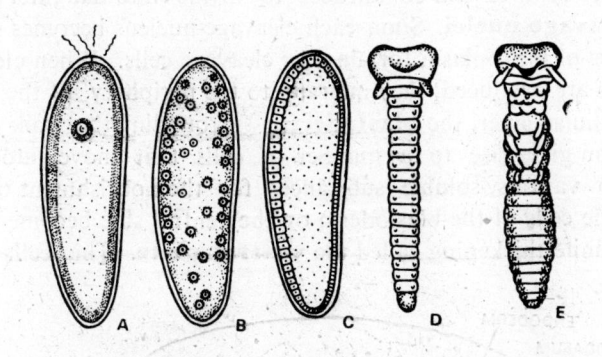

FIG. 17.28. Stages in the development of an insect egg. A—Fertilization; B—Cleavage; C—Blastoderm formation; and D and E—Advanced embryos.

chamber and in a way set them on end with their heads upwards. The eggs from the right ovary pass into the left side of the ootheca and *vice versa*. In *Periplaneta americana* some of the eggs are placed wrong, end up in the ootheca, and, though in such eggs, the development occurs normally the nymphs cannot emerge out of the egg case. The oothecae of *Periplaneta americana* and *Blatta orientalis* contain no chitin which is found in insect cuticle. The left colleterial gland secretes a water-soluble proiten and an oxidase while the right gland secretes a fluid containing a dihydroxyphenol. When the secretions from the right and left glands mix, the phenolic substance is oxidized, producing a quinonoid tanning agent. Interaction of the tanning agent with protein gradually hardens and darkens the ootheca. Both the protein and the tanning agent (quinone) are secreted by the colleterial glands. The walls of the vestibular organ have specialized cells the secretion of which coats the eggs as they pass through the vestibulum, and may well contribute to the substance which binds the eggs together within the casing of the ootheca. The alkali-resistant outer layer of the ootheca is formed by the secretion of the dermal gland cells lining the posterior end of the vestibulum.

Certain valvulae of the ovipositor of oviparous species are modified to mould the ootheca, especially the **crista** or **keel**. In *Periplaneta americana* the third valvulae of the ovipositor are modified to form the "horned die" which moulds the inner surface of the keel of the ootheca, providing respiratory chamber and respiratory duct for each egg.

The oothecae of different species of cockroaches are quite distinctive as they may vary in size, shape and number of enclosed eggs. Each egg cell in the ootheca is indicated externally by an evagination (forming half of the respiratory chamber) on each side of the upper part of the keel. Thus the number of respiratory chambers and their corresponding canals, which show clearly in the keels of certain oothecae, is often a good criterion for the number of eggs in the ootheca (*Periplaneta americana, Blatta orientalis*).

**Development.** The eggs of cockroach are enclosed in the egg case. Each egg is laterally compressed being concave on one side (future ventral side) and convex on the other (Fig. 17.25). After maturation and fertilization the **zygote-nucleus** passes inwards and commences to divide into daughter nuclei usually called the **cleavage-nuclei**. Soon each cleavage-nucleus becomes surrounded by a stellate mass of protoplasm forming the cleavage cells. When cleavage-cells in large numbers are produced, they migrate to the periphery of the egg forming a continuous cellular layer, the **blastoderm**, surrounding the yolk (Fig. 17.28C). The blastoderm gives rise to a number of cells that move into the yolk and convert it into various soluble substances for the nourishment of the embryo. Meanwhile, the cells of the blastoderm on the ventral side become columnar and produce a definite thickening called the **ventral plate**. The cells of the ventral

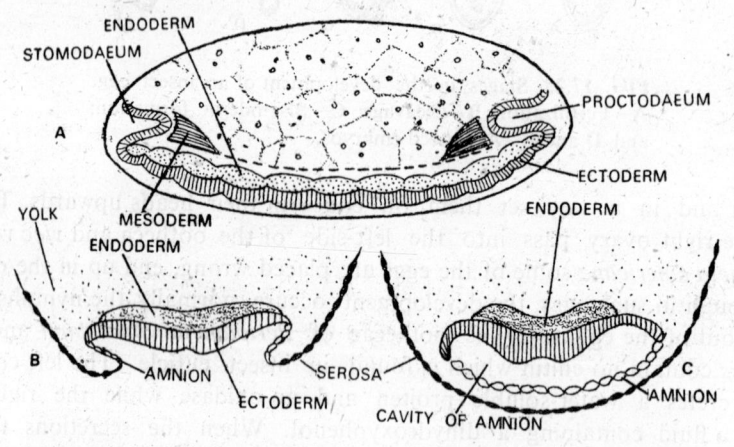

FIG. 17.29. A—Bipolar method of endoderm formation in an insect embryo; and B and C—Transverse section of an insect embryo showing the formation of amnion and serosa (*after* Korschelt and Heider).

plate multiply forming it several cells thick. It becomes broader anteriorly indicating the position of the head-lobes. At the opposite end also there is a specially-thickened area bearing a slight depression, perhaps the **blastopore**, on its surface. The **mesoderm** formation begins from this end and proceeds forwards. The mesoderm takes the form of a longitudinal band bifurcated in front in the position of the head lobes.

How the **endoderm** is formed in the cockroach is not known. Probably it is of bipolar origin. Soon after the stomodaeal and proctodaeal invaginations are formed, the endoderm cells are budded off directly from the apices of the two invaginations (Fig. 17.29). These two groups of cells approach each other as they grow, and ultimately meet to form a continuous layer that forms the wall of the mesenteron. By now a number of transverse lines appears in the ventral plate making it a segment (Fig. 17.30). As the development proceeds rudiments of appendages appear on the head, thorax and abdomen. The abdominal appendages disappear subsequently with the exception of the last pair, that forms the cerci. The segment giving rise to the antennae is at first postoral but it subsequently fuses with the preoral segment, (**prostomium**), thus giving secondarily the preoral position to the antennae. Six segments—the prostomial segment, the antennary segment, a segment devoid of appendages, another segment bearing rudimentary appendages and those bearing the two pairs of maxillae—unite to form the head.

The embryo now becomes surrounded by **embryonic membranes**. A fold of blastoderm arises on either side, the two folds grow inwards and eventually unite over the body of the embryo forming a two-layered covering. Of these the outer is called **serosa** and the inner **amnion** (Fig 17.29).

The mesoderm bands also become segmented, forming a number of somites. The somites become hollow and are then closely applied to one another. Finally all the somites fuse together with the result that the cavities of all the somites unite forming the coelom. The outer wall of the coelom lies close to the ectoderm and is called the **somatopleure** (somatic mesoderm), while the inner wall lying next to the endoderm is the **splanchnopleure** (splanchnic mesoderm). The body cavity of the adult is haemocoelic and is not derived from the coelom of the embryo. In fact a blood sinus replaces the entire body cavity. The heart is formed by special cells, the **cardioblasts**, that separate off from the two rows, of mesodermal somites where they become approximated dorsally.

Gradually the ventral plate grows upwards at the sides, the growing edges ultimately fuse with each other in the mid-dorsal line enclosing the entire amount of yolk. An ectodermal groove appears in the mid-ventral line forming the ventral nerve chain. This groove is cut off from the surface ectoderm forming the

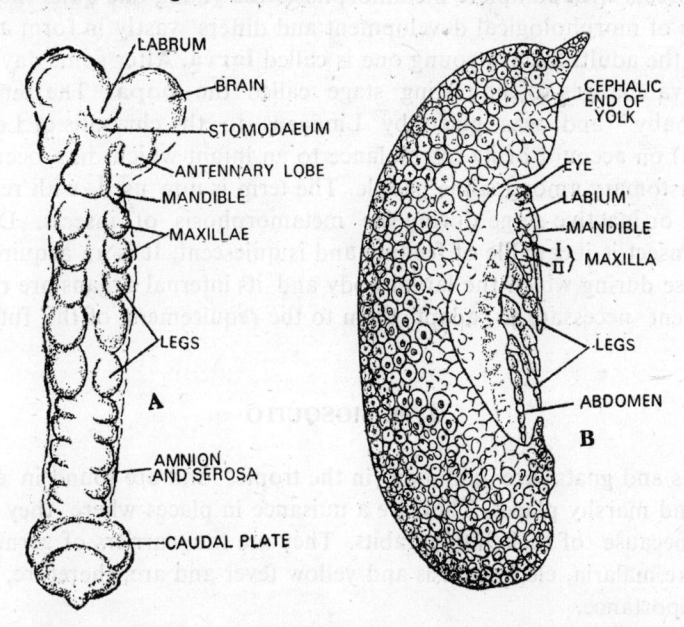

FIG. 17.30. A—Ventral plate of the embryo of cockroach, *Blatta germanica* isolated from the yolk; and B—Embryo of cockroach just after the rupture of amnion and serosa, lateral view of the entire embryo (*after* Wheeler).

chain. A pair of ectodermal thickenings on the anterior side forms the brain. A part of the brain, the **protocerebrum**, develops in pre-antennary region. This fuses with that developed in the following two segments to form the complete

**brain** or **syncerebrum**. In the case of the cockroach the embryo develops almost all the adult features, only it is without wings and smaller in size at the time of hatching.

**Metamorphosis**. At the time of hatching the cockroaches are white in colour and bear dark eyes. Soon the integument is acted upon by daylight and oxygen and gets dark and thickened. The young ones are devoid of wings but in other respects they resemble the adult. Such an immature young one that resembles the adult at the time of hatching is known as **nymph**. In fact the nymphs quit the egg in a relatively advanced stage of morphological development, only wings and genitalia are lacking or are ill-developed. The nymphs cast off their exoskeleton several times before they become adults. The process of casting off the skin is known as **moulting** or **ecdysis**. The adult characters are fully assumed during the last moult.

In majority of the insects the eggs almost always hatch in a condition morphologically different from that assumed in the imago. In order to reach the adult form and conditions the young ones have to pass through changes of form that are collectively known as **metamorphosis**. In the case of the cockroach practically there is no metamorphosis. The minor changes that take place in this case are known as **direct** or **incomplete metamorphosis**. In others the metamorphosis is **indirect** or **complete**.

In the insects with complete metamorphosis the young one quits the egg in an early stage of morphological development and differs vastly in form and habits, etc., from the adult. Such a young one is called **larva**. After some days of active life the larva changes to the resting stage called the **pupa**. The term "pupa" signifies "baby" and was given by Linnaeus to the chrysalis of Lepidoptera (butterflies) on account of its resemblance to an infant which has been wrapped up as is customary among some people. The term is now used with reference to the resting or inactive stage during the metamorphosis of insects. During this stage the insect is incapable of feeding and is quiescent. It is an acquired transitional phase during which the larval body and its internal organs are remodelled to the extent necessary to adapt them to the requirement of the future imago (adult).

## TYPE MOSQUITO

Mosquitoes and gnats are common in the tropics and are found in abundance in damp and marshy places. They are a nuisance in places where they occur in numbers because of their biting habits. They are the carriers of germs of many diseases like malaria, elephantiasis and yellow fever and are, therefore, of considerable importance.

**External Characters**

As in other insects, the body of a mosquito (Fig. 17.31) is composed of three regions—head throax and abdomen. A greater part of the **head** is covered by a pair of large **compound eyes**. The **epicranium** and **clypeus** strengthen the surface of the head, but the clypeus projects much forward. The head bears a pair of many (14) jointed antennae, clothed with silky hair, numerous and long in the case of male. This affords a ready means of distinguishing the sex. The

*Class Insecta* 695

other appendages of the head are the **mandibles, labium, hypopharynx** and two pairs of **maxillae** with the **maxillary palps**. Except the palps all the

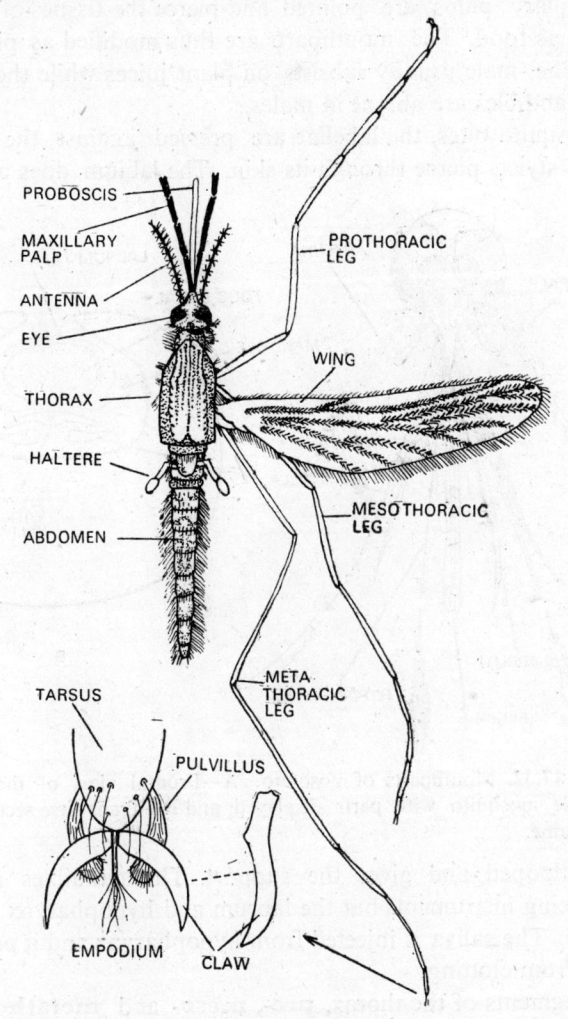

FIG. 17.31. Mosquito. A—Adult *Anopheles* female, wings and legs of one side removed; and B—Tip of tarsus enlarged.

appendages are modified to form a piercing and sucking **proboscis** (Fig. 17.32). In the female the labium (the fused second maxillae) forms a long gutter and has two lobes, **labella**, at the end. Mandibles and first maxillae are fine, paired rods or styles of chitin contained within the grooved labium. The incomplete tube formed by the labium is closed above by the labrum-epipharynx. The labrum is a pointed dagger-like structure, grooved below while the epipharynx is a projection from the roof of the mouth. It becomes completely fused with the labrum. The hypopharynx is a flattened plate fitting against the lower, grooved surface of the labrum-epipharynx. These two together form a tube for sucking the liquid food of the mosquito. The hypopharynx bears a salivary duct which opens at its tip. Two tactile maxillary palps remain outside the proboscis projecting from its base. These are long in the male mosquitoes (Fig. 17.33) and in the *Anopheles* female, but are very short in *Culex* female. The palps are pointed in the *Culex* male while they are club-shaped in the *Anopheles* male. All the mouthparts except the labium and the maxillary palps are pointed and pierce the tissues of the organism to suck up juices as food. The mouthparts are thus modified as piercing and sucking organs. The male usually subsists on plant juices while the female feeds on blood. The mandibles are absent in males.

When a mosquito bites, the labellae are pressed against the host and the enclosed pointed stylets pierce through its skin. The labium does not enter the skin

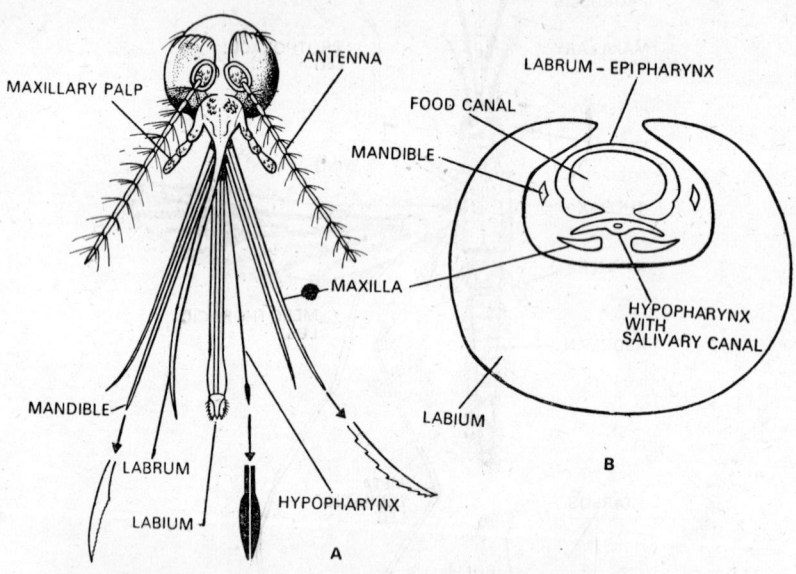

FIG. 17.32. Mouthparts of mosquito. A—Frontal view of the head of a female mosquito with parts displayed; and B—Transverse section through the same.

but becomes looped and gives the support. The mandibles and maxillae act merely as piercing instruments but the labrum and hypopharynx form a tube to suck up blood. The saliva is injected from hypopharynx and it prevents the blood of the victim from clotting.

The three segments of the thorax, **pro-**, **meso-** and **metathorax**, are easily recognized from sides and ventrally. Dorsally, however, only two terga are clearly visible. The pronotum can hardly be recognized as two small knobs known as **patagia**. The **mesonotum** is large and covers up a greater part of the thorax. Its posterior margin presents a lip-like structure, the **scutellum**. This is undivided in the *Anopheles*, but is three-lobed in the *Culex*. The mesothorax bears a pair of wings that are used for flight. The arrangement of veins (nervures) and scales on the wings is characteristic (Fig. 17.31). The metanotum is small and bears two club-shaped **halteres** or **balancers** attached to it. These are the rudimentary wings, useless for flight. Ventrally the thorax bears three pairs of legs. The abdomen consists of eight segments, each with a dorsal tergum and a ventral sternum.

**Life-history** (Fig. 17.34)

The sexes are separate; the pairing takes place while the insects are on the wing.

## Class Insecta

The female then finds out a suitable pool of stagnant water and lays 200 or 300 of her eggs on the surface. The female *Culex* lays her eggs in clusters in which they are cemented together to form a boat-shaped raft (Fig. 17.34). The *Anopheles* and *Stegomyia* lay their eggs singly. Each egg of the *Anopheles* is a boat-shaped oval structure (Fig. 17.34). Its flat upper surface is surrounded by frill of small air spaces and a pair of lateral floats make it more light. The culecine egg is elongated and broader at the base to which a rounded process is attached. It has neither frill nor float. Many eggs are cemented together to form a light boat-shaped raft to float on the surface of water.

**Larva.** In two or three days the eggs hatch out. From each egg emerges a small transparent larva that enters water. The larva is a small wriggling creature with three distinct regions, the head, thorax and abdomen (Fig. 17.34). The head is large and rounded bearing a pair of compound eyes, a pair of simple

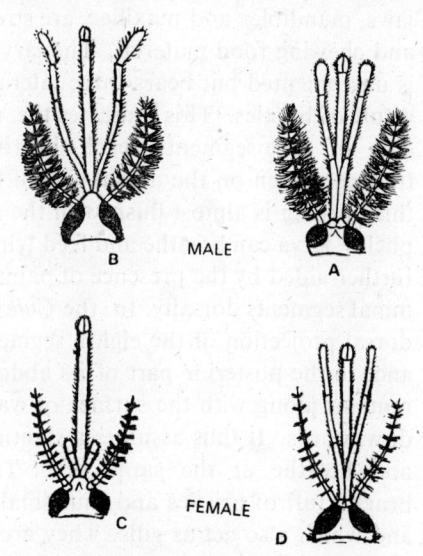

FIG. 17.33. Comparison of mouthparts of A—*Anopheles*; and B—*Culex*.

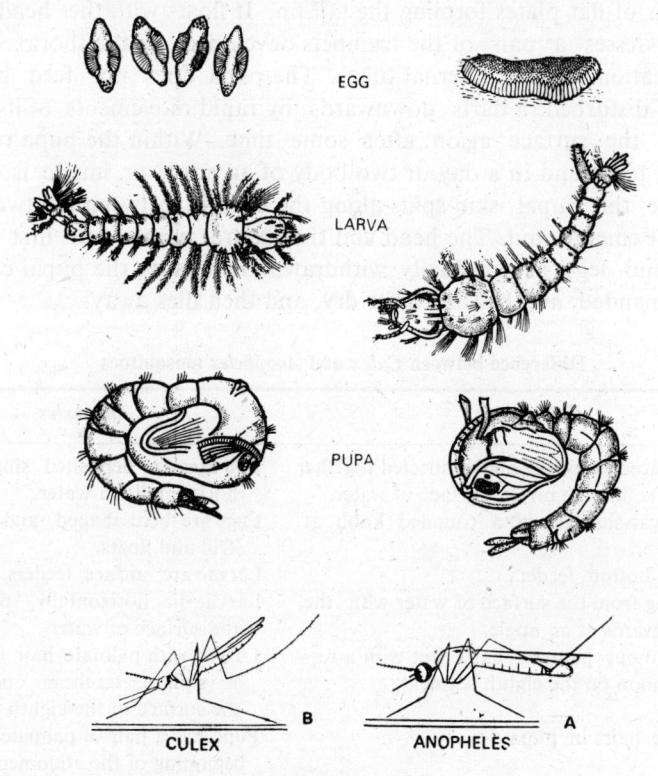

FIG. 17.34. Comparative study of the life-cycle of A—*Anopheles*; and B—*Culex*.

eyes or ocelli, feelers, jaws and two small feeding brushes. The movement of these brushes sweeps small organisms and other food particles into the mouth. The jaws, mandibles and maxillae, are strong as in cockroach and are meant for biting and chewing food material. The larva, thus, subsists on solid food. The thorax is unsegmented but bears three lateral pairs of unjointed tubercles, each bearing a tuft of bristles. This indicates the three-segmented thorax. The abdomen consists of nine segments, each bearing a tuft of bristles. The respiratory tubes, tracheae, open on the dorsal side of the eighth segment. In the *Anopheles* larva this opening is almost flush with the surface. In an undisturbed position the anopheles larva can breathe and feed lying parallel with the surface of water. This is further aided by the presence of palmate scales lying in pairs on some of the abdominal segments dorsally. In the *Culex* larva, however, the tracheae open on a long dorsal projection of the eighth segment, the siphon. This gives a forked appearance to the posterior part of its abdomen. While breathing the end of the siphon remains along with the surface of water and due to the weight, the head hangs downwards. It thus assumes a slanting position in water (Fig. 17.34). It can feed and breathe at the same time. The ninth segment in both types of the larva bears a tuft of bristles and four small leaf-like plates which contain the tracheae and which also act as gills. They are known as the **tracheal gills**.

**Pupa.** The larva casts off its skin three or four times before it reaches its full size. At the last moulting it passes into the next stage known as **pupa**. This is the resting stage. The pupa (Fig. 17.34) has a large rounded anterior part consisting of the head and thorax and a slender, bent posterior abdomen. The abdomen ends in a pair of flat plates forming the tail fin. It floats with the head end upwards. It possesses a pair of the trumpets developed on the thorax. These are in communication with the internal tubes. The pupa does not feed but is not quiescent. If disturbed, it darts downwards by rapid movements of its abdomen but rises to the surface again after some time. Within the pupa remarkable changes take place and in a day or two body of an adult or imago is perfected. At this time the pupal skin splits along the back in between the two trumpets and the adult emerges out. The head and thorax are pushed out first and then the wings and legs are gradually withdrawn. It rests on the pupal case till the wings are expanded, and became quite dry, and then flies away.

**Difference between *Culex* and *Anopheles* mosquitoes**

| *Culex* | *Anopheles* |
|---|---|
| 1. Eggs are placed vertically and connected together to form rafts floating on the surface of water. | The eggs are deposited singly and lie horizontally in water. |
| 2. Eggs are cigar-shaped with a rounded knob at the base. | Eggs are boat-shaped and possess the frill and floats. |
| 3. Larvae are bottom feeders. | Larvae are surface feeders. |
| 4. Larvae hang from the surface of water with the head downwards at an angle. | Larvae lie horizontally, parallel with the surface of water |
| 5. Larvae without palmate hairs but with a respiratory siphon on the eighth segment. | Larvae with palmate hair but without a siphon, tracheae open flush on the surface of the eighth segment. |
| 6. No palmate hairs in pupa | Pupa has a pair of palmate hairs at the beginning of the abdomen. |

| | |
|---|---|
| 7. The adult when at rest keeps the anterior and posterior ends of its body nearer the resting place; it thus appears humpbacked. | The adult when at rest keeps the body in a straight line making an acute angle with the resting place. |
| 8. The wings are not spotted. | The wings are spotted with scales. |
| 9. The maxillary palps in the female are very short. | The maxillary palps in the female are as long as proboscis. |
| 10. The maxillary palps in the male are pointed. | The palps in the male are club-shaped. |

## Indian Mosquitoes

*Anopheles, Culex, Stegomyia,* and *Nyssorhynchus* are common mosquitoes found in India. Malarial parasites are harboured in the *Anopheles* and *Nyssorhynchus*. At least 48 species of *Anopheles* are known; several species of *Culex* are also common.

## Economic Importance

The mosquitoes are definite nuisance as they inflict painful and irritating bites. Most of them carry germs of various diseases, such as **malaria, yellow-fever, dengue** and certain forms of **filariasis**. Female *Anopheles* spreads malaria while the common tropical gnat, *Culex*, is the carrier of larvae of *Filaria bancrofti* and causes filariasis in man and also conveys the virus of dengue fever. The tiger mosquito *Stegomyia* is the carrier of the yellow fever.

Being of medical importance mosquito naturally attracts a great deal of research, and it is probably true to say that more has been written about mosquitoes than about any other group of insects. Whilst the female mosquitoes are notorious for their biting, the most striking feature about the males is their habit of swarming. Large number of individuals, all of the same species, are found flying above, below or beside a landmark or marker which may be a small bush or post. While swarming insects show vertical movement so that the swarm appears as a more or less attenuated column of insects. Different species of mosquitoes often use rather different markers and they may fly at different distances from the marker. Male mosquitoes have the ability to detect and locate females for themselves from a range (in *Aedes*) of about 25 cm. The antenna of the male is equipped with many whorls of very long hairs, and its sub-basal is enlarged to house a battery of sense cells (Jhonston's organs) which is stimulated by the movement of antenna. In flight, the beating of the wings of the female mosquito makes a whining noise which causes the hairs on the male's antenna itself to vibrate and the sense organ is thus stimulated. Females of different species emit rather different notes.

Female mosquitoes respond to a variety of stimuli in locating their meal. Larman (1959) working with *Anopheles maculipennis atroparvus* found that smell is important in locating warm-blooded host; and that heat and moisture together are probably releasing stimuli for alighting upon the host. Wright and Kellogg (1962) write "it appears that *Aedes aegypti* are stimulated to fly in search of warm-blooded host by an alteration in the ambient carbon dioxide level, and that they are guided to the appropriate surface partly by colour, but mainly by warm, moist convection currents rising from it."

## TYPE HOUSE-FLY

No household pest is more annoying than the house-fly. They are in plenty during the rainy season, though they appear early in spring and continue till late in winter. Formerly they were regarded as **scavengers** and were not considered as a serious menace to our health. But it is now an undisputed and established fact based on evidences, both circumstantial and real, that the housefly is not merely an insignificant insect, but one that plays an important role in carrying and spreading certain diseases that are fatal to human life.

### Form of the Body

The adult house-fly (Fig. 17.35) is about 7 mm in length. The body is divided into the **head**, **thorax** and **abdomen** (as in cockroach and mosquito). It has only two thin membranous wings (Diptera), the second pair of wings is rudimentary and is represented by a pair of appendages called **halteres** or **balancers**

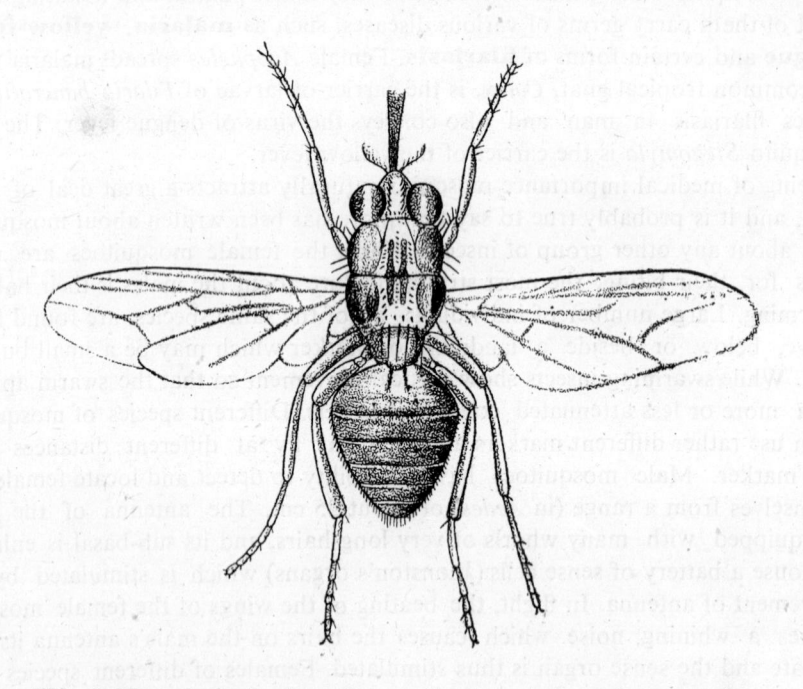

FIG. 17.35. A Cyclorrhaphous dipteran *Glossina palpalis*, male (tse-tse fly).

supposed to be balancing in function. There are several species of the housefly, the most common being *Musca domestica* in Europe and *Musca nebulo* in India, *Musca sorbens, M. ventrosa, M. vicina* and *M. conducens* are some allied flies frequently found about houses in India.

**Head**. The head (Fig. 17.36) bears a pair of large compound eyes occupying almost the whole of the hemispherical head and thus provides a wide field of vision of the insect. Each of the eyes consists of about 4,000 facets and consequently 4,000 ommatidia. The head bears a pair of **antennae** while the mouthparts are specially modified to facilitate its mode of feeding.

## Class Insecta

The hard chewing and biting jaws of cockroach are replaced by a greatly modified contractile tube-like structure called the **proboscis** the functions of which is to suck liquid food. The proboscis ends finally in a pair of soft cushion-like **oral lobes** (Fig. 17.36). When the house-fly sits on solid food the proboscis is protruded and the saliva is poured out on the object. Saliva dissolves even solid and makes it liquid ready to be sucked. The fly cannot swallow solid food.

The **labrum** is fused with **epipharynx** (cf. mosquito) and forms a narrow slender tube opening ventrally, into which fits the narrow blade-like **hypopharynx**. The mandibles and the first pair of **maxillae** are absent. Only **maxillary palps** are present. The labium is just like the half tube of the mosquito enclosing the labrum and hypopharynx Labium has no palps. The tip of the labium terminates in two fleshy heart-shaped lobes known as the labella or oral lobes (Fig. 17.36). Each lobe is traversed by over thirty narrow open channels. These channels open into a common channel which finally leads into the mouth opening. There are certain minute teeth-like structures on the undersides of the

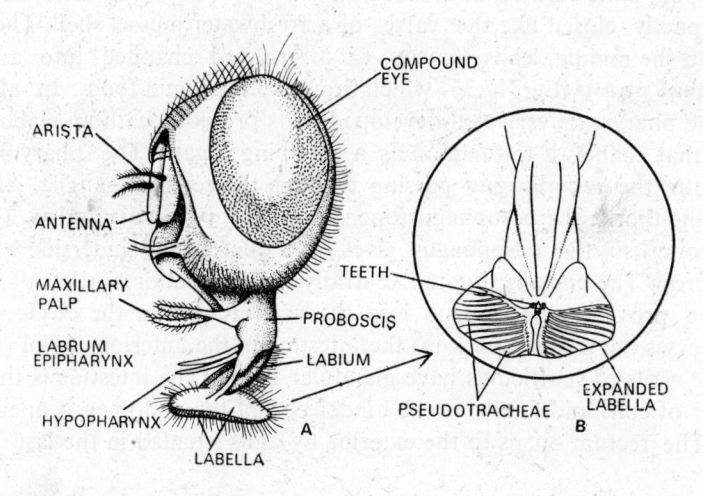

FIG. 17.36. A—Mouthparts of a house-fly; and B—Proboscis of the same, enlarged.

lobes to aid in rasping, breaking up of slids, so that they may be easily dissolved. Each of the oral lobes of the house-fly is provided with a large number of sense organs as to enable the insect to taste and smell. In a state of rest the proboscis is bent inwards.

**Thorax.** The thorax of the house-fly bears two membranous wings dorsally and three pairs of legs ventrally. In the adult stage the colour of the thorax is dusty grey dorsally having four dark and longitudinal stripes. The legs are covered with large number of hair-like bristles which are responsible for the carriage and transference of micro-organisms. Each leg is furnished with two sticky pads called **pulvilli** (singular **pulvillus**). Each pad or pulvillus is thickly beset with minute hairs which secrete tiny drops of a sticky liquid that literally sticks the fly to the object upon which it walks or sits. Unfortunately these sticky hairs from ideal organ for picking up all sorts of bacteria from the filthy materials upon which the fly mostly frequents. Thus, we see that the house-fly is suited, in many ways, for gathering and carrying the germs along with itself, but the relation of the insect with the germs it carries is a purely mechanical one.

**Abdomen**. The abdomen is distinctly marked out from thorax. In the adult it is the most pulpy bulging part of the body continuously held up from the ground on which it alights. Dorsally the abdomen is partially covered by the wings. The last four segments of the abdomen in the female fly form a peculiar tube-like structure called the ovipositor. When fully extended it is like long narrow tube, through which the eggs are conveyed to the places where laid. When not required for use the ovipositor is retracted entirely and telescoped within the abdomen. In the males the last four segments form the genital pouch and other structures associated with reproduction.

### Digestive System

The proboscis, as mentioned above, is continued into a pair of more or less heart-shaped soft cushion-like lobes, the **oral lobes**. When not in use these are compactly closed like the valves of a freshwater mussel shell. The **mouth**, situated in the middle, leads by way of a grooved channed into a powerful **pharyngeal pump** (Fig. 17.36) which draws up the liquid food. In all sucking insects the pharynx is very well-developed. It is provided with an elaborate musculature that enables it to function as a pumping organ. The pharynx is continued into the **oesophagus** passing through the cerebral ganglia. After entering into the thorax the oesophagus opens into the **proventriculus (gizzard)**, before doing so the oesophagus gives off a duct on the underside which leads into the crop. The **crop** is situated ventrally in the anterior portion of the abdomen. The proventriculus opens into the **ventriculus** or the **chyle stomach**. This traverses the thorax and joins the intestine in the anterior end of the abdomen. The walls of ventriculus have glandular lining. The intestine is thrown into a number of coils and finally at the hinder end of the abdomen it opens into the rectum. The rectum opens to the exterior by anus situated in the last segment.

### Respiratory System

The respiratory system of the house-fly is very highly developed. Altogether it occupies more space in the body of the fly than any other set of organs. It consists of three sets: the **spiracles** or **breathing pores**, situated on the sides of the body, **air-sacs** and **air tubes** or **tracheae**. A large pair of spiracles is situated over the bases of the first pair of legs. These spiracles supply air to the following: a set of air-sacs which fill up the vacant space in the head; a series of air-sacs in the thorax which give off tracheae to the muscles and legs, and two large air-sacs which occupy, in some cases, almost the whole of the front end of the abdomen and give off tracheae to the viscera. Above and behind the bases of the last pair of the legs is another pair of spiracles which supplies air to the large muscles of the thorax in that region. In addition to these thoracic spiraeles there are numerous pairs of spiracles situated at the sides of the abdomen, the abdominal spiracles. In the male there are seven pairs of abdominal spiracles but in the female only five pairs have been reported. All these spiracles communicate with tracheae which ramify among the intestinal organs of the abdomen.

## Circulatory System

The circulatory system of house fly is very simple. The body cavity forms a blood cavity or haemocoel so that all the organs and muscles, etc. are bathed by blood fluid which is colourless and contains fatty corpuscles. There is a muscular tube or "heart" lying in a cavity immediately under the dorsal surface of the abdomen. It extends from the posterior to the anterior end of the abdomen and is divided into four chambers, each having a pair of openings through which the blood is sucked, so to speak, from the pericardial cavity. The heart is continued as a dorsal vessel along the chyle stomach and appears to terminate in a mass of cells behind the proventriculus. Associated with the blood system is a diffuse structure known as the **fat body** which consists of a large number of very large fat cells well supplied with tracheae. The size of the fat body varies considerably, just before hibernation it seems to fill almost the whole of the abdominal cavity and after hibernation it is found to have shrunk almost to nothing. While it may have some excretory function, it would appear that it also stores up the products of digestion, which it obtains from the blood in which it is bathed.

## Reproductive System

Sexes are separate and are slightly different in size, the female being slightly larger than the male. The female reproductive organs consist of a pair of **ovaries, spermathecae** or **vesicula seminalis** and **accessory glands** and their ducts. The ovarioles in each ovary are much numerous than those in the cockroach.

The number of the ovarioles in each ovary is 70 and each consists of ova in various stages of development. The two ovaries occupy the major portion of the abdomen. The two oviducts unite to form a common duct, the **vagina**, which swells up posteriorly to form a **sacculus** or **ovipositor**. There are three **spermathecae,** two on the left side and one on the right. They all open on the dorsal side of the ovipositor. There is a pair of accessory copulatory vesicles.

The male reproductive organs consist of a pair of **testes, vasa deferentia, ejaculatory sac, ejaculatory duct**, and the terminal **penis**. The testes are always pyriform. The vasa deferentia are short and join each other distally to form a common ejaculatory duct which is a long winding tube. In association with the ejaculatory duct there is a muscular ejaculatory sac which contains a chitinous filiform **ejaculatory apodeme,** which aids in propelling the seminal fluid along the genital canal during copulation.

## Life-history

The life-history (Fig. 17.37) of house-fly provides an excellent example of complete metamorphosis. The insect begins its life-history after copulation followed by oviposition and then the eggs hatch into larvae which pass through a resting stage or pupal stage. The fecundity (rate at which the houseflies breed) is simply amazing. A single fly lays 100 to 150 eggs. While each fly is able to lay such 5 or 6 deposits in one season, the entire life-cycle takes only 10 days. Thus, if the progeny of a single pair of flies is allowed to develop without destruction the figures will go very high. C.F. Hodge and Malcolm Burr calculated the probable number of progeny for *Musca domestica* at the end of one season and came to the conclusion that a pair of flies and its progenies, if all were to live, will produce 191,010,000,000,000,000,000 (C.F. Hodge, *Nature and Culture*, 1911) or 5,598,720,000,000 (Malcolm Burr, *The Insect Legion*, 1939).

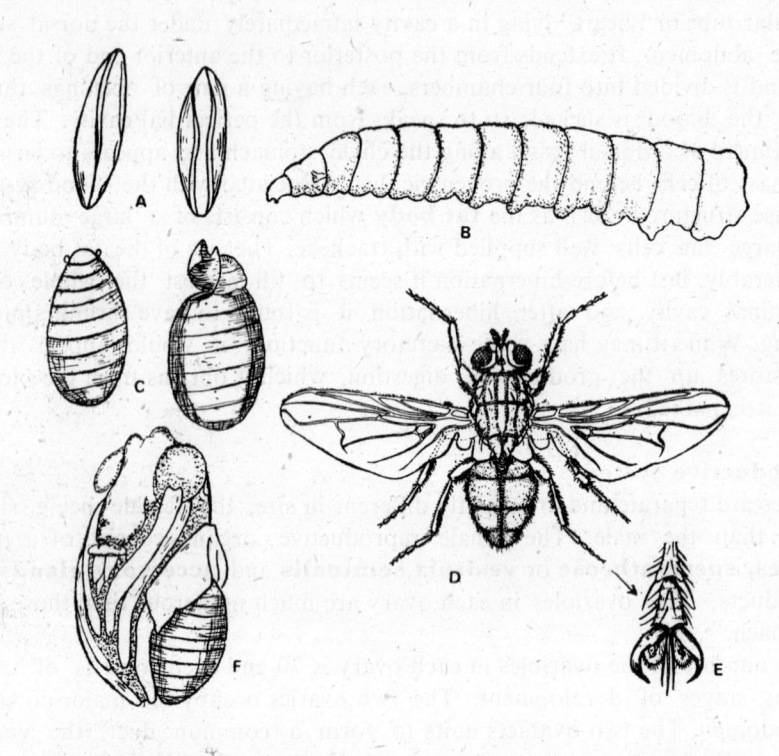

**FIG. 17.37.** Life-cycle of a house-fly. **A**—Eggs; **B**—Larva; **C**—Different stages of pupa; and **D**—Adult house-fly.

**Breeding season.** Normally the breeding season is form June to October, while the maximum activity is seen in the months of August and September. Temperature plays an important part in the development of the house-fly. A temperature of about 30-35°C will enable the fly to develop from egg to adult stage in five days.

**Breeding places.** Horse manure forms the principal place for this purpose but the flies also use cowdung, faeces, cesspool and other decaying vegetable and animal matter.

**Copulation.** The male alights on the back of the female, who extends her ovipositor and inserts into the genital pouch of the male. The two remain in position for some time, the female remains responsible for flight during the period.

**Oviposition.** The eggs are laid 6 to 8 days after copulation. The female extends her ovipositor into the manure dung or whatever substance she has selected as the nidus and lays about 100 to 150 eggs in groups at a time. The eggs are oval in shape about 1 mm in size and pearly white in colour and have two rib-like thickenings.

**Hatching.** In warm weather the egg takes 8 to 24 hours to hatch. The larvae, specially called maggots, emerge from a slit along one side of the egg; while growing up the larvae cast off their skin twice, thus dividing the larval period into three **stages** or **instars** (*i*) from hatching to the first moult that takes 20 to 30 hours, (*ii*) between the first and second moult taking another 24 hours, and (*iii*) from second moult till pupation that lasts three days ordinarily.

**Larva**. Full grown larva is creamy white in colour and measures about 12 mm in length. The body tapers from middle to anterior end which carries a pair of oral lobes (Fig. 17.37). The segment bearing the oral lobes is called the **cephalic segment**. Behind this are 12 trunk segments. Mouth is situated between the oral lobes on the ventral side. On the anterior border of the oral opening is situated a tooth-like black hook-like process called the **mandibular sclerite**. This is helpful both in locomotion and in tearing up the food. Larvae always consume semi-liquid food. For respiration each larva has four respiratory apertures or the **spiracles**, two anterior ones at the sides of the second segment, while the two posterior ones are located in the middle of the blunt posterior end. Ventrally each segment, except the first six, has got a crescentic pad beset with tooth-like processes. These are **locomotory pads** used in conjunction with the hook. The **anal** opening is situated on the ventral side of the last segment.

It is noteworthy that the larva is a bright shining creature whose life is greatly influenced by humidity, temperature and food supply. Scarcity of any of these retards the development and thus the flies grown under such conditions are undersized.

**Pupation**. When fully matured the larva usually crawls away from its original moist home and travels, sometimes many yards, to some dry and sheltered crevice. Here it rests for some time preparatory to pupal stage. The emergence of pupa is preceded by a general contraction of the body. Some of the anterior segments are even withdrawn within the body. The larva, which becomes almost barrel-shaped in a few hours time, changes its colour from creamy white to dark brown. Thus, we see that it is the larval skin that forms the pupal skin or the **puparium**. The ventral locomotory pads of the larva can still be recognized as rough crescentic areas on the ventral side of the puparium. Within the puparium the larval organs undergo degeneration, leaving only some groups of cells called the **imaginal discs**. These cells build up the organs and the body of the future fly anew. The whole of this process takes three to four days.

When the fly is ready inside the puparium it has a sac-like structure situated on the front of the head. This is known as the **ptilinum**. With the aid of this process the young fly pushes open the anterior end of the puparium, and comes out. The ptilinum is used only in the act of emergence of the fly from the puparium, thereafter it is withdrawn in the head. Its position is indicated externally by a crescentic slit called **lunule** situated above the base of the antennae.

At the time of emergence the flies appear plump, loose and whitish in colour. Soon, however, the exoskeleton darkens and becomes hard. The flies become sexually mature in about a fortnight's time and start laying eggs, four or five days after attaining maturity. Thus, from the limbless, headless, light avoiding larva develops an active, sun-loving aerial insect, the house-fly. Life of flies is not constant. Those bred in summer live for about two months, whereas, those bred before winter live through the winter months to summer.

## Economic Importance

House-fly has the habit of regurgitating a part of the food it takes in, and forms a vomit-spot on the surface it sits on. Such vomit-spots are often seen on glass panes after their visit. As seen earlier their mouthparts are adapted for sponging liquid foods. It is a known fact that houseflies visit all kinds of substances; they suck the substances on which they sit indiscriminately and much of the rubbish adheres to the hairs on their body. All kinds of bacteria find their way to the crop of the flies where they breed profusely. Many of these are passed undigested with their faeces to the different places they visit, and many are further thrown out with their vomit. Several organisms isolated from the body of the house-fly are such that cause dangerous diseases and epidemic such as cholera, typhoid, paratyphoid, anthrax, diarrhoea, dysentery, tuberculosis and a number of maladies caused by parasitic worms.

## Blood-sucking Flies

Among the mosquito (suborder Nematocera) blood-sucking is confined to the females, they alone being equipped with mandibles and maxillae which can pierce vertebrate skin. Males often feed at flowers and the proboscis was probably orginally an organ adapted for the taking of nectar. Among the suborder Brachycera the horse flies (Tabanidae) have some members that suck vertebrate blood. Among the higher (Cyclorrhaphous) Diptera there are two distinct groups which suck vertebrate blood. Firstly, there is a group of flies allied to the ordinary house-fly that spends most of its time away from the host and only visits the host for short feeding periods. And secondly there is a group of highly specialized ectoparasites, the horse flies and the bat-flies, which spend the greater part of their adult life, indeed sometimes all of it, on the body of the host.

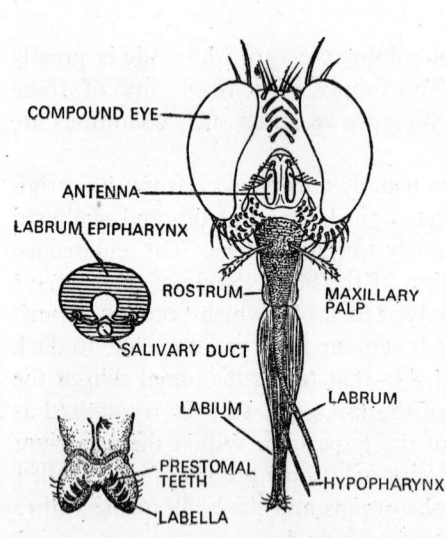

FIG. 17.38. Mouthparts of the stable fly, a blood-sucking dipteran.

Ordinarily the mouthparts of dipterans are soft having an elongated proboscis through which liquid food may be drawn. In the cosmopolitan stable fly (*Stomoxys calcitrans*) the proboscis itself is hardened into a stabbing organ with large prestomal teeth (Fig. 17.38) for penetrating vertebrate skin. Both man and his domestic animals are attacked by *Stomoxys*. *S. calcitrans* is, at first sight, easily mistaken for a house-fly but its bite at once reveals its true identity.

## TYPE HONEY-BEE

The domestic honey-bee, *Apis mellifera*, is the best known among the world's nectar gatherers. It has been kept under domestication for thousands of

*Class Insecta* 707

years and is now the common domesticated animal on almost every land. It was mentioned in Sanskrit writing as long ago as 3000 B.C. and was well-known in ancient Greece and Rome. Inscriptions on early Egyptian tombs prove that these industrious insects were kept for their honey at least 4,000 years ago. The Babylonians, like the Hindus, used honey for food and medicines and as offerings to their Gods.

Honey-bees are semi-tropical insects that have adapted themselves in temperate climates where they survive cold weather by forming themselves into winter clusters or balls within the hive, hollow trees or wherever the colony has taken up residence. These clusters are formed whenever the temperature within the hive falls to 57°F or lower. Within the cluster, a temperature of about 90°F is maintained by bees fanning their wings, thereby creating heat by muscular activity. This temperature is remarkably constant and is one of the rare instances among the cold-blooded insects where any effort is made to control temperature.

## Form of Body

The body of the honey-bee, like that of any other insect, is divided into three parts—head, thorax and abdomen—well separated from each other. The **head** carries the eyes, antennae or feelers and mouthparts; the **thorax**, the wings and legs; and the **abdomen** the **wax-glands** and **sting**.

**Head.** The flattened triangular **head** is widest crosswise through the upper corners, which are capped by the large compound eyes. It carries the antennae on the middle of the face (Fig. 17.39); the large compound eyes and three ocelli occur at the top of the face; and mouthparts at the bottom of the face. Each antenna is jointed, is sensitive to touch and contains the organs of smell. At the lower edge of the face is a loose flap, the **labrum**. On its underside is a small soft lobe on which are located the organs of taste. At the sides of the labrum are the two heavy **mandibles** which work sidewise. In the workers they are spoon-shaped as they lie, but sharp-pointed and toothed in the queen and drone, those of the queen being the largest, those of the drone the smallest. Behind the jaws is a bunch of long appendages usually folded back beneath the head, which together forms the **tongue**, that is actually modified labium. When the bee wishes to suck

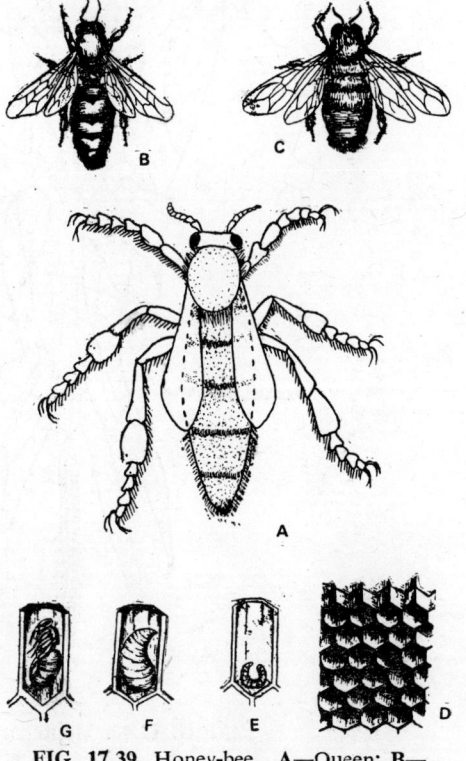

FIG. 17.39. Honey-bee. A—Queen; B—Worker; C—Drone of the honey-bee; D—A portion of the comb; E—A larva; F—Prepupa; and G—A pupa in cells.

up any thick liquid such as honey or syrup, the terminal lobes of the labium and maxillae are pressed close together as to make a tube between them. The food is then taken in by a sucking action of the pharynx. At the tip of the tongue there occurs a spoon-like lobe or **labella** and there is a groove running along the entire length on the ventral side. Within the tongue this groove extends into a double-barrel tube. A flexible chitinous rod (Fig. 17.40) lies along the back wall of this channel which is itself provided with a still finer groove along its ventral surface. This delicate apparatus is perhaps necessary for sucking up minute drops of nectar from the base of a flower. Even the smallest quantity of nectar finds its way up the groove by capillary action.

**Thorax.** The thorax comprises legs and wings. The wings are two, paired one on each side, and are united to each other by a series of minute hooks so that they work together, and the four wings are practically converted into two. Each wing is hinged at its base to the back, and pivoted from below upon a small knob on

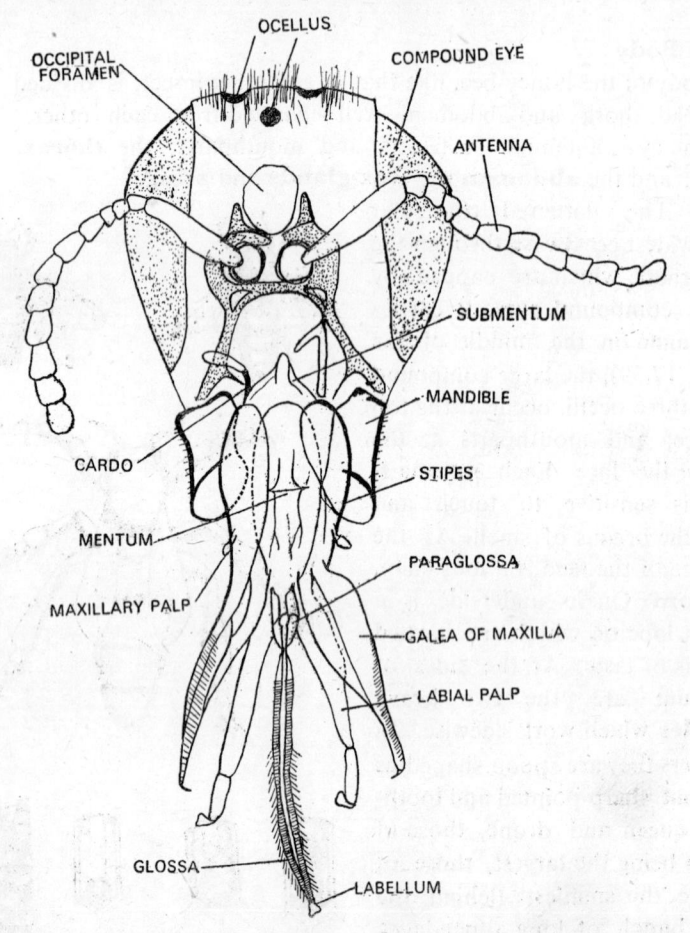

FIG. 17.40. Mouthparts of a worker bee.

the side wall of the thorax. The up and down motion of the wings is produced, not by muscles attached to their bases, but by two sets of enormous muscles, one vertical and the other horizontal, attached to the walls of the thorax whose contractions elevate and depress the back plates of the thorax.

## Class Insecta

There are three pairs of legs (Fig. 17.41)—pro-, meso- and metathoracic—each of which consists of coxa, trochanter, femur, tibia and a five-jointed tarsus. The tarsus terminates in a pulvillus and lateral claws. The legs are covered with bristles, and have other structures used in collecting and carrying pollen and in cleaning pollen from the body. An **antennae cleaner** and an **eye-brush** are present in the prothoracic legs. In the workers occurs the **pollen basket**, a concavity on the surface of the tibia of the metathoracic leg; and there is spur on the mesothoracic leg used to pry pollen from the basket.

**Abdomen.** The **abdomen** of the bee has only six visible segments and those of the head and thorax, but bears two important organs, viz. the wax glands and the sting. The wax glands are simply developed cells of the skin of the undersurfaces of the last four visible segments of the worker. The wax is secreted by the glands discharged through minute pores on the underside of each segment and accumulates in the form of a little scale in the pocket above the underlapping ventral plate of the segment next in front.

The **sting** is such a complicated organ that it is difficult to describe it clearly in a few words. The actual sting or **terebra** is a hollow organ formed of three

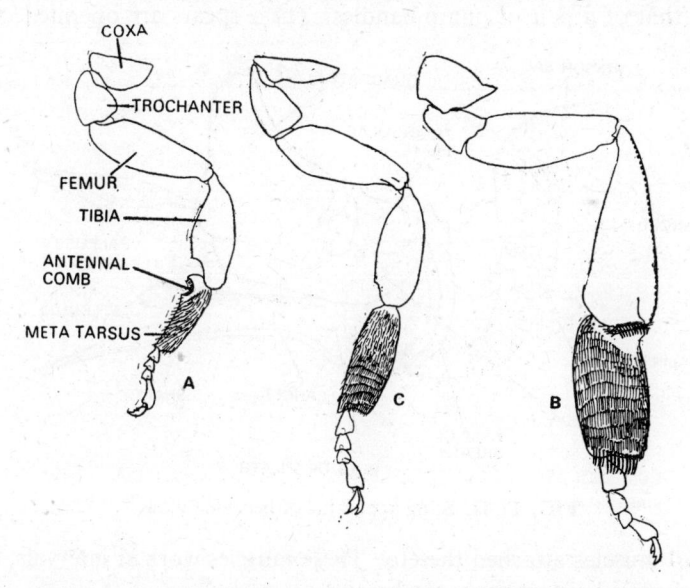

FIG. 17.41. Legs of a worker bee. A—Front leg; B—Middle leg; and C—Hind leg of a worker.

pieces bounding a central canal. The dorsal part of **stylet-sheath** has three functions: (*a*) to cut the wound; (*b*) to serve as the dorsal wall for the poison canal; and (*c*) to hold the stylets in position. The stylet-sheath (Fig. 17.42) expands at its base to form the **bulb** of the sting and the latter continues inwards as a pair of diverging arms. Within the sheath are two barbed **spears** or **stylets** or **lancets**, which are grooved along their entire lengths and along the sheath are two guide rails which fit accurately into the stylet-grooves (Fig. 17.42). The barbs of the stylets are much like those on a common fish hook. Associated with the sting are three pairs of plates. (*i*) The innermost pair is that of the **oblong plates** posterior in position and representing the divided ninth sternum. To each is attached the basal arm of the stylet sheath of its side. Distally each carries a palp-like appendage on its side. (*ii*) The two **triangular** or **fulcral plates** represent the reduced sternum of the eighth segment and to each is attached the corresponding arm of the stylet. (*iii*) The large **outer** or **quadrate plate** lies dorsally to the triangular plates at the posterior angle; the quadrate plates represent the sternum of the ninth segment. These three pairs of plates act as the sternum of the ninth segment and also function as levers. Two sets of **poison glands** are found associated with the ovipositor. A pair of filiform **acid glands** open either separately or by means of a common duct into a large **poison-sac** (Fig. 17.42), and their secretion has an acid reaction. The poison-sac discharges into the anterior end of the bulb of the sting and situated close to its aperture is the opening of an unpaired **alkaline gland** that secretes alkaline fluid.

The working of the poison gland is itself interesting. When the point of one spear penetrates far enough to get one barb under the skin, the bee has made a hold, and has no difficulty in sinking the whole length of the sting into the wound; the pumping motion at once commences, and the other spear slides down a little beyond the first, then the first beyond the second and so on with a motion like that of a pair of pump handles. These spears are operated by small

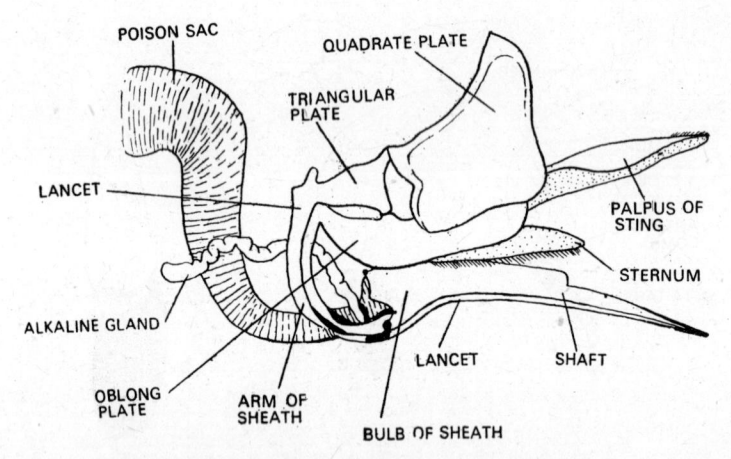

**FIG. 17.42.** Sting apparatus of bee, side view.

but powerful muscles attached thereto. These muscles work at intervals, for some time, after the sting is torn from the bee.

### Alimentary Canal

The **mouth** opens into the **pharynx**, a muscular widened tube, (Fig. 17.43) that narrows into a slender long **oesophagus**, running clear through the thorax into the front end of the abdomen, where it enlarges into a thin-walled sac, the **honey-sac** (the crop, in general). The sac narrows posteriorly and opens into the stomach proper. Then comes the slender small intestine with a circlet of slender tubes, **Malpighian tubes**. Finally forming the terminal part of the alimentary canal is the **rectum** of a varying size externally opening at the **anus**.

The honey-sac is of special interest in the worker. The nectar gathered from the flowers is held in it instead of being swallowed down to the stomach. From the honey-sac the nectar is **regurgitated** into the cells of the comb or given up first to another bee. The upper end of the true stomach sticks up into the lower end of the honey-sac as a small cone with an "x" shaped valve in the summit.

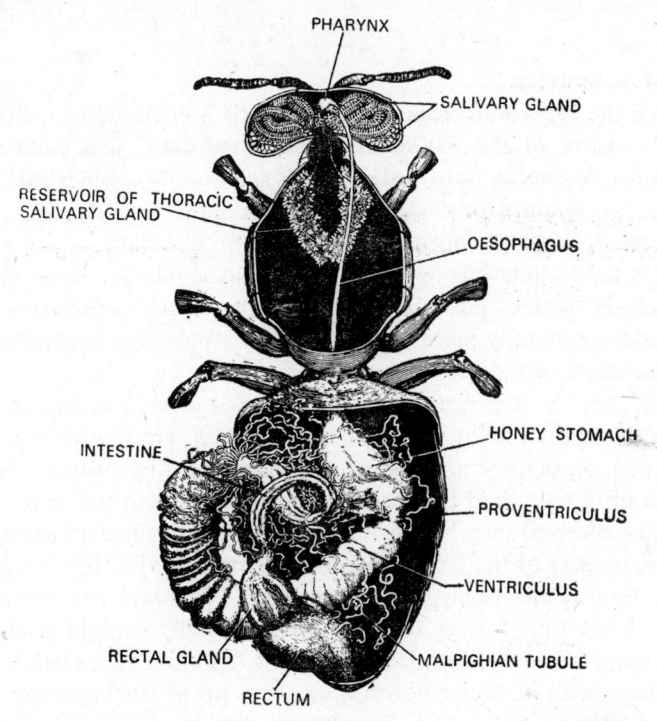

FIG. 17.43. Dissection of digestive organs of honey-bee (worker).

This opening is called the stomach mouth, guarded by four very active lips that take whatever food the true stomach requires from the honey-sac. The natural food of bees consists of pollen, nectar and honey. The **salivary glands** are located on the back of the head in the front part of the thorax and open upon the upper part of the labium. The saliva can thus affect the liquid food before the latter enters the mouth.

### Circulatory System

The circulation is maintained by a slender, dorsal contractile **heart** as in other insects (see cockroach). The heart in this case consists of four consecutive chambers, which are merely swellings of the tube, each having a vertical slit or **ostium** opening into each side.

### Respiratory System

The respiration is brought about by the tracheal system as found in other insects (see cockroach). In all there are 10 pairs of **spiracles**, none occurs in the head.

### Nervous System

The dorsal cerebral ganglion or "**brain**" communicates with the double ganglionated **ventral cord** by two commissures around the oesophagus. The organs of special sense are well-developed. The **antennae** bear end organs of smell, hearing and touch. Taste organs are situated near the mouth and on the tongue. There are two large **compound eyes** and three **ocelli**, all of the typical arthropodan type.

### Reproductive System

In the male the **testes** are separate. Each testis is enclosed in a double membrane and consists of 250-300 **seminiferous tubules**. The **vasa deferentia** enlarge to form **vesiculae seminalis** which are usually cylindrical or sac-like in form. The two **ejaculatory canals** which leave the vesiculae are rudimentary and the accessory glands open into the common ejaculatory duct. The ovaries in the females have numerous ovarioles. The two oviducts unite to form the vagina which is dilated posteriorly to form the **bursa copulatrix**. A median **spermatheca** is generally present together with a pair of colleterial glands which open into the ducts of the spermathecae.

The queen bee is impregnated by the male or drone but once in their lifetime, during the nuptial flight when the spermatozoa are received in the spermatheca and the copulatory organs of the male are torn away to remain in her genital bursa until recovered by workers after her return to the hive. The spermatozoa thus received may live for many years and fertilize all the eggs she will ever lay. The ovaries of the female begin to enlarge to fill the long abdomen, and in a day or two she begins to lay her eggs. The queen can control fertilization of eggs. Unfertilized eggs develop into genetically haploid drones or males (with 16 chromosomes) and fertilized eggs yield diploid females (with 32 chromosomes). In the season of nectar flow a queen lays up to 1000 eggs per day, gluing each to the bottom of a cell. The **larva** that hatches is tiny and worm-like devoid of legs or eyes. For two days all larvae are fed on "royal jelly" produced by the pharyngeal glands of young workers. Thereafter, drone, and worker larvae receive mainly honey and pollen, but queen larvae continue chiefly on royal jelly which makes them grow differently and they become larger. Each larva grows and moults several times then its cell is sealed with wax and the larva within spins a thin **cocoon** and changes into **pupa**. After complete metamorphosis it finally cuts the cell-cap with its mandibles and emerges as young bee. The time of development for each caste is standardized because of the temperature regulation in the hive. Queen takes 16 days—3 days as egg, $5\frac{1}{2}$ days as larva and $7\frac{1}{2}$ days as pupa; worker takes 21 days—3 days as egg, 6 days as larva and 12 days as pupa; whereas, the drone takes 24 days—3 days as egg, $6\frac{1}{2}$ days as larva, and $14\frac{1}{2}$ days as pupa.

Bees are very ancient creatures. They were fully developed in their present form before modern mammals had evolved from their primitive beginnings, long before man had become man. The honeye-bee is a symbol of industry and cooperation, gathering food in time of plenty against the needs of winter. When the warmth of spring brings early flowers, the workers gather nectar and pollen, the queen lays eggs rapidly, and new workers soon swell the population of the colony. When the hives become overcrowded, the old queen with several thousand workers emerges and flies to a new site previously located by worker scouts. This phenomenon is know as **swarming**. Before swarming is to take place some new queen larvae are started in the old colony. Later one of the queens emerges, stings the other queen larvae, is fertilized in a mating flight, and returns to serve the old hive in laying eggs and increasing the colony. The life of a queen may last from three to five seasons and she may lay a million eggs in her life-time. Drones are produced during active nectar flow, but, thereafter when brood production ceases they are mostly driven out to starve and die. Many of the workers live for 6-8 weeks; those hatched in autumn survive till spring.

## Comb

A beautiful work in nature is a piece of honeycomb with its snowy whiteness and its burden of sweetness. The walls of its cells are so thin that from 2000 to 3000 of them must be laid one upon another to make an inch in thickness. Each wall is so fragile as to crumble at a touch and yet so constructed that tons of honey stored in them is transported in safety to thousands of miles. The comb is made up of wax-scales of real bees-wax from the abdomen of a bee working at comb-building. These scales are somewhat pear-shaped, much more brittle than the wax, transparent, looking somewhat, like mica. Some say they are white, some say pale yellow, probably both are right, as the colour depends upon the pollen consumed.

Wax is nearly allied to fats and after being secreted it is converted into plastic material by prolonged chewing with secretions. It is with this material that the comb is made. A chief use for the honeycomb is to furnish cradles for the "baby" bees during their brood stages. The cells, therefore, are made to suit different types of "babies" they have to accommodate. The **worker cells** where workers are reared and honey and pollen is stored are 5 mm across; the **drone cells** are 5 mm across, serve to rear drones and for storage; and the **queen cells** are large peanut-like, open below and are built along the lower margins for queen rearing.

Honey-bees alone maintain necessary temperature within their cells, in other words, they are the only animals that achieve "air conditioning." In summer they ventilate the hive by fanning their wings vigorously and keep the temperature at about 33°C for brood rearing, and to evaporate excess water from honey in open cells. In hot dry weather they carry in water to increase the humidity of the colony and to dilute the honey if necessary. In chilly season, when the stored honey is used as food, they form a compact cluster and produce heat by the active movements of body and wings. Such actions can raise the temperature to 14°C or below to 24°C or 30°C even when outside the hive the temperature is below, at or near freezing point.

Bees have many enemies including spiders, ants, kingbirds, bee-martins, toads, skunks, and bears who prefer honey. Too serious diseases cause heavy losses in colonies. Bee colonies are also reduced by dearth of nectar and by the exhaustion of the honey stock in winter.

### TYPE BEDBUG

The bedbugs are domestic parasites of man that normally suck human blood. They belong to the order Hemiptera which includes plant-lice, frog-hoppers, white flies, scale insects and capsid-bugs, etc. It includes 30-40 species of wingless brownish blood-sucking insects, mostly parasitic on birds and bats.

There are two widely distributed species of bedbugs which attack man: one is the common cosmopolitan bedbug, *Cimex lectularius*, specially found in all temperate climates, the other is the tropical or Indian bedbug *Cimex hemipterus* (*rotundatus*) prevalent in many tropical countries including Southern Asia, Africa, and West Indies. The tropical bug differs from the common one only in minor details, such as greater length of body hairs, darker colour and more elongate abdomen. Another species *Leptoeimex boneti* attacks man in Guinea. Some species of the genus *Cimex* confine their attention to poultry and other birds, bats, etc.

The bedbugs are old pests of man even by historical standards. They are recorded from ancient Greece and China. It has been suggested that men gradually acquired these pests from bats and birds in prehistoric times while living in caves and shelters. Both species of bedbugs are less specific in their food requirements and are able to live and breed on fowls, rabbits, mice and rats, etc.

## General Characters

The bedbugs have flat oval bodies and are devoid of wings except for a pair of spiny pads which represent the first pair of wings. The prothorax has wing-like expansions at the sides which grow forward and partially surround the head. In the male the abdomen is quite pointed at the end, whereas, in the female it is evenly rounded. In unfed bugs the contour of the abdomen is almost a perfect circle. The eyes project prominently at the sides of the head; the antennae

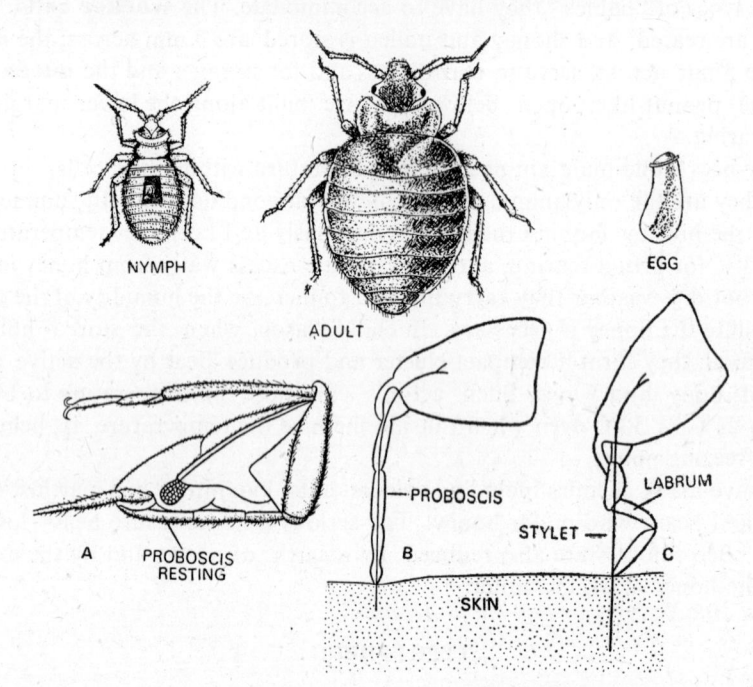

FIG. 17.44. First row centre—Adult bedbug (*Cimex*); Left—Nymph of the same; Right—Egg; A—Head and thorax with proboscis in the resting position; B and C—Dia-grammatic representation of stylet penetration

are flexible and four-jointed, and the jointed beak is folded under the head, so that it is not visible from above (Fig. 17.44). The greater part of the body is covered with bristles set in little cup-shaped depressions which are perforated at the bottom for the passage of muscles that moves the bristles. The bugs generally raise their bristles when they meet one another (Murray). The legs are also covered by a dense set of bristles. One of the most striking characteristics of the bedbugs is the peculiar pungent odour so well-known to all who have had to contend with pests. The odour is produced by a clear volatile fluid secreted by a pair of the glands of very variable size which open between the bases of the hind pair of legs.

## Habits

Bedbugs are normally night prowlers and exhibit a considerable degree of cleverness in hiding themselves in cracks and crevices during the daytime. They try to keep their body surface as much in contact with other surfaces as possible. Cracks and crevices provide this. They also tend to settle on surfaces where other bugs have lived even if these are in exposed situation. They, thus, are found in crowds, all stages cast skins and eggs, themselves fouled by black and whitish dots of excrement.

As a result of companionship with man for many centuries they have become quite experienced. They leave the bed during daytime and return again at night. In natural infestations active roving begins usually before midnight. Bugs seldom cling to the skin while sucking, preferring to remain on the clothing. In the course of 13 or 15 minutes a full meal is obtained and the bug, distended fully, retreats to its hiding place. A single meal is sufficient for about a week, although the bug may become ready to feed again in a day or two. The bite of bedbug seldom produces pain or swelling unless, rubbed or scratched. This indicates that their saliva is not irritating. Although human blood is their normal food, bedbugs are able to subsist on the blood of such animals as rats, mice, rabbits, cats, dogs and even chicken.

It is commonly believed that after roving the bugs habitually return to their resting places but experimental evidence to this is lacking. Because the number of resting places within quick and easy reach is limited it is natural that the same resting place may be used again and again. It is also believed that bugs take an unerring course to the host even though the latter may be yards away. Experiments, however, show that the perception of the host, which is a perception of warmth, not smell as far as known—does not occur beyond the range of 5 cm, and that bedbugs will wander at random rather than in a directional manner in quite small room containing the host and even then frequently fail to find it.

## Mouthparts

The mouthparts consists of four needle-like stylets lying in the long jointed lower lip or **beak (rostrum)**. The needles are grooved so as to fit together along their length. They consist of a pair of flattened, sharp-pointed **mandibles** and a pair of slightly shorter **maxillae** with serrated edges. The mandibles, worked by muscles in the head, slide independently on either side of the two maxillae which do not move independently of each other. The inner maxillary surfaces are grooved so as to form two exceedingly fine channels (Fig. 17.45) running along their entire length. One of these is the **feeding channel** and is connected to the pharynx. Blood is sucked through it. The other is the salivary channel, joined by ducts to the salivary glands, and is connected to a small pump

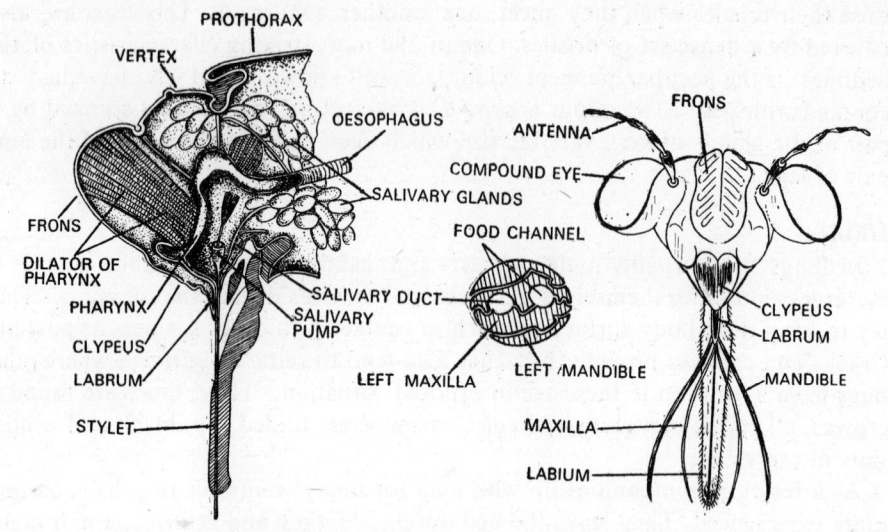

FIG. 17.45. Mouthparts of a generalised bug.

beneath the pharynx. All these are enclosed in the trough-like beak. The beak is grooved in such a way that the sides of the groove almost close together, thus forming a protective sheath for the stylets inside (Fig. 17.45).

When about to indulge in a meal the beak is bent back and the piercing organs gliding up and down past each other, are sunk into the flesh of the victim (Fig. 17.45). A strong sucking motion of the pharynx, into which a bit of salivary juice has already been poured, draws blood up through a tube made by the piercing organ.

## Alimentary Canal

A true mouth is absent, and the actual entrance into the digestive tract is the aperture of the suction canal which is situated at the apex of the maxillary stylet. The **suction canal** communicates with the **pharyngeal duct**, a continuation of the **pharynx** proper which is the principal organ of suction. The mid-gut is frequently divisible into serveral regions. The **hind-intestine** is not long and consists of a small bladder-like chamber, often quite extensive and opens at the anus. **Malpighian tubes** are four at their usual place, *i.e.*, beginning of the proctodaeum. The **salivary glands** exhibit a marked uniformity of structure among Heteroptera, the group to which the bedbug belongs. The principal gland is many-lobed and is drawn into main **salivary ducts**. There is filiform accessory gland also opening by a narrow duct into the salivary duct of its side. The two main ducts of opposite side converge to form a common canal opening into the **salivary pump**, a characteristic structure of the group, that opens into the **hypopharynx**. It propels the saliva down the ejection canal.

## Nervous System

The nervous system exhibits a very uniform and complete degree of concentration. The abdominal ganglia, to a large extent, are fused with the thoracic, though the connectives persist as the main single or paired abdominal nerve which gives off lateral segmental branches.

## Reproductive System

Each **ovary** has seven ovarioles which are composed of a small number of follicles. The two oviducts unite to form a common canal as in other insects

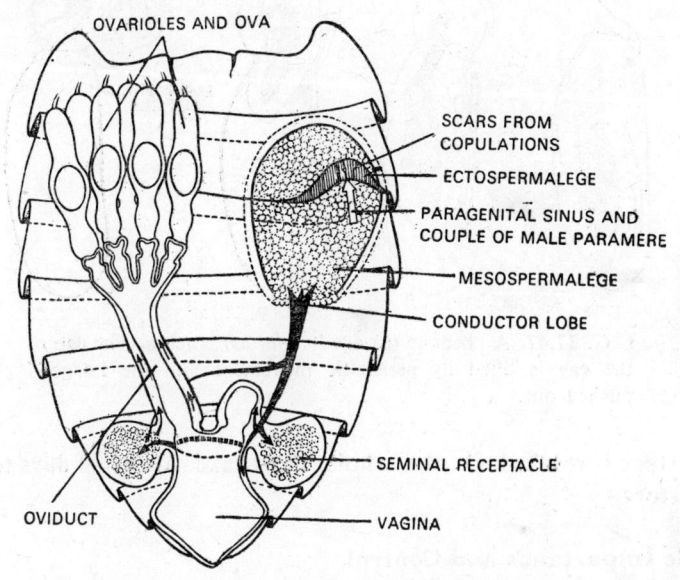

FIG. 17.46. Reproductive organs of bedbug.

(see cockroach). There is a small unpaired body known as the **organ of Berlese**, which is situated in the ventral region of the abdomen and functions as a copulatory pouch which receives the spermatozoa during coition. After that the sperms pass in large numbers through the haemocoel into the spermathecae which are unconnected with the common oviduct (Berlese).

The male organs consist of **testes, vas deferens, accessory glands** and **aedeagus**, etc., as usual.

## Life-history

The eggs of the bedbugs are pearly white oval objects, furnished with a little cap at one end which is bent to one side. The eggs are about 1/10 cm in length and are laid singly or in small batches. The ovaries hold about 40 eggs, at a time, all almost at the same stage of development. Often a female returns to lay more eggs in the same place so that batches of 40 or more may be found in the crevices where the adult insects hide.

The eggs hatch (Fig. 17.47) in from 6 to 10 days during warm weather but are retarded in their development by cold. The freshly hatched bugs are very small, about 1.5 mm long flat, active, delicate, and pale in colour. After a few hours it is able to pierce the skin and suck the blood, and if undisturbed it feeds to repletion in about three or four minutes. It may take a meal three or four times its weight and become globular and bright red. It needs shelter but no more food till it moults into the second stage. After their first hearty meal they have a much more robust appearance, and grow rapidly. After about five moults

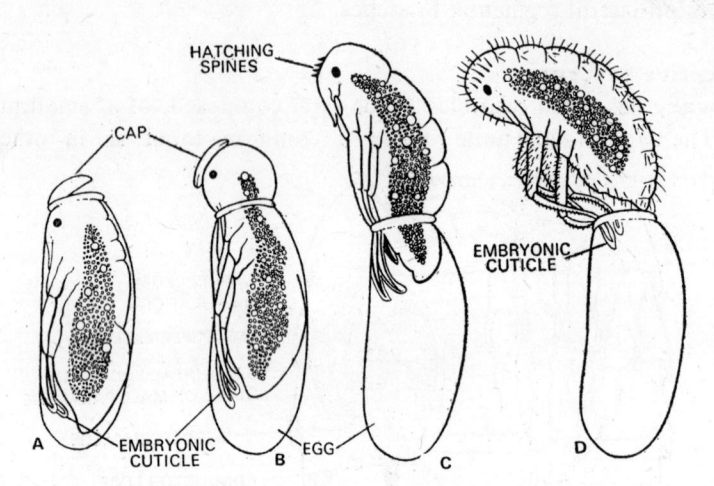

**FIG. 17.47.** A bedbug (*Cimex lectularius*) hatches. The cap of the egg is lifted by peristaltic movement and the imago is pushed out.

the adult stage is reached. On the whole they take about 29 days to complete the life-history.

### Economic Importance and Control

They feed on mammalian blood by piercing the skin, causing, in some cases only, pain and inflammation. It is one of the most disgusting pests of the bed. The bedbugs have been accused of the transmission of almost every form of epidemic disease. It is doubtful, however, if they are active or chief vectors of any major disease, but they appear to possess the organisms of plague, tularaemia and oriental sore and under exceptional circumstances they may become means of their dissemination and transmission. It is believed that the bedbug acts as a transmitting agent of **Kala-azar**, a very serious human disease, caused by protozoans in some parts of India, and also **Relapsing fever** (Spirochaetosis). When they are exceedingly numerous their bites prove extremely irritating, leading to scratching and secondary infections. The bug-bites are also responsible for disturbances in sleep which may become serious especially in children.

**Control** measures in this case should be directed at the cleanliness of rooms, mattresses, furniture, etc., by proper repairs, cleaning, etc. Necessary precaution to prevent entrance of bugs from infested beddings or luggage brought after a journey, should be taken. Badly infested furniture and rooms may be made bug-free by washing with insecticides which will penetrate the crevices and cracks in furniture, walls, etc., or even by fumigation. Boiling water when poured on cots kills them and makes the cots bug-free.

With the advent of DDT the control of the bed-bug by common man has become easier and effectual. It can be sprayed on the indoor surfaces so as to leave a deposit which remains lethal to bugs which crawl over it for many months. For small infested bedrooms all that is required is about 1/3 gallon of

## TYPE SILK-MOTH

The silk-moth is the insect that made the gift of silk to mankind. There are a number of moths and other insects that spin silk casings or cocoons to cover themselves when in the chrysalis stage. Of these the silkworm proper produces just the right quality of silk for making the thread. Nature invented the cocoon for the moth, but man has subordinated the moth to the cocoon with the result that the moth has degenerated and lost the power of flight. In India silk is produced by moths belonging to only two families, the Bombycidae and the Saturnidae. The silkworm par excellence, the mulberry worm (*Bombyx mori*) belongs to the family Bombycidae and the family Saturnidae includes the Tassar silk-moth (*Antheroea paphia*) and the Eri silk-moths (*Attacus ricni*). The description which follows is that of the mulberry silk-moth.

The mulberry silk-moth never occurs in the wild state, it is completely domesticated. As a result of domestication several strains of the moth have evolved and they produce silks of different varieties with respect to size, weight,

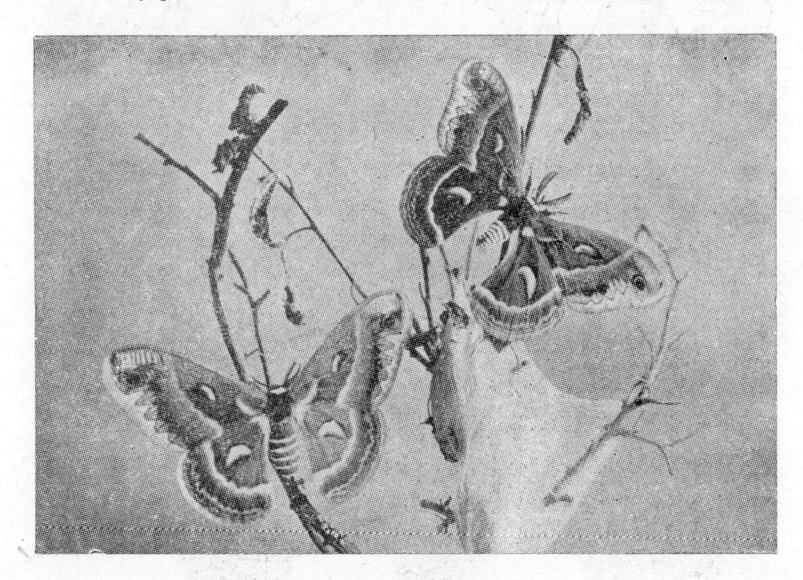

FIG. 17.48. Silk-moth.

colour, etc. *Bombyx mori* is extensively cultivated all over the world and usually produces a single brood in a year (**univoltine**). Some strains, on the other hand, produce two to seven broods a year (**polyvoltine**) and are cultivated in warm climates. In South Indian silk producing centres such as Mysore, Coimbatore and Salem such a strain is being extensively used.

### Adult

The adult moths are robust and creamy white in colour with several faint lines across the fore-wings. This medium-sized moth has a wing expanse of 40-45 mm (nearly two inches). Though both the pairs of wings are apparently properly developed the moths fly very rarely. The robust body is fatter in the females. The proboscis is absent, antennae are bipectinate in both the sexes and the body is densely covered with scales and hairs. The life-span of the adult is short usually ranging from two to three days. It does not take food during this period, only copulates, and the female lays from 300 to 400 seed-like whitish eggs (Fig. 17.49).

## Life-cycle

In the univoltine race overwintering takes place in the egg stage, as such the eggs may hatch in months. In others, however, the eggs hatch within a few days. Nowadays it is possible to control hatching artificially, if required. The newly hatched **larva** is rough and wrinkled, naked and whitish in colour. The larvae are highly gregarious and wander about rarely. They are phytophagous, having mandibulate mouthparts, and feed voraciously on the white and black mulberry

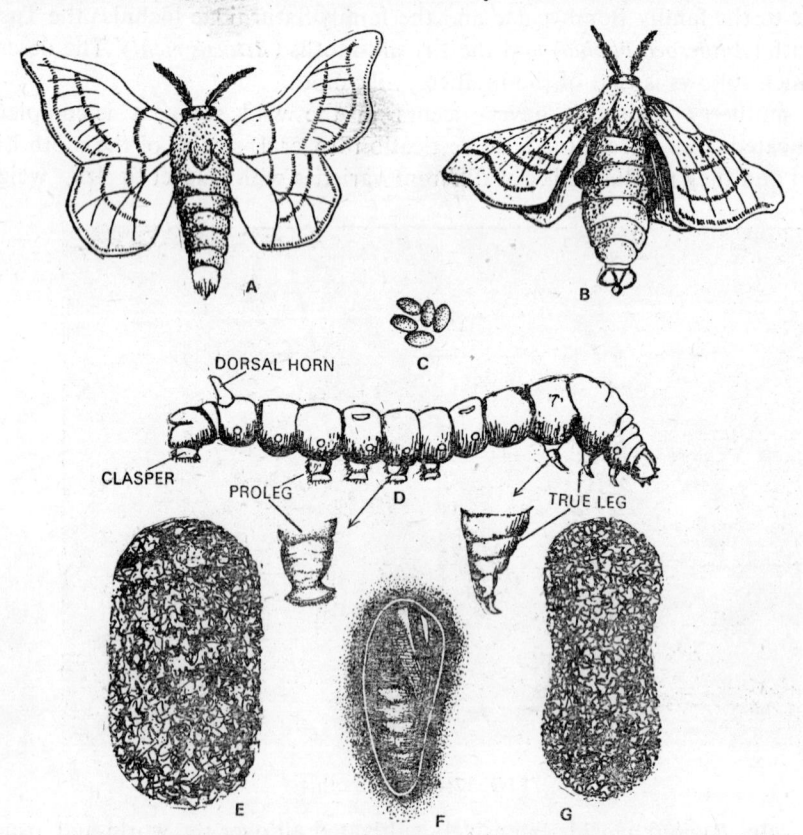

FIG. 17.49. Structure and life-cycle of a silk-moth. A—Male; B—Female; C—Eggs; D—Mature larva; E—Pupa in a cocoon; F—Pupa exposed; and G—Cocoon.

leaves. They need as many as five to nine feeds of chopped mulberry leaves per day. The larva outgrows the skin or cuticle which is periodically shed until it attains full size. The full-fed caterpillar is cylindrical, pale, yellowish-white in colour measuring about 8 cm. Behind the head there are thirteen segments. The first three thoracic segments following the head bear three pairs of 5-jointed legs

## Class Insecta

each ending in a curved claw. Four of the abdominal segments bear paired **prolegs** or **"cushion feet"** which are unjointed fleshy protuberances terminating in a cushion with a semicircular series of hooks by which the animal clings. The prolegs on the last segment turn backwards to form the **claspers**. It is provided with a small dorsal horn on the anal segement. The larvae have only three pairs of simple eyes or **ocelli**. The antennae, maxillae and labial palps of the larvae are rudimentary, while the mandibles, which are lacking in the adult, are large and strong biting organs here. Projecting from the labium is the **spinneret** by which the product of the paired **silk glands** is poured out. The silk glands are long sac-like structures occupying much of the body cavity. They are the longest in *Bombyx mori* being four times the body length and are folded so as to envelop the hinder region of the gut. Their fine ducts unite and open through the spinneret. The silk is a sticky fluid which hardens into the fine thread on exposure to air. The fine threads of silk are fairly strong and pliable. The larva respires through trachea, which open to the exterior by spiracles on the first thoracic and first eight abdominal segments. When fully grown the larva ceases to feed and eventually pupates.

### Pupa

The larval period lasts from three to five weeks after which pupation takes place. The larva stops feeding and the contents of the digestive canal are voided. The larva now spins its own silken cocoon around its body. The larva takes about three days to construct the entire cocoon. The outer or initial filaments of the cocoon are irregular, but the inner one that forms the actual bed of the pupa is one long continuous thread about 300-450 metres long, wound round in concentric circles. During the pupal stage there has been a tremendous breakdown and reorganization of organs, as well as the formation of new ones. The result is that the usual division of the body into head, thorax and abdomen is easily recognized in pupa.

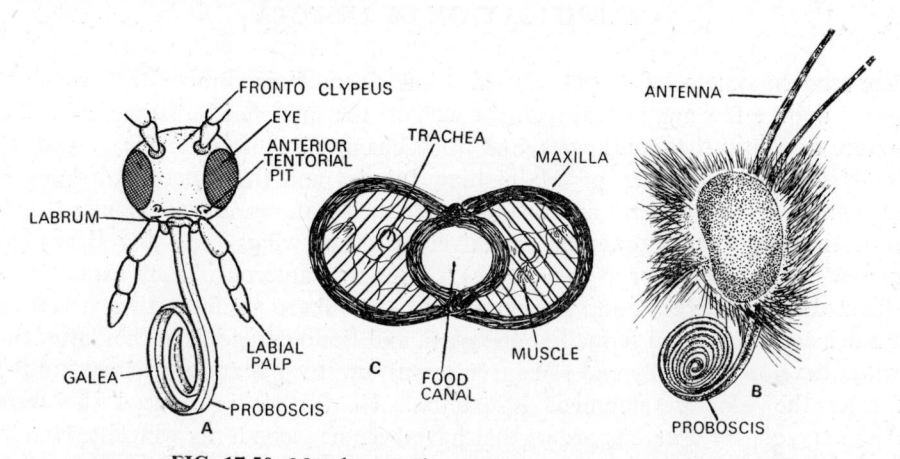

FIG. 17.50. Mouthparts of a moth, generalised structure.

**Emergence.** In the normal life-cycle pupation lasts from two to three weeks' time, after which period the adult or imago emerges. For emergence the pupa secretes an alkaline fluid that softens one end of the cocoon and after breaking the silk strands a feeble, crumpled adult squeezes its way out.

### Diseases

The silkworm larvae are subject to a severe hereditary disease known as **pebrine**, caused by a protozoan parasite, *Nosema bombycis*, and transmitted from generation to generation through the eggs. The caterpillars turn pale brownish and later on shrink and die. To protect the cultures from the disease the only remedy is to examine the body fluid of the mother moth. If it is free of the parasite, (**pebrine corpuscles**) the eggs may be reared, if not they may e discarded. Some other diseases such as **Flacherie** and **Grasserie**, etc. are caused by insanitary methods of rearing and feeding, but they are of minor importance.

### Preparation of Silk

About a day or two after pupation the pupae are killed in the cocoon by heat (drying in the sun) or by hot water and then the raw silk forming the cocoon is reeled out in skeins. This art requires some special skill which one learns by experience. Emergence is not permitted as the long silk thread will break, and the pierced cocoons can only be spun like cotton not reeled. A few cocoons which are necessary for the seeds of the next generation are kept to allow the adults to emerge and lay eggs.

### Economic Importance

Since the beginning of the written history the Chinese silkworm, *Bombyx mori*, has been the most commercially important and beneficial insect. It is the basis of large industries not only in China, Japan, India and European countries where the silkworm is reared and raw silk produced, but also in North America. Nowadays the intestine of the silkworm is used in the preparation of "**gut**" that is used in fishing and also for surgical works. This industry with good prospects is already growing in Italy, Spain, Formosa, Japan and India.

## CLASSIFICATION OF INSECTA

The modern system of insect classification dates from Brauer 1885 which is based upon a few anatomical features such as the presence, absence and the nature of wings; the mouthparts and their changes in ontogeny; the course of life-history, the number of Malpighian tubules and the structure of thoracic segments, wings, antennae and certain other such features. He recognized two subclasses: (*i*) **Apterygogenea**, primitively without wings, and (*ii*) **Pterygogenea**, having wings, or in some cases, secondarily apterous. These names later changed to **Apterygota** and **Pterygota**. In 1899 Sharp modified the system of Brauer and introduced terms Exopterygota and Endopterygota, in the latter the wings develop internally and remain internal up to pupation, whereas, in the former the wing development is external. He further introduced the term Anapterygota for apterous orders that have become secondarily wingless. Handlirsch (1908) related living with fossil forms, adding the dimension of geologic time and his revision of 1939 profoundly influenced the later workers. More significant contributions in this directions have been made by Martynov (1938), Essig (1942), Jeanel (1949), Brues, Melander and Carpenter (1954) and Carpenter (1961). According to the latest views the class Insecta is divided into four

## Class Insecta

subclasses and not two. This has been done with the idea that classification should illustrate phylogeny, reflecting both differences and similarities. Two orders of the Apterygota namely Collembola and Protura have been raised respectively to the status of subclass **Oligoentomata** and **Myrientomata** for reasons given with the descriptions of subclasses.

### I. Subclass Oligoentomata

This subclass includes the order (*i*) **Collembola** (*kolla*, glue + *embolon*, a peg.) commonly known as springtails (Fig. 17.51). Small, apterous insects without

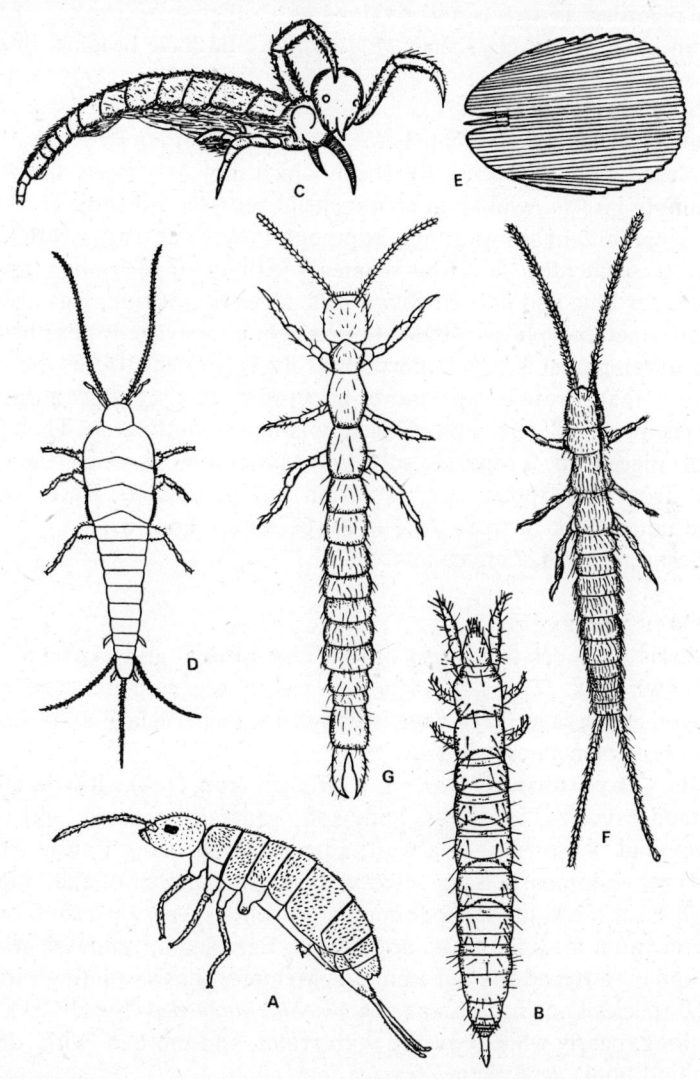

FIG. 17.51. A—A spring-tail *Axelsonia*; B—*Acerentomon doderi*; C—A proturan in nature; D—*Lepisma* sp.; E—Scale of *Lepisma*; F—*Campodea*; and G—*Heterojapyx*.

compound eyes. Mouthparts chewing; antennae large, 4-jointed; abdomen 6-segmented, first segment with an adhesive forked organ called ventral tube, fourth segment bears a forked springing organ; metamorphosis absent (ametabolic); occur under decaying vegetable matter. Cleavage of the zygote is combination type and postovarian development is apparently ametabolous metamorphosis. About 2000 species are known.

Examples: *Entomobrya, Hypogastrura* and *Sminthurus*, etc.

In 1961 Carpenter observed that all six-legged arthropods are not necessarily insects. He, therefore, removed Collembola from the class Insecta. It is true that Collembola are regarded as aberrant myriapods apparently related to the chilopods, but since their deviations are obviously in the insect direction they have been regarded as insects and retained as a subclass of insects. If they were placed in an independent class their *phylogeny* would have become obscure.

## II. Subclass Myrientomata

This subclass has been created to include the order **Protura**, which was formerly placed with the myriapods (from which they have been derived). Soft-bodied, minute insects, whitish in colour, head pear-shaped (Fig. 17.51). Mouthparts piercing, antennae wanting, compound eyes absent; prothorax short; abdomen 10 segmented, first four segments with a pair of leg-like appendages; first pair of legs long and carried forward in a raised position; legs also sensory in function, metamorphosis slight (ametabolic). Cleavage is peripheral (superficial) and development is by anamorphosis. It is evident from the characters given above that Protura are more insect-like as they show a differentiated thorax, three pairs of legs and Collembola-like mouthparts. Therefore, they have been placed in a separate subclass Myrientomota which indicates their derivation from myriapods and evolution towards insects. Rare, found under stones and under bark of trees. About 50 species are known.

Examples: *Campodea, Eosentomon*.

## III. Subclass Apterygota

This subclass includes two insect orders **Thysanura** and **Aptera** which are primitively wingless. The abdomen has eleven segments plus telson and may bear vestigial appendages. Cleavage is peripheral (superficial) and development is by ametabolous metamorphosis.

**1. Order Thysanura.** (*Thusanos*, a fringe+*oura*, tail). Bristle-tails. Adult smaller, moderate-sized, wingless, body soft, generally covered with scales or hairs; compound eyes present or absent; antennae long and filiform. Mouthparts of biting type; abdomen 11 segmented with varying number of abdominal appendages, styli on the 2nd to 9th abdominal segments. Anal tip bears one pair of jointed cerci and a median caudal appendage. Eggs small, yellowish white, oval. Nymph similar to the adults but smaller. Metamorphosis wanting (ametabolic). About 650 species known. *Lepisma* (*Thermobia*) *domestica* (Fig. 17.51), silver-fish is 1-4 cm long, pearly white, covered with scales, and mottled with dark spots. Found behind books and picture frames, etc.

**2. Order Aptera** (or **Diplura**). These are the insects that live in the soil, but also frequent decaying vegetable matter of various kinds; the largest form measures up to 5 cm long, but an average size is 3 to 5 mm. They are small (Fig. 17.51) eyeless mostly unpigmented insects with moniliform (beaded) antennae provided with segmental muscles. Mouthparts are of biting type. Abdomen

## Class Insecta

terminating in variably developed cerci or unjointed forceps, no medial tail filament, no ovipositor present. About four hundred species are known. They are popularly known as bristle tails.

Examples: *Japyx, Anajapyx*, etc.

### IV. Subclass Pterygota

Winged insects which may sometimes be secondarily apterous; metamorphosis very varied, rarely slight or wanting.

DIVISION 1. EXOPTERYGOTA (HEMIMETABOLA). Insects passing through a simple and sometimes slight metamorphosis, rarely accompanied by a pupal instar. The wings develop externally and the young are generalized nymphs.

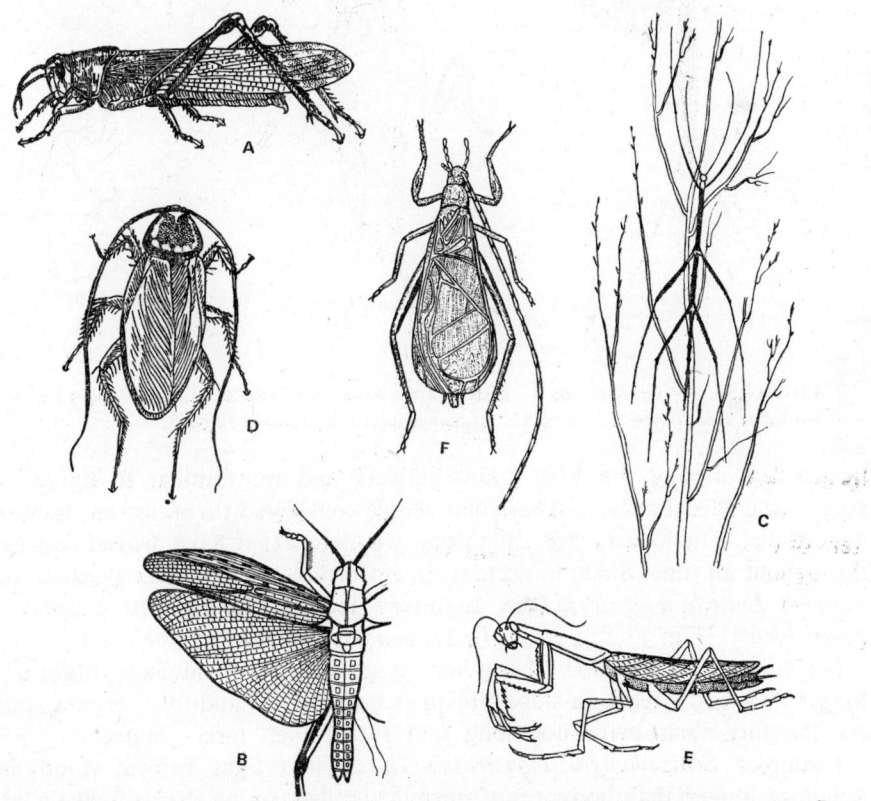

FIG. 17.52. A—Typical migratory locust, *Schistocerca*; B—A typical grasshopper with wings spread; C—A stick insect *Diapheromera arizonensis* male; D—A cockroach; E—A preying mantis; and F—A tree cricket (*Oecanthus niveus*).

1. **Order Orthoptera** (*orthos*, straight + *pteron*, wing). Crickets, grasshoppers, etc. Adults usually stoutly built (Fig. 17.52), terrestrial with chewing mouthparts; antennae variable; venation of a generalized type. Tegmina long and narrow, hindwings membranous, delicate, wingless forms also common. Abdomen usually 11-segmented, usually bears jointed cerci. Metamorphosis slight and gradual or absent. Nymphs resemble adults, but differ in size and in the absence of wings. Common in houses and in grassy lawns, or on plants and trees in the

fields. About 20,000 species are known. The important families of Orthoptera are:

(i) *Family Acrididae.* Short-horned grasshoppers. Medium-sized to large insects with antennae much shorter than the body, hind legs for jumping, tarsi 3-jointed, ovipositor small, inconspicuous, cerci 2-jointed; auditory organ on the first abdominal segment. Numerous grasshoppers are known. Grasshoppers and

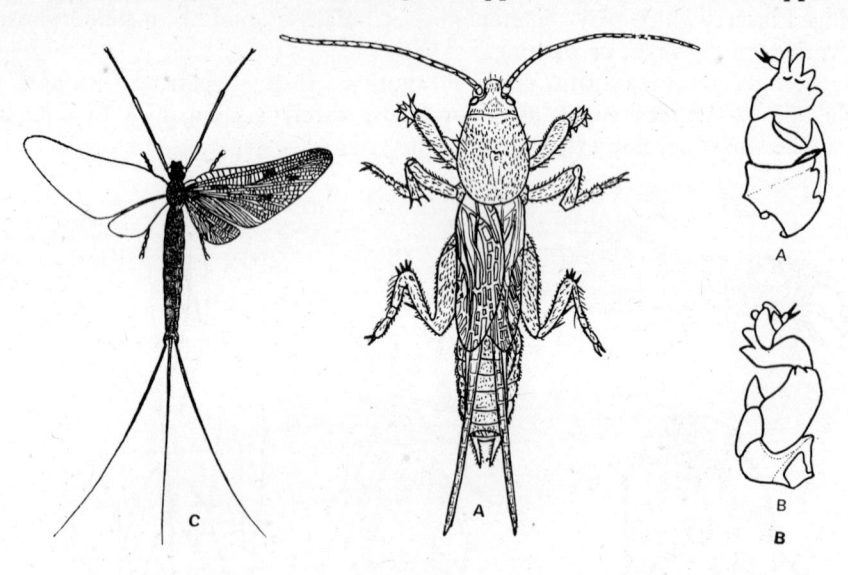

FIG. 17.53. A—*Gryllotalpa*; B—Limbs of the same: (a) Fore-leg outer side; (b) Fore-leg inner side (posterior); and C—Ephemeroptera, *Ephemera vulgata.*

locusts are among the best known insects and are familiar to almost every savage or civilized person. Aboriginal people considered them as an important item of diet. The locusts are migratory wanderers that have proved destructive throughout all time. Some important genera include *Chrotogonus* (surface grasshopper) *Locusta migratoria* (the migratory locust), *Schistocerca gregaria* (the desert locust) (Fig. 17.52); *Hieroglyphus banian*, (the rice grasshopper).

(ii) *Family Tettigonidae.* Long-horned grasshopper. Antennae filiform and longer than body; tegmina slope obliquely downwards; auditory organs situated on the fore-tibia; ovipositor long and sword-like; tarsi jointed.

Examples: *Schizodactylus monstrosus.* The adult is light brown; stoutly built; antennae longer than body; tip of tegmina rolled up on the back like a watch spring; with its anterior margin bent down, legs fussorial; ovipositor absent.

(iii) *Family Gryllidae.* Crickets. Antennae very long, filiform; hind femora enlarged for jumping; tegmina folded over abdomen; right one overlies the left; cerci long and unjointed, ovipositor slender and cylindrical, and stylets generally short, tarsi-jointed; auditory organs on fore-tibia.

Examples: *Gryllus domesticus* (the house-cricket), *Gymnogryllus erethrocephalus* (the field-cricket), *Gryllotalpa africana* (the mole cricket) (Fig. 17.53), etc.

2. **Order Grylloblattodea** (Lat. *gryllus*, a cricket+*blatta*, a cockroach). This is a small order including only six species. The members of this order show many primitive features. They are apterous with eyes reduced or absent. Ocelli are

also absent, filiform antennae are moderately long. Mouthparts are mandibulate. Legs are approximately similar to each other, tarsi 5-segmented. Females have well-developed ovipositor. Male genitalia are asymmetrical. Cerci long and 8-segmented.

Examples: *Grylloblatta campodeiformis, Gallosiana*, etc.

3. **Order Phasmida** (Gr. *phasma*, an apparition). This order is a group of predominantly tropical including the stick- and leaf-insects. They are large apterous or winged insects (Fig. 17.54), frequently with slender elongate body more

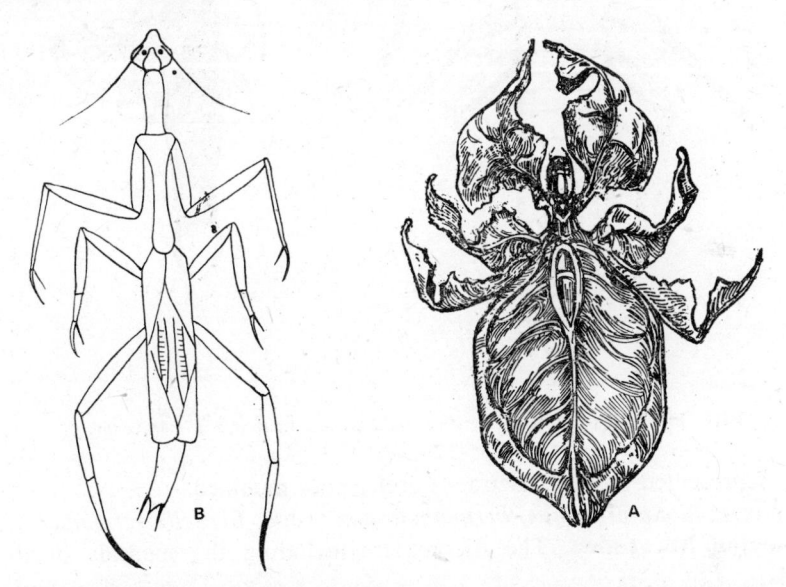

FIG. 17.54. Phasmida. **A**—Oriental leaf phasmid, a female *Phyllium siccifolium*; and **B**—Mantis female, dorsal view.

rarely depressed and leaf-like (Fig. 17.54). Mouthparts are mandibulate. Prothorax is short but meso- and metathorax usually elongate. Legs similar to each other, coxae, small and rather widely separated, tarsi almost always 5-segmented. Ovipositor concealed, external genitalia in male variable and asymmetrical, concealed. Cerci short, unsegmented. Metamorphosis slight. About 2,000 species are known.

Examples: *Phyllium crurifolium*, the leaf-insect of Oriental region; *Carausius morsus*, the stick-insect.

4. **Order Dictyoptera**. Cockroaches and Mantids. Medium or large-sized insects with almost invariably filiform many segmented antennae. Mouthparts of the biting type. Legs similar to each other, fore-legs raptorial coxae large rather closely approximated, tarsi 5-segmented. Fore-wings modified into more or less thickened tegmina. Ovipositor reduced and concealed in females. Male genitalia complex. Cerci many segmented; specialized stridulatory and auditory organs absent. Eggs contained in ootheca. They include two distinct homogeneous groups: the cockroaches (**suborder Blattaria**) and the mantids (**suborder Mantodea**). They are essentially terrestrial, though some semi-aquatic cockroaches are known. They do not fly well and the wings of many species are reduced or absent specially in the female.

**Suborder Blattaria.** The Dictyoptera in which the large shield like pronotum nearly or completely covers the head from above. Cockroaches. Two ocelli

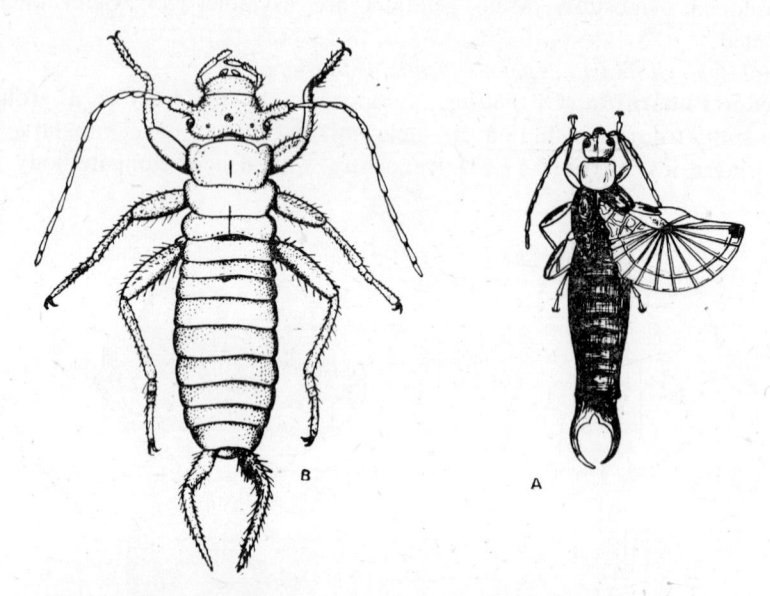

**FIG. 17.55.** Dermaptera. A—*Forficula auricularia*; and B—*Arixenia*.

usually represented by the fensetrae. Forelegs not modified.

Examples: *Blatta orientalis, Periplaneta americara, Blattella germanica*.

**Suborder Mantodea.** The dictyoptera including the mantids or praying mantids.

Example: *Paratenodera sinensis* (Fig 17.54)

5. **Order Dermaptera** (*derma*, skin+*pteron*, wing). Earwigs or Clipshears. Adult elongated (Fig. 17.55) with slightly flattened body; elytra truncate, apterous forms common, mouthparts chewing; antennae 10-15 jointed, half the length of the body, eyes circular; ocelli absent; tarsi 3-jointed; abdomen 11-segmented, last segment armed with two large, unjointed and horny forceps; ovipositor absent, wings four, shorter than the abdomen, fore-wings modified into leathery tegmina, hind-wings large, ear-shaped (Fig. 17.55) and membranous with veins radiating from the middle of the front margin; metamorphosis slight or wanting. Nymph wingless in younger stages, liable to be confused with Thysanura; advanced stages, however, similar to adult. Common in houses, fields, amongst decaying vegetable matter, etc. About 1,000 species are known.

Examples: *Nala lividipes*, and *Arixenia jacobsoni* (associated with bats), etc.

6. **Order Plecoptera** (*plekein*, to fold+*pteron*, wing). Stone flies. Body soft, antennae long and setaceous, compound eyes and three ocelli present, mouthparts chewing; wings four, membranous and held flat on the back during rest. Abdomen with long cerci. Nymph aquatic, with long antennae and cerci, with tufted tracheal gills. About 700 species are found.

Example: *Perla* sp.

7. **Order Isoptera** (*isos*, equal+*pteron*, wing). The termites or white-ants. Social or polymorphic insects (Fig. 17.56) living in colonies composed of (*a*)

# Class Insecta

**reproductive winged (macropterous)** castes; (*b*) (**reproductive brachypterous** or **complemental**) castes, nymph-like, with wings as wing-stumps incapable

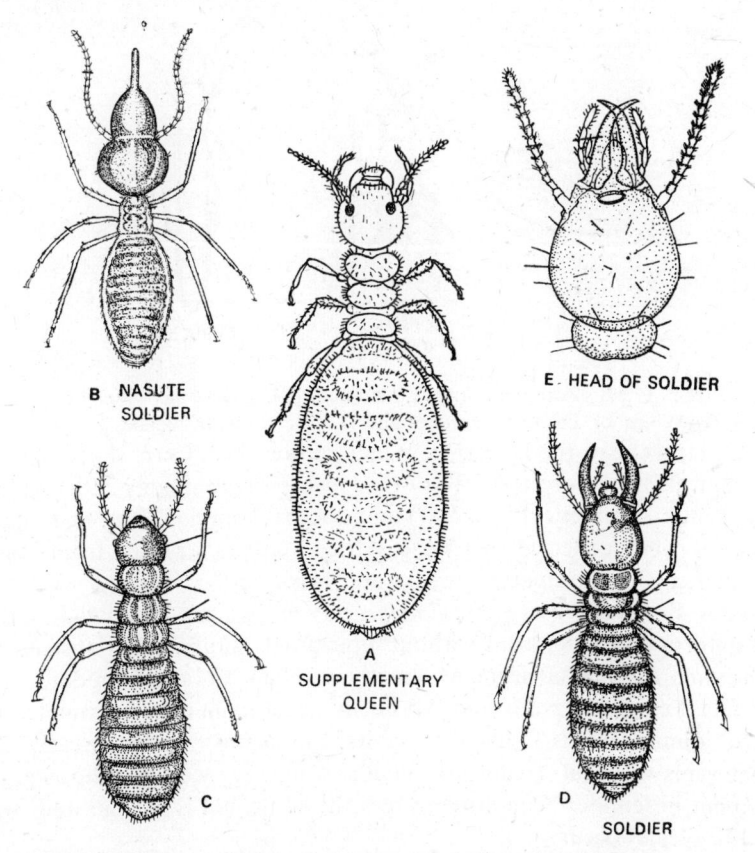

FIG. 17.56. Isoptera. **A**—*Reticulotermes flavipes*, apterous supplementary queen; **B**—*Nausitermes corniger*, nasute soldier; **C**—*Prorhinotermes simplex*, worker; **D**—*P. simplex* soldier; and **E**—Head of soldier of *Coptotermes ceylonicus*.

of flight; (*c*) **reproductive wingless** castes, worker-like and wingless; (*d*) **worker** castes; (*e*) **soldier** castes. In winged forms wings similar in shape, being long and narrow, and during rest are laid flat on the back, legs of the **ambulatory** type, abdomen 10-segmented, broadly-jointed to the thorax and at its caudal extremity carries two cerci, each 2 to 8-jointed, and two anal styli. Metamorphosis incomplete in winged forms but wanting in sterile forms. About 2,600 species occur. The termites are quite interesting, and destructive among insects. They construct burrows in wood and in the soil and build huge mounds of termitaria. Those in contact with man destroy portions of wooden buildings, fences, furniture, etc. causing heavy loss. Some are known to culture fungi for food in the specially prepared beds in their nests. Their nests are wholly underground or entirely or partly above ground and in some cases even aerial nests are constructed. *Microtermes obesi* and *Termes obesus, Hemitermes, Coptotermes, Leucotermes, Termopsis, Calotermes, Mastotermes*, etc.

8. **Order Zoraptera** (Gr. *zora*, pure + *apteron*, wing). This is a very small order of minute apterous or winged insects with biting mouthparts. Antennae

are segmented moniliform; tarsi two-segmented, eyes absent in apterous forms, winged forms possess compound eyes and ocelli, wings long and slender, fore

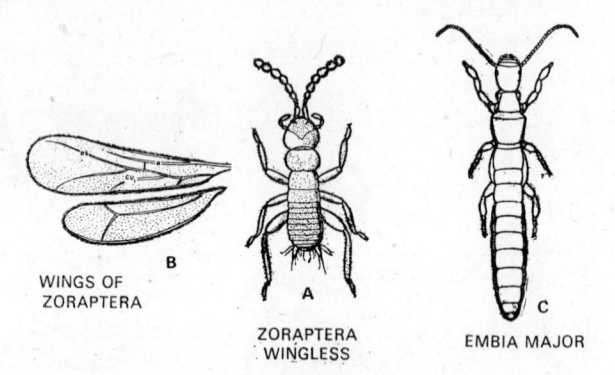

**FIG. 17.57.** Zoraptera and Embioptera. A—*Zorotypus* wingless; B—Wings of *Zorotypus snyderi*; and C—*Embia major*, female.

pair much larger than the hind one. Wings can be shed. Cerci short and unsegmented. Simple metamorphosis. The order has a single family Zorotypidae and as single genus *Zorotypus* (Fig. 17.57) and about 12 species. The first specimens all wingless were collected in West Africa. Others reported from North and South America, Java, Sumatra, Ceylon and Hawaii.

9. **Order Embioptera** (genus *Embia+pteron*, wings). Elongated with long, jointed filiform antennae, head without epicranial suture, ocelli absent, mouthparts chewing. First tarsal segment greatly swollen to accommodate silk spinning glands. Female apterous and larviform; genitalia in male asymmetrical, cerci small two-jointed. Wings in male two pairs, thin, smoky, radius greatly thickened, other veins vestigeal (reduced), female wingless. Metamorphosis gradual in male, absent in female. Nymph resembles the adult, but is smaller and wingless. About 140 species occur.

Example: *Embia* (Fig. 17.57 C).

10. **Order Psocoptera** (genus *Psocus+pteron* wing). Book-lice, having about 800 species; is divided into 2 suborders.

(*i*) **Zoraptera**. Small Psocoptera with moniliform antennae; prominent prothorax and short unjointed cerci; maxillae normal.

Example: *Zorotypus*.

(*ii*) **Psocida**, the Psocoptera with long, filiform antennae; prothorax much reduced; cerci absent. The head reduced with triangular frons, and very well-developed clypeus. The maxilla is provided with a rod-like rasping organ; maxillary palps fore-jointed; labium with reduced labial palps and glossae; hypopharynx with lingual glands; chitinized plates on ventral aspect and superlinguae on dorsolateral aspect. Nymph resembles the adult, but is smaller and wingless.

Examples: *Atropos pulsatoria*, *Psocus* (Fig. 17.58).

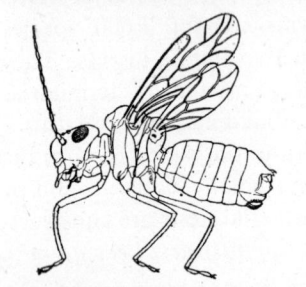

**FIG. 17.58.** Psocoptera. *Peripsocus phaeopteri*, a typical dsocopteran.

11. **Order Mallophaga**. Bird-lice (Fig. 17.59). Adult with mouthparts of modified chewing type; meso- and metathorax sometimes not distinct; tarsi

one or 2-jointed terminated by a single or paired claw. Thoracic spiracles ventral. Found mainly on birds but quite a number lives on mammals. They are small, flat-bodied, active insects entirely adapted for an ectoparasitic life. Legs fitted with claws to enable them to pass rapidly through hairs and feather of the host. Some cling to the feathers by jaws. They feed upon hair feathers, epidermal scales and often become very irritating to host. They are widely distributed and about 2,500 species are known.

Examples: *Menopon, Gliricola, Menacanthus, Myssidea, Trichodectes* (on dogs), *Felicola* (on cats), *Bovicola* (on bovine cattle).

12. **Order Siphunculata** (Anoplura by some authors). Sucking lice. Adult with mouthparts highly modified for piercing, sucking; retracted within the head

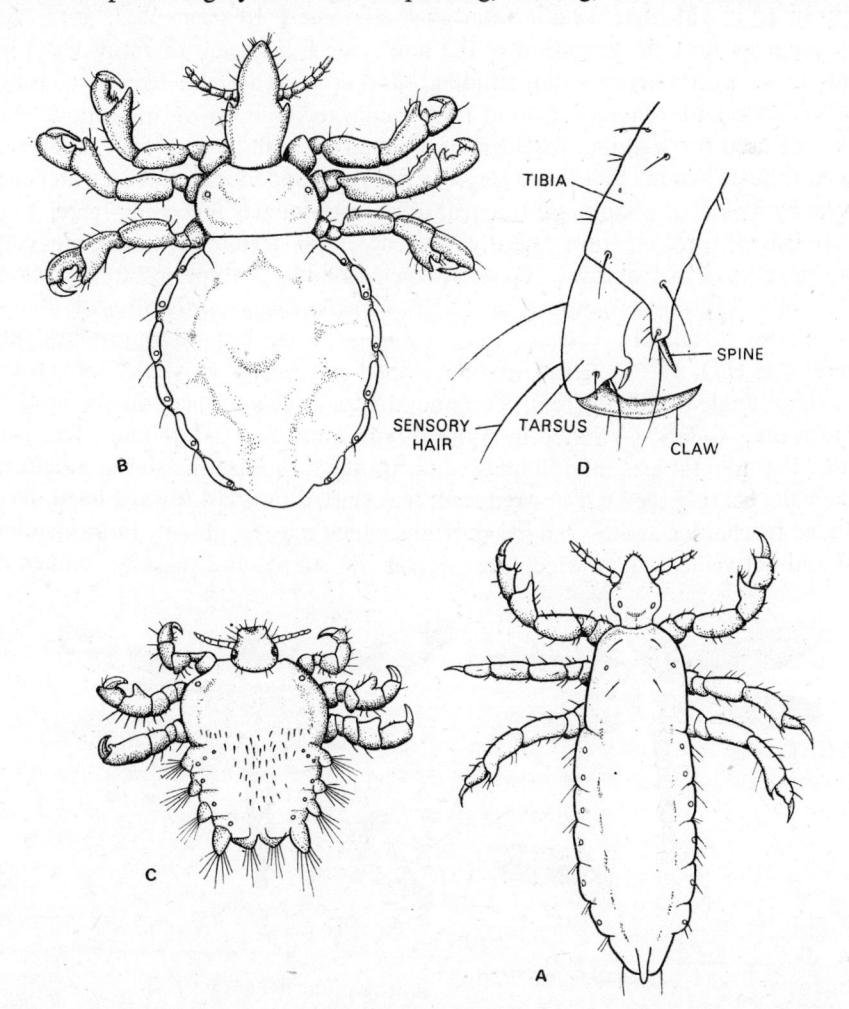

FIG. 17.59. Siphunculata. A—*Pediculus humanus*; B—*Haematopinus suis*; male from pig; C—*Pthirus pubis*; female of crab-louse of man; and D—Terminal part of front leg of *Pediculus humanus*, female.

when not in use; thoracic segment fused; tarsi one-jointed, claws single; thoracic spiracles dorsal. Occur on mammals. These are exclusively blood-sucking ectoparasites of mammals of which about 225 species are known. They are quite homogeneous and form a distinct group both in appearance and habits. Species infesting domesticated mammals have been taken to all parts of the world. They are certainly as old as the human race. The best known species *Pediculus humanus* (Fig. 17.59), is common louse of man. It has two races: *P. humanus capitis* is the head louse and *P. humanus corporis* (*vestimenti*), is the body louse. *Haematopinus asini*, the sucking horse-louse, *H. eurysternus*, the ox louse, *H. suis*, the hog louse, *Polyplax spinulosa*, the spined rat louse, also occur on cattle. *Linognathus ovillus*, the sheep louse, *L. stenopsis*, the goat louse, etc, are good examples.

13. **Order Ephemeroptera** (May-flies). Adult soft-bodied, shortlived insects that are attracted to light; moult once before maturity; mouthparts chewing (rudimentary); antennae small, setaceous; abdomen 10-segmented, cerci very long, usually with a prolongation of the same length between them; wings four, membranous with numerous longitudinal and cross veins, hind-wings considerably smaller than fore-wings, held at right angles to the body during repose; legs weak, not used for walking; forelegs longer than the middle and the hind pair; metamorphosis hemimetabolic. Nymph (naiad) campodeiform, live in water and breathe by means of abdominal tracheal gills, mouthparts chewing; cerci long, median caudal prolongation usually present, tarsi terminate in a single claw. Occur near lakes and streams. About 151 genera and 1,270 species are known.

Examples: *Ephemera vulgata* (Fig. 17.53), *Caenis*, *Ephoron*, *Campsurus*, etc.

14. **Order Odonata** (*Odous* gen., *odontos*, a tooth). Dragon-flies and Damsel flies (Fig. 17.53). Head well-developed and freely movable, can rotate through an angle of 180 degrees; compound eyes large and meet on the head in **Anisoptera**, widely separated in **Zygoptera**; antennae 3-7 jointed, flagellum bristle-like; mouthparts mandibulate and toothed; thorax; the sterna and terga of the wing bearing segment not reduced; legs shifted forward toward head, has a 2-jointed trochanter and a vestigial pulvillus which may be absent. Four membranous and net-veined wings which are similar in shape and basally stalked in

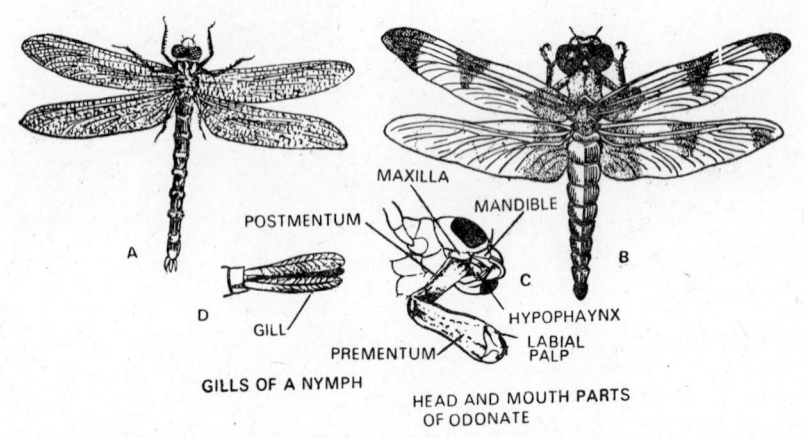

FIG. 17.60. Odonata. A and B—Two odonatans; C—Mouthparts of a larva; and D—Tracheal gills of a larva.

Zygoptera and dissimilar in shape without a basal stalk in Anisoptera. Wing venation very characteristic in the group. Nymph (naiad) has chewing mouthparts; labium in the form of an extensile and prehensile mask. They live in ponds, lakes, etc. and feed on mosquito larvae, pupae etc. which are seized by means of this mask. Adults occur near ponds, streams and stagnant pools, nymphs in water. About 500 genera and 4,500 species are known.

(a) **Suborder Anisoptera** (Dragon-flies). Sturdy built, powerful fliers, wing kept horizontal to the body during rest. Hind wings broader than fore wings near the base; compound eyes contiguous on the vertex of the head. Eggs laid on plants; nymphs breathe through rectal tracheal gills.

Examples: *Pantala, Anax, Aeschna.*

(b) **Suborder Zygoptera** (Damsel flies). Slender build; poor fliers, fore- and hind-wings of the same size and shape, with long narrow bases, wings lie flat or stand erect, back to back "like a butterfly" over the abdomen; compound eyes project from the sides of the head and do not meet at the vertex of the head, but are widely separated. Eggs inserted in the stem of water plants; nymphs breathe by three leaf-like caudal gills.

Examples: *Calopteryx, Lestes, Agrion,* etc.

15. **Order Thysanoptera** (*thusanos*, a fringe+*pteron*, a wing). Thrips. Small brownish or black, insects whose body is elongated, antennae 6-9 jointed, (Fig. 17.61), inserted close together, Mouthparts rasping sucking, cone-shaped,

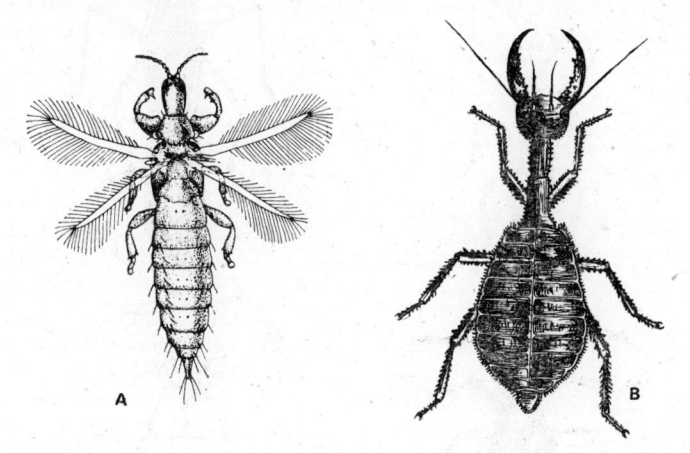

**FIG. 17.61.** Thysanoptera. A—A typical thysanopteran or thrips; and B—An antlion (Neuroptera).

asymmetrical, prothorax well-developed. Tarsi 1 to 2-jointed, each with a terminal protrusible vesicle. Maxillary and labial palp very small. Wings two pairs, narrow with greatly reduced veins and fringed with long marginal hairs. Cerci absent. Occur on leaves and flowers of plant. About 1,600 species occur.

Examples: *Thrips* (Fig. 17.61), *Taeniothrips.*

16. **Order Hemiptera** (*hemi*, half+*pteron*, a wing). Bugs. The order Hemiptera or Rhynchota (Fig. 17.62) includes about 50,000 species and is the largest among the Exopterygota. It is divided into two main groups that are often regarded as separate orders, viz., the **Heteroptera** and **Homoptera**.

(a) **Suborder Heteroptera**. Adult stoutly built, often coloured; mouthparts piercing and sucking, beak projecting from the front part of head, eyes compound, ocelli present; female genitalia conspicuous or concealed; pronotum large, sometimes spined, mesothoracic scutellum prominent; wings two pairs, fore-wings (hemi-elytra)—with their basal half horny and distal half membrannous—lie flat over the back during rest with tips overlapping; legs usually of the running type, metathoracic legs modified for swimming in aquatic forms, tarsi 1 to 2-jointed; abdomen 11-segmented, but usually 8 segments visible; metamorphosis incomplete. Adults occur on vegetation of all kinds, nymphs found along with their parents. There are as many as 43 families (according to Essig).

This group includes many important pests. *Blissus leucopterus*, the chinch bug (Family Lygaeidae), of the United States is a serious pest of wheat and other cereals.

*Dysdercus* (Fig. 17.62), the cotton strainers (Family. Pyrrhocoridae) includes common cotton pests of the tropics. The giant water bugs (Family Belostomidae); the water scorpions (Family Nepidae), the back-swimmers (Family Notonectidae); the water boatman (Family Corixidae) are some of the aquatic or semi-acquatic

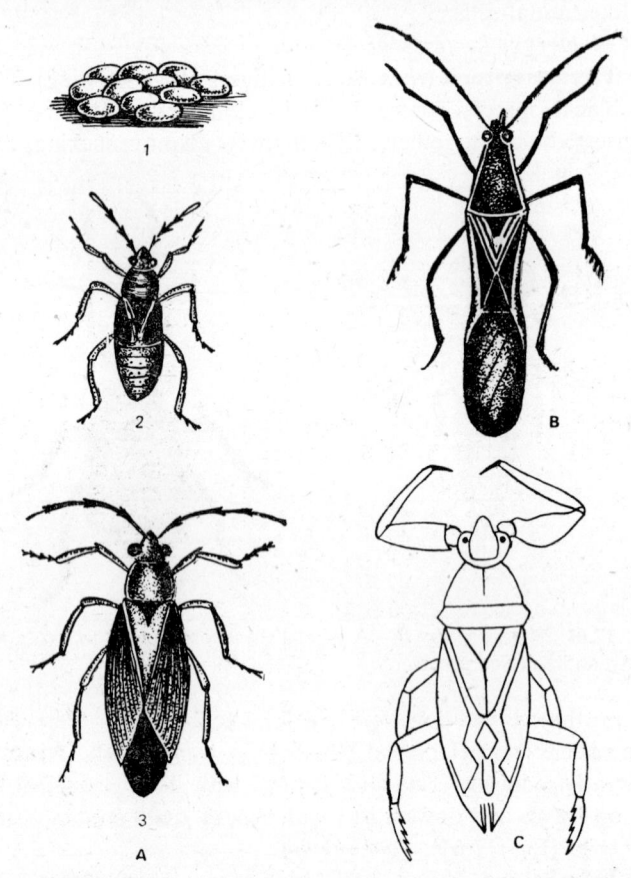

FIG. 17.62, Hemiptera (Heteroptera). A—Stages in the life-cycle of the red cotton bug, *Dysdercus singulatus*: 1—Egg; 2—Nymph; 3—Adult; B—*Leptocorisa varicornis*, a pest of paddy; and C—*Belostoma*, the water bug.

Heteroptera. Some (Family Reduviidae) have acquired the habit of sucking body fluids of insects and other small animals. The reduviid *Triatoma* is a blood-sucker of man. *Cimex lectularius* (Family Cimicidae) is another bug that sucks human blood. Other species of this family occur on bats, poultry and swallows. The famous stink bugs (Family Pentatomidae) also belong to this suborder (Fig. 17.62).

(*b*) **Suborder Homoptera.** Adults excepting "Cicadas" are mostly small insects; mouthparts piercing and sucking, the labium arises far back on the head or even between the fore coxae, eyes compound, ocelli present; antennae 2 to 5 jointed with a small terminal bristle; wings four, held sloping over the abdomen during rest, front pair larger and harder, hind pair wider, females of some forms apterous, some males have only one pair of wings; coxae of the front legs touch the head, tarsi 3-jointed; metamorphosis incomplete. Aleurodids (family Aleurodidae) and some male scales undergo complete metamorphosis. Adults found on vegetation of all kinds, nymphs generally found along their parents.

The Homoptera (Fig. 17.63) are known for three characteristics: they discharge sugary waste product or "honey dew" from the anus, notably in aphids; they excrete wax either in powdery form or as threads; and they lodge a peculiar tissue, the **mycetome**, in the abdomen, which harbours symbiotic microorganisms. Among the families of Homoptera the cicadas (Cicadidae) are well-known for the shrill sounds produced by the males, the lantern-flies (Fulgordiae) are large tropical insects known for brilliant colouration, the froghoppers

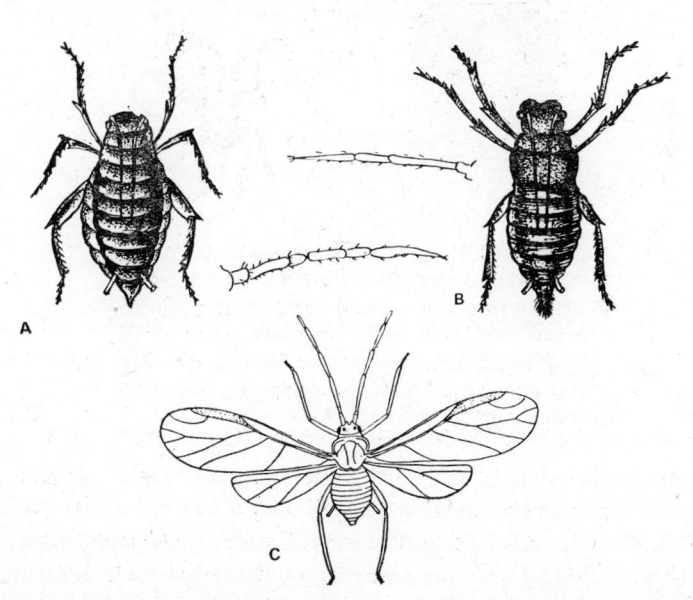

FIG. 17.63. Hemiptera (*Aphis fabae*). A—*Apterous viviparous* female; B—Fundatrix; and C—Winged male.

(Cercopidae) are small insects whose nymphs live within a frothy exudation, protective in nature; the leafhoppers (Cicadellidae or Jassidae) suck the sap of plants and some beet-leaf hoppers transmit virus causing plant diseases; and the white flies (Aleyrodidae) have their bodies and wings covered with powdery wax.

The largest family of Homoptera is that of the aphids (Aphididae) commonly called the plant lice or greenfly. They all pass through a more or less complex life-cycle on one or more plants. Alternation of host plants may also take place. Some forms reproduce parthenogenetically and some sexually. The aphids produce large amount of "honey dew" on which the ants feed. The grape phylloxera, *Phylloxera*, damages roots; the corn root aphis, *Anuraphis maidi-radicis*, destroys roots of corn and *Aphis gossypii*, the cotton aphis, is pest of cotton and other plants.

Another important family is the Coccidae including the scale insects and mealy bugs. They are the most modified among insects and show sexual dimorphism of

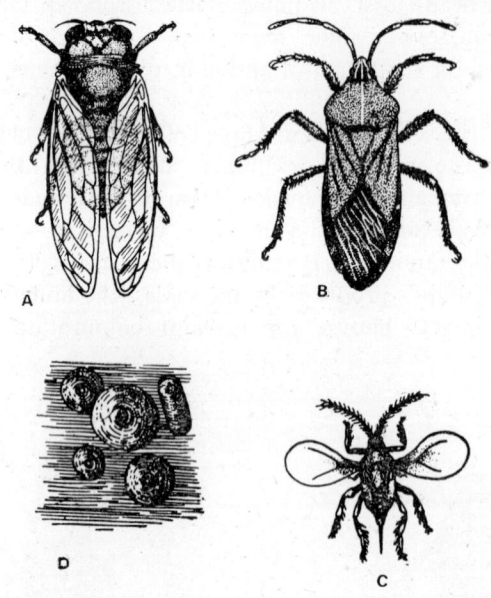

FIG. 17.64. Hemiptera (Homoptera). **A**—Typical cicada, dorsal view; **B**—Squash bug (*Anasa tristis*); **C**—San Jose' scale (*Quadriaspidiotus*), adult enlarged; and **D**—A group of scales enlarged; the large round ones cover mature females, the elongate ones cover males; and the smaller ones cover second instar nymphs.

extreme type. In the scale insects hard covering called "scale" covers the body while in the mealy bugs the insect is covered with a fine waxy exudation. Many of the coccids are important pests. The mussel-scale (*Lepidoaphis ulmi*), the San Jose scale (Fig. 17.64) (*Aspidiotus perniciosus*), the fluted scale (*Icerya purchasi*), and the citrus mealy bug (*Pseudococcus citri*) are the more important pests. Some of the scale insects on the other hand are useful. *Coccus cacti*, for instance, yields cochneal and the Indian lac insect *Laccifer lacca* is one of the most valuable insects known. It produces lac providing commercial shellac and thrives on the native fig and banyan trees. The bodies of the females are covered by copious exudation of lac. They are in such a large number that they cover the twigs of the host plant completely at places. This lac is pruned off the twigs, melted off in boiling water, refined and prepared for the market as shell-lac or shellac.

*Class Insecta*

DIVISION II ENDOPTERYGOTA. Insects passing through a complex metamorphosis are always accompanied by a pupal instar. The wings develop internally and the larvae are usually specialized.

17. **Order Neuroptera** (*neuron*, a nerve+*pteron*, a wing). Lacewings, aphid-lions, ant-lions etc. Adult small or large, mouthparts chewing, antennae generally long and filiform, wings four, leaf-like, net-veined, of nearly equal size, held sloping over the abdomen during rest, branches of veins prominently bifurcate at the margins, legs short; tarsus 5-jointed, metamorphosis complete; larva predaceous on aphis, coccids, white flies, etc., common on vegetation infested with aphids, coccids etc., *Chrysopa* sp (lacewing fly).

The Neuroptera have about 4,500 species and are divided into two suborders **Megaloptera** and **Planipennia**.

(*i*) **Suborder Megaloptera**. The Megaloptera (separated by some authors as an order) is relatively small suborder containing about 500 species including older flies (*Sialis*) and of allies that have aquatic larvae with seven or eight pairs of hair-fringed abdominal appendages. *Corydalis* (Fig. 17.65) occurs in Northern India and attains a wing expanse of 15 cm with gigantic mandibles in the male. The snake-flies (*Raphidia*) have a long neck-like prothorax and an elongated ovipositor, their larvae live under bark of conifers and allied plants.

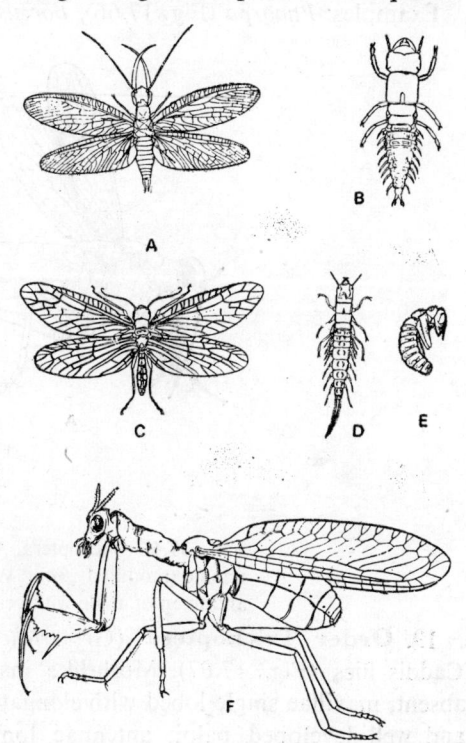

FIG. 17.65. Neuroptera. **A**—*Corydalus*, male; **B**—Larva of the same; **C**—*Sialis*, adult; **D**—larva; **E**—Pupa of the same; and **F**—*Mantispa*.

(*ii*) **Suborder Planipennia**. The majority of Neuroptera (about 4,000 species) belong to suborder Planipennia. This is divided into several families including the **Hemerobidae** (the brown lacewings) and **Chrysopidae** (the green lacewings). The larvae of these destory large number of Homoptera and other insects. The larvae of Myrmeleonidae are called ant-lions and commonly make pit-like snares for capturing their prey. The larvae of **Ascalaphidae** lurk under stones on leaves or on trees. The **Mantispidae** (Fig. 17.65) are predators and resemble mantids.

18. **Order Mecoptera** (*mekos*, length+*pteron*, wing). Scorpion-flies. Maxillae and labium produced into a broad snout at the end of which are the chewing mouthparts; antennae elongate, filiform, wings four, long and narrow, membranous with a large number of cross veins. In the male tip of abdomen swollen and curved, carried upwards like the sting of a scorpion; metamorphosis complete; larva cruciform; mouthparts chewing; three pairs of thoracic legs and prolegs present (up to eight pairs) or absent; pupa exarate. Occur amongst

herbage in cold, shady places. About 500 species are known. Examples: *Panorpa* (Fig. 17.66), *Boreus*, etc.

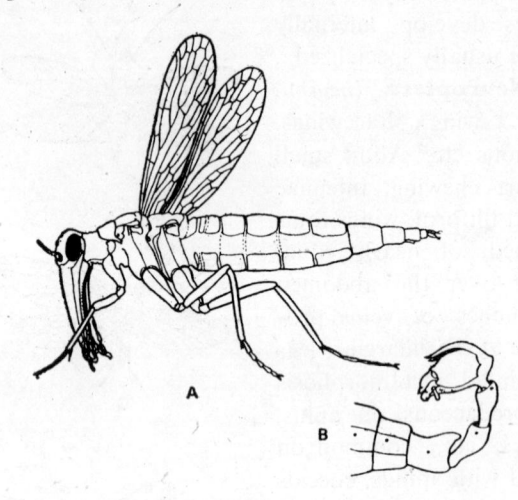

FIG. 17.66. Mecoptera. A—Scorpionfly, *Panorpa* female (modified *from* Webber); and B—Tip of abdomen of male with genitalia.

19. **Order Trichoptera** (Gr. *trichos*, a hair + *pteron*, wing). Hairy wing. Caddis flies (Fig. 17.67). Moth-like insects; mouthparts suctorial, mandibles absent, maxillae single lobed with elongate palpi, labium with a median glossa and well developed palpi; antennae long, setaceous, wings four membranous,

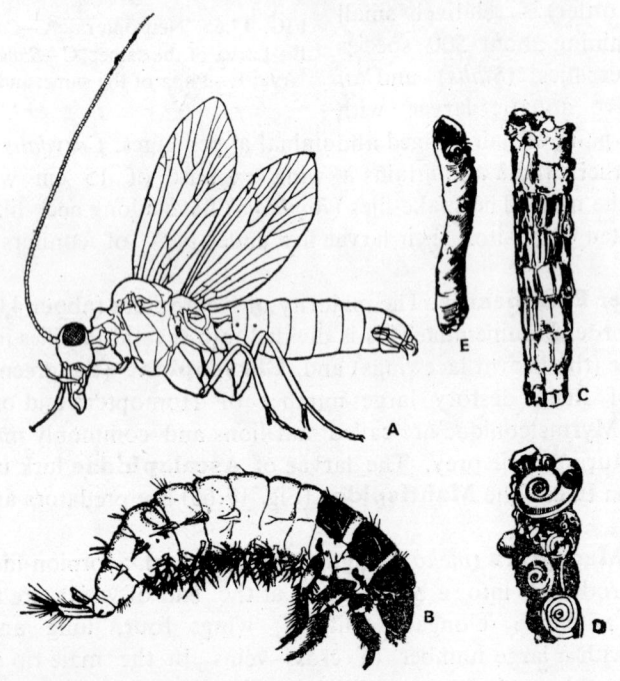

FIG. 17.67. Trichoptera. A—External features; B—Typical larva (Caddisworm); C—Larval case of *Phryganea*; D—Larval case of *Limnephilus*; and E—Larval case of *Coleophora*.

more or less densely hairy, held roof-like over the back in repose, fore-wings elongate, hind-wings broader with a folding anal area, tarsi 5-jointed, tibiae with spurs; metamorphosis complete, larva aquatic, more or less cruciform with three pairs of long thoracic legs, and one pair of prolegs; lives in a case; pupa exarate aquatic; also lives in a case. Found on plants near water generally on the hills.

Examples: *Hydroptila, Phryganea*, etc.

20. **Order Lepidoptera** (Gr. *lepidos*, a scale + *pteron*, a wing). Moths and butterflies. The Lepidoptera are familiar flying terrestrial insects of small to very large size. The body wing and other appendages are usually completely clothed by flat overlapping scales. The mouthparts modified into a specially coiled suctorial proboscis. Mandibles rarely present. Compound eyes large, ocelli two or absent; antennae variable. Wings, two pairs usually present, rarely vestigial, the fore pair often largest. Larvae terrestrial polypodous and phytophagous. Pupae, free or obtect usually in cocoons.

It is a large order having as many as 140,000 species with varied habits, and life-histories. Adult butterflies and moths are among the most delightful and

FIG. 17.68. Lepidoptera. A—Typical butterfly; B—Peried butterfly, *Pieris* sp.; C—Walnut moth; and D—Armyworm moth, *Leuconia unipuncta*.

pleasing objects of nature and have long been of great interest to mankind. The adults feed upon nectar, over-ripe fruit, honey dew, etc. while their larvae, feed on roots, seeds, wood. Most of these are quite destructive feeding on our crops, orchards and gardens. Some members (Family Saturniidae) yield silk of commercial value.

The Lepidoptera are divided into three suborders; Zeugloptera, Monotrysia and Ditrysia, which are divided into several superfamilies and families.

(*i*) **Suborder Zeugloptera**. The Lepidoptera in which the adults have functional mandibles, lacinia developed, galea not haustellate. These are the primitive Lepidoptera with one family (**Micropterygidae**) having a few species only.

Examples: *Micropteryx, Sabatinca*, etc.

(*ii*) **Suborder Monotrysia.** The Lepidoptera in which the female is provided with one, rarely two, genital apertures on sternites IX. Wings nearly always aculeate. This has four superfamilies (Imms) of which only the swift-moths (family Hepialidae) are important. The mouthparts in these are vestigial as in the genus *Hepialus*.

(*iii*) **Suborder Ditrysia.** The Lepidoptera in which the females have a copulatory pore on sternite VII and an eggpore on sternite IX. Fore-wings without jugum or fibula, hind-wings often with frenulum, its venation reduced. This suborder includes nearly 95 per cent of the lepidopterans, and includes a large number of families and superfamilies. Of these the **Cossidae** (goat-moths) have the most archaic venation; the **Pyraustidae** includes the genera *Acentropus* and *Hydreuretcis* whose larvae are aquatic; the **Pterophoridae** (or plume-moths) have deeply fissured wings and **Tineidae** are an extensive family of varied larval habits; this also includes the cloth-moth. Other notable groups are

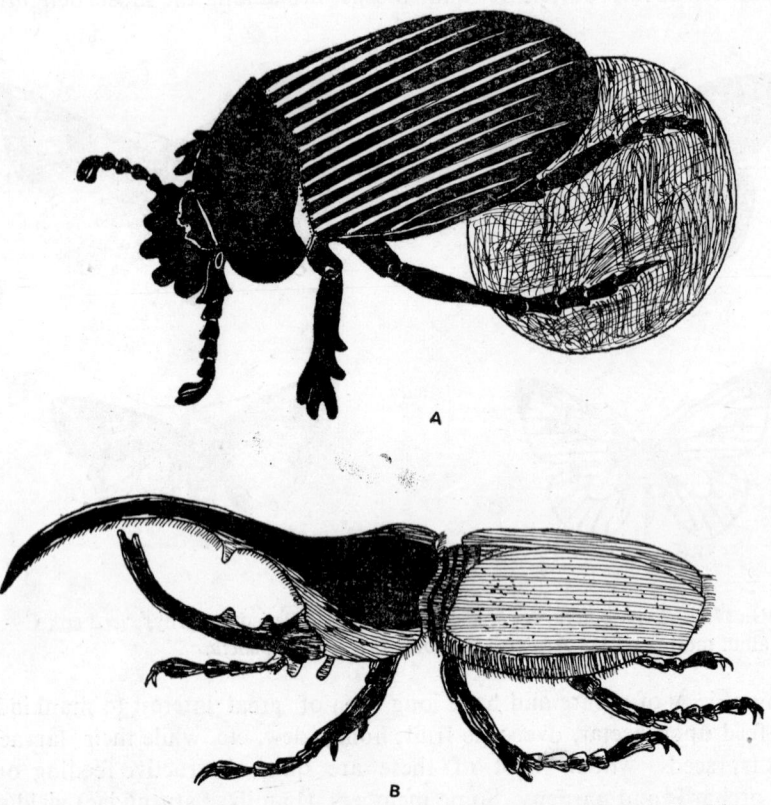

FIG. 17.69. Coleoptera. A—Sacred beetle, *Scarabus sacer* rolling a ball of dung (the ball may be of the size of a tennis ball); and B—The Hercules beetle, *Dynastes hercules* a huge beetle found in America.

the **Geometridae** whose larvae are loopers, the **Sphingidae** or the hawk-moth and the **Noctuidae** the great family of owlet moths. The family **Saturniidae** are characterized by dense silken cocoons, which are of commercial value in some cases and produce silk. The oriental atlas moths, *Attacus*, with a wing expanse of 25 cm is notable genus of this family. Another family, the **Bombycidae** includes *Bombyx mori* a native of China, whose larva is well-known as the "silk-worm." The super-family **Papilionoidea** includes butterflies distinguished by clubbed antennae and the absence of a frenulum.

21. **Order Coleoptera** (*koleos*, a sheath + *pteron*, a wing). Beetles (Fig. 17.69). Minute to large, robust with compound eyes, body highly chitinized. Mouthparts chewing; antennae simple, of medium length, generally 11-segmented, fore-wings veinless and modified into horny elytra which meet on the back to cover the hind-wings during rest, hind-wings membranous, and are of service in flight, abdomen tapering, one pair of cerci present, anal tube present or absent, female has seven visible ventral abdominal segments, while male has six. Metamorphosis complete. Grub campodeiform or cruciform with three pairs of thoracic legs. Anal prolegs when present without hooks, mouthparts chewing, ocelli present. The Coleoptera with over 25,000 species rank as the largest order in the animal kingdom.

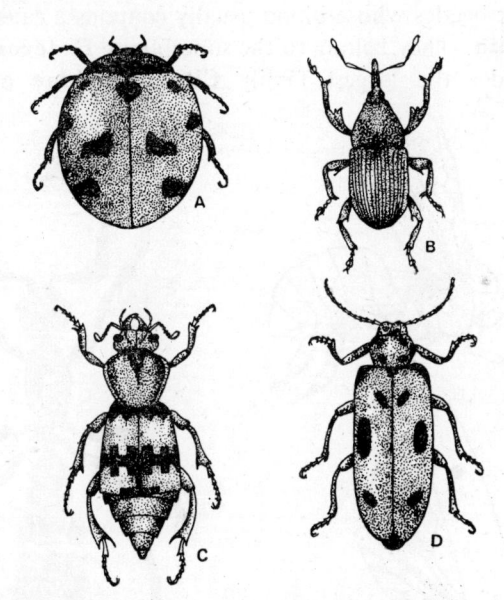

FIG. 17.70. Coleoptera. A—Ladybird bettle; B—Grain weevil; C—Burying bettle; and D—Milkweed beetle.

The beetles have successfully occupied the land and have more effectively met the diversified requirements of habitat and climate. They are found in every corner of the world excepting the polar caps and the oceans. The Coleoptera have been divided into three suborders—Adephaga, Archostemmata and Polyphaga.

(*i*) **Suborder Adephaga**. The **Adaphaga** are characterized with filiform antennae and wings with one or two cross veins. This suborder has a single superfamily, the Caraboidea that includes mainly predators both as adults and larvae. The large families **Carabidae** (the ground-beetles) and **Cicindelidae** (the tigerbeetles) include most of the terrestrial forms. Family **Dytiscidae** includes the aquatic caraboids and the family **Gyrinidae** includes the anomalous whirligig beetles.

(*ii*) **Suborder Archostemmata.** The **Archostemmata** are the Coleoptera in which the antennae are varied in character and hind-wings lack cross veins. It includes the majority of beetles. The larvae are of different types but are devoid of tarsi and have single claws. The rove-beetles (**Staphylinidae**), the carrion-bettles (**Silphidae**), etc. belong to the superfamily **Saphylinoidae**, The **Diversicornia** are ill-defined group of over 40 families. The **Coccinellidae** (lady bird-beetles) are mostly beneficial, their larvae and adults feed largely on aphids. The family **Dermestidae** or larder-beetles have densely hairy larvae. The **Hydrophilidae** includes aquatic or subaquatic beetles, as well as terrestrial forms, with greatly elongated maxillary palpi. The **Elateridae** include the tropical "fireflies" and the more numerous click-beetles, whose larvae, or "wire-worms" are destructive root feeders of crops. The **Buprestidae** are metallic green or blue creatures, their larvae are legless boarders living beneath the bark or trees and characterized by their greatly widened pro-thorax. Family **Tenebrionidae** includes the mealworms (*Tenebrio*) and flour beetles (*Tribolium*) and **Meloidae** includes the blister beetles whose blood usually contains a caustic blistering agent termed **cantharidin**. They belong to the superfamily **Heteromera** which is the largest and includes the largest family **Chrysomelidae** or leaf-beetles with

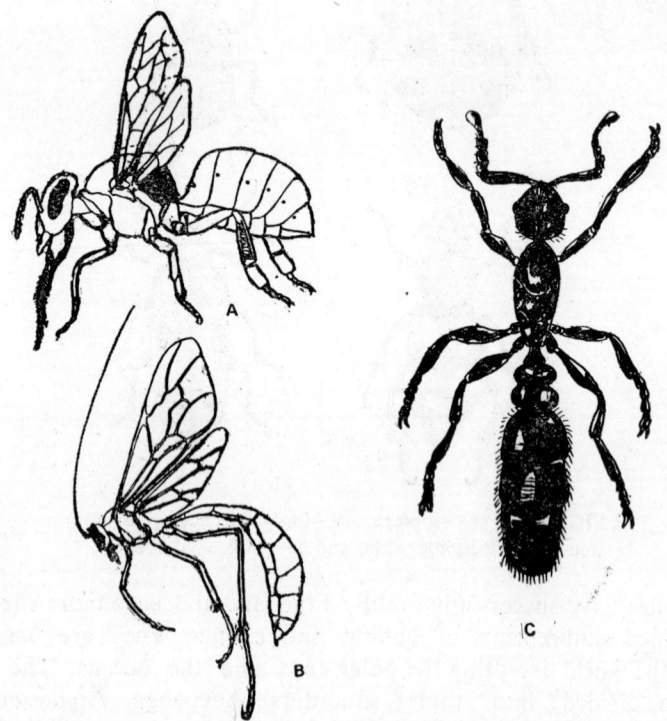

FIG. 17.71. Hymenoptera A—A Typical hymenopteran; B—Ichneumon, generalised; and C—Ordinary black ant, wingless female.

more than 30,000 species. Some members of this family, such as the asparagus-beetles (*Crioceris asparagi*), the flea-beetles (*Phyllotreta*) and the Colorado potato beetle (*Leptinotarsa decemlineata*) are destructive to crops. The Cerambycidae or longicrons include most other Phytophaga. They are forest insects whose larvae bore into the trees. The superfamily **Rhynchophora** includes

*Class Insecta* 743

beetles whose head is usually drawn out to form a rostrum. The family **Curculionidae** (weevils) includes over 70,000 known species. The rostrum is used to bore holes in the medium in which the eggs are deposited. The grain weevils (*Calandra*), the cotton boll-weevil (*Anthonomus grandis*) the pine-weevil

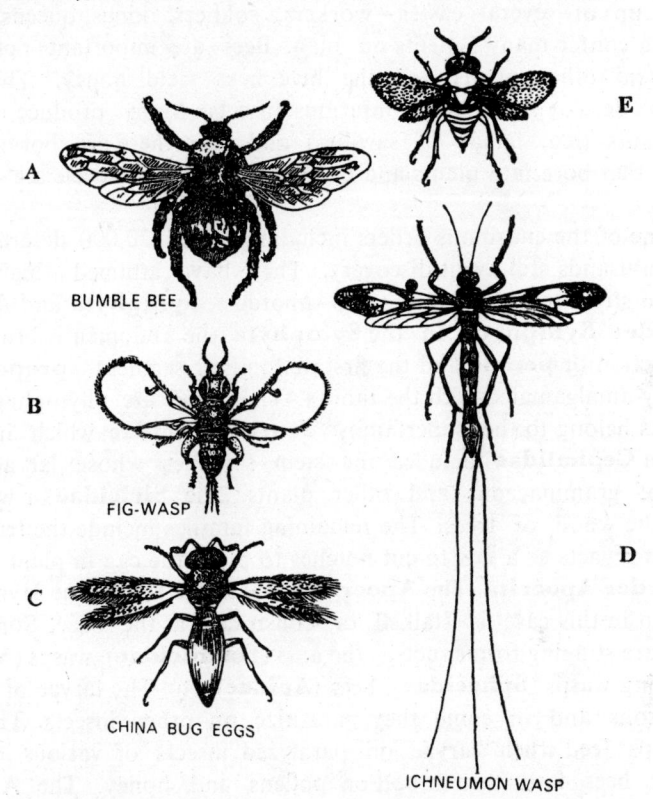

FIG. 17.72. Hymenoptera. A—Bumble bee; B—Figwasp *blastophaga*; C—*Eumicrosoma benefica*, parasite of chinch bug egg; D—Ichneurmon wasp; and E—A chalcid parasite.

(*Hylobius abietis*) belong to this group. The bark-beetles, **Scolytidae**, rank as major forest pests. The superfamily **Lamellicornia** includes beetles, that can be easily recognized by fossorial forelegs and an antennal club formed of plate-like components. The stag-beetles (**Lucanidae**), the chafers and dung beetles (**Scarabaeidae**) belong to this superfamily.

22. **Order Strpesiptera** (*strepsis*, a twisting + *pteron*, a wing). Male degenerate free-living with stalked eyes, flagellate antennae, mouthparts degnerate chewing; wings two pairs, fore-wings reduced to small clubbed structures called pseudohalteres, hind-wings large, fan-shaped; female larviform, endoparasitic, apodous with head and thorax fused, eyes and antennae atrophied; mouthparts vestigial, e.g. *Pyrilloxens competus*, parasitic on sugarcane *Pyrilla*.

23. **Order Hymenoptera** (*hymen*, a membrane + *pteron*, wing). Sawflies. ants, bees, wasps, etc (Fig. 17.71, 17.72). Minute to moderate-sized insects with membranous wings, the hind-pair is the smaller and connected with the fore-pair by hooklets. Venation of wings is specialized by reduction Mouthparts

are modified for biting and licking. Abdomen with first segment fused with thorax. A sawing or piercing ovipositor is present. Larvae usually with many legs, sometimes without legs and pupation takes place generally in cocoons. These are mostly social insects living in colonies, and individuals are polymorphic being made up of several castes—workers, soldiers, kings, queens, etc. The Hymenoptera confer many benefits on man. Bees are important pollinators of fruit trees and other plants and the hive bees yield honey. The parasitic Hymenoptera destroy myriads of injurious insects. Many produce larvae that defoliate plants (e.g. larvae of sawflies) and then there are boring larvae of wood-wasps that bore into plants and damage them (for life-cycle see description of honey-bee).

This is one of the enormous orders including about 120,000 described species and many thousands still await discovery. They have attained a high degree of specialization structurally. There are two suborders, **Symphyta** and **Apocrita**.

(*i*) **Suborder Symphyta**. In the **Symphyta** the abdomen is broad with no basal constriction or **petiole** and the first abdominal segment (**propodeum**) is only partially amalgamated with the thorax. The larvae are phytophagous. Most of the species belong to the superfamily **Tenthredinoidae** which includes six families. The **Cephalidae** includes the stem saw-flies whose larvae bore into the stems of graminaceous and other plants. The **Siricidae** or wood-wasps tunnel into the wood of trees. The remaining families include the true sawflies, their ovipositor acts as a saw to cut notches to place the egg in plant tissue.

(*ii*) **Suborder Apocrita**. The **Apocrita** include the rest of the Hymenoptera. The abdomen in this case is stalked or constricted at the base. Some of these (**Aculeata**) are stinging forms such as the ants (**Formicidae**), wasps (**Vespidae**), solitary digging wasps (**Sphecidae**), bees (**Apidae**), etc. The larvae of true wasps are carnivorous and in some they parasitize on other insects. The solitary digging wasps feed their larvae on paralyzed insects of various orders and spiders. The bees feed their brood on pollens and honey. The Apidae and Bombidae (bumble bees) are social insects.

Other members (**Parasitica**) of this suborder are ecto- or endoparasites of other insects. The ichneumon flies (superfamily **Ichneumonoidae**) are the largest of these and parasitize on caterpillars of Lepidoptera. Others are small to minute and some of these (**Cynipoidea**) produce galls and a number of these are parasitic on Diptera. Another superfamily, the **Chalcidoidae**, includes parasites and hyperparasites except for a small number that form galls. The **Prototrupoidea** include many minute egg parasites.

24. **Order Diptera** (*di*, two + *pteron*, a wing). Moderate-sized to very small insects with a single pair of wings, hind-pairs being modified into halteres. Mouthparts for sucking or for piercing also and usually form a proboscis. Larvae are worm-like, terrestrial, aquatic or parasitic, pupa formed without cocoon. Most flies are diurnal and many visit flowers for nectar, while numerous others feed upon decaying organic matter and diverse fluid substances. There is also a number of flies that are predators on smaller insects or have acquired blood-sucking habits.

This order includes over 64,000 described species. Structurally they are the most highly specialized among insects surpassing even Hymenoptera. The Diptera have three suborders. Nematocera, Brachycera and Cyclorrhapha.

*Class Insecta*

(*i*) **Suborder Nematocera.** The **Nematocera** have many jointed antennae and include such forms as the crane-flies (**Tipulidae**), midges (**Chironomidae**),

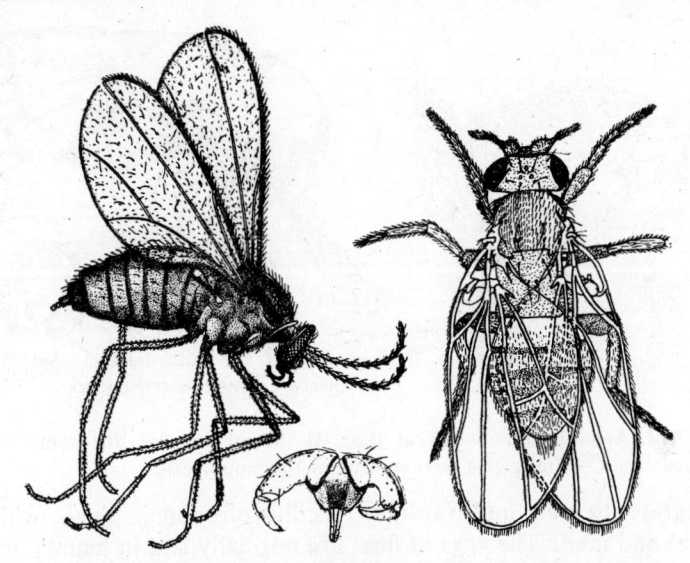

FIG. 17.73. Diptera. A—Cecidomyiid fly, *Erosomyia indica*, a pest of mango, female; B—Genitalia of male; and C—*Drosophila* sp.

the larvae of which (the blood-worms) have haemoglobin dissolved in the blood plasma, and mosquitoes (**Culicidae**) which are slender insects, the female of the majority of which can suck the blood through vertebrate skin. The buffalo-gnat (**Simuliidae**), the fungus-gnats (**Mycetophilidae**) and the gall-midges (**Cecidomyiidae**) are other members of this suborder.

(*ii*) **Suborder Brachycera.** The **Brachycera** include many families of stout-bodied flies among which the females of the **Tabanidae** (horse-flies) are blood sucking and the **Asilidae** (robber flies) are predators on other insects.

(*iii*) **Suborder Cyclorrhapha.** The **Cyclorrhapha** includes all other Diptera with 3-jointed antennae, each with a dorsal bristle-like **arista**. Most of the families possess a **ptilinum,** the **Syrphidae** (hover-flies), however, are without ptilinum. The **Trypaneideae** (fruit-flies) include destructive larvae that bore into fruits. The pomace-flies (**Drosophilidae**) have saprophagous larvae, while those of warble-flies and bot flies (Oesteridae) are endoparasites of mammals. The house-fly and its allies belong to the **Muscidae**. The blood-sucking stable flies (*Stomoxys*) and the tsetse flies (*Glossina*) also fall under this suborder. The **Hippoboscidae** are viviparous and live, as adults, as blood-sucking parasites on birds and mammals e.g. (*Hippobosca* and *Melophagus*).

25. **Order Aphaniptera** (*aphanes,* not apparent + *pteron,* a wing). Very small apterous laterally compressed insects whose adults are ectoparasites of warm-blooded animals. Mouthparts are of the piercing and sucking type. Larvae worm-like, pupa exarate in silken cocoons. The Aphaniptera (or Siphonaptera) are laterally compressed and not dorsoventrally flattened Eyes are present or absent and antennae are short 3-jointed organs reposing in grooves. Fleas are covered with a tough cuticle and the legs are adapted for clinging and leaping. The human flea, *Pulex irritans*, is known to leap vertically up to 20 cm. All fleas

are blood sucking ectoparasites of birds and mammals and rarely exceed 4 mm in length. Usually each species has its particular host. The rat-flea, *Xenopsylla*

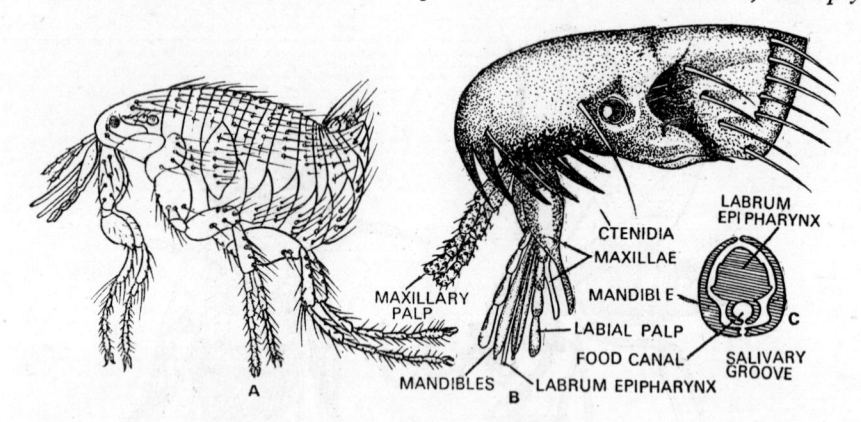

FIG. 17.74. Aphaniptera. A—A rat flea; B—Mouthparts of the same, highly magnified; and C—Transverse section through the mouthparts.

*cheopis* migrates to man and transmits bacillus of bubonic plague which affects both the rat and man. The eggs of fleas are normally laid in haunts or sleeping places of hosts. The larvae are whitish worm-like with a well-developed head, bearing biting mouthparts and 13 trunk segments. They feed upon organic matter found around the hosts. The pupae are exarate and are enclosed in silken cocoons. About 1,000 species of fleas are known.

## GENERAL NOTES ON INSECTS

### Origin of Insects

As the earliest known insects were fully winged they provide no clue to their origin. However, several theories have been proposed, based upon the studies on comparative morphology of insects, other arthropods and annelids, and these theories derive insect from different sources.

(*i*) The **Crustacea theory** derives them from forms resembling the zoea larva of the higher Crustacea with three pairs of appendages on both head and thorax; or from other Crustacea that resemble the young of some thysanurans. It is presumed that the crustacean forerunners migrated to land and there evolved into primitive wingless forms.

(*ii*) Some authors derive them from the **Trilobita (Trilobite theory)**, the extinct class of palaeozoic marine forms with three-lobed body. This view maintains that the earliest insects were winged and have originated from extinct Palaeodictyoptera which were derived from the trilobites.

(*iii*) The insects have also been derived from the polychaete worms (**Annelida theory**) because of the resemblance of the pleura and thoracic appendage of insects to the parapodia of polychaetes.

(*iv*) Finally there is the **Symphyla theory** that derives the insects from the Symphyla which possess the essential structural features required of an ancestor of the Thysanura. This theory is regarded as more probable because it is accepted by all, that the ancestors of insects were many legged animals like the Symphyla. This is supported by the existence of a polypod (many-legged) larva during the development of an insect.

# Class Insecta

## Number of Insects

It is estimated that 85 per cent of all the known kinds of living animals are insects. Their estimated number ranges between 2,50,000 and 10,000,000 species probably the latter being more authentic. Fully described species number 680,000 or more and more are being discovered, explored and established every day. The insects are, thus, the most abundant and widespread of all land animals, being the principal invertebrates that can live in dry environments and the only ones that fly.

## Habitat and Food of Insects

Insects abound in all habitats including the sea, various kinds are adapted to live in fresh and brackish water, in soil, on and about plants of all kinds and on or in other animals. Various sorts eat all kinds of parts of plants—roots, stems, or leaves, sepals or blossoms and seeds or fruits. Fungus feeding forms are also numerous. Some utilize the tissue fluids and excretions of animals and the scavenger insects consume dead animals and plants. Parasitic insects live in the eggs, larvae or adults of other insects and many other animals and plants.

## Fecundity

The rate of producing young ones varies in different insects from a single larva hatched at a time (viviparous flies) to the million eggs, more or less, laid by a queen bee. The actual number from any one female is less important than the rate of growth. In some species the rate of growth is very rapid, for instance, the pomace-fly (*Drosophila*) lays up to 200 eggs per female and the entire life-cycle is completed only in ten days at about 80° F. The house-fly completes its life-cycle in 8 to 10 days. Thus, in a very short time, the progeny of one generation matures and begins to produce fresh generations. Further the production of young ones is accelerated by the phenomenon of parthenogenesis (development of young ones from unfertilized eggs), which occurs in aphids (plant lice) and other insects. This leads to extremely rapid multiplication under optimum conditions of temperature, moisture and food supply. The offspring of a single aphid could cover the earth in one season if all survived.

## Flight

Insects are the only flying invertebrates. In the beginning the insects living on taller plants developed lateral extensions from the sides of their thoraxes enabling them to sail from leaf to leaf or from tall plants down to the ground. In time, these gliding vanes became larger and larger and more efficient, but as they glided through the air the vanes tended to vibrate or flutter because of aerodynamic forces. Slowly muscles were developed to actively move the wings up and down against the air, enabling the insects to sail farther. With further evolution powered flight became a reality. One of the first flying insects was *Stenodictya lobata*, a stone-fly-like insects that is known from fossilized remains found in the upper carboniferous strata. This early flying insect had well-developed wings attached to the two hind segments of the thorax. In addition it had short lobes extending out from the front segment. It is, therefore, assumed that earlier insects had three thoracic extensions of which only the last two developed into true wings.

The ability to fly is of great advantage. It makes it easier to evade enemies, enables predaceous species to capture prey, makes it possible for a plant eating insect to get to its food plant, helps insects to obtain mates and facilitates moving to new areas where food may be more abundant or where climatic conditions may be more favourable.

Wings are found in almost all insect orders except the most primitive. Apterygota. Some insects such as the Mallophaga and Anoplura have lost wings secondarily because of development of ectoparasitic mode of life. The sheep-tick (*Melophagus ovinus*) and the deer parasite (*Lipoitena cervi*) are wingless. Queen ants which have wings in the beginning lose them after mating. Termites have also wingless forms. Many species of cockroaches, among the most ancient of flying insects, have dispensed with wings. Although insect wings were apparently all derived from a common ancestor, the wings of various present-day insects have become greatly modified. The wings of butterflies and moths are thin, expanded vanes covered with scales, those of bees and many other insects are cellophane-like. Beetles have horn-like fore-wings and membranous hind wings that are folded under them when at rest. A somewhat similar arrangement is found in grass-hoppers. In housefly and mosquito only two wings (Diptera) are present, the hind wings have become modified into balancers or halteres.

Insect wings vary in size. Perhaps the greatest wing expanse was that of large dragonflies (*Meganeura*) that lived millions of years ago. The wing expanse was more than two feet across. Among the modern insects the largest wing expanse is found among the butterflies and moths. The fairyflies (*Allaptus*) has probably the smallest wings. It is interesting to know that the minute fairy-flies use their wings both in air and water.

The wings contain many hollow veins through which blood flows. The veins also serve as stiffening structures and along them especially in the basal region are found sensory organs. The veins in the front portions of the wings are heavier than those towards the rear margin so that as the wings beat up and down, the rear portions of the wings flex, causing a forward thrust. When an adult winged insect emerges from its pupal case, its wings are fluid-filled sacs that gradually expand and harden into the typical wing form.

The actual manner in which an insect's wings are caused to move up and down is very complicated. The thorax of an insect is a box-like structure having the wings attached to the upper corners. This box has large muscles extending from the top to bottom of the thoracic box (indirect muscles). The contraction of these muscles pulls the top of the thorax downward causing the wings to rise because of the way in which the wings are attached. There are other indirect muscles that extend in a lengthwise direction. The contraction of these muscles causes the wings to beat downward. This is because the longitudinal

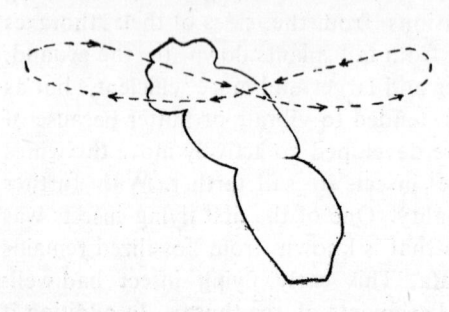

FIG. 17.75. Flight in an insect showing plane of vibration during forward flight. Figure of '8' is evident.

## Class Insecta

muscles, on contraction, cause the roof of the thoracic box to bend upward. The direct muscles attached to the bases of the wings bring other movements. Thus, on the downstroke the wing moves downward and forward and on the upstroke it moves upward and backward. As the wings move up and down they also rotate on their axes. When an insect wing rises, its hind portion is deflected downward and when it beats downward these deflections tend to drive the insects forward like the rotating blades of a propeller. Laboratory study of a flying insect's wing reveals that the tip describes a figure of eight (8) and not simple up and down arcs (Fig. 17.75).

In order to increase the efficiency of the wings many four-winged insects (bees, butterflies and moths) have evolved devices for hooking the two wings together so that they function as a unit. The two-winged flies have solved the problem by the elimination of the hind-wings. In the dragonflies the two pairs of wings beat alternately, i.e. the front pair beats downward while the hind pair rises.

The rate of wing-beat per second varies greatly. It is affected by age of insect, sex, temperature, fatigue and so on. Many methods have been used to study this. These include Kymograph, stroboscopic lights and electronic equipment. The flight speeds of different insects are given below:

| Insects | Wing-beat rate | Miles per hour |
|---|---|---|
| Beetle (Coccinella) | 75-91 | — |
| Butterflies | | |
| Papilio | 5-9 | — |
| Pieris | 12 | 5.7 |
| Monarch | — | 6.2 |
| Bumblebee | 250 | 6.4 |
| Dragonfly | | |
| Aeschna | 38 | 15.6 |
| Austrophlebia | — | — |
| Meganeura (prehistoric) | — | 4.3 |
| Honeybee | 250 | 5.6-13.9 |
| Hornet | 100 | — |
| Horsefly | 96 | 6-14 |
| House-fly | 190 | 4.4 |
| Hoverfly (Eristalis) | 190 | 7.8 |
| Hummingbird moth | 85 | 11.1 |
| Midgefly (Forcipomyia) | 988-1047 | — |
| Mosquito | | |
| Aedes | 587 | 2.5-5.5 |
| Culex | 278-307 | — |
| Noctuid moth | 30-50 | — |
| Hummingbird (for comparison) | 30-50 | — |

In asynchronous flight mechanism the wing stroke is not synchronous with the nervous stimulation of the muscle. Thus, the rate of stimulation does not bear any relation to the rate of the beat of wings, instead each stimulation causes a variable number of contractions. It is because of this that insects possessing an asynchronous flight mechanism can maintain such a remarkably high frequency of wing beat.

The mode of operation of an asynchronous flight mechanism is evident from the work of Boettiger on the fly *Sarcophaga bullata*. This analysis became possible

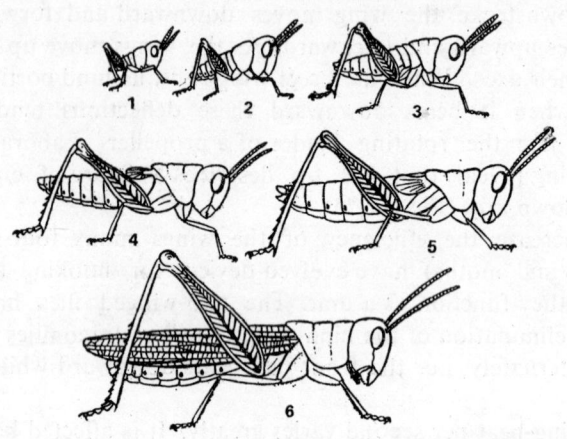

FIG. 17.76. Metamorphosis. Incomplete metamorphosis in grasshopper. 1-6—Different nymphal stages leading to adult (6).

by the fortunate circumstance that when the fly is exposed to fumes of carbon tetrachloride its wings become set in either up or a down position corresponding to the positions during the normal flight movements. Study of these positions and pushing of the wings from the position to another, thus, made possible the elucidation of the normal functioning of the relevant parts of the sketeton.

The high rates of muscular activity naturally demand a high rate of energy consumption. When the bee goes over from running to flying its oxygen uptake rises 50 times or more. In the fruit fly *Drosophila* it is increased by about 100 times. The honey-bee seems to be dependent on the store of sugar in its stomach. Sugar is present in the blood in the form of glucose at the surprisingly high concentration of 20 per cent. The bee weighing 100 mg uses this sugar at the rate of about 10 mg per hour and has a flying time of only about 15 minutes with flight range of about 6.5 km. Some insects can utilize fat. The fat content of the bee-leaf hopper *Eutettix* may be used as a measure of the distance over which it has flown from the breeding places. During flight of the desert locust *Schistocerca*, fat is consumed. In *Drosophila* glycogen is used during flight. During prolonged flight glycogen carried in the knobs of the halteres is consumed. The knobs of two halteres of *Drosophila* carry enough glycogen to provide for nine minutes' flight.

Insect flight muscles are very powerful. The fibrils are relatively large and the mitochondria are huge (about half the size of a human red blood cell). Insects are the only cold-blooded fliers. A low body temperature and a corresponding low metabolic rate impose limitations on mobility. On cold days many insects "warm up" before flying, i.e. they remain stationary on a tree trunk and move the wings up and down until sufficient heat is generated.

## Growth and Metamorphosis

Since an insect lives in an armoured exoskeleton, it can change form and increase in size only if its rigid armour is cast off (moulting), and none moults

## Class Insecta

after attaining the adult stage. The primitive orders Protura to Thysanura attain adult form and size by slightly graded changes and hence are called **Ametabola** (Gr. *a*, not +*metabola*, change). The **Hemimetabola**—Orthoptera to Thysanoptera—undergo a gradual change or **incomplete metamorphosis**. The young hatch as small nymphs, somewhat resembling the adults with compound eyes. In the successive stages called instars, the wings appear externally as small wing pads, that enlarge at successive moults and become functional in the adult called **imago**. Grasshopper is an example (Fig. 17.76). In others the young emerge as small worm-like segmented **larvae** with the head, thorax and abdomen much alike and with short legs, but no wings or compound eyes. The larvae moult several times and increase in size. Each finally enters a "resting" stage, the **pupa**

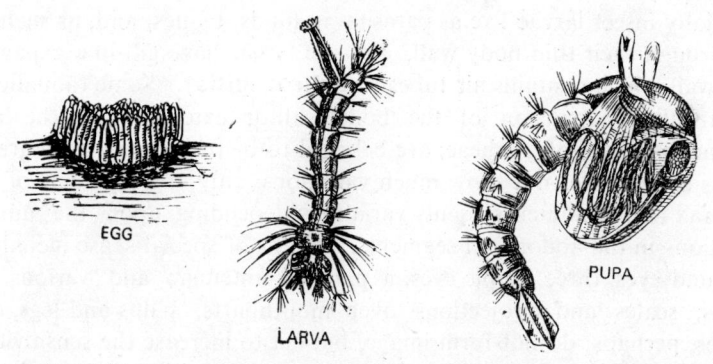

FIG. 17.77. Metamorphosis. Complete metamorphosis in a mosquito. Egg, larva and pupa mark different immature stages.

within the last larval skin, in a special puparium or in a cocoon. In this stage many larval organs break down and are absorbed by phagocytic cells while new structures for the adult arise concurrently. These profound changes occur before the adult or imago hatches out. This is known as **complete metamorphosis** (Fig. 17.77) as found in the honey-bee and animals in which it occurs are called **Holometabola**. In some cases during development an insect passes through two or more different larval instars, then it is said to undergo **hypermetamorphosis**. This process is characteristic of some parasitic groups.

### Variation in Structure

Variations in the structure of insects are correlated with the mode of life they lead. In the cockroach, for instance, the gizzard is well-developed and its lining is armed with hard plates and spines, but insects which suck juices have no gizzard. In the honey-bee the nectar is sucked up into a honey stomach which corresponds to the crop. The region between the honey stomach and the stomach is a valve which prevents the food in the honey stomach, designed for storage in the hive, from going into the stomach. The mouthparts are modified accordingly. Cockroach has **biting** mouthparts; the house-fly has one meant for **sucking**, the mosquito has **piercing** and **sucking** type and so has the bedbug. In most butterflies the mandibles are rudimentary and the maxillae are greatly elongated, each forming a half tube, so that when they are held together they form a long sucking proboscis, through which liquids are pumped up by the mouth pump. The proboscis is extended only when the insect is feeding, when not in use it is coiled under the head (Fig. 17.50).

The thoracic legs are modified in various ways, but basically all are made up of the same parts as figured in the case of cockroach. Land forms have walking legs, some have legs meant for digging, houseflies have sticky pads, the tips of which enable them to walk on smooth vertical surfaces. Water beetles have flattened legs for swimming. Legs of honey-bee are modified for collecting food. Thus, besides locomotion, the legs are modified for various other works.

The wings are also modified according to the need of the animal but do not show such pronounced variation as shown by other structures.

Respiration in the majority takes place by the tracheal system, but some, like most of the tiny collembolans, have no air tubes, and breathe through the body surface. Many insect larvae live as parasites in fluids, tissues, and, as such, the air diffuses through their thin body wall. Aquatic larvae have gill-like expansion of the body wall which contains air tubes (**tracheal gills**). Some aquatic larvae have thin-walled expansion of the body wall or extension of the hind gut, which, though devoid of tracheae, are believed to be respiratory structures.

Nervous system does not show much variations, only the number of ganglia in the thorax and abdomen presents variations depending upon the number of and variations in the abdominal segments. Organs of special sense include a pair of compound eyes, three simple eyes, a pair of antennae and various sensory hairs, pits, scales and projections over mouthparts, palps and legs, etc. The simple eyes, perhaps, do not form image, but act to increase the sensitivity of the brain to light stimuli from compound eyes. For, if the simple eyes are covered by some opaque paint, the insect does not react to light as rapidly as when the simple eyes were not shut off. If large eyes are covered, the insect does not respond to light. Insects are able to recognize colour. At least in the case of honey-bee it has been experimentally proved that it reacts to colour; of course the bees cannot detect minor differences in colour, such as blue and grey. Some insects respond to red. The antennae are the chief organs of touch and also carry organs for the sense of smell. The ability of the insects to smell has also been proved experimentally with bee as a type in most cases. That butterflies also possess sense of smell is proved by the fact that the males of some species emit scent to attract the females.

**Sensory receptors**. The sensory receptors maintain contact between the environment and the insect through its nervous system. The stimulus received by the receptor is changed into a nervous impulse. There are three types of stimuli that affect the receptors. These are (*i*) mechanical vibrations or pressures, (*ii*) contact with chemical, stimulating taste, the olfactory sense or the common chemical sense, and (*iii*) light rays. Sensory receptors may be external or exteroceptors (eyes, antennae, palpi and setae) or internal or interoceptors (proprioreceptors and the sympathetic receptors). Exteroceptors perceive external events, interoceptors react to internal stimuli. All external receptors develop from modifications of hypodermal cells, sometimes, as in the eye, to a striking degree, sometimes almost imperceptibly.

**Humidity and temperature**. Maintenance of acceptable levels of temperature and moisture is of prime importance for survival. Humidity receptors have been identified with several types of chemoreceptors in many different insects.

# Class Insecta

Some of these receptors appear to respond to dry, others only to moist air. In general the sense of humidity seems to depend upon the detection of water vapour as a chemical. Temperature receptors have not been identified. It is, however, felt that temperature receptors exist on the antennae, maxillary palps and tarsi. There are many experimental evidences that show the presence of temperature receptors.

**Tactile receptors.** Articulated tactile setae (Fig. 17.77) with sensory nerve connection are scattered generally over the whole body and are often dense on the antennae, cerci and tarsi. Movement of the seta in its membrane stimulates the nerve cell. Each sensory hair comprises the usual trichogen and tormogen cells which secrete respectively the hair and its socket.

**Campaniform sensilla** are a further development of the tactile setae. A modified hypodermal cell or cells support a dome-shaped area of thin cuticle with which a sensory cell is associated. Deformations of the dome, which may be above, below or on the level of the surrounding cuticle stimulate the neuron. They respond either to direct contact or to the stresses of the body movement.

**Chordotonal organs (scoloparia).** Chordotonal organs are subcuticular and more complex. Berlese considered them to be a further modification of campaniform sensilla, but recent investigators do not agree with this view. The unit of the chordotonal organ is the **scolopophore** (Fig. 17.78) composed of an

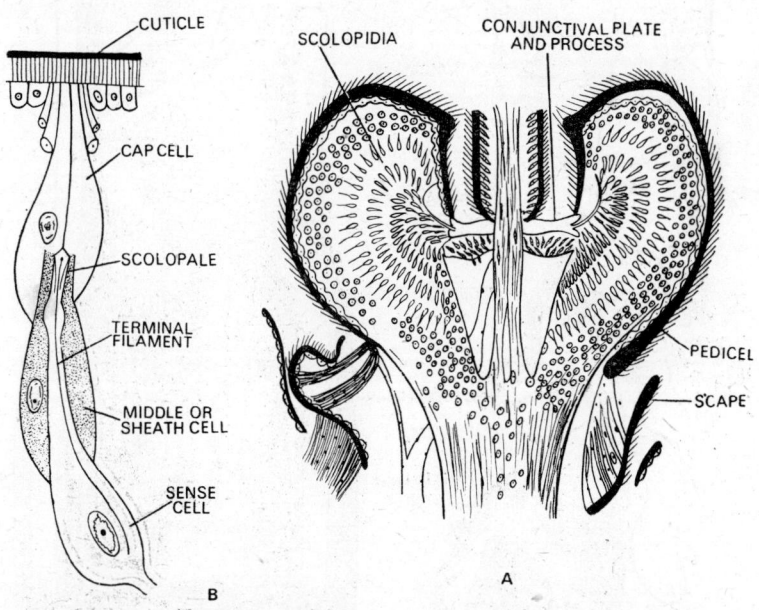

FIG. 17.78. A—The organ of Johnston in its complex form as in *Chaoborus*; and B—Scolopophore from such an organ.

apical **cap cell** attached to the body wall and an envelope cell; in the central part of the cap cell is a terminal ligament, which forms a terminal extension of the **sensory rod (scolops or scolopale)** in the **envelope cell**. The sensory rod is a vacuolated cylinder with walls strengthened by longitudinal ribs. The tip of the sensory dendron penetrates the envelope cell reaching the sensory rod. The organ may be made up of either one or a few scolophores, but usually a number of them is gathered together and enclosed in ensheathing membrane. In most cases a chordotonal organ is attached to integument at each end. Such organs are abundant in the abdomen and appendages and at the base of the wing. They serve sometimes as proprioreceptors, sometimes as receptors for perceiving internal pressures or external vibrations. They are essential parts of certain auditory receptors.

**Johnston's organ.** It is a mechanical sense organ situated in the second antennal segment of most insects (similar to Chordotonal sensillum in structure) and is sensitive to the movements of the antennae. The whirligig beetle (*Gyrinus*) extends its antennae forward near the water surface, apparently perceiving wavelets and irregularities through changes in the position of the flagella; The antennae of insects with well-developed Johnston's organs are quite sensitive tactile structures, even responding to light breaths of air. The auditory function is aided in some insects, as has been found in the mosquito *Aedes*. Sound vibrations set in motion the numerous setae along the flagellum, imparting to the whole segment a vibratory motion detected by Johnston's organ.

**Auditory receptors.** Sound detection is only a refined variation of sense of touch. The auditory receptors are, therefore, like ordinary tactile receptors. They

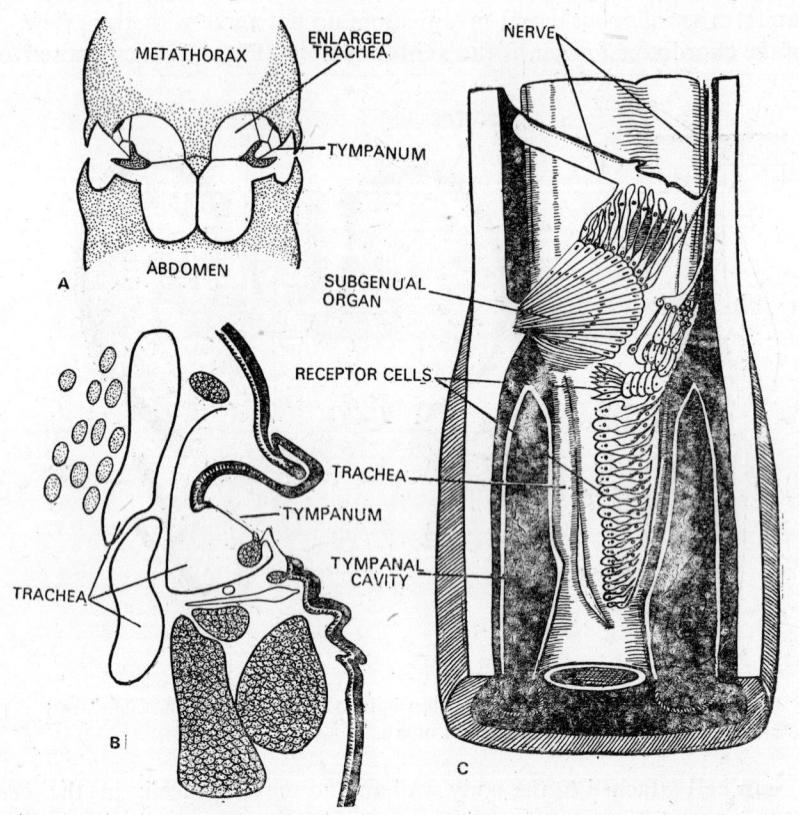

FIG. 17.79. A—Diagrammatic section through the thoracic tympanal organs of the moth *Catocala*; B—Section through abdominal tympanal organ of the orthopteran, *Oedipoda*: and C—The tibial organ of orthopteran *Decticus*.

are the simplest and at the same time least understood. Auditory setae are scattered all over the body. They probably occur in all insects. The insects continue to respond to sounds even when the obvious auditory structures have been experimentally removed or immobilized. This is an experimental evidence of the existence of auditory receptors. It may be pointed out here that low intensity disturbances of the air producing displacement of air particles along with a local pressure increase are interpreted by auditory receptors as sound. Sound may also be transmitted through solid and liquid media.

**Tympanal organs.** In some insects special auditory organs are present, especially in those insects that have elaborate sound producing organs. Many short-horned grasshoppers have a pair of ears or **tympana** on each side of the first abdominal tergum. The crickets and long-horned grasshoppers have auditory organs at the base of the fore tibiae. The cicadas have sound receptors at the base of the abdomen. Chordotonal tympanic organs occur in many Lepidoptera, placed on the first abdominal segment, on the metathorax or at the base of the wings.

The more complex auditory organs consist of tracheal air-sacs separated from the outside by thin cuticular tympanal membranes. Numerous chordotonal receptors are fastened to the tympanum or to the walls of the tracheal sac in such a way that they are stimulated by vibrations of the tympanum, or by the resultant compression of air within the sac. The tympanal membrane is exposed in short-horned grasshoppers as it lies on the surface but in the long-horned grasshoppers and crickets it is sunk into a pit.

The greatest complexity is reached in some of the long-horned grasshoppers, where two tympana, a trachea, three chordotonal organs and two nerves are involved. A pair of longitudinal slit-like openings may be seen near the proximal end of the fore tibia. Each slit leads into an invaginated cavity lined with thin cuticle. These two cavities have their inner walls closely applied to the wall of an enlarged trachea, the common walls forming an inner (medial) and an outer (lateral) tympanum. Between the tympana the widened trachea is divided by a medial septum. The supratympanal organ placed just proximal of the tympanal chambers, consists of a large number of scolopophores attached by their cap cells to the outer wall of the leg and at the other end to the tracheal wall; it receives sensory fibres from both the tympanal nerve, and the tibial nerve, both of which lead to the prothroacic ganglion. The small intermediate organ is placed just beyond the supratympanal organ and at the proximal extremity of the tympanal chambers. Its scolopophores serviced by the tibial nerve, are attached only to the tracheal wall and lie free in the haemocoel of the leg. The third chordotonal organ is the **crista acustica (Siebold's organ)**, composed of a long series of scolopophores which diminish in size distally. They are attached only to the tracheal wall, but are held against it and supported by ligamentous bands.

**Chemoreceptors.** Chemoreceptors in insects appear to be modification of tactile receptors, and in some, probably serve both functions. All chemoreceptors are characterized by the possession of cuticle which is at least partly thin-walled (though never perforated) and are innervated by one or a group of bipolar neurones, whose associated distal processes contain small refringent bodies. A number of different morphologic types have been described. The **trichoid** type **(sensillum trichoideum olfactorium)** is derived from some innervated articulate setae having very thin cuticular walls, either entirely or only around the base, and are frequently set up in cup-like depressions. If the hair become shorter and stouter the same is called **sensillum chaeticum.** Such receptors are found especially on the antennae, mouthparts and tarsi (in some Diptera and Lepidoptera). In some cases the setae assume a cone-like or peg-like appearance. Such a receptor is called **basiconic (sensillum basiconicum, sensillum styloconicum)** and principally occurs on the antennae and mouthparts. In some cases the setal cup is deepened so that the seta lies below the general surface. Such a receptor is termed **sensillum coeloconicum.** In some forms the seta lies in a chamber connected to the surface by a tube. This is **sensillum ampullaceum.** All these occur on antennae.

The excretory tubules vary only in number—from two to over a hundred, but all function almost in the same way.

**Reproductive organs.** The essential parts of the reproductive organs are almost similar. There is, however, difference in the mode of reproduction in different animals. One species of cotton-cushion scale is hermaphrodite in which the females are able to fertilize their own eggs. Unfertilized eggs give rise to males which are rare. Some species are without males and the eggs develop without being fertilized. Such a mode of development is known as **parthenogenesis.** In the vast majority of insects, sperms are stored in the receptacle of the female at the time of mating, and the eggs are fertilized at the time of laying. It has been established beyond all doubt that male honey-bees regularly develop from unfertilized egg and females (queens) and workers develop from fertilized egg. This also holds good for most Hymenoptera.

### Insect as Enemies

A great struggle between the insect and man began long before the dawn of civilization and has since then unceasingly continued. The insects destory or damage all types of growing crops and other important plants, by chewing leaves, buds, stem, flowers and fruits, by sucking the sap of plants, by boring into the interior of the roots, by laying eggs in the interior of their parts, and by disseminating various kinds of diseases. The Insects destroy or depreciate the value of stored products and possessions including food, clothing, drugs, animal and plant collections, paper, books, furniture, bridges, buildings, mines, timbers, telephone poles, telegraph lines, rail road tiers and the like. White ants devouring furniture or insect larvae ravaging warm clothing is a common feature.

The insects annoy and injure man and all other living animals, both domesticated and wild. They cause positive annoyance by their presence, e.g., the presence of housefly in our dining rooms. Sometimes their "buzzing" sound makes one irksome. They also bite and sometimes sting injecting venom. When crushed they leave poisonous caustic or corrosive body fluids on the skin. Some insects dwell on or in the body of animal as internal or external parasites hunting the host. Some insects disseminate diseases of a deadly nature.

### Beneficial Insects

Here and there the insects have established a relationship with man which has become definitely beneficial to us. The most obvious of the benefits is the utilization of things that insects make, collect or produce, such as silk, honey, beeswax, shellac, paints, dyes and medicine.

**Silk.** Everyone is familiar with silk. But does he know its origin? A creamy white moth (*Bombyx mori*) about 5 cm across the open wings lays 300-400 eggs which hatch producing larvae which spin cocoons of silk before pupation. These cocoons are picked up, animals inside killed, boiled and the silk obtained direct. For more than 35 centuries the silk-worm has toiled ceaselessly for man for countless generations laying their eggs, feeding on mulberry leaves provided for them in the larval stage, spinning their cocoons and then dying A perpetual sacrifice to adorn men and women (Fig. 17.48) indeed!

**Honey.** Plants lavishly produce nectar which the bees and wasps collect and store as food for themselves and their young. Man has domesticated honey-bees (*Apis mellifica*) and other species that collect nectar from the plants. This nectar undergoes chemical changes on mingling with saliva and transforms into honey rich in glucose (grape sugar) and fructose (fruit sugar) that are more readily assimilated by man. Honey-bees also produce bees-wax which is of immeasurable worth to man.

**Shellac.** A tiny species of scale insects, the lac insect, *Laccifer* (*Tachardia*) *lacca*, found in forests of India and Burma yields the substance from which shellac is made and is used in making varnishes and polishes, for hardening hot materials, as sealing wax in making phonograph records, airplanes, linoleum, buttons, shoe polishes, pottery, toys and bangles.

**Cochineal.** It is a beautiful carmine-red pigment or paint. It is the dried, pulverized bodies of a kind of scale insects, *Coccus cacti*, that lives on prickly pear. These days cochineal is used as a cosmetic or rouge, for decorating fancy cake, for colouring beverages and medicines and for treating whooping cough and neuralgia.

## Insects as Medicines

Many kinds of insects have been known to be of medicinal value. In the seventeenth century almost every insect was believed to possess some medicinal power or the other. In fact some insects do have real medicinal value. A medicinal substance called *cantharidin* is secured from the dried bodies of European blister beetle known as the Spanish fly (*Lytta vesicatoria*). The preparation known as "specific medicine Apis" is extracted from the bodies of honey-bee and used for the treatment of liver, diphtheria, scarlet fever, dropsy and urinary irritation. Homoeopathy also claims many medicines of insect origin (*Cimex, Cantharis, Blatta* etc).

## Insect Pollinators

For the growth and maintenance of plant life sexual reproduction is needed (much more than asexual reproduction). For this the essential male cell of plants called *pollen* is transferred to the female flowers. This is done by water, wind or insects, the latter being the most important. Flowers that depend on insects for pollination are generally beautifully coloured or are marked by odours. The wonderfully intricate mechanism of pollination is too long to be dealt here, but each citizen owes it to himself to know as to how the beneficial insects affect the complex currents of plant life.

## Social Insects

Some insects are **solitary**, each individual lives by itself. The sexes associate only to mate, and the female either deserts her eggs or dies after laying. Some species assemble in large numbers as in swarms of locusts and hibernating lady bird beetle. Such insects are called gregarious and the parents usually never see or live with their offsprings. But in some, parents and offsprings live in mutual cooperation. This is the real social instinct. This mode of existence is made possible owing to the lengthened life of the female parent which permits association with her offspring. About 600 species of insects exhibit **social instincts**, the female or both parents live cooperatively with their offsprings in a common shelter. True social life has been distinguished from sub-social relations in which the female guards her egg and later the young e.g., in earwig (order Dermaptera), cockroaches and crickets (order Orthoptera), in some beetles (Coleoptera) and bugs (Hemiptera), and in Embioptera. The true social life occurs in all termites, all ants, certain wasps and bees and in some caste differentiation becomes very pronounced.

Termites are social insects. They form a nest or colony in hollow wood on which most of them feed. Nests of termites are found in a number of varieties. Many are symmetrical. Some sort of ventilating system that regulates the oxygen supply to the termitaria is always present. The centre of the termitaria has the chamber of the queen (royal chamber). The centre receives fresh air by way of channels from a cellar cavity. Other channels convey stale air from the centre to the surface of the nest. In the nests of certain species the centre contains fungus gardens, which the termites maintain and utilize for food. Termites live in colonies composed of four distinct castes. The king and the queen form the **reproductive caste**. There are some winged termites which can become kings and queens of new colonies. They also belong to the reproductive caste. The enlarged and almost helpless queen produces thousands and thousands of eggs, most of these hatch into white blind **workers** who make up the second caste. **Soldiers** with large heads and jaws make the third caste. They defend the colony. Then there are nymphs which take over the task of reproduction should the king or queen die. At certain times of the year winged males and females develop and disperse. Following dispersion and pairing, the wings fall off and a male and a female begin excavating a new nest.

The social life of ants is similar to that of termites. A migrant winged male and a female mate in flight, on the ground or in bushes. The male dies and the female chews off her wings. She then establishes a colony which is generally constructed in soil. The eggs hatch into sterile wingless female workers, which soon take over the maintenance of the colony. The queen then assumes a purely egg-laying function. Many ant colonies are composed of only the queen and the worker castes. Others are much more complex, having a soldier caste or even including "slave" species.

Social wasps also have a caste of sterile workers but these possess wings. They are only slightly smaller than the queen and are similar in appearance. Only workers are produced in the spring and summer, but at the onset of autumn fertile males and females are produced. The male dies after mating. The female hibernates through the winter and forms a new colony in spring.

Honey-bees possess the same castes as wasps. They keep the colony alive during winter by storing enough food to last until spring. Food stored is in the form of honey and is stored in wax cells. Fertilized eggs develop into queens and workers depending upon the diet given to the larvae. Those feeding on **royal jelly** (secretion from the salivary glands of worker bees between 6 and 14 days old), develop into queens and those fed on **bee bread** (nectar and pollen) into workers. The larvae from unfertilized eggs fed like workers develop into male drones. Depending on their age workers perform different duties. A very young worker feeds the queen, the drones and the larvae on honey and pollen from the storage cells. They continue the process up to the age of two weeks after which they develop wax glands, construct cells and receive and store pollen and nectar brought by field bees. They also clean the cells and guard the entrance to the hive. The oldest workers develop large pollen baskets and are the field bees. New queens are produced by feeding some of the female grubs on royal jelly. This depends on the presence or absence of a queen. If a queen is present the workers cannot feed any larva on royal jelly. This is brought about by depriving the workers of a special pheromone called **queen substance**. This is done by a secretion produced by the mandibular gland of the queen and picked up directly by a few workers who disseminate it to others. When the old queen cannot secrete enough of the substance for all the workers, new queens are produced and swarming eventually occurs. When a queen is experimentally removed from a colony the worker bees transfer one of the queen's most recent eggs to a queen cell and feed royal jelly to the newly hatched larva. When the new queen matures she inherits the old hive and the old queen leaves the nest with part of the colony. After her departure the young queen takes wing on her nuptial flight with a train of drones following. She returns to the old hive and can lay fertile eggs for the next 3 or 4 years.

## Communication

Communication among insects involves a variety of stimuli and sensory channels, sound, visual stimuli, chemical stimuli, and mechanical contact, all serve as signals eliciting behavioural responses. Although many solitary insects display different forms of communication, the highest development of communication is found among social species.

Noises in most insects are produced by the rasping of one skeletal structure over another. Perhaps the simplest is the humming noise produced by extremely rapid vibrations of the wings during flight. Many Diptera and Coleoptera produce such noise. Some grasshoppers and crickets have special file-like sound-producing structures. For example, in some, the front margin of the hind wing scrapes over thick veins of the fore-wing, which begins to vibrate. Other grasshoppers and crickets have a file of teeth on the inner face of the hind femur. When the file is rubbed over the fore-wing sound is produced. Various crickets have files on one or both wings and produce sounds capable of being received up to one-quarter mile away.

Cicadas produce sounds by vibrating certain chitinous membranes called **tymbals**. These organs are usually found in pairs on the abdomen and are controlled by muscles. There are some insects that produce sounds by beating some part of the body against the substrate. There are still others that have a modified pharynx, which can produce a sound by the direct expulsion and vibra-

760                                                          LIFE OF INVERTEBRATES

tion of air.

The significance of sound production is still uncertain in many insects. It has, however, been suggested that sound is used in courtship as alarm signal and as

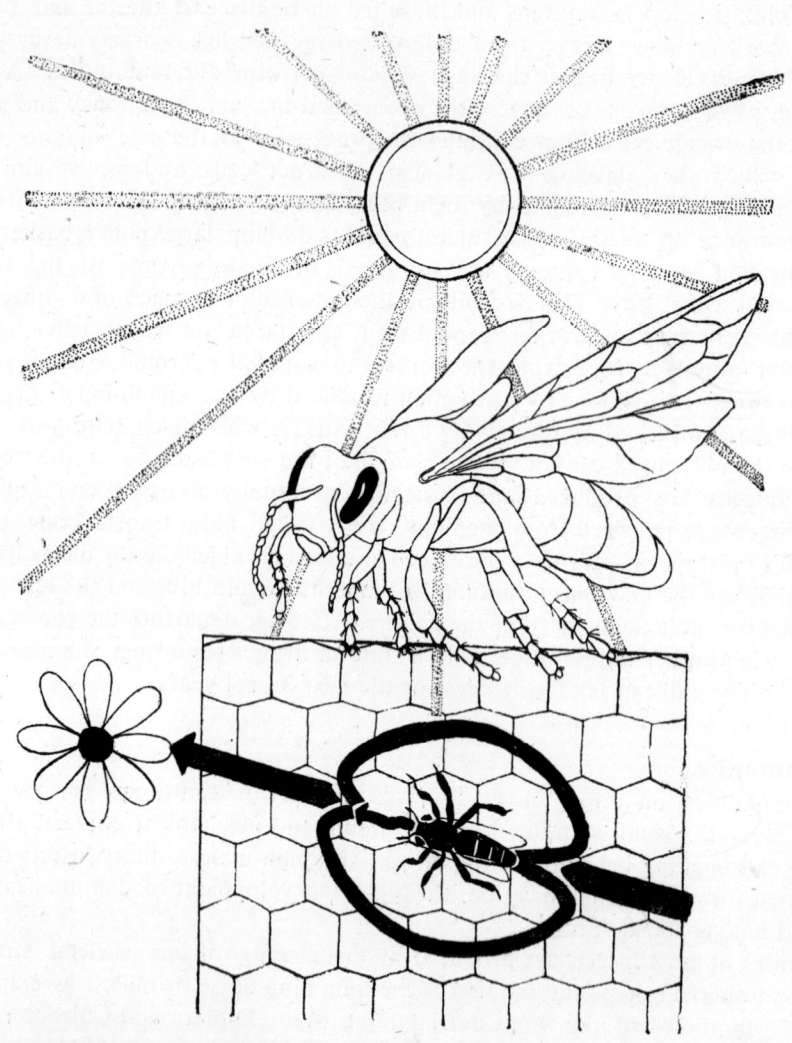

**FIG. 17.80.** Behaviour of honey-bee. The figure shows how the scouts among the worker-bees convey their message by dancing in a characteristic style.

a means to make their presence felt. The mating calls of cicadas and crickets are very well known.

Some insects communicate by body secretions, which may function as sex-attractions, trail-blazing juices, and irritants. This type of communication, found largely in the social insects, plays an important role in locating food and the nest, in attracting mates, in recognizing a member of the same colony, and in defence. In recent years the chemical sex-attractants, produced by insects, have provided considerable interest to insect physiologists. Some of these chemicals, known as pheromones, have been isolated and identified. The pheromones have been used in applied fields of insect control to lure males and females to their destruction. Many of the social insects blaze a trail from a food supply to their colony. The exact nature of this chemical secretion is unknown, but it is thought to be related to the citrals. Members of their own communities are also recognized by certain body secretions.

In 1956 Von Frisch pioneered very useful work on the language of bees and the later work of Wenner (1964) added considerably to our knowledge of communication in these insects. They literally decoded bee messages and found them to be quite explicit in giving the source of a food supply to other worker bees in the colony. The source of the food, its direction and distance from the colony and the richness of the source are all conveyed by a dance and sound vibration of the scout bee The successful foraging scout returns to the hive with nectar and pollen which she shares with the other field bees; this conveys them the kind and amount of food that has been found. The scout bee then begins an excited dance, circling to the right and to the left, with a straight-line run between the two semicircles traced by the dance. During the straight-line run, the bee wags her abdomen. Von Frisch discovered that the orientation of the circular movements shows the direction of the food, and that the frequency of the tail-wagging runs indicates the distance. The closer the food source is to the hive, the greater the frequency of tail-wagging runs. Bees use the angle of the sun and sky polarization as a means of orientation, and the dance of the scout bee indicates the location of the food in reference to the sun's position. If the tail-wagging run is directed upward, the food is located toward the sun; if the tail-wagging run is directed downward, the food is located away from the sun. The inclination of the run to the right or to the left of vertical, indicates the angle of the sun to the right or to the left of the food source An internal "clock" compensates for the passage of time between discovery of the food and the start of the dance, so that the information is correct even though the sun has moved during the interval. On cloudy days the sky polarization pattern acts as a reference point instead of the sun. If the food source is closer than 80 metres the clues provided by chemoreception are sufficient for finding the food, the tail-wagging dance is not performed by the scout bee.

The work of Wenner has shown that the foraging bee communicates to the other members of the hive not only by its wagging dance but also by sound. During the straight run the dancing bee produces vibrations emitted as a train of pulses. The vibrations are produced by wing movements. The surrounding bees apparently receive the dancer's vibrations through their antennae, which touch the dancing bee's thorax, or through their legs. By means of sound the foraging bee indicates the distance of the food source from the hive. The average number of vibrations is proportional to the distance of the food from the hive. There are indications that the sound signals of the forager may also provide clues as to the richness of the food source and perhaps other information.

The fireflies are known to produce flashes of light which serve as sex-attractants as well as for recognition of a member of the same species. The light producing organs are usually located close to the body surface behind a window of transparent cuticle.

# 18 PARASITISM

Parasitism is an intimate relation between two animals in which one is injured. Usually the parasite takes food from its host, but parasitism is not always concerned with food relations. Parasites may cause injury by occupying space within the bodies or dwellings of their hosts, by giving off poisons, and by other similar means. Throughout the animal kingdom parasites occur but mostly they are found among the protozoans, platyhelminths, nematodes, arthropods, etc.

**Types of Parasites**

Parasites which live outside animals (hosts) are called **ectoparasites**. Some parasites living within the body of their hosts are called **endoparasites**. Some species that are forced to lead parasitic life (they cannot live otherwise) are called **obligatory parasites** (filarial worms, cestodes, etc). Whereas the **facultative parasites** are those that can lead an independent life if opportunity comes (maggots of some flies). Some parasites must be continually associated with their hosts and are **permanent** (cestodes), while **transitory parasites** may come in contact with their hosts only at intervals or during certain phases in their life-cycles (*Gastrophilus, Haemonchus*). Then there are social parasites that invade societies and live at the cost of the community. **Accidental parasites** are such that are not normally parasitic but are forced to live as parasites due to some unusual circumstances. Some parasites are **periodic** and attack the host for short periods only, as for instance mosquitoes. **Erratic** parasites live in an organ which is not their normal habitat (*Fasciola hepatica* in the lungs). Parasitism leads to a high degree of specialization in their host requirements.

Some need only a single host while others need two or more specific hosts to complete their life-cycle. Therefore, a regular alternation of generations may occur in such cases. Parasites having associations with a single host are known as **monogenetic** parasites and have simple life cycles, but there are others that have relations with two or more hosts and have complicated life-cycles with **alternation of generations**. Such parasites are called **digenetic** or **trigenetic** depending upon the number of hosts.

There are numerous examples of parasitism. It is beyond the scope of the present chapter to describe even a few. Some diverse types are given here. Two shrimp-like crustaceans, *Synalpheus* and *Typton*, live in the canals off the walls. A midge larva of swift streams lurks under the wing-pads of mayfly nymphs and sucks nourishment from the bodies of its hosts. Order Isopoda includes a large number of parasites. No group of Crustacea presents more examples of parasitism than the Copepoda.

Among the protozoan parasites the most familiar example is furnished by the malarial parasite (*Plasmodium*), although there are many more like the dysentery amoeba (*Entamoeba histolytica*; *Giardia* that causes intestinal disorders; *Leishmania* (kala-azar); Trypanosoma (sleeping sickness), etc. The famous helminth parasites include the tapeworm (*Taenia*), the liver fluke (*Fasciola hepatica*), the bloodfluke (*Schitosoma haematabium*), the pinworm of children (*Enterobius vermicularis*), the hook-worm (*Ancylostoma duodenale*), the trichina worm (*Trichinella spiralis*) and *Ascaris lumbricoides*.

**Origin of Parasitism**

The parasites were known and discussed many centuries ago, for instance, by Aristotle and Hippocrates, but it is only during the last half a century that the subject has made great strides on account of the increasing realization of its importance and the amount of exact knowledge that has been gained. The origin of parasitic habits is associated with the attempt to secure food for itself or for its offspring. Sarcophagid flies usually deposit their eggs in dead flesh, but sometimes they lay eggs in open sores and their larvae invade living animals. Some animals become attached to certain others in reponse to their sessile habits and ultimately start drawing food from their hosts and become parasites. *Sacculina*, a crustacean, is an example of this type. After developing habits of attachment to a particular animal it has become parasitic and now depends on the absorption of nourishment from the host. Some (Clark, 1921) authors conclude that assumption of sessile life by a species is often the first step toward parasitism.

There is another view regarding the origin of parasitism. Some people believe that commensals have often become parasites. Ewing has given a very thorough discussion of the development of parasitism among mites. He concludes that among the "living forms we can trace out all stages of advancing, parasitism including the occasional or erratic parasitism, semiparasitism, facultative parasitism, even to the fixed and permanent type, and finally to endoparasites."

**Pathogenic Effects**

The parasites are responsible for many types of pathogenic effects. The pathogenicity of parasites varies greatly depending on their numbers, their habits, whether migration takes place in the host and especially the degree of adaptation that has developed between the host and the parasite. The following are the ways in which the parasites cause harm to their hosts: (*i*) by absorbing food intended for the host (tapeworms); (*ii*) by sucking blood or lymph (hookworms, ticks); (*iii*) by feeding on the tissues of the host (ascarids, maggots); (*iv*) by causing mechanical obstructions or pressure (ascarids, microfilariae); (*v*) by causing the growth of nodules and tumours which may be of malignant nature (*Gongylonema neoplasticum* in rats); (*vi*) by perforating vessels (*Habronema magastoma*); (*vii*) by causing wounds through which infection may enter (*Ascaris, Demodex*); (*viii*) by destroying the tissues (*Cysticercus temnicollis, Dioctophyme renale*); (*ix*) by irritation (lice, *Fasciola*); (*x*) by secreting toxins or otherwise harmful substances such as antidigestive enzymes (gastrointestinal worms), digestive enzymes (trichonemas, maggots of blowflies) and anticoagulatory and haemolytic enzymes (hookworms, leeches, blood-sucking arthoropods); and (*xi*) by transmitting causal agents of infectious diseases such as bacteria, blood protozoa, viruses and filariae.

## Host and Organ Specificity

Any species of parasite, generally speaking, becomes associated only with particular species of hosts or only a single species. If a parasite attacks more than one hosts species these hosts are usually closely related. The number of such parasites that can live in a large range of hosts is very little (e.g. *Trichinella spiralis, Sarcoptes scabiei*). Ordinarily same parasites of sheep cannot live in cattle and those of the donkey are not found in the horse. If the parasite is able to establish in a host, the latter is said to be **tolerant**, and if it is not able to establish in the host it is said to be **refractory**. An animal is said to be the natural host of a parasite when the latter is able to develop and live normally in it. If a parasite enters an unsuitable host it either dies very soon, or it may develop for some time, but does not reach sexual maturity, or in certain cases it may become encysted and be able to complete its development if it should eventually find its way into the suitable host.

The parasites usually attack a particular organ in the body of the host although the reasons underlying this organ specificity are not clear. Of the internal parasites the largest number inhabit the alimentary canal, especially the intestine. The lungs, body cavities and blood vessels are also attacked by the parasites frequently, even the heart, brain, eye and bones do not escape.

## Resistance and Immunity

Every host tries to put up a resistance to the effects of parasites. The parasites, on the other hand, adapt themselves in such a way that leads right to the state of commensalism, in which the host suffers no ill-effects. Ability of an animal to resist infection by a parasitic organism is called **immunity**. It is an essential requirement for survival, since most animals are perpetually menaced by viruses, bacteria and parasitic animals. Immunity of animals is due to many different mechanisms, such as impervious skin, antiseptic stomach (due to acid), activity of phagocytes, and chemical defence by antibodies. The immunity may be (**natural immunity**) or acquired (**acquired immunity**). Natural immunity to parasites varies from an absolute refractoriness to complete tolerance and it depends on age (**age immunity**) or on nutrition, vitamins and general conditions. In some cases a strong immunity results from the infection so that reinfection is unsuccessful. This is **acquired immunity**. In cases of continued reinfection the effect on the parasites may be stunting of growth or complete inhibition of development beyond a certain stage, or decreased egg-output or all these.

## Host-Parasite Relations

The association of the individuals may be in the form of parasitism, commensalism and symbiosis. A parasite lives partly or wholly at the expense of its host. It may be completely dependent on the host and definitely injurious to it. The most successful parasite is the one which does the least harm to its host. Example of such a parasite is provided by a crab (*Pinnotheres*) which lives in oysters, scrapes mucous string from the edges of the gills, without injuring the body, but steals its food all right. Certain chalcids lay their eggs upon the pupae of ants, the larvae hatch, live, and do not harm the pupae.

Hosts, on the other hand, try to avoid parasites by some means or the other. The common method is secretion of certain juices which do not allow the emergence of parasites from their cysts as in some protozoan parasites. Sometimes the body fluids may contain such qualities that may not allow the parasite to flourish. Then certain species become immune to the actions of some species but not to others. For instance, *Trypanosoma brucei* is killed by human blood, while *T. gambiense* is not, and causes the deadly disease, sleeping sickness.

The parasites are mostly sedentary creatures and their environment is usually constant or is not subject to much variations. This decreases the importance of sensory and nervous systems which are usually reduced or lacking in the parasitic animals. The eyes, for instance, are present in the free-living turbellarians, reduced in ectoparasitic trematodes, and are lacking in the endoparasitic trematodes and cestodes. The eyespot may be present in the larval stages, e.g., in the miracidium of *Fasciola*. In most of the endoparasites the locomotor organs are also lacking and so are the trophic organs. The extent to which the trophic locomotor, and neurosensory systems of a parasite are reduced depends to a large extent on the degree of dependence of the parasites on the host. In some cases they are reduced to such a degree that the parasites are referred to as "degenerate" which of course they are not. They are rather perfectly adapted. A good example of this type is furnished by *Argulus*, the carp-lice, in which the appendages become modified to form adhesive discs and hooks. Another crustacean *Sacculina* is parasitic on crab. In the adult condition it looks like a round swelling on the underside of the tail-end of the crab. It extracts the juices from the unfortunate host by means of branching root-like threads. The parasite is a little more than a bag of eggs which hatch out into larvae resembling those of the barnacles.

## Adaptations

The parasites develop remarkable means to adapt themselves to the various ways of life. Although the first major problem of the parasite is food, it must get it from the host and for this reason it must remain attached to or associated with the host. To attain this they often possess structures for attachment —

hooks and suckers—as in the cestodes, trematodes, mites and other insects like the anoplurans. The shape of the body may also be altered to suit the environment in which it lives. The tapeworm and round-worms have elongated bodies as they live in the alimentary canals of the host. Some mites that live inside the quills of the feathers of the fowl and other birds have elongated bodies; likewise the mite *Demodex* that lives in the hair-follicle of the host is also elongated.

The body of the flea is laterally flattened to enable it to move freely among the hairs and feathers of the host, the lice on the other hand are dorsoventrally flattened. The leaf-like bodies of the liver-fluke occupying spaces in the body and the elongated form of the blood-flukes, etc., are further examples of this type.

The structure of the parasites usually becomes simple. The tapeworm, for example, which lives in the food inside the host's food canal, has lost its own alimentary canal, and absorbs the surrounding food through its skin. Most of them lose their locomotor organs which they do not need. The parasitic insects—the fleas and and lice—have lost their wings as they live amidst the hairs and feathers on the surfaces of the bodies of their host.

There are numerous special adaptations for obtaining food. The mouthparts of the blood-sucking insects for instance, have developed into efficient blood-getting tools. Some structures are modified for piercing and other for sucking, all lying in a gutter formed by the lower lip when not in use. The mouthparts of the blood-sucking horseflies are built on the same plan, but they are shorter and coarser, that is why they inflict more painful and more irritating bites. The mouthparts of tsetse flies and the stable flies are also modified to cause painful wounds.

Some lice feed on solid tissues of the host body. They have their mouthparts adapted for chewing scales and epithelial debris present on skins of their hosts, so much so, that some of them can bite off bits of hairs or feathers and chew them. Similarly, there are some worms (e.g. hookworms) that live in the food canals and feed in this manner. They have powerful, bell-shaped mouths, behind which is a suctorial pharynx that enables the worm to suck in a plug of the lining of the host's food canal. There are teeth inside the bell-shaped mouth that rasp pieces off this plug of tissue. Hookworms combine this method with blood-sucking and may cause anaemia in the long run. It has been calculated that fifty dog hookworms can remove nearly two tablespoonfuls of blood every day.

After having obtained food and security the parasites have to fight the battle on which the survival of the species depends, i.e. **reproduction**. With the parasites firmly stationed within the body of the host its offsprings have to face the injurious effects of the world outside. The fertilized egg, for instance, is open to many risks. The climatic conditions such as rapid changes of temperature, damaging effects of the rays of the sun and drying, etc. may destroy large number of eggs. These risks are countered by the production of a large number of eggs. *Ascaris lumbricoides*, the roundworm of man and the pig, for example, produces 200,000 fertilized eggs a day. American workers estimated some time ago that in China 355 million people were infected with this worm and the total weight of the eggs produced each year was 18,000 tons, indeed a remarkable figure seeing that one egg is about 1/1500 inch long. The huge fish-tapeworm of man which may be 18 metres long may produce 36,000 to 1,000,000 eggs a day; the beef tapeworm may produce 50-150 million eggs a year.

To protect the eggs from climatic factors they are protected with resistant envelopes of various thickness and various designs. The eggs of *Ascaris* are so well protected that they can live outside the body of the host for about five long years. In some parasites the female lays the eggs in places where they get necessary warmth for development and the young parasites that hatch out have good chances of establishing themselves inside or on the surface of the host. The lice, for instance, glue their eggs to the hairs of man and their other hosts. One or more eggs of the warble flies are glued to a single hair of the cattle. On hatching they pierce their way through the skin of the host and live as parasites for some time. During the period of parasitic life, which lasts for months, the grubs of the warble flies grow up and wend their way to the skin alongside the vertebral column of the host and there they cause the formation of swellings, or warbles, inside which the grubs become mature.

Some insects whose grubs are parasitic lay their eggs actually inside the bodies of the hosts in which the grubs live and develop. The hosts themselves are often grubs of other insects. It is quite interesting to note that in some cases the parent of the parasitic grub tries to find out whether any other egg has already been laid inside the host. If so the mother will leave that host and try to find another.

These methods of protecting the offsprings are remarkable, but in still other cases it is found that the young ones develop inside the bodies of the mother. The tsetse flies nourish their larvae in their wombs and feed them on "milk glands" till they are ready to pupate. At birth she places the grubs in sheltered place, such as the bark of a tree or in warm moist soil where it pupates at once. The larvae of the tropical flies are open to many risks which are reduced by this method.

Apart from the production of large number of eggs, some parasites increase the number of infective stages, as it happens in the case of the liver-fluke. A single egg gives rise to a single miracidium which enters and parasitizes a certain species of snail. In the body of the snail each miracidium becomes a reproductive sac, the sporocyst, which produces several rediae, which are capable of producing more rediae. Then each redia produces several cercariae that infect the new host. From a single egg, in this way, several cercariae are produced. This helps the parasite to make a mass attack on the host whose resistance to the parasite may be relatively easily overcome. Many other parasitic animals use this kind of device for multiplying the number of potential adults derived from a single fertilized egg.

There are several types of methods developed by different parasites to enter the body of the definitive host. Special mention may be made of those parasites that work their way through the skin of the host. This capacity naturally increases their chance of reaching the right places to live in. For example, the young of the hook-worm of man, cattle and other animals, and those of some flukes are provided with structures that enable them to bore their way through the skin of their host.

From the above it is apparent that the parasites are the most interesting, remarkable and efficient animals. There is no ground for despising the parasite.

If the parasite causes incalculable illness and suffering in man and other animals, it is a danger signal for the parasite itself, because the death of the host means the elimination of the parasite. In fact all the pain and illness caused by parasite animals is due to faulty adaptation between the host and parasite and this fault is rectified some time by the process of evolutionary adjustment. The result is that there are many parasites and hosts that live together very harmoniously without disturbing the biological processes of each other.

# 19 PHYLUM MOLLUSCA

**General Characteristics**

The molluscan plan of animal structure is among the relatively few patterns (Fig. 19.1) which have been highly successful. This is true both in terms of the number of individuals and the number of species, and also in the more significant ecological sense. The ground plan of molluscs is a body divided into two functionally distinct regions: (*i*) the visceral hump, the upper part functions largely through mucous secretion and ciliary action; and (*ii*) the head and foot, largely muscular, also provided with cilia and mucous cells. This plan of constitution does not seem to be similar to that of any other phylum. Some biologists feel that such a structure might have evolved from a platyhelminth-like ancestral stock in which enlargement of the alimentary system led to the development of the characteristic molluscan visceral hump. Clearly, the Mollusca constitute a major phylum: there are probably about 110,000 living species of molluscs. That is, in terms of numbers of species, the Mollusca must rank along with the Nematoda, just after the arthropods. (Surprisingly, there are more than twice as many molluscan species as there are vertebrate species.) Most of the known species belong to the three major classes—Gastropoda, Bivalvia and Cephalopoda. The phylum name is derived from *mollis*, meaning soft, referring to the soft body within a hard calcareous shell which is usually diagnostic. Thus, most molluscs are readily recognizable as such. The Mollusca are bilaterally symmetrical (asymmetries occur in some in connection with certain organs) animals having unsegmented bodies sheltered in an external calcareous shell made up of 1, 2 or 8 parts. The shell is internal in some and reduced or wanting in others. The more obvious functional homologies in the group arise from the extensive use of ciliary and mucuous mechanisms in feeding, locomotion, and reproduction. The basic molluscan plan of structure and function is remarkably uniform throughout the group, but there is no single standard of molluscan shape. An extreme diversity

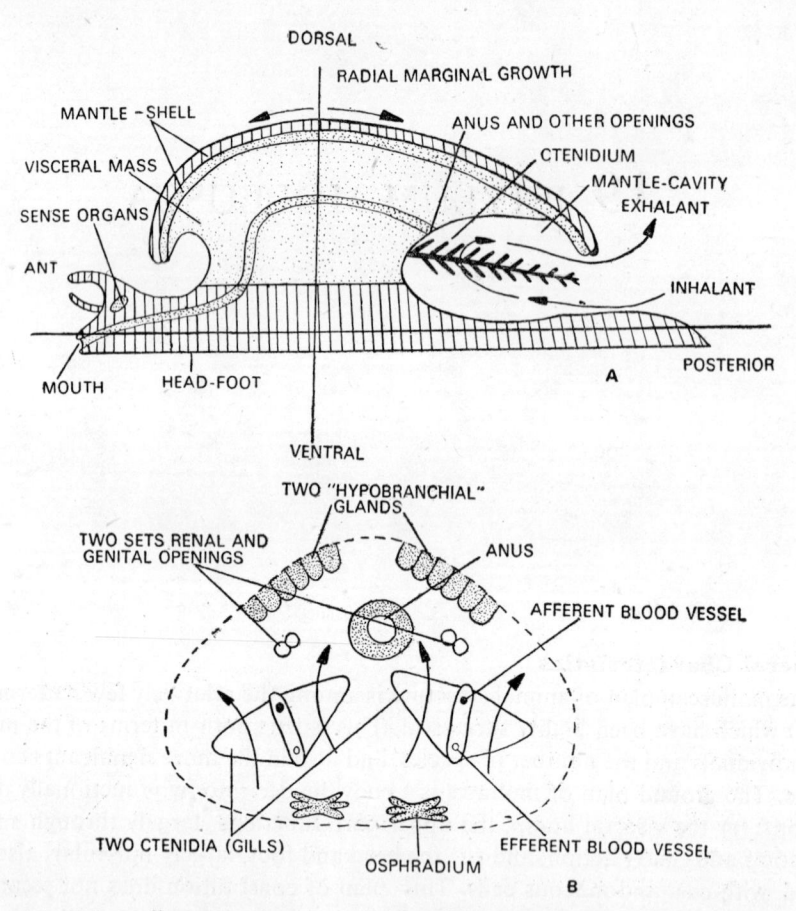

FIG. 19.1. Molluscan archetype and its pallial complex. Note these parts of molluscan body. A—Head-foot, visceral mass, and mantle shell; and B—Water circulation through the mantle cavity and pallial complex is from ventral inhalant pore to dorsal exhalant.

in external body form has been based upon this plan. A comparison between oyster, chiton, snail and octopus will reveal this diversity.

As is evident from the table of classification there are three major groups, and a number of smaller ones, among the molluscs. The **Gastropoda** constitutes a large, diverse group with the shell usually in one piece. This shell may be coiled as in typical snails, that is, helicoid or turbinate, or it may form a flattened spiral, or a short cone as in the limpets, or it may be secondarily absent as in the "slugs." Most gastropods are marine, but many are found in freshwaters and on land. The Gastropoda are the only successful non-marine molluscs. The

# PHYLUM MOLLUSCA

| | |
|---|---|
| **Classes:** | MONOPLACOPHORA (NEOPILINA) |
| | AMPHINEURA |
| **Suborders:** | APLACOPHORA |
| | POLYPLACOPHORA (CHITONS) |
| **Classes:** | SCAPHOPODA |
| | GASTROPODA |
| **Subclass:** | PROSOBRANCHIA |
| **Orders:** | Archaeogastropoda |
| | Mesogastropoda |
| | Neogastropoda |
| **Subclass:** | OPISTHOBRANCHIA |
| **Orders:** | Cephalaspidea |
| | Anaspidea |
| | Nudibranchia |
| | Thecosomata |
| | Gymnosomata |
| | Notaspidea |
| | Acochlidiacea |
| | Sacoglossa |
| **Subclass:** | PULMONATA |
| **Orders:** | Basommatophora |
| | Stylommatophora |
| **Class:** | BIVALVIA (or pelecypoda) |
| **Subclasses:** | PROTOBRANCHIA |
| | LAMELLIBRANCHIA |
| **Orders:** | Taxodonta |
| | Anisomyaria |
| | Heterodonta |
| | Schizodonta |
| | Adapedonta |
| | Anomalodesmata |
| **Subclass:** | SEPTIBRANCHIA |
| **Class:** | CEPHALOPODA |
| **Subclasses:** | NAUTILOIDEA (tetrabranchia) |
| | AMMONOIDEA (entirely extinct) |
| | COLEOIDEA (dibranchia) |
| **Order:** | Decapoda |
| **Suborders:** | Belemnoidea (extinct) |
| | Sepioidea |
| | Teuthoidea |
| **Orders:** | Vampyromorpha |
| | Octopoda |

**Bivalvia** are a more uniform group with the shell in the form of two calcareous valves united by an elastic hinge ligament. Mussels, clams, and oysters are familiar bivalves. The group is mainly marine with a few genera in estuaries and in freshwaters. There can be no land bivalves since their basic functional organization is as filter-feeders. The third major group, the **Cephalopoda**, includes the most active and most specialized molluscs. There is a chambered, coiled shell in *Nautilus* and in many fossil forms; this becomes an internal structure in cuttlefish and squids, and is usually entirely absent in octopods. Among the

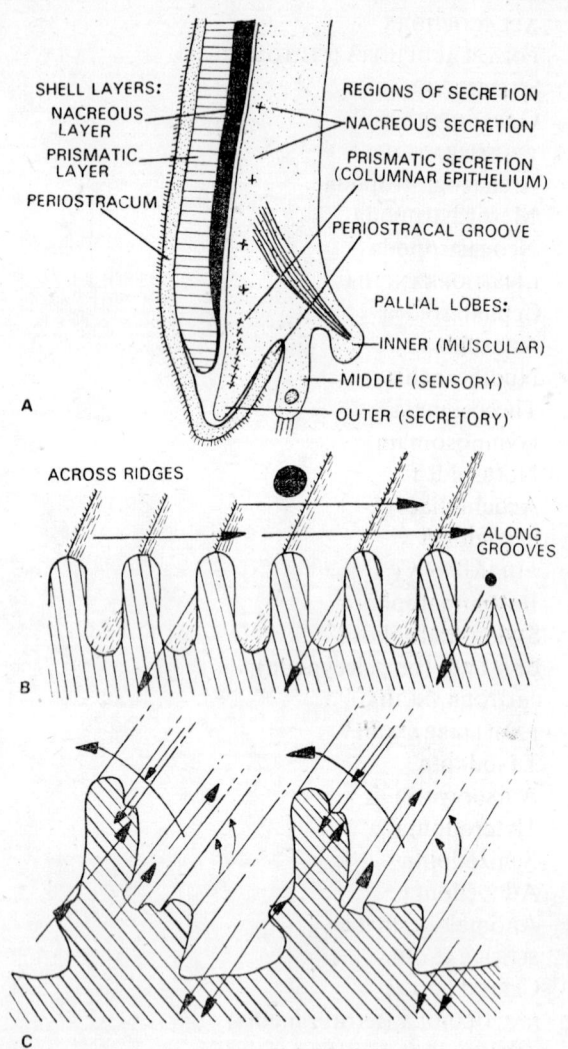

FIG. 19.2. A—The archetypic arrangement of the mantle shell margin in molluscs; B and C—Ciliated sorting surfaces, which are used externally and internally in molluscs for the mechanical separation of particles of different sizes. On a simple sorting surface (B) large particles are carried across the ridges, fine particles along the grooves; while on more complex sorting surfaces such as (C) five categories of particles are sorted in different directions.

minor groups of Mollusca comes the class **Monoplacophora**. It includes several fossil families and only one living genus *Neopilina*, an archaic mollusc recently discovered in deep-sea dredging from the Pacific Ocean. Class **Amphineura** has two types of molluscs included in two subclasses, **Aplacophora** and **Polyplacophora**. The latter includes about forty genera of chitons or "coat-of-mail shell" with eight plates. The worm-like Aplacophora includes only a few species. The other minor group, the Scaphopoda or "elephant's tusk shell", includes only a few species. Ecologically, bivalves and gastropods can make up an important part of the animal biomass in many natural communities. Certain snails of the seashore, such as *Littorina*, *Hydrobia*, and *Nassarius*, are known to occur in several parts of the world at densities of hundreds per square metre (Fig. 19.1), and one species of *Nassarius* on Cape Cod occurs in patches with up to 23,000 individuals per square metre. Even less obvious molluscs—nocturnal or cryptic ones—can be abundant. One English Zoologist, with a garden of a quarter of an acre in extent, removed four hundred slugs from it each night for many years without any obvious reduction in the population. Thus, even on land, molluscs can be of ecological importance.

## Archetypic Functioning

**Coelom**. The coelom in Mollusca has become typically small, although in the recently discovered *Neopilina* there are two extensive dorsal coeloms. In most molluscs the coelom comprises a pericardial coelom around the heart, a gonadal coelom, and paired coelomic ducts, which, together with the pericardial wall, serve as excretory organs. The origin and evolutionary relationships of the molluscan coelom is an example of biological problem that is not clearly understood as yet.

The unique features of molluscs depend on the modes of growth and of functioning of the three distinct regions of the molluscan body, namely (*i*) the **head-foot** with some nerve concentrations, most of the sense-organs, and all the locomotory organs; (*ii*) the **visceral mass** (or **hump**) containing organs of digestion, reproduction, and excretion; and (*iii*) the **mantle** (or **pallium**) hanging from the visceral mass and enfolding it and secreting the shell.

**Mantle**. Whatever the shape of shell it is always underlain by a fleshy fold of tissues, the **mantle**. In its development and growth, the head-foot shows a bilateral symmetry with an anteroposterior axis of growth. Over and around the visceral mass, however, the mantle-shell shows a biradial symmetry, and always grows by marginal increment around a dorsoventral axis (Fig. 19.2). A space is left between the mantle-shell and the visceral mass forming a semi-internal cavity, the **mantle** or **pallial-cavity**. The **ctenidia** or the typical gills of the mollusc develop in this cavity. The mantle-cavity is almost diagnostic of the phylum. Primarily it is a respiratory chamber housing the clenidia. The alimentary, excretory, and genital systems all discharge into it. Although its basic functional plan is always recognizable, it has undergone remarkable modifications of structure and of function in different groups of molluscs. In addition to its respiratory function, it provides the feeding chamber of bivalves and of some gastropods, a marsupial brood-pouch in some forms, and an organ of locomotion in few bivalves and in the most highly organized molluscs, the Cephalopoda.

**Ctenidia**. In the more primitive living snails and bivalves there are paired **ctenidia** of characteristic basic form. The **filaments** or **plates** always alternate on either side of an axis, which contains a dorsal afferent blood vessel bringing deoxygenated blood to the gill, and a ventral efferent blood vessel carrying the oxygenated blood back to the heart and then to the rest of the body. The flow of blood through the filaments is from dorsal to ventral.

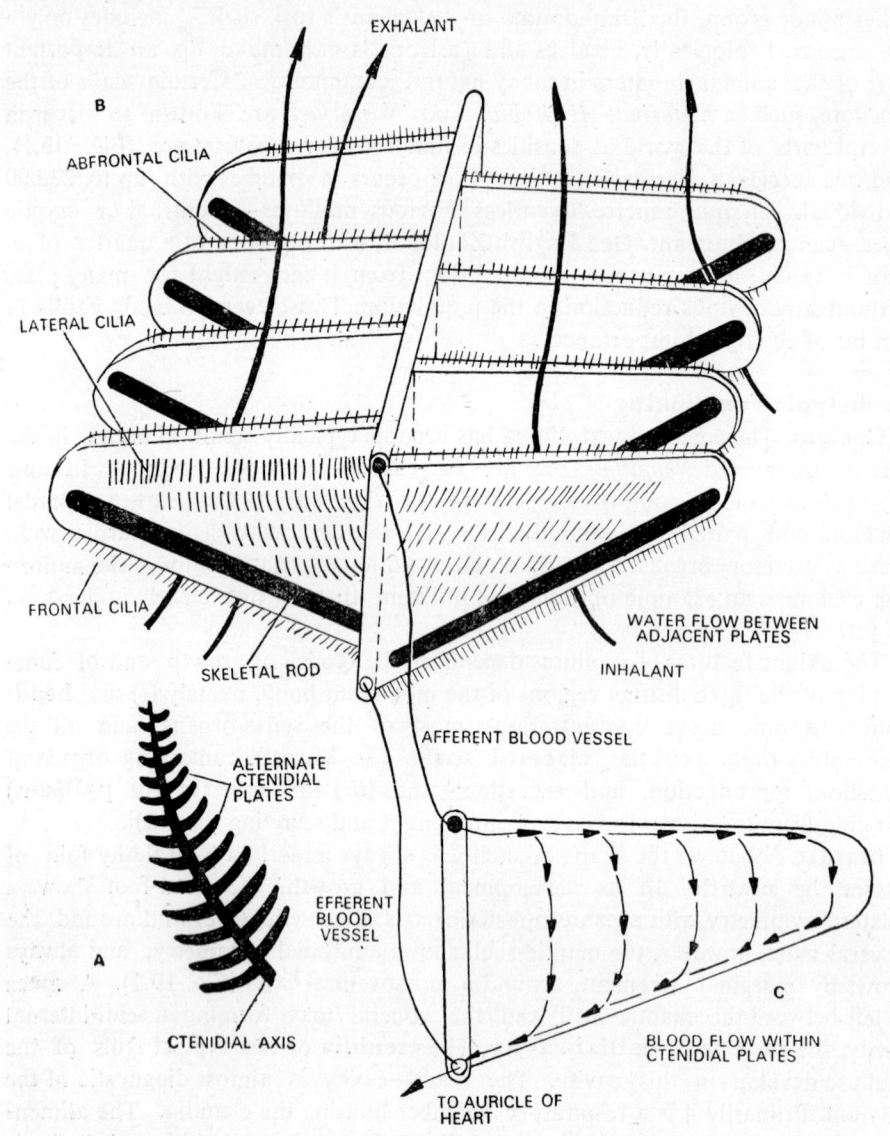

**FIG. 19.3.** The archetypic molluscan gill—the aspidobranch ctenidium as it is found in primitive gastropods and other groups. A—An aspidobranch or feather or plume-gill: B—Stereogram to show the water current from ventral inhalant pore to dorsal exhalant between adjacent gill plates; a current created by lateral cilia; and C—A single gill-plate to show the counter-current flow of blood within the plate which results in physiological efficiency of oxygen exchange.

The exposed surface of filaments has cilia that are so arranged as to produce a flow of water through the gill in the opposite direction to the flow of blood, which means that the water flow between adjacent filaments is from ventral to

*Phylum Mollusca*

dorsal. This counter-flow system in gills is universal among molluscs.

The arrangement of ctenidia in the mantle cavity displays a morphological and functional constancy. In almost all molluscs, the mantle cavity is functionally divided by the gill into an **inhalant** part which is usually ventral and an **exhalant** portion usually lying dorsal to the gills. The anus, the openings of the

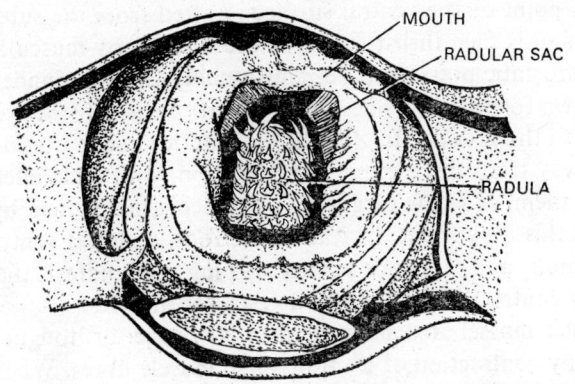

FIG. 19.4. The arrangement of radular apparatus in an archetypic mollusc.

kidney and genital ducts, all discharge into the exhalant current, that is they discharge dorsal to the bases of the ctenidia. More primitive forms of both snails and bivalves have the anus in the centre of the roof of the mantle cavity with genital openings and two kidney ducts symmetrically placed.

In the path of ihe inhalant current lies an **osphradium** or pair of osphradia, the pallial sense organ, which sample the incoming water current. Above the ctenidia in the exhalant region, adjacent to the anus lie the **hypobranchial** glands which are functionally integrated structures found in all Mollusca and are said to form the pallial complex.

**Movement.** The visceral mass and the mantle-cavity function slowly and continuously using mucous and cilia but the head-foot works largely by muscles which show many of the usual fast reflexes of bilaterally symmetrical animals. Functionally, it is possible to regard an unspecialized mollusc as being made up of two animals: a **muscular animal** responsible for locomotion and retraction into the shell, which carries about **ciliary animal** responsible for respiration, excretion and feeding.

The foot is the most powerful muscular organ having many potentialities. It operates in conjunction with the hydrostatic properties of the blood system. This is so because the blood of molluscs although conveyed from the heart in a system of arteries, passes eventually into a haemocoel, a complex of rather indefinite vascular cavities lying in spongy meshwork, but the hydrostatic properties of its haemocoel are not always greatly used. In many cases the foot is essential'y a solid muscular organ as in chitons and limpets. In such cases the foot is used as powerful adhesive sucker, bound to the substratum by a layer of mucous. By virtue of its adhesive power *Patella* is able to resist wave action on open shores (shape of shell also helps in this), to avoid desiccation when the tides leave it exposed, and yet to move around for grazing when the tide is in. This type of adaptation is clearly profitable, for a limpet-like form has been developed independently several times in the gastropod molluscs.

Many small molluscs glide by the action of cilia on the ventral surface of foot, very much as do the platyhelminths. In larger forms this is not practicable. Many larger forms such as *Helix* have a more flexible foot and employ waves of muscular contraction, which pass anteriorly over the sole of the foot in a steady rhythm. A given point on the ventral surface is lifted from the substratum as the wave reaches it, and is then thrust forwards presumably by muscular contraction coupled with hydrostatic pressure (Lissmann's analysis). The hinder part of the body is then drawn forwards, while the relaxed parts of the ventral surface remain adhered to the ground. In *Aplysia* the anterior and posterior ends of the foot are lifted away from the ground in alternation, with remainder of the body arching between them. It seems turgor pressure being maintained by the haemocoelic fluid helps this movement. In *Nucula* the foot, with flat ventral surface, is thrust into the mud and expanded into a plug or holdfast, the body is then drawn after it by contraction of the pedal retractor muscles.

In the freshwater mussel *Anodonta* the foot is wedge or tongue-shaped. It is forced forwards by contraction of its transverse muscle fibres. When these relax the foot swells under the pressure of the aemocoelic blood, the pressure being increased by the adduction of the valves of the shell which forces blood from the body into the vascular meshwork of the foot. The body is then drawn forwards by the contraction of the longitudinal muscles while the foot remains fixed. Such a progress has evolved into drilling mechanism in the shipworm (*Teredo*). In the pelagic forms the foot is drawn out into a highly mobile fin. In some cases (*Carinaria*) this is held upwards, the animal swimming upside down. There are some (sea-butterflies) that move by steady beat of lateral epipodia.

The cephalopods move by jet propulsion. The primitive form *Nautilus* draws water into its mantle cavity and expels it through a muscular funnel that is formed from the foot. In other recent cephalopods the funnel has been replaced by a complete tube, the shell has been reduced, with the result that mantle can contract under the action of circular muscles. Thus, the jet of water is expelled with great force. With the use of jet propulsion the cephalopods take us to the field of hydrodynamics.

**Sorting surfaces.** Cilia are used to create water currents for respiration and for feeding, to move materials through the gut, the kidneys and the genitalia, and to cleanse exposed surfaces. They are also used as **sorting surfaces**, (Fig. 19.2) which can segregate particles into categories of different sizes and send these to be disposed of in different ways in several parts of the organisms. The epithelium of sorting surfaces is thrown into a series of ridges and grooves (Fig. 19.2). The cilia in the grooves beat along them and the cilia on the crests of the ridges beat across them. Smaller particles are carried in the direction of the grooves while larger particles are carried at right angles. The labial palps of bivalves are used to separate larger sand grains that are rejected. They can be expanded to a varying extent by blood pressure. Molluscan sorting surfaces can segregate particles of four or more sized categories.

The alimentary canal of molluscs is extensively ciliated, with many internal surfaces organized as sorting areas. Digestion is both extracellular and intracellular. Although in a few highly specialized molluscs most of the digestion is extracellular, the typical molluscan gut is organized to deal slowly but conti-

*Phylum Mollusca* 777

nuously with a steady stream of finely divided plant material passing in from the feeding organs. Some feed by the filter-mechanisms, while others use the characteristic **rasping tongue** (or **radula**) of the grazing molluscs. Molluscs of all groups (with the sole exception of the bivalves) have this radula, a chitinous surface bearing teeth and covering a tongue supported by an odontophore (Fig. 19.5). The radula is moved by a variety of rhythmic strokes by the muscles forming the buccal mass. The radular apparatus has a two fold function: serving for rasping off food material (mechanically like an inverted version of the upper incisor teeth of a beaver), and for transporting the food back into the gut like a conveyor-belt.

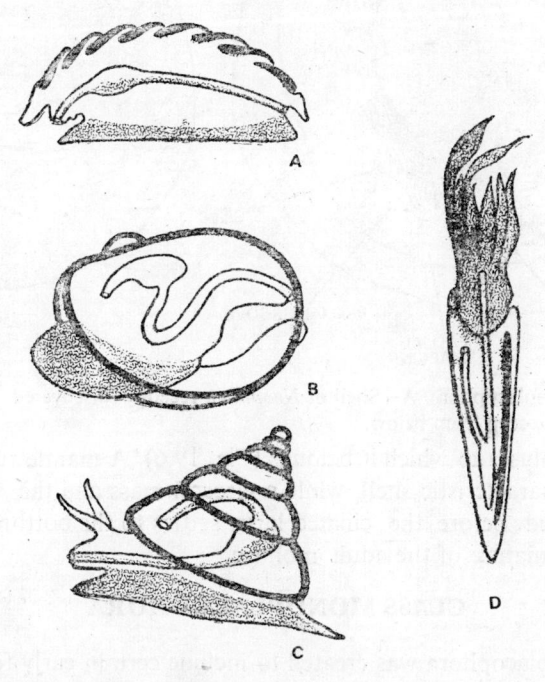

FIG. 19.5. Body plan of different classes of Mollusca. A—Chiton; B—Clam; C—Snail; and D—Squid.

A to-and-fro movement of the chitinous ribbon (at rates of about forty strokes per minutes) is achieved by alternate contractions of the antagonistic radular muscles attached to it (Fig. 19.4).

The molluscan nervous system shows a wide range in the degree of complexity, and in this, the phylum resembles the chordates. The nervous system of a chiton is not dissimilar to that of a turbellarian flatworm, whereas the nervous system and sense organs of a cephalopod (like an octopus) are equalled and exceeded only by those of the most highly organized vertebrates. In most typical molluscs, the nervous system is in an intermediate condition. The relative simplicity is probably related to the fact that, in molluscs other than the cephalopods, the main effectors controlled by the nervous system are cilia and mucous glands. In fact, apart from the muscles which withdraw it into its shell, the typical mollusc is a slow-working animal with little fast nervous control or quick reflexes.

In all primitive molluscs, the sexes are separate and fertilization is external after the spawning of eggs and sperms into the sea. The zygote undergoes spiral cleavage and eventually becomes a trochophore larva, with a characteristic ring of locomotory cilia. This later develops into a typical wheel-shaped velum bearing long cilia, and then this veliger larva develops some of the characteristics of

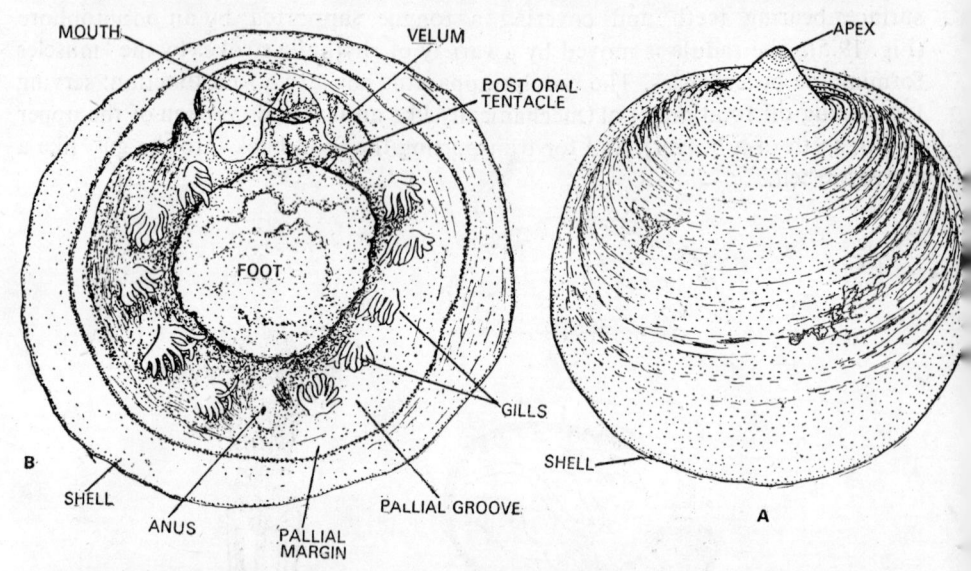

FIG. 19.6. Monoplacophora. A—Shell of *Neopilina*, recently discovered living mollusc; and B—The same seen from below.

the group of molluscs to which it belongs (Fig. 19.6). A mantle rudiment appears and secretes a characteristic shell, while a visceral mass and the head-foot have also differentiated before the ciliated larva settles to the bottom and metamorphoses into a miniature of the adult mollusc.

## CLASS MONOPLACOPHORA

The class Monoplacophora was created to include certain early fossil forms such as *Scenella* and *Pilina*, which had limpet-like shells and many muscle-scars indicating the attachment of pedal retractors. On the basis of certain Palaeontological data they were considered to be ancestral gastropods (pregastropods) prior to torsion. In 1952 the Danish oceanographic research vessel Galathea dredged some limpet-like animals from 3,570 metres in the Pacific Ocean off Costa Rica. This was found to be a living representative of the class and named *Neopilina galathea*. A preliminary report on the species appeared in 1957 and its morphology was described in 1959.

The genus *Neopilina* is an interesting survivor of an aberrant molluscan stock. It is almost perfectly bilaterally symmetrical, with a dome-shaped mantle and a flattened limpet-shaped shell having a spirally coiled protoconch (Fig. 19.6). The foot is a muscular disc attached by eight pairs of pedal retractor muscles. There are five pairs of gills, in the pallial grooves, on either side of the foot. The head bears elaborate branched postoral tentacular flaps, and contains a well-developed radula in a coiled radular sac. *Neopilina* contains two extensive dorsal coeloms. The alimentary canal is simple, stomach is provided with crystalline style-sac while the intestine is much coiled. The heart has two pairs of auricles, there are six pairs of nephridia and one or possibly two pairs of gonads. The nervous system comprises a ladder-like organization without much ganglionation involving ten sets of lateropedal nerve connections. It resembles that of *Chiton*.

*Neopilina* comes very close to chitons by many characters and also resembles primitive cephalopods. Some authors (e.g. Lemche) feel that in both *Neopilina* and the chitons there occur primitive characters that provide undeniable evidence of segmental origin of the phylum Mollusca.

The apparent metamerism of *Neopilina* is not represented externally, for there is a single shell and a foot without a vestige of segmentation. Internal organs are not disposed in a metameric pattern. Recently Morton has studied this aspect of *Neopilina*. He writes: "If *Neopilina* is the survivor of molluscs that were primitively metameric, it would be difficult to suggest from it what was the composition of a single generalized segment, or indeed to balance its segmental arithmetic at all." Morton further feels that "we fail to do it justice, if in our phylogenetic enthusiasm, we base upon it conclusions it will not easily carry."

## CLASS AMPHINEURA

The Amphineura are Mollusca with more or less elongated and symmetrical body with mouth and anus situated at opposite ends and their mantle is always provided with numerous spicules embedded in cuticle. They always possess a very large mantle that covers at least the dorsal surface and the sides of the body. The external symmetry of the body is repeated by some internal organs also. The nervous system comprises two longitudinal cords, one pedal, one pallial with ganglionic cells over their surfaces, and united with each other in front. The two pedal cords are united by anastomoses while the two pallial cords are united by a thick posterior commissure on the dorsal side of the rectum. Statocysts are lacking. Generally the buccal cavity is provided with radula (mandible in a species of *Chaetoderma*). The anal and renal openings are posterior. The heart is dorsally situated in the hind part of the coelom and its ventricle is more or less intimately united to the dorsal wall of the pericardium. The Amphineura are exclusively marine, found at all depths. They were found in the Ordovician era and are now divided into two orders: (*a*) the Polyplacophora or Chitons, and (*b*) the Aplacophora.

## TYPE CHITON

The chitons are by contrast the most successful of the molluscan minor groups. All are marine and live on rocks, mostly in shallow waters from the tide lines to moderate depths. Their body is characterized by a large foot occupying the whole of the ventral face of the body (Fig. 19.7), with eight calcareous plates on the mantle and a complete row of branchiae between the foot and mantle. Chitons cling tightly or creep slowly by foot, but if disturbed they curl up like a "pill-bug."

### Structure

The **body** is elongate, elliptical with a convex dorsal surface bearing eight overlapping calcareous valves or **plates** (Fig. 19.8) protective in nature. These are jointed to one another and covered at the sides (or entirely) by a thick fleshy girdle (part of mantle) containing bristles or spines. The **mantle** covers the dorsal

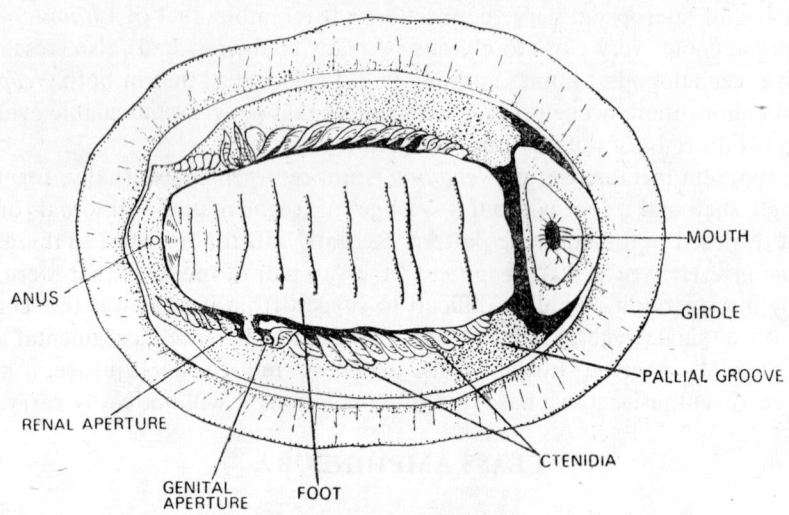

FIG. 19.7. Chiton, ventral view of *Chiton tuberculatus*.

and lateral surfaces, and the flat foot occupies most of the ventral surface (Fig. 19.7). Between the foot and the mantle there is a groove called the **pallial groove**. Beneath the anterior margin of the girdle there is the small more or less cylindrical head which is devoid of eyes and tentacles and contains the

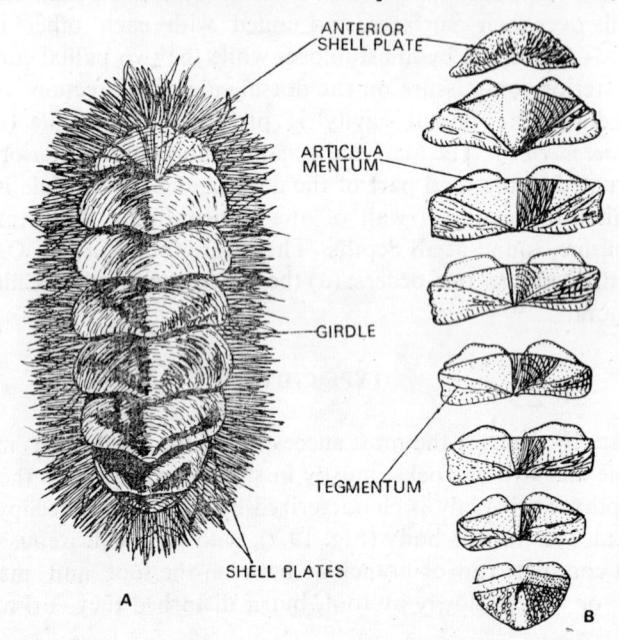

FIG. 19.8. A—*Chiton spinosus*, dorsal view; and B—*Chiton* shell-plates separated.

mouth, on either side of which there is a somewhat angular labial palp. The anus lies on the other side of the body (Fig. 19.9).

## Alimentary Canal

The **alimentary canal** extends from one end of the body to the other. The mouth leads into the buccal cavity in the floor of which lies the **radula** with many cross rows of 17 teeth of various shapes. A pair of salivary glands opens in the buccal cavity and lies at the sides. Two pairs of **mucous glands**, lying ventrally, also open into it. A short **pharynx** leads to the rounded **stomach** to which is connected the **digestive gland** or **liver**. The **intestine** is long and coiled and opens at the anus situated posteriorly in the pallial groove (Fig. 19.7). The food of *Chiton* comprises seaweeds and micro-organisms scrapped from the rocks with the help of radula.

The **heart**, enclosed in a large pericardium, occupies the posterior dorsal region of the body. It consists of two **auricles** and a median elongated **ventricle** connected to an anterior **aorta**, which carries blood to the various organs and intervisceral blood spaces. The venous blood from the different parts of the body is conducted back to a large sinus, on either side, near the line of union of the mantle with the body.

In each pallial groove are situated the **gills** 6 to 80 in number in different species. The gill rows may be either of **holobranchial** type, in which case they extend over the whole length of the body or of the **merobranchial** type confined to a limited space towards the posterior side of the body.

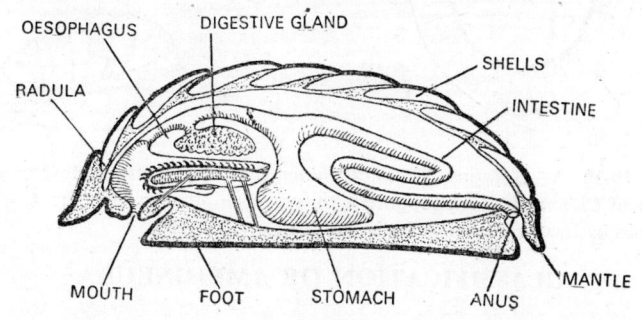

FIG. 19.9. Alimentary canal of *Chiton*, shown in a median vertical section.

## Excretory System

The **excretory organs** comprise two slender **kidneys** each of which consists of an elongated **renal canal**, situated on the lateral side of the visceral mass and folded on itself so that its two ends are posterior.

## Nervous System

The **nervous system** consists of a ring about the mouth that connects two pairs of longitudinal nerve cords, the **pedal** in the foot, and the **pallial** in the girdle. The cords are connected by many transverse connectives and are devoid of ganglia. Some chitons have **eyespots** or **eyes** in the integument over the shells. **Aesthetes** are tactile organs peculiar to chiton.

## Reproductive System

The **reproductive organs** are paired and symmetrical in one species *Nuttalochiton hyadesi*; in the others it is single and median lying on the dorsal side of the body, between the aorta and intestine, and extends from the anterior end to the pericardium. In *Chiton* the **ovary** is frequently of greenish colour and **testis** red. The ducts are paired, one running on each side to open posteriorly, in front of the renal aperture, between two of the posterior gills. Fertilization is internal and the eggs are laid singly or in strings consisting of 200,000 eggs. These strings readily break into fragments. The segmentation is total and regular in early stages. A gastrula is formed by invagination. The **blastopore** is placed on the vegetative pole of the ovum, and does not close but shifts to the anterior end of the embryo. A ciliated ring develops surrounding an apical tuft of cilia in the centre thus transforming the larva in a **trochophore** Fig. 19.10).

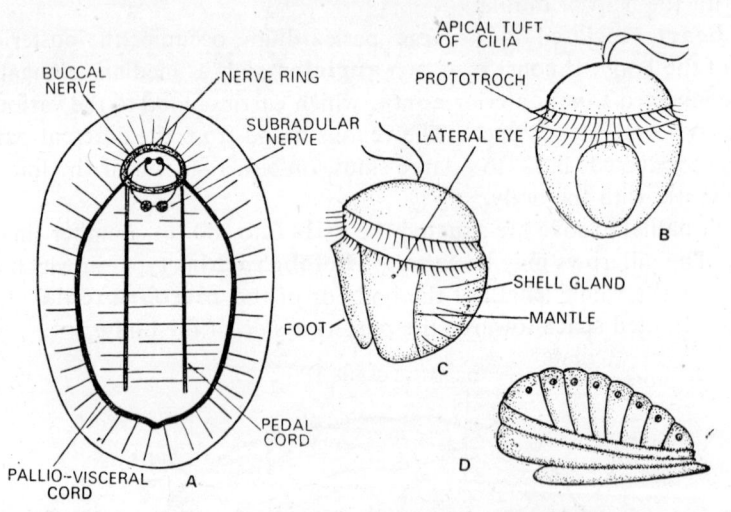

FIG. 19.10. A—Nervous system of *Chiton*; B to D—Different developmental stages of *Chiton*. B—Trochophore with foot beginning to appear; C—Larva attaches by foot; E—Shells clearly marked (*after* Heath).

## CLASSIFICATION OF AMPHINEURA

Class Amphineura is divided into two subclasses: Aplacophora and Placophora.

The **Aplacophora** or **Solenogasters** include small worm-like forms in which the mantle covers the entire body and contains five calcareous spicules representing the shell. The foot is a ciliated ridge in a ventral groove or is lacking. These animals live on corals or hydroids at considerable depths on the sea. Some authors (Parker and Haswell) have separated these two orders and raised them to subclasses.

(*i*) **Subclass Aplacophora (Solenogasters)**. The worm-like Amphineura with thick integument and calcareous spicules representing the shell. Foot is rudimentary. They live at the bottom of the ocean. *Neomenia* found in North Atlantic, is 2-3 cm long hermaphrodite form with two gonads. *Chaetoderma*, about 2.5 cm long, is another known example of the group found around North Atlantic. Sexes are separate, each having one fused gonad.

(*ii*) **Subclass Polyplacophora**. These are elliptical Amphineura with a shell made up of a mid-dorsal row of eight broad plates. This includes chitons found chiefly in shallow coastal water. 600 living and 100 fossil forms are recorded. The group consists of animals clearly adapted for life on hard and uneven surfaces in the littoral zone. *Chiton, Tonicella, Chaetypleura, Isnochiton,*

*Phylum Mollusca* 783

*Cryptochiton*, etc., form its examples.

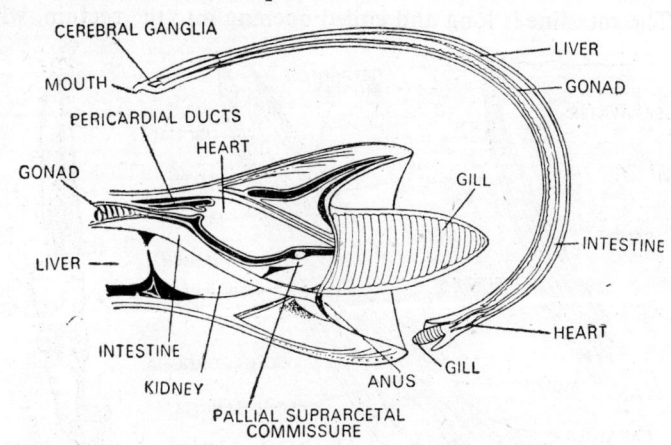

**FIG. 19.11.** *Chaetoderma nitidulum.* **A**—Median sagittal section; **B**—Enlarged view of the posterior, extremity (*after* Wiren).

## CLASS SCAPHOPODA

The **Scaphopoda** (Gr. *Skephe*, boat + *podos*, foot) or **tooth-shells** or **tusk-shells** are marine bilaterally symmetrical Mollusca in which the body and shell are elongated along the anteroposterior axis, are nearly cylindrical, and surrounded by a complete tube of mantle; head is somewhat rudimentary, without eyes, the foot is cylindrical adapted for digging, a radula is present but not ctenidium. They were formerly mistaken for tubicolous annelids but later were discovered to be molluscs. They are all marine animals living in shallow waters up to 4,560 metres partly buried in mud or sand allowing only the posterior extremity to project from the substratum. They feed on unicellular organisms such as Protozoa, Diatomacea, etc. About 200 living and nearly 300 fossil species are known. The fossil species extend back to the middle Silurian. *Dentalium* is the most known genus of the class.

### TYPE DENTALIUM

**Structure**

**Shell.** The shell of *Dentalium* (Fig. 19.12) is an elongate, slightly curved cone containing the entire animal. It is open at both ends. The mantle cavity extends continuously from one aperture to the other. The head is situated at the anterior end of the body on the concave or dorsal side. Laterally and posterioly it is provided with two pouches and quite at its posterior end, on either side of its dorsal surface are two broad symmetrical flattened tentacular lobes. About the mouth are several delicate ciliated and contractile **tentacles (capatula)** with expanded tips. These are sensory and prehensile, serving to capture the microplants and animals used as food. The **foot** has the form of an extensive elongated cylinder and protrudes from the larger ventral end of the shell serving as a digging organ for burrowing.

**Internal structure.** The non-invaginable **proboscis** leads directly into a true **buccal cavity** situated in the trunk at the base of the foot. The buccal cavity has in its interior the **radula** ventrally and **mandible** dorsally. A rather short **oesophagus** with two lateral pouches leads into the **stomach**, which receives the ducts of the **liver**, situated behind the stomach and the rest of the alimentary canal. The intestine is long and coiled opening into the rectum, which opens

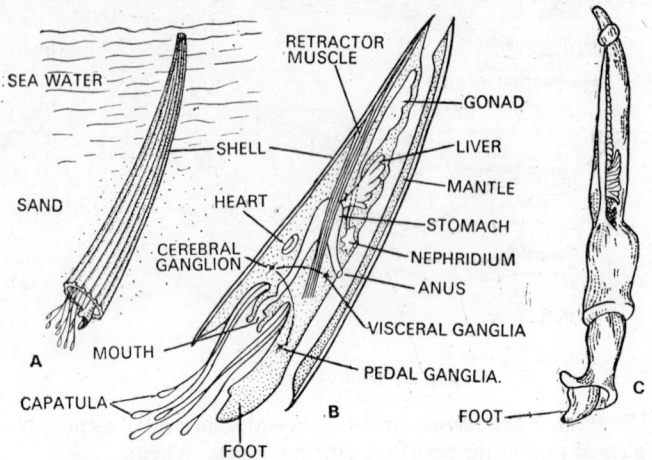

FIG. 19.12. Internal organization of *Dentalium*—B; A—Within shell; and C—With shell removed.

at the anus in the mid-ventral line. An **anal gland** lies on the right side of the rectum.

### Circulatory System

The circulatory system is simple devoid of differentiated vessels and ventricle, etc., and consists of blood spaces or sinuses. There is no specialized respiratory apparatus, respiration being effected by the internal surface of the mantle.

### Excretory System

**Excretion** takes place by two symmetrical kidneys situated in front of the gonad on the ventral side of the middle of the body. Each is a short but fairly wide sac with pleated walls, lying between the intestinal mass and the stomach (Fig. 19.12).

### Nervous System

The nervous system consists of the cerebral ganglia situated on the dorsal side of the oesophagus and connected with one another. Each cerebral ganglion is in close juxtaposition with the corresponding pleural ganglion. The cerebral and **pleural ganglia** are united to the **pedal ganglia** of the same side by a long connective. The two pedal ganglia are situated in the foot. The visceral ganglia are ill-defined ganglionic swellings lying on either side of the anus and connected to the pleural ganglia by visceral commissures. Among the organs of special sense are the capatula or tentacular filaments, statocysts and a subradular organ. The statocysts are situated in the foot and the **subradular organ** is a ciliated ridge on the ventral side of the

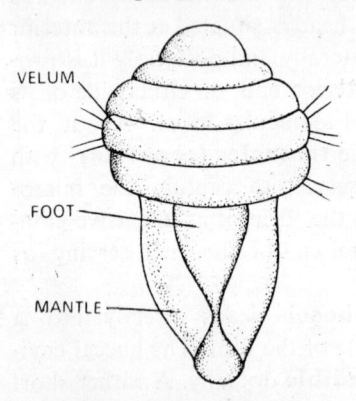

FIG. 19.13. Larva of *Dentalium*.

*Phylum Mollusca*

buccal cavity opposite to the mandible.

**Reproductive System**

The sexes are separate and the gonad is unpaired and median and is extremely long, occupying the whole of the posterodorsal region of the body. It is divided into symmetrical transverse lobes and its anterior extremity is contracted to form a duct, which diverges to the right and opens into the right kidney. The eggs are laid separately and undergo irregular segmentation immediately after fertilization and each leads to the formation of a ciliated larva (Fig. 19.13). After a short larval stage the young animal sinks to the bottom and transforms into the adult.

Shells of *Dentalium* kept on string were used as money by the Red Indians of the Pacific coast. Their trade equivalent ranged from 25 cents for 5 cm shell to five dollars for the rare shells of 6 cm length.

## CLASS GASTROPODA

The class Gastropoda (Gr. *gaster*, belly + *podos*, foot) is the largest and the most varied group of the molluscs. It includes a range from certain marine forms, which can be numbered among the most primitive of living molluscs to the highly evolved terrestrial air-breathing slugs and snails. The Gastropoda are Mollusca that possess a univalve (made of one piece) shell which is usually conical. The conspicuous anterior head and the long ventral foot present bilateral symmetry, but the visceral mass that is usually contained in a dorsal shell, is specially coiled and asymmetrical. It is presumed that the ancestral gastropods possessed bilateral symmetry throughout, but in living species the digestive tract, anus, heart, gills, kidneys and some nervous structures have been rotated or coiled to 180 degrees and certain parts have disappeared during the process. The head is distinct bearing eyes and tentacles. The foot is situated behind the head and has an extensive flattened ventral surface. The mantle is not divided into lateral portions, and the mantle cavity encloses two plume-like ctenidia, in some there may be only one ctenidinm and in the air-breathing forms the ctenidia may be lacking. Respiration takes place in such cases through the wall of the mantle cavity itself and through a pulmonary sac. Kidney is usually single. The nervous system contains distinct cerebral and pleural ganglia besides the pedal, visceral, abdominal and buccal ganglia. Sexes are separate or united, larva passes through trochophore and veliger stages. The Gastropoda occur everywhere from the tree tops to the depths of the sea. They included lands forms, water forms, and amphibious forms. They are of all sizes from less than that of a tiny pinhead up to a length of 60 cm, which is the size attained by the big horse conch *Fasciolaria gigantea*. They supply mankind with many useful articles like food, dyes, umbrella handles, etc. and indirectly also cause diseases. Some snails are restricted within almost unbelievably narrow limits. Thus, the Hawaiian tree snails (*Achatinella*) is confined to a single island and to a single valley in that island. Land snails and marine snails likewise have circumscribed habitat. This is the largest group of Mollusca including the snails, slugs, limpets, whelks, periwinkles, sea-hares and the like. Most gastropoda show all the chief molluscan features. The Gastropoda are divided into two subclasses—Streptoneura and

Euthyneura. The Streptoneura has its orders Diotocardia (Aspidobranchia) and Monotocardia (Pectinibranchia) and Stenoglsssa; whereas, the orders of Euthyneura include Opisthobranchia and Pulmonata. All these are divided into seven suborders.

## TYPE APPLE-SNAIL

The apple-snail or *Pila globosa* in a common example of the univalve molluscs found in freshwater ponds, pools, tanks, lakes, marshes and paddy fields, etc. They are the largest among the freshwater molluscs. The gastropods are very successful in water, but many like the snails and slugs have invaded the land. *Pila* is also a similar gastropod.

### Shell

The shell of *Pila* has the form of an elongate cone closely coiled in a spiral manner round a central axis. The apex of the shell is at the extreme top being

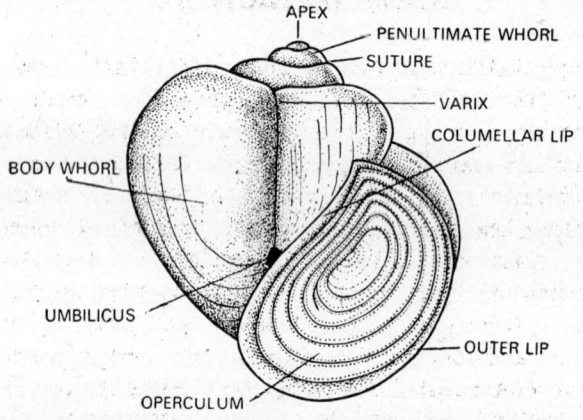

**FIG. 19.14.** *Pila.* Ventral view of entire animal to show external features.

the oldest part, the first to be secreted. It is followed by gradually increasing whorls, the last two of which are the largest and enclose the greater part of the body (Fig. 19.14). This is the **body whorl** having a wide opening called the **mouth** or **aperture** through which the body protrudes out. The margin of the aperture or **peristome** is smooth and continuous and its outer wall constitutes the outer **lip**, whereas, the inner one is the **columellar lip**. The remaining series of whorls is collectively known as the **spire** (Fig. 19.14). The last whorl of the spire that is in direct communication with the body whorl is often called the **penultimate whorl**. The various whrols of the shell coil around a central axis, the **columella** (Fig. 19.15) which is hollow and opens to the exterior by an opening called the **umbilicus**. Shells having an umbilicus are known as **perforate** or **umbilicated** shells. The coil is said to be **dextral**, when the shell held with the apex (summit) towards the observer has the mouth or aperture below and to the right. It is **sinistral**, when under the same conditions, the aperture is to the left. Dextral shells are much common than the sinistral ones. The line of contact between two successive whorls is known as the **suture**. The aperture is closed by a flat, calcareous plate, the **operculum**, formed as a cuticular secretion. External surface of the operculum shows a number of concentric rings of growth around a central circle, the **nucleus**. The gastropod body is attached to the shells by columellar muscles and withdraw themselves into their shells by their contraction. This muscle is symmetrical and horseshoe-shaped in species with conical shells (*Patella, Septaria*) but is asymmetrical in others. In forms without shells this muscle is absent (*Vaginula*).

Microscopical examination of the structure of shell reveals that it is made up of three layers. The outer layer is the thin chitinous layer, the **periostracum**, presenting an appearance of homogeneous membrane. The underlying calcareous layer in the adult is the **ostracum** consisting of parallel fibrils formed by the solidification of the secretion. The bundles of fibrils or plates run in the direction of the long axis of the shell. The **hypostracum** is the next layer following the ostracum. Essentially its structure is like that of the ostracum with the difference that its plates lie at right angles to the plates of the ostracum, in other words, they run parallel to the aperture.

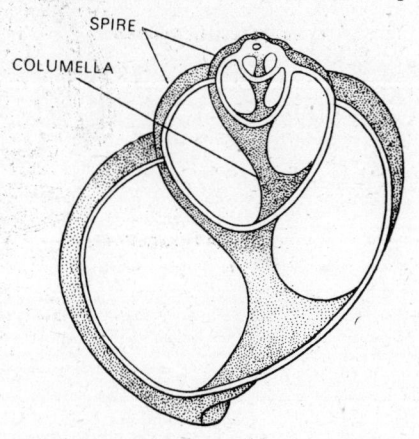

FIG. 19.15. Shell of *Pila* with half the portion ground off.

## Integument

The whole body is covered by a single layer of **epithelium** which, in parts not protected by the shell, may be more or less ciliated. In shell-less gastropods (Opisthobranchia) the whole surface of the body is ciliated. This layer is very rich in glands which are almost exclusively unicellular, some lying in the epithelium while others sunk in the subjacent tissue, their ducts, however, passing through the epithelium. The free edge of the mantle is particularly rich in **mucous, pigment** and **calcareous glands** as it plays the chief role in the secretion of the shell. The layer immediately beneath the body epithelium is called the **corium** and consists of connective tissue and muscle fibres. This layer is not clearly marked off from the tissues lying below. The remarkable colouring of the integument is caused by pigment cells which are more often found in the corium than in the epithelium. In several Nudibranchia stinging cells have been discovered in the integument.

## Mantle

This normally covers the whole of the visceral mass and projects all round it leaving only the head and foot projecting on the ventral side. On the anterior or on the dorsolateral aspect (or exceptionally on the posterior aspect) there is a large space called the **mantle cavity** or the **pallial cavity** This is an extensive cavity bounded dorsally by the mantle and laterally and ventrally by the dorsal surface of the animal posterior to the head. This space lodges a number of organs, the **organs of pallial complex**, which will be described below. In the gastropods the pallial cavity has a situation opposite to that which it occu-

788                                       LIFE OF INVERTEBRATES

pies in other molluscs due to the **torsion**, which the gastropod body undergoes towards the end of its development. The result of this torsion is that the anus opens anteriorly and lies above the head. The advantages of this arrangement are clear enough in an animal that lives in a shell with only one opening. The

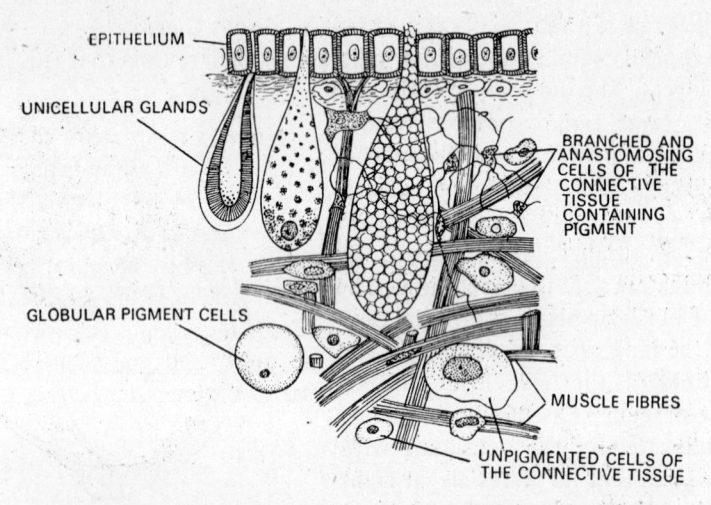

**FIG. 19.16.** Section of the integument of a mollusc (*after* Plate).

exact process, by which the torsion is brought about, is not clearly understood. What is known has been derived from the study of the development of the gastropod larvae.

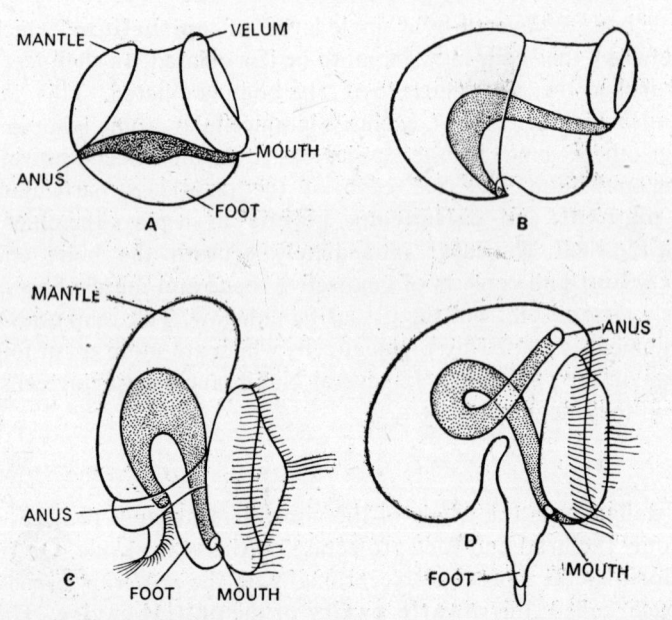

**FIG. 19.17.** Four stages in the development of a gastropod torsion. A—Embryo with straight alimentary canal; B—Flexure begins with ventral side; C—Embryo, with ventral flexure and an exogastric shell; and D—Embryo with lateral torsion and an endogastric shell (slightly modified *after* Roberts).

## Torsion

Torsion is a process of bending (**flexure**) on the ventral side of the larval gastropod. It takes place in an anteroposterior sagittal plane, about a transverse axis situated at right angles to the main antero-posterior axis of the animal. This brings about the two ends—mouth and anus—of the digestive tract close together. The approximation of mouth and anus occurs in some members of all classes of molluscs except the Amphineura. As a result of this the original saucer-shaped visceral mass and shell becomes cone-shaped and there is a simultaneous coiling up of these structures forming an exogastric coil (Fig. 19.17C) e.g. in *Patella*. The disposition of the shell in such cases is similar to that found in other molluscs with coiled shell but without **lateral torsion** (as in *Nautilus*). But what follows is peculiar to gastropods. During the completion of metamorphosis, there is a lateral torsion, following the primitive ventral flexure, with the result that the originally **dorsal** or **exogastric** shell becomes **ventral** or **endogastric**. It is, then, this lateral torsion that produces the adult condition of gastropods. While the head and foot remain stationary, the visceral mass is rotated through an angle of 180°, so that the anus and the mantle cavity are carried upwards and finally come to lie dorsal to the head. In addition the organs on one side of the body fail to develop. As a result of this unequal growth the visceral mass and mantle (and the shell secreted by the mantle) become spirally coiled. Torsion is thus brought about by (*i*) ventral flexure followed by (*ii*) lateral twisting. These activities are controlled by the contraction of the larval retractor muscle, followed by a rotation or coiling by differential growth processes. All this takes place quickly as torsion is a fairly rapid process.

The anterior position of the anus restricted the antero-posterior elongation of the body. This difficulty was solved by planospiral coiling. By this the bilateral

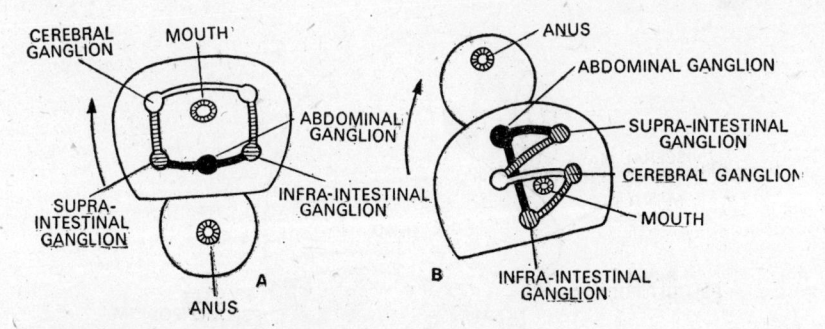

**FIG. 19.18.** Torsion of the visceral commissure in the *Streptoneura* seen from buccal side. A—Before torsion; and B—After torsion whose direction is indicated by the arrow (*after* Peleseneer).

symmetry of the visceral complex was not disturbed. In planospiral coiling each coil of the shell lay completely outside the preceding coil, rather the main disadvantage. To cover this a more convenient design was evolved. An asymmetrically coiled shell developed in which the coils formed a skewed helical pattern around a central axis (the columella) with each coil below the preceding one. This gave rise to a turbinate shell.

In most gastropods it is the organs of the original left side that degenerate. But since the visceral mass is rotated through 180° the adult appears to lack the kidney, the gill and one of the chambers of the heart (atrium) on the right side. The nervous system of these gastropods remains bilateral and uncoiled, but becomes twisted into a figure of 8 (Fig. 19.17) when the viscera rotates. In some gastropods there is reversal of the rotation of the viscera and the nerve cords get untwisted. In some such cases the coil of the visceral mass and spire may disappear in the adult, leaving the internal torsion and asymmetry unaltered but producing a secondary external symmetry as in **Patellidae**, etc.

**Advantages of torsion.** The majority of workers on molluscs have long felt that the twisting of the internal organs brought disadvantage to the adult gastropod. The views of recent workers are just the contrary: 1. The twisting is of immediate selective advantage to the larva. In the pretorsion condition, retraction of the larva into the developing mantle-shell involves withdrawal of the foot first followed by velum and head. After torsion, on the contrary, the velum and head are withdrawn first, affording them fastest protection, and the foot is withdrawn thereafter. The posterior part of the foot may develop a protective **plate** or **operculum**.

2. Morton feels that torsion has brought the mantle cavity anteriorly. This has certain advantages for gill breathers as it allows them to take the inhalant respiratory current from undistributed water ahead of the snail.

3. Further, torsion adds the mantle sense organs to those of the head region with the result that the osphradium, along with the head tentacles and eyes, etc., monitor the environment into which the snail is moving.

4. Torsion allows the snail to pull along coiled or elongate shell more easily. Although some stocks of primitive gastropods have simple cone-shaped or limpet-like shells the great majority have coiled shells. The turbinate shell, how-

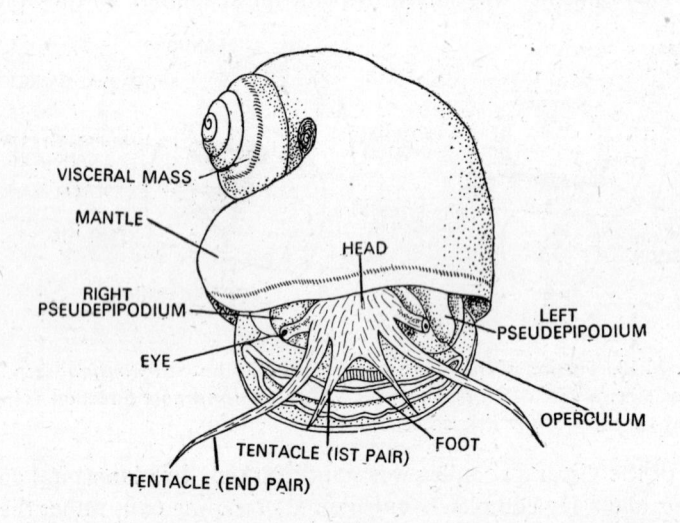

FIG. 19.19. *Pila*, front view after removing the shell (*after* Prashad).

ever, if its aperture were symmetrically placed over the head or foot, would be mechanically imbalanced. To bring about balance nature has made certain adjustments and produced two types of shell, **dextral** and **sinistral**. Most spiral shells are dextral, i.e., if the shell is held with the spire pointing upwards and the opening of the last whorl facing the observer, the opening is on the right. In sinistral spirals the opening is on the left. It may be pointed out here that in a few species both sinistral and dextral shells are known.

**Head.** The **head** of *Pila* is prolonged into a partly contractile snout and bears two pairs of filamentous **tentacles** and the **eyes**. The first pair of **tentacles** or **labial palps** are smaller filament-like anteriorly directed contractile structure (Fig. 19.19); the second **pair of tentacles (true tentacles)** are relatively longer structures lying on either side of the labial palps. The **eyes** are small situated on short or stumpy stalks, the **ommatophores**. Projecting anteriorly over the foot, on either side of the head there are two fleshy structures known as the **nuchal lobes** or **pseudepipodia**, of which the left is better developed and forms a long **respiratory siphon**. These are simple prolongations of the mantle.

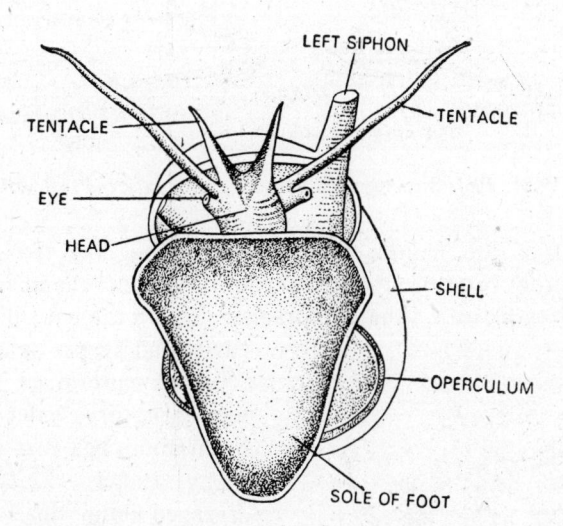

FIG. 19.20. Ventral view of the foot of *Pila* (*after* Prashad).

**Foot.** Normally and primitively the foot is formed by a powerful mass of ventral muscles with a more or less elongated ventral creeping surface. But this condition may be modified in relation to different conditions of existence. The smooth ventral surface of the foot is called the **sole** or better **creeping sole** which is often divided by a median longitudinal furrow. The foot of *Pila* that forms the main part of the fully expanded animal is roughly triangular in outline (Fig. 19.20). It is extensile and constantly changes its shape. It bears the operculum on its dorsal surface.

**Organs of the pallial complex.** Near the anterior edge of the right pseudepipodial lobe there is a prominent ridge that extends to the posterior end of the cavity. This is known as the **epitaenia** (Fig. 19.21) and is quite prominent in the living snail and separates the right branchial chamber from the mantle cavity. In preserved specimen it becomes reduced, due to shrinkage, and is about two millimetres in height. The branchial chamber lodges the gill or ctenidium, extending from the margin of the tentacle right up to the pericardium at the posterior extremity. The ctenidium hangs vertically downwards from the

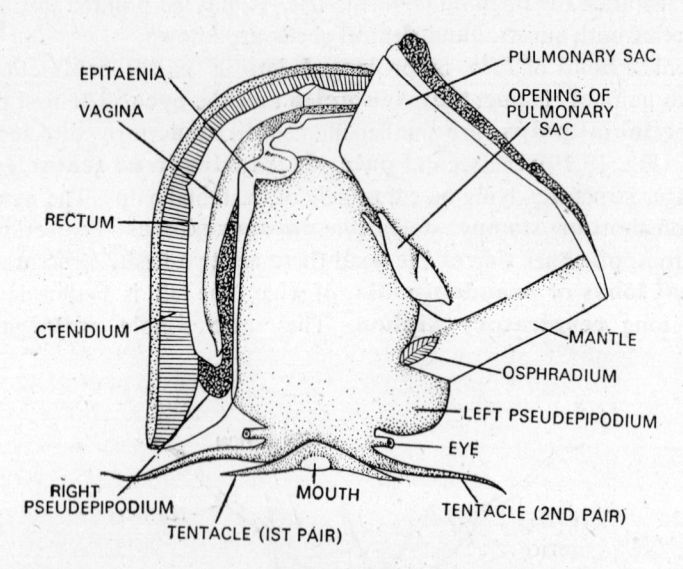

**FIG. 19.21.** *Pila.* Structure of the mantle cavity (*after* Prashad).

dorsolateral wall of the mantle cavity. A large bag-like structure, the **pulmonary sac** or the **lung** occupies the roof of the left chamber of the mantle cavity. The **osphradium** is situated dorsolaterally on the left side of the mantle cavity and keeps hanging from the roof-like curtain in the course of the respiratory water current. Each osphradium has two rows of thick fleshy leaflets 22-28 in number arranged along the two sides of a central axis. It is, thus, called a bipectinate structure. The leaflets are roughly triangular in shape and each is attached to the mantle wall along its base and to the central axis along its inner side. The rectum, the male or female ducts and the anterior chamber of the renal organ are other structures included in the organs of the pallial complex. The rectum is an elongate tubular structure lying on the left side of the ctenidium and extending from the posterior end to its external opening (Fig. 19.22), the anus, situated about quarter of an inch behind the anterior edge. The male or female duct (as the

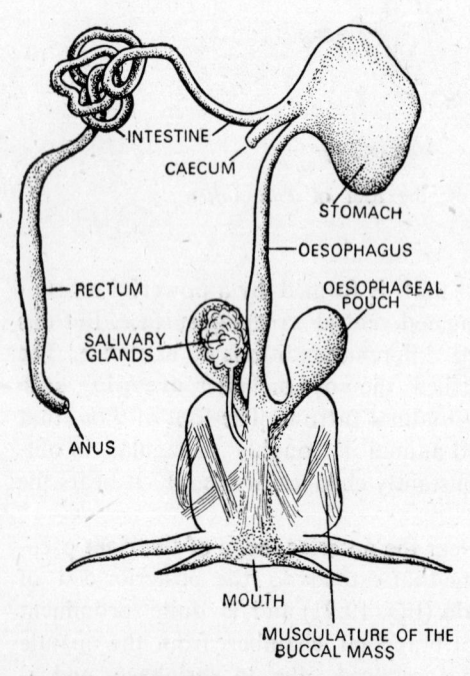

FIG 19.22. *Pila.* The alimentary canal.

case may be) lies on the left of the rectum (Fig. 19.22). In the male a special copulatory organ is met with. It arises as a separate appendage from the edge of the mantle and becomes associated with the end of the male duct during the act of copulation. At the base of the penis-sac there is a glandular thickening, the hypobranchial gland (found only in the Pilidae). The anterior chamber of the renal organ also emerges as a small elliptical projection of reddish colour, into the right of the ctenidial part of the pallial cavity.

## Digestive System

The **alimentary canal** comprises the **foregut** of stomodaeal nature, the **mid-gut**, and **hind-gut** of proctodaeal nature. The fore-gut includes the **mouth, buccal cavity** and **oesophagus**. The mouth (Fig. 19.22) is a vertical slit at the end of the snout. True lips are lacking, the plicate edges, however, serve as secondary lips. The mouth leads into the buccal mass (or pharynx) which is a thick-walled structure containing the radula internally for cutting the food. Its walls are strengthened by several pieces of cartilage. Externally the buccal mass is provided with special musculature, responsible for its movements. The **buccal mass** is a complex structure. Anteriorly it possesses a narrow space, the buccul cavity, divided by the jaws into an anterior and a posterior portion. The anterior portion is often called the **vestibule**. The jaws are local thickening of the culticle lining the cavity. They lie dorsolaterally, rather placed more towards the anterior side. Both the jaws are connected by a thin cuticular membrane. Seen from above, each jaw appears, more or less, elliptical, whose anterior edge is serrated forming a cutting surface. A fairly prominent ridge marks its dorsal surface. A more or less fleshy outgrowth, the **tongue mass** or **odontophore,** if formed by the elevation of the floor of the cavity, the greater part of which it occupies. Located on the roof of the buccal cavity are two elongated glandular areas, the **dorsal buccal glands,** that release accessory digestive fluids. About the middle of these areas are situated the openings of the **salivary glands**.

The **radula** or **lingual ribbon** is ribbon-like strip of horny basement membrane on which are situated many transversely arranged rows of minute

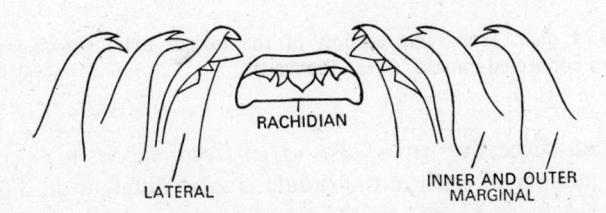

FIG. 19.23. *Pila*. One row of radular teeth.

recurved teeth. It runs longitudinally over the summit of the odontophoral mass. Each transverse row of *Pila* contains seven teeth. In the centre is a **median rachidian** tooth followed by one **lateral** and two **marginals,** on either side. It is secreted in a ventral diverticulum of the buccal cavity in which it is almost wholly contained, but its anterior extremity stretches out on the floor of the buccal cavity, where it forms a median projection. The radular ribbon is supported by a system of paired cartilaginous pieces furnished with protractor and retractor muscles (Fig. 19.24). The action of these causes the radula to move to and fro and work like a rasp on the prey seized by the animal. The teeth are secreted at the bottom of the diverticulum by a small number of matrix cells. As new teeth are being added, the old ones are pushed forward so that fresh rows of teeth come forward as the old ones wear off. The teeth are all simple devoid of secondary folds, ridges or basal processes. The rachidian tooth is roughly rectangular in shape with a broad posterior edge and slightly concave anterior edge, while the two sides slope backward. The anterior cutting edge is provided with a median and two or three lateral cusps on each side. The lateral teeth, as shown in Fig. 19.23 are elongated, greatly broadened at their toothed edges. Both the marginal teeth are similar in appearance, the outer being narrower than the inner. The number of teeth in any given transverse row is constant for species. It may, however, increase slightly with age (Aplysidae, Pulmonates). On the whole the number of teeth varies from group to group.

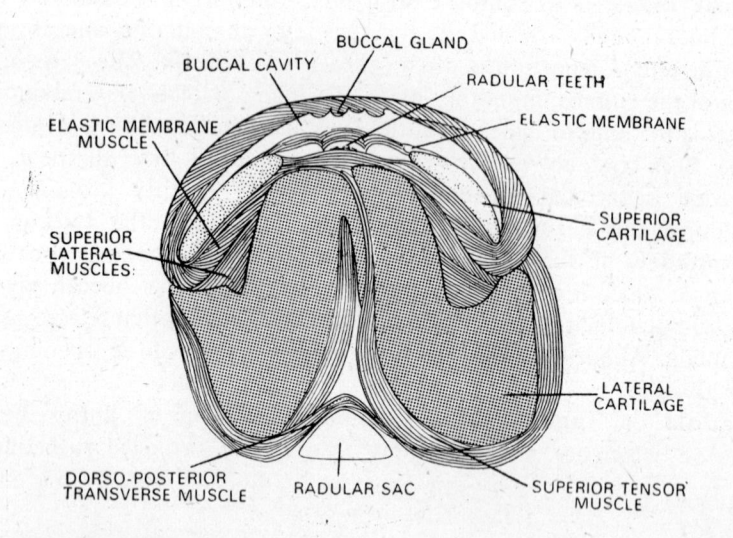

**FIG. 19.24.** *Pila.* Transverse section of the buccal mass showing cartilages and connected muscles (*after* Prashad).

The number of successive rows also varies from species to species and consequently total number of teeth in the radula is also different in different forms. The number varies from 10, one tooth in each row, in certain Eolidae, to 250 in *Buccinum undatum*, 1920 in *Patella vulgata*, 8343 in *Limnaea stagnalis*, 15,000 in *Helix aspersa*, and so on up to 750,000 in *Umbrella*.

The radula travels over the ends of the cartilages just like a band over its pulley, the cartilages being entirely passive in the matter. The radula is protruded from the mouth by the protractor muscles, and by the alternate action of the upper and lower sets of muscles into the tongue plate, a chain saw-like movement is communicated thereto, in consequence of which the teeth act as a rasp or saw upon the body with which they are brought in contact.[1] This movement, known as **chain-saw movement**, is common in gastropods and very limited in *Pila*. The buccal cavity of gastropods is provided with salivary glands of which there are two in *Pila* lying one on each side of the posterior limits of the buccal mass and partially cover the oesophagus as it comes out. The surface of the glands is rough giving it the appearance of a branched type of a gland. The ducts of the glands run anteriorly and open in its dorsal roof. These organs are generally simple mucous glands, without any digestive action, but in certain forms (*Dolium galea*) their secretion contains as much as four per cent sulphuric acid, which sets to dissolve the calcareous spicules of the animals eaten.

The **buccal cavity** is followed by **oesophagus**, a tubular structure, with plicated walls. For a little distance it runs posteriorly in the middle line, then turns to the left and enters into the visceral mass, where it opens into the **stomach**. At the junction of the oesophagus and the buccal cavity open two rounded **oesophageal pouches** which are intimately associated with the salivary gland only externally. Each opens by short duct at the junction of the buccal cavity and the oesophagus (as in the Aspidobranchs). They appear to be mere backwardly directed processes of the oesophagus, with unknown functions. The terminal part of the oesophagus sometimes becomes differentiated to form a **gizzard** with thick muscular walls and furnished internally with **masticatory teeth**. In *Pila* no such structure is found. The terminal part of the oesophagus opens into the stomach, a more or less rectangular mass of dark red colour lying buried in the liver mass and with its lower margin nearly straight. An internal examination reveals that the stomach encloses a broad U-shaped cavity of a rose-red colour. The oesophagus opens into its posterior limb known as the **cardiac chamber**, the walls of which are folded, the folds being very low. The anterior limb of the U-shaped cavity of the stomach forms the **pyloric chamber** of the stomach. From this arises the intestine. Its walls are thrown into transverse folds. A small bluntly rounded pouch, the **caecum**, arises from the outer wall of the pyloric chamber. It is elongated tube, but as major part of its length is embedded in the wall of the stomach it appears as a small protuberance, only for this reason its opening into the stomach is situated down below (Fig. 19.25). The caecum in other gastropods lodges a cuticular projection known as the **crystalline style**, but in *Pila* it is not found.

The **intestine** is narrow tube that originates from the pyloric chamber. It courses its way through the liver and on reaching beneath the posterior renal chamber it turns upwards and backwards, coils between the gonads (uterus in female and seminal vesicle in male) and the liver two or three times and then turns to enter the mantle cavity as rectum. The intestine is separated from the stomach by a sort of valve. In nearly all cases the intestine exhibits a well-marked longitudinal projection, the **typhlosole**, along certain parts of its course. The intestine is very long and coiled in herbivorous gastropods such as *Patella*, but it is very short, often straight, in carnivorous forms. The **liver** or **digestive gland** of *Pila* is a soft gland of brownish or dirty green colour. It is

---

[1] *Phil. Trans. Roy. Soc.*, London, CXLIII, page 31, 1853.

a more or less triangular plate, spirally coiled after the whorls of the shell. It is made up of two lobes, the separate ducts from which unite to form a common duct before opening into the stomach. The two lobes of the liver are equal and symmetrical in very few forms (*Neritina* and *Valvata*). More frequently the original left lobe is more deeply involved in the spire, and is larger from larval life onwards than the right lobe in dextral gastropods, the reverse is the case in sinistral forms. In some the right lobe may disappear and the left lobe may only persist. The liver discharges its digestive secretion into the stomach. In addition to their digestive properties the secretion exercises excretory function (in Euthyneura) and they also arrest the actions of poisonous substances. Lastly, the digestive glands take a share in intestinal absorption. The terminal part of the alimentary canal is a thick-walled tube which opens to the exterior through the **anus**, which is situated a quarter of an inch from the edge of the right pseudepipodial lobe. In some (*Murex*, *Purpura*) the rectal portion of the intestine is divided with a somewhat ramified gland, the **anal gland**, of unknown function.

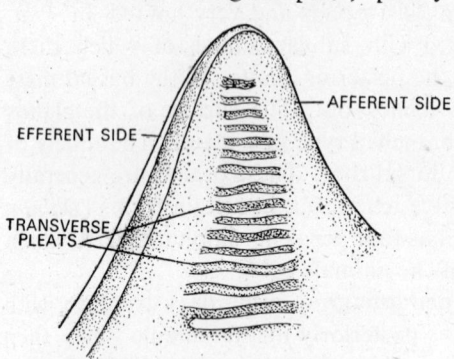

FIG. 19.25. *Pila*. A branchial lamella showing the ctenidial axis.

## Respiratory System

Primitively the respiration in gastropods is aquatic. The organs of respiration consist of a pair of leafy expansions of mantle situated in the pallial cavity and called **ctenidia** (Fig. 19.25). Each ctendium is the homologue of a single branchia of *Chiton*. Respiration in *Pila* takes place through two types of organs. In water it respires by ctenidium, whereas, on land it respires by means of a **pulmonary sac** and two **right** and **left nuchal lobes**.

The ctenidium is situated in the mantle cavity and consists of numerous thin **leaflets** or **lamellae** attached to the mantle wall by their broad bases, but have their apices free in the branchial chamber. The line of attachment is known as the **ctenidial axis**. A ctenidium with such an axis with a single row of leaflets is described as **monopectinate**. In the other types (*Patella*) the axis is free and bears two rows of leaflets. Such a ctenidium is called **bipectinate**. In the opisthobranchs ctenidium is a simple, flat and projecting tegumentary lamina transversely folded from its base to its extremity in such a manner that the ridges of one face correspond to the furrows of the other. Such a branchia is called **plicate**. The **lamellae** or **leaflets** are larger in the middle of the ctenidium and become smaller towards its ends. Each leaflet has a longer left side known as the **efferent side**, being on the side of efferent ctenidium sinus and the relatively smaller, right or **afferent side** lying on the side of afferent ctenidial sinus. The anterior and posterior faces of leaflets are provided with low transverse ridges, or **pleats** that are richly supplied with blood vessels. Each leaflet arises as a double-layer of epithelium internally supported by connective tissue and muscle fibres. The epithelial layer consists of three kinds of cells: (*i*) **non-ciliated columnar cells**; (*ii*) **ciliated columnar cells**; and (*iii*) a few **glandular cells**. Underlying the epithelium there is a thin basement

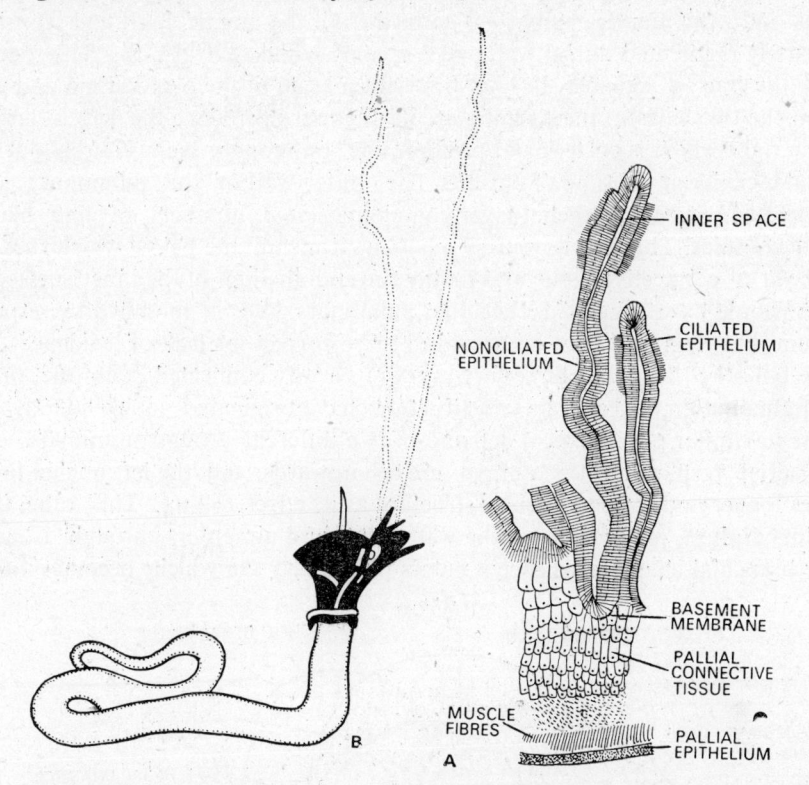

**FIG. 19.26.** A—Transverse section of the branchial lamellae with mantle; and B—Mucilage cards formed by *Vermetus* as a feeding device.

membrane below which lies connective tissue layer consisting of cells with a few scattered nuclei and large number of obliquely running muscle fibres (Fig. 19.26). The ctenidium of *Pila* is situated on the right side but morphologically it is the gill of left side which has been pushed to the right side by the extensive development of the pulmonary sac. Both the nuchal lobes are equally expanded forming shallow channels. The respiratory current enters through the left lobe. Passing underneath the osphradium the current reaches the posterior part of the pulmonary chamber. Thence it enters the right of the branchial chamber and flows through the leaflets where gases are exchanged. The current finally makes an exit from the right nuchal lobe. The osphradium helps the animal in selecting the suitable quality of water for breathing.

In certain gastropods various parts of the mantle serve as accessory respiratory organ. In forms in which the mantle has disappeared as a shell-forming organ, the dorsal envelope of the body acts as respiratory organ. In others a ctenidium coexists with secondary respiratory organs or pallial branchiae. If the ctenidium is atrophied and disappears altogether the respiratory function is taken over by the mantle. As a result of adaptation to a terrestrial life pulmonary respiration has developed in some gastropods. Some gastropods possess a special chamber known as the pulmonary sac or lung for this purpose.

In *Pila* **pulmonary sac** or **lung** is a sac-like structure hanging from the roof of the mantle cavity and is entirely closed except for a large opening with which it opens into the mantle cavity. It is derived from the mantle itself and consists of a densely pigmented dorsal wall and a creamy white ventral wall. The opening of the sac is situated in line with right margin of the osphradium and the flaps on the two sides of the opening are of unequal size. Since the left is larger than the right which is posterior in position, the pulmonary sac is entirely out of the mantle cavity at the time of closure. The outer wall of the pulmonary sac consists of an external epithelial layer containing large numbers of fine black pigment granules. This is followed by a basement membrane below which lies a thick layer of connective tissue and finally the endothelium of the inner surface.

Some people have suggested that the pulmonary sac is modified by second ctenidium but this has not been proved to be correct for lack of evidences. It may be remarked that the pulmonary sac is in no way connected with the lung of the pulmonate molluscs. The two structures have originated independently in response to similar physiological demand of two different groups of animals.

For aerial respiration the animal moves upwards and the left nuchal lobe becomes longer and forms a tubular structure as its edges roll up. This tube, the respiratory siphon, projects above the water level and air enters through it and fills the branchial chamber whence it enters pulmonary sac which becomes fully

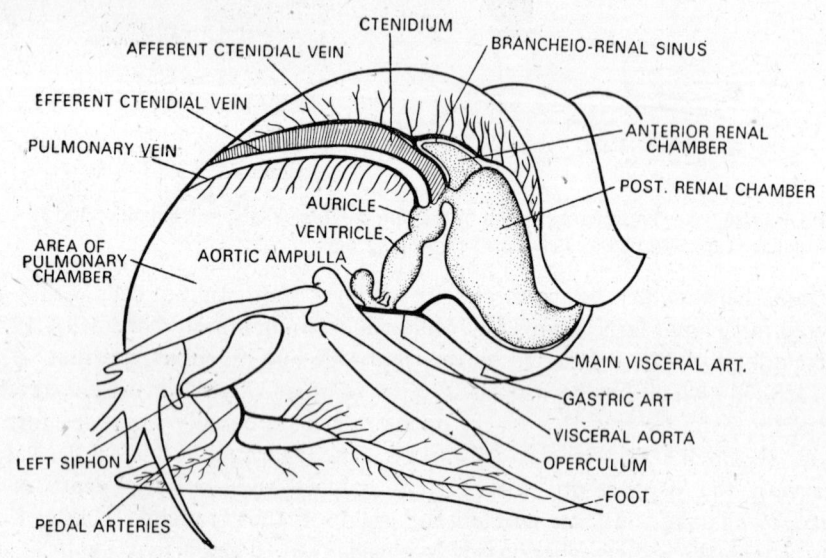

FIG. 19.27. *Pila*. Heart and principal vessels from the left side (*after* Prashad).

enlarged. At this time the opening of the pulmonary sac looks almost rounded and lies quite close to the posterior opening of the siphon. The inspiration of air is the result of suction caused by the expansion of the pulmonary sac. Alternating with the expansion the pulmonary sac contracts driving out air (expiration). Besides forming long tube, the siphon, sometimes, when water surface is clear, becomes broader at the tip, just like a funnel, and the animal lies near the surface. The rest of the process is the same. Thus, in *Pila* there are two types of aerial respiration depending upon the surroundings in which it lives. When in water, ordinarily *Pila* resorts alternately to an aquatic and aerial respiration rather

*Phylum Mollusca* 799

regularly. When moving on land *Pila* breathes directly with the pulmonary sac but no siphon is formed. Only the left nuchal lobe is expanded. Sometimes the snail remains buried in dry mud or at the bottom of tank with tightly closed operculum. Under these circumstances it respires through air stored in the pulmonary sac.

### Circulatory System

The **blood** of molluscs is a colourless liquid containing **amoebocytes**. In *Planorbis* it is red (except *P. albus*) in which haemoglobin is diffused in the plasma. In some gastropods the blood is of bluish tint because of the presence of **haemocyanin**, a copper containing albuminoid. In some molluscs the blood is coloured by pigments of extraneous origin absorbed by the amoebocytes (*e.g.* violet red colour in *Fasciolaria*).

The **heart** (Fig 19.27) is always dorsal and in the immediate neighbourhood of the respiratory apparatus. It lies enclosed in the **pericardial chamber** situated on the left side of the body-whorl wedged in between the pulmonary sac and the renal organ on the anterior side and the stomach and the digestive gland on the posterior side. It is a fairly deep oval cavity bounded by thin-walled **pericardium**. It communicates with the posterior renal chamber by a **renopericardial aperture**. This cavity respresents the true coelom in the molluscs. The pericardial cavity is the true coelom as the renal organs open into it and the reproductive cells are produced by its wall in the embryo.

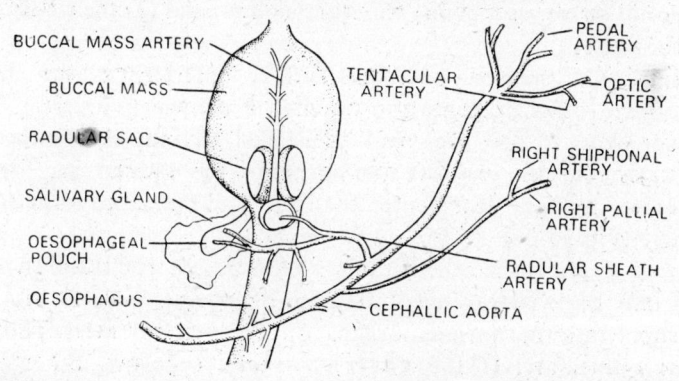

FIG. 19.28. Blood vessels of the buccal mass, diagrammatic (*after* Prashad).

The heart of *Pila* is two-chambered and lies enclosed within the pericardial cavity. The **auricle** is single like all other Monotocardia (except *Cypraea*). It is a thin-walled more or less triangular sac occupying the dorsal part of the pericardial cavity. The apex of the auricle points dorsally into which open veins from the ctenidium, the posterior chamber of the kidney and the pulmonary sac. The thin and transparent wall of the auricle are extensile though they have fewer muscular fibres. The **ventricle** is a thick-walled ovoidal sac slightly elongated situated vertically below the auricle. Its walls are spongy being made up of muscular strands. Between the two is a valvular opening, the **auriculoventricular** opening. The valves are semilunar in shape. From the lower end of the ventricle arises a very short muscular tube, the **aorta**, which bifurcates immediately into two main branches. The root of the aortic trunk is provided with two semilunar valves of unequal size. Of the branches the **cephalic aorta** is the anterior blood vessel which bulges out into bulbous outgrowth, soon after its origin forming a thick-walled bulb. The **aortic ampulla** is a contractile structure which aids circulation. The presence of aortic ampulla is one of the chief characteristics of the group (Pilidae) to which *Pila* belongs. The opening of the ampulla into the aorta is not provided with valves. The second branch of the aorta is known as the **visceral aorta** which proceeds backwards into the visceral mass.

**Arteries**. From these two branches a number of small arteries arise and conduct blood to various parts of the body. From the cephalic aorta arise the following important vessels: (*a*) the **cutaneous** artery, supplying the skin, a delicate vessel; (*b*) the **oesophageal artery**, which is relatively thicker supplying the oesophagus; (*c*) the **pallial artery,** stouter blood vessel, supplying organs of the left half of the mantle cavity including the osphradium, etc., These three branches arise from the aorta on the outer side; (*d*) The **pericardial artery** arises from the aorta from the inner side and supplies the pericardium; and (*e*) The **renal artery** arises from the aorta when it enters the posterior renal chamber.

Thereafter, the main limb, the **aorta**, runs along the left side of the oesophagus to the right where it appears quite prominent. It gives off many other branches. The prominent among them is one that branches into three arteries: (1) the **right pallial artery** supplying the right side of the mantle; (2) the **right siphonal artery** supplying the right siphon; and (3) the **penial arteries** supplying the penis.

The main trunk of the aorta continues further anteriorly as the **buccal** and gives rise to many other branches, the important of which are: (*i*) The **optic arteries** supplying to the eyes and eye-stalks; (*ii*) the **tentacular arteries** the two tentacles; (*iii*) the **radular sac arteries** the radular sac; and (*iv*) the **pedal arteries** the foot which enter the foot and branch to form an irregular network there (Fig. 19.28).

The **viseral aorta** is the other branch of the main aortic trunk that proceeds backwards into the visceral mass. During its course it gives rise to many branches supplying the various organs of the visceral mass. The prominent arteries among them are: (*i*) the **gastric artery** supplying the stomach; (*ii*) **intestinal arteries** supplying the intestine and the posterior renal chamber; and (*iii*) the **hepatic artery** supplying the gastric gland and gonad.

**Veins**. The blood distributed by the arteries is collected by certain capillary spaces called **lacunae**. The lacunar spaces are well marked, but they merge into large spaces called sinuses which are devoid of definite walls. There are four such sinuses; (*i*) The **anterior perivisceral** sinus lies in the floor of the mantle cavity surrounding the anterior part of the gut. Three channels conduct blood from this sinus to the renal chamber and the pulmonary sac. (*ii*) The **peri-intestinal sinus** runs along the coils of the intestine up to the junction of the anterior and posterior renal chambers. Blood from the viscera, the digestive gland and the greater part of the genital system is collected by this sinus which becomes tubular anteriorly forming two branches, the right or ventral which is known as the **afferent renal sinus**, and the left one forms the **afferent renal vein** and turns upwards to the roof of the posterior renal chamber. (*iii*) The **branchio-renal sinus**

lies dorsal the intestine along the right side of the anterior renal chamber. From the roof of both the renal chambers it receives many branches. Anteriorly it continues as the afferent ctenidial vein above the rectum. From this arise numerous blood vessels that take blood to the lamellae of the ctenidium. Into this sinus opens the efferent renal vein. The blood from the ctenidial leaflets is collected by efferent ctenidial vein which runs to the pericardial cavity and opens into the auricle. (*iv*) The **pulmonary** sinus lies in the walls of the pulmonary sac. It receives a large amount of blood from the anterior perivisceral sinus and returns the same through the pulmonary vein to the auricle after aeration.

**Circulation.** The cephalic and visceral aortae conduct blood to various organs of the body from where it is returned by many blood sinuses, mainly the peri-visceral and peri-intestinal sinuses. From these blood sinuses the blood has three courses open—it may go to the ctenidium, pulmonary sac or to the renal chamber. During aquatic respiration the major portion of the blood from the perivisceral sinus passes into the ctenidium for aeration; and during aerial respiration blood flows into the pulmonary sac. The blood returns to the auricle through the efferent ctenidial vein. The blood from the peri-intestinal sinus flows into the renal chamber where it gets rid of nitrogenous wastes and passes to the auricle without aeration. The auricle, thus, receives both aerated and non-aerated blood. It is this mixed blood that enters the *ventricle* for general circulation.

## Excretory System

The kidneys are the essential organs of excretion in gastropods on the whole, but in *Pila* the exact homologies of the renal organs are not established, the two renal organs are called (*a*) the **anterior renal chamber**, and (*b*) the **posterior renal chamber** (Fig. 19.27). The anterior chamber is ovoid sac opening into the mantle cavity in a deep crevice by an elongated opening that lies to the right of the epitaenia. There are numerous grooves on the dorsal surface of the chamber. Corresponding to these grooves a number of lamellae hang within the cavity of the chamber greatly reducing its lumen. The triangular lamellae are arranged on either side of a thick median axis of white colour running throughout the length of the roof. This is the **efferent renal sinus**. The floor of the chamber possesses a similar median ventral axis, the **afferent renal sinus**, furnished with lamellae on both sides. The afferent renal sinus is the right branch of the peri-intestinal sinus which breaks up into several branches to supply the numerous lamellae on both the sides. Posteriorly the anterior chamber communicates with the posterior renal chamber through an opening lying slightly above the point of entrance of the afferent sinus into the ventral axis.

The posterior chamber lies to the right of the pericardium and the digestive gland. It appears as a hook-shaped area of grey colour (may be brown sometimes). The cavity of the chamber is spacious and encloses a few coils of the intestine and a portion of the genital duct. Between the pericardium and this chamber there is only a thin partition, the **renopericardial septum** perforated by a slit-like opening, the **renopericardial aperture** surrounded by thick white lips, by which the cavity of the pericardium communicates with the posterior renal chamber. The roof of the chamber has a profuse blood supply. The floor of the chamber comprises a thin epithelium which separates its cavity from the organs that project into it. The two renal chambers are definitely excretory structures and separate nitrogenous wastes. The excretory material, collected in the posterior chamber, passes into the anterior, whence it is discharged into the mantle cavity. From the above it is evident that the kidney communicates with the coelom (pericardial cavity) at one end and with the exterior at the other. It, therefore, represents a true coelomoduct as in most other Mollusca.

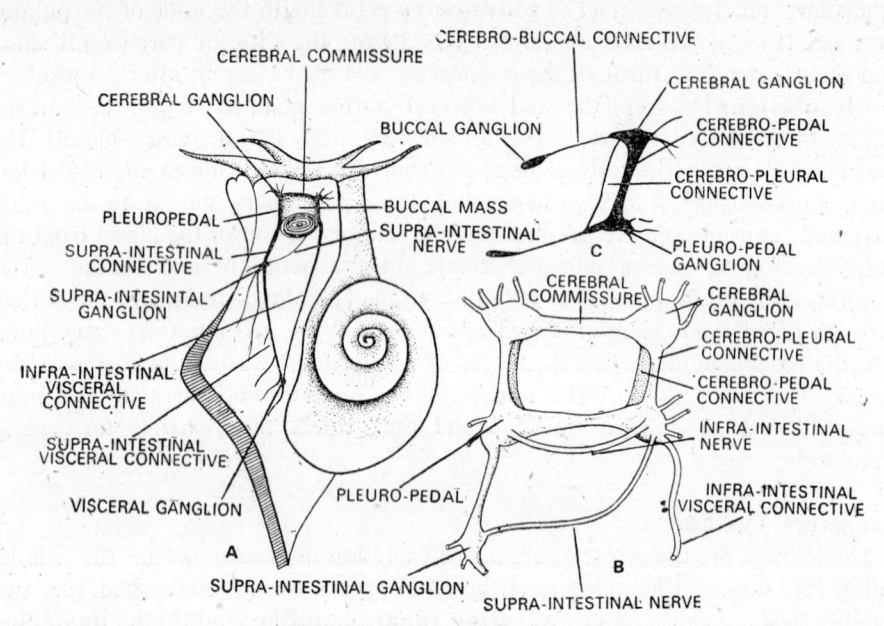

FIG.19.29. *Pila*. Structure of nervous system. A—Dissection of cerebral ganglia and the main nerves; B—Side view of the dissection showing the position of the buccal ganglia and also the cerebropleural and cerebropedal connectives; and C—The brain and the pleuropedal mass after removing the buccal mass.

In majority of other molluscs the **kidneys** are originally paired (a single pair being common) and the two kidneys open, one on each side of the anus. In gastropods their primitive symmetry is lost and the topographically left kidney is rudimentary and that of the right side alone is functional in almost every case. In some the right kidney no longer exists in the adult. Their primitive situation is wholly within the visceral mass and their migration outside the visceral mass is a specialization. The kidney is always a dorsal organ, situated in the neighbourhood of the pericardium with which it communicates by a ciliated aperture, which may be lost (Aspidobranchs) or may be multiple (*Elysis*). As a rule the external opening of the kidney is situated near the anus, and sometimes the two open together in a **common cloaca** (*Limax, Vaginula* but not in *V. willeyi*). The external renal opening is borne on a papilla in some (Aspidobranchs) but is a simple slit, shaped like a button hole in majority (Pectinibranchia and Tectibranchia). In *Pila* the exact homologies of the renal organs are not established.

**Nervous System**

The Gastropoda possess a well-developed nervous system in which the same

**cerebral, pedal, pleural, visceral** and **stomatogastric** nerve-centres and the **connectives** and **commissures** are to be found as in other molluscs. But the special character of the gastropod nervous system is the asymmetry of the visceral centres and of the nerves arising from them, an asymmetry resulting from that of the visceral organs themselves. *Pila* also possesses the nervous system of the same type comprising the **cerebral, buccal, pedal, pleural** and the **supra-** and **infra-intestinal ganglia** with their **commissures** (connections between similar ganglia of the opposite sides) and **connectives** (connections between dissimilar ganglia).

The **cerebal ganglia** of *Pila* (Fig. 19.29) are situated on either side of the buccal mass and connected with one another by cerebral commissure. Each ganglion is a flattened triangular structure from which arise various delicate nerves and connectives. The **cerebral commissure** is thick and band shaped running dorsally over the buccal mass. The **buccal ganglia** are small elongated ganglia embedded in the tissues of the buccal mass. Each ganglion is connected with the cerebral ganglia of its side by a **cerebrobuccal connective**. The **pleuropedal ganglionic** mass is situated ventrolaterally to the buccal mass on each side and is more or less rectangular in shape. Each is formed by the fusion of two ganglia—the **pedal** and **pleural**. The line of fusion between the two is indicated by a slight notch. In *Pila* the nerves that ordinarily arise from the sub-intestinal ganglion in other molluscs also arise from the right pleuropedal indicating the fact that the sub-intestinal ganglion is also fused with it. From the pleuro-pedal mass originate two connectives on each side (Fig. 19.29 B), and are connected with the cerebral ganglia, These connectives are (*i*) the **cerebropedal**; and (*ii*) the **cerebropleural connectives**, the former arising from the pedal part and the latter from the pleural part of the pleuropedal mass. The two pedal ganglia are connected by the pedal commissures of which, in fact, there are two, one lying above the other; as such, on seeing from the above only the first pedal commissure, which is relatively broader, is seen. A delicate nerve, the **infra-intestinal nerve**, is seen forming a loop over the first pedal commissure (Fig. 19.29) rather posteriorly, connecting the two pleuropedal masses. Another nerve, the **infra-intestinal-visceral** connective arises from posterior side of the right pleuropedal mass and continues backwards up to the visceral ganglion (see below).

The **supra-intestinal ganglion** is a single ganglionic mass lying behind the the pleuropedal mass of the left side. It is a slightly swollen ganglion connected to both the pleuropedal masses by the **pleuropedal-supra-intestinal** connective to the left and **supra-intestinal nerve** to the right. The supra-intestinal is also connected to the visceral ganglion by a stouter **supra-intestinal-visceral** connective. The **visceral ganglion** is formed by the fusion of two spindle-shaped masses and lies near the base of the visceral mass close to the anterior lobe of the digestive gland and to the right of the pericardium.

Several nerves arise from the different ganglia and innervate the various parts and tissues. The cerebral ganglia give off some anterior nerves supplying the skin of the snout, tentacles, ventrolateral margins of the buccal mass. Some nerves arise on the posterior side of the ganglia. These include the **tentacular**, the **optic** and a nerve to the statocyst. The buccal ganglia supply the muscles of the buccal mass, oesophagus, salivary gland and the oesophageal pouches, etc.

The nerves from the pleuropedal mass innervate the foot and adjoining structure. A stout nerve from the **supra-intestinal** supplies the mantle, ctenidium and another innervates the pulmonary sac. Nerves from the visceral ganglion supply the renal organ, genital system, pericardium, stomach, intestine and digestive glands, etc.

**Effect of torsion.** The narvous system in molluscs varies from simple as described in *Chiton* to complex as in the apple snail. The complexities are attained by various processes such as torsion, detorsion and abortion or exaggeration of certain parts. Looking at these complexities the nervous system of molluscs shows the following conditions:

*Protoneurous condition.* The nervous system is simple, undifferentiated as seen in *Chiton*, where the nervous system is without any definite ganglionic formation (Fig. 19.10 A). In the differentiated types ganglia formation begins. In the Solenogasters the nervous system is ladder-like with a single cerebral ganglion, but it becomes double in *Chaetoderma*. In Scaphopoda the nervous system is symmetrical. The pleuropedal connectives become fused with the cerebropedal connectives. There is no parietal ganglion. Figure 19.59A shows the nervous system of *Lamellidens*. It is similar to that of *Unio*. In this parietal ganglia are fused with the visceral ganglion forming visceroparietal ganglion.

*Streptoneurous condition.* This condition is produced by the lateral twisting of the nervous system as the whole pallial complex moves up. The original parietal

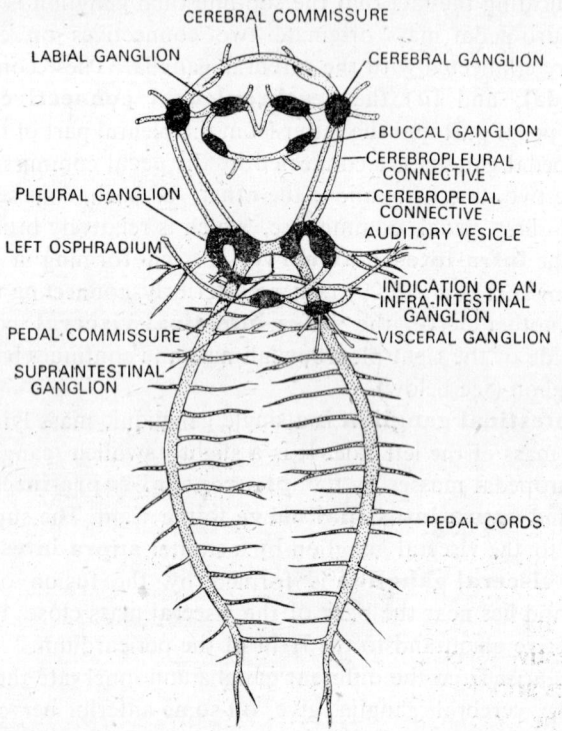

FIG. 19.30. Nervous system of *Patella* (*after* Pelseneer and Bouvery *from* Lang).

ganglia move to occupy a position above or below the level of the oesophagus. It is, therefore, called accordingly **supra-intestinal** or **infra-**

**intestinal**. The **pleurovisceral** connectives are crossed and form a figure of "8" (Fig. 19.18). The whole nervous system becomes asymmetrical. These changes occur in the Gastropoda especially prosobranchs. The streptoneurous condition or **streptoneury** in the gastropods exhibits many diversifications and complexities.

Interchange at corresponding points of the nervous system (as during torsion) is called **chiastoneury**, which may be **simple** or **complicated**. In those with simple chiastoneury the parietal ganglia are replaced by supra-intestinal on the left and infra-intestinal on the right. The pleurovisceral connectives are interrupted and the original right and left connectives cross each other as in *Pila* (Fig. 19.29A). The infra-intestinal ganglion fuses with right pleuropedal ganglionic mass, as is evidenced by the presence of the infra-intestinal nerve.

The circumenteric nerve ring remains unchanged in most gastropods, only there is shortening of the nerves between ganglia. In some the pleural ganglia give off **pallial nerves** to the mantle. In *Patella* (Fig. 19.30) a circumpallial nerve joins the two pallial nerves forming a complete nerve ring. The pleurovisceral loops, in most gastropods, cross with one another to maintain streptoneurous condition. In many cases there occurs gradual shortening of the nerves between the pleural and intestinal ganglia (pleurointestinal connectives). Sometimes this shortening is extreme with the result that the intestinal ganglion fuses with the corresponding pleural ganglion. In this way the original crossing of the pleurovisceral nerves is obliterated and only uncrossed portion of the pleurovisceral loop persists. In *Patella* the pleurovisceral loop is greatly reduced and is displaced to the right side. The intestinal ganglia are indistinct. The pleurovisceral loop gives nerves to gills and osphradia and makes the system more complicated.

The variations in the nervous system of gastropods include addition in the number of visceral ganglion, increase in the size of pedal ganglia, which in *Patella* and others give two pedal nerve cords (Fig. 19.30), absence of distinct intestinal and pleural ganglia as in *Haliotis*, and many similar changes.

In some cases (Diatocardia) the pallial nerve is connected to the pleural ganglion and a nerve runs from the intestinal ganglia into the mantle. The occurrence of secondary pleurointestinal connection has been named **zygoneury**. It is found in *Triton* and *Haliotis*. Such secondary connections exist in some cases on both sides **dialyneury**.

*Euthyneurous condition*. In some gastropods (many Opisthobranchs and pulmonates) the nervous system becomes secondarily symmetrical from the primary asymmetrical condition. Such a condition (euthyneurous condition) is the result of either detorsion or double torsion.

## Sense Organs

In addition to the sensory cells scattered over the whole surface of the body, gastropods possess special organs of sense called the **rhinophore, osphradium, eye**, and **statocyst**.

**Rhinophore**. The **rhinophore** or **olfactory organs** are constituted by the cephalic tentacles, especially by the posterior pair in the quadri-tentaculate types. The surface of the tentacles is covered by small ciliated papillae giving them a silky appearance and is innervated by fibres of the olfactory nerves.

**Osphradium.** The **osphradium** is unpaired in *Pila* situated dorsoventrally on the roof of the mantle cavity. In the types possessing two ctenidia the osphradia are paired. It is a bipectinate (Fig. 19.31) structure hanging like a

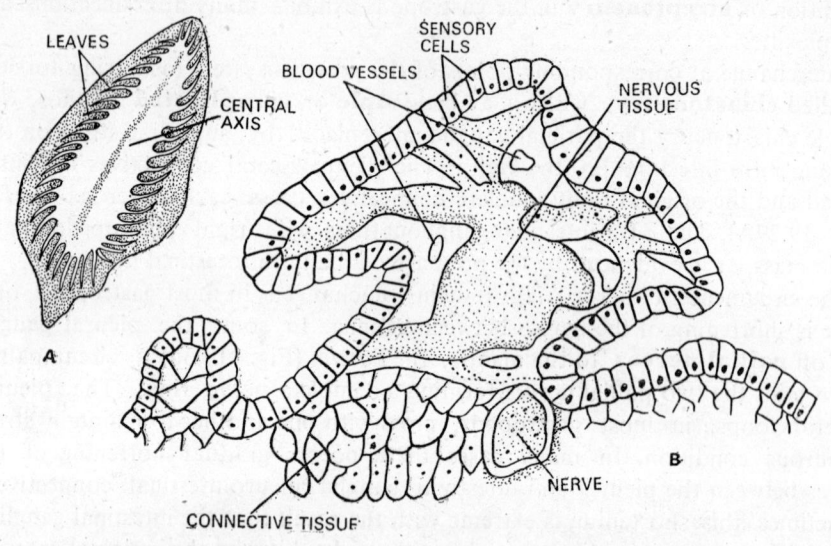

FIG. 19.31. A—Osphradium of *Pila*; and B—Transverse section through the same.

curtain in the course of the respiratory water current. An osphradium consists of a specialized and usually elevated and ciliated region of the epithelium in which there is an accumulation of the sensory cells. The leaves of the osphradium are arranged on both sides of a central axis. Each leaf is just like a double fold of the wall of the central axis. Formerly it was regarded as a rudimentary gill, but it has now been established that the osphradium serves to test the chemical and physical properties of water. Water passes beneath osphradium for respiration. The osphradium helps in the selection of food material (Spengel.)

**Eyes.** Two **eyes** are borne on short **stalks** or **ommatophores**, situated behind the second pair of tentacles. They lie slightly below the tips on their outer side and not on the extremity itself. The outer epithelium of the eye-stalk continues over the eye forming the **cornea** or **pellucida externa**. It consists of a row of low squarish cells devoid of pigments. Lying below the cornea is a thick layer of connective tissue made up of fibre and nuclei. This layer encloses the pyriform **optic vesicles** consisting of the following parts: the **pellucida interna** lying next to the connective tissue layer and forming the inner cornea (Fig. 19.32). The ovoidal structureless **lens** occupies the central space

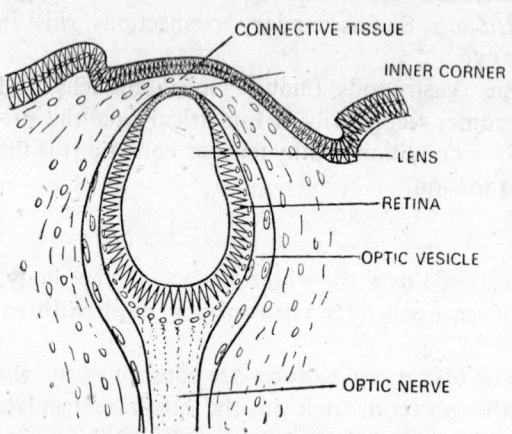

FIG. 19.32. Vertical section of the eye of *Pila*.

*Phylum Mollusca*

of the vesicle. The sensitive part of the eye of the **retina** forms the outer coat of the vesicle made up of large cells filled with pigment granules. The inner cornea is merely the anterior continuation of the retina, but there its cells become smaller and cubical. At the posterior end of the optic vesicle enters the optic **nerve**. A close examination reveals that the retina consists of two kinds of cells: (*i*) large broad **visual cells** having a brush of hair-like processes on their outer ends; and (*ii*) smaller **packing cells** wedged between them. No definite observations are available about the functions of the two kinds of cells (Fig. 19.32). The sense of sight of *Pila* is limited in range and it seems that the apple-snail does not rely on this sense very much.

**Statocysts.** The **statocysts** are two pyriform sacs of creamy white colour lying posterior to the pedal ganglion of each side. The internal walls are lined

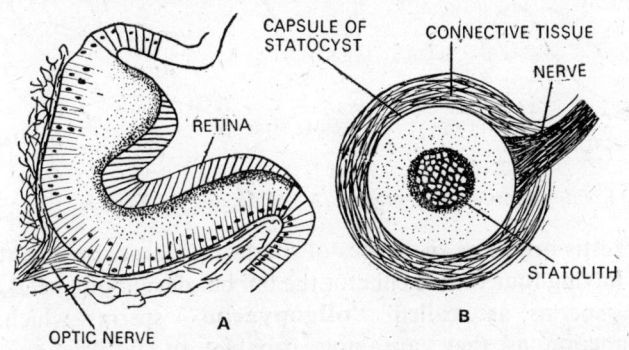

FIG. 19.33. A—Vertical section of eye of *Patella*; B—Statocyst of *Pila*.

by ciliated epithelium containing sense cells. The outer covering of the sac is leather-like and within the cavity lie a number of minute particles the **statoconia**, floating in the fluid contained therein. The statocyst is an organ that enables the animal to maintain equilibrium. The particles (statoconia) help in detecting its position in space. Formerly it was supposed to be an organ of hearing, hence it was called **otocyst**. But this nomenclature is now regarded misleading. In some cases the statocyst possesses only one particle, the **statolith**, in the cavity. In some cases the statoconia coexist with a statolith, (*Onicidium*, *Turitella*). Statocysts are absent in the adult *Vermetus*. In creeping gastropods the statoliths are situated in the foot, near the pedal ganglia, whereas, in swimming gastropods (*Glaneus*, etc.) they show a tendency to approach the cerebral centres.

**Reproductive Organs**

*Pila* is an example of dioecious Gastropoda in which sexual dimorphism is distinct, the shells of the males being generally much smaller than those of the females. There is a well-developed copulatory organ in the male which is rudimentary in the female. In other gastropods sexual differences have been found to occur in the aperture of the shell (*Cerithium*), in the radular teeth (certain Buccinidae) and in such other structures. The gonad is always unpaired even in the most primitive molluscs (Aspidobranchia). It is generally placed on the dorsal side and at the summit of the visceral mass.

In the male *Pila* the **testis** lies in the upper part of the first two or three whorls of the visceral mass closely associated with the digestive gland. It is cream-coloured in fresh specimens and can be easily distinguished from adjoining

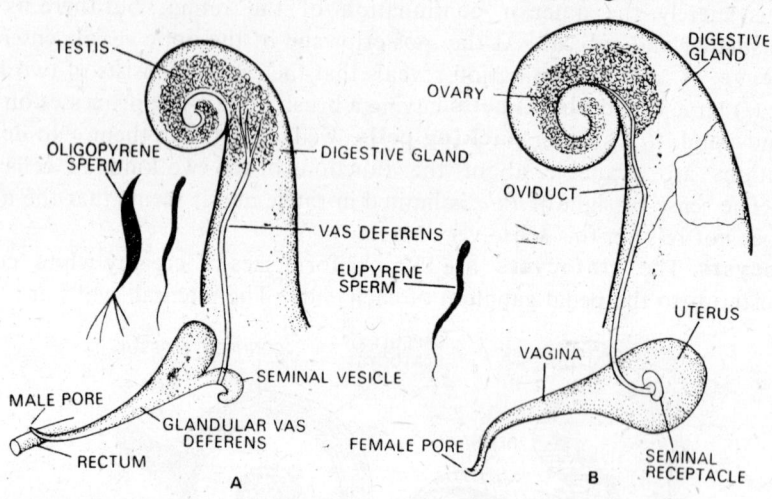

FIG. 19.34. *Pila*. Reproductive organs. A—Male; and B—Female.

tissues. The testis produces two types of sperms: (*i*) large **non-motile sperms** (Fig. 19.34) having four to five cilia for the tail but deviod of a head and a middle piece. Such sperms are called "**oligopyrene**" sperms which have some secondary function as they are not capable of fertilization, and (*ii*) the smaller hair-shaped second type of sperms having a head, a middle-piece and a tail and called "**eupyrene**" sperms which are capable of fertilizing eggs. From the testis arise several delicate ducts, **vasa efferentia**, that unite towards the posterior edge of the testis to form a common duct, the **vas deferens**, which emerges from the posterior end of the testis and runs along the inner edge of the posterior renal chamber. Morphologically the vas deferens shows three regions: the proximal thin tubular portion leading from the testis, the **vesicula seminalis**, and the thick terminal glandular portion (Fig. 19.34). The seminal vesicle is a swollen flask-shaped structure lying to the right of the pericardium immediately below the line of junction of the two renal chambers. The narrow duct-like portion of the seminal vesicle opens on the left into the glandular part of the vas deferens. The terminal part of the vas deferens that enters the mantle cavity is the glandular part, and it runs along the rectum to open at the tip of a claw-shaped protuberance, the **genital papilla** (or miniature second penis). This establishes connection with the true penis during copulation. The **penis** is a long extensile flagella-like structure arising as an outgrowth of the mantle on the right side. It is about one centimetre in length (normally increasing to about four centimetres during copulation) slightly curved, bearing a groove and lying within a sheath called the **penis sheath**, itself a sac-like outgrowth of the mantle wall at the base of the penis, lying a little to the left of the exreme limit of the mantle cavity. The **hypobranchial gland** is not deep-seated, its cells are tall having small basal nuclei and are filled with granular secretion or are empty. The secretions of the gland cells are directly poured out on the surface as the glands are not provided with ducts.

In the female the **ovary** occupies the same position as the testis in the male.

It is relatively smaller, much-branched structure, orange red in colour in younger specimens, and situated along the inner surface of the first 2 or 2½ whorls. A thin but stout coat covers it. As it is situated towards the inner surface the whorls have to be pulled apart to bring it to view. The branches of the ovary consist of single-layered, more or less, flask-shaped acini with their closed rounded ends directed outwards. The oviduct arises from the middle of the ovary and lies just beneath the skin. It is a tube that appears, more or less, transparent in fresh specimens. It runs along the margin of the gastric gland and turns downward about the level of the renal organ and finally enters the **seminal receptacle**, a bean-shaped structure lying within the posterior renal chamber, in close association of the **uterus**, which is a bluntly rounded sac of yellow colour (in fresh condition) lying to the right of the renal chambers. The sac narrows towards one side forming the **vagina** (Fig. 19.34) which is of lighter creamy colour in fresh condition and as such visible under the skin. It enters the mantle cavity at its right posterior corner and runs along the rectum to open by a narrow slit-like female **genital aperture**, situated on a papilla a little behind the anus. Some gastropods (Pulmonata) are hermaphrodite. The reproductive organs in these attain considerable complexity. There are **hermaphrodite glands** which produce both ova and sperms. This is followed by a convoluted **hermaphrodite duct**. Besides, there is an **albumen gland** in which the albumen of the relatively large eggs is formed and sometimes a separate **oviduct** and **sperm duct** leading to a common **genital** opening. Sometimes there is a single duct undivided throughout. With the oviduct a **receptaculum seminis** may be associated, besides a number of narrow oviducal glands. Frequently a **prostrate gland**, an **eversible sac** (sac of the dart) and a penis may also be present.

**Copulation.**[1] With the onset of rains the snails become active and begin breeding at once. Copulation takes place in water or on ground and lasts for two or three hours or more. At its base the penis becomes connected with the genital papilla and both the penis and its sheath are inserted in the mantle cavity of the female. The tip of the penis is inserted into the vagina and the spermatic fluid is transferred into the seminal vesicles. Fertilization takes place within the body of the female while the development occurs outside. The eggs are laid in large masses, 200-800 at a time, in some shallow sheltered crevice and are left to develop without incubation or other help of the mother. Each egg consists of a whitish shell, a double shell membrane, a thick layer of white solid albumen and a core of fluid albumen into which floats the embryo.

**Embryology.** The eggs may develop within the body of the mother or may be laid. In the oviparous species that do not copulate unfertilized ova are generally laid (*Patella*) one by one and are not united by accessory envelope, fertilization in such

FIG. 19.35. *Helix pomatia* in the act of depositing eggs in a specially dug-out crevice (*after* Neisenheimer).

[1]Bahl, K.N., *Mem. Ind. Mus.*, IV, 1928, pp. 3-8.

forms takes place after they are laid. In others the fertilization is internal and eggs are laid in masses, embedded in jelly, each having its own hyaline envelope. Sometimes large numbers of capsules are attached to rocks or sea weeds, etc. In aquatic species the shells surrounding the eggs may be in a single gelatinous mass, which may be ribbon-shaped and in some (Basommatophora and Opisthobranchia) the ribbon may be coiled. Some (Pulmonata) lay eggs in the earth. In the ovoviviparous gastropods the development takes place in the terminal portion of the oviduct (Streptoneura, Opisthobranchs, Pulmonates, etc.)

The **segmentation** of the ovum is always total and soon becomes irregular (except in those forms in which yolk is less). Throughout the class the cleavage pattern in similar (**typical spiral**) in early stages and resembles those of *Nereis* and other Chaetopoda. The first four **blastomeres** are usually equal or nearly so. These give rise to a new series or quartetes of smaller cells (**micromeres**).

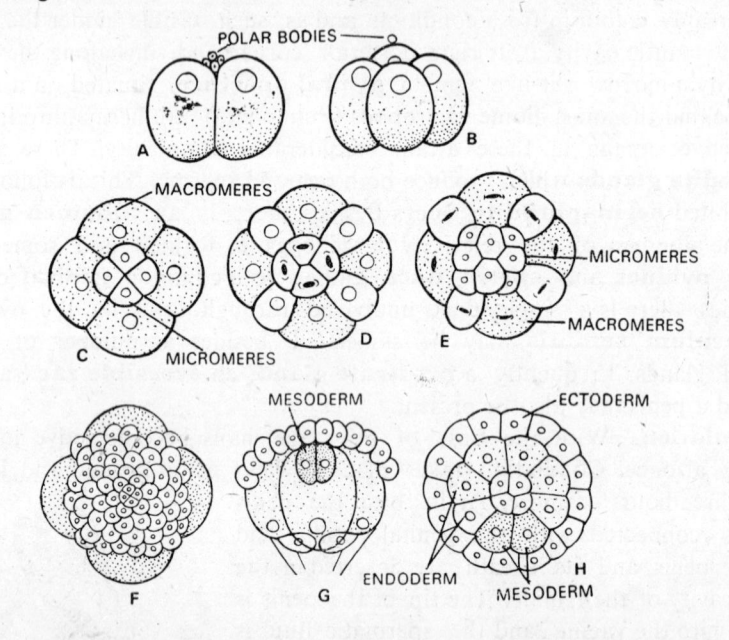

FIG. 19.36. Formation of germinal layers in Gastropoda (*after* Korschelt and Heider). A and B—Seen from side; C to F—Seen from upper pole; G—From lower pole, and H—An optical section.

The larger cells are known as **macromeres**. The micromeres multiply more rapidly than the macromeres and form a cap (Fig. 19.36) of small cells on the surface of the large cells. Of the four quarttetes formed, three give rise to ectodermal organs. Of the cells of the fourth quarttete, three become **endoderm cells** forming ultimately the endodermal lining of the mesenteron and the fourth becomes the parent cell of **mesoderm**. The **blastula** formed in this manner contains a blastocoel in the interior which is large in forms like *Patella, Planorbis, Limax*, but is much reduced in other gastropods. The **animal** or **formative pole** of the blastula is indicated by the presence of the polar bodies, whereas the **vegetative** or **nutritive pole** is opposite, greatly thickened owing to large macromeres forming it, and it is at this point that the blastropore will be formed.

# Phylum Mollusca

**Gastrulation.** In some forms (*Planorbis, Paludina*) the gastrulation is effected by invagination of the endoderm, formed by macromeres, into the ectoderm

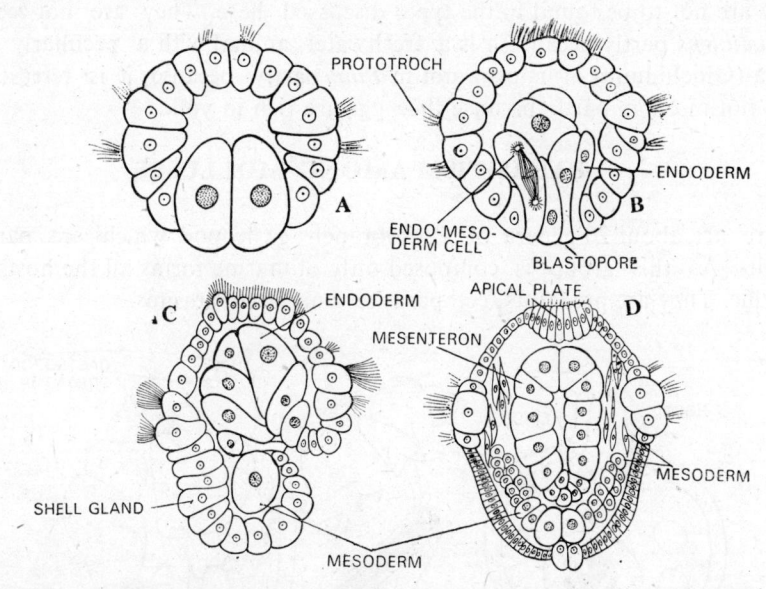

FIG. 19.37. Early development of *Patella*. A—Blastula; B—Gastrulation begins; C—Completed gastrula; D—Frontal section of a later stage.

formed by the micromeres. In others gastrulation is effected by **epiboly**, *i.e.* due to rapid multiplication, the ectoderm encroaches upon the much larger macromeres and the endoderm invaginates at a later stage forming small enteron. In most cases the blastopore closes and a definite mouth is formed by a new invagination at the point of closure (in *Paludina* a portion of the blastopore remains open and becomes the anus). The mesoderm is formed as two primary mesomeres from the fourth macromere, as mentioned above. The mesodermal organs (definitive kidney, heart, etc.) appear at a late period as during development first the provisional larval organs appear.

In molluscs generally two types of larvae occur in their life-cycle. The ciliated **trochophore** is one and the **veliger** another. The ciliated trochophore in its typical state appears in *Patella* and *Trochus* and in these it is formed at a very early period, even before the formation of the mesoderm, and becomes free at once. It bears a ring of cilia in front of the mouth, the prototroch or future velum. Soon the trochophore develops into a yet more efficient locomotor form, the veliger. Its head bears a ciliate area or "velum" often produced into retractile lobes; its body already shows the beginning of foot and mantle; on the dorsal surface lies the little embryonic shell gland. The young gastropods are bilaterally symmetrical up to trochophore stage, but afterwards torsion sets in as a result of which the asymmetry, characteristic of the gastropod, is established. Post-larval metamorphosis occurs in various cases. The

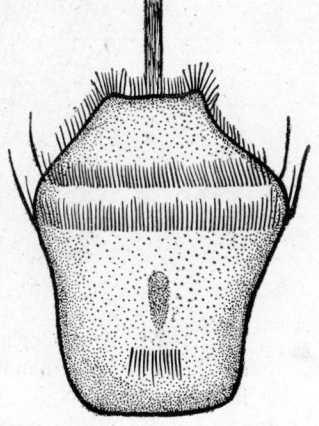

FIG. 19.38. Trochophore of *Patella*.

velum is absorbed by a process of phagocytosis and the adult organs appear.

Although trochophore and veliger occur in the development of most forms they are not to be found in the types discussed here. They are not seen in the *Lamellidens* partly because it is a freshwater animal with a peculiarly adhesive larva (Glochidium) of its own, not in *Pila* partly because it is terrestrial, and also not in *Sepia* partly because thee ggs are rich in yolk.

## PARASITISM AMONG MOLLUSCS

There are about 20 genera of Prosobranch gastropod which are parasitic as adults. As this group is composed only of marine forms all the hosts are also marine. They are predators, ectoparasites and endoparasites.

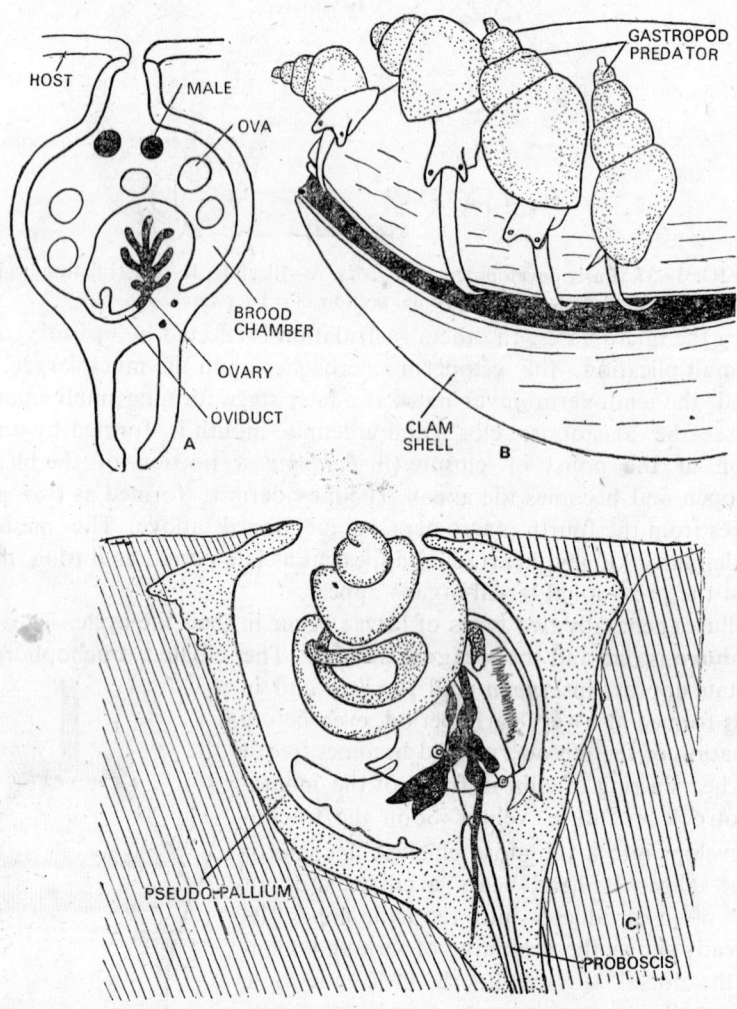

**FIG. 19.39.** A—Endoparasitic mollusc, *Entocolax*; B—*Odostomia scalaris*, attached to a bivalve; and C—Section of a starfish showing a mollusc established inside as a parasite.

The pyramidellids are predators possessing a long armed proboscis by means of which they perforate the tissues of their victims such us sedentary annelids and bivalve molluscs, which are always larger than themselves. For example *Odostomia scalaris* is always associated with *Mytilus edulis,* the edible mussel while *O. unidentata* is a predator of sedentary polychaete *Pomatoceros triqueter.*

Truly parasitic gastropods are associated with echinoderms; mainly holothurians, more rarely with sea-urchin and starfish. Some are ectoparasites which attach themselves to the surface of their host or bury into it to varying extents, others are endoparasites and live inside the body. In some genera like *Entocolax* and *Enteroxenos,* etc., parasitic in holothurians, the body is reduced, all parts lost and only the ovary remains. A brood-chamber is formed by the general body cavity of the parasite. This also encloses a neotenic male which implants itself in the wall where it was previously mistaken for a male gonad. The larva enters the intestine via the mouth of the holothurian. After casting off the shell the larva digs itself into the wall and as it grows it pushes the host peritoneum, before it which comes to invest the parasite very closely. The parasite is nourished by the host and does not need a digestive system of its own.

## CLASSIFICATION OF GASTROPODA

These are molluscs which always exhibit bilateral asymmetry and have a shell consisting of a single piece only (univalve). The shell is generally twisted into a spiral into which the animal can withdraw wholly or partially. The head is distinct bearing tentacles and eyes. The foot is well developed and as a rule flattened ventrally to form a creeping sole. Typically there are two plume-like ctenidia in the mantle cavity, but there may be one, or ctenidia may be wanting being replaced by secondary gills. In air-breathing gastropods the mantle wall is variously modified acting as a respiratory surface. The buccal cavity contains a radula. The larva is often of the trochosphere type. The nervous system consists of a cerebral, visceral and usually pedal ganglia with visceral loop.

(*i*) **Subclass Prosobranchia (Streptoneura).** Dioecious Gastropoda in which the visceral loop is twisted in the form a figure of "8" due to torsion. They are almost always with a shell and an operculum. The ctenidia lie in front of the heart and the mantle cavity opens anteriorly.

1. **Order Archaeogastropoda (Aspidobranchia).** Prosobranch gastropods in which the heart has two auricles and sometimes two aspidobranch (with two rows of gill-plates) ctenidia; secondary gills sometimes replace ctenidia (Patellidae) which may also be absent (Lepetidae). Kidney usually paired of which the right discharges the genital product; and osphradium not developed very much.

The order includes several families including Pleurotomariidae, Haliotidae (e.g. *Haliotis*); Fissurellidae, (Fig. 19.40, 8), Trochidae, Turbinidae, Acmaeidae (*Acmaea*), Patellidae (e.g. *Patella*), Lepetidae, Neritidae (*Nerita*), Helicinidae[2] *Helicina*).

2. **Order Mesogastropoda (Pectinibranchia).** Prosobranch gastropods in which the heart is with single auricle. The ctenidium is also single and is of the pectinibranch type, that is, with one row of leaflets. The kidney is single and gonads are with separate ducts opening far forwards in the male forming a

---

[2]Last two families are sometimes separated in a new order Neritacea.

a penis. The osphradium is well-developed. Radula normally possesses seven teeth in each row. Proboscis and siphon are generally absent. The buccal

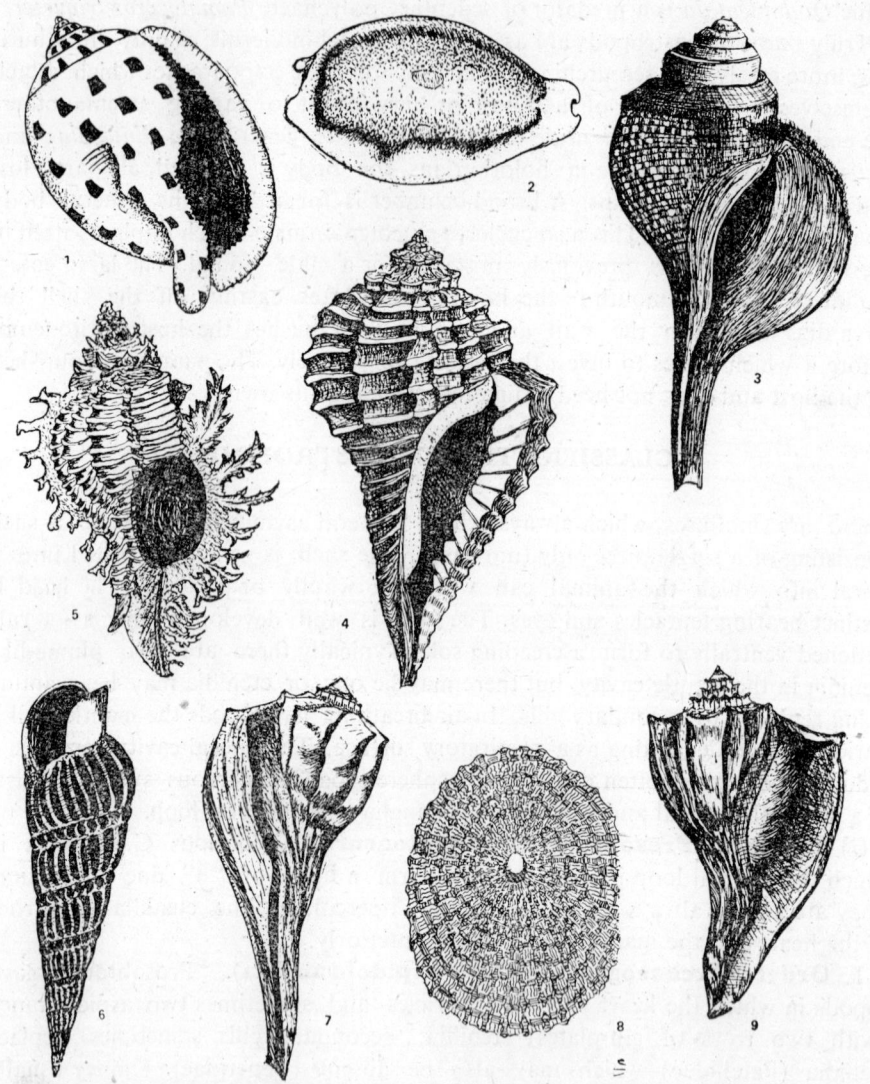

FIG. 19.40. Some gastropod shells. **1**—*Phalium granulatum* (the helmet shell); **2**—*Cypraea spadicea* (the Chestnut Cowire shell); **3**—*Busycon canaliculatum*; **4**—*Eupleura* (thick-lipped drill); **5**—*Murex florifer* (lace murex); **6**—*Terebra dislocata* (the auger shell); **7**—*Busycon contrarium* (the crown conch); **8**—*Diodora listeri* (key hole limpet); and **9**—*Busycon carica* (knobbed whelk).

ganglia are situated behind the buccal mass and united with the cerebral centres by connectives (e.g. *Pila*).

This is the largest order of gastropods including many common species such as *cypraea* (19.40, 2); periwinkles *Littorina* (Littorinidae); the pelagic *Janthina* (Janthinidae); slipper shells *Crepidula, Stombus, Lambis, Cerithium, Nassarius* (Fig. 19.41); the worms shell *Vermetus*; the freshwater snails *Viviparus, Campeloma* and many others.

3. **Order Neogastropoda (Stenoglossa).** The prosobranch gastropods with

Phylum Mollusca

highly concentrated nervous system, eversible proboscis and siphonate shell; osphradium large and bipectinate, usually carnivorous in habit; unpaired

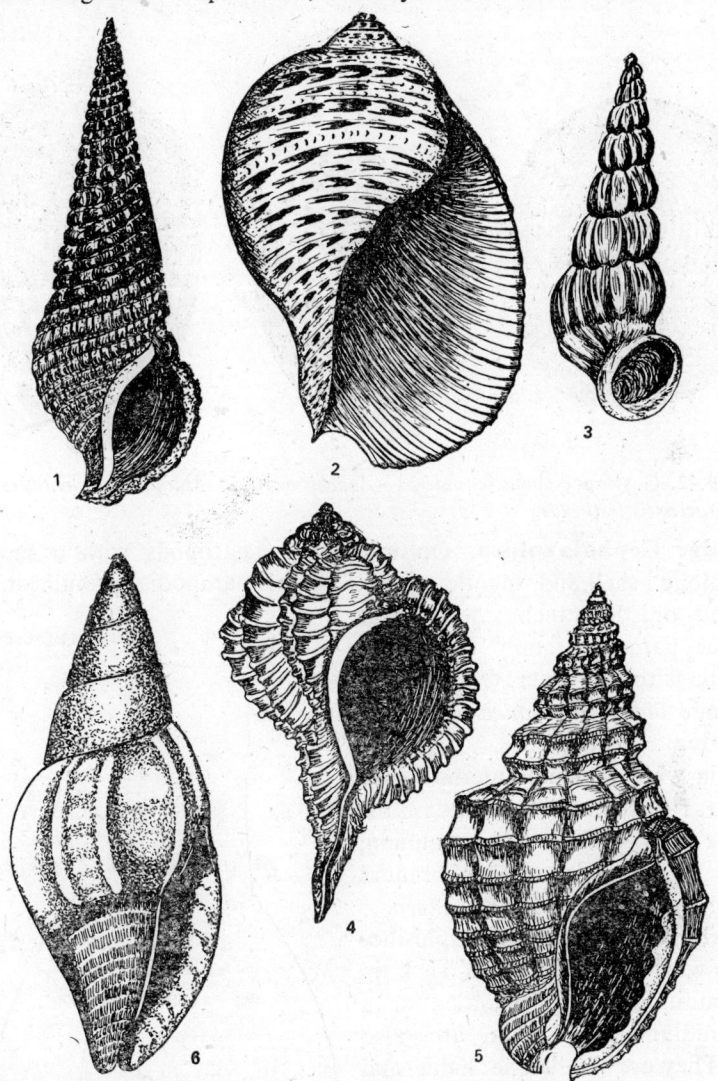

FIG. 19.41. Gastropod shells (contd). 1—*Cerithium floridanum*; 2 - *Tonna maculosa*; 3—*Opalia wroblewskii*; 4—*Forreria belcheri*; 5—*Nassarius actus*; and 6—*Arachis avara*, the greedy dove shell.

oesophageal gland, a pallial siphon and penis is always present. This order includes the boring muricids *Murex*, *Busycon* (Fig. 19.40, 3) *Fasciolaria*; *Buccinum Colus, Mitra*. Some genera with a poison gland associated with the radula are included in this order. Examples are *Conus, Terebra* (Fig. 19.40, 6).

(*ii*) **Subclass Opisthobranchia (Euthyneura)**. Hermaphrodite marine Gastropoda in which the visceral loop is untwisted, as a rule, the ctenidium lies dorsal to the heart; there is reduction or loss of shell. The mantle cavity tends to occupy the posterior position again due to detorsion of the visceral hump. The heart retains its primitive position and the auricle is situated behind the ventricle. The single ctenidium is often replaced by accessory respiratory organs such as secondary branchiae. The shell, if present, is very much reduced or internal.

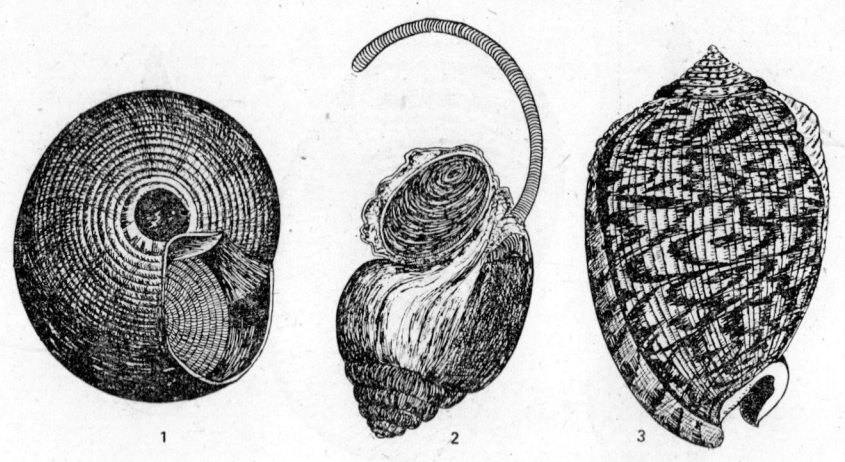

FIG. 19.42. Gastropod shells (contd.). 1 – *Anachisavara*; 2 – *Margarites lirulatus*; and 3 – *Cypraecassis testiculus*.

1. **Order Cephalaspidea.** Opisthobranch gastropods with a moderately well-developed shell and mantle cavity, lateral parapodia prominent, largely burrowing opisthobranch with large head. The parasitic family Pyramidellidae belongs to this order. Other examples include *Bulla, Seaphander, Philine*.

2. **Order Anaspidea.** Opisthobranchs in which a small internal shell is present, reduced mantle cavity on the right side and with a true ctenidium, crawling or swimming opisthobranchs including *Aplysia* (Fig. 19.43), *Akera*.

3. **Order Nudibranchia.** Ophisthobranchia which have neither a shell in adult condition, nor a mantle nor a true ctenidium. Osphradia are also lacking. They are slug-like in habit and have secondary branchiae usually arranged around the anus or in rows on the dorsal surface.

Examples: *Doris, Eolis* (Fig. 19.44).

4. **Order Thecosomata.** Opisthobranchs which are known as shelled pteropods or sea-butterflies (Fig. 19.44) and are planktonic with large parapodia.

Examples: *Limacina, Cavolina*.

5. **Order Gymnosomata.** Opisthobranchs that have no shell or mantle cavity. Naked planktonic species.

Examples: *Pneumoderma, Cliopsis*.

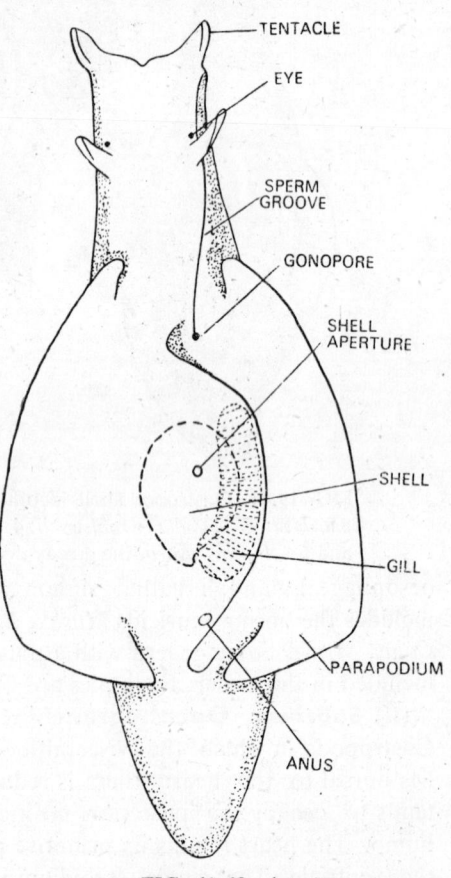

FIG. 19.43. *Aplysia*.

Phylum Mollusca

**6. Order Notaspidea.** Opisthobranchs with internal shell and no mantle cavity, skirt-like projection of mantle covers gill on right side.

Example: *Pleurobranchus*.

**7. Order Acochlidiacea.** Minute opisthobranchs with naked visceral mass or hump. Live in spaces between sand particles.

Example: *Acochlidium*.

**8. Order Sacoglossa.** Shelled slug-like opisthobranchs with modified radula and a pumping pharynx for suctorial feeding.

Examples: *Alisia, Alderia*.

(*iii*) **Subclass Pulmonata**. Hermaphrodite Gastropoda which show torsion and possess a shell without operculum. Ctenidia are absent, the mantle cavity

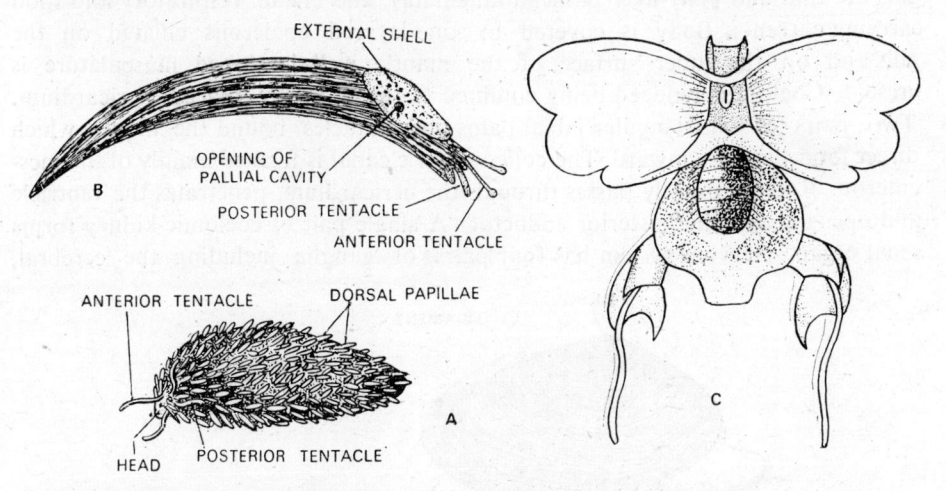

FIG. 19.44. A—*Eolis*; B—*Limax*; and C—Butterfly pteropod *Hyalaea limbata* with foot transformed into swimming wings.

acts as pulmonary sac. The nervous system shows symmetry which is secondarily acquired due to the shortening of the visceral connectives and the concentration of ganglia into the circumoesophageal mass.

**1. Order Basommatophora.** Mostly freshwater Pulmonata in which eyes are located at the base of the tentacles and an external shell is present. A few members are marine but these are all shore forms and breathe air.

Examples: *Limnaea, Planorbis*.

**2. Order Stylommatophora.** Pulmonata in which the eyes are at the tip of the retractile posterior pair of tentacles. Shell well-developed as in the land snails or reduced completely concealed in the mantle as in the slugs.

Examples: *Helix, Polygyra*, and the land slugs such as *Limax* (Fig. 19.44B).

## CLASS PELECYPODA

This class includes the clams, oysters, mussels and their numerous kins. They have neither head nor jaws nor teeth. Structurally speaking they are the humblest of the molluscs. Their class name, **Pelecypoda**, refers to the more or less hatchet-shaped foot common to most, but not all members of the group. The Pelecypoda (Gr: *pelekys*, hatchet+*podos*, foot) are bilaterally symmetrical molluscs with laterally compressed soft body, enclosed in a rigid shell of two parts (hence Bivalvia). The mantle consists of two thin tegumentary lobes attached to the trunk dorsally, and extending over the sides to the ventral surface, so that they can be brought together below the foot. The edges of the mantle lobes are generally free but in many Pelecypoda the lobes are partially united, at one or two regions. The ventral region of the body is differentiated into muscular foot of variable size and form, depending upon the habit of life adopted by the animal. The mass of the foot is frequently invaded by a part of the viscera. In some forms foot becomes rudimentary. Posterior to the foot there occurs a byssus gland secreting, in some cases, horny fibres, the **byssus**, by which the animals may be permanently attached (*Pecten, Ostrea*). The two gills are thin and plate-like (Lamellibranchiata) and create respiratory and food carrying currents. Body is covered by one-layered epidermis ciliated on the gills and on the inner surface of the mantle well-developed musculature is present. Coelom is reduced being confined to the dorsally placed pericardium. Two pairs of flat triangular labial palps and tentacles bound the mouth which direct food particles inward. The coiled enteric canal is formed mainly of the mesenteron. Rectum usually passes through the pericardium, penetrates the ventricle and opens above the posterior adductor. A single pair of coelomic kidney forms renal organ. Nervous system has four pairs of ganglia including the cerebral,

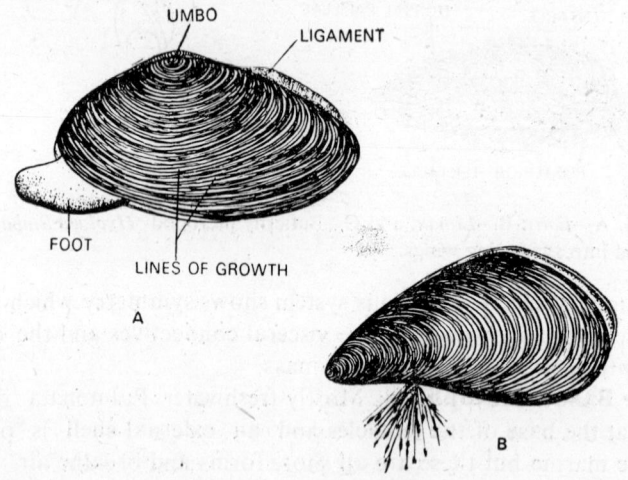

FIG. 19.45. A—Freshwater mussel, entire animal; and B—*Mytilus*.

pleural, pedal and visceral. Eyes and osphradia are chief organs of special sense. The sexes are separate. Development is accompained by metamorphosis. What all the members do have in common is a bivalve shell—the fortress and the skeleton of its possessor—and the animals that live in bivalve shells are built more or less upon a common plan. This applies equally to the tiny *Pisidium*, which may be no larger than a pinhead and which is so prolific that its progeny fairly line the beaches of some of our lakes and to the huge *Tridacna* of the Western Pacific, whose over one metre long shell is often used as a baptismal font.

The class is divided into three subclasses: (*i*) Protobranchia, (*ii*) Lamellibranchia, (*iii*) Septibranchia, and several orders.

# TYPE FRESHWATER MUSSEL

The freshwater mussels are found in streams, lakes and ponds all over the world. There are many genera and hundreds of species of these molluscs of which several are found in India. The large lamellibranch that is available here is found abundantly in the eastern districts of Uttar Pradesh and has now been named *Lamellidens marginalis*. These mussels live partially buried in mud or sand of the bottom and crawl about with the help of the protrusible muscular foot that protrudes between the two valves of the shell, *i.e.*, through the **gape**. From each side of the shell emerge two short tubes, the **siphons**, of which one has smooth edges and is the **exhalant siphon** and other has fimbriated margin and is called the **inhalant siphon**. On slight disturbance the foot and the siphons are withdrawn and the gape is closed. In size it varies from 5-10 cm in length.

## Shell

The shell of the lamellibranch consists of two **valves**, which are united along a straight **hinge-line** by a tough elastic substance that forms the **hinge-ligament**. The ligament passes transversely from one valve to another. Its elasticity is responsible to open and close the shell. Along the hing-line each valve has teeth-like projections that enable the shells to fit with each other. These are the **hinge-teeth**. The hinge is dorsal and the gape is ventral. In *Anodonta* the connection between the two valves is afforded only by a ligament, but in *Unio* each is provided with strong projections and ridges, the hinge-teeth, separated by grooves and sockets, and so arranged that the teeth of one valve fit into the sockets of the other. The same is the case with *Lamellidens*.

The valves are externally marked by a series of concentric lines, **lines of growth**. They lie parallel to the free edges of the shells. Alternate periods of

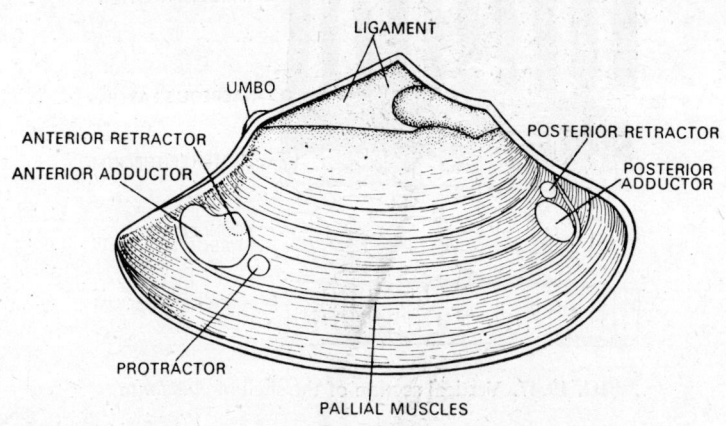

FIG. 19.46. Inner side of the right valve of the shell showing the scars and impressions.

slow and rapid growth can be marked out by their size. The **umbo** (plural **umbones**) represents the oldest portion of the valve. At this place the shell is the thickest This is the first part of the shell to be secreted along the edges of which subsequent layers are added. The inner surface of the shell also shows similar markings. A close examination reveals that the shell consists of three layers: (a) The **periostracum** is the outermost layer formed of a chitin-like chemical compound **conchiolin**. It is dark blue when wet and brown when dry. It is a thin and translucent layer partly responsible for some variations in the colour of the shell; (b) Beneath the periostracum lies the middle layer called the **prismatic layer**. It is composed of numerous calcareous prisms embedded in conchiolin. (c) The last layer of the shell (Fig. 19.47) is the **nacreous layer** or **nacre**. It is also called "**mother of pearl**" and is composed of alternate layers of calcium carbonate and conchiolin. The first two layers are secreted only by the edges of the mantle and has smooth lustrous surface. This layer may also secrete a pearl as a protection against foreign body as a parasite. When a parasite, say a larval stage of a fluke, enters the mantle it becomes enclosed in a sac formed by the growth of the mantle epithelium which later on, secretes thin concentric layers of pearly substance around the body (Fig. 19.48). Except for the concentric lines of growth the shell is smooth and there are no ridges or tubercles which are frequently found in the shells of other bivalve molluscs. The inner surface of the shell presents characteristic markings and structures: (i) the **hinge-teeth** that interlock the valves tightly are clearer on the inner side; (ii) the **scars** of the **adductor muscles** are other prominent structures. The scar of the **posterior adductor** muscles is larger and posterior. Posterior to the scar of the **anterior adductor** is the scar of the **anterior retractor muscle** of the foot. Ventral to this is the scar of the **protractor muscle**. Parallel to the margin between the adductor scars marking the attachment of the retractor muscles of the mantle is a fine line, the **pallial line** indicating the attachment of the mantle.

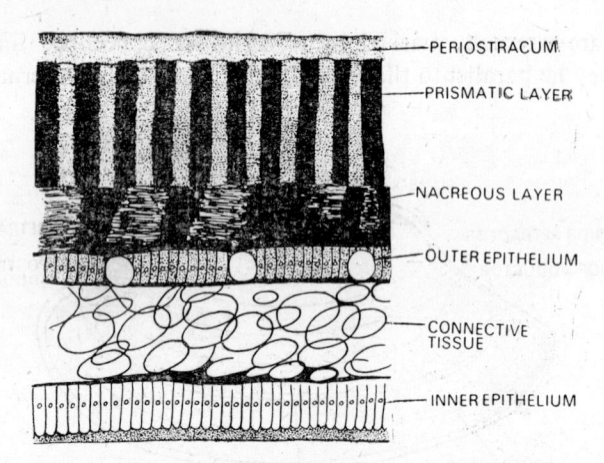

**FIG. 19.47.** Vertical section of the shell of *Anodonta*.

The mantle consists of two thin lobes attached to each other and the visceral mass dorsally. The mantle secretes two valves of the shell that lie over two lobes. Normally the mantle is creamy white in colour but preservation spoils it. Certain parts of the mantle edges are so thickened and rolled as to form a **dorsal exhalant aperture** and a **ventral inhalant aperture**. The space enclosed

*Phylum Mollusca* 821

between the two lobes of the mantle is known as the **mantle** or **pallial** cavity. Dorsally the pallial cavity is bound by the visceral mass.

Primitively the bivalve mantle consists of two large lateral lobes which, with the bivalve shell, completely enclose the body. The two mantle lobes are fused dorsally (below the hinge ligament) but their margins are free. The degree of fusion of the mantle margins of opposite lobes varies both in extent and area of fusion. The most common area of mantle fusion is below the posterior enlargement of the mantle to form a distinct and permanent exhalant aperture. Differentiation of the mantle edges below the fusion area and the apposition of the lobes below this region results in the formation of a functional, but not complete, development of inhalant siphon. Such a development occurs in superficially burrowing bivalve such as *Unio* and *Anodonta* (Fig. 19.49 B). Fusion below the inhalant region may form a complete inhalant siphon. With further complete fusion separate complete tubes are formed (Fig. 19.49 E). Further

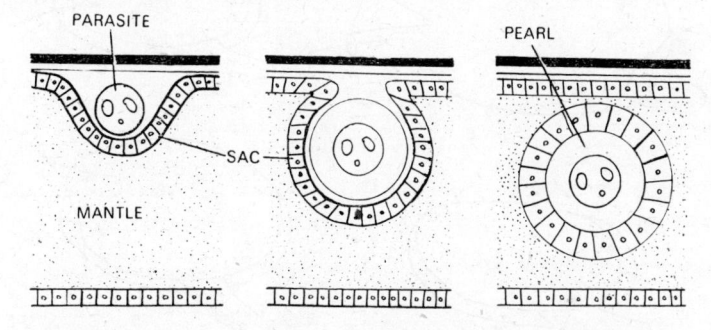

**FIG. 19.48.** Formation of a Pearl. **A**—A parasite (1) settles down between the shell and mantle; **B**—It is completely encircled in a sac, which secretes concentric layers of pearly substance; and **C**—A pearl of good cyst surrounding the parasite is formed. This prevents the parasite from harming the host and is protective device.

fusion of the mantle lob (Fig. 19.49) restricts the ventral opening to an anterior pedal aperture (Fig. 19.49 F). In some cases a byssal aperture is also set free. Some of the freshwater bivalves e.g., *Lamellidens* have an additional dorsal pallial or supra-anal aperture opening above and anterior to the exhalant siphon. Its function is not clear.

**Musculature**

The musculature of a clam consists of three main systems described below:

(*i*) The muscles of the mantle or the **pallial muscles** are those that connect the mantle lobe to the shell valve. In the bivalves with siphons, the **siphon retractor muscles** are developed from the neighbouring pallial muscles.

(*ii*) The muscles of the shell include the **adductor muscles** crossing transversely between the shell valves. Of these there are two: the **anterior** and **posterior adductor muscles,** which are large, cylindrical, transverse bands of muscles controlling the movement of the two valves. When there are two equal muscles the condition is called **dimyarian (isomyarian)**. Where the two muscles are of unequal size, the condition is termed **heteromyarian**. Many clams possess only one posterior adductor. This condition is known

**monomyarian**. Then there are the **anterior** and **posterior retractors** running from the foot to the shell. They are used in withdrawing the foot. A smaller band of similar muscles, the **protractor**, lies close to the anterior adductor. It serves to compress the visceral mass.

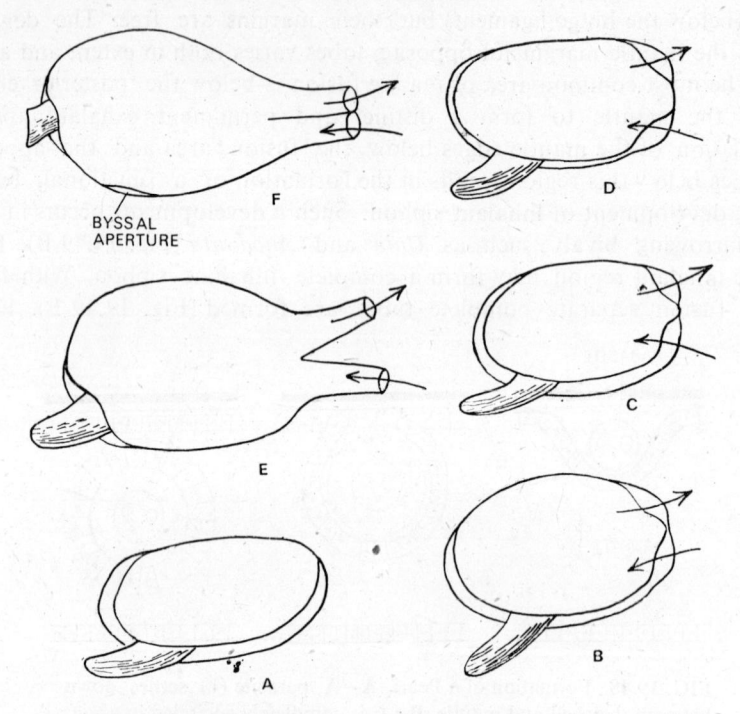

FIG. 19.49. Diagram illustrating various degrees of mantle lobe fusion. A—No mantle lobe fusion; B—Fusion between arrows at (1) to form an exhalant siphon and an incomplete inhalant siphon; C—Fusion below the inhalant region to form a complete inhalant siphon; D—Same as C; E—Marked development of siphons; and F—as in E, but with a byssal aperture.

(*iii*) The **intrinsic muscles** of the foot and visceral mass run in sheets below the foot epithelium, in various directions, and many run transversely across the foot itself. Their main functions seem to be to provide local tensions, which in conjunction with blood hydrostatic pressure, result in the turgidity and allow for fine control of foot movement.

The **visceral mass** comprises the foot, a large muscular organ, occupying approximately the middle of the mantle cavity along the anteroposterior line. It is composed entirely of a complex mass of intrinsic muscles, other organs of the visceral mass are four plate-like gills, two on each side, a pair of leaf-like palps on each side of the foot attached to the mantle near the anterior adductor muscles, a transverse mouth opening between the foot and the anterior adductor, and the anus which is situated on the posterior face of the posterior adductor muscle.

In a freshly killed animal some internal organs are also visible from outside. These include the **digestive gland**, which is greenish brown structure in the dorsal region near the anterior end, the **pericardial cavity,** the only remnant

*Phylum Mollusca*

of the true coelom containing the heart lying posterior to the digestive gland and **nephridium**, a dark coloured structure lying ventral and lateral to the pericardial cavity. The **coelom** is reduced to a single ovoidal chamber, the **pericardium**, lying in the dorsal region of the body and containing the heart and part of the intestine; it is lined by coelomic epithelium. In the remainder of the body the space between the ectoderm and the viscera is filled by the muscles and connective tissue.

**Foot**

The **foot** is a muscular projection from the ventral surface, its size and form are very variable depending upon the life of the animal. The mass of the foot is invaded by a portion of the viscera, at least the digestive canal, the liver and in some cases (superficially) the gonads. It is cylindrical in primitive forms.

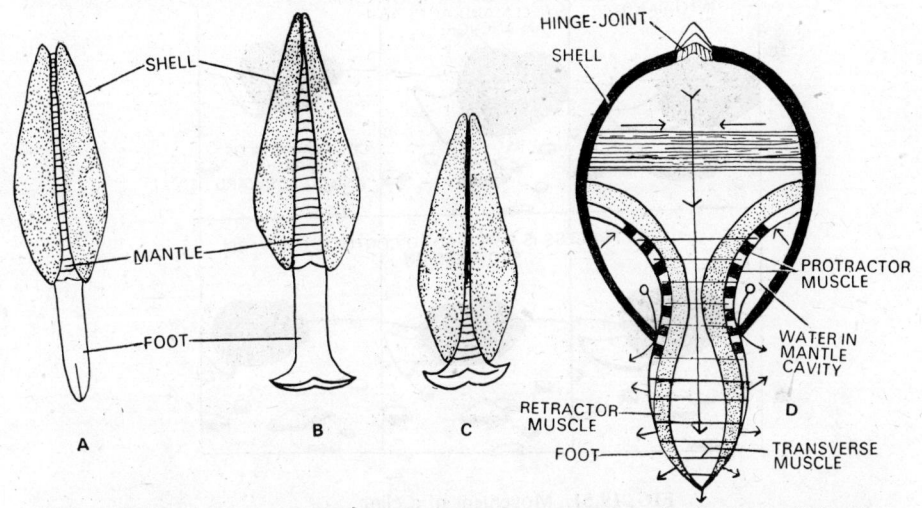

**FIG. 19.50.** Operation of foot of *Yoldia limatula*, a protobranch. **A**—End of the foot folded and extended into substratum; **B**—Foot opened and ancord; **C**—Body drawn into substratum; and **D**—Diagrammatic cross section of a bivalve showing hydrostatic forces that produce dilation of foot, central vertical arrow indicates flow of blood into foot.

Usually it is flattened from side to side and terminates below in a more or less elongated keel, which ends in a single point which is always anterior. In some (*Poromya*) the anterior portion is much elongated giving a tentacle-like appearance to the foot. In *Mytilus* a distinct posterior carinated projection may be seen behind the extensile pedal cylinder: this has been called the "Punch's hump." In some the foot is rudimentary as in the boring Pelecypoda (*Pholas, Teredo*) and in those that are attached by a byssus or by a shell (*Pecten, Ostrea*). The foot is an organ of locomotion. Figure 19.51 shows how a clam moves with the help of a foot.

The foot in primitive **Protobranchia** retains a small flattened sole. The two sides of the sole can be folded together producing a blade-like edge. The foot is thrust into mud or sand in this condition. The sole then opens and serves as an anchor. As the foot contracts the body is pulled downwards. A rapid mud-burrower such as *Solemya* can disappear under the mud surface with two thrusts of the foot. The sole has disappeared in other bivalves but the foot functions in burrowing essentially in the same way. The distal end of the foot dilates after being thrust in the substratum and the body is then pulled in following the foot.

The movement of the foot is brought about by the muscle action in conjunction with the hydrostatic properties of the blood system. Engorgement of the foot with blood plus the action of a pair of pedal protractor muscles produces extension. The protractors extend from each side of the foot to the opposite valve where they are attached to the shell as already noted.

It has been pointed out that the actual mechanism of burrowing involves a coordinated action of a number of forces. The protraction of the foot is initiated by contraction of the pair of pedal protractor muscles. The projecting

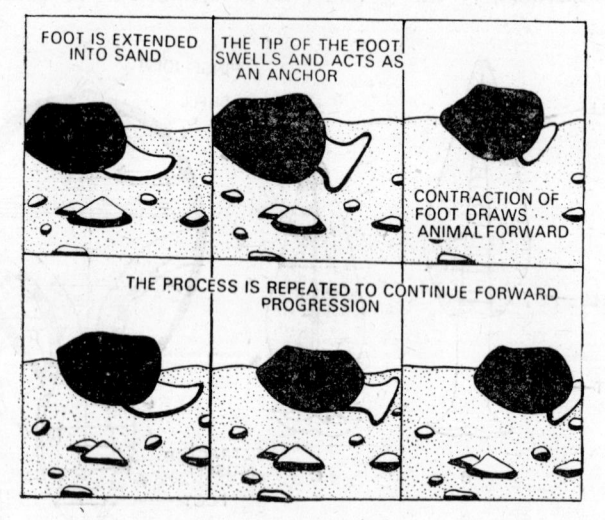

FIG. 19.51. Movement of a clam.

foot pushes into the surrounding sand. By now the valves begin closing by contraction of the adductor muscles. Some water from the mantle cavity is

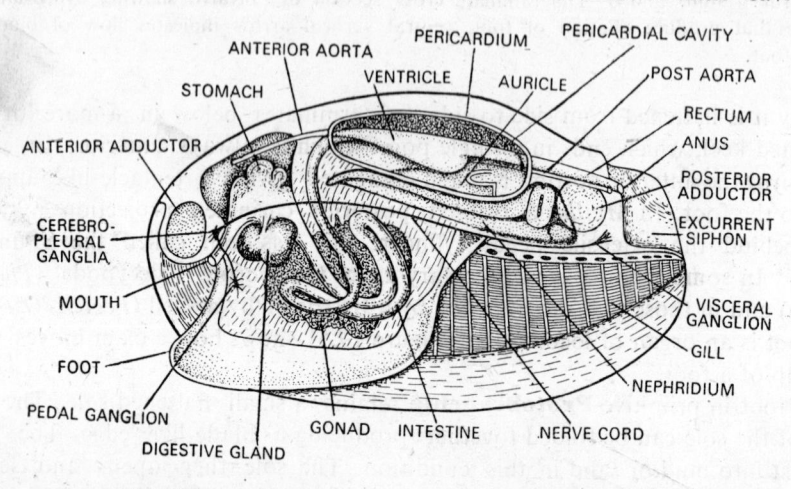

FIG. 19.52. Internal anatomy of the freshwater mussel.

*Phylum Mollusca*

expelled, which softens the sand or mud and facilitates the movement of the foot. The water remaining in the mantle cavity and the blood act as a hydrostatic skeleton. The pressure of these two fluids (water and blood) is increased by the closure of valves. Blood from the visceral mass is forced down into the pedal haemocoel, causing the foot to dilate and anchor into the substratum. Now the pedal retractors are contracted with the result that body is pulled downwards. Following pedal retraction relaxation occurs and valves gape.

This account applies to bivalves living in soft sand and mud bottoms, but some species have invaded harder substrate as wood or stone. In such forms adaptations for boring have developed. The shell becomes elongated, the anterior margins of valves have abrasive serrations for drilling, the hinge ligament is reduced and the hinge-teeth are either absent or poorly developed and byssus threads or a sucker-like foot usually provides for anchorage. Drilling is accomplished by rubbing the head of the bore with the serrated margins of the shell, but the mechanism varies in different groups.

There are many sessile bivalves which remain attached to the substratum by means of strong horny threads called byssal threads. *Mytilus* (Fig. 19.45B) is an example. In the formation of byssal thread, the foot is pressed against the stone or piling to which the bivalve is being attached. The glandular secretion (of a byssal gland) then flows down a groove along the back of the foot. The secretion hardens on exposure to seawater, the foot is withdrawn and a completely formed thread remains behind. Another thread is secreted in the same way in a new location thus producing a mass of threads attaching the clam to the substratum.

## Respiratory Organs

The ctenidium that performs only a respiratory function in gastropods is so expanded in the bivalves that it lines almost the entire surface of the mantle

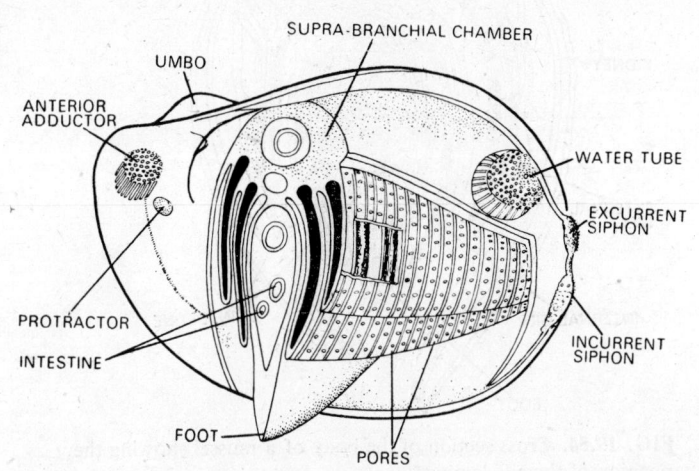

FIG. 19.53. A schematic section of the mussel showing the respiratory organs. Water tubes have been exposed, whereas, the pores are highly magnified.

cavity and does an additional function of feeding. The ctenidia or ciliated molluscan gills are so organized as to enable them to filter suspended and deposited material from a current of water, which had only respiratory role earlier. There are two **ctenidia** or **gills** hanging in the mantle cavity (Fig. 19.53), one on each side between the mantle and the visceral mass. Each **ctenidium** (Fig. 19.54) consists of a long, hollow axis, the **suprabranchial chamber**, which lies along the anteroposterior axis of the body. From the axis descend two **filaments** or **laminae** at right angles to it. Each filament is a hollow prolongation of the axis. It is made up of an outer **plate** or **lamella** and **inner lamella**, both united at the ventral tip. The cavity within these is divided by a series of vertical partitions, the **inter-lamellar junctions**, which separate many cavities, the **water-tubes**, of the gill. The entire outer surface of the laminae is covered by close set gill-filaments which impart striated appearance to them. At their bases the adjacent filaments are connected by horizontal **inter-filamentar** junctions imparting longitudinally striated appearance to it. The gills have a sieve-like structure being perforated by minute pores, the **ostia**, between the filaments and are covered with cilia. They connect the mantle cavity with the water-tubes. In filibranchs and eulamellibranchs the gills are folded forming ascending and descending limbs (Fig. 19.54). In the filibranchs (e.g. *Mytilus*) adjacent filaments are joined by ciliary junctions; in the eulamellibranchs (e.g. *Anodonta, Unio, Lamellidens*) they are joined to one another by vascular interfilamental junctions. In the primitive protobranchs the filaments are unfolded.

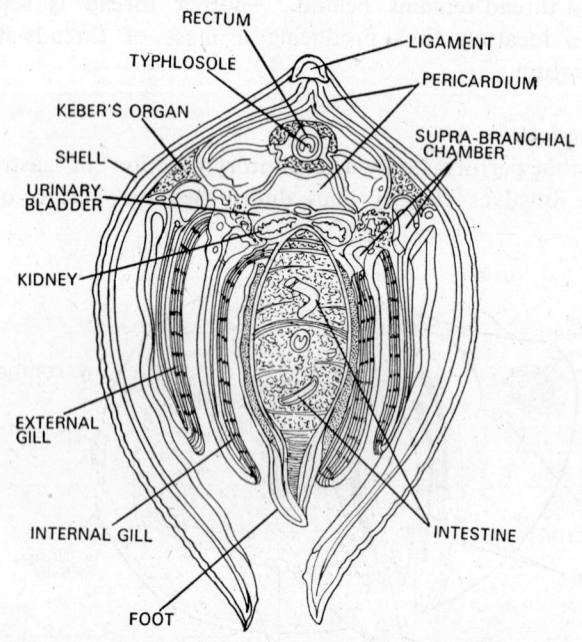

FIG. 19.54. Cross section of the body of a mussel showing the position of the gill laminae.

In Lamellibranchs the lateral cilia draw water into the mantle chamber and from there into the interlamellar and suprabranchial cavities. As the water passes between the filaments the laterofrontal cilia catch the food particle and throw them on to the frontal cilia, which then sweep them, entangled in mucous, over

# Phylum Mollusca

the surface of the gill lamellae. From here they pass either into a ventral marginal groove or into a dorsal groove along the axis of the gill at the base of the demibranchs. From here the food material is carried to the lateral palps. The sorting of particles partly occurs in the gills and partly on the palps.

**Course of water.** The water enters through the incurrent siphon into the mantle cavity. From here it passes into the water-tubes through the microscopic ostia. In the water-tube the water flows upwards into the suprabranchial chamber. From here it is directed to the posterior side and through the excurrent chamber and excurrent siphon it passes out. The water takes oxygen along with it and also food material which is strained as the water enters the minute ostia. The rate at which water passes through the mantle cavity has been recorded in the case of some oysters and marine clams. For an animal of average size the minimum rate averages about 2-5 litres (almost 3 quarts) an hour.

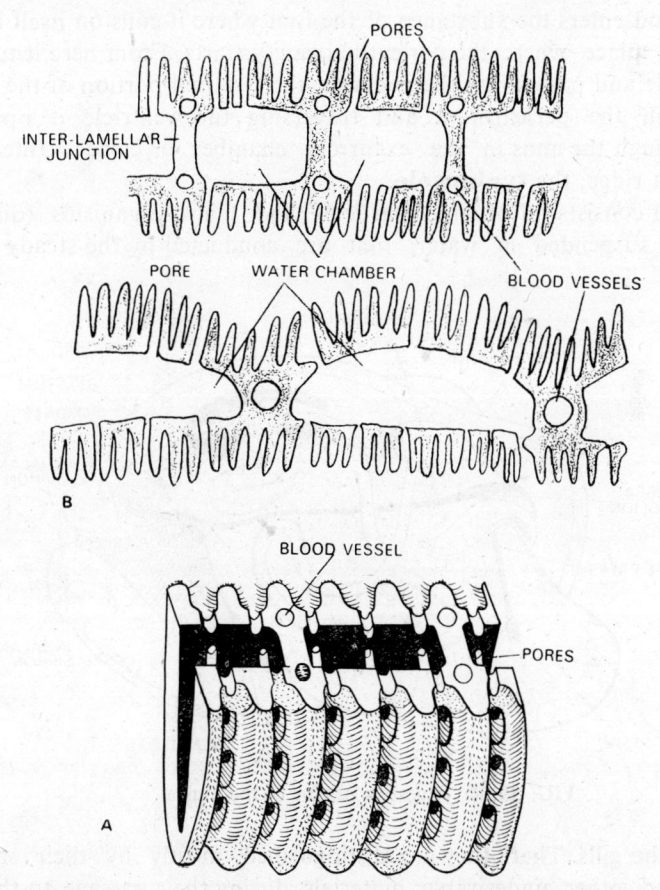

FIG. 19.55. A—Highly magnified portion of a gill; B—Transverse section of outer (upper), and of inner gill lamina.

## Digestive Organs

The **mouth** lies just below the anterior adductor. There are two triangular flaps on each side of the mouth. These are the **internal** and **external labial palps**, which unite to form **upper** and **lower lips**. The upper lip is formed by the union of the external palps in front of the mouth, whereas, the internal palps form the lower lip behind the mouth. Both the lips bear **cilia** on their outer surface. The mouth opens into a short, flattened tube, the **oesophagus**, (Fig. 19.56) which leads dorsally to the **stomach**, a sac-like enlargement of more or less oval shape embedded in the **digestive gland**. Into the stomach open wide ducts from the digestive gland or "liver." The lining of the stomach is thrown into folds. The stomach of well-nourished specimens contains a gelatinous rod, the **crystalline style**, which is supposed to help digestion. The style can be found in very fresh specimens only. In a starved individual it becomes reduced. The digestive glands are a pair of greenish brown structures of irregular shape surrounding the stomach. In mature specimens it closely intermingles with the **gonads** at its edges. It is made up of numerous blind diverticula. The gland produces fat and stores it. Some people associate it with excretion. It also produces digestive ferments. From the posterior end of the stomach arises the **intestine** and enters the substance of the foot where it coils on itself and again ascends the place where the pericardial cavity starts. From here it turns to the posterior side and passes into the **rectum**, the terminal portion of the gut which runs through the pericardium and traversing the ventricle it opens to the exterior through the anus in the excurrent chamber or cloaca. Internally the rectum has a ridge, the **typhlosole**.

Their food consists of organic particles and micro-organisms (diatom and protozoans) suspended in water, that are conducted by the steady stream of

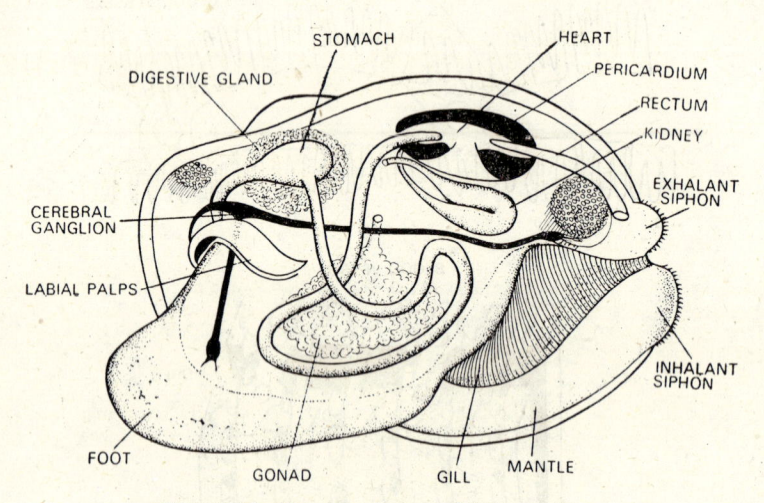

**FIG. 19.56.** Digestive organs of the mussel.

water up to the gills. There they are distinguished, mostly by their small size, from silt and other undesirable materials during their passage to the mouth. Heavy particles drop on the mantle and are carried backward by cilia of the mantle and are expelled posteriorly. The lighter particles become entangled in mucous secreted by the gills and are carried always by beating cilia, to the ventral edge of the gill and then forward until they meet the ciliary tracts on the palps. Here further sorting takes place. The larger particles are carried to the tips of the palps wherefrom they drop off into the mantle cavity from which they are removed. Digestible material is carried to the deep groove between the two palps. This groove leads directly into the mouth between the two lips or ridges that connect the palps of one side to the palps of the other. There is no radula in this case because it could be of no help to an animal feeding only on microscopic particles.

Nothing is known about the physiology of digestion. Whatever is known is based upon the study of marine clams. The digestion is mostly intracellular. Food from the stomach enters the digestive gland, the cells of which readily ingest and break down solid particles. Protein and fat digestion is exclusively intracellular, and the cells of the gland also absorb carbohydrates. The only extracellular enzyme is the carbohydrate digesting enzyme released in the stomach by the dissolution of the crystalline style. The working of the crystalline style is quite interesting. The style-pouch is lined with cilia, the beating of which rotates the style and moves it forward with the result that its free end constantly rubs against a particular portion of the stomach wall and its head is worn away getting mixed with the stomach contents.

### Excretory System

The excretory organs are **kidneys (urocoels)** lying beneath the pericardium, one on each side of the body, and the **Keber's organ** Each kidney (Fig. 19.56) is more or less U-shaped tube of which the lower arm is spongy and glandular and is known as the **glandular part** and opens into the pericardium or coelom. The upper arm is thin-walled and is known as the **urinary bladder**. The two bladders communicate with each other by an oral aperture. Each bladder communicates with the exterior by a minute aperture which is situated between inner lamina of the gill and the visceral mass. The kidney has the normal relations of a coelomoduct and is often called the **organ of Bojanus**, after its discoverer. The **pericardial gland** of **Keber's organ** is glandular mass of reddish brown colour anterior and lateral to the pericardial cavity. It discharges excretory material into the pericardial cavity. The kidneys extract waste products of metabolism from the blood and from the pericardial gland. The wall of the bladder is ciliated and maintains an outgoing current.

### Circulatory System

The vascular system is of open type. Many of the blood vessels are wide and capable of a great deal of variation in degrees of construction and distribution. **Venous sinuses** having connective tissue walls, of considerable dimensions, exist. **Lacunar tissue** is abundant. The blood consists of **leucocytes** floating in the colourless **plasma**. The leucocytes are of three types but they are simply developing stages of the same. The **heart** consists of a median muscular chamber, the **ventricle** wrapped around the rectum and two (right and left) thin-walled lateral **auricles**. The **auriculoventricular opening** has valves opening into the ventricle.

**Arteries.** From the anterior end of the ventricle arises the **anterior aorta**, runs anteriorly in the middle line along the anterior dorsal surface beneath the hinge ligament. After covering a little distance it bifurcates forming the **right and left terminal arteries**. During the course each terminal artery gives off several branches supplying the mantle and the digestive glands, etc., and finally terminates in the anterior pallial artery. Before this division the aorta gives rise to the following paired arteries: the **posterior pallial artery** supplying the

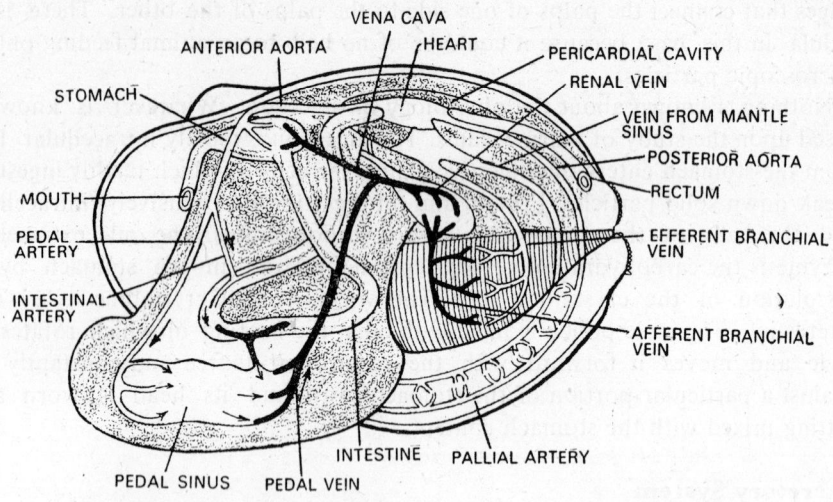

FIG. 19.57. Circulatory system of a bivalve mollusc (diagrammatic).

mantle; the **hepatic arteries** supplying the digestive gland; the **intermediate pallial artery** supplying the mantle in the intermediate region, the **anterior**

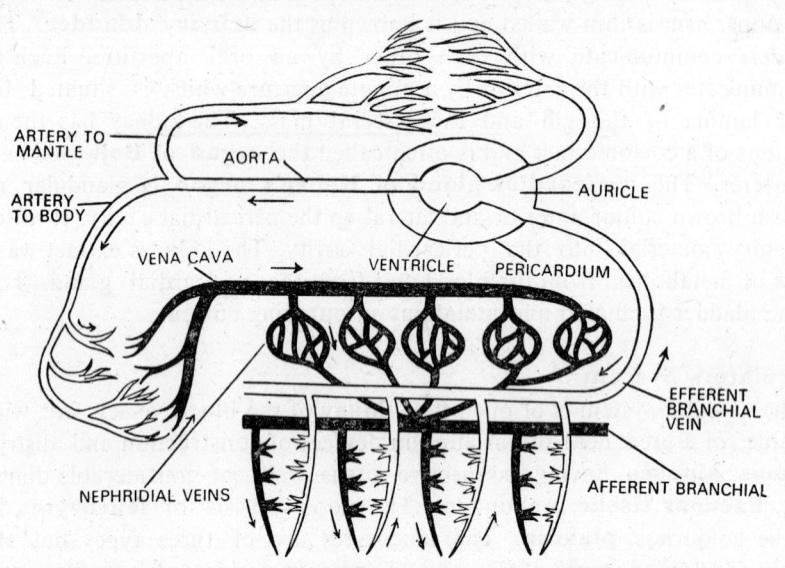

FIG. 19.58. The circulatory system of the mussel (diagrammatic).

**ventral artery**, an unpaired vessel arising from the aorta just before its bifurcation, its branches supply tentacles, the foot and the digestive glands. Below the rectum towards the posterior side runs the **posterior aorta** and gives off branches to the pericardium, ventral adductors, nephridia and other posterior organs.

**Veins.** The blood from various parts of the body is collected by many **sinuses** and passed into smaller veins, whence it is finally received by the **vena**

# Phylum Mollusca

**cava**, a long stout vein lying in between the kidneys. It pours all the blood into the kidneys through small **renal vein**. The **afferent branchial veins** collect blood from the kidneys after the elimination of nitrogenous waste, and conduct to the gills for aeration. The **efferent branchial veins** collect blood from the gills after aeration and carry to the **auricles** which also receive blood from the vena cava directly through a vein that arises before it enters the kidney. The **marginal sinus** is a fairly wide vessel which runs along the whole of the free margin of the mantle. Behind the posterior adductor muscle it communicates with the **horizontal veins**, a paired vessel lying in the mantle parallel with but below the gill axis. Numerous other blood vessels from the mantle pour blood into it. Blood from this blood vessel is returned to the vena cava through the **plicate vein**. The blood from the mantle is returned to the heart direct without going through the gills or even through the kidneys. It is so because the mantle also serves as an organ of respiration and as such aeration of the blood takes place while it is in the substance of the mantle.

## Nervous System

This includes the **cerebral**, the **pedal** and the **visceral ganglia** and their connectives. The **cerebral ganglia** are triangular structures of pale yellow

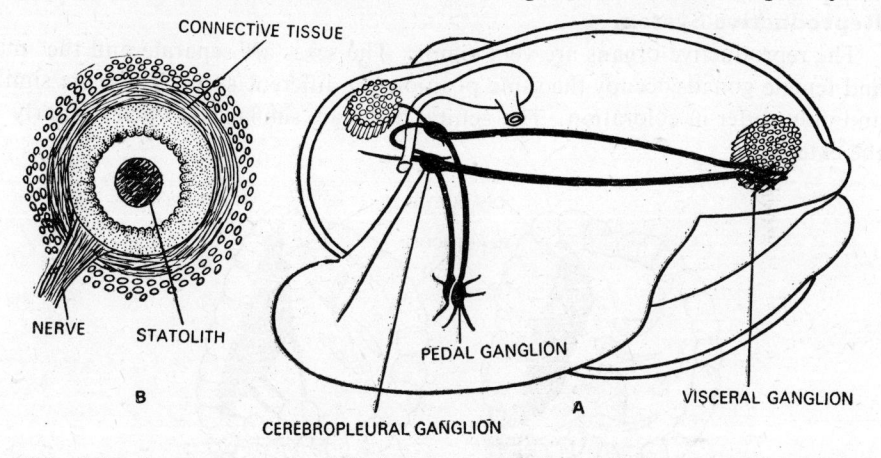

FIG. 19.59. A—Nervous system of a freshwater mussel; and B—Statocysts of *Anodonta*.

colour (sometimes orange red in fresh specimens) lying at the base of the palps. The two cerebral ganglia of two sides are connected with each other by a **cerebral commissure** that passes over the oesophagus. From the cerebral ganglia arise the cerebral connectives that pass downwards and backwards into the upper part of the foot. The **pedal ganglion** is rounded, coloured as cerebral. It is formed by the fusion of two ganglia. On close examination it appears to be bilobed. Many fine nerves originate from this ganglion and innervate the foot and the anterior retractor muscles, etc. The **cerebrovisceral connectives** are paired. Each arises from the posterolateral border along the cerebropedal and passes directly backwards. It runs just below the place of attachment of the gill laminae and passing through the substance of the kidney it emerges out and is connected with the visceral ganglion. During the course several fine branches arise and innervate the visceral mass. The **cerebrovisceral connective** is a delicate cord and is a bit difficult to find out as it lies in the connective tissue over the visceral mass. The **visceral ganglia** are triangular flattened structures lying on the ventral side of the posterior adductor muscle. In colour they are similar to other ganglia. These ganglia are more separated from one another (Fig. 19.56). Each ganglion gives off several nerves that innervate the posterior parts. The important nerves are: (*i*) the **posterior pallial nerve**, (*ii*) the **dorsal pallial nerve** (innervating the mantle, (*iii*) the **posterior renal nerve** (innervating the kidney), (*iv*) **the branchial nerve** (to the gills), and (*v*) the **posterior adductor nerve** (to the adductor muscles, etc).

**Sense organs.** The **sense organs** include the **osphradia**, the **statocysts** and scattered **epithelial sensory cells**. The osphradia are a pair of dark brown pigmented patches of sensory epithelium situated laterally or ventrally to the visceral ganglia and medial to the gill axis of that region. The statocysts are paired structures lying close to the **pedal** ganglia. Each statocyst consists of a round sac containing some granules or one granule (the **otolith**) within the cavity. The statocyst nerve too arises from the cerebropedal connective. The sensory cells occur in the outer epithelium of the body and inner border of the mantle. These cells are sensitive to touch. The ventral border of the mouth also contains certain pigmented cells sensitive to light.

### Reproductive System

The reproductive organs are very simple. The sexes are separate and the male and female gonads occupy the same positions in different sexes. They are similar and only differ in coloration. The genital ducts are similar and open similarly to the exterior.

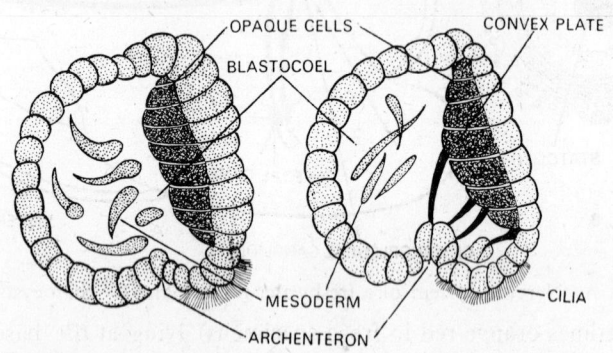

FIG. 19.60. Development of *Anodonta*. A—Blastula; and B—Gastrulation begins.

In the male the **testes** are paired racemose glands of white colour lying among the coils of the intestine. In the mature specimens the mantle lobes are almost filled with the genital tissue. The **genital duct (vas deferens)** is short in front of the testes and opens just in front of the opening of the kidney. The opening is called the **genital pore**.

In the female the **ovary** is also a racemose gland occupying the same place but is red in colour. The **genital duct (oviduct)** is likewise short and opens in front of the renal aperture. There are no accessory structures.

**Fertilization.** The spermatozoa are discharged in the breeding season through the genital pore into the cloaca (of the male), whence they pass out and

## Phylum Mollusca

enter, through the incurrent canal, into the cloaca of the female where the fertilization takes place. The zygote then passes to the water-tubes which act as brood-pouches. Here the larvae feed for some time on the mucous secreted by gills and each develops to a stage called **glochidium larva**, that hatches and passes out through the excurrent canal.

**Development.** The development has not been studied in the Indian freshwater mussel *Lamellidens*, but the development of *Anodonta* agrees with all, to some extent, differing only in certain details. These differences occur, perhaps, because of adaptations to freshwater conditions. Moreover, a temporary parasitism of the larva has complicated the later stages.

The egg-cell is surrounded by a **vitelline membrane** and is attached to the wall of the ovary by a minute stalk, the insertion of which is marked on the liberated ovum by an aperture (**micropyle**) through which the sperm enters during fertilization. Segmentation is total but unequal. A number of small, clear yolkless cells are rapidly divided off from a large yolk containing portion that

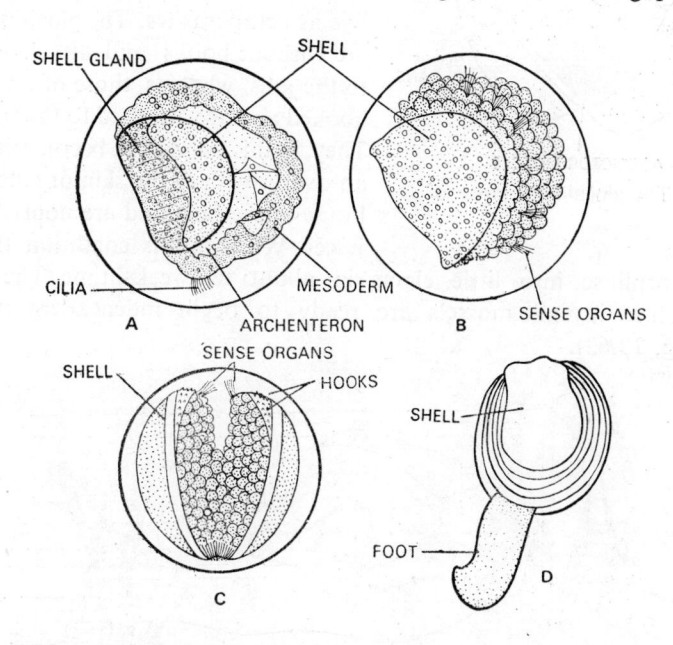

FIG. 19.61. Further stages in the development of young *Anodonta*. A—Gastrula with embryonic shell; B—Advanced embryo; C—Free glochidium; and D—Juvenile mussel.

multiplies slowly. Eventually a hollow ball of cells, the **blastula**, is formed. On the posterior dorsal regions of the blastula a number of large opaque cells form an internally convex plate (Fig. 9.60 A, B); this marks the beginning of the future **shell-sac**. A pair of large cells, pushed into the central cavity, produces the **mesoderm**. On the undersurface posteriorly there is a slight protrusion of ciliated cells forming ciliated disc. In front of this an invagination establishes the archenteron, but this happens at quite a later stage and the embryo becomes a **gastrula** (Fig. 19.61A). The shell-sac forms an embryonic shell and many of the mesoderm cells combine to form an adductor muscle extending from shell to shell. The mouth of the gastrula closes and a definite mouth is subsequently formed by an ectodermal invagination. Gradually a larva, peculiar to freshwater mussel, is built up. This is the **glochidium** and has two triangular delicate and porous shell-valves, each with a spiny incurved tooth on its free edge. The valves close together by the action of the adductor muscle, the mantle lobes are very small and their margins bear, on each side, three to four patches of glandular cells which secrete a long thread by the provisional byssus. Peculiar brush-like sense organs are also borne by the mantle lobes (Fig. 19.62).

The glochidia are produced in large numbers and fill the brood-pouch so much so that the gill-lamina appears as if packed with aggregated sand grains. Soon the glochidia are released in water where they slowly sink to the bottom and most of them die. To develop further, they must, within a few days, become attached to the fins or gills of a fish and live as ectoparasites. The glochidia of *Unio* are without hooks and attach themselves to the gills, whereas, those of *Anodonta* are hooked and are attached to the skin or fins. They, in this condition, become encysted by an overgrowth of the skin or mucous membrane of the host and are nourished by its juices. While in this condition the glochidia metamorphose into little clams in about ten weeks' time. The glochidia leave the fish as juvenile mussels are ready to begin independent life at the bottom (Fig. 19.63).

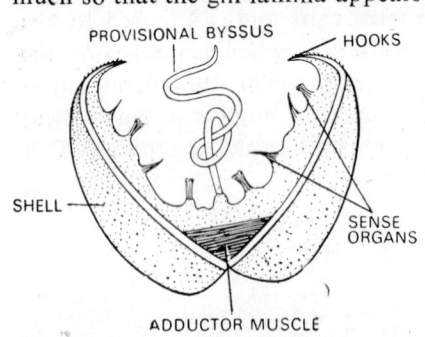

FIG. 19.62. The glochidium larva of *Anodonta*.

FIG 19.63. Life-cycle of freshwater clam. Fish with glochidia, Juvenile clam, and Matured clam producing new glochidia.

In some freshwater clams a parasitic stage is lacking, the young develop within the body of the mother. In marine Pelecypoda the eggs and sperms are shed into water where fertilization takes place and usual larval stages are produced—first a trochophore and then a veliger. The glochidia of freshwater clams correspond to the veliger of marine pelecypods.

## Other Pelecypoda

The pelecypoda are mostly sedentary bottom dwellers. *Mytilus*, the sea-mussel, is attached to solid objects by a "byssus" or thread-like structures secreted by reduced foot. The edible oyster, *Ostrea*, is permanently attached to rocks or shells. *Pholas* is a form that burrows in hard clay or soft rocks. The shipworms, *Teredo*, are slender bodied bivalve that burrow in the wood of ships. *Pecten* attaches by a byssus but can detach and swim by clapping its shells together. Some bivalves are commensals on echinoderms, some live in burrows of worms and crustaceans and some are embedded in sponges, and the tests of ascidians. One genus, *Entovalva*, is parasitic in the holothurians. Some bivalves are enormous in numbers. It has been calculated that 4,500,000,000 clams live on 1900 square kilometres of the Dogger Bank, east of England.

## Economic Importance

The Pelecypoda form food of man. For generations ancient people are known to have consumed molluscs sans shells. The shells are reported to have made "shell-mounds." An Indian "shell mound" on San Francisco Bay contained over 28,000 cubic metres of debris accumulated over an estimated 3,500 years. Annual consumption of molluscs in many countries is heavy. The soft internal parts of oysters are sucked out and consumed fresh, cooked or canned. Shell is used on road, and its lime is taken out and supplied to poultry, etc. Good natural oyster beds in shallow water are, therefore, highly valued. Oyster culture at various places has formed a good industry. Both Atlantic oyster and the Japanesae oyster (*Ostrea gigas*) have established on the Pacific coast. It is reported that United States takes 130,000,000 pounds of oysters and clams, etc. (valued at $37,000,000) annually. The shells also yield pearl buttons.

Natural pearls, so valued since ancient times, are produced by Pelecypoda. As stated earlier pearls are formed as a protective measure against foreign matter or parasite. Freshwater forms and oysters form some pearls. But the marine pearl oysters, *Magaritifera*, of Eastern Asia produce the best pearls. The Japanese artificially introduce small particles in the mantle of a bivalve, *Meleagrina*, and keep them in captivity sometimes for several years till the artificial pearls ("culture pearls") are produced.

The teredos are known for the damage they cause to wood in marine structures. Now as a measure of precaution, such constructions (wharves, etc.) are made of creosoted wood, concrete or metal.

## Parasitism

Parasitism occurs in freshwater mussels. Only larval forms are parasitic, adults are not. *Anodonta*, for instance, produces up to two million eggs. Instead of being laid the eggs are retained by the females and are incubated until hatched on the gills, which are modified as a brood-chamber. The larvae (glochidia) are ejected in spurts of several dozen at a time by means of exhalant siphon. The larvae reach the gills of a fish along with respiratory current and become attached to it. The larvae become inactive. Their presence, however, irritates the skin of the fish which reacts enveloping each glochidium in a minute cutaneous cyst, visible to the naked eye and found particularly on the fins. The glochidium lives

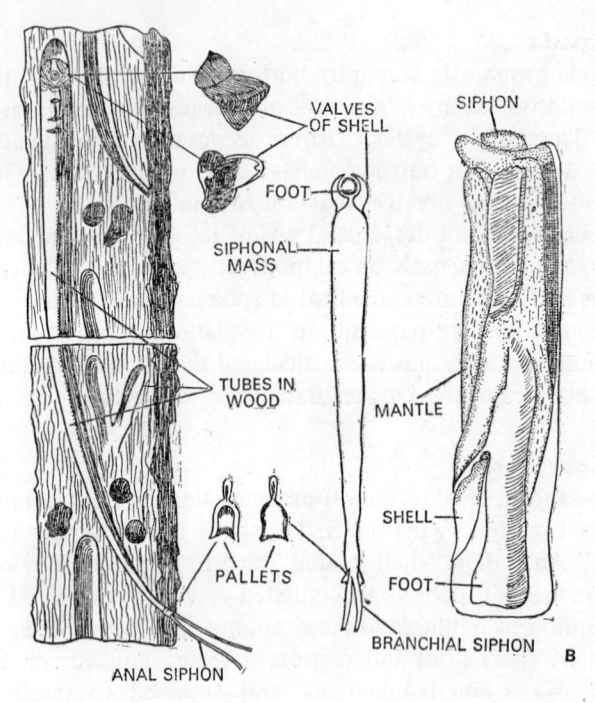

FIG. 19.64. A—*Teredo* (shipworm); and B—*Solen*.

as a temporary parasite developing into a small mussel inside. When fully formed the little mussel moves inside and breaks the wall by movements of its foot.

## CLASSIFICATION OF PELECYPODA

This class comprises typically bilaterally symmetrical Mollusca in which the body is very much compressed from side to side. The mantle is divided into two, right and left, equal lobes, each lobe secretes a valve of the shell. The two valves of the shell are joined together by a ligament and hinge. Head is not distinct whereas, the foot is more or less plough-shaped. Eyes, tentacles and radula are wanting. There are two ctenidia in the mantle cavity quite complicated in structure. The sexes are separate almost always. The larva is of trochophore type. More than ten thousand species have been described of which approximately four-fifths live in the ocean. They are classified according to the characters of the gills.

(*i*) **Subclass Protobranchia**. The Pelecypoda in which the gills consist of short, flattened leaflets and the foot has flat ventral surface acting as a sole. The shell is very characteristic having teeth on the hinge line. The labial palp is quite large. Marine species distributed along the Atlantic and Pacific coasts.

Example: *Nucula*.

(*ii*) **Subclass Lamellibranchia**. Gill-filaments folded and adjacent filaments attached by ciliary or tissue junctions.

*Phylum Mollusca* 837

1. **Order Taxodonta.** Hinge teeth numerous and similar; anterior and posterior adductor muscles approximately equal in size. Mantle margins unfused; gill filaments unattached, with no interlamellar junctions.
Example: *Arca*.

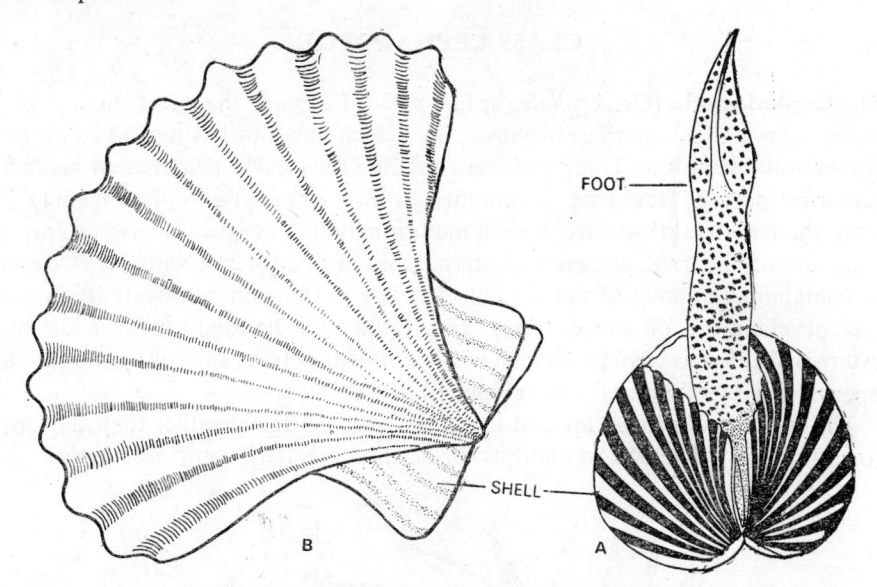

FIG. 19.65. A—*Cardium*; and B—*Pecten*.

2. **Order Anisomyaria.** Gills of the filibranch type with interlamellar junctions. Usually sessile; foot small; anterior adductor reduced or absent. No siphons.
Examples: Mussels—*Mytilus, Modiolus*; the oysters—*Ostrea, Crassostrea, Spondylus, Pinctada, Anomia, Lima*; the scallops—*Pecten,* the rock-boring *Lithophaga*.

3. **Order Heterodonta.** Hinge teeth consisting of a few large cardinal teeth with or without elongated lateral teeth. Gills of the eulamellibranch type; adductor muscles of equal size. Mantle fused to varying degrees; siphons commonly present.
Examples: *Cardium* (Fig. 19.65), the freshwater Sphaeriidae and Corbiculidae; the common edible clam *Mercenaria, Lasaea*; the borning clams—*Petricola, Tagelus*; the commensal clams—*Lepton, Dosinia, Donax, Tridacna*.

4. **Order Schizodonta.** Hinge teeth variable in size, shape. Gills of the eulamellibranch type.
Examples: Freshwater Unionidae—*Unio, Elliptio, Lampsilis, Anodonta,* and *Simpsoniconcha*, the freshwater Margaritiferidae, Mutelidae, and Aetheriidae.

5. **Order Adapedonta.** Hinge teeth reduced or absent; ligament reduced or absent and shell gapes. Gills of the eulamellibranch type; mantle margins sealed except for pedal gape; long siphons. Deep burrowing and boring clams.
Examples: Razor clams—*Ensis, Tanope, Mya*; boring clams—*Hiatella, Barnea, Martesia, Pholas*; the shipworms—*Teredo*. (Fig 19.64).

6. **Order Anomalodesmata.** No hinge teeth. Gills of eulamellibranch type but outer demibranch reduced and turned up dorsally. Hermaphroditic.

Examples: *Lyonsia, Pandora*; watering-pot shells—*Clavagella*.

(*iii*) **Subclass Septibranchia**. The Pelecypoda in which the gills consist of only a horizontal muscular partition and the adductors are two.

Examples: *Poromya, Cuspidaria*.

## CLASS CEPHALOPODA

The **Cephalopoda** (Gr. *kephale*, head+*podos*, foot) are the most highly developed symmetrical marine Mollusca in which the foot has moved by forming circumoral appendages completely surrounding the clearly demarcated head; the posterior part of the foot is modified to form a funnel or siphon leading out from the large mantle cavity; a shell may or may not be present; when present it is usually internal, sometimes external and in one group (Nautilli) is capable of containing the body of the animal. The key to the comparatively high nervous development of the octopus and squid can be found in their loss of an external shell. It is easy to understand how freedom from the confines of a shell opened new possibilities for development of the body.

The cephalopods have, around them, the aura of legend rather than the sober tone of reality. The octopus and its misdeeds share fame and fascination with

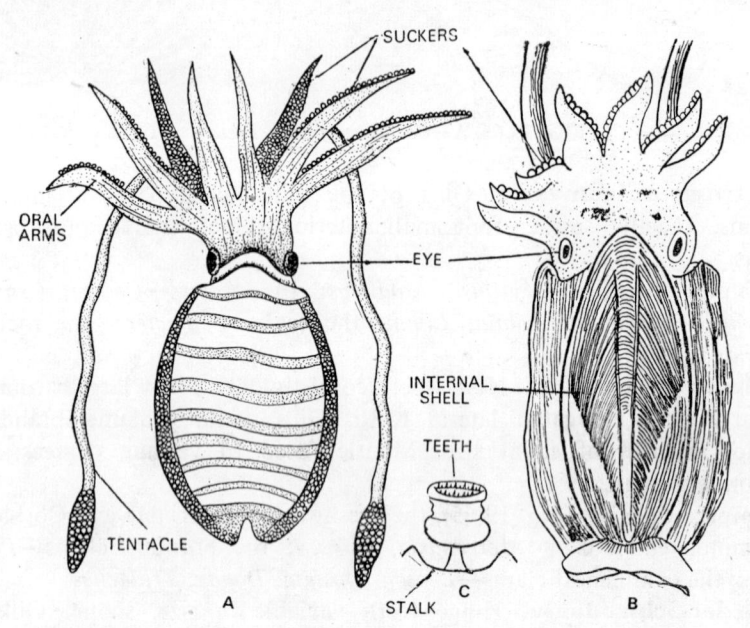

FIG. 19.66. A—*Sepia*, cuttlefish, entire; B—Its internal shell; and C—A stalked sucker.

such legendary beasts as the Minotaur and the sea-serpent (which may in fact be a giant squid's tentacle). Their size, strength, ferocity, and cunningness, their beauty, grace and speed are enough to excite curiosity and wonder in these animals. They are the stuff of which legends are made. Even students of natural history find them superior to other Mollusca. In the cephalopods the ventral surface is abbreviated because of the displacement of foot that forms a crown of appendages around the mouth. This is what is meant by the name Cephalopoda, i.e., "head footed." The members of this class include nautiluses, squids and

## Phylum Mollusca

octopuses. Like the gastropods all degrees of reduction of the shell are found in the cephalopods. While the nautiluses have a large, calcareous, external coiled shell, the squids have only a thin horny vestige of a shell embedded in the mantle, and the octopuses have no shell at all. In contrast to the snail and freshwater mussel the squid is very active and has adapted to predatory life. The eyes are like those of a fish, but unlike most bilateral animals, the body is elongated in an anteroposterior direction. The long axis of the squid is dorsoventral. The functional upper surface of a swimming squid is structurally anterior and the functional undersurface is structurally posterior; thus when a squid swims its ventral surface is forward, the dorsal surface is the hindmost, the anterior surface up, and the posterior down.

### TYPE CUTTLEFISH

The cuttlefishes, devilfishes, squids and nautili are the most highly developed among Mollusca. The cuttlefishes are marine molluscs living usually up to depth of a few fathoms, but often visit shallow water. They are quick swimmers and known for their strong arms. They have a pair of eyes (Fig. 19.66) that are as complex in structure as those of fishes.

### External Features

The cuttlefish has a distinct **head** devoid of appendages of its own. Its bears two well-developed eyes and is connected to the body by a constricted region,

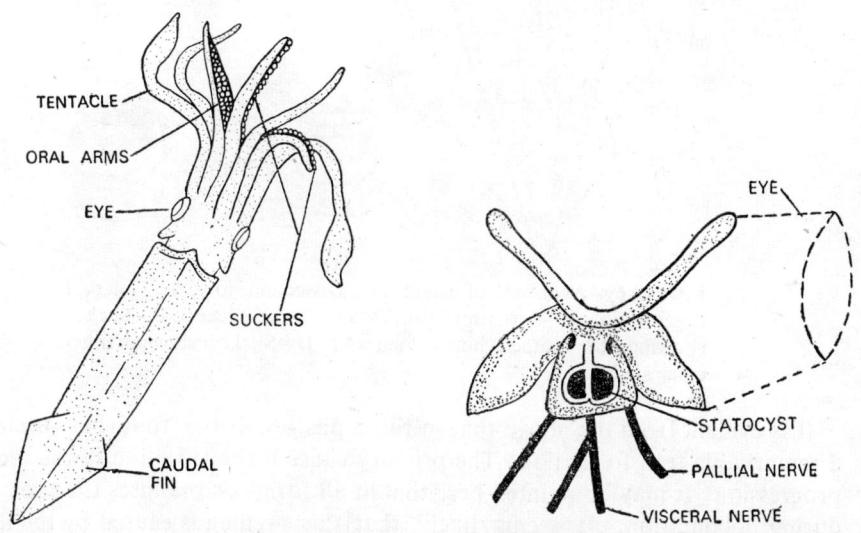

FIG. 19.67. *Loligo*, in a swimming posture.　　FIG. 19.68. Cranial cartilage of *Sepia*.

the neck. It is surrounded by appendages formed by the foot which consists of ten arms bearing suckers that are adhesive and prehensile. Of those two tentacular arms are different. They are more or less completely retractile within special pouches (*Sepia, Sepiola*, etc.), in some they are only very slightly retractile (*Loligo*). The arms are placed in pairs situated to the right and the left of the median plane. The arms are stout at the base and taper towards free end, their outer surfaces are convex and the flat inner surface bears cup-shaped suckers borne on short stalks (**Pedunculated**). Sessile suckers also occur in some forms. The tentacles are comparatively long and narrow and are provided with suckers only towards their free ends. Examined carefully the lip of the cup appears membranous, immediately within which their exists a horny rim (Fig. 19.65). The sucker works on the vacuum principle. When firmly applied to the objects the contractions of the muscles of the wall of the cup increase its cavity thereby creating a partial vacuum. Consequently the cup adheres to the object. The number of arms differs in different members of the group. In some (Dibranchia) there are four to five pairs of symmetrical arms but in octopuses there are only eight similar arms. The arms may be sessile or stumpy or variously reduced. In female *Argonauta* the two dorsal arms are enlarged to form a veil applied to the mantle and each secretes a protective calcareous shell. In most cases a single arm of the male is modified (**hectocotylized**) to form copulatory organ. The head region including the arms and the head proper is known as the **cephalopodium**. In *Amphitretus* the arms are united by web.

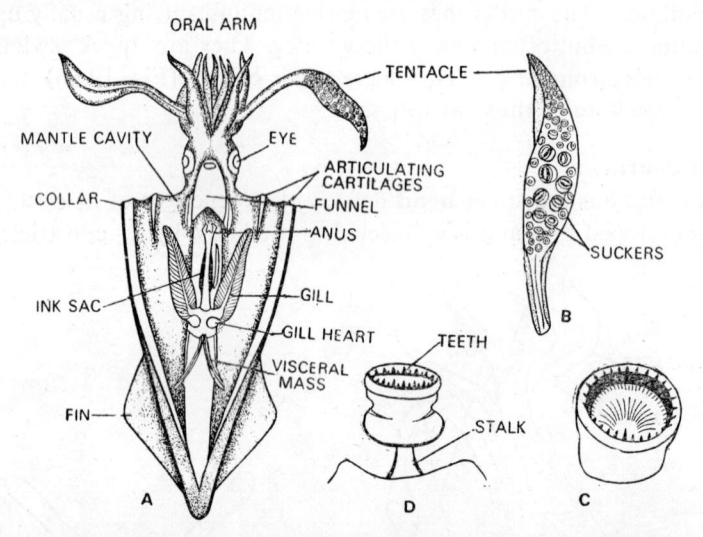

FIG. 19.69. Anatomy of squid. A—Dissection to show internal organization; B—An arm of squid studded with suckers; C—Sucker showing the toothed horny ring; and D—Suckers showing muscular stalk.

It is evident from the above that in the cephalopods the foot has undergone a remarkable transformation. The primary cause is the variation in the mode of progression. It may be pointed here that in all forms of molluscs the foot swells during locomotion. It seems likely that this swelling is caused by the flow of blood into the foot. This blood is not allowed to return to the body by special sphincter muscles.

The remaining part of the foot forms a muscular **funnel** or **siphon** for the egress of water. Some people have argued that the funnel alone should be regarded as the modified foot and arms as appendages of the head (hence the name **Siphonopoda**). This view does not appear convincing. The funnel opens to the exterior behind the neck internally communicating with the mantle cavity by a small aperture. Primitively the funnel has the form of two symmetrical

## Phylum Mollusca

lateral lobes which simply incline towards one another and overlap in *Nautilus*. In the Dibranchia the two lobes fuse together during development and form a complete tube. Through this tube are ejected the excrement, the secretion of the **ink-sac** and the generative products. In some (Nautilidae) the interior of the funnel is generally provided with a larger or smaller valve attached to its anterior or dorsal face.

The **mantle** is thick and muscular and covers the trunk. It contains both longitudinal and circular fibres. When the longitudinal fibres contract and the circular relax, the volume of the cavity is enlarged, the anterior collar of the mantle is drawn back from the funnel and water is forced in laterally and ventrally by the pressure of surrounding water. The mantle has taken over the protective function, which, in other molluscs, is served by the shell. Towards the oral

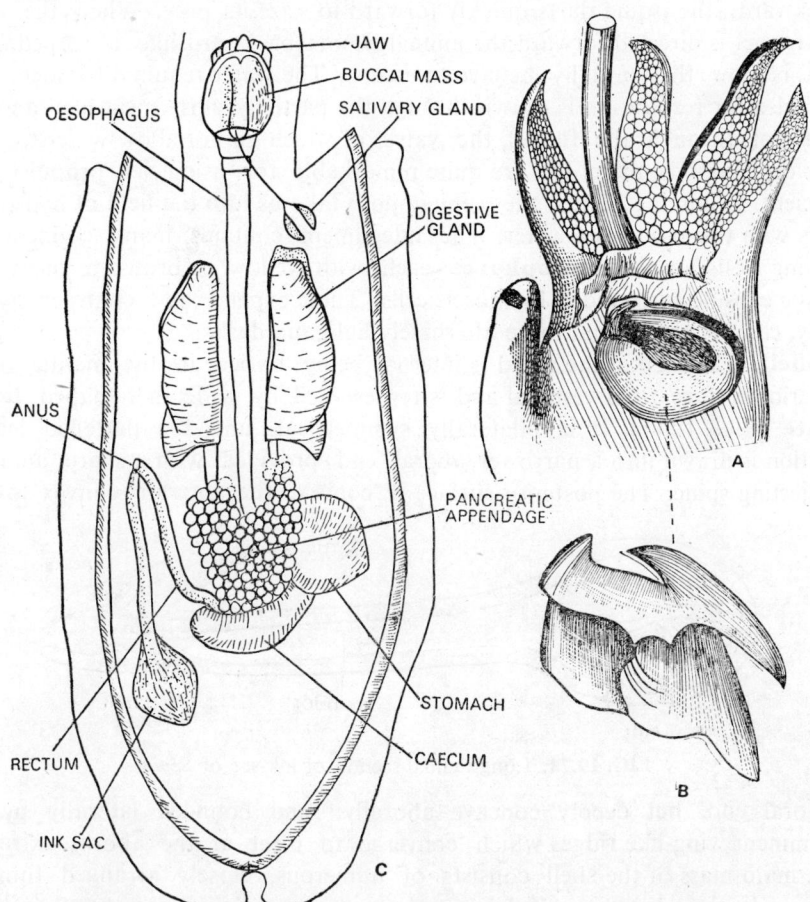

**FIG. 19.70.** Jaws of *Sepia*. A—*In situ*; B—The jaws taken out (*from* Parker and Haswell *after* C.N.H.); and C—Digestive organs of *Sepia*.

end it terminates in a ridge round the neck. Ventrally the mantle ends in free edge, the **collar**, which surrounds the neck between the head and the visceral mass. By three interlocking spaces the collar articulates with the visceral mass. These are in the shape of oval projections or ridges of the inner surface of the posterior mantle-wall near its oral border, which fits into the concave depressions or grooves on the facing surface. Thus, under ordinary circumstances, the funnel is the main outlet of the mantle cavity. The mantle is also the chief locomotory organ. At the dorsal end of the body, its anterior surface is extended into a pair of triangular folds or "fins" which can be undulated to move the animal slowly and to change the direction. The most important movements of squids are effected by jets of water emerging from the siphon, caused by the rhythmical contractions of the muscular walls of the mantle cavity. When the mantle is relaxed, water enters the mantle cavity around the edge of the collar. Now the mantle contracts, the edge is tightly sealed and the water is forced out through the opening of the funnel; when the mantle is contracted strongly a jet of water is expelled from the funnel. This pushes the animal in the direction opposite to that in which the jet is expelled. When the tip of the funnel is bent backward, the squid darts quickly forward to seize its prey. When the tip of the funnel is directed forward the animal shoots backward like a torpedo; and this is how they usually behave to escape. The water required for locomotion and also for respiration is drawn through the partially closed slit-like aperture, and not through the funnel, the valves of which do not allow water to get in. The cephalopods, therefore, are quite remarkable for using jet propulsion for efficient swimming. Thus the cephalopods take us into the field of hydrodynamics with this use of sea water. The integument contains many pigment containing cells or **chromatophores** each with yellow or brown pigment, in an elastic capsule surrounded by muscle cells. These expand and contract rhythmically, causing the animal to be alternately light and dark.

**Shell.** The shell in the squid is internal being buried in the mantle of the anterior surface. It is vestigial and is represented by a feather shaped **horny-plate** (Fig. 19.66). It is bilaterally symmetrical and the flattened leaf-like portion is drawn into a narrower aboral end provided with a sharp, anteriorly projecting spine. The posterior surface is convex, the anterior convex towards

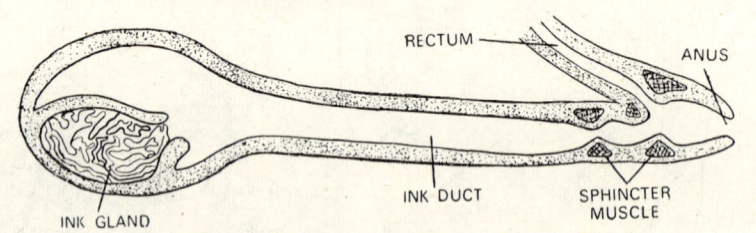

FIG. 19.71. Longitudinal section of ink-sac of *Sepia*.

its oral end, but deeply concave aborally; and bounded laterally by thin prominent wing-like ridges which converge to meet at the aboral extremity. The main mass of the shell consists of numerous, closely arranged thin calcareous laminae between which are interspaces containing gas. In *Spirula* the shell is partly covered. In *Nautilus* the shell is external, coiled (straight in some Palaeozoic Nautiloidea) and is provided with internal septa, disposed perpendicularly to the axis of the coil. The body occupies only the last chamber, but prolongation of the mantle runs back to the initial chamber through a calcareous tube, **shell siphuncle**, that penetrates all septa (Fig. 19.79). The female *Argonauta* also (Fig. 19.82) bears external shell which is different from other Cephalopoda in not possessing muscular attachments and is pedal in origin.

It is formed some ten or twelve days after birth.

The squids possess cartilaginous internal supports also. Prominent among these is a cartilage-like case, the **cranial cartilage**, surrounding the brain (Fig. 19.68). Besides this, there is a **nuchal cartilage** over the neck and a similar support of the siphon and fins. The mantle, fins, siphon and arms are all muscular. Powerful muscular bundles, originating from the cephalopedal mass and from the side of the funnel, unite together and are inserted symmetrically on the side of the shell. Other differentiated muscular bundles may be recognized; they are mostly due to the specialization of the funnel.

The **mantle cavity** is the sac enclosed by the mantle. The plume-shaped **cteninidia** lie in the mantle cavity, one on each side. The **anal apperture** lies in the middle line of the posterior surface close to the internal opening of the funnel. On either side of the rectum is a narrow projecting tube with a terminal opening, the excretory aperture. In the mantle cavity the gonoducts also open.

### Digestive System

The digestive tube of Cephalopoda comprises the **mouth**, a muscular **pharynx** (buccal mass) with a pair of horny **jaws** and **radula**, a long **oesophagus**, a muscular **stomach** with a **pyloric caecum** and a short **intestine**, which turns forward and opens in the middle line below the funnel. The mouth is surrounded by a thin **peristomial membrane** enclosing a circular lip beset with papillae. The jaws look like the inverted beak of parrot (Fig. 19.70). The jaws are for biting pieces which are swallowed rather rapidly, so much so that the radula (quite small in the squid) is probably seldom used. The mouth leads into the buccal cavity in which is situated the **odontophore** carrying numerous minute teeth. The oesophagus is a slender long tube which runs between the two halves of the liver towards the aboral side. The stomach into which opens the oesophagus is rounded thick-walled structure. A wide caecum opens into the stomach close to the opening which leads into the intestine. At this point the alimentary tube turns upon itself and runs almost parallel to the oesophagus to open at the **anus**. Two pairs of **salivary glands** connected to the pharynx are situated in the head behind the cranial cartilage. Whether they act like the salivary glands of other animals is not known. Extending from the neighbourhood of the salivary glands there is a large brown glandular mass, the **liver** or **digestive gland**. It consists of two (right and left) partly united portions. The ducts of the liver join the stomach. Another gland, **pancreas**, joins the stomach by minute ducts. The secretion of this has

FIG. 19.72. Radula of *Sepia*.

the property of converting starchy matter into sugar. The food of the squid comprises crustaceans, molluscs and fishes. It swims forward with the arms together, then darts at the prey by suddenly ejecting water from the siphon, the arms are spread, the prey grasped and brought to the mouth, crushed by the jaws and swallowed. Small bottom animals are quietly covered by its spreading arms and then gathered into the mouth.

Above the rectum occurs a characteristic organ the **ink-sac**. It is a pear-shaped body, a portion of the interior of which is glandular; this is separated from a reservoir that receives the secretion of the glandular part through a small orifice, connecting the two. The ink-sac develops as a diverticulum from the dorsal wall of the intestine and opens into the extreme terminal part of the rectum. It secretes a black substance or **sepia**. The Cephalopoda are able to expel the secretion contained in the reservoir at will. Thus, when attacked, it may emit the "ink" from the reservoir which mixes with water in the mantle cavity, and is ejected through the funnel as a black cloud which acts like a smoke screen, under cover of which the animal avoids attack and enemies. Some people feel that the ink forms a smoke-screen that gives the opportunity to avoid the attack and some people are of the view that it brightens dark objects that distracts the attention of the enemy while the squid goes off in another direction. Some squids emit a substance which, on coming in contact with water, starts shining brilliantly simulating a flash of light that protects the squid. The ink-gland does not occur in *Nautilus, Octopus arcticus, O. piscatorum* and some other species.

## Circulatory System

The Cephalopoda have a more complete and perfect circulatory system (Fig. 19.73) as compared to other Mollusca and the blood is nearly entirely

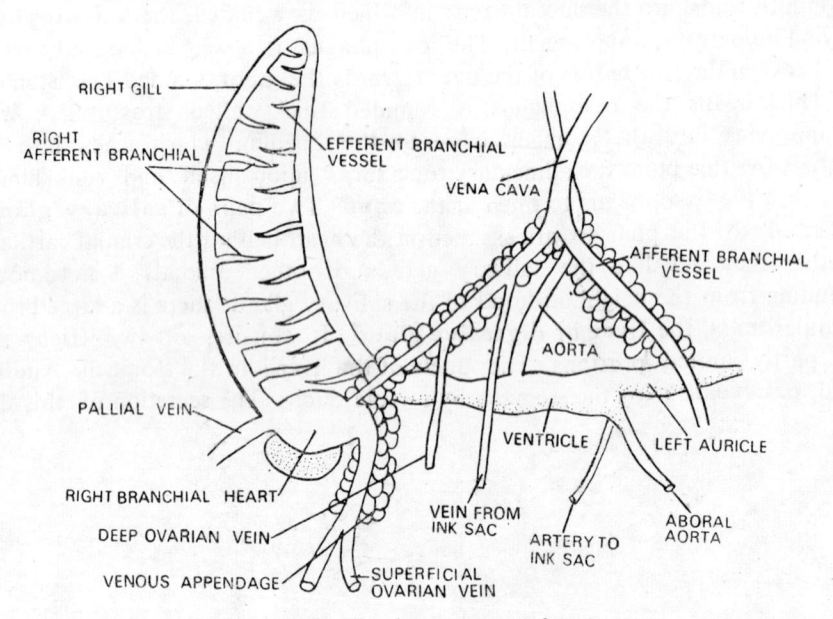

FIG. 19 73. Circulatory system of *Sepia*.

contained in true vessels. The **heart** of a squid consists of a ventricle and two **auricles**. The heart is situated somewhat superficially near the middle of the posterior or physiologically ventral surface. It lies in a **pericardial cavity** which is reduced very much in some (Octopoda). The essential part of the heart is the median ventricle which is divided into two lobes by a constriction, and is somewhat asymmetrical in position, other organs being quite symmetrical. The auricles, lateral and symmetrical, are nothing more than simple contractile expansions of the efferent branchial vessels (see below). From the ventricle arises the **oral aorta** at its oral end and smaller **aboral aorta** from its aboral end. The aboral aorta bends over the ink-sac and supplies the aboral portions of the body. From the aortae arise arteries that penetrate the tissues where they break into networks of fine vessels or capillaries. The capillaries are connected with a system of veins. Of these the important one is the **vena cava**, which runs from the head to the neighbourhood of the rectum, in front of which it bifurcates forming right and left **afferent branchial** veins, each running through the cavity to the renal organ of its side to the base of the gill, where it is joined by veins from the aboral region. At the base of each gill the afferent branchial vein becomes dilated forming a contractile sac, the **branchial** or **gill-heart**, which pumps the blood through the gill. It gives an impetus to the blood which passes through the gills at higher pressure. From the gills the blood is carried back to the ventricle by two dilated contractile vessels, the **efferent branchial veins**, one on each side. These bring the oxygenated blood to the heart that pumps it to various parts of the body. In *Nautilus* there are four auricles receiving blood from four branchial gills, the ventricle is transversely elongated and the circulation is partly lacunar.

**Coelom.** The pericaridum, as mentioned above, is the oral part of a pouch representing coelom. It gives off a pair of diverticula, each lodging the branchial heart of its side and communicating with the **renal sacs** (cavities of the kidney) by a pair of apertures. The aboral part of the coelom forms a capsule (**gonocoele**) which encloses the gonad.

### Respiratory System

The respiration is carried on by the paired, plume-shaped **ctenidia**, lying parallel with the long axis of the body. Throughout the greater part of its length the ctenidium is attached to the wall of the mantle cavity by a thin muscular fold. Each ctenidium is made up of numerous pairs of delicate lamellae, the surface of which is increased by the presence of a complex system of foldings. The branchiae are not ciliated, as they are in other Mollusca. The contractions of the muscular mantle are sufficient to produce a current of water enough for respiration. The exchange of gases takes place at the surface of the folding and the oxygenated blood is collected by capillaries and poured into efferent vessel which carries blood to the auricle. In *Nautilus* there are two pairs of branchiae (hence **Tetrabranchia**), all others have a single pair of branchiae (**Dibranchia**) and are free throughout their extent.

### Excretory Apparatus

In all cephalopods the two divisions of the coelom are in open communication with the exterior. This communication is effected through the **kidneys** which are a pair of thin-walled sacs, that open into the mantle cavity by means of conspicuous **excretory apertures**. On either side is an aperture through which the sac communicates with the pericardium. The right and the left

**excretory sacs** communicate with each other orally and aborally. Through each excretory sac runs the corresponding afferent branchial vein, formed by the bifurcation of the vena cava, and surrounding it are masses of glandular tissue, through which renal excretion is carried on. The product of the renal excretion is the nitrogenous excretory substance called **guanin**, which can be detected in the internal cavity.

### Nervous System

The Cephalopoda possesses very highly developed nervous system in sharp contrast with that of its slow moving relative clam. All the essential parts of the nervous system are centralized in the head round the initial part of the oesophagus. In *Nautilus* the nerve centres show much less concentration than other (Dibranchia). The **cerebral, pedal,** and **pleurovisceral ganglia** are all of relatively large size, are closely aggregated together in the head region and are supported and protected by the cranial cartilage. The **cerebral ganglia** are fused together into a rounded mass lodged in a hollow of the cranial cartilage. Anteriorly a strong fibrous membrane covers the brain. A pair of short thick processes, the **optic nerve** or **stalks**, arises from the sides and

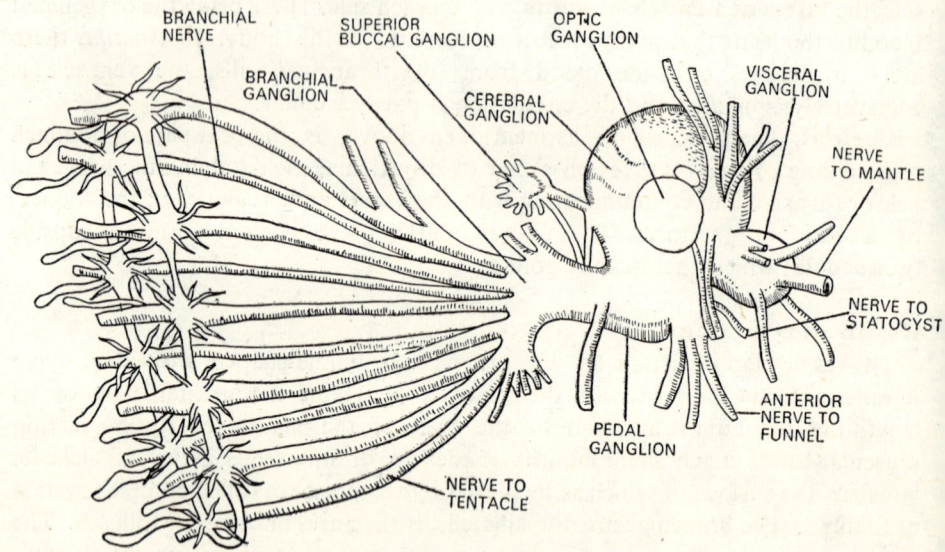

FIG. 19.74. Proximal part of the nervous system of *Sepia*.

soon expands forming large **optic ganglia** close to the eyes. Thick commissural bands of nerve-matter pass round the oesophagus to unite the pedal and pleurovisceral ganglia, lying behind. Like the cerebral, the pedal ganglia are also united into a single mass. This expands orally into a broad mass which gives rise to nerves to the arms called the **branchial nerves** (Fig. 19.74). In the same way the pleurovisceral ganglia unite to form a single mass, that lies in immediate contact with the pedal. From these ganglia arise various nerves supplying the various organs of the body.

**Brain.** By invertebrate standards cephalopods have enormous brain. Control of water pumping and, therefore, of locomotion involves the giant fibres whose conducting rate can be as great as 20 metres per second. In general the

*Phylum Mollusca*

brain of octopods and cuttlefish are larger than those of squids. According to J. Z. Young, the number of cells in the brain of a medium-sized *Octopus* is one hundred and sixty eight million. The greatest number of those cells are concerned in sensory integration and in "memory" functions. *Octopus* and cuttlefish are proving to be exceptionally useful experimental animals for neurophysiologists and physiological psychologists interested in localization of both innate and learned components on behaviour. Many scientists feel that the most significant advances in our future understanding of the cellular processes connected with memory will come from work presently being carried out on cephalopods.

The organs of special sense are very well-developed in the Cephalopoda. A pair of **ciliated pits** is found opening by slits on the surface behind the eyes. These are supposed to be olfactory organs. Numerous narrow sensory cells lining the pit are connected by nerve fibres at their bases. In the tongue there is a small elevation covered with papillae, probably an organ of taste. The **statocysts** are two embedded in the posterior portion of the cranial cartilage close to the pleurovisceral ganglion. A cartilaginous septum separates the cavities of the two organs. The inner surface presents a number of rounded and pear-shaped elevations and is lined with a flattened epithelium raised up on the posterior surface into a **ridge** or **crista statica** and **a macula statica** composed of cylindrical cells furnished with short hair-processes at their free ends and connected with nerve fibres on the other. A large statolith attached to the macula, is enclosed in the cavity. The statocysts are believed to maintain the equilibrium.

It is apparent that the sense organs are impressively complex and magnificently efficient among cephalopods. The statocysts are capable of detecting directions and angle of acceleration as well as static posture (compare the vertebrate labyrinth of the inner ear), and they are potentially sensitive to sound. There are olfactory pits on the head, and chemotactile as well as mechanotactile sensillae on various parts of arms and suckers.

**Eyes.** The **eyes** (Fig. 19.75) of the Cephalopoda are the most developed organs of special sense. They are supported by curved plates of cartilage forming a sort of **orbit**, connected with the cranial cartilage. The eyeball has a firm sclerotic supported by plates of cartilage. The sclerotic presents a **pupil** towards the outer side. The part of the sclerotic immediately around the pupil is known as the **iris**. The outermost transparent surface is the **false cornea**. The lens is a spherical body of dense glassy-looking substance. In fact it consists of two planoconvex lenses in close apposition. It is supported by an annular ciliary process projecting inward from the sclerotic. The lens divides the cavity of the eyes into two

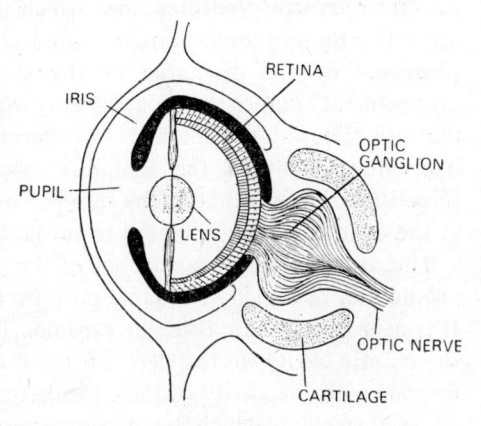

FIG. 19.75. Eye of a squid.

portions, the smaller outer filled with **aqueous humour** and the larger inner filled with **vitreous humour**. Somewhat complicated **retina**, the sensitive part of the eye, forms the inner lining of the large chamber of the eye (Fig. 19.75). Besides the optic ganglion, there occurs a body, the **optic gland** or the **white body**, of unknown function. This structure agrees with that of the vertebrate eye. When two similar structures having a similar function appear in two distantly related groups the structures are said to evolve independently and the process is said to be convergent evolution. Thus, the eyes of the squid and of a vertebrate, which have no possibility of a common ancestor, are said to have arisen by convergent evolution. Some squids also possess peculiar light producing organs.

The cephalopods have eyes with a retina, which is a little more efficient than that in the vertebrate eye since it is not developmentally inverted and thus the fibres of the optic nerve are collected together outside the eyeball. The lens of the eye is a complex structure which has a focal length of about two-and-a-half times its radius, although it is nearly spherical in form. This implies that the outer parts of the lens have a lower refractive index than the central regions, and in this cephalopods resemble fishes. The packing density of the retinal cells is similar to that of the higher vertebrates, and pigment movements allow light and dark adaptations. There are behavioural indications that cephalopods have colour vision. The eyes have power of considerable image formation and shape discrimination.

**Reproductive System**

The sexes are separate and in some forms sexual dimorphism is pronounced. Generally speaking the males are more slender or smaller than the female, but in *Nautilus* the cephalic hood and the aperture of the shell are wider in male. *Argonauta* presents maximum dimorphism, the males being much smaller and females have characteristic external shell. The gonad is single and median situated near the aboral extremity of the body in the coelom. In fact the gonad appears to be a projection into the wall of the coelom.

The **testis** (Fig. 19.76) is a compact mass of minute tubules enclosed in a capsule from which arises a single **spermiduct**, a convoluted tube that leads into the **seminal vesicles**, into which the **prostate glands** open. The sperms are rolled up into long narrow cylindrical chitinoid capsules, the **spermatophores**. For the discharge of the sperm, the spermatophore has a peculiar arrangement comprising an apparatus of the nature of a spring which ruptures the wall releasing the sperms. The spermatophores are stored in a wide sac-like enlargement of the seminal vesicle called the **spermatophoral sac (Needham's sac)** that opens into the mantle cavity by a small aperture located at the tip of the penis situated towards the left.

The ovary is simply a portion of the wall of the coelom, from which ova proliferate, occupying the same position as occupied by the testis in the male. It is also enclosed in a similar capsule. The oviduct is a wide tube opening into the mantle cavity to the left of the rectum. The lumen of the oviduct is continuous with the cavity of the capsule that lodges the ovary. The ovary presents an axial swelling which bears numerous follicles, containing ova in different stages of development. During the breeding season the ovary is packed with numerous ova so much so that it takes a polygonal form. Situated to the right

*Phylum Mollusca* 849

and left of the ink-duct, in the anterior wall of the mantle cavity, there occur **nidamental glands** of somewhat oval outline. The secretion of these glands is

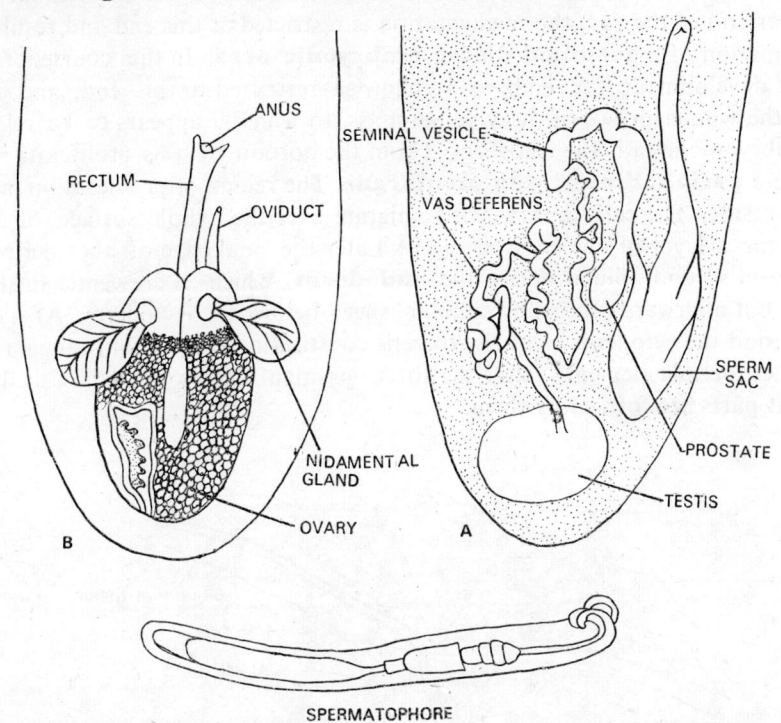

FIG. 19.76. Generative organs of *Sepia*. A—Male; B—Female; and C—A spermatophore.

viscid and glues the eggs in masses. Internally each gland possesses a median canal in its long axis. A glandular mass, the **accessory nidamental gland**, is at the sides and around the oral ends of the proper nidamental glands. The function of these is unknown.

Copulation is sometimes preceded by a courtship display by the male. The organ of copulation is the **hectocotylized** arm. In *Loligo* and *Sepia* it is the left fourth arm that is hectocotylized. In *Nautilus* the organ of copulation is the **spadix**. During copulation the hectocotylus withdraws a mass of **spermatophore** from Needham's sac and carries them at its specialized tip. It is inserted into the mantle cavity of the female where the tip becomes detached and remains. The eggs are laid, shortly after copulation, in masses or strings in a soft, gelatinous substance, usually attached to some foreign body. In some cases the egg has a long or short stalk. The segmentation is incomplete, at no

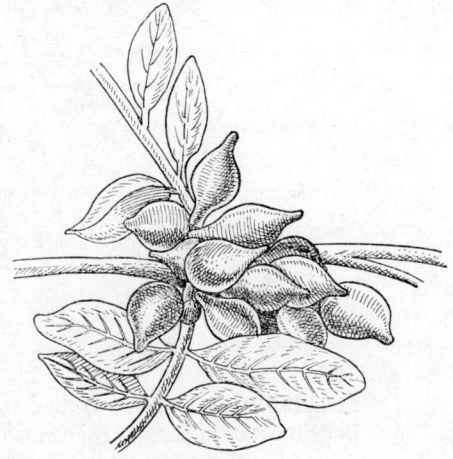

FIG. 19.77. Bunch of eggs of *Sepia* attached to a plant (*after* Jatra).

period does the ectoderm completely cover in the vitelline mass. There is no proper blastopore. As the formative protoplasm is located at the narrower end of the egg, the segmentation is restricted at this end and results in the formation of a **germinal disc** or **embryonic area**. In the course of subsequent development the embryo is likewise restricted to this end, and never covers the whole surface of the vitelline mass, to which it appears to be related. The embryonic area is the ectoderm. From the portion of this proliferate cells forming a **perivitelline** or **yolk membrane**. The region of proliferation marks the anal side. The cells thus formed migrate over the whole surface of yolk and form a layer of nuclei investing it. Later the anal edge of the periphery produces a second cellular layer, the **endoderm**, which is crescentic in shape at first, but afterwards forms a circular sheet below the ectoderm. At a still later period the ectoderm gives rise to cells constituting the genital rudiment and other mesodermic elements. After the three germinal layers are established the different parts are formed by them.

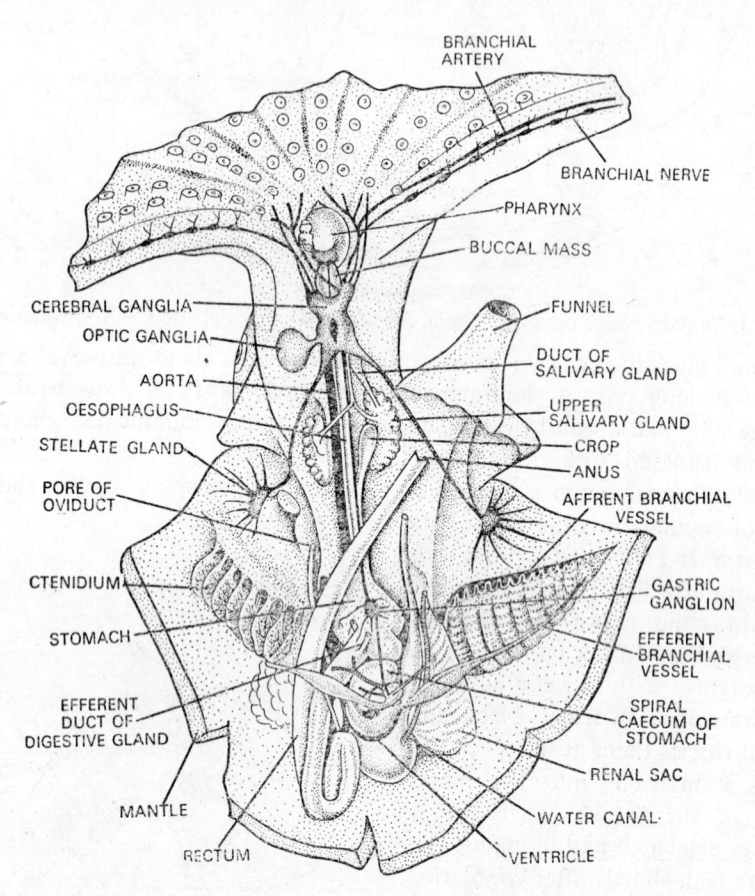

FIG. 19.78. Anatomy of *Octopus* (*after* Leucart and Milne Edwards). Body is cut open posteriorly and the liver is removed.

In their functional morphology the cephalopods differ from the rest of the molluscs in four principal ways: (*i*) The anteroposterior axis of the head-foot

*Phylum Mollusca*

has changed and followed the course of dorsoventral axis of the visceral mass so that the head lies in the same line with the elongate visceral mass, while the foot becomes the exhalant funnel of the mantle cavity. (*ii*) The importance of cilia is greatly reduced; the cilia of the mantle cavity and ctenidia being functionally replaced by powerful muscles which carry out pumping for the respiratory current and for locomotion by jet propulsion. (*iii*) The shell is either internal without any skeletal significance or absent (as in octopods). The shell of *Nautilus* is papery and different (see below). (*iv*) The course of water circulation, brought about by the rhythmic contraction of the mantle cavity (muscular action), is reversal from the usual molluscan pattern.

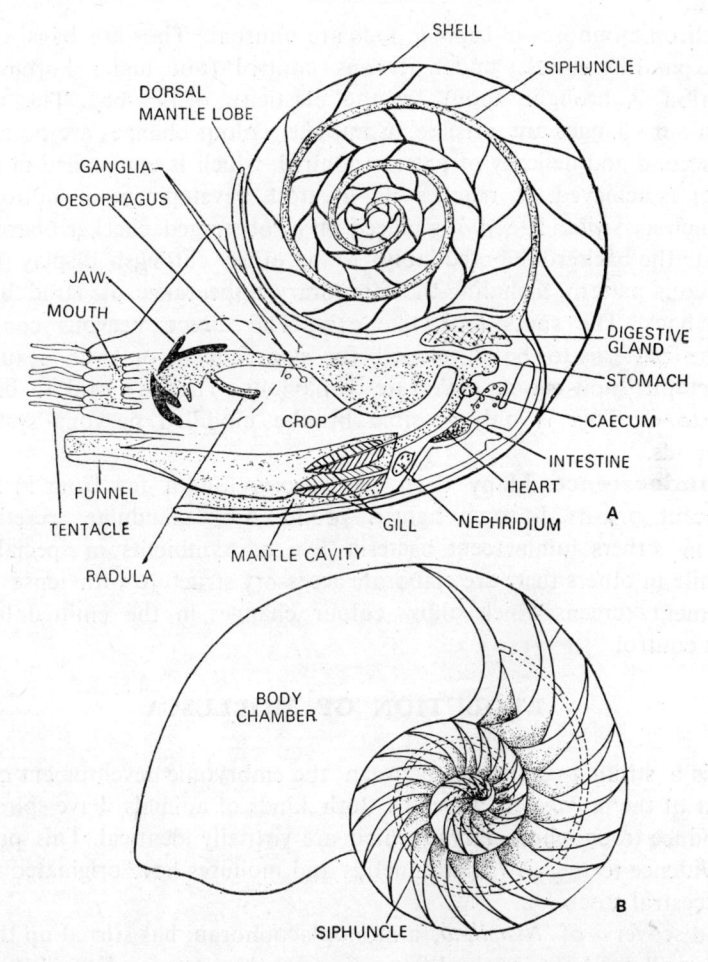

FIG. 19.79. *Nautilus*. A —Anatomy of a nautiloid; and B—Vertical section of the shell.

## Distribution

The cephalopods are marine animals. They are fundamentally a swimming group ranging from tidal limits to very deep waters. They are common in tropical and warm temperate seas. They are mostly carnivorous and their predators are fish, whales and sea birds.

**Behaviour**

The presence of an ink-sac has been mentioned earlier. The ink-sacs are used to lay down protective smoke screens as well as to produce smaller puffs. The true cuttlefish have the largest ink-sacs. Nearly all modern cephalopods are able to produce compact puffs under the cover of which they escape their predators. One small species of *Sepiola* reacts to danger by first darkening with its chromatophores, then emitting a compact puff of ink, then reverse the effect of darkening (i.e. whiten their colour) and jet away undetected by the predator who gets just a mouthful of ink-puff. These changes take place within less than a second.

The chromatophores of cephalopods are unusual. They are bags of pigment which expand by muscles under nervous control (not under hormonal). Their contraction is brought about by the elasticity of the bag. This means that rapid colour changes are possible. Complete colour changes are possible in less than a second and delicacy of pattern control, which is unequalled in the animal kingdom, is achieved. It reaches its greatest development in bottom-dwelling forms such as *Sepia*. If *Sepia* is placed on a submerged checker board it is able to imitate the background creditably. Many other cuttlefish display terrorizing, conspicuous pattern, including the temporary appearance of large black spots on the back. The spots look like eyes. The direct nervous control allows flickering patterns to be developed for concealment or even sexual display. Some octopus show more elaborate behaviour. They are said to build houses out of stones. This is only possible by the modified nervous system of all cephalopods.

**Bioluminescence.** Many deep sea species (and a few from inshore) have luminescent organs. In some light is produced by glandular secretion of the animal, in others luminescent bacteria live as symbionts in special ducts or sacs; while in others there are elaborate accessory structure with lenses, reflectors and pigment screens which allow colour changes in the emitted light under nervous control.

## EVOLUTION OF MOLLUSCA

There is a striking similarity between the embryonic development of molluscs and that of the polychaete annelids. Both kinds of animals have spiral cleavage and produce trochophore larvae which are virtually identical. This provides the main evidence to the view that annelids and molluscs have originated from common ancestral stock.

The discovery of *Neopilina*, a monoplacophoran, has stirred up the opinion that the molluscs have evolved from segmented worms, as *Neopilina* is believed to have segmented structure. It possesses eight pairs of retractor muscles, five pairs of gills and six pairs of nephridia, which, according to many zoologists, confirm the existence of metamerism and establish that molluscs have originated from annelids. But many authors feel that the evidence for metamerism, as found in *Neopilina*, is not very convincing (as discussed on page 779).

Even without *Neopilina* the affinity between molluscs and annelids seems to be

# Phylum Mollusca

probable, but it appears that the two groups diverged from some common ancestor prior to appearance of metamerism. The nature of such an ancestor is not clearly defined, but the likely steps taken by the ancestor are summed up below.

The molluscs have originated from metamerically segmented vermiform ancestors that employed hydraulic pressure for burrowing their way through the soil or for locomotion. The worm-design is such that animals can creep on hard soil badly because it is not able to flatten itself sufficiently due to the pressure of fluid-filled coelom. An improvement in crawling was possible only if the muscles of the ventral surface of the body wall were intensified so that they could have a thorough and better grip over the substratum. Such a need-based development of the musculature of the ventral surface of the body took place in the molluscs and a foot developed.

The formation of a foot, adapted for crawling, demanded a change in the feeding habit. Therefore, a rasping ribbon or radula developed in the mouth. The radula is a band-shaped structure carrying variously-shaped minute teeth which help the animal to rasp the food from a primordial permanent subsoil. Rasping of food on the substratum presupposes a firm hold of the body on the substratum otherwise the pressure needed for scraping on the subsoil could not be obtained. It is apparent, therefore, that the development of rasping organ in the mouth and foot for crawling are parallel and reciprocal.

In order to strengthen the crawling movement of the thick muscular-pad or foot on the ventral side the viscera of the body has been displaced more and more towards the dorsal side. With the displacement of viscera the shell, which was made up of several pieces, moved up and settled down on the dorsal side. The musculature of the dorsal body wall became thin and the viscera was protected by a firm shell. With further evolution the foot grew thicker, the visceral mass, covered by an uniform shell, moved upon the mantle cavity, became reduced, and so also the gills. Only a pair of enlarged gills persists.

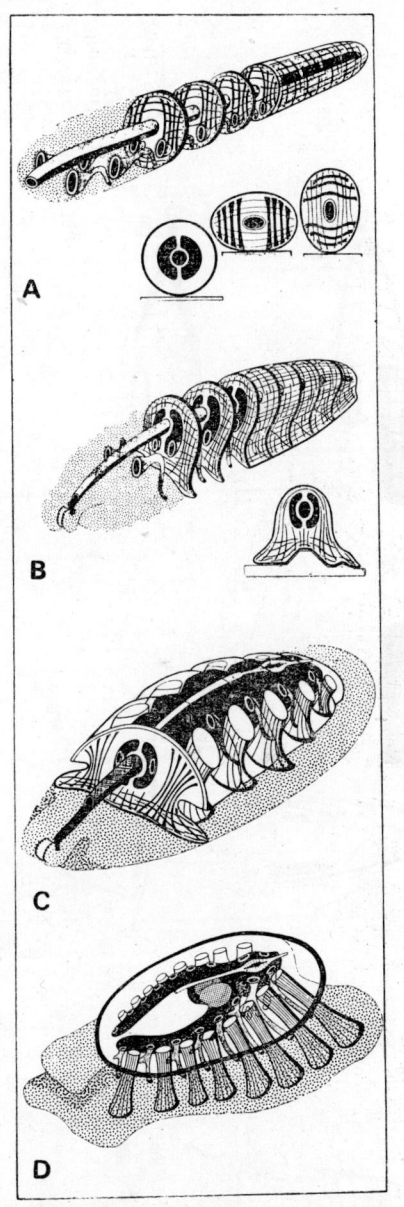

FIG. 19.80. Evolution of a mollusc from a segmented animal.

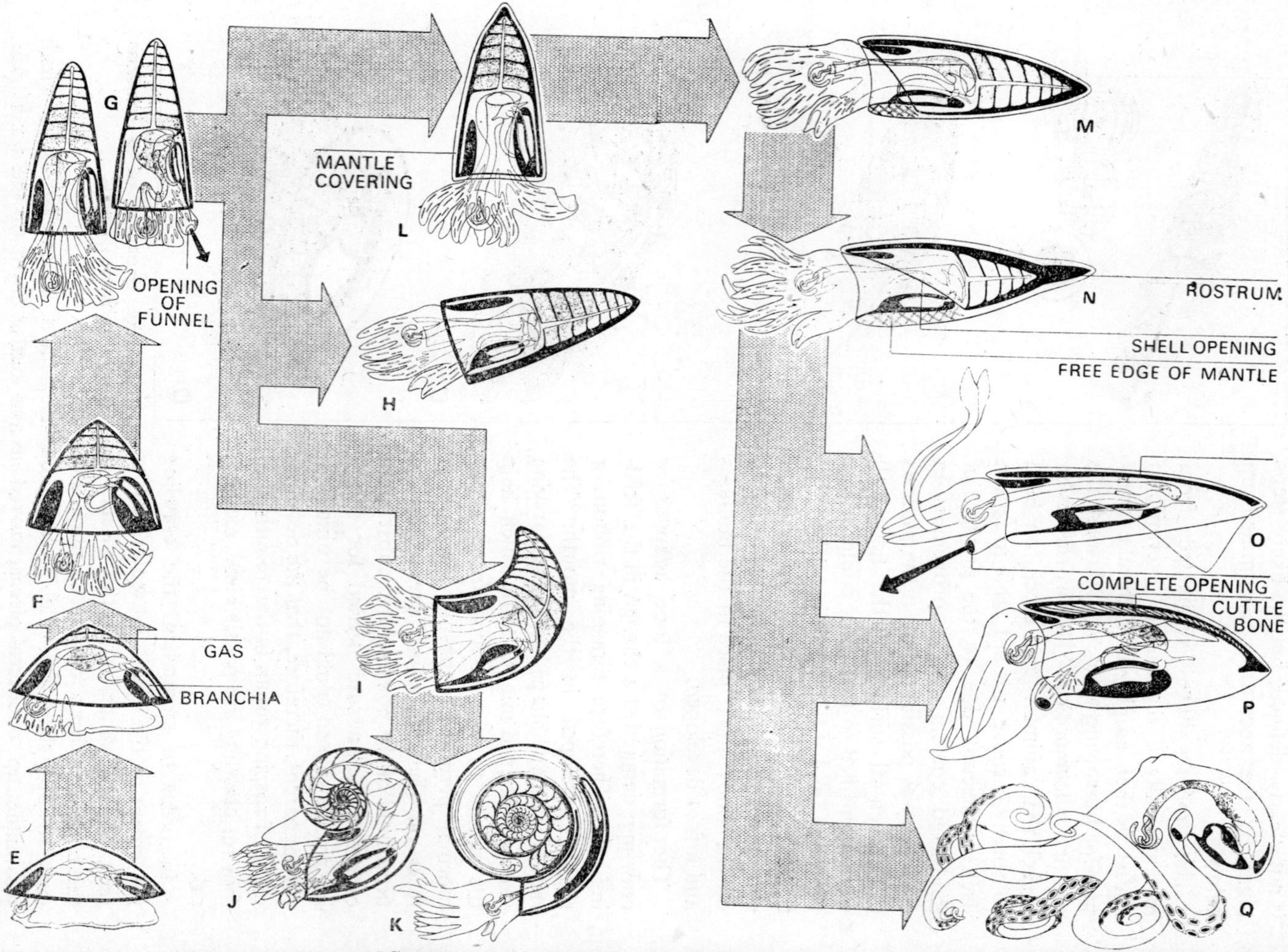


FIG. 16.81. Evolution of

*Phylum Mollusca* 855

The cephalopod or cuttlefish (Fig. 19.81) originated from fully developed molluscs, with uniform shell enclosing the visceral mass, which was not convoluted as in snail (E). A number of changes which ran more or less parallel paved the way for the evolution of cuttlefish. The first change is the appearance of a gas-filled chamber (F) for buoyancy. Due to the enlargement of the shell there appeared gas-filled chamber to help the animal to orientate in water. The number of gas-filled chamber varies with the weight of the visceral mass and shell.

Then developed the back-stroke drive (G). This developed form the retractor motion of the molluscs inside the shell. As it retracted inside the shell the corresponding quantity of water from the mantle cavity was ejected. This pushed the animal back (i.e. produced back stroke motion). This was further aided by the accumulation of gas in the body that made it lighter.

The development of predatory habit further added to the evolution of cephalopods. The foot that had served for crawling was also used to grasp the prey. As the animal began to float on water the foot performed the only function of catching the prey. The foot, thus, changed into a system of tentacle-like structures for seizing and holding the prey tightly. As a crawling forward motion was developed the mouth shifted in the centre of the grasping organ and the head became surrounded by the arms which originated from the anterior part of the foot. A suctorial disc, a zone of high adhesion, also developed from the surface of the foot. The posterior portion of the foot is transformed into a funnel through which water is ejected out during back-stroke movement. The water enters the covering during breathing and is expelled during movement.

Several groups of fossil cephalopods evidently retained the calcareous shell. In many the shell was straight (orthocone, H), in others shell was tilted a little. In modern cephalopods the external calcareous shell is lost and an internal shell may or may not be retained.

The spirally coiled shells of the nautiloids and ammonids are formed differently, they are not derivable from the slightly curved forms. The curvature of

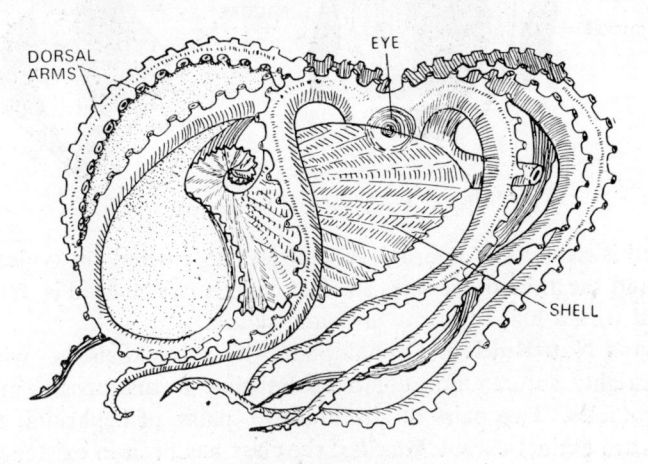

FIG. 19.82. *Argonauta.*

the anterior chamber shifted far in the middle and a form was moulded that does not hinder movement in water. Such forms were widely distributed during palaeozoic and mesozoic but now they are extinct except one genus *Nautilus*,

It is important that in all molluscs with external shells the forward movement is correlated with the drawing back of the soft body which may result in deformity. To avoid this situation develops a calcareous shell burried inside. It lies along the length of the body. The mantle is now free to grow and develop colour. It is assumed that the protection and repair of shell is easily possible in case of the internal shell. The mantle deposits calcareous matter on the shell. In order to maintain a balance in water a rostrum is formed (N) in Belemnite.

As a final step in cephalopod the outer wall of the living chamber, in which the viscera is lodged, is dissolved (N). The mantle thus becomes free from the reinforcement of the shell and can now contract or expand itself.

## CLASSIFICATION OF CEPHALOPODA

The cephalopods are bilaterally symmetrical Mollusca in which the foot moves near the mouth and is modified to form a series of long prehensile arms bearing suckers or tentacles. A part of the foot forms the funnel or siphon for **egress** of water from the mantle cavity. In most modern representatives of the

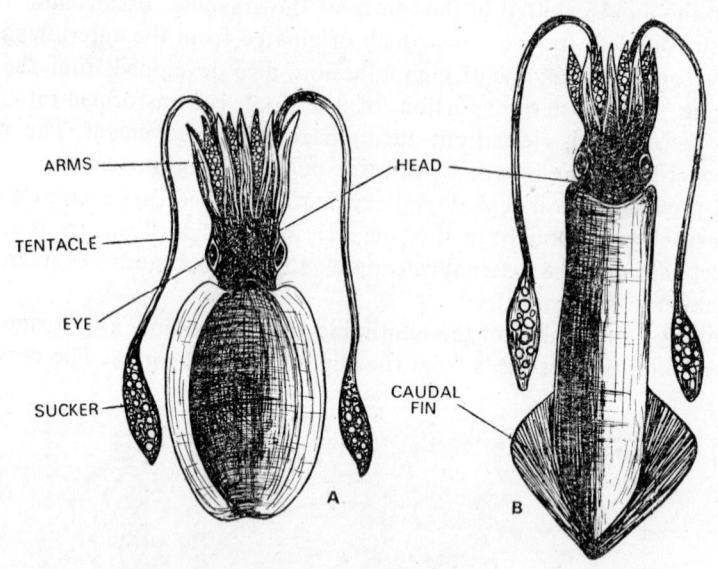

FIG. 19.83. A—*Sepia*; and B—*Loligo*.

class the shell is reduced or internal or entirely absent but in typical cases it is chambered and the animal occupies the last chamber. The head is well-developed and provided with a pair of horny jaws and radula.

(*i*) **Subclass Nautiloidea**. Cephalopoda with external shells, which may be coiled or straight; sutures not complex. Existing forms possess many slender suckerless tentacles. Two pairs of gills and two pairs of nephridia are present. All members are extinct except *Nautilus*; the class has been in existence from the Cambrian period to the present time.

Examples: *Endoceras* and *Nautilus*.

(*ii*) **Subclass Ammonoidea**. Extinct Cephalopoda with coiled external shells having complex septa and sutures. Existed from the Silurian period to Cretaceous period.

*Phylum Mollusca*

Examples: *Ceratites, Scaphites,* and *Pachydiscus.*

(*iii*) **Subclass Coleoidea.** Cephalopoda in which shells are internal or absent. A few tentacles bear suckers. One pair of gills and one pair of nephridia are present. Members have existed from Mississippian period to the present.

1. **Order Decapoda.** Coleoidea including the cuttlefish and squids. Two long tentacles and eight shorter arms present; body with lateral fins.

(*a*) **Suborder Belemnoidea.** Extinct species. Shell internal, chambered but with a posterior solid rostrum and a dorsal shield-like extension.

Examples: *Belemnites, Belemnoteuthis.*

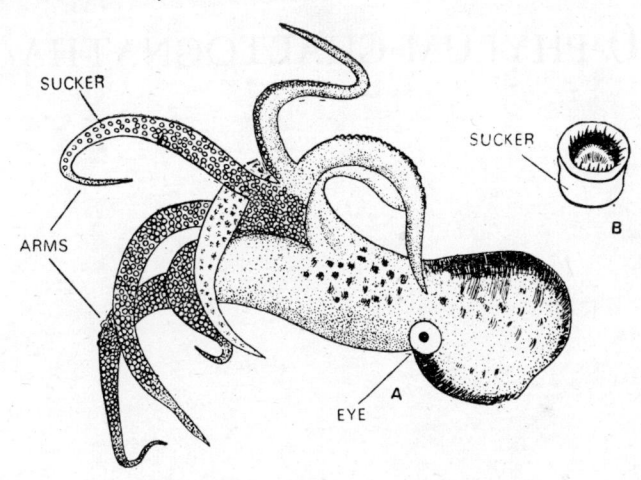

FIG. 19.84. A—*Octopus*; and B—Sucker.

(*b*) **Suborder Sepioidea.** Cuttlefish and sepiolas. Shell with septa or greatly reduced or lost. Body mostly short and broad or sac-like.

Examples: *Sepia* (Fig. 19.83), *Idiosepius, Spirula, Sepiola.*

(*c*) **Suborder Teuthoidea.** Squids. Shells often a flattened plate. Body mostly elongate with long tentacles.

Examples: *Loligo* (Fig. 19.83), *Architeuthis, Chiroteuthis, Onycoteuthis, Cranchia.*

2. **Order Vampyromorpha.** Vampire squids. Small deep-water coleoides, having octopod-like forms with eight arms united by a web, but two small retractile tentacles also present.

Examples: *Vampyroteuthis.*

3. **Order Octopoda.** The octopods. The coleoides possessing eight arms; globular body with no fins.

Examples: *Octopus* (Fig. 19.84), *Eledone, Eledonella, Vitreledonella, Amphitretus, Cirroteuthis, Argonauta* (Fig. 19.82).

# 20 PHYLUM CHAETOGNATHA

The **Chaetognatha** or the **arrowworms** (as they are popularly called) constitute an isolated group of transparent, pelagic animals having several genera (*Sagitta, Eukrohnia* and *Spadella*) and about 50 species. They are of great interest to naturalists inspite of their small number. The affinities of the group are very doubtful, therefore, it has been raised to the phylum status. The body is transparent, slender, usually 2.5-7.5 cm in length. In an open ocean they look like cellophane arrows as they dart after their prey. In certain seasons they appear in incredible numbers and form a part of the food of fish. The body is divided into three regions—head, trunk and tail, has fin-like projections, which probably act as balancers. The mouth is on the ventral surface of the head and is slit-like. The anus is situated at the junction of the trunk and tail, about a third of the way from the posterior end.

The **head** (Fig. 20.1) is slightly swollen. It is covered by a fold of skin which arises from the dorsal side of the head, and forms a kind of **hood** (the **prepuce**), more developed laterally than ventrally.

Around the head there is a slightly thickened area, where the epidermis is stratified. This is called **collarette**. The head carries groups of sickle-shaped setae, the **seizing jaws,** which are laterally placed. There are also present a number of spines or teeth. The head also bears a set of two eyes. Close behind the head there is an annular modification of the ectoderm, the cells of which are ciliated. This is supposed to be an olfactory organ. The main mass of head is composed of muscles having various functions.

The body possesses two pairs of cutaneous fins. The lateral fin-folds are on

# Phylum Chaetognatha

the sides and the caudal fin is situated around the tail. The fins consist of fold of ectoderm containing a gelatinous substance, on each surface of which lie some chitinous, rod-like rays beneath the ectoderm.

## Body Wall

Outermost epithelium, the **epidermis,** has thin flat cells with distinct oval nuclei arranged in one layer or more than one layers. The epidermis secretes structureless **cuticle.** This is followed by a structureless basement or basal membrane below which is a layer of longitudinally disposed muscle fibres which are all cross-striated. In the regions of the trunk and tail, muscles are arranged in

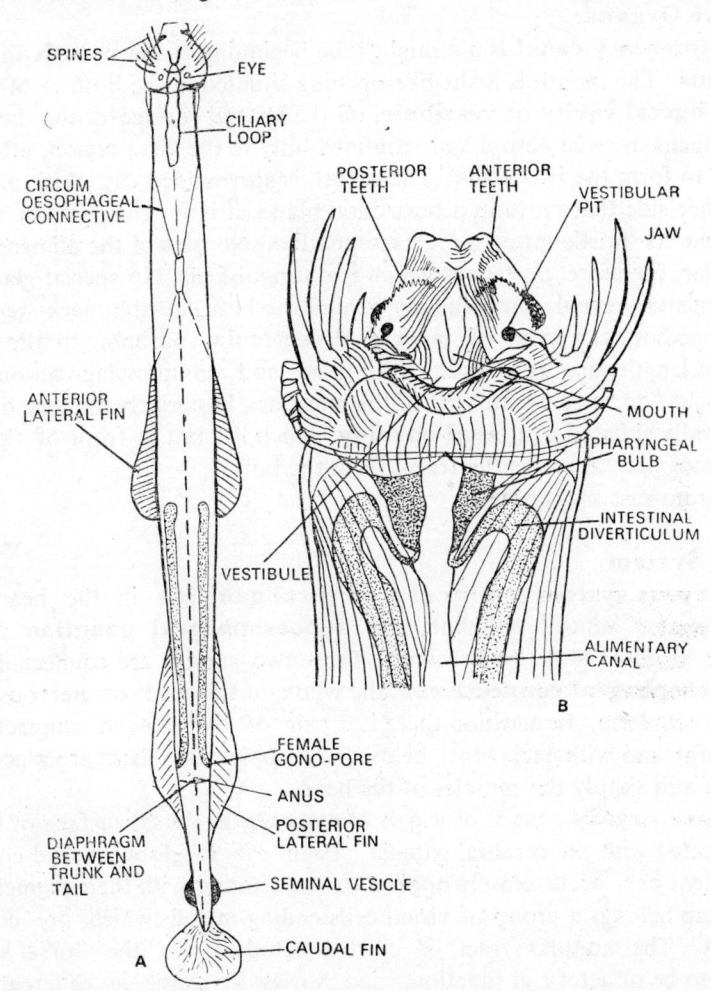

FIG. 20.1. A—*Sagitta elegans*, ventral view; and B—Head of *Sagitta elegans*, ventral view.

four bundles—two dorsal and two ventral. The muscle fibres are all cross-striped. The muscles, in their arrangement and appearance in transverse section, appear like those of the nematodes. In the head the muscles are broken up into bundles which work upon the jaws.

### Body Cavity

The coelomic body cavity is well-developed. It is lined by a layer of flat epithelium and is divided into two lateral halves by a longitudinal septum into which is situated the alimentary canal. Thus, the septum becomes divided into two parts—the **dorsal** and **ventral mesenteries.** Besides these, there are transverse partitions also. One is at the junction of the head and trunk and the other at the junction of the trunk and tail. Of these three chambers the posterior alone communicates with the exterior through the male generative ducts. The septa indicate segmentation.

### Digestive Organs

The **alimentary canal** is a straight tube beginning at the **mouth** and ending at the **anus**. The mouth is a slit-like opening situated at the bottom of a depression, the **buccal cavity** or **vestibule**, on the ventral surface of the head. The **mouth** opens into the oesophagus confined only to the head region, after which it widens to form the **intestine**. The mouth is surrounded by chitinoid hooks on its either side that move in a horizontal plane (it is for this reason that they are known as bristle-jawed or Chaetognatha). No part of the alimentary tract is glandular, therefore, no true stomach is distinguished. No special glands open in the alimentary canal. Immediately behind the head, in the neck region, the intestine possesses a pair of **lateral diverticula**, probably to allow certain amount of lengthening as a result of pulling and compressing actions of the head muscles, on the anterior part of the intestine. Posteriorly the intestine opens by a ventrally situated oval **anal opening**, which lies just in front of the transverse septum that separates the trunk from the tail.

There is no vascular system.

### Nervous System

The **nervous system** consists of a **cerebral ganglion** in the head and a large elongated ventral ganglion, the **suboesophageal ganglion** placed in about the middle of the body length. These two ganglia are connected by long **circumoesophageal connectives**. The whole of this part of nervous system lies in the ectoderm. In addition there is a pair of the ganglia connected with the cerebral and with each other below the oesophagus. These are placed in the mesoderm and supply the muscles of the head.

The **sense organs** consist of a pair of **eyes** on the dorsal surface of the head and connected with the cerebral ganglia. Each eye is globular and consists of five pigment-cup ocelli closely applied to one another with their pigment walls. To each cup belongs a group of visual cells ending in rods which are connected with brain. The annular tract of ciliated ectoderm on the dorsal surface is supposed to be olfactory in function, and varies in shape in different species. Besides these all over the surface there are **tactile papillae**. They are simply specialized ectoderm cells supplied with nerve endings.

### Excretory System

The **excretory system** is not well-developed. There are no segmental tubes nor any special organ of excretion. But there exists a pair of small glands situated dorsolaterally to the pharynx, one on either side. Some people were of the opinion that they were salivary glands, but they are not seen to open in the digestive tract, therefore, their digestive function is ruled out. Another view considers them excretory structures. They open to the exterior at the base of the hood (prepuce) and the base of the head and are known as **prepucial glands**.

## Reproductive Organs

The animals are hermaphrodite. The female organs or **ovaries** are contained in the middle segment of the body and the **oviduct** opens on each side of the anus at the junction of the middle and the caudal regions. The ovaries are placed one on each side. They are solid and are attached to the wall of the body projecting into the body cavity. The oviducts are narrow tubes extending along the outer sides of the ovaries. An inner sperm duct, which acts as sperm pouch, is found within the oviduct. The oviducts are blind anteriorly and open to the exterior in front of the posterior septum. The germ cells pass through the lining of the duct, it is believed. The exact methods have not been studied.

The **testes** are situated in the caudal chamber. They are in fact thickenings of the parietal coelomic epithelium in the anterior part of the chamber. Cells are released in the body cavity where they develop into spermatozoa. The male generative ducts are a pair of delicate tubes, the **vasa deferentia**, opening at one end in the coelom by a funnel-like opening. Each vas deferens, before opening, is dilated to form a **vesicula seminals**. The external male orifice is situated in the posterior regions of the tail.

**Development**. The ova are transparent structures which float on the surface of the sea, except in *Spadella cephaloptera* which attaches them to the sea-weeds. The mode of oviposition and place of fertilization is not known. For *Sagitta*, of course, it has been pointed out that fertilization takes place in the body cavity. The whole development takes place in the sea; there is no larval stage.

The segmentation is complete and leads to the formation of a gastrula by invagination resulting in the formation of a blastopore which closes soon. Mouth is a new formation. The archenteron gives off two folds from posterior end and thus becomes trilobed. The middle lobe later acquires an opening to the exterior by invagination at the front end, while posteriorly it opens into the archenteron besides the openings of the lateral pouches. Later the hind opening of the middle lobe is closed—it has only the external opening. This is the formation of enteron. The anus is formed later. Archenteron is separated off and so is the coelom. It consists of two lateral sacs opening into an unpaired portion behind. From the anterior end of the two lateral sacs a portion is cut off which apparently becomes the cephalic section of the adult coelom. The further changes of the hinder part of the coelom into the adult condition have not been followed.

In *Sagitta* the generative cells appear very early during segmentation. The two endoderm cells at the anterior end of the archenteron (opposite to the blastopore) soon increase greatly in size. Before long they migrate into the archenteron and divide forming a group of four cells, two of which subsequently become ovaries and two testes.

The gastrula gives rise to the deric epithelium of the pharynx through stomodeaum. From the same layer the nervous system arises at the later stage. The epithelium of the intestine arises from the mesial (inwardly turned) layers of the two endodermal folds. The muscular layer arises from the rest of the endoderm.

### Affinities

Slabber placed *Sagitta* with Vermes without any discussion, but since his time the subject has been much debated and several theories of its affinities have been put forward. O. Hertwig linked them with coelenterates on the basis of the similarity of the mesenteric cavity of Actinians. Gegenbaur, Haeckel and many others related them with the nematodes, because in both there is a tendency to develop cuticle, the alimentary canal is straight and the muscular system is similarly disposed. But the nematode body cavity is pseudocoelomate and that of the chaetognath is coelomic. Grassi compared the cerebral ganglia of the Arthropoda and Chaetognatha but this is not supported by embryological evidences. Seibold, d' Orbigny, Gunther, H. Milne Edward and many others linked them with Mollusca (Heteropoda). Meissner compared them with vertebrates and Metschnikoff with the Enteropneusta. In fact he formed one group Ambulacararia including the Echinoderms, Brachiopoda, Chaetognatha and Enteropneusta. Later works show that *Sagitta* cannot be compared with Enteropneusta.

The theory of their relationship with Annelida as supported by Kuhn, Hertwig, Kowalevsky, Huxley and others, appears attractive. Both Annelida (*Polygordius*) and Chaetognatha have a large body cavity and a longitudinal septum supporting the alimentary canal; have four groups of longitudinal muscles with striped fibres and pinnate four sections; circular muscles are lacking and epidermis is simple. The teeth and jaws are compared with setae of Chaetopoda and the ventral ganglion of Chaetognatha might be considered as a concentrated nerve cord. On the other hand, the Chaetognatha have no oblique septa, no nephridia, no blood vessels and embryology furnishes few points of comparison.

Development of *Sagitta* is very much abbreviated and as such it is difficult to interpret various structures in the light of embryology. The mesoderm develops by archenteric diverticula (and by pole cells in the annelida). A trochophore stage is lacking in the Chaetognatha and the segments of *Sagitta* arise differently.

Considering all these facts it seems more practicable to raise Chaetognatha to the status of an independent phylum with affinities with the echinoderms as the formation of coelom in both is similar (enterocoelic). MacBride and others believe that many years ago, there existed a stock of free-swimming forms, Protocoelomata, having a coelom still communicating with the gut. From this hypothetical ancestor have descended all the coelomates. The Chaetognatha have also evolved from the same stock and have remained unmodified.

# 21 PHYLUM ECHINODERMATA

The echinoderms are common and conspicuous marine animals. They have been known since ancient times. The name Echinodermata was first applied by Jacob Klein (1734), only to echinoids. Linnaeus in the 10th edition (1758) of his *Systema naturae* kept all invertebrates except insects in one class, Vermes. This was subdivided into the orders Intestina, Mollusca, Testacea, Lithophyta, and Zoophyta. Linnaeus placed the genera *Asterias, Echinus* and *Holothuria*, under the group Mollusca, which included a variety of naked, warty, or spiny animals, such as naked molluscs, some polychaetes, a few coelenterates and ctenophores, and *Priapulus*. He did not use the term echinoderm. The name Echinodermata was revived by Bruguiere (1791), who divided the Vermes into six orders: Intestina, Mollusca, Echinodermata, Testacea, and Zoophyta. Echinodermata was, thus, recognized as a distinct group of invertebrates. Bruguiere presented excellent figures of a number of echinoderms belonging to the asteriod, ophiuroid, and echinoid groups. He did not consider holothurians to belong to echinoderms. In 1801 Lamarck greatly improved the classification of the invertebrates, and recognized seven classes: Mollusca, Crustacea, Arachnida, Insecta, Vermes, Radiata, and Polypi. He placed echinoderms in the class Radiata, which was divided into two orders, Echinodermes and Mollasses. He included asteroids, echinoids, and holothuroids in the order Echinodermata, but Mollasses were medusoid coelenterates. This initiated the unfortunate association of the two main radiate groups (Coelenterata and Echinodermata) of invertebrates, which persisted for nearly fifty years. In 1847 Frey and Leuckart, in their textbook of invertebrate anatomy, separated Echinodermata as a group coordinate with the other main invertebrate groups. In 1854 Leuckart reiterated that the prevailing Radiata concept of Echinodermata was false and that they must be regarded as a separate main division of the animal kingdom, since their grade of structure is obviously higher than that of the coelenterates with which they had long been united. Since that time the Echinodermata have generally been regarded as a distinct phylum of invertebrates.

In 1875 Huxley proposed a group **Deuterostomata** for all the coelomate Bilateria, basing the name on the lack of relationship of the mouth to the blastopore. Deuterostomata was divided into three categories: Enterocoela for echinoderms, chaetognaths, and enteropneusts; Schizocoela for molluscs, polychaetes, and arthropods; and Epicoela for tunicates and *Amphioxus*. A little later Metschnikoff (1881) pointed out the striking resemblance between the larvae of echinoderms and enteropneusts and proposed to unite both groups under the name **Ambulacraria**,[1] a name that did not get further support. Apparently Goette (1902) created a [new category **Pleurogastrica** for the first time to include Chaetognatha, Enteropneusta, Echinodermata, and Chordata in one category. At the same time K.C Schneider (1902) presented the same assemblage of the four enterocoelous phyla under the name Enterocoelia. This has been supported by Grobben (1908) with minor changes here and there. He later created a phylum **Deuterostomia**. The terms Deuterostomia and Enterocoela or enterocoelous coelomates are now regarded as synonymous. They include the four phyla Chaetognatha, Echinodermata, Hemichordata, and Chordata. This relationship however, is too remote to justify the inclusion of these groups in a single phylum or even a superphylum. Hyman feels that the terms Deuterostomia or Enterocoela are covenient expressions to indicate certain similarities of embryological development but should not be employed as taxonomic categories.

The Echinodermata (*Gr. Echinos*, hedgehog+*derma*, skin) are "spiny-skinned" animals constituting rather a backward phylum of animals which are considered to have undergone a certain amount of retrogression in their features and structure. They are exclusively marine animals including starfishes (Asteroidea), brittle-stars (Ophiuroidea), sea-urchins (Echinoidea), sea-cucumbers (Holothuroidea) and sea-lilies (Crinoidea), etc. The echinoderm body plan is entirely different from that of other phyla, its relations however, are established by the larva. The animals are triploblastic, radially symmetrical usually with pentamerous arrangement of body parts in the adult stage, the larvae being bilaterally symmetrical. The body does not show any segmentation and there is no head. Most of the organs are ciliated. The body is covered by delicate epidermis over a firm mesodermal endoskeleton consisting of movable or fixed calcareous plates, usually in definite pattern, and with spines. In some (Holothuroidea) the skin is leathery and calcareous plates are microscopic. The digestive tube is simple usually having a second opening or anus (except in some). The coelom is large usually ciliated and is divided into several well-marked compartments carrying out different functions in life. One portion of the embryonic coelom forms the perivisceral coelom, a second forms the water vascular system and is connected with the exterior through a perforated plate (madreporite), a third part of the

---

[1] Also spelled Ambulacralia by all later authors.

## Phylum Echinodermata

coelom forms the haemal system, while the cavities from whose walls the gonads are developed are also coelomic. The water vascular system is provided with tube-feet serving for locomotion. There is no definite excretory system. Respiration takes place through minute gills or papillae protruding from the coelom, by tube-feet and in some (Holothuroidea) by cloacal respiratory trees. The nervous system comprises a circumoral ring and radial nerves arising from it. Sexes are separate. Rare exceptions also occur but sexes are alike externally. Gonads are large with simple ducts, ova are abundant and are usually fertilized in the sea. Larval life intervenes, the larvae being small ciliated transparent and usually free-swimming. Conspicuous metamorphosis follows. A few species are viviparous and some reproduce asexually by self-division. Many echinoderms are able to regenerate their lost parts rapidly.

The echinoderms present unique structural features such as the water vascular system, calcareous endoskeleton and the pedicellariae, etc., which isolate these widely from other invertebrates and give them an independent position. As a matter of fact they are quite ancient group differentiated from the Cambrian time.

Some biologists are of the opinion that the echinoderms have undergone retrogression from some type of more advanced and active ancestor. This idea is further supported by the fact that the larvae of echinoderms are free-living bilaterally symmetrical individuals that metamorphose into radial adults as they secondarily take to sessile life.

The echinoderm coelom is enterocoelic in nature. This places them in line with Chaetognatha, Chordata (and probably Phoronidea). But it is difficult to say whether it is point of affinity. MacBride and others believe that in the primitive

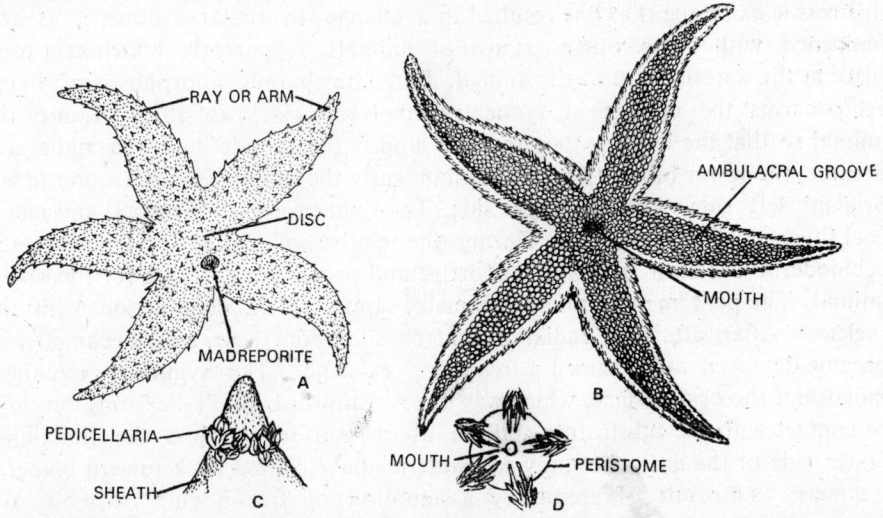

FIG. 21.1. A—Common starfish *Asterias*, aboral view; B—Oral view of the same showing ambulacral grooves; C—Tubercle of general surface of *Asterias* encircled with pedicellariae on a fleshy sheath; and D—Mouth armature of *Asterias*.

disposition of their coelomic sacs they present a certain resemblance to Chordata. These features include: (*i*) a mesodermal endoskeleton (not external or ectodermal as in other invertebrates); (*ii*) the retention of embryonic blastopore as the anus in the adults which in annelids and molluscs forms the mouth; (*iii*) the formation of mouth from stomodaeal inpushing that meets the endodermal archenteron; (*iv*) the appearing of mesoderm from outpocketings of the primitive gut; (*v*) the central nervous system never becomes separated by mesodermal tissue from the tracts of ectoderm from which it originates; (*vi*) special resemblance of the bipinnaria larva to the tornaria larva of Enteropneusta. All these points are concerned with fundamental and not superficial traits. These points of similarity suggest that the chordates and echinoderms may have had the same or similar ancestors.

The adult echinoderms are medium sized animals, the smallest being about one centimetre in diameter. *Pycnopodia helianthoides* is the largest starfish spreading to about 80 cm. Similarly *Echinosoma hoplacantha* is the largest of sea-urchins having a shell of 30 cm diameter. The spines of some tropical sea-urchins are about 22 cm long. *Synapta musculata* is the biggest sea-cucumber which attains a length of about two metres and diameter of five centimetres.

The echinoderms live on the sea-shore and sea-bottom, from the tide lines down to over 3,750 metres. They are slow-moving animals, mostly free-living and a few pelagic forms. None of them is parasitic. They do not form colonies though some may occur in large numbers. Some sea-lilies are permanently attached. Starfishes are of not much economic importance, some are known to damage commercial beds of oysters or clams. Some echinoderms are used as human food.

The echinoderms are believed to have originated from some coelomate ancestor which was bilateral and free-moving. The bilateral condition is perhaps recapitulated in the echinoderm larva. Some ancestral individuals of this type may have become attached to the bottom and assumed a sessile mode of life. It is this sessile existence that has resulted in a change to radial symmetry as has happened with some other groups of animals. Apparently attachment took place at the anterior end of the animal. Based on the metamorphosis of living echinoderms the change in symmetry involved a clockwise 90° rotation of the animal so that the left side became the upper (oral) side and the right side became the lower (aboral) side. Simultaneously the mouth moved around to the original left side, now the upper side. The right protocoel (axocoel) and mesocoel (hydrocoel) degenerated. During the course of these developments the echinoderm skeleton arose as supportive and protective structure for the sessile animal. The pentamerous radial symmetry appeared in conjunction with the skeleton. After attaining radial symmetry some of these sessile echinoderms became detached and resumed a free-living existence. The symmetry remained radial, but the oral surface, which was directed upwards in sessile forms developed contact with the substratum and the aboral surface became the functional upper side of the animal. Many echinoderms have shifted back toward bilateral symmetry as a result of secondary assumption of free-moving existence. All degrees between radial symmetry and distinct bilateral symmetry are illustrated by sand dollars, cake urchins, heart urchins and sea-cucumbers.

The phylum Echinodermata consists of the Echinozoa (sea-cucumbers), Homalozoa (fossil carpoids), Crinozoa (sea-lilies and various fossil groups) and Asterozoa (sea-stars). The older division of the phylum into subphyla Pelmatozoa and Eleutherozoa has now been given up. Of these the crinoids (sea lilies) are the most primitive class of living echinoderms, but in order to introduce the

*Phylum Echinodermata* 867

basic features of the echinoderm structure the more familiar asteroids (sea-stars) are being described first.

## CLASS ASTEROIDEA

The Asteroidea (Gr. *aster*, a star + *eidos*, form) includes free-living echinoderms belonging to the subphylum Asterozoa. The body is flexible and flattened and is in the form of a pentagonal or stellate (star-shaped) disc or often a disc continuous with usual five ray-like extensions or arms. The number of arms may be more. In some cases the adhesion of arms has resulted in the formation of pentagonal forms (*Pentaceros recticularis*). Each arm is provided with open ambulacral groove on the oral surface, provided with two or four rows of podia. Internally each arm contains gonads and a pair of digestive glands and the radial water canals located to the outer side of the ambulacral ossicles. The endoskeleton of the asteroids consists of separate calcareous pieces, bound together by connective tissue and usually bearing externally projecting knobs, tubercles or spines.

The asteroids are marine animals commonly called starfish or sea-stars (because the word "fish" is apt to mislead). The water vascular system is the most characteristic in *Asterias* and performs a number of vital functions. It starts with the madreporite situated on the aboral surface and gives off a system of tubes traversing the body. The pedicellariae are present. The larval forms are bipinnaria and brachiolaria.

### TYPE SEA-STAR

The sea-star lives on most sea-coasts and in the shore-waters from the tide lines to considerable depths in sand and mud. All the members of the genus *Asterias*

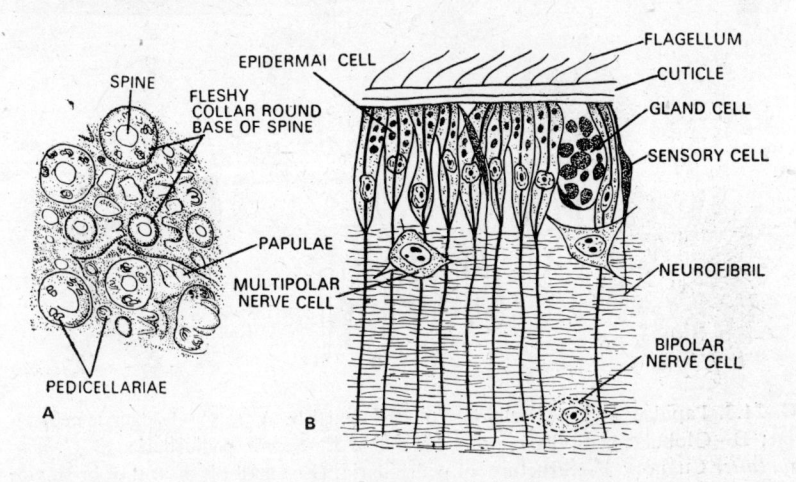

FIG. 21 2. A—Small area of aboral surface of *Asterias* (highly magnified); B— Section through the integument of *Asterias glacialis* (*after* Smith and Cuénot).

are benthonic animals as they inhabit the bottom of the sea. A few scattered individuals may be found on middle or sandy shores but they are quite scarce.

Due to food relationships they are usually found in the same area with marine clams, oysters and rock-barnacles. They are inactive during the day but at night they are much more active. Ordinarily they move about with the oral side next to the substratum and if turned over they set them right. There are little over one thousand species of sea-stars which differ in certain details but as a class are more similar in structure and habits. *Asterias rubens, A. vulgaris, A. forbesi, Pisaster ochraceus* and *Pentaceros* are some of the common species.

### Form of Body

The body (Fig. 21.1) of a sea-star is composed of a central disc from which radiate a number of arms. There are usually five **arms** or **rays** but some forms may have more, the sunstar (*Solaster*) for instance has 11-13 arms. The axes of the arms are termed **radii**, and the spaces between them on the disc are the **interradii**. The body has an **oral** or **actinal** surface on which the **mouth** is situated and an **aboral** or **abactinal** surface which is covered with spines of various lengths. In the centre of the convex aboral surface is situated a minute opening, the **anus**. It is doubtful whether the anus functions in egestion since it is an extremely small opening. On the aboral surface, which normally occupies upper position, there is an eccentrically placed sieve-like plate the **madreporite**. The two arms that lie adjacent to the eccentrically placed madreporite,

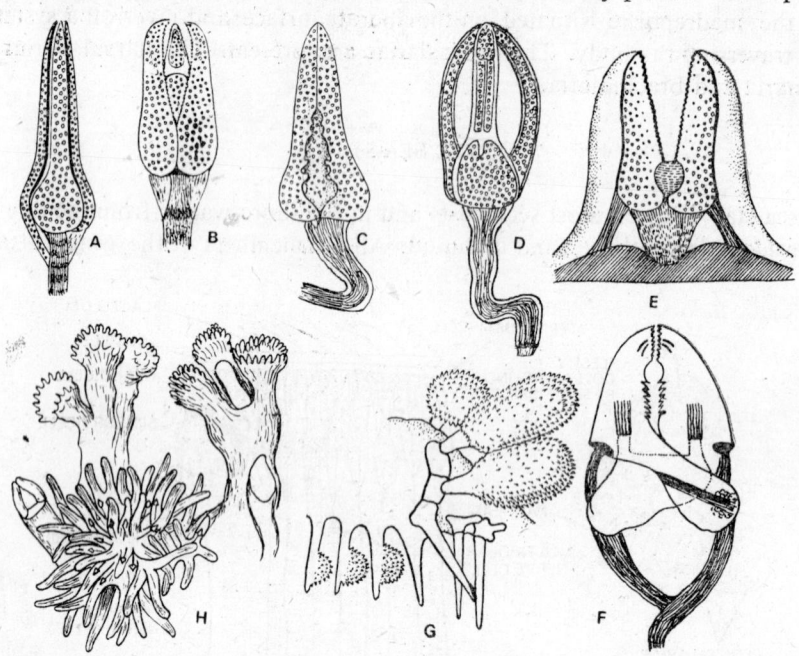

FIG. 21.3. Papulae and pedicellariae of the starfish. A to C—Tridactyle pedicellariae; D—Globular pedicellaria (*after* Koehler); E—Sessile pedicellaria of *Gymnatieria* (*after* Cuénot); F—Structure of pedicellaria; G—Small bit of arm of brisingid, *Odinella nuotrix* showing clusters of pedicellariae (*after* Fisher); H—Cluster of pedicellariae and branched papula of *Pycnopodia* (*after* Fisher).

with a portion of the latter are known to form the **bivium**. The **trivium**, likewise, is composed of the other three arms and the adjacent portion of the

*Phylum Echinodermata* 869

central disc. It is apparent from this that the presence of madreporite interferes with the radial symmetry. The two arms between which the madreporite is situated appear different from the other three. The sea-star can be said to be bilaterally symmetrical along one vertical plane passing through the middle of the madreporite and that of the arm opposite to it, but the symmetry is also slightly disturbed by the eccentric position of the anus.

On the aboral surface there are a number of short stout spines arranged, more or less, in rows on the arms and supported on irregularly shaped ossicles buried in the integument (Fig. 21.2). Some of the spines are fixed and others are movable. Between the ossicles there are numerous minute **dermal pores**, through which project small soft filiform cutaneous, retractile processes, the **papulae**, or **dermal branchiae** (gills). Their functions are respiratory and excretory. Peculiarly modified spines known as the pedicellariae occur in the space between the spines or in clumps around base of the spines. They are like small pincers meant to protect delicate breathing gills. Each is composed of a **basal plate** with which articulate two jaws or blades (looking like snaping scissor-blades mounted on a single soft handle (Fig. 21.3). The jaws are internally provided with muscles with the help of which they can be moved.

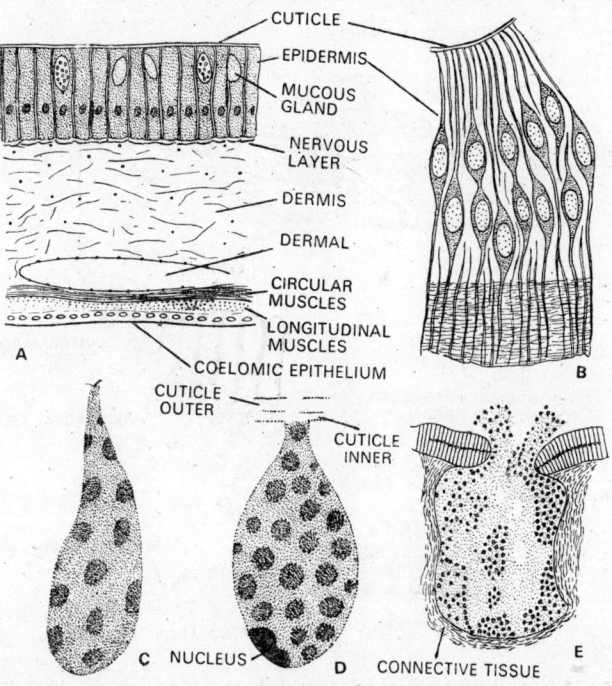

FIG. 21.4. Histology of the body wall. A—Schematized section through the asteroid body wall; B—Aboral epidermis of *Asterias*; C—Mucous type gland cell; D—Muriform gland cell; and E—Large gland cell of aboral wall of *Henricia*.

One family of asteroids (Porcellanasteridae), characteristically possesses **cribriform organs** (Fig. 21.5). When fully developed they consist of a vertical depression between adjacent marginal plates lined with a series of thin vertical plates like the leaves of a book or with vertical rows of papillae, supported by calcareous deposits and clothed with a tall ciliated epithelium. In *Ctenodiscus* the cribriform organs are reduced (Fig. 21.5) to a groove between successive marginal plates. The groups contain a few lamellae guarded by scalloped calcareous edges. The cribriform organs of *Ctenodiscus* serve primarily to sieve out coarser particles. The cribriform organs are found on the sides of the arms in the interradii between the marginal plates and vary in number from 1 (*Procellanaster*) to 14 (*Thoracaster*).

On the oral surface lies the five-rayed aperture called **actinosome** or **mouth** in the centre of the disc and surrounded by a **perioral** membrane, the **peristome**, and by oral spines. From the actinosome radiate five narrow grooves, the **ambulacral grooves**, of variable width, running along the middle of the arms to their extremities (Fig. 21.1). On either border of the ambulacral groove there are either two or three rows of movable calcareous spines, the **ambulacral spines**.

The groove also contains two (or four) rows of soft tubular bodies ending in sucker-like extremities called the **tube-feet** or **podia**, that act as locomotor organs. At the end of the ambulacral groove there is a sensitive reddish pigment spot, the **eye**. A median process, the **tentacle**, overhangs each eye. It is similar

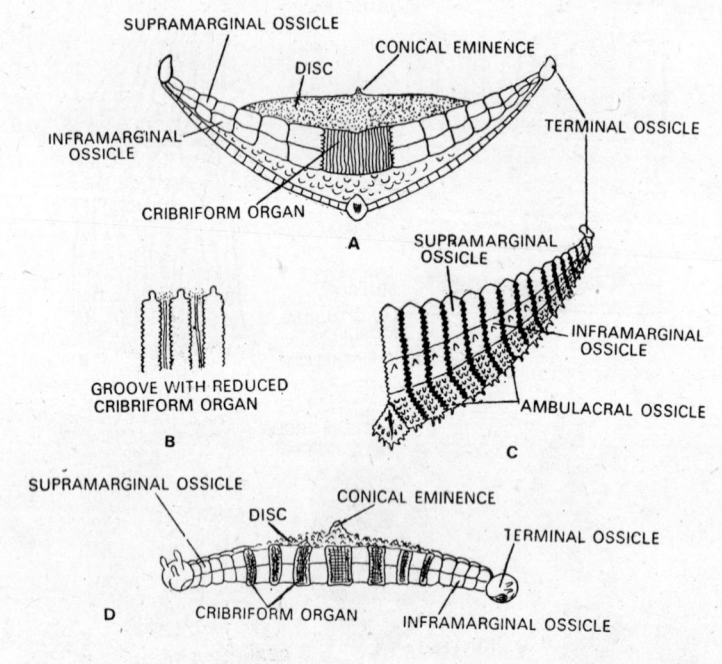

FIG. 21.5. Cribriform organs of starfish. A—Side view of *Procellanaster*, showing the single cribriform organ in the interradius; B—Three supramarginal plates of C, enlarged; C—Side view of an arm of *Ctenodiscus* with reduced cribriform organs along the length of the arm; and D—Two arms of *Hyphalaster*, side view with seven cribriform organs.

to the tube-feet but without sucker. The tentacle is believed to be tactile and also olfactory, as has been ascertained by some experiments. It is believed that the starfish is guided more by the olfactory sense of the tentacles than by sense of sight in capturing its food.

**Body wall.** The **body wall** is covered by a definite cuticle made up of two layers: an outer thicker homogeneous layer, and an inner layer marked with hexagonal impressions of the underlying epidermis. The cells of the epidermis are tall and flagellated, columnar throughout or attenuated at both ends or basally (Fig. 21.4). In the nervous regions the epidermal cells contain long elastic filaments that extend through the nervous layer. The epidermal cells are interspersed with neurosensory cells and with two sorts of gland cells (Fig. 21.4). These cells are goblet or mucous gland cells or muriform cells filled with coarse spherules. The epidermis contains pigment granules responsible for external colours. A nervous layer is found at the base of the epidermis. The epidermis is separated from the underlying thick dermis by a delicate basement membrane. The dermis is made up of fibrillar connective tissue which secretes and houses the endoskeletal ossicles and binds them together (Fig. 21.2). The dermis is followed by a layer of smooth muscles made of outer circular and inner longitudinal fibres. The muscle layer is weakly developed except in the aboral wall. The innermost layer of the body wall consists of coelomic epithelium (peritoneum). It is made up of cuboidal flagellated cells. Just to the outer side of the muscle layer in the dermis a ring canal runs around the base of each papula. (Fig. 21.11).

**Endoskeleton.** The endoskeleton is made up of calcareous ossicles that have a fenestrated structure. In the living condition the fenestrae are filled with connective tissue. The endoskeleton consists of the main supporting skeleton, embedded in the dermis of the body wall as well as of the more superficial skeleton of projecting **spines, tubercles, warts, granules**, etc. The main endoskeleton consists of discrete **ossicles** of various shapes bound by connective tissue. The ossicles may be rod-like, four-angled pieces or rounded, polygonal and squarish plates. The ossicles are so placed that they leave spaces for the emergence of groups of papulae or only papulae or in some case provision is made for special papular areas.

In some (Phanerozonia) the sides of the arms are made of two rows of large rounded, squarish or retungular plates often called **supra-** and **inframarginal plates.** The remaining body surface is supported by closely set plates that send up erect columns with expanded tops called **paxilla**. In some asteroids the aboral body surface with paxilliform ossicles looks like a field of flowers (of course under high magnification). The little tubercles or spinelets, forming the crown of the paxilla, are movable (by tiny muscles). They move horizontally foming a formidable surface when the animal is irritated. A larger spine or tubercle may be present in the centre of the crown.

In Phanerozonia the marginal plates are covered with little tubercles of spinelets as found in paxillae or they may produce a row of marginal spines. In some (Spinulosa) the endoskeleton is reticulated (Fig. 21.6). The endoskeletal ossicles are often poorly visible in the living animal as they are concealed by a thick leathery membrane. A definite arrangement of ossicles throughout the Asteroidea supports the ambulacral grooves. These are the ambulacral ossicles which never bear any spines, tubercles or other external appendages. Lateral to the outer ends of the ambulacral ossicles forming the edges of the groove is a row of adambulacral ossicles. In some asteroids there is present a row of small supra-ambulacral ossicles in the angle of the arms on the inner side. The ossicles are mesodermal in origin and lie in a flat position in the aboral positions of the body wall.

The **ambulacral ossicles** are elongated ossicles arranged in two rows meeting at the apex of the groove like the rafters supporting the roof of a house. These are, however (Fig. 21.6), movably articulated permitting opening or closing of the groove. At the end of the row the ambulacral ossicles terminate in a **median terminal ossicle**. Supporting the ambulacral spines and prominent tubercles at the edges of the grooves there is another row of ossicles on each side. Between

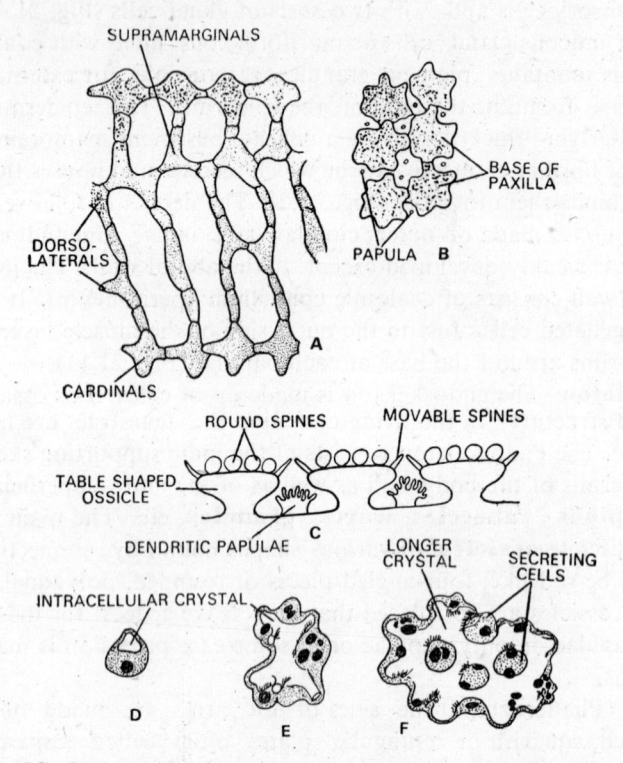

FIG. 21.6. Endoskeleton. A—Arrangement of endoskeletal ossicles of one of the Asteriidae from centre of ray laterally; B—Phanerozonic endoskeleton, seen from the inside, showing basal plates of the paxillae; C—Diagrammatic cross section through several paxillae of *Lindia*; D to F—Development of ossicles; D—Initial intracellular crystal; E—Larger; F—Still larger ossicle surrounded by numerous secreting cells.

the ambulacral plates there are a series of oval openings, the **ambulacral pores**, through which project the tube-feet. There are two rows of these openings, one row higher than the other, and the pores of the upper row alternating with those of the lower one. Five flat plates surround the mouth. These are known as the **oral ossicles**.

**Coelom.** The **body-cavity** or **coelom** is a large space within the body wall extending into the arms. It is lined by ciliated coelomic epithelium which also forms an investment for the organs of the cavity. The coelom in the living state is filled with **coelomic fluid** containing amoeboid corpuscles or **amoebocytes**. The coelomic fluid is similar in composition to sea-water and the amoebocytes collect particles of waste matter and when loaded with it pass it to the exterior

*Phylum Echinodermata* 873

by transversing the papulae. The amoebocytes also play an important part in removing foreign bodies from the organism acting as phagocytes as mentioned above. Developmentally the coelom originates from the enteron of the embryo as a single pouch which soon separates from the enteron and divides into a number of sacs, one (or one pair) of which is the **hydrocoel** and the others form **splanchnocoel**. The left hydrocoel gives rise to the water vascular system, the right is small (or lacking) and takes no part in the formation of the water vascular system. The splanchnocoel develops into the **perivisceral cavity** and its associated spaces (such as the haemal canals, aboral sinus, axial sinus, etc).

## Ambulacral System

The **ambulacral system** or **water vascular system** is peculiar to the Echinodermata. It is responsible for locomotion in the sea-star. The entire water vascular system is the modified part of the coelom. Essentially it consists of the **madreporite, stone canal, radial canal, Tiedemann's bodies, lateral canals** and **tube-feet**. The madreporite is the **sieve-plate** on the aboral surface

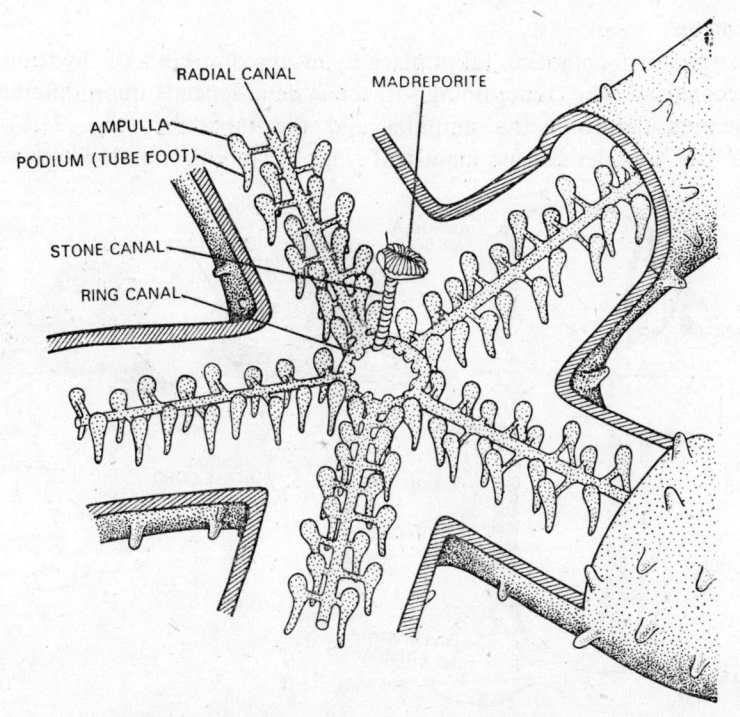

FIG. 21.7. Water vascular system of the starfish.

of the body, through which sea-water enters in. It (madreporite) leads into a S-shaped calcareous canal called the **stone canal** or **madreporic canal**. The walls of this canal are supported by a number of calcareous rings and its cavity is lined by cilia which are responsible for the movement of water through the canal. The madreporic canal opens into the **ring canal**, a pentagonal vessel around the mouth (Fig. 21.7). In most starfishes (except *Asterias*) from the ring-vessel arise a series of five pairs of thin-walled, pear-shaped bladders with long narrow necks. These are known as **Polian vesicles** and are interradially situated. On the medial surface of the ring canal there are nine small reacemose vesicles called **Tiedemann's vesicles**, the cavity of which is internally divided into many chambers, and which produce amoeboid cells or amoebocytes found in the fluid filling the system. Arising from the pentagonal vessel five **radial canals** or **ambulacral vessels** extend, one in each arm, above the ambulacral groove. From each radial canal arise paired lateral vessels each of which ends in a rounded muscular sac, the **ampulla**. Each ampulla extends ventrally as thin-walled hollow cylinder, the tube-foot, which ends in a sucker. Thus, a tube-foot is a closed cylinder with muscular walls having a sucker at outer or free end and a bulb-like ampulla (Fig. 21.8) at its inner end lying within the body cavity. The tube-feet also called **pedia** are organs of locomotion. Apart from locomotion, which is slow, the tube-feet also help the animals in the capturing and handling of food. In some they are tactile and respiratory. Internally the water vascular system is lined by a flat ciliated epithelium and contains albuminous fluid in which float the amoebocytes.

**Locomotion**

In the sea-star locomotion takes place by means of a kind of hydraulic pressure mechanism. The functioning of the system depends upon differences between the musculature of the ampulla and the tube-foot (Fig. 21.8). In the ampulla the muscles consist mainly of rings of smooth muscle fibres which are

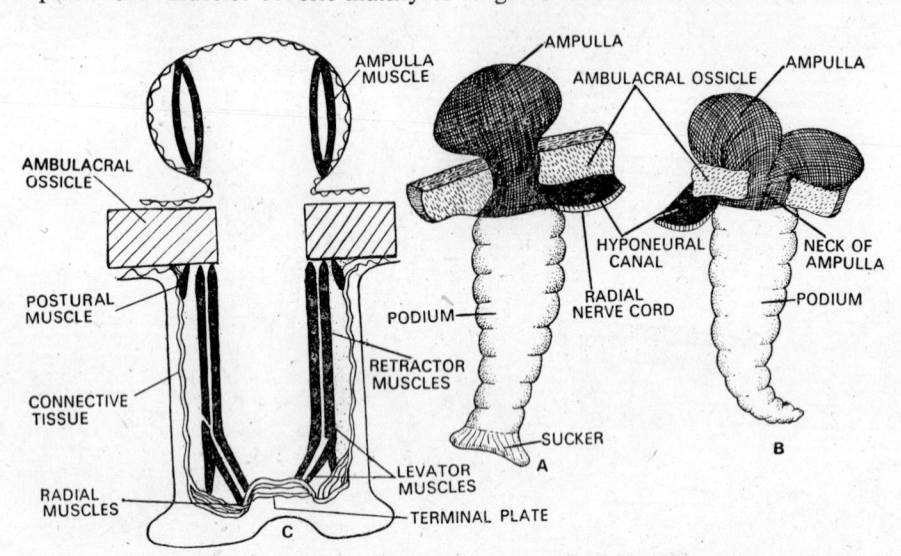

FIG. 21.8. Tube-feet. A—A scheme of ampulla and attached podium with a sucker in *Asterias rubens*; B—The same in *Astropecten irregularis* with podium without sucker (*after* J.E. Smith); and C—Arrangement of chief muscles in a longitudinal section through a foot and ampulla.

set vertically and which lie parallel to the long axis of the arm. Protraction of the tube-foot is brought about by the contraction of these muscles, the effect of this being to drive fluid out of the ampulla into the foot. The increase in pressure brings about the elongation of the foot, any wasteful lateral bulging of this organ being prevented by a collagenous sheath of connective tissue in its wall. This sheath allows extension but resists lateral pressure, so that it may be compared, from a functional point of view, with the circular muscle fibres in the tentacles of anemones. The musculature of the tube-foot (Fig. 21.8), in contrast to that of the ampulla, consists of longitudinal muscles, which are bounded interiorly by the ciliated epithelium of the coelom and exteriorly by the collagenous connective tissue, ectoderm and the cuticle. The hydrostatic pressure of protrusion brings about relaxation of these muscles (Fig. 21.9). Withdrawal of the foot is accomplished by their contraction. During this process

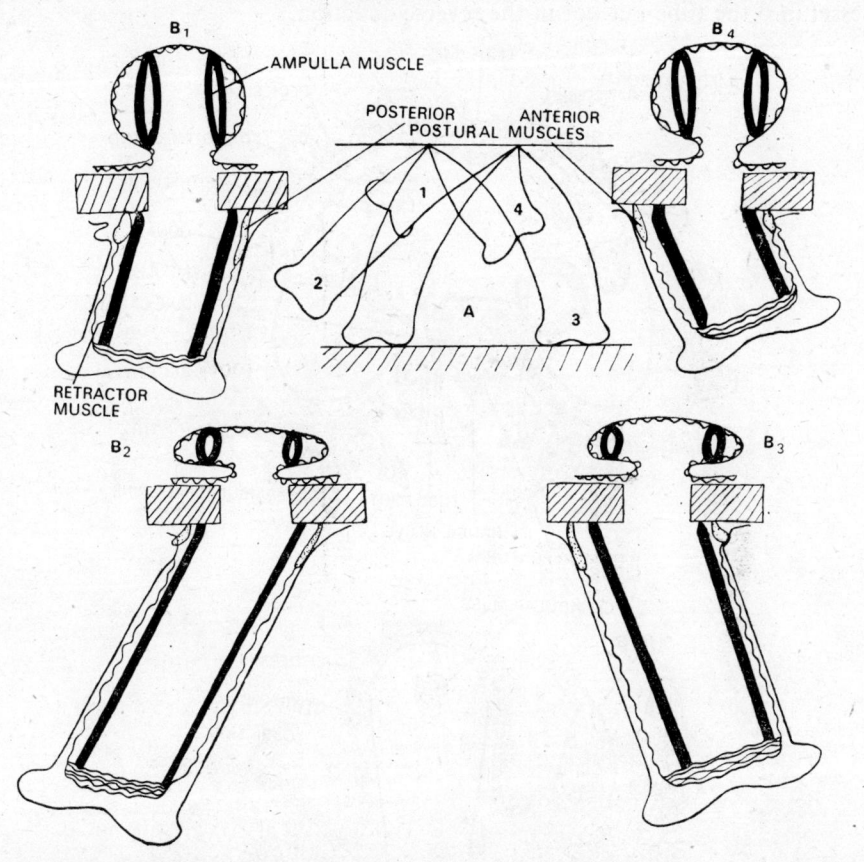

FIG. 21.9. Movement of tube-feet in *Asterias rubens*. A—1-4, the successive phases of the ambulatory step; B—1-4, show the conditions of contraction and relaxation of the protractor, retractor and postural muscles of the foot during the successive phases of static posture of the "ideal" step, the protractor and retractor muscles are shown in black, the orienting (postural) fibres are stippled. The anterior postural fibres orientate the foot in the forward direction, the posterior fibres in the backward direction of the step.

the connective tissue fibres become pushed together into layers. Bending of the foot is achieved by localized contraction of the longitudinal muscles, while postural muscles provide for its orientation. This makes possible the highly organized stepping movements by which these animals pull themselves along. The whole system is a striking example of the complexity of behaviour brought about by a **hydrostatic skeleton**, coupled with adequate specialization of the muscles and nervous system.

Protrusion is an entirely passive process, as far as the foot is concerned, thus in *Asterias rubens* the volume of the ampulla is about the same as the maximum extension attained by the foot. In this situation there must be no serious loss of fluid during protrusion and withdrawal, for the effect would be as disastrous as having a leaking hydraulic brake system. Loss is prevented by a valve at the junction of the tube-foot with the lateral vessel. This valve is essentially an extension of the lateral water vessel, arranged to permit fluid to pass from the vessel into the tube but not in the reverse direction.

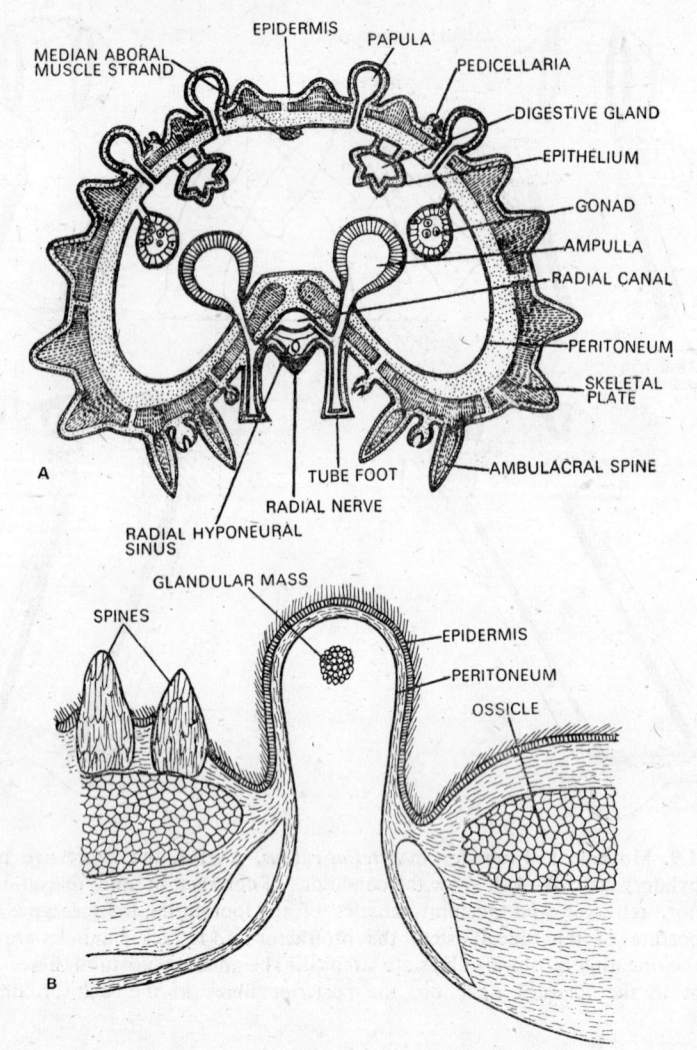

FIG. 21.10. A—A section through the arm of *Asterias*, nervous system black; and B—Section through an asteroid papula.

When the ampullae contract, the water is forced under pressure into the tube-feet, which extend as slender flexible processes that can be twisted about by muscles in their wall. Special valvular arrangement prevents the backflow of water in the ring canal. If the tip of a tube-foot touches the substratum or an object, the muscles of its wall may contract and return the fluid to the ampulla and the foot shortens. Withdrawal of the fluid lessens the pressure within the tip and causes it to adhere to the object because of greater pressure of the seawater or atmosphere outside the suckers of the foot, which, thus, work like vacuum cups. The tube-feet act either independently or in a coordinated manner. By the alternation of the activity (extension and contraction) of the tube-feet in different parts of the body the animal is able to move itself from one place to another.

## Haemal System

The **haemal system** is another derivative of the coelom comprising a system of cavities usually designated as **blood system** that is responsible for circulation in the absence of definite circulatory system. It is inconspicuous in asteroids, and has been studied mainly by means of serial sections. It is for the most part enclosed in coelomic spaces. The main vessel is the **oral haemal ring** that runs in the septum that subdivides the hyponeural or peribuccal ring sinus (Fig. 21.10). It gives off a **radial haemal sinus** into each arm. The radial haemal sinuses give off branches into the podia. From the oral haemal ring a haemal plexus ascends in the axial gland, and enters and **aboral haemal ring** which runs inside the **aboral** or **genital coelomic** sinus and gives off haemal branches to the gonads inside the coelomic branches to the gonads (Fig. 21.10). Close to its junction with the aboral haemal ring, the haemal plexus of the axial gland receives haemal strands, known as the **gastric haemal** tufts, which come from sinuses in the wall of the cardiac stomach and cross the general coelom to reach the axial gland. The gastric haemal tufts vary from one to four, their usual number, however, is two. The tufts are the only part of the haemal system not enclosed in a coelomic channel. There is a haemal channel at the base of each of the two mesenteries that attaches each pyloric caecum to the aboral wall of the arms, thus forming a total of 20 pyloric haemal channels. These communicate with the plexus in the stomach wall and so with the axial-gland haemal plexus by way of the gastric haemal tufts (Fig. 21.11). This system conveys products of digestion into the haemal system. It would appear that the haemal plexus of the axial gland is the centre of the haemal system, so that its original designation as heart is not surprising. As already indicated, the terminal process of the axial gland is enclosed in a contractile sac. Evidences confirm the contractions of the terminal process, the axial gland, the gastric haemal tufts, and the aboral haemal ring, indicating that there is little doubt that the fluid in the haemal system undergoes some movement.

The haemal channels have the usual histological structure essentially identical with the axial gland. Beneath the covering coelomic epithelium occurs a layer of connective tissue bounding internal channels containing coelomocytes.

The haemal system is well-developed in holothuroids and echinoids, less well-expressed in the other classes. As the channels of this system are not definite vessels with definite walls, the system may not be referred to as a vascular system, but rather as a lacunar system. The channels are best developed in relation to the digestive system and would seem to play a considerable role in the uptake and distribution of digested products. The fluid contained in the haemal system does not differ essentially from the coelomic fluid. A peculiarity of the haemal system of echinoderms is that most of the channels are enclosed inside tubular portions of the coelom. Old books describe these tubes of the coelom as the perihaemal system which is not correct. It may, however, be pointed out that all coelomic spaces contribute something to circulation, respiration, and excretion. The coelomic tubes surrounding the haemal canals have been termed by Hyman as the hyponeural canals because of their relation to the nervous system. The hyponeural system of canals appears functionally related to the nervous system, probably acting to cushion it against injury. But the mystery of enclosure of the haemal channels inside coelomic channels remains unsolved.

**The axial gland.** The **axial gland** is an elongated spongy body of brownish or purplish colour and is given several names as heart, ovoid gland, dorsal organ, septal organ, brown gland, etc. At its oral end the axial gland narrows and terminates in the septum that subdivides the hyponeural ring sinus while

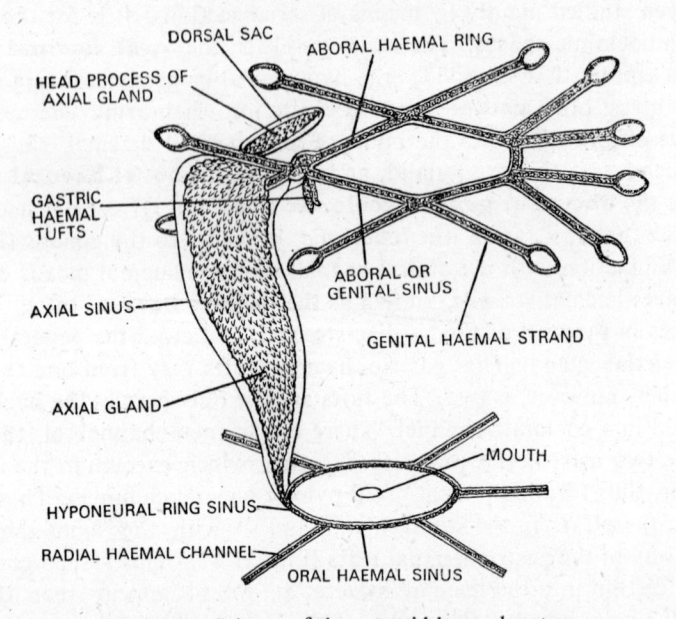

FIG. 21.11. Scheme of the asteroid haemal system.

on the aboral side the aboral extension of the gland is lodged in separate closed sac, the **terminal** or **madreporic vesicle**. The axial gland consists of connective tissue enclosing numerous spaces containing cells similar to coelomocytes. Externally it is bound by peritoneum. The axial gland is enclosed in a thin-walled tubular coelomic cavity, the **axial sinus**.

## Muscular System

The musculature is very variously developed in the different starfishes. A well-developed dermomuscular body wall occurs only in the holothurians. In all other echinoderms the skin is either slightly contractile or not contractile at all. In those with contractile skin the muscles are restricted to special bands acting on particular skeletal plates or other parts of the body, and to the muscular elements in the walls of tube feet and water vascular system and also in the viscera. In forms furnished with movable spines and pedicellariae muscles are attached to the base of these structures. The muscular tissue consists of contractile unstriped muscle fibres. The muscles of some pedicellariae and spines are striped.

## Digestive System

The digestive tract is a modified tube extending from the **mouth** on the oral side to the **anus** at the aboral surface (Fig. 21.14). The mouth leads into a short tube, the **oesophagus** and continues to the double pouched **stomach** with which it is not clearly demarcated. Of the two pouches one is the larger portion that receives the oesophagus and is called the **cardiac pouch**. It is a five-lobed structure, each lobe opposite to one of the arms. The walls of the sac are greatly folded. The structure is eversible. It everts out of the mouth by the contraction of the muscular fibres of the body wall. The other pouch, the **pyloric pouch**, is pentagonal in shape, each angle of which is drawn out to form a pair of large appendages, the **pyloric caecae**. Each pair of pyloric

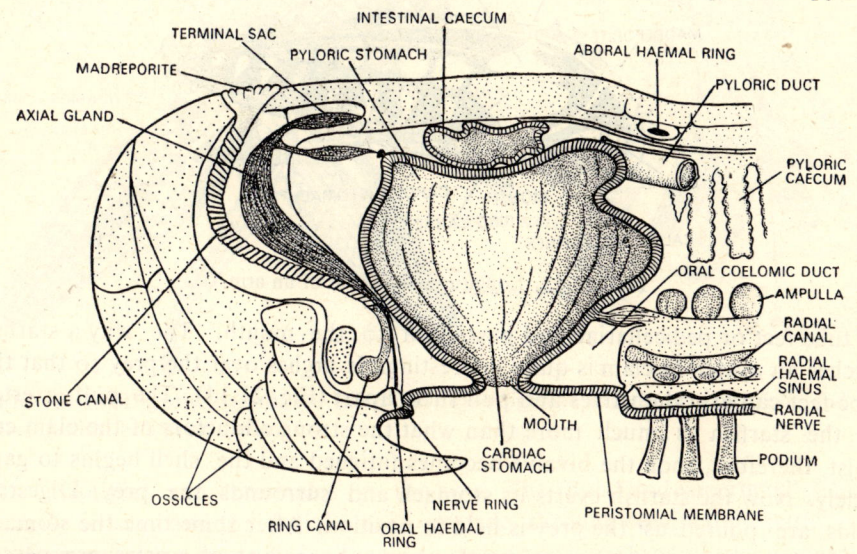

FIG. 21.12. Section of arm. Vertical section through the disc and base of one arm of an asteroid (*after* Cuénot).

caeca begins as a cylindrical canal or duct whose cavity is continuous with the cavity of the pyloric chamber. The duct bifurcates into two hollow arms that run almost to the end of the arm. Into each arm open several pouches, the walls of which are glandular and secrete digestive fluid. Hence they are also known as the digestive glands. From the pyloric pouch of the stomach a short tube, the **rectum** (or **intestine**) arises and opens aborally at the pore-like anus. Before opening it receives two brown branched pouches, the **rectal caeca** (or **intestinal caeca**). The buccal membrane is the part of the body wall around the mouth in the oral depression and is devoid of calcareous structures; the circular muscle fibres in it act as a constrictors and the longitudinal as dilators.

Echinoderms also have the ability to locate food by chemoreception. In the

On the other hand, if the circumoral nerve ring is cut between two radial nerves all movement is inhibited because coordination of podia in all of the arms is lost.

**Sense organs**. The organs of special sense are very poorly developed in the echinoderms. At the tip of each arm there is an **eye-spot** which can be easily located with the help of a hand lens. Each eye-spot presents a soft area having cup-shaped pockets. The lining of the wall consists of pigment cells containing orange pigment and rod-shaped visual elements. The animal also possesses an

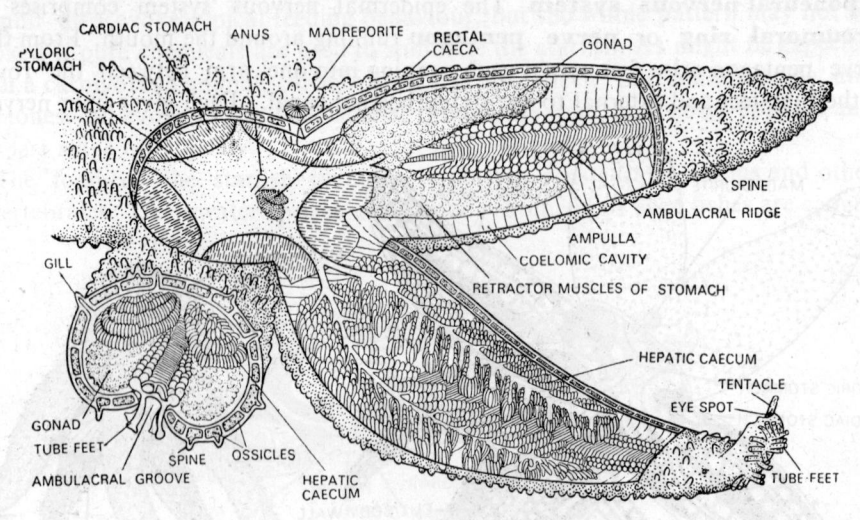

FIG. 21.15. General anatomy and reproductive organs of a starfish.

olfactory sense. The seat of this sense is not certain. It is either located in the tentacle or in the podia or at least in some of them, especially those located around the mouth. The sense of touch is scattered being located in the skin of the aboral surface. It is especially concentrated around the edges of the suckers of the tube feet.

**Behaviour**

The responses of the starfish to stimuli are too complex to be stated definitely. Starfishes can learn some performances. Professor Jennings taught individuals to use a certain arm in turning over. After 180 **"lessons"** extending over a period of 18 days one animal learnt the lesson and **"remembered"** it after an interval of seven days. Among these the young individuals were quicker to learn than the old individuals.

If placed on its aboral surface a starfish turns a sort of handspring by means of its arms, and performs the "righting reaction." The righting behaviour of inverted starfishes has been much studied as a piece of nervously coordinated activity. At the beginning of the movement the tip of each arm turns downwards, that is, towards the starfish's back when it is inverted. The arm which first makes contact "leads" the rest of the movement by turning further under, while the arms on the opposite side of the disc turn upwards and towards the leading arm. A starfish (*Astropecten irregularis* or *Asterina gibbosa*) will attempt to turn over if freely suspended in any position in water. But no attempt is made to right itself if its tube-feet are touched. Removal of much of the epidermis of the back has no effect, for it does not prevent attempts to turn when the starfish is put down on its back. Thus, contact on the back is not the stimulus, nor, indeed, does the stomach with its contents act as a statocyst-like gravity receptor, for its removal does not prevent the starfish from righting itself. Removal of the tube-feet does, however, prevent righting. But to remove the tube-feet, the ambulacral spines must also be removed and it remains to be proved whether the ambulacral spines or tube are responsible for tactile position reception.

When kept in aquarium tanks, starfishes, sea urchins and sea cucumbers can often be seen moving about on the vertical walls. Sea cucumbers and starfishes are as likely to go up as to go down. *Asterina gibbosa* for example, may creep up or down, according to the phase of behaviour it is in. If a cork float is attached to it, applying a strong upward pull, then the direction of its movement may be altered. Instead of moving upwards in the negatively geotactic phase it will go down, and *vice versa*. Thus, apparently, the pull of the body plays some part in the determination of the direction of gravity for this animal. Since it is supported on its tube-feet, it may be that an uneven stress is applied to the musculature of the feet which results in the perception of the direction. But this is not a highly sensitive response, for a slope of 5.6° fails to produce a response from the sea cucumber, *Thyone briareus*. It has been suggested that movement up the vertical faces may be simply continued movement in one direction, an animal crawling horizontally continuing to move when it reaches the wall of an aquarium, now moving upwards. This would account for the changes between going up and down, and remove these movements from consideration as true responses to gravity. But this is an unsatisfactory explanation, for it does not account for the behaviour of *Asterina*, for example, when it creeps towards the centre of a horizontal rotating turntable, behaviour is more consistent with a true gravitational sense.

**Reproductive System**

The starfishes are dioecious i.e., the male and female individuals are separate, but the male and female organs are morphologically alike and are situated in the same position in both; only microscopical examination reveals the difference. The genital system is remarkably simple. There are no organs of copulation, no accessory glands, no receptacles, etc, and the sex cells are directly discharged in the sea water where fertilization and development take place. In the axial organ (Fig. 21.10) there is a collection of cells, the **genital stolon**, which gives off sex cells at its aboral end. The **genital rachis** is a ring situated in the aboral sinus and continues below with the stolon. It is this genital rachis that becomes enlarged at intervals, corresponding to the arm, and gives rise to the gonads, located in the coelom at the sides of the arms near the disc. Each arm has a pair of them, of which each is a branched, lobed sac from which a duct opens to the outside through a perforated plate in the angles between neighbouring arms. This is the genital duct which is formed as the sex cells work their way through the walls of the arms for their exit.

**Fertilization**. The eggs and sperms released in the sea-water cannot survive long unless fertilized. Sperms are attracted to the egg and the first sperm that comes in its contact pushes its way through the membrane leaving its flagellum outside. On reaching in, it initiates physiochemical changes, the first of which is accumulation of fluid beneath the membrane which is elevated and thus, at least, new sperms cannot penetrate it. The sperm-nucleus migrates towards the egg nucleus and both undergo changes preparatory to the first cell division.

**Artificial fertilization.** Eggs and sperms of mature sea-stars and sea-urchins are quite easily obtained, as such they have been widely used by embryologists for experimentation. One of the most interesting phenomena is the development of the larva from an unfertilized egg when subjected to certain environmental conditions. This phenomenon has been called **artificial parthenogenesis** but since the development is initiated by subjecting the unfertilized eggs to various stimuli such as certain acids or concentrated sea-water it is preferred to call it **artificial fertilization**. Loeb reared normal larvae from unfertilized eggs of echinoderms by immersing them in solutions such as sodium chloride, potassium bromide, cane-sugar, etc. He attributed the cause of development to the increased

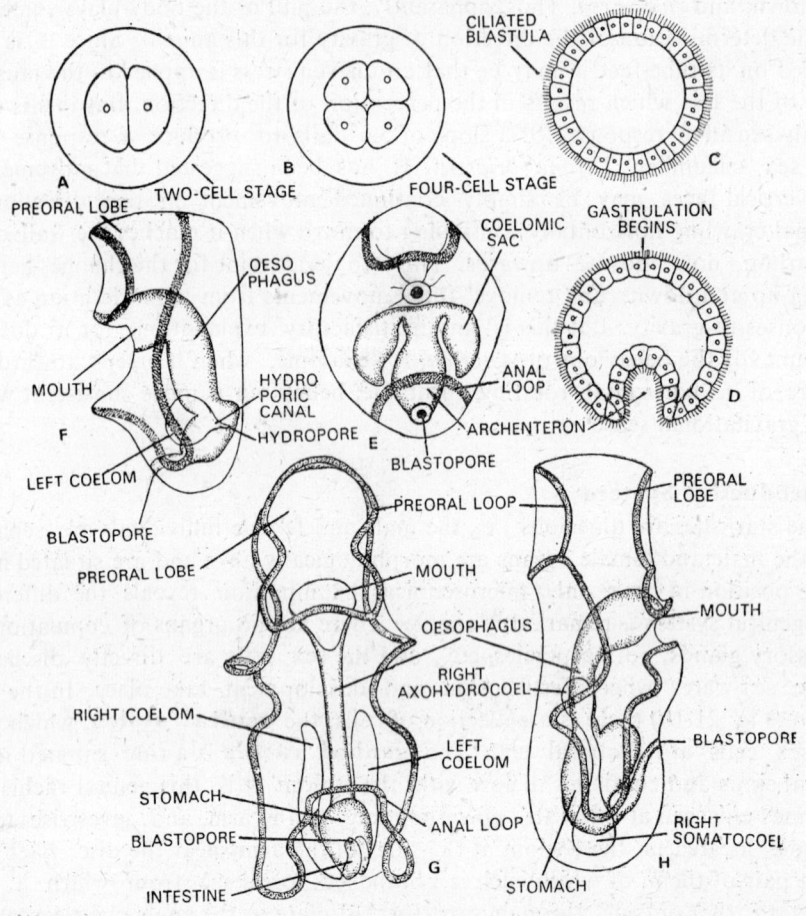

FIG. 21.16. Development of starfish. A to D—Early stages of embryology; E—Formation of coelomic sacs in *Astropecten auranciacus* with beginning of ciliary bands; F—Four-day stage of same; G—Ten-day larva; and H—The same seen from right side.

osmotic pressure and thought it probable that in ordinary fertilization the spermatozoon brings a solution with a high osmotic pressure into the egg, thereby causing the withdrawal of water. Sea-water concentrated to 70 per cent of its volume has a similar result. Change of temperature, mechanical agitation, etc. show similar results. The larvae that develop in these ways are perfectly normal. Strong centrifuging of unfertilized eggs causes them to become bell-shaped and ultimately break into two. The lighter halves contain the maternal nuclei; the heavier non-nucleated halves can be collected and fertilized by adding sperms to the dish. Such methods make it possible to isolate maternal and paternal characteristics. In the first the larva is normal without paternal characteristic and in the second it is without maternal characteristic. The echinoderm eggs, because of the ease with which they can be handled, have led to many important experiments on heredity and other similar topics.

**Embryology**

The development in starfishes takes place in two ways, (*i*) direct in which a free larval stage is omitted, and (*ii*) indirect with a free larval stage. The former mode is prevalent in asteroids with large yolky eggs that are usually brooded. Here only the indirect development is described.

After fertilization the zygote divides into two equal cells, the cleavage being total. In some cases the blastomeres towards the vegetal pole are slightly larger. The cleavage leads to the formation of **blastula** which sooner or later develops a blastocoel and is called **coeloblastula**. After a free-swimming existence of about a day the coeloblastula undergoes embolic invagination forming a two-layered **gastrula**, with an outer ectoderm and inner endoderm (Fig. 21.16). The archenteron formed as a result of gastrulation is narrow. When fully formed it gives rise to a coelomic sac, the hydroenterocoel, on each side. The two occupy the right and left of the archenteron and gradually elongate anteriorly in the plane of the long axis of the larva. From the left hydroenterocoel arises a dorsal evagination which forms the water pore. Ventrally a stomodaeal inpushing is formed. This meets the archenteron and completes the digestive tract. The blastopore forms the larval anus. The digestive tract is curved and differentiates into oesophagus, stomach and intestine, etc. The two hydroenterocoels continue to develop anteriorly where they fuse forming a U-shaped[2] coelom (Fig. 21.16). Finally the posterior parts of U are cut off forming right and left somatocoels.

The embryo develops cilia, hatches and leads a free-swimming life feeding[3] chiefly on diatoms. By the time the coelomic sacs separate in the interior the larva becomes angular in shape and the cilia concentrate to form a locomotory band, which takes up circuitous course forming loops anterioly and posteriorly. It forms a **preoral loop** around the ventral surface of the anterior end. Then it passes in front, about the mouth, proceeds backward along the sides and loops forward on the posterior part of the ventral surface, finally forming the **anal** or **post oral loop** anterior to the anus (Fig. 21.17).

The preoral loop separates and forms an independent loop (it is independent loop from the very beginning in many species). By now several slender project-

---

[2] In *Asterina gibbosa* the U-shaped sac is formed in a different way. The whole of the anterior half of the archenteron cuts off as single coelomic sac and from this arises a pair of extension, extending back alongside the digestive tract.

[3] In *Asterias rubens* and some other species the larva hatches in coeloblastula stage (after 24 hours) and begins feeding after developing the digestive tract.

ions of the body called arms are formed. These are bordered by the ciliary bands (Fig. 21.17). The arms grow longer as the larva develops and become very slender in some forms (*e.g. Asterias*). Such a larva with arms bordered by

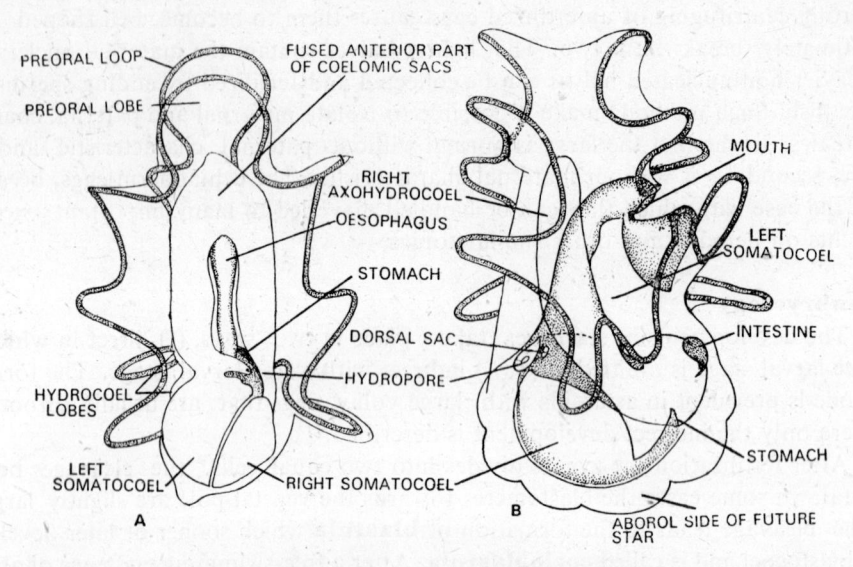

**FIG. 21.17.** A—Bipinnaria of *Astropecten auranciacus* dorsal view; and B—The same from right side (both *after* Hörstadius).

ciliary bands is known as the **bipinnaria larva** (Fig. 21.17), which can be distinguished after 4-6 days of development. The interior of the larva is occupied by the digestive tract and various coelomic compartments.

After some weeks of free-swimming life the bipinnaria may change into another larval stage, the **brachiolaria larva**. This develops three additional arms known as the **brachiolar arms**. One of them is anterior and median and the others are posterior and paired (Fig. 21.18). The brachiolar arms grow out from the anterior part of the ventral surfaces anterior to the preoral loop. The brachiolar arms are relatively short and carry the preoral loop along. Each arm has adhesive cells at its tip, and into each arm extends a prolongation of the coelom. An adhesive glandular area exists between the bases of the brachiolar arms, which constitute a device for attachment before metamorphosis.

The brachiolaria soon attaches to the substratum by the sucker and metamorphoses into the adult star. Its anterior part degenerates and is absorbed. The posterior part that contains the digestive tract and different body cavities changes into a baby star. It detaches and begins independent existence.

**Coelom formation.** As mentioned above the coelomic sacs form U-shaped cavity in the beginning. Near its posterior end on the left side the U bears the **hydroporic canal.** In the posterior side alongside the stomach lie the two somatocoels, the left one being larger than the right. Both somatocoels enlarge and extend forward. The left somatocoel gives off a process, the **ventral horn**, to the ventral side of the stomach in front of the intestine. This fuses with the dorsal axohydrocoel. The dorsal horn arises similarly and may fuse with the left hydrocoel and thus establishing a connection of left somatocoel with the left axohydrocoel. Thus, the left somatocoel with its derivatives surrounds the post-

## Phylum Echinodermata

erior part of the digestive tract and becomes the definitive coelom of the oral part of the disc. Part of the left axohydrocoel corresponding to the hydrocoel puts out similar lobes and forms coelomic lining of the podia. In the wall of the left hydrocoel appears a groove which forms the larval stone canal. The dorsal sac that lies beneath the madreporite in the adult originates in different ways in

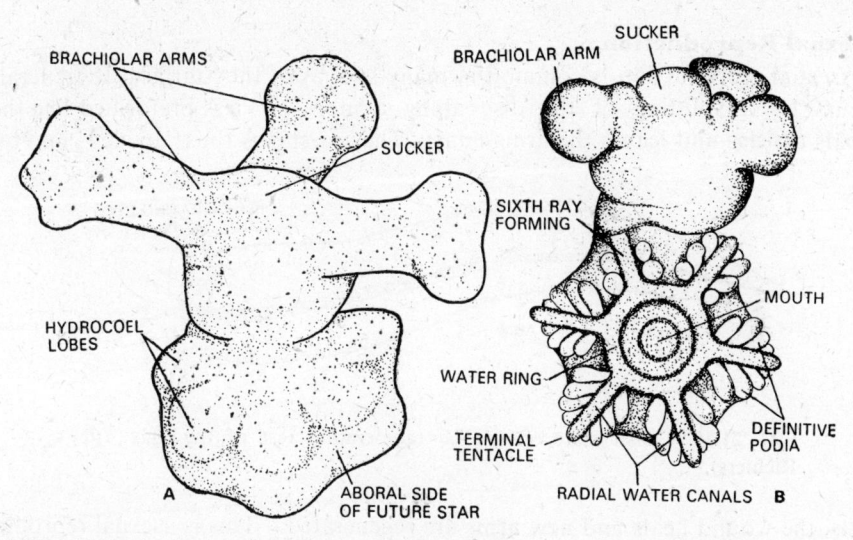

FIG. 21.18. A—Larva of *Leptasterias hexactis*, that develops in a brood-pouch and its metamorphosis into a star; and B—In which the attachment apparatus is degenerating.

different forms. In some asteroids it is formed by the rearrangement of mesenchyme cells, in others it is formed by an ectodermal invagination, while in still others it arises as an outgrowth of the right axohydrocoel. In the bipinnaria larva also the dorsal sac exists and it exhibits rhythmic contractions and is responsible for circulation of fluid in the larva.

### Metamorphosis

The larva is attached to the substratum by the sucker. The anterior half acts as a stalk and is absorbed later, while the posterior rounded portion forms the definitive star. As the anterior part of the larva degenerates, the lobulated hydrocoel makes a five-lobed disc on the left (Fig. 21.18). Ectodermal extensions cover these lobes forming five primary podia. Additional podia arise in pairs proximal to the primary podia. From the right (aboral) side five corresponding bulges are put out. These form the arms of the star. As soon as few podia are formed the star begins to attach itself to the surface and pull away from the sucker.

The internal structures also undergo many changes during metamorphosis. The most remarkable is the formation of new mouth and oesophagus. The old oesophagus breaks and most of it is absorbed. The portion of oesophagus that is left forms a projection of the stomach. After this the stomach undergoes a torsion with the result that the stump of the ruptured oesophagus is brought beneath the centre of the hydrocoel ring where it meets an ectodermal invagination with which it fuses forming a new oesophagus and mouth. The stomach also undergoes some changes with the result that it persists as a pyloric portion of the adult stomach. It gives off processes called pyloric caecae. Original intestine breaks away and the arms and distal portion of the intestine are absorbed. The remaining portion forms the proximal part of the adult intestine with intestinal caeca, distal part of the intestine is a new formation.

**Asexual Reproduction**

Asexual reproduction is common in many species of the starfish. It generally occurs by a splitting of the disc along a more or less preformed line that avoids ossicles and leaves the arms intact. The sea-star is thus divided into two

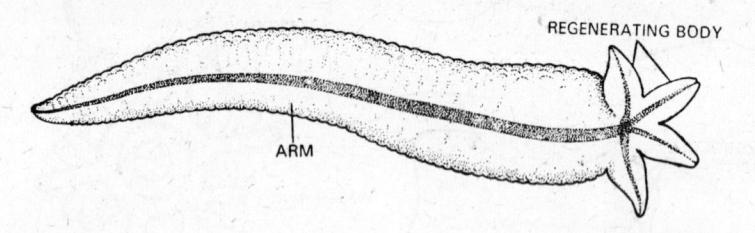

FIG. 21.19. Comet state of *Linckia* regenerating rest of the body (*after* Richters).

parts, the wound heals and new arms are regenerated. This is asexual reproduction by simple fission which does not take place during the season of sexual reproduction.

The genus *Linckia* has a habit of casting off arms, mostly at a more or less definite level (about two centimetres from the disc). This arm regenerates full sea-star. The factors that bring about autotomy are not known. The cast off arm generally regenerates four or five arms at torn surface. Such a regenerating arm is called a comet (Fig. 21.19).

**Regeneration**

The power of regeneration is remarkably developed in the starfish particularly with regard to its arms. Any or all of the arms of the starfish may be lost but the missing part is quickly regenerated. A single arm with a part of the disc will regenerate the entire body. The starfish is also capable of **autotomy**, i.e., **self-mutilation**. An injured or mutilated arm or one caught in the grip of some enemy is cast off by breaking at the point where it joins the central disc. The lost part is soon regenerated.

## CLASSIFICATION OF ASTEROIDEA

These are Echinodermata with starshaped body with a five rayed symmetry clearly indicated by the arms which radiate out from the central portion of the body. The arms are not distinctly marked off from the disc. The arms are hollow and enclose the prolongation of the coelom. The oral or aboral surface is distinct. The larva is either a bipinnaria or a brachiolaroia. There are about 1600 living species of asteriods. This class is divided into two subclasses.

(*i*) **Subclass Somasteroidea**. Extinct Palaeozoic starfishes in which the

## Phylum Echinodermata

skeletal structure of the arms shows a number of primitive and extinct features. The subclass has got only one living species *Platasterias latiradiata* that lives in deep water off the pacific coast of Mexico.

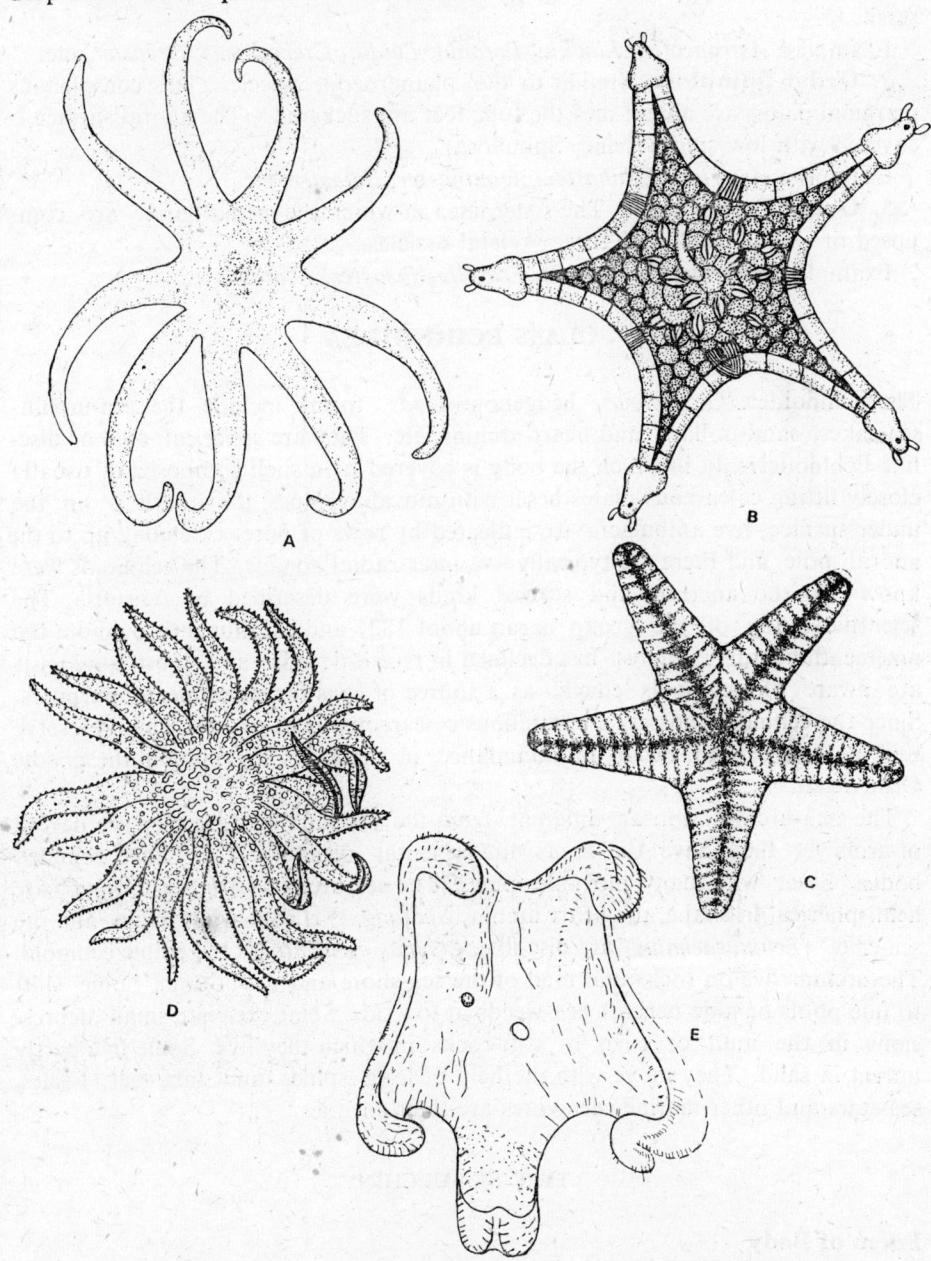

**FIG. 21.20.** Various sea-stars. **A**—*Solaster endeca*; **B**—*Albatrossater richardi*, most abyssal known asteroid; **C**—*Ctenodiscus crispatus*, oral side; **D**—*Heliaster kubiniji*; and **E**—*Tylaster willei*, high arctic star.

(*ii*) **Subclass Euasteroidea**. All the living species of asteroids, except *Plat-*

*asterias*, belong to this subclass which has three orders.

1. **Order Phanerozonia.** The Asteroidea with large marginal plates typically in two rows along the margin of arms and disc, in which the pedicellariae are sessile.

Examples: *Astropecten, Linckia, Porania, Luidia, Ctenodiscus, Oreaster,* etc.

2. **Order Spinulosa.** Similar to the phanerozone species, but conspicuous marginal p'ates are absent and the tube feet are suckered. The aboral surface is covered with low spines (hence Spinulosa).

Examples: *Asterina, Echinaster, Acanthaster, Solaster,* etc.

3. **Order Forcipulata.** The Asteroidea in which the pedicellariae are composed of a short stalk and three skeletal ossicles.

Examples: *Asterias, Pycnopodia, Freyella, Pisaster, Leptasterias.*

## CLASS ECHINOIDEA

The Echinoidea (Gr. *echinos,* hedgehog+*eiods,* form) include the sea-urchins sea-cakes, sand-dollars, and heart-urchins, etc. They are spherical, oval or disc-like Echinodermata in which the body is covered by a shell composed of usually closely fitting calcareous plates beset with movable spines, the mouth is on the under surface, five ambulacra are indicated by rows of pores extending up to the aboral pole and there are typically five inter-radial gonads. The echinoids were known to the ancients and several kinds were described by Aristotle. The scientific study of the group began about 1825 and continued throughout the nineteenth century. Interest has declined in recent decades and most zoologists are aware of echinoids chiefly as a source of eggs for experimental purposes. Since the echinoids possess a continuous calcareous theca or test they preserve well as fossils and occur in abundance in the rocks beginning in the middle Ordovician.

The sea-urchins appear different from the sea-stars because they are devoid of arms yet they have the same fundamental structure. They have rounded bodies beset with many movable spines. The sea-urchins (*Arbacia, Echinus*) are hemispherical in shape, the heart urchin (*Spatangus*) is somewhat ovoid, and the sea-cake (*Echinarachnius*) are disc-like. *Echinus esculentus* is the edible echinoid. The urchins live on rocks and mud of the sea-shore and sea-bottom. Some shift to tide pools or hide beneath sea-weeds at low tide. Some excavate small depressions in the mud or even in soft-rocks in which they live. Some live partly buried in sand. They move with the help of long spines and tube-feet. Fishes, sea-stars and other marine carnivores are their enemies.

### TYPE SEA-URCHIN

**Form of Body**

The body is covered over by a spine covered **external test** or **shell** composed of ten double rows of closely fitting calcareous plates, usually firmly sutured together (Fig. 21.21). The **mouth**, provided with five teeth, is in the centre of the oral surface, and is surrounded by a broad membranous area free from spines called **peristome**. Five pairs of branched **dermal branchiae** or "gills" occur around the margin of the peristome. There are five rows of long

*Phylum Echinodermata* 891

thin **tube-feet**, those on the oral surface being used in locomotion, but others in respiration. The anus opens in the centre of the aboral surface, within a central group of plates, the **periproct** (Fig. 21.22). This is surrounded by five genital plates, of which one (which is the **madreporite** also) is larger and finely perforated. The plates perforated for the tube-feet are the ambulacral plates corresponding to the rows of tube-feet on the five arms of the starfish. The intermediate rows which terminate in the genital plates are **interambulacral plates**. On the plates there are a series of rounded tubercles on which articulate the spines. Each spine has a cup-shaped base fitting over the tubercle (ball and socket joint) and can be moved by muscle appendages. Arround the basis of the large spines there are a number of very small **spinules**. Between the spines, here and there, occur three-jointed **pedicellariae**, borne on long flexible stalks. The pedicellariae in *Echinus* are of various types. Here and there occur small rounded bodies, the **sphaeridia,** which are perhaps spines modified into sense organs as they contain ganglia in the interior (Fig. 21.22).

**External appendages**. The most striking external feature of a sea-urchin is the covering of thickly placed spines, which may be approximately of one

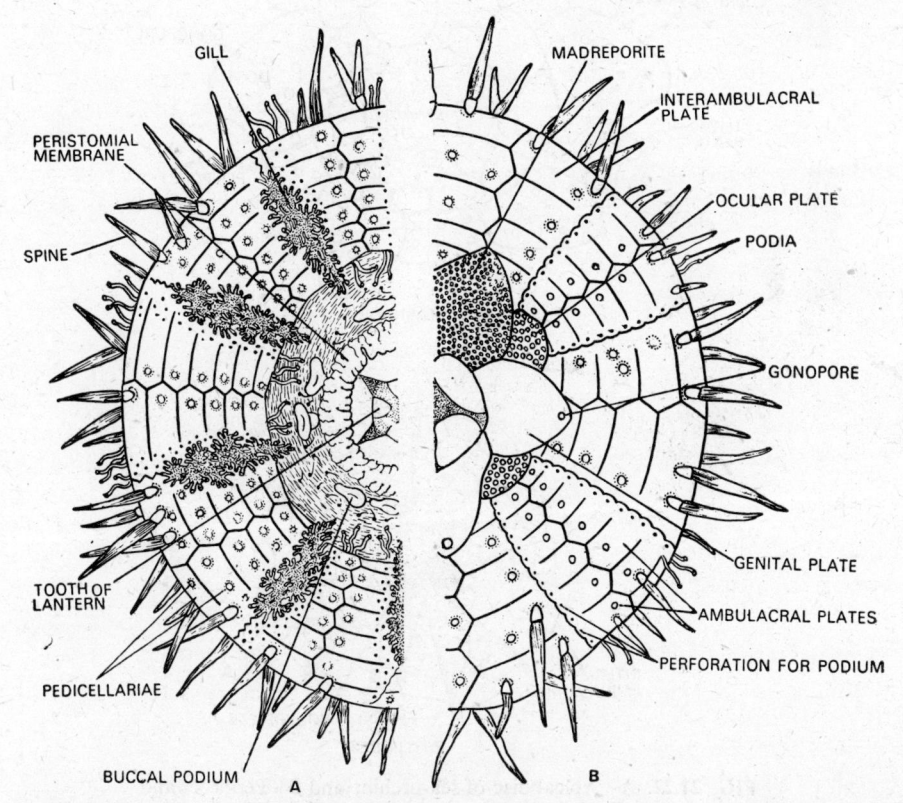

**FIG. 21.21.** External features of regular urchin (*Arbacea punctulata*) A—Oral view; and B—Aboral view.

size in some cases, but usually those of the oral and aboral areas are shorter than those around the sides. The spines are of various sizes and shapes. The larger spines are called **primary spines** (or radioles) and the smaller ones are the **secondary spines**. In the same way still smaller ones are called **tertiary** spines. In many echinoids there is no striking difference between primary and secondary spine and intermediate sizes may occur.

The typical spine is straight, circular in cross section tapering uniformly to a point or blunt tip. They are usually located on tubercles of the test, have an indented base that fits over the tubercle to form a ball-and-socket joint. Above the base, there is shelf-like projection, the **milled ring** (Fig. 21.23), marked with grooves (not unlike those on the milled edge of a coin). Above the milled ring a smooth band or **collar** may be present (Fig. 21.23). The main part of the spine known as the shaft lies distal to the collar. The shaft may be smooth or ornamented in some fashion or other.

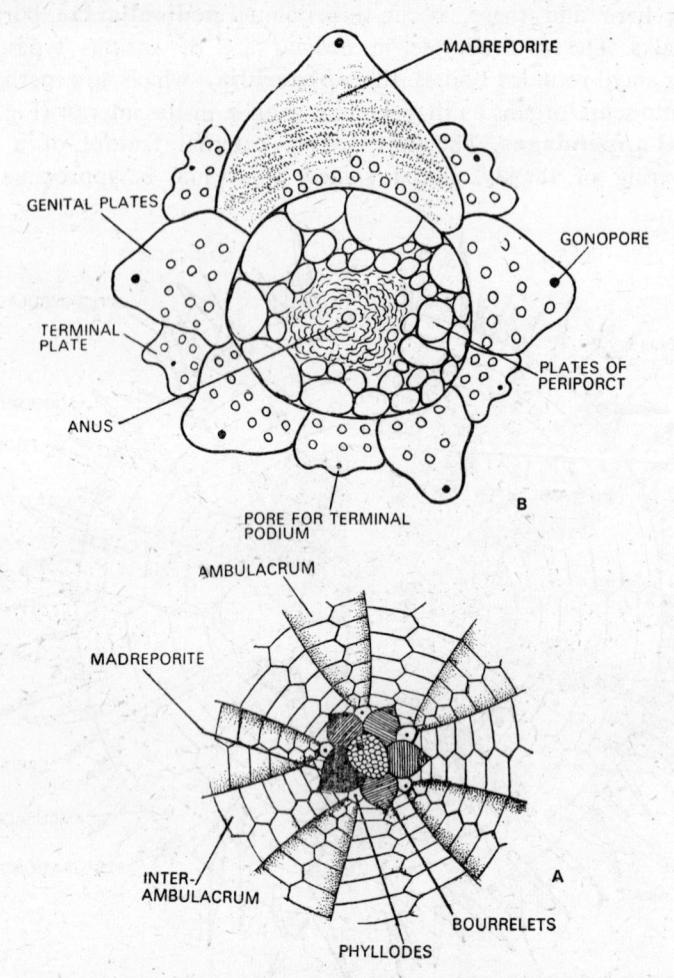

FIG. 21.22. A—Apical disc of sea-urchin; and B—Periproct and apical system of *Strongylocentrotus*, adult.

Other notable types of spines are sketched in (Fig. 21.23). The aboral spine of *Encope* is clubbed having a series of spines (Fig. 21.23 D); the aboral spines of *Goniocidaris clypeata* are mushroom-shaped (F). In *Asthenosoma varium* the aboral spines are poisonous. Each one carries a blue poison bag which makes

## Phylum Echinodermata

it conspicuous. The tip of the spine is surrounded by the poison bag. In some the primary spines of the oral surface have broad white ends, somewhat resembling hoofs (e.g. *Araeosoma*); on the other end the oral spine of *Laganum* is provided with crown end. The spines are composed of calcium carbonate intermingled with organic substance. The calcareous part is fenestrated, the fenestrae being filled with organic matter.

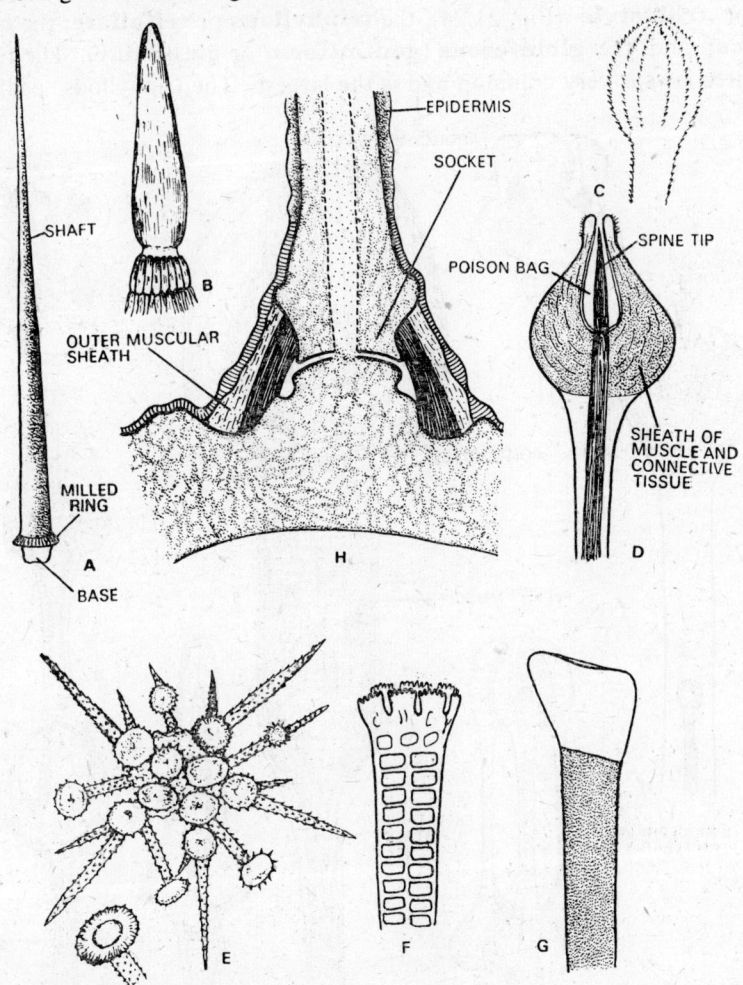

**FIG. 21.23.** Spine variations. A—Ordinary primary spine of a regular urchin; B—Primary spine of another echinoid encircled at base by secondary spines; C—Aboral spine of *Encope*; D—Aboral poison spine of *Asthenosoma varium*, an Indopacific spines; E—Mushroom-shaped aboral spines of *Goniocidaris clypeata*; F—Oral spine of *Laganum* with crown end; G—Primary oral spine of *Araeosoma* with hoof-shaped end; and H—Section through the base of *Cidaris* spine showing muscular sheaths.

**Pedicellariae.** The pedicellariae are minute appendages of the test peculiar to the echinoids and asteroids. They are always present in the echinoids and are always rather best developed here. They are borne abundantly on the test between the spines, and on the peristome and sometimes upon the periproct.

The pedicellariae consist of a head made up of three movable jaws (two, four or five may also be present) mounted on a stalk of varied length. Each jaw carries internal calcareous valves and the stalk carries internal skeletal rods (Fig. 21.24). Head is movable on the stalk and the jaws are also operated by muscles.

There are four general kinds of echinoid pedicellariae These include **tridentate** or **tridactyle** (Fig. 21.24), the **triphyllous** or **trifoliate;** the **ophiocephalous;** and the **globiferous** (**gemmiform** or **glandular**). The tridentate with three jaws is very common and is the largest. The triphyllous pedicellariae

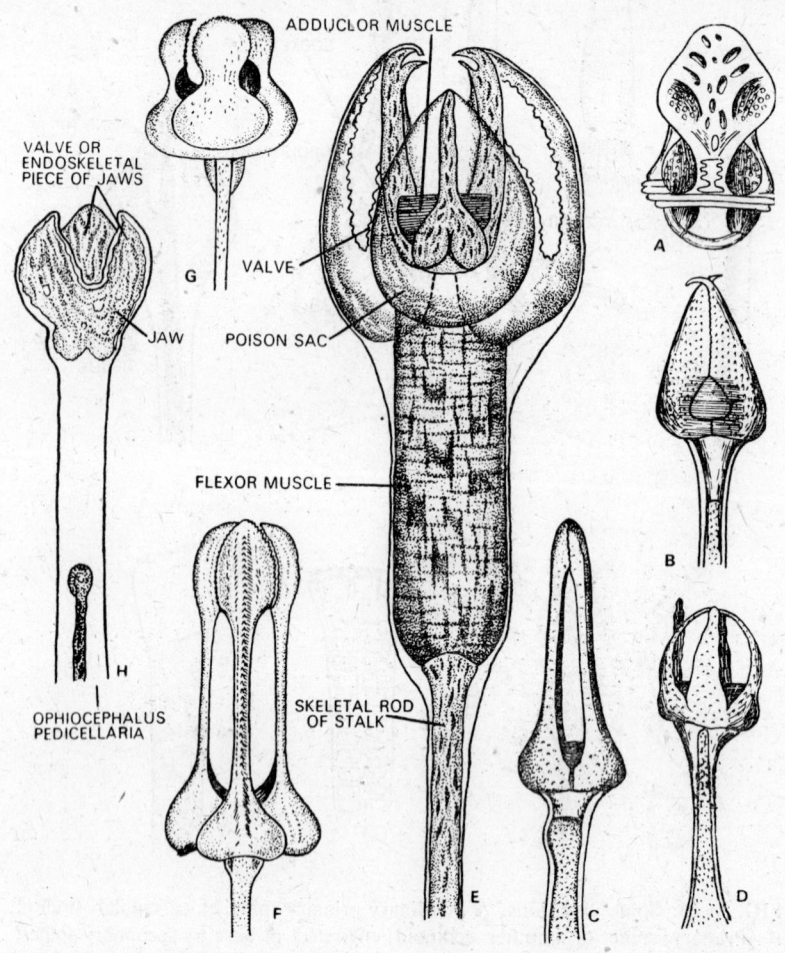

FIG. 21.24. Pedicellariae. A—Valve of ophiocephalous pedicellaria of *Tripneustes ventricosus* to show handle; B—Bidentate pedicellaria of sand dollar, *Echinarachnius parma*; C—Quadridentate pedicellaria of *Salenia*; D—Quinquedentate pedicellaria of laganid *Peronella* (*after* Mortensen); E—Globiferous pedicellaria of *Strongylocentrotus*; F—Tridentate pedicellaria of *Clypeaster rosaceous*; G—Ophiocephalous pedicellaria of *Clypeaster*; and H—Typical ophiocephalous pedicellaria of *Tripneustes ventricosus*.

are small with short broad jaws that do not meet distally. Some pedicellariae of this type have only two jaws (bidentate). The most interesting pedicellariae

## Phylum Echinodermata

are the globiferous ones provided with poison glands (Fig. 21.24E). Each jaw terminates distally in a sharp tooth bent inward, or a group of teeth and has a round plump appearance. The ophiocephalous pedicellariae have short stout inwardly concave jaws with blunt tips. Their valves are provided with a basal arc or handle whose interlocking holds the grip when the jaws are closed. Each of these has many variants. It is not possible to describe all here.

Besides these, there are minute glassy or transparent, hard, solid bodies of oval or spherical shape borne on the ambulacral areas of all echinoids except cidaroids. These are called **sphaeridia**.

The **podia**, in regular urchins, are arranged in five double rows on the ambulacra, extending symmetrically spaced from the peristome to the periproct. In some the rows of podia continue over the peristome to the edges of the mouth. In such forms buccal podia are wanting. The podia of regular urchin are mostly of locomotory type. They are long and slender and capable of great extension and are provided with a terminal disc or sucker supported internally by a circle of several calcareous pieces. The stalk of the podium is supported by calcareous spicules. The podia of the aboral side lack terminal discs and simply taper to a bluntly rounded end.

### Digestive System

Close to the edge of the peristome, on the inner surface of the shell project five processes, the **auricles**, within the ring of which lies a complex five-sided masticatory apparatus known as **Aristotle's lantern** (Fig. 21.28). It comprises

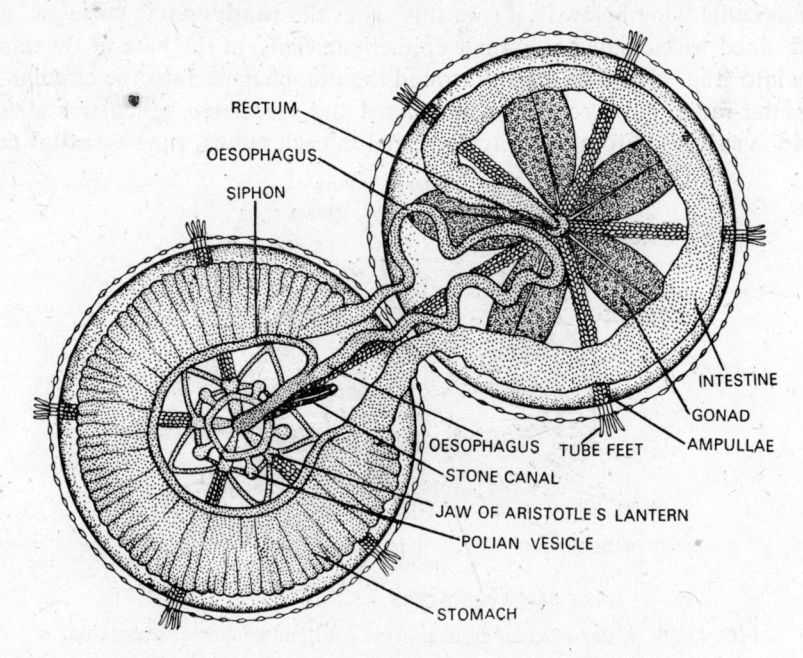

FIG. 21.25. Internal anatomy—Digestive organs of an echinoid.

five long, curved and pointed teeth, that project through the mouth, each supported proximally by a triangular ossicle, the **alveolus**. **Epiphysis**, a stout bar, is firmly united to the base of the alveolus. Adjacent epiphyses are in close contact with one another and from their inner margins arise **stout bars** or **rotulae**, the inner ends of which are bound together forming a circular aperture through which the oesophagus passes. With the inner end of each, rotula is movably articulated a more slender bar, **radius**, which runs parallel and closely applied to the rotula on its outside and is bifurcated at the free end. This is the masticatory apparatus which is worked by four sets of muscles and one set of ligament. The mouth leads into a five-rayed **pharynx**, each ray being attached by a pair of bands of connective tissue to the inner edge of the lateral face of adjacent alveolus. The pharynx opens into short **oesophagus** followed by a broad, lobed, coiled **stomach** which opens into a broad **intestine** coiled in the direction opposite to that of the stomach and finally there is a short rectum (Fig. 21.25). From the oesophagus to the posterior end of the stomach there leads a tube, the **siphon**, in the stomach wall. The food consists of sea weeds, but small organism and dead animal matter may be eaten. Chadwick, on the other hand, considers *Echinus* to be probably wholly carnivorous[4]. The occurrence of fragments of the shells of barnacles and serpulid worms attests the destructive power of the alveolar teeth.

### Water Vascular System

The water vascular system (Fig. 21.26) consists of tubular canals which communicate, on the one hand, with the exterior through the pores of the **madreporite** and, on the other hand, with the cavity of tube-feet and their ampullae. The madreporite, like that of the starfish, is perforated and opens into a madreporic ampulla lying below it. From this arises the **madreporic tube** or **stone canal** lined with ciliated columnar epithelium. This, at the base of the lantern, opens into the **circular canal** around the oesophagus. Into the circular canal open inter-radially five roughly pear-shaped and sacculated vesicular bodies, the **Polian vesicles**. From the circular canal in each radius, runs a **radial canal**,

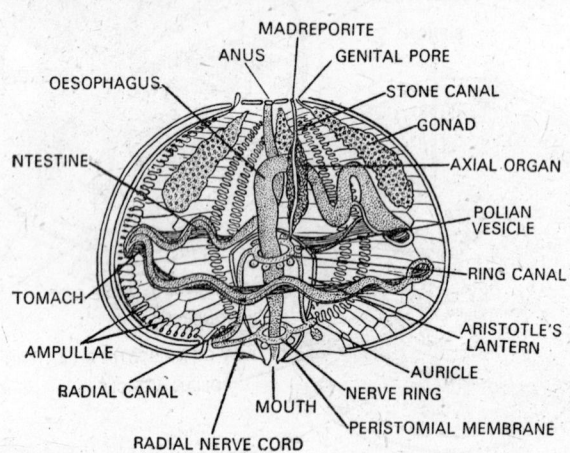

FIG. 21.26. Water vascular system, nerve cord and gonads of sea-urchin.

which passing beneath the rotulae down the outer surface of the lantern, traverses the whole length of the **ambulacrum**. On the inner surface of the test, it

[4] Chadwick, H., L.M. B.C. Memoir, *Echinus*.

## Phylum Echinodermata

runs to the apex, where it ends blindly in the pore of the radial plate. From this, during its courses, arise lateral branches opening into thin-walled ampullae of the tube-feet. The **tube-feet** are exceedingly mobile and can be extended far beyond the tip of the spines. Locomotion is brought about by the tube-feet with the help of spines. The fluid contained in the system is sea-water with traces of albumen. It is pale yellow or reddish in colour consisting of amoebocytes and corpuscles containing pigment.

### Nervous System

The nervous system consists of a ring of fibres and ganglion cells, the **nerve ring**, lying around the pharynx close to the mouth. From this arise five **radial nerves**, traversing the entire length of the ambulacra, upon the inner surfaces of the test, in close relation with radial canal. From the nerve ring nerves are given off to the integument and the wall of the pharynx in which they break up to form plexuses. Five lamellae of ganglion cells and nerve fibres lie in close proximity to the nerve ring at the points of origin of the radial trunks. From each lamella a pair of large nerve arises. These ascend along the edges of the five alveoli and are supposed to innervate their muscles. This has been described as **deep nervous system**. But for this, the deep and the aboral nervous systems are poorly developed. Organs of special sense do not occur. **Sphaeridia** and the **terminal tentacles** of the water vascular system are the only known sense organs of *Echinus esculentus*. The tube-feet are highly sensitive to external stimuli and those of *E. microtuberculatus* have a tubercle-like thickening of ectoderm close to the sucker disc, beneath which there is a ganglionic mass of nerve tissue.

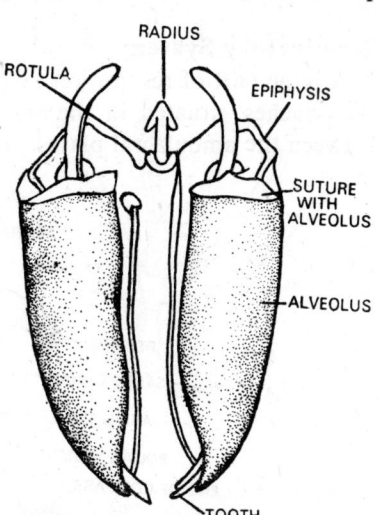

FIG. 21.27. Side view of Aristotle's lantern.

### Vascular and Coelomic System

Besides the radial canal of the water vascular system, two other canals traverse the **ambulacra**, in relation to the radial nerve trunks. Of these the **epineural canal** runs between the nerve and the test and the **pseudohaemal canal** runs along the inner axial face of the nerve trunk. Both these end blindly at the apex of the test in close proximity to the terminal tentacle and do not open into the circular epineural canal surrounding the pharynx. Little is known about the ontogenic history of the pseudohaemal system. The canals contain fluid similar to that found in the coelom and probably they provide nourishment to the nerve trunks with which they are so intimate.

The **blood vascular system** of *Echinus* and its allies has been subject of much discussion. There are undoubtedly two vessels which run alongside the inferior coil of the intestine in the mesentery which supports the latter. The **ventral vessel** runs along the inner axial side and the other along the outer

3. **Order Echinothurioida**. Flexible test with hollow spines. Simple ambulacral plates on peristomial membrane. Gills inconspicuous or lost.

Example: *Asthenosoma*.

(b) **Superorder Echinacea**. Sea urchins with rigid test and solid spines. Gills present.

1. **Order Arbacioida**. Periproct with 4 or 5 plates. Mid-Jurassic to Recent.

Example: *Arbacia*.

2. **Order Salenioida**. Anus located eccentrically within periproct because of the presence of large anal plate. Lower Jurassic to recent.

Example: *Acrosalenia*.

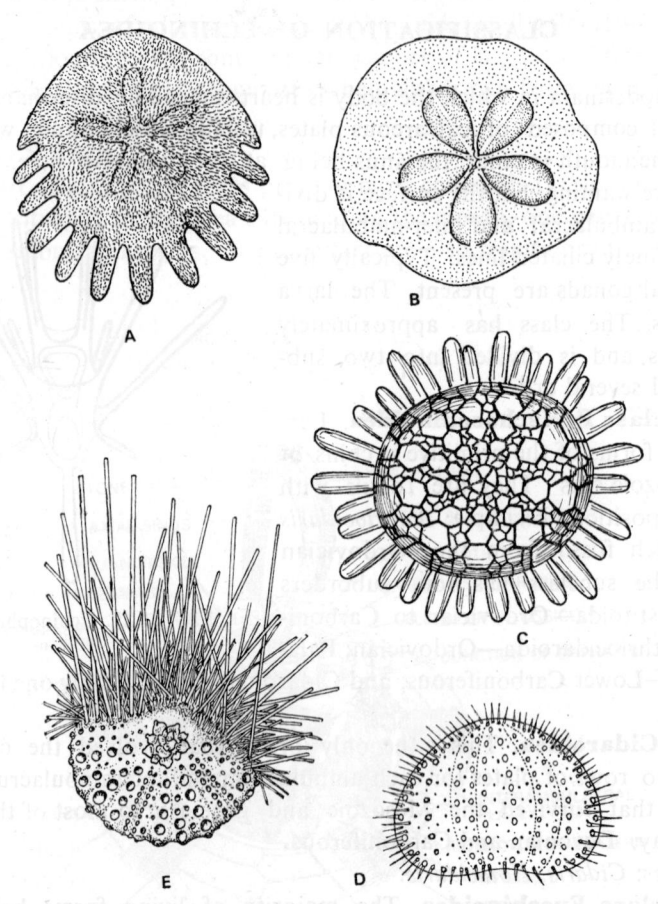

FIG. 21.30. Various types of sea-urchin. A—The African sand-dollar *Rotula orbiculus*; B—The West Indian sand-dollar, *Clypeaster roseaceus*; C—*Colobocentrotus atratus*; D—*Echinus*; and E—*Echinocidaris (Arbacea) pustulata*.

3. **Order Temnopleuroida**. Test usually sculptured. Camerodont lantern (large epiphyses are fused across top of each pyramid). Lower Jurassic to Recent times.

Examples: *Toxopneustes, Lytechinus*.

4. **Order Hemicidaroida**. Periproct without suranal plate, primary tuber-

cles perforate and lantern noncamerodont. Upper Triassic to upper Cretaceous.

5. **Order Phymosomatoida.** Characters similar to Hemicidaroida, but tubercles imperforate. Upper Triassic to Recent times.

Example: *Glyptocidaris*.

6. **Order Echinoida.** Test nonsculptured with imperforate tubercles; lantern camerodont. Lower Cretaceous to Recent.

Examples: *Echinus, Paracentrotus*.

(c) **Superorder Gnathostomata.** Irregular Echinoida in which mouth is in centre of oral surface but anus has shifted out of apical centre. Lantern present.

1. **Order Holectypoida.** Fossil members essentially regular in shape. No petaloids. Two living genera *Echinoneus* and *Micropetalon* are oval. Jurassic to Recent.

Examples: *Echinoneus, Micropetalon*.

2. **Order Clypeasteroida.** Test greatly flattened, no phyllodes present, petaloids present. True sand-dollars. Late Cretaceous to Recent.

Examples: *Clypeaster* (Fig. 21.30B), *Rotula* (Fig. 21.30A).

(d) **Superorder Atelostomata.** Irregular Echinoidea with lantern absent.

1. **Order Holasteroidea.** Oval or bottle-shaped echinoids with thin delicate test. Petaloids and phyllodes not developed. Deep water species. Arose in Jurassic and still survive in deep waters.

Example: *Pourtalesia*.

2. **Order Spatangoida.** Oval or elongate echinoids usually called heart urchins. Oral centre shifted anteriorly and anus shifted out of aboral apical centre. Petaloids present but may be sunk into grooves. Phyllodes present. Occur in low Cretaceous and is still found in existing seas.

Examples: *Spatangus, Echinocardium*.

3. **Order Cassiduloida.** Mostly extinct echinoids with round to oval test and a central or slightly anterior apical centre. Few existing species similar to sand-dollars.

Examples: *Nucleolites*, a fossil of the Mesozoic; *Apatopygus*, inhabit tropical seas.

4. **Order Neolampidoida.** Similar to Cassiduloida, but without petaloids or phyllodes.

## CLASS OPHIUROIDEA

The Ophiuroidea (Gr. *ophis*, a snake +*oura*, a tail+*eidos*, form) are flattened in which arms are distinct from the disc (Fig. 21.31). They are often called serpent-stars from the snake-like appearance of their long, slender arms. Since their arms have the tendency to break off readily they are also called brittle stars. The Ophiuroidea were separated from Asteroidea by Norman (1865). The knowledge of the Ophiuroidea was greatly enhanced by the study of the materials secured by the numerous dredging and collecting expeditions. They resemble the asteroids in having arms but in other respects they are quite different. There are no ambulacral grooves on the arms as they have been covered by the skeletal plates and converted into the epineural canal. There is no anus and the madreporite is on the oral surface. The arms do not possess the gonads or caecae. The ophiurans live in shallow to deep water. They bury themselves in mud or sand in the day isolated calcareous bodies; contractile tentacles surround the mouth, the apical system of plates is not developed and the water vascular pore is usually absent in the adult. The axial organ and generative axis are lacking. The holothurians have been known from ancient times as they are conspicuous animals of the sea shores. They are commonly called sea-cucumber. The common name has been apparently derived from Pliny's term *Cucumis marinus* that he applied to true holothurian. About 500 species of the class are known. The holothurians are

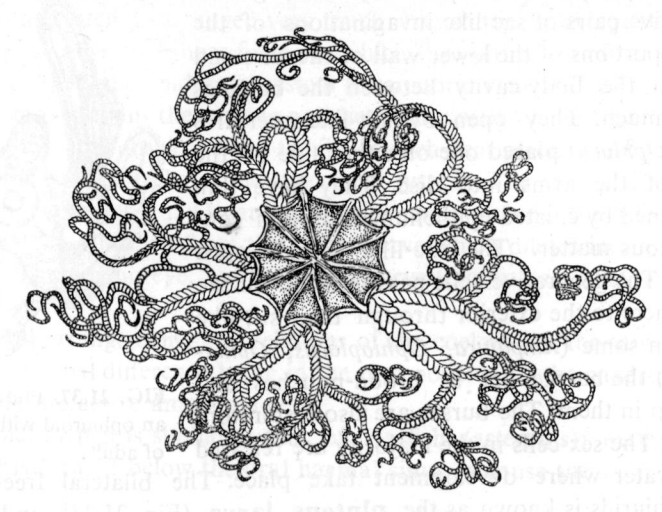

**FIG. 21.38.** *Astrophyton.*

elongated on principal axis at or near the two ends of which the mouth and anus are situated. The most striking external features include its muscular body wall almost devoid of large skeletal plates; its branching tentacles surrounding the mouth, and its lateral position when at rest or moving about on the sea-bottom. Sea-cucumbers have a soft sac-like elongated body with small calcareous plates in the skin. The common examples of these are *Cucumaria* and *Thyone* (Fig. 21.39) that occur on the sea-bottom or burrow in the surface mud or sand. They lie in lateral position when moving about or at rest, in which state they may be buried with only the ends of the body exposed. They are, however, quick to respond to the stimuli if disturbed. Most sea-cucumbers are coloured, black, brown, or olive green. Some may be rose coloured, orange or violet, striped coloured patterns are often found.

## TYPE SEA-CUCUMBER

### Structure

The most striking external features of the sea-cucumber are its muscular body wall almost devoid of large skeletal plates and its branching **tentacles** surrounding the mouth. The number of retractile tentacles varies from 10 to 30. They are hollow and extend by water pressure being connected with water vascular system (Fig. 21.42). They can, thus, be compared with the oral tube-feet of other echinoderms. The body is elongated along oral-aboral axis, of which one end is somewhat thicker than the other. The mouth opening is situated at the thicker

## Phylum Echinodermata

(oral) end and the cloacal aperture (anus) is at the posterior or aboral end. The body is five-sided and along each side extends a double row of **podia** or **tube-feet**. The three rows of tube-feet on one side (ventral) are broad and used for locomotion. This region with its three ambulacral areas corresponds

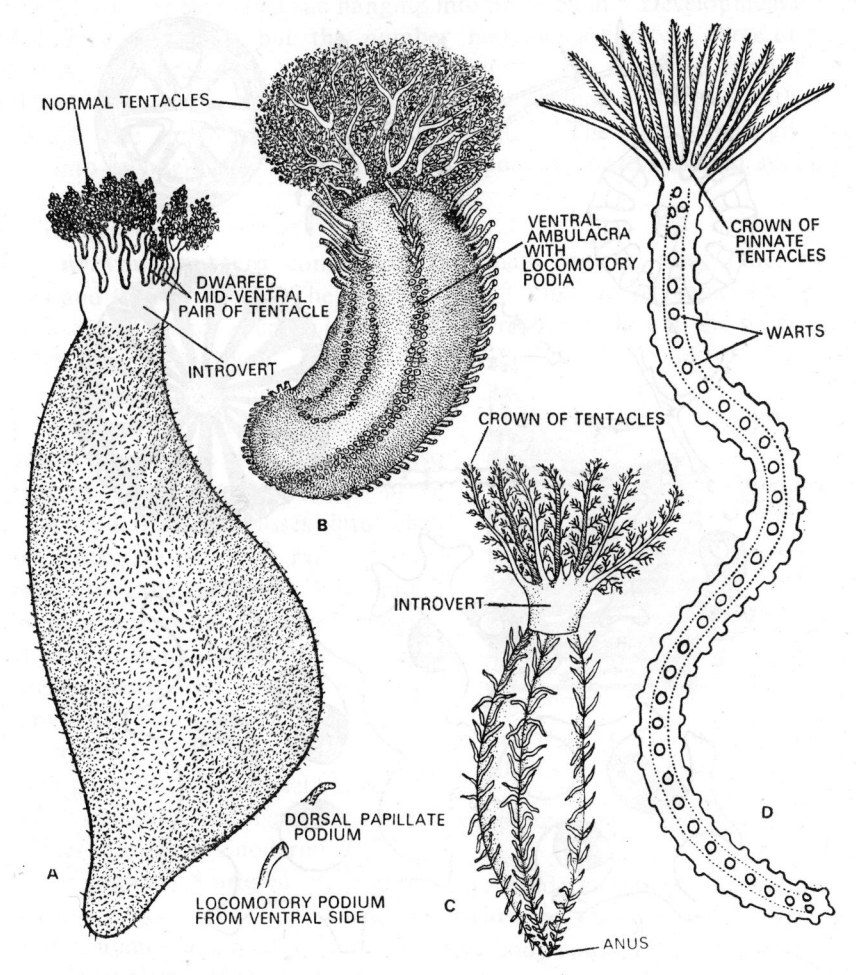

FIG. 21.39. Holothuroidea. A—*Thyone briareus*; B—*Cucumaria frondosa*; C—*Cucumaria planci*; and D—*Euapta lappa* with pinnate tentacles.

with the trivium of other echinoderm, the rest corresponds with the bivium and contains the two rows of thin and elongated tube-feet used for respiration. In some forms (*Cholochirus*) the ventral surface is distinct with the three ambulacral areas and is devoid of rows of tubercles that occur on the dorsal side with only two rows of ambulacral areas. In *Cucumaria* (Fig. 21.38) the ventral surface is not clear but its position can be determined by reference to the tentacles.

**Body wall.** A thin **cuticle** covers the **epidermis**, composed of non-ciliated columnar epithelium (Fig. 21.41). The basal ends of the epidermal cells merge with the dermis which is made up of a thick layer of connective tissue surrounding the microscopic calcareous ossicles. Hard calcareous plates are

## Locomotion

The animals move with the help of tube-feet but a wave of muscular contraction which travels from one end of the body to the other is of great help in locomotion which is also assisted by the tentacles.

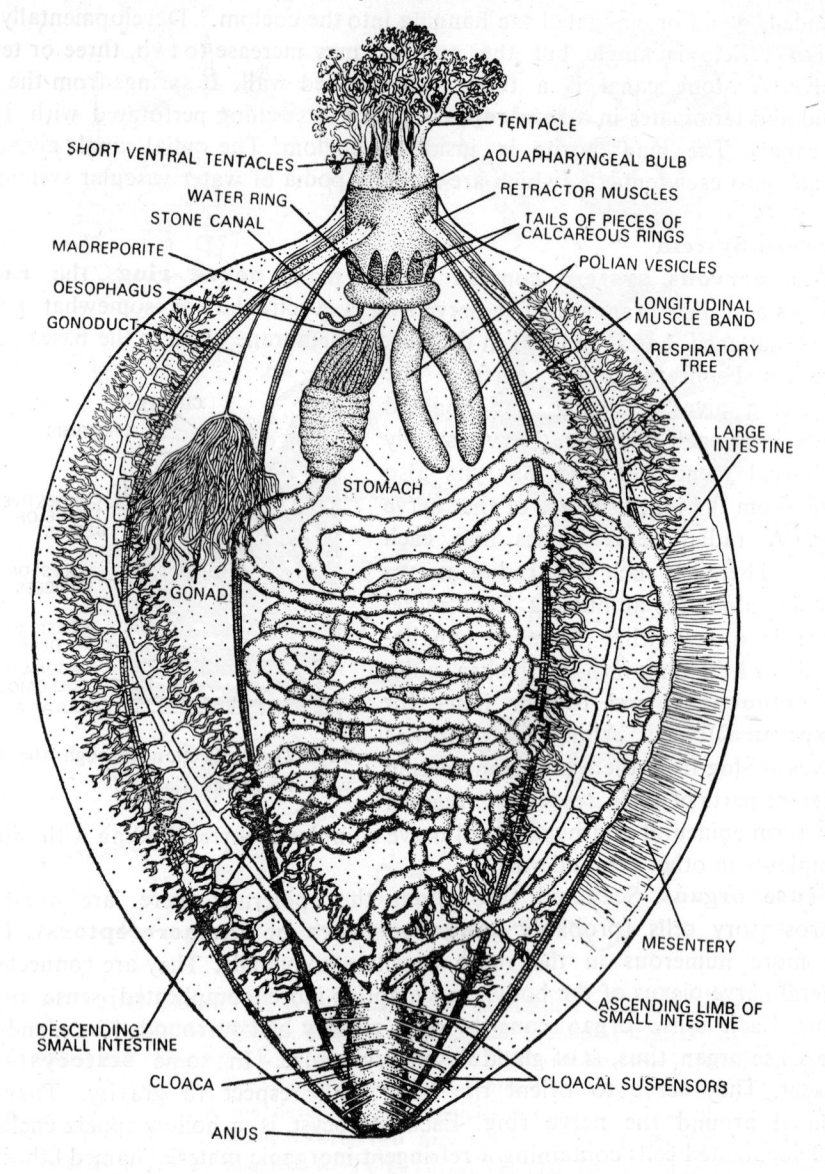

FIG. 21.42. Anatomy of a typical holothurian.

## Haemal System

The **haemal system** is mostly highly developed in the holothurians. It consists of a ring situated close to the nerve ring. From this arise radial haemal canals that run along the radial canals of the watervascular system. In larger species there are two main intestinal sinuses a dorsal and a ventral sinus running along the intestine and supplying it through a large number of smaller channels.

*Phylum Echinodermata*

Between the arms of first great loop of the intestine the dorsal sinus gives rise to a complex interconnecting network of channels containing many lacunae. A haemal network around the left respiratory tree also exists. Experimental evidence indicates that oxygen passes from the terminal vesicles of the left tree into the coelomic fluid and then into the haemal network.

The haemal channels do not have distinct lining, although muscle fibres and connective tissue are present. In the intestine the sinuses are merely spaces in the connective tissue layer. The coelomocytes are formed in the walls of

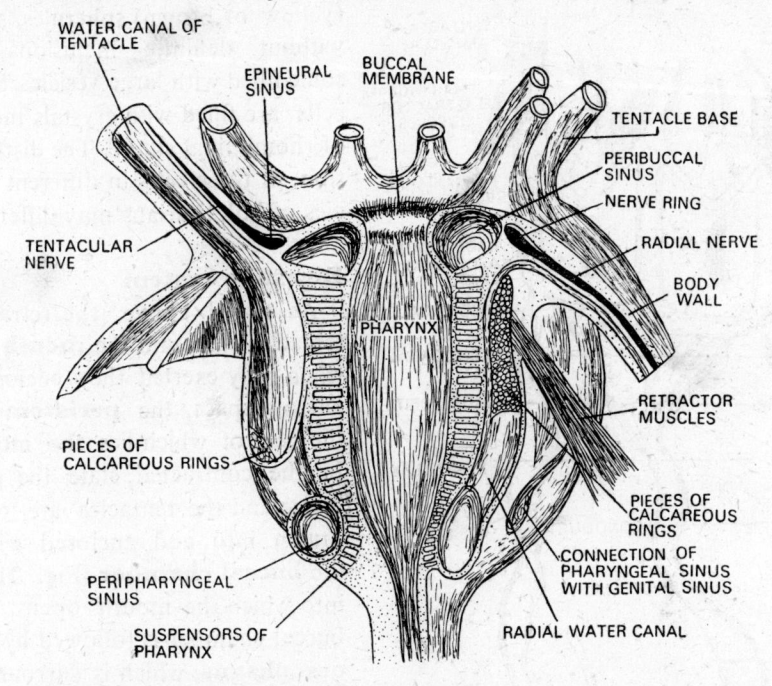

FIG. 21.43. Longitudinal section through the anterior end of *Cucumaria* showing location of the nerve ring.

the haemal channels. The haemal fluid is essentially coelomic fluid with similar coelomocytes. Apparently the function of the haemal system is the distribution of the food materials that are brought in the system by amoebocytes. The dorsal sinus is contractile and fluid is pumped through the intestinal sinuses into the ventral sinus and anteriorly into the haemal ring.

**Coelom.** The coelom is spacious occupying space between body wall and the digestive tract and expanding from the calcareous ring to the cloacal attachment. It is slightly subdivided by the mesenteries of the digestive tract mostly three in number. The mesenteries are incomplete and often permeated by holes. The coelom is lined by a flat to cuboidal ciliated epithelium. It is filled with a fluid into which float numerous corpuscles. The chemical nature of the coelomic fluid is similar to sea water except that its pH is less than that of sea water and is less alkaline than sea water. In the coelomic fluid float numerous free cells called **coelomocytes**. Some of these cells also occur in tissues, water vascular system and haemal system. They are of various types. Of these the **haemocytes** are found in the haemal system of all holothurians. They are circular or oval flattened biconvex nucleated cells containing pigment that looks red *en masse*. In some species they are so numerous that the haemal system, or the coelomic contents look red. Some people considered the pigment of haemocytes to be similar to haemoglobin, but it is not identical with haemoglobin of vertebrates.

Then there are amoebocytes of different types. Some are active amoebocytes phagocytizing degenerating cells; then there are amoebocytes with colourless spherules, some with coloured (yellow or brown) spherules, some without definite inclusions and some filled with large vesicles. Some cells are filled with crystals mostly of rhomboidal shape. The distribution of these cells in different species of holothurians may differ.

### Digestive System

As noted above, the retractile tentacles surround the **mouth** and when fully exerted they enclose a narrow space, the **peristome** at the base of which lies the mouth. In the contracted state the peristome and the tentacles are withdrawn into and enclosed within the **buccal chamber** (Fig. 21.42) into which the mouth opens. The buccal chamber is followed by the **oesophagus**, which is surrounded by a calcareous ring made up of ten **circumoesophageal ossicles** of which five are ambulacral in position and five inter-ambulacral. The oesophagus leads into a muscular **stomach** (not clearly marked out), which narrows into a very long convoluted intestine (Fig. 21.42). This expands into a short **muscular cloaca** which opens at the posterior end. The digestive tract is supported by a mesentery, mostly subdivided into three portions. The first part of **dorsal mesentery** is attached to the anterior part of the digestive tract (oesophagus, stomach and small intestine). The **left mesentery** is attached to the body wall in the left dorsal inter-radius, the **ventral mesentery** supports the large intestine. Histologically the digestive tract consists of five layers; the lining epithelium, the inner layer of connective tissue, the muscular layer (circular and longitudinal), the outer layer of connective tissue and the covering ciliated peritoneum. Two long branched **"respiratory trees"** open into the

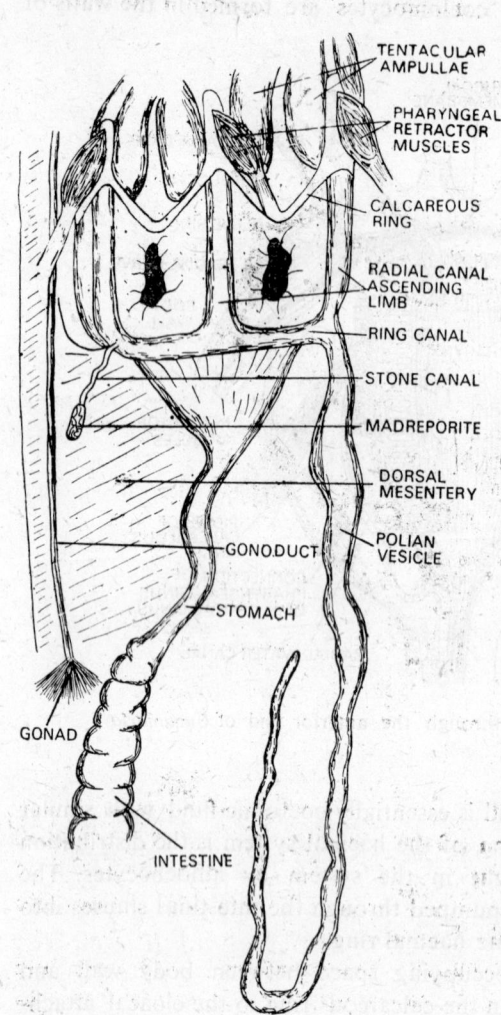

FIG 21.44. Aquapharyngeal bulb of *Cucumaria*.

cloaca. Each of these begins behind in a tubular stem, becomes elaborately branched in front, some of the branches reaching nearly to the anterior end of the body cavity. Each terminal branch is enlarged into an ampulla. The respiratory trees are permeated with coelomocytes, especially the type with coloured spherules and groups of granules of excretory nature. The function of these may be respiratory and excretory, but they are probably also hydrostatic regulators of the internal pressure. The food consists of very small animals and plants, in mud and sand, conveyed to the mouth by the tentacles.

In *Holothuria* and some other species a number of tubules are found attached to the common stem of respiratory tree. These tubules named **tubules of Cuvier** or **Cuverian tubules** may be white, pink or red in colour and, in some cases, may be attached to the bases of the respiratory tree, especially the left. It is evident from the description of many authors that the tubules are defensive in function. When irritated *Holothuria* turns its anal end towards the offending object, contracts in size and pushes the cuverian tubules out (Fig. 21.45A). The emerged blind ends of the tubules elongate, their tips swell and they dart about in all directions. The slender threads also become sticky. The long sticky threads wrap round irritating objects such as a lobster and render it incapable of movement. Threads break and the holothurian moves away quietly. In *Actinopyga agassizi*, according to Hyman the tubules are highly toxic to fish and other animals and serve as organs of defence in a different way.

**Axial complex.** The axial complex is poorly developed in the holothurians. In *Cucumaria* it is in the form of a spongy glandular ring around the pharynx and extends along the radial haemal lacunae. It is developed to different degrees in different species. In *Molpadia* it consists of a network of connective tissue towards the inner side of the water ring (Fig. 21.44) forming a tubular outgrowth that extends backwards along the dorsal wall of the oesophagus in the dorsal mesentery and also accompanies the gonoduct to the base of the gonoduct. According to Cue'not (1829) the part of the coelom that gives rise to axial sinus disappears during the embryology of holothurians. Hyman feels that there is some confusion between the true axial gland and sites of amoebocyte formation in the walls of the haemal sinuses, especially the haemal ring.

## Reproductive System

The sexes in the majority of holothurians are separate, but it is difficult to distinguish male from female (Fig. 21.45B) except in the case of females found brooding their young. Some forms are also hermaphrodite (e.g., *Cucumaria laevigata*, *Mesothuria intestinalis* and many synaptids). The gonad is single situated in the anterior part of the coelom and opens to the exterior in common with the external madreporite, when present. The gonad consists of numerous tubules united basally into one tuft attached to the left side of the dorsal mesentery. In some there may be two tufts, one on either side of the mesentery. The tubules may be long and numerous and short and few in number; they may be simple or branched in various ways. In synaptids the gonad usually consists of two branched tubules one on either side of the mesentery. In some others it takes the form of a pair of sacs. The gonad becomes very large at the time of sexual maturity and is made of numerous long tubules. The tubules open into a hollow tube (base of the gonad) attached to the mesentery, and from this common base the gonoduct proceeds in the mesentery to the gonopore. The gonopore may be located on a genital papilla or on the surface in the mid-dorsal line, between the tentacles or shortly behind them. In some cases the gonoduct may subdivide before opening with the result that a number of gonopores are formed. The length of the gonoduct varies in different holothurians. The lumen of the gonadial tubule is lined by germinal epithelium from which the sex cells originate. This is surrounded by a connective tissue layer, itself embraced by musculature (usually both circular and longitudinal) which is clothed with the coelomic epithelium made up of flat or columnar cells. Among the hermaphrodite forms the same tube produces both eggs and sperm; generally at the same time, often at different times. Histological structure of gonoduct is simple. It consists of a ciliated epithelium surrounded by connective tissue continuous with that of the dorsal mesentery.

Each species of holothurian breeds for one or more months at a definite time of the year, mostly spring and summer in temperate regions. They usually spawn in the late afternoon or evening or during the night apparently in response to dim light. Oviparous species release their eggs directly in sea-water where they are fertilized at once. The sex cells issue from the gonopore in a slow stream and are well disposed by tentacular movement. All the ripe sex cells are not discharged during one spawning. A holothurian may spawn at intervals.

Some eggs may develop in sea-water and pass through free-swimming larval stages, which are omitted in those form that are brooded. The free-swimming larva is called **auricularia** (Fig. 21.46), which soon transforms into a **doliolaria**, resembling doliolaria of crinoids. In some cases auricularia stage is omitted and doliolaria is developed directly. In some cases both stages may be omitted and a completely flagellated or ciliated oval larva may be formed as in *Cucumaria saxicola*.

In other cases the eggs develop and the young are held on the external surface of the mother. The young ones may be attached to the smooth part of the creeping sole (*Psolus antarcticus* and *Cucumaria curata*), tentacular crown or tentacles (*Bathyplotes natans*) or in specially formed pockets of the body surface of the the back (*Thyonepsolus nutriens*) or sole (*Psolus granulosus*). The spawning females place the eggs in surface depression in a very interesting way. They arch their expanded tentacles to form an interlocking mesh in which the extruded eggs are caught. The eggs are then transferred to the dorsal surface by the tentacles and by two usually extensile podia located near the gonopore. Other dorsal podia pass the eggs along the animals back into the preformed pits of the surface. Eggs are very adhesive and stick without difficulty. This applies to *Thyonepsolus nutriens* as recorded by Wootton in 1949. In some others definite incubatory pockets are formed. They are at first external and then become deeper.

Cleavage is typically equal, holoblastic and of radial type with tiers of cells in line with one another, very perfectly so in some species and irregularly so in others. Cleavage leads to the formation of a coeloblastula. In the eggs of incubating species cleavage is superficial and the blastula consists of a surface layer of cells enclosing the yolk. Gastrulation is typical embolic in both. Typically, the embryo soon acquires flagella and escapes the fertilization membrane. Mesenchy-

## Phylum Echinodermata

me is given off continuously from the tip of the archenteron. The mesenchyme fills the blastocoel in the form of endomesoderm. Further developments are similar to those in other echinoderms.

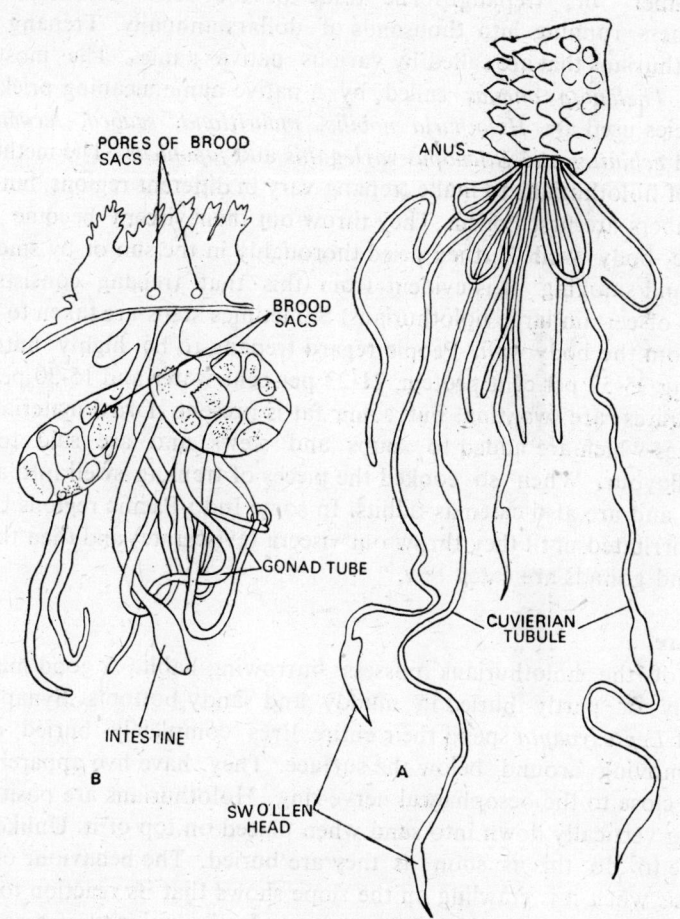

FIG. 21.45. A—Posterior part of *Holothuria forskali* with emerged cuvierian tubules; and B—Reproductive organs and brood sac of *Cucumaria glacialis*.

## Regeneration

Holothurians possess a remarkable power of regeneration. On irritation *Synapta*, for instance, violently contracts and fragments the body into two or more pieces (self-multilation), of which the anterior part soon regenerates its lost parts. This is a protective device. In some forms a portion or whole of the intestine is pushed out, the respiratory trees are the first to go, and in some the lower branches of these are emitted as tubules which swell up in sea-water and appear tough white sticky threads, the Cuvierian organs, in which the enemies are entangled. Enemies of the size of lobsters are rendered helpless by this technique. By all this the incovenience caused to the holothurian is only a temporary one, as the lost or injured part is soon regenerated.

## Economic Importance

Dried holothurians are used as food at many places. Among South Pacific Islands and on the coasts of Queensland in Southern China these are called "beche-de-mer" or "trepang." The trade of the dried holothurians forms a good business running into thousands of dollars annually. Trepang is made of large holothurians that are called by various native names. The most favoured species is *Thelenota ananas* called by a native name meaning prickly red fish. Other species used are *Holothuria nobilis, mauritiana, scabra, lecanora, argus, edulis* and *echinites* and *Stichopus variegatus* and *japonicus*. The methods of preparation of holothurians to make trepang vary in different regions, but in general the cucumbers are first boiled. They throw out their viscera become short and thick. The body wall is then dried thoroughly in the sun or by smoking or by both sun and smoking. It is evident from this that trepang consists of cured body wall of certain large holothurians. Sometimes steps are taken to remove the ossicles from the body wall. People regard trepang to be highly nutritious diet constituting 35-52 per cent protein, 21-23 per cent water and 15-30 per cent ash. Carbohydrates are wanting but some fat is present. Dried material is cut into small pieces which are added to soups and stews and are said to impart a delicate flavour. When so cooked the pieces of trepang swell into a gelatinous condition and are also eaten as tidbits. In some Indo-Pacific regions the holothurians are irritated until they throw out viscera (eviscerate) and then the curverian tubules and gonads are eaten raw.

## Behaviour

Many of the holothurians possess burrowing habit. A good many of them live wholly or partly buried in muddy and sandy bottoms. Synaptids such as species of *Leptosynapta* spend their entire lives completely buried in the soft bottom moving around below the surface. They have five apparent statocysts arranged close to the oesophageal nerve ring. Holothurians are positively geotactic, going vertically down into sand when placed on top of it. Unlike the worms they cease to do this as soon as they are buried. The behaviour of *Synaptula hydriformis* when it is crawling on the slope shows that its reaction to the gravity remains after the removal of the statocyst. In the majority of cases only those echinoderms which possesses statocyst show gravitational responses when tilted.

Sea-cucumbers and sea-stars are as likely to go up or to go down. *Asterina gibbosa*, for example, may creep up or down according to the phase of behaviour it is in. In these animals the pull of the body plays some part in the determination of direction of gravity for this animal.

Holothurian generally right themselves when turned over, that means that if the functionally ventral side of the holothurian is turned away from the substratum it will turn back to the right position. One of the behavioural defence mechanisms of some sea-cucumbers is the ability to throw out almost all the internal organs when disturbed or irritated (*i.e.* eviscerate). These organs soon regenerate. *Thyone* and *Stichopus* can be used for experiments to demonstrate evisceration. Place any of these Holothurians in 0.1 per cent ammonium hydroxide ($NH_4OH$) in sea water for 1 to 2 minutes and then remove the animal. If the organism is held up by the posterior end most of the internal organs will be extruded. Complete regeneration of the organs may take place once.

Holothurians contract in response to mechanical stimulation or contact. The tentacles contract, the body ends retract and the animal in general shortens. The contractions are accompanied by the emission of water from the cloaca so that the animal becomes much reduced. If the animal is suddenly handled by man this reaction is sudden and water is emitted as a jet. In *Holothuria surinamensis* the entire body surface responds to salts, acids, alkalies, alkaloids, essential oils and a variety of organic substances. When applied with a pipette held 5 mm away from the animal's surface the extended parts retract and local areas of body wall contract forming a depression.

## CLASSIFICATION OF HOLOTHUROIDEA

The Echinodermata in which the body is elongated, worm-like (cylindrical) or fivesided. The mouth and anus are at the opposite extremities of the body. The body wall possesses scattered and isolated calcareous bodies or many contractile tentacles surrounding the mouth and the water, vascular pore (madreporite) is usually absent in the adult. The larva is an auricularia.

1. **Order Aspidochirota.** The Holothuroidea in which the tube-feet are present, tentacle ampullae are distinct, tentacles are shield-shaped or leaf-shaped, without retractor muscles. Madreporite, internal respiratory tree present.

    Examples: *Holothuria, Actinopyga, Synallactes.*

2. **Order Elasipoda.** The Holothuroidea in which the body is flat ventrally, tube-feet present, tentacle ampullae are lacking and respiratory tree is wanting. Madreporite may be internal or external. Mostly deep-sea forms.

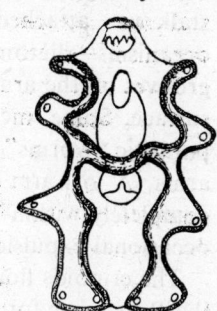

FIG. 21.46. Auricularia larva of holothuria.

    Examples: *Oneirophanta, Kolga.*

3. **Order Pelagothurida.** Pelagic Holothuroidea without tube-feet and with large tentacle ampullae.

    Example: *Pelagothuria.*

4. **Order Dendrochirota.** The Holothuroidea in which tube-feet, retractor muscles and respiratory trees are present. Tentacles arborescent, tentacle ampullae absent, madreporite internal.

    Examples: *Colochirus, Psolus, Cucumaria, Thyone* (Fig. 21.39), etc.

5. **Order Molpadonia.** Burrowing Holothuroidea without tube-feet, with respiratory trees and tentacle ampullae and tentacles unbranched. Madreporite internal.

    Examples: *Molpadia, Caudina,* etc.

6. **Order Synaptida** or **Apoda.** Burrowing Holothuroidea without tube-feet, without radial ambulacral vessels and without respiratory trees, tentacles pinnate without vestigial ampullae, retractor muscles present. Madreporite internal.

    Examples: *Synapta, Rhabdomolgus, Euapta* (Fig. 21.39), etc.

## CLASS CRINOIDEA

The **Crinoidea** (Gr. *krinon,* lily+*eidos,* form) are brachiate Echinodermata which remain attached during the whole or part of their lives by the aboral apex of the body; the arms are usually branched and provided with pinnules; the tube-feet are tentacle-like, without ampullae, and the water pore is always multiple. The oral surface carries interradially placed anus. They are commonly known as the feather-stars or sea-lilies. The feather-stars or sea-lilies differ from other echinoderms in being fixed permanent or temporarily by a jointed stalk. The arms arise from a corresponding number of **thecal plates** or **radials** below which there is a circlet of **basis** often with **infra-basal** plates alternating with them. The modern rosy feather-star (*Comatula* or *Antedon rosacea*) leave their stalk at a certain stage in life, but the other crinoids, e.g. *Pentacrinus* are permanently stalked like almost all the extinct stone-lilies or encrinites once so abundant. Most of them live in deep water and many in the great abysses. An anchorage is found on rocks and stones, or in the soft-mud, and great number grow together. The free comatulids swim gracefully by bending and straightening their arms and they have grappling cirri on the aboral side where the relinquished stalk was attached. By these cirri they move themselves temporarily. Small organisms—diatoms, protozoa, minute crustaceans—are wafted down ciliated grooves on the arms to the central mouth, which is of course on the unturned surface. Some members of the class e.g. *Comatula*, are infested by minute parasitic "worms" (*Myzostomata*) allied to chaetopods, which form galls on the arms. A lost arm can be replaced, and even the visceral mass be regenerated completely within a few weeks after it has been lost. It has been suggested that occasional expulsion of the visceral sac frees the crinoid from parasites (Dendy).

The crinoids flourished during the Palaeozoic era but by the end of the era the Palaeozoic forms died out. Only some living genera have continued to the present time. According to some estimates about 5000 species of crinoids have lived and died out in past geologic ages. At the present time there are about 630 living crinoids of which about 80 are stalked species (sea-lilies), evidently on their way out, and the rest have lost the stalk and have adopted a free existence (feather-stars).

### External Features

The stalked crinoids present jointed or scaly appearance similar to that found in extinct members. The body is called **crown** or **corona** or **calyx** and is attached by a stalk, the attachment being done by circular **discs** (Fig. 21.47) or root-like extensions or other similar devices.

The **stalk** (also called **stem** or **column**) is of cylindrical or polygonal contour and is supported by internal skeletal pieces, visible through the surface. The length of the stem in the present stalked crinoids is about 50 cm (about 2 feet), but in some extinct forms it was as much as 21 metres (or about 70 feet). The stalk is variously ornamented and bears many appendages called **cirri** (Fig. 21.48), in many cases, the stalk may be without cirri. In the feather-stars the stalk is lost during development but usually the cirri persist forming one to several circlets springing directly from the aboral surface of the crown (Fig. 21.47).

The cirri show great diversity in form and size. They are short, stout and curved in feather-stars living among arborescent growths; but are long, stout and rigid in those that hold on to rocks, while in those living on muddy bottoms the cirri are slender and straight. The number of cirri varies from 15 to 35. In some

## Phylum Echinodermata

cases they may be well over 80. The cirri usually terminate in a claw and may have distally a series of projecting spines.

The **crown** consists of a central rounded oval, hemispherical or discoidal mass that includes the viscera and a set of **arms** or **brachia**. The central mass is divisible into an aboral cup or saucer the **calyx** or **dorsal cup**. Roofing the calyx there is an oral membrane, the **tegmen**. The tegmen is pierced interradially by numerous minute pores (500 to 1500, the ciliated funnels of Chadwick 1907), leading into internal coelomic spaces and serve for the ingress of water into the interior of the crinoid. The calyx, strictly speaking, is the aboral body

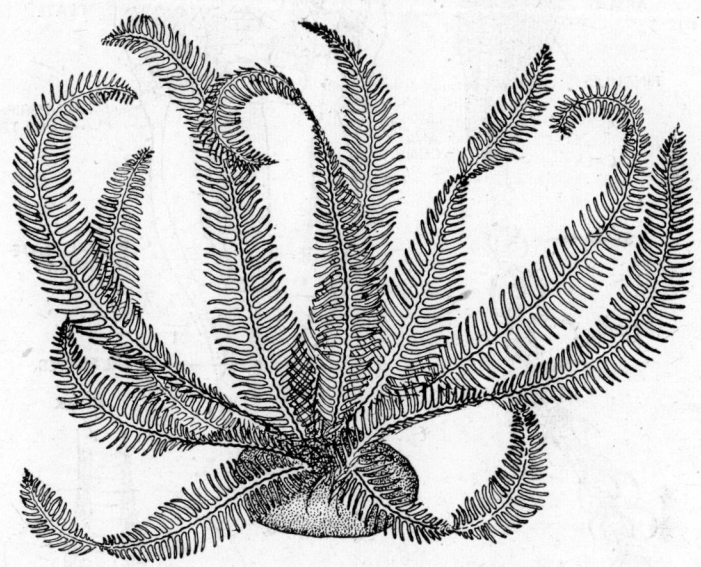

FIG. 21.47. A typical feather-star *Antedon*.

wall studded with skeletal plates. In extinct forms the calyx was high, but is greatly shortened in present crinoids. The oral body wall is the tegmen. In primitive existing forms it contains five deltoid plates (Fig. 21.47) encircled by small plates.

The **mouth** is usually located at or near the centre of the tegmen. When central, five ambulacral grooves arise from the mouth and run nearly symmetrically on the arm bases (Fig. 21.49). The anus is situated eccentrically at the tip of projecting **anal cone**, sometimes noticeably long.

The brachia grow out from the boundary between calyx and tegmen and extend freely into the water. They are jointed in appearance and covered over by scales. Primitively there are five arms, but the arms may divide and form 10 arms (Fig. 21.47). The arms may further divide to 40. Some feather-stars may have 60 arms while some have 80 to 200. The number of arms is probably in some way correlated with food conditions. Arms may be long and slender or short and broad. It is interesting to record that warm temperatures favour the evolution of long arms and cold water favours short arms. All the arms are typically bordered on each side by a row of short tapering side branches or **pinnules**. The pinnules of two rows of each arm alternate. The pinnules also appear jointed as the supporting endoskeleton is visible through thin covering tissue.

These pinnules have the same structure as the arms. The proximal pinnules are regarded as the **oral pinnules**, the middle ones are the **genital pinnules**

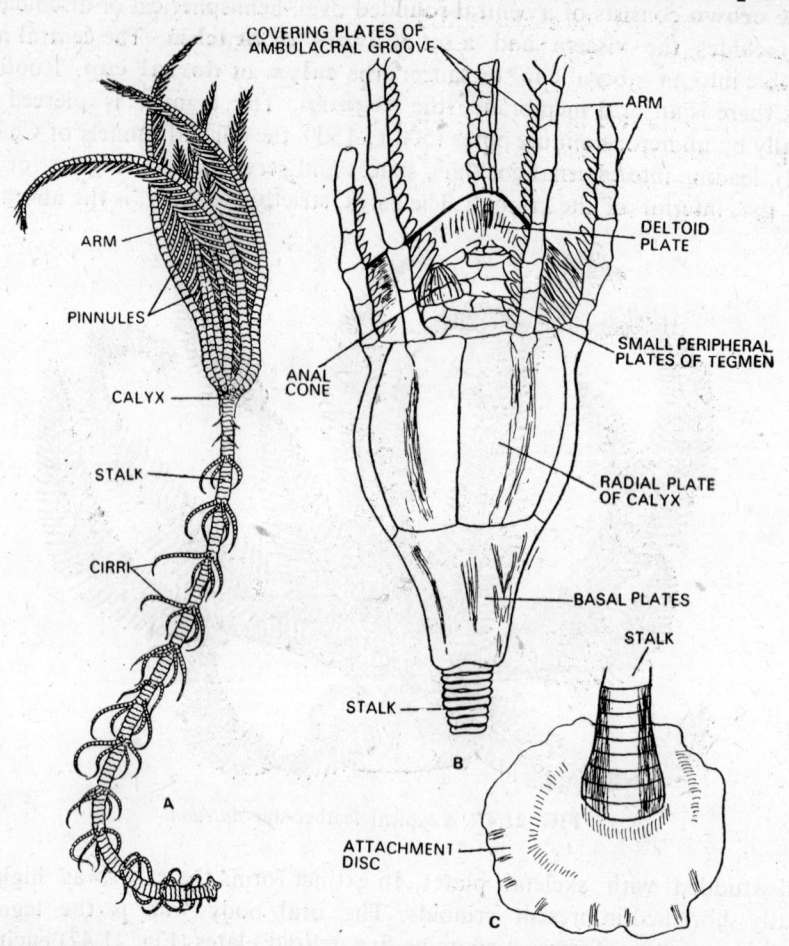

FIG. 21.48. A—*Metacrinus*, a stalked crinoid; B—*Hyocrinus*, a primitive existing crinoid with high calyx and deltoid plates; and C—Attachment disc of *Calamocrinus*.

the rest of them are referred to as **distal pinnules**. The oral pinnules are devoid of ambulacral grooves and podia and act as tactile and protective devices. The genital pinnules contain the gonads and swell during sexual maturity. The distal pinnules are mostly long and slender and become rapidly short near the arm end. Pinnules are generally more or less spiny distally.

The arms arise from a corresponding number of **thecal plates** or **radials** below which there is a circlet of **basis** often with **infra-basal plates** alternating with them. The ambulacral grooves on the arms are deep furrows that converge to the arm bases. Each ambulacral groove has raised edges which form regularly repeated **lappets**, alternating on the two sides. The grooves are accompanied by **podia** or **tentacles**. The podia and lappets become reduced along the terminal grooves. The podia along the mouth are termed oral or **labial podia**.

## Phylum Echinodermata

Swimming movements are made as in the ophiuroids, by means of the arms. In the ten-armed species every other arm sweeps downwards, while the alternate

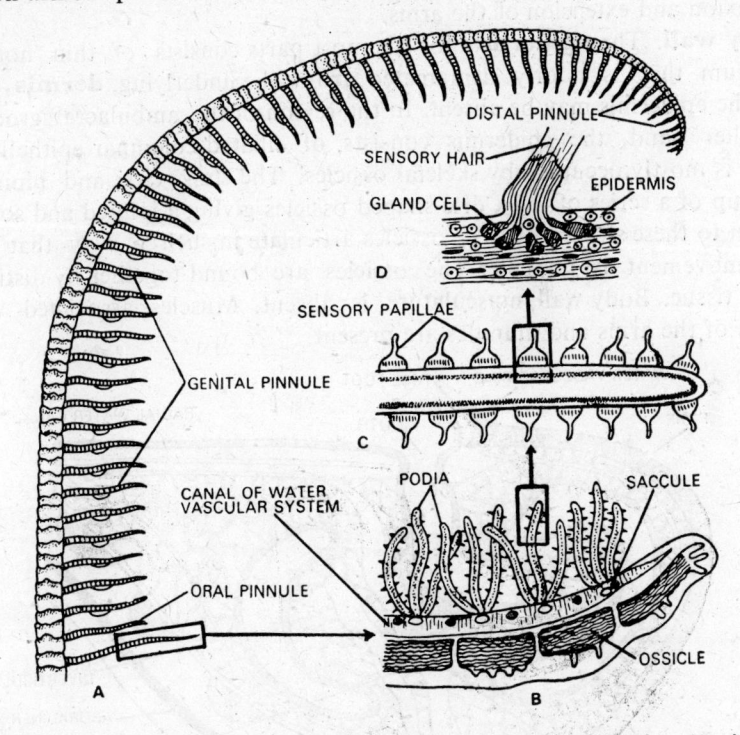

**FIG. 21.49.** *Antedon.* A—Detailed structure of an arm; B—Pinnules showing podia; C—A podium, magnified; and D—A sensory papilla in section.

set moves upward. In species with more than 10 arms, the arms still move in sets of five but sequentially. The arms are often used to pull the animal over

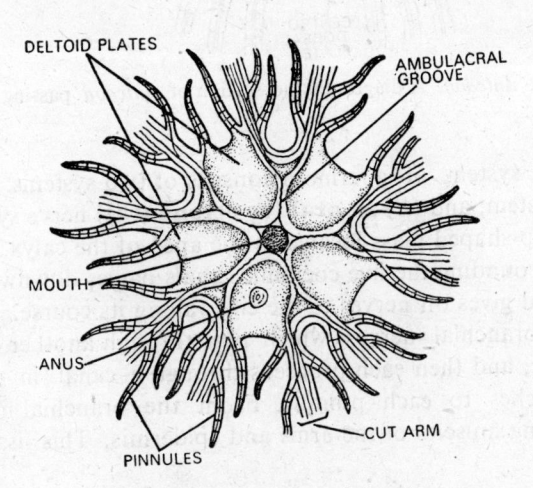

**FIG. 21.50.** Oral view of *Antedon.*

irregular and vertical surfaces. Feather-stars swim and crawl only for short distances and then rest on the bottom by holding them by grasping cirri. The movements in the sessile sea lilies are limited to bending movement of the stalk and flexion and extension of the arms.

**Body wall.** The **epidermis** for the most parts consists of thin non-ciliated syncytium that is poorly demarcated from the underlying, **dermis.** In some cases the epidermis may be absent. In the region of the ambulacral grooves, on the other hand, the epidermis consists of ciliated columnar epithelium. The dermis is mostly occupied by skeletal ossicles. The stalk, cirri and pinnules are made up of a series of thick disc-shaped ossicles giving a jointed and solid construction to these structures. The ossicles articulate in such a way that at least some movement is possible. The ossicles are bound together by distinct connective tissue. Body wall musculature is absent. Muscles associated with the ossicles of the arms and pinnules are present.

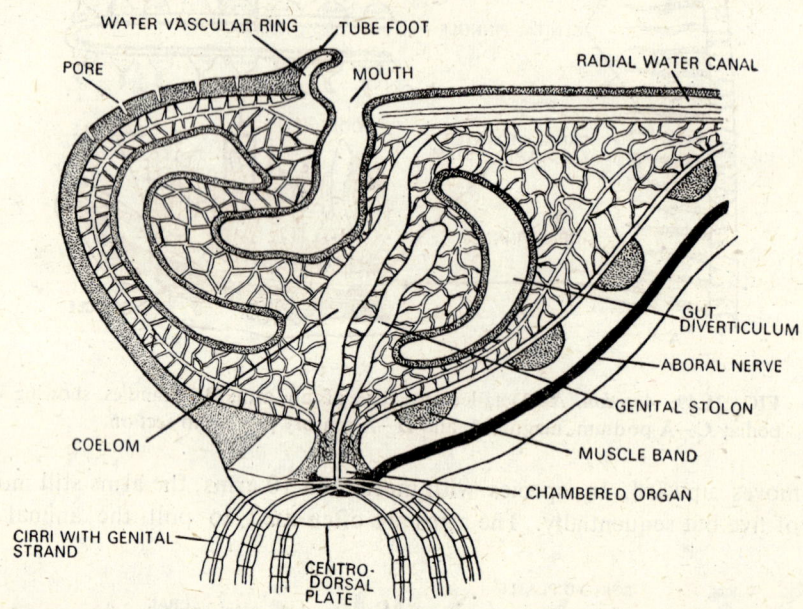

FIG. 21.51. *Antedon.* A diagrammatic section of *Antedon* passing through the central disc.

The nervous system of a crinoid consists of two systems: (*i*) an **aboral** (or **entoneural**) system; and (*ii*) an **oral** (or **ectoneural**) nerve system. The aboral system is a cup-shaped mass located in the apex of the calyx. From this a nervous sheath surrounding the five coelomic canals passes downward through the stalk ossicles and gives off nerves to the cirri, along its course. The aboral centre gives off five branchial nerves which pass through another concentric pentagonal nerve ring, and then each proceeds through a canal in the arm ossicles and send branches to each pinnule. From the branchial nerve arise smaller nerves innervating muscles of the arms and epidermis. This is a motor centre of the crinoids.

The oral nerve system consists of a subepidermal band of nervous tissue that runs just beneath the epidermis of the ambulacral groove in the arms. In the region of the mouth the nerve bands form a nervous sheath encircling oesophagus and the intestine. The ectoneural centre is sensosy in nature.

There are no organs of special sense. However, the ambulacral and podial epidermis are richly supplied with sensory cells. The podia (Fig. 21.49) bears numerous papillae each of which contains a gland cell and a few sensory cells with bristles. The crinoids are negatively phototropic and are known to possess righting reflex.

The **digestive organs** comprise the ciliated alimentary canal descending from the **mouth** (Fig. 21.50) into the cup, and curving up again to the anus, which is located on a papilla. The mouth leads into a short **oesophagus** which opens into an intestine. The intestine descends aborally and laterally, when it reaches the calyx it makes a complete round along the inner side of the wall. The terminal portion then passes upward into a short rectum that opens to the outside at the anus.

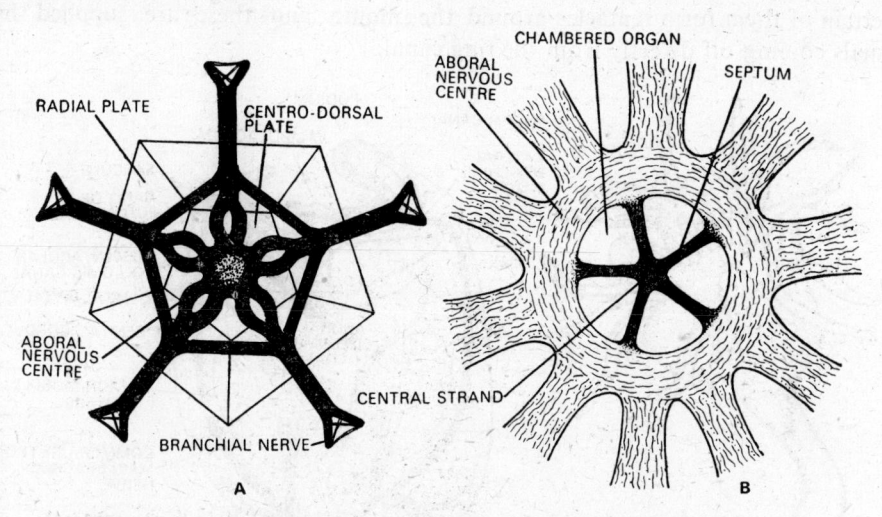

FIG. 21.52. A—Aboral nerve centre; and B—Chambered organ of *Antedon*.

The coelom is reduced to a network of communicating spaces as the cup is filled with connective tissue. In the oral side of the arms the coelom extends as five parallel coelomic canals. In the stalk five coelomic canals pass through the central perforation in the ossicles and give off one canal in each cirrus. The coelomic cavity contains several types of coelomocytes.

Crinoids are suspension feeders. During feeding the arms and pinnules are held outstretched and the tube-feet are stretched. The tube-feet bear long slender mucous secreting papillae. Any planktonic organism or detritus that comes in contact with tube-feet is trapped in mucous and tossed into the ambulacral grooves by a sudden whiplike action of the tube-feet. The cilia of the ambulacral grooves beat towards the oral aperture, carrying food down, arms into the mouth.

The **haemal system** is a network of spaces and sinuses within the connective tissue strands occupying the coelom. Around the oesophagus there is a definite plexus from which branches extend downward (aborally) through the centre of the crown into a spongy mass of cells, closely associated with the axial gland. The axial gland is an elongated tubular mesh of glandular tissue occupying the polar axis of the crown. Branches of the haemal system also supply the intestine. There is no haemal space in the stalk.

Respiration takes place through general body surface exposed to sea-water and through the tube-feet.

The **water vascular system** consists, as usual, of a **circumoral ring** and **radial vessels**, but in several respects it shows remarkable modification. The madreporite of other forms is represented by fine pores which open from the surface of the calyx directly into the body cavity, and which may be very numerous, they are said to be 1500 in *Antedon rosacea*. By these pores the water enters the body cavity, and from it enters the numerous stone canals which hang from the ring freely in the body cavity, and open into it near the pore canals. There are no **Polian vesicles** or **ampullae**, the tube-feet are small, arranged in groups of three, and are connected by delicate canals with the radial vessels. Certain of them form tentacles around the mouth and these are supplied by canals coming off directly from the ring canal.

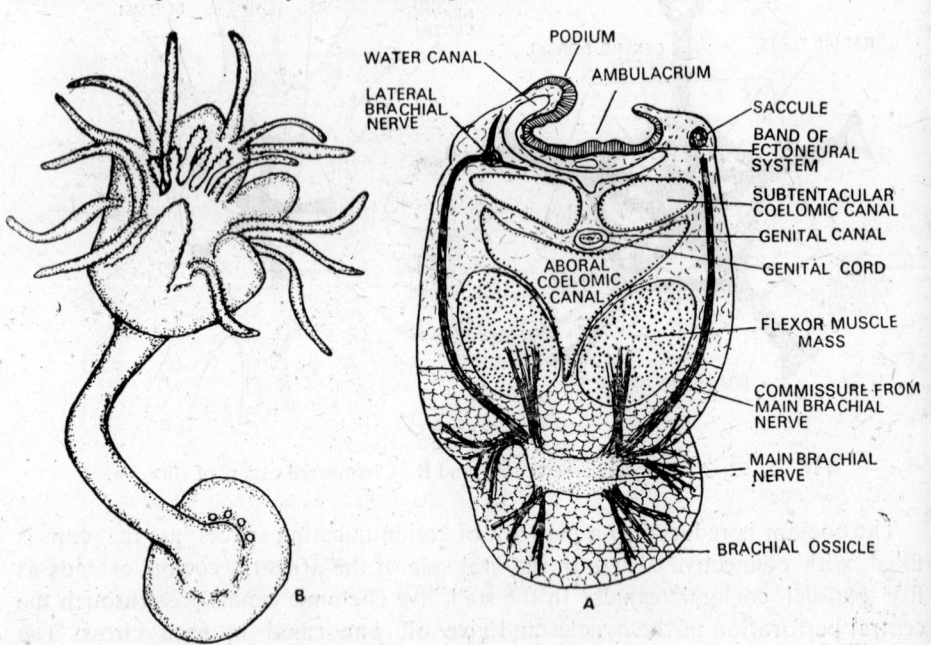

**FIG. 21.53.** A—Section through an arm showing the main brachial nerve and its branches; and B—Pentacrinoid larva of *Antedon*.

### Reproduction

The sexes are separate, but there is no sexual dimorphism. The reproductive organs extend as tubular strands from the disc along the arms, but are rarely functional except in the pinnules, from each of which the elements burst out by one duct in females, by one or two fine canals in males. Only those pinnules are involved in the formation of sex cells which occur along the proximal part of the arm length.

When the eggs and sperms are mature spawning takes place by rupture of the pinnule wall. Hatching takes place at the larval stage. Early development is similar to that of the asteroids and holothuroids. During the course of development the gastrula grows into a doliolaria larva without forming a bipinnaria. Hatching takes place when a doliolaria is formed. It leads a free-swimming life for a few days, then settles at the bottom, attaches to the substratum and metamorphoses into a minute sessile stalked crinoid. This is the pentacrinus larva. The pentacrinoid of *Antedon* is about 3 mm. The doliolaria develops into pentacrinus in about six weeks. It assumes the adult form after several months. In *Antedon* it becomes free-swimming, in others remains stalked.

Nothing is known about excretion in crinoids. Probably waste products are collected by coelomocytes which are then deposited in little saccules located in rows at the sides of the ambulacral grooves. It is likely that these saccules are thrown out periodically.

There are about 400 living species in twelve genera, but about 1500 species in 200 genera are known from the rocks. The class is obviously decadent. It is represented in the Cambrian, and attained its maximum development in Silurian Devonian, and Carboniferous times. The recent forms include the stalked **Pentacrinus, Rhizocrinus**, etc., and the free comatulids, which pass through a stalked *Pentacrinus* stage (Fig. 21.53), e.g., *Antedon*.

## CLASSIFICATION OF CRINOIDEA

Echinodermata in which the theca gives off five, usually branched, arms. The oral surface is directed upwards. The stalk springs from the aboral apex. Suckerless tube-feet and an open ambulacral groove present. No spines, no pedicellariae. This class is divided into four orders.

1. **Order Inadunata**. The extinct Crinoidea in which the calyx is made up of firmly fitted plates with conspicuous basals. The arms are typically free beyond the radials and may be three, five or more in number, branched or unbranched and with or without pinnules. They flourished from the ordovician to the Permian. According to Moore and Laxdon (1943) 53 families and over 300 genera existed during Palaeozoic.

Example: *Hyocrinus*, (Fig. 21.48B).

2. **Order Flexibilia**. The extinct Crinoidea in which the lower brachials (usually three) are flexibly incorporated into the calyx. The stalk is usually devoid of cirri. The arms are always branched but devoid of pinnules and typically curved inwards. They ranged from the Ordovician to the Permian. In 1943 Moore and Laxdon recorded 10 families and 47 genera.

Example: *Forbesiocrinus*.

3. **Order Camerata**. The extinct Crinoidea characterized by the rigid calyx of thick sutured plates into which is incorporated a variable number of the lower brachials as well as a number of interradial and interbrachial plates. The arms are branched and pinnulated. The camerate or box crinoids were limited to the Palaeozoic era. The group has about 29 families and 129 genera.

Examples: *Reteocrinus, Platycrinites, Technocrinus*.

4. **Order Articulata**. The existing Crinoidea having a pentamerous calyx the arms of which are uniserial, pinnulated and generally branched. The mouth and the ambulacral grooves are held open and exposed, although the grooves may be partially closed by the plates in the lappets. The order is divided into three suborders.

(*i*) **Suborder Isocrinida** includes the stalked crinoid.
Example: *Metacrinus*.

(*ii*) **Suborder Comatulida** includes those Articulata which in early life break from the stem and thereafter lead a free existence.
Examples: *Comatula, Heterometra, Antedon*.

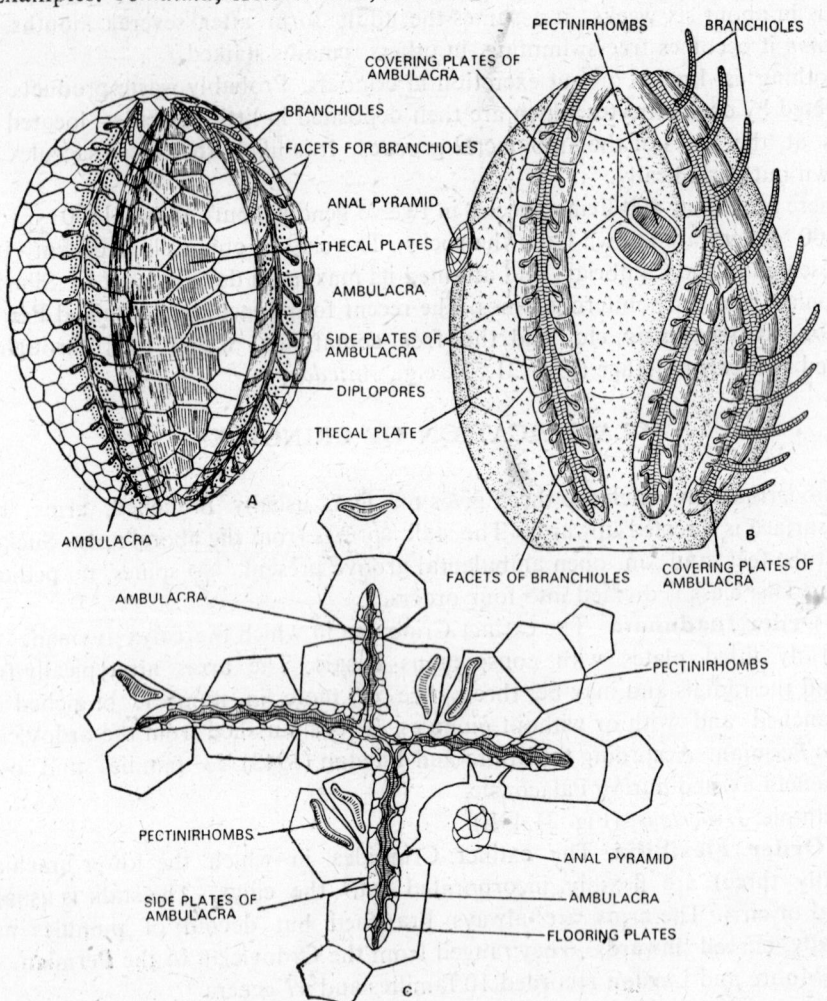

**FIG. 21.54.** Extinct cystidea. A—*Proteroblastus*, diploporite, with ambulacra reaching the aboral pole; B—*Callocystites*, rhombifer, also with long ambulacra, brachioles retained on one side; and C—*Lovenicystis*, oral region showing ambulacral grooves paved with little plates (*after* Regn'ell).

(*iii*) **Suborder Millericrinida** includes existing stalked crinoids with no or rudimentary cirri, limited to the attached end of the stem.
Examples: *Apiocrinus, Democrinus, Rhizocrinus*.

## FOSSIL CRINOIDS

The sea-lilies and the feather-stars are living members of the Crinoidea, which also includes many fossil forms. In fact the fossil Echinodermata seem to outnumber the living forms. The Homalozoa include all the fossil carpoids, while the majority of Crenozoa are extinct so are many genera in other groups. Among the Homalozoa there are forms with laterally flattened theca from the Cambrian to Lower Silurian (class Heterostelea). Some collected from the Ordovician to the Permian are of bud-like shape attached directly or by a short stalk in an upright position (class Blastoidea). Some are of discoid shape without stalks or arms found in the Lower Cambrian to the Carboniferous (class Edriasteroidea). Some extinct forms are oval or spheroidal (class Cystidea) attached directly or by a stalk in an upright position. They existed during the Middle Ordovician to Middle Devonian reaching their peak during Silurian.

The cystids or cystoids are a well-known group of extinct Crinoids mostly from Ordovician and Silurian eras. The cystids have an erect orientation with the oral surface facing upwards. Their form is base-like (Fig. 21.54). The theca is usually sessile, attached to the substratum directly. In some cases a long or short stem has also been found. The stem is made of a series of hollow pieces tapering to the attached end. For attachment the cystids had root-like extensions.

The oval or spheroidal theca is made up of closely fitting, immovable polygonal plates, both small and large. The small ones do not show any definite arrangement but the larger ones are disposed in cycles. The thecal plates of cystids are permeated with canals, which run inwards more or less at right angles to the surface. The function of the thecal canals of cystids is not clear.

The mouth is found near the centre of the oral surface. It is surrounded by numerous oral plates. The anus is located eccentrically in one of the interradii on a small anal cone (Fig. 21.54). Between the mouth and anus is found the gonopore, in the same inter-radius. There were originally three ambulacra, with one in the radius opposite that containing the anus. In *Proteroblastus* (Fig. 21.54) the ambulacra are long reaching the aboral pole. *Callocystites* also has long ambulacra. In *Lovenicystis* the grooves of ambulacra are lined by little plates cystids (Fig. 21.54).

# 22 PHYLUM HEMICHORDATA

The phylum **Hemichordata** has recently been removed from the Chordata and placed as an independent phylum of invertebrates. Many other names were suggested for the phylum by various authors but Hyman (1959) accepted the opinion of Van der Horst (1939) and retained the name Hemichordata as "suitable and just name of this phylum." This name does not imply the possession of a notochord by Hemichordates; it simply says that hemichordates are "part-chordates," that is related to chordate, a fact that cannot be denied. In 1948 Dawydoff proposed a new name **Stomochordata** for this group but there is no use in changing any appropriate name that has been in use for such a long time. Another name **Adelochordata** (*adelo*, hidden) given to the group, because notochord is not distinct, has been given up for the same reason.

The Hemichordata are solitary or colonial. The solitary forms are more or less vermiform, with the body and coelom divided into three unequal regions, the **proboscis, collar** and **trunk** (Fig. 22.2). These forms vary from moderate to great length, have a straight digestive tract and a terminal anus. Such forms belong to the class **Enteropneusta**. Then there are colonical or aggregated forms belonging to the class **Pterobranchia**. They are small or microscopic with a U-shaped digestive tract. Among these some form true colonies (Rhabdopleurida) and others are aggregation housed in common secreted encasement (Cephalodiscida). The members of the phylum differ so much in general characters that the inability of early workers to relate them properly is not surprising. The enteropneust *Balanoglossus*, commonly known as **acorn worm**, is being described here to introduce the phylum Hemichordata.

# TYPE BALANOGLOSSUS

*Balanoglossus*, the peculiar worm-like animal, lives an obscure life under stones, and burrows in the sand from between tide-marks down the abyssal region of the sea. The name refers to the tongue-shaped muscular proboscis by which the animal works its way through the sand. It was discovered in 1825 by J. F. Eschscholtz in Marshall Islands and was described as a worm-like Holothurian (Sea-cucumber). Several years later Kowalevsky (1866) discovered numerous pairs of gill-slits piercing the body wall. On this account Gegenbaur created a

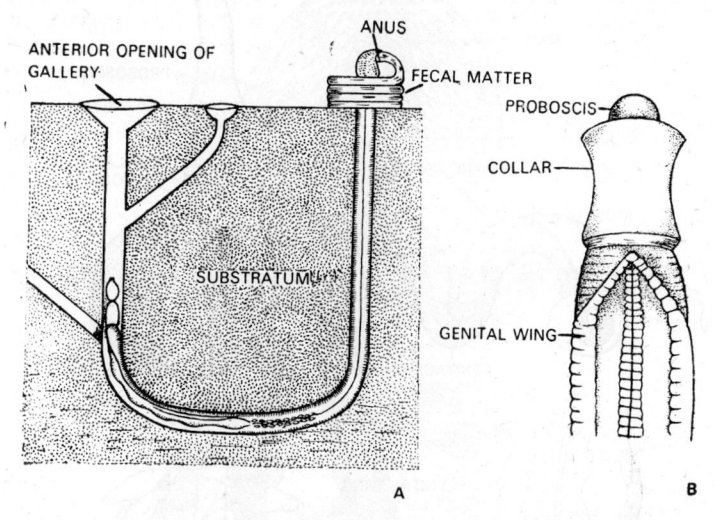

FIG. 22.1. A—*Balanoglossus* in its burrow in the sand; and B—*Balanoglossus numeensis*, anterior end (*after* Maser).

special class, the Enteropneusta. Bateson (1883-1886) showed by his embryological researches that the Enteropneusta exhibit chordate affinities in respect of the coelomic, skeletal, nervous and respiratory systems and that the gill-slits are formed on a similar plan to that of the gills-slits of *Amphioxus*, being subdivided by the tongue-bars which descend from the dorsal borders or the slits.

Some thirty species of *Balanoglossus* (Fig. 22.1) are known, distributed among all the principal marine provinces from Greenland to New Zealand. Some (the **Ptychoderidae** and **Spengelidae**) are predominantly tropical and subtropical, while others (**Balanoglossidae**) are predominantly arctic and temperate in their distribution. One species *Glandiceps chryssicola* (**Spengelidae**) was dredged during the "Challenger" expedition in the Atlantic Ocean off the coast of Africa at a depth of 2500 fathoms. Rao (1955 b, c, d) lists eight species from Madras and the Gulf of Mannar, of which five are not found elsewhere.

## External Characters

The balanoglossids range from 15-25 cm in length, some species, however, are as small as two centimetres and still others as long as 120 cm. The soft body of *Balanoglossus* is devoid of any hard structure and is divided into three parts, the proboscis (preoral lobe), collar and trunk. The **proboscis** or the collar and open to the outside by paired collar pores. The collar pores are remarkable for their constancy, this is probably owing to the fact that they have become adapted to a special function, the inhalation of water to render the collar turgid during progression. Like the collar coelom the proboscis coelom also gets distended by the intake of water through its pore. The proboscis and coeloms are occupied by muscle and connective tissue, produced by coelomic wall. Therefore, in the adult the coelom is greatly reduced and mostly devoid of a definite lining (peritoneum is absent).

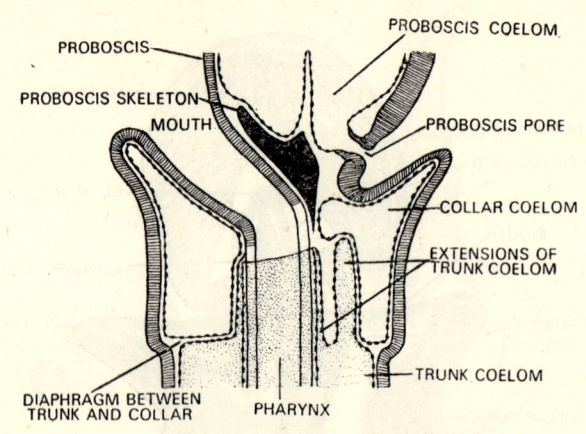

FIG. 22.4. Coelom of the collar region.

The trunk has got a pair of trunk cavities which do not open to the exterior. There are reasons for supposing that the trunk cavities were also provided with pore-canals and pores. Between the successive body cavities there is no connection, but in certain cases a communication between the two cavities of the same region occurs. Each of the paired cavities is, at one time, a closed lateral space between the skin and the alimentary canal. As these two spaces grow towards each other, both above and below the gut, they come into such a close apposition that they remain separate only by their conjoint walls which are sometimes referred to as **dorsal** and **ventral mesenteries**.

### Body Wall

The outermost layer consists of a single-celled layer of ciliated **epithelium** (as in *Amphioxus*). Numerous gland cells are scattered in the epithelium (Fig. 22.5). The secretion of these gland cells is very copious with the result that the animal is always covered with slime. Sand particles adhere to the slime and form a sort of temporary protective tube. The slime may also give **obnoxious** smell. The epidermis of the collar region is very glandular. It is divided into three to five histologically different transverse zones [heavily glandular zone, less glandular zone, etc. (Fig. 22.6)]. At the base of the epidermis lies a thick layer of nervous tissue (Fig. 22.6). It is traversed by the filamentous bases of the epidermal cells. Internally the nervous layer is bound by a strong basement membrane. Below the epithelium lies the musculature comprising an outer layer of circular muscle fibres and an inner layer of longitudinal fibres. The muscular sheath is not typically developed in this group. It is usually present in the proboscis and collarette, but is altogether absent in the main part of the collar, as a rule. In the trunk only the longitudinal fibres are generally present especially ventrally.

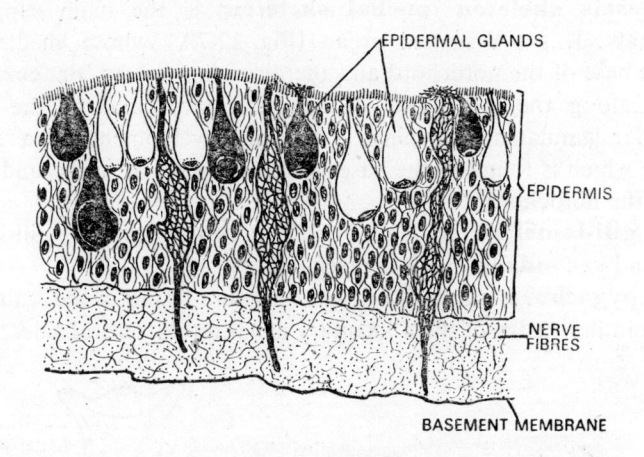

**FIG. 22.5.** Transverse section of the body wall of a balanoglossid *Glossobalanus minutus* (*after* Grasse).

The animal moves by pushing the proboscis and collar forwards in the sand first and then pulling the body behind. As the circular muscles are weak the protrusion of the proboscis cannot be vigorous. It is, however, brought about

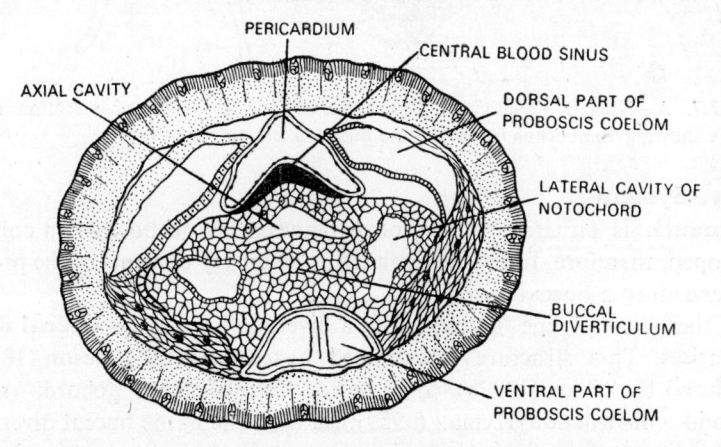

**FIG. 22.6.** Transverse section of the proboscis of *Balanoglossus misaksensis* to show the structure of the buccal diverticulum.

by ciliary action, distension of the coelom (hydrostatic skeleton) and contraction of the circular muscles. The cilia probably play main part in locomotion and are the chief burrowing organs.

## Skeleton

The **skeletal** structures include (*i*) the **proboscis skeleton,** (*ii*) the **gill-bars,** and (*iii*) **pygochord**. Now the notochord is not considered as a skeletal structure. The name notochord was given to hollow dorsal outgrowth of the alimentary canal of the collar region. It extends forwards into the basal (posterior) part of the proboscis through the neck, into the proboscis coelom, ending blindly in front (Fig. 22.3).

(*i*) **Proboscis skeleton (nuchal skeleton)** is the main support of the proboscis stalk. It is a Y-shaped organ (Fig. 22.7A) whose median part lies beneath the base of the notochord and the diverging **arms** or **cornua** extend backwards along the outer side of the alimentary canal of the collar. It is a non-cellular laminated thickening, a special development of a structureless membrane, which is found at the base of the layers of cells underlying that portion of the notochord.

(*ii*) The **gill-lamellae** are supported by two types of gill-bars called **primary** and **secondary gill-bars** (see below).

(*iii*) The **pygochord** is a hard rod-like structure found in the caudal region. It is a median derivative of the alimentary canal on its ventral side.

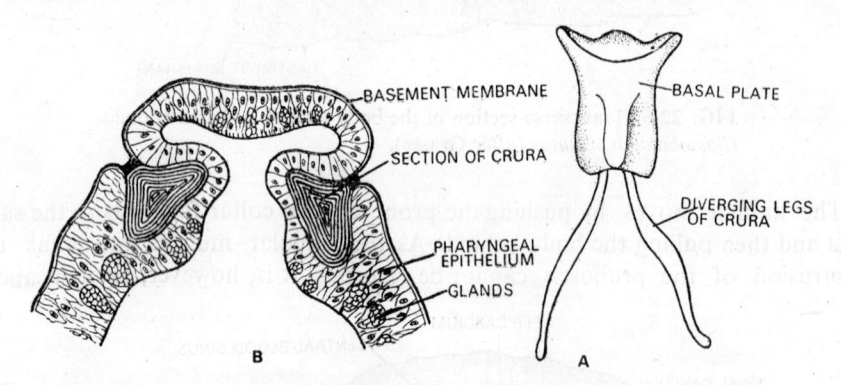

FIG. 22.7. A—Nuchal skeleton of *Balanoglossus*; and B—Transverse section of the pharynx showing the cornua (crura) of the nuchal skeleton cut in section.

## Digestive System

The **mouth** is situated in a groove between the proboscis and collar, and remains open, therefore, food is brought to it by ciliary currents of the proboscis. Mouth leads into a buccal cavity.

From the roof of the buccal cavity a diverticulum named **buccal diverticulum** arises. This structure was named **notochord** by Bateson (1885) and **stomochord** by Willey (1899b). Since words formed from "chord" seem to imply solid construction Hyman (1959) has used the name buccal diverticulum for it. The diverticulum begins as a narrow neck that passes anteriorly just above the main plate of the proboscis skeleton and continues in the posterior part of the proboscis. It often expands considerably and usually presents sacculations, because of the presence of ventrolateral pockets inside. The enlarged portion of the buccal diverticulum is called the **vermiform process**. The buccal diverticulum is hollow and opens into the buccal cavity. Its wall consists of tall, vacuolated epithelial cells similar to the epithelium of the buccal cavity. All recent workers on enteropneusts now agree that the buccal diverticulum does not represent a notochord and is preoral extension of the digestive tract (Silen 1950).

The buccal cavity leads into a wide pharynx which can conveniently be divid-

# Phylum Hemichordata

ed into two regions, upper respiratory region opening by a series of gill-slits, and lower digestive region with glandular lining. There is an incomplete parti-

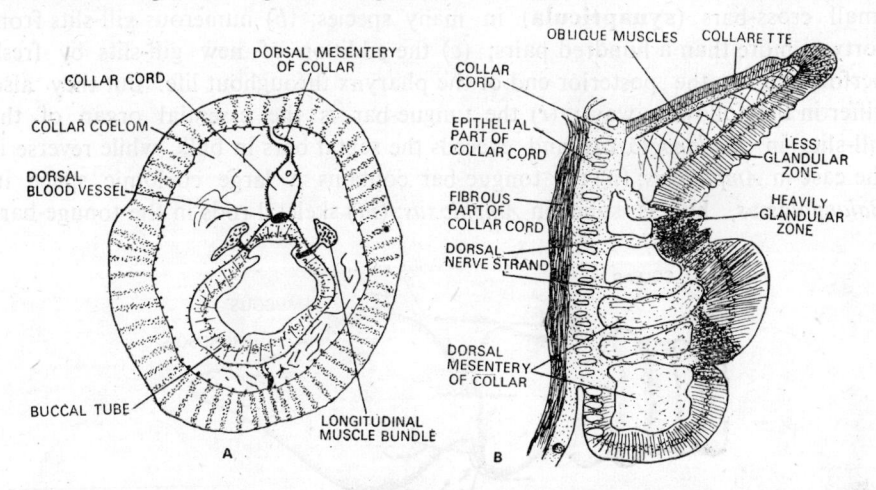

**FIG. 22.8.** A—Transverse section through the collar of *Protoglossus* (*after* Caullery and Mesnil); and B—Longitudinal section of the dorsal side of the collar of *Glossobalanus minutus* showing five epidermal zones.

tion between the respiratory and digestive regions (Fig. 22.13). The pharynx passes on into a straight intestine which opens to the exterior at the anus.

The gill-slits open freely and directly to the outside only rarely but usually open into branchial pouches or gill-sacs (Fig. 22.13) which open to the exterior by small pores, the **gill pores**, or **branchial pores** (Fig. 22.13) situated on the dorsal side, in a shallow longitudinal groove, not far from the middle line. Gill-sacs lie between the alimentary canal and body wall and communicate with the former by U-shaped slits (Fig. 22.10). Seen from dorsal side the gill-pores present a linear series of simple pores. The walls of gill-slit are thin and are

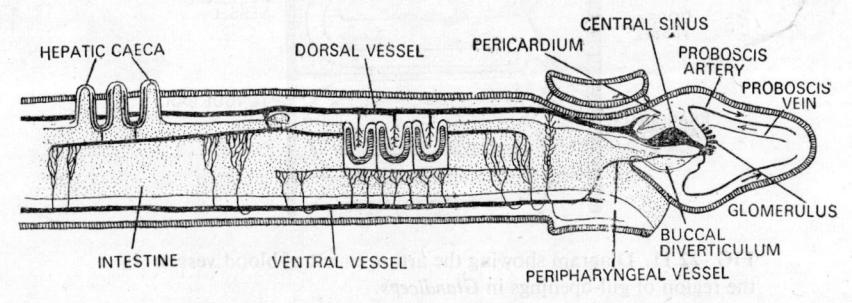

**FIG. 22.9.** Circulatory system of *Balanoglossus* (*after* van der Horst).

supported by gill-bars which are of two types, the **primary gill bars** surrounding the U-shaped opening and **secondary gill-bars** lying between the arms of "U." In some species the slits open into an atrium formed of lateral folds, usually turned upwards to leave a long mid-dorsal opening. The number of gill-slits is variable being about 700 pairs in some forms (*Balanoglossus aurapticus*). So far as the indefinite number of gill-slits is concerned *Balanoglossus* comes quite close to *Amphioxus*.

The gill-slits of *Balanoglossus* resemble those of *Amphioxus* in the following ways: (*a*) the presence of two types of gill-bars in all the species and also of small cross-bars (**synapticula**) in many species; (*b*) numerous gill-slits from forty to more than a hundred pairs; (*c*) the addition of new gill-slits by fresh perforation at the posterior end of the pharynx throughout life. But they also differ in the following ways: (*i*) the tongue-bar is the essential organ of the gill-slits in *Banlanoglossus* and exceeds the septal bars in bulk, while reverse is the case in *Amphioxus*; (*ii*) the tongue-bar contains a large coelomic space in *Balangolossus*, but is solid in *Amphioxus*; (*iii*) skeletal rods in the tonuge-bars

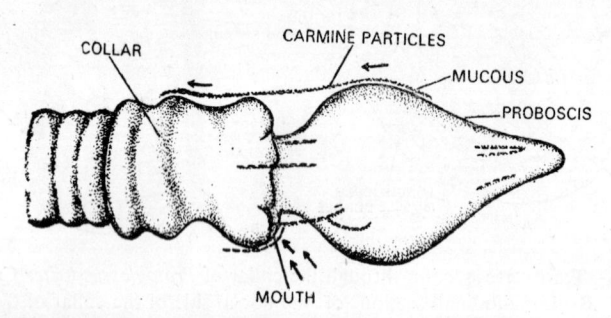

**FIG. 22.10.** Feeding-currents on proboscis of *Glossobalanus* (*after* Barrington).

are double; (*iv*) the tongue-bar in *Balanoglossus* does not fuse with the ventral border of the cleft, but ends freely below, thus producing a continuous U-shaped cleft; (*v*) in most species of *Balanoglossus* each gill-slit may be said to open into its own atrial chamber or gill-pouch, this in turn opens to the outside by minute pore, there are, therefore, as many gill-pouches as there are gill-slits and

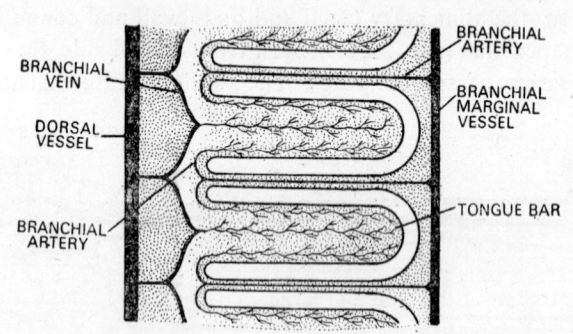

**FIG. 22.11.** Diagram showing the arrangement of blood vessels in the region of gill-openings in *Glandiceps*.

as many gill-pores as pouches. The respiratory current of water is, therefore, conducted to the exterior by different means from that adopted by *Amphioxus*.

The function of alimentation is closely associated with that of locomotion, somewhat as in the burrowing earthworm. As it excavates its burrow the sand is passed through the body, and the nutrient matter that may adhere to it is extracted during its passage through the intestine. The exhausted sand is finally ejected through the anus at the orifice of the burrow where it appears as worm-casting at low tide (Fig. 22.1). It is in response to this type of feeding that the

## Phylum Hemichordata

mouth remains constantly open. *Balanoglossus*, thus literally feeds its way through the sand. The considerable number of marine organisms living in the sand constitutes its food. A large amount of water also enters along with sand. This is passed through the dorsal respiratory part of the pharynx. Another method of feeding has been reported by recent workers. Water always remains circulating through the pharynx by ciliary currents of the proboscis. Food particles passing along with the currents are thrown down on the glandular surface and digestion begins. Formation of definite mucous strands for feeding has been reported for *Glossobalanus* (Fig. 22.11). The animal was placed in water containing carmine particles which were either taken directly along with the feeding currents or were caught in strands of mucous and passed backwards. There is no endostylar apparatus and the large amount of matter ingested passes directly to the intestine opening by terminal anus.

### Circulatory System

The circulatory system is quite peculiar consisting of sinuses or lacunae and blood vessels devoid of endothelium, situated within the thickness of the basement membrane of the body wall, of the gut wall and the mesenteries. The blood is a non-corpuscular fluid that circulates in the system (some authors even doubt the presence of blood). The main blood channels include a contractile dorsal vessel ending anteriorly in a sinus, a complex **glomerulus** at the base of

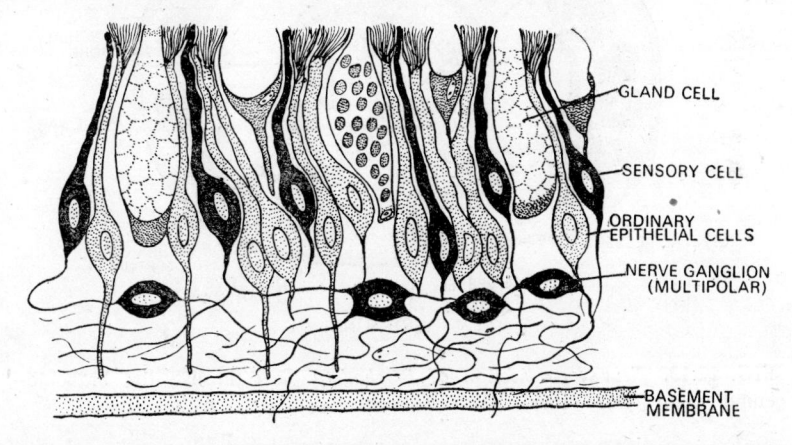

FIG. 22.12. Transverse section of the balanoglossid epidermis to show the nerve net (schematic figure from different sources).

the proboscis and a ventral vessel. Above the dorsal sinus lies a closed contractile vesicle the **cardiac sac** or **"heart"** (also called **pericardium** by some). The circulation of blood (if it occurs) is maintained by this vesicle. The dorsal and ventral vessels are connected together by a ring of vessels in the region of the glomerulus, which also sends blood vessels anteriorly in the proboscis. The blood is propelled forward by the contractile dorsal vessel and is collected in the **central blood sinus**. By the pulsation of the vesicle (best observed in the larva) the blood is driven into the glomerulus, from which it issues by efferent vessels which conduct it to ventral vessel. Details of the working of the circulatory system are not clearly understood. It is, however, clear that the circulatory system in *Balanoglossus* is built on a simpler plan than that of *Amphioxus*.

## Excretory System

Some authors (Young, 1955) maintain that there is no specialized structure for excretion in any part of the body. Whatever has been described under this heading is the **proboscis gland**, lying on one side of the notochord in the proboscis. It has been reported that this gland separates urea and uric acid from blood and passes on to the proboscis coelom wherefrom it is excreted through the proboscis pore. The proboscis peritoneum adjacent to the glomerulus or elsewhere in the proboscis may, in some species, present an appearance suggestive of excretion (Hyman, 1959).

## Nervous System

The nervous system of *Balanoglossus* presents interesting resemblances with that of the echinoderms. As in the latter it consists of a sheet of nerve fibres and cells lying beneath the epidermis all over the body. At various places this sheet is thickened to form a dorsal and ventral tract of nervous material in the trunk region, a **circular tract** connecting these two at the posterior edge of

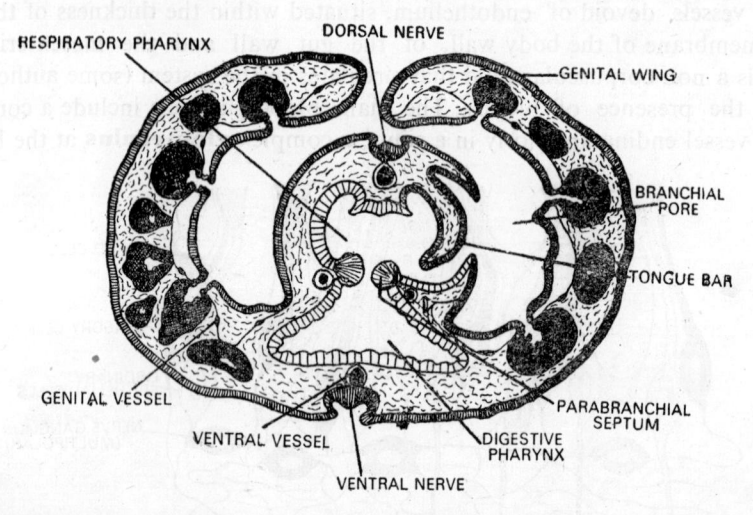

FIG. 22.13. Transverse section of *Ptychodera flava* through the branchio-genital region (*after* Willey).

the collar and a strong concentration of nerve tissue around the whole of proboscis stalk and the posterior end of the proboscis. In the region of the collar it is rolled up to form a hollow neural tube open at both ends. This part lying above the alimentary canal is known as the central nervous system. From this tube "dorsal roots" proceed to the skin of the back of the collar. Thus, it is evident that the nervous system of *Balanoglossus* presents unmistakable resemblances with the centralized subepithelial plexus of echinoderms on the one hand, and with the tubular dorsal nerve cord of the vertebrates on the other. Organs of special sense are apparently lacking. Some people have, however, described a patch of special ciliated cells on the collar as the only sense organs, probably chemoreceptors.

## Behaviour

The behaviour of *Balanoglossus* suggests the presence of only one clearcut reflex, that is a contraction of the longitudinal muscles in response to tactile stimulation. If separate pieces of the body are subjected to light or to tactile stimuli they react by moving away, suggesting, thereby, the presence of uncentralized nature of nervous system as that of the echinoderms. There are no special conducting pathways. This is proved by the stimulation of flaps of body partly cut off from the rest that produce a generalized contraction. Thus, it is evident that conduction can occur in all directions. The dorsal and ventral nerve cords act only as quick conduction pathways.

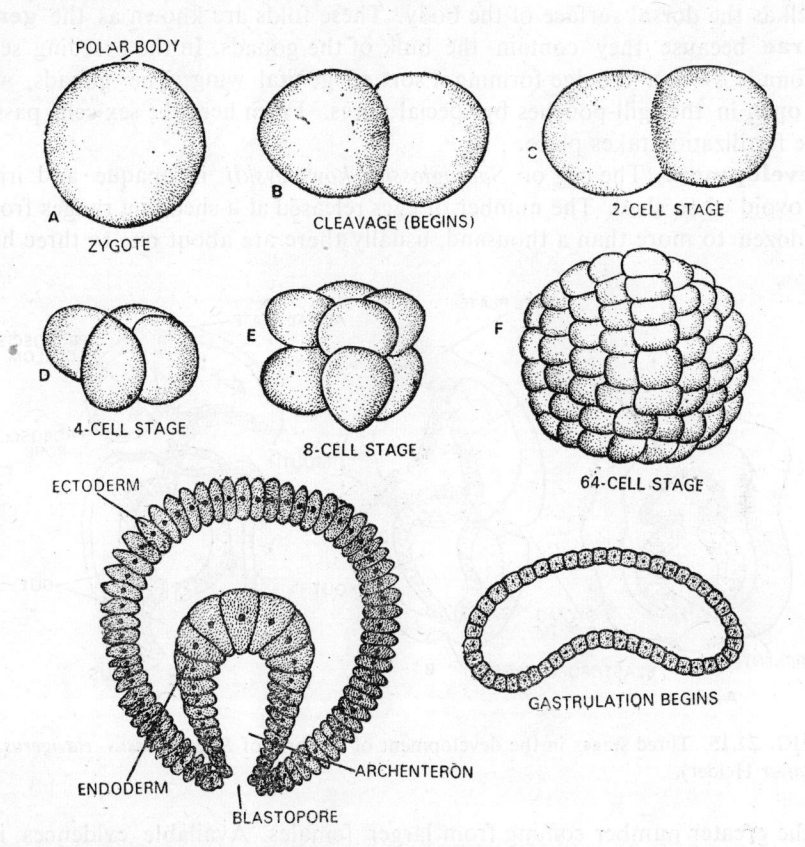

FIG. 22.14. A—A cleavage in *Saccoglossus*; and B—Gastrulation in *Saccoglossus*.

## Regeneration

*Balanoglossus* possesses great power of regenerating lost parts. If cut, the posterior part of the body is readily formed. The proboscis, collar and branchial parts can also regenerate the lost parts easily. It is reported that these parts can be regenerated from mere fragments of the regions.

## Reproductive System

The sexes are separate. When mature the colour of the genital region may be slightly different. Both male and female gonads consist of more or less lobulated hollow sacs connected with the epidermis by short ducts (Fig. 22.13). They may be either uniserial, biserial or multiserial in disposition. They occur in the branchial region, and also extend to a variable distance behind it. When arranged in uniserial or biserial rows the genital ducts open into or near the branchial grooves in the region of the pharynx (Fig. 22.13) and in a corresponding position in the post-branchial region. In some species the sides of the body form paired longitudinal pleural or lateral folds which are movable and can be approximated at their free edge so as to close in the dorsal surface, embracing both the median dorsal nerve-tract and the branchial grooves with the gill-pore. This, so to say, forms a peribranchial and meduallary tube opening behind where the folds cease. They can also be spread out horizontally so as to expose their own upper side as well as the dorsal surface of the body. These folds are known as the **genital pleurae** because they contain the bulk of the gonads. In the breeding season the gonads swell and bulge forming a sort of genital wing. The gonads, when ripe, open in the gill-pouches by special pores. From here the sex cells pass out where fertilization takes place.

**Development.** The egg of *Saccoglossus kowalevskii* is opaque and irregularly ovoid when shed. The number of eggs released at a shedding ranges from a few dozen to more than a thousand, usually there are about one to three hund-

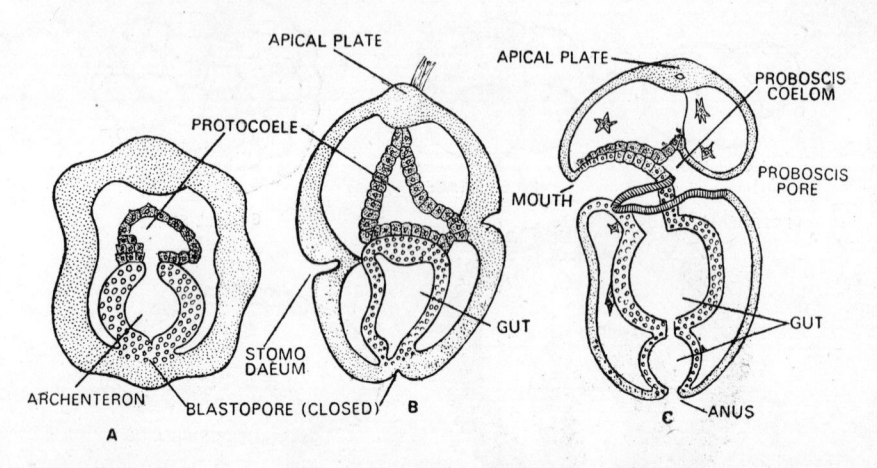

**FIG. 22.15.** Three stages in the development of tornaria of *Balanoglossus clavigerus* (*after* Heider).

red, the greater number coming from larger females. Available evidences indicate that maturation begins some hours before ovulation and that the egg is generally in the metaphase of the first meiotic division when shed. It is at this stage that the egg is fertilizable. Fertilization of eggs within 6 to 7 hours after shedding yields a high percentage of normal development. The spermatozoon is able to enter at any point over the entire surface After fertilization the egg undergoes an orderly series of successive changes in shape till cleavage sets in.

**Cleavage.** The first cleavage begins about two hours after fertilization and produces two generally but not invariably, equal cells. The second cleavage is also like the first and produces four usually (but not invariably) approximately equal cells. As a result of the third cleavage two tiers of cells are produced, a tier of four smaller animal cells and a tier of four larger **vegetal cells**. The cleavage continues till a multicellular **blastula** (of about 2456 cells) is formed.

## Phylum Hemichordata

As the cells multiply the volume of the blastula increases and the contour becomes more smooth. In the interior a blastocoel is formed. Next there comes a stage when the blastula is covered over with numerous indented areas, which later disappear. The approach of gastrulation is marked by a reduction in the volume in the animal-vegetal axis. Gradually the compact blastula begins to flatten at the vegetal or posterior end and the flattened end is pushed inwards forming a shallow concavity which deepens gradually and blastopore becomes reduced. As the blastopore decreases in size, a well-defined circular or elliptical rim develops around it. There is a considerable variation in the contour of the blastopore as it decreases and finally closes.

After about 18 hours the animal vegetal's axis begins to elongate again. This heralds the beginning of the establishment of the distinctive larval form. The proboscis, collar and trunk are delineated with the formation first of the anterior

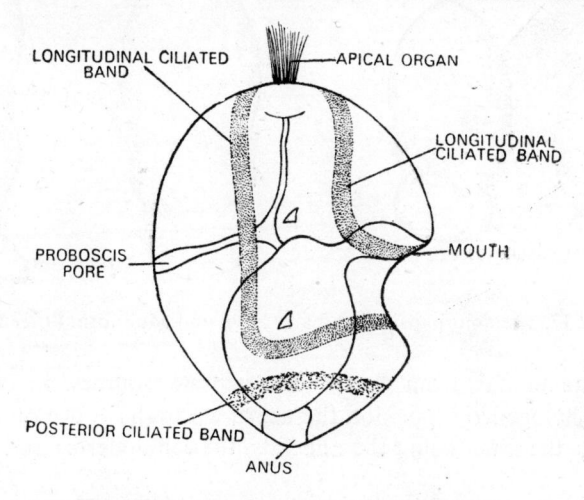

FIG. 22.16. Young tornaria larva, side view.

collar groove and then of the posterior collar groove (Fig. 22.17). With the increase in length the space within the egg membrane becomes smaller for the larva, as such it curves ventralwards.

Following this curving (sometimes called skewing) the "neural groove" appears as a longitudinal dent in the mid-dorsal line of the collar, which projects also in the proboscis and at least two-thirds the length of the trunk. Concurrently with these changes a small pit appears at the mid-ventral point of the anterior collar groove, this is the first external indication of the mouth. Various other larval structure including the cilia particularly the apical tuft, the transverse ciliated band, etc., appear before hatching that takes place usually on the seventh day. The arms appear on the posterior side in the newly hatched larva at the side of the former blastopore.

On hatching the larva (Fig. 22.16) is free-swimming and pelagic called **tornaria** (Fig. 22.16). The ciliated band of the larva has exactly the same relationship as found in the auricularia larva of sea-cucumbers (echinoderms). The band passes in front of the mouth, down the sides of the body, and in front of the anus. In later tornaria larvae in addition to the longitudinal bands, there is always present a posterior ring of stout cilia. The cilia of the posterior ring are purely locomotor, while those of the band set up feeding currents converging to the mouth. After leading a free swimming life for some time the larva sinks, the three parts of the body become clearly marked off, and it undergoes metamorphosis into the worm-like adult.

## Phylogeny

*Balanoglossus*, the shore-living acorn or barnacle worm, is more like invertebrates than the lower chordates. It is, however, believed to be related to the lower chordates. This is one of the mutually exclusive views which seeks to

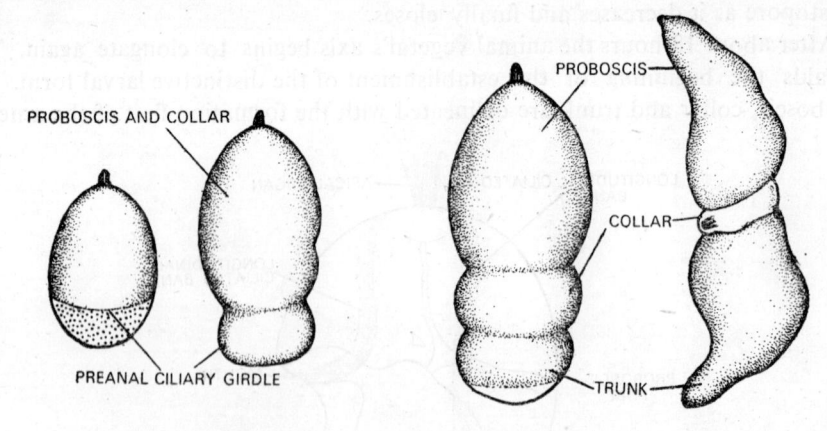

FIG. 22.17. Metamorphosis of tornaria larva into adult form (*after* Davis).

derive vertebrate animals from various invertebrate sources. In an attempt to establish its phylogenetic position the acorn worms have been linked to several diverse groups, the chief being the annelids, the echinoderms and the chordates.

## Annelidan Affinities

The habits of *Balanoglossus* (burrowing, food capture through permanently open mouth), the extraction of the nutrient matter in the long, almost straight gut, leaving off the refuse behind as well as its general form suggest it to be an earthworm. Naturally *Balanoglossus* was declared to be related to the Annelida. The tornaria larva of *Balanoglossus* was supposed to be a modified trochophore[1] of annelids and relationship was further strengthened. But now it has been established that, but for these superficial similarities between the acorn-worms and earthworms, there is nothing that can lead to the establishment of definite relationship.

The process of development of the modern annelids is entirely different from that of *Balanoglossus*. The cleavage by which the fertilized egg is divided into blastomeres follows in Annelida a "spiral" plan, in which many blastomeres arise in a regular way and the future fate of each can be exactly stated. In the chordates the cleavage is "irregular", the cells not forming any special pattern. The gastrulation in early chordates occurs by invagination, the folding in of one side of the ball of cells to form an archenteric cavity communicating with the exterior. In the annelid-worms this never occurs, the cells do not form any

[1]Ciliated, small free-swimming larva of Polychaete annelids.

special pattern. The gastrulation in early chordates occurs by invagination the folding in of one side of the ball of cells to form an archenteric cavity communicating with the exterior. In the annelid-worms this never occurs, the cells which will go to form the gut migrate inwards either at one pole or all round the sphere and only later form themselves into a tube, which opens to the outside secondarily. The method of the formation of mesoderm and coelom differs widely in the two groups. In the lower chordates the third layer is reduced by separation from the endoderm so that the coelom is continuous with archenteron and is said to be an **enterocoele**. In annelids the cells separate in many ways to form the mesoderm and the coelom that arises within this solid mass is a **schizocoele**.

In the light of these and many other points such as the organization of the nervous system, absence of gill-slits in annelids, etc. the annelidan affinities of *Balanoglossus*, have been give up.

**Affinities with chordates**. Strongest objection to the annelidan theory is presented by the presence of numerous paired gill-slits. The gill-slit apparatus of *Balanoglossus* show a rough analogy with that of *Amphioxus*. In fact it is this that stimulated a search of additional chordate characters in *Balanoglossus*, Bateson was the first to identify the structure that he regarded as notochord, the axial stiffening rod, characteristic of all chordates. Several objections have been put forward to emphasize that this structure is not a true notochord. The first main objection is that the true notochord occurs above the dorsal vessel and covers much more portion of the body while in *Balanoglossus* it occurs below the dorsal vessel and is shorter. The plate-like skeletal tissue contains large cells and bears a considerable resemblance to a notochord. A notochord extending throughout the length of the body would clearly be disadvantageous for an animal whose main movements are lengthening and shortening. Supporters of the chordate affinities feel that it is quite reasonable to accept the skeletal plate as a remnant of notochord produced in the same way in which it is formed in higher forms and serving as a fixed point by which the body is drawn forward on the proboscis. But the recent workers on enteropneusts do not support this view for the following reasons:

(*i*) *The Notochord*. It is evident from the account of the buccal diverticulum (page 936) that the name notochord was erroneously given to this structure by Bateson (1885) and his followers. The following differences support this view:

| *Buccal diverticulum* | *Notochord* |
|---|---|
| 1. Formed as a hollow forward evagination of the buccal wall | Formed as a solid rod-like elevation from the roof of archenteron. |
| 2. Resembles the anterior lobe of vertebrate hypophysis (in the manner of formation). | Does not resemble vertebrate hypophysis. |
| 3. Consists of large vacuolated cells and frequently of ordinary epithelial cells. | Consists of large vacuolated cells. |
| 4. Lacks sheath | A sheath is present. |
| 5. Not related to blastopore. | Related to the blastopore. |
| 6. Not supporting in function. | Supporting in function. |
| 7. Ventral to the mid-dorsal blood-channel. | Dorsal to the main dorsal blood vessel |
| 8. Some sort of preoral gut. | Axial stiffening rod. |

(*ii*) *The central nervous system.* The central nervous system of hemichordates and chordates shows a marked resemblance so far as the brain is concerned in structure and development. Its dorsal position, its occasionally hollow construction, its mode of formation from the dorsal epidermis and the occasional presence of a neuropore are more or less similar in the two groups. But the rest of the nervous system presents typical invertebrate pattern. It is intraepidermal in position, while the main nerve cord is ventral, additionally circumcenteric connectives (not found in chordates) are present.

(*iii*) *The gill-slits.* The gill-slits and the branchial apparatus in the two groups have the same general construction with tongue-bars, synapticulae, and an arcade of trifid skeletal supports. This identity suggests common ancestry. But Hyman (1959) says that this relationship does not, however, justify the inclusion of the hemichordates in the phylum Chordata. The few similarities are far outweighed by important differences which, from apart those given above, are listed below:

| Hemichordata | Chordata |
| --- | --- |
| Presence of coelomic sacs | Absence of coelomic sacs of the pattern of hemichordates |
| No trace of segmentation in these systems (at the most in incipient stage). | Segmentation expressed by the plan of muscular, nervous, circulatory and excretory systems. |

**Echinoderm affinities.** Showing through these special chordate features a plan is visible that resembles the echinoderms also. The similarity between a young tornaria and a young bipinnaria larva is so great as to have misled even Johannes Müller who regarded it (tornaria) as the larva of a starfish. Many other authors[2] have emphasized the relationship. The obvious resemblances between the two include: (*i*) the almost identical course of the longitudinal ciliated band in the young stages; (*ii*) the presence of a dorsal pore; and (*iii*) the cavities of *Balanoglossus* are represented in the echinoderm larvae which present trisegmental condition. As in *Amphioxus* the coelomic cavities are the archenteron, a method of coelom formation not found in most other invertebrates. The method of coelom formation presents greater phylogenetic importance than the resemblances between external features which might have developed due to similar ecological conditions. The proboscis pore of *Balanoglossus* compares well with the water pore of the echinoderms. It is for this reason that some authors have even visualized the beginning of the echinoderm water vascular system in the hemichords.

As mentioned above the larva of *Balanoglossus* is trisegmental and this character is retained in the adult inasmuch as the first segment becomes the proboscis, the second the collar and the third the trunk. Thus, it is concluded that *Balanoglossus* and its allies retain the larval organization throughout life. *Amphioxus* and vertebrates, on the other hand, constitute a group that started with trisegmental organization, but introduced metamerism secondarily.

---

[2]Willey, Metschnikoff, Masterman are prominent among these.

## Phylum Hemichordata

On the basis of the above discussions it cannot be concluded that *Balanoglossus* and its allies are chordates. They are assuredly below *Amphioxus* in the level of their organization, and are highly specialized. *Balanoglossus* on the other hand, links the chordates with echinoderms. On the basis of the surprising similarity between the larvae of the hemichords and echinoderms it is considered that they have descended from some common ancestor and in response to the different modes of life, they were called upon to lead, the echinoderm larva having metamorphosed into an adult with a kind of radial symmetry, while the adult balanoglossid has developed into a worm-like animal with bilateral symmetry.

## CLASSIFICATION OF HEMICHORDATA

**Tonuge worms** fall under this category. Peculiar marine worm-like animals in which the anterior end is usually provided with fleshy proboscis and collar,

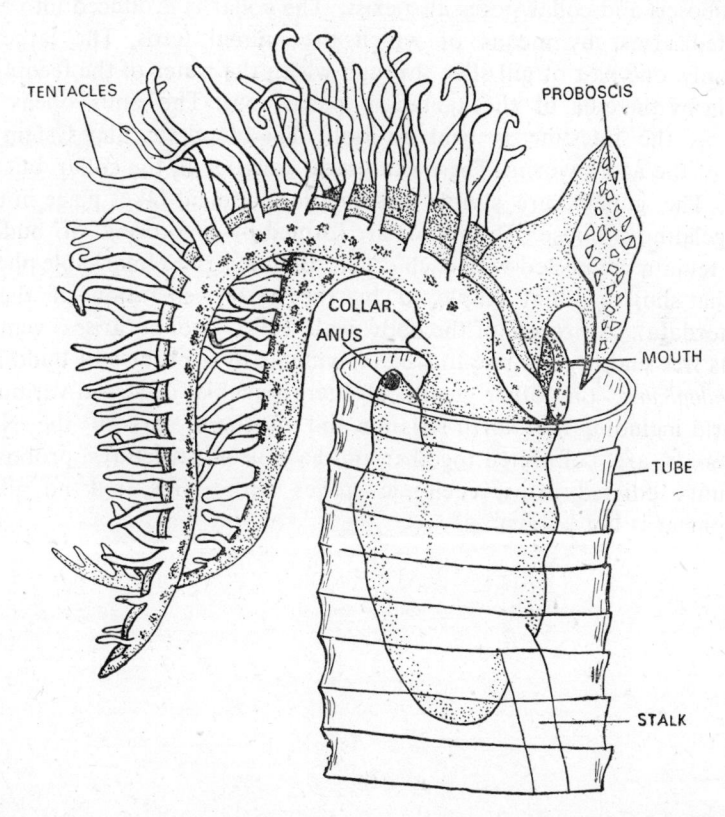

FIG. 22.18. *Rhabdopleura*, side view (*after* Delage and Herouard), a pterobranch hemichordate in its tube.

the body is not metamerically segmented, but is divisible into **preoral lobe** (proboscis), **collar** and **trunk**. Notochord is absent. There are many permanent gill-slits. Nerve tissues occur both dorsally and ventrally embedded in epidermis. There is a simple blood system and the direction of flow is the reverse of that in true chordates.

(*i*) CLASS ENTEROPNEUSTA. The Hemichordata with elongated worm-like burrowing animals with numerous gill-slits. The epidermis is ciliated and glandular.

*Balanoglossus, Saccoglossus* (=*Dolichoglossus*, Fig. 22.2).

(*ii*) CLASS PTEROBRANCHIA. The Hemichordata in which the preoral lobe bears ciliated tentacles which are meant to produce ciliary feeding currents of water, gill-slits are two or absent. They are sedentary, colonial animals, and show signs of the general echinoderm-chordate plan of organization.

*Rhabdopleura* (Fig. 22.18) and *Cephalodiscus* are two examples of this class. *Cephalodiscus* is the genus that has been found on the sea-bottom at various depths, mainly in the southern hemisphere. Many individuals live together in many chambered gelatinous house, therefore, they have been called colonial. Each individual or zooid has a proboscis, collar and trunk but these structures are visible only under careful examination. Ordinarily *Cephalodiscus* does not look like *Balanoglossus* at all. Each part of an individual possesses coeloms and proboscis and collar pores also exist. The collar is produced into a number of ciliated arms, by means of which the animal feeds. The large pharynx bears only one pair of gill-slits, through which the water of the feeding current, drawn in by the cilia of the tentacles, passes out. The anus opens near the mouth as the intestine is bent on itself. The blood vascular system is similar to that of the acorn worm. There is a dorsal ganglion in the collar but it is not hollow. The gonads are simple sacs and development takes place in the spaces of the gelatinous house. The zooids are formed by a process of budding, but do not remain connected with each other. Budding is a remarkable phenomenon somewhat similar to that employed by the nurse generation of the salpians (Urochordata). A process of the body wall called stolon arises ventrally and from its free end it gives rise to free individuals by a process of budding.

*Rhabdopleura*, the other genus of Pterobranchia, occurs in various parts of the world including the North Atlantic and northern part of the North Sea. The zooids are connected together and have usual structures, proboscis, collar and trunk, ciliated arms, coelomic spaces with pores, but no gill-slit. The development is not known.

# INDEX

Abdominal appendages 561
Aboral pole 244
Aboral surface 111, 868
Abductor muscles 607
Acantharia 82
Acanthocephala 315, 474
Acanthor larva 322
Acari 593
Accessory glands 703, 717
Acellular 1
Acetabula 285
Aciculum 405
Acid gland 710
Acineta 141
Acochlidiacea 817
Acoela 261
Acontia 231
Acorn barnacle 641
Acorn worm 931
Acrothoracia 642
Actiniaria 236
Actinomyxidia 107
Actinophrys 82
Actinosphaerium 83
Actinotrocha 516
Active phase 22
Adductor muscles 607

Adenosine triphosphate 6
Adhesive organ 265
Adhesive tube 343
Adhesive type 195
Adjustor neurone 443
Adoral zone 136
Adradial 222
Aeolosomatidae 452
Agamophilariae 377
Agamont 86
Agelenidae 595
Alcyonaria 229, 232, 235
Alcyonium 233, 235
Alkaline gland 710
Allocoela 263
Alveolar gland 562
Alveoli 111
Amastigote 36
Amblypygi 591
Ambulacral groove 870
Ambulacral spines 870
Ambulacrum 896
Amictic eggs 336
Ammonoidea 956
Amoebic dysentry 79
Amoebiasis 79
Amoebocytes 164, 171, 531, 799

Amoeboid movements 57
Amphiblastula 172
Amphids 358
Amphidiscophora 178
Amphilinidea 295
Amphineura 771, 779
Amphipoda 651
Amphiura 907
Ampulla 191, 447, 874
Anguina 367
Anisospores 85
Anal lacuna 312
Anal pore 245
Anaspidacea 646
Anisomyaria 837
Annelida 398
Annelida theory 523, 746
Anodonta 819
Anomalodesmata 837
Anomura 652
Anopla 313
Anostraca 635
Anoxybiotic respiration 272
Antapical spines 46
Anterior loops 438
Anthocodia 234
Anthozoa 228, 232
Anthozoans 228
Antipatharia 240
Antiplectic 120
Aphodal 162
Aphrodite 421
Apical field 325
Apical lobe 467
Apical organ 394
Apical papilla 275
Aplacophora 773, 782
Aplysia 816
Apoda 919
Apodemes 530
Apopyles 160, 165
Appendix interna 601
Appendix masculina 601, 626
Apple snail 786
Aporhynchus 298
Apostomatida 145
Arachnida 556
Arachnida Classification 587
Araeolaimoidea 381
Araneae 591
Arbacioida 900
Arcellinida 81
Arcella 82
Archaeocytes 164
Archenteron 50
Archigates 300
Archigregarinida 104

Archiannelida 475
Argas 594
Argonauta 855
Argulus 641
Arhynchobdellida 474
Aristotle's lantern 897
Arrow-worm 858
Articulata 524, 927
Artificial fertilization 884
Arthropoda 525, 532
Ascaris 349
Ascaroidea 385
Aschelminthes 324
Association cyst 89, 104
Asconoid canal 159
Aspidobothridia 283
Aspidochirota 919
Asplanchna 341
Asterias 889
Asteroidea 867, 889
Astomatida 145
Astrangia 240
Astropecten 889
Athalamida 83
Athrocytes 392
Atoll 238
Atoke 415
Atrium 265
Auricularia 916
Aurelia 221
Autogamy 83, 131
Autoheteroxenous 365
Autospasy 563
Autotomy 616, 888
Autozoids 234
Avicularium 504
Axial cell 152
Axial complex 915
Axopodia 59
Axial gland 878
Axial organ 878
Axial sinus 878
Axinellida 179

Balancers 246
Balanoglossus 931
Baroltaxis 129
Basal bodies 109
Basal plate 230
Barnacles 641
Barrier reef 236
Baccillariophyceae 4
Bdelloidea 340
Bdellonemertini 314
Bed bug 713
Beneficial insects 757
Beroe 248

*Index*

Beroida 248
Bicosoecida 48
Bilateria 250
Bioluminescence 851
Bipalium 252
Bipinnaria larva 886
Biramous type 601
Birth pore 277
Bivium 868
Bladder worm 291
Blastea 182
Blastocoel 202
Blastopore 450
Blastostyle 205
Blastula 172, 202, 222, 449
Blepharoplast 19, 21, 38
Blind gut 276
Bodo 43
Bolbosoma 312, 317
Bolinopsis 247
Bonellia 478
Book-lungs 568, 570, 581
Bothria 285
Bothridia 285
Botryoidal tissue 456, 458
Bougainvillia 215, 216
Brachiolaria larva 886
Brachionoidea 341
Brachiopoda 518
Branchellion 473, 474
Branchial crown 411
Branchipus 635
Bracts 213
Brittle stars 901
Brown bodies 506
Bryozoa 501
Buccal capsule 344
Buccal field 328
Buccal mass 792, 793
Buccinum 794
Budding 200
Bugula 502
Bursa 321
Bursaria 147
Buthus 573
Byssus 834

Caeca 266
Calandra 743
Calcareous ribbon 519
Calcareous spicules 159
Calcaronea 177
Calcinea 177
Calciferous glands 434
Calcoblasts 164
Calotte 152
Calthrops 166

Calyptoblastea 216
Calsispongiae 176
Calymma 84
Calyx 388, 920
Camerata 927
Campanularia 216
Campanularidae 216
Canal-fed vacuole 118
Cantharidin 742
Cantharis 757
Capatula 783
Capillary plexus 463
Carapace 558
Carchesium 138
Cardiac pouch 879
Cardiac stomach 612
Cardiopyloric aperture 613
Cardiopyloric strand 617
Carpus 604
Carybdea 226
Carybdeida 226
Caryotic 4
Cassiduloida 901
Cassiopeia 228
Catenula 262, 263
Caudae 331
Caudal appendage 481
Caudal vesicle 288
Caudo-vesicular ganglion 335
Caudina 919
Cell-mouth 4, 123
Cell-anus 117
Cell membrane 7
Cement gland 321
Centipedes 635
Centriole 21
Centrurus 587
Cephalic aorta 800
Cephalic lacuna 312
Cephalic papillae 357
Cephalic shield 547
Cephalic sucker 455
Cephalocarida 634
Cephalodiscus 947
Cephalopoda 856
Cephea 228
Ceractinomorpha 180
Ceratomyxa 107
Cercariae 277
Cerebral commissure 803
Cerebral ganglion 272, 443, 467, 860
Cerebro-buccal connective 803
Cerebro-pleural connective 803
Cerebro-pedal connective 803
Ceriantharia 240
Cerithium 807
Cervical gland 356

Cestida 247
Cestoda 284, 294
Cestodaria 295
Chaetoderma 779
Chaetogaster 401
Chaetognatha 858
Chaetognatha theory 523
Chaetonotoidea 345
Chaetopterus 419, 420
Cheilostomata 510
Chelae 605
Chelicerae 548
Chelecerata 458
Chelonethi 587
Chemoreceptors 257
Chemotaxis 128
Chemotropism 203
Chilaria 552
Chilomastix 43
Chilopoda 655
Chinese liver fluke 279
Chiropsalmus 226
Chiton 779
Chloragogen cells 434
Chloromonadida 47
Choanocytes 155, 159, 164
Choanoflagellida 48
Chondrioderma 86
Chondrotrichida 145
Choristida 179
Chromadoroidea 381
Chromosomes 13
Chrysaora 228
Cidaroidea 899
Ciliary movement 121
Ciliary rim 395
Ciliata 108
Ciliated chamber 181, 433
Ciliated funnel 467
Ciliated groove 258
Ciliated pharyngeal epithelium 433
Ciliated pits 335
Ciliated organ 415, 467
Cimex 714, 735
Cinclides 231
Cingulum 46
Circular nerve ring 209
Circumoral spines 480
Circumpharyngeal connectives 413
Circumvalation 60
Cirratulus 424
Cirripedia 641
Cirrus sac 279
Cladocera 636
Cladocopa 638
Clam movement 824
**Clathrinida 177**

Clathrulina 82
Clavularia 234
Clepsine 473
Cliona 179
Clitellar glands 457
Clitellar region 457
Clitellum 426, 457
Clone 130
Clypeaster 901
Clypeasteroida 901
Clytia 215, 216
Cnidaria 185
Cnidoblasts 191
Cnidocil 192
Cnidospora 1, 16, 105
Coccidae 736
Coccidia 104
Coccinellidae 742
Coccus 757
Cochineal 757
Cocolithophorida 45
Cocoon formation 448, 472
Coelenterata 187
Coeloblastula 173, 884
Coelodandrum 85
Coelodermatida 562
Coelogorgia 235
Coelomic septa 432
Coelomocytes 914
Coelomoducts 401
Coeloplana 248
Coenenchyme 234
Coenosarc 204
Coenosteum 216
Coenothecalia 235
Coleoidea 857
Coleoptera 741
Collar-cells 152
Collarette 859
Collencytes 164
Collosphaera 85
Collothecacea 341
Collozoum 85
Colochirus 919
Colpoda 145
Comatricha 86
Comatula 928
Comb-plates 242
Comidial 286
Compensation sac 503
Compensatrix 503
Commensal nemertines 306
Conchostraca 636
Conchiolin 820
Conjugants 130, 141
**Conjugation in Paramecium 130**
**Conjugation in Vorticella 141**

Contophora 4
Contractile bladder 268
Contractile vacuole 10, 25, 118
Convoluta 262
Copepoda 639
Copulatory sac 256
Copulatory spicules 356
Copulatory apparatus 274
Coracidium 300
Corallimorpharia 240
Corals 237
Coral reef 237
Corium 787
Cornua 932
Corona 326, 328, 920
Coronal sphincters 330
Coronary groove 226
Coronata 226
Costa 43
Cotylaspis 283
Cotylogaster 283
Cotylorhiza 228
Coxa 541, 547
Coxal gland 553
Coxopodite 601
Cradactis 230
Crested float 214
Cribriform organs 869
Crinoidea 920, 927
Crura 266
Crural gland 541
Crustacea 596, 634
Cryptomonadida 45
Cryptozoic schizonts 95
Crystalline cone 623
Ctenes 242
Ctenidial sinus 796
Ctenidium 773, 792
Ctenophora 187, 242
Ctenostomata 510
Ctenoplana 248
Culex and anopheles compared 697, 698
Cucumaria 909
Cunina 218
Cubomedusae 226
Cupelopagis 329
Curculionidae 743
Current producing 109
Cutaneous artery 800
Cutaneous habronemiasis 364
Cuticle 350
Cuticular teeth 409
Cuvierian tubules 915
Cyanea 227
Cyanophyta 4
Cyclic shell 71
Cyclophyllidea 300

Cyclops 639, 640
Cyclosis 123
Cyclostomata 504
Cydippid larva 243
Cydippida 247
Cylindrical gland 573
Cyphonautes 509
Cypraea 799
Cypris stage 634
Cysticercus 211
Cysticercosis 292
Cyst-passers 79
Cystosomal groove 43
Cytheresis 638
Cytogamy 123
Cytopharynx 116
Cytopyge 11, 110, 117
Cytostome 4, 19, 110
Cyzicus 636

Dactylopores 216
Dactylozooids 212
Daphnia 636, 637
Dead man's fingers 235
Decapoda 857
Deep nervous system 881
Deltidium 519
Demoscolecoidea 382
Demoscolex 382
Demospongiae 178
Dendrobranch 610
Dendroceratida 181
Dendrochirota 919
Dendrophylia 240
Dentalium 783
Dent articulaire 519
Deoxyribonucleic acid 7, 12
Dermal glands 562
Dermal membrane 181
Dermal pore 160
Determinate cleavage 497
Determinate spiral 337
Deutocerebrum 531
Deuterostomia 400
Deutomerite 104
Dexioplectic 120
Dextral shells 786
Diactines 165
Diadematoidea 899
Diaphragm 165
Diaplectic 120
Diastole 62
Diastylis 648
Didinim 144
Digenea 266, 283
Dileptus 144
Dilepis 300

Dibothriocephaloidea 298
Dibothriorhynchus 298
Dicranophorus 332
Dictyoceratida 181
Dictyosomes 9
Dicyemida 152
Dioecocestus 300
Digestive granules 139
Digitelli 221
Dinoflagellates 45
Dioctophymoidea 384
Diphyllidea 298
Diphyllobothrium 300
Diploblastic metazoans 185
Diplomonadida 49
Diplodal 162
Diplopoda 652
Diptera 744
Discomedusae 227
Directives 230
Ditrysia 740
Doliolaria larva 916
Dorsal light response 625
Dorsal vessel 411
Dorso-intestinal vessel 437
Dorylaimoidea 381
Dorylaimus 381
Double pentagon 881
Dracunculoidea 384
Drosophillidae 745
Dung beetle 382

Ebriida 47
Ecdysis 526, 616, 674
Echinobothrium 295
Echinococcus 285, 300
Echinoderes 348
Echinodermata 863
Echinoida 901
Echinoidea 890, 899
Echinopluteus larva 899
Echinorhynchus 316
Echinothuroidea 899
Echiuroidea 477
Echiurus 477
Ectocyte 88
Ectoneural nervous system 881, 924
Ectoplasm 56, 88
Ectoprocta 501
Ectosarc 54
Ectosomes 339
Edwardsia 236
Eimeria 104
Elasipoda 919
Elephantiasis 377
Eleutheria 211
Elytron 669

Embioptera 730
Embolus 584
Embryology 360, 508
Embryonic membranes 576
Embryophore 290, 389
Encystment 141
Endoderm 185
Endodermal lamella 208
Endodermal cell type 196
Endodermal core 205
Endogastric shell 789
Endomixis 134
Endophragmal skeleton 607
Endoplasm 55, 88
Endoplasmic reticulum 8
Endopleurite anterior 608
Endopterygota 737
Endsac 620
Endosome 26
Endosternite 558, 563
Enoploidea 380
Enopla 314
Entamoeba histolytica 77
Enteronephric nephridial system 441
Enterocoelous 399, 945
Entodiniomorphida 148
Entoneural nervous system 881, 924
Entoprocta 388
Ephelota 147
Ephemeroptera 732
Ephyrae 224
Epiactis 232
Epiboly 811
Epicyte 104
Epigynum 584
Epimastigote 36
Epimerite 104, 599
Epiostracum 562
Epiphanes 329
Epiplasm 112
Epipodite 547
Epipolasida 179
Epitaenia 791
Epitheloid membrane 164
Epitoke 415
Ergasilus 642
Eriophyes 592
Exconjugant 130
Endosarc 54
Excretory cytoplasm 27
Excurrent channel 139
Exflagellation 97
Exhalant siphon 819, 824
Exoerythrocytic 95
Exopterygota 725
Exogastric shell 789
Extracellular digestion 199

# Index

Exumbrella 207
Euasteroidea 889
Eucarida 651
Eucaryotes 4
Euglena 18
Euglenida 47
Euglenoid movement 18, 26
Eugregarinida 104
Eucestoda 296
Eumalacostraca 645
Euphausiacea 651
Euryalae 907
Eurypylous 161
Eurypterida 555
Eversible buccal region 402
External blood plexus 438
Eye-spot 26

Facultative 27
False scorpions 587
Fasciola 270
Fasciolopsis 284
Feather star 920, 921
Fecundity 747
Ferment cells 615
Fertilisation membrane 360
Fertiliun 415
Filaria worm 376
Filaroidea 384
Filipodia 59
Filosia 82
Filter feeding 411, 488
Filtration theory 62
Finger keel 560
Fire medusa 226
Flacherie 722
Flagellar movement 21
Flagellar pocket 36
Flagellata 17
Flagellated canals 160
Flame bulb 333
Flame-cells 254
Flexibilia 927
Flexor muscle 598
Flimmer filaments 21
Flosculariacea 341
Flukes 265
Flukes and man 279
Flustra 503
Food basin 552
Food vacuole 11
Foot 325
Foraminifera 60, 82
Foraminal tubule 171
Forcipate type 332
Forcipulata 889
Formation of a pearl 821

Formation of spicules 166
Founder cell 167
Fountain-zone hypothesis 58
Fringing reef 238
Frondicularia 82
Fulcral plates 710
Fulcrum 331
Fulgoridae 735

Galea 667
Galvanotaxis 128
Gammarus 651
Gametes 87
Gametocytes 74
Gametocyst 89
Gametogony 89
Gamont 73
Gangesia 298
Gas gland 214
Gastric cavity 207
Gastric mill 614
Gastric filaments 221
Gastric pockets 220
Gastropoda 785, 813
Gastropore 216
Gastrotricha 342, 345
Gastrozooids 212
Gastrulation 173, 202, 449, 629
Gastrea theory 182
Gastrovascular canals 221
Gastrovascular cavity 201
Gemmation 83
Gemmules 155, 170
Gemmule formation 171
Genae 547
Gene pool 132
Geniculate ganglion 335
Genital atrium 687
Genital chamber 255
Genital papillae 386, 808
Genital pleurae 942
Genital pores 898
Genital rachis 883
Genital sheath 321
Genital stolon 883
Genitointestinal canal 269
Geonemertes 307
Geophilus 658
Geotropism 204
Germ-balls 277
Germ mother cells 202
Germinal disc 350
Germinal epithelium 572
Germovitellarium 336
Gersemia 233
Giardia lamblia 41
Giardiasis 41

Gill-bars 936
Gill-heart 845
Gigantolina 296
Girdle of Venus 247
Gizzard 434, 702
Glabella 547
Glandular cells 197, 797
Glandular plexus 620
Glandula plicata 562
Glass sponges 177
Glass crab 633
Globiferous pedicellaria 894
Glochidium larva 834
Glomerulus 939
Glue cells 245
Glutathione 205
Glutinant nematocysts 195
Gnathobases 552
Gnathochilarium 652
Golgi bodies 9
Gonangium 207
Gonapophyses 688
Gonopalpons 213
Gonophore 207
Gonotheca 207
Gonozooids 206, 212
Gordiancean larva 323
Gordioidea 386
Gordius 386
Grantia 154, 177
Granuloreticulasia 82
Grasserie 722
Green glands 596
Gregarina 103
Gregarine movement 89
Gregarinida 103
Grillotia 298
Groove plate 613
Grylloblatodea 726
Guanin 846
Gubernaculum 359
Guinea worm 375
Gymnoblastida 144
Gymnosomata 816
Gynecophoric canal 279
Gyrocotyle 296
Gyrocotylidea 296

Hadromerida 179
Haemal system 877, 925
Haemoglobin 462
Haemocoel 526, 530, 617
Haemocoelomic canal 462
Haemosporina 86
Haemosporidia 105
Haemozoin 94
Hair-worms 385

Halistemma 213, 214
Hakenwurmer 323
Halichondrida 180
Halteres 696
Halteria 147
Hamanniella 320
Haplobothrium 298
Haplodiscus 262
Haplosclerida 181
Haplosporidia 108
Hastate plate 613
Hectocotylised 840, 849
Helicosporida 107
Heliopora 234, 235
Heliozoa 83
Hemicidaroida 901
Hemichordata 930, 947
Hemimetabola 751
Hemiptera 733
Hemixis 134
Hepatic caecae 553
Hepatozoon 105
Hermaphrodite 201
Hermocystis 103
Heteronemertini 314
Heterotrichida 147
Hexacanth 290
Hexacorallia 229, 236
Hexactinelliida 177
Hexactinomyzon 107
Hexasterophora 177
Heterochlorida 45
Heterocotylea 203
Heterodonta 837
Heteromyarian 821
Heterotricha 144
Heterotrophic 14
Heteroxenous 363
Hibernacula 506
Hirudin 461
Hirudinaria 455
Hirudinea 453
Hirudo 453, 454
Histoblasts 171
Holasteroidea 901
Hologamy 83
Holometabola 751
Holophytic 14
Holothuroidea 908
Holotricha 144
Homoptera 735
Homosclerophorida 178
Honey 757
Honey bee 706
Hoplocarida 646
Hoplonemertini 314
Horse hairs 385

*Index*

House fly 700
Hyaline cap 56
Hyalonema 178
Hyalospongiae 177
Hydra 189
Hydrachna 594
Hydractinia 216, 217
Hydranth 205
Hydratula 224
Hydrocaulus 204
Hydrocoel 873
Hydrocorallina 216
Hydroida 216
Hydroporic canal 887
Hydrorhiza 204
Hydrostatic skeleton 354
Hydrotheca 205
Hydrozoa 188, 215
Hydrula 211
Hymenoptera 743
Hymenostomatida 146
Hypermastigida 50
Hypobranchial gland 775, 808
Hypodermic impregnation 336
Hyponeural nervous system 881
Hypostome 189, 205
Hypostracum 562
Hypotrichida 150
Huxley's anastomosis 333

Idmonea 510
Imaginal disc 705
Imago 751
Immovable finger 560
Immunity 764
Import 60
Inadunata 927
Inanition 125
Inarticulata 524
Incurrent canal 139, 332
Incurrent pore 158
Incurrent siphon 827
Independent effectors 193
Infra-intestinal visceral connective 803
Infundibular canal 245
Infusorigens 153
Inhalant aperture 156
Inhalant siphon 819
Ink-sac 841
Insecta 662, 722
Insect pollinators 757
Integumentry nephridia 440, 442
Interambulacral plates 890
Internal blood plexus 438
Inter-radial 222
Intersternal membranes 599
Interstitial cells 191

Internal Sacculina 643
Inter-radii 208
Intestinal caecae 434, 879
Intracellular digestion 199
Introvert 502
Invagination 60
Inversion 172
Indeterminate cleavage 497
Iris sheath 623
Isogametes 90
Isomyarian 821
Isoptera 728
Isospores 84

Japyx 725
Johnston's organ 756
Jovan Hadzi 183
Julus 653, 654

Kappa particles 134
Katharobic 18
Kala-azar 36, 37, 718
Keber's organ 829
Kentrogon larva 643
Killer paramecia 134
Kidney worm 379
Kinetid 112
Kinetodesma 112
Kineties 4, 111
Kinetonucleus 138
Kinetoplast 36, 38
Kinetoplastida 36, 49
Kinetosome 112, 137
Kinoplasm 136
Kinorhynchya 346
Kircephalus 545
Koenenia 589

Labella 695
Labochirus 590
Labial papillae 357
Labidognatha 595
Labrum 559
Labyrinth 566, 567
Lac insect 736, 757
Lacinia 667
Lacrimyaria 144
Lacunar system 318
Lacoplectic 120
Lagena 75, 76
Lamellibranchia 836
Lamellidens 819
Lampshell 518
Lanceolate plate 613
Lappets 923
Larva 697
Larval forms 631

Lasso 245
Lateral hearts 438
Lateral oesophageals 438
Lateral subcerebral gland 330
Laura 644
Laurer's canal 274
Lecanicephaloidea 297
Leishmania 37
Lemnisci 318
Leishmania stage 36
Lepas 641, 643
Lepidoptera 739
Lepidurus 636
Leptomonas type 36
Leptostraca 645
Leucettida 177
Leucon type 161
Leuconid type 160, 161
Leucosolenia 157
Leucosolenida 177
Ligament sacs 320
Ligament strand 321
Ligia 650
Ligula 668
Limax type 60, 78
Limnaea 794
Limulus 554
Lineus 313, 314
Lingua 668
Linuche 227
Liriope 218
Lithistida 179
Lithocyte 209
Little-sail 214
Liver-fluke 270
Liver rot 270
Lobata 247
Luminiscence 494
Lobopodium 59
Lobose 60
Lobosia 81
Loligo 839, 856
Lopophore 513
Looping 198
Lophopus 507
Lophotaspis 283
Lorica 325
Loricate form 327
Lovenella 216
Loxosoma 388, 389, 390
Lucifer 631
Lumbricus 440
Lung Nematode of frog 369
Lunule 705
Lycosa 579, 591
Lymphatic glands 569
Lyriform organs 583

Lysosomes 11

Macna stage 645
Macrobiotis 546
Macroconjugant 140
Macrodasyoidea 345
Macrogametocyte 97
Macrura 652
Macula statica 847
Madreporaria 237
Madreporite 868, 896
Malacobdella 306, 314
Malacostraca 644
Malarial parasite 92
Maligant tertian 100
Malleate type 332
Malleus 331
Mallophaga 730
Malpighian tubules 565, 581
Mandibulata 595
Mantle 773, 841
Mantle cavity 773, 787
Mantle lobe 520
Manubrium 205, 207, 221
Marginal lappets 221
Marginal vesicle 209
Mastax 331
Masticatory teeth 795
Mastigamoeba 49
Malacocotylea 283
Mastigophora 17, 44
Mastigoneme 21
Maxillipede 604
Maxillary process 561
Maxillula 602
Mature proglottides 288
Maurer's clefts 99
Mecoptera 737
Median retrocerebral sac 329
Median sagittal plane 250
Medulla 256, 258
Medullary substance 351
Medusa 187, 207
Megalopa stage 633
Megaloptera 737
Megascleres 165
Megalospheric shells 70
Megamere 449
Megellania 519, 520, 521
Mehilis's glands 269, 274
Membrana quadripartita 116
Membranipora 510
Menipea 510
Menospora 103
Mermithoidea 381
Merostomata 549
Merozoites 87

*Index* 959

Mesenchyme cells 159
Mesenteric filaments 231
Mesenteries 231, 233
Mesogastropoda 813
Mesogloea 185, 191
Mesolamella 197
Mesothelium 315
Mesosaprobic 18
Mesosoma 556
Mesozoa 151, 182
Metaboly 26
Metacercaria 278
Metachronal rhythm 22
Metagenesis 211
Metamerism 400, 482
Metamorphosis 395
Metanauplius 631
Metaphase 13
Metasoma 556
Metraterm 275
Metridium 231
Microconjugant 140
Microfilariae 377
Microgametocyte 97
Microlinypheus 557
Micromere 449
Micropyle 171
Microscleres 165
Microspheric 74
Microspheric shells 70
Microsporidia 108
Microstomum 262, 263
Microtubular 11, 12
Millipedes 652
Miracidium 276
Mitochondria 9
Mitotic spindle 13
Mixotrophic nutrition 27
Molar process 601, 604
Mollusca 769, 852
Mollusca theory 523
Molpadia 919
Molpadonia 919
Monactinal monaxon 165
Monaxon 165
Monhysteroidea 381
Monocystis 87
Monogenea 266, 283
Monomyarian 822
Monoplacophora 778
Monopodal 56
Monothalamus 70
Monogononta 340
Monotrysia 740
Monoxenous 362
Morula 223
Mosaic image 685

Mosquito 694
Mother of pearl 820
Müller's larva 264
Multiple nucleus 68
Multiple fission 86
Multipodal 56
Muscle tails 189
Musculo-epithelial cells 189, 206
Mushroom gland 688
Mycetome 735
Mycetozoa 85
Mydocopa 637
Myocyte 88, 104, 165
Myonemes 88
Myrientomata 724
Mysidacea 648
Mysis 633
Mystacocarida 639
Mytilus 837
Myxosporidia 107
Myzophyllobthrium 285, 297
Myzostoma 422


Naiad 732
Naididae 452
Nais 425
Nacreous layer 820
Narcomedusae 217
Nassellaria 84, 85
Natantia 652
Nauplius 631
Nausithoe 226, 227
Nautiloidea 856
Nautilus 856
Neacter 384
Nebaliacea 645
Nectocalyx 213
Nectonematoidea 387
Nectonemertes 312
Nectosome 214
Needham's sac 848
Nemathelminthes 324
Nematoblasts 191
Nematocystis 87
Nematocysts 184, 191, 206
Nematoda 348, 362
Nematocera 745
Nematomorpha 384, 386
Nematotaenia 300
Nemertina 307, 313
Nemertopsis 314
Neodasys 345
Neogregarinida 104
Neolampidoida 901
Neomenia 782
Nephridia 412, 466
Nephridiopore 312, 389, 412, 467

Nephridiostome 440, 467
Nephroblast 451
Nephro-peritoneal sac 621
Nephrostome 412
Neuroblast 457
Nephrocytes 679
Nereis 403
Nerve cells 195
Nerve-net 195, 206
Nerve pentagon 881
Nesolecithus 296
Neuromotor system 112
Neuropodium 405
Neuroptera 737
Neutral red granules 139
Nidamental glands 849
Noctiluca 46, 47
Nodosaroid shell 71
Nodulus 427, 428
Non-ciliated columnar cells 797
Nosema 106
Notommatoidea 340
Notoplana 264
Notopodium 405
Notaspidea 817
Notopodial cirri 405
Notostigma 658
Notostraca 636
Notum 665
Nuchal cartilage 843
Nuchal lobes 791
Nuchal organ 404
Nuclear vacuole 118
Nuclear envelope 9
Nucleoli 13
Nucleus 12, 26, 38
Nuda 81, 248
Nutritive muscular cells 196
Nutritive pole 810
Nuttalochiton 781
Nyctiphanes 651
Nyctotherus 145
Nymph 694
Nymphon 594
Nyssorhynchus 699

Obligatory phototroph 27
Obligopyrene 808
Ocellus 208
Octocorallia 229, 232
Octopoda 857
Octopus 857
Ocular segment 456
Odonata 732
Odontostomatida 148
Odontophore 843

Odoriferous glands 654
Oesophageal glands 410
Oesophagus 116
Olfactory setae 625
Oligochaeta 424
Oligoentomata 732
Oligosaprobic 18
Oligotrechida 147
Olynthus 156
Ommatidium 685
Ommatophores 806
Onchobothrium 297
Onchosphere 290
Onicidium 807
Oniscus 650
Onychophora 535
Oocyst 87
Ooeciun 507
Oogenesis 584
Ookinete 87, 97
Ootheca 690
Opalina 50, 51
Ophiactis 907
Ophiopus 907
Ophiotaenia 298
Ophiurae 907
Ophiuroidea 901, 907
Opiliones 593
Opisthaptor 267
Opisthobranchia 815
Opisthopora 453
Opisthosoma 548
Optic organelle 444, 470
Oral-aboral axis 244
Oral arms 220, 221
Oral disc 230
Oral groove 116, 123
Oral lobes 701
Oral pole 224
Oral surface 111
Organ of Bojanus 829
Organs of pallial complex 791
Organ of Tomosvary 654
Organogeny 577
Orthognatha 594
Orthonectida 151, 153
Orthoptera 725
Oscular fringe 157
Oscular sphincter 159, 164
Osculum 157
Osmoregulation 125, 139
Osmotic theory 62
Osmotrophic heterotrophy 27
Osphradium 775, 792, 806, 832
Ossicles 871, 910
Ostia 175, 231
Ostracoda 636

# Index

Otolithic vesicles 311
Oxymonadida 49
Oxyspirura 384
Oxyuris 371, 383
Oxyuroidea 383
Ovarian balls 321
Ovary 359
Oviducal funnel 447
Oviduct 359, 687
Ovijector 370
Ovipositor 703
Ozobranchus 474

Packet glands 307
Palaeonemertini 314
Palamnaeus 558, 564, 572, 573
Palaemon 597
Palaeocaridacea 646
Pallial cavity 787
Pallial complex 787, 791
Pallial groove 509, 780
Palmella state 30
Palpifer 667
Palpigradi 588
Palpons 213
Paludicella 510
Pantala 733
Pancreatic gland 843
Papulae 869
Parabasal body 38
Paracyatholaimus 382
Paraeopoda 604
Paraflagellar 20
Paragastric cavity 155, 158
Paraglossa 668
Paramecin 134
Paramecium 109
Paramylum 26
Paranemertes 314
Paraoral cone 132
Parapodia 402, 660
Parapolar cells 115
Paraseison 340
Parasitic flagellates 35
Parasitism 762
Paraxial organ 574
Parazoa 175
Parazoanthus 240
Parenchymula 173
Parietal muscles 231
Parthenogenesis 756
Parthenogonidia 33
Patagia 696
Patella 552
Pathogenic trypansomes 38
Pauropoda 660
Paxilla 871

Pearl oyster 835
Pebrine corpuscles 722
Pectinal teeth 561
Pectinatella 510
Pedal disc 230
Pedal ganglion 784, 803
Pedalia 226
Pedicellariae 890, 812
Pedicellina 388
Pedipalpi 552, 560
Peduncles 225, 519
Pelagonema 381
Pelagonemertes 314
Pelagothurida 919
Pelecypoda 817, 836
Pellicle 17, 38, 111
Pellucida externa 806
Pellucida interna 806
Pelmatohydra 198
Penetrant 193, 195
Penial setae 350
Penis 255
Penis bulb 265
Pennatulacea 235
Pennicullus 116
Pentacrinus 927
Pentastomatida 543
Pericardial sinus 539
Perinephrostomial ampullae 464
Perionyx 440
Periostracum 787
Peripatus 536
Periplaneta 663
Periproctal sinus 896
Perisarc 205
Peristomial disc 136
Peristomial groove 136
Peristomial vacuole 118
Peristomium 426
Peritoneum 398
Peritrichia 146
Peritrichida 147
Peritrophic membrane 647
Perivitelline 850
Peroxysomes 10
Per-radii 207
Per-radial canal 222
Planula 220
Phaeophyceae 4
Phallomeres 689
Phanerozonia 889
Pharetronida 177
Pharyngeal glands 355
Pharyngeal nephridia 440, 442
Pharyngeal teeth 409
Phasmida 727
Phasmids 358

Pheretima 425
Philodina 340
Phoronidea 512
Phoronis 512, 513, 514
Phoronopsis 512
Photoautotrophic 14
Photoreceptor 443, 470
Phototaxis 28, 128
Phototrophic 27
Phototropism 64, 203
Phreodrilidae 453
Phylactolaemata 502
Phyllobranch 609
Phyllobothrium 297
Phyllocarida 645
Phylum Acanthocephala 315
—Annelida 395
—Arthropoda 525
—Brachiopoda 518
—Bryozoa (Ectoprocta) 501
—Chaetognatha 859
—Cnidaria (Coelenterata) 185
—Ctenophora 242
—Echinodermata 863
—Entoprocta 388
—Hemichordata 830
—Mesozoa 151
—Mollusca 769
—Nemathelminthes 249
—Phoronidea 512
—Porifera 154
—Protozoa 1, 5
Phymosomatoida 901
Phytomastigophorea 1, 44
Phytomonadida 47
Pila 786
Pilidium larva 307, 313
Pinacocytes 173
Pinnules 232, 922
Pin-worm 371
Piscicola 454
Piscicolidae 474
Placobdella 474
Plagiostomum 263
Planaria 250, 252
Planipennia 737
Planorbis 799
Plant parasitic nematodes 365
Plantulae 670
Planula 211, 220
Plasmalemma 8, 54
Plasmagel 54
Plasmodium vivax 98
Plasmasol 54
Plasmotomy 153
Plastida 10, 18
Platycopa 639

Platyctenea 248
Platyhelminthes 301
Plecoptera 729
Pleodorina 48
Pleopods 601
Pleura 428
Pleural ganglia 784, 803
Pleurobrachia 243
Pleurobranch 609
Plesiopora 452
Ploima 340
Pluteus larva 898
Plug 472
Pneumatocodon 213
Pneumatophores 213, 219
Pneumatosaccus 213
Podocopa 638
Podobranch 609
Podogona 592
Poecilosclerida 180
Poison claw 656
Poison glands 710
Poison sac 710
Polar plug 473
Polian vesicles 896, 911, 926
Pollen basket 709
Pollination 757
Polychaeta 402
Polynoe 421
Polysaprobic 18
Polysynergids 68
Pontobdella 454
Pork tapeworm 286
Porifera 154, 176
Porocyte 158, 164
Porpita 213, 219
Polyaxons 166
Polycladida 263
Polyembryony 278
Polygordius 475, 476
Polymorphism 212, 504
Polyp 187
Polypide 502
Polyplacophora 782
Polythalamus 70
Polyvoltine 719
Precoxa 547
Prepucial gland 861
Pre-patent period 100
Priapulida 480
Primary nematogen 152
Primitive endoderm 508
Prismatic layer 820
Proboscis 254, 309
Proboscis gland 940
Proboscis pore 309
Proboscis receptory 318

# Index

Proboscis sheath 309
Proboscis skeleton 935
Proctodaeum 410
Proloculum 70
Promastigote 38
Pro-prioceptors 684
Prosicula 219
Prosobranchia 813
Prosodus 162
Prosopora 453
Prosopyle 160, 164
Proteromyxidia 85
Protein coat 360
Protomerite 104
Protobranchia 836
Protonephridia 319
Prototaenia 300
Proteocephaloidea 298
Proteolytic ferment 461
Prostate glands 446
Protostomia 400
Protozoa 1, 5
Protozoea 631
Protozoea stage 632
Protrusible pharynx 402
Protrusible sac 660
Pseudocoelocytes 351
Pseudocoelomate bilateria 315
Pseudopipodia 791
Pseudoconjugation 89
Pseudopenis 689
Pseudophyilidea 298
Pseudopodium 3, 46, 57
Pseudonavicella 90
Pseudoscorpionida 587
Psoroptes 594
Pteripyles 567
Pterygota 725
Ptildinum 705
Ptygura 341
Pulmonata 817
Puparium 705
Pusule system 45
Puluillus 669, 701
Purine 15
Pycnopodia 866
Pygidium 406
Pygochord 936
Pyloric caeca 879
Pyloric pouch 879
Pyloric stomach 612
Pyrenoid 18, 26
Pyriform organ 509
Pyrimidine 15

Quadrants 351

Quadrate plate 710
Quadrinuclear stage 78
Quadriradiate 165
Quartan malaria 99
Queen cells 713
Queen substance 759
Quellkorper 115
Quinine 101
Quotidian malaria 99

Rachis 236
Radial canal 160, 222, 896
Radiolaria 84
Radula 777, 843
Ramate type 332
Raptorial 109
Recovery phase 22
Rectal caeca 879, 882
Rectal gland 355
Recurrent nerve 682
Redia 277
Reduction bodies 170
Reflex arc 443
Regeneration 201
Renal sacs 845
Renette 356
Reproductive cells 196
Reptantia 651
Retortomonadida 49
Reticulopodia 59
Retrocerebral organ 329
Retinal cell 259
Retractor muscles 231
Reunion masses 170
Rhabdiasoidea 382
Rhabditoidea 382
Rhabdites 253
Rhabdocoela 212
Rhabdom 623
Rhagon 162
Rhagon type 162
Rheotaxis 129
Rhinocricus 652
Rhinophore 805
Rhizomastigina 48
Rhizostomeae 228
Rhizocephala 643
Rhizopoda 52
Rhodophyta 4
Rhombogen 153
Rhombozoa 151
Rhopalioids 120, 222
Rhopalonema 218
Rhynchelmis 448
Rhynchobdellida 474
Rhynchocoel 307
Rhynchodaeum 309

Rhynchonella 521
Rhynchotremata 524
Ribonucleic acid 7
Ribosomes 7, 8
Ricinulei 592
Rictularia 384
Richtersia 382
Ring canal 207
Rostellum 286
Rostellar hook 286
Rostral lamellae 326
Rotifera 325, 340
Royal jelly 759
Rugosa 240

Saccocirrus 476
Saccule 553
Sacculina 643
Sacculus 703
Sacoglossa 817
Sagittocysts 184
San Jose scale 736
Salenioida 899
Salivary chamber 433
Salivary gland 711
Salivary pump 716
Sarcocyte 88, 104
Sarcoma 293
Sarcomastigophora 1, 15, 44
Sarcodina 81
Sarcosporidia 108
Scaphognathite 604
Scaphopoda 783
Schistocephalus 300
Schistocerca 726
Schistosomiasis 281
Schizochoerus 296
Schizocoeous 399, 945
Schizocystis 104
Schizodonta 837
Schizogregarinida 104
Schizogony 86, 95
Schizont 73, 96
Schuffner's dots 98
Scleroblasts 164
Sclerotium 86
Scolex 285
Scolopophore 753
Scopula 135, 580
Scorpionida 587
Scorpion 557
Scutellum 696
Scutigera 657, 658
Scyphistoma 221, 224
Scaphognathite 604
Scyphozoa 219, 225
Sea-anemones 230

Sea-cucumber 908
Sea-fans 235
Sea-mats 501
Sea-pens 236
Sea-urchins 890
Sea-wasps 226
Secondary mesenteries 231
Secretion theory 62
Seibold organ 755
Seisonacea 340
Semaeostomeae 227
Seminal vesicles 255
Seminal receptacle 256
Sensory cells 196
Sensory pits 222
Sepia 856, 857
Septal excretory canals 441
Septal nephridia 440
Septaria 787
Septibranchia 838
Septointestinal 438
Septonephridial vessel 438
Sergestes 632
Serialaria 506
Serosa 693
Sertularia 216
Setigerous sac 405
Sheath 20
Shells 202, 819, 842
Shellac 757
Shell glands 269, 289, 596
Shell plates 509
Shelly loop 519
Silicoblasts 164
Silicoflagellida 45
Silk-moth 719
Sinistral shells 786
Siphonoglyph 231
Siphonophora 219
Siphonosome 214
Siphonozooids 234
Sipunculata 731
Sipunculoidea 478
Skeletal axis 240
Skeleton of sponges 165
Slime glands 541
Sleeping sickness 36
Slipper animalcule 110
Sminthurus 724
Social insects 758
Soldier 729
Solenocyte 395
Solenogasters 782
Solifugae 588
Solmundella 218
Sol-gel theory 58
Somatic mesoderm 398

# Index

Somatoderm 153
Somersaulting 198
Somasteroidea 889
Sorting surfaces 776
Spadix 849
Spasmoneme 135
Spatangoida 901
Spathebothrium 285, 300
Spelaeogriphacea 648
Spermathecae 447, 687
Spermatocyst 172
Spermatogonia 202, 572
Spermatophore 575, 627, 849
Spermatozoa 88, 201, 572
Spermiducal vesicle 268
Sphaeridia 891
Spicules 155
Spider 578
Spider silk 585
Spinnerets 580
Spinulosa 889
Spiral cleavage 517
Spiral shell 71
Spire 786, 787
Spirostreptus 652
Spirotrichia 147
Spiruroidea 384
Splanchnic mesoderm 398
Splanchnocel 873
Spongocoel 158
Sporoblasts 90
Sporocyst 276
Spongilla 170, 180
Spongioblasts 164
Sporogony 97
Sporoplasm 105
Sporozoa 1, 16, 86, 103
Sporozoite 86, 95
Squama 603
Squid 838, 840
Starfish 687
Statoblast 506
Statoconia 807
Statocyst 209, 222, 246, 807
Statolith 209, 222, 246
Staurocalyptus 178, 181
Stauromedusae 225
Stegomyia 699
Steinaria 382
Stenoglossa 814
Stenostomum 263
Stercoral pocket 580
Sternum 528
Stephanoceros 339
Stereogastrula 173
Stereoline glutinant 195
Sterrasters 166

Stichocotyle 283
Stigma 26
Stigmata 535, 559, 676
Stolonifera 234
Stolons 204
Stomatopoda 646
Stomoblastula 172
Stone canal 873, 896, 911
Streptoneury 805
Strepsiptera 743
Streptoline glutinant 195
Strobila 285
Sting 707, 709
Stipes 667
Strobilation 221, 224
Strongyloidea 383
Stylaster 217
Stylet sheath 709
Stylommatophora 817
Stylonchia 150
Subdermal spaces 160
Sub-oesophageal ganglia 314, 571
Sub-neural vessel 437
Sub-radular organ 784
Subumbrellar surface 207
Sucker glands 457
Suctoria 141, 147
Suctorida 147
Suctorial pharynx 277
Sun-spiders 588
Superposition image 624, 625, 686
Suprabranchial chamber 825, 826
Supra-intestinal ganglion 803
Supra-pharyngeal ganglia 413
Supra-oesophageal ganglion 570, 621
Supra-oesophageal vessel 438
Surface tension theory 59
Surra disease 44
Swarm spore 74
Swimming bells 213
Sycettida 174
Sycon 155, 158, 161
Syconoid canal system 160
Syllis 422
Symphyla theory 746
Symphyla 659
Symphyta 744
Symplectic 120
Syncytial ectoderm 350
Syncarida 645
Syncitium 290
Syngamy 74
Syngens 132
Synkaryon 130
Synura 35
Systole 62
Syzygy 89, 103

Walking movement 59
Wandering nucleus 130
Water vascular system 904, 911
Wax glands 707
Wheel organ 328
Whip-worm 379
Worker caste 729
Wuchereria 370, 373, 376

Xiphosura 550
Xysticus 592

Yolk glands 256, 299
Yolk reservoir 269

Zeugloptera 739

Ziemann's dots 98
Zoantharia 229
Zoanthidea 240
Zoanthus 241
Zoea stage 632
Zonites 346
Zoochlorellae 83, 197
Zooecium 502
Zooids 187
Zonites 346
Zoraptera 729
Zygeupolia 310
Zygonemertes 311, 314
Zygoneury 805
Zygoptera 732, 733